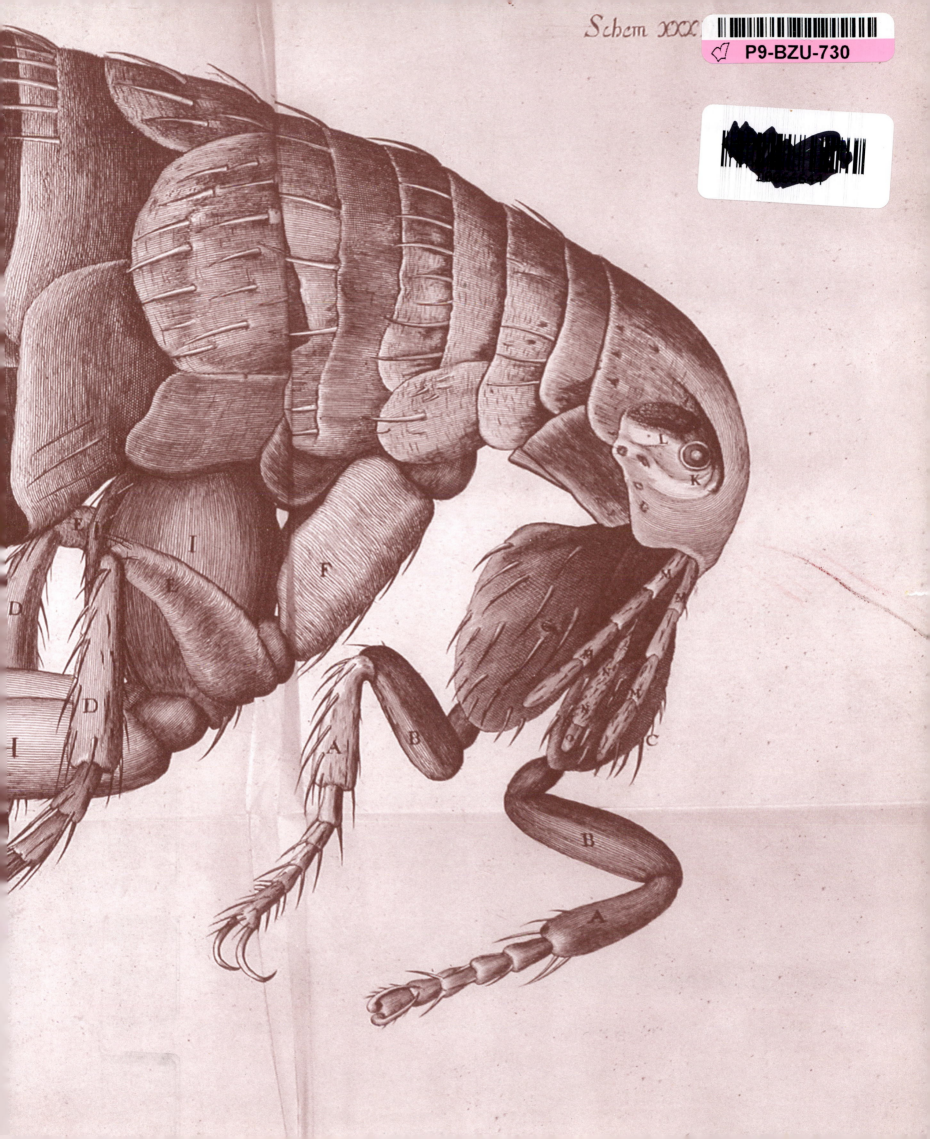

EDITOR-IN-CHIEF ADAM HART-DAVIS

# SCIENCE

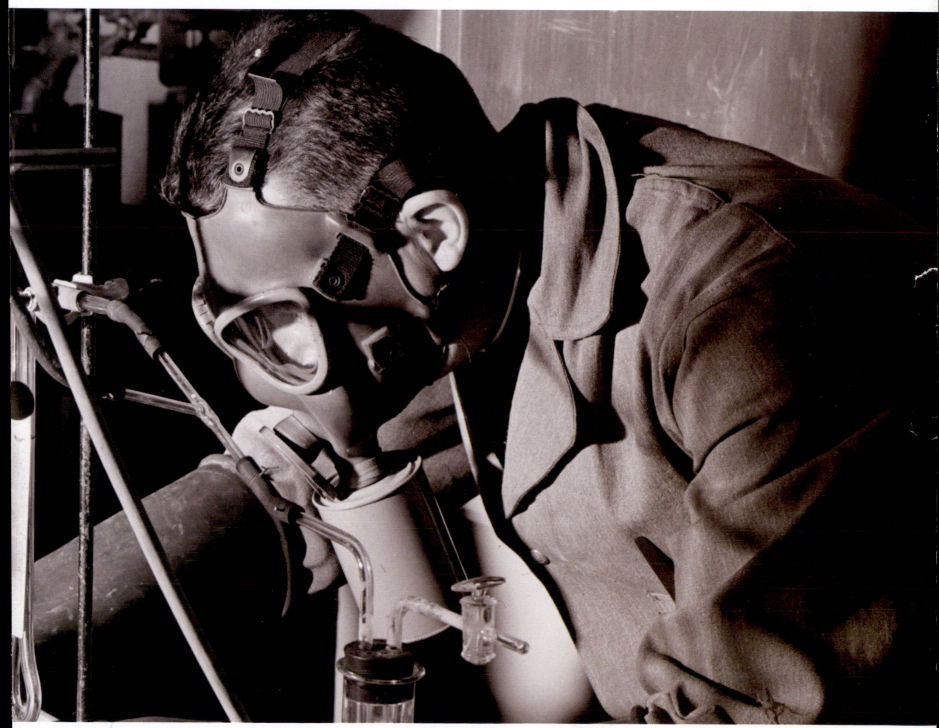

## THE DEFINITIVE VISUAL GUIDE

**LONDON, NEW YORK, MELBOURNE,
MUNICH, AND DELHI**

**Senior Art Editors**
Stephen Knowlden, Vicky Short

**Senior Editors**
Janet Mohun, Kathryn Wilkinson

**Section Designers**
Vivienne Brar, Paul Drislane, Mandy Earey,
Clare Joyce, Mark Lloyd, Heather McCarry,
Matthew Robbins, Alison Shackleton

**Section Editors**
Ann Baggaley, Kim Dennis-Bryan,
Jemima Dunne, Martha Evatt,
Nicola Hodgson, Neil Lockley,
Martyn Page

**Designers**
Keith Davis, Supriya Sahai, Silke Spingies

**Editors**
Katie John, Patrick Newman, Anna Osborn,
Frank Ritter, Nikki Sims, Sarah Tomley,
Angela Wilkes

**Illustration Visualizer**
Peter Laws

**Jacket Designer**
Duncan Turner

**US Editor**
Jane Perlmutter

**Production Controller**
Sophie Argyris

**Production Editors**
Phil Sergeant, Marc Staples

**Picture Researchers**
Ria Jones, Sarah Smithies, Louise Thomas

**Associate Publisher**
Liz Wheeler

**Managing Art Editor**
Louise Dick

**Managing Editor**
Julie Oughton

**Art Director**
Bryn Walls

**Publisher**
Jonathan Metcalf

**Illustrators**
Oliver Burston and Jurgen Ziewe at www.debutart.com
Tom Coulson and Martin Darlison at Encompass Graphics Ltd.
Adam Howard and Andy Kay at Invisiblecities
Tim Loughhead
Darren Awuah
Thomas Bayley, Robin Carter, Tom Connell, Barry Croucher,
Stuart Jackson Carter, Terry Pastor, and Mick Posen at The Art Agency

First American Edition, 2009

Published in the United States by
DK Publishing
375 Hudson Street
New York, New York 10014

08 09 10 11   10 9 8 7 6 5 4 3 2 1

[ID042—October 2009]

Copyright © 2009 Dorling Kindersley Limited
All rights reserved

Published in Great Britain by Dorling Kindersley Limited

A catalog record for this book is available from the Library of Congress

ISBN: 978-0-7566-5570-9

DK books are available at special discounts when purchased in bulk for
sales promotions, premiums, fund-raising, or educational use. For
details, contact: DK Publishing Special Markets, 375 Hudson Street,
New York, New York 10014 or SpecialSales@dk.com.

Printed and bound in China by Leo Paper Products Ltd.

Discover more at
**www.dk.com**

# THE DAWN OF SCIENCE

## PREHISTORY TO 1500

The ancient world saw the first breakthrough moments in science, as the coming of age of great civilizations from Egypt to Babylon proved a decisive spur to invention, people learned to write, and scholars from Aristotle to Zeno had time and space to think deeply about the world around them.

# THE DAWN OF SCIENCE

## PREHISTORY TO 1500

| 14000 BCE | 3000 BCE | 2000 BCE | 1000 BCE |
|---|---|---|---|

**14000 BCE**
First pots created in Japan.

**c.3000 BCE**
Acupuncture needles first used in China. Babylonians create a number system based on 60; it survives in the way hours and minutes are measured today.

◄ Early acupuncture chart ►

**c.2000 BCE**
First wheels with spokes made. Baked bricks used for important public buildings in Mesopotamia.

Ziggurat of Ur, Iraq ►

**c.670 BCE**
Hippocrates establishes medical school on the Greek island of Cos.

**c.600 BCE**
Oldest known world map drawn by Babylonians.

▼ Map by Herodotus, c.450 BCE

◄ Ancient Egyptian astronomical calendar

▲ Jomon bowl

**c.5500 BCE**
Copper smelting begins in the Balkans and West Asia.

**c.3500–3200 BCE**
In Mesopotamia solid wheels used for transportation.

**c.2700 BCE**
Abacus first appears in Mesopotamia and becomes widely used for calculations.

▲ 16th-century abacus

▼ Egyptian figurine

**1500 BCE**
Oldest known astronomical calendar created in Egypt.

Pythagoras of Samos ►

**500 BCE**
Babylonian astronomers note an eclipse cycle known as the Saros.

**495 BCE**
Pythagoras introduces the concept of mathematical proof.

▼ Mesopotamian chariot on the Standard of Ur

**c.3400 BCE**
Ring of standing stones erected and used as an astronomical calendar at Callanish, Scotland.

▲ Early plank wheel

▼ Callanish stones

**c.1300 BCE**
Vertical obelisks used as clocks and compasses in Egypt. Egyptians making objects in faience (ceramics).

**c.400 BCE**
Empedocles states that all matter is composed of four elements: fire, earth, water, and air.

**c.350 BCE**
Aristotle provides scientific evidence that Earth is a sphere.

▼ Earth's curved shadow cast on the Moon

▼ Chinese bronze vessel

**c.2500 BCE**
Chariots with solid wheels used in battle in Mesopotamia.

**c.1100 BCE**
Bronze cast in furnaces in China.

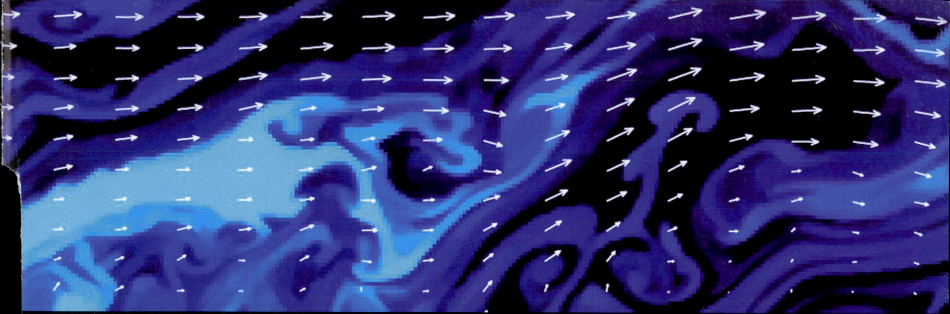

## Editor-in-Chief

# Adam Hart-Davis

Adam Hart-Davis is a writer, broadcaster, and photographer, and one of the world's most popular and respected "explainers" of science. His TV work includes *What the Romans*, *Victorians, Tudors and Stuarts*, and *Ancients Did For Us, Tomorrow's World, Science Shack*, *The Cosmos: A Beginner's Guide*, and *Just Another Day*. He is the author of more than 25 books on science, invention, and history.

## Main Consultants

### John Gribbin

**Physics**

Popular science writer, astrophysicist, and Visiting Fellow in Astronomy at the University of Sussex, UK.

### Jeremy Cherfas

**Biology**

Writer and broadcaster in biological subjects, with a PhD in animal behavior.

### Marty Jopson

**Biology**

Science communicator and TV broadcaster, with a PhD in plant cell biology.

### David Bradley

**Chemistry**

Science writer and editorial consultant, chartered chemist and member of the Royal Society of Chemistry, UK.

### Douglas Palmer

**Earth Sciences**

Science writer and lecturer for the University of Cambridge Institute of Continuing Education, UK, specializing in earth science and paleontology.

### Iain Nicolson

**Astronomy and Space Technology**

Formerly Principal Lecturer in Astronomy at the University of Hertfordshire, UK; writer, lecturer, and occasional broadcaster on astronomy and space science.

### Barry Lewis

**Math**

Formerly Director of Maths Year 2000 and now President of the Mathematical Association, UK.

## Contributors

**David Burnie** Biology and Medicine
**Jack Challoner** Physics
**Robert Dinwiddie** Earth Sciences and Physics
**Derek Harvey** Biology and Chemistry

**David Hughes** Astronomy
**Giles Sparrow** Physics and Space Technology
**Carole Stott** Astronomy
**Marcus Weeks** Math and Technology

**Other contributions** Ann Baggaley, Hayley Birch, John Farndon, Andrew Impey, Jane McIntosh, Sally Regan, Frank Ritter, Mark Steer, Amber Tokeley, Martin Toseland, James Urquhart, Diana Vowles

# Contents

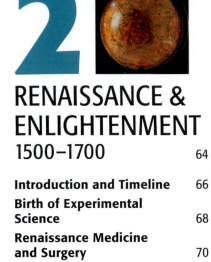
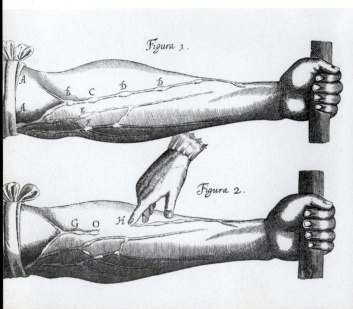

As empires rose and fell, from Egypt to China, the practical demands of the first great cities and armies stimulated a wave of inventions—bronze for making tools and weapons, wheels for moving loads and milling grain, gears for making machines, and the water and windmills to power those machines. Meanwhile, the administrative needs of the new rulers for everything from accurate calendars to tax and land inheritance calculations prompted the first great stirrings of science. At the same time early stargazers such as Hipparchus were mapping the night sky with astonishing precision, and brilliant scholars such as Euclid and Al-Khwarizmi were laying the foundations of mathematics.

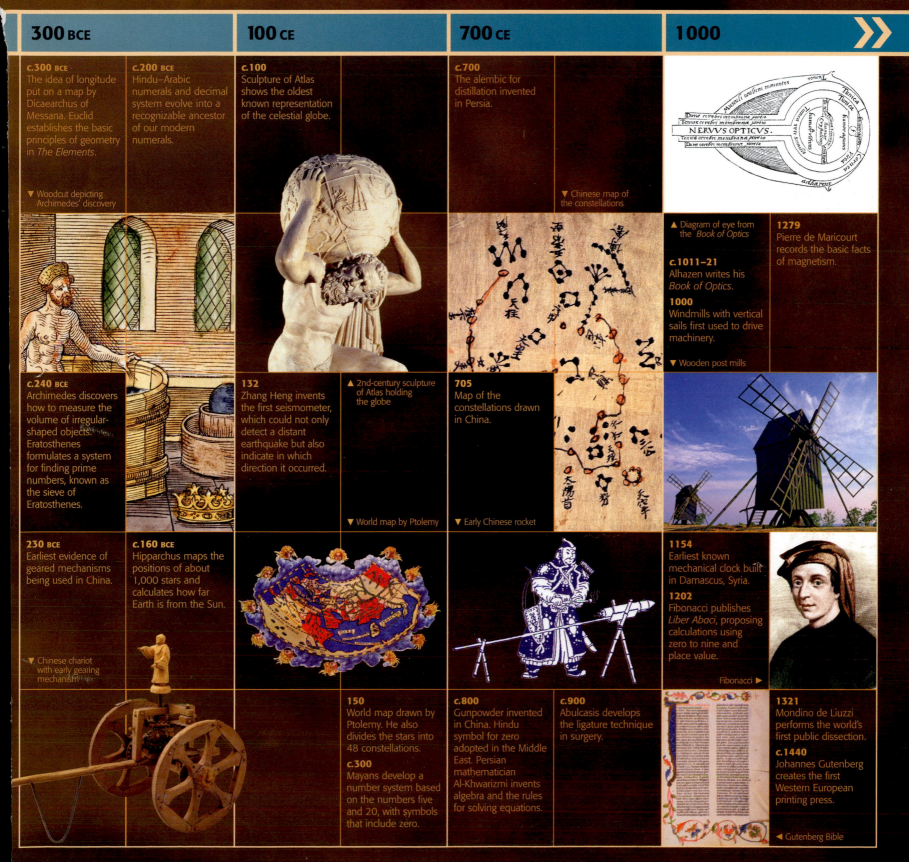

## 300 BCE

**c.300 BCE**
The idea of longitude put on a map by Dicaearchus of Messana. Euclid establishes the basic principles of geometry in *The Elements*.

**c.200 BCE**
Hindu–Arabic numerals and decimal system evolve into a recognizable ancestor of our modern numerals.

▼ Woodcut depicting Archimedes' discovery

**c.240 BCE**
Archimedes discovers how to measure the volume of irregular-shaped objects. Eratosthenes formulates a system for finding prime numbers, known as the sieve of Eratosthenes.

**230 BCE**
Earliest evidence of geared mechanisms being used in China.

**c.160 BCE**
Hipparchus maps the positions of about 1,000 stars and calculates how far Earth is from the Sun.

▼ Chinese chariot with early gearing mechanism

## 100 CE

**c.100**
Sculpture of Atlas shows the oldest known representation of the celestial globe.

**132**
Zhang Heng invents the first seismometer, which could not only detect a distant earthquake but also indicate in which direction it occurred.

▲ 2nd-century sculpture of Atlas holding the globe

▼ World map by Ptolemy

**150**
World map drawn by Ptolemy. He also divides the stars into 48 constellations.

**c.300**
Mayans develop a number system based on the numbers five and 20, with symbols that include zero.

## 700 CE

**c.700**
The alembic for distillation invented in Persia.

▼ Chinese map of the constellations

**705**
Map of the constellations drawn in China.

▼ Early Chinese rocket

**c.800**
Gunpowder invented in China. Hindu symbol for zero adopted in the Middle East. Persian mathematician Al-Khwarizmi invents algebra and the rules for solving equations.

**c.900**
Abulcasis develops the ligature technique in surgery.

## 1000

▲ Diagram of eye from the *Book of Optics*

**c.1011–21**
Alhazen writes his *Book of Optics*.

**1000**
Windmills with vertical sails first used to drive machinery.

▼ Wooden post mills

**1279**
Pierre de Maricourt records the basic facts of magnetism.

**1154**
Earliest known mechanical clock built in Damascus, Syria.

**1202**
Fibonacci publishes *Liber Abaci*, proposing calculations using zero to nine and place value.

Fibonacci ▶

**1321**
Mondino de Liuzzi performs the world's first public dissection.

**c.1440**
Johannes Gutenberg creates the first Western European printing press.

◀ Gutenberg Bible

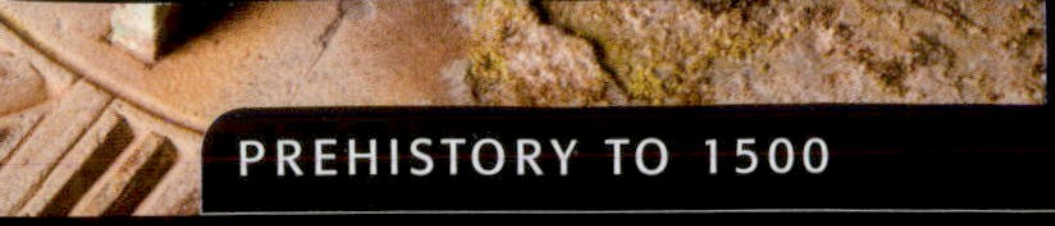

Fire is a chemical reaction between a fuel, such as wood, coal, or oil, and oxygen in the air; it usually gives off energy in the form of heat and light. This can be harnessed and used to positive effect. The reaction starts when the fuel is ignited, either naturally or by human effort by another fire, a spark, or heat. Particles of burning materials and ash that glow and discolor are visible as flames and smoke. A fire generally sustains itself until it has exhausted the supply of fuel or oxygen.

### Early humans and fire

Charcoal and burned bones from several African sites suggest people may have controlled and used fire as early as 1.5 million years ago and certainly by 400,000 years ago. For the last 100,000 years fire has been in common use. Control over fire brought many benefits, including the ability to live in colder regions and protection against predators. People's diet was improved by eating cooked food, roasted over a fire or baked in the embers. Fire also provided a social focus. Some groups used torches or lit fires to stampede animals into places where they could be trapped and killed.

From the earliest times when open fires were lit, people began to develop hearths, often ringed by stones. After 40,000 BCE more efficient hearths with clay surrounds and air-intake channels were invented to control and increase heat. Simple lamps were made from stones with a natural hollow: in this animal fat was burned, using a plant-fiber wick. Some groups began controlled burning to clear vegetation, facilitating hunting and the growth of useful plants.

**In certain circumstances fire occurs naturally. Humans in prehistory saw its potential and learned to control its power.**

**TECHNOLOGY**

By 2.5 million years ago, when they began making stone tools, early humans had understood that they could **alter the natural world** to their own advantage. By **employing tools** they could extend their abilities.

**FIRE'S POTENTIAL**

When early humans encountered naturally occurring bush fires, they came to appreciate not only fire's **power to destroy** the landscape and everything in it, but also its potential as a **source of heat, light, and defense,** and as a **tool for shaping the world**.

**Making fire**
Heat is needed to start a fire. One common method of creating heat is by friction: here, a stick is being rotated fast on a piece of wood until a glowing ember is created.

### Fired clay

Using clay in domestic hearths led people to discover the transformative power of fire on clay. This knowledge was used to create figurines of baked clay. Later, the technology was developed further, to produce pottery. Prepared clay was mixed with a temper, such as

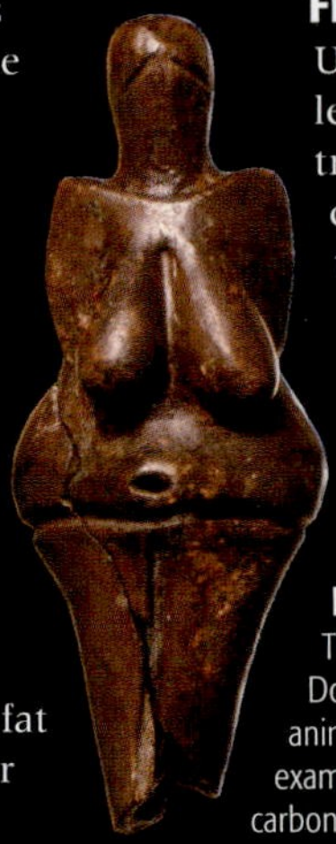

**Dolni Vestonice Venus**
The Palaeolithic inhabitants of the Czech site of Dolni Vestonice produced many human and animal figurines c. 20,000 years ago, including this example, modeled from a mixture of clay and carbonized bone before being baked.

# Fire Power

**Fire is a terrifying and potentially devastating natural phenomenon. Early mastery of fire offered many benefits to humankind including protection against wild animals, heat that enabled them to spread into cooler regions, and the ability to cook food.**

**IN PRACTICE**

## FIRED BRICKS

Clay, as daub or sun-dried bricks, was an important ancient building material. For more durable and impressive structures, the technology of ceramic production was used to make baked bricks. Hand- or mold-formed, the bricks were allowed to dry, then fired in large stacks encased in fuel. Fired-brick buildings and walls defended people from enemies and the elements. In the 3rd millennium BCE, the Harappans in the Indus Valley used baked bricks to protect against flooding, and for wells and bathroom floors, while the Mesopotamians used them for public buildings such as the ziggurat (stepped temple-mound) of Ur, Iraq (right).

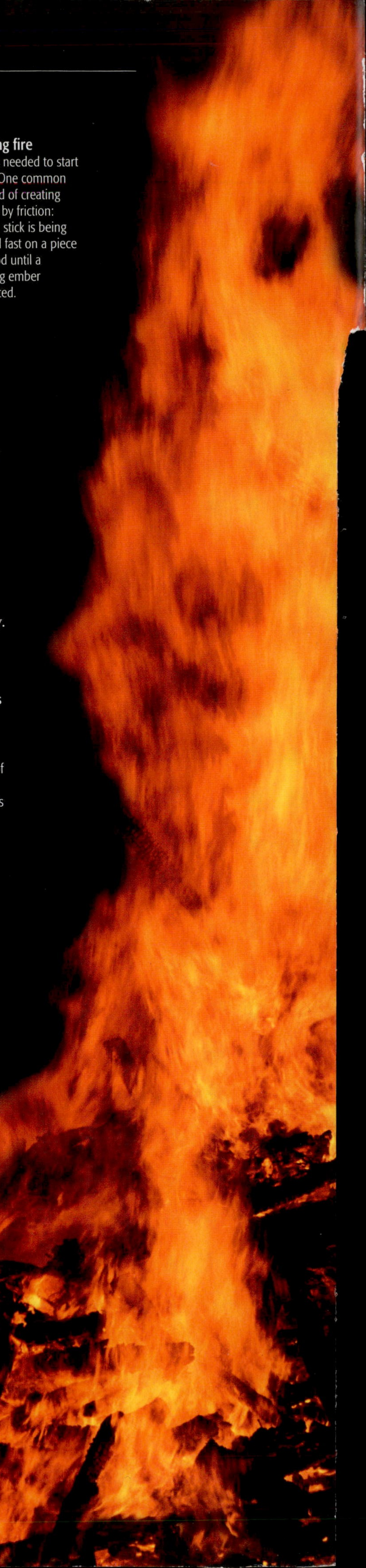

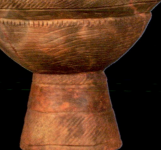

**Jomon pot**
The world's first pots were created in Japan around 14,000 BCE and soon developed into a well-made product, Jomon ware, of which this pedestaled bowl is a late example.

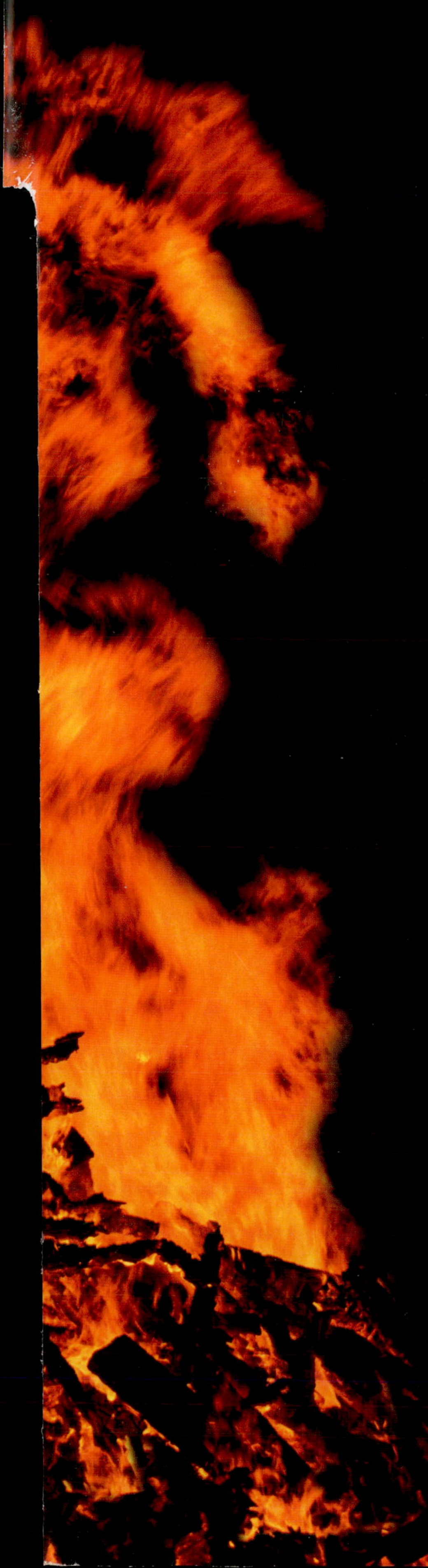

sand, to toughen it. After shaping, pots were left to dry and then were fired in a clamp (bonfire kiln). The pots were stacked on and covered with a layer of fuel, then sealed by a layer of clay. Holes in the top and around the sides allowed air to circulate so the fuel would burn. More permanent kilns appeared from *c.*6000 BCE in western Asia. A fired vessel's color depended on the type of clay and the firing conditions: circulating air produced reddish hues, while the absence of air turned the pottery black.

The pyrotechnical skills needed to control kiln temperature and air flow and to achieve high temperatures, coupled with a growing knowledge of other natural materials, were the prerequisites for the development of other transformational industries, including metallurgy and glassmaking.

## Cuisine

Fire made food easier to chew, more digestible and often more palatable, and improved human health by killing bacteria and parasites. Heat enabled foodstuffs to be preserved for later use by drying or smoking, and was used to remove or neutralize poisons present in foodstuffs. By 40,000 BCE some food was cooked in an animal skin or skin-lined pit containing water by adding heated stones. The invention of pottery made boiling easier: fire could be applied directly to the vessel of food and water. Unleavened bread was cooked on hot stones or pottery dishes over the fire. Clay ovens appeared from the sixth millennium BCE, heated by lighting a fire inside.

Through time, people experimented with different ways of cooking. Bread and cakes were leavened with yeast, which created $CO_2$, causing the dough to rise; cooking hardened the risen form.

◁ **Controlling fire**
Harnessing the power of fire was an important part of the development of civilization and has influenced every part of human life.

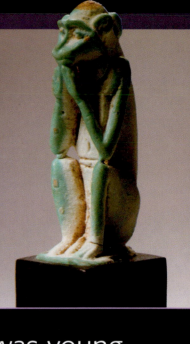

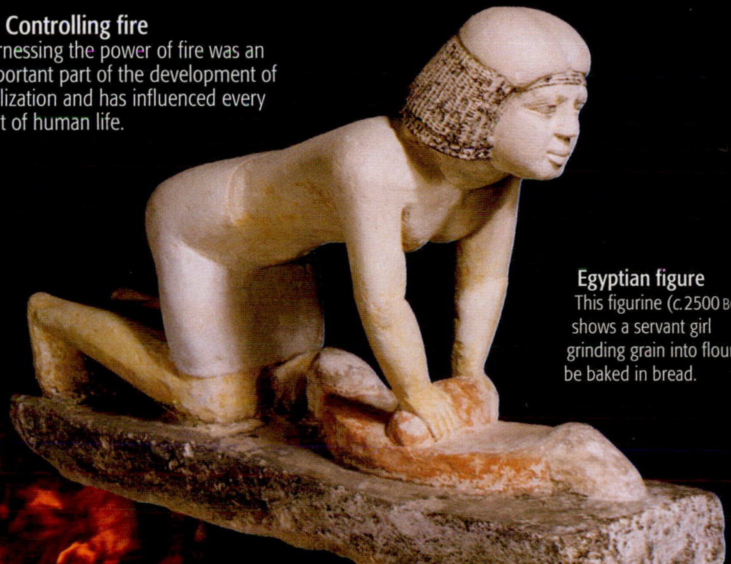

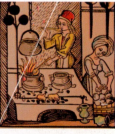

**Egyptian figure**
This figurine (*c.*2500 BCE) shows a servant girl grinding grain into flour to be baked in bread.

BREAKTHROUGH

### FAIENCE AND GLASS

Before 4000 BCE people in western Eurasia began producing faience objects, such as this Egyptian figurine (*c.*1300 BCE). This precursor to glass was made with silica, such as quartz sand, mixed with a stabilizer (lime) and an alkaline flux (often soda). The paste, colored with copper oxide, was shaped, then heated to about 1500–1800° F (850–1000° C), causing the grains inside the paste to sinter (adhere) and the object's surface to fuse as a thin glaze.

From *c.*1600 BCE, furnaces reaching higher temperatures enabled a similar mixture to be melted into a viscous liquid, true glass. During the first millennium BCE in the Levant, glassblowing was invented, revolutionizing the production of glass.

> "When the earth was young… **human beings**… did not know yet how to enlist the aid of fire."
>
> **LUCRETIUS (TITUS LUCRETIUS CARUS), ROMAN PHILOSOPHER**, *c.*100–*c.*55 BCE

## AFTER »

**Mastery of fire has enabled civilizations to transform the world, through technology and controlled destruction.**

**DIET**
Over the last 2,000 years people have developed ever more **elaborate ways of preparing food** and combining ingredients, using heat in increasingly sophisticated ways.

MEDIEVAL KITCHEN

**TRANSFORMING MATERIALS**
Heat has been used to transform many materials, including metal ores **18–19 »**. Advances in ceramics include **metallic glazes** and **kilns** capable of achieving higher and better controlled temperatures than before.

**HARNESSING ENERGY**
A broadening **range of fuels**, including coal and natural gas, have been used to create heat and power. Applications include sophisticated **ways of heating**, such as Roman hypocausts (underfloor heating). Steam power brought about Europe's **Industrial Revolution**.

**SEE ALSO »**
*pp.18–19* Early Metalworkers
*pp.132–33* Steam Power to Steam Engine

# Early Metalworkers

**Metallurgy started at different times across the globe, but by the 1st millennium BCE it was extensively practiced. The use of metals proved revolutionary: unlike stone, metal could be worked into any shape, and broken objects could easily be mended or recycled to make new things.**

**North American copper arrowheads**
Around 3000–2500 BCE many American communities made tools and ornaments by cold-hammering pieces of the native copper abundantly available around Lake Superior.

« **BEFORE**

At first metals were valued for their attractive appearance, but their other useful properties were quickly appreciated.

**TOOL MANUFACTURE**
During early times and, in some areas, until recently, **tools were shaped from stone, wood, and other nonmetallic materials**, by cutting or percussion (striking).

**METALS AS EXOTIC STONES**
When metals were first encountered, in their native (natural pure) form, they were **treated as attractive stones** that, instead of fracturing, changed shape when struck.

**PYROTECHNOLOGY**
The development of **pottery kilns** created the pyrotechnical skills (achieving high temperatures; controlling firing conditions) required for **smelting and melting metals**.

Native (naturally occurring pure) copper, gold, silver, lead, and tin were exploited first, to make small objects, including jewelry and daggers; copper was also used for tools. Copper could be worked by cold hammering but was easier to shape if heated until soft. Hammering made the metal brittle; its toughness could be restored by annealing (heating then cooling slowly), though this reduced its hardness.

## Smelting and casting
Copper was also extracted from ore using easily smelted copper oxide and carbonate ores (rock containing metal compounds mixed with other minerals). Smelting involved heating crushed ore and charcoal in a clay furnace. Air pumped in through a clay pipe using bellows helped raise the furnace temperature. Carbon displaced the metal from its oxide or carbonate and escaped as carbon dioxide ($CO_2$), leaving mineral impurities (as slag) and pure copper, which was denser and collected in the crucible base. When these ores were unobtainable, sulphide ores were used: before smelting, these required roasting in an open bonfire, to drive off the sulphur as an oxide gas.

For casting, smelted copper was melted (at 1982° F/1083° C) and poured into stone or fired clay molds or impressions in wet sand. At first, open molds were used, producing simple shapes, flat on one face. Later, two-piece molds enabled objects to be cast in the round.

## 5500–5000 BCE
**The time when copper smelting began in the Balkans and western Asia. Iron was not smelted until the 2nd millennium BCE.**

More elaborate objects could be made by lost-wax casting: a wax model was coated in clay, which was then fired, allowing the wax to run out, creating a mold. Metal was then poured into the mold, which was smashed to remove the finished object.

## Using alloys
Copper is relatively soft, so is unsuitable for heavy tools such as axes. Alloyed with certain other metals, however, it becomes harder. Early metallurgists often

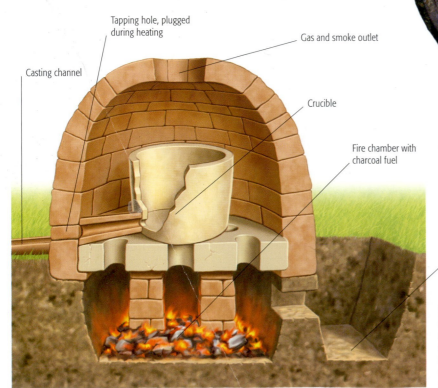

Tapping hole, plugged during heating

Casting channel

Gas and smoke outlet

Crucible

Fire chamber with charcoal fuel

Stoke-hole and fueling pit

**Bronze-casting furnace, China**
Bronze was melted in a crucible, from which it was poured into a mold. In sophisticated furnaces, such as this Chinese example (c.1000–800 BCE), the metal was tapped directly from the crucible along a channel into the mold.

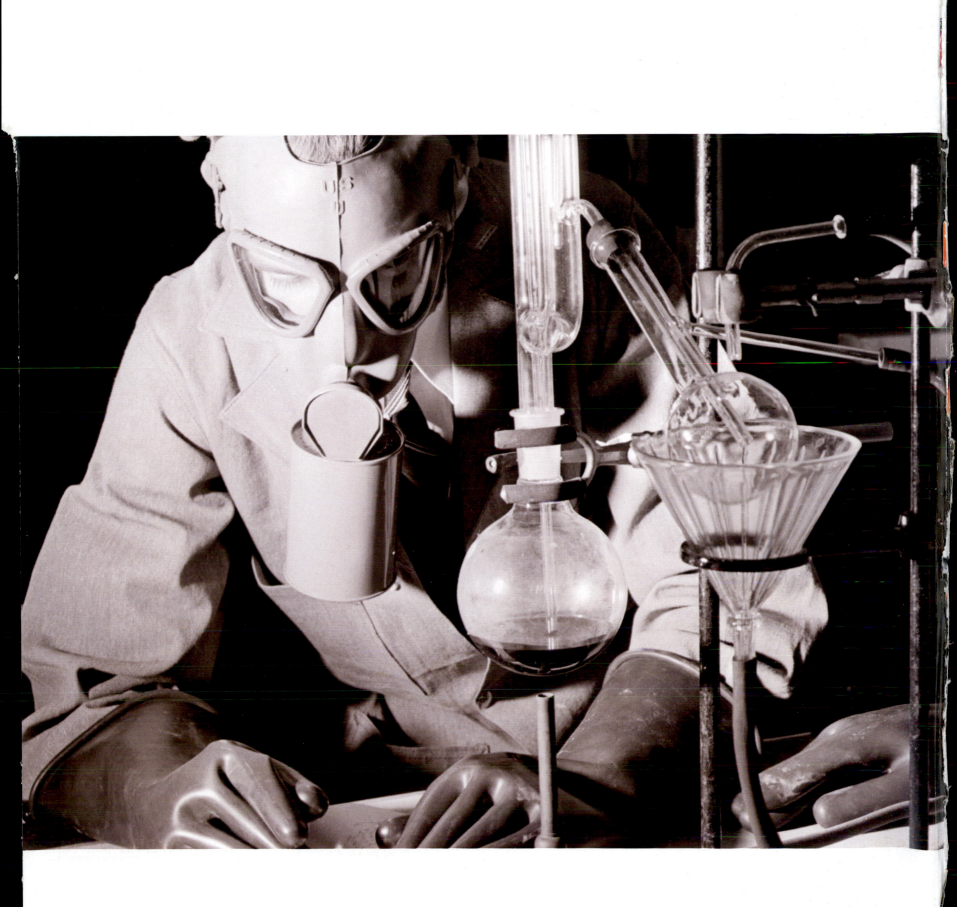

# SCIENCE

# 3
# THE INDUSTRIAL REVOLUTION
## 1700–1890

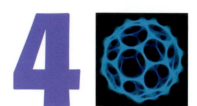

# 4

# THE ATOMIC AGE
## 1890–1970

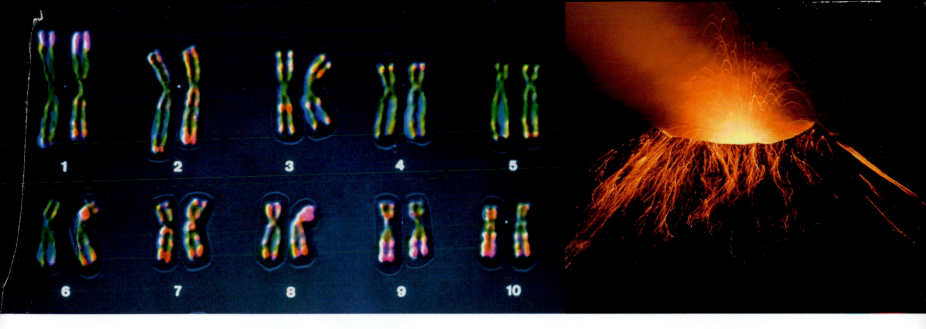

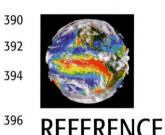

# 5
# THE INFORMATION AGE
## 1970 ONWARD    374

# REFERENCE    420

# Foreword

I have always been fascinated by science, and for the last 15 years by the history of science; so when I was asked to take part in the preparation of this book I jumped at the chance.

And here it all is; this superbly illustrated book paints a broad picture of the whole of science and its history. Arranged in chronological order, according to when a scientific principle was first laid out or when a process became technically possible, it begins with the ancient Babylonians, Chinese, and Greeks, with the idea of the four elements, and goes all the way through to string theory and space telescopes.

Science is not just a collection of answers, but an ongoing search for the truth about how the Universe works; it is not merely about the facts, but also about the struggle to discover them. One scientific idea often leads to another, and then another. This was especially true of the vacuum: theories and inventions followed one another rapidly in the mid-17th century; steam engines were a natural consequence in the 18th, cathode-ray tubes in the 19th, and today we have many more pieces of vacuum technology. The book is divided into five chapters, from the dawn of science through to the present day. Each chapter has its own timeline to help you find the various threads that make up that particular period.

Scientific ideas often occur to more than one person at a time, which has led to some disputes—over the invention of calculus, for example, or the discovery of oxygen. All these events are included. Alongside the ideas and theories in this book are the people who dreamed them up, from Pythagoras and Aristotle to Einstein and Marie Curie. There are double-page biographies of 19 major characters, and around 100 features on other great pioneers, from Eratosthenes to Richter. At the end of the book is a 54-page reference section, including brief biographies of all the major scientists, past and present, plus a plethora of scientific facts.

Because of its sheer size and complexity, this is the toughest book I have ever worked on, and it would never have been completed without a small army of writers, editors, designers, artists, and picture researchers. I thank them all, but particularly Janet Mohun and her team in the DK office.

Adam Hart-Davis

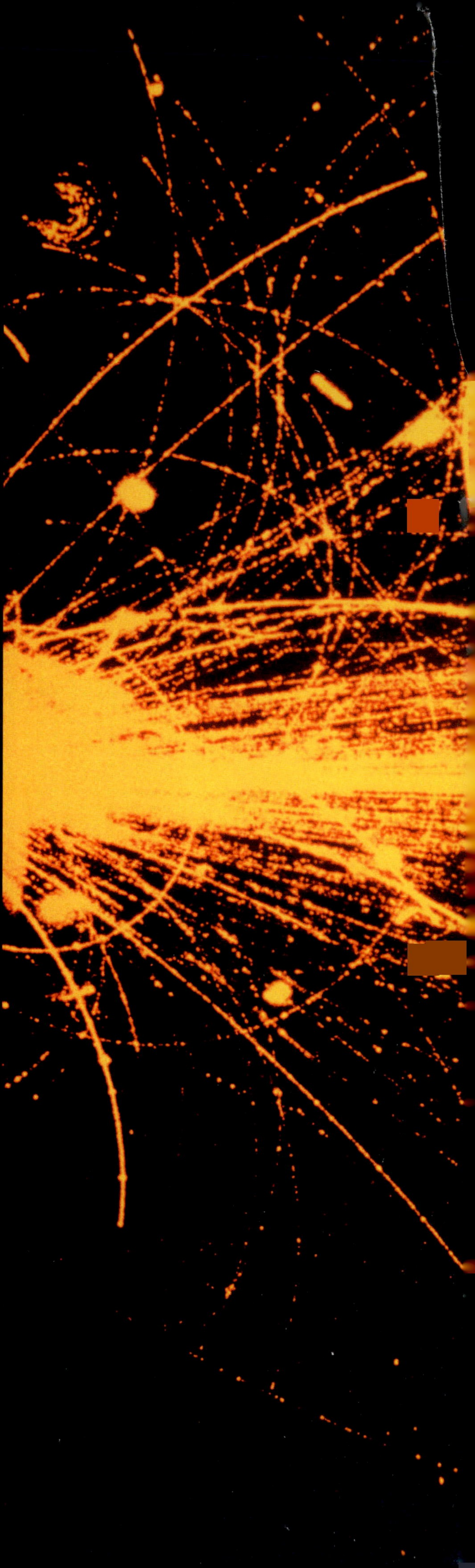

**Ancient Greek blacksmith**
On this 6th-century BCE Greek vase, a blacksmith is removing the bloom from a shaft furnace in which iron ore has been heated with charcoal.

stimulus to trade. While bronze was widely employed for tools and weapons, other alloys were also used. For example, adding lead lowered copper's melting point and increased its fluidity, producing an alloy suitable for casting complex shapes where strength was not required (such as ornaments).

preferentially exploited copper ores naturally containing arsenic, to produce a strengthened copper alloy. Generally alloyed in the proportion 5–10 percent, tin produced a superior metal, bronze, but was rare: once tin–bronze was developed (around 3000 BCE in western Asia) demand for tin became a major

### Iron

Ironworking developed late, since it presented technological challenges. Iron melts at a high temperature, around 2912° F (1535° C). In antiquity only the Chinese managed to achieve the temperature needed to produce cast (melted) iron: they built blast furnaces

Gold generally occurs as a native metal, often as electrum (gold-silver alloy, sometimes also containing copper). In South America and Turkey, golden objects were made from electrum, using gold's low reactivity: one method was depletion gilding, where acids removed baser metals from the surface. Gold's malleability makes it easy to work, using techniques such as punching (making holes for decoration), filigree (twisted threads of gold) and repoussé (creating a raised pattern), wiremaking, and hammering.

**GOLD HELMET, MESOPOTAMIA, _C._2500 BCE**

of good refractory clay capable of withstanding high temperatures, and created the necessary draft using water-driven piston bellows.

Elsewhere, only wrought iron could be produced. Iron ore and charcoal were heated in a furnace to around 2000–2100° F (1100–1150° C). At this temperature some of the impurities were given off, but others remained mixed with the iron in a spongy mass known as the bloom. Iron was laboriously extracted from the bloom by repeated hammering and heating to high temperatures. This process was also used to shape the resulting metal and to weld (join together) pieces of iron. Heated in a charcoal fire, the iron was converted into steel

(carbon-iron alloy), which is harder and tougher than wrought iron. Hardness was increased by quenching (plunging the white-hot object into water to cool it rapidly), although this also made it brittle. This was countered by tempering (reheating and cooling slowly), reducing both the brittleness and the hardness. Successful working required a balance between tempering and quenching.

Once these technologies were mastered, iron, which is far more common than copper and tin, rapidly became the main material for tools and weapons.

**Molten metal**
Metal was circulated in antiquity as ingots which could be melted. The molten metal was poured from the crucible into a mold.

**Chinese bronze vessel**
A Western Zhou (1100–771 BCE) bronze ritual vessel. The Chinese created elaborate bronzes in molds made of many interlocking pieces, which were disassembled after casting, for repeated use.

**AFTER**  »

For millennia metals have been a dominant material for manufacturing strong, durable tools and weapons and attractive ornaments.

**OTHER METALS**
Since antiquity, a wider range of metals, such as **zinc, aluminium, and tungsten**, have come into use and new alloys, such as **brass and pewter**, have been created.

**TECHNOLOGICAL ADVANCES**
From the 16th century onward the West also developed the technology to make **cast iron**. A more efficient blast furnace, using **coke as fuel**, was devised by Abraham Darby in 1709. Other technological advances included **stainless steel** and **tinned steel cans** for preserving food.

**BLAST FURNACE IN IRONBRIDGE, ENGLAND**

**NEW USES FOR METALS**
In recent centuries metals have been **put to new uses**, for example, to construct buildings, ships, airlanes, and rockets; also, in microcircuitry.

**SEE ALSO** »
_pp.178–79_ THE NATURE OF HEAT

## Development of the wheel

From their origins in ancient Mesopotamia, wheels have come to be used all around the world, and are still being developed and refined today.

**3000 BCE** Metal strips or nails are added to wheels to make **hard rims**; this protects the rims so that wheels last longer, even if it does nothing to improve the ride.

**c.1600 BCE** Spoked wheels first appear on Egyptian chariots. These wheels also seem to have developed independently in Europe some 200 years later.

**800–600 BCE** The Celts invent **pivoting front axles**, as found in burial mounds on archaeological sites. Such axles give vehicles much greater maneuverability than traditional fixed axles.

| 3500 BCE | 3000 BCE | 2500 BCE | 2000 BCE | 1500 BCE | 1000 BCE | 1400 CE |
|---|---|---|---|---|---|---|

**3500–3200 BCE** An unknown Mesopotamian innovator takes a solid potter's wheel and turns it through 90 degrees. Joining two solid wheels with an axle results in the first **transportation wheel**.

**2600 BCE** Thinner and lighter **plank wheels** (right) begin to become more popular than their solid counterparts.

**1800 BCE** The **crossbar wheel** (right) appears, although it is unclear if this is a step toward the first spoked wheel or a parallel innovation. The oldest known one was found in Italy.

**c.1400–1500 CE** Solid iron bands are used to reinforce wheel rims on carts and wagons. These bands are the first "tires."

« BEFORE

# Evolution of the Wheel

**Probably the most important mechanical invention of all time, the wheel has a long and varied history. For 5,500 years we have been thinking up ever more inventive ways of using wheels to improve our lives. From watermills to jet planes, a whole range of things that we use routinely rely in some way on the wheel.**

BEFORE

Even before the wheel had been invented, humans had devised various ways to move heavy objects around.

**BEASTS OF BURDEN**
Humans first domesticated animals between 9000 and 7000 BCE, and people began using **oxen** to draw **ploughs** around 4000 BCE. Camels, elephants, horses, llamas, yaks, and goats have all been used as beasts of burden.

**ALL-TERRAIN SLEDGES**
Known from at least 7000 BCE, **sledges** were used by **hunting** and **fishing** communities in **northern Europe**. Whether pulled by people, dogs, or deer, their long, thin runners spread the weight of a heavy load over almost any terrain, including snow or ice, sliding freely like skis.

Based on diagrams found on clay tablets, ancient Mesopotamian potters were the first people, as far as we know, to make use of wheels—for spinning clay to fashion their wares—as long ago as 3500 BCE. Wheels did not seem to catch on as a handy aid to transportation for another 300 years, when the Mesopotamians started to build chariots in 3200 BCE. Even with this intellectual barrier breached, the development of our most useful asset progressed at a very slow pace. It

would be another 1,600 years before the ancient Egyptians invented chariot wheels with spokes.

While the concept might seem simple to us today, there must, however, be something about the wheel that is conceptually difficult. Neither the Mayan, the Aztec, nor the Incan civilizations, all of which were highly developed, used the wheel. In fact, there is no evidence that the Americas ever saw a wheel before it trundled over with the Europeans.

In Europe, wheels evolved steadily through the ages until the early 19th century and the beginning of the Industrial Revolution, which saw rapid advances in their development.

▽ **Sumerian battle chariot**
The artefact known as the Standard of Ur, dating from c.2500 BCE, shows one of the oldest existing images of a chariot. This war chariot, from Sumer in Mesopotamia (now part of Iraq), is drawn by onagers (wild asses), and has solid wheels made from flat pieces of wood held together by pegs. The much lighter spoked wheels did not appear on chariots until around 1600 BCE.

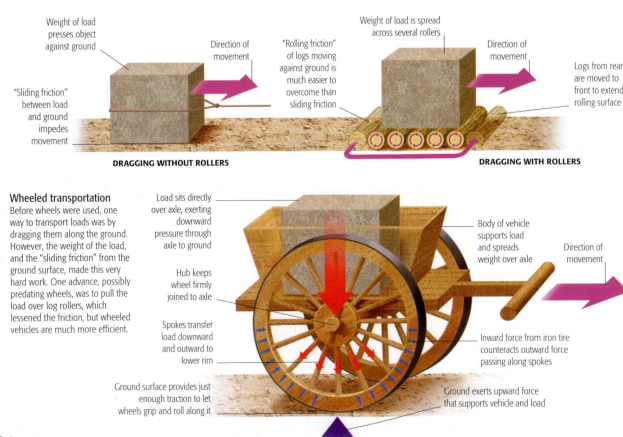

Weight of load presses object against ground

"Sliding friction" between load and ground impedes movement

Direction of movement

"Rolling friction" of logs moving against ground is much easier to overcome than sliding friction

Weight of load is spread across several rollers

Direction of movement

Logs from rear are moved to front to extend rolling surface

**DRAGGING WITHOUT ROLLERS**

**DRAGGING WITH ROLLERS**

## Wheeled transportation

Before wheels were used, one way to transport loads was by dragging them along the ground. However, the weight of the load, and the "sliding friction" from the ground surface, made this very hard work. One advance, possibly predating wheels, was to pull the load over log rollers, which lessened the friction, but wheeled vehicles are much more efficient.

Load sits directly over axle, exerting downward pressure through axle to ground

Hub keeps wheel firmly joined to axle

Spokes transfer load downward and outward to lower rim

Ground surface provides just enough traction to let wheels grip and roll along it

Body of vehicle supports load and spreads weight over axle

Direction of movement

Inward force from iron tire counteracts outward force passing along spokes

Ground exerts upward force that supports vehicle and load

**1820s** John Loudon Macadam and Thomas Telford build roads using compacted broken stone aggregate—the first **macadamized roads**.

**1846** The **pneumatic tire**, a hollow belt of inflated India rubber, is patented by Robert William Thomson. It is reinvented in 1888 by John Boyd Dunlop.

**1901** Edgar Purnell Hooley patents **tarmac**, a tar and aggregate mix spread out and rolled to make a much tougher road surface than just aggregate.

**1967** The first **alloy wheels** are made, for racing cars. Lighter than steel wheels, they improve steering and speed. Better heat conductors, they allow heat to dissipate from the brakes.

| 1800 | 1850 | 1900 | 1950 |

AFTER

**1820s** The **artillery wheel** (right), which has a metal hub, is invented. First used to move heavy steam vehicles without the spokes snapping, it is soon adopted for artillery pieces.

**1870** Improvements in metalworking enable the creation of **fine metal wheels with wire spokes** (right), invented by George Cayley in the 1850s. This allows for lightweight, nimble bicycles.

**2005** The **Tweel**, an experimental polyurethane tire-wheel hybrid, is invented by Michelin. Flexible spokes connect the hub to a thin, flexible rim and take on the shock-absorbing role of a traditional tire's sidewall.

Given the wheel's central place in everyday life, there is never likely to be a post-wheel period in human history.

**IMPROVING ROADS**
Wheels work much better when they can roll across a smooth surface. Rudimentary roads began to appear soon after the first chariots. Even today, a great deal of money is still spent on **developing even better road surfaces**.

**HARNESSING ENERGY**
Whether it is wind power, water power, or steam power that turns the blades, **wheels are at the heart of turbines**, which play a vital part in producing most of the electricity that we use around the world, in fact.

**SPREADING THE LOAD**
Narrow wheels sink into soft ground under the burden of a heavy load. To spread heavy loads, tractors have very wide wheels, while tanks and other extremely heavy armoured vehicles have **caterpillar tracks over wide wheels**.

**SEE ALSO »**
pp.40–41 SIMPLE MACHINES
pp.42–43 HOW GEARS WORK
pp.50–51 WATER AND WIND POWER

## How wheels work
There are three reasons why wheels make moving a load over the ground considerably easier than just dragging or pushing it. Most importantly, they greatly reduce friction. While the small part of a wheel that is in contact with the ground will not move, the rest of the wheel can roll on over the top of it, carrying the whole structure forward or back. Second, wheels make it much easier to change the direction of the pushing or pulling of the load. Third, wheels raise the load, so reducing the angle at which the force required to move the load must be applied. All three factors are evident when pushing a loaded wheelbarrow, for example.

## Importance of the wheel
Wheels have had a huge influence on the evolution of human society. The potter's wheel—the first known type of wheel—the watermill's turbine, and

the spinning wheel have all played a significant part in shaping history. Think also of the wheels on the chariots that armies used to subjugate enemies, or the wheels on the tractors that enabled the agricultural revolution to happen. One of the wheel's most important forms is the cogwheel. With a history dating back more than 3,000 years, to the first rudimentary gears made from wooden wheels with pegs driven into the rims, cogs have been central to the development of transportation and timepieces.

## Modern uses
Once you start to look closely at the modern world, you see that a host of things that we take for granted depend on wheels to function. Gas, diesel, and jet engines, disk drives, even the electric toothbrush—none of these everyday things could work without an internal wheel to keep them moving.

◁ **Early spinning wheel**
The first spinning wheels may have come from China, like this example from the Song dynasty (960–1279 CE). They were large, rimless, hand-cranked wheels, which were driven by a belt and turned a horizontally mounted spindle. Each revolution of the wheel produced several turns of the much smaller spindle.

« BEFORE

In ancient times even mundane happenings were seen to be part of huge cosmic events, far beyond human understanding.

To our ancient ancestors, the world was **dominated by powerful gods**, and Earth, the stars, and the sky were full of mystery. They wondered at what was going on around them, and attributed the events of the natural world to supernatural forces. The **ancient Greeks** brought organization, experimentation, and the search for **simple universal laws**. They began to realize that to understand the world one needed to know its nature (*physis*, hence the modern word "physics"), and that **natural phenomena** had logical explanations. This was a giant step from the assumptions of the old world that the supernatural determined almost everything.

The fundamental question of what the Universe is made of has occupied the minds of the greatest thinkers for thousands of years. The ancient Greeks, tireless and ingenious speculators about the Universe, turned their attention to the structure of the materials composing it. The Greek philosopher Thales of Miletus, who lived from about 625 to 546 BCE, is generally credited with being the first to speculate on the nature of the elements. He suggested that the basic principle of the Universe was water, and that all known natural substances were modifications of it. Water surrounded the land, and trickled through the soil—clearly, life

### Anaximander
A student of Thales, Anaximander thought that an invisible substance, *apeiron*, was the source of all things. He is known to have conducted the first recorded scientific experiment.

be universal and free from any specific characteristics. Anaximander identified an invisible, all-purpose plasma, which he called *apeiron*, from the Greek for "infinite." This mysterious substance could morph into all the materials on Earth.

Anaximander's own student, Anaximenes, rejected *apeiron*. He believed that everything had to be

# Elements of Life

The Universe is made from a relatively small number of naturally occurring elements. Our understanding of elements stemmed from what the ancient Greeks believed, and for over 2,000 years it was thought that everything was made from just four elements: water, fire, air, and earth.

could not exist without it. He described Earth itself as a flat slab, floating on top of an infinite mass of water.

### Fundamental substances
Thales's view of a water-based universe attracted fierce opposition. Anaximander, a pupil of Thales, rejected this theory and scoffed at the idea that Earth was like a floating log resting on a vast sea. He believed that Earth was curved, and that it dangled unsupported in space. Water could not be the main element because it was not versatile enough. According to Anaximander, if a particular element in nature, such as water, was the origin, a substance with an opposite nature, such as fire, could not emerge or co-exist. The origin must

composed of something, and proposed that the missing "something" was air. Air could be experienced, and turned into smoke and fire; when condensed, it became mist and water. The debate spread to Ephesus, which was home to Heraclites, who had a different theory. For him, fire was the key element, since it was dynamic and caused change in other materials. Xenophanes, from neighboring Colophon, favored earth. Although Earth changes slowly over time and its shape may shift, its fundamental essence stays the same.

### The Milesian School
Thales, Anaximander, and Anaximenes had much in common. They were all from Miletus and, despite their fierce disagreements, shared an intellectual approach. This later became known as the Milesian School. All three sought

to understand the true nature of reality, and to explain phenomena that most humans deemed terrifying, such as thunder, lightning, and earthquakes. The traditional view attributed such events to whims of the gods, but the Milesian School proposed natural explanations: lightning and thunder resulted from wind; rainbows were the result of the Sun's rays falling on clouds; and earthquakes were caused by the cracking of the ground when it dried out after being moistened by rain.

### Not one element, but four
The Greek philosopher Empedocles adhered to the teachings of Pythagoras (see pp.32–33). He, too, struggled with the problem of a fundamental substance, but concluded that the Universe was made of several elements: the water of Thales, the fire of Heraclites, the air of Anaximenes, and the earth of Xenophanes. These essential four accounted for all matter on Earth. Empedocles also proposed that two active principles united and divided the elements: love and strife. Love was the uniting principle that kept the elements together in different substances, and strife was the principle that divided them. No one believed that an empty space could exist, so love and strife were also counted as elements, filling the void between the other four. Empedocles called these substances *rhizomata*, the "roots" of matter. It was only after Aristotle, in the 4th century BCE, that they became known as elements.

### Thales of Miletus
The earliest of the natural philosophers of ancient Greece, Thales was a mathematician and astronomer. His assertion that the world started from water was the first to consider the elements of the Universe.

---

## EMPEDOCLES

Empedocles was a philosopher, physiologist, and religious teacher. A citizen of Acragas in Sicily, he was given to wearing a gold girdle, a laurel wreath, and bronze sandals. He was a famous healer and brilliant orator; Aristotle described him as the inventor of rhetoric. Empedocles is best known for his belief that all matter was composed of four

elements—fire, air, water, and earth—that were mixed and parted by the personified cosmic forces of love and strife. He believed in reincarnation, and insisted that he was a divine being. According to legend, he perished when he threw himself into the flames of Mount Etna, in an attempt to prove that he would return as a god.

"For from these [elements] **come all things** that are, or have been, or shall be…"

▽ **The four elements**
This illustration from a 1472 edition of Lucretius's *De Rerum Natura* ("On the Nature of Things") shows the four elements of Empedocles: air, fire, earth, and water. He explained the nature of the Universe by the interaction of love and strife on the elements.

▷ **Anaximenes**
The Greek philosopher Anaximenes believed that air was the most basic element of the Universe. The last of the important philosophers of Miletus, Anaximenes helped the transition from a mythological explanation of the world to a scientific one.

**Creatura**

**IGNIS Acutus. Tenuis. ac Mobilis**

**AQVA Crassa. Mobi= . et Obtusa.**

**AER Mobilis. Acu= tus. Crassus**

**TERRA Crassa. Ob= tusa. et Immobilis**

> "To the **elements** it came from Everything will return. Our **bodies** to **earth**, Our **blood** to **water…**"
>
> MATTHEW ARNOLD, "EMPEDOCLES ON ETNA," 1852

**Classical elements and their properties**
This diagram shows two squares, one inside the other. The corners of the larger square represent the classical elements, and the corners of the smaller square their properties.

FIRE

Hot — Dry

AIR — EARTH

Wet — Cold

WATER

**AFTER ≫**

The ancient Greek notion of four basic elements dominated Western scientific thought for more than 2,000 years.

**ARISTOTLE'S FIFTH ELEMENT**
Aristotle **36–37 ≫** reasoned that the heavens were composed of a **fifth element,** called **ether** (from a Greek word meaning "to glow"), which was **perfect, eternal, and incorruptible**. The remaining four were **confined to Earth**, and given the qualities of hot, cold, wet, or dry.

**THE WORKINGS OF THE UNIVERSE**

**ELEMENTS IN MODERN FORM**
The ancient Arab alchemists believed all metals were composed of two elements, **sulphur and mercury**. The Persians added **salt**. The 16th-century Swiss alchemist **Paracelsus 70–71 ≫** combined the four Greek and three Arab elements. In 1661 the Irish chemist **Robert Boyle 98–99 ≫** coined the **first modern definition of an element**: a substance that can not be broken down into simpler substances.

**SEE ALSO ≫**
*pp.232–33* THE PERIODIC TABLE

# Early Medicine and Surgery

The origins of medicine lie with the origins of civilization itself. People in early civilizations across the world attempted to explain the reasons for disease and create treatments. China has a strong tradition in using herbs in medicine and many of the earliest ideas still resonate today.

## ≪ BEFORE

**Prehistoric medicine, that is to say medicine before the written word, was probably characterized by a combination of primitive first aid and a belief in supernatural spirits.**

### TRIBAL MEDICINE

Clues about the **earliest medical practices** come from anthropological observations of indigenous peoples alive today and indicate that prehistoric societies probably used **herbs to treat simple ailments**. These would have been developed by a tribal shaman or medicine man.

### FIRE AND METALWORKING

Controlled fire that would have been used for sterilization purposes and cooking food may also have been used by early man for closing wounds. In addition, **mineral and metalworking tools ≪ 18–19** would have been used in crude surgical procedures. Some blades were made out of obsidian, a glass found in volcanic rock, and were remarkably sharp. The earliest surgical procedures included **skull trepanning**, which consisted of holes being bored into a person's skull in the hope of relieving headaches and epilepsy. Healed skulls (see right) indicate that some individuals even survived this practice.

**SURGICAL BLADE**

edical customs used to be handed down by word of mouth, but organized medicine began properly with the written word. The earliest known medical texts date from around 2000 BCE in China and Egypt.

## Origins of medicine and surgery

Medicine had almost certainly been established in China by the middle of the 3rd millennium BCE, when the Yellow Emperor purportedly composed the *Neijnh Suwen* or *Basic Questions of Internal Medicine*. This document (much expanded 3,000 years later) formed the basis of traditional Chinese medicine, which became mainstream throughout much of Asia. Practitioners diagnosed and treated disorders based on the interplay between humans and their environment, using techniques such as meditation and acupuncture.

In the same millennium, in *c.*2600 BCE in ancient Egypt, Imhotep (a great polymath and architect of the pyramids) was revered as a god of medicine and healing. The *Edwin Smith Papyrus*, from *c.*1700 BCE is arguably based on Imhotep's texts. The world's oldest surgical document, it is remarkably lacking in magical thinking with reference to diagnosis, treatment, and prognosis of disease.

Two centuries later in *c.*1500 BCE, a Babylonian text, the *Diagnostic Handbook* refers, perhaps, to the earliest

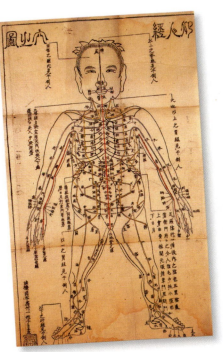

**Earliest acupuncture**
The first acupuncture needles date from 3000 BCE in China. Charts such as this identified points of the body where needles would bring maximum benefit.

physician—it suggests that a figure called Urlugaledin practiced primitive surgery way back in 4000 BCE.

The roots of the Ayurvedic (meaning "Life of Science") system of medicine are found in the Indus Valley even further back, possibly as early as 9000 BCE. This forms the basis of traditional Indian medicine today. It advocated a healthy lifestyle that prescribed herbalism, massage, and yoga. Its written records, such as the *Sushruta-samhita* text on surgery, appeared later, from *c.*500 BCE. This particular text refers to invasive practices that took place, including plastic surgery, cataract surgery, and even cesarean sections.

## Ancient Greek medicine

In 700 BCE the first Greek medical school opened at Cnidus. Ancient Greek medicine, like that in Egypt and India, placed emphasis on control of diet, lifestyle, and hygiene.

It was 300 years later that Hippocrates established his own medical school on Kos. He described many diseases for the first time and introduced lasting

medical terms, such as "acute" for an illness that is sharp and brief, and "chronic" for one that builds up slowly over time.

The Hippocratic School rejected supernatural causes in favor of seeking the physical causes of disease. The school emphasized care and prognosis, and encouraged thorough case studies, making it the forerunner of clinical medicine.

Hippocrates also championed the idea of

**Skull trepanning**
Holes were bored into patients' skulls in one of the earliest known surgical treatments dating back to 40,000 BCE.

## BREAKTHROUGH

### HIPPOCRATIC OATH

It is tradition that all physicians take the Hippocratic Oath, a guiding set of duties, which formed part of the *Hippocratic Corpus*, the texts from the Hippocratic School of ancient Greece. It is generally thought that Hippocrates himself wrote the oath, but it is likely that many contributed to the text. In the original version (see right), the oath swearer includes a debt of gratitude to their medical teacher, a promise that they will live a "pure" life and that they will preserve the confidentiality of the people in their care. Today the oath has been updated and modified in certain countries, for example, with the omission of clauses that forbid pregnancy termination.

AVICENA · ΥPOCRATES

## AVICENNA

Avicenna (also known as Ibn Seena) was a prolific Persian scholar who made important contributions in medicine, chemistry, astronomy, mathematics, psychology, and geology. His major work *The Canon of Medicine* became a standard text in European universities. He pioneered many medical practices, including quarantine to control the spread of disease (see pp.242–43), and clinical trials in the experimental use of drugs.

**Medical pioneers**
These three medical pioneers (who could never have met) were masters of their craft: Galen of ancient Rome, Avicenna from Persia, and Hippocrates of ancient Greece.

**AFTER »**

After the Middle Ages, the theory and practice of medicine advanced rapidly, but not always to the real benefit of the patient.

**BLOODLETTING**
This practice involved **withdrawing significant amounts of blood** for therapeutic reasons. William Harvey's circulatory theory discredited the practice in the 17th century. Today blood withdrawal is done chiefly for blood analysis or transfusion.

**BLOODLETTING**

**REFINEMENT OF SURGICAL TOOLS**
As **surgical techniques were refined**, tools evolved accordingly. During the Renaissance, saws for amputation were developed, but it was not until the discovery of stainless steel in the early 20th century that noncorrosive surgical tools were available for the first time.

**SEE ALSO »**
pp.70–71 RENAISSANCE MEDICINE AND SURGERY
pp.228–29 SAFER SURGERY
pp.312–13 THE DEVELOPMENT OF MEDICINES
pp.408–09 DISEASE CHALLENGES

humorism, which stated that the body contained four basic humors (fluids): black bile, phlegm, yellow bile, and blood. Moods and illnesses were attributed to imbalances of the humors. Hippocrates' student, the Greek physician Galen (129–c.216 CE), reinforced this idea with the belief that blood was continually made in the body and could stagnate. This encouraged the dubious practice of bloodletting, which involved the withdrawal of a large volume of blood in an effort to redress any imbalance in the humors. It wasn't until 1543 that a

**Egyptian surgical instruments**
Evidence suggests that bronze and copper instruments such as these were used in surgical procedures in ancient Egypt, such as drainage of sites of inflammation.

Flemish anatomist, Andreas Vesalius (see pp.72–73), disproved many of Galen's theories.

## The birth of scientific medicine
In the 5th century CE, as Greek scholars fled Byzantine persecution and settled in Persia, the Academy of Gundishapur emerged as a center for medical study. It later became the first teaching hospital.

The golden age of Islam (c.700 CE to 1200 CE) saw the very first pharmacies and free public hospitals in Baghdad. Medieval Islam also produced medical treatises from scholars such as Avicenna (see right). During this era, Islamic thinkers, including Avicenna, started to introduce experimental methods into their study of medicine, which had a lasting impact on the discipline.

# The First Astronomers

**To our ancestors, the sky and Earth, the humans and gods, and the animals and plants, were all parts of one, interacting environment. The Sun, Moon, and stars provided heat and light, allowing living creatures to thrive. Their regular patterns of movement and change meant they could be used as compasses, clocks, and calendars.**

The perceived "usefulness" of the stars and planets varied from place to place within the ancient world. People living by large rivers, such as the Euphrates and the Nile in the

## BEFORE

In the prehistorical period the night sky was thought to shed meaning and significance on human endeavor. Stories were weaved around the recognized constellations.

**EVIDENCE FROM CAVE PAINTINGS**
The fixed **constellations** were often represented on cave walls. The Pleiades feature in the prehistoric **cave drawings at Lascaux**, France. People noticed that certain **star groups** (such as Sirius and Orion's Belt) always **rose and set at the same places** on the horizon, and always occurred at the **same season** of the year. This would lead to the **first calendars** and celebrations of the seasons.

Middle East, were more influenced by rain and river levels than by the seasons. Around coastlines, however, temperature and daylight variations had a greater effect on human, animal, and plant life. The critical hunting and gathering seasons were ruled by a solar calendar, since the Sun's movement was recognized as causing the spring, summer, fall, and winter cycle. Its overwhelming importance led people to believe that it needed to be observed and worshipped.

### The importance of the Moon
In Mesolithic times, around 10,000 years ago, many people lived close to the water. In coastal areas, people noticed that tides were particularly high or low at certain times of the month and were connected with the lunar phases. These tides are now known as spring and neap tides (see pp.108–109) and would have affected fishing, shell collecting, and even opportunities for transportation.

The full Moon, acting as a "night light," would have affected how people lived, just as their daily tasks would have been governed by the number of hours of daylight. Their days were delineated by sunrise, noon, and sunset; and their activities determined by both a solar and a lunar calendar.

### Finding patterns
A long time ago people observing the phases of the Moon would have noticed a regular pattern occurred every 15, 30, 44, 59, and 74 days. The interval between successive full moons —29.5 days—became well known, establishing the concept that would later become known as a "month."

Importantly, the positions of heavenly bodies are predictable; both the Sun and the Moon rise in the east at regular times. Unlike both the vagaries of the weather and human relationships,

**Astronomical calendar**
Some of the Callanish standing stones in Scotland (c.1800 BCE) seem to have been set up to record the Metonic Cycle of 18.61 years (see p.29).

the sky is dependable. Observers would soon realize the time between one midsummer and another was 365 days.

### Using the Sun as a clock
At midsummer the Sun reaches its maximum noontime elevation outside the tropics. At midwinter, 183 days later, the Sun is at its lowest in the noon sky. The idea of using its movement as a clock began with observations of how the

shadows cast by objects such as a vertical stick or an obelisk—a tall thin stone—point toward the western horizon at sunrise and due north at noon. Every hour the shadow moves further around, acting as a sundial. Similarly, most stars travel across the sky. An imagined line from northern stars to the celestial pole acts like the hands of a gigantic clock. Early astronomical advances took place in Babylon and Greece in the 700 years BCE, spurred on by planetary »

**Egyptian calendar**
The world's oldest known astronomical representation lies in the Tomb of Senenmut, in Deir el-Bahri, Egypt. It dates back to 1500 BCE and depicts the decanal stars, which marked the ten-day Egyptian week.

### Phases of the Moon
Every month (every 29.5 days) the Moon goes through a complete set of phases from a full Moon to a new Moon (when it is dark). The shape appears to change as different parts of the illuminated Moon face Earth.

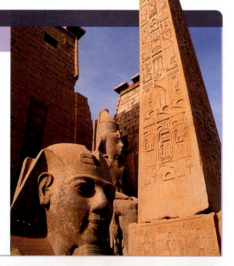

**BREAKTHROUGH**
## MARKING TIME

A vertical obelisk, like this 3,300-year-old granite monolith at Luxor in Egypt, was used as a clock. The movement of its shadow tracks the hours of the day, and the length of its shadow marks the progress of the year. At noon the shadow is cast due north, allowing the obelisk to act as a compass as well as a clock. The four sides of the obelisk's tapered stone shaft are embellished with hieroglyphs, typically praising Ra, the Sun god. The top tenth of the shaft was covered in gold to reflect the Sun's rays.

| WAXING CRESCENT | FIRST QUARTER | WAXING GIBBOUS | FULL MOON | WANING GIBBOUS | THIRD QUARTER | WANING CRESCENT |

**Pole star**
A photograph exposed all night shows the stars spinning around the pole of the sky. The elevation of the center of the circle above the horizon is equal to the latitude of the observing site.

ASTRONOMER (190–120 BCE)

## HIPPARCHUS

Hipparchus was a Greek astronomer who accurately mapped the positions of around 1,000 stars and divided them into six categories of importance. He measured the year length to an accuracy of 6.5 minutes and, by comparing his observations with those made centuries before, realized that the direction of the spin axis of the earth was changing. This caused the position of the Sun at the time of the equinox to move. He knew the seasons (intervals between solstices and equinoxes) were not of equal length and so calculated that the earth was not at the center of the Sun's orbit.

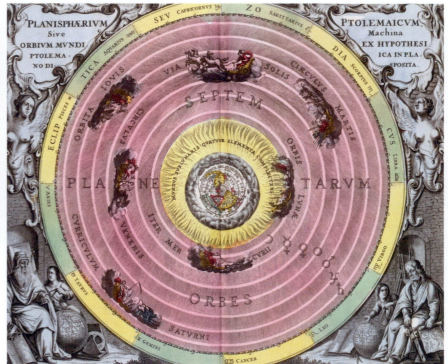

astrology. It was believed that the positions of the planets, the Sun, Moon, and stars, and the times of their risings and settings, influenced life on Earth and foretold the future.

Accurate observations of planetary positions over long periods of time were recorded on Babylonian clay tablets and used to predict the planets' future movement. In Greece, Hipparchus (see above) wanted to improve their accuracy, so he measured the position and brilliance of around 1,000 major stars. This led to the theory of spherical trigonometry, and the acceptance of the "degree" as a unit of angle.

By the 2nd century CE the Greek astronomer Ptolemy had divided the Mediterranean sky into 48 constellations in his famous treatise, *Almagest*. He created drawings of such positional accuracy that now—some 2,000 years later—they can still be used to detect the movement of stars that were previously thought to be fixed. Angles that had previously been estimated using the extended hand were now carefully measured with an adjustable cross staff or celestial compass.

### Movements of the planets

The future movements of the planets were predicted using complicated earth-centered models in which the planets moved at constant velocities around a series of theoretical circles. In the 6th century BCE the Greek astronomer Pythagoras

**Ptolemy's cosmos**
Ptolemy's vision of the universe with the known planets on spinning crystal spheres centered on Earth. Their order was determined according to their speed, relative to the fixed stars on the outermost sphere. This view of the universe was upheld for 1,400 years.

believed that the Earth was spherical. In the 3rd century BCE Aristarchus asserted that the earth revolves around the Sun, but this theory was not widely accepted at the time. He also used careful eclipse timings to calculate that the Moon was about 60 Earth radii away (within 1 percent of today's value). Around the same time, Eratosthenes accurately measured the earth's radius and the tilt of the spin axis (see pp.38–39).

### Marking time

Time was measured during the day using the Sun's shadow, and sundials became commonplace. During the night people relied on the use of stellar positions, and primitive water, sand glass, and candle clocks were used to keep

**Chinese constellations**
This North Polar sky map was found in Dunhuang, China, and is thought to date from *c*.705 CE. The constellation of Ursa Major can clearly be seen at the bottom.

check. This was important because many prayers had to be made at specific times of the day. Religious observance also meant that the Moon and the Sun were closely observed. Once the major religions were established many societies used both a lunar and a solar calendar to accurately set the dates of festivals. Even today, the Christian festival Easter is set on the first Sunday after the first full Moon after the spring equinox.

### Measuring the year

The length of the tropical year—the time interval between successive equinoxes—was estimated using the daily lengths of noontime shadows. The year length could also be found by

**Farnese Atlas**
This 2nd-century CE sculpture of the kneeling god Atlas includes the oldest known representation of the celestial globe. The sphere depicts 41 of the 48 classical Greek constellations described by Ptolemy.

" Astronomy **compels the soul** to **look up upward** and leads us from **this world** to another."

PLATO, GREEK PHILOSOPHER, *c*.400 BCE

**Astrolabe**
This astrolabe, used to measure time and position, was produced by Abu Ishaq Ibrahim al-Zarqali in *c.*1015 CE. If the latitude was known, the altitudes of the Sun or certain stars could be set to give the time. If the time was known, the Sun and star positions could be set to establish the latitude.

**Mater**
The base plate is called a mater. Its outer rim is marked with a degree scale.

**Tympan**
A rotating plate, or tympan, sits on top of the mater. The celestial sphere for one particular latitude is mapped onto this plate.

**Rete**
The rete is a cut-out plate that sits on top of the latitude tympan and rotates over it. This enables the user to line up the star pointers (on the rete) with the night sky for the particular latitude shown on the tympan beneath it.

**Ecliptic ring**
The ecliptic ring is part of the rete. It shows the annual path of the Sun across the sky.

**Ulugh Beg's mural sextant at Samarkand**
This gigantic 15th-century astronomical device has a 118 ft- (36 m) long meridian arc, which runs along the north–south plane. Stairs were carved out either side, to give Beg and his assistants access for measuring the altitude of celestial bodies as they crossed the meridian.

## AFTER »

Astronomical measurements of planetary years and orbit sizes spurred on physicists in their investigations of forces, leading to the discovery of gravitation.

### HELIOCENTRIC VIEW

The Polish astronomer **Nicolaus Copernicus** (1473–1543) started modern astronomy by proposing a **heliocentric system 74–75 »**, where the planets revolved around the Sun.

### INCAN EQUINOXES

The mysterious **Intihuatana stone** at Machu Picchu, dating from the 1400s, might have been a seat for a **priest-astronomer**, who observed the rising or setting of the Sun in certain valleys, with the timing of such events shaping the planting season and the harvest.

**INCAN INTIHUATANA**

### THE TELESCOPE

In 1609 **Galileo** turned a **telescope** to the sky, and began to **study** the **planets** and the **Moon** in greater detail. Astonishing new discoveries followed at an amazing rate.

**SEE ALSO »**
*pp.74–75* The Sun-centered Universe
*pp.76–79* Planetary Motion

measuring the time taken for the Earth to orbit the Sun using fixed stars to measure the movement. The dawn rising of stars such as Sirius or star groups such as Pleiades were used to do this. A breakthrough came when the Greek astronomer Meton (born *c.*460 BCE) discovered that 19 solar years was nearly equivalent in length to 235 lunar months. This became known as the Metonic Cycle, and it was used for setting the dates of religious festivals, and for predicting the times of solar and lunar eclipses.

The Julian calendar was introduced by the Roman Emperor Julius Caesar in 45 BCE. The year was taken to be 365.25 days long, and a leap year was added every four years. The inaccuracy was only one day in every 128 years, and this was corrected by Pope Gregory XIII in 1582 CE. The lunar calendar—with its month of 29.5 days—does not fit well with the solar calendar—whose year has 365.25 days. Chinese knowledge of astronomy had developed since the 12th century BCE but along independent lines. Chinese maps of the constellations (left) show a detailed knowledge of the night sky.

## The astrolabe

The latitudes of places on Earth could be measured using the elevation of the Sun. This was important in Islam for determining the direction of Mecca, and led to the development of the astrolabe (above).

# Ancient Number Systems

**From the time of the earliest civilizations various signs and symbols have been used to represent numbers. As recording and calculating became ever more complex, several numerical systems developed, eventually leading to the numerals and decimal number system we use today.**

The first way of recording numbers was probably making crude tally marks on stones or sticks, or, simply, piles of pebbles. Various numerical systems slowly evolved in the different cultures of the ancient world. Some proved inadequate for complex calculation, but others formed the basis of the system we are familiar with today.

One of the first ideas to emerge was the concept of counting in groups—what modern mathematicians call systems to a base. For example, eggs are still often counted in dozens, which can be seen as counting to

**Ancient Egyptian scribe**
In ancient Egypt scribes had detailed knowledge of writing and mathematics using hieroglyphs and, later, hieratics.

base 12. However, for most purposes, calculations now are usually done in a decimal system (that is, to base 10).

### Cuneiform and hieroglyphs

The decimal system was not universal in ancient cultures. The Babylonians, in around 3000 BCE, worked to base 60, and the legacy of this system can still be seen in the way we measure time in hours and minutes, and angles in degrees. The Babylonians inscribed numbers on clay tablets with the end of a pointed stick, which left behind a wedge-shaped (cuneiform) mark. These cuneiform symbols were the first true numerals.

In central America, between the 3rd and 9th century CE, the Mayans developed a sophisticated system based on 5 and 20, with symbols that included zero—a concept that did not appear in other cultures until many centuries later. The ancient Egyptians, working to base 10, assigned separate hieroglyphs for the numbers one, ten, a hundred, a thousand, ten thousand and a million. Over time these were simplified into simple brushstrokes known as hieratics. The ancient Chinese also worked in a decimal system, with characters representing the numbers one to ten, and multiples of ten, a hundred, and so on. As well as this system of writing numerals, they used a set of small sticks called counting rods arranged on a board that was divided into rows and columns for performing calculations.

« BEFORE

**The most ancient methods of representing numbers probably involved counting out small objects, such as stones, making marks—tally marks—on sticks or clay, or using counting rods.**

#### TALLY MARKS
A simple **way of recording numbers**, tally marks were used, for instance, in counting the number of days that elapse or the number of animals in a herd. These were sometimes **scratched** on stones or sticks or impressed on clay tablets (a precursor of the **Babylonian cuneiform numerals**).

**ISHANGO TALLY BONE, AFRICA**

#### COUNTING RODS
The system of using an object, such as a pebble, bone, or stick **to represent each unit** evolved into counting rods, which were used in ancient China. In the **counting rod system**, digits are represented by the number of rods. This system also enabled **fractions** and positive and **negative numbers** to be represented.

« SEE ALSO
*pp.26–29* THE FIRST ASTRONOMERS

**Algorist and abacist**
This woodprint is an allegorical representation of the fierce competition in medieval Europe between rival groups of mathematicians. Abacists (right) used a traditional abacus for calculation, while algorists (left) did their calculations on paper using the new numerals in a positional system.

## NUMBER SYMBOLS

Each of the ancient civilizations developed its own system of numeric notation. Some notation systems, such as the Mayan, had little or no influence on the evolution of modern numerals, and others, as in the case of Roman numerals, even hindered progress toward a universal positional notation system.

| Babylonian | Ancient Egyptian | Ancient Greek | Ancient Roman | Ancient Chinese | Mayan | Modern Hindu–Arabic |
|---|---|---|---|---|---|---|
| 𒁹 | \| | α | I | 一 | • | 1 |
| 𒈫 | \|\| | β | II | 二 | •• | 2 |
| 𒐈 | \|\|\| | γ | III | 三 | ••• | 3 |
| 𒐉 | \|\|\|\| | δ | IV | 四 | •••• | 4 |
| 𒐊 | \|\|\|\|\| | ε | V | 五 | — | 5 |
| 𒐋 | \|\|\|\|\|\| | ϛ | VI | 六 | •— | 6 |
| 𒐌 | \|\|\|\|\|\|\| | ζ | VII | 七 | ••— | 7 |
| 𒐍 | \|\|\|\|\|\|\|\| | η | VIII | 八 | •••— | 8 |
| 𒐎 | \|\|\|\|\|\|\|\|\| | θ | IX | 九 | ••••— | 9 |
| 𒌋 | ∩ | ι | X | 十 | ═ | 10 |

## Ancient Greek and Roman numerals

With the rise of Greek civilization and its interest in mathematics came the practice of using letters of the alphabet as numerals. The alphabet was also used by the Romans. In their system I represented one, and was simply repeated for two (II) and three (III); other letters were used for five (V), ten (X), fifty (L), one hundred (C), five hundred (D) and one thousand (M). Numbers in between were expressed by repeating these symbols, as in XXX for thirty, and simply adding progressively smaller numerals, for example CCLXVII for 267. If a numeral preceded a larger one, however, this indicated that it should be subtracted from it, so that IV was four (five minus one).

## Hindu–Arabic numeral system

The Roman system was difficult to use for calculation but still persisted in Europe until medieval times and the arrival of the Hindu–Arabic numerals

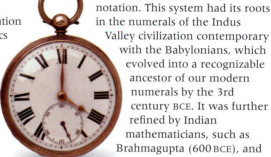

**Roman numerals**
Even today Roman numerals are still used in some contexts, such as copyright dates and on the dials of some clocks and watches.

and a decimal system in positional notation. This system had its roots in the numerals of the Indus Valley civilization contemporary with the Babylonians, which evolved into a recognizable ancestor of our modern numerals by the 3rd century BCE. It was further refined by Indian mathematicians, such as Brahmagupta (600 BCE), and spread to Persia and the Middle East, where it was adopted by Islamic scholars. It was at around this time, in the 9th

century, that a symbol for zero was adopted, rather than a gap left as a placeholder, making the positional notation system complete.

Translations of Islamic texts in the 12th century brought the Hindu–Arabic system (often referred to simply as "Arabic numbers") to Europe, where it gradually replaced Roman numerals. The ease with which the Hindu–Arabic system enabled calculations to be done, and the fact that it allowed numbers to be written down unambiguously, ensured its status as the universal mathematical language that has been used, with minor additions and modifications, to the present day.

**SEE ALSO »**
pp.48–49 ALGEBRA

## Mayan calendar

The Mayans used dots and bars to represent numbers, as can be seen on this calendar from the early-13th century Dresden Codex. The Mayan number system was based on the numbers 20 and 5 and included a symbol for zero—commonly represented by a shell—which enabled them to write very large numbers.

## BREAKTHROUGH

### ZERO

The adoption of a symbol to represent zero was a major turning point in the positional notation system. Previously in calculations a gap was left as a spaceholder in the appropriate column, but this led to ambiguity: out of context, it was practically impossible to distinguish between numbers such as 10, 20, 30, and 400 and simple 1, 2, 3, and 4. The gap representing zero was at first replaced with a small dot by Indian mathematicians, but this then evolved into the "0" symbol of Hindu–Arabic numerals. This symbol is still used today, although since the advent of computers, the symbol Ø is sometimes used for zero to distinguish it from the capital letter "O."

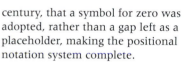

## AFTER »

The replacing of Roman numerals with Hindu–Arabic numbers in Europe made calculation simpler and helped accelerate progress in mathematics.

### MATHEMATICAL SYMBOLS

From medieval times onward, various symbols were introduced as abbreviations for **verbal instructions** in mathematical problems. The first to appear were symbols for operations such as **+ (plus)** and **- (minus)**, and the **= (equals) signs**, which were followed by conventions such as the superscript 2 for squares and √ for roots. Letters of the Greek alphabet were also used as symbols, for example π for pi, and later

typographical symbols such as ! (factorial), ∞ (infinity), and ≈ (approximation) became part of a universally understood shorthand.

### EXPONENTIAL NOTATION

One drawback of the positional notation system is that very large and very small numbers are difficult to read, especially when they include a long series of zeroes. This problem is overcome by the use of exponential notation, in which **large numbers** are **expressed as $a \times 10^b$** (a times ten to the power of b); for example, 100 is $1 \times 10^2$.

MATHEMATICIAN AND PHILOSOPHER **Born** *c.*569 BCE **Died** *c.*495 BCE

# Pythagoras

## "Number is the **ruler** of **forms** and ideas."

PYTHAGORAS, FROM "THE LIFE OF PYTHAGORAS" BY IAMBLICHUS OF CHALCIS, *c.*300 CE

Pythagoras, the most famous mathematician of the ancient world, is remembered now for Pythagoras's Theorem. The Babylonians and others were familiar with its principles about a thousand years earlier, but Pythagoras proved it to be true; indeed, he introduced the concept of proof, now fundamental to mathematics.

### Unclear beginnings

Pythagoras was such a towering figure that all sorts of stories were made up about him, and there is no way of proving which of them are true. He is known to have been born on the Greek island of Samos some time around 569 BCE. He may have traveled to Miletos; he may have been taught by Thales. It seems likely that he visited Egypt and Babylon, where priests may have taught the geometrical principles that underlie his theorem.

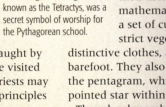

**The Tetractys**
This triangular arrangement, known as the Tetractys, was a secret symbol of worship for the Pythagorean school.

### Pythagoras's school

Returning in about 520 BCE, Pythagoras journeyed in 518 BCE to Croton, in present-day Calabria, Italy, where he founded a school that was partly mathematical, partly religious, and in part mystical. The mathematicians of the school regarded their life as a kind of exile, believing that the soul could leave the body, and that they would be reincarnated.

The school attracted about 2,600 young men, and probably women, who divided into two groups: an inner circle consisting of mathematicians, and their listeners. The mathematicians lived by a set of curious rules, were strict vegetarians, wore distinctive clothes, and went around barefoot. They also had a secret sign, the pentagram, which is a five-pointed star within a pentagon.

The school members lived their lives by mathematics, for Pythagoras thought that numbers were the essence of being. Odd numbers were thought to be male; even, female. For their arithmetical investigations they probably used pebbles in the

**The school of Pythagoras**
In southern Italy Pythagoras founded a mystical brotherhood, where "All is number." The mathematicians lived permanently in this unusual establishment, while listeners were permitted to attend during the day.

$$a^2 + b^2 = c^2$$

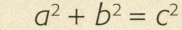

△ **Pythagoras's theorem**
In any triangle that contains a right angle, the square on the hypotenuse (c, the long side) is equal to the sum of the squares on the other two sides (a and b).

sand; in this way they learned about triangle numbers and square numbers. Especially important was the triangle with four pebbles on each side—the triangular number 10, which the

school called the tetractys (see opposite). The school held the tetractys to have many meanings: for example, its rows of one, two, three, and four points were held to represent dimensions, from zero to three respectively.

## Pythagoras's theorem

When he found a proof for the theorem that bears his name, Pythagoras is said to have sacrificed an ox at the school to celebrate. The theorem states that in any right-angled triangle the square on the hypotenuse (the long side) is equal to the sum of the squares on the other two sides (see above). In other words, if the sides of a right-angled triangle are of lengths a, b, and c, and c is the longest side, then $a^2 + b^2 = c^2$. There is an infinite number of integral solutions to this equation—values for a, b, and c, which are all whole numbers. The simplest examples of these "Pythagorean triples" are (3, 4, 5) and (5, 12, 13).

## Music and astronomy

Pythagoras enjoyed music. The story goes that he was walking past a blacksmith's shop and was intrigued by the notes as various hammers rang on the anvil. He investigated the notes made by plucking stretched strings, and so mathematically invented the musical scale. While this story is improbable, as he almost certainly studied music in Egypt, he may have experimented with stretched strings and gone on to formalize the musical scale mathematically.

Pythagoras believed that everything could be described in whole numbers, but in a right-angled triangle whose short sides are both one unit long, the length of the hypotenuse is given by the square root of two. Hippasus, one of the mathematicians in the school, managed to prove geometrically that the square root of two is an irrational number; it cannot be represented by a ratio, or fraction, of the form p/q, where p and q are whole numbers. Some say that Hippasus was thrown overboard and drowned; others that Pythagoras was so upset that he committed suicide. Whatever the truth of that, the square root of two is indeed irrational: written as a decimal it starts 1.4142135623730950488… but then goes on for ever, without any pattern.

## Pythagoras's legacy

In his work on whole, triangle, and square numbers, Pythagoras founded number theory, a current branch of mathematics that deals with whole numbers. Of more immediate influence, he introduced the concept of proof. Euclid's *Elements*, written in 300 BCE, is entirely about the subject of proof and became the most influential mathematical text in the world.

**Harmony of the spheres**
Pythagoras believed that the relationship between the planets in some way reflected his musical scale. Robert Fludd (1574–1637) produced this interpretation of Pythagoras's "harmony of the spheres."

**Mathematics of music**
According to legend, Pythagoras experimented with simple instruments and invented the musical scale, based on mathematical principles.

◁ **Pythagoras of Samos**
The most famous mathematician of the ancient world, Pythagoras was one of the first to establish the notion that Earth is a sphere. He founded a curious school or brotherhood for whom the study of mathematics was tantamount to a religion.

# TIMELINE

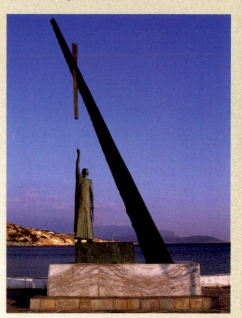

STATUE OF PYTHAGORAS AT PYTHAGORION, SAMOS, GREECE

■ **c.569 BCE** Pythagoras is born in Samos, on the island of Samos; his father was Mnesarchos, his mother Pythais. The town has since been renamed Pythagorion in his honor.

■ **c.550 BCE** Goes to Miletos and is taught by Thales, one of the earliest of the Greek philosophers, an astronomer who seems to have predicted an eclipse of the Sun in 585 BCE, and his pupil Anaximander, who was interested in cosmology and geometry.

■ **c.535 BCE** On Thales's advice he goes to Egypt, where there is already a sizeable community of people from Samos; they even have their own temple in Naucratis. Pythagoras studies astronomy and geometry.

■ **c.525 BCE** Reputedly taken prisoner by Cambyses II, the King of Persia, and taken to Babylon, where he studies arithmetic, music, and other disciplines with the scholars.

■ **c.520 BCE** Returns to Samos, where, after a visit to Crete to study its legal system, he forms a school. He is not treated well by the Samians, however, and travels on to mainland Greece, and from there to southern Italy.

■ **c.518 BCE** Settles in Croton, a Greek seaport in southern Italy, where he founds a school or brotherhood devoted to the study of mathematics, but also including a medical school. The Pythagoreans are sworn to silence and secrecy and are bound by a set of curious rules. In particular they are not allowed to eat meat, fish, or beans, nor to drink wine. They are not allowed to wear woollen clothes, because wool comes from animals. This may be connected with the fact that Pythagoras believes in reincarnation, and is worried about being reincarnated as an animal.

■ **c.510 BCE** The unusual brotherhood attracts hostility and distrust and is threatened with violence; Pythagoras escapes to Metapontion, another Greek city in southern Italy.

■ **c.495 BCE** Dies in Metapontion.

**BEFORE**

Mathematics before the ancient Greeks was largely unsystematized and emphasized practical applications.

**EARLY MATHEMATICS**

Mathematics in ancient Egypt was focused on practical issues, such as **counting ‹‹30–31**, **calendrical calculations** to help predict the flooding of the Nile, and **simple geometry** for dividing up land and **building large structures** such as the **pyramids**. Ancient **Chinese mathematics**, which developed independently, was concerned with many of the same subjects, and similarly **emphasized practicality**.

PYRAMIDS AT GIZA, EGYPT

‹‹ SEE ALSO
pp.30–31 Ancient Number Systems

**The five regular polyhedrons**
Also called Platonic solids, the five convex regular polyhedrons were first defined by Pythagoras. They are the only three-dimensional shapes whose faces are regular polygons and meet at equal angles.

**Regular tetrahedron**
The regular tetrahedron is composed of equilateral triangles. It has four triangular faces, six edges, and four vertices.

**Cube**
The cube is composed of squares. It has six square faces, 12 edges, and eight vertices.

**Regular octahedron**
The regular octahedron is composed of equilateral triangles. It has eight triangular faces, 12 edges, and six vertices.

**Regular dodecahedron**
The regular dodecahedron is composed of regular pentagons. It has 12 pentagonal faces, 30 edges, and 20 vertices.

**Regular icosahedron**
The regular icosahedron is composed of equilateral triangles. It has 20 triangular faces, 30 edges, and 12 vertices.

**MATHEMATICIAN (c. 325–265 BCE)**

**EUCLID**

One of the foremost ancient Greek mathematicians, Euclid may have studied at Plato's Academy and certainly taught at the Library of Alexandria during the reign of Ptolemy I. Although little else is known about his life, his work was widely translated and is well known, especially his major treatise, *The Elements*. As well as establishing basic principles of geometry with his axioms, Euclid wrote about number theory, and also on the subjects of optics, mechanics, and music.

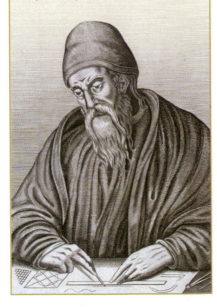

# Greek Mathematics and Geometry

**The mathematical thinking of the ancient Greeks marked a turning point in the development of the subject. Building on the empirical discoveries of the Babylonians and Egyptians, Greek mathematicians instigated a more scientific approach that is still the basis of mathematics today.**

The big innovation that distinguished ancient Greek from previous mathematics was the development of deductive logic, and with it the concept of proof. Where the Egyptians were content to accept that, for example, certain dimensions produced a right-angled triangle, early Greek mathematicians sought to understand the underlying laws of these discoveries, and applied logic to prove their theories.

**Early Greek mathematics**

The earliest known of these Greek mathematicians was Thales of Miletus, who probably acquired much of his practical knowledge from his travels to Egypt. Because he was an engineer, most of his studies were in geometry, which he is credited with introducing

**EUCLID'S FIVE AXIOMS**

1 Any two points can be connected by one and only one straight line.

2 Any line segment is contained in a full, infinitely long line.

3 Given a point and a line segment starting at that point, there is a circle that has the point as its center and the line segment as a radius.

4 All right angles are equal to one another.

5 Given a line and a point that is not on the line, there is only one line through the point that never meets the first line.

to Greece. Geometry—at that time, the study of lengths, areas, and volumes—therefore became the foundation for subsequent Greek mathematics.

Thales found new applications for his geometry: he calculated the height of a pyramid by measuring the length of its shadow and comparing this with the ratio of the length of his own shadow (measured at the same time) to his height. Theories attributed to him confirm his analytical as well as practical appreciation of geometry, and also established the principle of deductive proof in mathematics.

It is thought that Thales taught Anaximander, the philosopher who first asserted a scientific rather than supernatural order to the Universe and whose students included Pythagoras (see pp.32–33). From his school in Croton, Pythagoras taught that everything could be explained by mathematics, an idea that influenced later thinkers including Plato and Aristotle (see pp.36–37). Using Thales's notion of deductive logic, Pythagoras (or members of his school, the Pythagoreans) provided proofs for a number of theorems. These included Pythagoras's famous theorem for right-angled triangles—that the square of the hypotenuse (long side) of a right-angled triangle equals the sum of the squares of the other two sides.

Pythagoras's disciples included Hippasus, who discovered that the square root of 2 was what he called an "irrational" number—one that cannot be written as a simple fraction—and among later followers of his ideas was Hippocrates of Chios, writer of the first geometry textbooks to state Pythagorean theorems and proofs.

Mathematicians of this period were fascinated by how many geometric problems could be solved using only a straight edge and compasses. They found solutions for bisecting an angle, drawing parallel lines, and more. But three problems remained unsolved: trisecting an angle, doubling a cube, and squaring a circle (making a square with the same area as a given circle).

## Euclid's influence

Following the death of Alexander the Great in 323 BCE, the Greek empire seemed to be on the wane, but Greek mathematics flourished. The first, and perhaps greatest, mathematician of this period was Euclid (see opposite). His contribution was enormous, but at its heart was the notion of axioms, self-evident truths from which deductions could be made. In his treatise *The Elements*, probably the most

### NAVIGATION

The advances made by the ancient Greeks in geometry and mathematics were particularly useful for navigation, especially when combined with astronomical observations. By applying the principles of trigonometry (the study of triangles) to their observations, sailors were able to navigate by the stars with greater accuracy than before. For example, by measuring the angle between a celestial object, such as a star, and the horizon using a sextant, latitude can be determined.

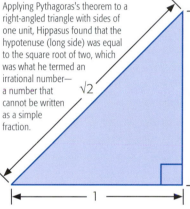

**Irrational numbers**
Applying Pythagoras's theorem to a right-angled triangle with sides of one unit, Hippasus found that the hypotenuse (long side) was equal to the square root of two, which was what he termed an irrational number— a number that cannot be written as a simple fraction.

$\sqrt{2}$

1

1

influential mathematics textbook ever written, Euclid laid down five axioms (see opposite) and proceeded from them to provide proofs for his mathematical theorems.

Building on Euclid's foundations, Archimedes proved that the area of a circle is pi ($\pi$) times the square of its radius, and managed to calculate $\pi$ (approximately 3.14) to a remarkable

degree of accuracy. Diophantus of Alexandria took Euclid's ideas to the next stage in his work on equations, pioneering algebra (see pp.48–49). A particular subject of interest was conic sections, the shapes made by slicing through a cone, such as parabolas and ellipses. Foremost in this field was Apollonius of Perga, who used conic sections to explain the apparent motion of the planets.

## Mathematics and astronomy

Astronomy dominated the last period of ancient Greek mathematics. Rather than simply mapping celestial objects, ancient Greek astronomers used geometric principles to understand their motions. The first comprehensive explanation of the motions of the planets was made by Ptolemy in the 2nd century CE. He devised a model that mapped their movements as seen from Earth. His treatise the *Almagest* (or *"Great Compilation"*) was as influential to astronomy (see pp.28–29) as Euclid's *Elements* had been to geometry.

AFTER ≫

Works of the Greek mathematicians were translated by Indian and Islamic scholars, influencing subsequent developments.

### CLASSICAL INDIAN MATHEMATICS

From around 400 CE the era of classical **Indian mathematics** followed on from both the Greek and Chinese traditions. Mathematicians such as **Aryabhata** and **Varahamihira** built on their discoveries, introducing **innovations** including a **symbol for zero** and **decimal numbers**.

### ISLAMIC MATHEMATICS

As the Islamic empire spread from about 700 CE **scholars translated** the works of ancient Greek mathematicians. **Islamic mathematicians** developed their work to make further **discoveries in geometry**, and in the emerging fields of **algebra 48–49 ≫** and **number theory**.

### MATHEMATICS IN MEDIEVAL EUROPE

From about the 12th century contact with Islamic scholars stimulated **interest in mathematics in Europe**, but there were no **major discoveries** until **Fibonacci 62–63 ≫** in the 13th century.

**MEDIEVAL GEOMETRY LESSON**

### SIEVE OF ERATOSTHENES

The sieve of Eratosthenes is a technique for finding prime numbers— whole numbers divisible only by themselves and by 1 (which, by definition, is not a prime). It consists of filtering out non-primes from a list of all numbers from 2 up to any top limit.

For example, to find all primes to 50:
**1.** List all numbers from 2 to 50.
**2.** Circle 2 then cross out all multiples of 2 (4, 6, 8, etc.).
**3.** Circle the next number that is not crossed out then cross out all multiples of that number.
**4.** Repeat step 3 until the end of the list is reached. The primes are the numbers circled.

Circled numbers are primes

Crossed-out numbers are non-primes

2 3 4̶ 5 6̶ 7 8̶ 9̶ 10̶
11 12̶ 13 14̶ 15̶ 16̶ 17 18̶ 19 20̶
21̶ 22̶ 23 24̶ 25̶ 26̶ 27̶ 28̶ 29 30̶
31 32̶ 33̶ 34̶ 35̶ 36̶ 37 38̶ 39̶ 40̶
41 42̶ 43 44̶ 45̶ 46̶ 47 48̶ 49̶ 50̶

PHILOSOPHER AND SCIENTIST **Born 384** BCE **Died 322** BCE

# Aristotle

## "The **whole** is *more than* the **sum** of its parts."

ARISTOTLE, FROM "METAPHYSICS," 335–323 BCE

One of the greatest of the Greek philosophers, Aristotle laid the foundations for scientific thought in the Islamic and Christian worlds. His studies of various subjects, including physics, mathematics, logic, and biology, have influenced many disciplines.

Aristotle was born in Stageira, Macedon, the son of a doctor, Nicomachus. Appointed personal physician to the King of Macedon, Nicomachus probably took young Aristotle with him, beginning a close association with the royal family.

Aristotle was orphaned at the age of 10 and brought up by his uncle, who complemented his early medical education with lessons in rhetoric and poetry. At the age of 17 he went to Athens to become a student at Plato's Academy. Aristotle spent 20 years at the academy, becoming one of a line of great Greek philosophers: Socrates had been Plato's mentor, and now Aristotle was an obvious choice as Plato's successor as head of the Academy.

### After Plato's Academy
Denied the job for political reasons, Aristotle left the Academy for Assos in Asia Minor (modern Turkey), where Hermias of Atarneus had assembled some philosophers. While at Assos, Aristotle explored the wildlife of nearby Lesbos, returning to the study of biology and zoology he had started with his father and beginning work on a systematic classification of living things. Assos was then invaded by the Persians;

▷ **Aristotle**
This bronze statue of Aristotle, created in 1915 by Cipri Adolf Bermann (1862–1942), is placed opposite one of Homer, also by Bermann, outside an entrance of the University of Freiburg in Breisgau, Germany.

ΑΡΙΣΤΟΤΕΛΗ

Aristotle fled, first to Lesbos, and then to Macedon at the invitation of King Philip. One of his pupils at the Macedonian court was Philip's son Alexander, who, after he succeeded the throne, wanted to send Aristotle to Athens in an ambassadorial role.

## The Lyceum and the "Peripatetics"

When Aristotle did not receive the headship of the Academy in Athens, Alexander encouraged him to set up his own establishment in competition: the Lyceum. This offered a broader range of disciplines than the Academy, with an emphasis on the study of nature as well as traditional philosophy, politics, and rhetoric. The Lyceum attracted students who were called the "Peripatetics" after the colonnades (*peripatoi*) of the school; the name may also derive from Aristotle's habit of walking while lecturing.

Aristotle prepared his lecture notes in the form of treatises, and these survived to become his most important

**Aristotle tutoring Alexander**
Aristotle taught at the Macedonian court before setting up the Lyceum in Athens. Among his pupils there was King Philip's 13-year-old son Alexander, later Alexander the Great, in whom he inspired a great love of learning.

works. In particular, he developed his "analytics," the system of deductive logic that underpins Western logic and the scientific method today. His teaching on the subjects of "natural philosophy" (physics), mathematics, and zoology showed that he regarded these just as highly as the philosophical subjects such as metaphysics, ethics, and politics covered by his treatises.

Aristotle's heyday at the Lyceum was sadly cut short by political wrangling after Alexander died. Athens once

again became fiercely anti-Macedonian, and Aristotle's connection with the deceased king made his position untenable. At the same time he was denounced for refusing to accept royalty as divine and was forced to leave Athens once more.

## Aristotle's legacy

Aristotle died of natural causes shortly after leaving the Lyceum. The works he had published during his lifetime are now lost, but much of his teaching was passed down by his students and a collection of writings was published posthumously. Translations into Arabic were later made by Islamic scholars, and subsequently into Latin by medieval European translators, preserving his extraordinary breadth of knowledge and clarity of intellect for posterity.

# "All men by nature desire knowledge."

ARISTOTLE, FROM "METAPHYSICS," 335–323 BCE

▽ **The School of Athens**
Raphael's fresco *The School of Athens*, painted in the Vatican, Rome, depicts many of the philosophers and scientists of ancient Greece. Deep in discussion at its center are Plato and his pupil, the young Aristotle.

---

### TIMELINE

- **384 BCE** Aristotle is born in Stageira, Macedon, on the Chalcidic peninsula of northern Greece. Shortly after the birth, his father, Nicomachus, is appointed personal physician to Amyntas III, King of Macedon.

- **374 BCE** Nicomachus dies, and Aristotle is brought up by his uncle, Proxenus of Atarneus.

- **366 BCE** Moves to Athens to study at Plato's Academy, staying for the next 20 years. After Plato's death, he leaves, with others including Xenocrates and Theophrastus, amid academic disagreements and anti-Macedonian feeling.

- **348 BCE** Philip of Macedon expands his territory and annexes Chalcidice.

- **346 BCE** Travels with Xenocrates to the court of Hermias of Atarneus in Assos, Asia Minor, and becomes leader of a group of philosophers. He visits Lesbos, where he studies the plant and animal life of the island with Theophrastus.

- **343 BCE** Assos is invaded by the Persians; Hermias is captured and later executed, and Aristotle flees to Lesbos.

- **342 BCE** Is invited by Philip of Macedon to Macedonia, and is appointed as tutor to students at the royal court, including the young Alexander the Great and Ptolemy I.

- **340 BCE** Aristotle marries Hermias's niece and adopted daughter Pythias. They have a daughter, also called Pythias.

- **339 BCE** After an unsuccessful bid to become head of the Academy (a post given to his friend and colleague Xenocrates), he loses the support of Philip and moves back to Stageira (now part of Macedon), but is still nominally employed by the Macedonian court.

- **336 BCE** Philip of Macedon dies and is succeeded by his son Alexander. Aristotle is sent to Athens to set up a rival to the Academy.

- **335 BCE** Establishes his own school in Athens, known as the Lyceum. Writes treatises, some in dialogue form, on the soul, physics, metaphysics, ethics, politics, and poetics during the next 12 years.

- **330 BCE** His wife Pythias dies. He later has a relationship with Herpyllis of Stageira, with whom he has a son, Nicomachus.

- **323 BCE** Alexander the Great dies, and anti-Macedonian feeling flares up again in Athens. Aristotle is denounced for seeing kings as merely mortal and flees Athens to his mother's estate in Chalcis on the island of Euboea; Socrates had died for just such views.

- **322 BCE** Dies of a stomach complaint in Chalcis, at 62, and requests in his will that he be buried alongside his wife.

**STATUE OF ARISTOTLE AT STAGEIRA**

« BEFORE

Clues to what people's ideas about the world may have been over 3,000 years ago come from examination of a small number of ancient texts and maps.

### EARLY LOCAL MAPPING

Long before they had any idea of the overall shape of the earth or the distribution of its landmasses, people were making primitive maps of their local areas. The **oldest known local map**, which was discovered in Mezhirich, Ukraine in 1966, is thought to date from approximately 10,000 BCE. Inscribed on a **mammoth tusk**, it shows several dwellings along the banks of a river.

### FOUR CORNERS OF THE EARTH

A common ancient theme was that **the earth was flat** and that it took the **form of a circle or square**, which was thought to be surmounted by a hemispherical sky. In ancient Egypt, which prevailed for nearly 3,000 years BCE, the sky was believed to be like a tent canopy, which stretched between great mountains that stood at the four corners of the earth.

---

MATHEMATICIAN (276–194 BCE)

### ERATOSTHENES

The Greek scholar Eratosthenes was born in Cyrene, in North Africa. As a young man he studied in Athens and Alexandria, becoming an expert in many different fields, including mathematics and astronomy. At the age of 40 he was appointed chief librarian at the library of Alexandria, the greatest center of learning in the ancient world. In addition to his first accurate determination of Earth's circumference (see right), Eratosthenes drew important early world maps, calculated the distance from Earth to the Sun, and made a key contribution to the mathematical theory of prime numbers. He also coined the term "geography" and was, arguably, the first geographer. Tragically, toward the end of his life he became blind and subsequently starved himself to death.

# Ancient Ideas of the World

**Over the course of several centuries BCE a succession of philosophers, mathematicians, and early geographers sought to understand the basic shape and size of our planet, as well as defining concepts such as latitude and climate zones. The conclusions that they reached demonstrated a high level of sophistication and accuracy.**

Greek and Roman scholars, including the first geographers, developed new concepts about the world, some of which are now known to be correct. Remarkably, some of them appear to have been based purely on philosophical or aesthetic grounds rather than scientific observation. Until around 2,500 years ago, the prevailing belief in the Mediterranean region and in nearby Mesopotamia was that these lands occupied the upper flat surface of a drum- or coin-shaped disk. Earth's surface at this time was thought to consist mainly of land, with a sea (the Mediterranean) at its center, and an ocean surrounding the land, extending to the edges of the disk.

#### Babylonian map
Part of this clay tablet (c.600 BCE) shows inhabited lands encircled by an ocean. It is considered to be the oldest known world map.

### A spherical Earth

Over a period of more than two centuries (c.550–300 BCE), Greek scholars made some key advances in understanding the world. The most important of these ideas, first advocated by the mathematician Pythagoras around 550 BCE, was that Earth is a sphere. Pythagoras reasoned that this shape is the most harmonious geometric solid and therefore the shape the gods would have chosen for the world when they created it. Accepting the possibility of a spherical Earth, around 360 BCE the philosopher Plato suggested that there was likely to be another landmass directly on the opposite side from the lands of the Mediterranean, which he called the antipodes. This idea, however, was based purely on a Greek belief in symmetry.

Aristotle (c.384–322 BCE) was the first scholar to put forward scientific evidence that Earth is a sphere. He argued that travelers to southern lands could see stars in the night sky, which were hidden below the horizon for those living further north, and this could only be explained if Earth's surface was curved. In addition, he pointed out that during lunar eclipses the shadow of Earth on the Moon has a curved edge. Aristotle's arguments were convincing and, by c.330 BCE, most educated people in the classical world had accepted that Earth was a sphere and not flat.

### Earth's circumference

The next challenge was determining how big this sphere was—the circumference of the Earth. Scholars realized they should be able to figure out the Earth's circumference. First they should find two places on a north–

#### Evidence for a spherical earth
One of Aristotle's arguments for a spherical earth was that its shadow, seen here cast on the Moon during a lunar eclipse, has a curved edge.

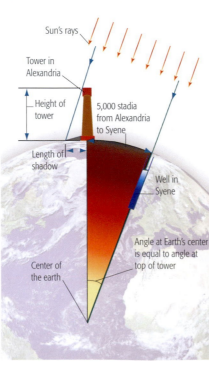

Sun's rays

Tower in Alexandria

Height of tower

5,000 stadia from Alexandria to Syene

Length of shadow

Well in Syene

Angle at Earth's center is equal to angle at top of tower

Center of the earth

#### Measuring Earth's circumference
By examining how the Sun's rays reached the earth at Alexandria and Syene, Eratosthenes figured out the difference in their latitudes (7° 12′), divided this into 360° (50), and multiplied it by the north–south distance between them (5,000 stadia) to give him the circumference of the Earth (250,000 stadia).

**World map by Ptolemy**
This 15th-century map is based on what was known about the extent of the world in Alexandria, Egypt, in about 120 CE, as written down by the Roman geographer and astronomer Ptolemy.

BREAKTHROUGH

## LATITUDE AND LONGITUDE

One of the most important advances of all time for the mapping of the world was the development of the concepts of latitude (distance north or south from the equator) and longitude (distance east or west from a meridian line drawn from top to bottom of Earth). Around 300 BCE Greek polymath Dicaearchus of Messana compiled a map of the world, as far as was known at the time (reconstructed below), and drew a line of latitude through it. The fact that the line accurately corresponded to what is now called 36° N shows that Dicaearchus must have fully understood the idea of latitude. Perpendicular to this line of latitude, and intersecting it in the vicinity of the island of Rhodes, Greece, he drew a single line of longitude.

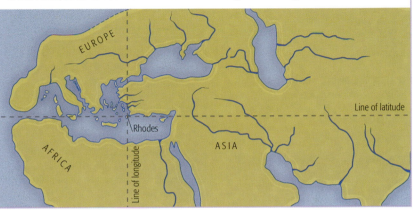

south line and measure the difference in latitude between them (in degrees). Then they had to divide this difference into 360 degrees (the "difference" around the whole sphere). Finally, they needed to multiply this by the distance between them on the ground, and the result should be the circumference of the earth. In c.240 BCE, Greek mathematician Eratosthenes estimated the earth's circumference to be 250,000 stadia (see left). The size of stadion he used in his calculation is uncertain, but if it was the Egyptian stadion, his measurement works out as 24,662 miles (39,690 km). This is an error of less than 1 percent from the

**STADION** Originally used to describe an ancient foot race, "stadion" was later used as a measure of length, equivalent to the distance over which the race was run. Slightly different versions were used in Greece, Asia Minor, and Egypt.

true value of the Earth's circumference around the poles.

### Climate zones

Scholars of the ancient world also sought to understand Earth's climate. In the 4th century BCE Aristotle had suggested that there were two cold climate zones near the earth's poles and a warmer region in the middle. Greek geographer Strabo proposed five zones—two frigid ones near the poles, a tropical or torrid zone near the equator, and two temperate zones either side of the tropical zone. He also suggested that there must be other unknown continents. Roman geographer Pomponius Mela wrote in c.43 CE that he believed that the torrid zone was so hot that people could not possibly cross it to reach the southern hemisphere, where there lived beings of whom nothing was known.

**Map by Herodotus**
This map by the historian Herodotus shows the extent of the world, as understood by the ancient Greeks, around the 5th century BCE. At this time the Greeks knew little about lands situated more than 600 miles (1,000 km) from the Mediterranean Sea.

> "the **shapes** which the **Moon** itself each month shows are of every kind…but in **eclipses** the outline **is always curved**."
>
> ARISTOTLE, ANCIENT GREEK PHILOSOPHER (384–322 BCE)

AFTER ▶▶

Ideas first proposed by the ancient Greeks continued to be developed in the Mediterranean world, Asia, and eventually medieval Europe.

**CALCULATIONS OF ARYABHATA**
The Indian mathematician and astronomer **Aryabhata** (476–550 CE) recalculated **Earth's circumference** to an accuracy of 99.8 per cent. He claimed that Earth rotates and determined the time for one rotation, relative to the stars, with astonishing accuracy, which was shown later to have an error of less than one in a billion.

**COLUMBUS'S MISTAKE**
In the 15th century Christopher Columbus utilized a **flawed ancient calculation of Earth's circumference** when planning his expedition in search of the Indies. This seems to have led him to believe that Asia is only about 3,100 miles (5,000 km) west of Europe. But some authorities think his "mistake" may have been intentional, in order to help him get funding for the expedition.

**1507** The year one of the first maps showing the New World was published—the *Universalis Cosmographia*. This was also the first known reference to name America.

SEE ALSO ▶▶
pp.78–79 PLANETARY MOTION
pp.190–91 DATING THE EARTH
pp.352–53 THE STRUCTURE OF THE EARTH

**BEFORE**

Archeologists have found many simple machines used by people who lived in the Stone Age, at least 5,000 years ago.

**PREHISTORIC TOOLS**
The earliest tools may have been **hand axes**, made from a single lump of flint by "knapping"—creating a sharp edge by knocking pieces off the flint using another stone.

**FLINT HAND AXE**

**PREHISTORIC MONUMENTS**
Transporting and erecting **monumental stones** weighing several tons each probably required the use of rollers, hand axes, wedges, and levers—as well as a large workforce.

**CALLANISH STONES ON THE ISLE OF LEWIS, OUTER HEBRIDES, SCOTLAND**

**« SEE ALSO**
*pp.20–21* EVOLUTION OF THE WHEEL

**IN PRACTICE**

**ARCHIMEDES SCREW**

This clever pump is said to have been invented by Archimedes during the 3rd century BCE. It has certainly been used for thousands of years for irrigation, and is still used today for pumping sewage and similar sludge. It is a simple helix—like a wood screw but of constant diameter—inside a close-fitting tube. As the handle is turned the liquid is pulled up the tube, each bucketful trapped in a compartment that seems to move steadily upward.

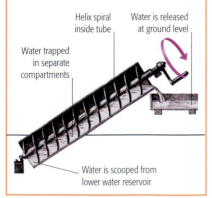

Helix spiral inside tube
Water is released at ground level
Water trapped in separate compartments
Water is scooped from lower water reservoir

# Simple Machines

**The first pieces of technology invented by primitive humans were simple machines that helped them solve physical problems. The simplest and most effective—the wedge, the lever, and the pulley—could multiply and change the direction of force. They are still in daily use in the 21st century.**

People have always wanted to do things that needed more than normal human strength—move heavy stones, cut wood, or kill animals for food or in self-defense—so they invented machines to increase the force they could exert. The simplest machines for multiplying force are the wedge, the lever, and the pulley. The wheel (see pp.20–21) was also helpful for moving things over smooth ground.

**The penetrating wedge**
A wedge is a piece of wood, metal, or stone that tapers to a thin edge so that it can be driven into or between objects to separate them. Because it gets gradually thicker, the wedge separates the material as it is pushed farther in. Knives, spears, axes, and arrowheads are all types of wedge: they use sharp edges to cut, penetrate, and kill.

When building the pyramids, the ancient Egyptians almost certainly used wedges in the form of earth ramps, to pull and lift heavy stones. We still use the wedge in many forms today, such

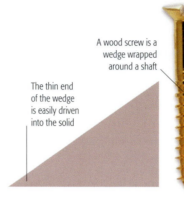

A wood screw is a wedge wrapped around a shaft

The thin end of the wedge is easily driven into the solid

**Simple and complex wedges**
A wedge has one sharp (or at least thin) edge which can be pushed or hammered into a solid to pry it apart—such as a knife into cheese, or an axe into wood. The effort is applied to the opposite end of the wedge.

as the spades and trowels we use for digging, and the axes and chisels used in woodworking. In each case the thin edge of the wedge is pushed into a resistant material to separate it.

Nails are types of wedges; they have sharp points that allow them to be hammered into wood. The woodscrew is a more complex type of wedge: the metal is twisted into a helix around the main shaft to form a thread. As the screw is twisted clockwise into wood with a screwdriver, the sharp wedge

bites its way into the wood and the thread pulls the whole screw forward.

**The empowering lever**
A lever is a rigid bar pivoted on a fulcrum (support point or pivot) which can be used to exert a force or effort on a load. Levers are generally used either to multiply the force or to apply it in a different place. People use levers all the time: a bottle opener is a lever; so are pairs of scissors, and some types of door and faucet handles.

The first person to describe levers in mathematical terms was probably Archimedes, the ancient Greek. There are three types or classes of lever, and Archimedes pointed out that to get maximum mechanical advantage (multiplication of force) it is best to use a lever of class 1 or class 2, and one

> "Give me a **lever long enough**… …and I shall **move** the **world**."
>
> ARCHIMEDES, MATHEMATICIAN, C.250 BCE

**Lever types**
There are only three different types of lever. They look similar but have quite different characteristics. All are used to extend the use of the muscles, especially the hands.

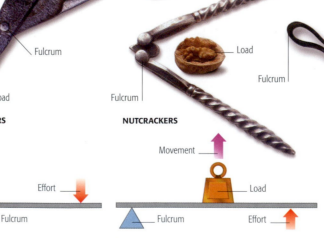

Effort

Fulcrum

Load

**SCISSORS**

Effort

Load

Fulcrum

**NUTCRACKERS**

Effort

Load

**TWEEZERS**

Movement
Load
Effort
Fulcrum

**Class 1 lever**
The most familiar kind of lever has the fulcrum (the pivot) between the effort and load. The closer the fulcrum to the load, the less the effort required.

Movement
Load
Fulcrum
Effort

**Class 2 lever**
With the fulcrum at one end and the effort at the other, Class 2 levers are immensely powerful if the load is close to the fulcrum.

Movement
Load
Fulcrum
Effort

**Class 3 lever**
Effort is exerted in the center of Class 3 levers, which are designed not for power but for precise action at a distance.

that is as long as possible, with the load as near as possible to the fulcrum or pivot of the lever.

In class 1 levers, the fulcrum is positioned between the effort (exerted by a hand, for instance) and the load. Examples include scissors and door handles. In class 2 levers, the load sits between the effort and the fulcrum, as in nutcrackers and wheelbarrows. Class 3 levers require the effort to be exerted between the load and the fulcrum, as in tweezers and fire tongs.

**The pulling machine**
Levers generally exert pushing forces and are most effective over short distances. Pulleys, on the other hand, exert only pulling forces, but can do so over long distances. A pulley consists of a grooved wheel mounted on a fixed axle; the groove allows the pulley to carry a length of rope.

Pulleys are more advanced than wedges and levers, since they depend on the existence of the wheel, which is itself an enormously useful machine.

The tension or pulling force is the same all along the rope; one pulley effectively pulls a load by two ropes of equal tension—so the force required to move the load is halved. If there are effectively three ropes (using two pulleys), the force required is only one third of the load, and so on.

An early use of a simple pulley with one wheel and a length of rope was to raise a bucket of water from a well. This allows the user to pull on the rope horizontally, rather than vertically.

△ **The power of the lever**
Archimedes was the first person to explain the power of levers mathematically. He is said to have claimed that levers were so powerful they could be used to raise Earth itself. This illustration assumes he would choose a Class 1 type of lever (see below) for the task, and would carefully position Earth as close as possible to the fulcrum, while standing far away from it himself, to gain maximum mechanical advantage.

**AFTER** »

All these simple machines are still in use today, although many have developed more complex mechanized variants.

**GEARING SYSTEMS**
Gears **42–43** »different size interlocking cogged wheels—are a direct development from pulleys, with additional advantages: they can exert **pushing forces**, and can be designed to **multiply force** or **increase speed**.

**PULLEYS AT TRAFALGAR**
Pulleys have always been used in **cranes and hoists** on building sites, and on sailing boats and ships. Admiral Nelson's flagship at the Battle of Trafalgar in 1805, *Victory*, had around 900 handmade **wooden pulley blocks**. These were essential for her crew to raise and lower her 37 large and immensely heavy canvas sails up and down the ship's tall masts in the shortest possible time—especially in the heat of battle.

**SEE ALSO** »
*pp.42–51* How Gears Work
*pp.50–51* Water and Wind Power
*pp.130–31* The Newcomen Engine

**Simple and complex pulleys**
A simple pulley with one wheel can halve the load. More complex pulleys with two, three, or more wheels reduce the load even more. Pulleys also have the great advantage of allowing you to vary the angle of pull on the rope.

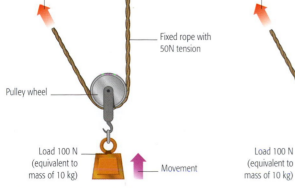

Effort 50 N

Fixed rope with 50N tension

Pulley wheel

Load 100 N (equivalent to mass of 10 kg)

Movement

Effort 33½ N

Load 100 N (equivalent to mass of 10 kg)

Movement

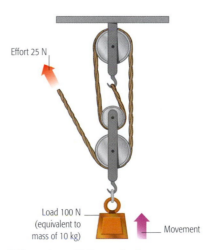

Effort 25 N

Load 100 N (equivalent to mass of 10 kg)

Movement

**Pulley system with one wheel**
Using one wheel, an effort of 50 newtons (N) is needed to raise a load of 100 N; the tension in the rope is 50 N and there are in effect two ropes raising the load.

**Pulley system with two wheels**
A two-wheel pulley effectively uses three ropes to raise a load of 100 N; so the tension in each rope has to be only one third of this—33⅓ N; this is the effort needed.

**Pulley system with three wheels**
With three wheels (four ropes) lifting the load, the tension in each, and therefore the effort, is reduced to a quarter of the load: 25 N.

# How Gears Work

The first gears were a natural development from the invention of the wheel. In early types of rotating machinery, the capacity of gears to shift and increase power from one rotating component to another made them a favorite device from the 3rd century BCE onward, and led to a multitude of practical applications.

Most gears consist of cogs (rotating wheels or cylinders bearing "teeth") that mesh with other cogs. They help movement to be transmitted from an input source—such as flowing water—to an output device, such as a pump, mill, or the hands of a clock. The position of the gears' teeth can vary, depending on the type of gear (see

**Large wooden gears**
Gears made of wood were commonly used in early water- or wind-driven mills, especially for changing the axis of rotation. As the gear teeth wore out from time to time, they could be replaced individually.

right). Sometimes the teeth are arranged along a linear rack or form a spiral along a shaft instead of being borne on a cog.

## Advantages of gears
The earliest gears were probably made from wood with cylindrical pegs for teeth. Various arrangements were used, depending on which of the gearing advantages was required.

Firstly, the direction of power can be changed from the input to the output gear. For instance, where a horse was used to pull a horizontal wheel, a gear arrangement could be used to transmit the power to another wheel that then raised water vertically from a river.

Secondly, gears can be used to change a slow input speed into a fast output speed. The rotational speed of horse-

or water-powered wheels was often too slow for grinding grain or hammering metals, so gears were used to increase the speed to a usable level.

Finally, gear sets can be used to change a small input torque (rotational force) into a larger output torque, effectively creating leverage. This is useful for purposes such as lifting heavy loads. The property of gears that allows the output torque and rotational speed to be different from the input is called the mechanical advantage. It is equal to the gear ratio: the ratio of the number of teeth in the driven gear to the number of teeth in the driving gear.

## Gears in ancient Greece and China
The earliest reliable accounts of the use of gears come from the classical world and early Imperial China. In Greece, Archimedes (c.287–212 BCE) is thought to have used gears in a number of constructions, including a milometer. Philon of Byzantium (died

**Antikythera mechanism**
Made by the ancient Greeks in around 100 BCE, the Antikythera mechanism contained more than 30 cogs. It was a mechanical "computer" that was used to predict the celestial positions of the Moon and planets.

### BEFORE

**Evidence of geared technology before 200 BCE comes from ancient manuscripts and a few archeological artefacts.**

**CHINESE DIFFERENTIAL GEARS**
Records indicate that a **differential gear** may have been developed in China as long ago as 800 BCE, for use in a device called a "south-pointing chariot." This was **a two-wheeled device** carrying a figure that always pointed in the same direction, acting as a mechanical **compass.**

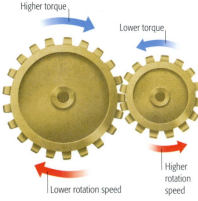
SOUTH-POINTING CHARIOT

**ANCIENT RATCHET MECHANISM**
The oldest archeological artefact showing evidence of a geared mechanism also comes from China, and dates from approximately 230 BCE. It consists of a **bronze ratchet wheel with 40 teeth**; its purpose is not yet known.

**SEE ALSO**
pp.20–21 EVOLUTION OF THE WHEEL
pp.40–41 SIMPLE MACHINES

**Improving performance**
In any pair of gears, the larger one slowly rotates with more torque (force), while the smaller one moves faster but with less torque. For this reason gears can be used to increase either the output torque or the output speed of a mechanism, but not both.

Higher torque / Lower torque / Higher rotation speed / Lower rotation speed

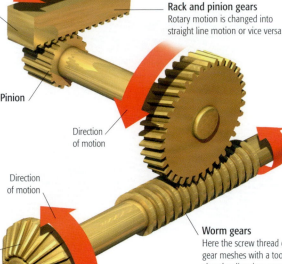

**Rack and pinion gears**
Rotary motion is changed into straight line motion or vice versa.
Rack / Pinion / Direction of motion

**Worm gears**
Here the screw thread on the worm gear meshes with a toothed wheel to alter the direction, speed, and force of motion. The worm gear can turn the toothed wheel, but the toothed wheel cannot turn the worm gear.

**Bevel gears**
These change the axis of rotation.
Direction of motion

**Types of gears**
Gears come in many different shapes and sizes, and their teeth may be straight, curved, or inclined. However, there are only four basic types of gear pairing, as shown here. Each changes the direction of motion in some way, and most also alter the speed and force of motion.

**Spur gears**
These gears intermesh in the same plane, rotate in opposite directions, and may change the speed or force of motion. The smaller gear turns faster.

Direction of motion

### INVENTION
#### MECHANICAL CLOCKS

Mechanical clocks differ from earlier clocks—such as those driven by water flow—in containing some type of oscillating mechanism. Gears are used in these clocks essentially as chains of counting devices, adding up the regular-timed pulses generated by the oscillator to mark the units of time. The final gears in the chain are attached to, and move, the hands of the clock. Mechanical clocks first appeared in Europe during the 13th century.

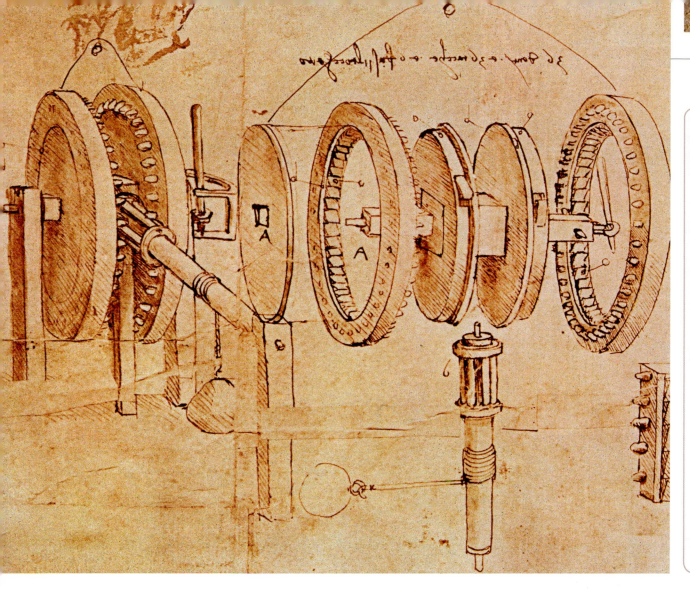

**Leonardo's designs**
This drawing shows Leonardo da Vinci's design for a two-wheeled hoist with a caged gear from 1503, with the assembled design (above left) and an exploded view (above right). Although probably never built, Leonardo's hoist could have been used for raising or lowering a weight. It incorporated safety ratchets to prevent the weight falling freely.

c.220 BCE) described the use of simple gears in water-raising devices. By around 100 BCE, the Greeks were using metal gears with wedge-shaped teeth to form complex calculating devices and astronomical calendars. We know this from the discovery of a device known as the Antikythera mechanism (see opposite), which is the oldest-known complex geared machine. Constructed around 100 BCE, it was found in a shipwreck off the Greek island of Antikythera in 1900. The mechanism contained over 30 cogs and was designed to calculate astronomical positions.

In China, artefacts and written records indicate that gears were being used in watermills, crossbow-arming mechanisms, and chain pumps during the period from 200 BCE to 250 CE.

### Middle Eastern technology
Whatever advances in gear technology were developed by the ancient Greeks, they were largely lost from Europe during the 1st millennium CE, and transferred to the Islamic world.

Several examples are known of geared calendars and astronomical devices from the Middle East dating from the 5th to 13th centuries CE. Parts have been found of a geared sundial calendar from around the 5th or 6th century that displayed the positions of the Sun and Moon in the zodiac. An early Persian astrolabe (see p.29) dating from the 13th century is the oldest geared machine in existence still in a complete state.

### Reintroduction to Europe
The knowledge of mechanical gearing that had been kept alive in the Islamic world was reintroduced to Europe via the Islamic conquest of Spain in the 12th century. It was most evident from the sudden advances in clockmaking. These included water-driven clocks that for the first time incorporated epicyclical or planetary gears (one or more small outer gears, revolving about a larger central gear).

In non-Islamic Europe, sophisticated gears for transmitting high torque first appeared in astronomical clocks in the 14th century, such as the "Astrarium" made by Giovanni de Dondi of Padua in Italy in around 1365.

Artefacts, writings, and drawings by luminaries such as Leonardo da Vinci and Albrecht Dürer indicate that all the main types of gears were known and used in Europe by the 15th century.

## AFTER ❯

Improvements in gear technology since 1500 have included new types of gears and more precise, harder-wearing metal gears.

### FROM ROTARY TO STRAIGHT LINE
In the 16th century, the Italian doctor and mathematician **Girolamo Cardano invented the Cardan gear**. It provided a means of converting rotary motion into back-and-forth, straight-line motion by means of a cog rotating inside a large-toothed wheel.

**19TH-CENTURY COTTON MILL**

### INDUSTRIAL REVOLUTION
The invention of **steam engines 132–33 ❯❯** and later **electric motors 168–69 ❯❯** during the Industrial Revolution of the 18th and 19th centuries led to an explosion in the use of metal gearing and further developments in the science of gear design.

**SEE ALSO ❯❯**
*pp.50–51* WATER AND WIND POWER
*pp.120–21* MEASURING TIME
*pp.168–69* THE ELECTRIC MOTOR

---

**IN PRACTICE**

## DIFFERENTIAL GEARING

Differential gearing allows power to be transmitted equally between the two wheels of a moving vehicle—such as a car—while allowing them to take paths of different lengths, and therefore turn at different speeds. This is necessary because the outside wheel of a car has farther to go when the car turns a corner. The differential consists of sets of bevel gears (see opposite) and pinions attached by a carrier to the crown wheel. When input torque (rotational force) is applied to the crown wheel, it turns the carrier, providing torque to the side gears. These drive the wheels via half-shafts. On a straight, the planet pinions do not rotate, and the wheels turn at the same rate. During a turn, the planet pinions rotate and as a result the side gears and wheels revolve at different speeds.

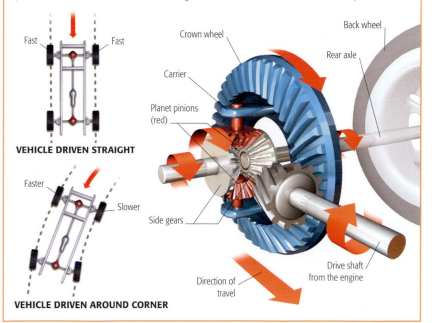

Fast    Fast
**VEHICLE DRIVEN STRAIGHT**

Faster    Slower
**VEHICLE DRIVEN AROUND CORNER**

Crown wheel
Back wheel
Rear axle
Carrier
Planet pinions (red)
Side gears
Drive shaft from the engine
Direction of travel

# "Eureka!"

**Many a great leap forward in science has come when a scientist has a sudden flash of inspiration. The first recorded case occurred 2,250 years ago in Syracuse and completely revolutionized the way people thought about measuring volume. The scientist in question was Archimedes, one of the finest minds of the time. His discovery came when trying to solve a tricky question of potential fraud.**

In 240 BCE, Syracuse, on the island of Sicily, was ruled by the tyrant Hieron II, who ordered an elaborate new crown as a tribute to the gods. It was to be made from pure gold, which the king weighed out very precisely. When it arrived, however, he suspected that the crown maker had cheated by mixing some cheaper metal with the gold. The king was furious at the thought of being tricked, and yet did not know how to prove it. To investigate, he summoned Archimedes.

Archimedes was perhaps the finest mathematician in the ancient world, and also a skilled engineer. He was still worrying about the question posed by the king when he went for one of his rare baths in the public bath house. He knew the weight of the crown because he had put it on his scale at home. If he could measure the volume of the irregularly shaped object, he could figure out its density—the weight per unit volume. Archimedes knew that metals have different densities. For gold it would be about 19 g/cm², while other common metals are much less dense: lead 11, copper 9,

silver 10. So from the density he would know immediately whether the gold was pure.

Archimedes' dilemma was how to measure the volume of the crown. Then as he sank down into the bath, some water slopped over the side, and Archimedes had his flash of inspiration and utterly changed the way people thought about measuring volume. His body sinking into the bath had pushed some water out of it—and Archimedes realized that any "body" pushed into water would displace its own volume of water. Overjoyed, he jumped out of the bath, shouting "Eureka," and ran all the way home, stark naked. He filled a bucket with water and lowered the king's crown in on a piece of string, slopping some water over the side. Then he pulled the crown out and measured how much water he needed to refill the bucket exactly to the brim—and that was the volume of the crown. From this he calculated the density, which turned out to be much less than that of pure gold; so the king really had been cheated. But, what mattered most was this new way to measure volume.

## A new breed of scientist
The story of Greek mathematician and inventor Archimedes' discovery of how to measure the volume of irregular objects is depicted in this 16th-century hand-colored woodcut.

# "Eureka! I have found it."

ATTRIBUTED TO ARCHIMEDES

**BEFORE**

For thousands of years people have found increasingly ingenious solutions to the challenges of transporting people and materials over water.

**ARISTOTLE**
The great Greek philosopher **Aristotle << 36–37** was a **keen observer** of science and technology. He noted that some boats which floated fully laden at sea sank when sailed up river. This is because fresh water is **less dense** than sea water, and is therefore less buoyant.

**EARLY BOATS**
The first **floating platforms** were probably rafts of logs, while some cultures made **quffas** (reed boats) or stretched animal skins over frames to make small craft such as **coracles**.

**QUFFA MADE WITH WOVEN REEDS**

**"EUREKA!"**
The Greek scientist and mathematician Archimedes **<< 44–45** is said to have shouted "Eureka!" ("I have found it!") when he noticed how his body **displaced** water from his bath. His understanding of the relationship between weight, volume, and water forms the basis of **Archimedes' Principle**.

# Floating and Sinking

**Whether things float or sink in water has always been important to people who live near rivers, lakes, and the sea. The ancient Greek inventor Archimedes was probably the first person to put buoyancy on a mathematical footing, and shipbuilders have made use of his ideas to manipulate buoyancy ever since.**

### The science of buoyancy
Items are buoyant if their weight (which includes the air they contain) equals or is less than the water displaced—the "upthrust." Ship designers can calculate the type and size of craft needed to carry heavy loads while remaining buoyant.

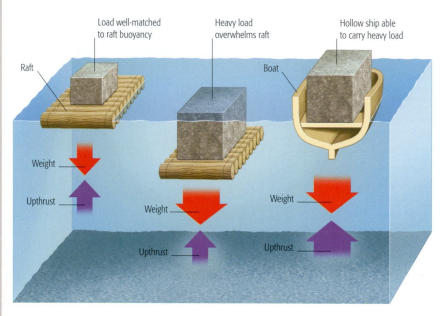

Load well-matched to raft buoyancy

Heavy load overwhelms raft

Hollow ship able to carry heavy load

Raft

Boat

Weight

Upthrust

Weight

Upthrust

Weight

Upthrust

**A buoyant raft**
The wooden raft can carry a small load provided the upthrust is greater than the combined weight of the raft and the load.

**A sinking raft**
With a heavier load the combined weight is greater than the upthrust; the water flows over the side, and the raft sinks.

**A buoyant ship**
Because a boat is hollow and deep, it displaces much more water than a raft, and the greater resulting upthrust can support the heavier load.

F or thousands of years people have wanted to travel over water, either in peace or in pursuit of war, and this has driven the gradual development of craft that operate either on the surface of water or under it.

### Buoyancy
Buoyancy is a force that pushes things upward; it holds up things that float. The ancient Greek mathematician Archimedes realized that objects immersed in water displace some of it, and that this diplacement is what causes the upthrust.

Archimedes' Principle states that any object immersed in water will experience an upthrust equal to the weight of water displaced. One consequence of this is that any object less dense than water will float, because if the object were to be submerged, it would displace a weight of water greater than

its own weight—and this upthrust would push it back up to the surface.

The density of an object is its mass per unit volume, normally measured in grams per cubic centimeter ($g/cm^3$) or kilograms per cubic meter ($kg/m^3$). Things that "feel" heavy for their size —such as lumps of metal or stone— have high density; while things that

"feel" light—such as table-tennis balls or cork—have a low density.

### The density of water
Pure water has a density of $1g/cm^3$; anything with a higher density than this will sink in water, and anything with a lower density will float. Cork floats high in the water, since it needs to displace only one quarter of its volume of water to feel an upthrust equal to its own weight. People and

> **ARCHIMEDES' PRINCIPLE** states that the upthrust on an object in water is equal to the weight of the water that the object displaces.

icebergs float low in the water because their densities are about $0.9g/cm^3$. Most kinds of wood have densities less than 1, so most wood floats. Boats float mainly because they are hollow and full of air.

Another consequence of Archimedes' Principle is that although dense objects sink, they feel lighter under water because they experience some upthrust. Marine mammals such as whales can grow larger than land mammals because water supports their weight; land animals could never grow legs strong enough to hold so much weight.

In pure water, no inanimate object can float halfway between the bottom and the surface. Any object with a density of 1.001 or more will sink; anything with a density of 0.999 or less will float. Fish have to keep swimming to maintain a constant depth.

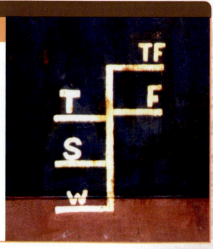

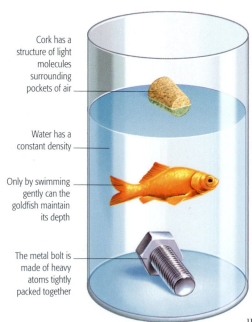

Cork has a structure of light molecules surrounding pockets of air

Water has a constant density

Only by swimming gently can the goldfish maintain its depth

The metal bolt is made of heavy atoms tightly packed together

### Will it sink or will it float?
Water is incompressible, so the density of fresh water is the same at all depths. Objects that are less dense than water will float, while denser items will sink. Seawater is more dense because it also contains salt.

**IN PRACTICE**

**PLIMSOLL LINE**

All large ships have to be marked with an International Load Line, or "Plimsoll Line" (after the British politician, Samuel Plimsoll, who helped introduce the law in 1876). This line is painted across a circle on a ship's hull, marking the position of the waterline when the ship is carrying the maximum safe load. The density of water varies with salinity and temperature, so various waterlines are marked: TF—Tropical Fresh Water; F—Fresh Water; T—Tropical Sea Water; S—Summer Temperate Sea Water; and W—Winter Temperate Sea Water.

## Submarine control

Submarines can be made to sink or rise to the surface through the use of their ballast tanks. Situated between the inner and outer hull, the tanks can be filled with air from compressed air tanks to decrease the vessel's density and make it rise; or with water to increase the vessel's density and make it sink.

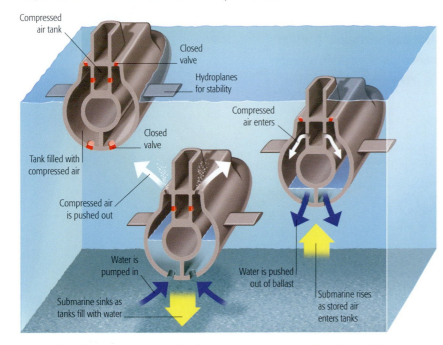

Compressed air tank

Closed valve

Hydroplanes for stability

Compressed air enters

Closed valve

Tank filled with compressed air

Compressed air is pushed out

Water is pumped in

Water is pushed out of ballast

Submarine sinks as tanks fill with water

Submarine rises as stored air enters tanks

**1 Submarine afloat**
When the ballast tanks are filled with air, the submarine's overall density is less than the water surrounding it, so it is able to float in the water.

**2 Submarine submerged**
For a dive, the ballast tanks are filled with water. This increases the submarine's weight, which become greater than the upthrust.

**3 Submarine surfacing**
Blowing compressed air into the ballast tanks drives out the water and lowers the weight of the submarine, so the upthrust takes it back to the surface.

## Controlling density

The first boats were probably simple rafts of wooden poles tied together, but then people found they could make lightweight boats by stretching animal skins over wooden frames. Wooden boats were the next step up, and ranged from the hollowed-out logs of Polynesian canoes to the birch-bark canoes of Canada, and the northern European "clinker-built" ships, which used overlapping wooden planks nailed to sturdy wooden frames. At the end of the 18th century the first iron boats appeared, and today the vast majority of large ships are made of steel.

All of these designs for boats and ships depend on the fact that the space inside the hull is filled with air, and the overall density of the vessel is equal to the total weight divided by the total volume—in other words, shipbuilders can control the density, assuming that water is unable to get inside. However, even huge ships can sink. The RMS *Titanic*, the largest passenger ship in the world in 1912, sank within three hours because the iceberg it collided with sliced through the ship's hull and allowed five of the buoyancy compartments to fill with water.

## Working with buoyancy

Boat and ship designers have learned to manipulate buoyancy, both in the materials and shapes of hulls and through the use of buoyancy tanks, while submarines are able to vary their buoyancy at will (see left).

Some marine creatures have gas bladders or swim bladders which fulfill the same function; they are able to fill them with oxygen to increase their buoyancy. But both supertankers and goldfish are subject to Archimedes' Principle—the upthrust is always equal to the weight of the fluid displaced.

### Boats for hunting

The Palestrina mosaic from the 1st century BCE is evidence of the ways in which mastery of the water developed. It shows Romans fishing or hunting with spears on the Egyptian Nile on a large oar-propelled boat, possibly specially built for the purpose.

**AFTER**

Great efforts have been made to reduce the number of deaths by drowning. Knowledge of density and buoyancy has influenced the design of boats and led to the development of personal flotation devices.

FLOTATION JACKET

**FLOTATION CHAMBERS**
Many recreational boats now have internal **flotation chambers** full of air, which prevent them from sinking even if they are swamped by water.

**BUOYANCY AIDS**
Life jackets have **saved many lives** by keeping their wearers afloat and their faces well above water. Buoyancy aids or life jackets are routinely made available not only to sailors, but also to all **passengers in commercial aircraft**. These buoyancy aids are designed to be easy to put on, secure, and durable, and they carry whistles and lights for attracting attention.

**SEE ALSO >>**
pp.144–45 LIQUIDS UNDER PRESSURE

# Algebra

**Early mathematicians lacked symbols for representing mathematical expressions and operations (such as addition and subtraction), which hindered progress significantly. With the introduction of symbols, algebra developed as a separate branch of mathematics and also led to major advances in other areas of the subject.**

Ancient Babylonian, Egyptian, and Greek mathematicians managed to describe and solve mathematical problems by expressing them in words and numbers (rhetorical algebra), but it was a long-winded process. A system was needed in which the answer being sought—the unknown—could be expressed in terms of the information known about it.

This remained a verbal process until the 9th century, when the Persian mathematician Al-Khwarizmi laid the foundations of what we now call algebra by setting out systematic and logical rules for solving equations. Later mathematicians built on his work by introducing symbols for operations, such as +, −, and =, and using the shorthand $x$ to represent an unknown and other letters for variables. The invention of specific mathematical

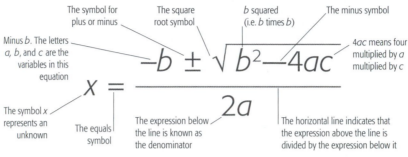

The symbol for plus or minus — The square root symbol — $b$ squared (i.e. $b$ times $b$) — The minus symbol

Minus $b$. The letters $a$, $b$, and $c$ are the variables in this equation

$4ac$ means four multiplied by $a$ multiplied by $c$

$$x = \frac{-b \pm \sqrt{b^2 - 4ac}}{2a}$$

The symbol $x$ represents an unknown — The equals symbol — The expression below the line is known as the denominator — The horizontal line indicates that the expression above the line is divided by the expression below it

**An algebraic equation**
The general formula for solving a quadratic equation $ax^2 + bx + c = 0$ shows some of the common mathematical symbols used in algebra.

signs and symbols simplified solving problems and established algebra as a field in its own right. It also enabled algebra to be used to solve problems in other disciplines, such as astronomy, physics, and economics, and to solve everyday problems too.

## Algebraic equations

Many problems can be expressed algebraically in the form of an equation, and unknown quantities can be calculated by applying the relevant rules of algebra.

As a simple example, suppose you paid a total of $28.75 for some stationery, and you know that the price included 15 percent tax. How much was the stationery before tax? To solve the problem algebraically, let the answer be $x$. Then $x$ plus 15 percent of $x$ is $28.75. In symbols, this

is represented as $x + 0.15x = \$28.75$, or $1.15x = \$28.75$. Therefore the price before tax, $x = \$28.75 \div 1.15 = \$25.00$.

It is also possible to use algebra to solve problems in which there are two (or more) unknowns, as in the following example. Two donkeys are walking along, each carrying bales of hay. The first donkey says to the other one: "If I had one of your bales, we would have equal numbers." The second donkey replies: "If I had one of your bales, I would have twice as many as you." How many bales was each donkey carrying? To solve the problem, let $x$ be the number of bales carried by the first donkey and $y$ the number carried by the second donkey. Then if one bale moves from donkey two to donkey one, donkey one has $x + 1$, donkey two has $y - 1$, and we know

▷ **Signs, symbols, and numbers**
A blackboard covered in algebraic expressions has become an immediately recognizable trademark of academic mathematics. Even when the equations are long and complex and contain unfamiliar symbols, the appropriate rules of algebra still apply.

« **BEFORE**

**Before the invention of algebraic symbols, problems in algebra were stated in a verbal form, known as rhetorical algebra.**

**ALGEBRA IN THE ANCIENT WORLD:**
From sources such as the Rhind papyrus we know that the **ancient Egyptians** could **solve linear equations** with several variables as long ago as 2000 BCE. The **Babylonians** of the same period were even more advanced, and **developed methods of solving quadratic** and possibly **cubic equations**. A Babylonian clay tablet (the Plimpton 322 tablet), dating from around 1600–1900 BCE, shows quadratic equations now known as Pythagorean triples.

**RHIND PAPYRUS**

**MATHEMATICIAN AND SCIENTIST (c.790–c.840)**

### AL-KHWARIZMI

Although Abu Ja'far Muhammad ibn Musa Al-Khwarizmi was the foremost Persian mathematician, astronomer, and geographer of his time, little is known of his life. He was probably born in Khwarizm, in present-day Uzbekistan, but spent most of his adult life as a scholar at the House of Wisdom (a library and intellectual center) in Baghdad. Sometimes known as the "father" of algebra, he established algebra as a distinct branch of mathematics, and translations of his writings later brought the idea to the West, along with Hindu–Arabic numerals. The word "algebra" is derived from *al-jabr*, part of the title of his book on the subject, *Hisab al-jabr wa'l-muqabala*.

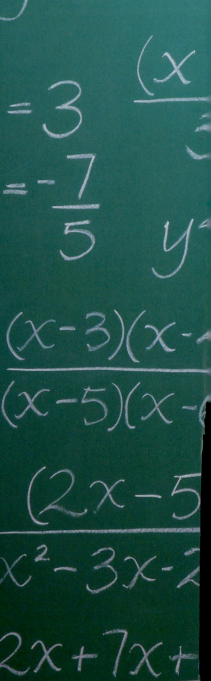

On the chalkboard (left):

$$1 = 5 - 6x - 3x^2$$

$$6x + 4y = 8 \qquad y =$$

$$\frac{6)^2}{\phantom{0}} - \frac{(y-8)^2}{64} =$$

$$x^2 - 2y + 4x - 4 =$$

$$< 0 \quad x^2 + (6-x)^2 =$$

$$x^2 - 6x + 8 =$$

$$\geq 0 \quad (x-2)(x-4) =$$

$$x =$$

$$> 0 \quad 1.96 = 0.5070$$

$$t \approx 8.1h$$

## TYPES OF ALGEBRAIC EQUATIONS

Equations can be classified into several different types according to the highest power to which one or more of their variables is raised. For example, in quadratic equations at least one variable is squared (raised to the power of two).

| Type of equation | Common mathematical form | Description/uses |
|---|---|---|
| Linear | $ax + by + c = 0$ | No variable is raised to a power greater than one. Used in simple problems of addition, multiplication, and division. |
| Quadratic | $ax^2 + bx + c = 0$ | At least one of the variables is squared. Used in calculations involving area. |
| Cubic | $ax^3 + bx^2 + cx + d = 0$ | At least one of the variables is cubed. Used in calculations involving volume. |
| Quartic | $ax^4 + bx^3 + cx^2 + dx + e = 0$ | At least one of the variables is to the power of four. Used in computer graphics. |
| Quintic | $ax^5 + bx^4 + cx^3 + dx^2 + ex + f = 0$ | At least one of the variables is to the power of five. Solution of quintic equations was important to the development of abstract algebra. |

they both now have the same number of bales, so $x + 1 = y - 1$. Adding 1 to each side gives $x + 2 = y$, or $y = x + 2$. Meanwhile if one bale moves from donkey one to donkey two, then donkey two has $y + 1$ bales, which is twice as many as donkey one has $(x - 1)$. In other words, $y + 1 = 2(x - 1)$, which is the same as $y + 1 = 2x - 2$. Subtracting 1 from each side of the equation gives $y = 2x - 3$. We already know that $y = x + 2$, so we can write $x + 2 = 2x - 3$. Subtract $x$ from each side to give $2 = x - 3$. Then add 3 to each side to give $x = 5$. And because $y = 2x - 3$, $y = 10 - 3 = 7$. So the answer is that donkey one had 5 bales and donkey two had 7 bales.

The same general principles can be applied to more complex equations. However, is not always possible to explicitly solve a given equation, although it is usually possible to obtain an approximate solution to a high degree of precision.

AFTER ≫

From the 19th century more complex fields of algebra developed, such as abstract algebra and Boolean algebra.

### BOOLEAN ALGEBRA
The philosopher and mathematician **George Boole** (1815–64) brought together **algebra and logic** in his book *An Investigation of the Laws of Thought* in 1854. In Boolean algebra, **arithmetical operations** (such as addition and multiplication) are **replaced with logical operations** such as AND, OR, and NOT, and numbers are dealt with in a binary form. Largely overlooked at the time, **Boolean algebra** later played a **major role in computing**—being used in search engines, for example.

### FERMAT'S LAST THEOREM
In the margin of a book the 17th-century mathematician **Pierre de Fermat** wrote "It is **impossible to separate** a cube into two cubes, or a fourth power into two fourth powers, or in general, **any power higher than the second into two like powers**." Known as Fermat's last theorem, this conjecture had confounded mathematicians for about 300 years until, using modern algebraic techniques, **Andrew Wiles** found a **proof in 1995**.

### IN PRACTICE
## COMPOUND INTEREST

Algebraic equations can be used to solve a huge variety of practical problems, including compound interest, which can be calculated using the equation:
$$A = P(1 + i)^n$$
Where **A** is amount of money accumulated, including interest
**P** is the amount of money deposited (or borrowed)
**i** is the interest rate per period (e.g. 5% per year, or 0.05 per year)
**n** is the number of periods for which the money is invested or borrowed

So if \$100 is invested for 2 years at an annual interest rate of 5% (0.05), the amount after 2 years is given by:
$$A = 100 \times (1 + 0.05)^2$$
Which gives the answer \$110.25.

**ANDREW WILES IN FRONT OF PART OF HIS PROOF OF FERMAT'S LAST THEOREM**

SEE ALSO ≫
pp.62–63 EAST MEETS WEST
pp.378–79 THE INTERNET

**BEFORE**

People's need for water led to some of the most ingenious and long-lasting engineering in the ancient world.

**ARCHIMEDEAN SCREW**
One early device for **raising water** to **irrigate fields** was the "Archimedes screw" **‹‹ 40–41**, which was attributed to **Archimedes** although it may have been in use since the 7th century BCE. It consists of an inclined spiral screw in a chute or cylinder, which scoops up water at its lower end, then carries it up as the screw is rotated.

**EARLY IRRIGATION**
**Agriculture** developed in the ancient world in areas where there was sufficient rainfall, and along the banks of large rivers, such as the Nile and the Euphrates. In order to **expand the area** available for **growing crops** in arid climates, **irrigation ditches** were dug to divert water from these rivers, in one of the earliest examples of **engineering** work.

**TRANSPORTING WATER**
As irrigation technology became more sophisticated, the irrigation channels developed into more **permanent structures**. Special water tunnels and bridges called **aqueducts** were built as early as the 7th century BCE, to carry water across long distances. These were perfected by the ancient Romans, who used them to supply both agricultural land and towns and cities.

IRRIGATION ALLOWED WATER CONTROL

**‹‹ SEE ALSO**
*pp.20–21* EVOLUTION OF THE WHEEL
*pp.40–41* SIMPLE MACHINES
*pp.42–43* HOW GEARS WORK

**Wind power from windmills**
Windmills were a common sight throughout the world until they were gradually replaced by steam, combustion, and electric-powered engines. Post mills such as these could be turned to face the prevailing wind, their rotating sails driving machines to grind corn or, more rarely, to pump water for irrigation.

**W**ater was an essential ingredient in early civilization: it was necessary for agriculture and for the growing towns. Early engineering focused on its transportation, and the flow of water in a river was soon recognized as a potential source of energy. By the time of the ancient Greeks, water was being used to drive simple machines. Wind power was also exploited in the ancient world by sailing boats.

**Harnessing power**
By the Middle Ages water and wind power were widely used to drive machinery. Water wheels—probably the earliest source of mechanical energy—powered the mills for grinding grain; later they were also used for sawing wood and clothmaking. They became more sophisticated in design: vertically mounted wheels were made possible by a system of gears that transferred power to the mill. Overshot wheels—where water flows over the top of the wheel—began to replace the less efficient undershot design, which relies on a fast supply of water under the wheel (see opposite).

The first windmills were built in the 7th century, in Seistan, Persia. Unlike the familiar modern windmill, these had sails mounted on a vertical shaft: wooden blades, sometimes covered with cloth, were mounted on arms to catch the wind. The sails were attached to a vertical axis driving a grinding stone.

**A global trend**
The idea soon spread to China and across the world. Windmills like those below soon began to appear in Europe in the 12th century; the sails were set at a slight angle to convert the

**Hurricane power**
The destructive power of the wind is seen at its most dramatic in tropical storms; hurricanes in the Atlantic and typhoons in the Pacific cause terrible damage on land and at sea every year. The enormous potential of wind power has been harnessed in minor ways since medieval times, and scientists today are researching the best ways of using this power to generate electricity.

# Water and Wind Power

**As early civilizations searched for alternatives to animal and slave power, two natural sources of energy were widely harnessed: water and wind. These remained the principal means of driving machinery until the discovery of steam and electrical power, and today they are seen as important sources of renewable energy.**

BREAKTHROUGH

## IMPROVEMENTS IN SAIL SHAPE

The rectangular sails of the earliest sailing ships were rigged perpendicular to the hull, efficiently catching the wind and propelling the boat forward when traveling in the direction of the wind, but they were not easy to handle against the wind. They were eventually replaced by triangular sails, which allowed the boats to sail closer to the wind.

**Rectangular sails**
These were hung from yards (crossbars), with one face to the wind. The yards could turn slightly around the mast.

**Triangular sails**
Fit lengthwise onto the ship on poles known as gaffs, triangular sails gave ships improved maneuverability.

**Chinese astronomical clock**
This mechanized armillary sphere and celestial globe demonstrates the technological sophistication of 11th-century China. Originally 37 ft (11.25 m) high, the water-powered clock also drove the sphere at the top of the tower, showing the positions of stars and planets.

Water and wind are the focus of today's scientific research into clean, renewable, and efficient forms of energy.

### WIND TURBINES
Windmill technology is now being used to **generate electricity**. Wind turbines **416–17 ≫** use huge propellers mounted on pillars to gather wind energy, which is converted into electricity. The turbines are often grouped in **wind farms**.

### WAVE AND TIDAL POWER
Experimental plants are being developed to generate cost-effective power from the **motion of sea waves**. The changing levels and currents of tides are also being exploited by dams across **tidal estuaries**, while underwater turbines offer a more reliable potential than wind power.

## 98–99
**percent of Norway's electricity was generated from hydroelectric plants in 2009, powered by high-level lakes.**

### HYDROELECTRIC POWER
Up to 20 percent of the world's electricity is produced from hydropower. This **harnesses the energy** of water, under **gravitational** force, to drive a turbine connected to an electrical generator. Most hydropower plants are built as a **dam across a river**, creating a reservoir with a constant store of water whose release through the **turbine** can be controlled according to demand and the availability of water.

horizontal force of the wind into rotational movement. These early windmills were built around a solid vertical post (hence the name "post mills") and could be turned to face into the wind. A system of gears derived from the design of vertical water wheels transferred the motion of the sails to the grinding wheels inside the mill.

Later medieval windmills improved upon this original design; the tower mill needed only the cap carrying the

sails to be turned into the wind. Windmills such as these were in common use even after the invention of steam engines, and only declined with the advent of the combustion engine.

### Understanding hydraulics
Wind and water power were also used in sawmills, clothmaking plants, and smelting metal furnaces. In the Islamic world the principle of the water wheel was developed into the *noria*, an undershot water wheel incorporating

containers, which raised water for irrigation. In medieval China windmills inspired by the Persian design were used to pump water. Hydraulic power was widely used in water-powered machines throughout China and the Islamic world.

Interest in water and wind power diminished during the 19th century as new forms of energy were discovered, but these natural powers are exciting interest again today in the search for renewable energy sources to replace non-renewable carbon-based fossil fuels. Wave power, tidal power, wind turbines, and hydroelectric plants are all increasingly being used around the world to generate electricity.

**ITAIPU HYDROELECTRIC DAM, BRAZIL**

**SEE ALSO ≫**
*pp.132–33* STEAM POWER TO STEAM ENGINE
*pp.252–53* GENERATING ELECTRICITY
*pp.414–15* GLOBAL WARMING

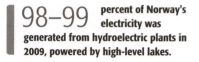

◁ **Overshot water wheel**
Water from a reservoir is directed through a channel on to the paddles or blades on the rim of a wheel, making it rotate. More efficient than the undershot water wheel, the overshot uses not only the flow of water, but also the weight of water carried in paddles on one side of the wheel.

The force of the water and gravity combine to turn the wheel

The wheel is turned purely by the force of the current beneath it

▷ **Undershot water wheel**
The earliest water wheels were slightly submerged in a stream or river, in an arrangement known as undershot. The flow of water acts on paddles at the bottom of the wheel to cause it to rotate, providing power for a mill.

**Medieval irrigation**
Widely used in the medieval Islamic world, *norias* use an undershot water wheel to raise water up to an aqueduct, where it can then be directed to a town or farmland. As the *noria* is rotated by the current, containers around its rim scoop water from the river and carry it up to the raised aqueduct.

## « BEFORE

The inspiration for alchemists of the Middle Ages and the Renaissance went back to ancient Egypt and ancient Greece.

### METALLURGY AND MYSTICISM
In the ancient world mystical significance was given to crafts, especially the craft of metalworking, where apparently **"magical" transformations occurred in the smelting process.** The occult learning of craftsmen was shared with the priestly class in cultures such as ancient Egypt, where the allegorical writings of Zosimos formed the basis for the "science" of alchemy.

### THE HERMETIC WRITINGS OF THOTH
Medieval alchemists based many of their ideas on what they believed to be **the writings of the Egyptian god Thoth**, which were in fact Hellenistic texts. Ancient Greeks had integrated the figure of Thoth with their god Hermes into one, Hermes Trismegitus. The *Hermetica*, supposedly the writings of this hybrid god, proposed a number of **"Hermetic Arts"** of which alchemy was considered to be one.

THOTH

## « SEE ALSO
*pp.18–19* EARLY METALWORKERS

# Alchemy

**Although regarded sceptically today as a mere pseudoscience, alchemy was an important and respected branch of learning from the time of the ancient Greeks to the Renaissance, and many of its techniques and discoveries formed the basis of the modern sciences of chemistry, biochemistry, and pharmacology.**

The central principle of alchemy, transmutation of matter, became the aim of philosophers in Hellenistic Greece, China, and India at much the same time, around the 3rd century BCE. Each of these cultures had sophisticated techniques of metal-working, often imbued with mystical significance. The processes of melting and mixing metals, essentially changing their character, became symbolic of the philosophical and religious ideas of the time. Alchemists saw them as the key to three distinct goals: the transmutation of base metal into gold; the discovery of a panacea or universal medicine; and creation of an elixir of life granting immortality.

From the empirical findings of these early metallurgists, which included the effects of fire on metals and elements such as mercury, sulfur, and arsenic, alchemy developed into a methodical discipline, involving experiments and the naming and classification of substances. Each of these was seen to have individual properties, which alchemists linked to mythical and astrological ideas. In addition, Greek alchemists proposed the possibility of creating a substance, the "philosopher's stone," which was thought to be both a panacea and a means of transforming metals to gold.

## Chinese and Indian alchemy
While Greek alchemists placed emphasis on the "perfect" metals—gold and silver—and the search for the philosopher's stone, in China the primary aim of alchemical experiments was to find the "Grand Elixir of

### Islamic alchemy
Rhazes, the most influential of the Islamic alchemists, discovered and classified a large number of natural substances in experiments using sophisticated apparatus invented by other Islamic alchemists.

COPPER CRYSTALS

METAL LIME

LIME

COPPER SAFFRON

URINE

**Alchemical symbols**
Alchemy's links with astrology and other occult beliefs were reflected in the symbols chosen to represent different materials, emphasizing their mystical significance.

Immortality." This quest was intimately associated with the search for the spiritual purity of Taoism and the martial arts. Discoveries of the medicinal uses of cinnabar (mercuric sulfide) and more importantly the mixture of sulfur, charcoal, and saltpeter (potassium nitrate) to make gunpowder (see pp.56–57), were in fact by-products of this experimentation.

Similarly, Indian alchemy of the same period stressed its pharmaceutical applications and was linked with Ayurvedic medicine and the Dharmic religions, even though its roots lay in the metallurgy of Indian craftsmen. In the course of their experiments, all three cultures developed techniques such as heating, burning, and most importantly distillation, which were later taken up by Persian and Arabic scientists in what was to be a golden age of alchemy in the Islamic world.

### The influence of Islam

The Islamic empire, which spread across the Middle East in the 8th century, boasted remarkable cultural and scientific achievements. Building

names given at this time: alchemy is from the Arabic *al-kimia*, "the art of transformation," and terms such as alkali, alkane, and alkaloid also have Arabic roots.

### The pursuit of gold

The great body of learning of the Islamic scientists, including alchemy, came to Europe in the 12th century through translations of Arabic and Greek texts. Medieval scholars such as the British Roger Bacon (see pp.56–57) took an avid interest in the art of alchemy, which fit well with the mystical Christianity of the time. The ability to transmute base metal to gold was also an attractive prospect in an increasingly commercial Europe, as was the possibility of eternal life. During the Renaissance, the emphasis in alchemy shifted back to the search for a panacea, with the Swiss Phillip von Hohenheim (1493–1541), who later took the name Paracelsus (see pp.70–71), leading the way in finding medicinal uses for chemicals. But with the advent of the so-called Scientific Revolution, from around the middle of

**Alchemist processes (c.1470)**
Alchemists tried many different processes, most of them involving heat or fire, in their attempts to make gold, and by chance made important chemical discoveries.

**Distillation by alembic**
Invented by Geber in the 8th-century, the alembic consists of a flask which is heated and from which vapor rises to be condensed and so transferred into a second flask. This woodcut shows a pair of intertwined alembics in action.

> "The **alchemists** in their **search** for **gold** discovered many other things of greater **value**."
>
> ARTHUR SCHOPENHAUER, GERMAN PHILOSOPHER, (1788–1860)

on the knowledge that had been passed down from the ancient Greek and oriental experimenters, alchemists in Persia such as Geber, Rhazes, and Avicenna treated the subject with a more rigorous scientific approach to experimentation. They refined and perfected the techniques of filtration, crystallization, and distillation using apparatus such as the alembic (a still for distillation invented by Geber), which allowed them to produce several new substances. Most important among the discoveries from their experiments were alcohol and the mineral acids (including hydrochloric, sulfuric, and nitric acids); they also found that a mixture of nitric and hydrochloric acids subsequently called *aqua regia* could dissolve gold—which for an alchemist had mystical significance.

Perhaps the most methodical of these Islamic alchemists was Rhazes. By documenting all his experiments, he was able to name and classify the many new substances that he and his predecessors had produced for the first time, almost free of mystical references. Many of the words used in chemistry today have their origins in

the 16th century, alchemy and its mystical associations fell out of favor—although it still had notable followers such as Robert Boyle and Isaac Newton in the 17th century.

The contribution of alchemy to scientific knowledge had, however, been huge. Stripped of its occult associations by the Enlightenment, it evolved into the science of chemistry as we know it today.

**AFTER**

As the science of chemistry emerged in the 17th century, ideas about the elements began to change.

**PHLOGISTON THEORY**
In 1667 the German scientist **Johann Joachim Becher**, put forward a theory that **in addition to the four classical elements, there was a fifth,** which he called phlogiston. This new element was supposed to exist inside combustible materials, and was released during burning. Although the theory was soon discredited, it had prominent adherents including the British **Joseph Priestley 146–47 »**, who was credited with **the discovery of oxygen**.

**CHEMICAL REVOLUTION**
The so-called **Chemical Revolution**, spearheaded by early chemists including the French **Antoine Lavoisier**, dismissed the phlogiston theory and its continued adherence to the notion of the classical elements. Lavoisier recognized and named hydrogen and oxygen, and made the first steps toward a complete list of the elements.

SEE ALSO »
pp.138–139 THE NATURE OF MATTER

**BREAKTHROUGH**

### DISCOVERING PHOSPHORUS

In the search for the mythical philosopher's stone, many important discoveries were made by alchemists. One of the most significant of these was made in 1669 by a merchant and alchemist from Hamburg, Germany, named Hennig Brand. By boiling and concentrating urine he managed to produce a substance that glowed with a pale green light, which he named phosphorus (from the Greek for "light-bearing"). This was a major turning point in the evolution of alchemy into chemistry.

ASTRONOMER **Born 78 CE Died 139 CE**

# Zhang Heng

> "**The Sun** is like **fire**, and the **Moon** like **water**. The fire gives out **light** and the water **reflects it**."

ZHANG HENG, FROM "LING XIAN" (MYSTICAL LAWS), 120 CE

**D**espite his modest and unambitious nature, Zhang Heng became an important official in the court of the Eastern Han emperors. He was also an influential astronomer, mathematician, inventor, and poet. His inventions and writings are among the most important of his time, and his significance in the field of astronomy is compared to that of his Greek contemporary Ptolemy (see p.28).

**Emergence of a poet**

Zhang was born in Nanyang, China— where his grandfather had been a governor—and showed a talent for poetry and literature from an early age. While studying in the provincial capital, Luoyang, he established a reputation as a literary scholar as well as a poet. He was offered several posts at the Imperial court, but turned these down to continue his literary career until eventually accepting a minor post in the civil service in Nanyang. It seemed that Zhang, while in his twenties, was destined for a career as an administrator and poet.

**Astronomer at the Imperial court**

At the age of 30 Zhang took an interest in science. He devoted his time to studying astronomy and mathematics, and was soon as well known for his writings on these subjects as he had been for his poetry. As word spread of his expertise, he was once again invited to serve at the Imperial court in Luoyang, and this time he accepted. After a few years, he was promoted to the post of chief astronomer—an important position. The Chinese at that time devoutly believed in the astrological influence of the stars over their fortunes, but Zhang's interest was more scientific; he made detailed observations of the night sky, mapping

**Water-powered armillary sphere**

An invention that originated in China, the armillary sphere was improved to a high degree of sophistication by Zhang Heng. Beginning with a bamboo model, he went on to construct a bronze globe that showed the position of all the known stars. The globe's movement of precisely one rotation each year was powered by a water clock standing beneath it.

**Lunar eclipse**
From his observations of the night sky, Zhang deduced that the Moon is a sphere that reflects the light from the Sun. His idea was proved during a lunar eclipse, when Earth comes between the light of the Sun and the Moon and creates a shadow on the Moon's surface.

## ZHANG HENG'S SEISMOMETER

Zhang Heng invented this device for detecting distant earthquakes in 132 CE. It consisted of an egg-shaped copper container, inside which was a sensitive pendulum. In the event of an earthquake, even at a great distance, the pendulum moved a mechanism that released a ball from the mouth of one of eight dragon heads fitted to the container. The ball would fall into the mouth of one of eight copper frogs representing points of the compass, indicating where the earthquake had occurred.

**TIMELINE**

- **78 CE** Born in Xi'e in Nanyang Commandery (north of modern Nanyang City, Henan province), China.
- **88 CE** His father dies, and he is brought up by his mother and grandmother.
- **94 CE** Already an accomplished writer at the age of 16, he leaves Nanyang to study in Chang'an and Luoyang, the Han capital. He soon becomes well known for his literary writings, and has several works published.
- **101 CE** Returns to Xi'e and works under Governor Bao De, with rank of officer of merit, as an archivist and writer of official documents.
- **106 CE** Emperor He dies and is succeeded by the infant Emperor Shang, then less than one year old. When Shang also dies later that year, 12-year-old An is declared emperor.
- **108 CE** Commences his studies in astronomy, and begins to work on treatises on astronomy and mathematics.
- **112 CE** His reputation as an astronomer and mathematician spreads, and he is summoned to Luoyang to work for the Imperial Secretariat in the court of Emperor An.
- **115 CE** Appointed chief astronomer to the Imperial court, and is made responsible for managing national documents and editing national history.
- **123 CE** Introduces reforms to the calendar to bring it into line with his astronomical observations, but his suggestions cause controversy and he loses favor in the court.
- **125 CE** Emperor An dies, at only 32, leaving a corrupt and failing government. He is succeeded by his son Shun, who is only 10 years old when he is declared emperor.
- **126 CE** Re-appointed as chief astronomer by Emperor Shun, and is later given the rank of palace attendant to the emperor.
- **132 CE** Presents his design for a seismometer to the court.
- **136 CE** Amid increasing political rivalry in the court, Zhang retires from office under Emperor Shun, and is appointed chancellor of Hejian (in modern Hebei).
- **138 CE** Zhang's seismometer detects an earthquake to the west of Luoyang, which is confirmed by messengers arriving a few days later.
- **138 CE** Retires completely and returns to Nanyang, but is called back to Luoyang to serve in the Imperial Secretariat after only a few months.
- **139 CE** Dies in Luoyang, at 61. His body is taken to be buried in his home town of Xi'e.

all the stars and planets. He stated in his treatise, *Ling Xian* (Mystical Laws):

"North and south of the equator there are 124 groups which are always brightly shining. 320 stars can be named. There are in all 2,500, not including those which the sailors observe. Of the very small stars there are 11,520."

He also came to some innovative conclusions about the nature of the Universe, observing:

"The sky is like a hen's egg, and is as round as a crossbow pellet, the earth is like the yolk of the egg, lying alone at the center. The sky is large and the earth small."

Zhang, unlike his predecessors, conceived a spherical Universe with Earth as a sphere at its center, and realized that the planets were also spherical, as well as the Moon, which he controversially suggested was not in itself a source of light, but merely reflected the light of the Sun.

## Mathematics

Zhang's mathematical work was no less innovatory. He studied "magic squares" of numbers—in which the rows, columns, and diagonals add up to the same number—and tried to find a method of calculating a value for pi ($\pi$), which until that time had been achieved only by practical measurement rather than theoretical calculation. In one of his mathematical treatises he provided a geometrical solution to the problem using circles drawn inside and around squares. It

**Figurine of Zhang Heng**
In China Zhang is commemorated by portraits, statues, and numerous porcelain figurines. A portrait of Zhang was featured on a Chinese postage stamp in 1955.

gave $\pi = \sqrt{10}$ (approximately 3.162), and working from this he calculated the volume of a sphere.

Unfortunately, his work was not always appreciated by the court. Among the court ministers there were power struggles; Zhang took no interest in these but became embroiled when he suggested alterations to the calendar based on his astronomical data. The idea conflicted with the traditional astrological basis of the calendar, causing a controversy that denied him any further promotion.

### Later achievements

Zhang nevertheless went on to devise some remarkable inventions. The first was a water-powered armillary sphere (see opposite), achieved through his detailed study of the night sky and thorough knowledge of the mechanics needed to make it rotate exactly once a year. Perhaps more famous was his seismometer (see above), which sensed distant earthquakes and indicated the direction from which they had been detected; presented to the court in 132 CE, it was not proved to be accurate until 138 CE.

Zhang died the next year, leaving a body of work that included some of the finest poetry of the age. His treatises on astronomy and mathematics were significant advances, and he is regarded as probably the most important Chinese thinker of his time. He is commemorated by his writings and inventions, and in the naming of later discoveries in his honor. These include the lunar crater Chang Heng, the asteroid 1802 Zhang Heng, and the mineral Zhanghengite.

### Zhang Heng and his seismometer

Mainly remembered today as the inventor of the seismometer, Zhang Heng was also a gifted astronomer, mathematician, poet, and literary scholar.

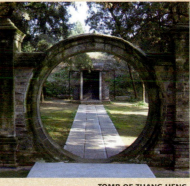

**TOMB OF ZHANG HENG**

« **BEFORE**

**So terrifying and devastating is fire that it has been used widely as a weapon for at least 3,000 years.**

**SECRET WEAPON**
"Greek fire" was a secret incendiary mixture brought to Constantinople by **Kallinikos** in 673 CE. It probably included **naphtha, or rock-oil**, a black, sticky material that oozes out of the ground in the region between the Black Sea and the Caspian Sea. This may have been distilled, and mixed with **sulphur or resin**. The material could be catapulted in tubs or pumped through blowpipes; it floated on water, and was easy to light, but extremely **hard to put out**, except, allegedly, with urine, vinegar, or sand. It was a major factor in the **defense of the Byzantine Empire** for 800 years.

**FLAMING ARROWS**
Arrows tipped with **burning oil-soaked rags** were fired at the wooden walls and buildings of **Athens** in 480 BCE. More than 2,200 years later, in 1806, **William Congreve** used his **rockets** to attack a French fleet in the harbor at **Boulogne**. The rockets all swerved off course, with the result that no ship was hit, but most of the town was **destroyed by fire**.

▷ **Early Chinese rocket**
The Chinese used bamboo for their rocket tubes; a firestick equipped with a nozzle and a guide stick to steady the direction made a powerful weapon.

# Gunpowder and Fire Weapons

**Invented by the Chinese in the 9th century, gunpowder rapidly became a crucial factor in battle. The secret spread slowly to the West and reached Europe by the 13th century. Incendiaries, materials that readily set fire to enemy property, predated gunpowder, and are still in use today.**

For hundreds of years gunpowder was a vital ingredient for such military hardware as bombs and guns, and in peacetime for mining and demolition. Gunpowder has its own built-in supply of oxygen, so that it burns extremely rapidly; in burning it produces a large quantity of gas. If it is ignited on an open surface it goes off with a fiery whoosh, but if it is contained then the gas generates enormous pressure in the container, and the result is an explosion.

## Invention of gunpowder
Gunpowder was invented in the 9th century by Chinese alchemists or monks looking for the elixir of life. They had used saltpeter (potassium nitrate) in various medicinal compounds for hundreds of years, and probably came across this highly inflammable material by chance. A Taoist text, which is thought to date from the mid-800s, includes the passage. "Some have heated together sulfur, realgar [an arsenic sulfide mineral], and saltpeter with honey; smoke and flames result, so that their hands and faces have been burned, and even the whole house where they were working burned down."

Popular legend has it that the Chinese used gunpowder for fireworks, as they do today, but in reality they quickly harnessed it for military purposes. Firesticks were bamboo tubes filled with gunpowder, lit, and pointed at the enemy at close range. Primitive guns were firesticks with pebbles packed in front of the gunpowder. Rockets were firesticks with a nozzle and a guide stick. Bombs were simply gunpowder in containers.

## Arrival in Europe
Knowledge of gunpowder reached the Arab world in the 13th century, and soon traveled through Europe, possibly via a German monk and alchemist called Berthold Schwarz, but certainly through an English monk in Oxford, Roger Bacon (see opposite).

Bacon wrote down what appears to be the formula of gunpowder in Latin and in code: "*Luru Vopo Vir Can Utriet.*" This can be interpreted as a mixture of charcoal, sulfur, and saltpeter, but it

is so low in saltpeter that it would only crackle and fizz; it would not explode.

Real gunpowder, or "black powder," is a mixture of the three ingredients in the proportions 15 percent charcoal, 10 percent sulphur, and 75 percent saltpeter. These should ideally be ground together to a fine powder, so that they are intimately mixed. When burned, gunpowder produces large amounts of gas, including carbon dioxide and sulphur dioxide.

Roughly speaking, a teaspoonful of gunpowder will burn to produce about 5 liters of hot gas in perhaps one hundredth of a second. If the gunpowder is contained this will create an enormous pressure surge, which may be used to cause an explosion, drive a bullet from a gun or a ball from a cannon, or push a rocket into the air.

Old-fashioned black powder produces lots of smoke, and because it burns rather than detonating is classified as a low explosive. However, it is still used in fireworks, where its behavior can be tailored to specific requirements, such as directed bursts, time delays, and

**Berthold Schwarz**
Commemorated by a statue in Freiburg, Germany, Schwarz is credited with inventing gunpowder and the gun in 1353. He may have been a Franciscan monk and an alchemist, but some authorities say he never existed and was merely a legendary character.

Charcoal 15%   Saltpeter 75%
Sulphur 10%

**The components of gunpowder**
Effective gunpowder contains 75 percent saltpeter (potassium nitrate), to provide the oxygen for the combustion; 10 percent sulphur; and 15 percent carbon. Sodium nitrate is a cheaper alternative to potassium nitrate but it tends to get damp.

AFTER »

secondary explosions, and where the addition of mineral salts produces all sorts of colors and effects.

## Flammable materials in warfare

Medieval alchemists produced their own recipes for incendiary materials, one of which was made "by taking petroleum, liquid pitch, and oil of sulphur. Put all these in a pottery jar buried in horse manure for 15 days… When the sun rises and before the heat has melted it the mixture will inflame."

Fire ships were often used to attack enemy vessels in protected harbors: allowing a burning ship to drift into a harbor packed with wooden vessels at anchor could have devastating results.

Flamethrowers were used in World War I, and even today the use of phosphorus bombs is common. White phosphorus catches fire when exposed to air, and continues to burn for a long time, starting fires and causing horrific burns on the skin.

### The siege of Orléans

An English army laid siege to Orléans on October 12, 1428, hoping to take the city and the rest of France. Despite the use of cannon and artillery, the English did not prevail and the siege was broken four months later by Joan of Arc.

### SCIENTIST AND PHILOSOPHER (1214–92)
### ROGER BACON

Born in 1214 at Ilchester in Somerset, Roger Bacon was a pupil of Robert Grosseteste in Oxford. He then went on to study in Paris, but returned to Oxford to become a Franciscan monk. Some deride him as a medieval necromancer (invoker of the dead); others hail him as the first experimental scientist. He was an advocate of the importance of observation and experiment, and said that "Nothing can be known except by experience." His views were too radical for some of his colleagues, however, and around 1278 he was placed under house arrest. He died at Oxford in 1292.

> "…of **saltpeter** take six parts, five of **young willow**, and five of **sulphur**, and so you will make **thunder** and **lightning**…"
>
> ROGER BACON, c.1252

Improvements in chemistry led to the development of high explosives and other advanced incendiary materials.

### THE USE OF NAPALM

During World War II American chemists devised a modern form of Greek fire, based on jellied gasoline. The thickener was a mixture of aluminium **NAphthenate and PALMitate**; hence the name. Modern napalm, napalm B, is mainly a mixture of **benzene and polystyrene**, which is more adhesive, and burns for up to 10 minutes. Napalm was used by the Americans in the **Vietnam War**, and has been a factor in many other conflicts. It is most commonly dropped from aircraft in incendiary bombs, but a thinner version can be used in flamethrowers by infantry.

**NAPALM IN THE VIETNAM WAR**

SEE ALSO »
pp.240–41 MASS PRODUCTION OF CHEMICALS

# The Printing Revolution

**The pace of development in the civilized world has always been associated with literacy. Thousands of years ago only nobles, scribes, and priests could read, but gradually more people became literate and there was a growing demand for information. Early printing methods were slow and expensive, but this changed in the 15th century with the invention of the Gutenberg press.**

Characters carved in stone, such as the hieroglyphics on ancient Egyptian pyramids, stand as testimony to humankind's desire to spread information—to make things known to a wider audience. The ambition to disseminate information in a more portable form is equally old: the ancient Egyptians wrote on scrolls made from the papyrus plant, and the Sumerians of Mesopotamia (now southern Iraq) pressed characters into clay from around 8000 BCE. But printing—the mass reproduction of text and images—only became a possibility after the Chinese invented paper in the 2nd century CE.

### Woodblock printing

Once in possession of thin, porous paper, the Chinese, Japanese, and Koreans carved whole pages of text into the surface of wooden blocks, reproducing complicated ideographs (graphic symbols) in reverse so that the printed image would read correctly. They "printed" the pages on one side only, by simply brushing paper on to the inked block using a special brush. Books were made by sewing the pages back to back, and then folding them to reach an accordion-like length.

### Movable type

Woodblock printing could not easily be simplified in Eastern languages, because of their thousands of ideographs, but a version of movable type characters did appear in China around 220 CE. Metal type followed in Korea around 1230.

These innovations were unknown in Europe, where woodblocks were still being used to make books, textiles, playing cards, and many other products. By the 15th century, however, many Europeans were looking for ways to speed up the printing process, and the true breakthrough came with the invention of movable type in about 1440.

German Johannes Gutenberg (see below) found a way to cast individual letters rapidly by pouring a molten alloy of lead, tin, and antimony into copper molds in which the characters had been stamped in reverse. Filed to

**The world's oldest printed book**
*The Diamond Sutra*, an ancient Buddhist text, was printed using woodblocks in China. It actually bears the date of production: the equivalent of May 11, 868 CE.

« **BEFORE**

**Before the advent of printing, writing was used in a variety of forms and media to record and spread information.**

**ANCIENT WRITING SYSTEM**
**Cuneiform script** was developed in Sumeria in the 3rd millennium BCE. The writing took the form of **pictographs** inscribed in vertical columns on clay tablets with the sharpened tip of a reed, known as a stylus. Later, the pictographs were read horizontally, from left to right.

**CUNEIFORM WRITING**

**PAPYRUS AND VELLUM**
The ancient Egyptians wrote on **scrolls made from the papyrus plant**, using an ink made from soot mixed with gum and water. The Romans wrote on **vellum** (stretched calfskin).

**ILLUMINATED MANUSCRIPTS**
From 400 CE onward text manuscripts were enhanced by **colorful, decorative elements.** The term "illuminated" refers to illustrations incorporating gold or silver.

« SEE ALSO
*pp.38–39* ANCIENT IDEAS OF THE WORLD

**Movable type**
Gutenberg created movable type characters by pouring molten metal alloy into individual copper molds, which had been shaped into letters using a hard-carved punch.

### Gutenberg press

No actual press has survived from Gutenberg's time, although illustrations of its use do exist. Most important was the use of a screw to press the paper firmly on to the inked type, ensuring a sharp and even impression.

**Tympan**
Paper or vellum is mounted on the tympan, which folds down on to the inked type.

**Platen**
When lowered the platen exerts even pressure over the closed tympan and inked forme.

**Forme**
The forme holds a page of type securely in the press, ready for inking and printing.

**Frisket**
When folded over the tympan, the frisket's slots allow the inked type to press against the paper.

**Coffin**
A wooden frame supporting the forme, which slides under the platen to make an impression.

**INVENTOR (c.1398–1468)**

## JOHANNES GUTENBERG

Born in the German city of Mainz, Gutenberg is thought to have first acquired a knowledge of goldsmithing through his father's work for the bishop of the city. It was probably his experience of molten metal that led to his realization, in a "ray of light," that individual type characters could be struck and molded many times over. His invention of movable type, revealed c.1440, was the innovation that led to the development of the first printing press in 1450. Five years later a dispute with a business partner bankrupted him, but in later life he was given a stipend by the archiepiscopal court.

**Printing techniques** ▷
The advancement of printing techniques has not made earlier methods almost obsolete. They are still used for specialized purposes, especially in arts and crafts, where very specific, minutely controllable effects are required.

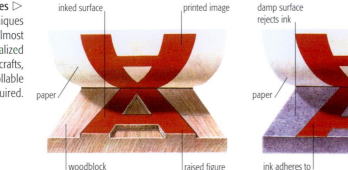

inked surface | printed image
paper | woodblock | raised figure

### Relief printing
This is essentially the method used in woodblock printing; the paper receives ink only from raised areas.

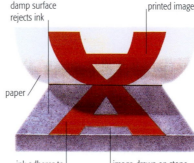

damp surface rejects ink | printed image
paper | ink adheres to greasy image | image drawn on stone with greasy medium

### Lithographic printing
The surface used for lithography is completely smooth. A greasy medium attracts ink that is offset onto paper.

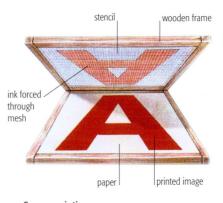

stencil | wooden frame
ink forced through mesh | paper | printed image

### Screen printing
An ink-blocking stencil is placed on a mesh screen and ink is sqeezed through unblocked areas onto the paper below.

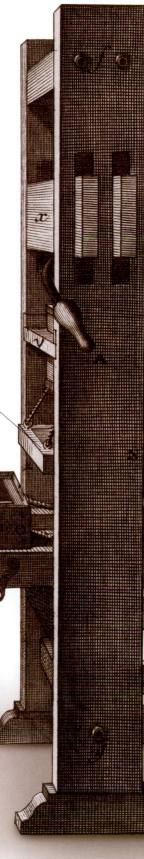

**Handle**
Pulling down on the handle screws the platen onto the back of the closed tympan.

**Frame**
The massive construction of the press ensures minimal slippage during printing.

a uniform depth, the characters could be set together to form blocks, and then pages, of text.

### The Gutenberg press
Gutenberg realized the potential of movable type for the mass production of printed matter. Borrowing large sums of money from friends, he secretly set about creating a press, taking as his model the screw presses used at the time for tasks such as crushing grapes. He also adapted existing inks for use with the new press. His large workshop was truly integrated, doing its own ink-making, engraving, and type-casting.

Printing presses did not change very much until the late 18th century, when steel components replaced the wooden structures. The next breakthrough—printing in a continuous vertical and horizontal motion—did not occur until the 19th century, when a German, Friedrich König, built the first steam-powered, flat-bed cylinder press for *The Times* newspaper in London, printing its first edition on November 29, 1814.

**378,000** The total number of new titles published in 2005 by the two leading book-producing countries alone: 206,000 in the UK and 172,000 in the US.

**Gutenberg Bible**
Printed in 1455 the Gutenberg Bible was the printer's crowning achievement; 180 copies were produced, 45 on vellum and 135 on paper. The illustrations were hand drawn, and each book took a year to finish.

## AFTER »

The 20th century saw the introduction of photography-based technology that offered cheap printing, initially on a small scale.

### PHOTOCOPYING
The office photocopier—Xerox—revolutionized the process of **producing paper copies** in the 1960s. The machine worked by a dry process using heat, called **xerography**, and quickly replaced other copying methods.

### COMPUTER PRINTERS
By the late 1970s office printers were linked directly to computers. Today, publishers can extend this direct link by sending digital files electronically to **computer-controlled presses** to be printed. Many people now have desktop printers at home and are able to print and publish their own work.

**OFFICE PRINTER**

**SEE ALSO »**
*pp.378–79* THE INTERNET

> "Like a new star it shall scatter the **darkness** of ignorance."
>
> JOHANNES GUTENBERG, INVENTOR, 1450

PHYSICIST AND ASTRONOMER **Born 965** Died 1039

# Alhazen

## "**Truth** is sought for **itself**, but the **truths** are immersed in **uncertainties**."

**ALHAZEN**, FROM "APORIAS AGAINST PTOLEMY," 1025–28 CE

Abu Ali al-Hasan ibn al-Haytham, whose name is usually Latinized to Alhazen or Alhacen, was born *c.*965 CE in Basra, in what is now Iraq. He is known as the "father" of the science of optics and, emerging from a thriving Islamic culture of curiosity and experiment, he was also one of the earliest experimental scientists. He was called "the second Ptolemy."

Alhazen was educated in Basra and Baghdad, where he also worked as a civil servant. He was interested in science and engineering and when, in 1011, he heard about the annual flooding of the Nile, he wrote to Caliph Al-Hakim bi-Amr Allah, offering to solve the problem. The Caliph received him with ceremony at the gates of Cairo, and he traveled south to build a dam.

Unfortunately the young man had no inkling of the vast size of the Nile. Even at Aswan, far from the coast, it is perhaps 1 mile (1.6 km) wide and has

**Alhazen's** *Opticae*
Translated into Latin in 1270, Alhazen's book was published in Basel in 1572 by Friedrich Risner with the title *Opticae Thesaurus: Alhazeni Arabis* (Encyclopedia of Optics by Alhazen the Arab).

multiple streams. To build a dam with the available technology was utterly impossible. However, he could not go back to the notoriously ruthless Caliph and admit defeat, since he would surely have been executed. So he feigned madness, was placed under house arrest, and remained "mad" until the Caliph died in 1021.

### Optical writings
Confined though he was, Alhazen did not waste his time under house arrest. He studied optics intensively, investigating shadows, the colors of sunsets, rainbows, and eclipses. From these empirical investigations came his *Book of Optics*, written in the form of seven volumes or scrolls. Entitled *Kitab al-Manazir*, it was translated into Latin in 1270 as *Opticae Thesaurus: Alhazeni Arabis*. This work had considerable influence on the English monk Roger Bacon, the German astronomer Johannes Kepler (see pp.78–79), and Francis Bacon (see pp.68–69), who wrote about experimental science in the 1620s.

### New understanding of vision
Among the topics covered was the way in which we see. Aristotle, and later Euclid and Ptolemy, had said that when we look at something, rays leave our eyes and bounce off it. Alhazen realized that this was wrong; the light is there anyway—during the day everything is lit by the Sun, with light bouncing off objects in all directions; at night, the Sun's light is reflected to us by the Moon. All we have to do is open our eyes, and the light pours in. In other words, vision is a passive rather than an active phenomenon, at least

**An impossible project**
Alhazen had hoped to be able to dam the Nile in this area near Aswan, but when he saw the sheer size of the river he realized it was beyond him.

**Alhazen**
One of the earliest experimental scientists, Alhazen showed how the eye works, and proved that light travels in straight lines.

### INVENTION

## CAMERA OBSCURA

*Camera obscura* is Latin for "dark room." In its simplest form this comprises a dark room with a small hole in one wall and a white wall opposite. Light from the outside world streams through the hole, and forms an image on the opposite wall. The image is upside down and faint, although it can be made brighter and more distinct by enlarging the hole and inserting a suitable lens to focus the incoming light.

A pinhole camera is a small version that can be used for taking photographs. The modern camera is a direct descendant, with a lens used in place of the pinhole.

### TIMELINE

- **965** Abu Ali al-Hasan ibn al-Haytham (the name is Latinized as Alhazen or Alhacen) is born in Basra, then part of Buyid Persia, now in Iraq.

- **1011** Goes to Egypt to dam the Nile, having proposed this to Caliph Al-Hakim bi-Amr Allah in ignorance of the river's size. Realizing he cannot dam the Nile and fearing summary execution for the failure, he pretends to go insane.

- **1011–21** Placed under house arrest by the Caliph, he uses the time to continue his researches and writes the *Book of Optics* during his ten years of captivity.

- **1021** Caliph Al-Hakim bi-Amr Allah dies.

- **1021** Alhazen is thought to have traveled to Spain in this year to further his knowledge of science with the Spanish Moors.

- **1025–28** At some time in this period Alhazen publishes *Doubts Concerning Ptolemy* or *Aporias Against Ptolemy*, in which he identifies flaws and contradictions in Ptolemy's writings on science, astronomy, and mathematics.

- **1027** In order to devote himself to his scientific researches, Alhazen resigns from his position as a civil servant.

until the light reaches the retina. Alhazen experimented with bulls' eyes and showed that the light enters a small hole at the front, the pupil, and is focused by a lens onto the sensitive surface, the retina, at the back of the eye.

### Investigations into light

Alhazen went on to experiment with mirrors and lenses, but perhaps the most fundamental he did was to show that light travels in straight lines. He probably

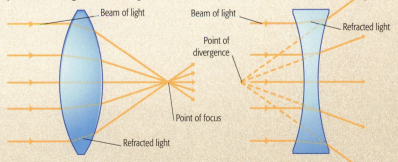

**Diagram of the human eye**
This diagram from Alhazen's *Book of Optics* is remarkably accurate and shows the cornea, aqueous humor, vitreous humor, and optic nerve as they are called today.

inferred this from the fact that solid objects cast sharp shadows in sunlight, but he proved it with the aid of the *camera obscura*, or "dark room."

Arranging for three lamps to be burning outside, he showed that three images appeared on the opposite wall inside. When he intercepted one of the paths with his hand, the corresponding image disappeared. He could "catch" the image anywhere on a straight line between the hole and the wall, and therefore the light that formed the image must be traveling along that straight line. He wrote, "The entering light will be clearly observable in the dust which fills the air." Further, he made a portable *camera obscura* in the form of a tent, and took it around on his travels. No doubt the moving images must have astonished all those who had the chance to see them.

Alhazen's groundbreaking *Book of Optics* remained the definitive book on the subject until the time of Isaac Newton, 650 years later. He also wrote

extensively on other subjects, ranging from astronomy to psychology. He tried to explain why the Sun and Moon look bigger when they are near the horizon than when they are high in the sky. He showed that twilight ends when the sun is 19 degrees below the horizon, and estimated the height of Earth's atmosphere as 9 miles (15 km); in fact, three-quarters of the atmosphere lies within 7 miles (11 km). He studied the attraction between masses, and the acceleration of masses due to gravity. He even suggested that a body will stay at rest or move at a constant speed unless acted on by an external force; 650 years later this came to be called Newton's first law of motion (see pp.104–105). In addition to

**The setting Sun**
Alhazen suggested that the Sun and Moon appear to be bigger when they are near the horizon because when they are high in the sky there is no foreground with which to compare them.

these achievements, Alhazen was a pioneer of the scientific method, the basis of scientific research today.

### The scientific method

The method or process begins with an observation, which is formulated into a statement of the problem to be solved. A suggested solution follows, called the working hypothesis. This is tested by experiment, ideally predicting the result expected. If successful, it can be refined into a theory; if not, another working hypothesis may be required and further experiments carried out to test it. When the problem is resolved, the work is written up and published.

Alhazen died in Cairo in 1039. His many and varied achievements in science are commemorated by a crater on the Moon, an asteroid, and by his portrait on Iraqi banknotes.

**AL-AZHAR MOSQUE AND UNIVERSITY, CAIRO**

- **1037** Almost certainly returns to Egypt, attending the Al-Azhar University and the Dar Al-Hekma (House of Knowledge) in Cairo. The latter's library is reputedly second only to that of Baghdad.

- **1038** Publishes *The Model of the Motions of Each of the Seven Planets,* the first model of the solar system since Ptolemy to be based on radical ideas; for example, he suggests, for the first time in science, that Earth is rotating around its axis.

**ALHAZEN ON IRAQI 10,000-DINAR BANKNOTE**

- **1039** Dies in Cairo at 74 years of age. In the course of his life he wrote more than 200 manuscripts, of which the modern world has knowledge of nearly 100 scientific works. About 50 of these are about mathematics, 23 on astronomy, 14 on optics, and others on medicine. Numerous further works are lost.

---

### BREAKTHROUGH

## SIMPLE LENSES

Alhazen investigated the optical properties of glass, and discovered that the ratio between the angles of incidence and refraction varies; the actual relationship was discovered by Snell 600 years later (see p.116–17). Alhazen studied mirrors and simple lenses, investigating their magnifying power. When used to view a close object, a convex lens produces a magnified real image; a concave lens produces a diminished virtual image.

Beam of light | Beam of light
Point of divergence
Refracted light
Point of focus
Refracted light

**Convex lens**
A convex lens is thicker in the middle than at the outside edge, and has the effect of converging incoming parallel rays of light and bringing them to a point of focus, where a real image is formed.

**Concave lens**
A concave lens is thinner in the middle than at the edge, and has the effect of diverging incoming parallel rays of light. A virtual image is formed at the point from which the rays appear to diverge.

## BEFORE

**Our view of the material world is thousands of years old, but science began in ancient Greece.**

### EARLY INDIAN SCIENCES
Scientific thought may have arisen as a necessity in India, where **medicine**, **metallurgy**, building, and crafts demanded an increasingly sophisticated understanding of the **physical processes** involved.

### EARLY EASTERN SCIENCE
Inscriptions from the 13th and 14th centuries BCE show evidence of scientific thought in China, Japan, Korea, and Vietnam. Early scholars explained nature in terms of **numbers, yin** and **yang**, the five **elements**, and other intellectual constructs.

### HELLENISTIC GREEK SCIENCE
Divisions of knowledge into subjects such as **mathematics and philosophy** were less distinct around 600 BCE. Thinkers such as **Thales ≪ 22–23** and **Pythagoras ≪ 32–33** began to observe and experiment, laying the basis of the modern **scientific method**. Much of what they discovered in the way of mathematics was included in **Euclid**'s *Elements*, written in *c.*300 BCE. At the same time, the Greeks drew ideas from other peoples, including the **Mesopotamians**, **Babylonians**, and **Egyptians**.

**≪ SEE ALSO**
*pp.38–39* ANCIENT IDEAS OF THE WORLD
*pp.52–53* ALCHEMY

# East Meets West

**After the collapse of the Roman Empire, science disappeared in the West. More alert to the achievements of the Greeks was the Islamic world, where many works were translated into Arabic, stimulating original thought in myriad areas. When scholars began to translate Islamic works into Latin after 1100 CE, Europe was transformed.**

Although Hellenistic Greece was the scene of an extraordinary flowering of philosophy and science between 600 and 300 BCE, that event did not occur in a vacuum—astronomy and mathematics were being studied in India, Persia, and among the Sabaeans in eastern Syria. After the fall of the Roman Empire in the 5th century CE, Western science all but disappeared, but in the East scholars, particularly the Nestorians of Syria, had established centers of Greek learning, from where Greek ideas were disseminated in Latin, Greek, Syriac, and Persian. At the same time, ideas were migrating westward from China, India, and Persia, creating fertile conditions for debate in centers such as Byzantium.

Between 622 and 750 CE Arab tribes, united under Islam, gained control of Syria, Persia, Egypt, Spain, and parts of Africa. There developed a renewed fascination with learning, and works of Greek philosophy, mathematics, and science were rapidly translated into Arabic; these were to become the basis of the Arabic sciences.

An important new development was the Arab emphasis on using experiment, the empirical test, to try to establish scientific truth. In the 10th century there began an era of original research that lasted until the 13th century, producing leaps of knowledge in astronomy,

**An Islamic view of the medieval world**
This map was created by the Muslim geographer Al-Idrisi in the mid 12th century. South is at the top of the map, and the Islamic empire is positioned in the center between the East and West. Modern-day Mongolia lies at the bottom left.

△ **Widely influential book**
In the 12th century the Italian scholar Gerard of Cremona translated into Latin (see above) the work *Book of Medicine for Mansur* by Abu al-Razi, known in the West as Rhazes or Rasis. Written in 903 CE for Mansur, the Persian prince, and divided into nine chapters, the work was highly influential in both the Islamic world and the West.

◁ **The Canon of Medicine**
Printed in Rome in 1593, the *Canon of Medicine* was one of the first Arabic works to be published in Europe. It was written as the *Qanun* by the great scholar Avicenna (980–1037) (see pp.52–53). The *Canon* was translated into Latin by Gerard of Cremona in the 12th century.

## LEONARDO FIBONACCI

The son of an Italian trader based in the port of Bugia, Algeria, Pisa-born Fibonacci learned the Hindu–Arabic numeral system while helping his father. He went on to study with Arab mathematicians. In 1202 he published *Liber Abaci*, in which he explained how using the digits 0–9 and place value (how the place of each digit has a value 10 times the place to its right) could benefit commercial activities such as bookkeeping. His book was hugely influential in Europe. Later works included *Practica Geometriae*, in which he addressed Arab geometry and trigonometry.

space were described in mathematical terms by Nicole Oresme, who based his work above all on Platonic concepts.

### The Renaissance

During the European Renaissance, from the 14th until the 17th century, thinkers responded to the stimulus of Greek and Islamic science with new developments. Among them Tartaglia, Cardano, and Ferrari's cubic and quartic equations; Stevin's decimal fractions; Viète's algebraic formulae; advances by Copernicus (see pp.74–75) and Galileo (see pp.82–83) in astronomy; and discoveries in physics, chemistry, and medicine.

▷ **Nicole Oresme and King Charles V**
A 14th-century vellum manuscript shows the philosopher, bishop, polymath, and popularizer of science, Nicole Oresme, presenting his work to King Charles V of France. Beneath the two illustrations is the text of "Human Happiness" from Aristotle's *Ethics*.

mathematics, and medicine. Arab alchemy was doomed in its bid to transform base metals to gold but its intense researches yielded empirical physical, chemical, and biological information. Important, too, were the instruments and apparatus introduced by the Arabs for scientific experiments, and the notion of the laboratory as a place dedicated to such research.

### Translation into Latin

In Europe, scholars such as Boëthius and Cassiodorus in the 6th century had tried to preserve Greek learning, although many of their books and manuscripts were lost. From the 11th century, however, Arabic scientific works began to be translated into Latin by scholars such as Gerard of Cremona and Adelard of Bath. The ideas they contained, both early Greek and Arab, had a profound effect on European thought.

Direct contact with Arabic culture was also bringing rewards in the West. In the 13th century Fibonacci, also known as the Leonardo of Pisa (see above), founded his very influential approach to numbers and mathematics on his tuition as a youth in Algeria; his work included the Fibonacci series for calculating the offspring of two rabbits: 0, 1, 1, 2, 3, 5, 8, 13, 21, and so on, each figure being the sum of the preceding two figures. Fibonacci drew on not only Euclid (see pp.34–35) and Arab ideas but also Diophantus of Alexandria, who is credited as the "father of algebra."

In the field of astronomy, the recovery of Aristotle's beliefs about the Universe challenged medieval thinking that had become mired in superstition. The Arabs, too, had been preoccupied by astronomy and astrology; in the 1080s, observers in Toledo and Córdoba, Spain, had compiled tables of the stars.

### The infinity of space

In the medieval period, God and the heavens were seen as inextricable, yet Arabic texts discuss an infinity of empty space. In the 14th century Bradwardine of Oxford used mathematics to attempt to prove God's presence in the infinity of space, and theories of motion within

AFTER »

In many branches of science, proposals first made in the ancient world, East and West, had to wait centuries before full development.

#### CALCULUS

In the 5th century CE **Zu Chongzhi** used calculus to find the volume of a sphere; 1,000 years later, Bonaventura Cavalieri formulated what is known as **Cavalieri's principle** to achieve exactly the same end. **Pierre de Fermat** applied calculus (see pp.102–103) to **curved lines**. **Isaac Newton** subsequently applied calculus to **physics** in general, while his contemporary **Gottfried Leibniz** devised much of the **notation** used in calculus.

#### COPERNICUS AND GALILEO

For **scientific understanding** to advance, it was sometimes critical for **religious misconceptions** to be overturned. **Copernicus** observed the stars and concluded that Earth **revolves** around the **Sun** (see pp.74–75), while **rotating** about its axis. **Galileo** (see pp.82–83) agreed after observing moons crossing the face of **Jupiter**. Both men were thought heretical, yet the notion of such a **heliocentric** solar system was first presented 1800 years before Galileo by the Greek astronomer **Aristarchus of Samos**, in *c*.250 BCE.

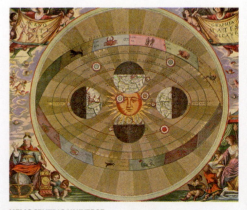

**HELIOCENTRIC UNIVERSE**

SEE ALSO »
*pp.68–69* BIRTH OF EXPERIMENTAL SCIENCE

# 2

# RENAISSANCE AND ENLIGHTENMENT

## 1500–1700

From 1500, the reemergence of classical and Islamic scientific ideas in Europe's Renaissance galvanized scientific thinking. Brilliant minds such as Galileo and Newton learned about the world by observation and experiment, and thinkers such as Descartes ushered in a new age of Enlightenment governed by the power of human reason.

# RENAISSANCE AND ENLIGHTENMENT 1500–1700

## 1500

## 1550

**c.1503**
Leonardo da Vinci creates a design for a two-wheeled hoist with a caged gear.

**1525**
Galen's works on the human anatomy published in the original Greek.

**1526**
Paracelsus pioneers the use of minerals and poisons in medicine.

**1537**
Ambrose Paré revives the practice of ligature for gunshot wounds.

**1540**
Robert Recorde introduces the mathematical symbols + and—in his book *The Ground of Artes*.

Paracelsus ▶

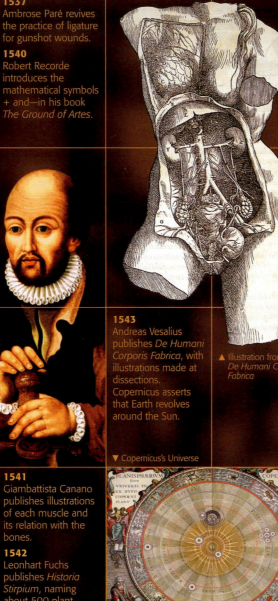

▲ Illustration from *De Humani Corporis Fabrica*

**1543**
Andreas Vesalius publishes *De Humani Corporis Fabrica*, with illustrations made at dissections. Copernicus asserts that Earth revolves around the Sun.

**1530**
Father of botany, Otto Brunfels publishes the first part of *Herbarum Vivae Eicones* with realistic illustrations by Hans Weiditz.

**1541**
Giambattista Canano publishes illustrations of each muscle and its relation with the bones.

**1542**
Leonhart Fuchs publishes *Historia Stirpium*, naming about 500 plant species.

**1535**
Berengario de Carpi produces *Anatomia Carpi*, the first anatomical text with illustrations based on dissections of human bodies.

**1546**
Girolamo Fracastoro proposes that diseases are contagious and are spread by seedlike entities. Georgius Agricola publishes *On the Nature of Fossils*.

▼ Copernicus's Universe

**1551**
Leonard Digges invents the theodolite, used in surveying to measure angles.

**1553**
Miguel Seveto states that blood travels from one side of the heart to the other through the lungs.

**1557**
The equals sign = introduced in mathematics.

**1559**
Matteo Colombo states that the pulmonary vein carries blood and not air as previously believed.

**c.1562**
Gabriel Fallopio describes the ovaries, the uterus, and the tubes connecting them.

**1564**
Julius Aranzi discovers that the blood of a mother and fetus do not mix.

▼ The heart and the pulmonary artery

**1572**
Tycho Brahe reports sighting a new star, revolutionizing the idea that the heavens never change.

▼ Brahe's sextant

**1577**
Geronimo Mercuriali suggests that disease is spread by flies.

**1579**
Fabricius gives the first accurate description of the valves in the veins.

**1583**
Galileo discovers that the oscillations of a swinging pendulum take the same amount of time regardless of their amplitude. Felix Platter proposes that the retina is the part of the eye stimulated by light.

**1586**
Simon Stevin shows that water pressure increases with depth.

**1590**
Hans and Zacharias Janssen make the first compound microscope.

**1593**
*The Canon of Medicine* is one of the first Arabic works published in Europe.

▼ *The Canon of Medicine*

Robert Hooke's ▶ compound microscope

The Renaissance placed humanity at the center of things and in Italy, where it began, anatomists such as Vesalius made the first systematic studies of the human body. As Renaissance ideas spread across Europe, the emerging belief in the power of human reason gave new impetus to mathematics and its application to science. And as the conviction grew that the best way to learn about the world is to study it directly, the invention of the telescope and microscope helped reveal things too distant or too tiny to see with the naked eye. The telescope enabled Galileo to confirm Copernicus's insight that Earth is not the center of the Universe, while Leeuwenhoek and Hooke were startled by the teeming life under their microscopes.

## 1600

## 1650

**1600**
William Gilbert publishes *On the Magnet, Magnetic Bodies and the Great Magnet of the Earth*, proving that magnetism could penetrate and attract through solid objects.

William Gilbert ▶

**1605**
Michal Sedziwój proposes the existence of a "food of life" in the air.

**1608**
Willebrord Snellius publishes *Eratosthenes Batavus*, laying out the foundations of modern geodesy.

**1609**
Galileo makes his first telescope. Johannes Kepler publishes his first two laws of planetary motion. (The third one is published in 1619.)

◀ Galileo's telescope ▶

**1610**
Galileo discovers the mountains on the Moon, that the Milky Way consists of innumerable stars, Jupiter's four largest moons, the phases of Venus, and sunspots.

▼ Galileo's drawings of the Moon

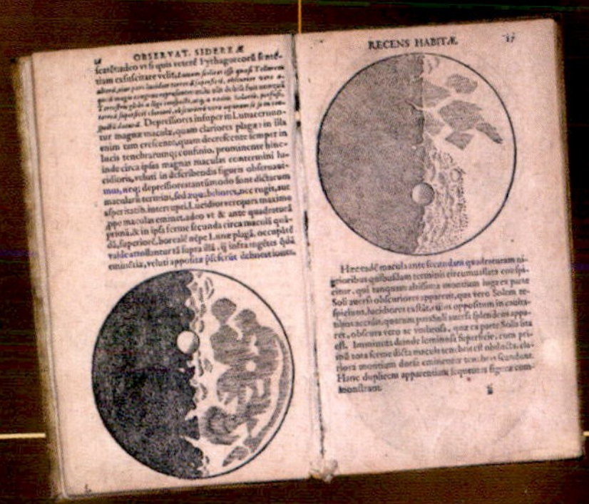

**1614**
Logarithms devised and published by John Napier.

**1621**
*The Doctrine of Signatures* advocates the medicinal use of plants that resemble the human body.

▼ Napier's bones; an early calculating tool

**1622**
William Oughtred invents the slide rule.

**1628**
William Harvey proposes that blood is pumped around the body by the heart.

**1630**
Pierre de Fermat and Blaise Pascal develop a probability theory.

**1631**
The Vernier scale invented. Pierre Gassendi publishes the first data on the transit of Mercury.

**c.1632**
Galileo proposes the concept of inertia.

**1637**
René Descartes states that the heart is just a mechanical pump.

**1643**
Evangelista Torricelli creates a vacuum.

▼ Evangelista Torricelli

**1648**
Von Helmont measures the water taken up by a tree.

**1649**
Pierre Gassendi states that the properties of atoms depend on their shape and that they may be joined together to form molecules.

**1651**
William Harvey states that young animals develop from eggs.

**1653**
Blaise Pascal publishes *On the Equilibrium of Liquids*, setting out Pascal's law of pressure, the basis of all hydraulics.

**1654**
Von Guericke gives a public demonstration of the vacuum.

**1657**
Christiaan Huygens builds the first mechanical clock with a pendulum.

**1658**
Jan Swammerdam observes red blood cells under a microscope.

**1660**
Robert Boyle discovers that air is necessary to carry sound, feed a flame, and sustain life.

**1661**
Richard Towneley and Henry Power note air pressure decreases with altitude. Marcello Malpighi discovers capillaries. Robert Boyle publishes *The Sceptical Chymist*, trying to bring scientific ideas to alchemy.

**c.1665**
Calculus developed by Isaac Newton and Gottfried Leibniz. Robert Hooke publishes *Micrographia*, and uses the term "cell" to describe cells seen in a slice of cork.

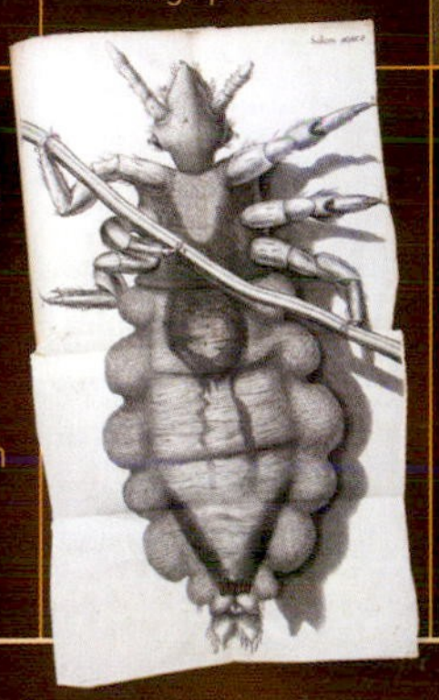

**1669**
Nicolas Steno proposes that fossils are the remains of extinct animals. Hennig Brand makes phosphorus, isolating a new element.

**1671**
Newton builds the first reflecting telescope. Marcello Malpighi publishes *Anatomia Plantarum*.

**1675**
Newton publishes *Hypothesis of Light*.

◀ Newton's telescope ▶

**1676**
Olaus Roemer calculates the speed of light. Robert Hooke publishes Hooke's law. Antonie van Leeuwenhoek observes single-celled organisms under a microscope.

Spring demonstrating ▶
Hooke's Law

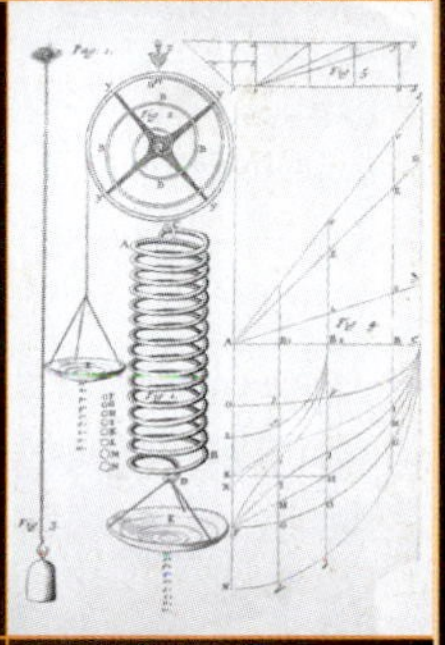

**1681**
In *Sacred Theory of the Earth*, Thomas Burnet suggests that Earth contains subterranean voids full of water.

**1684**
Isaac Newton writes a paper about the orbits of comets.

**1690**
Christiaan Huygens publishes *Treatise on Light*. Paul Hermann divides seed plants into angiosperms and gymnosperms.

**1691**
Jacob Bernoulli introduces polar coordinates.

▲ Hooke's air pump

▼ Illustration from Hooke's *Micrographia*

▲ Isaac Newton

**1687**
Isaac Newton's *Principia* published, stating Newton's laws of motion and the law of universal gravitation.

## BEFORE

Ancient Greek and early Islamic scientists and philosophers asked many questions about the nature of the Universe.

### ANCIENT GREEKS
Philosophers in Ancient Greece believed that knowledge would be found through argument and reason. The idea of **performing experiments** rarely occurred to them.

### PLATO AND ARISTOTLE
**Plato** argued that if enough men of sufficient intelligence thrashed out an idea for long enough, they would reach the "truth" about any subject. **Aristotle**, a student of Plato's, allowed that science could be studied but denied any place for experimentation.

### ISLAMIC SCIENTISTS
For about 800 years after the great Greek philosophers, scientific study was led by the Arabs, notably **Alhazen**, **Al-Jahiz**, and **Al-Farabi**. While there was an emphasis on **investigation**, the discipline was still hampered by mysticism.

« SEE ALSO
pp.36–37 ARISTOTLE
pp.52–53 ALCHEMY

Until the mid-15th century, the ideas that governed European learning were a mixture of beliefs handed down from ancient Greece and the teachings of the Catholic Church. Then, with the start of the Renaissance, scholars began to break away from traditional methods, proclaiming their belief in reason as the measure of all things.

### The Scientific Revolution

We now know that the physical sciences start with observations, followed by experimentation under carefully controlled conditions. This was not always the case.

Galileo Galilei (see pp.82–83) was the first great scientist of the modern era to rely on observation and experiment. Among his spectacular discoveries were the principle of the pendulum (see p. 83) and the law on the behavior of falling bodies (contradicting the views of Aristotle, which had held for almost 2,000 years). Legend has it that he dropped objects from the tower of Pisa in order to prove that all objects fall at the same rate, whatever their mass. Similar experiments had been carried out by scientists such as the Flemish Simon Stevin, who in 1586 used two lead balls, one ten times the weight of

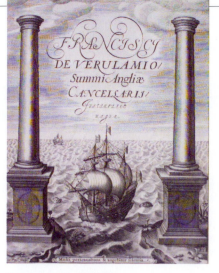

**Novum Organum (1620)**
The title page of Francis Bacon's book shows people sailing through the Pillars of Hercules—the limits of knowledge—into the future.

the other, which he dropped 30 feet from the church tower in Delft.

Galileo also refined the telescope (see p.83). He gathered powerful evidence to show that the Sun, not Earth, was the center of the solar system. This observation clashed with the accepted wisdom of theologians. He was pressed by the Church to retract, standing trial in 1633 for heresy. Found guilty, he was forced to issue a public denial.

Galileo was part of a new wave of scientists using experimentation to make sense of the world around them. Englishman William Gilbert, Queen Elizabeth I's physician, studied the effects of magnetism (see pp.80–81). He determined that the ordering of the heavens was not mythical: the magnetic nature of "attraction" held the planets on their course. William Harvey developed ingeniously simple experiments to study the circulation of the blood (see pp.90–91). His ideas were considered so heretical that he had his work published by an obscure press in Germany.

### A new optimism

Despite the opposition of the Church, scientific thinking continued to flourish and men developed a new optimism about what could be achieved by it.

The French mathematician and philosopher René Descartes (see pp. 102–03) was an admirer of Galileo. He claimed that he had learned nothing in school that wasn't subject to doubt, and stated that knowledge had to be "clear and distinct." When he came to write his *Regulae* (Rules) in 1628, and then the *Discours de la Methode* (Discourse on Method) in 1637, his purpose was to

# Birth of Experimental Science

Science as we know it today is no more than 500 years old. It is firmly based on certain rules of procedure that scientists must follow to obtain accurate knowledge. These rules were formulated during a revolution in scientific thinking—the birth of experimental science—and are upheld today by the world's scientific societies.

▽ **Recording observations**
Good science needs precise recording of information. These notes by the French mathematician and physicist, André-Marie Ampère, concern the magnetic properties of electrical current.

▷ **Testing theories**
Scientists must design experiments that are open—not to prove a preconceived theory but if possible to disprove it. Objectivity and repeatability are vital.

set down a method not simply for solving mathematical or physics problems, but for reaching the truth about everything. He started with the first principle, "I think, therefore I am," and believed he could use this principle to construct a new body of knowledge.

A contemporary of Descartes, the English philosopher Francis Bacon (see opposite), was equally contemptuous of false knowledge, or what he called "idols," based only on a few untested observations. Like Descartes, Bacon demanded a new standard of precision. He proposed a scientific method that required the patient accumulation of data in three steps. The scientist makes his observations, then formulates a theory that might explain them, and finally he tests his theory by rigorous experiment.

### System of logic

In 1620 Bacon published *Novum Organum*, or "New Instrument," in which he set out his belief in

**AFTER** »

Scientists still rely on the scientific method as a regular and generally accepted procedure for conducting research.

**DESIGNING EXPERIMENTS**
Scientists must first form a **hypothesis**, and then look for evidence to support it. The method requires **careful recording of data** and other information. **Experiments** must be performed under **controlled conditions**, and must be repeatable by others.

**PEER REVIEWS**
Scientists often hold conferences with their peers to **report on their findings**. The results are passed on for analysis. Only when the findings have been **rigorously tested** will they be **published**, usually in a scientific journal.

**SEE ALSO** »
pp.82–83   GALILEO GALILEI
pp.90–91   CIRCULATION OF THE BLOOD

"**Many** will pass through and **knowledge** shall be **increased**."

FRANCIS BACON FROM "NOVUM ORGANUM," 1620

this scientific method. His work had a direct influence on scientists Robert Boyle, Isaac Newton, and Robert Hooke.

One of the greatest legacies of Bacon's philosophy, which he stressed in his *Novum Organum*, was the concept of establishing an organization within which scientific methods and research could be shared and explored. By the end of the 17th century this had become a reality.

### Scientific societies

The first scientific society that published any proceedings was the *Accademia die Lincei* (Academy of

**Birth of scientific societies**
The first scientific societies were founded in the 17th century. This engraving depicts King Louis XIV of France visiting the Academy of Sciences in Paris.

Lynxes), so called because of the members' sharp observational skills. Founded by Duke Frederico Cesi in 1603 in Rome, Italy, it welcomed Galileo into its ranks in 1610 following his astronomical discoveries, but then got into trouble with the Church and closed after the Duke died. The *Accademia degl' Investiganti* (Academy of Investigators) was founded in Naples in 1650, and the *Accademia del Cimento* (Academy of Experiment) in Florence in 1657, although it ended 10 years later.

Also during the 1650s a group of "natural philosophers" had begun meeting in England, sometimes at Gresham College in London, and later in the rooms of Robert Boyle in Oxford. They called themselves the "Invisible College," but when Charles II

was restored to the monarchy in 1660 they became the Royal Society of London for the Improvement of Natural Knowledge, generally abbreviated to Royal Society. Robert Hooke was appointed the first curator of experiments, which meant that he was expected to give practical demonstrations at every meeting. The English example was swiftly copied on the continent. In 1666 Louis XIV accepted the founding of the French Royal Academy of Sciences, and by 1700 similar organizations were established in Naples and Berlin. In turn they spawned many societies that are still in existence today.

These scientific societies provided not only a focus and a meeting place for discussion, by which scientific ideas thrive, but also a conduit for information. When Volta invented the electric battery in 1799, he wrote from Italy to Sir Joseph Banks, the President of the Royal Society in London, and told him about it. Banks immediately wrote to all those who he thought would be interested, and within weeks batteries were being made all over Britain, with the result that studies in electricity gathered rapid momentum. Even the Napoleonic wars were no barrier to such dissemination of information through the Royal Society.

PHILOSOPHER AND POLITICIAN (1561–1626)

## FRANCIS BACON

Although not a scientist, Bacon established the principles of the Scientific (or Inductive) Method, based on observation and experiment, which has been a basic tool of scientists ever since. Bacon rose to be Lord Chancellor under James I, but in 1621 he was convicted of bribery and driven out of public life. He then devoted himself to writing. He died after catching a chill while stuffing a chicken carcass with snow, to see if the meat would be preserved.

# Renaissance Medicine and Surgery

The 14th century saw a flowering of the arts and sciences around the world, and a new intellectual freedom that produced an upwelling and dissemination of new ideas as never before. Medicine at this time grew out of a fusion of ideas from ancient Greece and the Middle East. However, a combination of science and supernatural belief meant that trained physicians worked alongside practitioners of folk medicine.

## « BEFORE

**Previous knowledge passed on from the ancient world had a huge impact on the development of medicine and surgery.**

### DISEASE AND HYGIENE
The 13th-century Italian surgeon **Gugliemo da Saliceto** recognized that **good hygiene** was vital to health, and suggested that pus formation was bad for wounds (indicating an infection).

### PHARMACOPEIAS
Some of the first books on the preparation of medicines, called **pharmacopeias**, were written in antiquity and added to by physicians and alchemists of the Middle Ages. In ancient Greece, **Dioscorides** wrote a text that **influenced pharmacological thinking** for 1,600 years.

**« SEE ALSO**
*pp.24–25* EARLY MEDICINE AND SURGERY

From the 14th century Europe saw great advances in understanding the human body and the scientific basis of medicine —including surgery, anesthesia, and pharmacy. The new universities at Paris, Bologna, and Padua became key centers of learning and trained some of the finest physicians of the time.

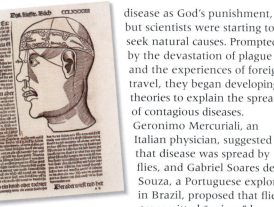

**Early textbooks**
This 1512 book included an illustration of phrenology or craniology—the study of the shape of the skull.

## Health and disease
Disease is a disorder of the body, and in the 17th century a new science of disease, pathology, was becoming a discipline in its own right. The Christian world still tended to see disease as God's punishment, but scientists were starting to seek natural causes. Prompted by the devastation of plague and the experiences of foreign travel, they began developing theories to explain the spread of contagious diseases. Geronimo Mercuriali, an Italian physician, suggested that disease was spread by flies, and Gabriel Soares de Souza, a Portuguese explorer in Brazil, proposed that flies transmitted "poison" by sucking it from the sores of their victims. The first specialist of tropical medicine was the Jewish physician Garcia de Orta. He left Portugal, perhaps fleeing the Inquisition, to settle and practice in India, where he studied the effects of cholera. Increasingly, biological factors were being blamed for disease, even though the details were misguided. In 1600 mass hysteria in Italy brought on by fear of spider bites spawned a "therapeutic" dance, called the Tarantella.

## Balance and lifestyle
The ancient Greeks had proposed that good health relied upon maintaining a balance between the four body fluids, or humors: blood, black bile, yellow bile, and phlegm. The idea that disease resulted from an imbalance of body constituents was central to later medicine, too. It was championed by Paracelsus, a Swiss physician–alchemist who considered that the imbalance in the body associated with disease was caused by outside influence—"poison." For Paracelsus, disease could be treated by finding another external substance that would redress the imbalance caused by the poison, and so bring about a cure. He believed that there was a remedy for every disease in the minerals and herbs of the natural environment.

◁ **Surgery in progress**
This 1563 woodcut depicting a surgical operation on a man's head is from one of the original works of Paracelsus (see below).

# "The dose makes the **poison**."

**ATTRIBUTED TO PARACELSUS,**
PHYSICIAN–ALCHEMIST

Pharmacopeias—catalogues of medicines—had existed in ancient China and Mesopotamia. However, Paracelsus was perhaps one of the first scholars to put forward a theory that explained the role of drugs in curing disease. More importantly, Paracelsus recognized the significance of dose. In fact, he articulated something that is still a principle of pharmacology today: "All things are poison and nothing is without poison, only the dose permits something not to be poisonous." Even something as seemingly innocuous as water can harm if taken in sufficiently large quantities. Substances such as arsenic are simply more toxic because they are lethal at much lower doses.

Many physicians thought they knew the appropriate cures for various diseases. However, in spite of the increasing influence of experimental medicine, some ideas were less than scientific. The Doctrine of Signatures, which probably originated in ancient Greece and was popularized by the German mystic Jakob Böhme, became especially influential. Böhme suggested that cures for disorders could be found by looking for the Creator's "signature" in the natural world. According to the Doctrine, plants that resembled the human body, or parts of it, harbored cures for disease—such as mandrake roots, thought to resemble human limbs.

While many of these teachings were at best ineffectual and at worst harmful, the herbalist origins of pharmacology did produce some things of genuine worth. For example, the painkilling properties of willow bark eventually led to the production of aspirin, while digitalins extracted from foxglove would later be used to treat heart problems.

**Doctrine of Signatures**
The Doctrine advocated the medicinal use of plants with a similarity to the human body. These illustrations depict a plant with roots that mimic the shape of a hand (left) and flowers resembling the eye (below).

▽ **Tongue depressor**
By the 17th century medical instruments such as this one were routinely used by physicians as part of their clinical examinations.

## Advances in surgery

Surgery would always be risky as long as there was pain, blood loss, and infection. Nevertheless, the Italian surgeon Theodoric Borgognoni refined a mixture of opiate-based substances that formed the basis of anesthesia (see pp.228–29) until the 19th century. In the mid 16th century, the French surgeon Ambroise Paré improved techniques of ligature (tying limbs or blood vessels to reduce blood loss) during amputations, which had been developed in the 10th century by Abulcasis, a physician in Spain. Paré also revived an ancient Roman potion of turpentine, egg yolk, and oil extracts from roses for healing wounds, which proved more effective than the prevailing treatment of pouring boiling oil on the injury. Despite these efforts, many patients died after surgery as a result of a factor that was to evade even the finest minds of the time: infection.

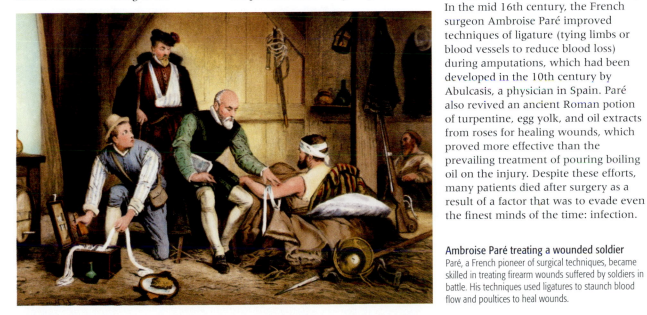

**Ambroise Paré treating a wounded soldier**
Paré, a French pioneer of surgical techniques, became skilled in treating firearm wounds suffered by soldiers in battle. His techniques used ligatures to staunch blood flow and poultices to heal wounds.

**AFTER** »

The modern study of biochemistry grew out of European medicine between the 14th and 17th centuries. Scientists began to understand the chemical effects of drugs on the body with greater precision.

**SAFER DRUGS**
In the 17th century **Sylvius**, a medical professor, suggested that all processes of life and disease were based on **chemical actions**, ushering a new age of biochemical study. Most **traditional therapies**, such as using mercury to treat syphilis, had been developed by **trial and error**, without knowing how or why they worked. In time, **modern pharmacology** would replace these remedies with safer, more effective treatments.

**OCCUPATIONAL DISEASES**
When **Paracelsus** wrote a treatise on the illnesses of metalworkers, he sparked an interest in **occupational diseases**. With the rise of anatomical pathology, physicians began to see links between **some diseases and particular occupations**, such as the high levels of lung disease in miners.

**SEE ALSO** »
pp.228–29 SAFER SURGERY
pp.312–13 THE DEVELOPMENT
 OF MEDICINES
pp.406–07 MODERN
 SURGICAL PROCEDURES

**METALWORKERS**

**ALCHEMIST (1493–1541)**

## PARACELSUS

Paracelsus was born in Switzerland but traveled widely in Europe, Africa, and the Middle East. His many interests included alchemy, astrology, medicine, psychology, and toxicology. A prolific writer, Paracelsus had a reputation for being arrogant. Born Phillip von Hohenheim, he adopted the name Paracelsus later in his life: it means "equal to or greater than Celsus" (a Roman scholar of medicine). In 1526 he recorded his discovery of zinc.

# The Human Body Revealed

**Rejecting ideas of old, European scientists of the 14th–16th centuries explored the body as geographers explored the world, dissecting the human form with a high degree of expertise to discover complex anatomical systems.**

Anatomists have been dissecting the body to learn about its structure since the days of ancient Greece and Egypt.

### Anatomy by direct observation

The world's first public dissection took place around 1315: Mondino de Liuzzi, an Italian Professor of Medicine, used the body of an executed criminal. European laws concerning human dissection at this time varied and scholars relied heavily on the availability of executed criminals' bodies. Liuzzi's purpose was to

discover, to educate, and even to entertain. His publication, *Anothomia*, set out a basic body plan, but it supported old Greek errors too, especially those of the physician Galen 129–200 CE (see pp.24–25), the established authority of anatomy at that time.

Denying the wisdom of antiquity was risky because the Catholic Church used to condemn challengers as traitors, but many remained undaunted. In 1535 the Italian physician Jacopo Berengario da Carpi produced *Anatomia Carpi*, the first anatomical text to be extensively illustrated and based on dissections

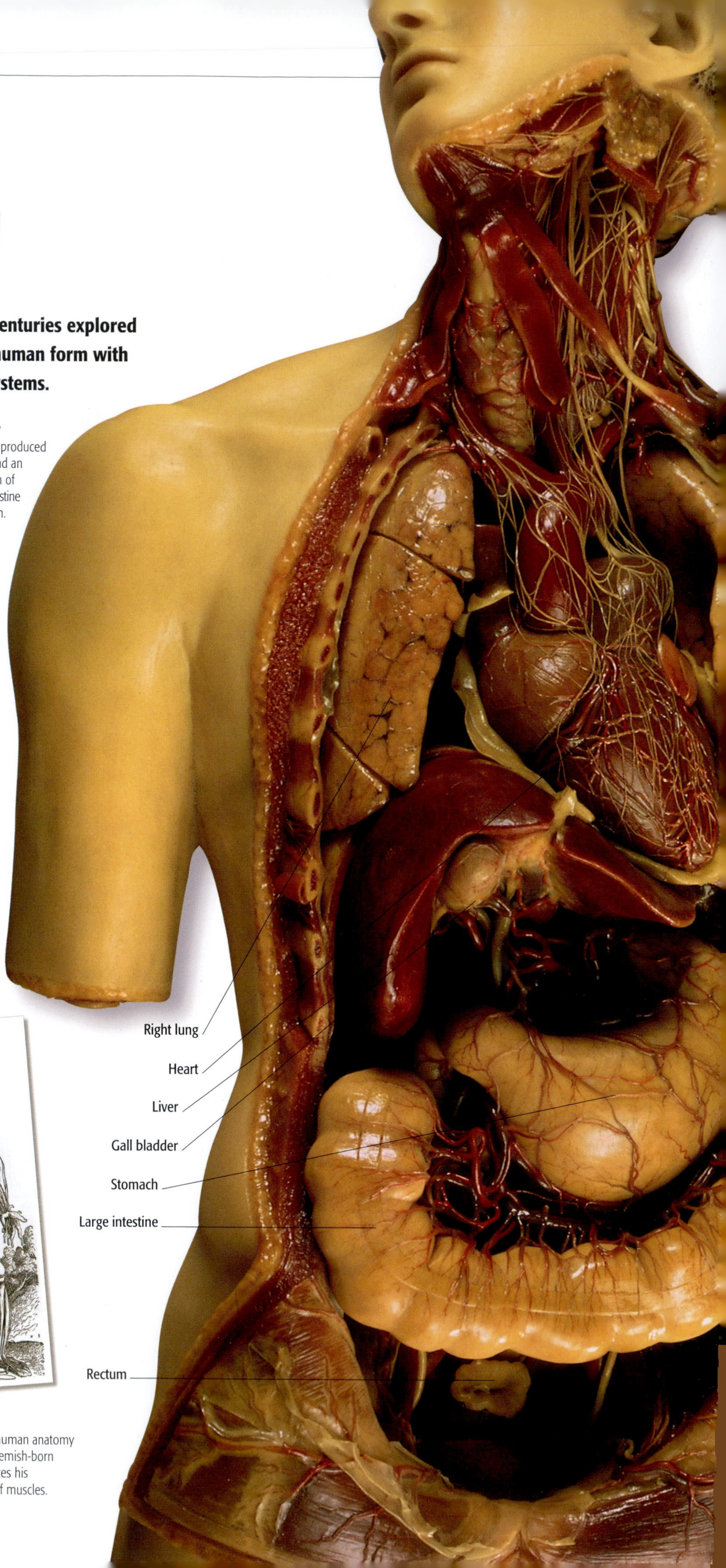

### Understanding the human body

This Italian anatomical wax figure was produced in the 18th century, when scientists had an accurate understanding of the position of most internal organs. Here, the small intestine has been removed to show the rectum.

Right lung

Heart

Liver

Gall bladder

Stomach

Large intestine

Rectum

## BEFORE

Many early cultures regarded the human body as sacred and would only dissect animals. This eventually changed, first under Egyptian law and later under Christian law.

### EARLY DISSECTIONS

The first anatomical text, written in the 6th century BCE by ancient Greek theorist **Alcmaeon**, is believed to have been based on **animal dissection**. Some 300 years later, pagan Rome forbade human dissection, which meant that Galen, the city's most famous physician, **proposed many mistaken theories** about human anatomy based on dissecting animals.

### GALEN DISCREDITED

Human dissections were sanctioned in medieval Islam and **Avenzoar** was thought to be the first physician to perform **autopsies** in the 12th century CE. Many Arabic scholars doubted the Greek idea that disease was caused by an imbalance in the four humors (body fluids) and **disagreed with Galen's anatomical theories**, such as the belief that blood was continually made in the body. However, they had little impact on Renaissance Europe where Galenic wisdom dominated.

**FOUR HUMORS**

« SEE ALSO
*pp.24–25* EARLY MEDICINE AND SURGERY

**Vesalius's muscle man**
This illustration from the key work on human anatomy *De Humani Corporis Fabrica*, by the Flemish-born physician Andreas Vesalius, demonstrates his detailed understanding of the shapes of muscles.

of human bodies. Carpi's work undermined Galen's theories, including the description of a network of blood vessels below the brain, which were found not to be present in humans. In doing so Carpi was heralding a golden age of anatomy that was dominated by four men: Eustachi, Colombo, Fallopio, and Vesalius.

In the 16th century Italian anatomist Bartolomeo Eustachi made accurate diagrams of almost all the main organs, though his findings were not published until 1714. Italian surgeon Matteo Colombo studied blood circulation at the University of Padua and went on to

become possibly the first Renaissance physiologist. Like others who followed him, he experimented on animals to answer his questions. Gabriele Fallopio an anatomist from Italy studied the bones, muscles, and reproductive organs, describing the tubes that link the ovaries to the uterus, which now bear his name.

However, the most famous anatomist of the 16th century was Flemish-born Andreas Vesalius (see below)—a pioneer for understanding anatomy in medicine. His superbly illustrated *De Humani Corporis Fabrica* from 1543 documented observations made from human dissections, and his work finally disproved many of Galen's theories.

### Microscopic anatomy

By the 1600s anatomists saw that organs were made of different types of tissue, such as muscle, bone, and lining. The body was defined as an organized structure – tissues made up organs, and organs interacted in systems. Around this time a new type of anatomist – micro-anatomist – was routinely using the microscope to look at the body in even closer detail.

Many organ specialists of this period were physicians who were eager to improve their diagnoses and treatments. One such was Marcello Malpighi, an Italian doctor, who scrutinized the skin and found it to be a multilayered organ —the Malpighian layer has dividing cells that generate the layers above it, and bears the doctor's name today.

### The science of pathology

Scientists also sought to find out more about diseased anatomy. In 1761, Italian anatomist Giovanni Morgagni published

a treatise based on his observations of autopsies to study the causes of disease. It was this that established anatomical pathology as a science. With the rise of medical teaching and as executions became scarcer, demand for bodies exceeded supply. In 1832 the Anatomy Act was introduced in England, allowing licensed anatomists to dissect unclaimed and donated bodies. By the 1800s, when English surgeon Henry Gray wrote *Anatomy*, physicians such as Carl Ernst Bock, a German professor, routinely performed autopsies, and dissection was mainly used to discover cause of death.

**Peripheral nerve in arm**

**Left lung**

**Diaphragm**

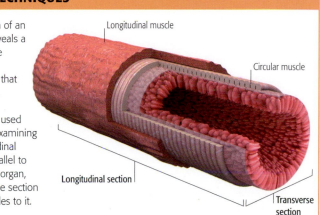

**Anatomical illustrations**
Vesalius's *De Humani Corporis Fabrica* was illustrated with a ground-breaking style by artists who were present at dissections.

Liver

Kidney

Fallopian tube

Ovary

Bladder

"I profess to **learn** and to **teach anatomy** not from books but **from dissections**." WILLIAM HARVEY, ENGLISH PHYSICIAN 1628

**AFTER »**

From the 1700s, anatomists were looking at bodies in new ways, using new methods.

#### COMPARATIVE ANATOMY
Anatomical studies comparing species underscored **humans' place in the animal kingdom**. In 1698 British scientist **Edward Tyson** dissected a chimpanzee and found it was closer to humans than to monkeys. In the mid-18th century Dutch physician **Petrus Camper** showed that the orangutan was a different species and not another type of human.

#### ANATOMY FOR EXHIBITION
Anatomists developed techniques for preservation and **display of body parts**. Today, collections are used for research and teaching. New interest in anatomy came with the *Body World* exhibits of German anatomist Gunther von Hagens in the 1990s.

**HUMAN BRAIN SPECIMEN**

#### IMAGING TECHNIQUES
Modern anatomy uses techniques to create images of the body without dissection. **X-rays** can reveal bone fractures and lung tumors, while magnetic resonance imaging (**MRI**) and **ultrasound** can show soft tissues.

**SEE ALSO »**
pp.294–95 THE DISCOVERY OF X-RAYS
pp.404–05 BODY IMAGING
pp.440–53 REFERENCE: BIOLOGY

---

ANATOMIST (1514–1564)

### ANDREAS VESALIUS

Vesalius was born in Brussels, studied medicine in Paris, and later moved to Italy, becoming Chair of Surgery and Anatomy at the University of Padua. He dissected executed criminals to create anatomical drawings, correcting errors of earlier anatomists, including Galen. In 1544 he became court physician for Charles V, the Holy Roman Emperor. He died after being shipwrecked on the Greek island of Zakynthos.

---

IN PRACTICE

### DISSECTION TECHNIQUES

When a thin section of an organ is taken, it reveals a great deal about the arrangement of the microscopic tissues that make up the organ. There are two techniques that are used for dissecting and examining an organ. A longitudinal section is sliced parallel to the long axis of the organ, whereas a transverse section is taken at right angles to it.

Longitudinal muscle

Circular muscle

Longitudinal section

Transverse section

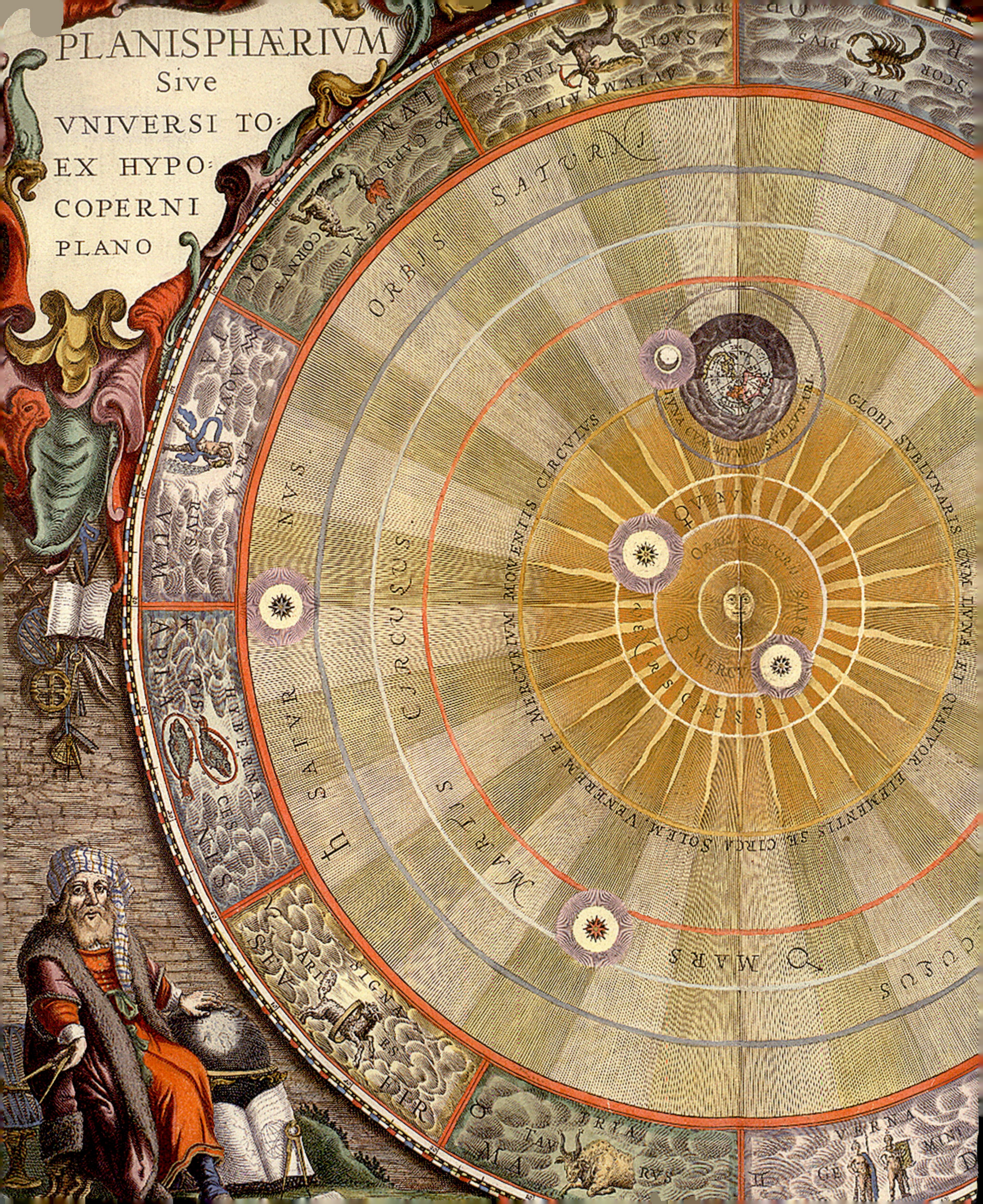
PLANISPHÆRIVM
Sive
VNIVERSI TO:
EX HYPO:
COPERNI
PLANO
ORBIS SATURNI
SATVRNVS
MARTIS CIRCVLVS
MARS
ORBIS MERCURII
GLOBI SVBLVNARIS

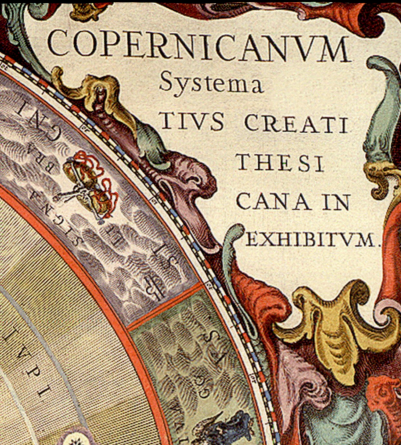

# The Sun-centered Universe

**The ancients believed that the Sun, Moon, and stars spun around the earth, and—to the casual observer—this still feels true today. But in 1543 a Polish scholar named Nicolaus Copernicus suggested otherwise, placing the Sun firmly at the center of the Universe.**

Almost 1,800 years before Copernicus, Aristarkhos of Samos suggested that the Sun was the center of the Universe. But this view, together with his assertion that the Universe was therefore much larger than anyone had imagined, was ignored. The generally accepted view was that Earth lay at the center of the Universe, and this belief became crystallized in Ptolemy's great work *Almagest*, a 13-book treatise written in the 1st century CE. It was to remain the accepted model for more than a thousand years, until after the death of Copernicus in 1543.

Copernicus was a keen astronomer who came to the conclusion that it was unlikely that thousands of stars were spinning rapidly around the earth every 24 hours. Apart from anything else, the distant ones would have to travel at impossible speeds. Surely it was much more probable that they were all stationary, while the earth was spinning?

He then developed his own theory of the Universe, in which Earth and all the other planets orbit around the Sun. He described these planetary motions in his book, *De revolutionibus orbium coelestium* (*On the Revolutions of Heavenly Bodies*). This was the first scientific study of cosmology, and it challenged both the views of his peers and the authority of the Church. It was not published until 1543, when—according to legend—Copernicus was on his deathbed. His fears may have caused him to delay publication until he was beyond persecution; the book's dedication to the Pope certainly seems to be an attempt to soften the Church's response.

What worried Copernicus was that if the Sun really was the hub of the Universe, then Earth must be moving. This was contrary to various statements in the Bible, and was to cause immense trouble to the Italian scientist Galileo, when he confirmed it some 90 years later (see pp.82–83).

In fact the Sun is the center of our solar system, but not of our galaxy—the Milky Way (see pp.340–41)—nor the Universe. Nevertheless, Copernicus's theory was an adequate approximation, and it slowly revolutionized the way people thought about the Universe.

**The Copernican cosmos**
This 1660 painting by Andreas Cellarius shows "The system of the entire created universe according to Copernicus," including the six known planets and the moons of Jupiter, which were discovered in 1610.

> "Finally we shall place the **Sun** himself at the center of the **Universe.**"
>
> COPERNICUS, *c.*1540

## « BEFORE

Modeling the paths, speeds, and brightness of planets had spurred astronomic endeavor for two millennia.

### ANCIENT IDEAS ON PLANETS

Ancient Greek astronomers **« 28–29** noticed that five bright "stars" were not "fixed" but **wandered about the ecliptic zodiac** of the sky. Two of these—**Mercury** and **Venus**—stayed relatively close to the Sun, while the other three—**Mars, Jupiter,** and **Saturn**—did not.

### OLD ECLIPSES

Around 500 BCE Babylonian astronomers noted an **eclipse cycle known as the Saros.** They realized that 18 years, 11 days, and 8 hours after an eclipse of the Sun and Moon, these bodies, together with Earth, returned to similar geometries and **another eclipse** took place.

### CHARTING THE STARS

One of the oldest maps of the stars is Chinese. **Su Song's** *Celestial Atlas* dates back to 1092 CE, and marks the position of **more than 1,350 stars,** including the correct position of the Pole Star.

**« SEE ALSO**
pp.26–29 THE FIRST ASTRONOMERS
pp.74–75 THE SUN-CENTERED UNIVERSE

During the course of the 16th century, the Copernican hypothesis that the Sun lies at the center of the planetary system became more widely accepted. If true, this also meant that Earth moved across its orbit every six months, and so the nearby stars should be seen to move against the background of more distant stars.

Astronomers wanted to detect this movement, so they began to make more accurate instruments. They were correct in their assumption, but unfortunately the huge distances involved—even the nearest bright stars are at least a quarter of a million times farther away from Earth than the Sun—meant the movement they hoped to see was too fine for their instruments.

### Mapping the skies

Tycho Brahe (1546–1601) was Europe's first really great observer. This Danish nobleman turned his back on a political career to concentrate on astronomy. The amazing sight of the 1560 solar eclipse first hooked his interest, and when astronomers failed to predict accurately the conjunction of Jupiter and Saturn three years later, he decided to devote his life to astronomy. In 1572 he reported a supernova (see right), and became

**Brahe observing the skies ▷**
A 16th-century depiction of Tycho Brahe in his observatory on the Danish island of Hven. He sits beside his mural quadrant—a huge brass arc used for measuring celestial positions.

**BREAKTHROUGH**

### BRAHE'S SUPERNOVA

Observing the constellation of Cassiopeia in mid-November 1572, Tycho Brahe noticed that a new star had appeared. Rivalling Venus in brightness, the new star remained visible until March 1574. Its appearance revolutionized astronomy by destroying the stale Aristotelian view that the heavens never changed. The exploding star is about 7,500 light years away, and it is shown below as it appeared 400 years after its diskovery.

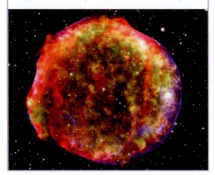

# Planetary Motion

**The Ptolemaic system, which placed Earth at the center of the universe, had held sway for more than 1,500 years, from the ancient Greek era to the 16th century. But then one astronomer accurately mapped the celestial sky and carefully charted planetary motion, laying foundations for the laws of planetary motion.**

**Brahe's sextant**
This large floor-mounted wood and brass instrument used two sighting arms to accurately measure the distance between planets and stars that were closer than 60 degrees.

famous. The Danish king, Frederick II, gave Brahe the small island of Hven and provided him with funding for his work. Brahe built two great observatories there—Uraniborg and Stjerneborg—equipping them with a series of extremely well-made and stable quadrants and sextants for measuring celestial movements. These were used with great care, to minimize experimental errors.

Brahe showed that the comet of 1577 was farther away than the Moon, measured the length of the year to an accuracy of one second, and proved that the rate of precession (a change in direction of Earth's rotational axis) was not irregular. His greatest legacy was a long, accurate series of observations of the positions of the planets. These were consistently accurate to within a minute of arc (equal to $^1/_{60}$ of one degree).

### Joining forces

Brahe's quarrelsome nature led him to leave Denmark for Bohemia, in what is now the Czech

> " Man makes a **great fuss** about **this planet**, which is only a **ball bearing** in the hub of the **universe.** "
>
> CHRISTOPHER MORLEY, AMERICAN AUTHOR, 1890–1957

ASTRONOMER (1571–1630)

## JOHANNES KEPLER

The German astronomer Johannes Kepler was assistant to Tycho Brahe and Imperial Mathematician to Emperor Rudolf. His belief in the ancient philosophical concept of the music of the spheres, used to describe proportions of the solar system, and sometimes credited to Pythagoras, led to Kepler's own ideas of ordered geometric spacing of planetary orbits, as explained in his *Mysterium Cosmographicum*. He also wrote three more very influential books: *Astronomia Nova*, *Epitome Astronomia Copernicanae*, and *De Harmonices Mundi*. He is best known for the three laws of planetary motion.

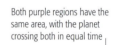

Planet moves faster when it is closer to the Sun

The sun at one focus of the elliptical orbit

Both purple regions have the same area, with the planet crossing both in equal time

Orbiting planet

Planet moves more slowly when it is farther from the Sun

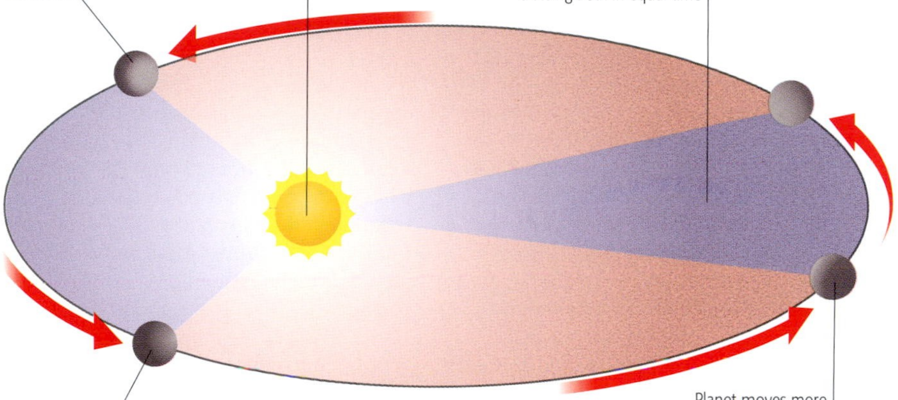

### Elliptical planetary orbit
The first of Kepler's laws of planetary motion showed that the orbit of a planet forms an ellipse around the Sun. The second law showed that a planet moves faster when closer to the Sun and slower when farther away.

Republic, in 1597. Johannes Kepler also moved to Prague in 1600, suffering from religious persecution, and became his assistant. Unfortunately Brahe died just one year later, but Kepler inherited his observations and used them to work out the true nature of the planetary orbits. He was a great mathematician and developed three laws of planetary motion still in use today.

### Kepler's three laws
The first law states that every planet orbits around the Sun in an elliptical orbit. The second law concerns the speed of a planet around its orbit, and notes that an imaginary line between the planet and the Sun would sweep out the same area in the same given time (shown above).

Both these laws were published in *Astronomia Nova*, 1609. His third law was in the last chapter of *De Harmonices Mundi*, 1619. It states that the square of the time a planet takes to go around its orbit is proportional to the cube of the

orbit's semi-major axis (half the diameter of the planet's orbit). But not all of Kepler's ideas were correct. One of his theories stated that planets were swept along by radiated magnetic influence from the Sun; later disproved by the Newtonian theory of gravitation (see pp.106–107). »

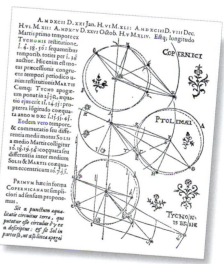

Kepler's calculations in *Astronomia Nova* This book demonstrates Kepler's mathematical workings, as he abandoned the Copernican and Ptolemaic views on the circularity of planetary orbits.

## BREAKTHROUGH
## PLANETS IN TRANSIT

A transit occurs when one astronomical body travels across the face of another. Only three major bodies pass between Earth and the Sun: Mercury, Venus, and the Moon. During a transit of Venus, the planet moves directly between Earth and the Sun, appearing as a dot against the Sun's disk. Venus transits the Sun four times every 243 years, so this is one of the rarest predictable astronomical phenomena. This image shows the movement of Venus over a five-hour period during the 2004 transit.

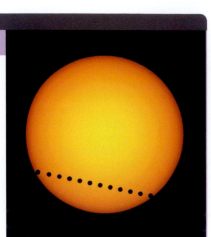

increasing precision led to a greater understanding of the stability of the Solar System and the actions of one planet on another. It also improved our knowledge of the lunar orbit and our ability to predict eclipses. During the solar eclipse of 1715 Edmond Halley predicted its path across the earth to an accuracy of less than half a mile (1 km).

### Predicting eclipses

Eclipses themselves had always been hard to predict: Thales of Miletus (640–546 BCE) was only able to estimate their timing to within a year. Luckily, the eclipse he did predict passed over the River Halys in Turkey during a war between the Lydians and the Medes, and the soldiers on both sides were so awestruck that they immediately lowered their weapons and declared peace. The eclipse allows us to date the battle precisely, to May 28, 585 BCE, demonstrating the importance of these events to astronomers and historians.

Solar eclipses occur only when the New Moon is close to a point (the node) at which the Moon's orbit crosses Earth's. If

>> Kepler's third law revealed the relative sizes of the planetary orbits but not their absolute sizes. The ancient Greeks had estimated that the Sun was 1150 Earth-radii away, but Kepler's Mars calculations convinced him that the real value was far greater. There was much confusion in the 17th century—the 1687 and 1713 editions of Newton's *Principia* show a doubling of his estimate. Newton also increased the estimated mass of the Sun by an astonishing factor of eight.

In 1716 Edmond Halley suggested a way of measuring the distance from Earth to the Sun, using the timing of the transits of either Mercury or Venus, measured by observers over a wide range of latitudes. This method was used in 1769 to observe the Venus transit. Today we know that the Sun is about 93,000,000 miles (150,000,000 km), or 23,455 Earth-radii, away.

### A day on Mars

The 1609 introduction of the telescope (see pp.84–85) revolutionized planetary astronomy. It led to the discovery of the moons of Jupiter, and revealed the

## 250 MILLION YEARS

The time it takes for the Sun, and our Solar System, to orbit once around the Milky Way, traveling at 155 miles (250 km) a second.

planets to be physical disks with observable surface features, rather than bright points of light. The sizes, spin rates, and spin-axis orientations could all now be measured. It was realized, for example, that Mars had a "day" that was 24 hours, 37 minutes long, and a spin-axis inclination of 24°, which implied that Mars had seasons, just like Earth. Saturn's inclination was 26.7°, which led to changes in the appearance of its rings.

### Calculating the Earth-path

Around 135 BCE, Hipparchus (see p.28) had measured planetary positions to an accuracy of 60 minutes of arc. By 1586 Brahe had improved this to 1 minute, by 1700 Flamsteed to 0.3 minutes and by 1856, Argelander was working to an accuracy of 0.015 minutes. This

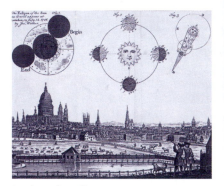

### London solar eclipse
This print from 1748 was published to inform Londoners of the cause and appearance of a solar eclipse that would be visible from the capital on July 14, of that year.

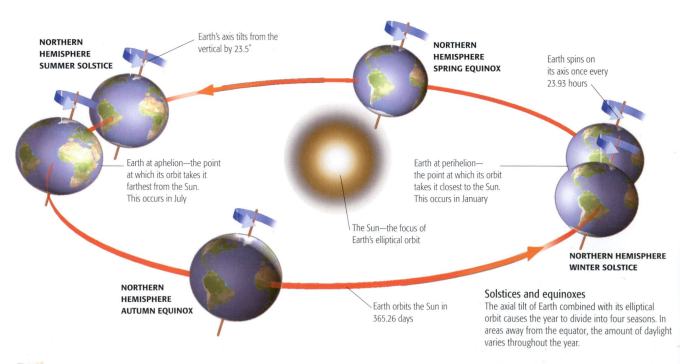

NORTHERN HEMISPHERE SUMMER SOLSTICE

Earth's axis tilts from the vertical by 23.5°

NORTHERN HEMISPHERE SPRING EQUINOX

Earth spins on its axis once every 23.93 hours

Earth at aphelion—the point at which its orbit takes it farthest from the Sun. This occurs in July

Earth at perihelion—the point at which its orbit takes it closest to the Sun. This occurs in January

The Sun—the focus of Earth's elliptical orbit

NORTHERN HEMISPHERE WINTER SOLSTICE

NORTHERN HEMISPHERE AUTUMN EQUINOX

Earth orbits the Sun in 365.26 days

### Solstices and equinoxes
The axial tilt of Earth combined with its elliptical orbit causes the year to divide into four seasons. In areas away from the equator, the amount of daylight varies throughout the year.

| **PERIGEE** The point on the Moon's orbit where it is closest to Earth.

| **APOGEE** The point on the Moon's orbit where it is farthest from Earth.

the Moon is near perigee (at its closest to Earth), its shadow cone intersects Earth's surface. Anyone standing in that shadow experiences a brief solar eclipse. The shadow can be as wide as 135 miles (217 km) across, sweeping Earth at a speed of 1056 mph (1700 kph). Totality can last for as long as 7.7 minutes.

Total solar eclipses are rare. Even though they occur somewhere on Earth every 18 months or so, a total eclipse only occurs at a fixed locality every 370 years. This rarity has given them a special place in mythology and folklore.

## Eclipse science

When the Moon completely covers the solar disk, the Sun's coronal atmosphere springs into view. The corona is a gaseous region around the Sun, extending more than 600,000 miles (1,000,000 km) from its surface, with temperatures exceeding one million degrees Celsius. Closer to the solar limb (the apparent edge of the Sun in the sky) you can see large arches of red gas confined by magnetic fields: these are the solar prominences (cooler clouds of gas within the hotter coronal ones).

Kepler correctly identified these as solar phenomena, not lunar volcanic activity, as previously thought. But it took another 200 years for astronomers to realize that the size of the corona varied with the phase of the solar cycle.

The interest in stellar bodies, rather than planetary ones, increased hugely during the 19th century as telescopes improved.

### STELLAR DISTANCE
Astronomy was revolutionized by the measurement of the **distance to the star 61 Cygni** in 1838. Stellar energy outputs could then be calculated, and stellar lifecycles debated.

### EINSTEIN
The stellar energy problem was solved in essence when **Albert Einstein** realized that $E = mc^2$ in 1905. **The theory of general relativity 304–05** » also explained the extent to which starlight bends as it passes close to the Sun.

### SOLAR CORONA SCIENCE
Eclipses are now observed mainly for pleasure. Coronal observations are obtained using spacecraft **396–97** », which can view the **solar outer atmosphere** continuously, not just during the brief occasions of an eclipse.

**SOLAR CORONA**

SEE ALSO »
pp.182–83 THE SOLAR SYSTEM
pp.394–95 INSIDE THE SOLAR SYSTEM

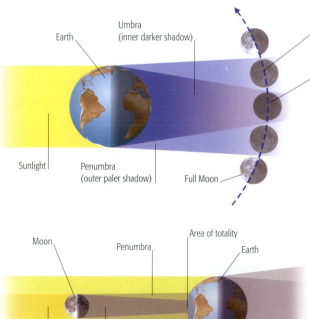

Only a slight darkening of the Moon occurs in the light outer shadow.

Earth — Umbra (inner darker shadow)

The Moon is darkest within the umbra.

**Lunar eclipse**
If the Full Moon is near an orbital node, it can pass through Earth's shadow cone and be eclipsed. People inhabiting the half of Earth that faces the Moon would see the Moon as darkened.

Sunlight — Penumbra (outer paler shadow) — Full Moon

Moon — Penumbra — Area of totality — Earth

**Solar eclipse**
The shadow cast by the New Moon just intersects Earth's surface. People standing in this small shadow region would see the solar disk totally covered by the Moon.

Sunlight — Umbra — Area of partial eclipse

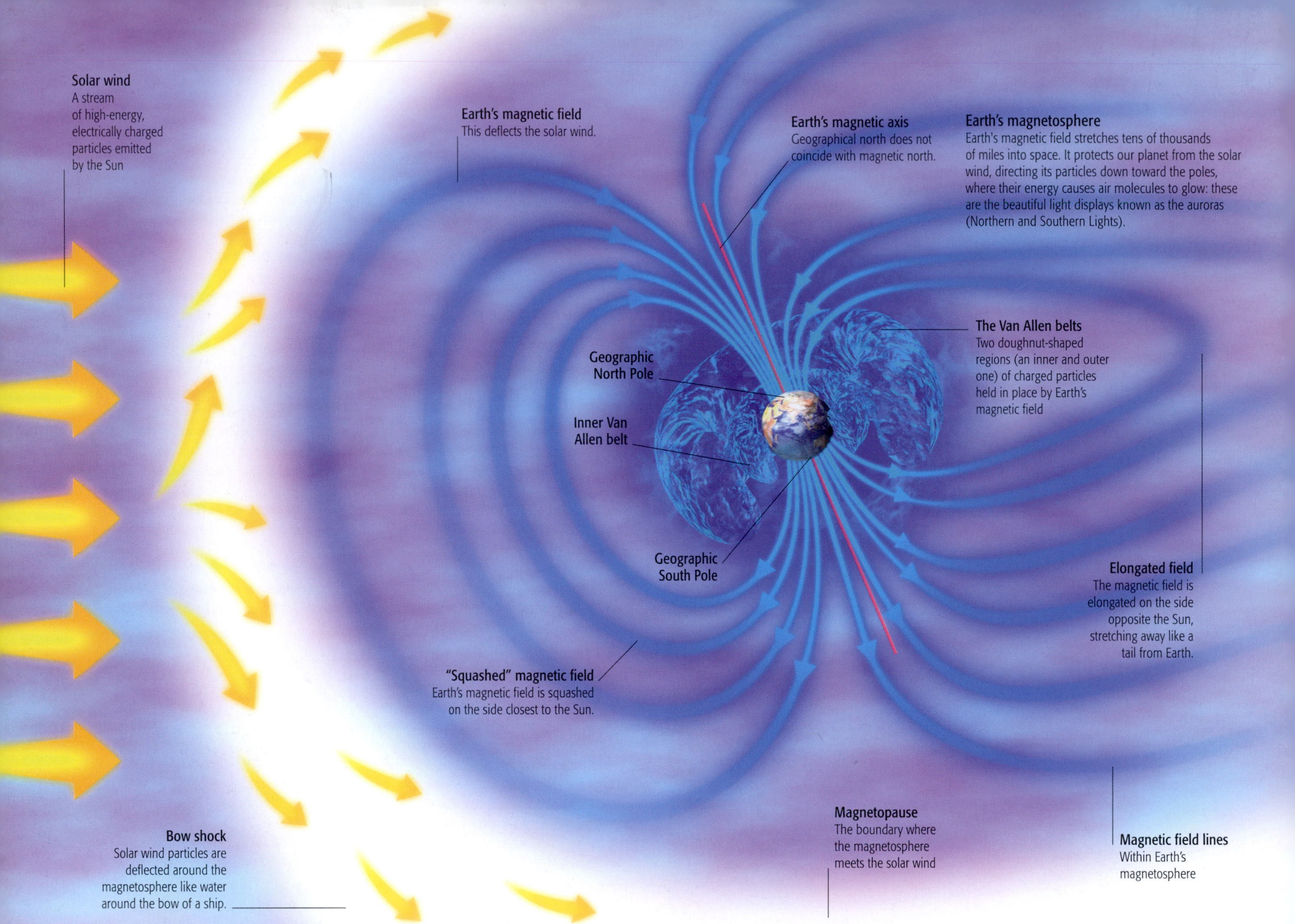

## Diagram labels

**Solar wind**
A stream of high-energy, electrically charged particles emitted by the Sun

**Earth's magnetic field**
This deflects the solar wind.

**Earth's magnetic axis**
Geographical north does not coincide with magnetic north.

**Earth's magnetosphere**
Earth's magnetic field stretches tens of thousands of miles into space. It protects our planet from the solar wind, directing its particles down toward the poles, where their energy causes air molecules to glow: these are the beautiful light displays known as the auroras (Northern and Southern Lights).

**The Van Allen belts**
Two doughnut-shaped regions (an inner and outer one) of charged particles held in place by Earth's magnetic field

Geographic North Pole

Inner Van Allen belt

Geographic South Pole

**Elongated field**
The magnetic field is elongated on the side opposite the Sun, stretching away like a tail from Earth.

**"Squashed" magnetic field**
Earth's magnetic field is squashed on the side closest to the Sun.

**Bow shock**
Solar wind particles are deflected around the magnetosphere like water around the bow of a ship.

**Magnetopause**
The boundary where the magnetosphere meets the solar wind

**Magnetic field lines**
Within Earth's magnetosphere

« **BEFORE**

People first became aware of magnetism through finding naturally magnetic rocks.

**ACCIDENTAL DISCOVERY**
No one is sure of the true origin of the name "magnet." One of the most popular stories centers on a shepherd in Magnesia, Greece, who found nails sticking to a black rock. Until the 15th century, **magnetic rocks** were thought to have magical powers, which could be destroyed by garlic or diamonds.

**THE WORK OF PEREGRINUS**
French scholar Pierre de Maricourt (also known as Peter Peregrinus) was the first person to record the basic facts of magnetism, noting them in a letter written in 1269. He wrote in detail about magnetic poles, forces, and **compass** needles.

« SEE ALSO
pp.18–19 EARLY METALWORKERS

# Magnetic Fields

The area of influence of a magnet is called a magnetic field. Earth itself has a vast magnetic field emanating from its core and extending far out into space. The fields around much smaller, more familiar magnetized objects follow similar "lines of force" flowing from opposite poles.

Although magnetic fields are invisible, they can be visualized using tiny pieces of iron called filings. When iron filings are sprinkled around a magnet, they immediately arrange themselves along the field's "lines of force," illustrating the magnetic field.

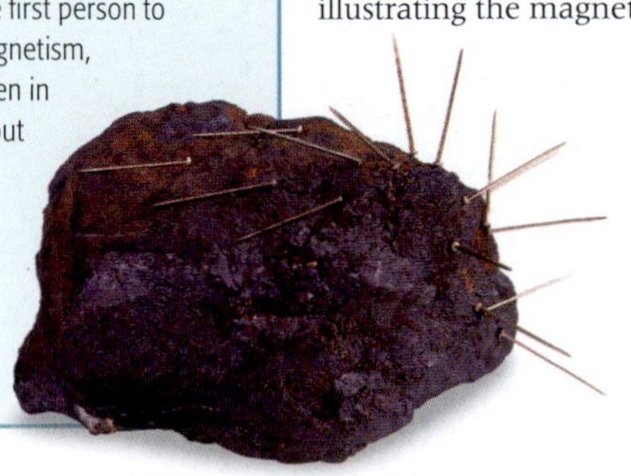

**Lodestone**
Lodestones are a form of magnetite that were recognized as adopting a north–south alignment, and exercising repulsion and attraction. The ancient Chinese called them *tzhu shih* ("loving stones"), because they liked to "kiss."

Every magnet has two ends, or poles: one that is north seeking (N), and one that is south seeking (S). If you put together two magnets with like poles facing each other (N–N or S–S), they push each other apart, while if you hold the unlike poles facing each other (N–S), they pull together.

Materials that can be magnetized, such as iron, are composed of tiny regions called domains, each of which also has a north and south pole. When iron is unmagnetized, the domains are randomly oriented within the metal. But when the metal is magnetized, the domains all align—forming dense lines with all north-seeking ends pointing in

the same direction. If a metal paper clip is held near to a magnet, for instance, it becomes magnetized—its domains align, and it is pulled toward the magnet.

**Earth's magnetism**
Any magnet that is free to move and carefully balanced on a frictionless bearing will swivel so that the north-seeking end always points northward and the south-seeking end always points southward (a compass needle is just a small magnet on a pivot). This is because Earth itself has a magnetic field, caused by the powerful electric currents that run through the liquid metal of its outer core. This magnetic field—the magnetosphere

## Creating a magnet

A piece of iron or steel can be made into a magnet by rubbing it repeatedly with another magnet.

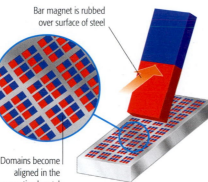

Bar magnet is rubbed over surface of steel

Domains become aligned in the magnetized metal

Domains randomly oriented

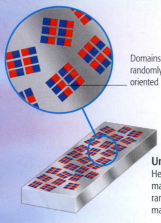

**Unmagnetized steel bar**
Here the tiny regions of magnetism—domains—are randomly oriented. The overall magnetic field is zero.

**Magnetized steel bar**
Rubbing steel with a magnet causes the domains to line up in the same direction. The metal is now a magnet, even after the other magnet is taken away.

---

ENGLISH PHYSICIAN AND SCIENTIST 1544–1603

# WILLIAM GILBERT

English natural philosopher William Gilbert graduated as a medical doctor, and began practicing in London in 1577. By 1599 he was the president of the College of Physicians. In 1600 he published his principal work, *On the Magnet, Magnetic Bodies, and the Great Magnet of the Earth*, which had taken 17 years to write. He proved that magnetism could attract through solid objects, not just air. He became the Royal Physician in 1601, but died—possibly of the plague—two years later. A second work, *A New Philosophy of our Sublunar World*, was published posthumously.

---

—extends into space, where it protects Earth from a stream of harmful particles emitted by the Sun called the solar wind.

### Shifting magnetic fields

Although compasses point northward, they do not point to Earth's geographical north pole, defined by the planet's axis of spin. Compass needles actually point toward "magnetic north," which at present lies a few hundred miles away from the geographical north pole.

Earth's magnetic north pole gradually shifts position, as the planet's liquid outer core is free to swirl somewhat separately from the rest of the planet's spin. Occasionally, Earth's magnetic field flips—possibly as a result of turbulence in the liquid metals—so that the magnetic north ends up near the geographical south pole (in Antarctica). This "field reversal" occurs at irregular intervals, from a few hundreds of thousands to many millions of years.

### Discovering magnetism

Magnetism was first recorded as a phenomenon around 600 BCE, when people realized that some minerals had magnetic properties. Naturally occurring forms of magnetite, known as lodestones (meaning "leading stones"), were observed always to align themselves in a north–south direction, and were used in the earliest compasses, from around 1000 CE onward.

The first person to study magnetism in a systematic way was English philosopher William Gilbert. In 1600, he published the results of his investigations, and suggested that our planet acts as a magnet. He also investigated static electricity (see pp.158–9). Like everyone else at the time, he considered electricity and magnetism to be completely separate. But in 1820, Danish natural philosopher Hans Christian

◁ **Magnetite compass**
Early compasses had a piece of magnetite attached to the bottom of a movable dial, with the north-seeking pole of the magnetite aligned with the North pointer on the dial.

Ørsted discovered that electric currents produce magnetic fields when the needle of a compass is deflected by a current going through a nearby wire. Physicists quickly came to realize that magnetism and electricity were very closely interrelated, and not separate.

---

**AFTER**

The study of magnetism led physicists to discover the nature of light, and ultimately radiowaves, X-rays, and infrared radiation.

**MAGNETISM AND ELECTRICITY**
English scientist Michael Faraday **170–71 »** discovered **electromagnetic induction** (the inverse of Ørsted's discovery) and showed that light is affected by magnetic fields.

**ELECTROMAGNETIC RADIATION**
In the 1860s Scottish theoretical physicist James Clerk Maxwell **167 »** produced a set of theorems—**Maxwell's Equations**—that unified the sciences of magnetism, electricity, and optics, and predicted the existence of electromagnetic waves.

**SEE ALSO »**
pp.166–67 ELECTROMAGNETISM
pp.170–71 MICHAEL FARADAY

---

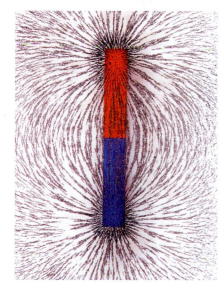

**Iron filings around a magnetic field**
A magnetic field can be made visible by placing iron filings around a magnet. Each tiny piece of iron lines up with the field, along the "lines of force." The closer each line is to the next, the stronger the force of the field.

### Magnetic forces

A magnet always attracts unmagnetized objects made of iron or steel. But two magnets close together can either attract or repel, depending on whether like or non-like poles are placed closest to each other.

Magnetic field lines

**Attraction and repulsion**
Two poles of opposite types—north/south—always attract each other.

South-seeking pole

North-seeking pole

**Force direction**
Attraction runs toward the south-seeking pole

**ATTRACTION**

North-seeking pole

North-seeking pole

**Like poles repel**
Two poles of the same type—north/north or south/south—always repel each other.

**REPULSION**

MATHEMATICIAN AND ASTRONOMER **Born 1564** Died 1642

# Galileo Galilei

## "**Science** is written in the language of **mathematics**."

GALILEO GALILEI

**G**alileo has been called the "father of modern science." He was one of the first thinkers to challenge the doctrines of Aristotle (see pp.36–37) and test his ideas by experiment. Galileo applied mathematics to the real world. He got into trouble with the church over his views on the solar system, and he was eventually summoned for trial in Rome and put under house arrest.

### From medicine to mathematics

As a teenager Galileo was tempted to go into a monastery, but his father persuaded him to study medicine; he therefore attended medical school at the University of Pisa. Galileo did not complete his medical degree; instead, he spent most of his time studying mathematics. He went on to teach that subject and later became professor of mathematics at the University of Pisa.

While at the university, Galileo experimented with pendulums, studied the fall of objects, and devoted himself to astronomy. Sitting in the cathedral, he noticed the great bronze lamp swinging to and fro on its chain in the

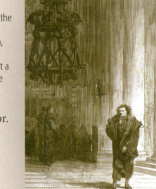

**Galileo observing a lamp in Pisa cathedral**
Watching the swinging of the great bronze lamp in the *duomo* (cathedral) in Pisa, and timing it against his pulse, Galileo realized that a pendulum makes a simple and reliable timekeeper.

draft from the door. He timed the swings, using his pulse, and was surprised to find that each swing took the same time, regardless of whether the lamp swung only a few centimeters or much more. Realizing that the pendulum is a regular natural timekeeper, he invented a little pulse meter (see left), and toward the end of his life he also designed a pendulum clock.

While at Pisa, Galileo performed

### INVENTION
### PULSE METER

While he was a medical student in Pisa, Galileo invented a simple pulse meter—a pendulum on a stand some 8 in (20 cm) high, constructed so that the length of the string was variable. The doctor would feel the patient's pulse, adjust the length of the string until the swing of the pendulum was just in time with the pulse, and then read the pulse rate off the scale on the stand, which probably just said "slow," "medium," and "fast." The medical authorities stole the idea, and Galileo received no credit for it.

## Trial for heresy
Galileo's discovery of the moons of Jupiter was praised by the church, but when he suggested that Earth orbits the Sun, he was accused of heresy and summoned for trial by the Inquisition in Rome.

**Galileo Galilei**
Seen here in a 1636 portrait by the Flemish painter Justus Sustermans, Galileo is credited as the originator of modern science. He made fundamental contributions to mathematics, physics, and astronomy.

various experiments in motion and inertia (see pp.112–13). He challenged Aristotle's assertion that heavy objects fall faster than light ones, in proportion to their mass. According to legend, Galileo tested the theory by dropping balls of various weights from the Leaning Tower of Pisa. In reality he tested Aristotle's theory by rolling balls down an inclined plane and ascertaining their speed of descent (see pp.104–105).

In 1592 Galileo moved north to Padua, and settled into a long-term relationship with Marina Gamba; the couple never married, but had three children. His two daughters were sent to a convent; his son, Vincenzo, was selfish and often a trial to his father.

### Further researches
In 1602 Galileo returned to the study of falling objects and trajectories; in due course he showed that a projectile, such as a cannonball, will follow a path close to a parabola as it is acted upon by gravity.

In 1609 Galileo heard about the spyglass invented by the Dutch spectacle maker Hans Lippershey. Although they never met, Galileo constructed his own instrument, and then increased the magnification from three to 30 times, creating the first astronomical telescope. Galileo made some money, or obtained some promise of favors, by offering spyglasses to the military, and then used his own to look at the heavens. First he

noted that the Moon was not the smooth and perfect sphere that Aristotle had claimed, but had a rough and irregular surface, with mountains and craters.

On January 7, 1610 Galileo looked at Jupiter. He could see colored bands on the planet, but there were also curious points of light in line—two on the right of the planet and one on the left. He thought they were distant stars, but on the following night all three seemed to have moved to the left of Jupiter. On January 13, he saw one more point of light. He then realized that they were not stars at all, but four small worlds—moons— orbiting the planet. This was a shock. Everyone "knew" that Earth was the center of the Universe, but here were moons orbiting Jupiter, not Earth. The moons, named Io, Europa, Ganymede, and Callisto, are now referred to as Galilean moons in his honor.

### Punishment for heresy
The Jesuits in Rome were fascinated when Galileo showed them the moons of Jupiter; he was welcomed with enthusiasm by the cardinals and even by the Pope. However, these moons prompted Galileo to turn toward the idea that the Sun was the center of the Universe, as proposed by Copernicus in 1543 (see pp.74–75). This in turn meant that Earth must move around the Sun, which directly contradicted the Bible.

Galileo was sentenced to life imprisonment by the Inquisition in Rome, though this was reduced to permanent house arrest at his villa in Arcetri. He died there in 1642. In 1992, 359 years after his imprisonment for heresy, Galileo was pardoned by Pope John Paul II.

### Pisa's leaning tower
According to legend, Galileo dropped balls of various weights from the Leaning Tower of Pisa to disprove Aristotle's assertion that heavy objects fall more quickly than light ones.

### TIMELINE

**1564** Galileo di Vincenzo Bonaiuti de Galilei is born in Pisa, Italy, the eldest child of musician Vincenzo and Giulia Ammannati.

**1581** Encouraged by his father, he enrolls to study medicine at the University of Pisa, and invents a pulse meter. Later he switches to the study of mathematics.

**1585** Begins teaching mathematics in Florence and then in Siena.

**1589** Becomes professor of mathematics at the University of Pisa.

**1590** Begins study of motion, rolling balls down inclined planes and measuring their velocity and acceleration.

**1592** Moves to the University of Padua as professor of mathematics; the post is much better paid than the one in Pisa. He teaches geometry, astronomy, and mechanics, and describes this as the happiest time of his life.

**1600** His first daughter, Virginia, is born to Marina Gamba.

**1601** His second daughter, Livia, is born. She and Virginia are to enter a convent in 1614, remaining there for the rest of their lives.

**1606** His son, Vincenzo, is born.

**1609** Hears about the spyglass invented by Hans Lippershey, makes his own, improves it, and sells it to the militia.

**GALILEO'S TELESCOPE**

**1610** Looks at the Moon and then at Jupiter, discovering four moons. He publishes the results in *Sidereus Nuncius*, or *Starry Messenger*.

**1611** Travels to Rome to show the Jesuits the moons of Jupiter and is made a member of the Accademia dei Lincei. At the celebratory dinner a Greek mathematician coins the word "telescope" for Galileo's instrument.

**1614** His views on the motion of Earth in a sun-centered universe are denounced as dangerous and close to heresy by Father Tommaso Caccini.

**1623** Publishes his first book, *The Assayer*.

**1632** Publishes his second book, *Dialogue*, in Florence. He is summoned to appear before the Holy Office in Rome.

**1633** Undergoes a papal trial in which he is accused of heresy. He is forced to recant and deny his own teachings and is then confined to house arrest in his villa at Arcetri.

**1638** Loses his sight. Proposes a method for measuring the speed of light.

**1641** Devises an escapement mechanism to govern the timekeeping of a pendulum clock; the idea is taken up 15 years later by Christiaan Huygens.

**1642** Dies at Arcetri.

**GALILEO'S TOMB IN FLORENCE, ITALY**

## BEFORE

For thousands of years astronomers had observed and recorded heavenly bodies. By the 2nd century BCE the effect of lenses on light had been documented in the East.

**VIEWING BY EYE**

The **Babylonians** (c.1800 BCE) named the constellations, while the ancient **Chinese** recorded supernovae and comets in the 4th century BCE. **Arabian and European astronomers ‹‹26–29** made catalogues of the stars as long ago as the 2nd century BCE.

**EARLY WORK ON OPTICS**

In about 1010 Arab mathematician **Alhazen** working in Egypt, laid down the basic principles of optics, describing how lenses work in his *Book of Optics*.

**‹‹ SEE ALSO**
*pp.60–61* ALHAZEN
*pp.82–83* GALILEO GALILEI

# Exploring the Skies

**People have always been fascinated by the stars and planets in the night sky, but until 1609 they had to rely on just the naked eye to make them out. The invention of the telescope opened up entirely new views into the Universe.**

The first telescope may have been made by English polymath Leonard Digges, around 1551, but there is no evidence that he used it to study the stars. Because his work was not widely known, the credit for inventing the telescope usually goes to the Dutch.

### Early designs

The first designs for a telescope were produced by a Dutch glasses maker called Hans Lippershey. He discovered that if two lenses were installed in a tube, distant objects could be made to appear closer. The Italian scientist Galileo Galilei heard about Lippershey's telescope; he made his own versions, improved the magnification to 30 times, and sold the idea to the local prince. Then, during the last month of 1609, he turned his telescope toward the sky and was the first person ever to see a up-close view of the Moon. He also discovered four of Jupiter's satellites, and his instrument was the first to be given the name "telescope." In 1611 the

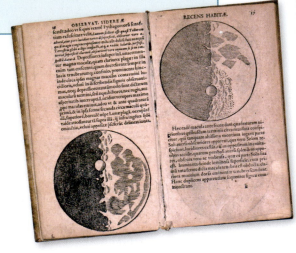

**Galileo's drawings of the Moon**
Galileo observed that the Moon was not the smooth sphere described by Aristotle—instead, it was covered with mountains, valleys, and craters. His illustrations of the Moon were published in *Sidereus Nuncius* in 1610.

astronomer Johannes Kepler improved on Galileo's design by using a convex lens for the ocular (eyepiece) instead of a concave one. This gave him a wider field of view and potentially higher magnification. One problem, however,

**40 TIMES was the approximate magnification of Newton's telescope, even though it was only about 8 in (20 cm) long, with a 1¼-in (3-cm) mirror.**

was that to achieve high magnification the objective (the lens at the front) needed a long focal length. As a result telescopes grew ever longer—some reaching 33 ft (10 m).

### Refracting and reflecting light

Another problem associated with these first telescopes came from the way in which the lenses work. In 1671 Newton wrote to the Royal Society about his experiments with prisms, his explanation of the colors of the spectrum, and his theories about the nature of light. Newton explained that

lenses, like prisms, refract (bend) light as it passes through, and refract blue light more than red; that is why a prism produces a spectrum. When light was refracted in a "refracting" telescope it caused all the stars to have colored fringes around them.

Almost in passing, Newton described his invention of a reflecting telescope, which used mirrors rather than lenses to focus the light. Mirrors do not refract light, and all colors are reflected equally; so a reflecting telescope produces no colored fringes. A Scottish astronomer, James Gregory, had described a similar instrument in 1663, but could not find anyone to grind the mirrors for him. Newton was lucky—he managed to get his design made, and presented it to the Royal Society.

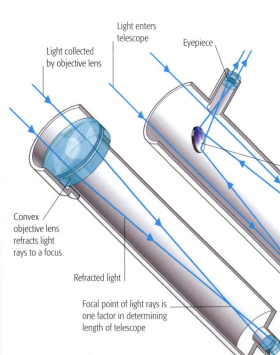

◁ **How a reflecting telescope works**
The primary mirror reflects the incoming light back up the tube. The secondary mirror in a Newtonian reflector is at 45° and reflects the light through the side of the tube to the eyepiece; in a Gregorian reflector it reflects the light back down the tube through a hole in the primary to an eyepiece behind.

Light enters telescope

Light collected by objective lens

Eyepiece

Secondary flat diagonal mirror reflects light to the eyepiece

Convex objective lens refracts light rays to a focus

Refracted light

Concave primary mirror

Focal point of light rays is one factor in determining length of telescope

Eyepiece magnifies image

△ **How a refracting telescope works**
Rays of light entering the objective lens at the top of the tube are brought to a focus near the lower end of the tube. They then pass through the eyepiece to the observer who sees a clearer, magnified, virtual image. Generally, both lenses are convex.

## BREAKTHROUGH

### THE ACHROMATIC LENS

In the mid-18th century Chester Moore Hall and John Dollond independently solved the problem, highlighted by Isaac Newton, of chromatic aberration and the resulting color fringes around stars viewed through refracting telescopes. Hall and Dollond made lenses from a sandwich of two different materials: crown glass and flint glass. These have different refractive indices—meaning that they bend light waves to different extents. The resulting combined lens focused light of different wavelengths (and colors) at the same point, creating a sharp image with no fringe.

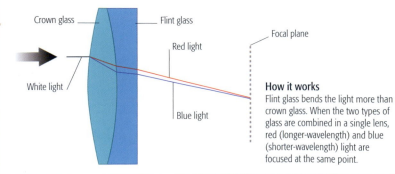

Crown glass

Flint glass

Focal plane

Red light

White light

Blue light

**How it works**
Flint glass bends the light more than crown glass. When the two types of glass are combined in a single lens, red (longer-wavelength) and blue (shorter-wavelength) light are focused at the same point.

**Eyepiece**
The horizontal eyepiece made the telescope easier to use than refracting telescopes, whose users had to peer awkwardly upward at the sky.

▽ **Telescope development**
Technological advances allowed telescopes to become more and more powerful.

**1611 Kepler** introduces the convex eyepiece, improving clarity, magnification, and field of view.

**1763** The **apochromatic lens**, or triple achromat, focuses red, blue, and green light together.

**1857** Advances in chemistry allow the manufacture of **silver-coated mirrors**, which are more reflective.

**1990s Adaptive optics** are used to remove atmospheric distortion.

| 1600 | 1700 | 1800 | 1900 |

**1609** Galileo makes the first **refracting telescope**, with a concave eyepiece.

**1663** The **reflector telescope** is described by James Gregory and later built by Isaac Newton.

**1733** The invention of the **achromatic lens** allows red and blue light to be focused together.

**1932 Aluminized mirrors** are introduced; these are cheaper to produce and more durable.

**1990** The **Hubble Space Telescope** avoids atmospheric distortion.

**Focus adjustment control**
By turning this screw, the user can shorten or lengthen the telescope tube.

△ **Paranal Observatory, Chile**
The Very Large Telescope (VLT) at the Paranal Observatory comprises four units that can be used alone or combined to make one of the world's most powerful telescopes.

◁ **The remains of a star**
This picture, taken by the VLT, shows the remains of a star after a massive supernova explosion. This body, called the Crab Nebula, lies close to the constellation Taurus.

**Sliding focus**
The telescope tube was made of leather, in two parts, so that one could slide over the other to adjust the focus.

A replica of it is shown here. They were delighted, and invited him to become a fellow.

## Further developments
In the centuries that followed, telescopes were used not only for star gazing but also for spying on enemy forces and for navigation at sea. Both refracting and reflecting telescopes were built with increasing precision. Today almost all the major astronomical telescopes are Gregorian reflectors—even the Isaac Newton Telescope on the island of La Palma.

**Replica of Newton's telescope**
The Royal Society in London keeps a replica of Newton's original reflecting telescope. It had a 1¼-in (3-cm) mirror and was surprisingly compact and simple, but it changed the way astronomers study the stars. Most modern telescopes are reflectors.

Telescopes have grown considerably since Newton's instrument with its 1¼ in (3 cm) mirror. One example is the Very Large Telescope (VLT) at Cerro Paranal in Chile. This array is actually a combination of four main telescopes, each with a mirror 27 ft (8.2 m) in diameter. The point of such huge mirrors is not for greater magnification, but for more effective light gathering. Double the diameter of the mirror and you can collect four times as much light, and so see objects that are much fainter, which would be invisible with a smaller instrument. In 2005 the VLT photographed a planet in orbit around a star in the constellation Hydra, 230 light years from Earth—more than one and a quarter million billion miles (two million billion kilometers) away.

After its light-gathering capacity, the next most important feature of a telescope is its resolving power; that is, its ability to pick up fine details and produce sharp images. A modern large telescope could form a sharp image of a dime from several miles or kilometers away.

Magnification is much less important; indeed, many astronomical observations have been made with ordinary binoculars, which are, in fact, mini-telescopes.

**AFTER** »

Today telescopes look at stars from Earth and from space. Some gather radio waves, gamma rays, or X-rays rather than light.

**GATHERING RADIO WAVES**

**RADIO TELESCOPES**
Grote Reber built the first **radio telescope** in 1937. It worked like a reflecting telescope, but picked up **radio waves** rather than light waves.

**THE HUBBLE TELESCOPE**
Launched in April 1990 the **Hubble telescope** operates in space, **above the atmosphere**. As a result it has taken some of the most **spectacular** pictures ever seen **414–15**»

SEE ALSO »
*pp.110–11* Isaac Newton
*pp.116–17* Splitting and Bending Light
*pp.396–97* Spaceprobes and Telescopes

**« BEFORE**

For a long time, even thinkers such as Aristotle were misled about the nature of motion by common-sense experience.

**EXTERNAL FORCES**

**Aristotle** argued that things only keep moving because they are **continually pushed or pulled** by an **external force** and slow down as soon as the force is removed, just as a cart slows down when the ox stops pulling.

" ...A body in motion can **maintain** this motion only if it remains **in contact** with a **mover**..."

ARISTOTLE, 384–322 BCE

**ANTIPERISTASIS**

**Aristotle** stated that an arrow keeps flying after it has left the bow because of **a natural reaction** or **antiperistasis** with the air, which generates **eddies** or **vibrations** that drive it on.

**« SEE ALSO**
*pp.36–37* **ARISTOTLE**

# Motion, Inertia, and Friction

One of the great scientific breakthroughs of the 17th century was to understand how and why things move and stop moving. The key to the problem is that objects have inertia, which means that they slow down and stop only when something, typically friction, forces them to do so.

For almost 2,000 years people subscribed to Aristotle's assertion that, unless things are continually propelled onward by some force, they soon slow down naturally to rest. It seemed only common sense. Yet there were problems with this view. How, for example, does an arrow continue in its flight long after it has left the bow? In the Middle Ages the Islamic scholar Avicenna and the French priest Jean Buridan spoke of moving objects having "impetus." At first sight, this idea seems close to the modern idea of momentum—the natural tendency of an object in motion to keep on moving —but for Avicenna and Buridan, impetus was an active internal motive force pushing things onward against the natural tendency to slow down.

## Galileo's experiments

The great insight that Italian scientist Galileo Galilei achieved was to realize that common sense is not always

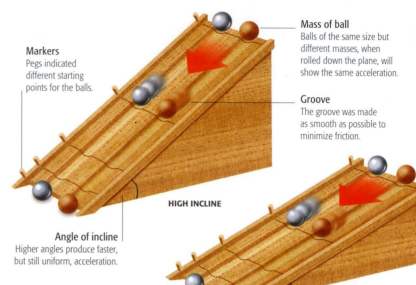

**Markers**
Pegs indicated different starting points for the balls.

**Mass of ball**
Balls of the same size but different masses, when rolled down the plane, will show the same acceleration.

**Groove**
The groove was made as smooth as possible to minimize friction.

**HIGH INCLINE**

**Angle of incline**
Higher angles produce faster, but still uniform, acceleration.

**LOW INCLINE**

△ **Galileo's inclined plane experiment**
By rolling balls down sloping wooden grooves and timing them as they traveled over different distances, Galileo proved that the acceleration at a given angle was uniform, no matter how far the ball traveled.

enough to explain different phenomena—careful observation and experiment are also needed to help people arrive at the truth.

During the 1630s, while under house arrest for championing Nicolaus Copernicus's heliocentric view of the Universe (see pp.74–75), Galileo carried out a series of experiments, the most famous of which involved rolling balls down a pair of inclined wooden grooves.

Rolling a ball down a slope allowed Galileo to slow the acceleration of

**Inertia in action**
When a ridden horse stops suddenly in front of a fence, inertia will often propel the rider over the horse's head—as in this case, where the rider is being thrown right over the jump.

falling objects to a measurable rate. By timing how long the ball took to travel a particular distance down the slope, he could figure out its acceleration, and by altering the slope, he could change the acceleration. To study deceleration, he rolled the ball down one slope and allowed it to roll up a slope opposite. Galileo had no clock, but he devised ingenious ways to time events, such as using his pulse or weighing how much water spurted from an urn as the ball traveled a certain distance.

### Discovering inertia

It took Galileo to figure out that the role played by friction in slowing things down is crucial. Things don't have a

**Galileo's demonstration**
Galileo used his inclined plane experiment to demonstrate the law of falling bodies. He apparently exhibited it to groups of clergy and aristocrats, as shown in this 1839 fresco by Florentine painter Giuseppe Bezzuoli.

natural tendency to slow down at all. The opposite is true: a moving object will keep moving at the same speed unless something slows it down, and that slowing force is usually friction.

In fact, as Galileo realized, there is no real difference between an object moving at a steady rate and one that is not moving at all—both objects are unaffected by forces. A force is needed either to make the object speed up or slow down or to start it moving. This is the principle of inertia—the tendency of an object to resist a change in motion, whether it is moving steadily or stationary.

Galileo also saw that it is impossible to distinguish between a steadily moving object and a still one without some point of reference; this idea formed the foundation of Einstein's special theory of relativity (see p.302) in the early 20th century. Galileo further realized that the rate of acceleration depends on the strength of the force and the heaviness, or mass, of the accelerating object.

### Projectiles

The concept of inertia solved the problem of the arrow. An arrow keeps flying while inertia is dominant, but gravity is continually pulling it down. Galileo showed that the combination of inertia and gravity would mean that the path of a projectile, such as an arrow, through the air must be a curve called a

parabola (see below). The effect of drag, or air resistance, is to slow the arrow down, so that it falls short.

Galileo's ideas enabled a giant leap in the understanding of force and motion, and laid the groundwork for Isaac Newton (see p.104) to bring them together in his three laws of motion. Newton's first two laws are, in fact, restatements of Galileo's discovery of inertia and his finding that acceleration depends on the relative size of the

## IN PRACTICE
### BICYCLE V–BRAKES

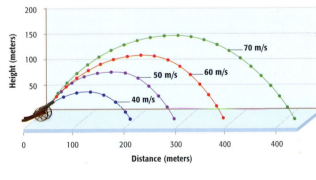

Brakes work by exploiting friction to counteract inertia. When a cyclist brakes, the extra friction as the brake pads are squeezed on the wheel rim slows the rotation of the wheel and reduces the bicycle's forward inertia.

Cable acts on end of short lever called a caliper when lever on handlebars is squeezed

Calipers are pivoted so that tension of cable squeezes brake pads against wheel

Brake pads are squeezed against rim, increasing friction

Brake pads are made of a rubberized material that causes maximum friction with minimum damage to wheel rim

## 3.2 SECONDS
**The time it would take for a ball or body to fall to the ground from the top of the leaning tower of Pisa.**

force and the object—or, rather, the object's mass. In this way, Galileo's discoveries underpin our knowledge of virtually every movement in the universe, from the motion of atoms to the rotation of galaxies.

**The flight of projectiles**
The discovery of inertia enabled Galileo to explain the trajectories of objects such as cannonballs. Their flight traces a parabola. The higher the angle at which a cannonball is fired (up to a maximum of 45°), the farther the ball travels.

## AFTER

**The concept of inertia is central to our understanding of motion.**

### STREAMLINING FOR SPEED
A race car relies on friction to make its wheels turn, but is held back by friction in its moving parts and by friction with the air, known as **drag**. To minimize drag, designers **streamline** the car's body so that it slips through the air with the **minimum of disturbance to air molecules.**

**FORMULA ONE RACING CAR**

### SWIMSUITS
Champion **swimmers** are **held back by drag** as they race through the water. Swimsuit manufacturers use supercomputers in the **design of swimsuits** to minimize the **"skin friction."**

**SEE ALSO »**
pp.104–105 NEWTON'S LAWS OF MOTION
pp.110–11 ISAAC NEWTON
pp.112–13 SPEED AND VELOCITY

« BEFORE

An ancient yet sophisticated system, used in China from the 5th century BCE, involved "counting rods" (*suan chou*).

### CHINESE COUNTING RODS
Counting rods were usually made from lengths of bamboo. They were used according to the **base 10 number system** « **30–31**. To form numbers, rods were set out in groups vertically or horizontally. The rods could be used for basic arithmetic, finding **square and cube roots**, and solving **complex equations**. **Negative numbers** could be formed by using black rods or laying rods diagonally across numbers.

« SEE ALSO
*pp.30–31* ANCIENT NUMBER SYSTEMS

## The abacus
The abacus as we know it, a frame with beads on wires, originated in China. Different forms developed in Japan, Korea, and Russia, and are still sometimes used in stores and schools. Shown here is a 16th-century Russian abacus.

---

### MATHEMATICIAN (1550–1617)
## JOHN NAPIER

Born in Edinburgh, Scotland, John Napier entered St. Andrew's University at 13, but it is thought that he acquired most of his mathematical learning while studying in Europe. In addition to inventing logarithms and creating "Napier's bones," Napier popularized the use of the decimal point and introduced decimal notation for fractions. His powers of invention went beyond mathematics—he also experimented with the use of fertilizers to enrich farmland and designed machines for warfare.

# Methods of Calculating

**Arithmetical calculating—adding, subtracting, multiplying, and dividing numbers—has always been important in commerce, industry, and everyday life. The development of more effective calculation methods, and even of mechanical calculating aids, enabled huge advances in scientific and practical knowledge.**

In the civilizations of the ancient world, traders learned through practice the basic addition, subtraction, multiplication, and division that they required for their business, but any calculations more complicated than these required the work of an expert mathematician.

### Early methods of calculating
One way to make calculation easier was to use mechanical aids. The abacus, for example, was a visual counting aid that first appeared in Mesopotamia in 2700–2300 BCE, and became widely used in the ancient world for calculating with large numbers in trade and engineering. The earliest versions were boards or tables with grooves, or even lines in the sand, into which people placed pebbles. (The Latin name for a pebble, *calculus*, has given us our word "calculate.") In China, people used counting rods on a board arranged in rows and columns. These devices were adequate for simple, quick calculation until a good system of writing numbers was formulated.

The development of new techniques also made calculation easier and more accurate. A breakthrough came when Hindu–Arabic numerals and a place-value system of notation (see p.31) were adopted in the West, replacing the old Roman numerals. Because digits were arranged in columns according to their value in the decimal system, addition and subtraction were made more straightforward. However,

> "...by **shortening** the **labors** [Napier's logarithms] **doubled** the **life** of the astronomer."
>
> PIERRE-SIMON LAPLACE (1749–1827), FRENCH MATHEMATICIAN AND ASTRONOMER

multiplication and division were still difficult, labor-intensive jobs. Advances in the sciences, especially astronomy and navigation, demanded ever more complex calculations, and mathematicians sought ways to lessen this work. One method devised in the 16th century was prosthaphaeresis, in which formulas from trigonometry were used for multiplication and division, but this was short-lived due to its complexity and inaccuracy. Then a discovery was made that would revolutionize calculation.

### Development of logarithms
In 1614 Scottish mathematician John Napier (see left) published *Mirifici logarithmorum canonis descriptio* ("Description of the Marvellous Rule of Logarithms"), in which he explained the principle of calculation using logarithms. What Napier had found was that any number can be expressed as a power of 10, which he called a logarithm, and that when the logarithms of two numbers were added together the result was the logarithm of the product of those two numbers (the numbers multiplied together). Similarly, division could be achieved by subtracting the logarithm of one number from that of another.

With his friend Henry Briggs, Napier worked out logarithms of numbers to the base 10, which Briggs published as a set of tables. Using these tables, it was simple to look up the logarithms of numbers, add or subtract them, and consult a table of antilogarithms to find the result. Napier and Briggs also discussed logarithms to a base other than 10 and discovered the natural

### "Napier's bones"
In this late-17th century set, the "bones" are cylinders marked with 0 to 9 and columns of multiples of the same. You turned the cylinders then read off the digits along a specific row.

## Pascal's calculator
This replica of Blaise Pascal's original calculator shows the dials, labeled for units, tens, hundreds, thousands, and so on. Numbers are "dialed in" as on an old-fashioned telephone. The windows at the top show black figures for addition and red for subtraction.

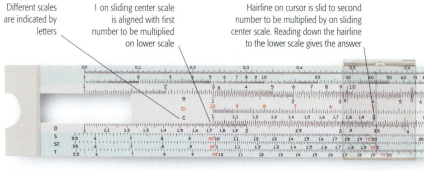

Different scales are indicated by letters

1 on sliding center scale is aligned with first number to be multiplied on lower scale

Hairline on cursor is slid to second number to be multiplied by on sliding center scale. Reading down the hairline to the lower scale gives the answer

**Slide rule**
To use a slide rule, a mark on the sliding central scale is aligned with a mark on either the top or bottom fixed scale, and the sliding cursor moved along to read off the answer. Here a simple multiplication is shown.

logarithm (see right), also called the Napieran logarithm, which became important in later mathematical work.

Logarithms made it easier for users to do tasks such as figuring out square and cube roots, and trigonometrical functions. This, in turn, facilitated advances in many scientific fields.

## Mechanical calculating tools
It was a small step from the tables of logarithms and antilogarithms to mechanical devices based on similar principles. In 1617 Napier invented a

kind of abacus called "Napier's rods." This comprised a set of rods inscribed with columns of numbers; the user arranged the rods on a special board, then read along one of the resulting rows of numbers, adding or subtracting digits to perform multiplication or division. Variations on this device soon appeared, usually in the form of rotating wooden or ivory cylinders ("Napier's bones") in a box. Another calculating machine was invented by French mathematician Blaise Pascal in 1645, using an elaborate mechanism of rotating wheels to perform additions and subtractions (see opposite).

Most important of all, however, was an ingenious and relatively simple device—the slide rule. Invented in

1622 by William Oughtred, it had logarithmic scales (in which the gradations become closer together as the value increases) inscribed on a set of connected rulers: the user slid the rulers to realign the scales, then read off the answer to a calculation. Slide rules were so effective that scientists, engineers, and technicians continued to use them until the advent of the electronic calculator in the 1970s.

## NATURAL LOGARITHMS

Natural logarithms are logarithms to the base *e*. Discovered by John Napier and Henry Briggs, their significance was only recognized by Nicholas Mercator in 1688. The number *e* is irrational (it cannot be written as a simple fraction), and has been calculated to hundreds of decimal places; to five places, it is 2.71828. It provides a "baseline" for complex calculations such as the rate of decay of a radioactive isotope (see pp.298–99). The number also appears in nature, in objects or systems that undergo a constant, exponential rate of growth, such as a spiraling Nautilus shell (right).

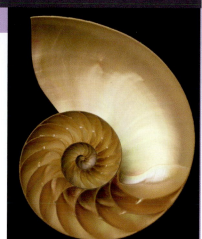

## AFTER »

Further advances in calculating techniques made mathematics an increasingly powerful tool in all areas of science.

**CALCULATING MACHINES**
**Tables of logarithms** continued to be used for calculations, not just in science but in everyday fields such as navigation and banking. There was, however, a constant worry about **inaccuracies** in the figures. This prompted inventor **Charles Babbage 174 »** to design his **Difference Engine** and **Analytical Engine**, to perform these calculations **mechanically** and print the results. In this way, he hoped to prevent errors.

**MATHEMATICAL NOTATION**
Swiss scientist **Leonhard Euler** was one of the greatest mathematicians of all. His innovations included **new symbols** for major concepts, such as $\Sigma$ for "sum," $i$ for the square root of -1, and e for the base of natural logarithms. He also popularized the use of $\pi$ for the ratio between the circumference and diameter of a circle.

SEE ALSO »
pp.102–03 GRAPHS AND COORDINATES
pp.174–76 CALCULATING AND COMPUTING

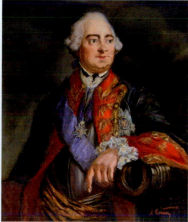

LEONHARD EULER

« **BEFORE**

Ancient Egyptians never dissected humans and so proposed the theory that arteries carried air, an idea that persisted for centuries.

### CHANNELS OF BLOOD AND AIR
In the 3rd century BCE ancient Greek physicians **Herophilus** and **Erasistratus** dissected blood vessels and saw heart valves. They distinguished **arteries** (which carry blood away from the heart) from **veins** (which carry blood toward the heart) by observing that it was only the severed arteries of a living pig that pulsed blood. However, in dead bodies, arteries empty into veins, so (like the Egyptians) they thought that **intact arteries carried air**.

### BLOOD FLOW
In the 2nd century CE Roman physician **Galen** knew that arteries carried blood, but **proposed erroneous theories**, which endured for centuries. For example, he thought that arterial blood was made **in the heart** and venous blood made **in the liver**, and that the body consumed both.

GALEN

« SEE ALSO
*pp.72–73* THE HUMAN BODY REVEALED

By the 18th century scientists were faced with the challenge of reassessing blood production and blood flow in the body. They set out to disprove the established idea that large amounts of blood were constantly produced in the body in order to put forward the theory that we all possess a fixed quantity of blood, which is pumped around the body by the heart.

### Blood in circulation
In 1553 a Spanish theologian named Michael Servetus proposed that blood circulated through lungs (pulmonary circulation). But his bold assertion conflicted with ancient Greek anatomy (see BEFORE) and so cost him his life. He was burned at the stake as a heretic. In fact, the medieval Arabic physician Ibn al-Nafis had said the same as Servetus back in the 12th century, but his theory was lost,

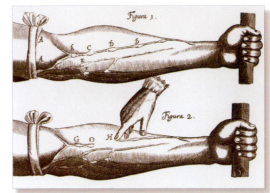

**Studying blood flow and valves**
William Harvey found that blood pooled in the surface veins of a ligatured arm, and that valves—marked here by swellings—stopped blood inside from flowing back toward the hand.

### Capillaries fill a gap
There remained one part of the story of circulation to complete: how blood passed from arteries to veins. Harvey predicted that blood passed through minute vessels too small to see and, in 1661, Italian doctor Marcello Malpighi discovered these microscopic tubes, called capillaries, in a frog's lung. Capillaries permeate all body tissues, bringing food and oxygen to respiring cells and removing waste carbon dioxide.

### Life-giving blood
Malpighi may have been the first microscopist to see red blood cells, and English physician John Mayow

and so deduced that a fixed volume was circulated instead. He confirmed Colombo's idea of the heart as a pump, but went further in describing that blood flows away from the heart in arteries and returns to the heart in veins. He proposed that as well as pulmonary circulation there was another, more extensive, one through the rest of the body—the systemic circulation. Although some

# Circulation of the Blood

**Blood is a life-giving source on which all cells depend. It is pumped around the body through a complex system of vessels by the heart muscle. But although blood is such an obvious commodity in the body, its function and means of passage were not fully understood until the end of the Renaissance.**

PHYSICIAN (1578–1657)

## WILLIAM HARVEY

William Harvey was born in England and studied at the University of Padua. He had a deep conviction in the importance of experimental science and a particular interest in blood. In 1628 he published the results of 30 years of inquiry that established that blood circulated around the body through the heart's pumping action. His reputation as an accomplished doctor meant that he was favored in high quarters: he was physician to both King James I and King Charles I. He also tended the victims of the English Civil War.

or ignored. Proof of their theory of pulmonary circulation came in 1559 with the experiments of Italian physician Matteo Colombo, who found that the pulmonary vein (which leads from the lungs to the heart) carried blood and not air as was previously believed. This was confirmed in 1564 by Italian anatomist Julius Aranzi, who also discovered that the blood of a mother and her fetus do not mix.

### The pumping heart
Once scientists had ascertained that blood circulated around the body, their next challenge was to determine how this circulation took place. The ancient Greeks had thought that the heart worked by suction and that the pulse was due to arteries pumping blood. But it wasn't until the mid 16th century that Colombo suggested that the heart itself was the pump.

In 17th-century England the physician William Harvey (see left) had a special interest in blood and was profound believer in the importance of experiments in science. In 1628 he established once and for all that blood flow was entirely circulatory. He experimented on animals to quantify the daily volume of blood pumped by the heart. He realized that such a great amount could not be continually made,

Renaissance scholars, such as French anatomist Jean Riolan the Younger, insisted on adhering to older Galenic ideas, by 1700 Harvey's doctrine was established science.

In the late 17th century Danish polymath and physician Nicolas Steno found that the heart wall was composed of muscle; he was also the first to show that the heart consists of two relatively independent pumps. Around the same time, French anatomist Raymond Vieussens examined the left ventricle in detail, noting its extra thick muscle, necessary to pump blood all around the body.

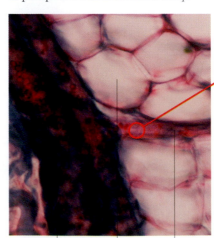

Arteriole | Body cells | Capillary

went as far as suggesting that blood carried oxygen from the lungs. However, it was many years before scientists could prove that it was hemoglobin (a protein respiratory pigment) in the red blood cells that carried oxygen from the lungs, where blood is oxygenated, to the body tissues (see pp.270–71).

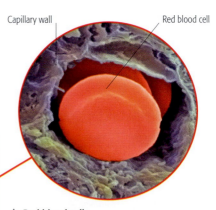

Capillary wall / Red blood cell

△ **Red blood cells**
These cells contain hemoglobin, which delivers oxygen around the body. Their disk form bends as they squeeze through tiny capillaries, bringing the oxygen close to capillary walls so it can be released into other cells.

◁ **Capillaries supply blood to organs**
As blood flows through capillaries such as these (known as exchange vessels), nutrients and oxygen pass from the blood across the ultra-thin capillary walls into neighboring cells. Carbon dioxide and metabolic wastes diffuse inward.

**Aorta**
The body's main artery, the aorta carries oxygenated blood away from the heart.

◁ **The human heart**
A wall (septum) divides the heart: deoxygenated blood from the venae cavae enters the right side and is pumped to the lungs, where it is oxygenated; the oxygenated blood enters the left side of the heart via the pulmonary vein and is pumped out through the aorta to the rest of the body.

**Pulmonary artery**
Artery that carries deoxygenated blood from the right ventricle to the lungs.

**Semilunar valves**
After the ventricles relax, these valves close to prevent backflow of blood from the arteries.

**Pulmonary veins**
Veins that carry oxygenated blood to the left atrium from the lungs.

**Superior vena cava**
Vein that carries deoxygenated blood to the right atrium from the upper body.

**Right atrium**
Thin-walled chamber that pumps blood into the right ventricle below.

**Tricuspid valve**
After the ventricles contract, this valve closes to prevent backflow of blood into the right atrium.

**Right ventricle**
Chamber with a thick wall to pump blood to the lungs.

**Inferior vena cava**
A large vein that carries deoxygenated blood to the right atrium from the lower body.

**Descending aorta**
Carries oxygenated blood away from the heart to the body tissues.

**Left ventricle**
Chamber with an extra thick wall to pump blood all around the body.

**Left atrium**
Chamber with a thin wall to pump blood into the left ventricle.

**Mitral valve**
After the ventricles contract, this valve closes to prevent backflow of blood into left atrium.

Direction of blood flow

Valve fully open

Muscle contracts

Muscle contracts

Blood

Vessel wall

Valve closed preventing backflow of blood

Direction of blood flow

△ **How valves in veins work**
Big veins run between limb muscles, which help squeeze deoxygenated blood upward against gravity. Valves within the veins prevent backflow as the blood returns to the heart. Portuguese physician, Amato Lusitano, was the first person to discover the role of valves in the mid-1500s by showing that air could be blown only one way through a dissected vein.

## AFTER ≫

Once the basic structure and function of the heart was understood, 19th-century scientists made advances that became important in cardiovascular medicine.

### THE ARTIFICIAL PACEMAKER
In 1839 a Czech anatomist called **Jan Purkinje** discovered specialized muscle fibers (known as Purkinje fibers) in the heart. The heart beat is coordinated by the fibers as they carry an **electrical impulse** that causes the muscle tissue of the heart's ventricles to contract. Malfunctions of this system are today treated with an artificial pacemaker, developed by Australian scientists **Mark Lidwell** and **Edgar Booth** in 1928. This device **stimulates the heart** with an implanted electrode.

**SEE ALSO ≫**
pp.228–29 SAFER SURGERY
pp.406–07 MODERN SURGICAL PROCEDURES
pp.448–49 REFERENCE: BIOLOGY

## How the heart beats
The cardiac cycle is the series of heart movements associated with a single heart beat. Contraction of heart muscle is called systole, relaxation is called diastole. Blood always flows from high to low pressure, with valves preventing backflow.

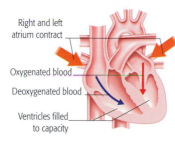

Right and left atrium contract

Oxygenated blood

Deoxygenated blood

Ventricles filled to capacity

### Atrial systole
As the thin walls of the atria contract, they pump blood the short distance into the ventricles, ensuring they are filled to capacity with blood.

Semilunar valves open

Tricuspid and mitral valves forced shut

Right ventricle contracts

Left ventricle contracts

### Ventricular systole
The thick walls of ventricles pump blood out of the heart; at the same time the triscupid and mitral valves shut, causing the first heart sound.

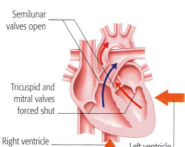

Semilunar valves close

Tricuspid and mitral valves open

Ventricles relaxed

### Diastole
As the whole heart relaxes, blood flows from the veins into the atria and then partly fills the ventricles, causing the semilunar valves to shut, which produces the second heart sound.

PHILOSOPHER AND POLYMATH **Born 1635** Died 1703

# Robert Hooke

## "The science of **Nature**…should return to the plainness and **soundness** of observations."

ROBERT HOOKE, "MICROGRAPHIA," 1665

Robert Hooke has been described as "England's Leonardo." An inventor of instruments such as the compound microscope, he was the scientist who helped form the Royal Society, as well as Christopher Wren's equal partner in London's reconstruction after the Great Fire of 1666. His *Micrographia* was so enthralling that Samuel Pepys recorded in his diary how he read it until 2 am, describing it as "the most ingenious book that ever I read in my life."

Hooke came from a modest home on the Isle of Wight, where he was born in 1635, the youngest child of John Hooke, the curate of All Saints Church in Freshwater. A sickly child, he was too poorly to attend school at first, but was very smart, with an aptitude for making things. Given an old brass clock to play with, he took it apart to see how it kept time, and made a wooden copy of it that actually worked. After his father died in 1648, Hooke left the island with an inheritance of £50 to attend Westminster School, London, where he studied music and became an excellent draftsman.

### The first professional scientist

At age 18 Hooke gained a place at Oxford University as a choral scholar, although he was really studying what we would now call science. By this time he had developed a curvature of the spine that gave him a pronounced stoop, probably due to a condition called Scheurmann's kyphosis, aggravated by poor diet as a child.

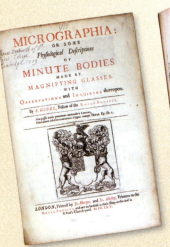

**Title page from *Micrographia***
Robert Hooke's *Micrographia* astonished readers such as Samuel Pepys with its detailed drawings of tiny creatures, but the work also dealt with the nature of gravity, fossils, craters on the Moon, and many other topics.

This foldout, detailed illustration of the underside of a louse is typical of Hooke's illustrations.

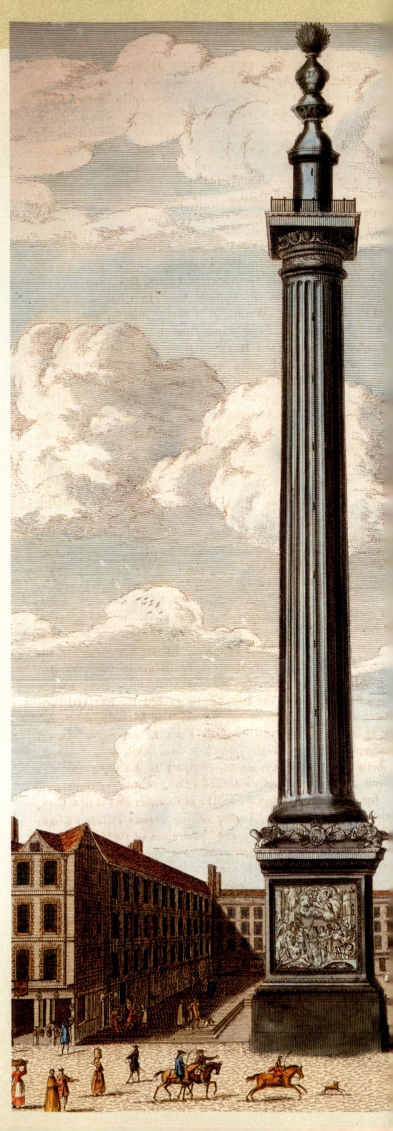

INVENTION

## COMPOUND MICROSCOPE

While lenses had previously been used together for magnification, Hooke invented the first practical compound microscope ("compound" meaning it has more than one lens). He used an oil lamp as a light source and a concave mirror to direct the light onto the object via a lens. Inside the microscope itself were four extendable concentric tubes containing an object lens, a middle lens, and the eyepiece. To focus, Hooke raised and lowered the entire tube to change the distance between the objective lens and the object. Fine adjustments were made by turning the end of the tube.

Lacking money to support himself, Hooke worked as a part-time assistant of an Oxford doctor, Thomas Willis, who had his own chemical laboratory, and then got a similar job with the great Robert Boyle. It worked out so well that he never completed his degree, becoming instead the first full-time, paid, professional scientist.

Boyle was a wealthy aristocrat, but he always treated Hooke as a scientific equal, not as a servant. While they were working together, Hooke discovered the law that relates the pressure and volume of a gas. Boyle wrote about this in a book published in 1661, and was careful always to say that it was Hooke's discovery. But because the discovery appeared in Boyle's book, to this day it is known as "Boyle's Law," although Hooke did get his own name on the law that describes how springs stretch.

### Birth of the Royal Society

In the same year, 1660, that Charles II was restored to the throne, a group of men, including Robert Boyle, formed a society "for the promoting of Experimentall Philosophy" under the King's patronage. This became the Royal Society. Hooke became its first "Curator of Experiments." His job was to demonstrate and test new discoveries, either those made by the Fellows of the Society, or by himself, or reported from other places.

In his own right, Hooke worked a lot on optics; his book *Micrographia* was published in 1665. Containing many engravings of creatures he had seen under the microscope, it was the first truly popular book about science.

#### London Monument

As City Surveyor, Hooke designed the Monument to the Great Fire of London, a hollow pillar 202 ft (61.5 m) high, although Wren is unjustly credited for it. The pillar was completed at the end of 1676.

astronomer and a Fellow of the Royal Society, as well as being an architect) were given the job of planning the reconstruction of the city. Hooke now had a salary as a City Surveyor, with the opportunity to accept commissions. Hooke was Wren's equal partner in this work and many of the "Wren" churches in the City of London were actually designed by Hooke.

Hooke knew poverty for many years, but when he died in 1703 he left a fortune of £8,000 in a great iron chest, earned through his many achievements.

BREAKTHROUGH

## HOOKE'S LAW

Hooke's Law describes the behavior of a stretched spring or other elastic object. It states "As the extension, so the force" —the amount of stretch is proportional to the force pulling on the spring. When the force is doubled, the spring stretches twice as much; when it is removed, the object reverts to its original size. This is only true for any elastic object up to its "elastic limit." If stretched beyond this limit, the springy object will never return to its original size, and may break.

Hooke's life was greatly changed after the Great Fire of 1666 destroyed much of London. He and Wren (who was an

TIMELINE

- **July 18, 1635** Hooke is born at Freshwater on the Isle of Wight, but is not expected to live.
- **1648** His father dies, and at the age of 13 Hooke goes to Westminster School in London.
- **1653** The headmaster at Westminster, Richard Busby, gets Hooke a place at Oxford University.
- **1654** Taken on as Robert Boyle's assistant.
- **1660** Joins a group of "experimental philosophers" who meet in London to found what becomes the Royal Society.
- **1661** and **1662** Attends meetings of the Royal Society as Boyle's assistant.
- **1663** Becomes Curator of Experiments at the Royal Society; in June is elected as a Fellow.
- **May 1664** Studies a dark spot on Jupiter's surface through a telescope and watches it move—the first evidence that Jupiter rotates.
- **December 1664** Hooke and Christopher Wren make observations of a comet and confirm that comets follow curved paths.
- **January 1665** Publishes *Micrographia*.
- **March 1665** Becomes Gresham Professor of Geometry, with rooms in Gresham College, London. He continues to live there for the rest of his life.

**CHRISTOPHER WREN**

- **April 1665** Leaving London with colleagues to escape the plague, Hooke develops his ideas about gravity. He speculates that the Sun influences the planets through a force obeying an inverse square law.
- **1667–1676** Although continuing as Curator of experiments at the Royal Society, Hooke is distracted from his own scientific work by his role in the rebuilding of London after the Great Fire.
- **1672** Isaac Newton presents his theory of light to the Royal Society, without giving due credit to the ideas of Hooke.
- **1676** Publishes Hooke's Law.
- **1677** Elected to the Royal Society's Council.
- **1680** Suggests the idea of an inverse square law of gravity to Isaac Newton.
- **January 1684** Hooke, Wren, and Edmond Halley agree that orbits seem to obey an inverse square law. Halley then asks Newton to prove the law.

**ST. HELEN'S CHURCH, BISHOPSGATE**

- **June 1702** Makes his last recorded contribution to a Royal Society meeting.
- **March 3, 1703** Dies, to be buried at St. Helen's Church, Bishopsgate, although his remains were removed to north London in the 19th century.

## ANTONY VAN LEEUWENHOEK

Van Leeuwenhoek first used a magnifying glass during his apprenticeship with a textile merchant in Amsterdam, Holland, and later made his own microscopes. He was not trained in science and never wrote a book, but academic organizations published his findings. The Royal Society was initially sceptical about his discovery of single-celled organisms, but he proved their existence in 1680, so securing his reputation.

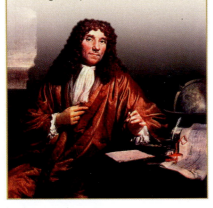

# Microscopic Life

**The invention of the microscope opened up a new world of discovery for biologists of the Renaissance. Not only could they study anatomy in far more detail than they had previously thought possible, but they found whole communities of new life forms in a single drop of pond water.**

Studies of medieval scholars were limited by simple magnifying glasses. Microscopy opened up life in miniature and eventually led to the development of ideas critical to biology.

### First views of a miniature world

The identity of the first inventor of the microscope is debatable, but the Dutch glasses makers Hans and Sacharias Jansen are likely contenders. Around 1590 they discovered that a hollow tube containing two lenses, one behind the other, could produce more enlarged images than a simple magnifying glass. They had made a compound microscope. In 1609 Italian astronomer Galileo (see pp.82–83), inspired by the invention of the spyglass by Dutch spectacle maker Hans Lippershey, constructed his own instrument for observing the skies and increased the magnification from three to 30 times. Around this time German botanist Giovanni Faber coined the word "microscope"—from the Greek words *micron* (small) and *skopein* (to look at)—to distinguish it from the telescopes that had been developed using similar technology (see pp.84–85). It was in 1665 that English philosopher Robert Hooke (see pp.92–93) coined the term "cell" to describe the microscopic units he found in a section of cork.

#### Early simple microscope
Van Leeuwenhoek fixed a tiny spherical lens in the hole in a brass plate. His specimen was held behind the lens on the tip of a needle, its position adjusted by turning a screw to focus the image.

In the same year Hooke published his book *Micrographia*, an exquisitely illustrated miscellany of the microscopic world (see below).

### The father of microbiology

After seeing *Micrographia* Dutch merchant Antony van Leeuwenhoek developed a technique of making lenses that would have allowed for more cost-effective production of his microscopes without compromising on the quality of the lens. However, he kept his lens-making technique a secret, and it was 200 years before comparable models were made. His microscope had a single spherical lens, made by drawing out a thread

◀◀ BEFORE

The idea that a microscopic world existed beyond the naked eye is an ancient one, but it wasn't until the first lenses were made that this world could be seen.

**THE FIRST MAGNIFIERS**
The writings of **Seneca the Younger** of ancient Rome include perhaps one of the earliest explicit references to a **device for magnification**: "letters, however small and indistinct, are seen enlarged and more clearly through a globe or glass filled with water." In the Middle Ages, **Al-hazen's** *Book of Optics* ◀◀ **60–61** was the foundation of an area of science that led to the **invention of eyeglasses** in 13th-century Italy, and the **microscope** in 16th-century Holland.

**MAGNIFYING GLASS**

◀◀ SEE ALSO
pp.72–73 THE HUMAN BODY REVEALED
pp.84–85 EXPLORING THE SKIES

#### Microscopic illustrations
Robert Hooke's *Micrographia*, published by The Royal Society, illustrated microscopic life with a degree of detail not seen before. It is known for its foldout plates, such as this example depicting a flea, but Hooke also included scientific ideas about light and astronomy.

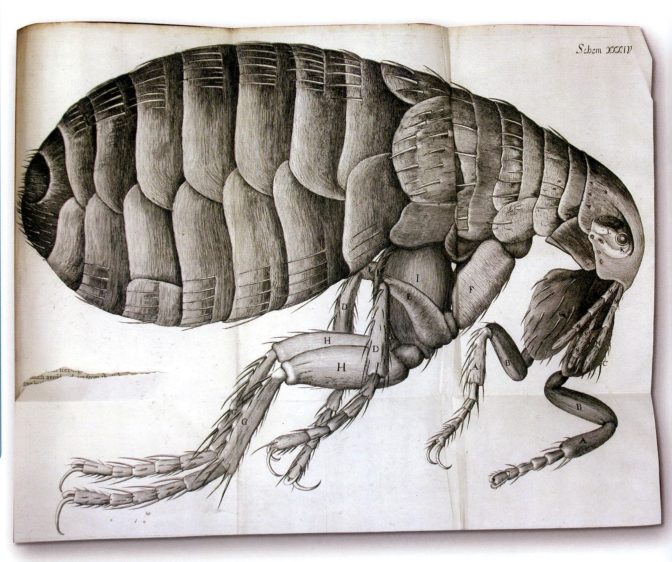

Schem XXXIV

## Modern microscope
Light reflected from a mirror through the specimen passes through different lenses in the microscope's tube to produce a magnified image. Focusing is achieved by knobs on the side of the instrument.

**Fine focus**
Function that allows small movement of the tube for use at high magnifications.

**Eyepiece lens x10**
Together with objective lenses, total possible magnifications are x400, x100, and x40.

**Coarse focus**
Function that allows large movement of the tube for use at low magnifications.

**Tube**
Light passes through here from the objective lens to the eyepiece lens.

**Revolving nosepiece**
Piece that holds different objective lenses above the specimen.

**Stage**
Part of the microscope that supports the specimen.

**Objective lens**
Magnifications are one of three: x40, x10, and x4.

**Specimen**
A specimen would be placed on a glass slide held over a hole in the stage by clips.

**Light**
Light is reflected up to the specimen by the surface of the mirror.

**Pivot**
Allows microscope to be angled for ease of use.

**Base**
The base is weighted to provide stability.

of glass, melting and rounding off its tip, and filing one side flat once it had cooled. He showed that pond water was teeming with miniature life forms, which he called "animalcules." He was seeing single-celled microbes, and—unlike Hooke's static cork cells—they were buzzing with life. Light microscopes rely on light rays being directed through a very thin transparent specimen. Leeuwenhoek's lenses magnified up to 270 times and he was the first to see some of the smallest things that can be viewed by a light microscope, including human sperm and even, possibly, bacteria.

### Improving the view
The light microscope was routinely used in science after the Renaissance. Today, liquids, such as blood or pond water are drawn out into a fine film, while solid objects, such as wood or stems, must be cut into slices of around 0.02 mm thick, using an instrument called a microtome.

In the mid-19th century, German physicist Ernst Abbe made a breakthrough in understanding the critical position of the lenses in the microscope for sharper image quality. But the light microscope is limited by its resolution due to visible light available. Two objects closer than half the wavelength of light (around 0.004 mm) appear as one, which means that certain tiny particles, such as viruses, are entirely beyond the microscope's resolution. Three centuries after Hooke and Leeuwenhoek, the electron microscope (see right) solved this problem and produced images of the miniature world that they would scarcely have thought possible.

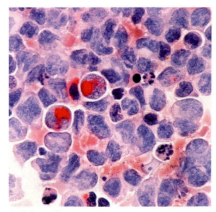

**Staining cells**
These blood cells have been stained and photographed through a microscope to reveal their structures more clearly. Here, white blood cells have been stained purple from a patient with myeloid leukaemia, a disease characterized by the large numbers of white blood cells.

AFTER ≫

**Emerging theories dominated a new kind of biology, which relied on the microscope.**

#### MICROSCOPIC THEORIES
In 1850 the **microscope** was key to uncovering the theory that all living organisms were **composed of cells 214–15 ≫**. In the 1800s microbiologists used microscopes to show that **decomposition** could be caused by **microorganisms 242–43 ≫**.

#### ELECTRON MICROSCOPE
First built by German physicist **Ernst Ruska** in 1931, the electron microscope resolves **more detail** than the light microscope since **electrons** have a **smaller wavelength** than light.

**SEE ALSO ≫**
*pp.214–15* ANIMAL AND PLANT CELLS
*pp.392–93* NANOTECHNOLOGIES
*pp.404–05* BODY IMAGING

### Microscopic findings
Early microscopes could magnify 30 times, revealing objects around 0.1 mm in size, such as plant cells. Advances in design allowed greater magnifications, so revealing smaller objects inside cells.

**1676 Leeuwenhoek** observes protozoa, such as *Amoeba* and *Paramecium*; diameter 50 micrometers.

**1876 Robert Koch** observes rods of anthrax bacteria in blood of infected animals; length 3 micrometers.

**1890 Richard Altmann** publishes description of mitochondria (energy-generating cell structures); length 1 micrometer.

**mid 1950s George Palade** observes ribosomes (protein-generating cell structures) with an electron microscope; diameter about 0.02 micrometer.

| 1650 | 1700 | 1750 | 1800 | 1850 | 1900 | 1950 |

**1663 Robert Hooke** observes cork cells; diameter 40 micrometers.

**1683 Leeuwenhoek** possibly observes bacteria; diameter 2 micrometers.

**1674 Antony van Leeuwenhoek** observes "giant coiled" chloroplasts in *Spirogyra*; length around 60 micrometers.

**1831 Robert Brown** discovers nucleus in cells of plant leaves.; diameter 2 micrometers.

**1848–49 Wilhelm Hofmeister** sees clusters of chromosomes in plant leaves coiled in nucleus; cluster diameter 2 micrometers.

**1939 Gustav Kausche, Edgar Pfankuch, and Helmut Ruska** view Tobacco Mosaic Virus using a transmission electron microscope; length 0.3 micrometer.

The Magdeburg hemispheres
In 1650 Otto von Guericke invented the air pump and four years later gave a public demonstration of the vacuum. Having pumped the air from two copper hemispheres, he showed that two teams of horses could not pull apart "nothing at all."

« BEFORE

**Before the 17th century, people believed that there was no such thing as a vacuum.**

**"WHAT IS" AND "WHAT IS NOT"**
The ancient Greek philosopher and astronomer **Thales of Miletus** said that everything in the universe must be "Thing" or "No-Thing." A **vacuum** was "No-Thing" and therefore, by definition, did not exist.

**FOUR ELEMENTS**
The Greek philosopher **Empedocles** asserted that all matter is made of only **four elements**: earth, air, fire, and water, in different proportions. He ingeniously showed that **air is not nothing** by pushing an upturned bucket into the sea and showing that the inside did not all get wet.

**ESSENTIAL PARTICLES**
The Greek philosopher **Democritus** believed that everything is made of tiny particles (which he called "**atoms,**" from the Greek word for "**indivisible**"). A **vacuum** would be "nothing"—an **absence of particles**. One of Democritus's successors, **Aristotle**, disagreed with the theory of atoms, stating that a **vacuum was impossible**, and so could not exist. "Nature abhors a vacuum" is an Aristotelian idea. This view was later supported by the Catholic Church, as they said the absence of everything implied the absence of God.

ARISTOTLE

# Discovery of the Vacuum

**The ancient Greek philosophers were convinced that a vacuum was nothing, and therefore an impossibility: "Nature abhors a vacuum." In the 17th century, however, vacuums were discovered and their properties explored in detail. Scientists were quick to realize their potential value.**

A vacuum is literally nothing; it is empty space. For thousands of years philosophers thought such a thing could not exist, but in the early 1600s the Italian scientist Galileo Galilei (see pp.82–83) became involved. He had been told by pump makers that they could not pump water out of a well more than 30 ft (10 m) deep. Their pumps could push water higher than this, but they could not suck it up.

He discussed this observation with a pupil, Evangelista Torricelli, who suggested experimenting with mercury instead of water. Mercury is 13 times denser than water and might show the same phenomenon at a lower height, making it much easier to investigate.

## Torricelli's experiments

Torricelli took a 3 ft (1 m) tube, closed at one end, and filled it completely with mercury. He dipped the open end in a bowl of mercury and raised the closed end. Suddenly—when the height of the mercury column reached about 30 in (76 cm)—a gap appeared in the tube above the mercury. It could not be air that had been dissolved, because when he lowered the end of the tube the gap closed immediately; air cannot dissolve so quickly. The gap reappeared when he raised the end of the tube again. The gap was a vacuum, and it became known as a Torricellian vacuum.

Torricelli realized that the mercury must be rising because of atmospheric pressure, which pressed down on the

**BOTTOM OF MOUNTAIN** — Mercury column 28 in (71 cm) high; Torricellian vacuum in tube above mercury; Tall, thick-walled glass tube (closed at top); Open end of tube immersed in dish of mercury; Atmospheric pressure pushes down on mercury surface

**TOP OF MOUNTAIN** — Vacuum above mercury; Mercury column 25 in (63 cm) high; Atmospheric pressure lower at mountaintop

Pascal's experiments with atmospheric pressure
Blaise Pascal's brother-in-law Florin Périer measured a height of 28 in (71 cm) of mercury at the bottom of the 3,300 ft (1,000 m) mountain Puy de Dôme, then carried the apparatus up, and measured 25 in (63 cm) at the top.

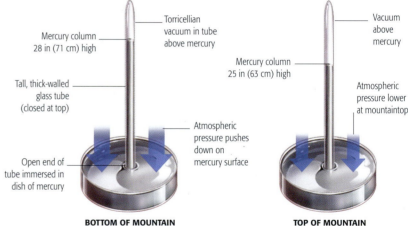

PHYSICIST (1608–47)

## EVANGELISTA TORRICELLI

Born in Faenza, Italy, of a poor family, Torricelli was fortunate in being able to study mathematics, and in becoming assistant to Benedetto Castelli, who taught at the University of Sapienza in Rome. Through Castelli, Torricelli made contact with his hero Galileo, and was taken on as his assistant in October 1641. When Galileo died three months later, Torricelli was appointed to succeed him as court mathematician to the Grand Duke Ferdinando II of Tuscany.

**Creating the unbreakable seal**
Otto von Guericke pumped the air out of two copper hemispheres of around 20 in (50 cm) diameter sealed with a greased leather ring. Because this created a vacuum, there was no air pushing outward from the insides of the sphere, but there was pressure (atmospheric) pushing on them from outside, which forced the spheres tightly together.

mercury in the bowl, forcing the heavy liquid up the tube. The atmospheric pressure could only push the mercury up by 30 in (76 cm), because this figure corresponds to normal atmospheric pressure. But water is 13 times less dense than mercury, so atmospheric pressure can push water up 13 times higher (about 33 ft/10 m).

### Varying pressure
French philosopher Blaise Pascal predicted that the height of a mercury column would be lower on top of a mountain, where the pressure of the atmosphere should be lower. Experiments by his brother-in-law, Florin Périer, in 1648, proved him right. He also predicted that the atmosphere must therefore gradually thin out, and at some height above the earth give way to empty space—in other words, a vacuum.

Robert Boyle and Robert Hooke repeated Torricelli's experiments in Oxford, UK, from the 1650s onward. They noticed that the height of the mercury column varied from time to time. At first they thought that it depended on the phases of the Moon, like the tides, but then realized that it depended on the weather: when the mercury was rising there was good weather to come; when it was falling there was bad weather on the way. In effect, they invented the barometer.

### The air pump
By 1659 Hooke had made an effective air pump for Boyle, and they used it in a series of experiments. They discovered that a piece of red-hot iron inside could be seen undimmed—so light does travel through a vacuum—but a bell rung inside could not be heard, which shows that soundwaves need air (or another medium) to carry them. They discovered that a candle would not burn in a vacuum, and also that small animals died if left inside the flask. Later Joseph Priestley and others showed that oxygen is necessary both for combustion and for breathing (see pp.146–147); the air pump removed the air, including the oxygen.

In practice, making a perfect vacuum is impossible. Von Guericke (see top), Boyle, and Hooke probably reduced the air pressure by three-quarters but no more; even in deepest space there is on average about one hydrogen atom in every cubic yard (or cubic meter).

Glass flask full of air

Flame is extinguished

Air is pumped out of flask

**Flame is extinguished in a vacuum**
A candle burns in a flask full of air, but when the air is pumped out the flame is extinguished, showing that air (or specifically, oxygen) is needed to support combustion.

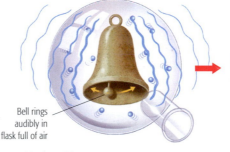

Bell rings audibly in flask full of air

Air is pumped out of flask

Bell still rings, but is now inaudible

**Sound is muted in a vacuum**
A bell ringing in a flask full of air can be clearly heard from outside the flask. But when the air is pumped out, no sounds escape the flask. This is because sound is carried by air, and cannot travel through a vacuum.

**AFTER** »

Scientists soon realized they could harness a vacuum's power.

**ATMOSPHERIC STEAM ENGINES**
The first effective **atmospheric engine 130–31** » was built by **Thomas Newcomen** in 1712. By generating a vacuum, it used the power of atmospheric pressure.

**THE LIGHT BULB**
**Joseph Wilson Swan** demonstrated the first useful **light bulb** in the UK, in February 1879, just ahead of **Thomas Edison 264–65** » in the US. The bulb's vacuum ensured that the carbon filament did not burn out.

**THE CATHODE RAY TUBE (CRT)**
Built for research in atomic physics, CRTs became useful as **television tubes**. They made use of an internal vacuum.

**SEE ALSO** »
pp.98–99 ROBERT BOYLE
pp.292–93 VACUUM TUBES

CHEMIST AND PHYSICIST **Born 1627** Died 1691

# Robert Boyle

> "There is a **spring**, or **elastical power**, in the air we live in."

**ROBERT BOYLE,** FROM "NEW EXPERIMENTS PHYSICO-MECHANICALL," 1660

Robert Boyle has been called the "father of chemistry"; he broke away from the old ideas of alchemy and defined what is meant by an element. With the help of Robert Hooke, he performed extensive research into the properties of the vacuum, and the relationship between pressure and volume in gases.

Robert was born in 1627 at Lismore Castle, in County Waterford, Ireland. He was the seventh son and 14th child of Richard Boyle, who had become the Great Earl of Cork in 1602 when he bought Lismore from the adventurer Sir Walter Raleigh (then in prison). Boyle watched as his father married off all his older siblings, mainly for financial or social advantage, and this

> **BOYLE'S LAW** Robert Boyle referred to the relationship between pressure and volume as "Mr. Towneley's hypothesis."

sorry spectacle may have put Robert off marriage. At the age of eight he was sent to Eton College, the well-known English private boarding school. He loved the academic work but hated the compulsory games. Four years later he and his brother Francis were taken by a French tutor on a tour of Europe, winding up in Italy, where they spent the winter of 1641–42 in Florence. The great Galileo (see p.82–83) died there in 1642, and the public reaction inspired Boyle to pursue similar ideas.

### Reduced circumstances
Returning to England in 1644 Boyle found the country in civil war, his father dead, and himself broke, since all the income that had come from the Irish estates had been stopped. He first settled at Stallbridge in Dorset, but visited his favorite sister, Katherine, in

**Pendle Hill**
On this hill in Lancashire, once famous for its witch trials, Richard Towneley and his physician Henry Power first noted, in 1661, that a trapped sample of air increased in volume as they went up the hill.

**Gresham College**
Founded in London in 1597 by Sir Thomas Gresham, Gresham College provided lectures that were free to the public. It became a meeting place of the "invisible college," and later of the Royal Society.

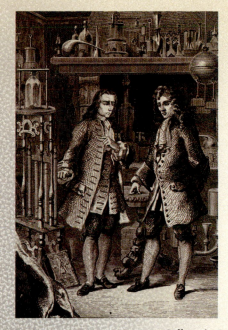

**Boyle and Papin in Maiden Lane Laboratory**
French scientist Denis Papin, a friend and colleague of Boyle's, invented some of the first steam engines. In this laboratory Boyle's technician, Ambrose Hanckwitz, was set up to make phosphorus.

London. Full of enthusiasm for science, the "new philosophy," Boyle met at Katherine's house a group of like-minded people who called themselves the "invisible college." They included John Wilkins, diarist John Evelyn, Robert Hooke, and William Petty, inventor of the catamaran. They met sometimes at Gresham College in London, and later in Boyle's rooms in Oxford—safely distant from Oliver Cromwell's base—and exchanged ideas by letter, or by sending one another books with notes in the margins. In 1652 Boyle visited Ireland, then torn apart by rebellion, and, with the help of William Petty, restored some order to his estates, which meant that money started to flow again. Thereafter he had a substantial income for life.

**A scientific partnership**
In 1654 Boyle hired the ingenious Robert Hooke as an assistant, and although Hooke was of a lower social class they seem to have worked together as equal partners. Boyle had lots of ideas, and Hooke was the right person to put them into practice. They investigated the Torricellian vacuum (see p.96), and it may well have been Boyle and Hooke who discovered that the height of Torricelli's column of mercury varied with the weather, and thus invented the barometer. When Hooke built an air pump, which they used and demonstrated extensively, it came to be called the Machina Boyleana.

Two years after Cromwell's death in 1658 Charles II returned from exile and was

**Robert Boyle**
Johann Kersebook painted this portrait around 1689, when Boyle was 62, although he barely looks so old. Boyle was uncomfortable about having his portrait painted, which may be why he looks so awkward.

> "I could be content **the world** should think I had scarce looked upon any other book than that of **nature**."
> ROBERT BOYLE

crowned king, and the pursuit of science was once again safe for gentlemen. In the same year the invisible college became the Royal Society of London for the Promotion of Natural Knowledge, abbreviated to the Royal Society. Boyle was one of the founding Fellows.

**Discovery of "Boyle's Law"**
In 1661 Boyle received a letter from Richard Towneley (see opposite), an astronomer and the first person in England to measure rainfall systematically. On April 27 he and his physician, Henry Power, had climbed Pendle Hill, 1,827 ft (557 m) high, near Towneley Hall in Lancashire, carrying a barometer. They found that the air pressure at the top of the hill was lower than at the bottom, and that trapped air occupied a greater volume when the pressure was lower. After getting Hooke to check the measurements, Boyle published the results in 1662, but then Power published the findings himself in 1663.

Although Boyle referred to the work as "Mr. Towneley's hypothesis" and said that Hooke had done the checking, the relationship is known as "Boyle's Law" (see p.93). In Europe the relation is often attributed to Edme Mariotte, but he did not publish it until 1676. With the large income from his Irish estates, Boyle was able to do scientific research, publish his own books, and pay assistants, but he remained a modest man. He refused offers of a peerage and a bishopric, and turned down an invitation to become President of the Royal Society.

**Air pump**
Built by Hooke, this device was a lot like a bicycle pump working in reverse, pulling the air out of a flask. There were great technical difficulties in making airtight seals, however, and it is unlikely that a complete vacuum was ever achieved with it.

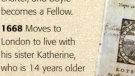

**TIMELINE**

**January 25, 1627** Robert Boyle is born at Lismore Castle in County Waterford on the south coast of Ireland. He is the 14th child and the seventh son of the First Earl of Cork.

**1635** Sent to Eton College, near Windsor, England.

**1639** Travels to Europe with his brother Francis (who had been married only days before) and a French tutor. They go to France, settle in Geneva for a while, and then go on to Italy.

**1641** Spends the winter in Florence, where the aging Galileo dies the following year. Boyle is inspired by the public interest in him to read Galileo's work and to follow in his footsteps.

**1644** Returns to England and begins to work on the "new philosophy," namely the ideas of experimental science as advocated by Francis Bacon (see pp.68–69) some 20 years before. He becomes one of the leading exponents of science based on observation and facts, rather than mere guesswork.

**1652** Visits Ireland, with William Petty, in order to recover his estates, ravaged by the Irish rebellion. Thereafter he has a substantial income.

**1654** Moves to Oxford and rents rooms in Cross Hall on the High Street. A plaque now hangs there, on the outside wall of University College. He hires Robert Hooke as assistant.

**1659** Boyle and Hooke make an air pump, following the example of Otto von Guericke (see p.97). They use it to carry out a series of experiments, showing that air is necessary to carry sound, to sustain a flame, and to sustain life.

**1660** Publishes the results of his experiments with the air pump: New Experiments Physico-Mechanicall, Touching the Spring of the Air, and its Effects. The work is criticized by the English scientist Franciscus Linus. There is much philosophical argument about the nature of the vacuum (see p.96).

**1661** Publishes The Sceptical Chymist, an attempt to bring scientific ideas to alchemy, but also the beginning of chemistry as we now understand it.

**1662** In the second edition of New Experiments Physico-Mechanicall, Boyle publishes "Boyle's Law," although he does not claim credit for the idea for himself. The Royal Society receives its charter, and Boyle becomes a Fellow.

**1668** Moves to London to live with his sister Katherine, who is 14 years older than him, but his favorite sibling. He never marries, nor seems to have had a long-term partner.

**1669** Publishes Paradoxa Hydrostatica Novis Experimentis, describing hydrostatical paradoxes.

**1691** Katherine dies on December 24; Robert Boyle dies on December 30.

**PARADOXA HYDROSTATICA**

# The **Behavior** of **Gases**

**Before the 17th century no serious scientific work was done on gases—they were all thought to be "just air." However, the pioneering work of scientists studying the weather led to the discovery that air and other gases behave according to a set of mathematical laws. This led to a greater understanding of atoms and molecules.**

## « BEFORE

**Gases were not studied in detail before the 17th century, although the ancient Greeks included "air" as one of the four elements from which everything is made.**

**ISLAMIC INGENUITY**
In 9th century Baghdad the **Banu Musa brothers** produced a book of **ingenious devices**, some of which used clever control of air pressure – including a jug that would not pour unless you removed your thumb from a hole near the handle.

**THE WORK OF TORRICELLI**
**Torricelli « 96–97**, inspired by his mentor **Galileo « 82–83**, investigated and discovered the vacuum. His work inspired other scientists to start experimenting on air held at low pressure.

**« SEE ALSO**
*pp.44–45* "EUREKA"
*pp.96–97* DISCOVERY OF THE VACUUM
*pp.98–99* ROBERT BOYLE

During the 17th century science flourished in England, and several natural philosophers became interested first in the behavior of the air and, later, of other gases. The result of their investigations was a set of "gas laws" that mathematically describe the interrelationships between pressure, volume, and temperature.

### Pressure and volume
Richard Towneley of Lancashire, England was a keen meteorologist interested in the characteristics of the atmosphere. In 1661 he collected a sample of "Valley Ayr" (probably trapped by mercury in a J-tube) and took it up Pendle Hill, where he found that the volume of the air had increased. Then he collected a sample of "Mountain Ayr" and took it down the hill, to find its volume had decreased. Knowing that the atmospheric pressure must be lower at the top of the hill, because of the experiments done by Florin Périer in France 13 years

earlier, he realized that for a fixed amount of air, lowering the pressure increased the volume, and vice versa. He told his friend Robert Boyle (see pp.98–99) about this, who in turn asked his colleague Robert Hooke (see pp.92–93) to confirm the findings, using the air pump that they had

**25** **MM (1 in)** is about the distance mercury falls in a barometer for every 1,000 ft (300 m) increase in height, reflecting the fall in air pressure.

developed to create a vacuum. In 1662 Boyle published the results, and the relationship came to be called "Boyle's Law," although Boyle himself referred to it as "Mr. Towneley's hypothesis."

### Bouncing gas
Robert Boyle thought air particles were like tiny coils of wool, which could be squashed, but would bounce back. In fact, gases are made up of millions of

Single weight produces pressure (P) in gas

Two weights produce double the pressure (2P) in gas

Low pressure in flask

High pressure squeezes particles into half the original volume

### Boyle's Law
The pressure and volume of a fixed mass of ideal gas at constant temperature are inversely proportional to one another. If the pressure is doubled, the volume is halved. This relationship, which can be written as the formula $PV = constant$, was first formulated by Robert Boyle in 1662.

Two weights produce equal pressure (P) in each flask

High pressure inside

Hot particles move faster and expand volume while maintaining pressure

Heat

### Charles's Law
At constant pressure, the volume of a fixed mass of ideal gas is directly proportional to its absolute temperature. If the temperature is doubled, the volume is doubled. This can be written as $V/T = constant$. The relationship was formulated by Jacques Charles in around 1787.

One weight produces pressure (P) inside

Two weights produce double the pressure (2P) inside

Low pressure inside

Pressure increases because hot particles move faster

Heat

### Gay-Lussac's Law
For a fixed mass of ideal gas at constant volume, the pressure is directly proportional to the absolute temperature. If the absolute temperature is doubled, the pressure is doubled. This can be written as $P/T = constant$, and it was first published, along with Charles's Law, by Joseph Gay-Lussac in 1802.

◁ **The dawn of hot-air ballooning**
The first hot-air balloon flight carrying people was made in France in 1783. Scientists soon realized that hot-air balloons could provide them with a platform for studying the atmosphere at various altitudes.

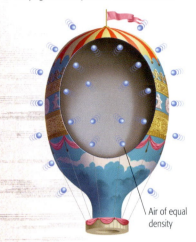

Air of equal density

**Balloon steady**
The air becomes gradually less dense with increasing altitude. When the air's density matches the balloon's density, the balloon stops climbing.

Air inside is less dense than the air outside

Direction of movement

**Rising balloon**
The air inside the balloon is at the same pressure as the atmosphere; when it is heated, its volume increases, and some of the balloon's air escapes. This lowers the average density of the balloon, and causes it to rise.

individual atoms or molecules, all flying about at great speed. The pressure on the walls of any container they are in is the same in all directions, and it is caused by these particles crashing into the walls and bouncing off them. If the volume of a container is halved, any particle inside will hit the walls twice as often, and so the pressure doubles.

## The effect of temperature

In the 18th century French scientists Jacques Charles and Joseph Gay-Lussac investigated the effect of temperature on gases. They found that when a fixed amount of gas is heated and its pressure is kept constant, its volume increases. This became known as Charles's Law (see opposite). On the other hand, if its volume is kept constant, its pressure increases (Gay-Lussac's Law).

When a fixed amount of gas is heated, the particles travel much faster—their speed is directly related to their temperature—so they hit the walls of a container much harder. This maintains the pressure if the volume is allowed to expand, or increases it if the volume is kept constant. However, these theoretical gas laws are only approximate, because the atoms and molecules of real gases are all somewhat "sticky," which tends to decrease their volume and their pressure slightly.

## Independent gases

One of the most enthusiastic early weather watchers was Englishman John Dalton, who realized that gases in a container do not sit in the bottom, as liquids do, but occupy the whole space. This means that if there are two gases in a container, both will occupy the entire volume, and their particles will behave independently. The total

pressure exerted by the gases is therefore the sum of the partial pressures of all the gases inside.

In 1858 the Italian chemist Stanislao Cannizzaro applied Dalton's Atomic Theory and Avogadro's Hypothesis (for both, see pp.138–39), and the experimental methods of Gay-Lussac to deduce relative atomic masses for

**Pioneering the pneumatic tire**
The first practical air-filled rubber tires were made by the Scot John Boyd Dunlop, who described them as "pneumatic." He made them in response to his son's complaint that a bicycle ride was "too bumpy."

various elements. Thus, what started for some scientists as an interest in the weather gradually became more mathematical and led to real insights into the behavior and properties of atoms and molecules.

## AFTER

Understanding how gases behave started a whole train of inventions and discoveries in both science and technology, ranging from the gas laws to the steam engine and the household vacuum cleaner.

**AIR COMPRESSION**
Compressed air is used to drive brakes in trucks, to spray paint, and to power many **industrial machines**. It is also used in hovercraft and in divers' air tanks.

**AIR PRESSURE**
The discovery that air flows from areas of **high pressure** to areas of **low pressure**, causing winds and weather, was the **beginning of meteorology**.

**COMPRESSED AIR TANK**

SEE ALSO ≫
pp.104–05 NEWTON'S LAWS OF MOTION
pp.132–33 STEAM POWER TO STEAM ENGINE
pp.138–39 THE NATURE OF MATTER

---

PHYSICIST AND CHEMIST (1778–1850)

## JOSEPH GAY-LUSSAC

Frenchman Joseph Louis Gay-Lussac studied chemistry in Paris before working in Claude-Louis Berthollet's prestigious laboratory as a research assistant. In 1802 he published his groundbreaking work on the relationship between gas volume and temperature. Two years later he and a colleague took a hot-air balloon to a height of 21,000 ft (6,400 m) to collect samples of air in order to establish that the composition of the atmosphere remains the same with increasing altitude. Gay-Lussac died in Paris in 1850, and is buried in the Père Lachaise cemetery.

# Graphs and Coordinates

**Graphs present statistical information visually so that anything from monthly rainfall to rising crime rates can be seen at a glance. However, underpinning graphs and coordinate systems is a fundamental link between algebra and geometry that places them at the heart of mathematics and science.**

## BEFORE

The roots of the coordinate system lie far back in the geometry of ancient Greece and the mathematical brilliance of ancient Islam.

**ANCIENT GREEK GEOMETRY**
As long ago as the 3rd century BCE, ancient Greek mathematicians such as **Dicaearchus** and **Eratosthenes** were **pinpointing locations** on maps using **grids** that were **precursors** to the modern system of **latitude** and **longitude** ≪ 38–39.

**KHAYYAM'S CUBICS**
In the 11th century CE the Islamic scholar **Omar Khayyam** found an ingenious way of using **geometry to solve cubic equations** ≪ 48–49. "No attention should be paid," he wrote, "to the fact that **algebra and geometry** are different in appearance."

**OMAR KHAYYAM**

≪ **SEE ALSO**
*pp.38–39* ANCIENT IDEAS OF THE WORLD
*pp.48–49* ALGEBRA

The connection between algebra and geometry was established in the 17th century, principally by two French mathematicians, Pierre de Fermat and René Descartes. What they realized is that algebra can be converted into geometry, and geometry into algebra, using coordinates—two or more numbers that pinpoint a location in relation to reference lines such as the axes of a graph. Through coordinates all geometry can be expressed as algebraic equations—and, in theory, every algebraic equation can be represented on a graph.

## Development of coordinate systems

What interested Fermat and Descartes was curves, which are involved in everything from the trajectory of missiles to the orbits of planets. All but the simplest curves were hard to deal with using the basic geometric approach developed 2000 years earlier by Euclid. Fermat and Descartes realized independently that the key to working with curves is coordinates.

Coordinates had been used in maps for thousands of years. Descartes' and Fermat's breakthrough was to see how coordinates bring geometry and algebra together. In studying the geometry of curves in 1629, Fermat realized that a single equation could be used to pinpoint every point along the line of the curve according to its distance from two other fixed lines. In *La Géometrie* of 1637, the appendix to his great book *Discours de la Méthode*, Descartes expressed this idea more fully, describing two coordinate lines intersecting at right angles called the *x* and *y* axes. This system later became known as the Cartesian coordinate system, after Descartes.

## The algebra of graphs

Coordinates make it possible to convert curves and shapes into equations, which can then be manipulated using algebra and, conversely, algebraic equations can be shown visually on a graph. Using Cartesian two-dimensional coordinates, a straight line is simply $y = mx + c$, where *y* is how far up, *x* is how far along, *m* is the slope of the line, and *c* is where the line crosses the *y* axis (see above).

Equations can also be found for much more complex lines and shapes. In 1691 the Swiss mathematician Jacob Bernoulli introduced polar coordinates (see opposite). With polar coordinates simple equations could describe curves as complex as the cycloid, the path traced by a point on the rim of a wheel rolling along the ground. Adding extra axes to the basic *x* and *y* ones or basing graphs on different shapes (such as a cylinder or sphere) allowed points to be specified in three or more dimensions, and equations could be created for shapes in multiple dimensions.

The connection between algebra and geometry opened up a whole new kind of mathematics called analytic geometry, which could analyze how things move and change. It also gave birth to calculus, created principally by the English scientist Isaac Newton (see pp.110–11) and the German philosopher and mathematician Gottfried Leibniz. Descartes saw his ideas as the key to a mechanical view of the Universe in which everything can be explained in terms of motion and extension (length, breadth, and height). However, there were two missing elements: rates of change and quantity. For example, motion and extension show the distance an object travels in a particular time, but not its velocity and acceleration at a particular instant. Calculus filled this gap (see box, left).

## The power of coordinates

Separately, algebra and geometry can deal well with static relationships but together—in analytic geometry and calculus—they can explore dynamic relationships, which is why they are essential tools for the scientist today. Without this combination, scientific breakthroughs such as Newton's law of gravity, quantum physics, and the Big Bang theory would not have been possible. Coordinates also have a huge number of practical uses. They are employed in everything from computer graphics to satellite positioning, and for plotting statistical data (see pp.196–97).

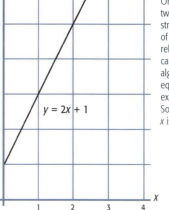

**Straight line graph**
On a Cartesian two-dimensional graph, a straight line means values of *y* always have the same relationship with *x*, and can be represented algebraically by the simple equation $y=mx+c$. In this example, *y* is always 2*x*+1. So when *x* is 2, *y* is 5, when *x* is 3, *y* is 7 and so on.

$y = 2x + 1$

**BREAKTHROUGH**

## CALCULUS

The development of calculus by Gottfried Leibniz and Isaac Newton answered two fundamental questions: how to figure out the tangent to a curve (which on a graph gives rates of change) and how to calculate the area under a curve (which gives quantities). For example, in a graph of speed against time, applying differential calculus (differentiation) gives the slope of the tangent at a point, which is the acceleration (rate of change of speed with time) at that point; applying integral calculus (integration) between two specified times gives the distance traveled in that period.

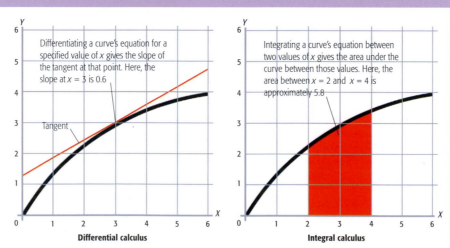

Differentiating a curve's equation for a specified value of *x* gives the slope of the tangent at that point. Here, the slope at *x* = 3 is 0.6

Tangent

**Differential calculus**

Integrating a curve's equation between two values of *x* gives the area under the curve between those values. Here, the area between *x* = 2 and *x* = 4 is approximately 5.8

**Integral calculus**

## Coordinate systems

Using coordinates, any point can be located precisely in space using the distance along reference lines, or axes, from the point where they intersect, known as the origin. Many coordinate systems have been developed; shown here are three of the basic ones.

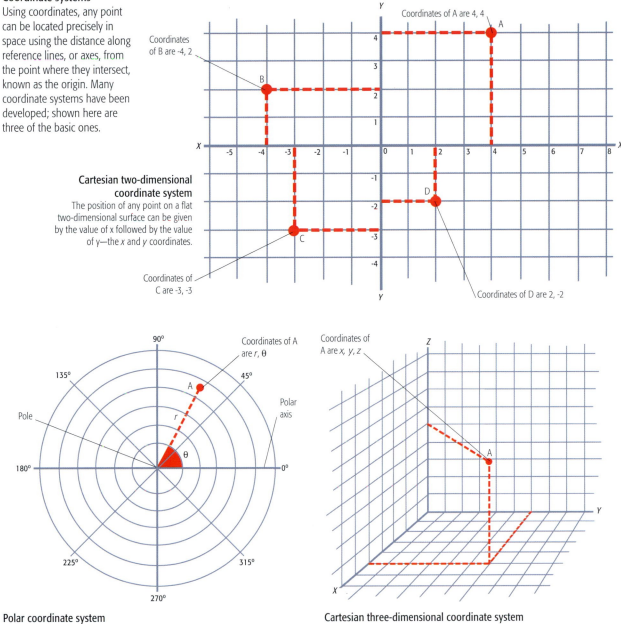

Coordinates of A are 4, 4

Coordinates of B are -4, 2

### Cartesian two-dimensional coordinate system
The position of any point on a flat two-dimensional surface can be given by the value of x followed by the value of y—the x and y coordinates.

Coordinates of C are -3, -3

Coordinates of D are 2, -2

## MATHEMATICIAN (1596–1650)

### RENÉ DESCARTES

The work of René Descartes marked a turning point in Western philosophy and science. Born in Touraine, France, he moved to Holland in 1628. There he set out a system for thinking, famously starting with the premise "Cogito ergo sum" (I think, therefore I am), the only thing he believed he could be certain of. As well as philosophy, he made key contributions to mathematics and optics. He moved to Stockholm, Sweden, in 1649 but died of pneumonia shortly afterward.

Coordinates of A are $r$, $\theta$

Pole

### Polar coordinate system
This system pinpoints location in two dimensions by the distance $r$ from the origin or "pole" and the angle $\theta$ measured anticlockwise from the polar axis.

Coordinates of A are $x$, $y$, $z$

### Cartesian three-dimensional coordinate system
By adding a vertical axis ($z$ axis) to the horizontal $x$ and $y$ axes on a Cartesian two-dimensional graph, a point in three dimensions can be specified using $x$, $y$, and $z$ coordinates.

### AFTER

As well as playing a key role in advancing areas of mathematics such as analytic geometry, the development of coordinates was also crucial to our modern understanding of space, time, and gravity.

#### CURVED AND MULTIDIMENSIONAL SPACE
The coordinate system opened up new ways of thinking about **space** in purely mathematical terms. One effect was the realization that **Euclid's geometry** works only for **flat surfaces**. For instance, according to Euclid, only one line can be drawn parallel to another through a given point, and the **internal angles** of a triangle always add up to 180 degrees. But on a **curved surface**, many other parallel lines can be drawn through a given point, and the internal angles of a triangle add up to **more than 180 degrees**. Even further complications arise with the **fourth dimension: time**. As a result, the German mathematicians **Carl Gauss** and **Bernhard Riemann**, and the French mathematician **Henri Poincaré**, among others, developed **new geometries** for curved and multidimensional space. It was Riemann's geometry for curved surfaces that led the German-born US scientist **Albert Einstein** to his understanding of **gravity** as a curvature in **space-time**.

**SEE ALSO >>**
*pp.302–303* THEORIES OF RELATIVITY

### IN PRACTICE

## GEOGRAPHICAL COORDINATES

The coordinate system has long been vital for making accurate maps. On topological maps, coordinates known as grid references are used for locating places, while on world maps places are located by the worldwide grid of latitude and longitude. Latitude gives the position north or south of the equator, and lines of latitude run around the world in bands parallel to the equator, which is why they are sometimes called parallels. A place's latitude is the angle in degrees a line drawn from the place to the center of the earth makes with the plane of the equator. Because the earth is not a perfect sphere, a degree in latitude is slightly longer at the poles than at the equator. Longitude gives the position east or west, and lines of longitude or "meridians" are circles that run from pole to pole, dividing the world into segments like an orange. A place's longitude is the angle in degrees that a meridian through it makes with the prime meridian, which runs north and south through Greenwich in London, England.

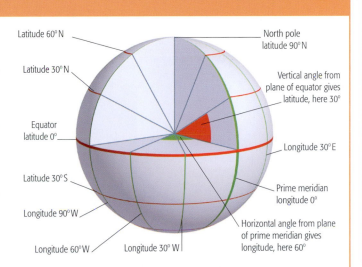

Latitude 60° N

North pole latitude 90° N

Latitude 30° N

Vertical angle from plane of equator gives latitude, here 30°

Equator latitude 0°

Longitude 30° E

Latitude 30° S

Prime meridian longitude 0°

Longitude 90° W

Longitude 60° W

Longitude 30° W

Horizontal angle from plane of prime meridian gives longitude, here 60°

## BEFORE

Forces and motion were concepts that had puzzled philosophers since antiquity.

**ARISTOTLE'S IDEAS**
The Greek philosopher **Aristotle ‹‹ 36–37** thought force was needed to **maintain motion**. He said that an arrow continues to fly only because the air continuously rushes around behind it and **pushes it forward**. However, he did not do experiments.

**GALILEO'S IDEAS**
**Galileo Galilei**, one of the world's first experimental scientists, disagreed with Aristotle, but thought "**natural motion**" was in circles. He rolled balls down slopes of various inclines, and found that **friction ‹‹ 86–87** was a retarding force. He also showed that the distance the balls rolled was proportional to the square of the rolling time.

**ROBERT HOOKE**
In his book *An attempt to prove the motion of the Earth* (1674), **Robert Hooke ‹‹ 92–93** wrote, "all bodies whatsoever that are put into a direct and simple motion, will so **continue to move forward** in a streight line, till they are by some other effectuall powers deflected." This came to be Newton's first law, but was clearly well understood at the time.

**‹‹ SEE ALSO**
*pp.86–87* Motion, Inertia, and Friction
*pp.92–93* Robert Hooke

# Newton's Laws of Motion

**In his great book *Principia*, published in 1687, Isaac Newton included a set of mathematical equations that are now called the Laws of Motion. They explain how objects come to move or to stay at rest, and how they react with other objects and forces. Newton's laws form the basis of all classical mechanics.**

Newton clearly began thinking about forces, motion, and gravity during the two intensely creative years of 1665–66. However, he might never have written down the details if he had not been persuaded to do so by Edmond Halley (see p.119) when he went to visit Newton in 1684.

### Newton's first law
Newton's first law—which says that moving objects carry on moving at the same speed and in the same direction unless they are stopped—is counter-intuitive: experience tells us that moving objects come to a stop unless there is a force pushing them forward. However, this fails to take into account the "hidden" forces of gravity (see pp.108–09) and friction (see pp.86–87). On Earth, all objects are being pulled down by gravity, and all moving objects are subject to friction, or drag, the force opposing movement.

For Italian scientist Galileo Galilei, the resistance of bodies to changing velocity, their inertia, was an important idea (see pp.86–87). It is why people on

**Newton's third law**
The force pushing a rocket upward is equal and opposite to the force pushing the exhaust gases downward—this shows Newton's third law in action.

### NEWTON'S LAWS OF MOTION

1  Unless it is disturbed by a force, an object will either stay still or travel at a constant speed in a straight line.

2  The force acting on a body is equal to its rate of change of momentum, which is the product of its mass and its acceleration.

3  To every action there is an equal and opposite reaction.

the surface of Earth feel stationary, even though it is hurtling through space at enormous speed, because their velocity is barely changed. Newton said that all objects have "forces of inertia." These forces oppose change in velocity, and cause objects to continue in their motion or to remain at rest. In other words, inertia acts to oppose acceleration.

### Second law
Newton's second law states that the acceleration of an object depends upon the net force acting upon it, and on the

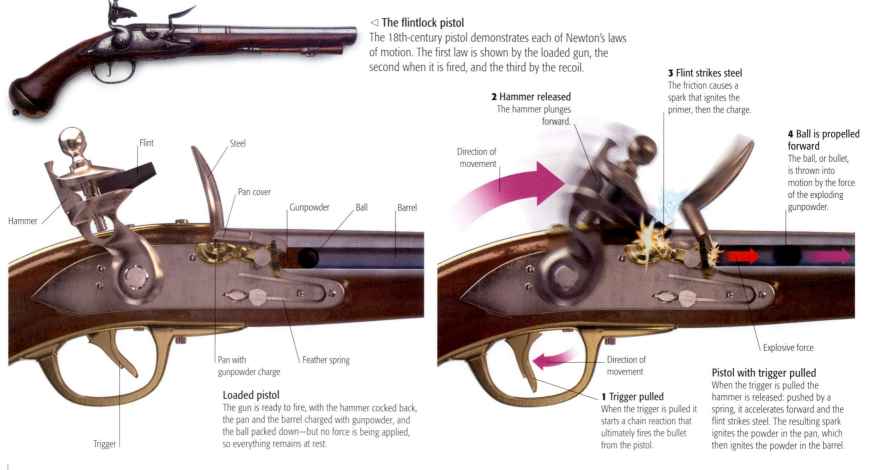

◁ **The flintlock pistol**
The 18th-century pistol demonstrates each of Newton's laws of motion. The first law is shown by the loaded gun, the second when it is fired, and the third by the recoil.

Flint

Steel

Hammer

Pan cover

Gunpowder    Ball    Barrel

Pan with gunpowder charge

Feather spring

Trigger

**Loaded pistol**
The gun is ready to fire, with the hammer cocked back, the pan and the barrel charged with gunpowder, and the ball packed down—but no force is being applied, so everything remains at rest.

**2 Hammer released**
The hammer plunges forward.

Direction of movement

**3 Flint strikes steel**
The friction causes a spark that ignites the primer, then the charge.

**4 Ball is propelled forward**
The ball, or bullet, is thrown into motion by the force of the exploding gunpowder.

Explosive force

Direction of movement

**1 Trigger pulled**
When the trigger is pulled it starts a chain reaction that ultimately fires the bullet from the pistol.

**Pistol with trigger pulled**
When the trigger is pulled the hammer is released: pushed by a spring, it accelerates forward and the flint strikes steel. The resulting spark ignites the powder in the pan, which then ignites the powder in the barrel.

**1 Car at rest**
Equal and opposite forces are acting and so are balanced.

**2 Car accelerates in a straight line**
Forces are unbalanced, because the accelerating force (red arrow) must exceed friction and drag (purple arrow) in order for the car to move.

**3 Car moves at a constant speed in a straight line**
Equal and opposite forces are acting, so are balanced and the car neither speeds up nor slows down.

Friction and drag

Accelerating force

**4 Car travels around a bend**
An equal and opposite force acts on a car as it goes around a bend, the wheels push sideways on the road, slightly distorting it, and the road pushes back. However, for the car, this is an unbalanced force that carries it around the corner.

Sideways accelerating force

**Balanced and unbalanced forces**
When all the forces acting on an object are balanced, it remains at rest or continues moving in a straight line. Unbalanced forces cause a change of velocity—acceleration when the force pushes forward—and a change of direction when the force pushes sideways.

**AFTER »**

Newton combined the study of mechanics and mathematics in a revolutionary way, which was to have a profound effect on scientific thinking and practical inventions.

**RELATIVITY**
Einstein's theories of relativity **302–303 »** showed that **Newton's laws** are only an **approximation**, but luckily they are a very good approximation in the real world—good enough for normal tasks on Earth, and to **transport people to the Moon 372–73 »** and back.

**SPACE TRAVEL**
There is **no friction** or drag **in space**, so Newton's first law ensures that spacecraft keep moving. **Voyager 1**, launched in 1975, is still traveling 1,000,000 miles (1,600,000 km) every day, leaving our solar system behind.

**VOYAGER 1**

Even at this colossal speed, Voyager will take something like 40,000 years to reach the vicinity of the nearest star. Future spacecraft may sail using the **solar wind « 80–81**, or be propelled by **ion engines**, which push charged atoms from the back of the rocket to achieve forward thrust. More powerful engines are needed because using those that are available now, a spacecraft would take around eight months to travel between Earth and Mars.

**SEE ALSO »**
*pp.108–09* GRAVITATIONAL FORCE
*pp.110–11* ISAAC NEWTON
*pp.112–13* SPEED AND VELOCITY

mass of the object. This is easier to accept: push a toy car and it goes faster. The acceleration is proportional to the force and the mass, so the harder you push it, the faster it goes; and the larger the car, the less acceleration your hand will achieve using equal force.

The same law applies to negative acceleration, or deceleration—catching a ball or stopping a car requires a force that opposes the motion in order to reduce the velocity. Put simply, forces cause a change in speed.

**Third law**
Newton's third law—for every action there is an equal and opposite reaction—explains why a rocket moves forward: it pushes hot exhaust gases with great force out of the back, and there is an equal force from the exhaust gases

pushing the rocket forwards. The driving force does not depend on pushing against the air behind—there is no air in space—but simply from the rocket and its exhaust pushing one another apart.

If you stand on a wooden floor, your weight exerts a downward force on the floorboards; at the same time the floorboards are holding you up—by pushing upward with a force exactly equal to your weight. A weighing machine is therefore actually measuring the force it has to exert to hold you up.

**Changing direction**
Anything that changes direction requires acceleration. If an object is moving in one direction and is going to change direction, it needs an external force to accelerate it in the

new direction. For example, if you are riding a bicycle and you want to turn to the left, you turn the handlebars to the left. What happens then is that the outside of the front tire cannot continue to roll in the original direction, but instead pushes sideways against the road. By Newton's third law the road pushes back, and this sideways push makes the bicycle and you accelerate toward the left, increasing your speed in that direction.

At the same time (or, preferably, a fraction earlier) you should lean to the left, so that your weight counteracts the centrifugal force you feel throwing you outward on the curve. Another way of considering this is to say that you lean "into the curve" so that the sideways force on the bicycle does not push it out from under you as the cycle turns.

**5 Recoil action**
The gun gives off a backward "kick" (force), felt by the person firing the pistol.

Bullet propelled down barrel

**Bullet fired**
The rapidly expanding gases from the ignited gunpowder push the ball with great force down the barrel. By Newton's third law an equal and opposite force pushes the gun back—this is the recoil.

**IN PRACTICE**

## CENTER OF MASS

Gravity is a force of attraction between all the mass in an object and all the mass of Earth, but Newton pointed out that mathematically this is equivalent to all the object's mass being concentrated at one point—its center of mass. This is equally true for Earth.

The center of mass is important for balancing; an irregular object placed on the edge of a table will not fall off as long as its center of mass is over the table.

The center of mass of an object may not lie within the solid body of the object; in a mug, for example, it lies somewhere in the middle of the air inside it.

**How to find the center of mass**
The tankard is suspended from two points with string and a line extended vertically downward from each. Where the lines intersect is the center of mass.

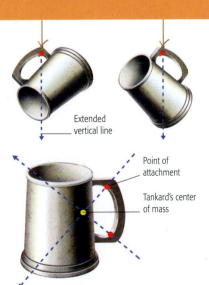

Extended vertical line

Point of attachment

Tankard's center of mass

# Newton's Idea of Gravity

**One of the greatest breakthroughs in the history of science was Isaac Newton's realization that the laws of gravity apply to the Moon and other celestial bodies just as they do to small objects near Earth's surface. Prompted by Edmond Halley and Robert Hooke, he eventually figured out the mathematics and wrote it all down in *Principia*, the most important science book ever published.**

**N**ewton apparently had the first glimmerings of an idea about universal gravity in 1665–66, his most prolifically creative years. He claimed, many years later, that the notion of gravity applying equally to objects on Earth and to the stars and planets was inspired by seeing an apple fall from a tree.

According to Newton's biographer William Stukeley, they had lunch together in Kensington on April 15, 1726, and "the weather being warm, we went into the garden, & drank thea under the shade of some appletrees, only he and myself. Amidst other discourse, he told me, he was just in the same situation, as when formerly, the notion of gravitation came into his mind."

Newton said he had seen an apple fall, and asked himself "Why should that apple always descend perpendicularly to the ground: why should it not go sideways, or upwards? Assuredly, the reason is, that the earth draws it; there must be a drawing power in matter, and the sum of the drawing power in the matter of the earth must be in the earth's centre."

Newton wondered how far the pull of gravity extends—clearly it reaches from the center of the earth up to the top of the apple tree, but could it go as far as the Moon? If so, it would surely affect the orbit of the Moon. Indeed, could it control the orbit of the Moon? He did some calculations, and they "seemed to answer, pretty nearly."

This theory overthrew the notions of Aristotle, who thought the celestial bodies were quite different from Earth and unconnected with it. It also ran counter to the theories of philosopher and mathematician René Descartes (see pp.102–103), who thought the stars and planets were whirled around in vortices.

It was a revolutionary new idea, but whether it actually occurred to Newton in a blinding flash in 1665–66 is open to question; apparently he did not mention the incident with the apple until 1726—60 years after the event. The theory may have been the result of many year's work, owing as much to his synthesis of the work of other great scientists—such as Copernicus, Kepler, Galileo, and Hooke—as to his own natural brilliance.

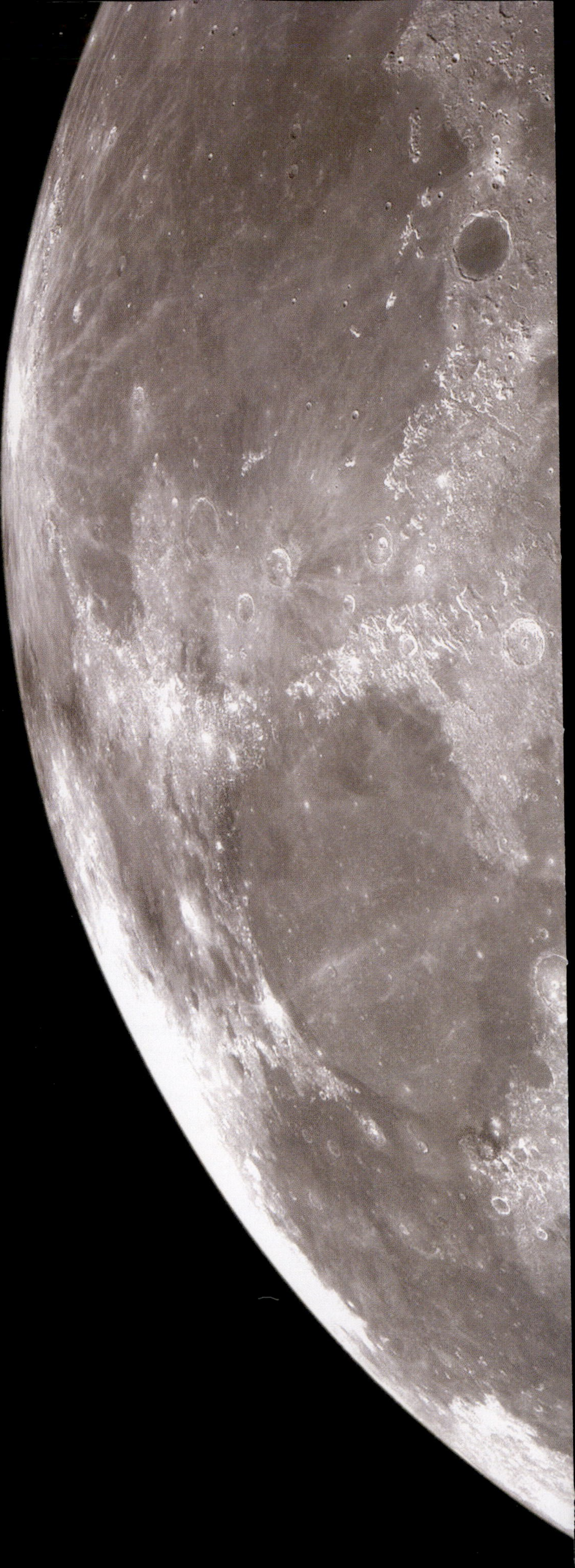

▷ **Celestial gravity**
In *Principia*, Newton argued that the Moon must be "perpetually drawn aside towards the Earth", in the same way that a projectile, "by the force of gravity, may be made to revolve in an orbit."

> "Nor could **the moon** without some such **force**, be retained **in its orbit**."

ISAAC NEWTON, PHYSICIST, FROM *PRINCIPIA* BOOK I, 1687

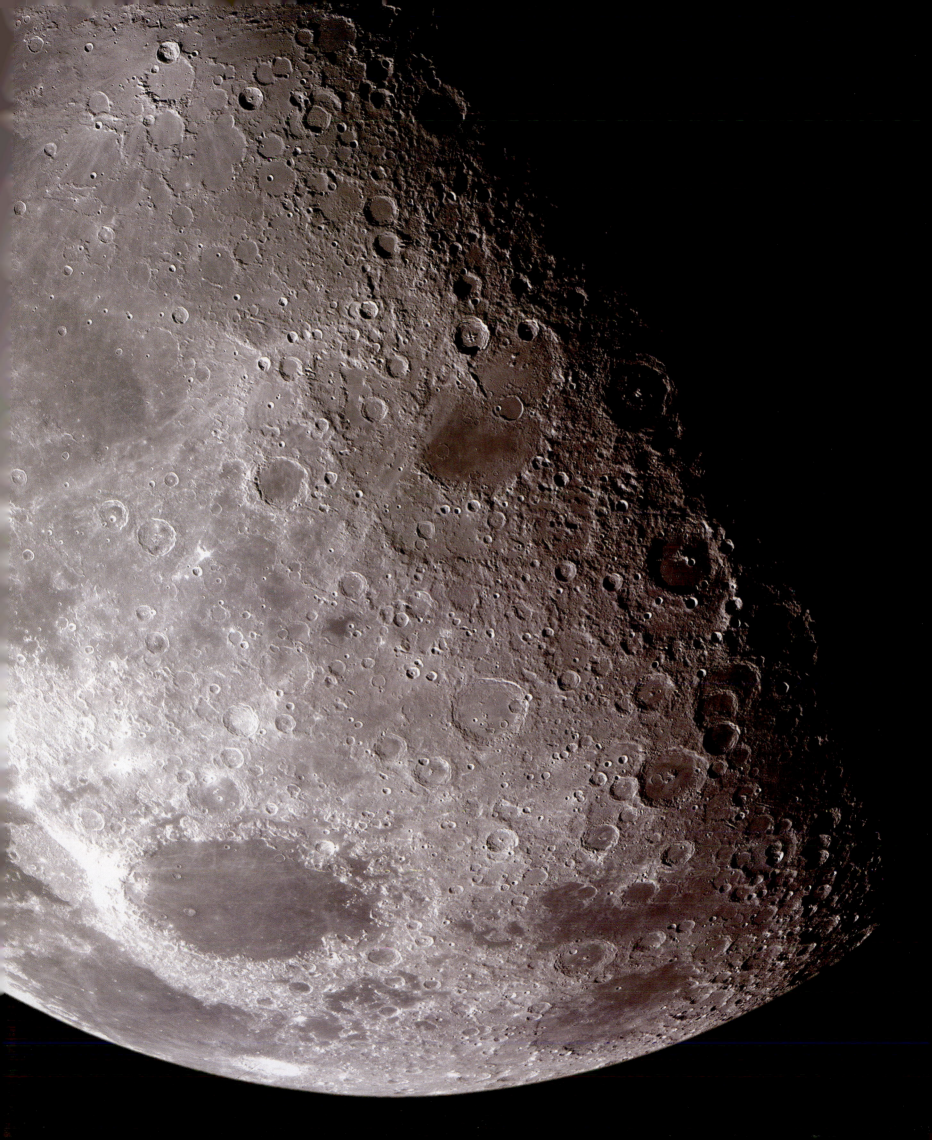

## « BEFORE

The fact that things fall to Earth was obvious to the earliest humans—but not why they fell.

### ANCIENT GREEK THEORIES
Aristotle believed in the four elements of earth, air, fire, and water « 22–23, and thought that things want to **return to where they belong**; so stones fall toward Earth, and they accelerate because as they get closer to home they move more jubilantly.

### GALILEO'S EXPERIMENTS
In the 16th century **Galileo « 82–83** conducted experiments to try to find a **mathematical** description of falling objects. He rolled balls down inclined planes rather than dropping them, so their **rate of fall** was slower, and measurable. But **he never solved the mathematical equations**, nor extended the idea to the heavens.

« SEE ALSO
*pp.86–87*
MOTION, INERTIA, AND FRICTION
*pp.106–07* NEWTON'S IDEA OF GRAVITY

# Gravitational Force

Why things fall to the ground has always been a mystery. Isaac Newton did not invent gravity, but he was the first to realize that things do not just fall, but are attracted to one another. He saw that this force affects everything in the Universe, from stars and planets to people and apples.

Gravity is a force of attraction between masses. It holds us on Earth and holds Earth together; it holds the solar system and galaxies together. It also determines weight and the speed at which things fall.

## Newton's discovery
Legend has it that Newton witnessed the fall of an apple and spontaneously realized how gravity works, but it seems that in reality it was the result of years of deliberation and calculation. He was aware of the theories of the French philosopher Renée Descartes and the Italian scientist Galileo Galilei, and he knew about the German astronomer Johannes Kepler's laws of planetary motion (see p.77), but apparently he did not apply himself to the problem seriously until the 1680s, when he finally figured out the mathematics, and

explained the concept of universal gravity. This was "a drawing power" and Newton further recognized that "If matter thus draws matter; it must be in proportion of its quantity." This means the attraction is proportional to mass; larger objects exert a stronger attractive force. An astronaut on the Moon weighs only a sixth as much as at home, because the Moon is much less massive.

## Weight and mass
A person's weight is the attractive force between the mass of the person and the mass of Earth. The mass of an object is the amount of matter in it: a 2.2 lb (1 kg) bag of flour contains 2.2 lb (1 kg) of flour, and this amount does not change whether the bag of flour is on the Moon or weightless in deep space. But the weight of that bag of flour is the force with which it

is pulled down by gravity—this would be 9.8 newtons on Earth, or about 1.6 newtons on the Moon. (A newton, usually expressed as a metric measure, is the force needed to accelerate a mass of 1 kg by 1 m per second every second.)

Fellow scientist Robert Hooke (see pp.92–93) had figured out that the force decreases with distance, and

Newton proved that this was in accordance with an inverse square law—that the strength of the force is inversely proportional to the square of the distance from the source of the force. Therefore an object twice as far away from the source has only one quarter of the attraction, while an object three times as far away has only one ninth of the attraction.

### IN PRACTICE

## SPRING TIDES AND NEAP TIDES

The ocean tides are caused mainly by the gravitational pull of the Moon. As it passes over any point, the Moon attracts Earth, but it attracts the ocean more strongly because it is slightly closer. This draws up a bulge—high tide. When the Sun is in line with Earth and the Moon,

the Sun's pull is added to the Moon's, and so the tides are more extreme; these are spring tides. When the line from Earth to the Moon is at right angles to Earth–Sun line, the two pulls cancel each other to some extent; these are the smaller neap tides.

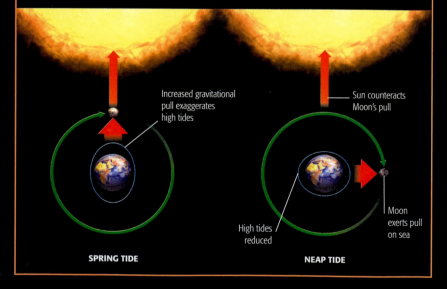

**Increased gravitational pull exaggerates high tides**

**Sun counteracts Moon's pull**

**High tides reduced**

**Moon exerts pull on sea**

SPRING TIDE                    NEAP TIDE

## The relationship between mass and distance
Gravitational attraction depends on the product of objects' masses divided by the square of their separation. If the masses are doubled, the attraction is four times stronger; but if the distance is doubled, the attraction is four times weaker.

**Attraction**
Two bodies are attracted by a gravitational force.

**Doubling the mass**
If each body is twice the mass, then the gravitational force is four times as strong.

**Increasing the distance**
If the distance between the masses is doubled, the gravitational force is four times less.

AFTER »

## Saturn: the ringed planet

Saturn is the sixth planet from the Sun, and the second most massive planet in our solar system. It is a gas giant, with no firm surface, orbited by at least 61 moons. Its main rings (below) measure around 124,000 miles (200,000 km) from side to side, but are only about 33 ft (10 m) thick. They consist mainly of millions of pieces of ice, all held in individual orbits by Saturn's gravity.

Since Einstein's groundbreaking theories, scientists have been attempting to learn more about the gravitational force, including the speed of gravity itself.

### THE FOURTH DIMENSION

Einstein's theory of relativity 300–301 » introduced the new concept of **four-dimensional spacetime** and redefined gravity, but did not quite explain how it works.

### GRAVITATIONAL WAVES

According to Einstein, objects with mass moving around can cause **fluctuations or ripples** in spacetime 302–303 »; these are gravitational waves and **should be detectable**, but so far no-one has succeeded.

### THE SEARCH FOR THE GRAVITON

The **graviton** is a theoretical fundamental particle proposed by **quantum field theory** 314–17 ». It has no mass, nor any electrical or magnetic charge, but is thought to carry the **gravitational force**. The existence of gravitons has not yet been confirmed experimentally.

SEE ALSO »
pp.300–01 EINSTEIN'S EQUATION
pp.368–69 ARTIFICIAL SATELLITES
pp.400–01 GRAND UNIFIED THEORY

Newton's great leap forward was to realize that the force of gravity extended out into space, and that it controlled not only the fall of the apple in front of him, but also the orbit of the Moon. Today Earth's gravity holds the Moon, the International Space Station, and hundreds of satellites in orbit.

Gravity is the weakest of the four forces that operate in the universe: electromagnetism, the strong nuclear force, the weak nuclear interaction, and gravity. However, only gravity and electromagnetism work over long distances.

## Spacetime distortion

Einstein (see pp.304–305) had a different explanation. He said that there is no attraction between masses; what is happening is that objects with mass distort spacetime around themselves, as weights would stretch a horizontal rubber sheet. Because an object that has mass makes a dent in spacetime, other objects with mass will tend to "roll downhill" toward it, making it appear that such objects attract one another.

Einstein also said that gravity acts at the speed of light. If this is true, something must be "carrying" the gravity—possibly gravitational waves.

### Changing weight

Astronauts and spacecraft in orbit around Earth have the same mass as on the ground, but are weightless because they are in "free fall," with the centrifugal force counterbalancing the force of gravity.

# 6,000 MILLION

**million million tonnes is the total mass of planet Earth, as first estimated by Isaac Newton, and as subsequently confirmed by Henry Cavendish's ingenious torsion balance experiment in 1797 (see below).**

## Michell-Cavendish experiment

Below is a modern representation of Henry Cavendish's experiment of 1797 in which he used a torsion balance apparatus, designed by John Michell, to measure the mass of Earth. From this scientists were able to calculate a value for the gravitational constant, $G$, which is part of Newton's Law of Gravity.

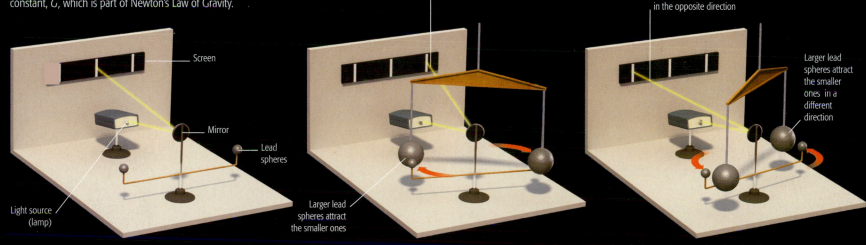

Mirror moves with the rod, and deflected light beam indicates amount of attraction

Light beam is deflected in the opposite direction

Screen

Mirror

Lead spheres

Light source (lamp)

Larger lead spheres attract the smaller ones

Larger lead spheres attract the smaller ones in a different direction

**1** Two spheres are mounted at either end of a free-swinging beam that has a fixed mirror in the middle. When the rod is still, light shone at the mirror is reflected to the center of the screen.

**2** Two large lead spheres are brought up close to the small ones, one in front and one behind, and their gravitational attraction on the small spheres turns the rod and mirror clockwise, moving the light to the right.

**3** When the large spheres are introduced on the other side, the rod and mirror turn counterclockwise and the light moves to the left. Using the relative attraction of the large spheres and Earth, Earth's mass can be calculated.

PHYSICIST AND MATHEMATICIAN **Born 1642 Died 1727**

# Isaac Newton

> "**Truth** is ever to be found in **simplicity**, and not in the **multiplicity** and **confusion** of things."

ISAAC NEWTON, FROM "RULES FOR METHODIZING THE APOCALYPSE"

Perhaps the greatest mathematician and scientist of all time, Isaac Newton was a contentious genius who made extraordinary strides in mathematics, optics, mechanics, and gravitational attraction. He invented calculus, he explained the colors of the rainbow, and he realized that the same gravitational force that gives us weight also controls the movements of the Moon and planets. In later life he became Master of the Mint, and President of the Royal Society.

### Childhood without parents

Isaac Newton was born in 1642 at Woolsthorpe Manor, near Grantham, England. His father had died three months earlier, and when Newton was three his mother married a rich clergyman, leaving him in the care of her parents. She returned when her second husband died, but that was not until 1653; it is possible that Newton's difficult relationships may have begun with the absence of his parents. After

education at the King's School in Grantham and some time assisting on his mother's farm, Newton went to Cambridge University.

### Works of genius

As an undergraduate he showed no great promise, but in 1665 the University was closed because the Plague was rife, and Newton returned home to Woolsthorpe, where, during the next 18 months, he produced his best work in mathematics and physics. Some time later, in 1671, he made a reflecting telescope, and presented it to the Royal Society.

His first scientific paper explored the nature of light. Robert Hooke (see pp.92–93) disagreed with some of Newton's conclusions in this paper. Unfortunately Newton took this as a personal affront, and vowed not to publish anything else about optics during Hooke's lifetime. He had disputes with others, too: with Gottfried Leibniz about who

**Woolsthorpe Manor**
Isaac Newton was born in this house in Woolsthorpe, near Grantham, England. His grandparents were entrusted with his care.

Profl I

*(handwritten note)*

◁ **Isaac Newton**
This portrait was painted by Sir Godfrey Kneller in 1689, when Newton was 46 years old, and just after the publication of *Principia* had made him famous.

◁ **An optical experiment**
In this color engraving Newton is shown using a prism to experiment with focused rays of light. He was the first scientist to observe and document the components of white light, and later to publish his findings.

**Brachistochrone curve**
The brachistochrone curve, or curve of shortest time, is the curve between two points that is covered in the least time by a body that travels under the action of constant gravity. It was suggested by Newton in 1696. Today his solution is called a cycloid curve.

invented the calculus, and with John Flamsteed over his astronomical catalogue of stars.

In the late 1670s Newton spent most of his time studying alchemy (the hidden forces of nature) and writing religious tracts. Having embraced the Christian sect of Arianism around 1672, he went to great lengths to try to show that Jesus, although more than just a man, was not God. In 1678 he had a nervous breakdown, and when his mother died in 1679 he shut himself away. He suffered another breakdown in 1693.

### Publication of *Principia*
In 1684, persuaded by his friend Edmond Halley, Newton wrote a paper about the orbits of comets. He revised this, and then expanded it into *Philosophiae Naturalis Principia Mathematica*—usually called just *Principia*—which contained Newton's laws of motion (see pp.104–105), and the law of universal gravitation (see pp.108–109). Both were to transform our understanding of the world.

In 1696 Newton moved to London to become Warden of the Mint, and two years later Master of the Mint. Elected

President of the Royal Society, he ruled it with a rod of iron, remaining in the post for 24 years. During this time the only known portrait of Robert Hooke mysteriously disappeared.

### A falling apple
On April 15, 1726 William Stukeley had lunch with Newton. He later described how Newton told him that the notion of gravitation was suggested by the fall of an apple as he sat in his garden. This seems to have been the first time that Newton mentioned the apple, which is odd, because he said it had happened 60 years earlier. He may have made up the story, to "prove" that he had the great ideas about gravity before Hooke.

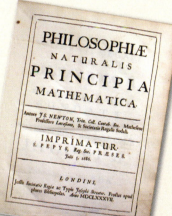

**Principia**
Newton's *Principia*, the most important science book ever written, was published in 1687 at the expense of his friend Edmond Halley, because the Royal Society had spent their budget on a history of fish.

**TIMELINE**

**December 25, 1642** Isaac Newton is born at Woolsthorpe Manor, Lincolnshire, England. He is not expected to survive the night.

**1646** His mother marries a rich clergyman and leaves Newton to be cared for by her parents.

**1652** Sent to King's School in Grantham; his name, carved by him into a windowsill, can still be seen in what is now the library.

**1659** Works on his mother's farm.

**1661** Goes to Trinity College, Cambridge.

**1665** Cambridge University closed for fear of the Plague. Newton goes home to Woolsthorpe for 18 months—his "annus mirabilis," in which he solves various mathematical problems, invents calculus, investigates the colors of the rainbow, and has his first great insight into gravity.

**1667** Returns to Cambridge, lectures on optics, and is made a Fellow of Trinity College.

**1669** Elected Lucasian Professor of Mathematics.

**1671** Makes the first reflecting telescope and presents it to the Royal Society.

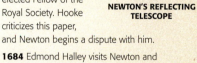

**NEWTON'S REFLECTING TELESCOPE**

**1672** Writes his first scientific paper, about light and the colors of the rainbow, and is elected Fellow of the Royal Society. Hooke criticizes this paper, and Newton begins a dispute with him.

**1684** Edmond Halley visits Newton and persuades him to write down his thoughts on mathematics, gravity, forces, and the orbits of planets and comets.

**1687** Publishes *Philosophiae Naturalis Principia Mathematica*, generally known as *Principia*, the most important science book of all time. After this he devotes much time to alchemy and writing religious tracts.

**1689** Elected Member of Parliament for Cambridge University, but is not very active.

**1692** Suffers a nervous breakdown, expressed by him as having "lost his reason."

**1696** Becomes Warden of the Mint, and reforms the currency.

**1698** Becomes Master of the Mint.

**1701** Resigns his Fellowship of Trinity and his Lucasian Professorship.

**1703** Elected President of the Royal Society.

**1704** Publishes a book, *Opticks*, the year after Hooke dies.

**1705** Knighted by Queen Anne, ostensibly for his work at the Mint, but probably in reality for political favors.

**1727** Dies, and is interred in Westminster Abbey.

**MONUMENT TO NEWTON IN WESTMINSTER ABBEY**

# Speed and Velocity

**People have an innate understanding of speed: it is simply how fast an object is moving. Velocity is speed together with the direction of movement. Along with acceleration and momentum, speed and velocity are fundamental scientific concepts that affect almost everything we do, from driving to mowing the lawn.**

## BEFORE

Early ideas about motion were based on common sense, but from the 17th century scientists analyzed it more systematically.

**EARLY RELATIVITY**
**Galileo Galilei** showed in his 1632 book *Dialogue Concerning the Two Chief World Systems* that **all motion is relative**—a precursor to Albert Einstein's **special theory of relativity**.

**THE WORK OF GASSENDI**
French philosopher **Pierre Gassendi** was one of the first thinkers to attempt a **scientific understanding of motion**. Through experiments he realized that **falling objects accelerate as they fall**.

**« SEE ALSO**
*pp.86–87* MOTION, INERTIA, AND FRICTION
*pp.104–05* NEWTON'S LAWS OF MOTION

**SCALAR** A quantity that is given only by a number (the magnitude) is called a scalar quantity. Speed is a scalar.

**VECTOR** A quantity with a magnitude and a direction is called a vector quantity. Velocity is a vector.

Speed is distance traveled divided by time taken. In everyday life, people usually measure it in miles per hour (mph) or kilometers per hour (kph). Scientists normally measure it in meters per second (m/s).

## Speed is relative

An object's speed is usually measured in relation to its surroundings; so, for example, an aircraft has an airspeed and a separate speed over the ground. Speed can also be defined as relative to the speed of another object. Two cars traveling in the same direction at 60 kph and 40 kph have a relative speed of 20 kph (60 – 40). If the two cars are traveling in opposite directions at 60 kph and 40 kph, their relative speed is now 100 kph (60 + 40).

### Relative speed

The relative speed of two objects moving independently depends on two factors: the difference between their individual speeds, and their individual directions of motion.

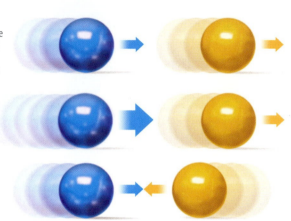

**Relative speed zero**
If two objects are traveling at the same speed in the same direction, they will always stay the same distance apart—as if they were both stationary.

**Catching up**
If one object is traveling at a greater speed than another and in the same direction, it will catch up and eventually overtake the other.

**Heading for collision**
If two objects are moving in opposite directions, their relative speed is the sum of their two speeds.

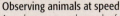

**Observing animals at speed**
A racehorse can reach speeds of up to 40 mph (65 kph). When scientists such as Newton were figuring out the laws of motion, horses were among the fastest things they could observe.

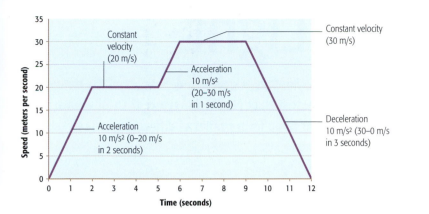

Constant
velocity
(20 m/s)

Constant velocity
(30 m/s)

Acceleration
10 m/s²
(20–30 m/s
in 1 second)

Acceleration
10 m/s² (0–20 m/s
in 2 seconds)

Deceleration
10 m/s² (30–0 m/s
in 3 seconds)

### Speeding up and slowing down
Plotting velocity against time on a simple graph like this one allows ready calculation of mean acceleration and deceleration of an object moving in a straight line. A horizontal line indicates no acceleration or deceleration.

Velocity is not the same as speed. Velocity is an object's speed in a particular direction. It is defined in terms of both speed and direction, as in "60 kph north."

People normally measure speed relative to the ground. But the ground is not fixed: Earth is traveling through space. Measuring speeds relative to the Sun is equally valid, because the Sun, too, is moving. English scientist Isaac Newton (see p.104) assumed that all speeds could, in principle, be measured relative to space, which was fixed or "absolute." However, in 1905 German physicist Albert Einstein published his special theory of relativity (see p.302), which describes the consequences of all speeds being relative in a universe where the speed of light is the same for all observers.

### Acceleration and momentum
Although Newton was mistaken about "absolute" space, he did pioneer our understanding of how things move. In his laws of motion, he proposed that an object's velocity can be changed only if a force acts on the object. The rate of change in velocity—in speed, direction, or both—is called acceleration. If a car goes from 0 m/s to 30 m/s in 6 seconds, and continues in the same direction, then its mean acceleration is 30 m/s divided by 6 s, or 5 m/s/s, or 5 m/s². Every second, it goes 5 m/s faster.

### Energy transfer
During a collision, energy is transferred from one object to another. The energy of a fast-moving pool cue transfers to the cue ball, which accelerates instantly. When the cue ball hits the target ball it transfers the energy, causing the target ball to accelerate. After impact the cue ball stops dead, its energy gone.

Direction of
movement

Cue ball gains
velocity from cue

Energy transfers to
target ball on impact

### Angular momentum
Any spinning object has what is called angular momentum, which is proportional to the mass of the object, its rotational speed, and the average distance of the mass from the centre of spin. The angular momentum must be conserved

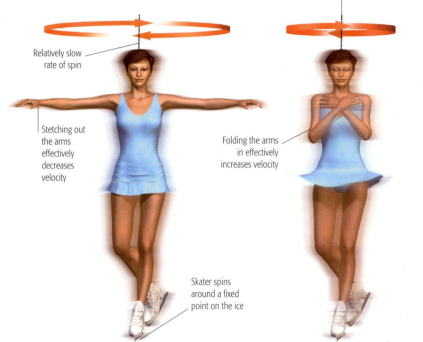

Relatively slow
rate of spin

Stetching out
the arms
effectively
decreases
velocity

Skater spins
around a fixed
point on the ice

Rate of spin increases
to maintain angular
momentum

Folding the arms
in effectively
increases velocity

1 A skater spins on the spot with her arms outstretched, and has a particular angular momentum, proportional to mass, rotational speed, and average distance of mass from the centre.

2 By bringing her arms in she decreases the average distance of her mass from the centre of spin, and so to conserve her angular momentum her rotational speed must increase.

If the car maintains the same speed but changes its direction, this is also acceleration. An object moving in a circle is accelerating, even when its speed is constant, because it is always changing direction. Without a force acting on it, the object would move in a straight line. Newton was the first to realize that the force keeping Earth in orbit around the Sun is gravity.

Newton also helped define momentum—the tendency for a moving object to keep moving—as velocity multiplied by mass. A heavy object traveling slowly may have the the same momentum as a lighter one traveling much faster.

### Rotational motion
Scientists refer to the rate of an object's movement through space as its translational speed. If the object is spinning, it also has a rotational speed, normally measured in revolutions per minute (rpm) or per second (rps). Rotational speed is also sometimes measured in degrees per second—the angle through which an object turns each second. In this case, it is normally called angular speed.

Our understanding of rotational motion was made complete in the 19th century by French mathematician Gaspard-Gustave Coriolis. His work shows how the laws of motion take effect on our rotating planet—for example, in the motion of large air masses and ocean currents.

## AFTER

Thanks to technological advances we have built ever faster machines in the past 100 years, from cars on the ground and boats on water to airplanes in the sky.

### BREAKING SPEED RECORDS
The first official **land speed record** achieved by a car, in Paris in 1898, was 39 mph (63 kph). In 1997 Andy Green set a new mark of 763 mph (1,228 kph) in the jet-propelled *ThrustSSC*. Donald Campbell set seven **water speed records** in *Bluebird K7* between 1955 and 1964, peaking at 276 mph (445 kph). Ken Warby set the current record of 317 mph (511 kph) in *Spirit of Australia* in 1978. In 1903 Wilbur Wright reached only 7 mph (16 kph) in his *Wright Flyer*. In 1976 Eldon Joersz flew a Lockheed SR-71 Blackbird jet at 2,194 mph (3,530 kph), an **air speed record**.

BLUEBIRD K7, ULLSWATER, UK, 1955

SEE ALSO »
pp.120–21 MEASURING TIME
pp.300–01 EINSTEIN'S EQUATION
pp.302–03 THEORIES OF RELATIVITY

## TERMINAL VELOCITY

Drop an object through the air, and the force of gravity will cause it to accelerate. But another force called air resistance, or drag, also acts on it. The faster the object moves, the stronger the drag—and it always acts in the opposite direction to the motion. At a certain speed, the drag exactly balances the force of gravity. This is why skydivers in free fall—with arms and legs outstretched—soon reach a "terminal velocity" of about 120 mph (195 kph), at which the drag will not allow them to fall any faster.

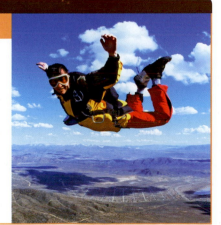

**BEFORE**

The dominant theory during the Classical era was that light emerges from the human eye to illuminate objects around it.

**LIGHT AS A PARTICLE**
In a long poem called *Re Rerum Natura*—*On the Nature of Things*, or *On the Nature of the Universe*—the Roman poet and philosopher **Lucretius** (c.99–55 BCE) deviated from the common view that light is generated by the human eye by suggesting instead that light is a **particle** emitted by luminous objects.

**REFLECTION OF SUNLIGHT**
In the 10th century, while under house arrest, Islamic scientist **Alhazen** wrote the influential *Book of Optics*, in which he argued, based on empirical observations, that light is a particle that travels **faster than sound**, and that objects are illuminated by **the reflection of sunlight**.

**≪ SEE ALSO**
*pp.60–61* ALHAZEN

# The Nature of Light

From the late 1600s to the early 1800s advances in optical technology triggered a debate about the nature of light. Was it made of particles, as the eminent English physicist Isaac Newton suggested? Or of waves, as Dutch astronomer Christiaan Huygens insisted? The answer came from studies of diffraction and refraction.

Isaac Newton (see pp.110–11) began experimenting with light and optics from around 1666. He showed that color is intrinsic to light and that white light is in fact a mix of many different colors that can be separated ("dispersed") and reunited using prisms.

Newton's studies convinced him that lens-based optical devices could never be free of color dispersion, so he set out to build a reflecting telescope, which used mirrors, to avoid this problem. His investigations of the nature of reflection, and in particular the fact that light moves in straight lines and casts sharp shadows (rather than spreading out like a wave), helped to convince him that light was a stream of "corpuscles," or particles.

Newton published his theories in a letter to the Royal Society in 1672. He elaborated on them further in *Opticks* in 1704, by which time he had a formidable reputation, and they were rapidly accepted.

## The wave-front theory
The first person to propose a wave theory of light was René Descartes (see pp.102–03), around 1630. In 1665 Newton's rival Robert Hooke (see pp.92–93) published a theory of light waves, based on studies using the recently invented microscope. His theory was corroborated by Christiaan Huygens in the late 1670s. Huygens's initial evidence was based on the way in which beams of light could cross each other without being affected. Particles, he argued, would inevitably collide and bounce away in different directions. Huygens published his *Treatise on Light* in 1690, and while Newton's theory held sway through much of the 18th century, the weight of scientific opinion slowly began to swing toward a wave theory. The clinching evidence came in the early 1800s, when Thomas Young showed the interference between two light waves diffracting after passing through two narrow slits (see below).

In the 20th century Einstein (see p.304–05) and Planck (see p.314) showed that light could act both as a particle and a wave. The quanta of electromagnetic radiation they recognized later became known as photons (see p.287).

**MATHEMATICIAN (1629–95)**

### CHRISTIAAN HUYGENS

Astronomer, physicist, mathematician, and instrument maker Christiaan Huygens was born to a wealthy Dutch family. The French philosopher René Descartes guided him in his early studies, and he studied law and mathematics at the University of Leiden before turning to science. Aside from his brilliant theory of light, he made many other important scientific contributions, such as developing the tools of calculus and writing the first book on probability theory. He built the most ambitious telescopes of his day, discovered Saturn's largest moon, and invented the pendulum clock.

**Water waves**
Dutch astronomer Christiaan Huygens's observations of ripples in water convinced him that light behaved in a similar way.

**Double slit experiment**
By demonstrating wavelike properties such as diffraction and interference, Thomas Young's ingenious experiment appeared to show for once and for all that light was a wave and not a stream of particles.

**Single slit panel**
Blocks out all light except that traveling at a narrow range of angles.

**Light source**
Light spreads out in every direction.

**Diffracting light**
Light spreads out or diffracts after passing through the first slit.

**Double slit panel**
Different parts of the light wavefront travel through two parallel slits.

**Interference of waves**
As the light diffracts again from the two slits, the waves interfere with each other. If the light were a diffracted stream of particles, they would simply overlap.

**Banded pattern**
A pattern of light and dark bands, corresponding to peaks and troughs of interference, forms on a screen.

**AFTER ≫**

Throughout the 1800s the wave theory was refined and eventually absorbed into a new field of science, but problems remained.

**TRANSVERSE WAVES**
French physicist **Augustin-Jean Fresnel** studied light waves and showed that their vibrations must be "**transverse**" (perpendicular to their motion). He used this to explain the **polarization** of light **116–17 ≫** which is caused by different vibrations.

**MAGNETIC ATTRACTION**
**Michael Faraday's** 1845 discovery that light could be affected by magnetic fields led to James **Clerk Maxwell's** realization that light is a form of **electromagnetic wave 258–59 ≫**.

**SEE ALSO ≫**
*pp.116–17* SPLITTING AND BENDING LIGHT
*pp.258–59* ELECTROMAGNETIC SPECTRUM
*pp.300–01* EINSTEIN'S EQUATION

**Light in the darkness**
Sun beams, seen here filtering through small spaces between the leaves, illuminate and warm the forest floor. The reality of light, however, is highly complex as beams of light behave sometimes like waves and sometimes like particles.

**« BEFORE**

People have been making mirrors and lenses since ancient times, for decoration and practical use, such as starting fires.

**POLISHED SURFACES**

The ancients used mirrors of polished **bronze or iron** that reflected up to half of the light that fell on them. **Archimedes** is alleged to have used mirrors to set fire to Roman ships that were invading Syracuse.

**ARABIAN OPTICS**

In his 984 treatise *On Burning Mirrors and Lenses*, Arabian mathematician **Ibn Sahl** anticipated **Snell's Law of refraction** by some 640 years. This knowledge allowed him to figure out the optimal shapes for mirrors and lenses.

**ANCIENT MIRROR**

**EARLY TELESCOPES**

Early reflective **telescope mirrors « 84–85** were made of "speculum metal," which is an alloy of one part tin to three or four parts of copper.

**« SEE ALSO**
*pp.60–61* ALHAZEN
*pp.114–15* THE NATURE OF LIGHT

---

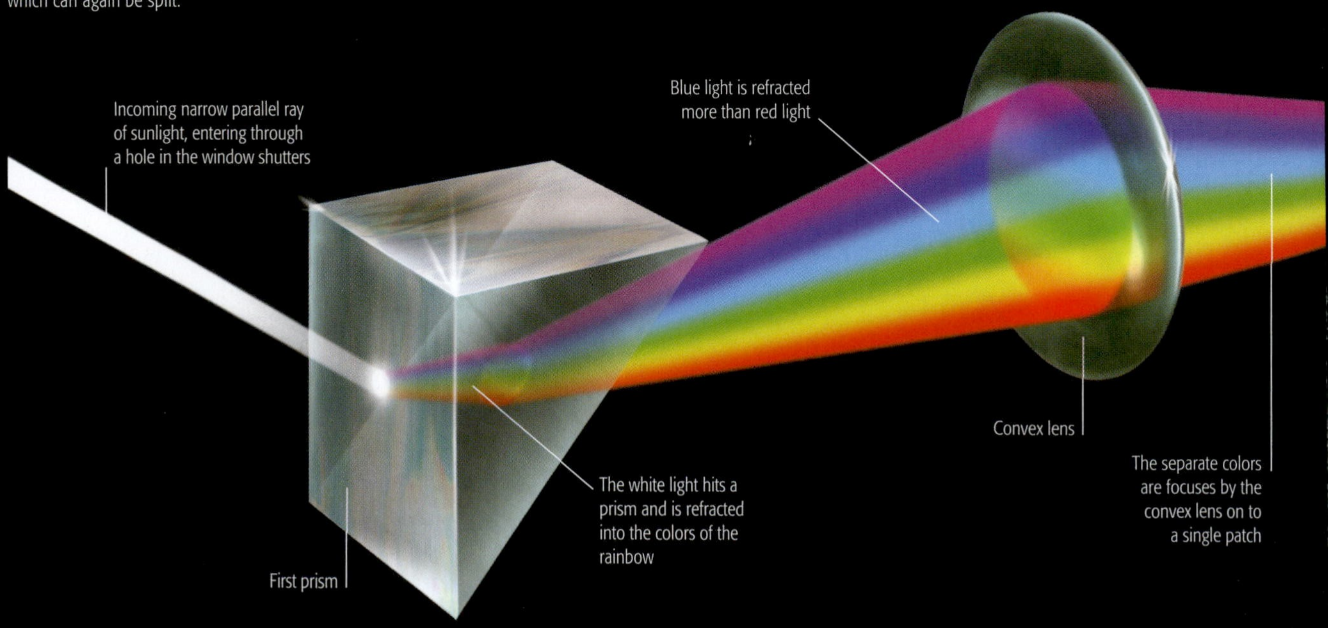

MATHEMATICIAN (1580–1626)

## WILLEBRORD SNELL

Snell initially studied law at university in his home town of Leiden, the Netherlands, before changing to mathematics. From around 1600 he traveled around Europe, meeting key scientists such as Kepler and Brahe, before returning to Leiden to marry Maria de Lange in 1608. His first major work, *Eratosthenes Batavus*, formed the foundations of modern geodesy (methods for measuring the earth). He also published studies on astronomy and navigation, but he is best remembered for his law on light refraction.

---

# Splitting and Bending Light

**The light-reflective quality of mirrors has made them useful for thousands of years. But it was not until the 17th century that scientists really began to understand the properties of light—formulating equations for describing reflection and refraction, and recognizing that sunlight is made up of many different colors.**

Light is crucial for life: it enables many animals to see, and is used by plants during photosynthesis (see pp.154–55) to convert carbon dioxide and water into carbohydrates. However, the behavior of light itself has taken many centuries to understand.

Light rays have three important characteristics: they are absorbed by dark rough surfaces, reflected by light shiny surfaces, and bent when they pass from one material to another.

## Reflection

Rays of light bounce symmetrically from mirrors; the angle of reflection is equal to the angle of incidence (see right). The surface of water or any liquid in a bowl or pool will act as a mirror. Many ancient tales, such as those of Aesop (*c.*600 BCE), include stories of people and animals seeing reflections in pools of water.

Reflection has long been useful scientifically. The ancient Egyptian Eratosthenes (*c.*240 BCE) was told by a traveler about the reflection of the Sun in a deep well at Syene (now Aswan) one midsummer's day. He realized this must mean that the Sun was directly overhead. He then used this information to help calculate the size of the Earth (see pp.38–39).

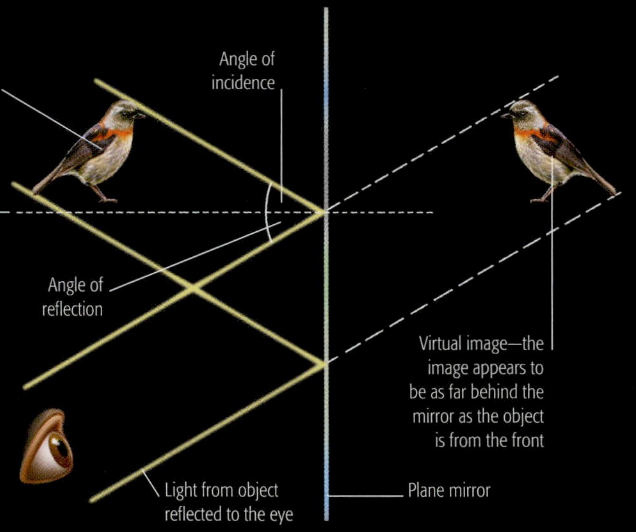

### Newtonian prism experiment

This experiment is similar to one described in Newton's book, *Opticks*. It is designed to show that white sunlight can be split by a prism into the colors of the rainbow; that these can be recombined to give white light, which can again be split.

### The principles of reflection

The angle at which light is reflected from a surface is the same as the angle at which light strikes that surface. In scientific terms the angle of incidence equals the angle of reflection.

Object

"Normal" line (imaginary line drawn at 90° to mirror's surface)

Angle of incidence

Angle of reflection

Virtual image—the image appears to be as far behind the mirror as the object is from the front

Light from object reflected to the eye

Plane mirror

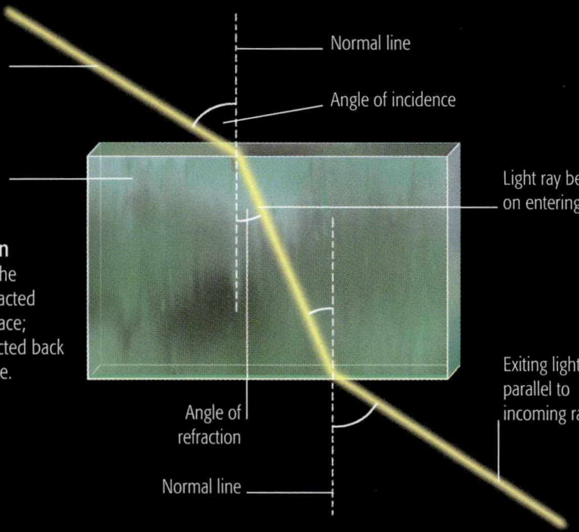

### The principles of refraction

A ray of light passing through the surface of glass or water is refracted downward, away from the surface; as it re-enters the air, it is refracted back to an angle closer to the surface.

Light ray enters

Normal line

Angle of incidence

Glass block

Light ray bent on entering glass

Exiting light ray parallel to incoming ray

Angle of refraction

Normal line

Incoming narrow parallel ray of sunlight, entering through a hole in the window shutters

Blue light is refracted more than red light

Convex lens

The white light hits a prism and is refracted into the colors of the rainbow

The separate colors are focuses by the convex lens on to a single patch

First prism

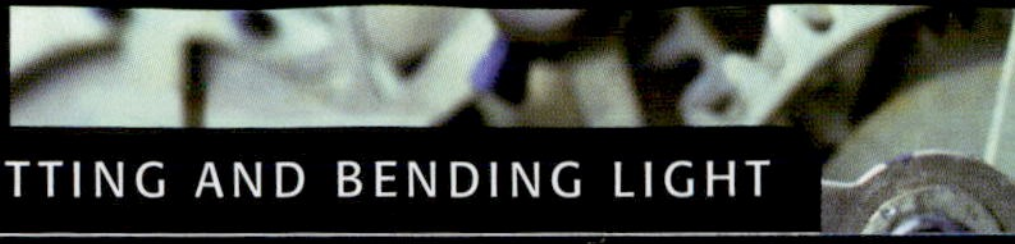

However, a sheet of polished metal is more convenient than a pool of water, and metal mirrors have long been used by men and women for personal grooming. Modern telescopes and domestic mirrors use a sheet of glass backed with a coating of silver or aluminum; the clean metal surface reflects the light, while the glass protects it from dirt and tarnishing.

## Light-bending materials

When light passes from one material into another, for example from air into glass or water, it changes direction; light is refracted (bent) as it goes through the interface. This refraction makes ponds look shallower than they really are, and can make a spoon appear bent when it stands in a glass of water.

Light travels more slowly through denser materials, such as water and glass, than it does in air, and it refracts to minimize the time it takes to travel between two points. In around 1620 the Dutch scientist Willebrord Snell discovered that the amount of bending depends on the refractive indices of the two media. He created a mathematical formula, now known as Snell's Law, that can be used to calculate the refraction of light when traveling between two materials of differing refractive index.

## Splitting light

In 1666 Isaac Newton (see pp.110–11) acquired a glass prism. He cut a hole in his window shutters, so that a narrow beam of sunlight shone through, then he intercepted this beam with his prism so that a spectrum formed on the opposite wall. He noted that the spot

### IN PRACTICE

## POLARIZATION OF LIGHT

Light consists of packets of electromagnetic waves that oscillate at right angles to the direction of travel. A wave going from left to right oscillates up-down, front-back, and all other directions in between. Polarizing filters act by absorbing the waves oscillating in one direction only, leaving a majority oscillating at right angles. In the diagram below, vertical waves travel through the polarizing filter; while horizontal waves are absorbed. Photographers use polarizing filters to reduce surface reflection, and to deepen the blue of the sky.

Light beam — Polarizing filter — Vertical waves are transmitted through the filter — Photographic sensor

UNFILTERED IMAGE

POLARIZED IMAGE

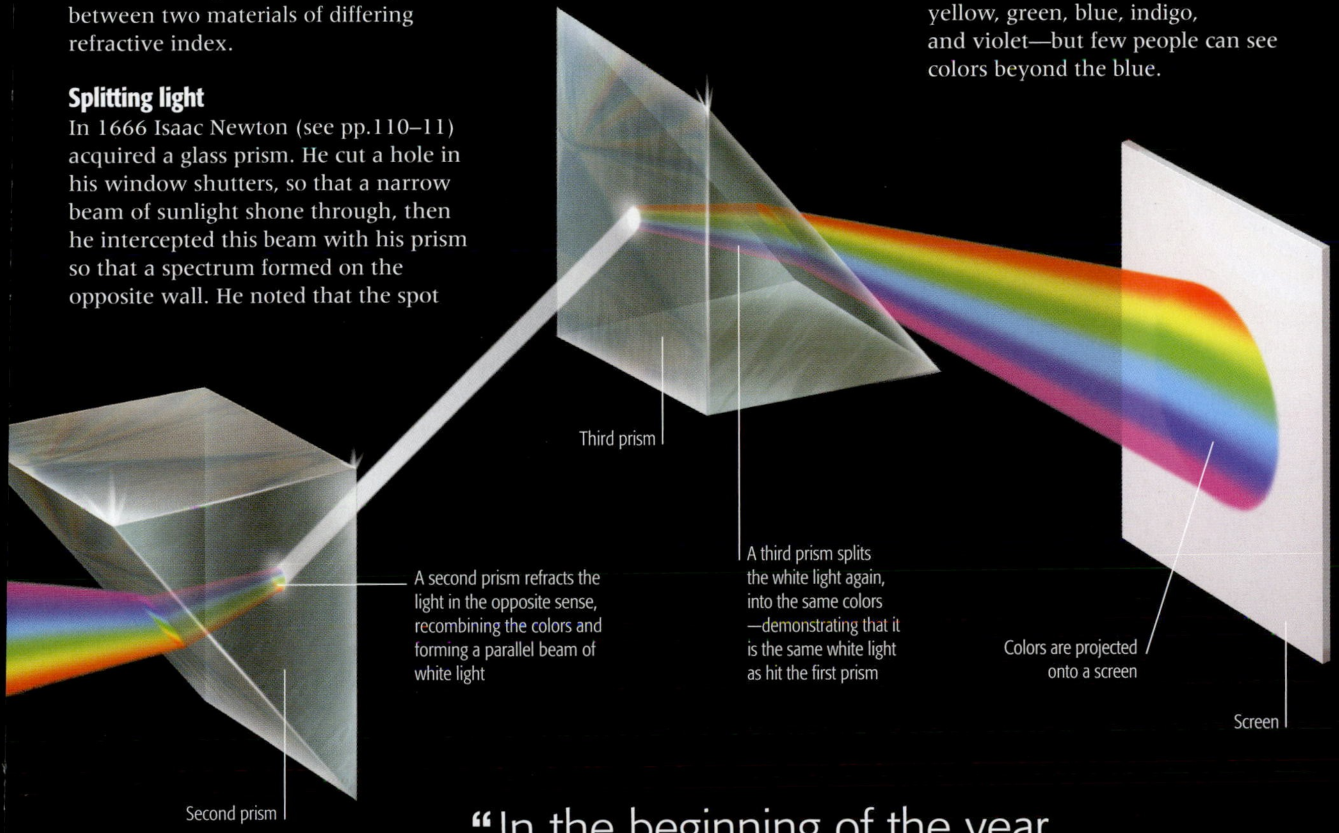

of light which came into the room spread out into a long patch of different colors: blue (which refracted the most) lay at one end and red (which refracted least) at the other. Newton showed that white sunlight is a mixture of all the colors of the rainbow. It can be split into its many colors using a prism, while a second prism will recombine them to make white light again (see below). Newton himself claimed, probably for mystical reasons, that he could see seven colors—red, orange, yellow, green, blue, indigo, and violet—but few people can see colors beyond the blue.

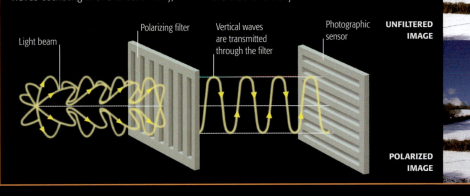

Third prism

A second prism refracts the light in the opposite sense, recombining the colors and forming a parallel beam of white light

A third prism splits the white light again, into the same colors —demonstrating that it is the same white light as hit the first prism

Colors are projected onto a screen

Screen

Second prism

"In the beginning of the year 1666… I procured me a **triangular glass–prisme**, to try therewith the **celebrated phaenomena of colors**."

ISAAC NEWTON, LETTER TO THE ROYAL SOCIETY, FEBRUARY 6, 1672

AFTER »

Intricate optics—lenses and mirrors—are crucial components within many modern scientific instruments. Some optics are used with infrared and ultraviolet light.

### VISIBLE AND INVISIBLE LIGHT
The light we see is only a small fraction of the **electromagnetic spectrum 258–59 »**, which ranges from gamma rays to radio waves. Most of these waves are invisible to the human eye.

### SPECTROSCOPY
A crucial development of the splitting of white light into its constituent colors was the development of **spectroscopy 259 »**, which has become a major analytical tool. Scientists can identify many elements and compounds by looking at the **unique spectrum** of light they emit or absorb.

### HOLOGRAPHY
In 1947 Dennis Gabor invented a way to make **pictures appear three-dimensional**. However, **holography** did not become a practical possibility until some ten years later when the laser had been invented.

### APOLLO 11
The first astronauts to visit the Moon left a special mirror on the **lunar surface**: the "laser ranging retroreflector array." This has allowed scientists to discover that the **Moon is receding** from the Earth by $1\frac{1}{2}$ in (4 cm) each year.

SEE ALSO »
*pp.258–59* ELECTROMAGNETIC SPECTRUM
*pp. 300–01* EINSTEIN'S EQUATION
*pp. 302–03* THEORIES OF RELATIVITY

## « BEFORE

Until 1577 comets were a complete mystery. They were regarded with fear and their occurrence was rare and startling.

### BAYEUX TAPESTRY EVIDENCE

An embroidered comet hovers above King Harold, noting **Comet Halley's 1066 appearance**, and is included as a symbol of death and disaster.

### ENSISHEIM METEORITE

On November 7, 1492 **a meteorite landed near the church of Ensisheim**, Alsace, France. Some of the meteorite was taken by locals, but a large chunk of it is still exhibited in the village.

« SEE ALSO

*pp.76–79* PLANETARY MOTION

**Comet Hale-Bopp**
This comet was seen in 1997. Its tail was 60 million miles (100 million km) long,

# Comets and Meteors

**Huge numbers of comets were formed at the dawn of the solar system some 4.6 billion years ago. Comets are largely made up of dusty rock and ice. Meteoroids are smaller fragments of rock in space. When they enter our atmosphere they become meteors or "shooting stars".**

Comets are found mainly in the outer regions of the Solar System, beyond Neptune's orbit. At the center of each one is an icy, dusty lump of kilometric proportions, the nucleus. As the comet travels across the solar system, inside the orbit of Mars, for example, solar radiation heats the surface of the nucleus enough to convert its water-ice into gas. The escaping gas begins to push away from the nucleus, until eventually the expanding gas and dust form a huge spherical cloud called a coma, and two tails (see right). This is what we see in the night sky. The comet's tails extend away from the coma, in the opposite direction from the Sun. They can extend for many millions of kilometers into space.

## A comet's life

A comet nucleus can range in size up to 25 miles across (40 km). Halley, for example, has a nucleus that is around 9 miles (15 km) long and 5 miles (8 km) wide. Halley takes about 76 years to travel around its medium-sized orbit

and loses about 6 ft (2 m) from its surface every time it passes close to the Sun. This is how comets die: Halley will last around another 3,000 orbits.

## Orbiting bodies

The major breakthrough in our understanding of comets came in 1687 when Isaac Newton showed how to calculate the orbit of a comet using three positional sightings taken a few weeks apart. He used the Great Comet of 1680 as an example. Unfortunately he had to assume that the comet had an open, parabolic orbit; so it seemingly came from infinity

## The comet's tail

As a comet nears the Sun, it develops two tails, which extend away from the Sun as the comet orbits. The curved tail is formed of dust, pushed away by solar radiation. The straight tail consists of ionized gas that has been blown away from the coma by the solar wind.

and returned there. Newton passed his accumulated positional data on to his friend Edmond Halley, who then started to calculate cometary orbits. By June 1696 Halley realized that the comets that had been seen in 1607 and 1682 had very similar orbits, and were

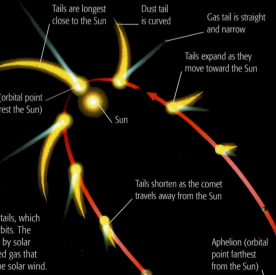

Tails are longest close to the Sun

Dust tail is curved

Gas tail is straight and narrow

Tails expand as they move toward the Sun

Perihelion (orbital point nearest the Sun)

Sun

Tails shorten as the comet travels away from the Sun

Aphelion (orbital point farthest from the Sun)

Naked nucleus

ASTRONOMER (1656–1742)

## EDMOND HALLEY

This English astronomer, mathematician, and geophysicist proved that some comets were permanent members of the solar system. As a Royal Navy sea captain Halley plotted magnetic deviation in the north and south Atlantic, and showed how the transits of Mercury and Venus could lead to an assessment of the Earth–Sun distance. He was at the heart of English Renaissance science: he became Professor of Geometry at Oxford, was the Astronomer Royal, and helped to publish Newton's *Principia*.

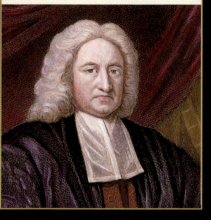

### Shooting stars
About 40 meteors, or shooting stars, could be seen every minute at the height of the Great Leonid Storm on 18 November 2001. During this yearly event, tiny dust particles from Tempel-Tuttle Comet hit Earth's upper atmosphere at 55 miles (72 km) per second.

most likely the same celestial body, returning to the Sun every 76 years. This made them part of the Sun's family. Halley predicted that this same comet would return again in 1758, and when it did, it was named after him.

## Two types of comet
By the mid-19th century two types of comets were recognized. One group travels in the plane of the planets and in the same direction, going around the Sun every five to seven years. The second group has random orbital inclinations—the comets can come in towards the Sun from any direction—and some are retrograde; they go backwards relative to the planets. These have much longer periods (the lengths of time it takes them to orbit the Sun).

Some long-period comets take up to 10 million years to complete an orbit, traveling far out into the freezing Oort cloud, which extends halfway between the Sun and the nearest star.

Bright comets visible to the naked eye are seen every 20–25 years, and for a few months only, while they are within the inner solar system. Around 180 short-period comets have been found,

### Meteorite crater
Meteorites are chunks of rock or metal—fragments of asteroid collisions—large enough to survive plunging through the atmosphere. Barringer Crater in Arizona, was formed when a massive iron–nickel meteorite hit Earth's surface at high speed some 50,000 years ago.

while each year about 20 long-period comets are discovered. Huge numbers of faint comets pass without being seen.

## Meteor showers
Large particles, up to $^2/_5$ in (1cm) in diameter, leave a comet nucleus at low speeds and then overtake or fall behind the comet depending on the direction of their emission.

### Earth-bound meteoroids
Incoming meteoroids produce bright fireballs in the earth's atmosphere. The surface of this meteorite remnant melted and boiled away.

After around 20 cometary orbits, the particles form a complete meteoroid stream along the comet's orbit. When the earth passes through one of these streams, the night skies are full of meteors all coming from a specific region of the sky, called the radiant. It may take the earth hours or weeks to pass through one stream.

A meteor forms when a dust particle hits our atmosphere, and its surface temperature rises to boiling point. Atoms from the boiling meteoroid collide with air molecules and ionize them (convert them from molecules to ions by the removal of electrons). The excited atoms then decay by emitting light, and the ionized electrons scatter radar waves. After traveling tens of kilometers in about one second, the meteoroid has boiled away completely, leaving a luminous train which quickly decays.

AFTER »

Four comets have been visited by spacecraft in the 20th century. The Rosetta spaceprobe was launched in 2004 to begin a 10-year mission that includes landing on a comet.

## COMET NUCLEUS
In 1951 American astronomer **Fred Whipple** predicted that a **comet's nucleus** would turn out to look like a "dirty snowball." Pictures taken by the **European Space Agency's Giotto spaceprobe** of Halley in 1986 proved him right.

## GIOTTO IMAGES
In 1986 the Giotto camera showed that the nucleus of Comet Halley is shaped like a potato; it is about **9 miles (15 km) long and 5 miles (8 km)** wide. Approximately 10 percent of the surface was actively emitting gas and dust.

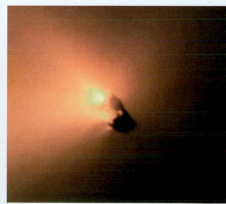

**NUCLEUS (CORE) OF COMET HALLEY**

**‹‹ BEFORE**

Ancient people took their time from the Sun. The day was the basic unit, and the period of daylight was divided into hours.

### SUNDIALS

A **sundial** is a simple instrument for measuring time. A stick—**the gnomon**—in the center casts a shadow on the face, and **the position of the shadow** indicates the hour. Sundials are accurate, but of course are of no use at night or on cloudy days.

**SUNDIAL**

### WATER CLOCKS

The earliest **water clocks** were probably made 4,000 years ago. The **klepshydra** ("water thief") used by the ancient Greeks measured a fixed amount of time **from water flowing out of a hole** in the bottom of a jar, and was improved by **Ktesibios** in the 3rd century BCE. Ancient Indians used a **ghati**—a bowl with a hole in the bottom that sank in a bigger bowl of water in 24 minutes (their "hour"). The **conical water clock** shown here (left) is a replica of an ancient Egyptian one. It is marked with **10 columns of 12 indentations on the inside** to indicate the time as water seeped from the small hole in the base. The outside is decorated with reliefs of astronomical subjects. It is similar to water clocks described in 1206 by the Persian inventor **Aljazari**, who also made more elaborate ones.

**WATER CLOCK**

### CANDLES

**Aljazari** also made **candle clocks**; one with elaborate caps and counterweights, another with a dial. More often candle clocks were just **candles with the hours marked** on them. They were usually shielded from drafts so that they burned at a constant rate.

**‹‹ SEE ALSO**
*pp.26–30* THE FIRST ASTRONOMERS

# Measuring Time

Early peoples measured time in days and noticed that the full Moon appeared every 29 or 30 days, and that winter arrived every 12 or 13 full moons. Measuring hours and minutes gradually developed in the ancient world and the Middle Ages, and today's atomic clocks are accurate to one second in several million years.

There are many "clocks" in nature, such as the daily rotation of Earth and the yearly orbiting of Earth around the Sun. The division of each day into hours, minutes, and seconds, however, is an arbitrary, human one.

### Mechanical clocks

The earliest known mechanical clock was the striking clock in Damascus, Syria, built in 1154. This clock, like many of its successors, had no time display; it only sounded bells to call people to prayer. Clocks like this one were known as turret clocks or tower

**Early pendulum clock with anchor escapement**
Made in London by William Clement—originally for King's College, Cambridge—this clock was in use from 1671. It was one of the first pendulum clocks to have an anchor escapement, like the one in the diagram (right).

clocks, and were controlled by a verge-and-foliot escapement—or speed regulator—a verge being a shaft, and a foliot a weighted iron crossbar. In Europe, Verge-and-foliot escapements were first described in France, by Villard de Honnecourt, in 1237.

Later came large tower clocks with faces and hands, so that people could tell the time. But with or without hands, verge-and-foliot escapement clocks probably gained or lost at least 15 minutes a day, mainly because of fluctuations in air temperature. When it was hot the foliot expanded and lengthened, so the clock lost time. When it was cold it contracted and shortened, so the clock gained time. Then in the late 16th century an Italian medical student named Galileo

**Galileo's pendulum inspiration**
Galileo used his pulse to time the great lamp that swings in Pisa Cathedral, and realized it always takes the same time for each swing, giving him the idea for a pendulum clock.

**Anchor escapement**
The anchor escapement allows one cog of the escape wheel to pass with each swing of the pendulum.

**Escape wheel**

**Drum**
As the drum turns it turns the mainwheel, the first in a series of gears.

**Pendulum**
The pendulum swings back and forth, taking one second to swing from side to side.

**Weight**
Its rate of fall controlled by the anchor escapement, the weight takes eight days to unwind the cord from the drum, at which point the clock stops and the weight needs to be wound up again with a key.

(see pp.82–83) noted the regularity of the swing of a pendulum and suggested it would make a better timekeeper for a clock than a foliot. Toward the end of his life he did indeed design a clock with a pendulum, but by then he was blind, so he never built it. His son built one based on his design, but it did not work very well. The first successful

## How a longcase pendulum clock works

A longcase pendulum clock is powered by the fall of a weight on a cord wrapped around a drum. As the drum turns it drives gears that turn the hour, minute, and second hands. An anchor escapement connected to the gears and to a pendulum regulates the speed at which the weight falls and the drum turns. Setting the pendulum swinging starts the clock. A second weight connects to a chiming mechanism (not shown here).

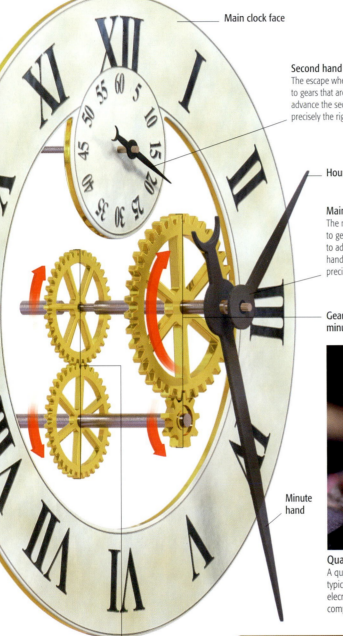

Main clock face

**Second hand**
The escape wheel connects to gears that are scaled to advance the second hand at precisely the right rate.

Hour hand

**Main hands**
The main wheel connects to gears that are scaled to advance the minute hand and hour hand at precisely the right rates.

Gears that drive the minute hand

Minute hand

Gears that drive the hour hand

mechanical clock with a pendulum was patented and built by the Dutch scientist Christiaan Huygens in 1657.

The period of swing of a pendulum depends on its length. This can be controlled with precision, so a pendulum clock can be made to keep excellent time. A typical longcase clock has a 39 in (1 m) pendulum. This swings from one side to the other in exactly one second. In the early 20th century standard US time was based on Riefler pendulum clocks, which were accurate to 10 milliseconds (one hundredth of a second) a day.

One early alternative to the pendulum was an improvement of the foliot—the balance wheel, a weighted wheel that rotates back and forth, controlled by the coiling and uncoiling of a balance spring. John Harrison used balance wheels in his four marine chronometers, the last of which was the most accurate timekeeper in the world at the time (see pp.134–35). This development allowed miniaturization and hence the first accurate watches.

## Modern timepieces

The search for ever more precise timekeeping led to the invention of the quartz clock by J.W. Horton and Warren Marrison at Bell Telephone Laboratories in New Jersey in 1927. Quartz watches followed in 1969. In a quartz clock or watch, a precision-made quartz crystal is set in a circuit connected to a battery. The crystal vibrates precisely 32,768 times a second, and every second the circuit generates a pulse. This powers either a digital display, or a tiny motor to drive gear wheels that turn second, minute, and hour hands.

A good quartz watch is accurate to 10 seconds a year. The first atomic clock, built by Louis Essen in 1955—based on the frequency of microwave radiation emitted by a transition between two energy states of caesium atoms—was accurate to one second in 300 years. Now, atomic clocks are millions of times more accurate even than that.

**Quartz crystal oscillator**
A quartz crystal oscillator in a clock or watch is typically Y-shaped. Quartz is "piezoelectric": electricity makes it vibrate, and conversely when compressed it generates a tiny electric current.

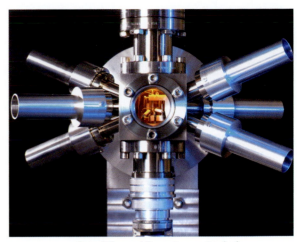

**One of the world's most accurate clocks**
This strontium optical atomic clock at the National Physical Laboratory at Teddington in London counts the oscillations of strontium ions between two energy levels stimulated by a laser beam. It is accurate to within one second in 650 million years.

### AFTER

In the mid-20th century most clocks and watches were analogue, with the time indicated by the positions of hands on a dial, but today many have digital displays.

#### DIGITAL TIME

Digital clocks and watches **display the time in numbers.** Today digital clocks are often incorporated into radios, stoves, and television sets. The **first ever digital watch** was made in the US **in 1972** by the Hamilton Watch Company, Lancaster, Pennsylvania; upon pressing a button on the side, faint light-emitting diodes displayed the time in glowing red.

**DIGITAL CLOCK**

#### ORBITING ATOMIC CLOCKS

The **Global Positioning System** (GPS) is an array of some 30 satellites. Each satellite **carries an atomic clock,** and continually transmits radio signals that include the time and its position in space. When a receiver on the ground picks up signals from four satellites, it can calculate its own position to within about 70 ft (20 m).

#### PULSAR PRECISION

Discovered by astronomy student **Jocelyn Bell** and her supervisor **Antony Hewish** in 1967, pulsars are **neutron stars** that emit pulses of radio waves with astonishing precision. Some pulsate almost as precisely as the most accurate atomic clocks. The period of one pulsar, for example, is 0.0148419520154668 seconds.

**SEE ALSO »**
*pp.134–35* HARRISON'S CHRONOMETER
*pp.298–99* RADIATION AND RADIOACTIVITY
*pp.368–69* ARTIFICIAL SATELLITES

### IN PRACTICE

## RAILROAD TIME

The building of the first railroads in England had a profound effect on time coordination. Until the 1840s English people lived by local time. Then in 1841 the railway from London reached Bristol and it became clear that Bristol time was 11 minutes behind London time. This is because Bristol is farther west than London. This small difference caused great confusion—not least for the railroad timetable—and soon all England was using London time, or "railway time," as standard. Railway time was eventually adopted as standard time everywhere railroads were built. For example, Italy adopted Rome time as standard in 1866, and India adopted Madras time as standard in 1870.

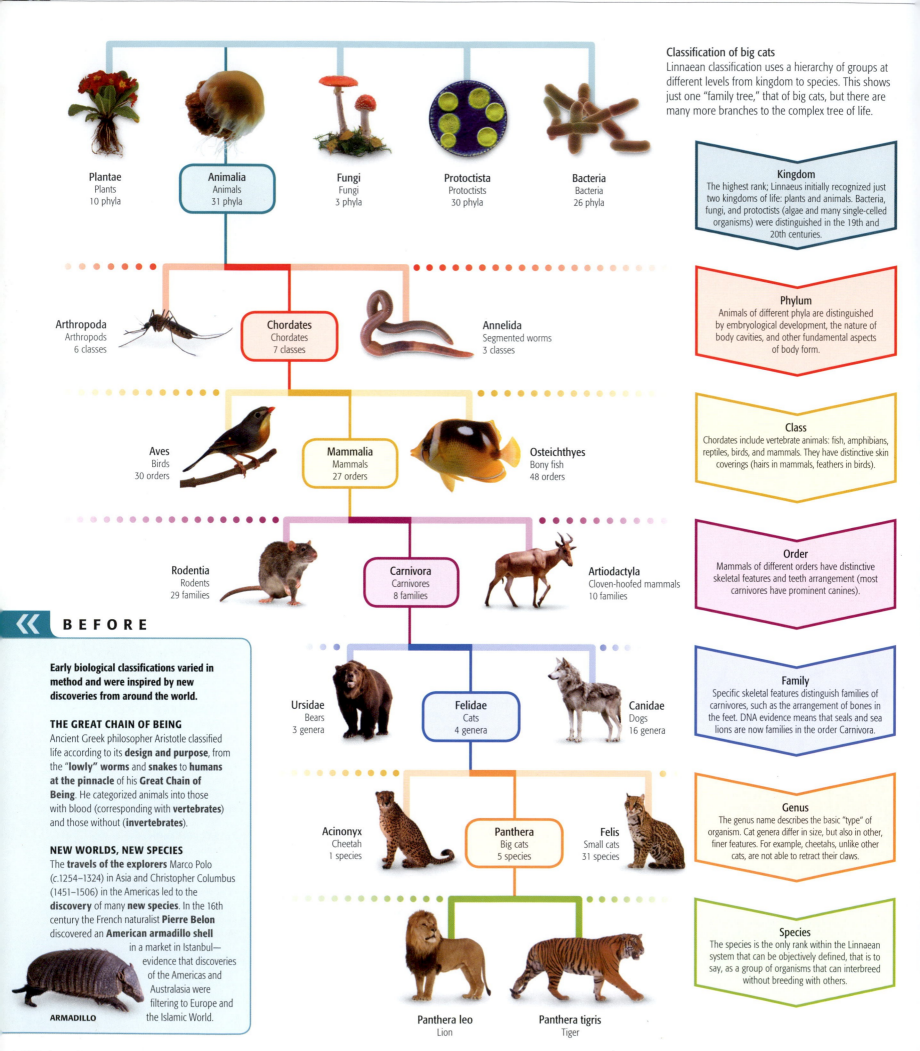

**Classification of big cats**
Linnaean classification uses a hierarchy of groups at different levels from kingdom to species. This shows just one "family tree," that of big cats, but there are many more branches to the complex tree of life.

**Plantae**
Plants
10 phyla

**Animalia**
Animals
31 phyla

**Fungi**
Fungi
3 phyla

**Protoctista**
Protoctists
30 phyla

**Bacteria**
Bacteria
26 phyla

**Kingdom**
The highest rank; Linnaeus initially recognized just two kingdoms of life: plants and animals. Bacteria, fungi, and protoctists (algae and many single-celled organisms) were distinguished in the 19th and 20th centuries.

**Arthropoda**
Arthropods
6 classes

**Chordates**
Chordates
7 classes

**Annelida**
Segmented worms
3 classes

**Phylum**
Animals of different phyla are distinguished by embryological development, the nature of body cavities, and other fundamental aspects of body form.

**Aves**
Birds
30 orders

**Mammalia**
Mammals
27 orders

**Osteichthyes**
Bony fish
48 orders

**Class**
Chordates include vertebrate animals: fish, amphibians, reptiles, birds, and mammals. They have distinctive skin coverings (hairs in mammals, feathers in birds).

**Rodentia**
Rodents
29 families

**Carnivora**
Carnivores
8 families

**Artiodactyla**
Cloven-hoofed mammals
10 families

**Order**
Mammals of different orders have distinctive skeletal features and teeth arrangement (most carnivores have prominent canines).

**Ursidae**
Bears
3 genera

**Felidae**
Cats
4 genera

**Canidae**
Dogs
16 genera

**Family**
Specific skeletal features distinguish families of carnivores, such as the arrangement of bones in the feet. DNA evidence means that seals and sea lions are now families in the order Carnivora.

**Acinonyx**
Cheetah
1 species

**Panthera**
Big cats
5 species

**Felis**
Small cats
31 species

**Genus**
The genus name describes the basic "type" of organism. Cat genera differ in size, but also in other, finer features. For example, cheetahs, unlike other cats, are not able to retract their claws.

**Panthera leo**
Lion

**Panthera tigris**
Tiger

**Species**
The species is the only rank within the Linnaean system that can be objectively defined, that is to say, as a group of organisms that can interbreed without breeding with others.

« **BEFORE**

**Early biological classifications varied in method and were inspired by new discoveries from around the world.**

**THE GREAT CHAIN OF BEING**
Ancient Greek philosopher Aristotle classified life according to its **design and purpose**, from the **"lowly" worms** and **snakes** to **humans at the pinnacle** of his **Great Chain of Being**. He categorized animals into those with blood (corresponding with **vertebrates**) and those without (**invertebrates**).

**NEW WORLDS, NEW SPECIES**
The **travels of the explorers** Marco Polo (c.1254–1324) in Asia and Christopher Columbus (1451–1506) in the Americas led to the **discovery** of many **new species**. In the 16th century the French naturalist **Pierre Belon** discovered an **American armadillo shell** in a market in Istanbul—evidence that discoveries of the Americas and Australasia were filtering to Europe and the Islamic World.

**ARMADILLO**

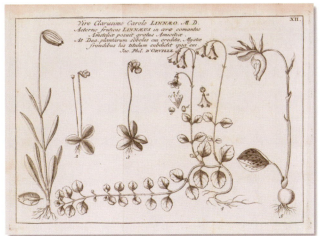

During the Renaissance scientists started to look carefully at species classification (taxonomy), and experimented with different aspects of structure in order to set out a coherent and consistent grouping system.

### Origins of scientific classification
Italian physician and botanist Andrea Caesalpino was the first to deviate from tradition, in the 16th century, by classifying plants not according to medicinal use, but by the structure of their fruit and seeds. Later, in the late 17th century, German physician and botanist Augustus Rivinus classified plants by flower structure. Both

**Naming plants**
*Linnaea borealis,* the twinflower on the left, was Carl Linnaeus's favorite plant and is named after him. Like all other plants, its two-part name indicates its genus and species in a system devised by Linnaeus.

# Classification of Species

**There are around a million and a half species of living things known to scientists, and undoubtedly many more awaiting discovery. The classification of organisms emerged from a desire to find order in the world and developed into sophisticated methods of reconstructing the evolutionary relationships of life on Earth.**

Caesalpino and Rivinus had seen that the reproductive structures of plants were more consistently useful in classification than whether they produced trunks and grew into trees. In England, naturalist John Ray (1627–1705) classified both plants and animals, no longer accepting the wisdom of ancient scholars, but drawing conclusions from first-hand observations.

### Improving the names
Although classification systems were improving, botany and zoology were still burdened with cumbersome names. New discoveries meant that these names, designed to be unique and descriptive, kept getting longer. Then, in 1735, a young botanist from

Sweden, Carl Linnaeus, published a pamphlet called *Systema Naturae*—the "System of Nature," in which he proposed a hierarchical system for classifying the natural world (see opposite). In 1753 he produced a companion work just on plants, *Species Plantarum*, in which his Latin descriptions of species were supplemented with single-word references. From this emerged the binomial system of biological names: the first part denoting the genus and the second part denoting the species, for example: *Panthera tigris* (see opposite). Linnaeus had made plant nomenclature consistent, and in the 10th edition of *Systema Naturae* in 1758 he applied the same

system for animals. His 1753 and 1758 publications mark the beginning of recognized scientific names of plants and animals respectively.

### Revealing patterns of evolution
With the rise of evolutionary thought in the 19th century, classification became a search for genealogical relationships. Some characteristics can evolve independently in groups that are not closely related, such as the wings of birds and insects. Scientists needed to classify on the basis of homologous characteristics: features that may be superficially different, but were so similar internally or developmentally that it was likely that they had evolved from a common ancestor. In this

respect, the wing of a bird has more in common with the hand of a mammal than with the wing of an insect. In 1950 German biologist Willi Hennig used this principle to devise a new method of classification—cladistics. Today cladistics is the dominating principle of taxonomy and may use evidence from homologous sections of DNA to map evolutionary relationships.

**Cladistic classification**
According to cladistic methods (based on ancestral similarities), birds share characteristics with reptiles. The diagram shows that each new characteristic relates to a point of divergence from the group to the left. Here, the birds diverge first, followed by the bipedal predatory dinosaurs, and so on.

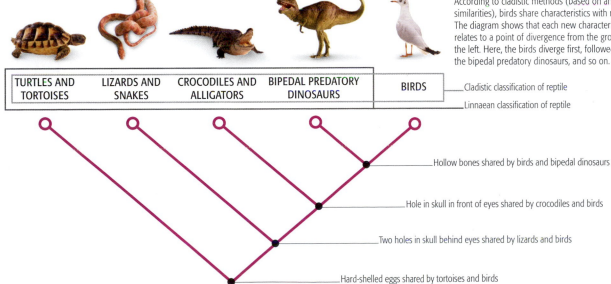

| TURTLES AND TORTOISES | LIZARDS AND SNAKES | CROCODILES AND ALLIGATORS | BIPEDAL PREDATORY DINOSAURS | BIRDS |
|---|---|---|---|---|

Cladistic classification of reptile
Linnaean classification of reptile

Hollow bones shared by birds and bipedal dinosaurs

Hole in skull in front of eyes shared by crocodiles and birds

Two holes in skull behind eyes shared by lizards and birds

Hard-shelled eggs shared by tortoises and birds

123

# 3

# THE INDUSTRIAL REVOLUTION

## 1700–1890

As the Industrial Revolution began to transform the world with factories and urban sprawl, so science truly came of age. It was no longer just the intellectual pastime of a few gentlemen "natural philosophers" but a career pursued by professionals in academic and scientific institutions.

# THE INDUSTRIAL REVOLUTION

## 1700–1890

### 1700

**1701**
Giacomo Pylarini gives the first smallpox inoculations in Europe.

**1709**
Daniel Gabriel Fahrenheit invents the alcohol thermometer.

**1712**
Thomas Newcomen builds the first atmospheric steam engine, to pump water out of mines.

**1717**
Sébastien Vaillant proposes that plants reproduce sexually and have male and female parts.

▼ Newcomen's engine

**1704**
Isaac Newton publishes *Opticks*, exploring theories on light. Final volume of John Ray's *History of Plants* published.

**1714**
Bartolomeo Eustachi's *Tabulae Anatomicae* published posthumously, giving an accurate description of the human nerve map. Fahrenheit invents the mercury thermometer.

▲ Isaac Newton

**1705**
Edmond Halley predicts that the comet of 1682 will reappear in 1758. The comet is later named after him.

**1715**
During a solar eclipse, Edmond Halley predicts the totality path to within less than half a mile (0.8 km).

▼ Comet in night sky

▼ London solar eclipse

**1710**
John Ray's *History of Insects* published posthumously.

**1735**
John Harrison produces his first marine chronometer. Carl Linnaeus publishes *Systema Naturae*, proposing a system for classifying the natural world.

### 1725

**1727**
Stephen Hales demonstrates that plants transpire in his essay *Vegetable staticks*.

**1742**
Anders Celsius develops the centigrade (Celsius) temperature scale.

**1744**
Pierre-Louis de Maupertuis proposes the principle of least action.

**1746**
Leonhard Euler describes the refraction of light mathematically.

Stephen Hales ▶

**c.1730**
Charles du Fay discovers that there are two types of electricity (positive and negative charges).

**c.1740**
Benjamin Franklin starts to keep regular weather records.

**1733**
Stephen Hales takes measurements of blood pressure.

**1741**
Stephen Hales presents the first ventilator to the Royal Society, UK.

**1747**
James Lind discovers that citrus fruits prevent scurvy.

◀ James Lind

**1745**
Pieter van Musschenbroek invents the Leyden jar, in which a static electric charge can be stored. Pierre Maupertuis states that offspring are made from components derived from their parents' bodies.

▲ John Harrison's first chronometer

**1749**
The Comte de Buffon publishes the first of 44 volumes of *Histoire Naturelle*, covering the entirety of natural history.

◀ Leyden jar

The rise of industry spurred scientific discovery both directly and indirectly, as scientists explored the nature of heat and energy, and of chemicals and their processes, leading to the discovery of the laws of thermodynamics, the atomic theory, and the periodic table. The power promised by science became intoxicating, and no discovery was more exciting than that of electricity and its links with magnetism. Yet in some ways the most profoundly disturbing discoveries were made, as natural historians left the factories and cities behind to explore the natural world. They revealed the shocking truth that we humans are just recent arrivals in an ancient world that has seen many creatures, including dinosaurs, come and go.

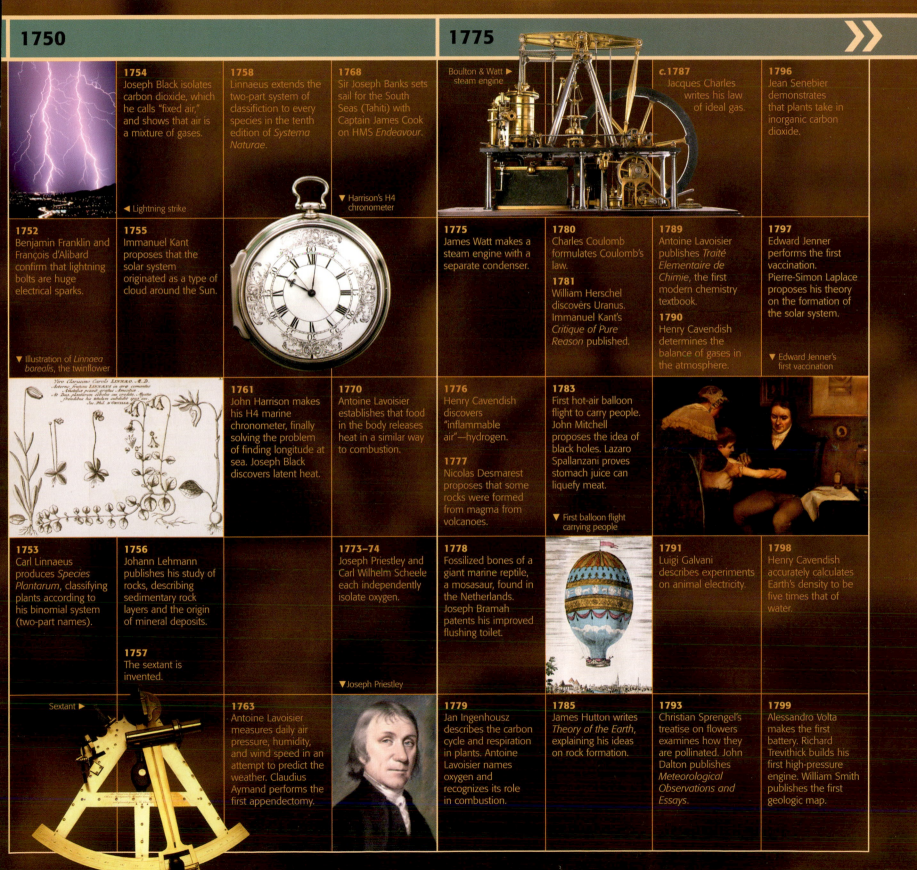

## 1750

**1754**
Joseph Black isolates carbon dioxide, which he calls "fixed air," and shows that air is a mixture of gases.

**1758**
Linnaeus extends the two-part system of classification to every species in the tenth edition of *Systema Naturae*.

**1768**
Sir Joseph Banks sets sail for the South Seas (Tahiti) with Captain James Cook on HMS *Endeavour*.

◀ Lightning strike

▼ Harrison's H4 chronometer

**1752**
Benjamin Franklin and François d'Alibard confirm that lightning bolts are huge electrical sparks.

**1755**
Immanuel Kant proposes that the solar system originated as a type of cloud around the Sun.

**1775**
James Watt makes a steam engine with a separate condenser.

**1780**
Charles Coulomb formulates Coulomb's law.

**1781**
William Herschel discovers Uranus. Immanuel Kant's *Critique of Pure Reason* published.

**1789**
Antoine Lavoisier publishes *Traité Elementaire de Chimie*, the first modern chemistry textbook.

**1790**
Henry Cavendish determines the balance of gases in the atmosphere.

**1797**
Edward Jenner performs the first vaccination. Pierre-Simon Laplace proposes his theory on the formation of the solar system.

▼ Illustration of *Linnaea borealis*, the twinflower

▼ Edward Jenner's first vaccination

**1761**
John Harrison makes his H4 marine chronometer, finally solving the problem of finding longitude at sea. Joseph Black discovers latent heat.

**1770**
Antoine Lavoisier establishes that food in the body releases heat in a similar way to combustion.

**1776**
Henry Cavendish discovers "inflammable air"—hydrogen.

**1777**
Nicolas Desmarest proposes that some rocks were formed from magma from volcanoes.

**1783**
First hot-air balloon flight to carry people. John Mitchell proposes the idea of black holes. Lazaro Spallanzani proves stomach juice can liquefy meat.

▼ First balloon flight carrying people

**1753**
Carl Linnaeus produces *Species Plantarum*, classifying plants according to his binomial system (two-part names).

**1756**
Johann Lehmann publishes his study of rocks, describing sedimentary rock layers and the origin of mineral deposits.

**1757**
The sextant is invented.

**1773–74**
Joseph Priestley and Carl Wilhelm Scheele each independently isolate oxygen.

**1778**
Fossilized bones of a giant marine reptile, a mosasaur, found in the Netherlands. Joseph Bramah patents his improved flushing toilet.

**1791**
Luigi Galvani describes experiments on animal electricity.

**1798**
Henry Cavendish accurately calculates Earth's density to be five times that of water.

◀ Boulton & Watt steam engine

**c.1787**
Jacques Charles writes his law of ideal gas.

**1796**
Jean Senebier demonstrates that plants take in inorganic carbon dioxide.

Sextant ▶

▼ Joseph Priestley

**1763**
Antoine Lavoisier measures daily air pressure, humidity, and wind speed in an attempt to predict the weather. Claudius Aymand performs the first appendectomy.

**1779**
Jan Ingenhousz describes the carbon cycle and respiration in plants. Antoine Lavoisier names oxygen and recognizes its role in combustion.

**1785**
James Hutton writes *Theory of the Earth*, explaining his ideas on rock formation.

**1793**
Christian Sprengel's treatise on flowers examines how they are pollinated. John Dalton publishes *Meteorological Observations and Essays*.

**1799**
Alessandro Volta makes the first battery. Richard Trevithick builds his first high-pressure engine. William Smith publishes the first geologic map.

## 1800

## 1825

**1800**
William Herschel discovers infrared light. Henry Maudslay invents a screw-cutting lathe.

**1801**
Guiseppe Piazzi finds the first asteroid.

**1802**
Joseph Gay-Lussac formulates a law stating that the pressure of gas is proportional to the absolute temperature.

Joseph Gay-Lussac ▶

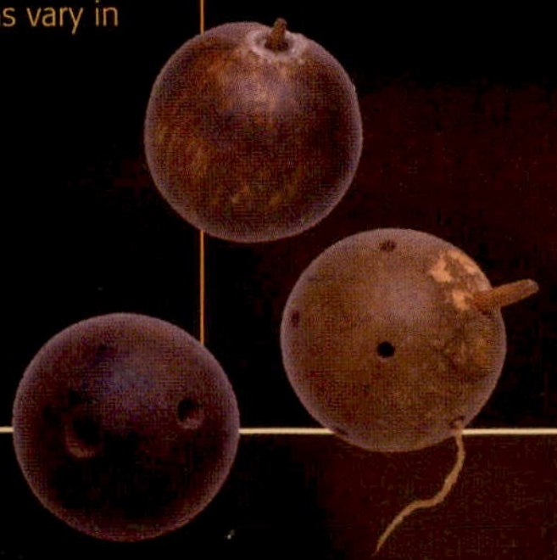

**1804**
Nicolas-Théodore de Saussure confirms photosynthesis as a chemical process.

▼ Section of leaf showing internal structure

**1803**
John Dalton proposes that atoms vary in weight.

▼ James Dalton's models of atoms

**1808**
John Dalton publishes *A New System of Chemical Philosophy*, suggesting that all elements are made of indivisible atoms.

Dalton's symbols ▶

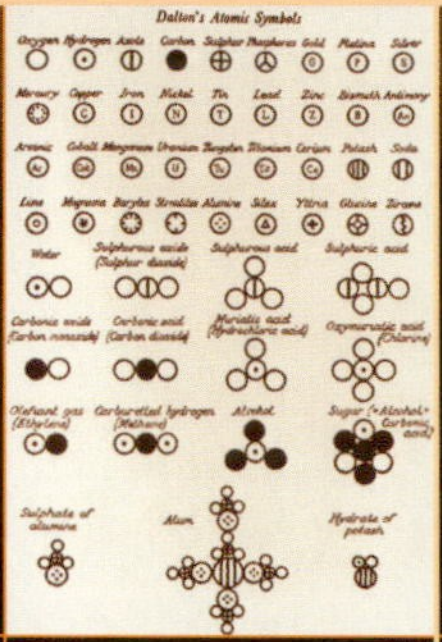

**1810**
Humphrey Davy invents the arc lamp and isolates and names chlorine.

**1811**
Amedeo Avogardo states that equal volumes of different gases contain the same number of molecules.

**1820**
Christian Ørsted discovers that electric currents produce magnetic fields. Jean Fourier proposes a formula for estimating the age of Earth.

**1821**
Michael Faraday builds the first electric motor.

Faraday's ▶ original experiment

**1822**
Iguanodon fossils discovered, strengthening the case for extinction. Charles Babbage proposes the first design for his Difference Engine—predecessor of the computer.

▼ Difference Engine

**1824**
Joseph Fourier discovers what is later known as the greenhouse effect. Nicolas Carnot publishes *Reflections on the Motive Power of Fire*, in which he explains how heat can make things move.

**1825**
William Sturgeon invents the electromagnet. Jöns Berzelius publishes a list of the atomic weights of 43 elements.

**1827**
William Prout classifies nutrients in food categories. Robert Brown defines Brownian motion.

**1828**
Christian Ehrenberg observes bacteria.

**1830**
Charles Lyell publishes the first of three volumes of *Principles of Geology*, in which he argues that Earth must be several hundred million years old.

**1831**
Robert Brown identifies the nucleus of a cell. Michael Faraday discovers electromagnetic induction. HMS *Beagle* sails with William Fitzroy and Charles Darwin.

The Coriolis effect ▼

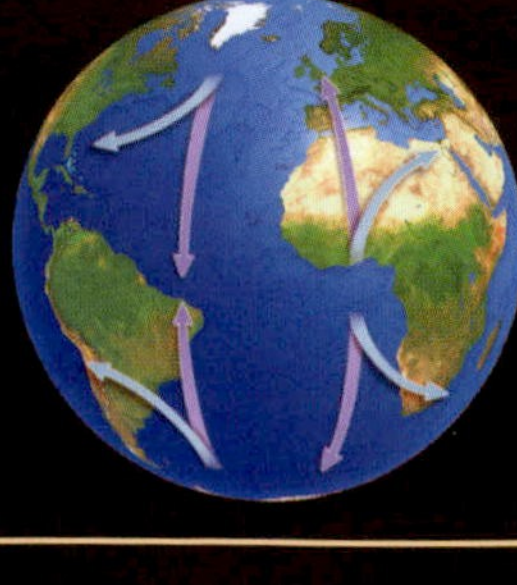

**1833**
Karl Gauss and Wilhelm Weber build the first operational telegraph system. Daguerre produces the daguerrotype, an early form of photograph.

▲ Early telegraph machine

**1835**
Gustave Coriolis describes the movement of winds around the globe, now known as the Coriolis effect. Robert Anderson invents the first electrically powered carriage.

**1837**
Louis Agassiz proposes that Europe went through an ice age. Samuel Morse and Alfred Vail devise the Morse Code.

▲ Anemometer

**1846**
The anemometer introduced. Ignaz Semmelweis tries to introduce antiseptic procedures into hospitals. Johann Galle discovers the planet Neptune.

**1839**
Linus Pauling publishes *The Nature of the Chemical Bond*, establishing chemical reactions.

▼ Neptune

**1840**
William Fox-Talbot develops the calotype, enabling multiple reproduction of the image—the photograph. The British government substitutes inoculation with vaccination.

**1847**
William Budd suggests that typhoid is transmitted through infected water. James Joule introduces the law of conservation of energy.

**1842**
Justus von Liebig discovers body heat is produced by chemical processes. Christian Doppler predicts the Doppler effect, based on sound waves.

**1845**
William Armstrong invents the hydraulic crane.

**1848**
William Thomson shows that there is an absolute zero of temperature. James Joule estimates the average speed at which gas molecules move.

"We must, however, acknowledge… that man, with all his **noble qualities**… still bears in his bodily frame the indelible stamp of his **lowly origin**."

CHARLES DARWIN, *The Descent of Man*, 1871

## 1850

## 1875

**c.1850**
Claude Bernard finds that the pancreas releases digestive juices. Rudolf Clausius proposes the conservation of energy as the first law of thermodynamics. Matthew Maury suggests there is a Mid-Atlantic Ridge.

**1857–58**
Friedrich Kekulé proposes his theory of the chemical structure of carbon atoms, the basis of organic chemistry.
**1859**
First edition of the *Origin of Species* published by Darwin.

**1864**
Louis Pasteur demonstrates that living airborne microbes make nutrient liquids ferment.

**1869**
Dimitri Mendeleev creates the periodic table. John Wesley Hyatt makes celluloid.
**1870**
Louis Pasteur and Robert Koch develop germ theory.

**1875**
William Crookes produces Crookes tubes. Francis Galton publishes the first newspaper weather map in *The Times*.

**1877**
Ernst Mach publishes a paper on the sound effects observed during the supersonic motion of a projectile. Thomas Edison produces the first phonograph.

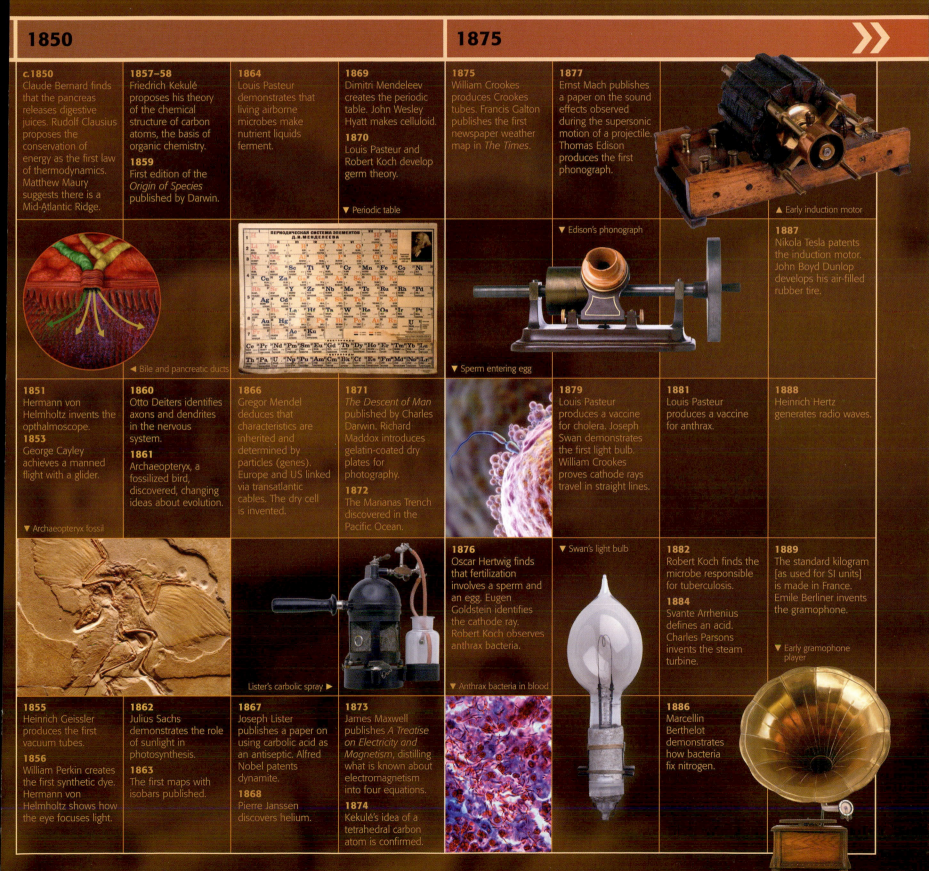

▲ Early induction motor

▼ Periodic table

▼ Edison's phonograph

**1887**
Nikola Tesla patents the induction motor. John Boyd Dunlop develops his air-filled rubber tire.

◀ Bile and pancreatic ducts

▼ Sperm entering egg

**1851**
Hermann von Helmholtz invents the opthalmoscope.
**1853**
George Cayley achieves a manned flight with a glider.

**1860**
Otto Deiters identifies axons and dendrites in the nervous system.
**1861**
Archaeopteryx, a fossilized bird, discovered, changing ideas about evolution.

**1866**
Gregor Mendel deduces that characteristics are inherited and determined by particles (genes). Europe and US linked via transatlantic cables. The dry cell is invented.

**1871**
*The Descent of Man* published by Charles Darwin. Richard Maddox introduces gelatin-coated dry plates for photography.
**1872**
The Marianas Trench discovered in the Pacific Ocean.

**1879**
Louis Pasteur produces a vaccine for cholera. Joseph Swan demonstrates the first light bulb. William Crookes proves cathode rays travel in straight lines.

**1881**
Louis Pasteur produces a vaccine for anthrax.

**1888**
Heinrich Hertz generates radio waves.

▼ Archaeopteryx fossil

**1876**
Oscar Hertwig finds that fertilization involves a sperm and an egg. Eugen Goldstein identifies the cathode ray. Robert Koch observes anthrax bacteria.

▼ Swan's light bulb

**1882**
Robert Koch finds the microbe responsible for tuberculosis.
**1884**
Svante Arrhenius defines an acid. Charles Parsons invents the steam turbine.

**1889**
The standard kilogram [as used for SI units] is made in France. Emile Berliner invents the gramophone.

▼ Early gramophone player

Lister's carbolic spray ▶

▼ Anthrax bacteria in blood

**1855**
Heinrich Geissler produces the first vacuum tubes.
**1856**
William Perkin creates the first synthetic dye. Hermann von Helmholtz shows how the eye focuses light.

**1862**
Julius Sachs demonstrates the role of sunlight in photosynthesis.
**1863**
The first maps with isobars published.

**1867**
Joseph Lister publishes a paper on using carbolic acid as an antiseptic. Alfred Nobel patents dynamite.
**1868**
Pierre Janssen discovers helium.

**1873**
James Maxwell publishes *A Treatise on Electricity and Magnetism*, distilling what is known about electromagnetism into four equations.
**1874**
Kekulé's idea of a tetrahedral carbon atom is confirmed.

**1886**
Marcellin Berthelot demonstrates how bacteria fix nitrogen.

# The Newcomen Engine

In 1712 Thomas Newcomen built an "atmospheric" engine to pump water up from the depths of a coal mine at Dudley in the West Midlands of England. This was the first time that reliable power had been made available when and where it was needed. The Newcomen engine was arguably the most important invention of all time—one that marked the beginning of the Industrial Revolution.

The demand for coal in England increased steadily from 1709, when Abraham Darby discovered how to smelt iron using coke rather than charcoal. By 1712 coal miners were digging deeper and deeper to supply the demand, and flooding of coal mines was an ever-increasing problem. Manual and horse-powered pumping was no longer enough. Then Newcomen, an ironmonger from Dartmouth in Devon—today he would be called an engineer—built the world's first successful engine to pump out the water.

The engine was a huge lumbering beast. Coal was burned to boil water and make steam, which was fed into a vertical cylinder; this forced a piston inside to rise. When the piston reached the top of the cylinder, cold water was squirted into the cylinder to condense the steam and create a partial vacuum. The pressure of the atmosphere then pushed the piston down, pulling down in turn a huge beam above, and so pulling up pump rods at the other end of the beam. The whole enormous engine was protected from the weather by an engine house some 33 ft (10 m) high.

Because they were inefficient, Newcomen engines were used mainly in coal mines, where there was a plentiful supply of fuel. Each engine cost around £1,000 to build—a vast some of money in the early 18th century. Nevertheless the engines were so important that some 75 of them were built in the 17 years before Newcomen died, and by the end of the century the figure had risen to more than 1,000. They were exported to Sweden, Germany, Belgium, France, Spain, and even to America. They were rugged and reliable, and some of them lasted for more than 100 years.

The first Newcomen engine had a brass cylinder 21 in (53 cm) in diameter and 8 ft (2.4 m) long; it ran at about 12 strokes a minute and could lift 10 gallons (45 liters) of water from 150 ft (45 m) down with each stroke. Later versions had much larger cylinders of cast iron, and so were more powerful. Improvements were made to the basic engine in the middle of the 18th century, and in the late 1770s James Watt's improved version with a separate condenser (see pp.132–33) moved the Industrial Revolution into the next phase.

### Newcomen's historic engine
No Newcomen engine survives today, but the Black Country Living Museum at Dudley has a full-size working replica of the original 1712 engine. The replica is based on this early drawing, which shows all the main working parts, and is occasionally operated under steam for the benefit of the visiting public.

> "In the whole history of **technology** it would be difficult to find a greater single **advance** than this, nor one with a greater significance **for all humanity**."
>
> L.T.C. ROLT, FROM *"THOMAS NEWCOMEN: THE PREHISTORY OF STEAM,"* 1963

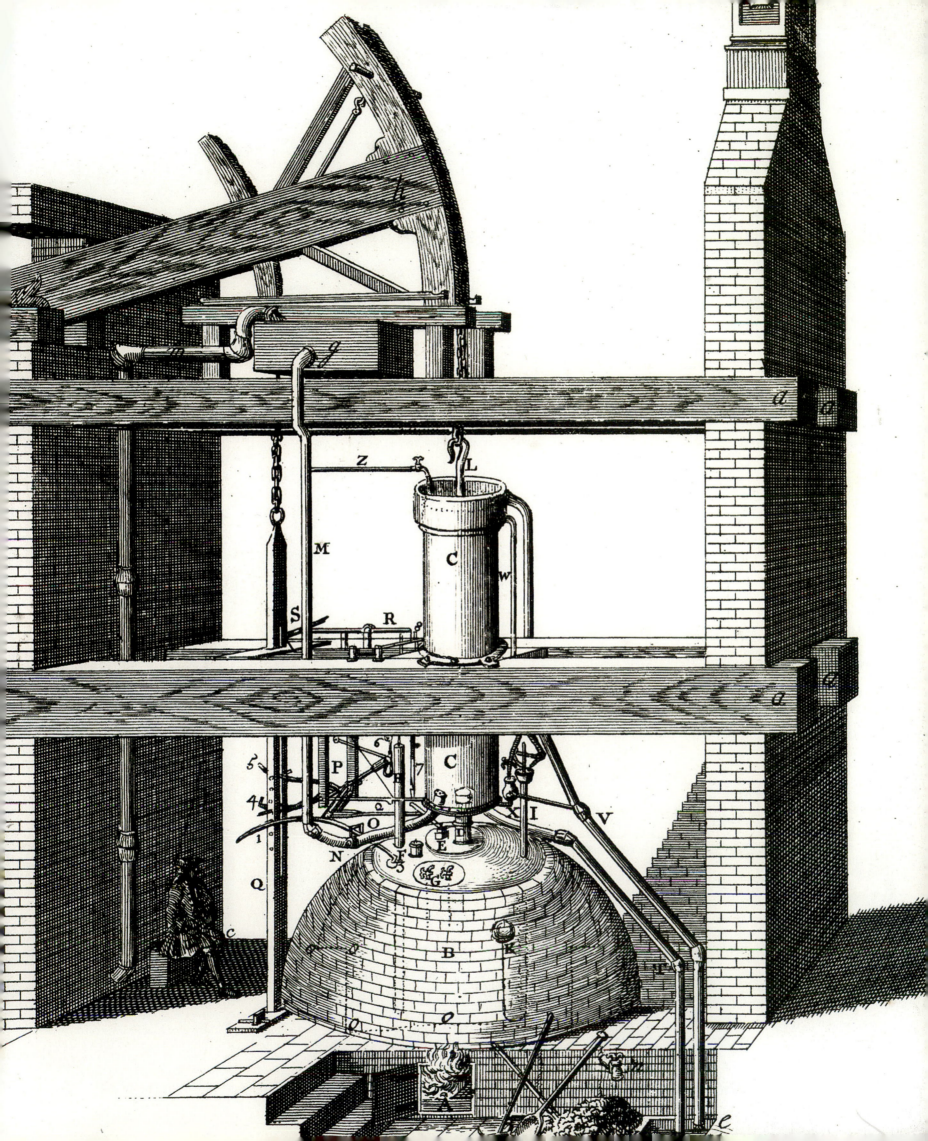

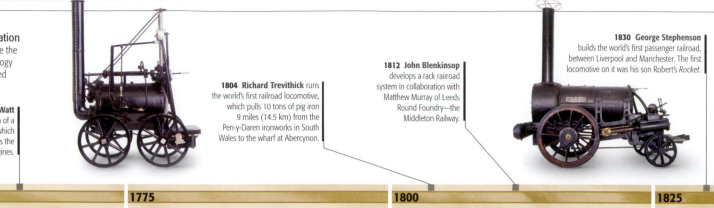

**Development of steam transportation**
James Watt and Richard Trevithick made the breakthroughs in steam engine technology that allowed a century of steam-powered transportation on both land and sea.

**1765 James Watt** conceives the idea of a separate condenser, which greatly improves the efficiency of steam engines.

**1804 Richard Trevithick** runs the world's first railroad locomotive, which pulls 10 tons of pig iron 9 miles (14.5 km) from the Pen-y-Daren ironworks in South Wales to the wharf at Abercynon.

**1812 John Blenkinsop** develops a rack railroad system in collaboration with Matthew Murray of Leeds Round Foundry—the Middleton Railway.

**1830 George Stephenson** builds the world's first passenger railroad, between Liverpool and Manchester. The first locomotive on it was his son Robert's *Rocket*.

| 1750 | 1775 | 1800 | 1825 |

# Steam Power to Steam Engine

**The "common engine," or atmosphere engine, invented by Thomas Newcomen in 1712 marked the start of the Industrial Revolution, and by the end of the 18th century it had been greatly improved by James Watt, whose more efficient double-acting steam engines were the first to power machinery. Watt's innovations paved the way for steam-powered locomotion.**

The discovery of the vacuum in the middle of the 17th century showed philosophers the immense power of atmospheric pressure, which pushes with a force of about 2.2 lb (1 kg) on every square centimeter of Earth's surface. One way to use this force was to create a vacuum by condensing steam, then allow the atmosphere to eliminate the vacuum. The Newcomen engine worked in this way, and for decades pumped water from coal mines, allowing miners to go deeper.

### Improving the efficiency
In 1764 a model Newcomen engine was brought for repair to a young Scottish engineer called James Watt (see p.149) at the University of Glasgow. After fixing it, he discovered that it needed loads of coal to run at all, because it wasted so much heat. Watt had heard about latent heat—the heat needed to turn water into steam—from his friend and mentor Joseph Black (see pp.148–49). He realized that the Newcomen engine was inefficient because for every stroke the entire cylinder had to be heated above boiling point to fill it with steam, then cooled below it for the steam to condense.

Watt pondered the problem for a few months, then one Sunday morning in May 1765 the answer came to him in a flash of inspiration as he was walking on Glasgow Green. He realized that it would be far better to keep the cylinder above boiling point all the time, and have a separate condenser.

In 1769 Watt went to London to acquire a patent, and when his wife died and his Scottish backer went bust he moved to Birmingham to go into partnership with Matthew Boulton. Boulton gave him money and skilled craftsmen as his assistants. Watt's engine was right at the edge of what was technically possible, but in 1775 he finally got it

**Lots of coal**
Early steam engines burned vast quantities of coal and so were initially used only at mines where coal was available as fuel.

**Steam engines in coal mining**
While the first steam engines were used to pump water out of deep coal mines, later ones were used to raise the mined coal itself up the deep shafts to the surface.

**The inspiration for Watt's big idea**
The idea of a separate condenser—Watt's great step forward, and why he became rich and famous—came to him after fixing a model Newcomen engine.

to work. From then until about 1800 Boulton & Watt steam engines were the best in the world.

Watt made a double-acting engine by letting in steam at both ends of the cylinder, instead of steam at the bottom and air at the top. This was a real steam engine, as opposed to an atmospheric engine; both up and down were power strokes (see opposite). Installed with a flywheel it could drive machinery in carpenters' shops, mines, cotton mills, and factories of every kind.

The next big step forward was the high-pressure steam engine, pioneered by fiery Cornishman Richard Trevithick

in 1799. This used the force from the steam to drive the piston. In 1801 Trevithick built a steam carriage, but it quickly came to grief after getting into a rut and turning over. Then in 1804 he successfully ran another engine on wheels along a track. The age of steam transportation had begun.

**1841 The railroad**
from London reaches Bristol, after which it is clear that local time differences and railroad timetables are not compatible, so London time, or railroad time, is adopted as standard.

**1883 Mark Twain**
publishes *Life on the Mississippi*, detailing his experiences as a paddle steamer pilot in the years before and after the American Civil War. Steamboats and steamships gradually replace sailing vessels throughout the 19th century.

**1897 The Stanley twins**
Francis and Freelan make the first Stanley Steamer, which goes on to become the world's most popular steam-powered car.

1850

1875

## AFTER

Internal combustion and electricity replaced steam for powering engines and machinery in the 20th century, but today we depend on steam turbines for most of our electricity.

### ELECTRIC MOTORS
In 1834 an American blacksmith named **Thomas Davenport** proved the **potential of electric motors** by using one to power his own tools.

### INTERNAL COMBUSTION
In 1885 Karl Benz made the **Benz Patent Motorwagen**, the first ever motor vehicle to be powered by an internal combustion engine.

### THE STEAM TURBINE
In 1884 **Charles Parsons** invented the **steam turbine**: a series of fans driven by super-high-pressure steam, like propellers in reverse. This is enormously powerful, and steam turbines now generate 75 percent of the world's electricity.

**SEE ALSO »**

pp.168–69 THE ELECTRIC MOTOR
pp.252–53 GENERATING ELECTRICITY
pp.254–55 THE INTERNAL COMBUSTION ENGINE

▽ **How a double-acting cylinder works**
Steam is exhausted to the separate condenser from the bottom of the cylinder, creating a partial vacuum inside. Steam from the boiler then enters the cylinder from the top end, allowing the piston to move down. The process is then reversed, so that both up and down are power strokes.

Steam inlet from boiler

Steam outlet to condenser

Piston moves down

Piston moves up

Steam outlet to condenser

Steam inlet from boiler

**1** Down stroke. Steam from the boiler enters the cylinder at the top, allowing the piston to move down, while steam from the previous stroke escapes to the separate condenser.

**2** Up stroke. Steam escapes from the top to the condenser, while steam from the boiler enters below, allowing the piston to rise.

**Boulton and Watt engine**
In this working model of a Boulton and Watt steam engine the double-acting cylinder is near the left side and the large flywheel on the right. The parallel motion above the cylinder was Watt's proudest invention.

**Parallel motion**
The parallel motion converts the vertical movement of the piston rod to the curved movement of the end of the beam.

**Beam**
The beam moves up and down in an arc.

**Governor**
The speed of the engine is controlled by a centrifugal governor, which cuts off the supply of steam if the engine goes too fast, and increases it if the engine goes too slow.

**Flywheel**
Cranks or sun-and-planet gears convert the up-and-down motion of the beam to turn the flywheel.

**Connecting rod**

**Piston**

**Valve gear**
The entry of steam into the cylinder is controlled by the valve gear.

**Double–acting cylinder**

**Crankshaft**

**Cistern**
Inside a cistern full of cold water are a condenser and a pump. The condenser cools the steam back into water, which is removed by the pump.

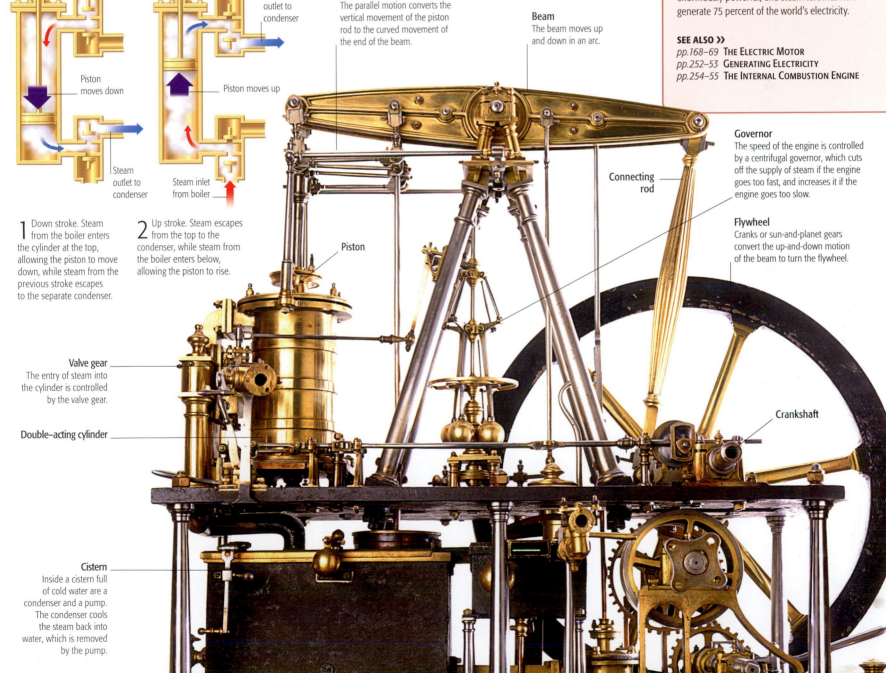

Longitude is a measurement of the position east or west of a fixed location, such as the home port of a vessel. Since the late 19th century this location has been along the prime meridian, at Greenwich, England. The key to finding longitude was deceptively simple. This is because a vessel's location on Earth's surface can be determined by the time at which it experiences local noon in relation to the time on a clock that has been set from noon at a fixed location.

In 1714 a prize of £20,000 was offered by the British parliament for a solution to the problem, and a "Board of Longitude" was established to sit in judgement on the ideas proposed.

Most of the suggestions offered to the Board were astronomical in nature, and all relied on precise observations that would require a steady deck, a fine telescope, and an experienced observer. But Yorkshire-born clockmaker John Harrison invented a mechanized alternative—a precise, robust, maritime timekeeping machine or "chronometer," which was capable of keeping track of the time in the vessel's home port.

In 1730 Harrison traveled to London and presented the Astronomer Royal, Sir Edmond Halley, with plans for his revolutionary marine chronometer. Halley referred him to the renowned clockmaker George Graham, who personally financed the construction of Harrison's first prototype, now known as "H1," over the next five years.

While pendulum timepieces were unreliable timekeepers at sea because the pendulum would not always swing accurately, H1's main technological advantage was that its spring-driven mechanism was unaffected by the pitch and roll of a ship's deck. The clock would only run for a day at a time, but Harrison's later timepieces incorporated an ingenious mechanism that allowed them to continue running while being wound. H1 caused a sensation when it was presented to members of the Royal Society, and proved its accuracy at sea on its voyage from England to Portugal and back. But despite his development of three further chronometers (named H2 to H4, see pp.136–37), bias against a mechanical solution meant that Harrison was never awarded the longitude prize.

## Harrison's first chronometer

When H1 was put to a sea trial onboard *HMS Centurion* in 1736 it lost only a few seconds during the voyage to Lisbon, Portugal. Decorated with eight cherubs and four crowns, the brass timepiece is almost 2 ft (0.6 m) tall, wide, and deep, and weighs 75 lb (34 kg). The brass balls on the top of the clock are part of the oscillating mechanism that replaced the pendulum.

**"One of the most exquisite movements ever made."**

WILLIAM HOGARTH, ON H1, "ANALYSIS OF BEAUTY," 1753

## « BEFORE

Through the ages, people have devised a variety of methods for finding their way at sea.

**STICK CHART**

### USING STICK CHARTS

Between 2300 BCE and 1500 CE, the **Polynesians** and **Micronesians** gradually **colonized a large part** of the **Pacific**, without any physical compass. Their methods included noting prevailing directions of waves and winds, and the use of **early types of maps** called **stick charts**.

### HIPPARCHUS' CONTRIBUTION

Though it had already been in use for decades, the system of **latitude** and **longitude** for locating a point on Earth's surface was formally described by the Greek astronomer Hipparchus in about 150 BCE **« 38–39**. It has continued, almost unchanged, to the present day.

### VIKING SUN COMPASSES

In the 8th and 9th centuries CE, the **Vikings** used **non-magnetic compasses**, based on the Sun's position, to provide crude **estimates of latitude** and the **direction of true north**. When they thought they were near land, they released ravens to find out the direction of the coast.

## « SEE ALSO

pp.26–29 THE FIRST ASTRONOMERS
pp.134–35 HARRISON'S CHRONOMETER

In order to navigate the seas with any degree of certainty, seafarers needed to ascertain not only what direction they were sailing in, but also where they were at any one time in terms of latitude and longitude (see pp.38–39).

### The magnetic compass

Until around 1100 CE the main way of navigating at sea was to stay in sight of land, or to observe the Sun and stars. The invention of the magnetic compass was a big step forward, since it allowed seafarers to orient themselves and steer a course even in cloudy weather or fog. The first definite reference to a magnetized device "for finding south"—a tiny iron fish in a water bowl—comes from China in about 1044. By 1120 this device had become a magnetized needle. These early Chinese compasses were "wet"—they consisted of needles floating in bowls of water. The true mariner's compass, using a pivoting needle in a dry box, was in use in Europe by c.1300.

By the end of the 15th century seafarers realized that compass readings require a degree of correction, because the directions in which a compass needle points (toward the magnetic poles) are not exactly the same as the directions to the true or geographic poles.

### Determining latitude and longitude

Though useful, the compass was not much help when determining position at sea. To find out where they were, and to create maps, navigators needed to work out their latitude and longitude. Latitude is easy to

**Mariner's compass**
In this 16th-century compass, the magnetized needle is attached to the underside of a pivoting circular card, which displays the compass points.

determine by measuring the angle of elevation of the Sun or a particular star in relation to the horizon (see opposite). To make such a measurement, a variety of instruments such as the astrolabe, cross-staff, and backstaff were used until the mid-18th century when these became extinct following the invention of the sextant (see opposite).

Between the 15th and 18th centuries, although voyagers could determine latitude, they had only a vague idea of where they were in the East–West direction. This was because longitude was much harder to measure. Compasses gave virtually no clue, and instruments such as quadrants (predecessors of sextants) were also of no use on their own, since the celestial position and paths of the Sun and stars appear exactly the same at different longitudes. To find their way back to home ports after voyages to the Americas, European seafarers relied on sailing north or south until they reached the same latitude as their home port, then sailing directly east.

# Navigating the Oceans

**Until instruments such as the magnetic compass came into use around 1300 CE, navigating the world's oceans was extremely difficult. Precise navigation only became possible several centuries later, following the invention of the sextant (an angle-measuring device) and an accurate seagoing clock.**

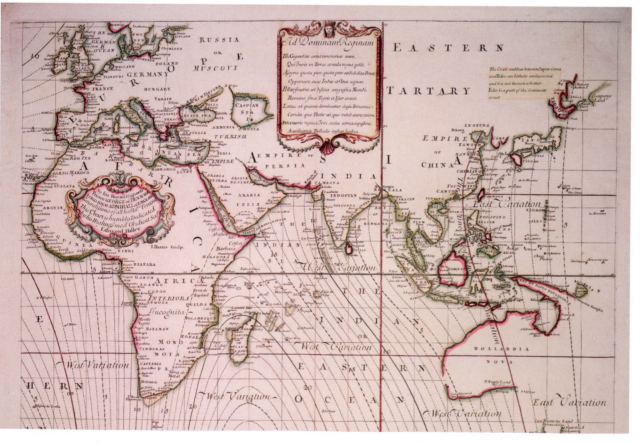

**Navigator's chart**
The lines on this chart, which dates from 1702, indicate the correction that had to be applied to compass readings to adjust for the difference between Earth's magnetic poles and the true (geographic) poles.

However, this method resulted in frequent shipwrecks in storms or fog. By the 18th century, the military and economic importance of solving the longitude problem was a major concern for some European governments.

### Solving the longitude problem

The scientists of the day already knew, in general terms, the solution. Ships' navigators needed a "universal clock," which would always tell them the exact time at a particular location, such as the navigator's home port. Wherever they went, they could then calculate their longitude by comparing the time on this universal clock to their local time (determinable using a sextant). Essentially, each hour of disparity would be equivalent to a 15 degrees' difference in longitude (see opposite). Two different types of clock were possible. The first would be a physical clock on board a ship, which would accurately keep track of the time at the

## Using a sextant

To measure the angle of the Sun above the horizon, the observer looks through the telescope and moves the sextant's arm until an image of the Sun lines up with the horizon. He or she then reads the angle of elevation of the Sun from the scale at the bottom. Measurements were made around local noon. At night, specific stars were used instead.

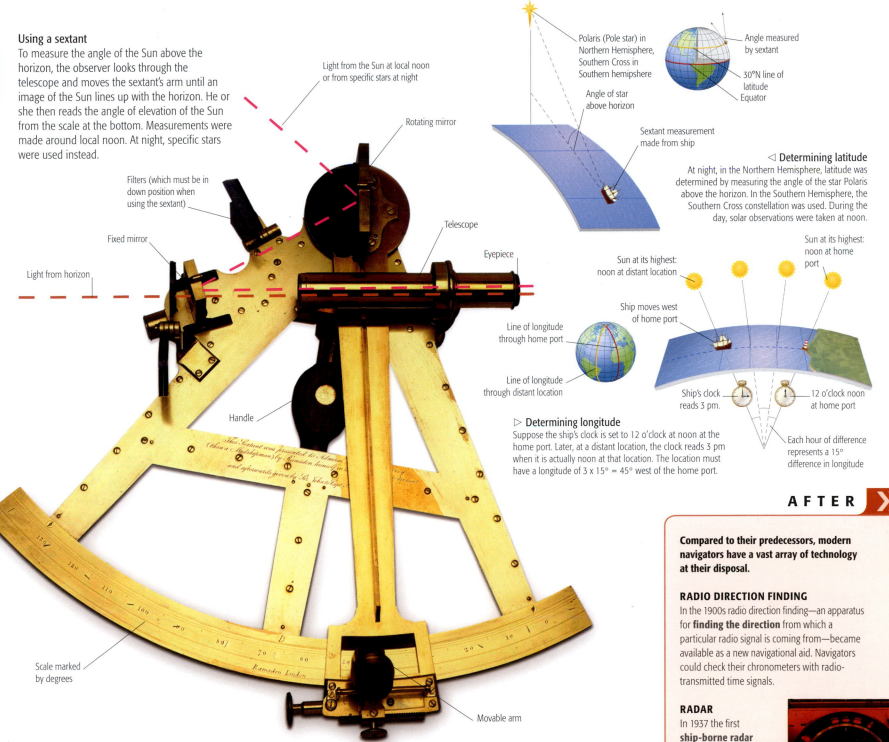

Light from the Sun at local noon or from specific stars at night

Rotating mirror

Filters (which must be in down position when using the sextant)

Fixed mirror

Telescope

Eyepiece

Light from horizon

Handle

Scale marked by degrees

Movable arm

Polaris (Pole star) in Northern Hemisphere, Southern Cross in Southern hemisphere

Angle of star above horizon

Angle measured by sextant

30°N line of latitude

Equator

Sextant measurement made from ship

◁ **Determining latitude**
At night, in the Northern Hemisphere, latitude was determined by measuring the angle of the star Polaris above the horizon. In the Southern Hemisphere, the Southern Cross constellation was used. During the day, solar observations were taken at noon.

Sun at its highest: noon at distant location

Sun at its highest: noon at home port

Ship moves west of home port

Line of longitude through home port

Line of longitude through distant location

Ship's clock reads 3 pm.

12 o'clock noon at home port

Each hour of difference represents a 15° difference in longitude

▷ **Determining longitude**
Suppose the ship's clock is set to 12 o'clock at noon at the home port. Later, at a distant location, the clock reads 3 pm when it is actually noon at that location. The location must have a longitude of 3 x 15° = 45° west of the home port.

ship's home port—but such a clock, capable of coping with the rough seas of a long voyage, was considered impossible to make. For many years, scientists looked instead for a second clock based on astronomical observations, and settled on a method relating to the Moon's position relative to other celestial objects.

In 1759, however, an English clockmaker called John Harrison produced a physical clock that performed the task with astonishing precision, and effectively solved the problem of longitude. In 1772 English explorer Captain Cook used a copy to navigate successfully all around the Pacific and produce maps that were far more accurate than previous efforts.

**AFTER**

Compared to their predecessors, modern navigators have a vast array of technology at their disposal.

### RADIO DIRECTION FINDING
In the 1900s radio direction finding—an apparatus for **finding the direction** from which a particular radio signal is coming from—became available as a new navigational aid. Navigators could check their chronometers with radio-transmitted time signals.

### RADAR
In 1937 the first **ship-borne radar** systems became available. These proved invaluable for navigating near **uninhabited coasts** at **night**, and **avoiding collisions** with other vessels **in fog**.

RADARSCOPE

### SATELLITE NAVIGATION
The first navigational systems based on **global positioning system (GPS) satellites 368–69 »** became available in the 1980s. Use of these systems is now standard for vessels making long ocean voyages. A GPS system can be linked to a vessel's steering system to produce **nearly automatic navigation**.

### SEE ALSO »
pp.212–13 STUDYING THE OCEANS

**IN PRACTICE**
## HARRISON'S MARINE CHRONOMETER

In 1759 the English clockmaker John Harrison (1693–1776) produced his masterpiece for solving the problem of longitude. It was a large watch (right), called H4, some 5 in (12 cm) in diameter, and weighing 3 lb (1.4 kg), which had taken four years to construct. In two trials to the West Indies, the chronometer proved astonishingly accurate at keeping track of the time in London, losing only five seconds on its first sea voyage, a two-month trip to Jamaica. With this, and a means of determining local time (which could be done by means of sextant readings), navigators could determine their precise longitude.

## « BEFORE

For perhaps 3,000 years philosophers have argued that it is impossible to cut something up into smaller and smaller pieces for ever; there must be a limit.

### MATTER AS INDIVISIBLE

Indian philosophers in the 6th century BCE first suggested that matter is made of **small indivisible particles**. In the 5th century BCE the Greek philosophers **Leucippus** and **Democritus** concurred, stating that the Universe is made of tiny particles, which were *atomos* ("indivisible").

**DEMOCRITUS**

### ISLAMIC ATOMISM

The Asharite School was a center of early Muslim speculative theology, founded by **Abu al-Hasan al-Ashari** in the 10th century. Al-Ashari believed in **atomism**, and taught that Allah created every particle of matter. His best-known disciple was **Al-Ghazali**, a Persian polymath who supported **scientific observation** and **mathematics**, and moved Islamic philosophy toward an idea of cause and effect determined by God. Al-Ghazali said that **atoms** are the only things that **continue to exist**; everything else is "accidental" and fleeting.

### « SEE ALSO
*pp.22–23* ELEMENTS OF LIFE
*pp.52–53* ALCHEMY

**ATOM** The smallest possible particle of an element.

**MOLECULE** Two or more atoms joined together.

**ELEMENT** A substance composed of only one kind of atom.

**COMPOUND** A substance composed of two or more elements.

# The Nature of Matter

Philosophers have long argued that the Universe is made of tiny indivisible particles. But there was no evidence for this until the 17th century, when groundbreaking experiments proved the existence of atoms, elements, and compounds. By the 19th century scientists thought "matter" had been completely explained.

The world is made of atoms. The word "atom" comes from the Greek *atomos* meaning "indivisible," and atoms are the smallest possible particles of elements (the chemical substances from which everything is made). A few atoms exist on their own, and are not bound to any others—such as the gas helium—but most atoms exist in tightly bound groups.

A molecule consists of two or more bonded atoms. Hydrogen gas, the lightest of all the elements, consists of two atoms of hydrogen linked by a chemical bond. Single hydrogen atoms can exist, but they are extremely reactive, and so do not remain single for long.

### A new set of elements

The 17th-century Anglo-Irish scientist Robert Boyle (see pp.98–99) saw the principal role of chemistry as finding out what things are made of. He studied the behavior of gases and promoted the idea of atoms, suggesting that all matter is composed of tiny particles that join together in various ways.

Boyle rejected the old idea of the world being composed of the four Aristotelian elements—air, earth, fire, and water (see pp.22–23)—because his observations suggested that there were many more than this. In his book *The Sceptical Chymist* (1661) he described elements as substances that could not be broken down into anything simpler by chemical processes. He also introduced the notion of compounds—materials made of more than one element—although he was unsure whether some materials were elements or compounds.

**Dalton's atomic models**
John Dalton imagined the atom as a hard, consistent sphere. He used specially-made wood balls to represent atoms when demonstrating his atomic theory.

### Dalton's atomic theory

English scientist John Dalton first suggested the idea of studying atomic weights in 1803, and elaborated on his atomic theories in his book *A New System of Chemical Philosophy* (1808). He suggested that all elements are made up of indivisible atoms, which cannot be divided or destroyed; that the atoms of each element are identical but differ from those of other elements; that atoms can be distinguished by their atomic weights; and that compounds are formed when atoms of various elements combine in definite proportions.

### Dalton's symbols
Dalton represented each type of atom by a pictorial symbol (today we use abbreviations: H for hydrogen, He for helium, and so on). Molecules were shown by adding the symbols together.

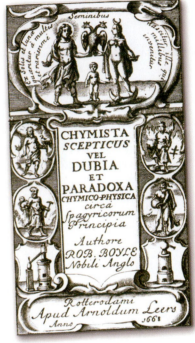

**Robert Boyle's The Sceptical Chymist, 1661**
This was the first chemistry book, as opposed to the many existing works on alchemy. For the first time, Boyle effectively defines what is meant by an element.

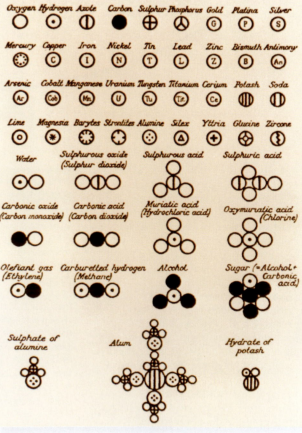

Dalton's Atomic Symbols

### CHEMIST (1726–1824)

## JOHN DALTON

Born in northern England to a Quaker family, Dalton started teaching in his brother's school at only 12. He was guided by two older teachers—Elihu Robinson and John Gough—who were enthusiastic amateur meteorologists. He learned the practicalities of meteorology from them and began seriously pursuing the subject, eventually making more than 200,000 notes in his weather journal. His observations, experiments, and publications were largely responsible for the acceptance of meteorology as a science. Best known for his atomic theories, Dalton also wrote the first scientific description of color blindness, which for many years was known as "Daltonism." In 1822 he was made a fellow of the Royal Society. When he died following a series of strokes, his body was placed in Manchester Town Hall, where 40,000 people filed past it to pay their respects.

Dalton invented symbols for the atoms of the 36 elements he knew about (more than 100 elements are now known), and calculated the atomic weight of each element by comparing it with hydrogen, the lightest element. Unfortunately he assumed that elements combine in the simplest proportions; so he assumed water was HO (rather than $H_2O$). Although this meant that many of his atomic weights were wrong, the principle was correct.

## Law of Definite Proportions

Around the same time, the French chemist Joseph Louis Proust and Swedish chemist Jöns Berzelius independently formulated the Law of Definite Proportions: atoms always combine in simple, fixed, and constant ratios. So "fixed air" is always $CO_2$: one atom of carbon combined with exactly 2 of oxygen. In 1825 Berzelius published a list of the atomic weights of the 43 elements that were then known; most of his weights were in good agreement with modern values.

### Giant molecules
Some molecules are so large that the material becomes solid. This is the structure of diamond, which is made entirely of carbon atoms. Each atom is bonded to four others in an infinite lattice.

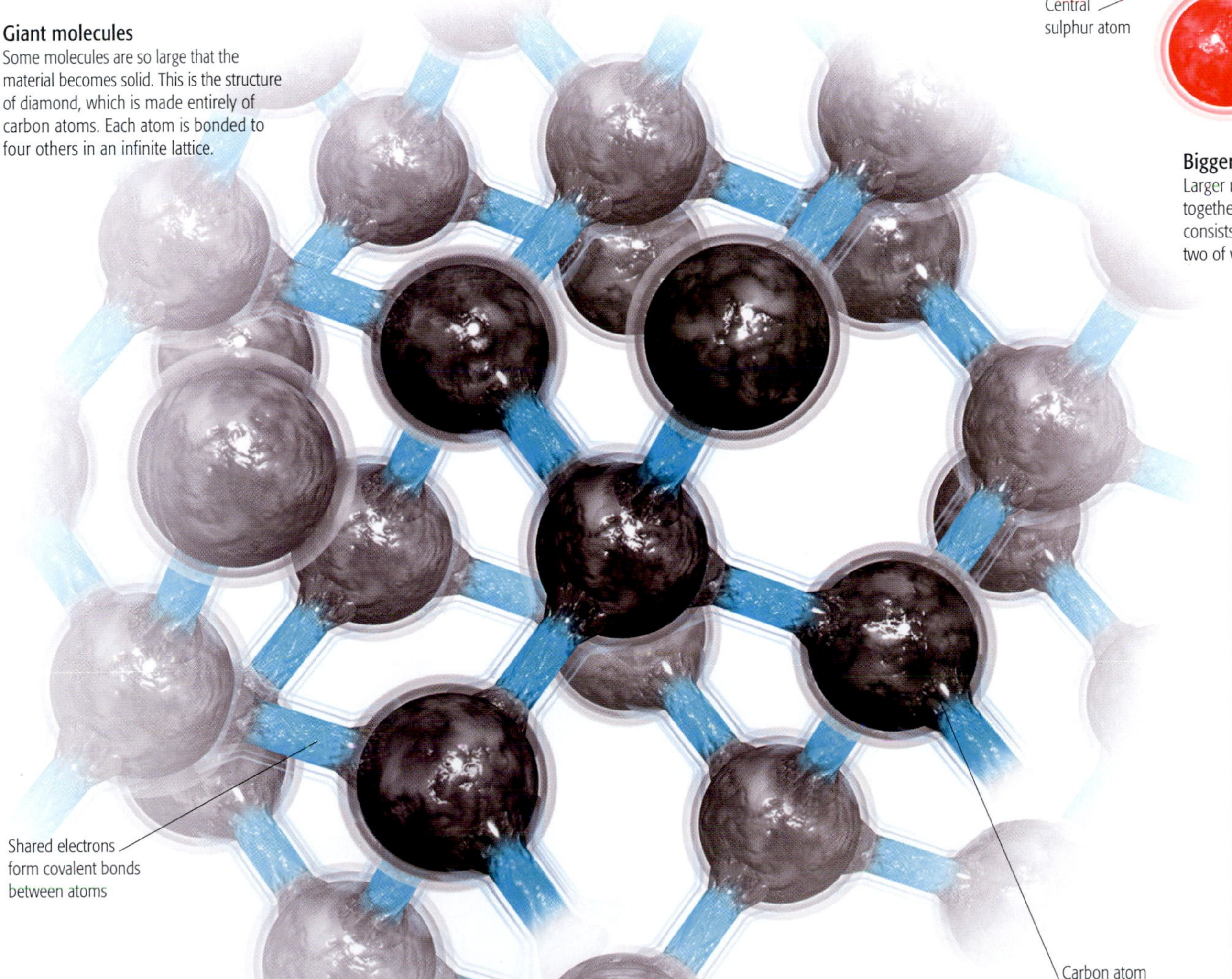

### Single atoms
Some elements form molecules only with extreme difficulty and therefore exist as single atoms at normal temperatures and pressures. Examples include noble gases (so called because of their very low reactivity) such as helium, argon, and krypton.

Complete shell of electrons allows no possibility of bonding under normal conditions

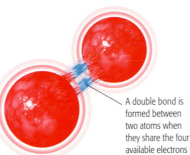

A double bond is formed between two atoms when they share the four available electrons

### Small molecules
Light elements may form small molecules by bonding with themselves—for example oxygen—or with other elements, as in water ($H_2O$—hydrogen bonded with oxygen).

**BREAKTHROUGH**

## AVOGADRO'S NUMBER

In 1811 Italian scientist Amedeo Avogadro stated that equal volumes of different gases (at the same temperature and pressure) contain equal numbers of molecules. Known as Avogadro's number, it is equal to $6.022145 \times 10^{23}$ (602,214,150,000,000, 000,000,000). The molecular weight of any substance (in grams)—a mole—contains Avogadro's number of molecules; thus 2 g of hydrogen and 32 g of oxygen both contain this number of molecules.

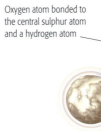

ONE MOLE QUANTITIES OF VARIOUS CHEMICALS

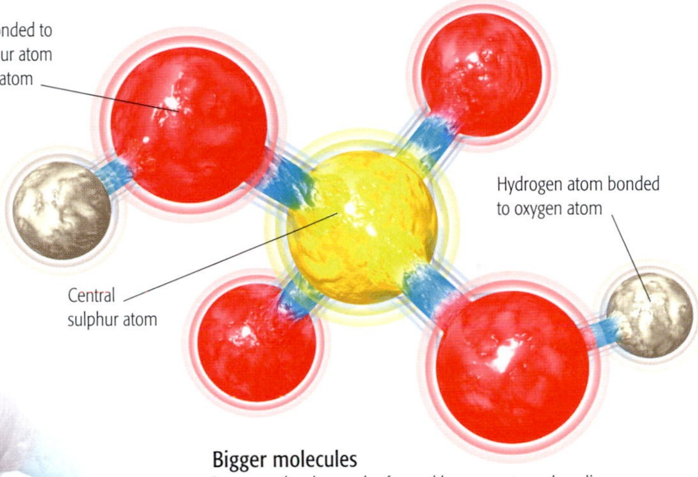

Oxygen atom bonded to the central sulphur atom and a hydrogen atom

Hydrogen atom bonded to oxygen atom

Central sulphur atom

### Bigger molecules
Larger molecules can be formed by many atoms bonding together. The molecule of sulphuric acid ($H_2SO_4$, shown here) consists of a central sulphur atom bonded to four oxygen atoms, two of which are also bonded to hydrogen atoms.

Shared electrons form covalent bonds between atoms

Carbon atom bonded to four other carbon atoms

**AFTER »**

During the 19th century chemists became gradually more definite in their understanding of the nature of matter.

**ACCURATE WEIGHTS**
In 1860 **Stanislao Cannizzaro** laid down the differences between **atoms and molecules**. He also provided much more accurate values of atomic weights.

STANISLAO CANNIZZARO

**THE PERIODIC TABLE**
In 1869 **Dmitri Mendeleev** constructed his **periodic table 230–31»**, the foundation of all the chemistry of the elements.

**ATOM SMASHING**
During the late 19th and early 20th century scientists discovered that atoms are made of **smaller particles 286–87»**; some are **radioactive 298–99»**; and some can be smashed apart **382–83»**.

SEE ALSO »
pp.140–43 STATES OF MATTER
pp.288–89 CHEMICAL BONDS

"No new **creation** or **destruction** of **matter** is within the reach of **chemical agency**."

JOHN DALTON, CHEMIST, 1808

Ideas about the nature of matter are rooted in ancient Greece, but the particulate nature of matter was made more meaningful by the emergence of atomic theory in the 19th century.

### PARTICULATE MATTER

The Greek philosophers **Leucippus** and **Democritus** proposed that matter was made up of **indivisible particles** now recognized as **atoms**. However, **Aristotle** had a different view. He believed substances could be understood in terms of **matter and form** and a theory of **five elements** (fire, earth, water, air, and ether), a concept that lasted for the next millennium.

### MATTER OF THE ENLIGHTENMENT

In the 17th century philosopher, physicist, and chemist **Robert Boyle** pioneered modern chemistry by establishing **ideas concerning atoms and molecules**. At the same time, **Jan Baptista van Helmont's** posthumously published *Ortus Medicinae* contained the first conceptualization of **gaseous matter** as well as an early expression of the fundamental scientific idea that **matter cannot be created or destroyed**.

**« SEE ALSO**

pp.22–23 ELEMENTS OF LIFE
pp.100–101 THE BEHAVIOR OF GASES
pp.138–39 THE NATURE OF MATTER

PHYCICIST (1856–1940)

## JOSEPH JOHN THOMSON

J.J. Thomson trained in engineering and mathematics, and rose to the position of Regius Professor of Physics at Cambridge University. It was here that he made his groundbreaking discovery during experiments on the electrical conduction of gases: that atoms contained very light, negatively charged particles. These particles later came to be called electrons. For this he received the Nobel Prize in Physics in 1906. He also developed mass spectrometry, which led to the discovery of isotopes (varieties of an element with different atomic masses).

# States of Matter

**Matter is more than a solid that can be held in the hand, a liquid than can run between the fingers, or a gas that we breathe or feel as a jet of air or gust of wind. Science has revealed more exotic forms of matter, such as plasma, which exist only under special conditions of temperature or pressure, or both.**

All naturally existing matter on Earth is made of atoms; groups of atoms bonded together to form molecules; or ions—atoms or molecules that have gained or lost one or more electrons (see pp.288–89). In everyday life the matter around us exists in three states—solid, liquid, and gas—and it is the arrangement and movement of its constituent particles that determine its state.

### Particles in motion

In a solid the particles are fixed in position by chemical bonds (see pp.288–89), and as a result can do no more than vibrate. Heat gives them energy and makes them vibrate more, but they remain in position, giving the solid its rigidity. Its shape can be changed only by physical force, which is more easily done with some solids than others.

In liquids and gases the particles have enough energy to move, meaning that liquids and gases can flow. In liquids there are still attractions between the particles that keep them together and the volume fixed (at a given temperature) but the inter-particle attractions are weaker than those in solids. Liquids, unlike solids, have no fixed shape and adopt the shape of any container that confines them. In gases the inter-particle attractions are

**Particles of matter**
The state of matter depends upon the motion of its particles (atoms, molecules, or ions) and the distances between them.

Particles of gas move freely and independently of one another. This enables them to scatter, so their volume is not fixed

Particle movement not limited by inter-particle attraction

**GAS PARTICLES**

Ice is made up of molecules of water bound rigidly together in a crystalline arrangement

Particle movement limited by inter-particle attraction

Liquid particles are more widely spaced than solid particles and change position, but inter-particle attraction keeps them in a fixed volume

**LIQUID PARTICLES**

Solid particles are bonded in a fixed position, but are capable of vibrating

**SOLID PARTICLES**

> "The **phenomena**... reveal to physical science **a new world**—a world where matter may exist in **a fourth state**..."
>
> SIR WILLIAM CROOKES, FROM "PHILOSOPHICAL TRANSACTIONS," 1879

**Three states of matter**
Ice and snow are solid forms of water that melt to form liquid. This hot volcanic spring turns liquid to gaseous water vapor, which condenses back to water to form clouds of tiny droplets.

## AMORPHOUS SOLIDS

An amorphous solid is not crystalline, which is to say there is no regular repeating pattern to the position of the atoms (see p.142). Many substances form amorphous solids when they are cooled so rapidly that the particles do not have a chance to become arranged in an organized way. Glass is an amorphous solid—the common belief that it is really a supercooled liquid arises from the fact that its solidification from liquid form does not happen in the same abrupt way as the phase transition exhibited by other types of solids.

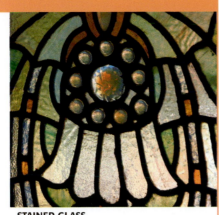

**STAINED GLASS**

neglible. As a result the particles are free to move independently of one another, so that neither the shape nor volume of a sample of gas is fixed.

### Phase transitions

When matter changes from one state to another, the change is known as a "phase transition." For example, a solid melting, a liquid boiling, a vapor condensing into a liquid, and a liquid solidifying are all phase transitions. Metal becomes more malleable when heated so its atoms become more energized, but they are still fixed by bonds: heated metal is still solid. But when it finally melts its particles move freely in the new liquid phase: melting is the abrupt phase transition.

A pure substance melts or freezes, and boils or condenses, at characteristic temperatures. These temperatures can be used as a gauge of the purity of substances, since any contamination alters their values. For example, pure water freezes at 32°F (0°C), whereas the salt dissolved in seawater depresses its freezing point to about 28.4°F (−2°C).

In order for a solid to be melted into liquid, the particles need to gain sufficient energy to escape the bonds of their rigid confinement. They must acquire yet more energy to become gaseous. In each case the amount of heat that is required to change the

state (but without changing the temperature) is called the latent heat, and this varies from substance to substance (see pp.178–79).

A few substances have anomalous properties that mean phase transition occurs directly from solid to gas, with no liquid state in between: this is called sublimation; for example, dry ice (solid carbon dioxide) sublimes directly into gaseous carbon dioxide.

### The fourth state of matter

In 1879 Sir William Crookes discovered the fourth state of matter by means of experiments on gases. He called it "radiant matter." Later, the English physicist J.J. Thomson showed that a proportion of the particles in the gas had lost their electrons, with the result that this state of matter was composed of flowing negative electrons and positive ions. It was named plasma in 1928 by US scientist Irving Langmuir.

Plasma resembles gas but differs from it in conducting electricity and responding strongly to magnetic fields, often forming beams or filaments when it does so. It is the most abundant state of matter in the Universe, being a component of stars—including the »

◁ **Sublimation**
Solid carbon dioxide (also known as dry ice) dropped into a tube of water turns instantly into gas. Dry ice is one of relatively few substances that sublime—change directly into gas when it is heated, bypassing a liquid phase.

Rising temperature gives bound water molecules more energy, turning them to liquid

Stream of light emitted from ionized gas

The bonding of water molecules keeps them in pools

Metal ball in the center of the sphere is charged with electricity

**Water's changing states**
As ice melts its molecules escape from their fixed positions in the ice crystals, but—unlike most other substances—come closer together. As a consequence ice, being less dense than water, floats (see p.142).

▷ **Liquid mercury**
Mercury is one of only two liquid elements (the other being bromine), and the only metal that is liquid at room temperature. Its high surface tension makes it form rolling spheres that do not wet surfaces as water droplets do.

Mercury drops on a surface are more spherical than water drops

**Plasma globe**
An electrically charged metal ball in a glass sphere filled with low-pressure gas discharges electricity through the sphere to the glass wall. This forms plasma by making particles of gas ionized along the path of discharge, emitting light in the process.

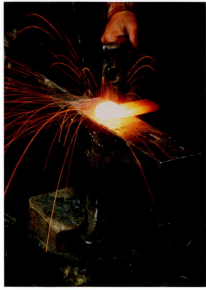

**Malleable materials**
Metals are highly malleable because their atoms can slide past one another so that the metal changes shape without fracturing. Reshaping by application of force is more easily done when the metal is red hot.

### Mohs' Hardness Scale

Ten common minerals are listed in Mohs' Hardness Scale, which enables the hardness of a solid to be determined by finding the hardest mineral listed that is capable of scratching it. For example, if a solid is scratched by feldspar but not by apatite, its Mohs hardness is determined to be between 5 and 6.

| Mineral | Mohs hardness |
|---|---|
| Talc (softest) | 1 |
| Gypsum | 2 |
| Calcite | 3 |
| Fluorite | 4 |
| Apatite | 5 |
| Feldspar | 6 |
| Quartz | 7 |
| Topaz | 8 |
| Corundum (e.g. ruby) | 9 |
| Diamond (hardest) | 10 |

### Ductility

Glass consists mainly of quartz sand (silica) with other additives, such as sodium carbonate, which lowers its melting point. When heated, glass is immensely ductile and can be stretched almost indefinitely; yet when cold it is brittle.

Solids also vary in their response to different types of stresses, which may be pulling (tensile) or pushing (compressive). Some are highly ductile, meaning that they can be pulled without breaking. Others are highly malleable and can be compressed without breaking; for example, copper and gold can be hammered without breaking whereas cast iron cannot.

### Fluidity and wetness of liquids

When a solid melts to form a liquid the particles become separated from their neighbors in such a way that the volume usually increases. However, the melting of ice into water is a notable exception: ice shrinks to form water that has higher density, with the result that ice floats.

Water owes its bizarre property to its bonding. In ice, the molecules are held rigidly apart in an "open" lattice arrangement by hydrogen bonds (see pp.288–89). On melting at 32°F (0°C), the molecules come slightly closer together, and the lattice begins to disintegrate. As the temperature rises, this disintegration continues and is complete at 39.2°F (4°C), when the cold water has its maximum density. As the temperature rises more the density of water drops a little as the molecules move faster and farther apart, but its density in liquid form never becomes as low as that of ice.

A liquid's resistance to stress is called viscosity; the more viscous a fluid, the less easily it flows. The physical principles of viscosity were first formulated by Isaac Newton (see pp.110–11). Today we know that viscosity—just like the hardness of solids—depends upon the size of the molecules and forces between them. Water has very low viscosity, while in olive oil it is higher, and in syrup higher still. The lack of rigidity of liquids also means that many can be mixed, though certain mixtures—such as water and oil—will spontaneously separate, repelled by the intermolecular forces that preferentially engage water

Sun—and is formed naturally by lightning. In the home, artificial plasma is used in fluorescent lamps and TV plasma displays.

### Hardness of solids

In many crystalline solids the particles naturally adopt a specific repeating pattern as a result of the highly organized ways in which the particles bond with their neighbors. The nature of the crystal arrangement has important consequences for the properties of the solid. Some solids—such as glass—lack a crystalline structure entirely and are called amorphous solids (see p.141).

In 1812 Friedrich Mohs came up with a scale of hardness based on the ability of one solid to scratch another. Mohs' Hardness Scale is still used by geologists today. Thomas Young later devised a way of measuring a property of solids which he described as elasticity. Some solids are highly elastic, meaning they can revert to their original shape when deformed, or plastic, meaning that they cannot. Young's expression of elasticity, published in 1807 and now known as Young's Modulus, has important applications in mechanical engineering since it quantifies how materials behave when under a load.

**Reinforced concrete**
Concrete on its own is a brittle material, but its strength can be improved for building purposes by incorporating bars of non-brittle steel.

with water and oil with oil but keep the different molecules apart.

Intermolecular forces also account for the fact that liquid particles at the surface pull inward to create a "skin"—a phenomenon called surface tension. The cohesive molecular forces responsible for surface tension plus the repulsive forces that keep oil and water apart explains why perfect rolling drops of water form on oily surfaces, such as

> **PLASTICITY** The characteristic of a solid to change shape without recovering.
>
> **ELASTICITY** The capacity of a solid to change shape and recover.
>
> **MALLEABILITY** The ability of a plastic solid to be reshaped without breaking.
>
> **DUCTILITY** The capacity of a plastic solid to be stretched without breaking.
>
> **VISCOSITY** The tendency of a liquid to resist flow.

a waxy leaf or a duck's back. However, on something that can be more easily wetted—such as clean glass—the attractive forces between the water and glass molecules are stronger than those between the individual water molecules. In 1804 Thomas Young explained that the attraction between water and glass is so strong that it enables water to creep up the sides of a narrow tube—a phenomenon known

### Surface tension

Water molecules at the surface are under tension, being pulled inward by their neighbors. The result is a thin surface "skin" that can support the weight of certain insects with water-repelling hairs on the ends of their legs, such as this pond skater.

## GOLD LEAF

Gold is extremely malleable and can be hammered into thin sheets without breaking. These sheets are called gold leaf and can be as thin as a thousandth of a millimeter (about 0.00004 in). Traditionally this was done by pounding it with mallets between layers of parchment, but today the process is mechanized. Gold does not rust or tarnish, and gold leaf has been used extensively in the decoration of buildings, artworks, and even food. Pure gold is nontoxic because it is very unreactive in the body.

as capillary action—to produce a distinctive concave meniscus (downcurved upper surface). In contrast, mercury has a convex meniscus (upcurved upper surface) because the attraction between mercury molecules is greater than between glass and mercury molecules.

### Nebulous matter

The idea of an ideal fluid (liquid or gas) in which there are no intermolecular forces—and therefore zero viscosity—is a useful theoretical concept because it simplifies understanding of the behavior of fluids. However, in real life there are forces between molecules of fluids, although in gases these forces are so small that

**Viscosity**
The attraction between molecules of olive oil mean that they are less free-moving than those in water, giving the oil a higher viscosity and thus making it flow more slowly.

"I call this **Spirit,** unknown hitherto, by the **new name** of **Gas**, which can neither be constrained by vessels, nor reduced into a visible body."

JAN BAPTISTA VAN HELMONT, *"ORIATRIKE: OR, PHYSICK REFINED,"* 1662

### Compression
Bubbles of gaseous carbon dioxide emerging from carbonated drinks cause a buildup of the pressure that is responsible for the explosive popping of a champagne cork.

they can usually be ignored and most real gases behave as if the forces were absent. As result, gases can flow and disperse freely, limited only by the container that holds them. The nebulous nature of gases also means that they can be compressed into smaller volumes, something that cannot be achieved with solids and liquids. Compression forces the gas particles closer together until, at a critical point, the gas undergoes a phase transition and becomes liquid.

### Mixtures and solutions

Gases and some liquids mix completely because their particles can combine. But interactions between one type of particle and another can result in perfectly blended mixtures called solutions. In a solution one type of substance, called a solvent, dissolves another, called a solute. Liquid solvents, such as water, can dissolve other liquids, solids, or gases. When a solid dissolves in water, the attractive forces that otherwise keep the solid particles in fixed positions become disrupted and the solid particles separate as they engage with water molecules instead. Some solids, such as common salt, engage with water very effectively and so lose their rigidity when they dissolve. Others, such as sand, cannot bond with water in this way so do not dissolve.

**AFTER**

Continuing research into states of matter led to the production of entirely new forms of matter during the 20th century.

**IDEAL LIQUID IN PRACTICE**
An ideal liquid with **zero viscosity** was made in 1937 by a science team that included **Pyotr Kapitsa, John Allen,** and **Don Misener.** They achieved it by supercooling helium, creating a **"super-fluid"** whose properties served to further the understanding of quantum physics **314–17».**

**SUPERCRITICAL FLUIDS**
Described as the **"fifth state of matter,"** supercritical fluids have properties of both **gases and liquids.** They are formed when gases are **compressed and heated** simultaneously and are used as a substitute for organic solvents in various processes, such as the decaffeination of coffee.

**SEE ALSO »**
*pp.286–87* STRUCTURE OF THE ATOM

**‹‹ BEFORE**

Water pressure has been used in various devices since ancient times, but the physical laws governing liquids under pressure were not known until the 16th and 17th centuries.

### EARLY APPLICATIONS

The ancient Greeks used a klepshydra, or **water thief ‹‹ 120–21** to determine how long a defendant could speak in court. Around 1,900 years ago Chinese inventor **Zhang Heng** built a **water-powered model of the movements of the stars ‹‹ 54–55**. In the 9th century the Persian Banu Musa brothers' *Book of Ingenious Devices* described **water wheels** and many other devices using water pressure.

### DIVING BELLS

A diving bell is a **diving chamber**, a bell-shaped hollow structure that is lowered into water so that a person can go down, breathing the **air trapped inside**, to do work under water. Simple diving bells have been **used since ancient**

**EARLY DIVING BELL**

**times**. In more advanced, later diving bells, **compressed air** was pumped down from the surface to replenish the fresh air inside.

**‹‹ SEE ALSO**

pp.54–55  ZHANG HENG
pp.100–01  THE BEHAVIOR OF GASES
pp.132–33  STEAM POWER TO STEAM ENGINE

# Liquids Under Pressure

**In 1586 Dutch scientist Simon Stevin showed that pressure in a liquid increases with depth: hence the enormous pressures in the deepest oceans. Then in the 1650s French mathematician Blaise Pascal showed that liquids cannot be compressed, a fact that is the basis of all hydraulics.**

Water power was used by the Greek inventor Ktesibios in the 3rd century BCE, and clever water clocks were described by various people from that time, including Arab inventor Al-Jazari 1500 years later. The first people to study the question scientifically were the French polymath Blaise Pascal in the middle of the 17th century and the English engineer Joseph Bramah (see opposite).

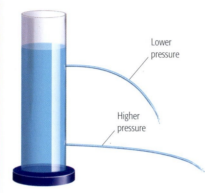

Lower pressure

Higher pressure

**Proving that pressure increases with depth**
If you fill a tall container with water and make one hole near the top, and one near the bottom, water comes out of the bottom hole under higher pressure.

At the surface of a liquid such as the sea, the pressure in the water is the same as the pressure of the atmosphere above it—15 lb/in² (1 kg/cm²), or one atmosphere. Because seawater is dense (1 liter has a mass of just over 2.2 lb or 1 kg), the pressure increases rapidly with depth: it reaches 2 atmospheres at a depth of 33 ft (10 m), 3 atmospheres at 66 ft (20 m), and so on.

One result of this is that snorkel tubes cannot be longer than about 39 in (1 m) because our chest muscles are not strong enough to expand against the pressure much below that depth. Below about 39 in (1 m) we have to hold our breath, or carry tanks of compressed air or oxygen.

### Under pressure

Submarines have to be immensely strong to cope with the pressure of the sea at great depths; at 1,650 ft (500 m) the pressure on a submarine is some 735 lb/in² (50 kg/cm²). The "crush depth" of a German World War II U-boat—the depth at which its hull would collapse under the pressure—was some 820 ft (250 m). That of a modern-day nuclear submarine is about 2,460 ft (750 m).

In deep ocean trenches the pressures are many hundreds of times that at the surface; so specialized vehicles are needed to withstand them: unmanned Remotely Operated Vehicles (ROVs) lowered by cable, free-roving unmanned Autonomous Underwater Vehicles (AUVs), and Manned Underwater Vehicles (MUVs) (see left). In 1960 a two-man MUV named the *Trieste* became the first and last MUV to descend to and return from the deepest point in the world's oceans: Challenger Deep, almost 36,000 ft (11,000 m) down in the Mariana Trench in the Pacific. Even at this depth, the crew saw small fish.

A more maneuverable and much better equipped MUV than the *Trieste* is the US Navy's *Alvin*, which can take a pilot and two scientists down 14,750 ft (4,500 m) on an eight-hour dive. Over the years *Alvin* has been used for much scientific research. In spite of the huge pressure—not to mention the cold, the lack of light, and the low levels of

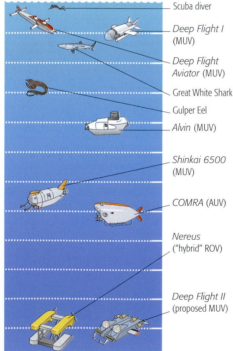

**Sea level**

|  | |
|---|---|
| 3,000 ft (1,000 m) | Scuba diver |
| | *Deep Flight I* (MUV) |
| 6,500 ft (2,000 m) | *Deep Flight Aviator* (MUV) |
| 10,000 ft (3,000 m) | Great White Shark |
| | Gulper Eel |
| 13,000 ft (4,000 m) | *Alvin* (MUV) |
| 16,500 ft (5,000 m) | *Shinkai 6500* (MUV) |
| 20,000 ft (6,000 m) | |
| 23,000 ft (7,000 m) | *COMRA* (AUV) |
| 26,000 ft (8,000 m) | *Nereus* ("hybrid" ROV) |
| 29,500 ft (9,000 m) | |
| 33,000 ft (10,000 m) | *Deep Flight II* (proposed MUV) |
| 36,000 ft (11,000 m) | |

**Exploring the depths of the world's oceans**
No Manned Underwater Vehicle (MUV) since the *Trieste* in 1960 has visited the deepest point in the world's oceans—Challenger Deep, in the Pacific—but Hawkes Ocean Technologies' *Deep Flight II* is planned to do so.

**Deep-sea diving**
Specialized deep-sea diving suits allow descents to 2,000 ft (600 m), at which depth the pressure is 61 atmospheres, or 900 lb/in² (61 kg/cm²).

### CHERRY PICKERS

Hydraulically raised platforms known as "cherry pickers" are commonly used to repair and maintain street lights and overhead cables. They are also used for tree surgery, high camera shots, and perhaps even occasionally for picking cherries. In transit the platform and folding "arms" rest on top of the back of the truck. In use the truck has stabilizing "legs," and the hydraulic pistons or "rams" can be operated either from the truck or from the platform itself.

**Capping an oil well**
When drilling contractors strike oil in natural reservoirs deep underground, the oil sometimes gushes out because of the enormous pressure exerted on it by the weight of the overlying rock. The well is then temporarily capped.

## JOSEPH BRAMAH

Born into a farming family in Yorkshire, Bramah injured his leg when young. Unable to work on the farm, he was apprenticed to a carpenter. In 1778 he patented an improved water closet—with a hinged valve below the bowl—that remained in use for many decades. He married a local girl, Mary Lawton, in 1783, then invented an almost unpickable lock, improved versions of which are still obtainable. After moving to London he applied Pascal's Law (see below) to invent first a hydraulic engine to pump beer up from the cellar of a tavern, then a hydraulically operated printing press.

Imagine a container full of liquid with two pistons in vertical cylinders connected to it: one cylinder with an area of 0.15 in² (1 cm²), and one with an area of 1.5 in² (10 cm²). Because the liquid cannot be compressed, applying a force of 2.2 lb (1 kg) to the smaller piston increases the pressure throughout the liquid by 15 lb/in² (1 kg/cm²). So the force pushing up on the larger piston is multiplied to 22 lb (10 kg).

Following Bramah, 19th-century engineers used hydraulics in a host of ways. William Armstrong invented the hydraulic crane in Newcastle, England in 1845, and filled his house with all sorts of hydraulic devices, including a lift and a rotating spit in the kitchen.

### High-pressure water in use

Hydraulics also works in reverse. If you remove the smaller piston and push down hard on the larger one, liquid squirts out off the smaller cylinder at great speed. This is how syringes and water pistols work, but high-presssure water has been used for all sorts of other purposes. Set up in 1883 the London Hydraulic Power Company built high-pressure water mains all

over London. The water was used to raise elevators, theater curtains, and even Tower Bridge.

High-pressure water jets are used to clean everything from cars to sewers. When abrasive grit is added to the jet it can cut stone and concrete. Some Scandinavian fire services use such jets to gain rapid access to buildings by cutting through walls. Firefighters also use high-pressure sprays or mists to cut off the supply of air from the source of the fire, reduce the temperature, and clear the air of lethal smoke.

**Mining tin ore with a high-pressure water jet**
Breaking up tin ore with high-pressure jets of water—a process known as gravel pumping—is the most common method of mining tin in Southeast Asia. The method is highly economic for small, scattered deposits of ore.

dissolved oxygen—a variety of animals, including several types of squid, the pelican eel, and other strange-looking fish, thrive in the deep sea. They tend to be small and few, because of the scarcity of food, apart from near hydrothermal vents, or "black smokers," where there are dense populations of clams, shrimp, and giant tube worms. They are not crushed by the pressure as we would be because they have the same pressure inside their bodies.

### Pressure and hydraulics

In the late 18th century, English engineer Joseph Bramah became the first person to apply the principles of hydraulics—as laid down by Blaise Pascal—for practical purposes. Pascal's Law states that pressure in a liquid in a small closed system is equal in all directions and constant: unlike gasses, liquids cannot be compressed.

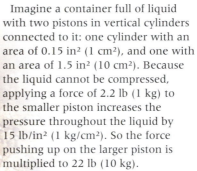

Small force
Small piston
Large piston
Large force

**How hydraulics works**
A liquid cannot be compressed. Therefore, in a hydraulic system the application of a small force by a small piston increases the pressure throughout the liquid. As a result, the force on a large piston is multiplied.

### AFTER

Hydraulics is used in a range of machines, from lifts to robots, while high-pressure water is widely used to generate electricity.

**HYDROELECTRIC POWER**
When water at the bottom of the dam of a deep reservoir is channeled into a pipe **the pressure is high enough to spin a turbine** and so generate electricity. Hydroelectric power supplies some 20 percent of world electricity.

**HYDRAULIC BRAKES**
All cars today have **hydraulically operated brakes**. When the driver's foot presses on the brake pedal,

pressure is increased in the brake fluid in a cylinder. This pressure is transferred by pipes to disk brakes, which clamp against the wheels.

SEE ALSO »
pp.212–13 STUDYING THE OCEANS
pp.416–17 RENEWABLE ENERGY

THREE GORGES DAM, CHINA

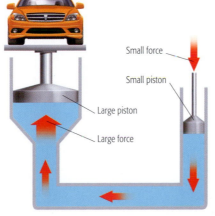

<< BEFORE

The ancients often confused air with "spirit" or "soul," because these disappeared when a person stopped breathing and died.

### EMPEDOCLES

**Empedocles**, a Greek philosopher, put forward the idea that **everything is made up of four elements**—earth, air, fire, and water << 22. He proved that **air is not "nothing"** by upending a vessel and pushing it down into water. The water did not fill the vessel—it did not wet all the inside—so something must have been keeping the water out; that "something" was air.

### PHLOGISTON THEORY

In the early 18th century inflammable materials were thought to contain the substance **phlogiston** (from the Greek word for "inflammable"), which was released as they burned. Scientists showed that **metals gained weight** when they burned, so they could not be losing phlogiston.

<< SEE ALSO
*pp.22–23* ELEMENTS OF LIFE
*pp.100–101* THE BEHAVIOR OF GASES

# Discovery of Gases

In the middle of the 18th century scientists realized that there were various kinds of "air," and each had its own chemical properties. Within 50 years scientists had isolated and identified a dozen new gases, which gave a considerable boost to the developing subject of chemistry.

The ancient Greeks had understood that air was not "nothing," and indeed was necessary for life, but they did not appreciate that air is not the only type of gas. This was at least partly because they lacked the equipment and skills needed to investigate gases.

Later, when these techniques and skills became available, scientists were able to show that the atmosphere is a mixture of gases, and that some gases are produced in chemical reactions, while other reactions involve an uptake of gases from the atmosphere itself.

## Quantitative techniques

In the research for his doctoral thesis of 1754, the Scottish chemist Joseph Black (see pp.140–41) was careful in all his experiments to record the weights of both the starting materials and the products of chemical reactions. This was the first time that anyone had made such meticulous records, and with the results Black was able to show that carbon dioxide, which he named "fixed air," was emitted when limestone was roasted or treated with acid. He discovered that the same gas was contained in the out-breath of animals, and also formed a small part of the atmosphere. Black called it "fixed air" because the gas had been fixed in the limestone, proving for the first time that there were different types of "airs."

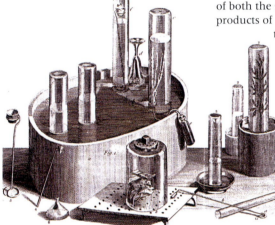

**Priestley's experimental apparatus**
Much of Priestley's apparatus was destroyed when his house was attacked in 1791, but this illustration from one of his books shows the apparatus he used for investigations of gases.

## The discovery of oxygen

One man who followed up Black's pioneering work was Englishman Joseph Priestley. In 1767 he lodged next to a brewery in Leeds, and enjoyed watching the wort (malty liquid) fermenting in the stone tanks known as Yorkshire squares. Above the froth of the yeast was a layer of carbon dioxide, as Priestley demonstrated by lowering a candle down into the square. Some 8–12 in (20–30 cm) above the froth the candle flame went out, and the smoke drifted above the froth along the interface between the fresh air and the "fixed air."

Leaning over the side, Priestley poured water from one jug to another within the layer of carbon dioxide, which gradually dissolved in the water, and so made soda water. While this discovery interested the admiralty as a possible treatment for scurvy, its real importance lay in the fact that it spurred Priestley on to investigate other gases. He developed a crucial piece of apparatus—the pneumatic trough—to collect samples of gas over water or mercury, and used it to identify 10 more gases, including ammonia, hydrogen chloride, nitrous oxide (laughing gas), and sulphur dioxide. His greatest discovery was without doubt oxygen.

Priestley probably discovered oxygen in 1771, while he was still in Leeds, although he moved to Calne in Wiltshire in 1773, and a plaque on the wall in Bowood House states that he discovered oxygen there in August

1774. He made oxygen by using a burning glass to focus sunlight on "calx of mercury" (mercuric oxide) in a flask. Mercuric oxide decomposes when heated to produce oxygen.

**1 Dilute acid**
Acid trickles down from funnel into the round flask

**2 Airtight stopper**
Cork prevents escape of gas

**3 Chemical reaction**
Acid reacts with limestone to produce carbon dioxide gas

Hydrochloric acid is added to the magnesium carbonate

**How gases are isolated**
Gas formed by a chemical reaction in the round flask (right) passes through the angled glass tube and bubbles up through the water (or mercury) in the pneumatic trough, to be collected in the receiving jar.

---

CHEMIST (1733–1804)

## JOSEPH PRIESTLEY

Born in 1733 on a farm in Yorkshire, Priestley was educated at Daventry, and became a teacher at Warrington, where he wrote an important 700-page book entitled *The History and Present State of Electricity*. He was the minister of Mill Hill Chapel in Leeds from 1767 to 1773, Librarian to the Earl of Shelburne from 1773 to 1780, and a part-time minister in Birmingham from 1780 to 1791. His open support for the French Revolution made him so unpopular that in 1791 a mob burned his house to the ground. In 1794 Priestley emigrated to America, where he died in 1804.

## GAS LIGHTING

The Flemish scientist Jan Baptista van Helmont found that heating coal without air produces a "wild spirit," which he described as "gas" in his book *Origins of Medicine* (1609). This "gas" is in fact a mixture of inflammable gases—including methane, hydrogen, and carbon monoxide—and burns with a smoky yellow flame. It is commonly known as coal gas. A few rooms were illuminated by coal-gas flames in the 1780s and 1790s; by the 19th century many factories, houses, and streets had coal-gas lighting (see pp.250–51), including in London in 1813 and in Paris in 1820.

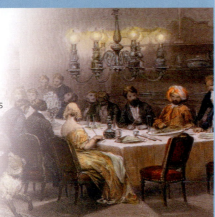

At roughly the same time that Joseph Priestley isolated oxygen, the German-Swedish chemist Carl Wilhelm Scheele also obtained the gas in Uppsala, Sweden, although he did not publish his results until 1777.

Unfortunately, Priestley called his new gas "dephlogisticated air," and it was left to the French scientist Antoine Lavoisier, to whom Priestley had shown his experiments, to make his own gas and call it "oxygène."

### Inflammable air

The English physicist Henry Cavendish (1731–1810) was a quiet genius who worked alone in his south London house. He wrote in the Royal Society's *Philosophical Transactions* (1776) that the "air" given off when metals react with acids is a distinct entity, different from anything we breathe. He called this gas "inflammable air," but we now know it as hydrogen. When Cavendish exploded it with oxygen, he found that the only product was water (or rather steam, since it was hot), and that the formula for water is $H_2O$. The reaction is:

$$2H_2 + O_2 = 2H_2O.$$

Hydrogen proved to be much less dense than air; in fact it is the least dense of all gases. This made it an obvious choice for lighter-than-air balloons. It proved a dangerous choice, however, since hydrogen is extremely flammable and explodes with oxygen in any proportions. After numerous disasters, helium replaced hydrogen in balloons and airships.

**Burning of the *Hindenburg***
On May 6, 1937 the German airship *Hindenburg* had crossed the Atlantic and was about to land in New Jersey, when its hydrogen ignited. The airship was engulfed within 37 seconds, killing 36 people.

> " On completing one **discovery** we **never fail** to get an **imperfect knowledge** of others of which you could have no idea before…"
>
> JOSEPH PRIESTLEY, 1786

**5 Trapped carbon dioxide**
Carbon dioxide rises up through the hole in the top of the "beehive"

**6 Gas builds up**
Carbon dioxide collects at top of receiving jar, allowing water back into trough

**4 Flow of gas**
Carbon dioxide gas is released through tube to pneumatic trough

Pneumatic trough containing water

AFTER ≫

The pioneering work of the 18th-century scientists led to the discovery of many more gases in the decades that followed.

**THE DISCOVERY OF HELIUM**
In 1868 the French scientist **Pierre Janssen** noticed an **unusual yellow line in the Sun's spectrum** (the pattern of lines produced by splitting sunlight with a prism). He said this must come from a **new element**, but no one believed him until the English astronomer **Norman Lockyer** verified the observation. Lockyer called the element **helium**, after *helios,* the Greek word for "Sun." Helium is the **second most abundant element** in the Universe; it is formed in the Sun and other stars by the **nuclear fusion** of hydrogen **324–25 ≫**.

**HYDROGEN FLAMES**
Hydrogen reacts vigorously with oxygen to make water, $H_2O$. A hydrogen flame is **pale blue** and difficult to see, which makes it dangerous. Its **high temperature** allows the production of limelight **251 ≫**, first used in the early 19th century for various forms of public lighting.

SEE ALSO ≫
pp.230–31 MENDELEEV'S TABLE

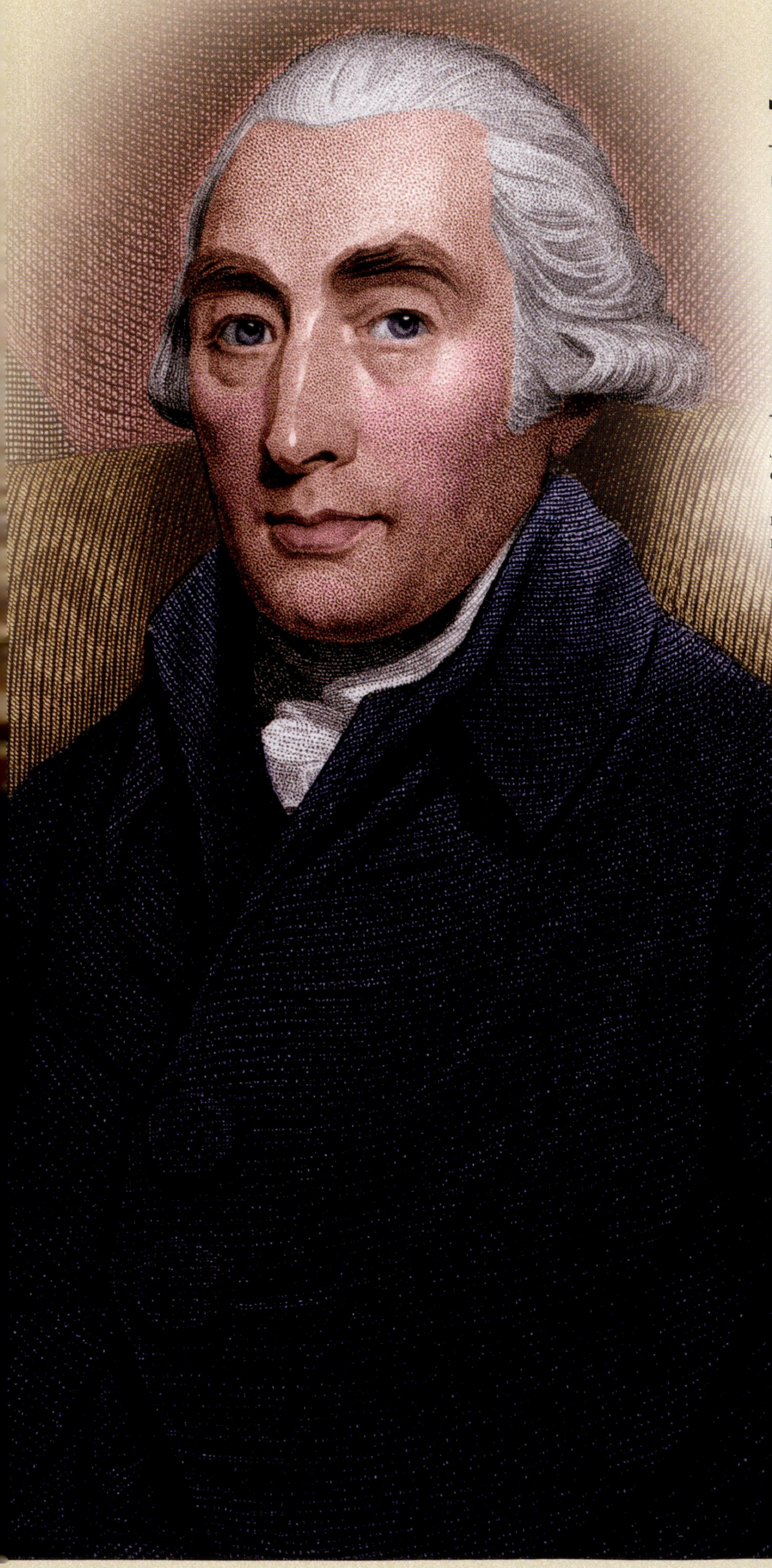

PHYSICIAN AND CHEMIST **Born 1728 Died 1799**

# Joseph Black

## "A very **great** quantity of **heat** is necessary to the production of **vapour**."

**JOSEPH BLACK,** FROM "LECTURES ON THE ELEMENTS OF CHEMISTRY," 1803

Joseph Black was the "father" of quantitative chemistry; his thesis of 1754 laid the foundation for a whole new approach to the subject, based on measurement rather than surmise and guesswork. His curiosity about the behavior of liquids during boiling and freezing opened the door to a new understanding of heat and molecular behavior.

Black was born in Bordeaux, where his father was a wine merchant, and was sent to school in his father's home town of Belfast. He spent his adult life at the universities of Glasgow and Edinburgh. At Glasgow he studied the arts for a few years, but was then drawn into science. Realizing that it was important to know how much material he was dealing with in each experiment, he developed an analytical balance that allowed him to weigh samples very accurately; this soon became a vital tool in all chemistry laboratories.

DISSERTATIO MEDICA
INAUGURALIS,
DE
HUMORE ACIDO
A CIBIS ORTO,
ET
MAGNESIA ALBA:
QUAM,
ANNUENTE SUMMO NUMINE,
*Ex Auctoritate Reverendi admodum Viri*
D. JOANNIS GOWDIE,
ACADEMIAE EDINBURGENAE PRAEFECTI;
NEC NON
*Amplissimi* SENATUS ACADEMICI *consensu*,
*Et nobilissimae* FACULTATIS MEDICAE *decreto*;
PRO GRADU DOCTORATUS,
SUMMISQUE IN MEDICINA HONORIBUS ET PRIVILEGIIS
RITE ET LEGITIME CONSEQUENDIS,
ERUDITORUM EXAMINI SUBJICIT
JOSEPHUS BLACK GALLUS.
Ad diem 11 Junii, horâ locoque solitis.
EDINBURGI:
Apud G. HAMILTON ET J. BALFOUR
ACADEMIAE TYPOGRAPHOS.
M,DCC,LIV.

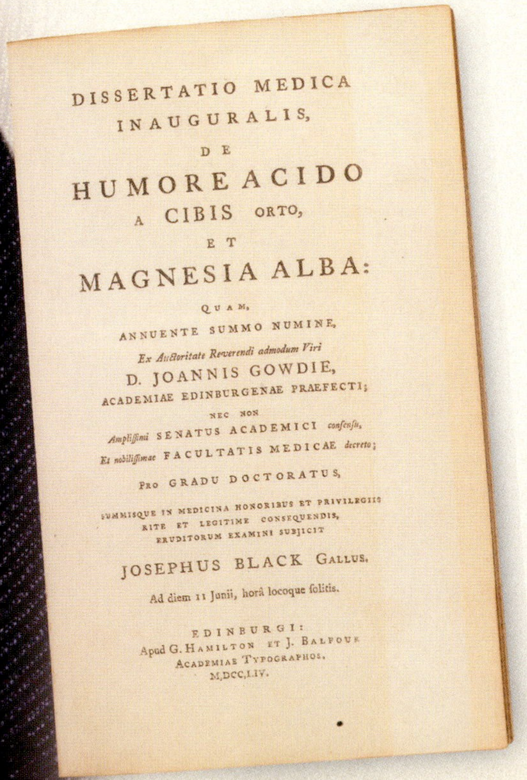

△ ***Magnesia alba***
Black's doctoral thesis of 1754 on the chemistry of magnesium carbonate has a significance in the history of science comparable to that of Isaac Newton's *Opticks*.

**Joseph Black**
Always quiet, refined, courteous, and beautifully dressed, Black was a skilled chemist and a superb lecturer. Throughout his life he enjoyed the reputation of a perfect gentleman.

## INVENTOR AND ENGINEER (1736–1819)

# JAMES WATT

Born in Greenock on the banks of the Clyde, Watt studied instrument-making in London, and set up a workshop in the University of Glasgow. Asked in late 1764 to repair a model Newcomen engine (see pp.130–31), he figured out how to improve its efficiency by using a separate condenser; the idea came to him while he was walking on Glasgow Green in May 1765: "I had not walked further than the golf-house when the whole thing was arranged in my mind."

It took him 10 years to get a full-sized engine working, by which time his wife had died and he had moved to Birmingham to go into partnership with Matthew Boulton. From 1775 until 1800 Watt's steam engines were the world's best. Meanwhile, Watt also invented a copying machine, sun-and-planet gears, a governor to control an engine's speed, and—his proudest invention—"parallel motion" to connect an engine's piston rod to its rocking beam (see pp.132–33).

---

Black began to study what happened when he heated "magnesia alba," a form of magnesium carbonate. With hindsight the chemistry is simple: heating the carbonate drives off carbon dioxide, leaving magnesium oxide. This solid product is quite caustic, and it was thought that when the carbonate was heated over a burner, some caustic material got into it from the flame; therefore the product should be heavier than the starting material. Black carefully weighed his starting material and his product, and showed that weight was lost in the process. No caustic material had been absorbed; on the contrary, something had escaped.

## Isolation of carbon dioxide

Since Black could not see what had escaped, he reasoned that it must be some sort of gas. This he managed to capture, calling it "fixed air" because in those days all gases were called airs, and this had been fixed in the magnesia alba. He was the first person to isolate any pure gas; he then showed that the same gas is produced by reacting limestone with acid, by burning charcoal, and in fermentation, and appears in the breath and in the atmosphere.

This was the first ever example of anyone making careful measurements in chemistry, and doing real analysis.

▷ **Temperature gauge**
Careful measurement of weights and temperatures was central to Black's methodical work. This 1784 sketch of a "temperature gauge" is attributed to Black, and was found with notes about his lectures on heat.

Black wrote up his results in 1754 for his MD thesis, which has been compared in importance to Newton's *Opticks*, for by conducting a beautiful sequence of simple experiments and observations he laid out the basis for quantitative chemistry.

Black used to lecture five times a week, often basing his lectures on his latest research results, and produced spectacular demonstrations to entertain his audience. He received no salary and his income came from students' fees.

## Latent heat

Among Black's students were sons of whisky distillers. They were eager to make their businesses more efficient, and in particular to save money in the distillation process; they wanted to know why they had to spend so much money boiling the wash, and then more money cooling down the vapor to condense the spirit. He pondered this question, and came up with the notion of latent heat (see pp.178–79): heat was needed to boil the wash even when it was at boiling point. The principle also applied to melting solids.

When, in 1758, James Watt obtained a job at the University of Glasgow, Black almost certainly explained the idea of latent heat to Watt, who went on to improve the efficiency of the steam engine.

**University of Edinburgh**
The "Old College," Edinburgh University, faces on to South Bridge Street. Black was appointed as professor of chemistry at Edinburgh in 1766.

Without Black's insight into latent heat, the Industrial Revolution might have been stalled in its tracks.

Black never married, although it seems that he was fond of female company. He performed on the flute and enjoyed going to clubs with friends, including political economist Adam Smith, geologist James Hutton, and philosopher David Hume.

Joseph Black died on December 6, 1799, while at dinner with his family. He set his bowl gently down upon his knees and died, without spilling a drop, somehow contriving to die as precisely and as peacefully as he had lived.

---

## TIMELINE

- **April 16, 1728** Joseph Black is born in Bordeaux, France. Both his parents are of Scottish descent, although his father John was born in Belfast, Northern Ireland.

- **1740** Sent to school in Belfast, where he learns Latin and Greek.

- **1746** Goes to the University of Glasgow to study philosophy and languages, but later, because his father wants him to do something useful, switches to medicine and anatomy.

- **1747** Attends lectures in chemistry by William Cullen, who later employs him as assistant in his laboratory. He goes on to complete his medical degree, but does his best research in chemistry and physics.

- **1752** Goes to the University of Edinburgh to further his studies, and develops his analytical balance to study "magnesia alba" (white magnesia). No one understands what happens when this material is heated, and Black pioneers a new type of analytical research.

- **1754** Publishes his doctoral thesis, a widely praised piece of science writing. The thesis is mainly about the chemistry of magnesia alba, but because this is for a medical degree he includes a section about how it can be used as a purgative and antacid.

- **1756** Returns to Glasgow as professor of anatomy and botany, although he is an expert in neither field. The following year he manages to change jobs.

- **1757** He is appointed Regius Professor of Medicine at Glasgow. He has qualified as a doctor, and is personal physician to the philosopher David Hume, but actually spends his time on chemistry and physics.

- **1757–58** Meets James Watt after the latter takes a job as instrument-maker in the university. The two become friends, and Black becomes Watt's mentor on heat-related matters.

- **1761** Develops the idea of latent heat. He is intrigued by the fact that melting ice stays at the same temperature while it turns into liquid water. He is the first to make a clear distinction between heat and temperature, and introduces the term "specific heat" (see pp.178–79).

- **1766** Returns to Edinburgh as professor of chemistry and medicine, succeeding Cullen, and continues to lecture there for 30 years, although by the end of this time his health is failing.

- **1795** Appoints a former student, Thomas Hope, as joint professor of chemistry. Hope gradually takes over his lecturing duties.

- **December 6, 1799** Dies in Edinburgh, and is buried in the famous Greyfriars churchyard. He leaves an impressively large financial legacy, which he stipulates is to be transformed into 10,000 shares so that it can be divided easily and fairly between his legatees.

**MEMORIAL LOCKET**

**BEFORE**

The first chemists of the Islamic world recognized substances belonging to two classes of organic compounds: alcohol and organic acids.

**ORIGINS OF ORGANIC CHEMISTRY**

The **fermentation process** that turns **fruit juice into wine** was undoubtedly known to people of antiquity, once they found that overripe fruits produced a pleasing beverage. However, it was **Geber ≪53** in the 8th century CE who recognized the **intoxicating component**, released as vapor when wine was boiled. Later, the Iraqi polymath **Al-Kindi** distilled wine to produce **ethanol**—the **first organic compound** isolated by science.

AL-KINDI

**SOUR GRAPES**

Wine can turn into **vinegar** as a result of the reaction of **alcohol** with **oxygen** in air to produce organic acids, predominantly **acetic acid**. Although Geber did not understand the nature of the acid's chemical production, he recognized the chemical affinity of the acidic product with **sour components** of grape residue and lemons— **tartaric acid** and **citric acid** respectively.

**≪ SEE ALSO**
*pp.52–53* ALCHEMY

**CHEMIST (1800–82)**

## FRIEDRICH WÖHLER

Born near Frankfurt, Germany, Friedrich Wöhler began his scientific experiments while still a schoolboy working in his bedroom. As well as achieving the first artificial synthesis of a biological molecule, Wöhler isolated the elements beryllium, aluminium, and silicon. He had an eclectic range of interests, becoming an authority on the chemistry of meteorites and later in life helping to set up a factory for the purification of nickel.

# Organic Chemistry

**Carbon is the foundation of our food, our fuel, and even our bodies, and it can combine with itself and other elements to form a huge variety of natural and synthetic molecules; as a result, organic chemistry—the study of carbon compounds—is a major branch of the subject.**

The term "organic chemistry" was coined in 1808 by the Swedish chemist Jöns Berzelius, originator of the modern method of depicting chemicals by formulae. Berzelius held that organic molecules such as urea could not be made artificially—they could only be made by life. However, German chemist Friedrich Wöhler, during the course of his research in 1828, accidentally created urea, a natural constituent of urine, prompting him to announce: "I can make urea without thereby needing to have kidneys."

Up to that point, the chemistry of organic substances was understood as being based on a mysterious life force. Encouraged by Wöhler's achievement, the German chemist Justus von Liebig later asserted that chemists would one day be able to synthesize sugar and the painkillers salicin and morphine. Liebig maintained that the secret of organic chemistry lay in knowing that the properties of substances containing just carbon and hydrogen (and sometimes oxygen and nitrogen) were as variable as all the other elements put together.

### The carbon atom partnership

The German chemist Adolphe Kolbe was a strong supporter of the idea that not only could organic substances be made, but their precise molecular structures were decipherable too. After the British chemist Edward Frankland

suggested in 1852 that atoms of each element formed a fixed number of bonds, the German chemist Friedrich Kekulé proposed that four bonds around each carbon atom projected into three-dimensional space and that they could link to form chains or even rings. When Dutch scientist Jacobus van't Hoff and French chemist Joseph La Bel independently confirmed the idea of Kekulé's tetrahedral carbon atom in 1874, the age of structural organic chemistry had truly arrived. Organic molecules had definite, sometimes very complex, arrangements of atoms.

### Different groups

Structural theory gave a scientific basis to analysis, as many chemists applied it to dissecting organic molecules. Even the simplest combination of carbon and hydrogen could make long branched

chains, components of combustible fuels. Hydrocarbons with single carbon–carbon bonds are called alkanes; those with more rigid double bonds (and therefore containing fewer hydrogen atoms) are alkenes.

By the 1870s, largely thanks to the pioneering work of Wöhler and Liebig, organic molecules could be classified according to their functional groups: carboxylic (organic) acids, alcohols, and aldehydes contained oxygen; amines contained nitrogen. The affinities of more complex molecules of life were being established too. The German chemist Emil Fischer discovered that proteins were built up from acids with amine groups, and that sugars were aldehydes; he even managed to synthesize sugar artificially. With

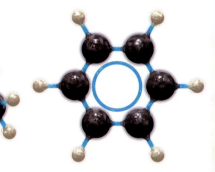

**Carbon compounds**

Carbon atoms have a tetrahedral (pyramidal) arrangement of bonds and can link together to form chains or rings, as shown in the examples below.

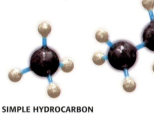

SIMPLE HYDROCARBON (METHANE)     CHAINS OF CARBON (BUTANE)     RINGS OF CARBON (BENZENE)

### Organic molecules

The categories of organic molecules are determined by the functional groups (atoms or groups of atoms) attached to the carbon atoms, as exemplified here by six kinds of two-carbon molecules. Carbon atoms are shown black, hydrogen white, oxygen red, and nitrogen blue.

> "The structural theory of Kekulé has been the **growth hormone** of organic chemistry."
>
> GERALD EYRE, THE THEORY OF ORGANIC CHEMISTRY, 1941

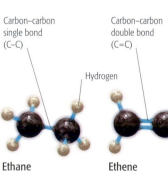

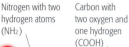

Carbon–carbon single bond (C–C)

Carbon–carbon double bond (C=C)

Hydrogen

Oxygen with hydrogen (OH)

Carbon with two oxygen atoms and hydrogen (COOH)

Carbon with one oxygen and one hydrogen (COH)

Nitrogen with two hydrogen atoms (NH₂)

Carbon with two oxygen and one hydrogen (COOH)

**Ethane**
The "-ane" ending of the name denotes the presence of a carbon–carbon single bond.

**Ethene**
The "-ene" ending of the name denotes the presence of a carbon–carbon double bond.

**Ethanol**
The "-ol" ending of the name denotes the presence of an oxygen–hydrogen group.

**Ethanoic acid**
This has a carbon double-bonded to one oxygen and single-bonded to an oxygen–hydrogen group.

**Ethanal**
The "-al" ending denotes a carbon double-bonded to oxygen and single-bonded to hydrogen.

**Amino-ethanoic acid**
The presence of both an amino (NH₂) and an acid group (COOH) makes this an amino acid.

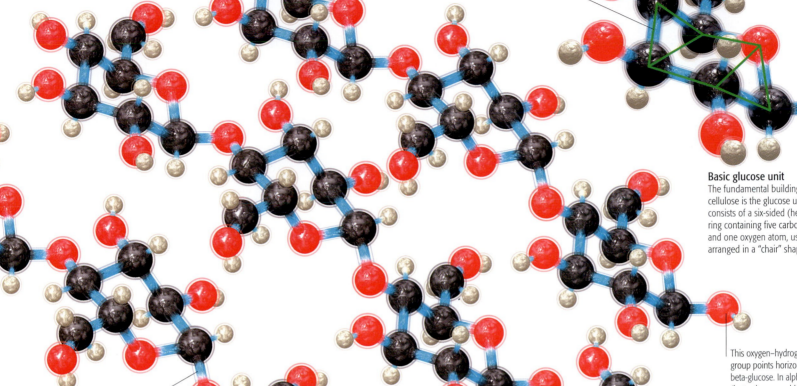

Hexagonal ring arranged in a "chair" shape

**Basic glucose unit**
The fundamental building block of cellulose is the glucose unit, which consists of a six-sided (hexagonal) ring containing five carbon atoms and one oxygen atom, usually arranged in a "chair" shape.

This oxygen–hydrogen (OH) group points horizontally in beta-glucose. In alpha-glucose (better known as blood sugar) the OH group points downward.

Glycosidic bond is a strong covalent bond between adjacent glucose building blocks.

The beta-glucose unit is a form of glucose unique to cellulose. The orientation of its atoms allows parallel chains to link together by hydrogen bonds.

**Structure of cellulose**
Cellulose is the most abundant organic molecule on Earth and makes up the cell walls of all plants. Its molecules consist of billions of glucose (sugar) building blocks, bonded to form long chains, each linked to other chains by hydrogen bonds.

French chemist Charles Gerhardt's synthesis in 1853 of an analgesic using salicylic acid (now best known as aspirin, its original trade name), Justus von Liebig's prediction had come true.

**Cotton grass**
The natural fibers of cotton are virtually pure cellulose. Because of its resilience it is useful for making fabric.

## Organic reactions

Back in 1819 the French chemist Henri Braconnot found that adding sulphuric acid to straw produces sugar. The acid digested the organic cellulose in straw to its sugar building blocks by a reaction called hydrolysis. Living organisms make proteins in their cells by bringing together smaller building blocks (amino acids) and combining them by reacting the acid group from one with the amino group from another – effectively a reversal of hydrolysis. Then in 1903, 75 years after Wöhler made urea, Finnish chemist Gustaf Komppa made camphor. This became the first ever commercially viable product of organic synthesis. The idea that organic chemical reactions were based on a mysterious life force had finally been laid to rest.

### IN PRACTICE

## FRACTIONAL DISTILLATION

Mixtures of liquids can be separated by boiling off each of the components sequentially. As each component is boiled off, its vapor is condensed into liquid—a process called fractional distillation. In the laboratory, sequential separation is done using a fractionating column to collect the vapor from a mixture while it is being heated. Vapor from the component with the lowest boiling point rises to the top of the column, where it passes into a tube and is condensed into liquid. The temperature is then increased and the component with the next highest boiling point is then distilled off. Fractional distillation is done in oil refineries to separate crude oil into more useful components.

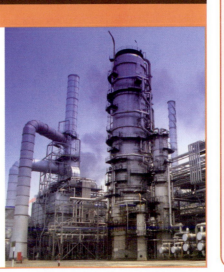

**AFTER ≫**

**Principles of organic synthesis led to ever more ambitious developments during the course of the 20th century.**

### THE AGE OF PLASTICS

In 1907 the Belgian chemist **Leo Baekeland** reacted together the organic molecules **phenol** and **formaldehyde** to produce a hard, moldable material that his company named **Bakelite**. This was the first plastic that retained its shape after being heated, and it heralded a new age of useful synthetic substances.

### ARTIFICIAL LIFE

In 1953 **Stanley L. Miller** and **Harold Urey** showed that **biological molecules** could be **synthesized** from a "soup" of simple molecules. Yet it was not until 2002 that the first **artificial virus**—polio—was made, and the artificial creation of a living cell still eludes scientists.

**SEE ALSO ≫**
*pp.336–37* THE AGE OF PLASTICS

# Plant Life Cycles

**Plant reproduction encompasses everything from flamboyant flowers that entice pollinating insects, to magnificent mosses and ferns whose free-swimming sperm are transported by rainwater or are blown by the wind. Scientists of the 18th and 19th centuries were forced to rely on the microscope—and a great deal of patience—to understand the secret sexual life of plants.**

Sexual reproduction occurs in both plants and animals when sperm and eggs fuse in a process called fertilization. The sperm and eggs each contain half the number of chromosomes and are produced by a special type of cellular division, called meiosis.

## Sexual reproduction

In 1729 Carl Linnaeus (see pp.122–23)—the great pioneer of biological classification—offended the sensibilities of 18th-century society by publishing a treatise on flower biology that was necessarily all about sex. He was articulating something that botanists had only then proved: flowers contained sexual parts. Their pollen-releasing stamens were distinctively male, and their carpels were female.

Seeds developed inside the carpel only after pollination: there was a clear parallel with animal reproduction. For most plants pollination is only

« **BEFORE**

The earliest studies on plant life cycles focused on seed plants that produced flowers, and later on cone-bearing plants.

**FLOWERS AS SEXUAL STRUCTURES**
At the end of the 17th century the German botanist **Rudolf Camerarius** demonstrated the sexuality of plants by showing that **the removal of stamens** from the flowers of the castor oil plant **prevented seed formation**. Previous scientists had suspected that stamens were male organs, but the work of Camerarius provided proof.

**FLOWER CLASSIFICATION**
In 1690 the German-born botanist **Paul Hermann** divided seed plants into **angiosperms** ("seeds within receptacles") and **gymnosperms** ("naked seeds"). This system was adopted by **Linnaeus**, who **divided plants** still **further according to flower structure**. In 1827 the British botanist **Robert Brown** reclassified the term "gymnosperm" to refer mainly to the cone-bearing conifers, whose ovules lie externally on cone scales.

**LINNAEAN SYSTEM**

### Life cycle of a flowering plant
Pollen grains of a flowering plant contain the male gametes (sperm). Pollination brings these male gametes to the female parts of flower, which contain the eggs (female gametes). Fertilization happens within the ovary. The new embryonic plants develop in seeds before germination makes them independent of their parent.

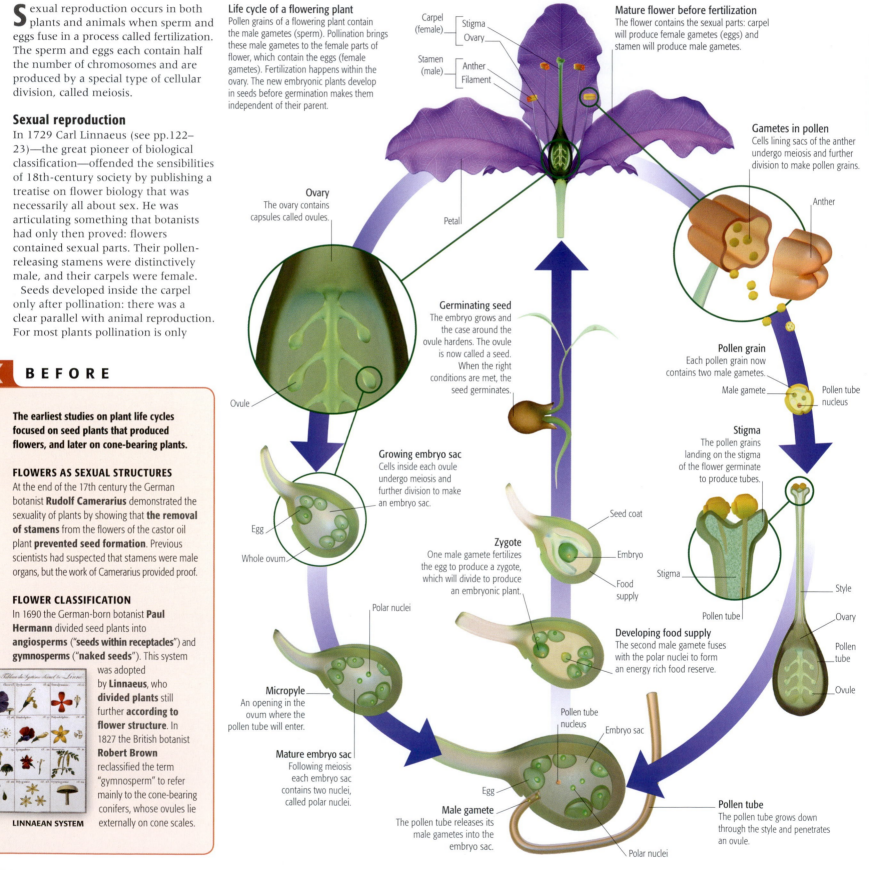

Carpel (female)
Stigma
Ovary
Stamen (male)
Anther
Filament

**Mature flower before fertilization**
The flower contains the sexual parts: carpel will produce female gametes (eggs) and stamen will produce male gametes.

**Gametes in pollen**
Cells lining sacs of the anther undergo meiosis and further division to make pollen grains.

Anther

**Ovary**
The ovary contains capsules called ovules.

Petal

Ovule

**Germinating seed**
The embryo grows and the case around the ovule hardens. The ovule is now called a seed. When the right conditions are met, the seed germinates.

**Pollen grain**
Each pollen grain now contains two male gametes.

Male gamete
Pollen tube nucleus

**Stigma**
The pollen grains landing on the stigma of the flower germinate to produce tubes.

**Growing embryo sac**
Cells inside each ovule undergo meiosis and further division to make an embryo sac.

Egg

Whole ovum

Seed coat

**Zygote**
One male gamete fertilizes the egg to produce a zygote, which will divide to produce an embryonic plant.

Embryo

Stigma

Food supply

Pollen tube

Style

Polar nuclei

**Developing food supply**
The second male gamete fuses with the polar nuclei to form an energy rich food reserve.

Ovary

Pollen tube

Ovule

**Micropyle**
An opening in the ovum where the pollen tube will enter.

**Mature embryo sac**
Following meiosis each embryo sac contains two nuclei, called polar nuclei.

Pollen tube nucleus
Embryo sac

Egg

**Male gamete**
The pollen tube releases its male gametes into the embryo sac.

**Pollen tube**
The pollen tube grows down through the style and penetrates an ovule.

Polar nuclei

## Life cycle of a fern
Spores from the mature adult fern are released to the air and germinate to form gametophyte plants. These plants produce gametes, which fuse to form sporophyte plants and grow to adult ferns. Ferns rely on rainwater for transporting their sperm.

Adult fern

**Fern produces spores**
Mature fern produces spores by meiosis in tiny spore-producing pods called sporangia that sit in clusters beneath small scales on the underside of the leaf.

**Zygote develops into young fern**
A young fern begins to grow from the zygote, while the prothallus starts to wither away. This marks the start of the sporophyte generation.

Prothallus withers

Spore-producing pod

**Zygote after fertilization**
A zygote is formed after sperm from the antheridia swim to eggs in the archegonia and fertilize them.

Female sex organ (archegonia)

Egg

Spore

Mature pod breaks open, releasing spores

**Sex organs mature**
The prothallus develops both male (antheridia) and female (archegonia) sex organs.

Male sex organ (antheridia)

Sperm

Rhizoid

Mature prothallus

**Spore develops into prothallus**
Spores released from the spori germinate on moist ground to form a small structure called a prothallus. This is the gametophyte generation of a fern.

**BOTANIST (1817–1911)**

## JOSEPH DALTON HOOKER

As a young man Hooker was part of James Clark Ross's expedition that confirmed the existence of Antarctica. During the trip he discovered many new species of plants on the islands of the Southern Ocean, and established his reputation as a botanist. A later Himalayan expedition resulted in a major publication on rhododendrons, and in 1865 he succeeded his father as director of the Royal Botanical Gardens, Kew, UK. He became Darwin's closest friend and was an ardent supporter of evolutionary theory.

**Devil's claw**
Hooks on the pods of this plant catch on to the fur of passing animals to help seed dispersal. Black seeds are released from the pods as the hooks split apart.

made possible through the important relationship between insects and flowers, as German botanist Christian Sprengel realized in the 18th century.

### Regenerating through seeds
Sprengel discovered that pollen is transferred—by wind or insect—to the stigma, and fertilization begins (see opposite). The resulting seeds may be scattered from a drying cone or flower, or helped by parachutes that get caught on the breeze. Flowering plants also encase their seeds within a sweet fruit, encouraging their dispersal when an animal eats the fruit.

In all cases the seeds must take their chances: they will only germinate to form a new plant if they land on soil that provides the right conditions for the species. In flowering plants,

fertilization happens safely within the protection of the flower. Mosses and ferns, on the other hand, rely on a two-stage process that occurs outside the plant. The moss life cycle was first fully understood when Johann Hedwig, a botanist from Romania, revealed its secrets in the late 18th century.

### Alternating generations
Hedwig studied mosses under a microscope, and discovered that they too had minute sex organs. Ferns defeated him, but in fact mosses and ferns (see above) go through a similar two-stage reproduction process: the gametophyte generation (whose "haploid" plants carry a single set of chromosomes) and the sporophyte generation (whose "diploid" plants carry the full, double set of chromosomes). In mosses, the sporophyte plant sends up a tall stalk topped by a large cap full of spores. These are released into the air, germinate in the soil, and bud into gametophyte mosses. The gametophytes have female and male organs, and the sperm from the male organs swim through rainwater to reach the female organs and fertilize the eggs. These eggs

form the new generation of sporophytes—they grow within the gametophyte until it bursts open, releasing a new stalk and spore pod, to restart the cycle.

### Two generations
In the mid-19th century a German scientist called Wilhelm Hofmeister first realized that alternation of generations is a characteristic of all plants. Female gametophytes are reduced to structures that are contained within the flower, whereas male gametophytes are within pollen grains. Fertilization happens after the

**AFTER**

**SEE ALSO >>**
pp.154–55 How Plants Work
pp.204–05 Laws of Inheritance
pp.214–15 Animal and Plant Cells

**Animal pollination**
Colorful scented flowers attract insects, which get coated in pollen that sticks because of the grains' spiky coats.

Pollen is of particular interest to research scientists, as well as to those working in agriculture and horticulture.

**POLLEN ANALYSIS**
In 1916 the Norwegian **Lennart von Post** developed the idea of studying pollen grains (palynology), which preserve well, to record the **geographical distribution of plant species**.

**CONTROLLED POLLINATION**
Today agriculturalists rely mainly on insect-pollination for crops to produce fruit, but **hand-pollination** (using a brush) is done

where pollinators are inadequate—such as in greenhouse situations—or where fruit formation is unreliable. Plant breeders also manage pollination, to **control the parentage** and genetics of seeds.

**POLLEN GRAIN**

▷ **Transportation in plants**
Plants have two different types of vessel involved in the movement of water and sugar. Xylem vessels carry a continuous upward stream of water (and dissolved minerals) due to both the "pulling" effect of evaporation from the leaves and the "pushing" effect of water absorbed at the roots. Phloem vessels transport a solution of mixed foodstuffs made in the leaves to many other plant parts—both upward to growing shoots and flowers and downward to roots.

**Palisade mesophyll**
These elongated cells, packed with chloroplasts, are the site of most of the leaf's photosynthesis.

**Xylem vessel**
Water and dissolved minerals are carried by the xylem vessel into the leaf.

**Phloem vessel**
This vessel carries sugar and other food away from the leaf.

**Spongy mesophyll**
Along with the palisade mesophyll, these cells make up the tissue between the upper and the lower epidermis. They fit together with many air spaces between them.

**Lower epidermis**
The cuticle on the lower of the leaf is thinner since it faces away from the sun's drying effects.

**Stoma**
Tiny pores (stomata) allow gases to exchange.

**Upper epidermis**
Epidermal cells produce a wax coating (the cuticle), which helps prevent too much water from being lost by evaporation.

**1 Movement of water by transpiration**
Water and dissolved minerals move from the xylem vessels of the leaf vein into the spongy mesophyll cells.

**2 Evaporation**
Water evaporates from the moist thin walls of the spherical mesophyll cells to create water vapor in the air spaces.

**Air space**

**3 Diffusion**
Water vapor diffuses through the stomata (pores) in the undersides of the leaves into the air.

# How Plants Work

**Plants carry out many functions in order to stay alive. Photosynthesis enables them to make food to sustain themselves and, ultimately, all animals. An efficient transportation system brings raw materials from the soil to the leaves, and moves the food to where it is needed.**

Today we understand that in order to sustain itself and grow, a plant converts the inorganic substances carbon dioxide and water into organic components, such as sugar. This process is known as photosynthesis and utilizes the ever-abundant energy from the sun. Knowledge of photosynthesis, and how both the raw materials and the products move around the plant, grew from experiments that began in the 1600s.

### Growth from air and water
Jan van Helmont, a Flemish chemist, showed through experiments on a willow tree over a five-year period in the early 1600s that plants grew by absorbing water from the soil. But the importance of air (or carbon dioxide, to be exact) to plant growth was not recognized until over a century later by the Dutch biologist Jan Ingenhousz. His studies clarified the basic requirements for photosynthesis—light, water, and carbon dioxide. Carbon dioxide provides the carbon that is key to making organic food, such as sugars, fats, and proteins.

Photosynthesis was confirmed as a chemical process by Swiss chemist, Nicolas-Théodore de Saussure, in 1804. Further studies revealed details of this process, such as the role of the green pigment chlorophyll in chloroplasts (see p.215) in harnessing the energy of the sun.

### Transporting water and minerals
An understanding of how water and food substances are transported in plants was also built up gradually. In 1727 the British physiologist Stephen Hales showed that the water that evaporates at the leaves of plants is replaced by uptake from the soil. This evaporation results in a stream of water (known as the transpiration stream) being pulled up through plants' xylem vessels (see above). There is also a push from water entering the roots via the osmosis.

### Transporting food
It was not until the 1930s that a German biologist, Ernst Münch, proposed a theory to explain the movement of food through a plant—a process known as translocation.

He stated that sugar travels through phloem vessels (see above) from high-pressure sources, such as the leaves, to places of low pressure such as the roots.

### The pathway to sugar
In 1961 the American chemist Melvin Calvin received the Nobel Prize for Chemistry for working out the photosynthetic reactions that resulted in the formation of glucose from carbon dioxide. The process is known as the Calvin cycle.

◁◁ **B E F O R E**

**The first breakthroughs in the understanding of plant function were due to the efforts of experimental chemists and physiologists in the 17th century.**

**THE SECRET OF THE SOIL**
In 1627 the English scientist **Francis Bacon** published *Sylva Sylvarum*, describing experiments where he grew plants in water. He concluded that soil merely supported plants and water alone contributed to growth.

**THE EXISTENCE OF GASES**
The **modern concept of gas** was developed by the Flemish chemist **Jan van Helmont**. He showed that air is composed of different gases, one of which was emitted by burning charcoal. This gas later became known as **carbon dioxide**. Van Helmont was never to appreciate its role in photosynthesis.

**PLANT PHYSIOLOGY**
**Marcello Malpighi**, an Italian physician, carried out studies on the structure of plants and in 1671 published the first significant contribution to **plant function**, *Anatomia Plantarum*.

**ANATOMIA PLANTARUM**

◁◁ **SEE ALSO**
*pp.140–41* **STATES OF MATTER**
*pp.152–53* **PLANT LIFE CYCLES**

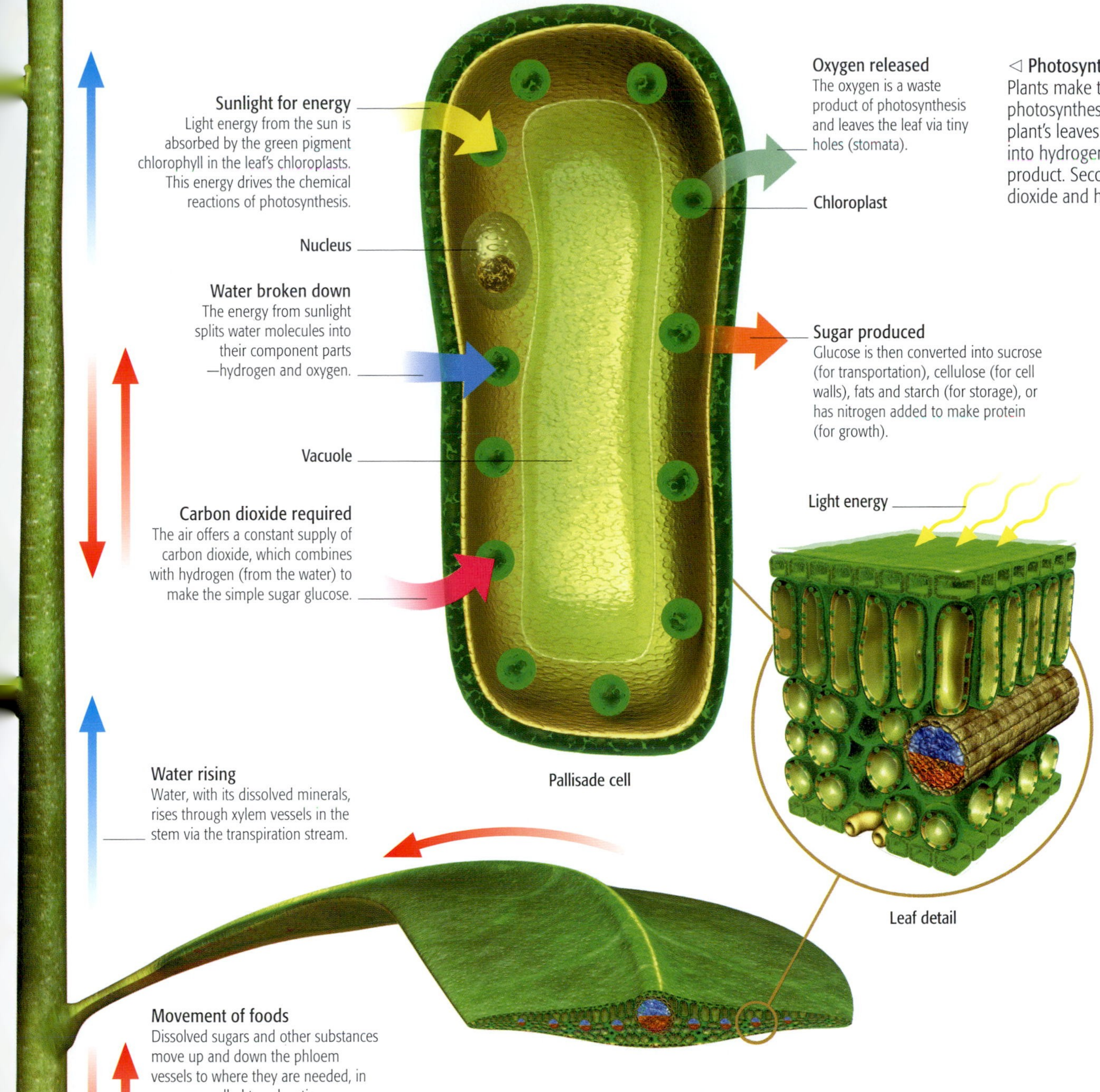

### Sunlight for energy
Light energy from the sun is absorbed by the green pigment chlorophyll in the leaf's chloroplasts. This energy drives the chemical reactions of photosynthesis.

Nucleus

### Water broken down
The energy from sunlight splits water molecules into their component parts —hydrogen and oxygen.

Vacuole

### Carbon dioxide required
The air offers a constant supply of carbon dioxide, which combines with hydrogen (from the water) to make the simple sugar glucose.

### Oxygen released
The oxygen is a waste product of photosynthesis and leaves the leaf via tiny holes (stomata).

Chloroplast

### Sugar produced
Glucose is then converted into sucrose (for transportation), cellulose (for cell walls), fats and starch (for storage), or has nitrogen added to make protein (for growth).

◁ **Photosynthesis**
Plants make their food in a two-stage process called photosynthesis, which takes place inside chloroplasts in the plant's leaves. First, light energy is harnessed to split water into hydrogen and oxygen, which is released as a waste product. Second, a series of chemical reactions brings carbon dioxide and hydrogen together to make glucose.

Light energy

Pallisade cell

Leaf detail

### Water rising
Water, with its dissolved minerals, rises through xylem vessels in the stem via the transpiration stream.

### Movement of foods
Dissolved sugars and other substances move up and down the phloem vessels to where they are needed, in a process called translocation.

▽ **Absorption of water and minerals**
The roots of a plant not only help to anchor it in the soil but also absorb water and dissolved minerals. Tiny root hairs increase the roots' surface area for absorption. Water moves from the soil (an area of low solute concentration) into the root (an area of high solute concentration) by the process of osmosis; this water flow forms a "pushing" force to drive water through the root and up the stem.

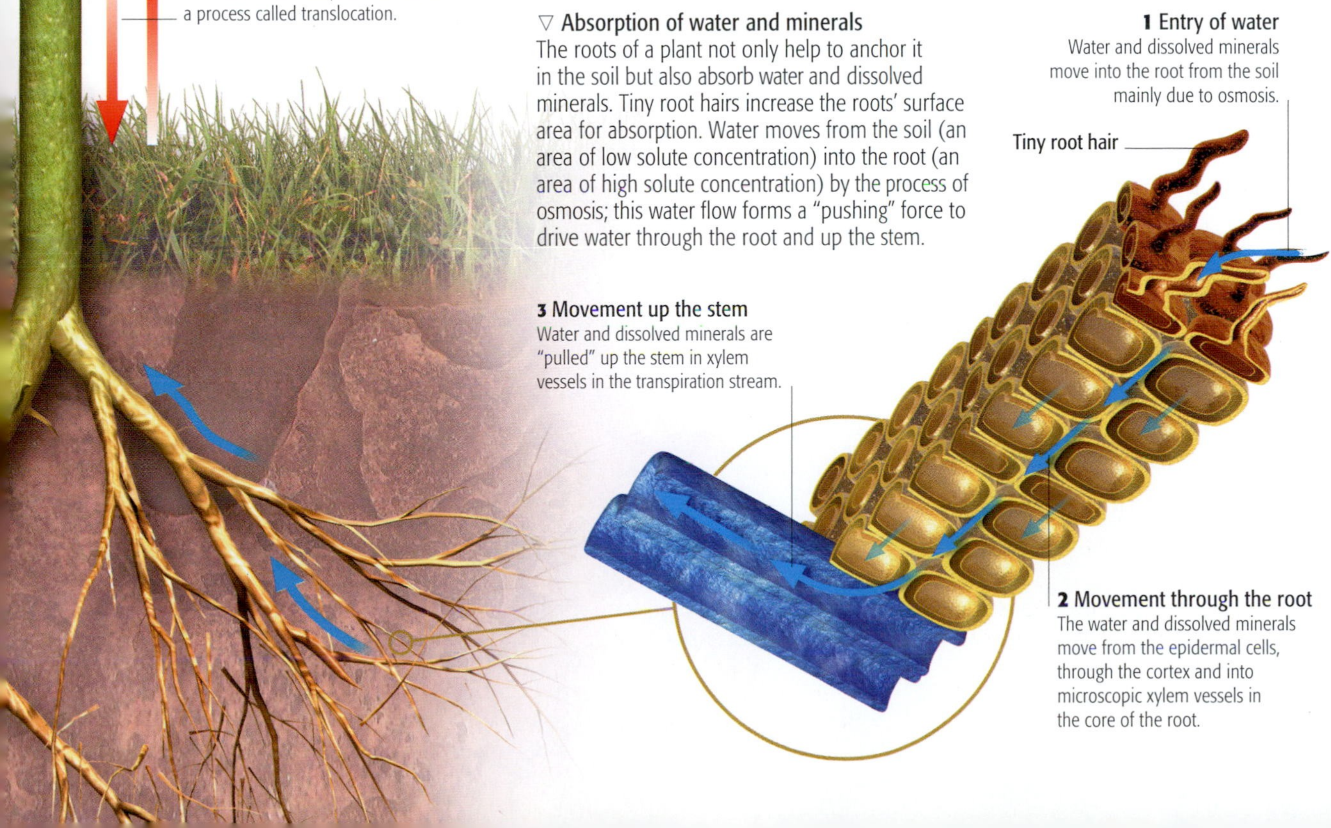

**3** Movement up the stem
Water and dissolved minerals are "pulled" up the stem in xylem vessels in the transpiration stream.

**1** Entry of water
Water and dissolved minerals move into the root from the soil mainly due to osmosis.

Tiny root hair

**2** Movement through the root
The water and dissolved minerals move from the epidermal cells, through the cortex and into microscopic xylem vessels in the core of the root.

**AFTER** »

The rate of photosynthesis is affected by three factors: light intensity, carbon dioxide concentration, and temperature—any of these can be limiting factors.

### MAXIMIZING GROWTH
Farmers can **increase crop growth in greenhouses** by using their knowledge of factors that control the rate of photosynthesis. They may use artificial light so that photosynthesis can continue beyond daylight hours. Use of **paraffin lamps** increases the rate of photosynthesis by producing carbon dioxide, and heat too.

### CHEMICAL PLANT GROWTH REGULATORS
In 1928 the Dutch biologist **Frits Warmolt Went** found that plants contain **minute quantities of auxin**, a chemical that controls the growth of plants. It is responsible, for example, for shoots growing toward the light. Auxins and similar chemicals are now widely used in horticulture.

**SEE ALSO** »
*pp.214–15* ANIMAL AND PLANT CELLS

PHYSIOLOGIST (1677–1761)
### STEPHEN HALES
An English parson and fellow of the Royal Society, Stephen Hales made significant advances in experimental physiology, furthering the understanding of animal and plant function. His essay, *Vegetable Staticks*, published in 1727 showed the role of plant transpiration; *Haemostaticks*, 1733, concerned transport in animals. He was the first to figure out blood pressure, blood flow rate, and the heart's capacity. He invented the ventilator, which was first presented to the Royal Society in 1741.

# The First Vaccination

**In every year of the 18th century nearly half a million people in Europe—and countless others around the world—died from a disease that had blighted humanity since antiquity: smallpox. This disease had defeated the best medical minds, until an English doctor made a discovery that would change the course of medical history.**

In 1773 Edward Jenner, newly trained in London, set up a medical practice in the village of Berkeley in England's Gloucestershire. Jenner had been mentored by John Hunter—a surgeon noted for his ground-breaking experiments—and the young country doctor had every sign of being faithful to Hunter's motto, "don't think, try." Jenner was passionate about addressing the inadequacies of his profession, and, perhaps inevitably, his thoughts turned to smallpox.

The disease was traditionally controlled by deliberate inoculation: scratching material from scabs of a milder form of smallpox into the skin to boost immunity. This had been widely practiced in Asia since antiquity, and by the early 1700s it was being adopted by the elite of Europe. Evidently the prospect of protection from smallpox was reason enough to risk infection. However, medicine needed a safer alternative. Jenner approached the problem as a scientist and noted a pattern: milkmaids who visited his medical practice having caught cowpox from their cows—a far more benign disease—did not contract smallpox. In 1796, when a milkmaid suffering with cowpox visited Jenner as a patient, he saw his opportunity to test the protective potential of the disease against smallpox. On May 14, 1796 he transferred blister fluid from the milkmaid to both arms of James Phipps, the eight-year-old son of his gardener: this was the world's first vaccination.

Next Jenner needed to find out whether his vaccination had worked and on July 1 he did the unthinkable —he deliberately attempted to infect Phipps with smallpox. To Jenner's relief the boy not only survived, but further tests continued to prove that he was unaffected by smallpox exposure for the rest of his long life.

In 1840 Jenner's early trials finally persuaded the British government to substitute inoculation with his new vaccination. Jenner was admired by government and queen alike. Smallpox vaccination would go on to save billions of lives and ultimately precipitate a global initiative that would be the first—and still the only—program to eradicate an infectious disease.

**Jenner performing his first vaccination**
Edward Jenner vaccinated James Phipps using a vaccine made from the cowpox blisters of a milkmaid. The source of the cowpox was a Gloucester cow called Blossom and her hide now hangs in St. George's Hospital in London, where Jenner trained to be a doctor.

**"It is a privilege to make a real difference in your own lifetime."**

EDWARD JENNER (1749–1843)

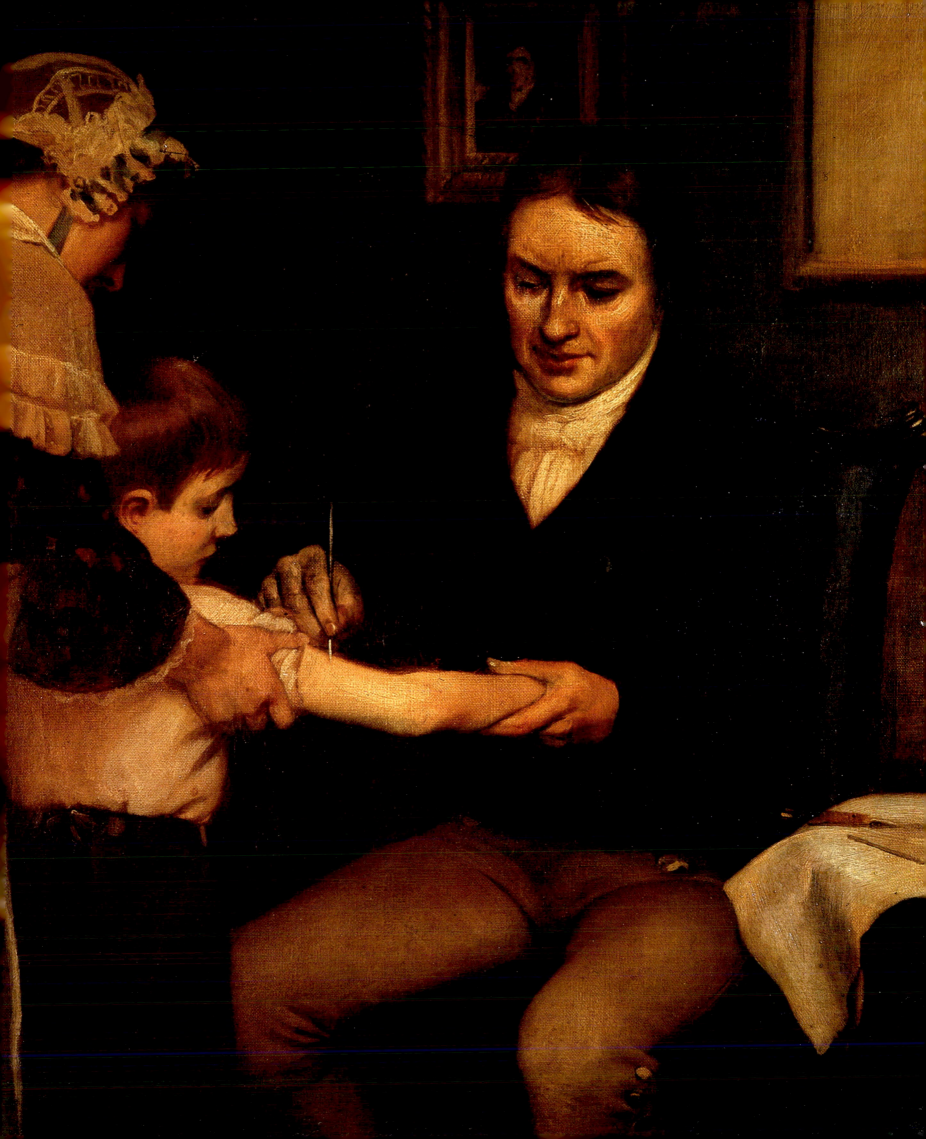

# Static Electricity

**From the dramatic flash of a lightning bolt to the gentle, invisible force that makes a rubbed balloon stick to a wall, the effects of static electricity are everywhere. They are all due to a simple imbalance of electric charge, and the fact that opposite charges attract one another while like charges repel.**

## BEFORE

People probably observed the effects of static electricity in ancient times, by rubbing objects with fur, for example, and seeing lightning.

### DISCOVERY OF STATIC

In 600 BCE the Greek philosopher **Thales of Miletus** noted that rubbing fur on the natural material **amber** (fossilized tree resin) gave the amber the ability to attract feathers and other small objects. Amber has an especially high **electrical resistance**; its own **electrons** are bound very tightly to their atoms, but amber does allow electrons from other objects to **migrate** onto it, so that it becomes **negatively charged** overall. The electrons in materials such as fur, feathers, hair, wool, silk, and paper are not so tightly bound, which allows them to have an **electrical attraction** to materials such as amber. The word for amber in Greek is "elektron," which gave us the word "**electricity.**"

**AMBER ATTRACTING FEATHER**

Its usage can be traced back to the British scientist **William Gilbert ‹‹ 81**, who used the word "**electricus**" to describe the force that exists between charged objects.

Electric charge comes in two varieties, either positive (+) or negative (–). All everyday objects contain these two charges, since they are all made up of atoms—and every atom contains both positively charged protons (within its nucleus) and negatively charged electrons (which orbit around the nucleus).

Generally, the negative charges balance the positive ones; atoms are electrically neutral. But the orbiting electrons can move from one atom to another in certain circumstances—such as when some materials are rubbed together—which confers an overall negative charge on the atoms they migrate to, and a net positive charge on the ones they leave. For example, when rubbing a balloon on a sweater, friction causes the negative charges to move from the sweater to the balloon, making it negatively charged.

Any two oppositely charged objects (+ and –) attract each other, and similarly a large imbalance of negative charges in one object will attract the positive charges of a second object. Sometimes, when two objects come near each other a stream of electric charges passes across the gap between the objects, causing a spark, which equalizes the imbalance of electric charges—as with lightning.

### Experimenting with static

Investigations into static electricity became popular during the 17th century. At that time, scientists thought that heat, magnetism, and electricity were made of weightless, invisible fluids, like gases or liquids. When they charged objects with static electricity, they believed they were filling them with electrical fluid—in English the verb "charge" originally meant "fill."

Early experimenters were able to make small sparks by charging objects and placing them close to each other. In 1745 Dutch scientist Pieter van Musschenbroek invented a device called a Leyden jar, which made it possible to store static electric charge. In 1752 American statesman and scientist Benjamin Franklin proved that lightning bolts are just huge electrical sparks. In his classic experiment, Franklin—after insulating himself to avoid electrocution—flew a kite up into a thundercloud and successfully charged up a Leyden jar by drawing electrical charge down the wet string from the cloud. He also made sparks fly from a metal key attached to the string.

### Electric force fields

About 20 years later French physicist Charles-Augustin de Coulomb was measuring the forces between objects that were charged with static electricity, known as electrostatic forces. Coulomb formulated an equation that relates the strength of the force, amount of charge, and the distance between two charged objects. The equation proves that doubling the distance between objects, for example, reduces the force to one quarter of its original magnitude. The unit of electric charge, the coulomb, is named after him.

**Detecting static electricity**
The gold-leaf electroscope, an instrument for detecting static electricity, relies on the fact that inside metals electrons are free to move around. A charged object just above an electroscope causes the metal parts to become charged, and this causes the foils to separate. The greater the charge, the higher the foil lifts.

**1 Charge introduced**
A positively charged resin rod is moved over the metal plate of the electroscope to induce a charge.

**3 Negative charge**
The metal plate becomes negatively charged due to the arrival of free electrons.

**2 Movement of electrons**
The free electrons from the gold-leaf foil pass up the metal rod to the metal plate.

Escaped free electrons move to metal plate

Metal plate

Glass case protecting gold leaves from damp

Metal rod

Gold-leaf foil strips which hang together when uncharged

Gold atom

Negatively charged electron

**4 Gold leaves part**
As the gold leaves become positively charged, they move apart.

**BEFORE CHARGE INDUCED**

**AFTER CHARGE INDUCED**

## AFTER

By the end of the 19th century physicists were able to make sense of static electricity and how it related to atoms.

### THE CONNECTION WITH MAGNETISM

In the 17th century scientists believed **electrical** and **magnetic forces** were completely separate until, in the 1870s, British physicist **James Clerk Maxwell** proved they were both forms of a single phenomenon: **electromagnetism 166–67 ››**.

### THE FORMS OF ELECTRIC FIELDS

In 1839 British scientist **Michael Faraday 170–71 ››** visualized the shapes of electric fields and proved that **static** and **current electricity** were different aspects of the same phenomenon.

### ATOMS AND IONS

The **electron** was discovered by physicist Joseph John (J.J.) Thomson in 1897. In due course **quantum physics 314–17 ››** was able to work out the behavior of electrons in detail.

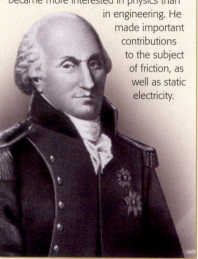

## PHYSICIST (1736–1806)

### CHARLES COULOMB

Like any other child of a wealthy French family in the 18th century, Charles-Augustin de Coulomb received a good education in literature, philosophy, and the sciences. He particularly excelled at mathematics, and after further study in applied math he became a qualified civil engineer. He worked for 20 years on large engineering projects for the French Crown, in several different countries. After winning a prize in a competition organized by the French Academy of Sciences, Coulomb became more interested in physics than in engineering. He made important contributions to the subject of friction, as well as static electricity.

**Storing electric charge**
The Leyden jar was the first capacitor, "storing" electricity between two electrodes.

**Lightning strike**
In a thundercloud, violent updrafts cause ice crystals to collide with each other, creating positively and negatively charged fragments. The negative fragments usually collect at the bottom of the cloud; objects on the ground then develop the opposite charge, which allows the huge amount of electricity that has built up in the cloud to discharge itself in the form of a lightning bolt.

SCIENTIST, POLYMATH, AND POLITICIAN **Born 1706** Died 1790

# Benjamin Franklin

## "He seized the **lightning** from **Heaven** and the **sceptre** from the hands of **tyrants**."

ANNE ROBERT JACQUES TURGOT, BARON DE LAUNE, 1778

Benjamin Franklin had what would be described today as a portfolio career—he was a printer, author, publisher, diplomat, librarian, inventor, philanthropist, scientist, and one of the leaders of the American Revolution. Famous for flying a kite during a thunderstorm, he was a pioneer in disproving the myth that electricity had anything to do with the gods. Due to the bravery and curiosity of Franklin and scientists like him, electrical power is now an everyday reality.

### Modest beginnings

Born in Boston, Massachusetts, Franklin was soon to gain a reputation for being industrious and something of a genius, but unfortunately at the age of just 10 his parents' funds ran out and his schooling was cut short. For the following two years he worked in his father's chandlery business, but it

was clear that a career of filling molds and cutting wicks for candles would not satisfy him. Desperate for a change and driven by his fascination for and love of books, Franklin at 12 years old became an apprentice to his elder brother, who was a printer.

### Life as a printer

Books took over Franklin's life and he would do anything he could to further his passion. He stopped going to church to secure more time to read, and he became vegetarian to save money—for the purchase of more books. However, at the age of 17 he became unhappy with his life and ran away to Philadelphia, and then to London, in search of new discoveries. In London he gained employment with a printer, but his new life was not all he had imagined and he soon returned to Philadelphia, determined to set up his own printing house.

In Philadelphia Franklin published *The Pennsylvania Gazette,*

**Benjamin Franklin**
Seen here in a *c.*1745 portrait by Joseph Siffred Duplessis, Franklin was one of the most influential Founding Fathers of the USA. His work in science took place relatively late in his varied career.

### Printing and communication

Franklin worked in printing until 1757, for almost four decades of his life. He was a central figure in the creation of networks that facilitated communication between the British colonies in North America.

## INVENTION

## BIFOCAL SPECTACLES

In the early 1760s Benjamin Franklin, who was far-sighted, hit upon the idea of incorporating lenses for both distant and close viewing within a single spectacle frame. He had what he called his "double spectacles" made by opticians in London and France. Four full lenses had to be cut in half and shaped to provide the four half-lenses. A pair of bifocals could be expensive, especially since lenses tended to break during cutting.

issued the famous *Poor Richard's Almanack*, founded the first public library in Philadelphia, and created the first American volunteer firefighting company. Public affairs became an important part of Franklin's life and in 1743 he set in place a plan for an academy; opening eight years later, this would eventually become the University of Pennsylvania.

With a thriving business, and printing partnerships in other cities, Franklin retired from his business activities, deciding instead to concentrate on scientific experimentation and invention. This was not wholly unexpected because in 1743 Franklin had already invented a heat-efficient stove, and had devised swim fins and a musical instrument

called the glass armonica. But his most remarkable achievements were to be in the field of electricity.

In 1746 Franklin saw some electrical demonstrations and became obsessed with the entire subject. Aided by crude apparatus sent to him by a friend, Peter Collinson, he set about investigating the nature of electricity.

### Electrical experimentation

Greek philosophers had discovered that rubbing amber against cloth would cause lightweight objects to stick to the amber. Franklin wanted to prove that electricity was a mysterious force that diffused through most substances: the static electrical charge was not created by rubbing, it was merely transferred from the amber to the cloth.

During his experiments Franklin was well known for playing pranks on his neighbors; he gave shocks to people by touching or kissing them, sent a current through water to ignite alcohol, caused fake spiders to move mysteriously, and even produced artificial flashes of lightning.

Franklin's amusing experiments had a serious purpose, however. In the 18th century beliefs about the natural world were steeped in superstition and folklore; lightning, for example, was considered a sign of God's wrath. Enlightened thinkers such as Franklin wanted to prove objectively that there was nothing supernatural involved.

### Franklin's kite

With the help of a simple kite made from a silk handkerchief and two sticks, Franklin proved that lightning was actually a form of electricity. He attached the kite to a length of twine, tying a metal key to the other end. Franklin then flew the kite into a thunderstorm. This was actually a very dangerous thing to do; a Russian professor died trying to repeat the experiment. Franklin was careful to

### Pillar of government
Franklin was the first to propose that Congress be split into two houses, one for senators and the other for representatives. The idea was eventually adopted, which is why the Capitol has two wings, one for each house.

insulate himself to avoid electrocution, but when electricity was conducted down the wet twine he was able to draw electrical sparks from the key.

Franklin's experiment may have seemed crazy at the time, but from the information it yielded he was able to develop the lightning rod, which he believed would save buildings struck by lightning from burning to the ground. He knew that lightning always struck at the highest point, so he placed his metal rod at the top of each structure.

### The nature of lightning
By drawing lightning down the rain-soaked twine of his kite to a metal key, Franklin proved that it was a natural form of electrical current. Today lightning is classified as plasma, the fourth form of matter.

The lightning hit the rod and the electricity was conducted down the side of the building to discharge harmlessly into the ground, leaving the building undamaged. Franklin considered the lightning rod to be his most valuable invention. Today lightning rods are hardly changed from Franklin's original design.

Franklin's legacy to science does not stop with his list of inventions. He developed the theory of positive and negative electricity, and during his many experiments he also coined the terms "battery," "charge," "condenser," and "conductor."

### TIMELINE

- **January 17, 1706**. Born in Boston, Massachusetts.
- **1715** Forced to leave Boston Latin School by his parents' lack of funds.

**BOSTON LATIN SCHOOL**

- **1718** Becomes apprentice to his brother James, a printer.
- **1723** Runs away from his apprenticeship because of bullying from his jealous brother. He goes to Philadelphia and then to London.
- **1724** Returns to Boston. Tries to borrow money from his father to start his own print shop. His father refuses, so he returns to Philadelphia.
- **1727** Suffers first pleurisy attack. The same year Franklin has an affair with a woman that results in the birth of his illegitimate son, William.
- **1730** Elected the official printer for Pennsylvania. Takes a common-law wife, Deborah Read Rogers.
- **1732** Birth of his son, Francis Folger. He begins to print *Poor Richard's Almanack*, and America's first German-language newspaper, *Philadelphische Zeitung*, which soon fails.
- **1734** Elected Grand Master of the Grand Masonic Lodge of Masons of Pennsylvania.
- **1735** Brother James dies.
- **1736** Son Francis (Franky) Folger dies of smallpox at four.
- **1737** Appointed Postmaster of Philadelphia.

**FRANKLIN'S ALMANACK**

- **1741** Advertises the "Franklin Stove."
- **1743** Publishes *A Proposal for Promoting Useful Knowledge* for the prototype American Philosophical Society. Daughter Sally is born.
- **1744** His father, Josiah Franklin, dies.
- **1747** A friend from London, Peter Collinson, sends him a Leyden jar (electrostatic capacitor); Franklin embarks on a journey of extensive electrical experiments.
- **1752** Proves his theories of electricity with his famous kite experiments. Awarded the Copley Medal by the Royal Society, London.
- **1762** Maps postal routes in the colonies. Invents the glass armonica.
- **1764–65** Charts Gulf Stream.
- **1774** His partner, Deborah, dies in Philadelphia.
- **1785** Begins four-year term as president of the Supreme Executive Council of Pennsylvania.
- **1787** Signs the United States Constitution.
- **1789** Becomes President of the Society for Promoting the Abolition of Slavery.
- **April 17, 1790** Dies in Philadelphia. More than 20,000 mourners attend his funeral.

**Volta's demonstration**
Volta successfully generated electricity from his wet cell battery when demonstrating it to the Institute of France in November 1801, in the presence of Napoleon Bonaparte.

# The First Battery

**When the battery was invented more than 200 years ago it marked a true breakthrough. Luigi Galvani had demonstrated "animal electricity" in the 1780s, but in 1791 Alessandro Volta showed that a circuit of two different metals and brine produces an electric current. After much experimentation, in 1799 he made a "Voltaic pile"—the first battery.**

In 1791 Volta had generated an electric current by separating copper and zinc with paper soaked in brine. Before then it had only been possible to generate a few sparks in a laboratory—but Volta's discovery made it possible to produce a steady supply of electrical power.

In 1799 Volta found that by stacking disks of brine-soaked paper, zinc, and copper in multiple layers, the amount of electricity generated was greatly increased. The Voltaic pile was the first practical method of generating electricity, and would later power some key scientific discoveries.

Volta first reported his electric pile in a letter to the Royal Society of London in 1800, but it was a demonstration in Paris one year later that secured his fame. At that time French politicians were very eager to support any new ideas or inventions that might give them an edge on their political rivals, so Napoleon Bonaparte himself invited Volta to address the Institute of France with his findings. He even helped out with the demonstration, by melting a steel wire and discharging an electric pistol.

Napoleon rewarded Volta by making him a count and a senator of the kingdom of Lombardy. But the Italian scientist would have been most proud of the fact that in 1881, in recognition of all his research in the field of electricity, the standard unit of electrical potential was named after him. It has been known ever since as the "volt."

The Voltaic pile was a "wet cell" battery. After Volta, various wet cell batteries were developed, with acid instead of brine as the electrolyte. All were prone to leakage and corrosion. Many also used glass jars, which were easily broken. Not until the development of "dry cell" batteries—which replaced the liquid electrolyte with a paste—were portable electrical devices practical. Today we have a variety of dry cell batteries, including rechargeable ones, and take Volta's invention very much for granted.

**A Voltaic pile**
Volta stacked zinc, copper, and brine-soaked paper disks in layers—zinc, paper, copper, zinc, paper, copper, and so on. When a wire was connected to both ends of the pile, an electric current was generated. Adding more disks increased the amount of electricity generated.

## "It is the **difference in the metals** that does it."

ALESSANDRO VOLTA, PHYSICIST, 1792

## BEFORE

Scientific investigations of static electricity in the 18th century laid the foundations of much of our present understanding of it.

### STATIC ELECTRICITY
Scientists experimented with producing and capturing **static electricity ≪ 158–59** extensively in the 18th century.

### THE BAGHDAD BATTERY
In 1936 archaeologists working near Baghdad, Iraq found a pot dating from around 200 BCE. Its contents suggested that it had been used in the production of an **electric current**. It contained a copper cylinder and an iron rod, around the end of which were traces of pitch, possibly used to **insulate** the iron from the copper. Residues suggested that an acidic liquid, such as lemon juice, was used to start an **electrochemical reaction** between the two metals.

### ≪ SEE ALSO
*pp.158–63* STATIC ELECTRICITY
*pp.162–63* THE FIRST BATTERY

# Electric Current

**Electricity is unusual in that it is a form of energy that can move along wires, in a flow called "electric current." This manageability has made it one of science's most useful and popular discoveries, from providing the power for lighting and heating, to running computers and cell phones.**

During the 1780s Italian physicist Luigi Galvani discovered that a dead frog's leg would twitch whenever it was hung from an iron hook and its nerves were touched by a copper hook. Galvani had also observed the same effect when frogs' legs were shocked by a spark of static electricity. He believed that "animal electricity" from the muscles was causing the legs to twitch.

### Development of the battery
Another Italian physicist, Alessandro Volta, suggested that the electricity was being produced because two different metals were present (iron and copper). In 1799 he built a device that could produce a constant and reliable electric current. Volta's invention, the "Voltaic pile," was the world's first battery (see pp.162–63), and gave other scientists a convenient way to produce electric current on demand.

### Electromotive force
Volta realized that the pile of metal disks was somehow driving the electric current. Scientists call this effect "electromotive force" (emf). The stronger the emf, the greater the current. Emf is also referred to as "voltage," and the unit of emf is called the "volt" in Volta's honor.

In 1810 English scientist Humphry Davy showed that it was not simply the presence of two different types of metal that was responsible for the emf. Instead, it was a chemical reaction between the two metals (the electrodes) and the liquid solution (the electrolyte). Each copper-cardboard-zinc unit is an "electrochemical cell."

**VOLT** The unit of electromotive force (emf). The greater the emf—the driving force—the greater the current.

**AMP** The unit of current, used to measure the rate of flow around a circuit.

**OHM** The unit of electrical resistance within a circuit. The greater the resistance, the lower the current.

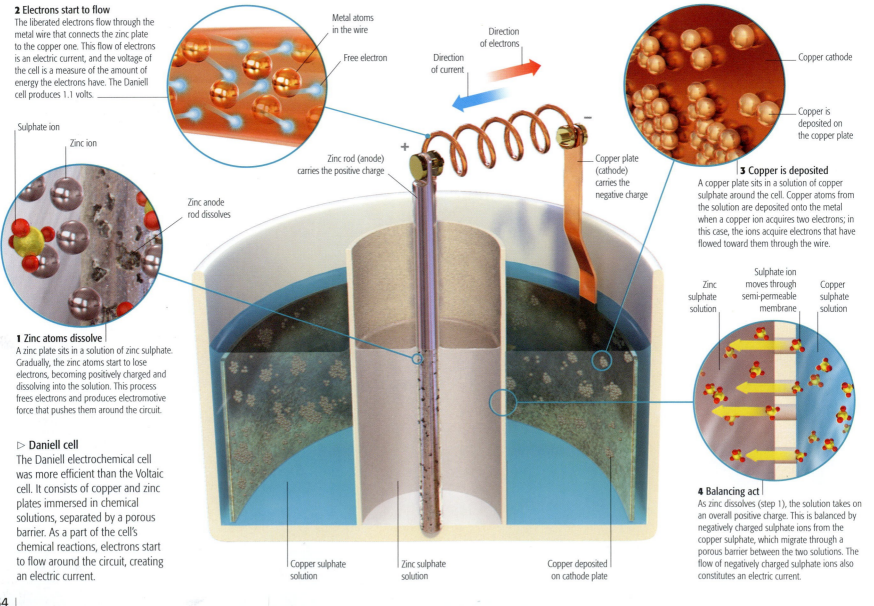

**2 Electrons start to flow**
The liberated electrons flow through the metal wire that connects the zinc plate to the copper one. This flow of electrons is an electric current, and the voltage of the cell is a measure of the amount of energy the electrons have. The Daniell cell produces 1.1 volts.

Metal atoms in the wire

Free electron

Direction of electrons

Direction of current

Copper cathode

Copper is deposited on the copper plate

**3 Copper is deposited**
A copper plate sits in a solution of copper sulphate around the cell. Copper atoms from the solution are deposited onto the metal when a copper ion acquires two electrons; in this case, the ions acquire electrons that have flowed toward them through the wire.

Sulphate ion

Zinc ion

Zinc anode rod dissolves

Zinc rod (anode) carries the positive charge

Copper plate (cathode) carries the negative charge

**1 Zinc atoms dissolve**
A zinc plate sits in a solution of zinc sulphate. Gradually, the zinc atoms start to lose electrons, becoming positively charged and dissolving into the solution. This process frees electrons and produces electromotive force that pushes them around the circuit.

▷ **Daniell cell**
The Daniell electrochemical cell was more efficient than the Voltaic cell. It consists of copper and zinc plates immersed in chemical solutions, separated by a porous barrier. As a part of the cell's chemical reactions, electrons start to flow around the circuit, creating an electric current.

Copper sulphate solution

Zinc sulphate solution

Copper deposited on cathode plate

Zinc sulphate solution

Sulphate ion moves through semi-permeable membrane

Copper sulphate solution

**4 Balancing act**
As zinc dissolves (step 1), the solution takes on an overall positive charge. This is balanced by negatively charged sulphate ions from the copper sulphate, which migrate through a porous barrier between the two solutions. The flow of negatively charged sulphate ions also constitutes an electric current.

## DRY CELL

While "wet" cells such as the Daniell cell (see opposite) use liquid chemical solutions, dry cells contain their chemical solutions in the form of a paste and are therefore far more convenient. First invented in around 1866 by the French engineer Georges Leclanché, the dry cell is also extremely portable, since all its materials are sealed tightly into a metal case. The most common type of dry cell today is the alkaline cell, which contains zinc powder and a paste of manganese dioxide to provide the electric current.

Steel cap is the positive terminal

Brass or steel pin acts as a current collector

Zinc powder (anode)

Manganese dioxide paste (cathode)

Carbon powder is mixed with manganese dioxide to increase conductivity

Steel base is the negative terminal

There are many other types of electrochemical cell, with different electrolytes and different electrodes but they all produce an emf as a result of chemical reactions.

### Electrical generators

Soon after Christian Ørsted's discovery of electromagnetism in 1820 (see p.80), scientists began to develop other ways of driving current around a circuit: electrical generators and dynamos. Alternators produce an emf that continuously changes direction, creating a current that flows back and forth (an alternating current, or "AC"). Dynamos and batteries produce current that flows in one direction only (direct current, or "DC").

### Types of circuit

Components connected in a single, continuous path are said to be in "series." If the circuit branches off, the components are said to be in "parallel."

### Electric circuits

Electric current is simply the movement of electric charge. This charge can be positive or negative. In most cases, electric current is the movement of negatively charged particles called electrons, but the direction of the current is defined as the direction in which positive charge moves.

The current can flow only if there is a complete circuit. Metals are the best conductors, because the electrons inside are free to move. The circuit can be broken by a switch; when the switch is open the current stops flowing, even if the emf is still present.

The energy provided by an emf can produce sound, as in a loudspeaker or a buzzer; it can make light in a lamp; it

can make heat in a toaster, hairdryer, or electric fire; or it can make things move, as it does in an electric motor, used to drive a wide variety of domestic and industrial appliances.

### The work of Georg Ohm

In the 1820s German physicist Georg Ohm figured out the relationship between emf and current. He found that the same voltage will not always produce the same current, and that it depends on what the current has to flow through—the circuit "resists" the flow of current. The resistance of a circuit, or a particular part of it, depends upon the material of which it is made, its length and thickness, and upon its temperature.

Ohm had sufficient skill to be able to manufacture his own wire for his experiments, creating it in specific

**Ohm's apparatus**
Georg Ohm figured out the scientific law that bears his name using this type of equipment. Junctions of copper and zinc connected together but held at different temperatures produced an electromotive force that drove a current around the circuit. Ohm observed how the current varied as he inserted different lengths of wire into the circuit.

Glass cylinder prevents any air currents from affecting the galvanometer needle

Copper plate

Cold copper and zinc junction

Magnetized needle swings in response to electric current

Zinc plate

Eyepiece enabled Ohm to view how much the needle swung

Hot copper and zinc junction

Pots filled with mercury into which Ohm dipped his test wires to make the circuit connection

lengths and thicknesses. In recognition of his important discovery the unit of electrical resistance, the ohm (represented by the symbol $\Omega$), was named after him.

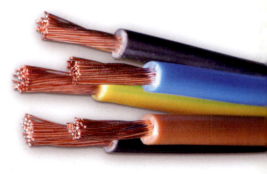

**The importance of insulation**
In practical, everyday circuits, most wires are coated in flexible plastic, such as plastic or PVC. This is to avoid a "short circuit," where current flows along a different path from the intended one.

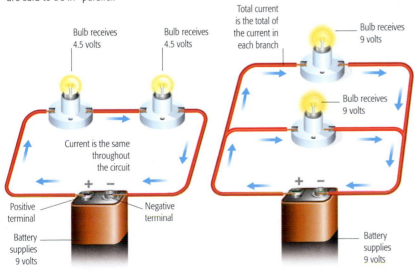

Bulb receives 4.5 volts

Bulb receives 4.5 volts

Total current is the total of the current in each branch

Bulb receives 9 volts

Bulb receives 9 volts

Bulb receives 9 volts

Current is the same throughout the circuit

Positive terminal

Negative terminal

Battery supplies 9 volts

Battery supplies 9 volts

**Series circuit**
The voltage of the battery is shared between both bulbs, and the amount of current flowing depends upon the voltage and the total resistance of the bulbs.

**Parallel circuit**
Here each branch has just one bulb. Each bulb receives the full voltage of the battery, and shines more brightly than in the comparable series circuit.

AFTER »

**Electric current helped scientists understand the nature of matter, and it enabled engineers to construct the modern world.**

### ELECTROLYSIS AND ELECTROPLATING

Electric current can **drive chemical reactions**. For example, it can make metal ions "come out of solution" and coat an object, in a process called electroplating. It can also make chemicals split apart, in a process known as **electrolysis**.

### ELECTRICITY AND MAGNETS

Mastery of electric current made possible the discovery of **electromagnetism** and Michael Faraday's invention of the **electric motor** and the **dynamo**.

SEE ALSO »
pp.166–67 ELECTROMAGNETISM
pp.170–71 MICHAEL FARADAY
pp.252–53 GENERATING ELECTRICITY

**‹‹ BEFORE**

Electricity and magnetism were hot topics of scientific investigation during the Scientific Revolution that began in the 16th century.

### GILBERT'S RESEARCH
The first person to study both **electricity and magnetism** in a scientific way was the English physician William **Gilbert**. He is famous for writing De Magnete (*On the Magnet*), in 1600, in which he conceived of the Earth as a giant magnet, and distinguished between magnetism and electricity. He worked with **static electricity**, however, so did not know how **electric currents flow**.

### INVENTION OF THE BATTERY
The battery, invented in 1799 by Italian physicist **Alessandro Volta** (1745–1827), was the first **man-made source of electric current**. It gave scientists the opportunity to study currents, and made the discovery that an **electric current produces magnetism** possible.

**‹‹ SEE ALSO**
pp.80–81 MAGNETIC FIELDS
pp.162–63 THE FIRST BATTERY
pp.164–65 ELECTRIC CURRENT

△ **Ørsted's announcement**
In July 1820 Ørsted published this pamphlet setting out his discovery of electromagnetism. The pamphlet caused a sensation, and prompted a surge of activity as Ørsted's peers explored the phenomenon.

# Electromagnetism

**Electricity can produce magnetism, and magnetism can generate electricity. Electric motors, electrical generators, electromagnets, and a host of other devices that are integral to our modern appliances depend on this interconnection—a phenomenon called electromagnetism.**

On April 21, 1820 Danish natural philosopher Hans Christian Ørsted made a startling discovery while he was setting up some apparatus for a lecture demonstration.

Among the apparatus that he was about to use was a magnetic compass needle, some wires, and a battery. At first the compass needle was aligned north–south, with Earth's magnetic field (see pp.80–81). But when Ørsted made an electric current flow in the wires, he noticed that the compass needle moved. Ørsted realized at once that the electric current was producing a magnetic field.

## Investigating electromagnetism
After three months of experimenting with his new discovery Ørsted published his findings to the rest of the scientific community. His work was a revelation—until then everyone had assumed that electricity and magnetism were two completely separate forces of nature. A magnetic field can be thought of as a series of field lines—the field strength can be defined as the density of these lines. Ørsted's name has now been adopted as one of the internationally recognized units of magnetic field strength.

Scientists around the world were quick to start performing their own investigations into electromagnetism. Within six months of Ørsted's publication, French mathematician and physicist André-Marie Ampère had demonstrated that two wires carrying electric currents can be made to attract or repel—just as ordinary magnets

## How a solenoid works
A solenoid is a coil of wire through which an electric current is passed to produce a uniform magnetic field like that of a bar magnet. There are several solenoids in a car's remote locking system. Changing the direction of the current in the coil changes the direction of the fields, making magnetized bars inside the fields move back and forth, locking or unlocking the car.

**Current direction**
The arrow shows the direction of the electric current flowing into the coil. If the current reverses, so does the magnetic field.

Negative terminal

Negative terminal of electric supply

Magnetic field

Positive terminal of electric supply

Electric current

◁ **Magnetic field around a wire**
The flow of electric current in a wire produces a magnetic field that loops around the wire. The direction of the field depends on the direction the current flows.

▽ **Ørsted's experiment**
The original experiment that Hans Christian Ørsted conducted can be replicated very simply, by passing an electrical current through a wire laid on top of a magnetic compass.

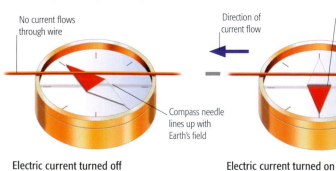

No current flows through wire

Direction of current flow

Needle lines up with magnetic field produced by current

Compass needle lines up with Earth's field

**Electric current turned off**
A magnetic compass needle normally aligns in a north–south direction, in line with Earth's magnetic field.

**Electric current turned on**
Electric current flowing through the wire produces a magnetic field that wraps around the wire. The compass needle swings to line up with that instead.

do—depending on the direction in which the current is flowing in each wire. Ampère also devised an equation to demonstrate how the strength of the electromagnetic field varies according to the strength of the electric current and the distance from the wire.

## Motors and solenoids
Within a year of Ørsted's discovery, English chemist and physicist Michael Faraday had found a way to exploit electromagnetic forces to make a wire rotate continuously—and, in so doing,

created the first rudimentary form of electric motor. Ampère intensified the forces by winding wire into coils, which he called solenoids. The magnetic field around a solenoid is the same shape as the field around a bar magnet and is stronger than that around a single wire. In another development English physicist William Sturgeon placed an iron bar inside a solenoid; the coil's field magnetized the iron, intensifying the effect. Sturgeon's "electromagnet" could lift 9 lb (4 kg). In 1830 American scientist Joseph

## Faraday's induction ring

Two wire coils are wound around an iron ring, with small gaps separating them. Switching an electric current on and off in one of the coils generates a changing magnetic field in the ring, which in turn produces an electric current in the other coil.

## Combined field

The magnetic fields from all the loops of wire combine to form a powerful field of nearly uniform strength, which is concentrated in the center of the coil.

## The field of a solenoid

The field of the wire is similar to that of a bar magnet, with north-seeking and south-seeking poles. This end of the coil is the north-seeking pole. The more turns of wire in the coil, the stronger the magnetic field.

**Positive terminal**

## Magnetic fields

The strength of magnetic fields varies in their proximity to the coil.

PHYSICIST AND MATHEMATICIAN (1831–79)

### JAMES CLERK MAXWELL

The Scottish theoretical physicist and mathematician James Clerk Maxwell did groundbreaking work on electromagnetism and the molecular behavior of gases. Taking Faraday's work on electromagnetism further, he formed a set of equations (see below), first published in his textbook *A Treatise on Electricity and Magnetism* in 1873. He also showed that Saturn's rings must consist of numerous small particles, and developed a statistical means to describe how atoms and molecules move in a gas. In 1861 Maxwell was the first person ever to create a true color photograph.

electricity supplied. This is how transformers work—devices that change high-voltage electricity from power lines to lower-voltage electricity for the home.

## Maxwell's equations

During the rest of the 19th century engineers improved their mastery of electromagnetism and built ever more powerful electromagnets, transformers, motors, and generators. Scientists were striving to understand this phenomenon. In the 1860s James Clerk Maxwell (see above) distilled everything that was known about electromagnetism into four equations, now known as Maxwell's equations. These described the properties of the electric and magnetic fields and demonstrated that light was an electromagnetic wave. This finding opened the door for the discovery of radio waves in 1888, by German physicist Heinrich Hertz.

**Electromagnetic power**

Hundreds of yards (meters) of wire carrying a strong electric current is wrapped thousands of times around a large piece of iron to create an electromagnet that can lift extremely large weights, such as scrap metal.

**AFTER**

Henry made an electromagnet with many more turns of wire; it could lift an impressive 750 lb (340 kg).

## Electricity from magnetism

In 1831 Faraday made a discovery that mirrored Ørsted's. He found that moving an ordinary magnet near a circuit would make electricity flow in the circuit. The faster the movement, the stronger the current. When the magnet was not moving, no current flowed. Again the effect was stronger when the wire was coiled, so that more of it was exposed

to the magnetic field. This effect is called electromagnetic induction, and is the basis of electrical generators.

Faraday also found that a changing magnetic field has the same effect as a moving magnet—so switching an electromagnet on and off, or varying the current flowing through it, produces an electric current in a nearby circuit. He wound two separate coils around an iron ring, and by connecting one coil to a battery and turning it on and off, he made current flow in the other coil.

By using different numbers of turns of wire in each coil, Faraday found he could "transform" the voltage of the

Electromagnetism plays a crucial part in the modern world and in our understanding of the Universe.

### TELECOMMUNICATIONS

With the invention of the **telegraph** in the 1840s and the **telephone 260–61 ≫** in the 1870s, people started using electromagnetism to send and receive messages.

### ELECTRIFICATION

Electromagnetic generators and **transformers** were used in electrification programs in many countries in the first half of the 20th century, making it possible to **generate** and distribute **electric power** on a huge scale.

### DRIVING TRANSPORTATION

**Electric trains** and **electric cars** use heavy duty motors. **Maglev trains** have electromagnets underneath that

**repel** electromagnets in the rails, making the trains **hover** above the track.

### MODERN PHYSICS

Physicists now know that the electromagnetic force is **one of the four fundamental forces 400–01 ≫**; the other three are strong nuclear forces, weak nuclear forces, and gravity.

**SEE ALSO ≫**
pp.168–69  THE ELECTRIC MOTOR
pp.170–71  MICHAEL FARADAY
pp.252–53  GENERATING ELECTRICITY

**MAGLEV TRAIN, SHANGHAI**

# The Electric Motor

As with so many inventions, the development of the electric motor was largely a collective effort over many years. Most sources credit Michael Faraday with building the first motor in 1821, but recognition should be given to Hans Ørsted, William Sturgeon, Joseph Henry, André Marie Ampère, and Thomas Davenport.

## « BEFORE

**The Industrial Revolution of the 18th century saw great changes in manufacturing and transportation, but it took some time for electric motors to play a major role.**

### STEAM AND ELECTRICITY
Water power and muscle power (mainly horses) were replaced with **steam power**, brought about largely by **James Watt's** modifications to **Thomas Newcomen's** steam engine of 1712.

The demand for **electricity**, which provided lighting to both homes and factories, gradually increased, giving the incentive to create electric generators. It was not until **dynamos** were used to run the first lighthouses that a real demand was created for constant power, and as with most technological advances, this demand fueled the **development of electric motors**.

### « SEE ALSO
*pp.162–63* THE FIRST BATTERY
*pp.164–65* ELECTRIC CURRENT
*pp.166–67* ELECTROMAGNETISM

### PHYSICIST (1797–1878)
## JOSEPH HENRY

Considered one of America's greatest scientists after Benjamin Franklin, Joseph Henry was a pioneer in the field of electromagnetism, where he discovered the phenomenon of self-inductance—voltage produced by a varying current. His work was crucial to the invention of the telegraph, the electric motor, and the telephone. Henry was the first Director of the Smithsonian Institution in 1846, and a founding member of the National Academy of Science. In 1893 his name was given to the standard electrical unit of inductive resistance, the "henry."

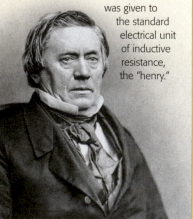

A simple definition of an electric motor is any apparatus that converts electrical energy into mechanical energy. The earliest motors didn't look anything like what we'd consider a motor today; if anything, they resembled a toy fishing game.

By the early 1800s the basic principles of electromagnetic induction were being studied, and by 1820 Danish physicist Hans Christian Ørsted, together with the French physicist and mathematician André-Marie Ampère, discovered that an electrical current produces a magnetic field (see p.166). When the current was switched on in a straight piece of wire the magnetic field seemed to encircle it, like a sleeve around an arm.

### Movement with magnetism
In London, English scientist Michael Faraday was intrigued by the news of these results, and wondered whether the phenomenon could be put to practical use. In 1821 he devised an experiment to see whether a magnetic field around a wire carrying a current could be used to generate motion

Faraday built an apparatus with a metal wire that was free to move and was suspended in a cup of mercury. (This experiment can easily be repeated using saltwater in place of the highly poisonous mercury.) Standing upright in the middle of the cup was a permanent bar magnet. The ciruit was completed via the mercury: the current flowing through the wire produced an electromagnetic field, which interacted with the existing magnetic field from the permanent magnet. The effect of this interaction was to make the suspended wire rotate around the magnet—in essence, forming the first ever electric motor.

To the uninitiated a spinning wire may not have seemed that exciting, but Faraday and many of his fellow scientists quickly saw the potential in mechanical energy. In 1822 Faraday's motor was developed further by English mathematician and physicist Peter Barlow. The dangling wire was replaced by a suspended spur wheel with spokes, and the cup was replaced by a trough of liquid mercury. Barlow's wheel was lowered so that the tip of one of the spokes just dipped into the mercury, and when voltage was applied the electromagnetic forces caused the wheel to rotate. Just by adding a wheel

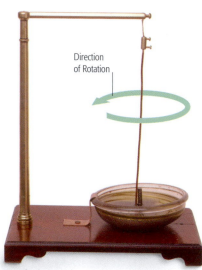

Direction of Rotation

**Faraday's experiment**
The original apparatus comprised a wire hung from a metal arm, and a bowl with a magnet standing upright in the center. Faraday filled the bowl with mercury to complete his circuit, but any conducting liquid will do, such as salty water.

instead of a wire, he made it much easier for people to grasp the concept of the ways in which an electric motor could be put to various uses.

An American blacksmith named Thomas Davenport improved the operating principles in 1834 by using four magnets—two fixed and two revolving. Crucially, he was able to prove the potential of motors by using his own model to operate his drills and woodturning lathes.

Despite improvements, the primary battery power was so expensive in the 1830s that electric motors proved commercially unsuccessful. At the time there was no electricity distribution, and many developers, including

**DC Electric Motor**
This cutaway shows the main components of a simple electric motor—a rotating armature held within a permanent magnet. Direct current is passed through the coil via brushes and the commutator—a device which switches the direction of the current every half turn, ensuring the electromagnetic field of the coil is always pushed the same way by the permanent magnet.

### IN PRACTICE
## THE ELECTRIC CAR

The Scottish inventor Robert Anderson is credited with inventing the first electrically powered carriage in about 1835, using non-rechargeable primary cells. However, the idea didn't catch on until the development of the lead-acid storage battery by Camille Fauré in 1881.

Great Britain and France were the first nations to embrace the electric car. Then, in 1899, a Belgian-built electric racing car set a world land speed record of 68 mph/h (110 km). The first electric taxis hit the streets of New York in 1897 and by 1900, 28 percent of all cars produced in the US were powered by electricity.

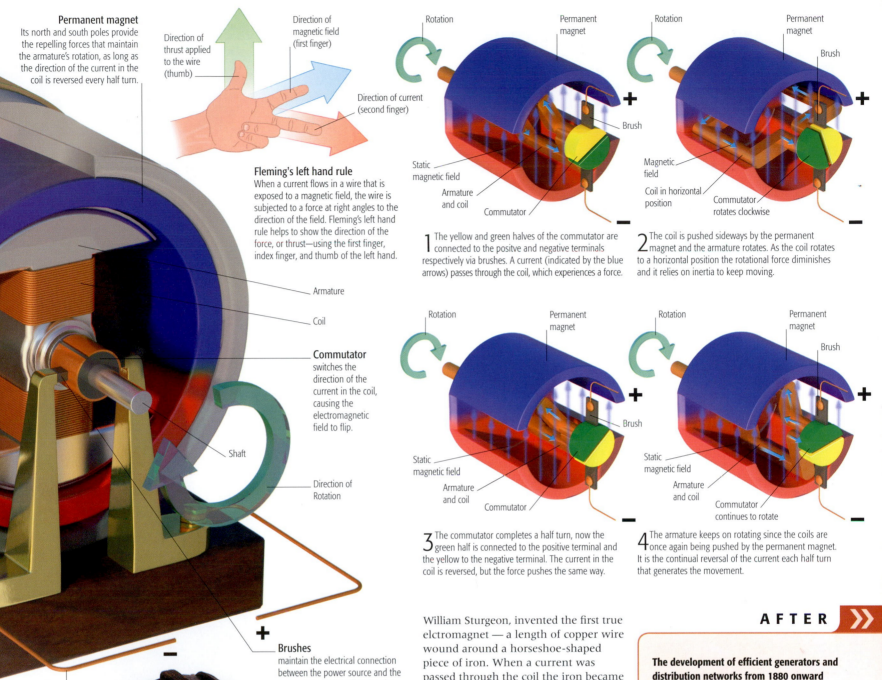

**Permanent magnet**
Its north and south poles provide the repelling forces that maintain the armature's rotation, as long as the direction of the current in the coil is reversed every half turn.

Direction of thrust applied to the wire (thumb)

Direction of magnetic field (first finger)

Direction of current (second finger)

**Fleming's left hand rule**
When a current flows in a wire that is exposed to a magnetic field, the wire is subjected to a force at right angles to the direction of the field. Fleming's left hand rule helps to show the direction of the force, or thrust—using the first finger, index finger, and thumb of the left hand.

Armature

Coil

**Commutator**
switches the direction of the current in the coil, causing the electromagnetic field to flip.

Shaft

Direction of Rotation

Power source

**Brushes**
maintain the electrical connection between the power source and the commutator while still letting the commutator rotate.

**Early induction motor**
The machine shown here is an example of the induction motor patented by Nikola Tesla in 1887–88. The induction motor, which uses alternating current (AC), has since become the most widely used form of electric motor.

Rotation | Permanent magnet

Static magnetic field

Armature and coil

Commutator

Brush

1 The yellow and green halves of the commutator are connected to the positve and negative terminals respectively via brushes. A current (indicated by the blue arrows) passes through the coil, which experiences a force.

Rotation | Permanent magnet

Brush

Magnetic field

Coil in horizontal position

Commutator rotates clockwise

2 The coil is pushed sideways by the permanent magnet and the armature rotates. As the coil rotates to a horizontal position the rotational force diminishes and it relies on inertia to keep moving.

Rotation | Permanent magnet

Static magnetic field

Brush

Armature and coil

Commutator

3 The commutator completes a half turn, now the green half is connected to the positive terminal and the yellow to the negative terminal. The current in the coil is reversed, but the force pushes the same way.

Rotation | Permanent magnet

Brush

Static magnetic field

Armature and coil

Commutator continues to rotate

4 The armature keeps on rotating since the coils are once again being pushed by the permanent magnet. It is the continual reversal of the current each half turn that generates the movement.

William Sturgeon, invented the first true elctromagnet — a length of copper wire wound around a horseshoe-shaped piece of iron. When a current was passed through the coil the iron became magnetized. Sturgeon's 0.4-lb (200-g) lump of iron could lift an 8.8-lb (4-kg) weight using the current from a single cell battery. What's more, he could regulate the magnetization by switching the electrical current on and off.

This was the innovation that put in place the final building blocks for another electromagnetism pioneer, American scientist Joseph Henry (see left). By developing an even stronger magnet, Henry was able to build prototypes of the first telegraph system and the first electric motor. These early motors used direct current (DC) that flows in only one direction, but later alternating current (AC) motors became the more popular since they were safer and more powerful.

Efficient and durable, electric motors are found in many devices, from disk drives to electric trains, and since they contain relatively few moving parts, they can remain in use for many years.

Davenport, went bankrupt in their attempts to prove that motors were a viable option.

**From failure to success**
However, there was further development on both sides of the Atlantic. In 1823 British electrician,

**AFTER**

The development of efficient generators and distribution networks from 1880 onward finally enabled widespread use of electric motors. Today, motors are used in all areas of life, and the technology is still developing.

**PEOPLE POWER**
Most analog **quartz wristwatches** run on a battery that powers a **"stepper motor"** to drive the hands. Seiko Watch Corporation of Japan have produced **"kinetic"** watches, **powered by the wearer's own arm movements**; energy from movements is converted into a magnetic charge, which is turned into electrical power.

**NANOMOTORS**
Scientists at the Univeristy of California, Berkeley have built **molecule-sized motors** around carbon nanotubes **393 >>**. Suggested future applications include acting as switches in communications systems and optical computers.

**SEE ALSO >>**
pp.254–55 THE INTERNAL COMBUSTION ENGINE

PHYSICIST AND CHEMIST **Born 1791** Died 1867

# Michael Faraday

## "**Nothing** is too wonderful to be true if it be consistent with the **laws of nature**."

MICHAEL FARADAY, FROM HIS DIARY, MARCH 19, 1849

Although he received little formal education, Michael Faraday became one of the finest and most influential scientists of his time. He was Director of London's Royal Institution for many years, and delivered many superb lectures there. He invented the electric motor and the dynamo, and laid the foundations for the science of electromagnetism.

Born just south of London Bridge in 1791, Faraday was the son of a poor village blacksmith from the north of

England. His family belonged to the Sandemanian sect of Christianity, and Michael was to remain devoutly religious all his life. At the age of 14 he was apprenticed to a bookbinder in the west end of London, where he took time to read many of the books.

### First scientific appointment

Faraday attended lectures given by Humphry Davy (see left) at the Royal Institution (RI) in Albermarle Street, only 15 minutes' walk from where he worked. Captivated, he wrote, illustrated, and bound a book from his lecture notes and sent it to Davy, asking for a job. A year later Davy took

---

CHEMIST (1778–1829)

#### HUMPHRY DAVY

Humphry Davy was born in Penzance in southwest England. He learned some chemistry from Gregory Watt, son of James Watt (see p.149), and in 1798 he was hired by Thomas Beddoes as medical superintendent at the Pneumatic Institution, Bristol, where he investigated the effects of many gases, including nitrous oxide, or laughing gas.

In 1801 he was invited to join the Royal Institution (RI) in London as lecturer and then professor of chemistry, and was also elected a Fellow of the Royal Society. He did brilliant research at the RI, isolating several elements, including potassium and sodium. But when asked what his greatest discovery had been, his answer was "Michael Faraday."

**Christmas Lecture**
Faraday gave 19 sets of Christmas Lectures in what is now called the Faraday Lecture Theatre at the Royal Institution in London. He originated the lectures to introduce complex scientific ideas to the public, especially to young people.

**Electromagnetic rotation**
On September 3, 1821 Faraday performed the experiment that was to result in the world's first electric motor. He wrote out the details in this laboratory notebook.

**Michael Faraday**
An intuitive physicist, Faraday had an exceptional gift for visualizing lines of magnetic force and how they interacted with electric current, when both phenomena were poorly understood. In addition he was an exceptional experimental chemist.

him on, first as his secretary, and then as chemistry assistant. When, in 1813, Davy decided to make a tour of Europe with his new wife, a grand heiress named Jane Apreece, Faraday was asked to go with them as scientific assistant. Davy's valet did not wish to go, and Faraday was given his duties, too. Jane treated him as just another servant and he had a miserable time.

For five years after they returned, Faraday worked on various projects in chemistry. Then, in 1821, he started on what was to become his greatest work, in electricity and magnetism.

## Visionary power
Lacking formal education, Faraday had almost no mathematics. Instead he had an extraordinary ability to visualize things. He looked at a magnet and could "see" the magnetic field lines sprouting from the ends and curling around from one pole to the other; he called them "lines of force," and we still use the expression today. This visionary power enabled him to make strides in several branches of chemistry and physics. His ideas were also crucial to physicist James Clerk Maxwell, who

was to continue in electromagnetism what Faraday left unfinished.

In 1820 the Danish scientist Hans Christian Ørsted showed that around a wire carrying a current there is a magnetic field, almost like a sleeve around an arm. Faraday talked about this with Davy and William Hyde Wollaston, and they speculated about whether this sort of magnetism could be used to generate motion.

## Ground-breaking invention
On September 4, 1821 Faraday made the world's first electric motor—a wire dipping into a pool of mercury and revolving around a magnet. His invention paved the way for others to develop the idea (see pp.162–63). Faraday failed to acknowledge the contribution of Davy and Wollaston sufficiently, and they argued. Davy then gave Faraday a useless project investigating various types of glass, which in effect wasted six years of his life. Only when Davy died in 1829 was he able to return to electromagnetism; Faraday then discovered the process of electromagnetic induction, the basis of transformers and dynamos.

## Later interests
Faraday's broad scientific curiosity was not confined to the laboratory. He investigated the spiritualist movement, which began in 1848, although he was convinced it was all nonsense. He devised experiments to show that, during table-tipping séances, the sitters themselves were unconsciously moving the table.

Michael Faraday also was a brilliant lecturer and demonstrator. He initiated the Christmas Lectures at the Royal Institution in 1825. And yet he was humble and self-effacing. He twice refused to be President of the Royal Society, and he even refused a knighthood for religious reasons, saying he preferred to remain plain Mr. Faraday to the end of his life.

### BREAKTHROUGH

## LAW OF INDUCTION

Faraday's law of induction, recognized independently in the US by Joseph Henry, is a basic law of electromagnetism and applies to electrical generators. The law states that the voltage induced in a coil of wire is proportional to the rate at which it cuts through a magnetic field, multiplied by the strength of the field. In practical terms, a higher voltage can be obtained by spinning a generator more quickly, by using a stronger permanent magnet, or by using more turns of wire in the input side of an induction ring or transformer.

**FARADAY'S INDUCTOR**

### TIMELINE

- **1791** Michael Faraday is born on September 22 in Newington Butts, south London, England, the son of a poor Yorkshire blacksmith.

- **1805** He is apprenticed to a bookbinder, George Riebau, in Bloomsbury, London.

- **1812** Goes to a series of lectures given by Humphry Davy (see opposite) at the Royal Institution (RI). He sits in the seat above the clock, and is entranced. He then writes, illustrates, and binds a 300-page book based on his notes from the lectures, and sends it to Davy, asking for a job.

- **1813** Davy employs Faraday first as secretary and then as chemical assistant at the RI.

- **1813–1815** Davy takes Faraday as assistant on his honeymoon tour of Europe, during which he investigates volcanic action. Davy's valet refuses to go, so Faraday has to act as valet as well. Davy's wife treats him as one of the servants, and Faraday becomes so miserable that he is tempted to return home and give up science.

**VOLTAMETER USED BY DAVY**

- **1821** Marries Sarah Barnard on June 2. She is a fellow member of the Protestant Sandemanian sect, which opposes the Church of Scotland.

- **1821** Invents the first electric motor.

- **1823** Liquefies chlorine.

- **1824** Elected a Fellow of the Royal Society.

- **1825** Elected Director of the RI. Initiates the Christmas Lectures, which are still running today.

- **1827** Gives his first of 19 sets of Christmas Lectures; one of his best-known was called "The Chemical History of a Candle."

- **1829** Invents the induction ring—the basis of the transformer and the dynamo.

- **1831** Observes that a moving magnet induces an electric current.

**DANIELL AND FARADAY**

- **1833** Elected Fullerian Professor of Chemistry at the RI.

- **1836** Inspired by Faraday, John Frederic Daniell succeeds in making the first battery (see pp.162–63) to produce electrical current over a long period.

- **1847** In what could be called the birth of nanoscience (see pp.392–93), Faraday reports that the optical properties of gold colloids differ from those of the bulk metal.

- **1855** Writes a strong letter to *The Times* newspaper about the Great Stink, caused by foul pollution of the River Thames.

- **1867** Dies on August 25 at his grace-and-favour house in Hampton Court, Surrey.

# Accurate Measurement

**Nearly every scientific experiment involves some kind of measurement. Today scientists can measure to extreme accuracy, using a wide range of precise measuring devices. They use a logical and clearly defined system of units called the International System of Units, or SI units.**

Accurate measurement in scientific investigations became much more important in the 17th century, when people began to use the scientific method (see pp.68–69). Experimenters became aware that only with reliable measurements could their experiments support or disprove their theories.

Along with the measurement of time (see pp.120–21), the most frequent and important measurements that scientists make are mass (or weight) and length (or distance).

## Mass and weight

In everyday language, people tend to confuse the terms "mass" and "weight" —both words seem to refer to how heavy something is. But to a scientist

there is an important difference: an object's mass is the amount of "stuff" it contains, while its weight is the force of gravity (see pp.108–109) on that object.

In 1678 the English scientist Robert Hooke (see pp.92–93) invented the spring balance, which measures weight directly from the extension of a spring. Although Hooke's invention is convenient, it is not generally very precise. This is because the force of gravity varies from place to place. An object weighs slightly more at the poles or near a large mountain range than at the equator or in the middle of a desert, for example.

In contrast, the lever-arm balance gives the same reading wherever you take it. It works by comparing standard weights or masses on one side of the pivot with the weight of an object on the other. A traditional balance would even give the same readings on the Moon, where an object's mass is unchanged, but its weight is only one-sixth of that on Earth.

Modern chemical balances are accurate to less than millionth of a gram (about a 30-millionth of an ounce), while a device called a mass spectrometer can select individual atoms by mass.

## Length or distance

In 1631 French physicist Pierre Vernier invented the Vernier scale—an add-on to existing measuring devices to help scientists measure small distances very accurately. Vernier attached his scale to

« **BEFORE**

All the ancient civilizations had systems of weights and measures, which were essential for many different activities, including building, making items such as clothes, agriculture, and trade.

### ANCIENT MEASUREMENTS

The units of measurement in early civilizations were often based on the **human body** or other natural quantities. For example, the cubit, which originated in ancient Egypt, was the **length of the forearm**, and in several civilizations the weight of precious stones was often given in terms of grains of wheat. However, human bodies and wheat grains vary; so it is impossible to measure consistently and accurately using these. As a result, even in ancient times governments and rulers made **"standard" weights and measures** against which people could compare their own measuring devices. Nevertheless, systems of weights and measures remained complicated and confusing, and the measurements people made were probably inaccurate.

**ANCIENT ASSYRIAN WEIGHT**

**ENGINEER (1771–1831)**

## HENRY MAUDSLAY

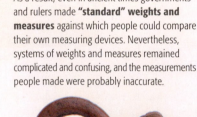

English toolmaker and inventor Henry Maudslay gained early experience of engineering when he began filling artillery shells in the Royal Arsenal, Woolwich, at the age of just 12. Later he was apprenticed to the English lockmaker Joseph Bramah, who was obsessed with precision. In 1791 Maudslay set up on his own, and in 1800 invented a type of lathe that enabled standard screw threads to be mass produced. He is known as the "father" of precision engineering.

Rack-and-pinion links spring to pointer

Pointer indicates object's weight of 7 lb (equivalent to 3.2 kg)

110 0 100 90 80 70 60 50 40 30 20 10

Dial shows weight in pounds

Oval spring deforms into a more circular shape when a weight is hung on the hook

## Spring balance
Inside this 18th-century spring balance is an oval spring, which deforms when an object is hung from the hook. The greater the mass of the object, the greater its weight and the more the spring deforms. As the spring deforms, a ratchet connected to it moves a pointer around a calibrated dial.

Weight (7 lb) suspended on a hook, which is linked to the spring by a metal bar

## VERNIER SCALE

The Vernier scale was invented in 1631 by French mathematician Pierre Vernier. Here, a Vernier scale is fitted to a calliper, a device whose jaws close around an object (or open into a space to make internal measurements). In the example shown, the approximate reading on the fixed main scale is somewhere between 18 mm and 19 mm. The point at which the markings on the fixed scale and Vernier scale coincide give greater precision. The actual reading below is 18.64 mm.

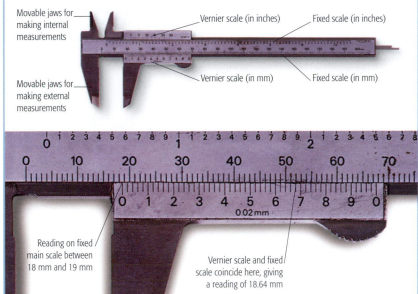

Movable jaws for making internal measurements

Vernier scale (in inches)

Fixed scale (in inches)

Movable jaws for making external measurements

Vernier scale (in mm)

Fixed scale (in mm)

Reading on fixed main scale between 18 mm and 19 mm

Vernier scale and fixed scale coincide here, giving a reading of 18.64 mm

a quadrant—a device for measuring angles—but it was later used on the calliper, a device that enables dimensions of objects to be measured.

Around the same time, English astronomer and mathematician William Gascoigne invented another device that would revolutionize accuracy: the micrometer. Gascoigne's device was based around a screw: as you turn a screw, it advances very slowly, and incorporating this into a measuring device makes it possible to measure small distances accurately. Gascoigne used his micrometer in telescopes, but the principle was soon being used to engineer tools.

In 1800 Henry Maudslay invented the screw-cutting lathe, which made it possible to cut standard screw threads. Until then nuts and bolts were made in pairs, and only a matching pair could be guaranteed to work together, but Maudslay's invention enabled engineers to mass produce screws, nuts, and bolts.

Maudslay used his expertise with screw threads to produce an extremely accurate micrometer that was capable of measuring to an unprecedented accuracy of one ten-thousandth of an

### Standard kilogram
Of all the SI units, the kilogram is the only one defined in terms of a physical object—the meter, for example, is defined by the distance light travels in a certain time. The International Prototype Kilogram (IPK) was made in 1889, and is held under controlled conditions in a vault in Sèvres, France.

inch (about 0.0025 mm). One of Maudslay's ex-employees, Joseph Whitworth, did even more to promote precision in the 19th century. In 1833, having learned the technique from Maudslay, he produced a new way of making extremely flat surfaces. These "surface plates" are used to set up accurate measuring devices, and Whitworth's technique enabled him to make a measuring device that was accurate to one millionth of an inch (about 0.000025 mm).

### SI units
By the end of the 19th century great accuracy was commonplace in science and engineering. However, scientists were still using traditional systems of measurement, with the result that different systems were being used in different countries.

With science becoming increasingly international during the 20th century, it became important for there to be a single measurement system. The Système International, or SI units system (see pp.422–23), was introduced in 1960, enabling scientists to share the results of their measurements and calculations meaningfully in international journals.

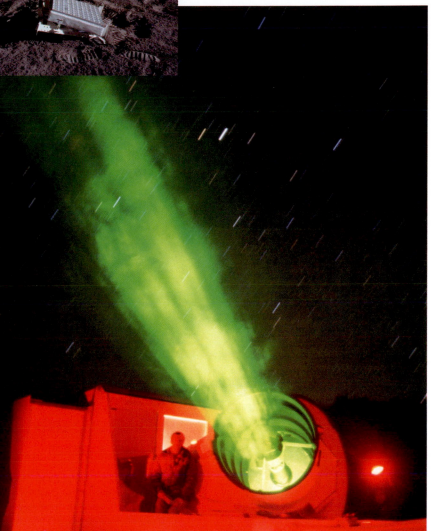

### Lunar laser ranging
The measurement of the distance to the Moon is one of the most precise ever made, accurate to better than one part in ten billion. It is made possible by reflectors placed on the lunar surface (left) in 1969 by the crew of Apollo 11. Laser beams from observatories on Earth (below) bounce off the reflectors, and the distance is calculated from the time taken to receive the reflections.

**AFTER** »

Modern science and technology is so sophisticated that measurements of unprecedented precision are necessary.

**EXTREME PRECISION**
It is now possible to measure the distances **between individual atoms** using an instrument called an **atomic force microscope**. This works by measuring the tiny forces between atoms on the microscope's probe and those on a surface, enabling the **surface atoms to be mapped**. Microelectronics is another area where extreme precision is needed, with components measured in nanometers (millionths of a millimeter). A 2004 satellite-based experiment, called Gravity Probe B, used nearly perfect spheres, **accurate to within 40 atoms' thickness**, to measure effects predicted by Einstein's general theory of relativity.

**GRAVITY PROBE B**

# Calculating and Computing

**From the last years of the 18th century onward the demands of science and industry prompted the development of techniques that ultimately gave us the computer. These techniques included the forerunners of apparently modern concepts such as programs and computer languages.**

As the Industrial Revolution spread across Europe, a demand arose for more automated machinery. At the same time scientific discovery called for increasingly complex calculations.

## BEFORE

For millennia, people have used mechanical aids for calculation. Tools such as the abacus, Napier's bones, and the slide rule are the direct ancestors of our computers.

### LOGARITHMS AND NAPIER'S BONES

First set out by Scottish mathematician **John Napier**, **logarithms** make calculations easier by allowing **multiplication** and **division** to be **simplified** to addition and subtraction. Napier also invented **Napier's bones**, a set of **rods** or **cylinders** marked with **multiplication tables** for numbers 1 to 9, as a calculating tool.

**NAPIER'S BONES**

### SLIDE RULE

The slide rule is a **mechanical computer** invented in 1622 by English clergyman **William Oughtred**. The device comprises two sliding bars marked with **logarithmic scales**, plus a marker, or **cursor**. Deceptively simple in appearance, **slide rules** enabled users to do highly complex **calculations**. They were essential in science and engineering until the 1970s—even being used in the **Apollo** missions.

**SEE ALSO**
*pp.30–31* ANCIENT NUMBER SYSTEMS
*pp.88–89* METHODS OF CALCULATING

Although various calculating machines had been invented, they were limited in capability and were not always accurate. A more useful tool was the slide rule, which was reasonably accurate and easy to use, but for large and difficult calculations logarithms provided the best solution. The main problem with logarithms, however, was that their use depended on sets of tables—which themselves had to be calculated.

In 1792 Napoleon Bonaparte appointed the French mathematician Gaspard de Prony to compile a set of more comprehensive and accurate logarithmic tables. To do so, de Prony employed a hierarchy of workers. Mathematicians devised the formulae, assistants broke the calculations down

### INVENTOR (1791–1871)

## CHARLES BABBAGE

Born in London, England, Charles Babbage was already widely read in mathematics by the time he went to Cambridge University. He was Lucasian Professor of Mathematics from 1828 to 1839, and was an eminent spokesman for science. To his dismay, none of his calculating machines was built during his lifetime. He did, however, manage to build part of his Difference Engine for demonstration purposes; this is recognized as the first automatic calculator.

**Babbage's Difference Engine**
The Difference Engine consisted of columns of brass cogwheels, and was operated by cranking a handle. The machine was never completed in Babbage's time; it was not until the 1990s that his design was realized, for the Science Museum in London.

into instructions for simple addition and subtraction, and these basic tasks were then carried out by workers known as "computers."

## Babbage's calculating engines

Even de Prony's tables contained inaccuracies. The absence of a reliable set of tables prompted Charles Babbage to start thinking about a machine that could do calculations mechanically. In 1822 he proposed his first design for a "Difference Engine" to the British Royal Astronomical Society. The idea was granted government funding, but amid arguments with the engineers and bickering about money, the machine was not completed. Undeterred, Babbage produced an improved design (his "Difference Engine No. 2"), but more importantly also devised what he called an "Analytical Engine."

**Electronic desktop calculator**
Electronic calculators, like this 1973 Hewlett-Packard HP-46, utilized integrated circuits. They operated as "electronic slide rules," programmed to perform specific mathematical functions with single keys.

For operating the Analytical Engine Babbage took an idea from the textile industry—the use of punched cards to program patterns for looms, developed by the French inventor Joseph Marie Jacquard around 1804. If the Analytical Engine had ever been finished, it would in effect have been the first programmable computer. It was described (and improved on) by Babbage's follower Ada Lovelace. The daughter of English poet Lord Byron, Ada Lovelace wrote programs for the cards and is often regarded as being the first computer programmer.

Computers are still evolving rapidly. The machines of the future are likely to function very differently from those in use today.

### QUANTUM COMPUTING

Computer **microprocessors** are getting smaller and yet more powerful all the time. It is predicted that the next stage will be **quantum computing**, in which processing uses **atoms** or **subatomic particles 382 »**. The technology is still in its infancy, but quantum computers are expected to be **many times faster** than existing machines.

### ARTIFICIAL INTELLIGENCE

We are still far from having **robots** with the full range of human abilities, but computers already use some of the **skills** linked with intelligence, such as **problem solving**, in a variety of fields. Examples include systems to aid **medical diagnosis**, automatic **gearboxes** in vehicles, and **search engines** and **spam filters**.

**SEE ALSO »**
pp.342–43 CODES AND CIPHERS
pp.366–67 MICROCHIP TECHNOLOGY
pp.380–81 ARTIFICIAL INTELLIGENCE AND ROBOTICS

**Modern supercomputer**
One of the world's fastest and most powerful computers is the Columbia supercomputer, built in 2004 and used by NASA. It has a processing power of 50 trillion operations per second.

Around the same time, English mathematician George Boole was working on a new approach to algebra and logic. Although Boole's ideas went largely unnoticed at the time, Boolean algebra later revolutionized computing, as the basis for digital computer logic.

## Modern computers

During World War II research led to the first electronic computers. Two landmark computers were designed: in Britain, Colossus was developed from the work of code breakers such as Alan Turing, and in the US ENIAC was built to calculate the trajectories of shells. From these machines, with punched-card programs and vacuum tubes, improved designs quickly followed. Analogue computing changed to digital. Improvements in electronic circuits led to the first stored-program computers, such as EDVAC in the US and ACE in Britain, and the first commercial computer, UNIVAC.

These early electronic computers filled whole rooms but had limited memory. Transistors improved computers' capabilities in the 1960s, but the turning point was the invention of the microchip. As microchips became smaller, cheaper, and more powerful, they were added to desktop and pocket calculators, and drove the boom in personal computing. Today, PCs have huge memories and can perform numerous functions.

### IN PRACTICE

## BINARY NUMBERS AND COMPUTING

Computers are constructed from assemblies of electronic circuits. Each circuit can switch between "on" and "off." These two states can be written in code using the binary (base 2) number system, which consists of 1s (representing "on") and 0s ("off"); the binary equivalents of decimal numbers 1 to 10 are shown here. Sequences of binary digits, or bits, form the instructions for all data processing and storage.

| Decimal number | Binary visual 16s 8s 4s 2s 1s | Binary number 16s 8s 4s 2s 1s |
|---|---|---|
| 1 | ☐ ☐ ☐ ☐ ■ | 0 0 0 0 1 |
| 2 | ☐ ☐ ☐ ■ ☐ | 0 0 0 1 0 |
| 3 | ☐ ☐ ☐ ■ ■ | 0 0 0 1 1 |
| 4 | ☐ ☐ ■ ☐ ☐ | 0 0 1 0 0 |
| 5 | ☐ ☐ ■ ☐ ■ | 0 0 1 0 1 |
| 6 | ☐ ☐ ■ ■ ☐ | 0 0 1 1 0 |
| 7 | ☐ ☐ ■ ■ ■ | 0 0 1 1 1 |
| 8 | ☐ ■ ☐ ☐ ☐ | 0 1 0 0 0 |
| 9 | ☐ ■ ☐ ☐ ■ | 0 1 0 0 1 |
| 10 | ☐ ■ ☐ ■ ☐ | 0 1 0 1 0 |

# Energy Conversion

**One of the fundamental laws of physics is that energy cannot be created or destroyed, but it can be converted from one form to another. During the 19th century scientists figured out what energy is and the forms it can take, and since then engineers have been finding more ways to convert it from one form into another.**

Energy is the capacity for doing work, and so the fact that energy can be converted from one type to another is useful in many ways.

The most versatile form of energy is electricity (see pp. 252–53), which is easy to transmit over long distances and can be used to generate motion energy in a range of machines, from toothbrushes to trains.

## Using energy for heat

Electricity can also be converted directly into heat energy in electric heaters and stoves, light energy in light

**Newton's cradle**
A simple toy can demonstrate the conservation of energy. Let the ball at one end swing into the rest and its potential energy is converted to kinetic energy (movement) as it falls; this is passed on to the ball at the other end, which swings out in turn.

## « BEFORE

Energy was being converted for various uses long before electric current was available.

### ENERGY FROM FOOD

Green plants use **photosynthesis « 154–55** to convert light energy from the Sun into carbohydrates. These are a form of **potential energy** used by the plants to grow, and by the animals that eat them for **heat and motion**.

### WIND AND WATER ENERGY

For centuries windmills and water wheels have converted **potential** energy to **kinetic** energy. Where that energy has been used to grind corn, for example, the energy has been converted to more potential energy (the grain for food) and to **heat** and **sound** energy.

**« SEE ALSO**
*pp.50–51* WATER AND WIND POWER

bulbs, gravitational energy through elevators and escalators, and sound in loudspeakers. Most forms of energy tend to become converted to heat. Friction converts kinetic (moving) energy to heat; for example, when a car driver applies the brakes, some of the kinetic energy of the car is converted to heat in the brake pads, which then heats the atmosphere. Much of the food eaten by mammals is used to provide energy to keep them warm, and all that heat is eventually lost to the atmosphere.

## Converting potential energy

The various forms of energy can be broadly divided into two classifications: energy by virtue of movement—or motion energy; and energy "waiting" to be collected—or potential energy.

Roman soldiers often defended their forts by throwing fist-sized stones down from the walls at their attackers. These stones fell perhaps 16 ft (5 m), giving them enough kinetic energy to inflict lethal damage. The amount of potential energy depends on the mass and the height, and is calculated by multiplying mass (m) by height (h) and by the acceleration due to gravity (g), to give mgh (m×g×h).

This is why falling down stairs can be so dangerous, and how hydroelectric projects are able to harness the potential energy of mountain water as it falls. The greater the amount of water (m) and the further it falls (h), the greater the potential energy and, ultimately, electricity generated.

Potential energy can also be locked up in chemicals; explosives have the power to destroy structures because the chemical energy contained in them can be released very rapidly. Coal has the

In the boiler the heat from the burning coal turns water into high-pressure steam: another carrier of potential energy

Sound energy is produced when the high-pressure steam escapes through the whistle, vibrating the air

Coal carries potential energy, which is converted into heat energy through burning

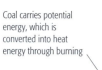

When the brakes are applied, friction converts motion energy into heat

High-pressure steam in the cylinders pushes the pistons back and forth, and they push the cranks to turn the wheels

**Steam locomotive**
The heat energy from burning coal is used to boil water and make high-pressure steam, which makes kinetic energy in the cylinders and sound energy in the whistle.

## Potential Energy

**Magnetic**
The north pole of one magnet is strongly attracted to the south pole of another.

**Gravitational**
A skydiver is pulled toward Earth, losing potential energy and gaining kinetic energy.

**Chemical**
Once fireworks are ignited they release their potential energy extremely rapidly.

**Electrical**
Many of our domestic gadgets run on electricity, as do certain trains and types of automobiles.

**Elastic**
Elastic energy is stored in such things as springs and trampolines, or can be generated in catapults.

Smoke emitted from the smokestack is how the train loses heat energy to the atmosphere

Light energy comes either from a dynamo producing electricity, or from the burning of kerosene

### ASTRONOMER (1818–89)
## JAMES PRESCOTT JOULE

Born in Salford, James Prescott Joule was a pupil of John Dalton (see p.138) at the University of Manchester. He firmly believed that heat was a form of energy and that other forms of energy could be converted to heat. Joule spent years investigating the connection, and even spent most of his honeymoon trying to measure the temperature difference of water at the top and bottom of a waterfall. He was finally proved correct in his beliefs, and the unit of energy—the joule—is named after him.

potential to create heat and so is burned in power stations to generate electricity. Gasoline has the power to drive combustion engines (see pp.254–55), and the chemicals in a battery can produce electricity directly (see pp.162–63). The food we eat—especially the carbohydrates—has the potential to create kinetic energy in muscles.

### Energy in motion
Any moving object has kinetic energy, and this is proportional to its mass and the square of its speed. Hence the kinetic energy of a car moving at 30 mph (48 kph) is more than twice that at 20 mph (32 kph). The asteroid that is traditionally thought to have wiped out the dinosaurs 65 million years ago had a huge impact despite having a relatively small mass; it hit the ground at around 19 miles per second (30 kilometers per second), giving it as much energy as a million express trains. Heat is motion

energy, caused by the motion of the atoms or molecules that make up an object. In a hot object the atoms are moving much faster than in a cold one. Atoms have tiny masses, but move extremely fast and carry plenty of heat.

Similarly, light is motion energy, since it is made of small particles called photons (see pp.286–87), which are always moving. These are easily converted to heat and can be converted directly to electricity by photovoltaic cells. Like light, sound energy comes in waves, but, in this case, of compressed air. Telephones, for instance, convert it into electrical energy for transmission.

Nuclear energy has two forms. Fission is the splitting of heavy nuclei, such as uranium; fusion is the combination of two nuclei, such as hydrogen. Both of these processes have the potential to liberate vast amounts of energy.

## AFTER

Although traditional methods of generating energy are still dominant, scientists are actively researching alternative methods.

### SUN AND TIDES
**Solar light energy** falling on Earth's surface can be converted directly to heat or electricity. **Movement of seawater 417 》** as waves and tidal currents carries immense amounts of energy; research into the possibilities for tapping this has been going on for decades.

### NUCLEAR POWER
Einstein's famous equation $E = mc^2$ **300–301 》** tells us that the loss of a small amount of mass (m) yields a colossal amount of energy (E). This is the principle behind nuclear weapons and **nuclear power stations 324–25 》**. Conventional nuclear power stations use the fission of **uranium nuclei**. They already provide much of the world's electricity, and may become more important, while research suggests that the fusion of **hydrogen nuclei** could yield huge amounts of energy.

### SEE ALSO 》
pp.180–81 LAWS OF THERMODYNAMICS
pp.324–25 FISSION AND FUSION

## Motion Energy

**Nuclear**
The Sun produces mass energy from the nuclear fusion of hydrogen atoms to make helium.

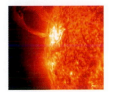

**Heat**
The individual atoms or molecules of hot objects are moving around or vibrating at high speed.

**Light**
Visible light is part of the electromagnetic spectrum, which is a range of mainly invisible moving waves.

**Sound**
Vibrations of the air are caused by sound waves; their energy can move eardrums so we can hear.

**Kinetic energy**
All moving objects—live or inanimate—have kinetic energy, simply because they are moving.

**Solar radiation**
The temperature of the surface of the Sun varies
but is about 5,780 K (5,500° C) on average. It takes
eight minutes for heat radiated from the surface of
the Sun to reach Earth in the form of sunlight.

**BEFORE**

The concepts of "hot" and "cold" were obvious to the ancients, but what caused things to turn from cold to hot was not clear.

**FOUR ELEMENTS**
The 5th-century BCE Greek philosopher **Empedocles** thought that everything was made from **four elements**: air, earth, water, and fire.

**CALORIC THEORY**
**Antoine Lavoisier** introduced the idea of a fluid named **"caloric."** He claimed that hot items contained high amounts of caloric, which **flowed into cold things** that they touched.

**« SEE ALSO**
pp.16–17 FIRE POWER
pp.176–77 ENERGY CONVERSION

# The Nature of Heat

People have known how to make things hot since they first tamed fire, but not why their methods worked. In the early 18th century a German scientist, Daniel Fahrenheit, invented a thermometer with a reproducible scale, and shortly afterward an American spy made a breakthrough in the understanding of heat itself.

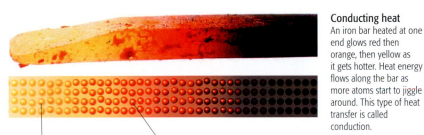

**Conducting heat**
An iron bar heated at one end glows red then orange, then yellow as it gets hotter. Heat energy flows along the bar as more atoms start to jiggle around. This type of heat transfer is called conduction.

Atoms move more vigorously as they gain heat energy from the fire

Atoms adjacent to the vigorously moving ones also start to move more actively as the heat energy flows between them

**PHYSICIST (1753–1814)**
## COUNT RUMFORD

Benjamin Thompson became a spy during the American Revolution and eventually fled to England. From there he went to Bavaria, reorganized the army, invented a stove and a coffeepot, and was made a Count of the Holy Roman Empire. Returning to England in 1799, he founded the Royal Institution before going to Paris to marry the widow of the guillotined Antoine Lavoisier.

Heat makes things get hotter; it raises their temperature. But the nature of heat puzzled not only the ancients but also the scientists of the 17th and 18th centuries. The solution came from a spy and serial philanderer, Benjamin Thompson, Count Rumford.

At the end of the 18th century cannons were made by casting the metal in one solid piece and then drilling or boring out the barrel to make an accurate cylindrical bore. While he was making cannons in Bavaria, Rumford realized that not only did the barrel always become hot

**Convection currents**
As water is heated, the warm water closest to the heat source becomes less dense and rises. It is replaced by cooler, denser water sinking down from above. The convection currents continue until the water boils.

**IN PRACTICE**
## KEEPING COOL

Many mammals have fur, which traps air and is a good insulator. It keeps them warm in cold weather, but can be a problem in higher temperatures. To avoid overheating, some animals have big ears which radiate heat, while others animals pant heavily—the evaporation of water from their mouths and tongues takes away the latent heat. Kangaroos lick their forearms to make them wet, and this water also evaporates to help keep them cool.

during the boring, but also that when the borer became blunt, the barrel became much hotter than it did when the borer was sharp.

This surprised him, because people then believed that heat was a fluid called caloric, which lay hidden within substances. The blunt borer was removing less iron, so it should be generating less heat—but instead it was generating more. Thompson came to the conclusion that heat is nothing to do with caloric, but is in fact a form of motion. The borer was causing vibration of the iron atoms—the longer he ground away with his blunt borer, the more he was making the particles vibrate.

### Three types of heat transference
During the cannon boring, both cannon and borer became hot; heat was being transferred from one to the other. It is now known that heat can be transferred in three different ways: by radiation, conduction, and convection. Anyone lying on a beach can feel the radiant heat from the Sun, traveling straight through space and the atmosphere to the skin. Conducted heat is soon felt by anyone putting the end of an iron poker into a

fire—the fire end may get red hot, but heat is also conducted along the poker to the other end. Convection is the movement of air or other fluids caused by heat. If a radiator is turned on at one end of a room, it heats the air in front of and above it. This hot air rises (because it is less dense than cool air), and moves across the ceiling, then slowly drifts down a wall as it cools. In due course the whole room becomes warmed by convection.

### Inventing the scale
People have long wanted to measure how hot things are. In 1714 Daniel Fahrenheit proposed a scale of temperature that included the freezing point of water being 32° and the boiling point being 212°, but this has generally been superseded by the scale devised by Swedish scientist Anders Celsius in 1742. He originally suggested that the boiling point of water should be 0° and the freezing point 100°, but after he died in 1744 this was reversed, and the scale used today has the freezing point of water at 0° C and its boiling point at 100° C.

These temperatures were essentially relative and arbitrary, but in 1848 William Thomson (later Lord Kelvin) showed that there is an absolute zero of temperature at −273° C. This means there is also an absolute scale of temperature, ranging from 0K (kelvin) to water's freezing point at 273 K and boiling point at 373 K (see pp.180–81).

### Specific heat and thermal capacity
Some materials are easier to heat than others. The "specific heat" of a substance is the amount of heat needed to raise the temperature of 1 g of it by 1° C. Water has an unusually high specific heat, which means it can carry more energy than other common liquids, making it useful for heating systems.

The "thermal capacity" of an object is the amount of heat needed to raise its temperature by 1° C; a large pan of water takes much more heat to reach boiling point than a small pan. Thermal capacity depends on the mass of the object and on its specific heat. Joseph Black (see pp.148–49) discovered that even after being heated to 100° C a pan of water needs yet more heat to make it boil. He called this heat "latent" (from the Latin, meaning "hidden"), because it seemed just to disappear.

**AFTER »**

Heat measurement has produced surprising results, including the quantum revolution.

**BLACK BODY RADIATION**
A heated black body (perfect emitter) glows, and the higher the temperature the shorter the wavelength. The exact relationship could not be explained without **quantum theory 314–17 »**.

**TEMPERATURE OF THE UNIVERSE**
The temperature in space is not absolute zero, but 2.725K, due to the background radiation—the heat echo of the Big Bang **320–21 »**.

**SEE ALSO »**
pp.180–81 THERMODYNAMICS

# Laws of Thermodynamics

**The area of science known as thermodynamics originated in efforts to understand and improve the steam engine early in the 19th century. Thermodynamics is the study of heat and energy. Its findings are set out in four laws, which provide deep insight into the way the Universe works.**

The origins of thermodynamics go back to 1824, when Nicolas Sadi Carnot, a French engineer, published an influential book, *Reflections on the Motive Power of Fire*. In this groundbreaking work Carnot explained how steam engines work as he understood it, and how heat can make things move.

### Motive power

Carnot used the term "motive power" to describe the ability to do work or to make things move. For example, steam engines have the motive power to lift heavy weights or pull trains. The source of this motive power is heat.

Carnot's book explained how heat could drive any engine. Whenever heat flows from a hot to a cold place, some of the heat can be "extracted" to do work. Carnot proposed that the most efficient way to extract heat was via a cycle of changes in the temperature and pressure of a gas. Known as the

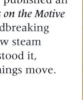

PHYSICIST AND MATHEMATICIAN (1847–1907)

### WILLIAM THOMSON, LORD KELVIN

Born in Belfast and raised in Glasgow, William Thomson published more than 600 scientific papers during his life. He made important contributions in many areas of science, including electricity and chemistry. Thomson also invented a device that predicted ocean tides, and he supervised the laying of the first successful transatlantic cable in 1858. He became known as Lord Kelvin after he was given the title Baron Kelvin in 1892, in recognition of his accomplishments (the River Kelvin runs past Glasgow University, where he was a professor for 50 years).

Carnot cycle, this idea paved the way for the development of the internal combustion engine (see pp.254–55) toward the end of the 19th century. In his book, Carnot relied on the caloric theory—the (incorrect) idea

### Bouncing ball

A bouncing ball rises a little less high with each bounce, because some of the ball's motion energy is shared out among the atoms and molecules of the ball and with those of the floor. Each impact also produces a tiny amount of sound energy.

that heat is a physical substance. By 1830 he had rejected this notion, and realized that heat is related to the movement of the particles of matter: the hotter something is, the more vigorously its atoms and molecules move. This is the dynamic theory of heat ("dynamic" means moving).

### Conservation of energy

With the recognition that heat is related to the motion of atoms and molecules, scientists began to understand heat as a form of energy. When an object slows down because of friction, for example, the energy of the moving object is not lost—it is merely transferred as heat to the atoms and molecules of the object and its surroundings. In other words, the energy of motion, kinetic energy, changes to thermal (heat) energy.

During the 1840s scientists realized that energy can neither be created nor destroyed, an idea investigated by the English scientist James Joule and expressed mathematically by the German physicist Hermann von Helmholtz. By 1850 Rudolf Clausius, another German physicist, had suggested that this idea, called the conservation of energy, should be the first law of thermodynamics.

### Entropy and absolute zero

Clausius suggested a second law, which concerns the fact that not all energy is usable. When a steam engine cools down, for example, the heat energy still exists, but it can no longer do any work since it has dissipated into the air. Heat produced by friction dissipates, too. The second law of thermodynamics states that the amount of unusable,

« BEFORE

Benjamin Thompson's work in the late 18th and early 19th centuries was fundamental to the understanding of heat as energy.

### CALORIC THEORY

During the 17th and 18th centuries, physicists believed that **heat was a physical substance**, which they called caloric. They suggested that **caloric passed from hot to cold objects**, explaining why hot things cool down. The word "caloric" comes from the Latin word for heat, *calor*, and is the origin of the word "calorie."

### BORING CANNONS

The American-born English physicist **Benjamin Thompson** (Count Rumford) **threw the caloric theory into doubt** after observing the boring

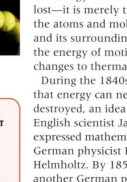

THOMPSON AT A CANNON WORKS

out of cannons. He noticed that when a drilling machine bored into a cannon barrel the cannon grew extremely hot. The **heat was being produced by friction** between the drill and the cannon, not released from within the gun metal as the caloric theory suggested.

« SEE ALSO
pp.176–77 ENERGY CONVERSION
pp.178–79 THE NATURE OF HEAT

### LAWS OF THERMODYNAMICS

**0** The "zeroth law" states that if two thermodynamic systems are each in thermal equilibrium with a third, they will also be in thermal equilibrium with each other.

**1** Energy can neither be created nor destroyed.

**2** The total entropy of an isolated system increases over time.

**3** There is a minimum temperature, at which the motion of the particles of matter would cease.

dissipated energy can only stay the same or increase; it can never decrease. In 1865 Clausius coined the term "entropy" to refer to this unusable energy. Entropy is therefore a measure of disorder in isolated systems, and as time passes the Universe is becoming increasingly disordered.

Ship's engine room
Here, seamen stoke the furnace in the engine room of the USS *Mississippi*. Heat from the fire boiled water into steam, which was used to turn the ship's propellers.

# "Nothing in life is certain except death, taxes and the second law of thermodynamics."

SETH LLOYD, PROFESSOR OF MECHANICAL ENGINEERING AT MIT, 2004

The third law of thermodynamics concerns temperature. As an object cools, its atoms and molecules vibrate less energetically. Scottish physicist William Thomson—Lord Kelvin (see opposite)—realized that there must be a temperature at which all motion would cease. He called it "absolute zero," and he even figured out that its value would be −459.67° F (−273.15° C). The kelvin temperature scale (K), which scientists use today, is named after him. Absolute zero is 0 K.

In 1852 Thomson came up with the the idea of a device he called a "heat pump." Such a device would be able to move heat from a cold place to a hot one—the opposite of what occurs naturally. With his heat pump idea,

Thomson had outlined the principles behind the refrigerator and air conditioning. The first cooling device based on Thomson's ideas was built in 1855, by an Austrian mining engineer, Peter von Rittinger.

## The "zeroth law"
The fourth law of thermodynamics is also known as the "zeroth law." (The name comes from the idea that it is more trivial than the first law and

should come before it.) It states that if two objects are in thermal equilibrium with a third object—in other words, at the same temperature—they will also be in thermal equilibrium with each other. This basically means that using the same thermometer to measure the temperature of different objects will give meaningful results. The term "zeroth law" was coined by the British physicist Ralph Fowler in the 1920s.

The laws of thermodynamics give scientists a better understanding of the Universe, including the world of living things.

**BLACK HOLES AND QUANTUM PHYSICS**
The **mathematics of thermodynamics** play a **key role in modern physics**. For example, it helps physicists to understand the relationship between **black holes 330–31 »** and **quantum physics 314–17 »**.

**ENTROPY AND LIFE**
The second law of thermodynamics states that the entropy of the Universe is increasing, but there are isolated processes in which **entropy is decreasing**. The 20th-century German physicist **Erwin Schrödinger** looked at the role of entropy in the processes of life. He realized that **living things produce order from disorder**, like "islands of negative entropy," although they need an input of energy to do so.

**SEE ALSO »**
*pp.314–17* QUANTUM REVOLUTION
*pp.328–31* THE LIFE CYCLE OF STARS

**AFTER** »

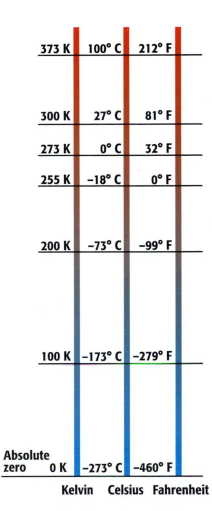

| | | |
|---|---|---|
| 373 K | 100° C | 212° F |
| 300 K | 27° C | 81° F |
| 273 K | 0° C | 32° F |
| 255 K | −18° C | 0° F |
| 200 K | −73° C | −99° F |
| 100 K | −173° C | −279° F |
| **Absolute zero** 0 K | −273° C | −460° F |
| **Kelvin** | **Celsius** | **Fahrenheit** |

Temperature scales compared
The Kelvin scale takes absolute zero as its starting point. On the Celsius scale, zero (0° C) represents the freezing point of water, while the Fahrenheit scale starts from the freezing point of saturated salt water (0 °F).

## BEFORE

Every ancient culture watched the Sun, Moon, and planets progress across the sky, assuming they revolved around the earth.

### FROM EARTH- TO SUN-CENTERED
In 1543, Nicolas Copernicus described a **sun-centered planetary system << 74–75**, overturning centuries of belief that the earth was at the center of the universe. In the early 1600s Johannes Kepler gave this theory mathematical form and structure.

### THE INVENTION OF THE TELESCOPE
Telescopic observations made from 1609 by Galileo Galilei **<< 82–83** provided **proof of the heliocentric view** for the first time. Although Galileo did not invent the telescope, he used a description of the first telescope to make a much more powerful version, allowing the skies to be more thoroughly studied than ever before.

**<< SEE ALSO**
*pp.74–75* THE SUN-CENTERED UNIVERSE
*pp.84–85* EXPLORING THE SKIES

**Formation of the solar system**
About 4.6 billion years ago, a vast spinning cloud of gas and dust contracted to form a central star: the Sun. This was surrounded by a slowly rotating flattened disk of material. Over millions of years, tiny pieces fused and grew, until eventually eight large lumps remained; these were the planets. Rocky metallic material near the Sun formed Mercury, Venus, Earth, and Mars. The four giants—Jupiter, Saturn, Uranus, and Neptune—formed at a greater and cooler distance out of snow, ice, rock, metal, and gas.

**2 The Sun takes shape**
Gravity pulls material into the center of the nebula, where it becomes concentrated and heats up. The central young Sun is surrounded by a flattening disk of material.

**1 Solar nebula forms**
Gas and dust collect together to make a huge cloud many times larger than the present-day solar system. The cloud spins and contracts.

**4 Rocky planets**
Close to the Sun, where it is hot, rocky and metallic material forms planetesimals. These join together to form the rocky planets.

**5 Giant planets**
In the outer, colder regions of the disk the planetesimals are of rock, metal, snow, and ice. These form together into bodies which become the cores of the giant planets. These cores attract large amounts of gas.

**6 Unused debris**
Remaining chunks of dust, rock, and snow are drawn into the Sun and destroyed or thrown out of the planetary solar system. Others form the vast Oort Cloud of comets.

# The Solar System

**Earth is part of the solar system, which consists of the Sun and the large number of objects that orbit around it. Five planets had been noticed in the skies since antiquity, but the invention of the telescope brought with it the discovery of new planets and asteroids, leading to theories about the origin of the solar system itself.**

The majority of the known objects within the solar system are too small and distant to be seen easily from Earth, and they remained undiscovered until the invention of the telescope in the early 17th century. By the end of the 19th century, astronomers had discovered two major planets, approximately 20 planetary moons, and more than 400 asteroids.

### A larger universe
In the mid-18th century astronomers believed the solar system consisted of six planets: Mercury, Venus, Earth, Mars, Jupiter, and Saturn. Astronomers were convinced that Saturn marked the outer limit of the system. So it came as a complete surprise when a seventh planet, subsequently named Uranus, was discovered. Even more

surprisingly, this discovery meant that the known solar system was double the size anyone had previously thought.

### New planets
The discovery of Uranus was made on March 13, 1781 by the British astronomer William Herschel. He was making a routine observation of the sky as part of a longer-term survey, when he noticed an unexpected object. He recorded this as "a nebulous star or perhaps a comet," but further observation confirmed that he had, in fact, discovered a planet.

Irregularities in Uranus' orbit led astronomers to the prediction of another, more distant planet. Urbain Le Verrier and John Adams used Newton's gravitational theory to pinpoint the planet's position, and on

September 23, 1846 Neptune was discovered by Johann Galle, using Le Verrier's calculations.

Meanwhile, others had discovered smaller but more numerous objects. In the early 17th century Johannes Kepler had come to the conclusion that there should be something between Mars and Jupiter. In the 1770s this idea was strengthened by the Titius-Bode Law, which states that the spacing of the planets follows a mathematical sequence. When the newly discovered Uranus fit into this theory, it confirmed the likelihood of its truth.

It was late 1800 before a group of astronomers, organized by Franz Xaver von Zach, and known popularly as the "Celestial Police," started to look in the Mars–Jupiter gap. In 1801 Giuseppe Piazzi found the first asteroid here as

**Jupiter**
The largest planet and the fastest spinner. It rotates in less than 10 hours.

**Belt of asteroids**
This is the remains of a planet that failed to form. Ceres, the largest asteroid, was the first to be discovered.

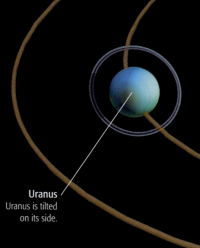

**Uranus**
Uranus is tilted on its side.

**The solar system**
The solar system has existed for around 4.6 billion years. The Sun is the most massive member; its gravity keeps the system together. All the other objects orbit around it, and rotate as they travel.

**3 Planetesimals form**
The disk of material continues to spin and flatten. Grains of material collide to form planetesimals: rocky objects that are the precursors of planets.

ASTRONOMER (1749–1827)

# PIERRE-SIMON LAPLACE

This Frenchman was one of the most influential mathematicians and astronomers of the 18th and 19th centuries. His mathematical ability was recognized at an early age, and by the time he was 20 he was working with Jean d'Alembert, the famed mathematician. Laplace's scientific legacy includes his work on positional astronomy, the stability of the solar system, and the theory of its formation. He survived the French Revolution, and was made chancellor of the Senate by Napoleon.

he was updating a star catalogue. The asteroid—Ceres—was joined by three more in the next six years; within 70 years, 112 were known.

## Theories of the system's birth

The excellence of William Herschel's telescopes and skills enabled him to discover two moons of Saturn and two of Uranus. He also produced extensive catalogues of star clusters and nebulae. The term "nebula" then referred to

indistinct starlike objects that we now know to include a range of objects, such as galaxies, as well as true nebulae (clouds of gas and dust).

Such discoveries opened up new areas of research, such as the idea that the planets had formed from a nebulous ring of material that once surrounded the Sun. This proposal, made in 1796 by Pierre de Laplace, remains the basis of our understanding of the formation of the solar system.

The size and content of the known solar system continues to grow. We also know that our solar system is not unique; we have seen others in the process of formation.

### OORT CLOUD OF COMETS

In the early 1930s **Ernst Opik** studied the paths of **long-period comets** (comets taking more than 200 years to return) and concluded there is a reservoir of them, far beyond Neptune. This is now known as the **Opik-Oort**, or Oort Cloud.

### SPACECRAFT EXPLORATION

From the 1950s onward, robotic spacecraft **visited all seven planets other than Earth**, as well as the Sun, planetary moons, comets, and asteroids. Between 1969 and 1972, 12 astronauts landed on the Moon and returned to Earth.

### PRE-PLANETARY DISKS

In 1983 the **infrared space telescope** IRAS provided the first evidence of a star with a flattened disk of pre-planetary material.

### DATING THE MOON AND EARTH

In 1956, a study of iron meteorites and ocean sediments **dated Earth to about 4.55 billion years**. Dating of moon rocks brought to Earth in 1969–72 shows that **the Moon is 4.527 billion years old**.

SEE ALSO »
pp.394–95 INSIDE THE SOLAR SYSTEM
pp.396–97 SPACEPROBES AND TELESCOPES

**Mercury**
The smallest planet and the closest to the Sun, Mercury also has the widest range of temperatures of any planet.

**Earth**
Earth is the only planet known to have liquid water and to support life.

**Mars**
The outermost of the four rocky planets, Mars is rust-red in color.

**Venus**
Just a little smaller than Earth, Venus is shrouded by a thick and permanent atmosphere, and is the slowest-spinning planet.

**Saturn**
Surrounded by a spectacular system of rings, Saturn is the most distant planet visible to the naked eye.

**Neptune**
The most distant and coldest planet, Neptune is 30 times farther from the Sun than Earth is.

# How Rocks Form

**Until the mid 18th century few scientists had investigated the composition of rocks or tried to figure out how they had formed. By around 1770 two distinct theories of rock formation had been proposed and by the beginning of the 19th century it had become clear that there were three main rock types.**

Early geologists came up with a variety of theories about rock formation, but it was not until the late 18th century that Scottish geologist James Hutton's observations indicated that there are three main rock types.

**≪ BEFORE**

A small number of scientists before the 18th century helped lay the foundations of modern petrology (the study of rocks).

**AVICENNA'S THEORY**
11th-century Persian scholar **Avicenna** proposed that **rock formation** occurs through processes that have **operated uniformly throughout Earth's history**—uniformitarianism.

**EARLY STUDIES OF ROCKS**
**Johann Lehmann's** 1756 study of rocks in southern Germany was a pioneering work. In addition to **describing layers of sedimentary rock**, he explained the **origin of many mineral deposits** within them.

**LEHMANN'S GEOLOGICAL DRAWINGS**

**≪ SEE ALSO**
*pp.38–39* ANCIENT IDEAS OF THE WORLD

These all form part of a cycle, in which rocks are constantly altered and recycled in different forms.

### Neptunists versus plutonists

Before the 18th century Earth's rock layers were widely believed to have been deposited during the biblical flood. But, by the 1770s, two rival theories of rock formation became established. The German geologist Abraham Werner thought that an ocean had once covered the whole planet, and that granites and other hard rocks, which he believed to be the oldest (because in his local area they mostly lay at great depth) had formed as minerals crystallized out from this sea. As the fluid level lowered, continents began to emerge. Werner's theory, termed "neptunist" due to its emphasis on the ocean, was highly influential for a time.

Opposing Werner were vulcanists and plutonists, who looked to volcanoes and magma (molten rock from inside the planet) solidifying on top of, or

beneath, Earth's surface as the main cause of rock formation. Thus, in 1777 French geologist Nicolas Desmarest proposed that columnlike basalt formations in the Auvergne in France and off Ireland (the Giant's Causeway) were the result of magma from local volcanoes crystallizing as it cooled.

### Hutton's theories

James Hutton leaned toward the vulcanist and plutonist view. He was sure that some rocks were formed from magma—he saw fingers of granite cutting through sandstones in a way that indicated the granite had once been molten. And he identified some geological formations as rocks that had been altered by heat (metamorphic rocks), confirming the importance of heat in rock formation. However, he also accepted sedimentation (the process whereby solid matter is deposited from air, water, or glacial ice) as important. He thought that a high proportion of rocks originally

**Geological tools**
A geologist uses a rock hammer (right), and sometimes chisels (bottom center and left), or a pocket knife (top center) for collecting rock samples. Many of these tools haven't changed in centuries.

**SEDIMENTARY PARTICLES**

Compaction and cementation

Uplift, weathering, erosion, and sedimentation

Uplift, weathering, erosion, and sedimentation

**Sedimentary rock layers**
These multi-hued layers of sedimentary rock—sandstones and mudstones—are in Arizona. They were sedimented between 200 and 230 million years ago in a river environment.

**Giant's Causeway**
In the late 18th century, this 60 million-year-old formation of basalt in Northern Ireland was the subject of a dispute: vulcanists (correctly) thought it had a volcanic origin; neptunists that it was altered sedimentary rock.

**IGNEOUS ROCK**

**GEOLOGIST (1726–97)**

## JAMES HUTTON

Scottish geologist James Hutton qualified as a medical doctor at age 23, but then immersed himself in studying chemistry and experimental farming. His agricultural interests led him to start investigating soils and rocks, and from the 1760s he made a series of geological tours of Scotland. These led him to develop his theories of rock formation. Often referred to as the father of modern geology, his *Theory of the Earth*, 1785, was highly influential, despite being written in a somewhat unintelligible style.

formed from sediment at the bottom of the sea. Later, these rock layers were uplifted and eroded, sometimes tilted and deformed, subsided again, and had more sedimentary layers added. He believed that this cycle had occurred many times over in the past.

## Three rock types
By 1800 it had become clear that there are three main rock types. The first type is igneous: rocks formed from cooling and crystallization of magma. These include granite, basalt, and obsidian. Secondly, there are sedimentary rocks, such as sandstone, shale, and limestone, formed from compaction of sedimented particles of mineral or organic material, or precipitated chemicals. Third are metamorphic rocks, such as gneiss, schist, and marble. These occur whenever existing rocks of any type

become buried under kilometer-thick layers of overlying rock or when rocks are affected by movements of Earth's crust (see pp.352–53). The pressure and heat in these circumstances cause physical and chemical changes, which result in the formation of completely new types of rock.

## The rock cycle
Also arising from Hutton was the idea of a rock cycle. Thus, when any rock is exposed at the surface, it is weathered and eroded, turning it into sedimentary particles. Over time, these are compacted and cemented to make sedimentary rocks. Both these, and igneous rocks, may be heated and compressed to form metamorphic rock, and any rock may melt to form magma, which cools to create new igneous rock. Rock may also be uplifted and exposed to start a new cycle.

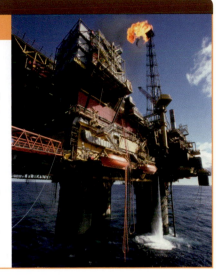

### OIL EXTRACTION
Some sedimentary rocks are oil bearing—they can hold a mixture of hydrocarbon chemicals (crude oil or petroleum) in their pores. These substances are derived from the remains of organisms, which settled millions of years ago in the mud or sand at the bottom of a lake or sea. This material was then compacted into sedimentary rock and, as more rock layers formed above, pressure and heat altered the organic matter into petroleum. When searching for oil, geologists look for promising permeable rocks trapped between layers of impermeable rock. Oil rigs, such as this one in the North Sea, enable oil to be extracted from beneath the seabed.

## " A **stone** is **ingrained** with **geological** and **historical** memories."

ANDY GOLDSWORTHY, BRITISH SCULPTOR, 1956

**AFTER** ≫

### ▽ Rock cycle
This illustrates how rocks are formed and altered by a variety of different processes to make new rock types.

Uplift, weathering, erosion, and sedimentation

Heat and pressure

**METAMORPHIC ROCK**

Melting

**SEDIMENTARY ROCK**

Melting

Heat and pressure

Cooling

Melting

**MAGMA**

Metamorphic rock
These crumpled layers of metamorphic rock, mainly schist, in the Sierra Nevada mountains of California, were originally deposited as sedimentary rock but were later altered by subjection to heat and pressure.

Over the past 200 years, the details of rock formation have been further elaborated by laboratory and field studies.

#### ARTIFICIAL MARBLE
In 1901 Canadian geologist **Frank Adams** subjected marble to great pressures in hydraulic presses, **simulating processes that produce metamorphic rocks**. In doing so, he showed how thin, leaflike layers, such as those seen in natural marbles, can be produced artificially.

#### PREDICTING MINERAL FORMATION
During the 1920s and '30s the Swiss–Norwegian geochemist **Victor Goldschmidt** developed methods for **predicting** what **minerals** would form in a rock given specific combinations of **chemical elements** and **geological conditions**.

**CROSS SECTION OF QUARTZ**

#### PLATE TECTONICS
The plate-tectonic theory of the 1960s modified ideas on rock formation. Spreading **mid-ocean ridges 212–13 ≫** (underground mountain ranges) emerged as sites of rock formation, and **subduction zones 358–61 ≫** (where one section of Earth's crust moves below another) as regions where rocks melt to form magma.

SEE ALSO ≫
pp.356–57 **PLATE TECTONICS**

Igneous rocks from volcanoes
The cooling of lava flowing from volcanoes such as Arenal in Costa Rica (one of the most active in the world) produces igneous rocks including basalt and andesite.

« BEFORE

The study of fossils and what they reveal has a history going back over 2,500 years.

**MARINE DISCOVERIES**
The 6th-century BCE Greek philosopher **Xenophanes** described **imprints of fish and seashells** he had found **in rocks** far from the sea. He recognized these as animal remains that must once have been under water.

**RENAISSANCE VIEWS**
In 16th-century Florence **Leonardo da Vinci described fossil shells** he had observed in the foothills of the Alps and proposed that they were the **remains of once-living organisms**.

**FOSSIL DRAWINGS**
Swiss naturalist **Konrad Gesner** produced the **first drawings of fossils** in his book compiled in 1558, **AL**though it is unclear whether he realized that they were the remains of animals.

« SEE ALSO
pp.184-85 How Rocks Form

## Geological succession
Each geological period has characteristic fossils, which lie in sedimentary rocks formed during that period. An example fossil from each period is shown in the sequence below, and the start date for each period is given.

# The **Fossil Record**

**Fossils record the history of life on Earth, revealing the remains (or evidence) of plants, animals, and other organisms preserved in rocks. Although the study of fossils, called paleontology, is a field of biology, its development has been closely linked to efforts to understand the history of Earth itself.**

Fossils have been the key to understanding extinct life forms and informing us about evolution. They have also helped us to identify the geological periods in Earth's history.

### Reality of extinction
The scientific study of fossils (the preserved remains of animals, plants, and organisms) arguably started in the mid 17th century with English scientist Robert Hooke (see pp.92–93), who observed fossilized seashells and wood. He was the first to formulate a reasonably accurate explanation of how fossils form: they had been soaked with "water impregnated with stony and earthy particles." In 1667 Danish geologist Nicolas Steno deduced that objects known as tongue stones (often found in certain rock layers in northern

Italy) were, in fact, fossilized shark teeth. In 1778 giant bones and a skull were found near the River Meuse in the Netherlands, which were later identified as the relics of a giant marine reptile, a mosasaur. Although many found the concept of extinction difficult to accept on religious grounds, most scientists recognized that such fossils were the remains of once-living

PALEONTOLOGIST (1769–1832)
### GEORGES CUVIER
Born in France, Georges Cuvier studied in Stuttgart, Germany, then made his reputation as a naturalist. By 1799 he was Professor of Natural History at the Collège de France in Paris. He had a brilliant ability to reconstruct animals and their lifestyles from fragments of fossil remains. In addition, he was appointed a state councilor, remaining in this position through the reigns of Napoleon and three successive kings of France.

**Cambrian period 542 mya**
Trilobites (*Paradoxides*) were marine arthropods that left many seabed fossils.

**Ordovician period 488 mya**
*Didymograptus* was a marine animal that left fossils in seabed shales.

**Silurian period 433 mya**
*Monograptus* was a graptolite, a fossil common in Silurian slates and shales.

**Devonian period 416 mya**
Brachiopods like this *Mucrospirifer* were hard-shelled marine animals.

**Carboniferous period 359 mya**
Plants like *Asterophyllites* grew in swampy areas and fossilized in coal layers.

**Permian period 299 mya**
Marine reptiles such as *Stereosternum* left fossils in some limestone shales formed on the seabed.

## Fossilization
Animal and plant remains will be preserved as fossils only if they are protected from destruction by being quickly buried after death. Even then, usually only the hard parts of the organism will be fossilized.

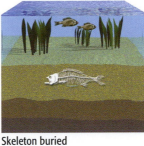

**Organism dies**
To become a fossil, an organism must die in an environment (such as on the seabed or near a river mouth) where it will be buried by sediment before it disintegrates.

**Skeleton buried**
After burial, the soft parts of the organism soon decay and dissolve, leaving just a skeleton in the case of an animal and a cellular framework in the case of a plant.

**Rock forms**
As the skeleton or plant remains are compressed by overlying rock, organic substances are replaced by minerals, although the remains retain their shape.

**Fossil revealed**
Earth movements slowly lift up the sedimentary layers, and weathering and erosion then wear them down, eventually exposing the fossil at Earth's surface.

species that had died out. In the early 19th century French naturalist Georges Cuvier (see left) established the reality of extinction, noting that fossils such as those of mosasaurs and pterodactyls bore no resemblance to anything living.

A series of fossil discoveries in England in the 1800s, including the first fossil dinosaur (a massive lizard that they called *iguanodon*), strengthened the case for extinction. Since many of the species found were restricted to specific rock layers, it seemed that they had existed for only a part of Earth's history.

### Rock layers and evolution
During the early part of the 19th century growing knowledge of the fossil record played an increasing role in the development of stratigraphy: the identification of particular rock layers of different ages. By 1880 nearly all the currently recognized geological periods (see pp.190–91) had been named and described using fossils as evidence.

The second half of the 19th century saw much paleontological activity. One remarkable discovery, in 1861, was of a fossilized bird, *Archaeopteryx* (see pp.188–89). In the 1870s and 1880s two American fossil hunters, Edward Drinker Cope and Othniel Marsh, engaged in "bone wars" as they directed fossil searches in the American West, leading to discoveries of the first pterosaur (flying reptile) fossils, in 1871, and the first skeleton of a *diplodocus* (a gigantic plant-eating dinosaur) a few years later.

In 1909 the Burgess Shale fossil site was found in British Columbia, Canada, showing an explosion of sea life in the Cambrian period, around 530 million years ago. During the 1920s new dinosaur and mammoth fossils were found in the Gobi desert, and in the 1940s, the discovery of the fossils of soft-bodied animals in Australia showed that complex organisms have existed on Earth for well over a billion years.

**AFTER**

Since the 1950s, there has been an ongoing search to find even older fossils.

**PRECAMBRIAN FOSSILS**
In 1954 American paleontologist **Elso Barghoorn** and geologist **Stanley Tyler** discovered **microfossils of bacteria** in a **2 billion-year-old** rock formation found in the Canadian Shield.

**FOSSILS IN CHINA**
Since the 1980s intriguing fossils of fish, dinosaurs, birds, and mammals have been found in China, such as **feathered dinosaurs**, which used feathers for insulation rather than flight.

**FEATHERED DINOSAUR**

SEE ALSO »
pp.188–89 FINDING *ARCHAEOPTERYX*
pp.190–91 DATING THE EARTH
pp.200–01 HOW EVOLUTION WORKS

**Triassic period 251 mya**
*Dicellopyge* was a freshwater fish that left fossils in rocks formed from lake or river sediments.

**Jurassic period 199 mya**
*Ichthyosaurus* was a streamlined marine reptile that is found as a fossil in limestone shales on the seabed.

**Cretaceous period 145 mya**
Sea urchins such as *Micraster* were marine animals of which the test, or shell, is a common fossil in chalk beds.

**Tertiary period 65 mya**
*Heliobatis* was a freshwater stingray and is found as a fossil in some rocks formed from lake sediments.

**Quaternary period 1.6 mya**
Hominins—modern humans and extinct relatives—have left fossils in locations such as rock formed from lakeside mud.

**BREAKTHROUGH**

## MASS EXTINCTION
Mass extinctions are sudden cataclysmic decreases in the number of living species on Earth. During the 19th century it became clear from the fossil record that a severe mass extinction happened at the end of the Cretaceous period when huge numbers of species died out, including the dinosaurs and, from the sea, ammonites (below).

# Finding *Archaeopteryx*

**Discovery of a "missing link" is a rare event in the palaeontological world. Few such finds have been more significant than *Archaeopteryx*, an extraordinary fossil skeleton that came to light in a Bavarian quarry in 1861. Combining the characteristics of a dinosaur with the feathered wings of a bird, the fossil was the first example of evolution in action. *Archaeopteryx* marks the midway stage between birds and reptiles.**

In the 19th century the limestone quarries of Solnhofen, in southern Germany, yielded a number of interesting fossils. It was not unusual for the quarry workers to come across relics from Jurassic times within the 170-million-year-old rocks. One notable find in 1860 was a fossilized feather that predated any previously discovered fossil feathers by millions of years. But this was nothing when compared to the sensational discovery made the following year.

The splitting of a slab of rock exposed the almost entire skeleton of a small animal surrounded by the outlines of feathered wings, and at the time of this discovery any animal with feathers was classified as a bird. Named *Archaeopteryx* (meaning "ancient feather"), the fossil quickly acquired the status of "first bird," although it had some distinctly unbirdlike features. The long, bony tail and claws on the wing bones were definitely those of a reptile. After the discovery of the first specimen, a number of complete fossils of *Archaeopteryx* were found, which showed that this particular bird also had teeth.

Karl Haberlein, a local doctor and fossil collector, bought a fossil from the quarrymen and, scenting a large profit to be made from avid collectors, offered it for sale. The prize was bought for £700 by Dr. Richard Owen, superintendent of the British Museum's natural history collection.

*Archaeopteryx* was a priceless gift to the supporters of Darwinism. At the time of the fossil's discovery, Darwin's still-new theories on evolution (see pp.200–03) and the divergence of species were not universally accepted. His opponents had been demanding evidence that one life form could evolve into another—and here the evidence was, set in rock. Owen himself was no evolutionist, though he was correct in his understanding of *Archaeopteryx* as a bird at a primitive stage of development. But it was left to his Darwinist arch rival Thomas Henry Huxley to recognize *Archaeopteryx* for what it was—the intermediate stage between reptiles and modern birds.

Recently discovered Chinese fossils show that birds are effectively avian dinosaurs, having evolved from small feathered theropods.

**Snapshot in time**
This specimen of *Archaeopteryx* has been almost perfectly preserved in a slab of limestone. The imprints of feathers are clearly visible splaying out from the animal's wing bones and around the long, whiplike tail.

## "If **evolution has taken** place, there will its **mark be left**."

THOMAS HUXLEY, "HOW PALAEONTOLOGY CAN HELP TO UNDERSTAND LIFE," 1860

# Dating the Earth

**From the mid 18th century, a growing body of scientific evidence made it clear that the earth was much older than previously believed. Discoveries in geology contributed to this revolution, but ultimately the key to dating the earth was supplied by advances in physics.**

The first person to try to figure out Earth's age using a reasoned scientific approach is thought to have been the 18th-century French naturalist, the Comte de Buffon. He argued that Earth must have started as a hot, molten orb, and that its age could be calculated by figuring out how long it had taken to cool to its present state. He estimated Earth's age at about 75,000 years—far older than conventional opinion of the time. After pressure from the Church, Buffon was forced to retract this claim, but belief that the earth was just a few thousand years old was already fading.

### Rock layers and fossils
Observation of fossils and rock layers (see pp.186–87) led to an understanding of the relative sequence of rock deposition. Geologists were starting to realize that layers of sedimentary rock were laid down at different ages and that very thick layers of rock must have taken millions of years to deposit. By 1800 it was realized that rock layers from different sites could be regarded as the same age if they contained the same fossils. However, geologists still had only a vague idea of the absolute age of each rock layer.

One of the most important 19th-century believers in the view that Earth is extremely ancient was Charles Lyell (right). In the 1830s, his book *Principles of Geology* produced a convincing argument that Earth must be several hundred million years old. His ideas of gradual change influenced Charles Darwin's Theory of Evolution (see pp.200–201).

### Cooling Earth and the salt clock
In 1862 a new figure entered the arena: the respected Irish-born physicist Lord Kelvin. Like Buffon, Kelvin maintained that it must be possible to calculate Earth's age from the time taken for it to cool from a molten state. Using new theories in physics, Kelvin first estimated that Earth must be between 24 and 400 million years old, but later refined his estimate down to 20–40 million years. His findings dismayed most geologists and biologists, few of whom thought that this time frame fit with their observations in the natural world.

In 1899 the Irish physicist John Joly suggested a novel approach to dating the earth, known as the salt clock solution. He argued that the age of Earth's oceans could be ascertained from the time it had taken for them to reach their current salinity, which results from salt added each year by rivers. He came up with a range of between 80 and 150 million years. However, Joly hadn't realized that salt does not continue to build up but is constantly being removed from the oceans by geological processes, and so his calculated age was too low.

### The radioactivity revolution
At the end of the 19th century a major advance in physics occurred that subsequently proved key to establishing the age of the earth. This was the discovery of radioactivity in 1896 by the French scientist Henri Becquerel (see pp.302–303). One early finding was that Earth has an internal heat source—in the form of radioactive elements such as uranium. This finding cast doubt on Kelvin's calculations, which assumed that no such heat source existed.

More importantly, advances in the early 1900s demonstrated that radioactivity supplies a type of natural "clock" that can be used to establish the absolute ages of some rock types (see below). Taking up this idea, the American chemist Bertram Boltwood tried dating a variety of rock samples and came up with ages for the different samples ranging from 265 million to over 2 billion years old.

For a while, the scientific community was sceptical of the new technique of "radiometric dating," but following further isotope work by Alfred Nier in the US and British geologist Arthur Holmes the method came to be accepted as valid. Holmes's 1947 timescale of Earth's history is close to accepted modern values.

This was not quite the end of the story. Because Earth may have existed for some time before any rocks currently

GEOLOGIST (1797–1875)

## CHARLES LYELL

Born in Forfarshire, Scotland, Charles Lyell started his career as a lawyer, but soon turned to geology. In 1828 he became convinced of Earth's immense age from his examination of the enormously thick layers of volcanic rock produced by Mount Etna in Sicily. A great traveler and observer of natural phenomena, Lyell popularized the idea that geological history could be explained by the operation of the same slow, gradual processes that can be seen on Earth today. His *Principles of Geology* was one of the most influential scientific books of the Victorian era and, in 1848, he was knighted.

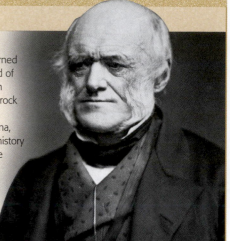

# 4.55 BILLION
**years old is the scientifically accepted age of the earth today.**

> "...we find **no vestige** of a beginning, **no prospect** of an end."
>
> JAMES HUTTON, GEOLOGIST, 1788

## BREAKTHROUGH

### RADIOMETRIC ROCK DATING

In 1902 the New Zealand-born physicist Ernest Rutherford suggested that the rate at which radioactive material decays can be used to date rocks. Granite, for example, can be dated by comparing the concentration of the radioactive isotope, uranium-235, with its decay product, lead-207, in minerals within the rock. When granite forms, grains of the mineral zircon usually crystallize within it. Inside each grain of zircon, atoms of uranium-235 are trapped and decay into atoms of lead-207. The rate of this decay can be accurately predicted—every 704 million years half the remaining uranium atoms in the mineral grain decay into lead-207 atoms. The ratio of the two isotopes allows the rock to be dated.

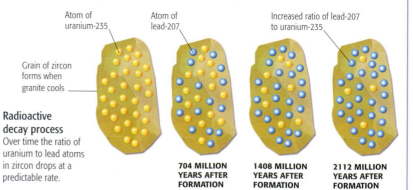

Atom of uranium-235

Atom of lead-207

Increased ratio of lead-207 to uranium-235

Grain of zircon forms when granite cools

**Radioactive decay process**
Over time the ratio of uranium to lead atoms in zircon drops at a predictable rate.

**704 MILLION YEARS AFTER FORMATION**

**1408 MILLION YEARS AFTER FORMATION**

**2112 MILLION YEARS AFTER FORMATION**

present in its crust came to be formed, scientists looked at materials, such as meteorites, that formed at the same time as Earth elsewhere in the solar system. In 1955 an American geochemist, Clair Patterson, used radiometric dating to date a piece of a meteorite whose chemical makeup matched Earth's. Patterson dated this rock at 4.55 billion years old, a figure that is still widely accepted as the true age of the Earth.

## Geological timescale
This geological "clock" shows the start dates for the presently accepted divisions of the geological timescale (mya—million years ago). It is an updated version of a timescale proposed by the geologist Arthur Holmes in 1947 as a result of his work on radiometric rock dating.

TERTIARY **65 MYA**
CRETACEOUS **145 MYA**
JURASSIC **199 MYA**
TRIASSIC **251 MYA**
PERMIAN **299 MYA**
CARBONIFEROUS **359 MYA**
DEVONIAN **416 MYA**
SILURIAN **433 MYA**
ORDOVICIAN **488 MYA**
CAMBRIAN **542 MYA**

ORIGIN OF EARTH **4554 MYA**
PRECAMBRIAN

1138 MYA
3415 MYA
2277 MYA

Since the 1950s rock dating methods have increased in number and sophistication and in the range of materials that have been dated.

### THE OLDEST MINERALS
In 1972 a rock found in the lunar highlands was brought back to Earth by the Apollo 17 astronauts, and has subsequently been dated at 4.5 billion years old. This is the **oldest moon rock** found so far. Its age meets with the currently accepted theory of how the Earth–Moon system formed.

ZIRCON CRYSTAL

### DATING OF SEDIMENTARY ROCKS
Since the 1980s techniques have developed for direct **radiometric dating** of some sedimentary rocks, as well as igneous (volcanic) and metamorphic rocks. This is done by measuring the age of tiny crystals on mineral grains during the sedimentation process.

### WORLD'S OLDEST ZIRCON CRYSTAL
In 2001 a zircon crystal was radiometrically dated as 4.4 billion years old. This crystal holds the current record for the **oldest known object** of terrestrial origin.

SEE ALSO »
pp.298–99 RADIATION AND RADIOACTIVITY

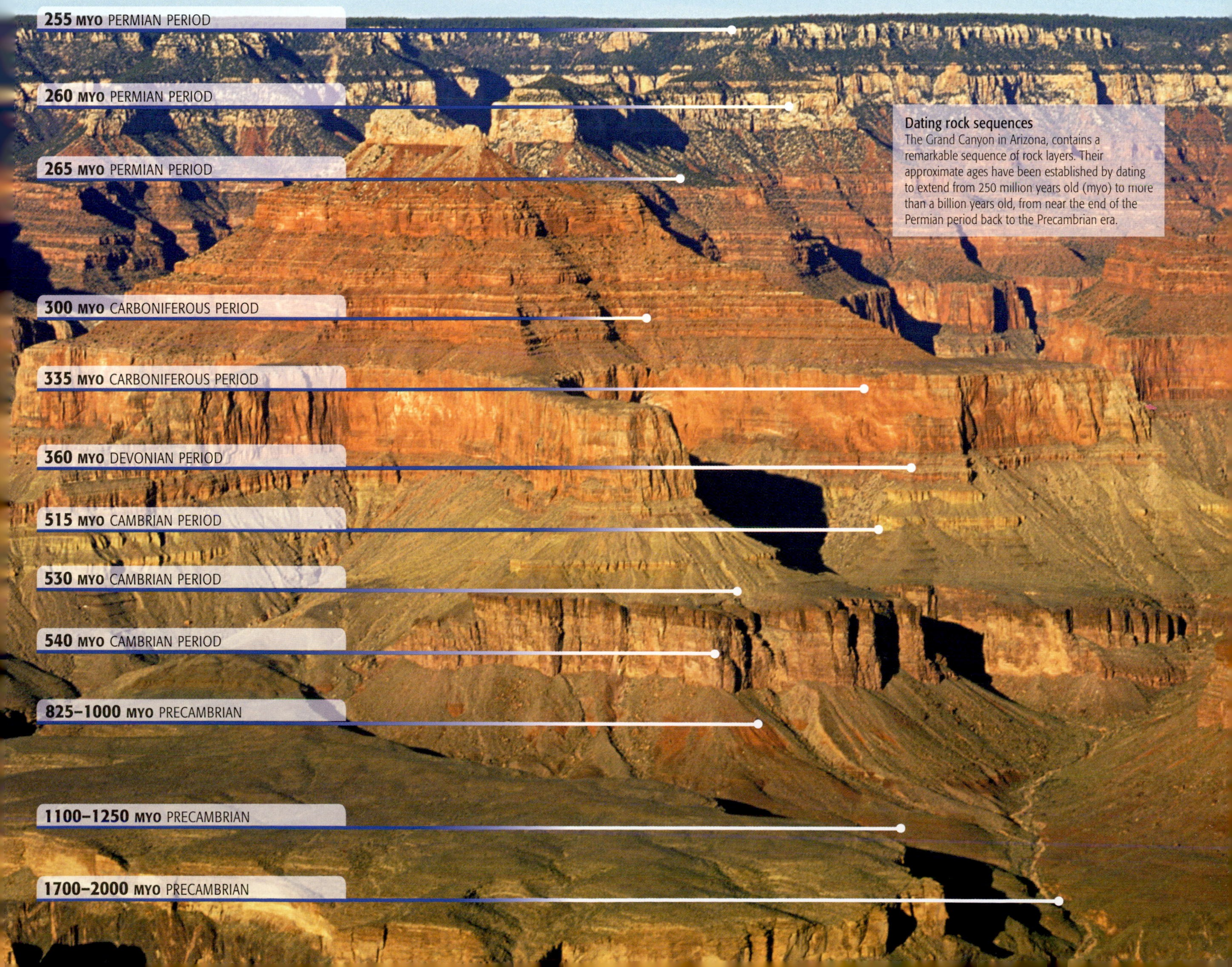

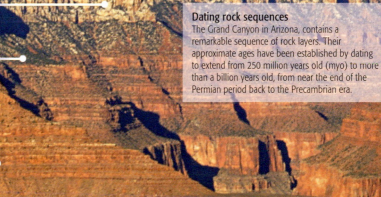

255 MYO PERMIAN PERIOD

260 MYO PERMIAN PERIOD

265 MYO PERMIAN PERIOD

### Dating rock sequences
The Grand Canyon in Arizona, contains a remarkable sequence of rock layers. Their approximate ages have been established by dating to extend from 250 million years old (myo) to more than a billion years old, from near the end of the Permian period back to the Precambrian era.

300 MYO CARBONIFEROUS PERIOD

335 MYO CARBONIFEROUS PERIOD

360 MYO DEVONIAN PERIOD

515 MYO CAMBRIAN PERIOD

530 MYO CAMBRIAN PERIOD

540 MYO CAMBRIAN PERIOD

825–1000 MYO PRECAMBRIAN

1100–1250 MYO PRECAMBRIAN

1700–2000 MYO PRECAMBRIAN

**《 BEFORE**

Some explanations of the formation of common features, such as hills and valleys, were first put forward over 1,000 years ago.

**THE EFFECTS OF WATER**
The 11th-century Persian scholar **Avicenna** stated that land may be shaped by "the effect of water, which, cutting itself a new route, has **denuded the valleys**, the strata being of different kinds, some soft, some hard."

**FOUNDATION OF GEOLOGY**
After studying local landscapes, the Chinese scientist **Shen Kuo**, in his *Dream Pool Essays* of 1086, outlined the principles of **uplift, erosion**, and **sedimentation**.

**VALLEY FORMATION**
Danish scientist **Nicolas Steno** proposed in 1669 that some **valleys** in Italy may have originated from the **collapse** of sections **of crust into vast underground chambers**. Although this was wrong, the idea that **gravity** was the only force powerful enough to create mountains and valleys seemed rational then.

**《 SEE ALSO**
*pp.184–85* HOW ROCKS FORM

# Shaping the Landscape

**Earth's land surface has an extremely varied relief, with landscapes ranging from extensive flat plains, craggy peaks, and deep canyons to delicately sculpted valleys and hummocky hills. Explanations for the formation of such features have varied from the effects of past catastrophes to the results of weathering and erosion.**

The formation of landscape involves three basic processes: uplift, weathering and erosion, and sedimentation. Land is built up through uplift, caused by plate movements (see pp.356–57) and volcanic activity. Uplifted land is then lowered, primarily by the processes of weathering and erosion. Weathering is the breakdown of rocks by physical or mechanical action (these include extremes of temperature which can break rocks apart), chemical attack, or biological activity (such as effects of lichens and plant roots). Erosion is the carrying away of the loosened material by wind, water, glaciers, and landslides.

## Early explanations
From the 16th to the 18th centuries reports of explorers made it plain that Earth had undergone great physical

IN PRACTICE

## UNRAVELING PAST CLIMATES

Some periods in Earth's past have had a colder or warmer climate than today, which has affected the shaping of the landscape. Evidence is found in the geographical distribution and chemical makeup of fossils, particularly of climate-sensitive species, such as beetles, some plants, and tiny marine organisms such as foraminifera (right). Other evidence comes from fossil tree rings and gases trapped in ice sheets.

changes in the past, which had altered its landscapes. Examples included evidence of massive historic landslides, and the existence of huge solidified sheets of volcanic lava in places such as Iceland and India. In America immense

canyons had been discovered, which had multiple layers of thick sediments forming their walls.

There were two main theories as to the cause of these phenomena. Most geologists thought that sudden

**Karst formation**
The dramatic
known

### Processes of weathering and erosion

Weathering and erosion are the main processes whereby rocks that are exposed to the air at Earth's surface are broken down and lowered. Four common types are shown here.

catastrophic events, such as massive earthquakes, floods, and volcanic eruptions were responsible. This idea came to be known as catastrophism.

But, in his 1785 book, *Theory of the Earth*, Scottish scientist James Hutton outlined a different view. He proposed that most geological features could be explained by slow, gradual processes taking place over very long periods of time, such as weathering, erosion, and deposition (process whereby solid matter is deposited from air, water, or glacial ice). His philosophy came to be known as uniformitarianism (see p.184). It can be summarized by the phrase, "the present is the key to the past," which means that past geological changes are explainable by the same uniform events that can be seen operating today. During the 19th century uniformitarianism gained in popularity within the scientific community. It was supported, in

**Wind erosion**
Yardangs such as this one in the central Sahara are shaped by the cutting action of wind-blown sand, which is strongest near ground level.

**Freeze-thaw weathering**
If water gets into cracks in rocks and then freezes, it can split the rock as it expands to form ice. When it thaws, the rock may disintegrate or new cracks may appear.

**Water erosion**
Antelope Canyon, in Arizona, was sculpted by rainwater picking up sand in an upstream basin, then rushing into the canyon, forming "flowing" shapes in rocks.

**Biological weathering**
This occurs where plant roots penetrate and force apart cracks in rocks. Some plants also release organic acids, which dissolve the rock and widen the spaces.

particular, by Scottish geologist Charles Lyell. Today, it is still widely used to explain how Earth's landscapes evolved.

### Wind erosion and deposition

Weathering and erosion by wind occurs mainly in desert regions with little or no vegetation. There are two main aspects to it. First, is the actual transportation of material (grains of sand and dust) in the wind, either directly or by a process called saltation, in which flying and bouncing grains help move the heavier ones. In this way, a persistent wind can

transport loose material over huge distances. Second, is the abrasive effect of wind-blown particles sculpting rock formations into sometimes striking shapes. Where the wind drops, the

particles are deposited to form sand dunes or loess (blankets of fine silt). The size of these deposits can be considerable, including sand dunes as high as 6,500 ft

»

# "Running water on the land has ever had the same power of wear and transportation."

JAMES DANA, AMERICAN GEOLOGIST, 1867

**Glacial erosion**
Glaciers such as the Fox Glacier in New Zealand erode the bedrock in high mountainous areas by "plucking" (removing) material from it and, lower down, by grinding trapped rocks against it.

Ice becomes heavily fractured as glacier moves downhill

Bedrock

Boulder is "plucked" from bedrock after freezing to glacial ice

Trapped boulder scours bedrock as it is carried along by glacier

**PLUCKED ROCK**

Kettle lake
Small circular lake formed by ice blocks burying themselves in ground and melting as glacier retreats.

Glacier

Erratic
Glacier has carried this large boulder down mountain.

**Erosion by a waterfall**
A waterfall typically occurs where a layer of hard rock lies over a softer layer. As the soft rock is eroded, an overhang of hard rock develops. This eventually collapses due to lack of support and the waterfall erodes its way upstream leaving a steep-sided gorge, as in Lower Yellowstone Falls, Yellowstone National Park, US.

A layer of hard rock is undercut by erosion of softer rock

Waterfall

Soft rock is worn away by splashback and abrasive action of rock particles in water

A plunge pool is formed by force of water hitting soft bedrock at its base, and is deepened by stones rubbing against base

**WATERFALL FORMATION**

Gorge
Deep valley carved from the landscape by river or waterfall.

Floodplain
Flat region next to a river that has formed from sediment deposited by river.

Meander
Bend in river caused by erosion of one bank and deposition on opposite bank.

(2,000 m) in Peru, while in China there are accumulations of loess more than 300 ft (100 m) deep, covering hundreds of square miles.

## Glacial erosion and retreat

As a glacier flows through mountains, it can reshape them, producing features such as cirques (hollowed-out depressions) and U-shaped valleys. At the lower end of a glacier, where it melts, features such as moraines (piles of rock debris), erratics (large boulders), and lakes can occur.

Retreating glaciers, such as those at the end of the last ice age 13,000 years ago, left other marks on the landscape. As ice melted, parts of the world depressed by the weight of the ice sheets, such as Scandinavia, Scotland, and Canada rebounded upward. In Washington State, a glacial lake (Lake Missoula) unleashed cataclysmic floods, creating deep, rough channels—the Channelled Scablands.

## Water erosion and deposition

Erosion by water flowing downhill, in streams and rivers, is the main cause of landscape features such as gullies,

**10 IN (25 CM)** The rate at which the Himalayas, in Asia, are being eroded down every century. They are also being uplifted at about 24 in (60 cm) a century.

wadis (dried-up river channels), gorges, and V-shaped valleys. Even slow, smooth flows of water wash away small particles of rock, but where the flow is fast and turbulent, due to a steep gradient, it has much greater erosive power. Typically erosion is both downward, deepening the gorge or valley, and upward, extending the gorge or valley into the hillside. As water flow continues downstream, it picks up more and more solid material (usually sand and rock) and it is the abrasive effect of this material that causes most of a river's erosive power.

As a river reaches its base level (the lowest point to which it can flow) and its gradient decreases, the erosive activity switches from vertical to horizontal, the stream carries away rock and soil from its banks by friction, suction, and dissolution. This widens the valley floor and creates a floodplain. Where the river slows (as a result of widening, for example) and its carrying capacity decreases, sediment begins to be deposited, producing features such as point bars (ridges of sand or gravel on the inside bends of meandering rivers—see above

## Erosion and deposition in drainage basin

From the glaciers that are the source of many rivers, along the course of the rivers themselves, and around the coast into which they discharge, a variety of erosional and depositional processes have their effects on the landscape, ranging from U-shaped glacial valleys and canyons, to floodplains and deltas.

right), deltas (triangular regions of deposited sediment), and sand bars (low, partially submerged ridges), where the river meets the sea.

A different pattern of erosion occurs when rainwater eats away at regions dominated by layers of limestone rock, producing karst landscapes (see main picture, pp.192–93). Rainwater trickles down along the vertical joints that divide limestone blocks. Since rainwater is usually slightly acidic, it dissolves the limestone, widening the joints, forming chimneys, and creating underground chambers linked by passages.

## Coastal erosion and deposition

Coastal cliffs undergo continual weathering and erosion through the action of waves (see above right).

Weathering occurs through abrasion, as waves hurl beach material against the cliff. As waves impinge on a cliff they also compress air in cracks in rocks. As the air then decompresses it can shatter the rock. Depending on the cliff's rock composition, wave action may produce a cave, or the cliff above the eroded part may collapse.

Like coastal erosion, deposition is primarily caused by wave action though tidal currents also play a part. Sand and silt in the water may come from a river, eroded cliffs, or offshore. Waves that hit the coast at an angle can shift this material along shorelines, sometimes for long distances, by a mechanism called longshore drift. Where the sediment settles out and

**U-shaped valley**
Typical shape of valley with rock erosion from both sides and base by moving ice of glacier.

**End moraine**
A broad, rounded ridge of rock debris marks the farthest point of glacier's advance.

**River deposition**
A river meandering through a floodplain such as the Bynoe River in Queensland, Australia, shifts sedimentary material downstream by eroding the outer parts of bends and depositing the material a little farther along on the inner banks of bends.

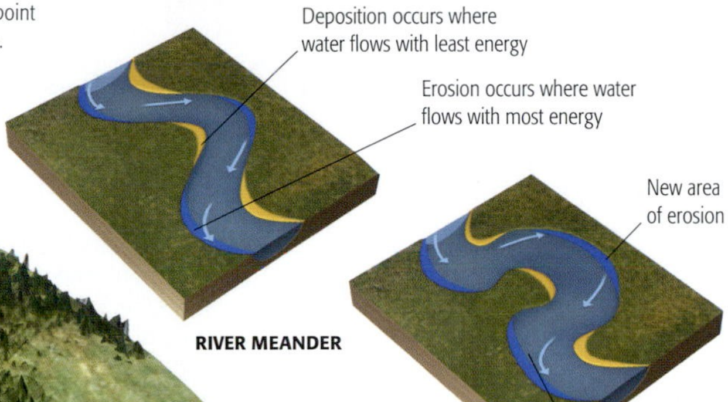

Deposition occurs where water flows with least energy

Erosion occurs where water flows with most energy

**RIVER MEANDER**

New area of erosion

**RIVER DEPOSITION**

Course of river changes, reworking sediments across whole floodplain

GEOLOGIST AND GLACIOLOGIST (1807–73)

## LOUIS AGASSIZ

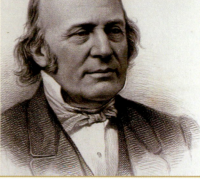

Swiss scientist Louis Agassiz was one of the first to describe the erosive effects of glaciers and the marks they leave on landscapes. He originally qualified as a medical doctor, but soon launched himself into the study of geology and biology. He was the first person to propose, in 1837, that Europe had once been through an ice age. In 1848 he moved to America, where he achieved great fame, but also expressed controversial racist views about human origins that have tarnished his scientific reputation.

**Coastal marine erosion**
Where waves encounter an indented cliffed coast, their energy is concentrated on to the sides of headlands and carves out caves, rock arches, and sea stacks, as at the Twelve Apostles off southeastern Australia.

Arch forms as waves erode both sides of headland, finally penetrating through it

**ARCH FORMATION**

Collapse of arch leaves sea stack, which will eventually be worn down to a stump

**STACK FORMATION**

**AFTER** »

Sea stack

Arch

**Headland**
An area of land formed of erosion-resistant rock protruding out to sea.

**Beach**
Material eroded from headlands is moved along shore by currents and waves.

**Sand spit**
Projecting deposit of sand is formed from settling of sediment.

**Delta**
A region where a river has deposited its load of sand and silt on meeting the sea.

builds up, it can form coastal features such as spits (sandy projections) and sand bars.

## Mass wasting
Mass wasting is an erosional process in which soil or rock moves downhill under the influence of gravity. It includes events such as rock falls, landslides, and mudflows. These events are often triggered by heavy rainfall (which makes the moving material more fluid) and an earthquake or volcanic eruption. Frequently, the sliding or falling material has previously been loosened by some form of weathering process. Satellite observations have shown a high frequency of such events in mountainous regions with moderate to high rainfall, and some earthquake or volcanic activity. These include parts of Southeast Asia and California.

**Knowledge of erosional processes is continually updated by new discoveries.**

**ROCKS ON THE MOVE**
In 1991 **unique granite blocks** from the island of Ailsa Craig, in Scotland, were found 150 miles (240 km) farther south in England, indicating they had been **moved** there **by glaciers.**

**EROSION OF MARS**
Studies of the **surface of Mars** in 2001 concluded that the planet's surface has been heavily **eroded** in the past **by water flows** similar to those on Earth.

**SALTATION AND ELECTRICITY**
In 2008 researchers found that during **saltation** (a process in which wind moves sand and dust) **static electricity is generated** as particles rub against each other. As this electricity may play a part in the maintenance of atmospheric aerosols (dust in air), which affect climate, these findings may help to **improve climate modeling.**

**SEE ALSO** »
pp.356–57 PLATE TECTONICS
pp.358–61 ACTIVE EARTH

« BEFORE

**Information gathering has taken place for thousands of years but data analysis is a relatively recent development.**

### EARLY CENSUSES

The **earliest known census** dates from **ancient Babylonian** times, about 3800 BCE, and recorded the human population and agricultural data. Many of the other ancient civilizations also regularly **recorded population numbers**, often for the purpose of taxation. In the Middle Ages, probably the best-known census is the *Domesday Book*, which was instigated by William I of England in 1086 to tax the recently conquered population. These early censuses were simply records of numbers because **mathematical techniques for analyzing data** had not yet been developed.

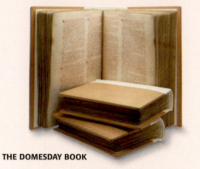

**THE DOMESDAY BOOK**

LAWYER (1601–65)

### PIERRE DE FERMAT

Born near Toulouse, France, into a wealthy family, Fermat graduated in law from Orléans University in 1631. After graduation he moved to Toulouse where he spent the rest of his life, although he died at Castres where he had a second home. Fermat married, had five children, and a busy professional life as a lawyer and government official. He did mathematics only in his spare time but still managed to make fundamental contributions to probability, number theory (including his famous "last theorem," see p.49), analytical geometry, calculus, and optics.

# Probability and Statistics

**We live in an information age in which modern technology makes it easy to gather large amounts of information on almost every aspect of our lives. However, on its own this information is of only limited value—it needs to be organized and analyzed to be of practical use.**

Even though information about population numbers has been collected since ancient times, the science of analyzing and making sense of data—statistics—is relatively recent. Although now not usually considered to be a branch of mathematics, statistics relies on mathematical analysis to interpret information and is closely linked to the area of mathematics known as probability theory.

## Chance and probability

The beginnings of probability theory came from the fascination that two 17th-century French mathematicians had with games of chance. Blaise Pascal and Pierre de Fermat discussed, in a series of letters, a method of calculating the chances of success in gambling games, and they were the first to give the subject of probability a scientific treatment.

What they discussed was a mathematical way of determining the probability of a particular outcome occurring in a random event, such as tossing a coin or throwing a die. When a coin is tossed there are two possibilities: heads or tails. Each is equally likely: there is one chance in two that the coin will come up heads (or tails), or in other words the probability is $1/2$. The six faces of a die give one chance in six of throwing any particular number, a probability of $1/6$. In games using more than one die, or a deck of cards, or a roulette wheel, the calculation becomes more complex but is essentially the same. From this discussion of gambling games, a theory of probability evolved.

The idea was further developed by the next generation of mathematicians. French mathematician Abraham de Moivre discovered a pattern to the probability of outcomes, now known

### Roulette

First played in its modern form in the late 18th century, roulette is a simple game of chance in which players bet on what numbered slot the ball will land in. After bets have been placed, the wheel is spun and the ball is spun in the opposite direction around the outside of the wheel. Eventually the ball comes to rest in one of the numbered slots—the winning one.

### CHANCE OF WINNING AT ROULETTE

The American roulette wheel has 38 slots: numbers 1–36 (colored red or black), 0 (green), and 00 (green). It is possible to bet on one or more specific numbers, on various combinations of numbers, such as any odd or even number, or on any red or black number. The table below gives the probability of winning for various bets.

| Bet placed | Chance of winning |
| --- | --- |
| 00 | 1/38 |
| 0 | 1/38 |
| Any single number | 1/38 |
| Any even number (2, 4, 6…36) | 18/38 |
| Any odd number (1, 3, 5…35) | 18/38 |
| Any red number | 18/38 |
| Any black number | 18/38 |
| Any number from 1 to 18 | 18/38 |
| Any number from 19 to 36 | 18/38 |

**Florence Nightingale**
Working as a nurse for British troops during the Crimean War, Florence Nightingale kept records of troop deaths and later used the information to create what are now called "coxcomb" graphs, which highlighted the number of deaths that were not directly caused by combat but by factors such as wound infection and disease.

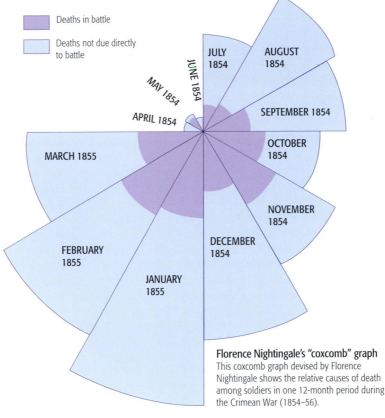

Deaths in battle

Deaths not due directly to battle

JULY 1854

AUGUST 1854

MAY 1854

JUNE 1854

SEPTEMBER 1854

APRIL 1854

OCTOBER 1854

MARCH 1855

NOVEMBER 1854

FEBRUARY 1855

DECEMBER 1854

JANUARY 1855

**Florence Nightingale's "coxcomb" graph**
This coxcomb graph devised by Florence Nightingale shows the relative causes of death among soldiers in one 12-month period during the Crimean War (1854–56).

The development of probability and statistics gave scientists new ways to analyze and conceptualize the physical world.

**COMPUTERIZED MODELING**
Many natural systems are influenced by numerous factors and **exhibit chaotic behavior 350–51 »**. For example, influences on the **weather** include air, land, and sea **temperatures, winds, sea currents, humidity**, and **amount of sunlight**, and minute changes in these factors can have a profound effect on the weather. Because of this, **weather forecasting** relies on numerical models in which **statistical methods** are used to arrive at forecasts that have various **degrees of probability** of being correct.

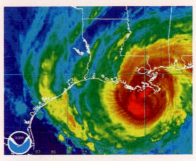

**COMPUTER MODEL OF A HURRICANE**

**QUANTUM THEORY**
The currently accepted theory of the nature and behavior of matter at the subatomic level, **quantum theory 314–17 »** uses **probability** as one of its fundamental concepts. For example, according to quantum theory it is impossible to know precisely the location and momentum of subatomic particles such as electrons "orbiting" the nucleus of an atom; it is only possible to specify **regions**—known as clouds—where particles may be located with the **highest probability**.

**SEE ALSO »**
pp.400–401 GRAND UNIFIED THEORY
pp.402–403 STRING THEORY

as normal distribution and represented graphically as the bell curve (see below). British mathematician and clergyman Thomas Bayes took de Moivre's ideas further with his theorem of conditional probabilities, which makes it possible to calculate the probability of a particular event occurring when that event is conditional on other factors and the probabilities of those factors are known. Bayes' work was further developed by Pierre-Simon Laplace, a French mathematician and astronomer whose application of Bayes' theorem to real cases led to a new field of study: statistics.

### Detecting patterns
The pioneering work in statistics was done by de Moivre, who used data about death rates and rates of interest to devise a theory of annuities, which enabled insurance companies to compile tables of risk for life insurance based on scientific principles. This application of mathematics to data in records was at first known as "political arithmetic," and as patterns emerged in collections of data, research began into their underlying statistical laws. To begin with, statistics was concerned with social issues, and advances in sociology and criminology were

made by the Belgian mathematician Adolphe Quetelet, who introduced the concept of the "average man." He also believed that mathematics lay at the heart of every science, and statistical

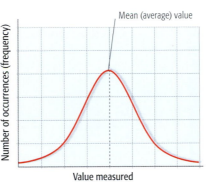

Mean (average) value

Number of occurrences (frequency)

Value measured

**Normal distribution**
When certain values (such as height) are plotted against the number of occurrences of that value (how many people are of a specific height), the result is often a bell-shaped curve—the normal distribution. The most common value, at the peak, is the mean (average).

analysis could be applied to data of all kinds. Perhaps the area where this had greatest effect was medicine, where an important new study, epidemiology (occurrence of disease in populations, see pp.242–43), developed from medical statistics. As more practical use was made of probability theory and statistics, the mathematics behind them was developed by various mathematicians, including the Frenchman Adrien-Marie Legendre, the German Carl Friedrich Gauss, and the Russian Andrey

Nikolaevich Kolmogorov, whose systematic approach to the subject forms the basis for much of modern probability theory.

### Modern statistics
Statistics plays a key role in much of modern life. Governments collect and analyze a wide range of personal data to detect patterns that can help shape policies. Businesses use market research to gather information about potential customers, and statistical methods to analyze the data. In science, statistics and probability are central to subjects such as quantum theory and are also essential to many other subjects, from psychology and economics to information science.

**IN PRACTICE**

## DATA SAMPLES

The data used for statistical analysis must be sound for analysis to produce useful results. The data must be collected using a valid method that measures what is intended, and the data must be accurate. It is also essential that the set of data is large enough and constitutes a representative sample. For example, in general public opinion polls the right questions must be asked in an unambiguous, neutral way; sufficient numbers of people must be polled; and, as a whole, the respondents must be representative of the population (for instance, in age and gender).

# Darwin's Theory of Evolution

**In the 19th century many scientists suspected that species were not fixed at creation but could change over time. A theory emerged from a quietly respectable man, which made such a change—evolution—an irresistible likelihood and the most powerful idea in biology.**

In September 1838 acclaimed English naturalist Charles Darwin read *An Essay on the Principle of Population* by British economist Thomas Malthus, which had a pivotal influence on his developing—but still very secret—ideas about the evolution of life. Malthus expressed his concerns about the burgeoning human population: he predicted starvation and destitution as a direct result of finite resources being spread ever more thinly among a growing number of people.

Darwin had recently returned from a voyage around the world and now started to amass evidence for a theory of evolution. He was convinced that organisms arose by descent from a common ancestor and Malthus's essay proposed a possible mechanism. Darwin realized that overpopulation of plant and animal species was prevented because many individuals did not survive. He had discovered a disturbing truth: there is a high mortality rate in nature and the individuals with less favorable characteristics are the ones that die. He thought that in this struggle for existence the favorable characteristics of the surviving

individuals would be inherited, so gradually altering the features of succeeding generations, and accounting for evolutionary change. Darwin called his proposed mechanism natural selection.

On June 18, 1858 Darwin received a letter from British naturalist Alfred Russel Wallace who had devised the same theory. They agreed to a joint paper on natural selection, which was read to the Linnean Society (British natural history society) on July 1. Remarkably, it passed with little notice. Darwin's *On the Origin of Species* which expanded on his theory, was published on November 22, 1859 and sold out immediately.

As expected, its general release caused a furor. At first even some members of the scientific establishment objected to the way that Darwin's theory conflicted with the biblical story of creation. Nevertheless, natural selection became accepted as a driving force of evolution, largely through the efforts of British biologist Thomas Henry Huxley and botanist Joseph Hooker. More than 150 years later it is still recognized as the principle that underpins all life on Earth.

### Darwin's finches
On his voyage to the Galapagos Islands, Darwin collected many specimens of finches (see right) and mockingbirds from the different islands. In finches he noticed variations that he thought may have descended from a common ancestor. He kept detailed notes on the diversity he saw in mockingbirds and concluded that the differences had arisen by natural selection.

> **"Nothing in biology makes sense except in the light of evolution."**
> THEODOSIUS DOBZHANSKY, UKRANIAN GENETICIST AND
> EVOLUTIONARY BIOLOGIST, 1973

<< BEFORE

**Fossil evidence before the 1800s was occasionally used to challenge traditional views that species were fixed.**

### THE EARLIEST EVOLUTIONARY IDEAS

The idea that organisms had natural origins began with ancient Greek philosophers such as **Empedocles**, but **Plato's** belief that organisms had a fixed "**essence**" hampered acceptance of evolution. Nevertheless, **fossils** were often taken as evidence of past life: Roman author **Pliny** described insects preserved in **amber** and medieval Chinese scholar **Shen Kuo** noted fossils of marine life inland.

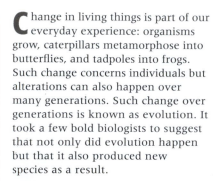

SPIDER IN AMBER

### HISTORY IN THE ROCKS

In 1669 Danish scientist **Nicolas Steno** proposed that **fossils** preserved in rocks in different layers were **remains of extinct animals**, which occurred in different ages of Earth's history, laying the foundations for **stratigraphy**, the study of rock layers.

<< SEE ALSO
pp.122–23 CLASSIFICATION OF SPECIES
pp.186–87 THE FOSSIL RECORD
pp.198–99 DARWIN'S THEORY OF EVOLUTION

Change in living things is part of our everyday experience: organisms grow, caterpillars metamorphose into butterflies, and tadpoles into frogs. Such change concerns individuals but alterations can also happen over many generations. Such change over generations is known as evolution. It took a few bold biologists to suggest that not only did evolution happen but that it also produced new species as a result.

### The first ideas of evolution

Following on from the work of French philosophers in the 18th century, Erasmus Darwin (Charles's grandfather) published *Zoönomia* in 1794, proposing that higher forms of life had emerged from primitive ancestors. Geologists were soon documenting a great deal of fossil evidence to support the view that life on Earth had changed (see p.187). The first comprehensive theory of evolutionary change came from the French scientist Jean Baptiste Lamarck. He proposed that living things acquired characteristics in their desire to gain food, shelter, and mates, by changing their body parts through use or neglect. Although he was correct that structures, such as muscle, could be modified, Lamarck also proposed that such acquired traits are passed on. Modern genetics tells us that this is impossible because characteristics are determined by DNA (see pp.308–309).

# How Evolution Works

**The idea that all creatures on Earth evolved from the same primitive ancestors was rejected in the past for religious reasons. From the 19th century scientists proposed various theories to explain the processes behind evolution; currently, natural selection, genetic drift, and mutation are the main protagonists.**

**Yellow-and-green lorikeet**
*Trichoglossus flavoviridis* has a distinctive green body and yellow-brown head.

**Rainbow lorikeet**
*Trichoglossus haematodus haematodus* has a yellow-green nape with a red border and black-blue head.

**Ornate lorikeet**
*Trichoglossus ornatus* has a distinctive red chin and black barring.

Banda Sea

GREATER SUNDA ISLANDS

NEW GUINEA

Arafura Sea

LESSER SUNDA ISLANDS

Timor Sea

Coral Sea Basin

**Flores lorikeet**
*Trichoglossus weberi* is almost entirely green with a distinctive lighter green breast and neck.

### Evolution of new bird species

Geographical separation of lorikeets led to the evolution of different species, illustrated by their diversely colored plumage. As islands separated from the mainland, the interbreeding bird populations were fragmented into smaller populations.

**Marigold lorikeet**
*Trichoglossus capistratus* has a yellow nape and breast.

**Red-collared lorikeet**
*Trichoglossus rubritorquis* is the largest of the lorikeets and has a distinctive red nape and varying shapes of blue over its head.

AUSTRALIA

»

## NATURALIST (1823–1913)

### ALFRED RUSSEL WALLACE

Wallace was an acclaimed British naturalist and explorer. He formulated explanations for geographical distribution of organisms, and became known as the "father" of biogeography. He communicated his ideas on natural selection to Charles Darwin, resulting in a joint publication of the theory in 1858 (see p.198). Later in life Wallace became an advocate of various causes, including socioeconomic reform and—controversially—spiritualism. Unlike Darwin, he rejected the idea that humans were also products of evolution.

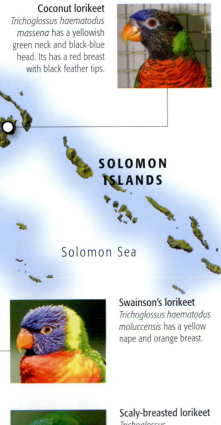

**Coconut lorikeet**
*Trichoglossus haematodus massena* has a yellowish green neck and black-blue head. Its has a red breast with black feather tips.

**SOLOMON ISLANDS**

Solomon Sea

**Swainson's lorikeet**
*Trichoglossus haematodus. moluccensis* has a yellow nape and orange breast.

**Scaly-breasted lorikeet**
*Trichoglossus chlorolepidotus* has a distinctive green body and yellow breast "scales."

### Modes of natural selection
Natural selection happens when individuals with certain inherited characteristics are more likely to survive and breed. These characteristics then predominate in the next generation.

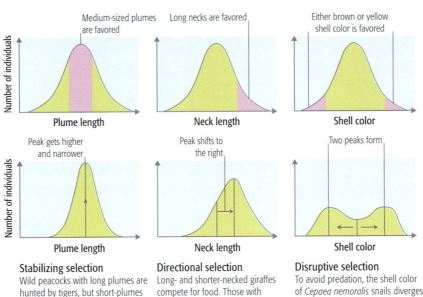

*Medium-sized plumes are favored* — Plume length (Number of individuals)

*Long necks are favored* — Neck length

*Either brown or yellow shell color is favored* — Shell color

*Peak gets higher and narrower* — Plume length (Number of individuals)

*Peak shifts to the right* — Neck length

*Two peaks form* — Shell color

**Stabilizing selection**
Wild peacocks with long plumes are hunted by tigers, but short-plumes don't attract a mate, so the range of length is reduced.

**Directional selection**
Long- and shorter-necked giraffes compete for food. Those with long necks are more successful so will predominate.

**Disruptive selection**
To avoid predation, the shell color of *Cepaea nemoralis* snails diverges to the extremes, leading to two or more variants over time.

### Darwinism
In 1858 British naturalist Charles Darwin (see pp.202–203) published his theory of evolution (see pp.198–99) with fellow naturalist Alfred Russel Wallace, who had come up with the same idea. They proposed that organisms usually produce more offspring than can be supported by resources in the environment. The offspring vary by virtue of inherited characteristics, and those individuals with characteristics favorable for their environment will reproduce and prevail. As Charles Darwin put it "It is not the strongest species that survive, nor the

*Forest fire wipes out most purple butterflies in population*

*Next generation will contain only genes of the survivors*

*A series of chance events can lead to the total loss of the purple population*

### Environmental influences on genetics
Random death—such as from a natural disaster—can drastically alter the genetic makeup of a population. When combined with the randomness of genetic drift (see right), an entire species can be wiped out.

> ## "Endless **forms** most **beautiful** and most **wonderful** have been, and are being, **evolved**."
>
> CHARLES DARWIN, "ON THE ORIGIN OF SPECIES," 1859

most intelligent, but the ones most responsive to change."

### The importance of genetics
A lack of understanding of inheritance meant that the Darwin–Wallace theory was deficient in certain areas. Then in 1900 the discovery of work carried out 50 years earlier by Gregor Mendel showed that the inheritance of genetic traits follows predictable patterns (see pp.204–205) and there was no blending. However, it was not until the 1930s that the mechanism behind natural selection became fully understood; the fusion of evolutionary and genetic ideas became known as Neo-Darwinism. Evolution then began to be understood in terms of changes in the genetic

makeup of groups. Today other mechanisms are also believed to influence evolution. These include genetic drift, which involves random changes in inheritance of certain characteristics; and mutation, which creates new genes by errors of DNA replication. Both exert influences independently of the environment.

### Production of new species
By the 1940s German biologist Ernst Mayr defined a species as a group that could interbreed among itself but not with others. It seemed that new species could evolve as a result of geographical splits or spontaneous change.

**Colorful camouflage**
The shell color of the *Cepaea nemoralis* snail (left) reflects its environment. Brown snails escape predation by thriving in woodlands while green snails survive by living in grassland.

---

Advances in understanding evolution in the 20th century have relied on discoveries in geology and chemistry.

#### CONTINENTAL DRIFT
In 1912 **Alfred Wegener's** theory of **continental drift 356–57 »** partly explained the distribution patterns of organisms that had been documented by biologists such as **Alfred Russel Wallace**. The **changing geography** of the planet is now known to have had a profound influence on the history of **life on Earth**.

**CONTINENTAL DRIFT**

#### MOLECULAR CLOCKS
The rise of **molecular biology** in the 1950s and 1960s enabled the **evolutionary relationships** of organisms to be established by comparing differences in their **DNA and proteins**. Scientists think that these differences have built up over time and that the amount of change reflects the **evolutionary time** elapsed.

#### SEE ALSO »
pp.202–03 CHARLES DARWIN
pp.204–05 LAWS OF INHERITANCE
pp.308–09 CHROMOSOMES AND INHERITANCE
pp.356–57 PLATE TECTONICS

NATURALIST **Born 1809** Died 1882

# Charles Darwin

## "The **mystery** of the **beginning** of all things is **insoluble** by us."

CHARLES DARWIN, "THE AUTOBIOGRAPHY OF CHARLES DARWIN," 1887

**Young Darwin**
Darwin's prolific correspondence and experimentation ensured that he was able to accumulate a great deal of evidence for his theory of evolution by natural selection, which he formulated in his late 20s.

There can be few biologists who have left as lasting an influence as Charles Darwin. His ideas not only explained the way by which all life has come to be, but he also showed that our own species sits side by side with others in the tree of life. He was such an astute observer of the natural world and amassed such a wealth of facts that his arguments became irresistibly persuasive. But he was also a cautious man and fully anticipated the controversy that his most famous book—*On the Origin of Species*, 1859—would cause.

**Collecting natural specimens**
Darwin collected many specimens, which laid the foundations for his ideas that species were not fixed, but constantly evolving. Above, a specimen of a brown butterfly with a blue spot was stored in an old pill box.

### Early life and voyage on the *Beagle*
Darwin's early life was marked by change. He aborted a medical career at Edinburgh University after being appalled by the trauma of 19th-century surgery. Then he went to study for the church in Cambridge. He passed his exams, but was more interested in beetles than theology, and his professor John

Henslow recommended him for a position as "gentleman naturalist" on HMS *Beagle*; in December 1831, with the reluctant consent of his father, he set sail with Captain Robert Fitzroy.

The *Beagle* was at sea for five years to chart coastlines of the southern hemisphere. Working as a naturalist, Darwin wrote assiduously about what he saw, collecting specimens and making discoveries that steadily

---

**BREAKTHROUGH**

## CORAL ATOLL FORMATION

Darwin made contributions that were important in geology as well as biology. In 1842 he wrote *The Structure and Distribution of Coral Reefs*, which included an accurate explanation of the formation of coral atolls (pictured below), based on his observations on board the HMS *Beagle* of islands in the Pacific. He suggested that a fringe of coral forms around the base of a sinking volcanic island, eventually persisting as a ring enclosing a lagoon: the atoll.

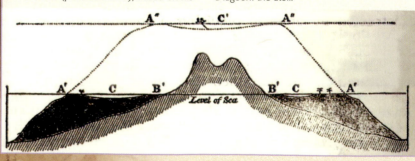

established his reputation back in England. During the voyage he read Charles Lyell's *Principles of Geology* 1830, which suggested that Earth had changed gradually over very long periods of time (see pp.190–91) and Darwin saw evidence for this all around him. But Lyell also suggested that species had originated in "centres of creation" and, based on observations on the distribution of species in places such as the Galapagos Islands, Darwin was harboring doubts about this view.

## Developing his ideas

Returning to England in 1836 Darwin began to organize his specimens with the help of experts. British ornithologist John Gould pointed out that the slightly varying mockingbirds he had collected in the Galapagos Archipelago were actually three unique species individual to different islands. This, and many examples like it, served to foster Darwin's growing conviction that species were not fixed. After writing the journal of his voyage he began a secret notebook on the subject, known then as "transmutation."

In 1838 Darwin read an essay by British economist Thomas Malthus, that had a profound influence on his

**Voyage of the *Beagle***
Charles Darwin spent five years aboard HMS *Beagle*. The *Beagle*'s task was to map southern coastlines, as seen here, but with Darwin on board as naturalist, it also inspired a huge breakthrough in biology.

**Rural retreat: Down House**
Darwin's seclusion in rural Kent, England meant that he was isolated from much of the furor caused by the publication of his evolutionary theory.

ideas. It made stark predictions about human overpopulation. Darwin began to consider whether the same theory would apply to plants and animals— that population growth would lead to competition over resources.

## Substantiating his theory

In 1839 Darwin was elected Fellow of the Royal Society. He also married his cousin, Emma Wedgwood, and published his journal of the *Beagle's* voyage, which was popular and made him well known.

In the years that followed he settled with his growing family in Down House in Kent. He established a routine of prolific correspondence,

experimentation, and research on topics ranging from barnacles to the behavior of plants. He gathered facts about the variation that could be produced by animal breeders, such as pigeon fanciers, and explored human behavior too. During this time Darwin's Christian faith was dwindling— much to the consternation of his wife, Emma—and there can be little doubt that the death of his beloved

**Darwin's tree of life**
Darwin penned a sketch that illustrated his idea that life evolved and diversified like the branches of a tree.

"No work on **Natural History** Science has made so **great** an **impression** on me."

BRITISH BIOLOGIST THOMAS HENRY HUXLEY, DISCUSSING *ON THE ORIGIN OF SPECIES*, 1859

daughter Annie, from scarlet fever in 1851, was the final blow.

In July 1858 he produced a joint paper with British naturalist Alfred Russel Wallace (who had formed the same idea independently) describing the theory of evolution by natural selection (see pp.198–99). A year later the first edition of Darwin's "summary" of his evidence and ideas appeared in the famous *On the Origin of Species*.

Amid the furor that followed publication— condemned by many for contradicting biblical creation—Darwin continued with his work at Down House. He subsequently produced work that made it clear he considered humans to be part of this natural world too. When he died in 1882 he was honored with a state funeral and buried in Westminster Abbey in London.

## TIMELINE

- **February 12, 1809** Charles Robert Darwin is born in Shrewsbury, England, son of wealthy physician, Robert Darwin, and grandson of Erasmus Darwin and Josiah Wedgwood.
- **September 1818** Attends Shrewsbury School.
- **1825** Starts to study medicine at University of Edinburgh.
- **1827** Abandons medicine to train for the clergy at Christ's College, Cambridge University.
- **1831** Performs well in final examinations.
- **1832** Studies geology and natural history of South America and Galapagos Islands on board HMS *Beagle*.
- **October 2, 1836** Returns to England.
- **January 4, 1837** Publishes his first scientific paper on the geology of South America.
- **February 17, 1837** Elected to council of Geographical Society.
- **June 20, 1837** Finishes writing *Beagle* journal.
- **March 1838** Becomes Secretary of the Geological Society.
- **January 24, 1839** Elected Fellow of the Royal Society.
- **January 29, 1839** Marries his cousin, Emma Wedgwood.
- **May 1939** Publishes *Journal and Remarks* about *Beagle* voyage.
- **May 1842** Publishes *The Structure and Distribution of Coral Reefs*.
- **November 1842** Family moves to Down House in Kent.
- **1846** Starts a six-year study of barnacles, using the idea that all organisms are descended from a common ancestor.
- **1853** Awarded the Royal Society's Royal Medal for work on barnacles.
- **1856** Investigates whether eggs and seeds could survive sea water and spread over oceans.
- **July 1, 1858** Darwin-Wallace paper presented to the Linnean Society (British natural history).
- **Nov 22, 1859** First edition of *Origin of Species* released and sells out on the first day.
- **1868** Publishes *Variation of Plants and Animals under Domestication*, first part of his "big book."
- **1871** Publishes second volume, *The Descent of Man*, dealing with sexual selection.
- **1872** Publishes *The Expression of the Emotions in Man and Animals*, concerning human psychology; smaller books follow on insectivorous plants, climbing plants, and earthworms.
- **April 19, 1882** Darwin dies at Down House, at 73. He is given a state funeral.

**BEETLE SPECIMEN**

**DARWIN'S COMPASS**

# Laws of Inheritance

**Family resemblances have long been recognized, but a scientific explanation for them was slow in coming. Then in the middle of the 19th century an Augustinian monk began to breed peas and so discovered the answer to this mysterious question.**

Most biological characteristics are inherited, and early efforts to explain how came from French philosopher Pierre Maupertius who concluded in 1745—as English naturalist Charles Darwin (see pp.201–202) did 100 years later—that offspring are made from components derived from every part of their parents' bodies. In a way they were correct: DNA in every cell of each parent determines the outcome in the progeny. But that alone could not explain why some traits skip generations.

### Mendel's experiments
In 1866 an Austrian monk published the results of his experiments in breeding peas. Gregor Mendel (see below) had found pea plant characteristics, such as flower color, were passed down in fixed proportions. He called this the Law of Segregation (see opposite). He deduced that each characteristic was determined by a particular particle (today called a gene) and that they exist in different variables, such as color. His theory suggested that characteristics could be inherited intact. This differed from popular notions of the time, which suggested that a purple-flowered plant crossed with a white-flowered plant, for example, would produce a blend of the color. Importantly, Mendel's work explained generation-skipping traits; these recessive traits can remain hidden in generations when "masked" by dominant traits.

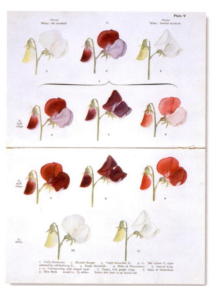

**Rediscovering Mendel's laws**
Geneticist William Bateson publicized Mendel's laws in his 1909 book *Mendel's Principles of Heredity*. Here, however, he shows how sweet peas do not follow the laws.

Mendel also looked at what happened when two characteristics, such as seed color and shape, were inherited together. This resulted in his Law of Independent Assortment (see opposite), which stated that different characteristics are inherited independently of one another. These findings established the distinction between phenotypes (the observable characteristics of organisms) and genotypes (the genetic characteristics of organisms that cannot be seen).

### Human heredity
Pea plants clearly showed Mendel's patterns because any two parents can

« BEFORE

**Early attempts at figuring out meaningful explanations for inheritance were flawed by basic misconceptions.**

**EARLY THEORIES OF HEREDITY**
Early understanding of heredity was hindered by a belief in preformation: **individuals were preformed in sex cells**. This was championed by scholars such as French philosopher René Descartes **« 102–03** in the 17th century.

**THE SCIENCE OF BREEDING PLANTS**
Plant breeders had long since experimented with **selective breeding** of crops but it was 18th century German botanists Christian Sprengel and Joseph Kölreuter who established the role of **pollination** in plant breeding **« 152–53**.

**« SEE ALSO**
*pp.152–53* PLANT LIFE CYCLES

**Pea plant: understanding inherited traits**
This snow pea (*Pisum bijou*) has a number of variable features such as flower color, plant height, and seed shape. The fact that these traits are inherited in simple ways, and that it is easy to cultivate, made this plant the ideal experimental organism for Gregor Mendel.

**PRIEST AND SCIENTIST (1822–84)**
## GREGOR MENDEL

As a young Augustinian monk, Gregor Mendel studied at the University of Vienna, Austria, and then returned to his abbey in Brno (in what is now the Czech Republic) as a science teacher. His mathematical skills proved useful in establishing his laws of inheritance (see opposite) in his breeding experiments—between 1856 and 1868 he cultivated and tested more than 29,000 pea plants. His work was not recognized until 1900. His scientific work came to an end in 1868 when he became an abbot, and his papers were burned by his successor after his death.

## Mendel's Law of Segregation

The inheritance pattern of a single characteristic, such as color, suggests that it is determined by a pair of alleles (gene varieties), some of which are dominant and some recessive (always "masked" by a dominant gene).

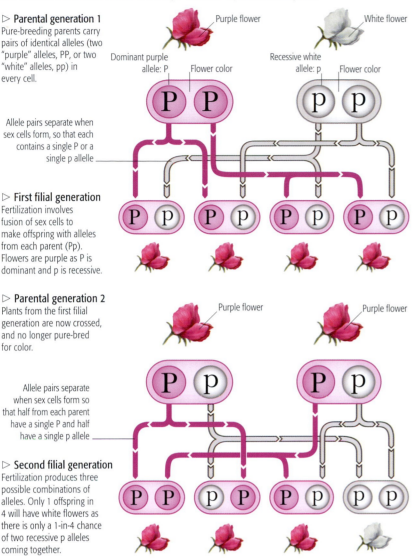

▷ **Parental generation 1**
Pure-breeding parents carry pairs of identical alleles (two "purple" alleles, PP, or two "white" alleles, pp) in every cell.

Purple flower
White flower

Dominant purple allele: P
Flower color
Recessive white allele: p
Flower color

Allele pairs separate when sex cells form, so that each contains a single P or a single p allele

▷ **First filial generation**
Fertilization involves fusion of sex cells to make offspring with alleles from each parent (Pp). Flowers are purple as P is dominant and p is recessive.

▷ **Parental generation 2**
Plants from the first filial generation are now crossed, and no longer pure-bred for color.

Purple flower
Purple flower

Allele pairs separate when sex cells form so that half from each parent have a single P and half have a single p allele

▷ **Second filial generation**
Fertilization produces three possible combinations of alleles. Only 1 offspring in 4 will have white flowers as there is only a 1-in-4 chance of two recessive p alleles coming together.

produce a very large number of offspring but the principles apply to other organisms too. Mendel's laws do not dictate the actual proportions of offspring born with characteristics, but rather the chance of each one carrying a particular trait.

## Mendel's Law of Independent Assortment

The inheritance pattern of two characteristics together, such as seed color and seed shape, suggests that their allele pairs are inherited independently of one another.

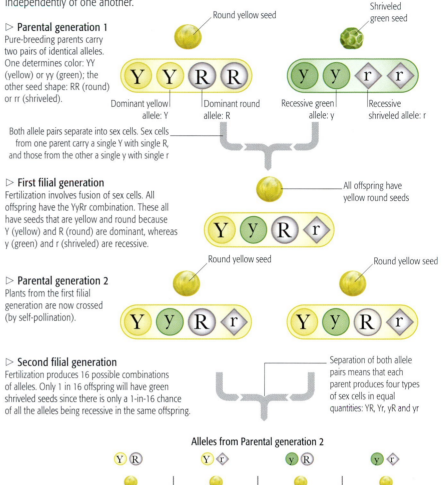

▷ **Parental generation 1**
Pure-breeding parents carry two pairs of identical alleles. One determines color: YY (yellow) or yy (green); the other seed shape: RR (round) or rr (shriveled).

Round yellow seed
Shriveled green seed

Dominant yellow allele: Y
Dominant round allele: R
Recessive green allele: y
Recessive shriveled allele: r

Both allele pairs separate into sex cells. Sex cells from one parent carry a single Y with single R, and those from the other a single y with single r

▷ **First filial generation**
Fertilization involves fusion of sex cells. All offspring have the YyRr combination. These all have seeds that are yellow and round because Y (yellow) and R (round) are dominant, whereas y (green) and r (shriveled) are recessive.

All offspring have yellow round seeds

▷ **Parental generation 2**
Plants from the first filial generation are now crossed (by self-pollination).

Round yellow seed
Round yellow seed

▷ **Second filial generation**
Fertilization produces 16 possible combinations of alleles. Only 1 in 16 offspring will have green shriveled seeds since there is only a 1-in-16 chance of all the alleles being recessive in the same offspring.

Separation of both allele pairs means that each parent produces four types of sex cells in equal quantities: YR, Yr, yR and yr

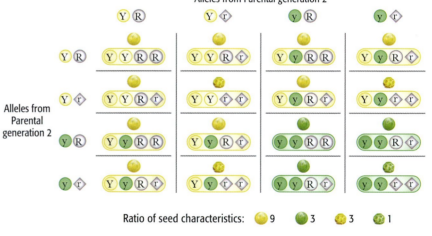

Alleles from Parental generation 2

Alleles from Parental generation 2

Ratio of seed characteristics: 9  3  3  1

## Exceptions to Mendel's laws

Mendel's work was only recognized posthumously in 1900, when his laws were rediscovered by botanists Hugo de Vries and German Carl Correns. Mendel's conclusions were initially doubted by some scientists—especially once it became known that not all patterns of inheritance followed his laws. But subsequent studies showed the reasons why. In 1902 American geneticist Walter Sutton and German biologist Theodor Boveri independently established that genes occur on chromosomes. In 1911 the work of American geneticist Thomas Hunt Morgan showed how some genes are linked on the same chromosomes and so are inherited together, contravening Mendel's Law of Independent Assortment. The discovery of genes on sex chromosomes explained why some characteristics—known as sex-linked ones—were more common in males than females (see pp.308–309).

AFTER ≫

**Twentieth-century genetics flourished after the rediscovery of Mendel's work and new discoveries about chromosomes and DNA.**

### GENETIC COUNSELING
In humans, Mendelian genetics is a powerful tool in **genetic counseling**. By careful study of their family history, a professional counselor can allow a couple to make an **informed preconceptual decision** by calculating the risk of an **inherited disease** being passed on to their children. It predicts, for example, that if both parents are normal but carry the **allele**

(gene variety) for cystic fibrosis, then there is a 1 in 4 chance of the child being born with this disease.

### THE DOUBLE HELIX
In 1953 geneticists James Watson and Francis Crick showed that **DNA** was a **double helix molecule**. Later studies showed that the structure of DNA was the key to carrying genetic information. **346–47 ≫**

**SEE ALSO ≫**
pp.306–07 HOW CELLS DIVIDE
pp.308–09 CHROMOSOMES AND INHERITANCE
pp.346–47 THE STRUCTURE OF DNA
pp.348–49 THE GENETIC CODE

### HUMAN INHERITED CHARACTERISTICS

Some characteristics are inherited simple Mendelian ways, but others are determined by more than one gene (see below).

**Cleft/smooth chin**
Cleft chin dominant, smooth chin recessive.

**Dimples**
Dimples dominant, lack of dimples recessive.

**Freckles**
Freckles dominant, lack of freckles recessive.

**Detached/attached ear lobes**
Detached is dominant and attached is recessive, but may involve more than one gene.

**Right/left handedness**
Right handed is dominant, and left handed recessive, but this may be due to two genes.

# Atmospheric Movement

**Scholars have long been aware that the earth's atmosphere is not static but is constantly in motion, producing areas of ever-changing pressure around the globe. But from 1700 onward, scientists began to discover how and why large-scale atmospheric movements occur and were able to explain the effect that they have on the earth's surface winds.**

Since the trade winds had blown Columbus to the Americas in 1492, scientists had been seeking to explain the earth's surface winds by examining atmospheric patterns, paving the way for modern weather forecasting.

## Hadley's model

In the early 1700s English amateur meteorologist George Hadley figured out a basic atmospheric circulation model. At the equator, strong solar heating causes air to rise, creating a belt of low pressure beneath it. This air moves north and south at height, then cools and sinks, producing belts of higher pressure at higher latitudes away from the equator. The air is then propelled by pressure differences back to the equator, creating surface winds. But scientists still couldn't explain why these winds moved east to west. In 1735 Hadley proposed that Earth's rotation, from west to east, modifies the direction of air movement, causing it to veer east when moving away from the equator and west when moving toward it. In 1835 the same proposal was made more precisely by a French mathematician, Gustave Coriolis. Today this is known as the Coriolis effect.

## Ferrel and polar cells

During the 19th century Hadley's model of atmospheric circulation gradually gained acceptance. But, by the middle of the century, scientists realized that it only explained the trade winds of subtropical regions. In higher latitudes there were winds known as westerlies, and in polar regions there were winds known as polar easterlies. To account for these, in 1856, an American teacher William Ferrel (see opposite) improved Hadley's model by proposing the existence of mid-latitude circulation cells. Later a third set of cells, polar cells, were added to the model to account for the prevailing winds in polar regions.

## Moving pressure systems

From the 1870s the attention of scientists, such as the English polymath Sir Francis Galton, shifted from studying prevailing winds at the Earth's surface to examining much less constant, but occasionally strong, winds

## Global atmospheric circulation

The part of the atmosphere shown here is called the troposphere, in which there is constant circulation of air. The pattern of movement is produced by a combination of basic circulation cells (Hadley, Ferrel, and polar cells) and the Coriolis effect. This pattern produces three sets of prevailing winds at Earth's surface—trade winds in low latitudes, westerlies in mid-latitudes, and polar easterlies near the poles.

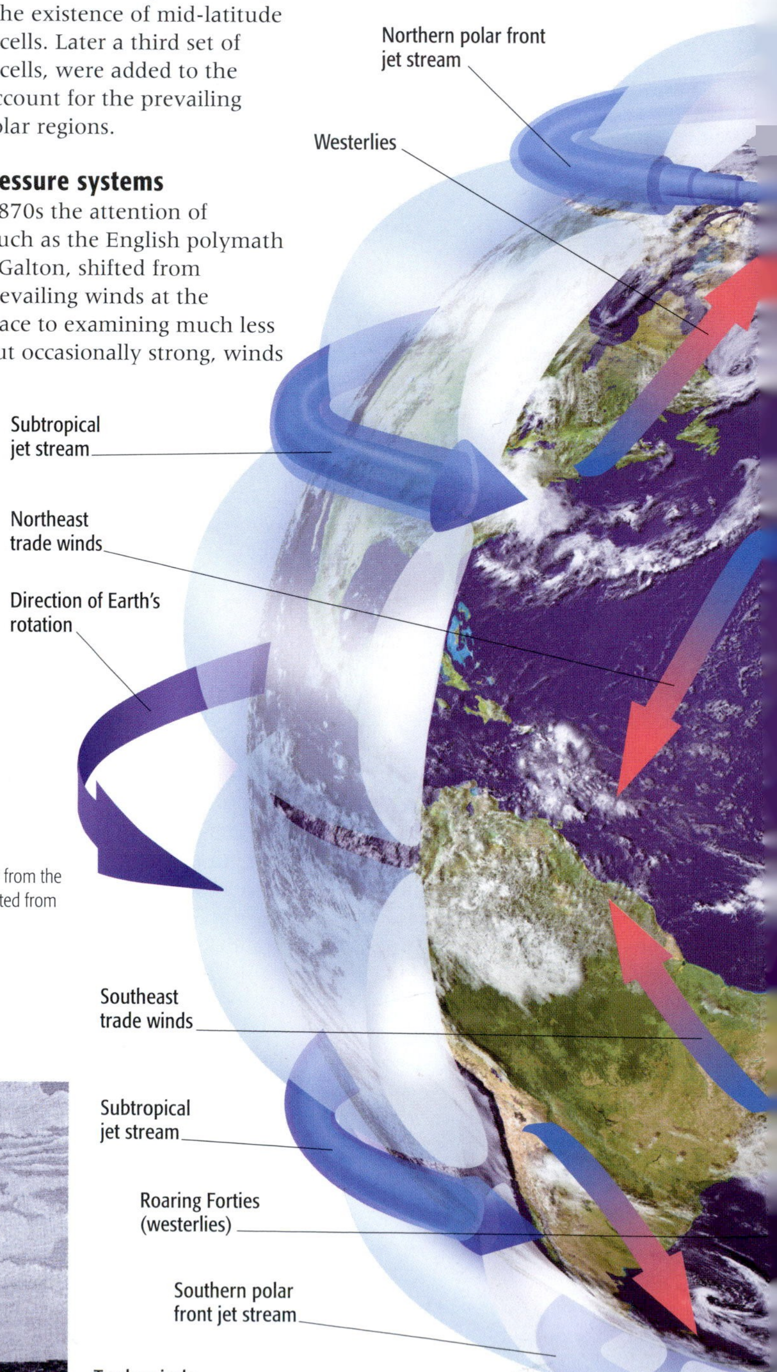

<< BEFORE

People have been using the force of the wind for over 5,000 years, but there is no evidence that its cause was understood until the time of the ancient Greeks.

### ANAXIMANDER EXPLAINS WIND

The Greek philosopher **Anaximander**, writing in c.550 BCE, gave a nearly correct explanation of **the nature and cause of wind**. He described it as a "flow of air, occurring when its finest elements are **set in motion by the Sun**."

### HALLEY'S TRADE WINDS

In 1686 after an expedition to the South Atlantic, the English astronomer Edmond Halley produced a **chart and description of trade winds**. He identified **solar heating** as their cause, correctly supposing that, as heated air rises in the tropics, it is replaced by air flowing in from higher latitudes, creating winds.

HALLEY'S CHART OF TRADE WINDS

Rotation of Earth from west to east

Air moving toward the equator is deflected from east to west

Expected path of air moving away from the equator

Air moving away from the equator is deflected from west to east

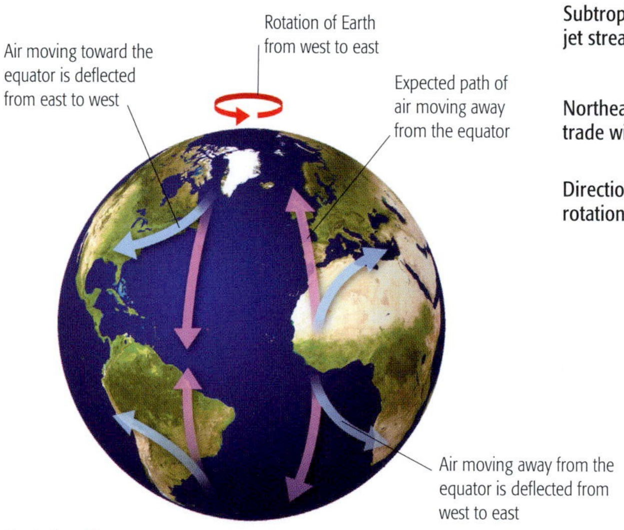

**Coriolis effect**
This effect describes the observed deflection to the west for air moving toward the equator, and to the east for air moving away from the equator.

**Trade winds**
In his flagship *Santa Maria*, shown here with companion ships *Pinta* and *Niña*, Christopher Columbus made full use of the trade winds during his voyage of 1492 across the Atlantic.

Northern polar front jet stream

Westerlies

Subtropical jet stream

Northeast trade winds

Direction of Earth's rotation

Southeast trade winds

Subtropical jet stream

Roaring Forties (westerlies)

Southern polar front jet stream

## OCEANOGRAPHER AND METEOROLOGIST (1817–1891)

### WILLIAM FERREL

William Ferrel is best known for developing a theory of how air circulates in the atmosphere in middle latitudes, modifying an earlier model developed by George Hadley. Born in Pennsylvania, Ferrel received only limited schooling but taught himself enough to become a schoolteacher. In his forties, he switched to working for *The American Ephemeris* and *Nautical Almanac* and became an expert on tides. At the age of 65 he finally took up professional meteorology on joining the US Army Signal Service, which later became the US Weather Bureau.

Warm air rises

**Cyclone**
Rising warm air leaves a surface area of low pressure. The surrounding air spirals inwards.

Cool air sinks

**Anticyclone**
Sinking cool air produces a surface area of high pressure.

**Cyclone and anticyclone**
For every low-pressure center (cyclone or depression, left) there is always a neighboring center of high pressure or anticyclone (right). Air flows from the anticyclone toward the cyclone, but the Coriolis effect modifies this flow to produce spiraling wind patterns.

Polar easterly

**Polar cell**
In polar regions warm air rises, moves toward the poles, cools, and then falls. It then flows back toward the equator and is deflected, producing polar easterly winds.

**Ferrel cell**
Air in Ferrel cells rises at about 60° N and 60° S, moves toward the equator, cools and then falls at around 30° N and 30° S. It then moves away from the equator and is deflected as a result of the Coriolis effect to the east to produce westerly winds.

**Hadley cell**
Hot air in Hadley cells rises and moves north and south, cools and then falls. As it flows back toward the equator, it veers to the west producing trade winds at low altitudes.

**Intertropical convergence zone**
Hot air rising in this region produces a belt of pressure.

Hadley cell

Ferrel cell

Polar cell

affecting localized areas. Barometers showed that these localized winds were associated with centers of low and high pressure, which move over the earth's surface. Low-pressure areas came to be known as cyclones or depressions, and the high-pressure centers as anticyclones. In each case, the air moves in a spiral pattern, which results from pressure differences and the Coriolis effect. Thus, in the northern hemisphere air moves in a clockwise fashion around an anticyclone (high pressure) and in an counterclockwise fashion around a cyclone (low pressure). In the southern hemisphere, the directions are reversed. By the late 19th century the study of cyclones and anticyclones was used in weather forecasting (see pp.208–209).

### Jet streams

In the mid 20th century, understanding of atmospheric circulation was further improved with the discovery of jet streams—narrow air currents flowing eastwards above the junctions between circulation cells. They result from a combination of pressure differences at these junctions causing air movements away from the equator, and the Coriolis effect deflecting these movements to the east. Monitoring the northern polar front jet stream is important in modern weather forecasting (see pp.208-209).

## AFTER

In recent years, research into atmospheric circulation has ranged from low-tech approaches to the use of sophisticated satellite technology.

### TRADE WIND DEMONSTRATION
In 1970 Norwegian adventurer **Thor Heyerdahl** showed that **trade winds can blow an unpowered sailing vessel** made of reeds from Morocco in Africa to Barbados in the Caribbean.

### SATELLITE WIND SURVEYOR
Since 1999 the **pattern of surface winds** over Earth's oceans has been monitored by **NASA satellite QuikScat**. It has proved useful for monitoring tropical cyclones and hurricanes.

### POLAR JET STREAMS IN MOTION
In the early 21st century it was discovered that the **northern polar front jet stream has been moving northward** for about 20 years, at a rate of about 1¼ miles (2 km) a year. The southern polar front jet stream has also been drifting toward the South Pole. The cause of this is not known, but, if it continues, it may contribute to future climate change.

**SEE ALSO ≫**
*pp.208–09* PREDICTING THE WEATHER
*pp.210–11* STRUCTURE OF THE ATMOSPHERE

### IN PRACTICE

## JET STREAMS AND AIR TRAVEL

First noticed by pilots and meteorologists in the 1920s and 1930s, the wind speed at the center of a jet stream is typically about 65 mph (105 kph), so by flying within a jet stream on an eastbound trip, an airliner can cut time off its flight and make fuel savings. This is the main reason why, for instance, eastbound flights across the Atlantic or North America often take around an hour less than westbound flights.

<< BEFORE

People have been concerned with the weather, and ways of predicting it, throughout history.

**ANCIENT GREEK WEATHER PREDICTION**
In *Meteorologica*, which was published around 340 BCE, the Greek philosopher **Aristotle** wrote down everything that was known at the time about weather. It was already appreciated that **water evaporates** from Earth's surface **to form clouds**, and that rain comes from those clouds. Greek scientist **Theophrastus** wrote a **treatise on winds** and "on the signs of rain, wind, storms, and fair weather."

**LEONARDO'S HYGROMETER**
During the 15th century in Florence, Italian polymath **Leonardo da Vinci** was interested in weather measurement and drew a design for a **crude hygrometer** featuring a sponge and a stone on a balance (to measure air humidity) as well as one for a **basic anemometer** (to measure wind speed).

<< SEE ALSO
*pp.206–07* ATMOSPHERIC MOVEMENT

LEONARDO'S HYGROMETER

METEOROLOGIST (1862–1951)
## VILHELM BJERKNES

Born in Norway, Vilhelm Bjerknes studied mathematics and physics at Stockholm University and collaborated with German scientist Heinrich Hertz in work that influenced the development of the radio. From 1895 to 1932 he held university posts at Stockholm, Leipzig, and Oslo. His major contribution to meteorology was to devise a basic model of atmospheric processes affecting weather, and equations to describe the model, that are still used today in weather forecasting.

# Predicting the Weather

**In the 18th century, many new instruments were invented for measuring aspects of the weather, and improvements made to existing ones. These innovations gave a boost to the new science of meteorology, setting a trend that has continued right up to today's computer-assisted forecasts.**

Swift progress from the 18th century boosted the new science of meteorology, setting the foundations for modern computer-assisted forecasts.

### Early observers of the weather
Some of the most important advances in weather measurement occurred in the decade up to 1724, as German physicist Daniel Fahrenheit first developed the mercury thermometer, then devised the temperature scale that bears his name. In 1742 Anders Celsius devised the alternative centigrade (Celsius) temperature scale. Other notable innovations included an improved hygrometer (for measuring humidity) and mercury barometer (for measuring atmospheric pressure), and widespread use of rain gauges.

American polymath Benjamin Franklin maintained regular weather records from the early 1740s, and tracked hurricanes on the Atlantic coast from the 1770s. In Europe, French scientist Antoine Lavoisier measured daily air pressure, humidity, and wind speed from 1763, while in

"It is almost possible to predict **one or two days** in advance… **what the weather** is going to be… it is not impossible to publish daily forecasts, which would be very useful to society."

ANTOINE-LAURENT DE LAVOISIER (1743–94)

CUMULONIMBUS

HEAVY PRECIPITATION

**Cold front clouds**
Distinct lines of tall, dense cumulonimbus clouds (top) form along a cold front. This type of cloud is often associated with heavy downpours of rain (above).

England, scientist John Dalton (see p.138) introduced a rain gauge network, and published *Meteorological Observations and Essays* in 1793.

### First weather maps and centers
In the 1820s the first crude maps were drawn of the weather from the previous few days. Soon it was realized that by drawing maps, storm systems could be tracked. New telegraph networks in the 1840s and 1850s made it possible to gather weather data simultaneously from several locations and map it quickly enough to put the data to practical use in "real time."

In the 1850s, staff at the Paris Observatory began to collect weather data, mainly to provide storm warnings for the French Navy. In the UK, British sea captain and meteorologist Robert Fitzroy was appointed chief of a new department to deal with collection of weather data at sea, the forerunner of the UK Meteorological Office.

In 1863 French meteorologist Edme Hippolyte Marié-Davy published the first modern-style maps, with lines (isobars) to connect areas of equal

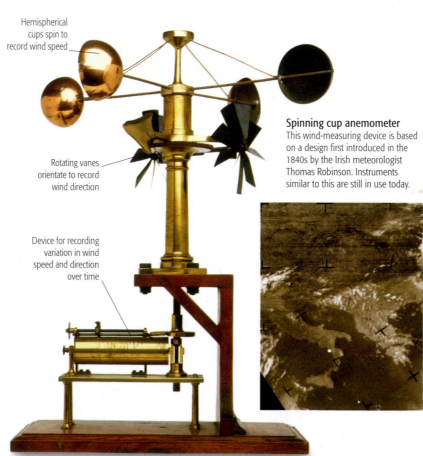

Hemispherical cups spin to record wind speed

**Spinning cup anemometer**
This wind-measuring device is based on a design first introduced in the 1840s by the Irish meteorologist Thomas Robinson. Instruments similar to this are still in use today.

Rotating vanes orientate to record wind direction

Device for recording variation in wind speed and direction over time

**Early weather satellite image**
This image of part of southern Europe was one of the first ever to be taken by a Nimbus satellite, which was one of the earliest models. Modern satellites are able to gather a great deal more data.

## Frontal systems

Many weather phenomena result from the movement of warm and cold air masses, which are separated by boundaries called fronts. Here, a warm front (caused by an advancing mass of warm air) is pushing to the east, followed closely behind by a cold front.

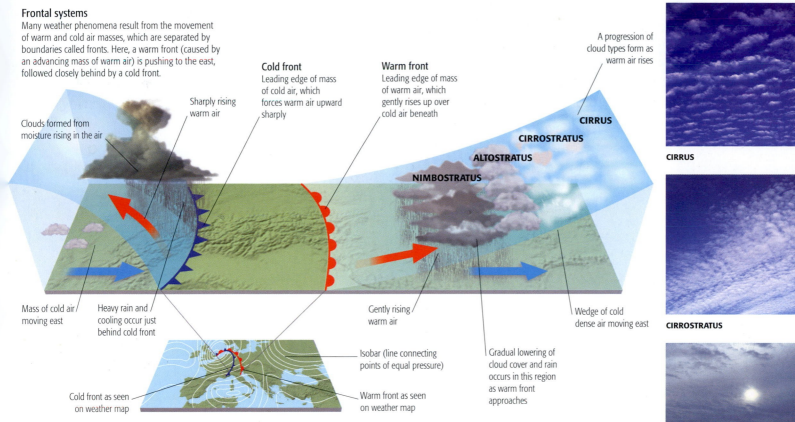

Clouds formed from moisture rising in the air

Sharply rising warm air

**Cold front**
Leading edge of mass of cold air, which forces warm air upward sharply

**Warm front**
Leading edge of mass of warm air, which gently rises up over cold air beneath

A progression of cloud types form as warm air rises

CIRRUS

CIRROSTRATUS

ALTOSTRATUS

NIMBOSTRATUS

Mass of cold air moving east

Heavy rain and cooling occur just behind cold front

Gently rising warm air

Wedge of cold dense air moving east

Cold front as seen on weather map

Isobar (line connecting points of equal pressure)

Warm front as seen on weather map

Gradual lowering of cloud cover and rain occurs in this region as warm front approaches

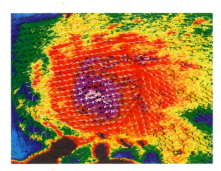

**Satellite storm tracking**
Satellite tracking of tropical storms and hurricanes, and prediction of their future behavior, has acquired a huge economic importance in the modern world, because of the impact on the lives of millions of people.

**CIRRUS**

**CIRROSTRATUS**

**ALTOSTRATUS**

**NIMBOSTRATUS**

barometric pressure. In the same year, English polymath Francis Galton's book *Meteorographica* was the first systematic attempt to gather, chart, and interpret weather data across Europe. He was also first to describe an anticyclone (a moving high-pressure area, see p.207), and in 1875 he published the first newspaper weather map in *The Times*.

In the US, the US Weather Bureau was set up in 1870 and produced daily weather forecasts. Meteorologist Cleveland Abbe sent out telegraphic weather reports, daily weather maps, and weather forecasts from 1871.

## Early 20th-century meteorology

In the early 20th century meteorologist Vilhelm Bjerknes (see opposite) and his son Jacob organized a system of weather observation in Norway, paving the way for numerical methods of predicting the weather. They were the first to describe the role of weather fronts (see above) and, along with anticyclones and depressions, this forms the basis of the model still used to predict the weather.

The development of radar and rockets (see pp.338–39) in World War II aided data collection from the late 1940s—radar locates and shows the intensity of precipitation. Computers were developed in the 1950s that could "crunch" the data and complex equations used in weather forecasting. The first computerized 24-hour weather forecast was put out in 1950 by US scientist John von Neumann.

**Warm front clouds**
An approaching warm front is signaled by a succession of cloud types, starting with high cirrus and cirrostratus, which are followed by altostratus (clouds of medium height, often forming a featureless sheet), and then finally nimbostratus (low rain clouds).

**AFTER** ⟩⟩

## WEATHER BALLOON

Weather or sounding balloons were the first method used to collect data from parts of the atmosphere far above the ground. The first were launched at the turn of the 20th century and, by the 1940s, they provided detailed three-dimensional views of the atmosphere. Modern balloons collect data from heights up to 25 miles (40 km). Each has two parts—the balloon itself, which is filled with hydrogen or helium gas, and an instrument package, which takes regular measurements of temperature, humidity, and pressure, as well as wind speed and direction. Since the 1930s they have carried radios to transmit data back to the ground.

**Since the 1950s forecasting has improved, but its limitations have been recognized.**

### WEATHER CHAOS

In 1961 American meteorologist **Edward Lorenz**, using a computer model, found that small changes in initial weather conditions can result in large changes in the final outcome, so concluded that **weather is inherently unpredictable**, conforming to what mathematicians call **chaos 350–51 ⟩⟩**.

### METEOROLOGICAL SATELLITES

The world's **first weather satellite**, TIROS I—a Television Infrared Observation Satellite—was launched in 1960. Now there are now more than 30 operating weather satellites in use worldwide, bringing about almost-accurate five-day forecasts.

**SEE ALSO ⟩⟩**
*pp.210–11* STRUCTURE OF THE ATMOSPHERE
*pp.212–13* STUDYING THE OCEANS
*pp.368–69* ARTIFICIAL SATELLITES

<< **BEFORE**

Early scholars made significant contributions to our understanding of the atmosphere.

**HEIGHT OF THE ATMOSPHERE**
In 1027 **Alhazen << 60–61** estimated the **height of the atmosphere to be 9 miles (15 km),** based on light refraction and the fact that twilight ceases or begins when the Sun gets to an angle of 19 degrees below the horizon.

**WEIGHING AIR**
In 1613 Italian scholar **Galileo Galilei** estimated that **the density of air is about 0.22 percent that of water**. The correct figure is 0.12 percent. Nevertheless, Galileo's estimate was a remarkable achievement for his day.

<< **SEE ALSO**
pp.206–07 ATMOSPHERIC MOVEMENT
pp.208–09 PREDICTING THE WEATHER

**Earth's atmosphere from space**
From a spacecraft, the atmosphere appears as a delicate blue shroud. The colour is due to gas particles scattering blue light.

**Investigating the composition of air**
In 1804 French scientists Joseph Gay-Lussac and Jean Biot made one of the first ascents to 2½ miles (4 km) to measure air composition and temperature.

# Structure of the Atmosphere

**Together with the lithosphere (rocky crust and mantle), hydrosphere (watery parts), and biosphere (living organisms), the atmosphere is one of the four main components of our planet. Our knowledge about its complex composition and structure has progressed steadily since the 18th century.**

The atmosphere consists of layers surrounding Earth. Each layer has its own temperature and gas composition that alters with height. Earth's atmosphere contains our weather systems and has played a key role in allowing life to evolve and survive on the surface.

**Troposphere and stratosphere**
In the 19th century scientists such as French scientist Joseph Gay-Lussac ascended high into the atmosphere in balloons. They established that temperature, pressure, and oxygen concentrations decrease steadily up to a height of 6 miles (10 km). Manned exploration to greater heights seemed

too dangerous, so 90 years later French scientist Léon Teisserenc de Bort used unmanned balloons. Between 1892 and 1896 he launched a series of meteorological instrument packages, carried by hydrogen balloon, to heights of up to 11 miles (17 km). In 1902 de Bort proposed that there were two layers in the atmosphere: a lower layer, the troposphere, where weather occurs, and a higher, more uniform and placid layer, which he called the stratosphere. This latter layer is now known to extend up to 30 miles (50 km).

**The upper atmosphere**
Above the stratosphere are more layers of very thin air. Clues about this region arose in 1904, after the first transatlantic radio transmission (pp.268–69). Physicists deduced that there must be a layer of plasma (electrically charged particles) at high altitude that reflects back the radio waves so that they follow Earth's curvature instead of shooting off into space. In 1926 this electrically charged region, caused by the effects of ultraviolet radiation

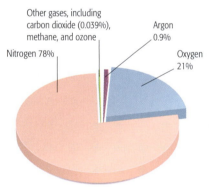

**Proportions of gas in the atmosphere**
This chart shows the relative amounts of different gases in dry air. There is also water vapor in the atmosphere, which makes up 1–4 percent of air at the surface of the earth and about 0.4 percent of the whole atmosphere.

Other gases, including carbon dioxide (0.039%), methane, and ozone

Argon 0.9%

Nitrogen 78%

Oxygen 21%

ionizing the air, was named the ionosphere. In the late 1940s probing of the region above the stratosphere commenced with rockets as well as balloons. This showed the existence of at least two more layers: the mesosphere, in which temperature drops with height, and the thermosphere (above 50 miles/85 km), in which the reverse occurs. The ionosphere was found

**CHEMIST (1731–1810)**

## HENRY CAVENDISH

Born into a British aristocratic family, Henry Cavendish was a shy and reclusive eccentric, who only ever spoke with his scientific friends. He was the first person to determine the composition of the atmosphere, and to show that this composition was the same everywhere. He also discovered hydrogen (which he called "inflammable air") and made an astonishingly accurate determination of Earth's density. A private income enabled him to devote his life to unpaid scientific investigation, and it is said that he made many important discoveries that he never published.

**"We live submerged at the bottom of an ocean of air."**

EVANGELISTA TORRICELLI, PHYSICIST, 1644

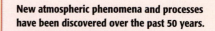

**Layers of the atmosphere**
Each layer is defined by the way temperature varies within it, and is separated by distinct boundaries.

## Thermosphere
This layer of rarefied gas extends from about 55 miles (85 km) up to between 300 miles (500 km) and 600 miles (1,000 km) depending on solar activity. Much of the gas here is ionized (consists of charged particles) and this layer of ionized gas is known as the ionosphere. It is found in the thermosphere, and the exosphere higher up.

## Mesosphere
The mesosphere extends from 30–50 miles (50–85 km) above the surface. Its boundary with the thermosphere (the mesopause) is the coldest part of Earth's atmosphere, with temperatures as low as –130° F (–90° C). Most meteors burn up here.

## Stratosphere
This layer extends up to a height of about 31 miles (50 km). The ozone layer is found between 12 miles (20 km) and 25 miles (40 km).

## Troposphere
All weather occurs in this layer, which extends up about 10 miles (16 km) at the equator and 5 miles (8 km) at the poles.

80 miles / 130 km
74 miles / 120 km
68 miles / 110 km
62 miles / 100 km
56 miles / 90 km

**MESOPAUSE**

50 miles / 80 km
43 miles / 70 km
37 miles / 60 km

**STRATOPAUSE**

31 miles / 50 km
25 miles / 40 km
18 miles / 30 km
12 miles / 20 km

**TROPOPAUSE**

6 miles / 10 km
Sea level

**HEIGHT ABOVE SEA LEVEL**

Aurora borealis

Meteor burning up as a result of air friction

Noctilucent clouds

Thin layer of ozone gas

140° F / 60° C
14° F / –10° C
–112° F / –80° C
–130° F / –90° C
–112° F / –80° C
–58° F / –50° C
–22° F / –30° C
14° F / –10° C
–4° F / –20° C
–40° F / –40° C
–76° F / –60° C
–76° F / –60° C
59° F / 15° C

**AVERAGE TEMPERATURE**

**Aurora borealis**
This occurs when charged solar particles collide with atoms and molecules in the thermosphere.

**Noctilucent clouds**
These ice crystal clouds, which form in the mesosphere, reflect light after the Sun has set.

**Nacreous clouds**
Forming in the lower stratosphere, these bright clouds, occasionally seen in polar regions, contain water and sometimes nitric acid.

**Rainbows**
Rainbows are caused by water droplets in the troposphere, which refract and reflect the light from the Sun.

**New atmospheric phenomena and processes have been discovered over the past 50 years.**

### BOUNDARY WITH SPACE
There is **no clearly defined boundary** between the atmosphere and space, although in the 1950s, the International Aeronautic Federation accepted a **height of 62 miles (100 km)**, called the **Kármán line**, as a boundary above which the air is too thin for aeronautical purposes. The level of harmful radiation also increases above this line.

### DISCOVERY OF OZONE DEPLETION
In the mid 1970s scientists warned that chemicals called chlorofluorocarbons (CFCs) were destroying Earth's ozone layer, resulting in increased ultraviolet radiation at the surface. In 1978 several countries banned the use of CFC-containing aerosol sprays. But, in the **mid 1980s**, a seasonal ozone "hole" (serious depletion) was detected above Antarctica. From 1987 an international treaty sharply reduced CFC production, which was phased out by 1996. By 2003 the **depletion of the ozone layer was slowing**, but the hole still reappears annually from September to December.

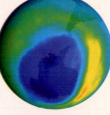

**OZONE HOLE OVER ANTARCTICA**

SEE ALSO ≫
pp.368–69 Artificial Satellites
pp.414–15 Global Warming

**IN PRACTICE**

## LIGHTNING: AN ELECTRICAL DISCHARGE

The idea that lightning might be an electrical discharge was first suggested in the early 18th century and, in 1752, American Benjamin Franklin (pp.160–61) and Frenchman Thomas-François Dalibard confirmed this view. They noted that clouds are electrically charged: negatively at the base and positively on top. A bolt of lightning can carry a current of over 10,000 amps, generating temperatures higher than 36,000° F (20,000° C), which is hotter than the surface of the Sun.

mainly to occupy the thermosphere. Later, an even higher layer, the exosphere, was defined, so rarefied that its gas particles hardly ever collide.

## Atmospheric chemistry
In the 20th century scientists began to consider how the concentrations of gases in the atmosphere are maintained and how they have changed over time. This has included the study of where the oxygen in the atmosphere came from—it is now thought to have been produced almost entirely by the activities of plants and other photosynthetic organisms over the course of Earth's history. Another example was the investigation of the ozone layer in the stratosphere, discovered by French physicists Charles Fabry and Henri Buisson in 1913. Its essential role in protecting life at Earth's surface was uncovered in 1930 by British physicist Sydney Chapman—molecules of ozone ($O_3$) and ordinary oxygen ($O_2$) are continuously converted into each other in a cyclical process that also absorbs harmful ultraviolet radiation and produces heat.

Scientists now study the chemistry of both the oceans and the atmosphere when trying to solve problems such as global warming.

**« BEFORE**

People have contemplated oceanic phenomena ever since they first went to sea.

**EARLY IDEAS ON TIDES**
The 2nd-century BCE Babylonian astronomer **Seleucus** of Seleucia was the first to state that **tides** are linked to the **phases of the Moon**.

**ORIGIN OF OCEAN BASINS**
In 1644 the French philosopher **René Descartes** suggested that in the distant past, **parts of Earth's crust had collapsed** into underground chambers, which had been formed by ancient gaseous outpourings. He thought this had **created the ocean basins** with the remaining crust forming continents.

**THE GULF STREAM**
Originally discovered in the 16th century by Spanish sailing expeditions, the strong, warm ocean current known as the **Gulf Stream**, was **first studied and mapped** in the 18th century by **Benjamin Franklin**.

**« SEE ALSO**
pp.108–09 GRAVITATIONAL FORCE
pp.136–37 NAVIGATING THE OCEANS
pp.206–07 ATMOSPHERIC MOVEMENT

# Studying the Oceans

**The roots of oceanography go back centuries. By around 1800 certain aspects—such as winds and tides—were already well understood because of their importance to trade and naval warfare. But it was not until later in the 19th century that marine science first began to take off as a recognized field of study.**

From the early 1800s scientists began to gather immense amounts of data about marine life, paving the way for future knowledge about the importance of the planet's oceans. In the 1840s and '50s American scientist Matthew Maury (see opposite) began to analyze data on currents and the depths of ocean basins. Among other accomplishments, Maury noted that the Atlantic is shallower in the middle (the first indication of the mid-Atlantic ridge, see p.356). The lowest point in the oceans—the Mariana Trench of the western Pacific, at over 35,800 ft (10,900 m) deep—was discovered in 1872.

## Ocean currents

Strong surface currents in the oceans, some warm, others cold, were known about from the 16th century onward.

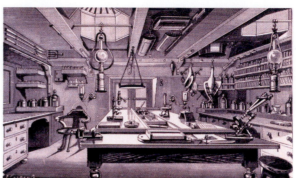

**HMS Challenger**
The natural history laboratory on board HMS *Challenger*, where some 4,000 new animal and plant species were identified.

During the 19th century scientists gradually realized that many of these currents are linked in gyres—these are large-scale circular patterns of currents that operate over whole oceans (see left). In 1905 the Swedish oceanographer Walfrid Ekman published a theory about the effect of prevailing winds blowing over ocean areas, which demonstrated that these winds were driving the gyres.

Today it is known that ocean currents play a major role in affecting localized climate. Warm currents carry heat away from the tropics and subtropics, giving some temperate regions, such as the British Isles and Scandinavia, a warmer climate than they would otherwise

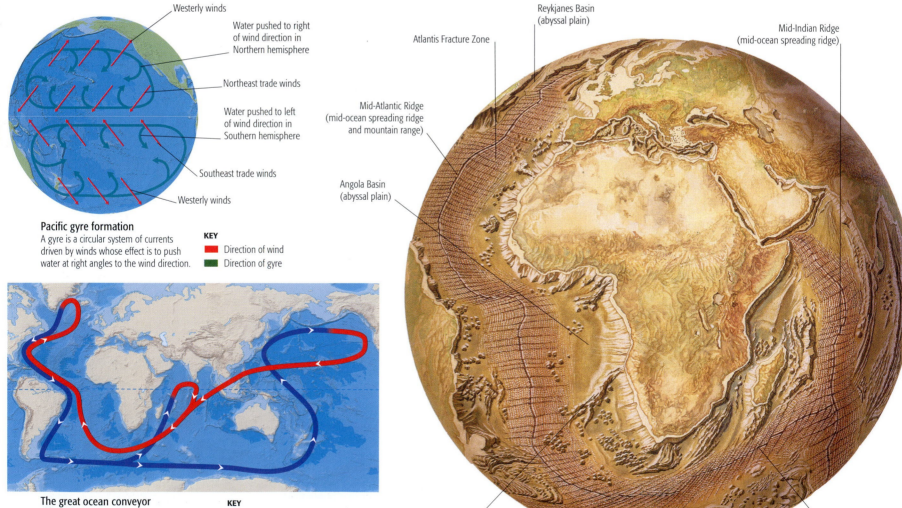

Westerly winds
Water pushed to right of wind direction in Northern hemisphere
Northeast trade winds
Water pushed to left of wind direction in Southern hemisphere
Southeast trade winds
Westerly winds

**Pacific gyre formation**
A gyre is a circular system of currents driven by winds whose effect is to push water at right angles to the wind direction.

**KEY**
🟥 Direction of wind
🟩 Direction of gyre

**The great ocean conveyor**
This continuous global flux of seawater is driven by water cooling becoming more dense in Arctic waters, then sinking to the seabed.

**KEY**
🟥 Warm surface current
🟦 Cold deep-ocean current

Reykjanes Basin (abyssal plain)
Atlantis Fracture Zone
Mid-Indian Ridge (mid-ocean spreading ridge)
Mid-Atlantic Ridge (mid-ocean spreading ridge and mountain range)
Angola Basin (abyssal plain)
Walvis Ridge (volcanic ridge caused by a hot spot)
Atlantic-Indian Ridge (mid-ocean spreading ridge)

## OCEANOGRAPHER (1806–73)

## MATTHEW MAURY

Often called the "father of oceanography," Matthew Maury was born in Virginia, and spent some time in the navy before an accident ended his seagoing days. Between 1842 and 1861 he was the first superintendent of the United States Naval Observatory. He devoted himself to caring for the Navy's navigational equipment, and to charting the winds and currents of the North Atlantic by analyzing old ships' logs. In 1855 he published *The Physical Geography of the Sea*, now credited as the first oceanography textbook.

enjoy. Conversely, cold currents give some coastal areas a cooler climate than they would otherwise have.

### The great ocean conveyor

In the 19th century scientists found that, while the surface temperature of the oceans varies with latitude, water

**50–80 PERCENT** The estimated amount of life on Earth that lives in the oceans. Less than 10 percent of this has been explored so far.

hauled up from great depths anywhere is always cold. Research in the 20th century established that all deep water throughout the oceans has originally sunk to the bottom in either the Arctic or Southern oceans and flowed toward the equator at depth. In the 20th century a slow, three-dimensional flux of water was identified—the great ocean conveyor (see opposite). This starts when warm water is carried into the North Atlantic, where it cools, becomes denser, and sinks. It then flows south along the bottom of the Atlantic and into the lower reaches of

the Southern Ocean, where it is joined by more cold, dense water sinking close to Antarctica. The cold water branches into the Indian and Pacific Oceans, and returns to the surface by mixing with warmer waters above. Finally, surface currents return the water to the Atlantic

### INVENTION

## SONAR

Sonar (an acronym for sound navigation and ranging) is a technology that uses sound waves to measure distances and locate objects underwater. A sonar device sends out sound waves and listens for echoes. The data is then usually displayed on a monitor. Sonar's uses range from mapping the seafloor to detecting fish schools and submarines. Many people contributed to the invention of sonar, but arguably the key figure was French physicist Paul Langévin, who developed submarine-detecting devices during World War I.

to start the cycle again. There are concerns that global warming and ice melting might disrupt the conveyor and bringing a cooler climate to some areas.

### Mapping the ocean floor

Sonar (see above) revolutionized mapping of the seafloor, revealing seamounts and guyots (submarine volcanoes), abyssal plains (flat areas), deep-sea trenches, and fracture zones (bands of linear cracks). A pattern of gently sloping shallow areas (continental shelf) was identified, linked to the deeper seafloor by steeply shelving areas (continental slope). In the mid 1950s a system of mid-ocean ridges was charted (see pp.356–57).

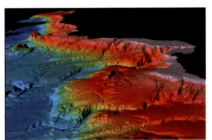

**Sonar image of seafloor**
This area off California, shown in black, is color-coded by depth. The dark orange shows continental shelf; the orange, yellow, and green, continental slope; and the blue abyssal plain.

**AFTER**

Research over the past 40 years has focused on knowledge of the seafloor.

**DEEP-SEA DRILLING PROJECTS**
Between 1968 and 1983 the Glomar Challenger research vessel, designed by the American National Science Foundation, undertook deep-sea drilling projects. Atlantic samples proved **seafloor spreading**, while **Mediterranean** samples showed that the sea had **dried up 6 million years ago**. A key find was that **ocean-floor rocks are no older than 180 million years**.

**HYDROTHERMAL VENTS**
In 1977 scientists discovered **seafloor vents gushing warm, mineral-rich fluids** on the bottom of the Pacific. Clustered around the vents was an **extraordinary abundance of life**. Many more vents have since been discovered.

SEE ALSO »
pp.354–55 MOVING CONTINENTS
pp.356–57 PLATE TECTONICS
pp.358–61 ACTIVE EARTH

Emperor Seamounts
Aleutian Basin (abyssal plain)
Aleutian Trench (deep-sea trench)
Mariana Trench (deepest deep-sea trench)
Gilbert Seamounts

Hawaiian Ridge (volcanic ridge caused by hot spot)

Mid-Pacific Mountains (seamounts)

East Pacific Rise (mid-ocean spreading ridge)

Tuamotu Ridge (volcanic ridge caused by hot spot)

Agassiz Fracture Zone

Tonga Trench (deep-sea trench)

**Mapping the ocean floor**
Mid-ocean ridges, fracture zones, abyssal plains, deep-sea trenches, and seamount chains were some of the most obvious features first revealed by sonar mapping of the ocean floor in the 20th century.

Although life is astonishingly varied, its cellular building blocks are surprisingly uniform. When biologists examined cells through microscopes, many began to see that to understand a cell is to understand life itself.

## Cell Theory

At the beginning of the 19th century, German botanist Johann Moldenhawer loosened the glue between plant cell walls and teased them apart, showing they were discrete (separable) units. At the same time botanist Matthias Schleiden and biologist Theodor Schwann combined their observations to develop Cell Theory, stating that all organisms are made up of cells and cell products. Before Schwann and Schleiden, the prevailing view was that living material could form spontaneously. This was finally laid to rest in 1850, when German physician Rudolf Virchow proposed that cells could arise only from other cells.

Schleiden speculated that the cell's nucleus may have something to do with the production of new cells, but the idea that DNA in the nucleus was the cell's genetic material did not become clear until the 20th century. DNA is copied and then the cell divides to create new cells with the same genetic makeup as the parent cell (see pp.306–07).

### The role of cell membranes

Toward the end of the 19th century, a botany student in Zürich named Charles Overton suggested that the

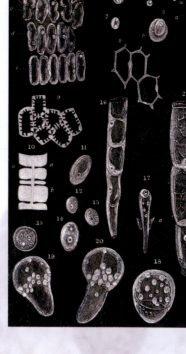

**Schwann's microscopic research**
German biologist Theodor Schwann laid the foundation of Cell Theory, which asserted the cellular nature of all living organisms, as demonstrated here in his microscopic image from 1847 of plant and animal cells.

# Animal and Plant Cells

**The cell is the smallest living thing. Some organisms, such as bacteria, consist of a single, independent cell which carries out all living processes. But the bodies of plants and animals consist of trillions of cells, which all interact together, making them the building blocks of life.**

## ◀◀ BEFORE

**Cells were first seen in the 1600s, but it was not fully appreciated for another 200 years that cells are the basic units of all life.**

### SPONTANEOUS GENERATION OF LIFE
Before Cell Theory, biologists thought there was a gradual transition from nonliving to living matter, coming from ancient Greek philosopher **Aristotle**'s idea of "Scala Naturae" (scheme of nature). It prompted belief that **microscopic life could spontaneously develop** in water, and encouraged the search for life's smallest unit.

### THE FIRST OBSERVER OF THE CELL
In 1665 English philosopher **Robert Hooke** coined the word "cell" to describe the **microscopic structures** he saw in **cork bark**. This was an allusion to the **boxlike cells** in **monasteries**. Hooke's cells were dead, but, soon after that, Dutch scientist **Antony van Leeuwenhoek** recorded seeing living, moving, single-celled life forms in pond water.

### ◀◀ SEE ALSO
*pp.94–95* MICROSCOPIC LIFE
*pp.154–55* HOW PLANTS WORK

**Mitochondrion**
Structure that generates the cell's energy. Oxygen releases energy stored in nutrients, such as those derived from glucose.

**Nucleus**
The nucleus contains the cell's genetic material, DNA, which directs the cell's activities.

**Cytoplasm**
The fluid between the nucleus and the cell membrane, and site of many chemical processes, such as buildup of stored food.

**Centrioles**
A pair of structures that are essential to cell division.

**Nucleolus**
A dense region of the nucleus used for making ribosomes (see below).

**Ribosome**
The site where the cell makes proteins.

## Structure of an animal cell
The animal cell is a eukaryote cell, which means it has a nucleus that controls all its activities. The nucleus is found in the watery cytoplasm, which is enclosed by the flexible cell membrane. There are additional membranes around the nucleus and other internal structures.

**Golgi apparatus**
A system of membranous sacs used for processing and secreting substances.

**Lysosome**
Enzymes for breaking down worn out parts of the cell and viruses are contained here.

**Cell membrane**
The permeable outer layer through which substances must enter and leave the cell.

**Endoplasmic reticulum (ER)**
A network of internal membranes for transporting and assembling large molecules, such as proteins and lipids.

## THEODOR SCHWANN

Born in Prussia, Schwann was a biologist whose belief in Cell Theory (that all organisms are made up of cells and cell products) debunked existing ideas of spontaneous generation of life and vitalism (that life was due to a mysterious vital force). His studies led to an understanding of the roles of enzymes in digestion and metabolism, and the Germ Theory of Disease, later developed by Louis Pasteur (see pp.244–45).

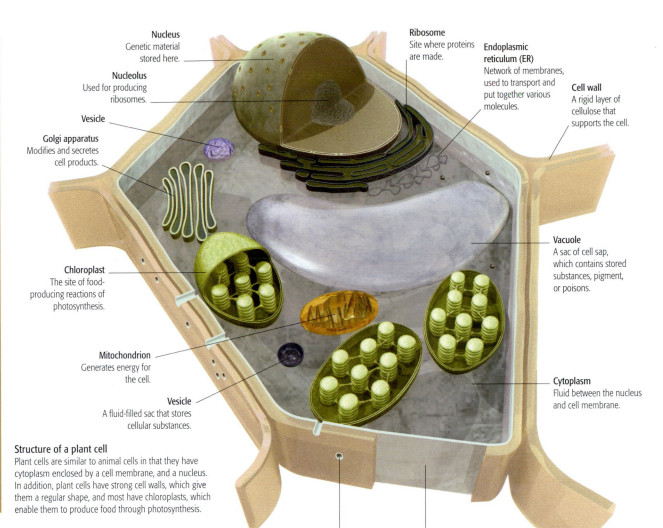

Nucleus
Genetic material stored here.

Nucleolus
Used for producing ribosomes.

Vesicle

Golgi apparatus
Modifies and secretes cell products.

Ribosome
Site where proteins are made.

Endoplasmic reticulum (ER)
Network of membranes, used to transport and put together various molecules.

Cell wall
A rigid layer of cellulose that supports the cell.

Chloroplast
The site of food-producing reactions of photosynthesis.

Vacuole
A sac of cell sap, which contains stored substances, pigment, or poisons.

Mitochondrion
Generates energy for the cell.

Vesicle
A fluid-filled sac that stores cellular substances.

Cytoplasm
Fluid between the nucleus and cell membrane.

### Structure of a plant cell
Plant cells are similar to animal cells in that they have cytoplasm enclosed by a cell membrane, and a nucleus. In addition, plant cells have strong cell walls, which give them a regular shape, and most have chloroplasts, which enable them to produce food through photosynthesis.

Cell membrane
Permeable layer that lets substances enter and leave.

Plasmodesma
Fine threads pass through adjacent cell walls, allowing substances to pass between cells.

cell membrane was made of a substance similar to olive oil, and that oily substances could pass through it. All cells are bound by an oily membrane, which forms the exposed surface of an animal cell, and lies under the cell wall of plants. Structures inside a cell called organelles—such as the nucleus, chloroplast, and mitochondria—are bound by membranes too. A membrane is critical in determining what passes across it

because it is selectively permeable— it lets some things through more easily than others, depending on their chemical properties and concentration differences (see below). Membranes are now known to be important in a range of functions, from the passage of electrical impulses along nerve cells to the responses of immune system cells.

## Acquiring food for energy
Although animal and plant cells share common features, animal cells depend on food consumed by the body while green plant cells are able to generate their own food. In the 17th century English botanist Nehemiah Grew observed plant chloroplasts,

which are now known to be the site where the plant cell generates sugar through photosynthesis (see pp.156–57).

Animal and plant cells both release energy in the same way. In 1890 German pathologist Richard Altmann found mitochondria in animal and plant cells—these granular structures release usable energy from food, such as sugar (see pp.272–73).

▽ In and out of cells
Movement of particles across cell membranes can occur in three main ways, depending upon the nature of the particles moving, concentration differences, and whether energy is used in the process.

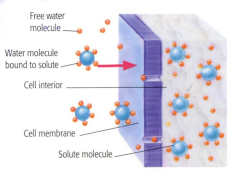

Free water molecule

Water molecule bound to solute

Cell interior

Cell membrane

Solute molecule

### Osmosis
Osmosis is the diffusion of molecules of water from an area of low solute concentration to high concentration through a semipermeable membrane. This is a passive process (no external source of energy is required).

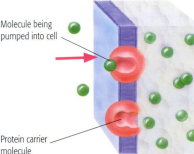

Molecule being pumped into cell

Protein carrier molecule

### Active transportation
Some molecules are pumped across the membrane from low to high concentration by a protein carrier molecule. This process requires energy.

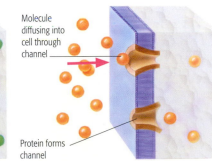

Molecule diffusing into cell through channel

Protein forms channel

### Diffusion
Molecules move from an area of high to low concentration by diffusion, sometimes across a cell membrane. The process is passive (no energy required).

AFTER ≫

Advances in the 20th century revealed that cell membranes are more than just oily layers.

### BILAYERS
In 1925 Dutch pediatrician **Evert Gorter** found that the oil extracted from red blood cells covered an area that was about twice the surface area of the cells. He concluded that **cell membranes** were exactly **two oil molecules thick**.

CELL MEMBRANE BETWEEN TWO ANIMAL CELLS

### FLUID MOSAIC MODEL
In 1935 British biologists **James Danielli** and **Hugh Davson** suggested that oily membranes were coated with protein. This model was modified, in 1972, by American biologists **S.J. Singer** and **Garth Nicolson**. It is accepted today that **proteins are interspersed in the oil** like a mosaic.

SEE ALSO ≫
*pp.244–45* BACTERIA AND VIRUSES
*pp.270–71* BREATHING AND RESPIRATION
*pp.306–07* HOW CELLS DIVIDE

PHYSIOLOGIST (1837–1900)

## WILHELM KÜHNE

Born in Hamburg, Germany, Kühne was appointed Professor of Physiology at Amsterdam University in 1868, and at Heidelberg University in 1871. He was active in many areas of research, including vision, muscles, and the chemistry of digestion. A notable teacher, Kühne initially refused to let women in his classes until a female physiology student proved her ability and he awarded a new degree for female graduates.

**3** Chemical digestion: After leaving the stomach, partly digested food enters the duodenum, the first part of the small intestine. Ducts deliver bile from the liver and complex secretions from the pancreas, enabling many digestive processes to take place here.

**4** Product absorption: As the partly digested food passes through the rest of the small intestine, a lining of tiny projections called villi provide a big surface area for the absorption of soluble products. The villi are each no more than a couple of millimeters high and filled with blood capillaries.

## BEFORE

Knowledge of the digestive system began with the first dissections in ancient Greece.

### STRUCTURE OF THE DIGESTIVE SYSTEM
During the 3rd century BCE, ancient Greek philosopher **Europhiles** made many important anatomical discoveries, including the fact that the **liver** and **intestines** were **connected by a large vein** (now known as the hepatic portal vein). He coined the word "**duodenum**" for the first part of the small intestine just beyond the stomach, and, in 200 BCE, Greek physician **Herophilus** identified the **pancreas**.

« SEE ALSO
pp.72–73 THE HUMAN BODY REVEALED

**2** Food reaches the stomach from the mouth via the esophagus. Strong muscles in the stomach walls churn the food and it can expand to hold 1¾–2½ pints (1–1.5 liters) of food at any one time. Gastric juice present in the stomach is highly acidic to kill pathogens and to provide the right conditions for enzymes to work.

**Gastric gland**
Some cells here secrete hydrochloric acid and others secrete pepsinogen. The acid converts inactive pepsinogen into pepsin, which digests protein.

**Gastric pit**

STOMACH

**Bile duct**
Releases bile from the liver.

**Pancreatic duct**
Releases pancreatic juice made in the pancreas.

**Pancreatic juice**
Enzymes present in this juice break down starch, protein, and fat.

**Bile**
The size of fat droplets in liquefied food is reduced by bile, making it easier for digestive enzymes to work.

DUODENUM

**Villus**
A tiny projection where soluble products are absorbed.

**Lacteal**
The lymph capillary of the villus for absorbing digested fat.

**Blood capillaries**

**Artery branches**
These carry blood to the intestine.

**Direction of blood flow**

**Vein branches**
These carry nutrient-rich blood away from intestine.

SMALL INTESTINE

**5** Final absorption: After leaving the small intestine, the partly digested food moves into large intestine, which consists of the cecum, colon, and rectum.

**Vitamin K**
Obtained from bacteria in feces with other vitamins.

**Bicarbonate and potassium**
These move into feces from blood to help salt balance.

**Amino acids**
Made by bacteria for use in protein production in body.

**Chloride and sodium**
Salts needed in the body are absorbed into the blood.

**Water**
Large intestine absorbs water from feces into the blood.

LARGE INTESTINE

### The human digestive system
Food passes from the mouth to the stomach, then through the small and large intestines. As it passes through the complex tubular system, food is churned, broken down by enzymes from glands, and then the digested products are absorbed.

**1** The first stage of the digestive system takes place in the mouth. Chewing breaks down food into small particles, so that enzymes can work more easily. Saliva lubricates food to help it pass through the system, and contains the enzyme amylase, which digests some starch into sugar.

MOUTH

Esophagus

# Digestion

**The idea that the gut is somehow involved in processing food is a very old one, but understanding how digestion splits large molecules into smaller ones requires the more recent science of biochemistry.**

All living things need food to give them energy to grow and survive. Food is digested (broken down) by the body in two ways: by physical pounding and chemical breakdown.

### Physical breakdown of food

The gut is a muscular tube that runs through the body. Back in the Renaissance, anatomists used microscopes to determine that all parts of the gut have an outer muscle layer and an inner glandular layer—the mucosa. The pulsating contractions of the muscle layer are capable not only of churning food, but also moving it through the system. The churning pulverizes the food, producing a nutrient-rich liquid. Some of the nutrients, such as sugar, dissolve easily in the liquid and are absorbed directly into the bloodstream. Others, such as starch, have molecules that are too large to be absorbed. No amount of pounding will change starch into sugar, so instead the gut has to break down such molecules chemically.

### Chemical breakdown of food

In 1783 Italian physiologist Lazzaro Spallanzini performed experiments—using himself as a guinea pig—to show that stomach juice can liquefy meat, even without the churning effects of the stomach. In 1833 French chemist Anselme Payen discovered a natural product (diastase or amylase) that changes starch into sugar. Then, in 1836, German physiologist Theodor Schwann found an extract from stomach lining that has the same sort of effect on protein. Wilhelm Kühne (see opposite), another notable German physiologist, called these products "enzymes."

The pancreas, a separate structure below the stomach, is the biggest gland of the entire digestive system. In 1850 Claude Bernard (see p.274) found that it released juices capable of working on all the main nutrient groups. Together, the stomach and pancreas ensure that the chemical digestion of food releases as much soluble nutrient as possible before the liquid passes through the intestines. Once in the intestines, more enzymes are released from fissures in the intestine wall (called the crypts of Lieberkühn), discovered in 1874 by the German physician Johann Lieberkühn. The released enzymes help further digestion of proteins and sugars.

### Absorbing nutrients

The intestines are lined with millions of villi—tiny projections just a few millimeters high. Each villus is packed with blood capillaries, which collect nutrients for distribution around the body in the blood. Products that the human body cannot digest are expelled, for example, tough plant fibers. These can be digested by herbivores such as cows (see below).

◁ **Human gastric experiment**
In 1783 Italian physiologist Lazzaro Spallanzani extracted gastric juices from himself, and demonstrated that they could liquefy meat inside or outside the body.

> "The fertility of **an idea** can be **proved** only **by experiment**."
>
> CLAUDE BERNARD, PHYSIOLOGIST 1927

## EXTRACTING NUTRIENTS

The digestive process takes, on average, 12–20 hours. Food is in the stomach for 1–2 hours, and in the small intestine for 1–5 hours. The final stages of digestion in the large intestine may take 12 hours. The vertical arrows below indicate where nutrients are absorbed. A number of digestive enzymes are available as food passes through the system—these catalyze (speed up) the breaking of chemical bonds, converting big molecules into small ones.

| | MOUTH | STOMACH | SMALL INTESTINE | LARGE INTESTINE |
|---|---|---|---|---|
| **PROTEINS** | | Pepsin breaks protein into peptide chains. | Peptides snipped into amino acids. | |
| **CARBOHYDRATES** | Salivary amylase breaks starch into maltose (sugar). | | Maltase breaks down maltose to glucose. | |
| **FAT (LIPIDS)** | | | Pancreatic lipase products enter lacteals. | |
| **FIBER** SOLUBLE INSOLUBLE | | | | Soluble fiber broken down—not absorbed. |
| **WATER** | | Small amount absorbed by lining. | Absorbed by small intestine lining. | Most water absorbed by large intestine. |
| **FAT-SOLUBLE VITAMINS (A, D, AND K)** | | | Bile from liver helps split droplets. | Bacteria make vitamin K. |
| **WATER-SOLUBLE VITAMINS (B, C, A, D, AND K)** | | | | |
| **MINERALS** IRON SODIUM CALCIUM | | | | |

### Digestion in ruminants

The gut of plant-eating cattle has four chambers, one of which is called the rumen. Here, a community of microbes releases an enzyme that can break down the tough cellulose of plant cell walls, making its nutrients available to the animal.

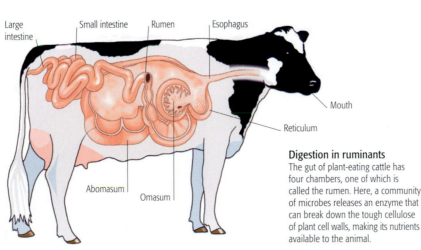

Large intestine — Small intestine — Rumen — Esophagus — Mouth — Reticulum — Abomasum — Omasum

**AFTER** ⟫

**With 20th-century knowledge of molecular biology, the digestive system was revealed to be a precisely controlled chemical system.**

### ENZYMES ARE PROTEINS

In 1926 American chemist **James Sumner** found that an enzyme, urease, was a protein. More research showed that **all enzymes are proteins**, and that their **complex structure** enables them to **accelerate chemical reactions**.

### ADMINISTERING DRUGS

An obvious way to administer a drug is by mouth. But the **digestive juices** that break down food **do the same to certain drugs**. Insulin, for example, is a protein, so is digested by the stomach. Recent research, however, has developed a way of encapsulating insulin so that it reaches the intestine intact, offering the hope of **needle-free treatment for diabetics**.

**SEE ALSO** ⟫
*pp.218–19* FOOD AND HEALTH
*pp.274–75* REGULATING THE BODY
*pp.448–49* REFERENCE: BIOLOGY

# Food and Health

**There is some truth in the saying "you are what you eat": the chemical substances in digested food are transported to cells in the bloodstream, where they are used in the vital processes of life. As such, too much or too little of certain types of food can have dire consequences for the health of the human body.**

Nutrition is the way that any living organism provides food for its cells. Animals—including humans—must eat food, then digest it until it becomes a complex soup of chemical nutrients, which are then distributed to body cells in the blood (see pp.216–17).

## Balanced and imbalanced diets

A balanced diet is one that contains the right combinations of nutrients in the appropriate quantities to maintain health (see opposite). Malnutrition results from any imbalance. People of developed countries enjoy a life where fat and sugar are relatively cheap and in plentiful supply. Overconsumption of foods containing such nutrients is the cause of some of the most common diseases that blight these people, including cardiovascular (heart) disease

**Early vitamin deficiency**
In the 18th century Scottish doctor James Lind was the first to recognize the importance of citrus fruits in preventing scurvy on board ships. The routine use of limes in naval nutrition prompted British sailors to be nicknamed "limeys."

and Type 2 diabetes. In developing countries, in contrast, where these expensive nutrients are in short supply, these diseases are rare.

## Role of nutrients

In 1827 English chemist William Prout classified nutrients in food into categories still recognized today: carbohydrates, fats, and proteins. In 1840 German chemist Justus Liebig found that these substances were made up of the chemical elements that occur in all other organic substances, notably carbon. Large amounts of carbohydrates and lipids (fats and oils) are required to provide energy, which is released in cells in the chemical process of respiration (see

> " Let **food** be your **medicine** and **medicine** be your **food**."
>
> HIPPOCRATES, ANCIENT GREEK PHYSICIAN *c.*460–370 BCE

pp.270–71), usually in the presence of oxygen. The nutrients are digested first in the gut and are then supplied to cells in their simplest forms: carbohydrates as sugars and lipids (fats) as fatty acids. Amino acids are also vital nutrients— they are assembled into proteins in cells, and used for cell growth and division (see pp.306–307), as well as other diverse cell functions.

Some nutrients are classified as "essential" because they cannot be made in the body and so must be obtained from food. Of the 20 required kinds of amino acids, about half need to come from the diet. Some fatty acids are essential, too, in regulating a range of functions, including blood pressure and blood clotting. Perhaps the most well known

of the essential nutrients are vitamins. Their key role was first recognized at the turn of the 20th century by English biochemist Frederick Gowland-Hopkins and Dutch physician Christiaan Eijkman. The importance of vitamins to a person's health is seen most vividly when these micronutrients are absent, resulting in a deficiency disease.

## Nutritional deficiencies

Perhaps the first explicit demonstration that a nutrient could reverse the effects of a skin disease occurred in 1747, when Scottish naval physician James

**Starchy foods**
Complex carbohydrates include starch and cellulose (fiber), found in foods such as potatoes and pasta. Starch is broken down to release sugar for energy. Fiber aids healthy movement of food through the digestive system, reducing the risk of bowel cancer.

**‹‹ BEFORE**

Early medicine recognized the importance of a balanced diet for health, while scientific experiments during the 1700s established its role in providing fuel for the body.

**DIET AND HEALTH**
Some of the earliest physicians noted **the effect of diet on health**. In ancient Greece a balanced diet was seen as an integral part of a healthy lifestyle. The Greek philosopher **Anaxagoras** was one of the first scholars to appreciate the fact that **nutrients in food** consumed were **absorbed into the body**.

**THE "FATHER" OF NUTRITION**
In 1770 French chemist **Antoine Lavoisier** established that food in the body **releases heat** in a way similar to combustion **270–71 ››**. Lavoisier devised a piece of apparatus called a **calorimeter**, in which he was able to **measure the amount of heat** given off by a small animal, such as a guinea pig. Lavoisier's work laid the foundation of many aspects of modern nutritional biology.

**ICE CALORIMETER**

**‹‹ SEE ALSO**
*pp.150–51* ORGANIC CHEMISTRY
*pp.216–17* DIGESTION

**DOCTOR (1716–94)**

## JAMES LIND

Lind was a Scottish physician who worked in the Royal Navy and was the pioneer of many practices that improved the health of sailors, including better ventilation and hygiene. His discovery that citrus fruits could alleviate the symptoms of scurvy (a major cause of death on long voyages, now known to be due to vitamin C deficiency) improved nutrition on board ship too. His systematic methods to prove this makes his study possibly the first ever clinical trial. He also suggested that a constant supply of fresh drinking water could be obtained on board ships by distilling seawater.

**Dairy products**
Milk, soft cheese, and yogurt are valuable sources of protein, and are also rich in calcium—needed for proper development of bone.

## A balanced diet

When a diet is described as "balanced" it contains all the appropriate nutrients from the different food groups in amounts that maintain good health for the body.

Lind (see left and below left) treated scurvy (which initially presents as spots on the skin, spongy gums, and bleeding from the mucous membranes) in sailors with citrus fruits, although the active ingredient—vitamin C—was not discovered until the 1930s. Vitamin C is needed to make collagen, which strengthens the skin, so when vitamin C levels are low the skin is weakened. Deficiencies of macronutrients cause

fundamental problems. In 1816 French physiologist François Magendie first demonstrated the effects of protein deficiency in dogs. In humans extreme protein deficiency causes the condition kwashiorkor (whose symptoms include swellings and enlarged liver; it can even lead to coma and death). The damaging effects of such deficiencies are most notable in children who require protein for proper growth and development.

## FOODS THAT BENEFIT HEALTH

Some specific foodstuffs are beneficial to health because of the abundance of one or more key nutrients or micronutrient—although other aspects of the diet and individual needs must be considered when judging their actual impact on a person's health.

| | |
|---|---|
| **Bananas** | Rich in antioxidants and potassium, which helps lower blood pressure. |
| **Black currants** | Contain a high vitamin C content, essential for healthy skin. |
| **Broccoli** | Rich in vitamin C and fiber and has potent cancer fighting properties. |
| **Cranberries** | Have cancer-reducing antioxidants and prevent urinary tract infections. |
| **Lentils and legumes** | Packed with vegetable protein and fiber, reducing the risk of bowel cancer. |
| **Salmon** | Contains omega-3 fatty acids, which prevents depression and heart disease. |
| **Sunflower oil** | Has a high vitamin E content and unsaturated fatty acids for a healthy heart. |
| **Yeast extract** | Rich in vitamin B, which helps maintain the nervous and immune systems. |
| **Yogurt** | Contains *Lactobacillus* bacteria, which promotes a healthy digestive tract. |

### Fruit and vegetables
All fruit and vegetables provide a certain amount of sugar and fiber, but some types are good sources of other essential nutrients, such as protein, vitamins, and antioxidants.

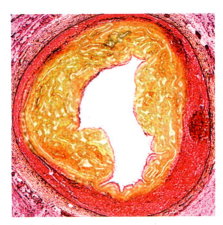

### Hardened artery
A high-fat diet can cause fatty substances in the blood to build up (in yellow) on the inside of artery walls; if such deposits form within the heart's own arteries, blood flow may be restricted and a heart attack results.

**AFTER »**

In the 20th century improved understanding of metabolism has shown the value of taking certain vitamin supplements.

### FUNCTIONS OF VITAMINS
Some vitamins **improve the functioning of enzymes** (which speed up reactions of metabolism). Vitamins of the B complex are needed in respiration (release of energy from food), which explains the lethargic effects of deficiency diseases such as beriberi. Some vitamins are now available as supplements: **folic acid**, for instance, is taken to reduce the incidence of spina bifida in unborn children.

**VITAMIN SUPPLEMENTS**

### ANTIOXIDANTS
Many fruits and vegetables, such as red peppers and tomatoes, contain antioxidant chemicals that prevent a type of metabolic reaction called **oxidation**, which is associated with **cancer**. Some antioxidants are vitamins.

### SEE ALSO »
pp.270–71 BREATHING AND RESPIRATION

### Meat and fish
Animal protein provides the essential amino acids that are needed by the body. Fish—in addition—is a good source of beneficial oils, from which essential fatty acids are derived.

### Fatty and sweet foods
Nutrients that are used in respiration to release energy. Excessive consumption of this food group, which includes butter, oils, hard cheeses, and cakes, can cause obesity, especially if the body is not exercised.

**« BEFORE**

Ancient Greeks distinguished nerves from blood vessels by dissection, but thought both had substances running through them.

**MAPPING THE NERVOUS SYSTEM**

In ancient Greece the philosopher **Alcmaeon** and physician **Herophilus** revealed how **nerves connected** to **brain, sense organs**, and **muscles**. The skills of Renaissance anatomists went further and Italian **Bartolomeo Eustachi's** *Tabulae Anatomicae* (published posthumously in 1714) gave the most **accurate description** of the **nerve map** then known.

**BODILY SPIRITS**

Crude vivisection experiments gave some clue about the roles of the nervous system. The Roman physician **Galen** found that **severing** the **spinal cord** of a living animal resulted in **paralysis**. Like many scholars of Greek antiquity, Galen **« 24–25** adopted the theory of the physician **Erasistratus** to explain this: muscles were animated by an "**animal spirit**," carried there in hollow nerves, and made in the brain from the "**vital spirit**" of the heart.

**« SEE ALSO**
*pp.72–73* THE HUMAN BODY REVEALED

## Nerve impulses

Nerve impulses pass down long nerve fibers (axons) of single neurones (nerve cells) and terminate at a synapse: a minute gap between the nerve fibers of adjacent neurones. The nerve impulses stimulate the release of neurotransmitter chemicals from vesicles (fluid-filled sacs) in the presynaptic bulb (swelling at the end of nerve fiber). These diffuse across the gap to stimulate a new nerve impulse in the next neurone fiber.

**Myelin sheath**
Layer of fatty material that insulates the electrical charges of the nerve fiber and makes the nerve impulse travel faster.

**Node of Ranvier**
Gaps between adjacent myelin sheaths where electrical activity is concentrated: nerve impulses effectively "jump" from node to node.

**Dendrite**
Short, branching nerve fiber that carries nerve impulses toward the cell body. Many dendrites enable this neurone to communicate across the synapses with neighboring neurones.

**Cell body**
Contains the neurone's nucleus and other organelles (cell structures).

**Axon**
The main nerve fiber that carries nerve impulses away from the cell body.

**Cranial nerves** connect the brain to sense organs

**Brain** is the site of sophisticated processing of information, such as conscious thought

**Spinal nerves** split into two roots: dorsal root (back) carries sensory nerve fibers into the spinal cord; ventral root (front) carries motor nerve fibers out of the spinal cord

**Spinal cord** is the site of simple processing of information, such as unconscious reflex actions

**Nerves** carry bundles of nerve fibers, some sensory and some motor

**How nerve impulses travel through the body**
Nerve impulses from sense organs travel through sensory nerves to the central nervous system (brain and spinal cord) and send impulses back out through motor nerves to muscles and glands.

# The Nervous System

**Animals can respond to a stimulus at a phenomenal speed, in just a fraction of a second. Understanding how this happens was one of the major challenges of biology, and it wasn't until the late 19th century that scientists were finally able to reveal the astonishing truth about how the nervous system works.**

All animals have a nervous system, made up of a network of branching fibers, which allow different parts of the body to communicate with each another at remarkable speed.

## An electrical system

In vertebrates the central nervous system consists of the brain and spinal cord. The peripheral nervous system is made up of nerves: sensory nerves carry nerve impulses from receptors in sense organs toward the central nervous system; motor nerves carry impulses from the central nervous system to effectors (so called because they put the impulses into effect) in muscles and some glands. Every nerve and muscle

can propagate an electrical charge with great speed, and establishing this was the key to understanding neurology—the science of the nervous system.

In the 17th century Dutch microscopist Jan Swammerdam had a breakthrough when he removed a frog's leg together with the nerves attached and made the muscle of the leg contract by pinching the nerves. In the late 18th century, Italian physician Luigi Galvani carried out an experiment where the exposed nerve of a dissected frog made contact with a metal scalpel and its legs went into convulsions—the legs were "galvanized" (so named after Galvani). He thought that the nerves were conducting electricity, but could not

prove it (see pp.162–63). A century later, German physician Emil du Bois-Reymond proved the theory and discovered tiny voltages pass along the lengths of nerves. In the 20th century the speed of these impulses was realized: they travel along nerve fibers insulated

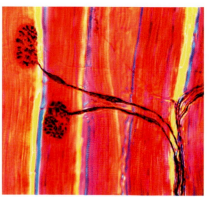

**Light micrograph of neuromuscular junctions**
A neuromuscular junction is a site of communication between nerve fibers and muscle fibers. A nerve impulse stimulates the release of a chemical neurotransmitter at the junction, which causes contraction of the muscle fiber.

> **"Problems** that **appear small** are **large problems** that are not **understood."**
>
> SANTIAGO RAMÓN Y CAJAL, SPANISH PHYSICIAN, 1897

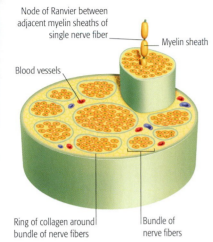

**Inside a nerve**
A single nerve is made up of a bundle of nerve fibers, which are held together with a layer of material called collagen. Larger nerves may carry blood vessels, too.

*Node of Ranvier between adjacent myelin sheaths of single nerve fiber*

*Myelin sheath*

*Blood vessels*

*Ring of collagen around bundle of nerve fibers*

*Bundle of nerve fibers*

by a myelin sheath (see opposite) at 230–240 feet (70–120 m) per second, and along uninsulated fibers at 7 feet (2 m) per second. In reality, reaction times are delayed by complex processing in the brain, which is itself influenced by age, health, reflexes, and practice.

## Nerve cells and the gaps between
In 1860 German scientist Otto Deiters identified extra-long fibers in the nervous system (axons) and shorter, branching fibers (dendrites). By the end of the century Spanish microscopist Santiago Ramón y Cajal proposed that the entire nervous system was cellular and that nerve fibers were not fused together. This proposition marked the origins of a new Neuronal Theory, according to which there is a narrow gap, called a synapse, between adjacent nerve cells and between nerve cells and

muscles. In the 20th century German pharmacologist Otto Loewi identified the first neurotransmitter—the chemical released at the ends of nerve fibers that triggers an impulse on either side of the gap. This kind of chemical signaling is now known to dominate the workings of synapses.

**PHYSICIAN (1737–98)**

## LUIGI GALVANI

Born in Italy, Galvani trained as a physician at the University of Bologna, but is most well known for his theories about animal electricity, and his accidental discovery that electricity could make the leg of a dead frog twitch (which inspired Mary Shelley's *Frankenstein*). He was made Public Lecturer in Anatomy at the University of Bologna in 1762 and became President of the university in 1772.

## AFTER »

Research on synapses (gaps between nerve cells) has led to recent medical advances and further understanding how drugs can affect the nervous system.

### PHARMACOLOGY
Advances in pharmacology have shown how **certain chemicals** can interfere with the function of synapses by **amplifying** or **blocking** effects of **neurotransmitters** (chemicals that relay impulses between nerve cells).

### NERVOUS DRUGS
In 1957 Swedish scientist **Arvid Carlsson** identified dopamine as a neurotransmitter in the brain, which led to a **dopamine precursor** being used in the treatment of **Parkinson's disease**.

**SEE ALSO »**
pp.222–23 THE BRAIN
pp.224–25 MUSCLES, BONES, AND MOVEMENT
pp.274–75 REGULATING THE BODY
pp.312–13 THE DEVELOPMENT OF MEDICINES

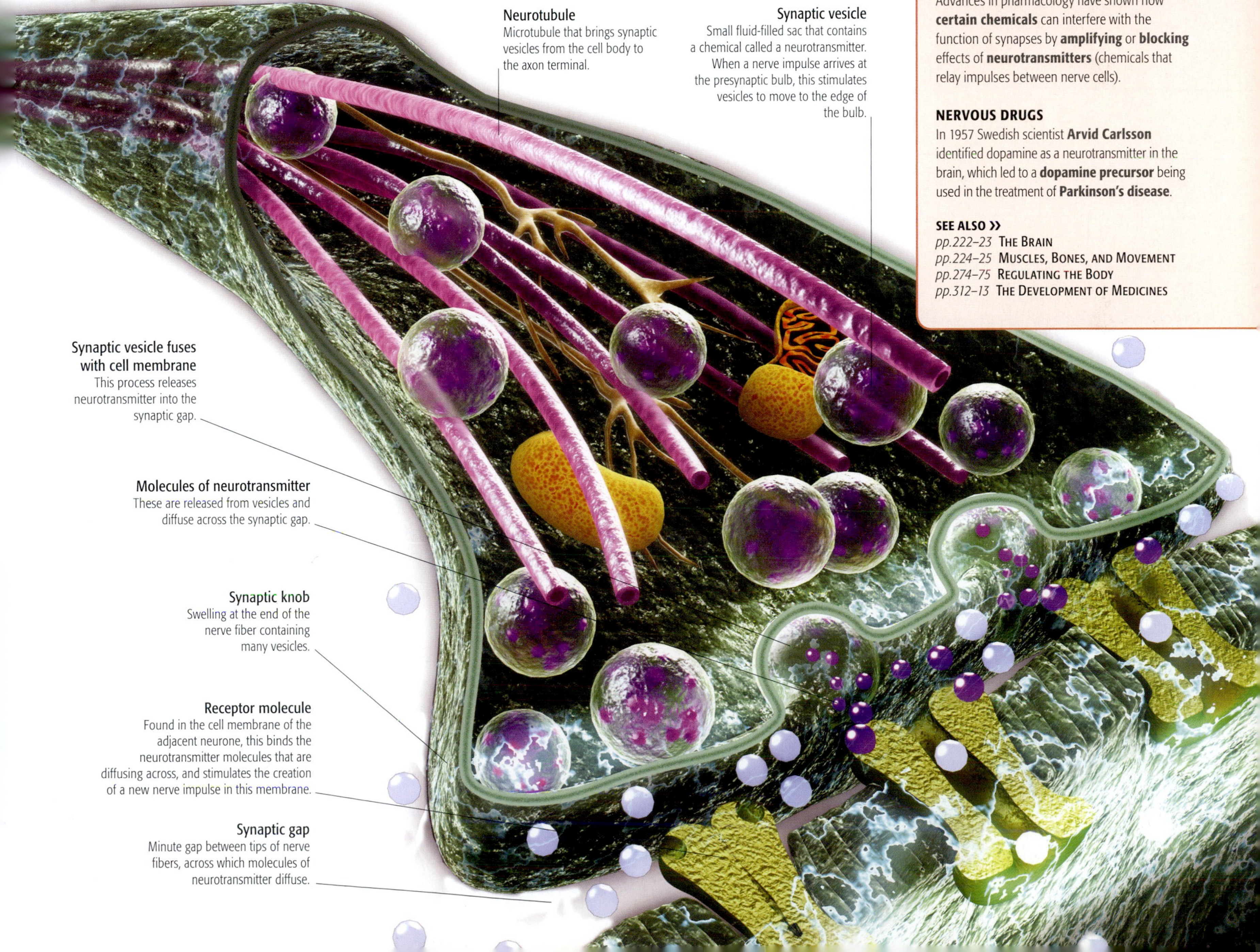

**Neurotubule**
Microtubule that brings synaptic vesicles from the cell body to the axon terminal.

**Synaptic vesicle**
Small fluid-filled sac that contains a chemical called a neurotransmitter. When a nerve impulse arrives at the presynaptic bulb, this stimulates vesicles to move to the edge of the bulb.

**Synaptic vesicle fuses with cell membrane**
This process releases neurotransmitter into the synaptic gap.

**Molecules of neurotransmitter**
These are released from vesicles and diffuse across the synaptic gap.

**Synaptic knob**
Swelling at the end of the nerve fiber containing many vesicles.

**Receptor molecule**
Found in the cell membrane of the adjacent neurone, this binds the neurotransmitter molecules that are diffusing across, and stimulates the creation of a new nerve impulse in this membrane.

**Synaptic gap**
Minute gap between tips of nerve fibers, across which molecules of neurotransmitter diffuse.

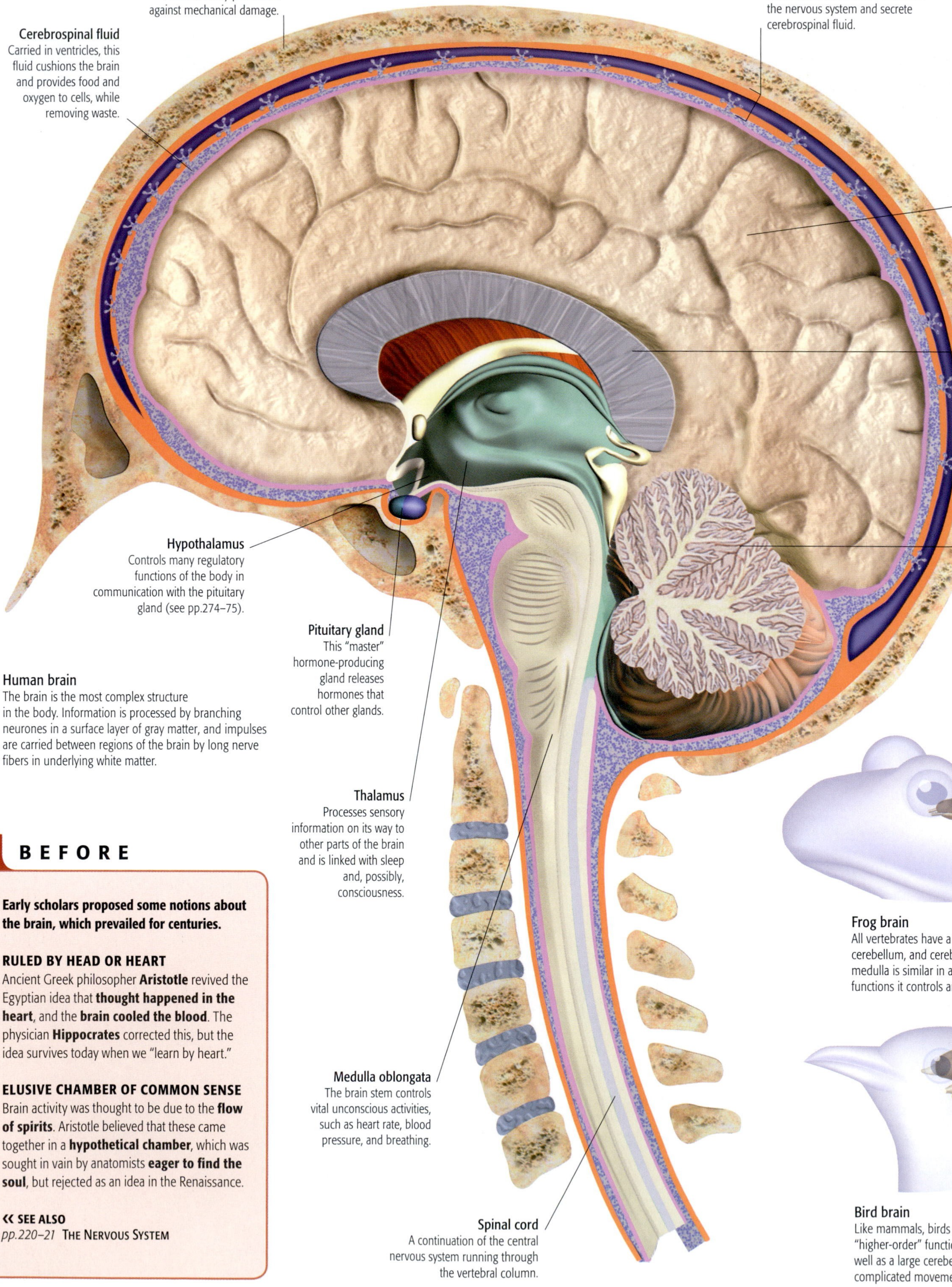

**Cranium (skull)**
Provides bony protection against mechanical damage.

**Cerebrospinal fluid**
Carried in ventricles, this fluid cushions the brain and provides food and oxygen to cells, while removing waste.

**Meninges**
These three membranes surround the nervous system and secrete cerebrospinal fluid.

**Cerebrum**
The cerebrum consists of two main layers. The cerebral cortex, often known as "gray matter," is highly folded to maximize the area for better processing of information. Beneath the cerebral cortex is the paler "white matter" of the cerebrum's interior, composed mainly of nerve fibers.

**Corpus callosum**
Connects the left and right hemispheres (sides) of the brain, controlling passage of nerve impulses between the two.

**Hypothalamus**
Controls many regulatory functions of the body in communication with the pituitary gland (see pp.274–75).

**Pituitary gland**
This "master" hormone-producing gland releases hormones that control other glands.

**Cerebellum**
Maintains balance and coordinates movements of large muscles during voluntary motions, such as walking. Information is processed by nerve impulses in the cerebellar cortex.

**Human brain**
The brain is the most complex structure in the body. Information is processed by branching neurones in a surface layer of gray matter, and impulses are carried between regions of the brain by long nerve fibers in underlying white matter.

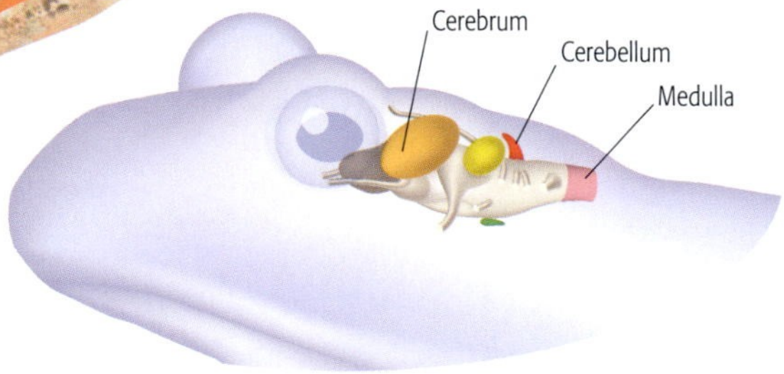

**Thalamus**
Processes sensory information on its way to other parts of the brain and is linked with sleep and, possibly, consciousness.

## BEFORE

**Early scholars proposed some notions about the brain, which prevailed for centuries.**

**RULED BY HEAD OR HEART**
Ancient Greek philosopher **Aristotle** revived the Egyptian idea that **thought happened in the heart**, and the **brain cooled the blood**. The physician **Hippocrates** corrected this, but the idea survives today when we "learn by heart."

**ELUSIVE CHAMBER OF COMMON SENSE**
Brain activity was thought to be due to the **flow of spirits**. Aristotle believed that these came together in a **hypothetical chamber**, which was sought in vain by anatomists **eager to find the soul**, but rejected as an idea in the Renaissance.

« SEE ALSO
pp.220–21 THE NERVOUS SYSTEM

**Medulla oblongata**
The brain stem controls vital unconscious activities, such as heart rate, blood pressure, and breathing.

**Spinal cord**
A continuation of the central nervous system running through the vertebral column.

Cerebrum
Cerebellum
Medulla

**Frog brain**
All vertebrates have a brain divided into medulla, cerebellum, and cerebrum. The relative size of the medulla is similar in all vertebrates, since the vital functions it controls are the same in all groups.

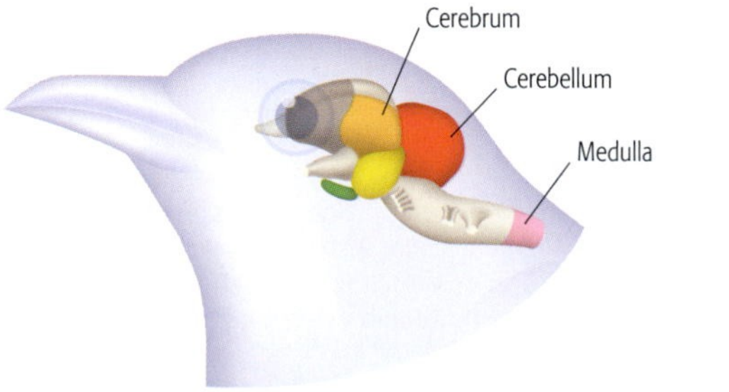

Cerebrum
Cerebellum
Medulla

**Bird brain**
Like mammals, birds have a large cerebrum to perform "higher-order" functions, such as raising offspring, as well as a large cerebellum for controlling the complicated movements of flight.

# The Brain

**Early ideas about the brain varied from the ancient Egyptians, who thought it was little more than skull stuffing, to anatomists of the Middle Ages, who expected to find the soul inside it. By the start of the 1800s its structure had been documented in detail, but its functions were only just then beginning to be understood.**

The brain is the most sophisticated part of the nervous system. Its complex network of neurones (nerve cells) engages in a frenzy of chemical and electrical activity, which controls a wide range of body functions, from steadying the heart beat to eliciting feelings of happiness or hunger.

## Understanding the brain

By 1700 the main parts of the brain were established: the medulla where it connects to the spinal cord, and, above this, the cerebellum and cerebrum (see left). Unlike the heart, the dead brain

**10 PERCENT** of brain cells are neurones (nerve cells) —the rest are glial cells. The function of glial cells is to supply nutrients to neurones, insulate them, or defend them from infection.

gave little clue as to its function, so scholars drew conclusions by observing disability instead. In 1710 French physician François Pourfour du Petit observed that soldiers wounded on one side of the head retained mobility on the opposite side of the body.

In the 19th century French physician Paul Broca made a classic study on an individual called Tan—so called since this was the only word he could say. Tan had syphilis and, after his death in 1861, Broca found that it had damaged a part of his brain on the left side, so he deduced that this was the part of the brain that was involved in speech. At the end of the century understanding of brain function

advanced further as techniques allowed electrical brain stimulation in living animals. By 1909, German neuroscientist Korbinian Brodmann defined 52 areas of the cerebral cortex in terms of function, many of which are still recognized today (see below).

### Areas of the cerebral cortex
The sensory regions of the cerebral cortex receive impulses, association regions process them, and motor regions send them to muscles.

**Premotor cortex**
Creates the intention to move.

**Motor cortex**
Controls coordinated muscle movements.

**Somatosensory cortex**
Receives and analyzes nerve impulses from touch receptors.

**Prefrontal cortex**
Involved in determining personality and thought processes.

**Sensory association cortex**
Processes sensory information.

**Broca's area**
Associated with the production of language.

**Visual association cortex**
Integrates visual data with memories and other senses.

**Primary auditory cortex**
Receives and analyzes nerve impulses from the ears.

**Auditory association cortex**
Integrates auditory data with memories and other senses.

**Wernicke's area**
Associated with the understanding of language.

**Primary visual cortex**
Receives and analyzes nerve impulses from the eyes.

## The functions of the brain

The cerebrum is involved in the so-called higher functions of the brain: decision making, problem solving, emotion, and language. Particularly big in humans, it receives nerve impulses from sense organs into its sensory areas

and transmits them to muscles from its motor areas. The regions in between (association areas) process the information. Behind the cerebrum is the smaller, but equally convoluted, cerebellum. This takes over as a kind of "automatic pilot" for complicated

movements that become routine in life, such as walking.

The medulla, or brain stem, contains nerves that form the autonomic nervous system, which controls internal functions, such as heart rate, blood pressure, and breathing. For

example, the medulla detects when blood has too much carbon dioxide, and speeds up heart and breathing rate to get rid of it. Some of its regulatory functions are controlled by neurotransmitter chemicals produced at the ends of the autonomic nerves. The medulla's job remains the same with different species, and includes no conscious decision making, and so its relative size varies little in fishes, birds, and mammals.

Modern techniques in neuroscience have advanced understanding of the brain.

### DIAGNOSIS WITHOUT DISSECTION
Techniques such as positron emission topography **(PET) 404–05 »** and electroencephalography **(EEG)** allow the brain to be studied without opening it up. PET is a neuroimaging technique used **to diagnose brain disorders, dementia,** and **tumors**. EEG measures electrical activity using electrodes on the scalp. First used by the Russian physiologist **Vladimir Neminsky** on a dog in 1912, EEG **allows diagnosis of neurological** (nerves and nerve disorders) **disorders**, such as **epilepsy**, but is also used in **psychology** and **cognitive science** (study of the mental processes of thinking and reasoning).

**3 TIMES** as much radiation as normal background radiation is required for a PET scan.

### BRAIN–MACHINE INTERFACES
In the 1970s scientists set up direct **electrical connections** between the brain and an external device, which can be used to **restore hearing** in the deaf and **vision in the blind**.

#### SEE ALSO »
pp.272–73 THE FIVE SENSES
pp.274–75 REGULATING THE BODY
pp.414–15 BODY IMAGING
pp.440–53 REFERENCE: BIOLOGY

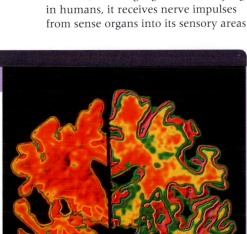

### BREAKTHROUGH
## ALZHEIMER'S DISEASE

In 1901 German neurologist Alois Alzheimer described the first case of the type of dementia that bears his name. The brain of the Alzheimer's patient (see left) is considerably shrunken, due to the deterioration and death of nerve cells. Today we know that the brains of people with Alzheimer's produce fewer neurotransmitter chemicals, but that some drugs can slow the natural breakdown of this limited supply.

Alzheimer's patient's brain | Normal aged brain

### PHYSICIAN (1824–80)
## PAUL PIERRE BROCA

Broca, the French physician and anthropologist, is best known for identifying the region of the brain involved in speech production. In Paris he founded the Anthropological Society, the Revue d' Anthropologie, and the School of Anthropology. He devised instruments for measuring skull dimensions and believed that brain size reflected intelligence; he also claimed that men were more intelligent than women. A great supporter of Darwin's theory of evolution, his own radical ideas often brought him into conflict with authorities.

## ‹‹ BEFORE

Bones and muscle inspired the earliest anatomists to propose theories, which were subsequently debunked.

### BONES FROM EARTH
Early scientists drew conclusions about the **development of bone** from its appearance. In ancient Rome physician **Galen** thought that its pale color meant that it was **formed from semen**, an idea that was adopted by physicians of the Renaissance. In the Middle Ages the Persian physician **Avicenna ‹‹ 24–25** asserted that **bones were composed of earth** because they shared its cold, dry quality.

### BALLOONIST THEORY
Galen also thought that **muscles contracted** as **fluid flowed** into them **from nerves**, an idea that became known as the **Balloonist Theory**, which remained popular until the Renaissance.

**‹‹ SEE ALSO**
*pp.72–73* The Human Body Revealed
*pp.220–21* The Nervous System
*pp.222–23* The Brain

# Muscles, Bones, and Movement

**Early Renaissance anatomists laid the foundations for today's understanding of how muscles and bones work together around a system of joints. This enables the living body to be a moving form.**

Early anatomists illustrated the human body with varying degrees of accuracy, but it was Flemish-born Andreas Vesalius, in the 16th century, who corrected anatomical errors passed down from antiquity. Renaissance anatomists recognized that bones varied in hardness and appreciated their role in protecting critical organs, such as the brain. However, they understood little about how bones in the skeleton are formed or how the contraction of muscles around a sophisticated system of joints make movement of the body possible.

## The skeleton
The skeleton supports all other parts of the body and protects the organs. In the animal kingdom there are different types of skeleton, including internal endoskeleton and external exoskeleton (see opposite). In the vertebrate skeleton the two main types of tissue

**Thigh muscle**
Contraction extends lower leg; contraction of muscle behind thigh flexes knee.

**Tendon**
Tough fiber that connects muscle to bone.

**Artery**
Blood vessel that brings oxygenated blood to muscle, bone, and joint.

**Vein**
Blood vessel that takes deoxygenated blood back to heart and lungs.

**Patella (kneecap)**
Protective bone of the knee joint.

**Ligament**
Fibrous tissue encapsulating the synovial joint, and connecting bones.

**The knee**
The knee can flex because muscles pull on the bones around a synovial joint— a capsule containing fluid. Tendons connect muscle to bones, and internal and external ligaments connect bone to bone.

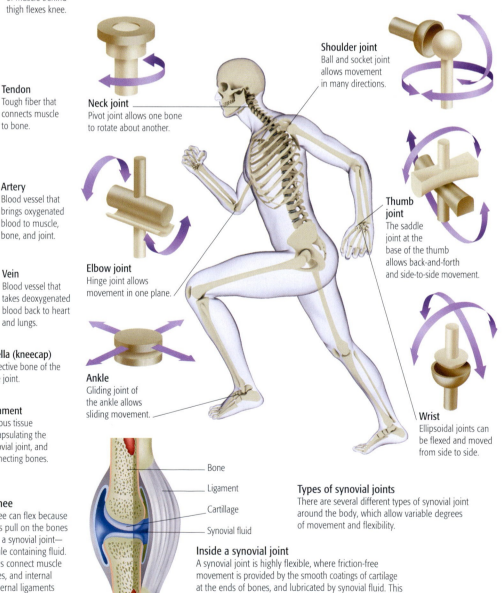

**Neck joint**
Pivot joint allows one bone to rotate about another.

**Shoulder joint**
Ball and socket joint allows movement in many directions.

**Elbow joint**
Hinge joint allows movement in one plane.

**Thumb joint**
The saddle joint at the base of the thumb allows back-and-forth and side-to-side movement.

**Ankle**
Gliding joint of the ankle allows sliding movement.

**Wrist**
Ellipsoidal joints can be flexed and moved from side to side.

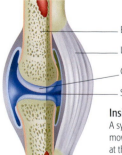

Bone
Ligament
Cartilage
Synovial fluid

**Types of synovial joints**
There are several different types of synovial joint around the body, which allow variable degrees of movement and flexibility.

**Inside a synovial joint**
A synovial joint is highly flexible, where friction-free movement is provided by the smooth coatings of cartilage at the ends of bones, and lubricated by synovial fluid. This also nourishes the cartilage and protects it from infection.

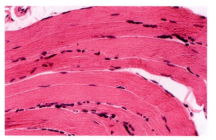

**Striated muscle**
Contraction of striated muscle moves one bone with respect to another. The body has voluntary control over this—stimulation of the nervous system causes nerve impulses to travel via motor nerves to muscles.

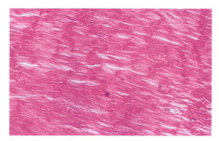

**Smooth muscle**
Smooth muscle occurs in organs, such as the gut, and in blood vessels; contraction of this muscle is important in internal processes. It is stimulated by a part of the nervous system over which we have no conscious control.

that make up the skeleton are bone and cartilage. Bone contains important minerals for the body and is harder than cartilage because it contains calcium phosphate. A joint occurs where bone meets bone. Some joints, such as those in the skull, are fixed, but others allow movement of bones—synovial joints (see opposite). Some, such as the elbow and knee joints, can only move in one direction, acting like hinges. Others allow for more rotation, such as the ball-and-socket joints of the shoulder and hip.

**Muscles**
Muscles are fibrous tissues, that twitch when stimulated, pulling their ends together to create tension. Muscle

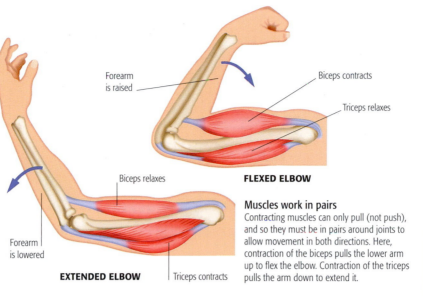

Forearm is raised

Biceps contracts

Triceps relaxes

**FLEXED ELBOW**

Biceps relaxes

Forearm is lowered

**EXTENDED ELBOW**

Triceps contracts

**Muscles work in pairs**
Contracting muscles can only pull (not push), and so they must be in pairs around joints to allow movement in both directions. Here, contraction of the biceps pulls the lower arm up to flex the elbow. Contraction of the triceps pulls the arm down to extend it.

contracts when it receives an impulse from the nervous system via nerves (see pp.220–21). This contraction is caused by the interaction of two muscle proteins: myosin, which was first isolated in 1864 by German physiologist Wilhelm Kühne, and actin, which was found by British physiologist William Halliburton in 1887. It was not until 1954 that two more British scientists, Andrew Huxley and Jean Hanson, showed how individual filaments of the two proteins slide over one another to contract muscle with energy provided by a substance called ATP (see right).

Voluntary contraction of muscles connected to the skeleton can move one bone in relation to another bone (see above). This type of muscle is sometimes found in pairs around joints, which allows movement in both directions (see above). Other muscles are found in the walls of organs, and their involuntary contraction is important in continuous vital functions, such as pulsation of the gut (see pp.216–17).

Thin actin filaments

Thick myosin filament

**RELAXED MUSCLE**    **CONTRACTED MUSCLE**

**How muscle contracts**
In relaxed muscle, thick myosin filaments overlap slightly with thinner actin filaments. When the muscle is stimulated to contract, an energy-rich substance called ATP causes the myosin to crawl along the actin filaments, shortening the muscle.

"**Muscles** are in a most **intimate** and **peculiar** sense the **organs of the will**."
GRANVILLE STANLEY HALL, AMERICAN PSYCHOLOGIST, 1907

**AFTER**

Study of the chemistry of muscles has helped to understand neuromuscular diseases, offering hope for new treatments.

**MUSCULAR DYSTROPHY**
In 1987 American geneticist **Louis Kunkel** identified a protein called **dystrophin**, which **stabilizes muscle fibers**. Its **deficiency** causes **Duchenne muscular dystrophy**. Current research into possible treatment is focused on the use of **stem cells 390-91 》**.

**FRAGILE BONES**
**Osteoporosis** is a bone disease where the thinning of bones increases risk of fracture. The American biologist **Fuller Albright** found that the disease is most common in **postmenopausal women** after their estrogen levels fall. This led to the development of **treatment with hormones**.

**SEE ALSO 》**
pp.274–75 REGULATING THE BODY

Cranium encases the brain

Ribs form a cage to surround and protect heart and lungs

Vertebral column encases spinal cord

**Endoskeleton**
The internal endoskeleton of a mammal, such as this dog, is composed of bones articulated at joints. As well as providing a hard framework, bone also acts as a reservoir of calcium for the body and contains marrow for production of blood cells.

Limbs consist of upper parts (single limb bones) and lower parts (two parallel limb bones)

**Exoskeleton**
Many invertebrates have a hard casing around the body, with flexible material at joints. An exoskeleton restricts growth so it must be periodically molted.

**Hydroskeleton**
Many animals, such as this starfish, are supported by an internal skeleton of water. This is effective in invertebrate animals in aquatic environments.

**BIOLOGIST (1729–99)**

# LAZARRO SPALLANZANI

Born in Italy, biologist Lazarro Spallanzani spent much of his career as Director of Natural History at the University of Pavia. He refuted the prevailing idea that life came spontaneously from inanimate matter rather than reproduction and used frogs to prove that eggs only develop into embryos after contact with semen. He was the first scientist to perform artificial insemination, on a dog. He also experimented with chemical digestion in the stomach (see p.217).

## Conception to embryo

About once a month an egg is released from the ovary and funneled down through the fallopian tube, where, if sperm are present, fertilization will take place. Over the next few days, the fertilized egg divides into an embryo, initially just a ball of cells, which becomes implanted into the blood-rich lining of the wall of the uterus (womb).

**2 Earliest embryo stage**
Enclosed within a coat, the embryo forms when the zygote undergoes its initial cell cleavage (division) in the first 24 to 36 hours. The earliest divisions are not accompanied by cell growth, so individual cells shrink to the size of normal body cells.

**Fallopian tube lining**
The lining has a coating of hairlike cilia to waft the embryo down the tube.

**First cleavage (division)**

**Nucleus of embryo cell**

**Egg cell**
The egg cell contains 23 maternal chromosomes.

**Fallopian tube**

**Follicle cell**
A cell that helps nourish the egg when it is in the ovary.

**Sperm tail**
Part of the sperm that lashes to propel sperm toward the egg.

**Sperm head**
The head of the sperm contains 23 paternal chromosomes.

**Acrosome**
Tiny bag of chemicals in the sperm head that digests the egg coat to penetrate it.

**1 Fertilization**
Fertilization takes place in the upper third of the fallopian tube, and occurs when the sperm cell penetrates the much larger egg to form a zygote.

# Human Reproduction

**All human life starts as a fertilized egg. Unraveling the events that produce this egg and its subsequent development relied on the microscope and bold experiments. Today scientists are not only able to monitor the health of a baby in the uterus, but can even manipulate the entire reproductive process.**

« BEFORE

Before fertilization was observed through a microscope, many scientists doubted the roles of eggs and sperm in reproduction.

**LIFE FROM EGGS**
**William Harvey**, discoverer of blood circulation and physician to King Charles I of England, studied the reproductive systems of deer in the royal parks, proposing in 1651 that **young animals** could **only develop from eggs**, refuting the theory of spontaneous generation (see above).

**MYSTERY OF SPERM**
In 1678 Dutch scientists **Antony van Leeuwenhoek** and **Nicolas Hartsoeker** made the first **microscopic observations** of **sperm**. Other microscopists of the period reported seeing **tiny human beings** (homunculi) inside **sperm heads** (since condemned as hoaxes), which fueled the idea of **preformation**—that embryos are preformed in sex organs.

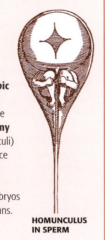

**HOMUNCULUS IN SPERM**

« SEE ALSO
*pp.72–73* THE HUMAN BODY REVEALED

Sexual reproduction involves the fusion of the gametes (sperm and egg) in a process of fertilization to make a single cell called a zygote. Today we know how hormones control the monthly release of human eggs. But until the early 1800s, scientists were not even convinced that sperm was needed to make a baby.

### Identifying the sperm and egg

In 1835 celebrated English zoologist Richard Owen classified the tadpolelike microbes in semen as parasites. In reality, they were sperm. A decade earlier, in France, physician Jean-Louis Prevost and chemist Jean-Baptiste Dumas had shown that sperm were in the testes of many different kinds of animals. Later, in 1852, English physician Henry Nelson claimed to see fertilization down his microscope—in a worm called *Ascaris*—and others saw sperm penetrating eggs of sea urchins and starfishes. But all these animals naturally fertilized outside their bodies. Finding fertilization inside the body of a human being proved to be much more elusive.

Sperm are produced from the walls of tiny tubules in the testes in a process that involves meiosis (see pp.306–07)—a type of cell division that halves the genetic content of the cells to prevent genetic doubling at fertilization. Meiosis also happens in ovaries to produce eggs. In 1826 a German embryologist Carl Ernst von Baer found eggs inside the ovaries of a dog. Swiss physicist Pierre Prévost and French chemist Jean Dumas studied the ovaries of mammals too. They recorded the growth of swellings in ovaries—the follicles that contain eggs —and noted the presence of embryos in the fallopian tubes after mating, suggesting fertilization happened there.

### Understanding female hormones

Two of the hormones controlling reproduction (estrogen and progesterone) are produced in the ovaries themselves. In the late 1920s a group of American scientists found that

### The menstrual cycle

The hormone FSH stimulates follicle growth (swelling in the ovary). Estrogen is released from the follicle and stimulates the uterus's wall lining to thicken and fill with blood. The hormone LH stimulates ovulation (the release of an egg from the swollen follicle). Without fertilization the empty follicle (corpus luteum) shrinks, progesterone levels drop, and the uterus lining is shed in menstruation.

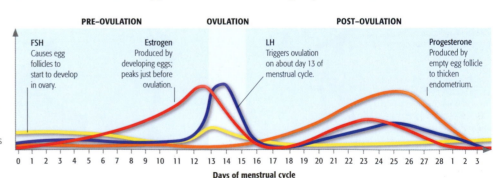

| PRE-OVULATION | | OVULATION | POST-OVULATION |

**FSH** Causes egg follicles to start to develop in ovary.

**Estrogen** Produced by developing eggs; peaks just before ovulation.

**LH** Triggers ovulation on about day 13 of menstrual cycle.

**Progesterone** Produced by empty egg follicle to thicken endometrium.

0 1 2 3 4 5 6 7 8 9 10 11 12 13 14 15 16 17 18 19 20 21 22 23 24 25 26 27 28 1 2 3

**Days of menstrual cycle**

**3 Morula stage**
The term "morula" is used when the embryo is still a ball of cells with a protective coat. It leaves the fallopian tube to enter the uterus three to four days after fertilization.

**Morula**
A compact ball of 16 to 32 cells.

**Coat around embryo**

**Fimbria**
A fringe of tissue at the opening of the fallopian tube that catches and guides the released eggs from the ovary.

**Ovary**

**Ovarian ligament**

**Blastocoele**
A fluid-filled cavity.

**Blastocyst**

**Inner cell mass**
Mass of cells that will develop into the embryonic body.

**Trophoblast**
An outer layer of cells that branches into the uterus lining, forming the placenta.

**Endometrium**
Blood-filled uterus lining.

**4 Blastocyst stage**
Six days after fertilization, this structure has a fluid-filled hollow—a liquid formed from the embryonic cells. The inner mass of cells becomes the maturing embryo; the outer layer of cells will form part of the placenta as it implants in the uterus lining, two days later.

**Maternal blood vessels**

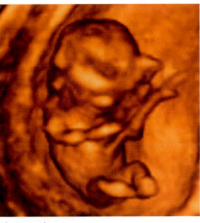

**Ultrasound scan**
Twelve weeks into pregnancy, the fetus is revealed by a technique that uses high-frequency sound waves, which reflect from tissue to give a three-dimensional image.

Mammalian in vitro (literally "in glass") fertilization (IVF)—in a hamster—was first achieved by Japanese clone specialists five years later. Then, in the 1960s, British physiologist Robert Edwards and physician Patrick Steptoe achieved IVF with human sex cells, heralding new hope for thousands of infertile couples (see pp.388–89).

## Stem cells

By the mid 1800s, scientists had discovered that embryos develop from shapeless balls of cells. In 1892 German biologist Hans Driesch found that each cell in the ball had all the information needed to form any body part. Today, we call these stem cells.

**Understanding the role of hormones in reproduction led to developments, which are considered routine today.**

**PREGNANCY TEST KITS**
In the 1930s American scientist **Georgeanna Jones** discovered that a **hormone** called human chorionic gonadotropin (**hCG**) was produced by the **placenta**. Early pregnancy tests detected hCG in urine. Later this method was replaced by a more direct analysis of hCG concentration using an **antibody-binding reaction**, which became the basis of the home pregnancy test kit in the 1970s.

**BIRTH CONTROL**
**Contraceptive methods** involving suppositories or barriers can be traced back to ancient Egypt, but the development of a **contraceptive pill** only became possible with 20th-century **understanding** of **hormonal control** of the menstrual cycle. "**The Pill**" contains hormones that **suppress ovulation** (the release of an egg from the ovary), reducing the chance of fertilization.

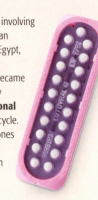

**CONTRACEPTIVE PILLS**

**SEE ALSO »**
pp.348–49  THE GENETIC CODE
pp.386–87  GENETIC TECHNOLOGY
pp.388–89  IN VITRO FERTILIZATION (IVF)
pp.390–91  HOW CLONING WORKS

hormones from the pituitary gland, at the base of the brain (see p.275), also caused changes in the ovaries: follicle-stimulating hormone (FSH) triggers growth of the follicle and luteinizing hormone (LH) causes it to burst at ovulation (see opposite). Today, drugs that mimic FSH are important in treating infertility.

## On the path to IVF

In 1951 Australian biologist Colin Austin and Chinese scientist Min Chueh Chang discovered that sperm cells need to be "activated" by chemicals before they can fertilize an egg—a process known as capacitation. Austin found how capacitated sperm penetrated the egg coat by releasing digestive enzymes from their heads.

# 20 MILLION
**The average number of sperm cells in a milliliter of human semen (fluid produced by the testes).**

**Placenta**
An organ made up of tissues of the fetus and the mother, this controls uptake of food and oxygen from mother to fetus.

Space filled with maternal blood bathes the chorionic villi, bringing the mother's source of food and oxygen close to the fetus's blood circulation.

Chorionic villus contains blood capillaries of the fetus's umbilical blood circulation.

**Endometrium**
The inner blood-filled lining of the uterus.

**Umbilical cord**
The structure that carries the oxygenated blood to the fetus from the mother and deoxygenated blood from the fetus back to the mother.

**Amniotic sac**
The fluid in which the fetus is suspended is contained in this sac, providing physical support.

**Cervix**
The neck of the womb.

**EXCHANGE ACROSS PLACENTA**

**Myometrium**
An outer muscular layer of the uterus that allows birth contractions.

**Maternal blood vessels**
Vessels that carry oxygenated blood from the mother's heart and lungs to the uterus; also transports deoxygenated blood away from the uterus.

**Fetus in the uterus**
Food and oxygen pass from mother to fetus along a vein in the umbilical cord, and waste passes back along an artery. This circulation stops at birth, when the baby's lungs and gut begin to function.

« BEFORE

Early surgeons attempted to address problems of blood loss and pain, but infection was a major issue.

**HERBAL PAIN RELIEF**

Early attempts to numb sensation used **plants with chemicals** that affect the nervous system —alkaloids. Poppies contain alkaloids called **opiates** and were used as anesthetics in Egypt and Asia. Islamic physicians pioneered the use of **inhalant anesthesia** with drug-soaked sponges, using a combination of different opiates.

OPIUM POPPY

**BLOOD TRANSFUSIONS**

Early **blood transfusions** were seen as a way of boosting "vitality." In 1665 an English physician Richard Lower used it to restore the health of dogs after substantial blood loss. Jean Denys, a French physician, tried to transfuse **sheep blood into humans**: animal blood was thought to be "impure of passion or vice." But a fatal incident led to its widespread condemnation.

« SEE ALSO
pp.24–25 Early Medicine and Surgery
pp.70–71 Renaissance Medicine and Surgery

# Safer Surgery

**Surgery is a form of medicine concerned with treating injuries and diseases and curing life-threatening conditions. The history of surgery is a triumph of medical science over pain, blood loss, and infection. Some of the biggest surgical breakthroughs took place during the 19th century.**

Surgery is highly invasive. It manipulates the body, exposing the patient to severe pain and blood loss in its aim to cure disease. And, for those who survive the ordeal, postoperative infection is an ever-present threat.

### Reducing infection

At the start of the 19th century many physicians subscribed to the theory that disease was caused by "bad air"—the miasma theory. Florence Nightingale advocated a regimen of fresh air in her hospital wards. Some physicians advised improving hygiene, but they were largely ignored. Then in 1864 the French microbiologist Louis Pasteur (see p.244) demonstrated that organisms in the air caused nutrient liquids to ferment, and German physician Robert Koch found bacteria in animals

suffering from a disease called anthrax. This gave new focus to the control of infection (see pp.242–43). If bacteria— and other "germs" (pathogens)—could spread from person to person they could just as easily invade an open wound. But germs could be suppressed or killed by chemicals, and could not develop in a germ-free— sterile—environment.

Foremost in the campaign to eliminate microbes using chemicals was Joseph Lister (see opposite). He treated both wounds and surgical instruments with carbolic acid (also known as phenol—a chemical that had been used to deodorize sewage), establishing an antiseptic technique that significantly reduced infection.

### Numbing the pain

Chemicals that deaden pain are called analgesics; those that eliminate all sensation are called anesthetics, a term coined by American physician Oliver Wendell Holmes in 1846. The earliest anesthetics were herbal and usually used on small areas of the body (locally). General (whole-body) anesthesia was too risky, since standardizing doses of plant extracts was difficult. The first recorded use of general anesthesia using plant chemicals was by Japanese surgeon Seishu Hanaoka in 1804. Soon after, the German pharmacist Friedrich Sertürner isolated morphine, the active ingredient of the opium poppy. Because of its purity, doses could be consistent.

**Lister's carbolic spray**
This instrument was developed by Joseph Lister to deliver a fine spray of the antiseptic solution over the wound during surgery, drastically reducing the incidence of infection.

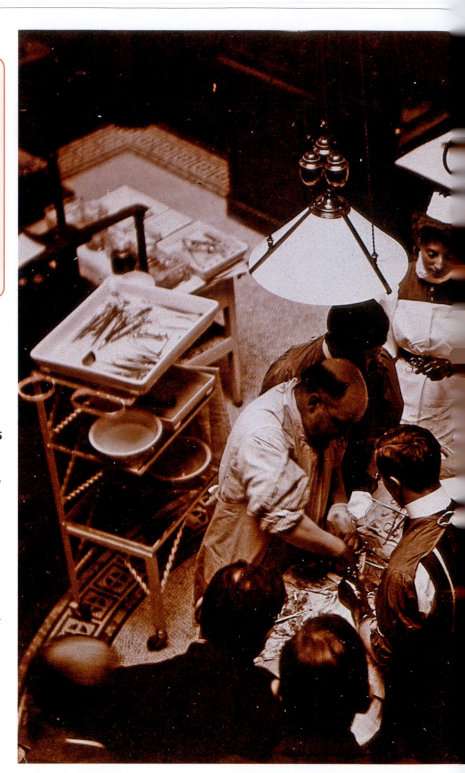

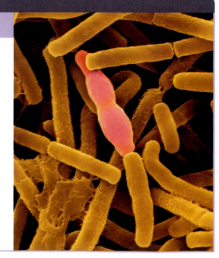

BREAKTHROUGH

### GERM THEORY OF DISEASE

Louis Pasteur demonstrated in 1864 that microorganisms did not arise spontaneously (as previously thought), but were present because of infection. Robert Koch applied this idea to explain the cause of infectious disease and, in 1890, arrived at a set of criteria known as Koch's Postulates that determine if a disease is caused by microbes. These microorganisms could be isolated from infected individuals and grown in a laboratory for identification, so that an appropriate treatment regimen could be devised. Today many disease-causing microbes have been identified.

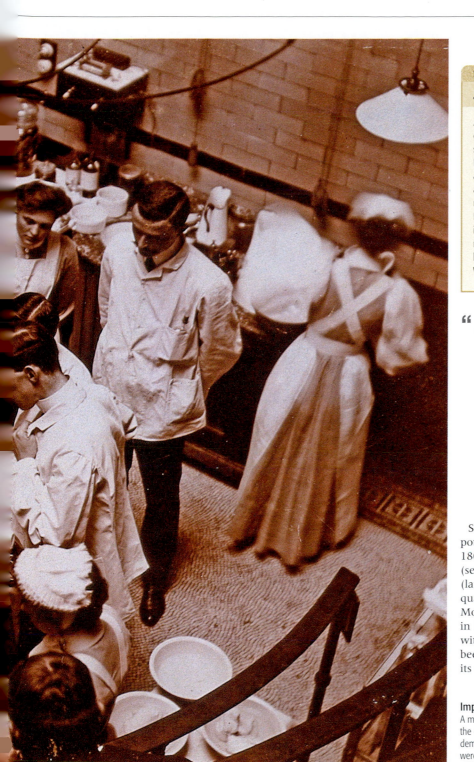

## JOSEPH LISTER

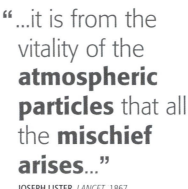

Joseph Lister was a skilled surgeon. He developed his ideas of sterile practice in surgery when working at Glasgow Royal Infirmary, Scotland. He promoted the use of antiseptics, recommended the use of gloves and the replacement of instruments with nonporous materials to minimize infection. Lister became President of the Royal Society in London but, in spite of his achievements, remained a modest character. He was honored with a funeral service in Westminster Abbey.

> "...it is from the vitality of the **atmospheric particles** that all the **mischief arises**..."
>
> JOSEPH LISTER, *LANCET*, 1867

hemorrhage after childbirth. The descriptions of patients' reactions afterward, which Blundell attributed to air bubbles in the blood, suggest that many had received incompatible blood. This issue was resolved in 1901 after Austrian biologist Karl Landsteiner discovered that there were four blood groups. Each group contains different antigens (chemicals that cause immune response). If compatible blood is transfused, see below left, immune response can be prevented.

Scientists also began to investigate the potential of gases as anesthetics. In 1800 English chemist Humphry Davy (see p.170) recorded that nitrous oxide (laughing gas) had pain-numbing qualities. American dentist William Morton used it for a tooth extraction in 1846, but a month later had success with a potent vapor: ether. Ether had been used privately in the US, but now its use spread to Europe. The English

**Improvements in infection control**
A medical team working in an operating room in the early 1900s at London's Charing Cross Hospital demonstrates that, although Lister's antiseptic principles were in practice, certain precautions—such as the use of face masks—were yet to be adopted.

surgeon James Simpson advocated the use of another gas, chloroform, in 1847. Chloroform posed a greater risk of overdose, so ether remained popular throughout the 19th century.

## Dealing with blood loss
Bungled attempts at blood transfusion in the 17th century had meant that the practice had all but disappeared. A revival of interest came with the work of English physician James Blundell. In 1818 he performed the first human-to-human blood transfusion on a man with what would now be called stomach cancer; the man died 56 hours later. His first successful human blood transfusion was some years later in a woman suffering from a

The 20th century saw important scientific breakthroughs in anesthesia and hematology (the study of blood).

**MODERN ANESTHESIA**
In 1951 British chemist **C.W. Suckling** made the anesthetic **halothane**, a nonflammable replacement for ether. Later it was supplanted by less toxic, ether-based drugs: **isoflurane** and **enflurane**. Today, faster-acting **desflurane** and **sevoflurane** are being introduced.

**BLOOD GROUPS AND BLOOD BANKS**
By 1907 **routine blood-typing** for transfusions was recommended. World War I brought a new urgency for transfusions—and with it developments that enabled long-term storage of blood and blood products.

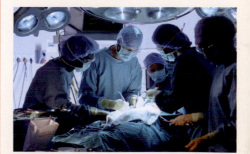
MODERN OPERATING ROOM

SEE ALSO >>
pp.406–07 MODERN SURGICAL PROCEDURES

### ABO blood groups
Safe blood transfusions are possible only when chemicals called antigens on the surface of donated red blood cells do not stimulate an immune reaction in the recipient. There are two types of antigens. These can be present on their own (blood groups A and B), together (blood group AB), or both absent (blood group O).

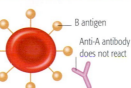

**Blood group A**
This blood cannot be donated to blood groups B or O since the A antigens react with their anti-A antibodies.

**Blood group B**
This blood cannot be donated to blood groups A or O since the B antigens react with their anti-B antibodies.

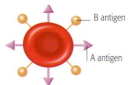
**Blood group AB**
This can be donated only to a recipient of the same blood group, to avoid anti-A and anti-B antibodies in others.

**Blood group O**
This blood can be donated to anyone since the lack of antigens prevents the possibility of an immune response.

# Mendeleev's Table

**No chemistry textbook, classroom, or laboratory wall today is complete without a copy of the periodic table—recognized by anyone who has studied science as an exceedingly useful framework for classifying and comparing the elements. But what is perhaps more surprising is that the original version of the table was literally "dreamed up" by an eccentric Russian professor while taking an afternoon nap.**

Over the first 65 years of the 19th century a bewildering variety of new chemical elements were discovered, many of them rare and obscure metals, but also substances that are well known today such as silicon and aluminium. By 1866 a total of 63 elements were known, but although these fell into a few obvious categories, such as metals and nonmetals, overall there seemed to be no rhyme or reason to them, nor indeed an explanation for why there were so many or why new ones kept turning up. In short, chemists at the time were in desperate need of a new way of understanding and classifying the elements.

By early 1869 a Russian chemist, Dmitri Mendeleev, had been trying to make sense of the elements for some months. He wondered whether if they were ordered by their atomic weights, or in some other way, a pattern might emerge. One February day, he became so engrossed with the problem that he canceled all engagements and began writing down the names of the elements on the backs of playing cards. He then tried arranging the cards, with the elements ordered by atomic weight, in a grid of rows and columns. This met with mixed success—there seemed to be some "periodicity," or recurring pattern, to the elements, but also many misalignments. In the midst of these efforts, he fell asleep and, while dreaming of playing patience, an idea came to him. What if spaces were left in his grid for elements that were undiscovered? When he awoke, Mendeleev found that a few adjustments to his grid produced a layout that worked brilliantly to show the relationships between all the known elements. It was Mendeleev's very own "Eureka" moment—the very first version of the periodic table had been born.

**A table of elements**

This version of the periodic table, with a photograph of Mendeleev (top right), comes from a textbook that was published in Russia during the Soviet era. It is somewhat different from the modern version of the table, particularly in its placing of the transition elements (the ones displayed here with black or dark blue element symbols).

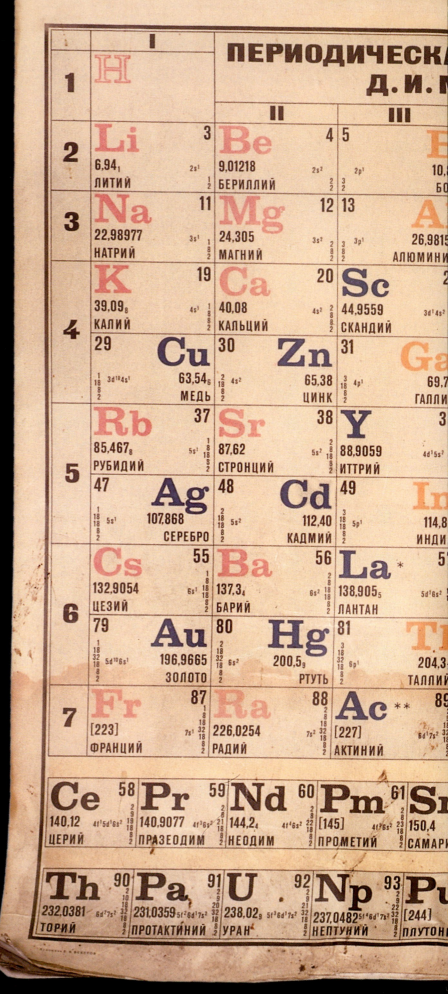

> "It is the **function** of **science** to discover the **existence** of a general reign of **order in nature** and to find the **causes** governing this order."
>
> DIMITRI MENDELEEV, LATE 19TH CENTURY

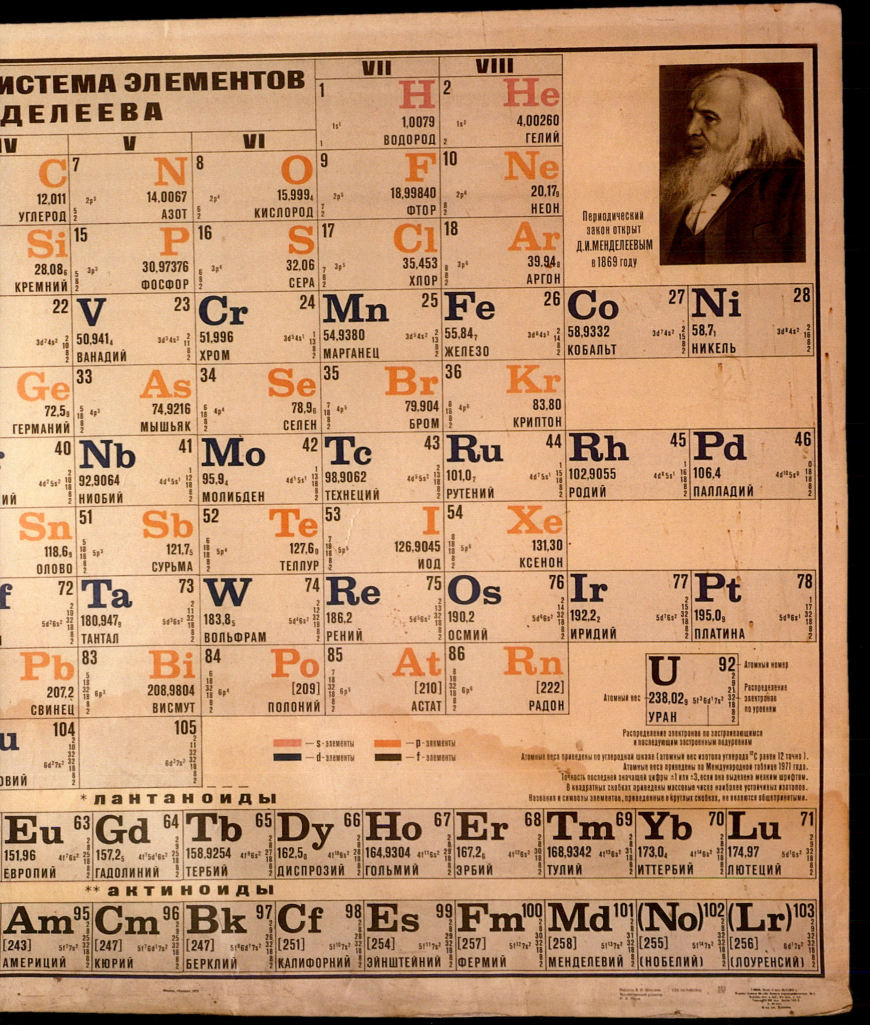

**BEFORE**

A key step in the development of the table was the determination of the atomic weights of elements in the early 19th century.

### DÖBEREINER'S TRIADS
In 1829 German chemist **Johann Döbereiner** noted **several triplets of similar elements**, where the middle element had an atomic weight equal to the average of the other two. This **introduced the concept of repetition** in the properties of elements.

**JOHANN DÖBEREINER**

### LAW OF OCTAVES
In 1864 English chemist **John Newlands proposed the "law of octaves."** He noticed similarities between every eighth element, ordered by atomic weight, **but the pattern petered out** after the first 17 elements.

**‹‹ SEE ALSO**
*pp.150–51* ORGANIC CHEMISTRY
*pp.230–31* MENDELEEV'S TABLE

# The Periodic Table

**The periodic table currently contains 117 elements, of which around 90 occur naturally on Earth. The position of an element in the table is determined by its atomic number (the number of protons in the nucleii of its atoms), while the properties of each element depend on the configuration of the electrons in its atoms.**

The periodic table is structured in order of increasing atomic number. It starts with hydrogen or H (atomic number 1) positioned top left, followed by helium or He (atomic number 2) at top right, then proceeds row by row through the rest of the elements in sequence by their atomic numbers. (Before 1914 the elements had been ordered according to their relative atomic masses, or weights.) Two rows of "inner transition elements" are set aside at the bottom of the table to avoid making their rows too long to fit on the page.

Since the 1920s chemists have known that the electrons within atoms are arranged in "shells," which occupy regions of space at differing distances from the nucleus (see pp.286–87). Progressing through the elements in order by atomic number, new shells are added and become sequentially filled by electrons, to produce each element's unique electron configuration (see right). This pattern forms the basis of various features of the table.

### Periods in the table
The periods of the table are its rows. The first period contains just hydrogen and helium. Atoms of these elements have just one electron shell, which can hold either one electron

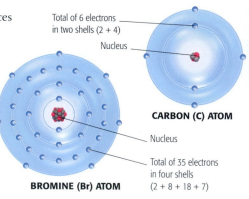

Total of 6 electrons in two shells (2 + 4)
Nucleus
**CARBON (C) ATOM**
Nucleus
Total of 35 electrons in four shells (2 + 8 + 18 + 7)
**BROMINE (Br) ATOM**

### Electron configuration in atoms
Atoms of different elements vary in their electron configurations—that is to say, the total number of electrons they hold and the way that these are arranged in shells around the nucleus.

> **"The elements,** if arranged according to their **atomic weights,** exhibit an apparent **periodicity."**

DMITRI MENDELEEV, RUSSIAN CHEMIST, 1869

| Group 1 | Group 2 | Group 3 | Group 4 | Group 5 | Group 6 | Group 7 | Group 8 | Group 9 | Group 10 | Group 11 | Group 12 | Group 13 | Group 14 | Group 15 | Group 16 | Group 17 | Group 18 |
|---|---|---|---|---|---|---|---|---|---|---|---|---|---|---|---|---|---|
| 1 H Hydrogen 1 | | | | | | | | | | | | | | | | | 2 He Helium 4 |
| 3 Li Lithium 7 | 4 Be Beryllium 9 | | | | | | | | | | | 5 B Boron 11 | 6 C Carbon 12 | 7 N Nitrogen 14 | 8 O Oxygen 16 | 9 Fl Fluorine 19 | 10 Ne Neon 20 |
| 11 Na Sodium 23 | 12 Mg Magnesium 24 | | | | | | | | | | | 13 Al Aluminium 27 | 14 Si Silicon 28 | 15 P Phosphorus 31 | 16 S Sulphur 32 | 17 Cl Chlorine 35 | 18 Ar Argon 40 |
| 19 K Potassium 39 | 20 Ca Calcium 40 | 21 Sc Scandium 45 | 22 Ti Titanium 48 | 23 V Vanadium 51 | 24 Cr Chromium 52 | 25 Mn Manganese 55 | 26 Fe Iron 56 | 27 Co Cobalt 59 | 28 Ni Nickel 59 | 29 Cu Copper 64 | 30 Zn Zinc 65 | 31 Ga Gallium 70 | 32 Ge Germanium 73 | 33 As Arsenic 75 | 34 Se Selenium 79 | 35 Br Bromine 80 | 36 Kr Krypton 84 |
| 37 Rb Rubidium 85 | 38 Sr Strontium 88 | 39 Y Yttrium 89 | 40 Zr Zirconium 91 | 41 Nb Niobium 93 | 42 Mo Molybdenum 96 | 43 Tc Technetium 98 | 44 Ru Ruthenium 101 | 45 Rh Rhodium 103 | 46 Pd Palladium 106 | 47 Ag Silver 108 | 48 Cd Cadmium 112 | 49 In Indium 115 | 50 Sn Tin 119 | 51 Sb Antimony 122 | 52 Te Tellurium 128 | 53 I Iodine 127 | 54 Xe Xenon 131 |
| 55 Cs Caesium 133 | 56 Ba Barium 137 | | 72 Hf Hafnium 178 | 73 Ta Tantalum 181 | 74 W Tungsten 184 | 75 Re Rhenium 186 | 76 Os Osmium 190 | 77 Ir Iridium 192 | 78 Pt Platinum 195 | 79 Au Gold 197 | 80 Hg Mercury 201 | 81 Ti Thallium 204 | 82 Pb Lead 207 | 83 Bi Bismuth 209 | 84 Po Polonium 210 | 85 At Astatine 210 | 86 Rn Radon 220 |
| 87 Fr Francium 223 | 88 Ra Radium 226 | | 104 Rf Rutherfordium 267 | 105 Db Dubnium 268 | 106 Sg Seaborgium 271 | 107 Bh Bohrium 272 | 108 Hs Hassium 277 | 109 Mt Meitnerium 276 | 110 Ds Darmstadtium 281 | 111 Rg Roentgenium 280 | 112 Uub Ununbium 285 | 113 Uut Ununtrium 284 | 114 Uuq Ununquadium 289 | 115 Uup Ununpentium 288 | 116 Uuh Ununhexium 293 | (117) (Uus) (Ununseptium) (not known) | 118 Uuo Ununoctium 294 |

**Periods** 1, 2, 3, 4, 5, 6, 7

**LANTHANOIDS**

| 57 La Lanthanum 139 | 58 Ce Cerium 140 | 59 Pr Praseodymium 141 | 60 Nd Neodymium 144 | 61 Pm Promethium 145 | 62 Sm Samarium 150 | 63 Eu Europium 152 | 64 Gd Gadolinium 157 | 65 Tb Terbium 159 | 66 Dy Dysprosium 163 | 67 Ho Holmium 165 | 68 Er Erbium 167 | 69 Tm Thulium 169 | 70 Yb Ytterbium 173 | 71 Lu Lutetium 175 |
|---|---|---|---|---|---|---|---|---|---|---|---|---|---|---|
| 89 Ac Actinium 227 | 90 Th Thorium 232 | 91 Pa Protactinium 231 | 92 U Uranium 238 | 93 Np Neptunium 237 | 94 Pu Plutonium 244 | 95 Am Americium 243 | 96 Cm Curium 247 | 97 Bk Berkelium 247 | 98 Cf Californium 251 | 99 Es Einsteinium 252 | 100 Fm Fermium 257 | 101 Md Mendelevium 258 | 102 No Nobelium 259 | 103 Lr Lawrencium 262 |

**ACTINOIDS**

**Atomic number**
The number of protons in the nuclei of the element's atoms.

**Chemical symbol**
Used as a form of shorthand in chemical equations (in this case for aluminium).

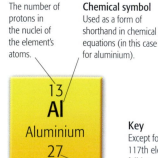

13 Al Aluminium 27

**Relative atomic mass**
The average number of nucleons (protons and neutrons) in the nuclei.

**Key**
Except for the as-yet-unsynthesized 117th element, all the elements fall into several color-coded categories or blocks containing similar elements. Nine categories are shown here, some of them corresponding closely or exactly to groups (columns) of the table.

**Alkali metals**
A group of soft metals that are extremely reactive.

**Alkaline-earth metals**
A group of silver-colored, moderately reactive metals.

**Transition metals**
A varied group of metals, many with highly valued properties.

**Inner transition elements**
(Also known as lanthanoids and actinoids) Reactive metals, some rare or synthetic.

**Other metals**
A group of mostly rather soft metals with low melting points.

**Metalloids**
Elements with properties halfway between metals and nonmetals.

**Other nonmetals**
A collection that includes the most critical elements in living matter, notably carbon.

**Noble gases**
Colorless, odorless gases that have very low chemical reactivity.

**Halogens**
Highly reactive nonmetals that comprise two gases, a liquid, and two solids.

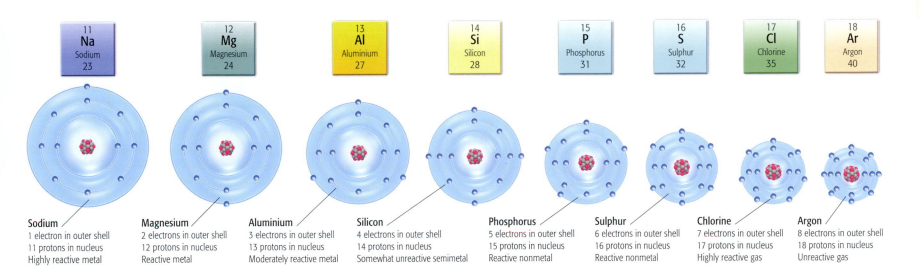

| 11 Na Sodium 23 | 12 Mg Magnesium 24 | 13 Al Aluminium 27 | 14 Si Silicon 28 | 15 P Phosphorus 31 | 16 S Sulphur 32 | 17 Cl Chlorine 35 | 18 Ar Argon 40 |

**Sodium**
1 electron in outer shell
11 protons in nucleus
Highly reactive metal

**Magnesium**
2 electrons in outer shell
12 protons in nucleus
Reactive metal

**Aluminium**
3 electrons in outer shell
13 protons in nucleus
Moderately reactive metal

**Silicon**
4 electrons in outer shell
14 protons in nucleus
Somewhat unreactive semimetal

**Phosphorus**
5 electrons in outer shell
15 protons in nucleus
Reactive nonmetal

**Sulphur**
6 electrons in outer shell
16 protons in nucleus
Reactive nonmetal

**Chlorine**
7 electrons in outer shell
17 protons in nucleus
Highly reactive gas

**Argon**
8 electrons in outer shell
18 protons in nucleus
Unreactive gas

▽ **A group (column) of the table**
The elements in group 1 all have one electron in their outermost shell, making most of them very similar in their properties: each of the elements shown here is a soft, highly reactive metal. Working down the group, the size of the atoms steadily increases, as with each element an extra electron shell has been added.

△ **A period (row) of the table**
The elements in period 3 have three electron shells. The number of electrons in the outer shell increases left to right, causing a change in properties. The atoms also get smaller, since at each step there is an extra proton in the nucleus, making it more positively charged so it pulls the negatively charged electrons in with increasing force.

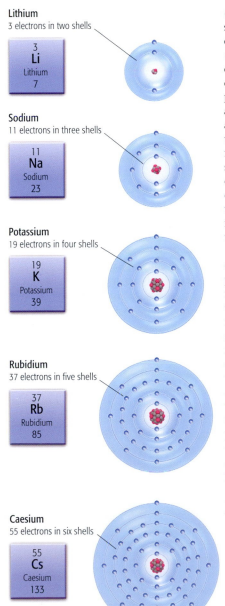

**Lithium**
3 electrons in two shells

3 Li Lithium 7

**Sodium**
11 electrons in three shells

11 Na Sodium 23

**Potassium**
19 electrons in four shells

19 K Potassium 39

**Rubidium**
37 electrons in five shells

37 Rb Rubidium 85

**Caesium**
55 electrons in six shells

55 Cs Caesium 133

(hydrogen) or two (helium). The second period consists of the next eight elements, lithium (Li) to neon (Ne). Progressing through these, electrons are slotted into a second electron shell until it is full. The third period also contains eight elements and represents the sequential filling of a third shell (see above). Subsequently the makeup of each period becomes more complex, as extra electrons are slotted into new shells before previous ones are completely full. Most elements in the block called the transition metals possess an incompletely filled shell adjacent to their outermost shell.

Across each period is a progressive change in the physical and chemical properties of the elements, from reactive metallic elements on the left, through less reactive metals to reactive nonmetals, and finally on the far right, a barely reactive, or "noble," gas. The main factor causing this progressive change is the number of electrons in the outermost shells of the elements' atoms. This affects the atoms' tendency to lose or gain electrons in chemical reactions (see pp.286–87), as well as properties such as melting points and electrical conductivity (see pp.164–65). Another trend across each period is a progressive decrease in size of atoms.

**Uniform groups**
The groups are the columns of the periodic table, which are collections of elements that resemble each other in their physical and chemical properties because they contain the same number of electrons in their outermost shells (see left). The most uniform are groups 1, 2, 17, and 18.

Group 1 (excluding hydrogen) and group 2 contain reactive metals, which easily give up the one or two electrons in their outer shells during chemical reactions. Group 18 consists of noble gases. These have full electron shells and are nearly unreactive since they have little tendency to lose, gain, or share electrons. Group 17 is the halogens —highly reactive nonmetals that tend to gain or share electrons in reactions.

**Less uniform groups**
Outside these four groups there is less uniformity, the elements falling instead into similar blocks and categories. An example is the metalloids (semimetals), including silicon (Si) and germanium (Ge), which are widely used as semiconductors (see pp.164–65). In the middle of the table is the large block of transition metals, which includes gold (Au), copper (Cu), and iron (Fe). The category classified as "other nonmetals" contains many critical elements, including hydrogen—the most common element—as well as carbon (C), nitrogen (N), and oxygen (O), which along with hydrogen and a little phosphorus (P) and sulphur (S) form 99 percent of all living matter.

**DMITRI MENDELEEV**

Born in Siberia, Russia, Mendeleev was Professor of Chemistry at the University of St. Petersburg. When formulating the first periodic table in 1869 (see pp.230–31), he left spaces for three undiscovered elements, whose properties he predicted. By 1888 these three elements—gallium (Ga), scandium (Sc), and germanium (Ge)—had all been discovered, with exactly the properties Mendeleev had predicted. Mendeleev also helped to develop Russia's oil and vodka industries, and introduced the metric system there. In 1955 the synthetic element mendelevium (Md) was named in his honor.

**AFTER**

Over the past 70 years a number of new elements have been synthesized.

**EXTENDED PERIODIC TABLE**
In 1969 an **extended periodic table** was suggested by the American scientist **Glenn Seaborg**. It was a logical extension of the principles behind the standard modern periodic table to include possible undiscovered elements up to an element with **atomic number 218**. Chemists believe it may be possible to synthesize relatively stable elements—"islands of stability" in a sea of unstable elements—far heavier than any made so far.

**LATEST MEMBER ELEMENT**
A joint team of Russian and American scientists claimed in 2004 to have briefly synthesized four atoms of the previously unsynthesized 115th element **ununpentium** (Uup), by fusing calcium (Ca) and americium (Am) nuclei. The atoms decayed in less than a second.

SEE ALSO »
pp.234–35 CHEMICAL REACTIONS
pp.286–87 STRUCTURE OF THE ATOM
pp.288–89 CHEMICAL BONDS

« **BEFORE**

The nature of chemical reactions was revealed as early chemists established that the elements combined together in fixed ways.

**THE BIRTH OF MODERN CHEMISTRY**
The foundation for understanding chemical reactions lies with the work of those such as the French chemist **Antoine Lavoisier**. In 1789 he surveyed the elements known at the time and established that **matter cannot be created or destroyed** « 138–39, which was important in showing that chemical reactions involve the **rearrangement** of chemical components.

**PROUST'S LAW**
In 1797 French chemist **Joseph Proust** proposed his **Law of Definite Proportions**, which stated that chemical elements combine **in simple ratios** to form **compounds**.

**JOSEPH PROUST**

« **SEE ALSO**
pp.52–53 ALCHEMY
pp.140–43 STATES OF MATTER

# Chemical Reactions

The gradual rusting of iron and the explosive response of gunpowder show that chemical reactions can take years or be over in seconds. At the heart of any chemical reaction is a substance (or mixture of substances) that changes its character so that, by the end of the process, a completely new chemical substance is formed.

Substances change their form for many reasons. In some cases, there is a change in state (see pp.140–43) but the chemical identity stays the same, such as when ice melts into water. Chemical reactions, however, involve a complete overhaul of chemical identity. In 1803 English chemist John Dalton (see p.138) showed that when chemicals react, they do so in fixed proportions because their particles (atoms) come together in set combinations—an idea later developed into Atomic Theory (see pp.286–87).

## From reactants to products
The identity of a chemical substance is determined not only by the elements it contains, but also the precise way the atoms of these elements are bonded together (see pp.288–89). Even the most subtle alterations in the arrangement of atoms can radically alter the properties of a substance, and therefore its identity. Chemical reactions can involve small shifts in position or complete rearrangements. The bonds that hold the atoms together in the original structure must be broken apart and new ones formed as the initial reactants are converted into products.

## Types of chemical reaction
In the 19th century scientists identified most of the types of chemical reactions recognized today. At this time Swedish chemist Svante Arrhenius (see below) also found that reactions in solution could involve charged particles called ions (see pp.288–89), as well as atoms and molecules.

Some reactions involve combining small particles to make bigger ones—synthesis—while others break particles down—decomposition. Sometimes there is displacement: an atom, or

**REACTANT** A chemical that undergoes a change during the process of a chemical reaction.

**PRODUCT** A chemical that is formed from the change during the process of a chemical reaction.

group of atoms, is ousted from its position by more reactive ones. There are also redox reactions, which involve the transfer of electrons (see pp.286–87), the negative parts of atoms. Some elements, such as oxygen, have atoms that pull electrons away from other

particles. The particle that loses electrons undergoes a process known as oxidation; the particle that gains electrons undergoes reduction.

Chemical reactions can also be characterized in terms of energy changes. All substances contain chemical (potential) energy. If there is more chemical energy in the reactants than the products, then this is given off as heat during the reaction, making it exothermic. Heat-giving reactions are the basis of fuels, such as gasoline. If products contain more energy than reactants, then energy must be taken in by the reaction, making it endothermic. In 1840 Swiss-Russian chemist Germain Hess found that the amounts of energy released or taken in

**Endothermic reaction**
This reaction is endothermic because it takes in some of the energy from the heat applied. Heating blue crystals of hydrated copper sulphate causes them to release water, forming white anhydrous copper sulphate.

**SCIENTIST (1859–1927)**

## SVANTE AUGUST ARRHENIUS

Svante Arrhenius initially studied physics at the University of Uppsala in his native Sweden, but later turned to chemistry, where he first made important studies on electrolysis (the use of electricity to drive chemical reactions). He became Professor of Physics at the University of Stockholm, and later helped to set up the Nobel Prizes. After 1900 he made contributions to biology, astronomy, and geology. He supported a theory of panspermia (that life originated by spores carried from outer space), and devised an early version of a theory that explained the greenhouse effect.

**Changing chemical bonds**
During this chemical reaction of combustion, a molecule of methane reacts with two molecules of oxygen to form carbon dioxide and water; the bonds between atoms must be broken and new bonds formed.

"**Chemistry:** that most **excellent child** of **intellect** and **art**."
CYRIL NORMAN HINSHELWOOD, BRITISH CHEMIST (1897–1967)

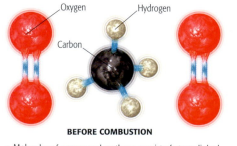

Oxygen    Hydrogen
Carbon

**BEFORE COMBUSTION**

**1** Molecules of oxygen and methane consist of atoms linked by strong covalent bonds (where electrons are shared). The bonds are double in oxygen and single in methane.

**BONDS BREAK APART**

**2** Combustion causes the oxygen and methane molecules to collide and react chemically. The hydrogen atoms are torn off sequentially.

Carbon dioxide molecule

Water molecule

**AFTER COMBUSTION**

**3** Two hydrogen atoms combine with oxygen to create water, while carbon forms double bonds with oxygen to create carbon dioxide.

**Exothermic reaction**
Potassium metal added to water reacts vigorously, giving off a great deal of heat, typical of an exothermic reaction. Hydrogen gas is released and a solution of the alkali potassium hydroxide results.

## Types of chemical reaction
Chemical reactions can be classified according to the way in which the particles behave when their atoms are rearranged as a result of the reaction.

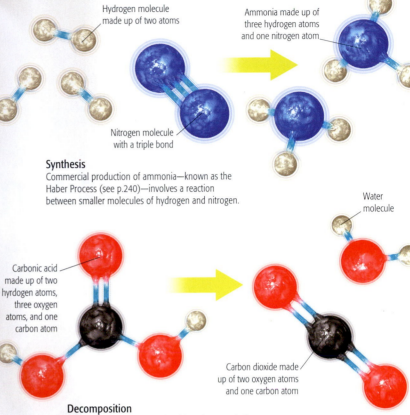

Hydrogen molecule made up of two atoms

Ammonia made up of three hydrogen atoms and one nitrogen atom

Nitrogen molecule with a triple bond

### Synthesis
Commercial production of ammonia—known as the Haber Process (see p.240)—involves a reaction between smaller molecules of hydrogen and nitrogen.

Water molecule

Carbonic acid made up of two hyrdogen atoms, three oxygen atoms, and one carbon atom

Carbon dioxide made up of two oxygen atoms and one carbon atom

### Decomposition
Carbonic acid is a very weak acid, and automatically breaks down into carbon dioxide and water when warmed in water.

by reactions was determined only by the reactants and products and not on how the reaction takes place—this had important implications in industry.

## The role of electrons
The behavior of all chemical reactions depends on electron transfer, and redox reactions are explicitly defined by this process. Electrons form the cloud of negative charge around every atom, but the nature of this cloud—its size and organization—determines how an atom interacts with others. The atoms of some elements can engage by sharing electrons in covalent bonds (if they are both nonmetals), while others can only form ionic bonds. When a chemical reaction occurs, there is a redistribution of electrons as new bonds are made and old ones destroyed (see pp.288–89).

The electron configuration of an element's atom also determines how it engages with other elements. Some substances react well together, and when they do so the new bonding patterns they form are very restrictive, making the products formed very predictable. This is the case when methane and oxygen are mixed together and heated (see left).

**AFTER**

In the 20th century the behavior of elements during chemical reactions was explained in terms of the way electrons allowed the formation of bonds.

**NATURE OF THE CHEMICAL BOND**
In 1939 American chemist **Linus Pauling** published a book on chemical bonding, which explained the difference between **ionic** (electron exchange) and **covalent** (electron sharing) bonding—and so established the basis of the modern understanding of chemical reactions. He was awarded the **Nobel Prize** for his work.

**MOLECULAR ORBITALS**
Chemical bonds involve **interaction between electrons** of atoms. In 1965 American chemists **Robert Woodford** and **Roald Hoffmann** used mathematics to describe the positions of **electrons** in oribitals around **molecules**, which helped to further understanding of changes in the position of atoms during chemical reactions.

**SEE ALSO »**
*pp.236–37* SPEEDING UP REACTIONS
*pp.238–39* ACIDS AND BASES
*pp.286–87* STRUCTURE OF THE ATOM
*pp.288–89* CHEMICAL BONDS

« BEFORE

Understanding of rates of chemical reactions relied on advances that encouraged chemists to think in terms of particles.

**HIGH-PRESSURE GASES**

In 1662 Irish polymath **Robert Boyle** quantified the way that **decreasing the volume of a gas increased its pressure**. In modern terms, he had noted that the particles of a **high-pressure gas** are brought close together.

**UNDERSTANDING CHEMICAL PARTICLES**

In 1803 English chemist **John Dalton** showed that when chemicals **react**, they do so in **fixed proportions** because of their particles « **234**, heralding the arrival of **atomic theory**, which laid the foundations for **collision theory** to be developed a century later.

« SEE ALSO
pp.100–01 The Behavior of Gases
pp.140–43 States of Matter
pp.234–35 Chemical Reactions

Some chemical reactions go faster than others because the particles interact more readily with each other: for example, combustion of methane is faster than rusting of iron. The speed of a chemical reaction is determined in complex ways, and the rate of any type of reaction can vary greatly depending on prevailing conditions.

In the latter half of the 19th century chemists, such as Jacobus van't Hoff in Holland, were building theories to account for these observations, and laying the foundation of physical chemistry today.

There are various ways of measuring the speed of a chemical reaction: in some cases it is easier to measure how much reactant (chemical undergoing change) is used up in a period of time; in others, it is easier to measure the rate at which the product (chemical produced by change) is made.

## Reaction processes

All chemical particles are more or less dynamic in their behavior: those in solids vibrate, while those in liquids and gases move around. For a chemical reaction to occur, particles must collide

effect makes them collide more frequently and with greater force, again increasing the rate of reaction. Light levels have a similar effect on particles that can absorb light energy. In 1889 Swedish chemist Svante Arrhenius (see p.234) proposed that there was a minimum amount of energy that a particle must have in order to react—the activation energy. When temperature or light increases, more particles reach this activation energy and so react when they collide. In 1900 Henry Le Chatelier (see below) established his law of equilibrium:

# Speeding up Reactions

**Although some chemical reactions are naturally more vigorous than others, all chemical reactions can be made to go faster or slower by changing conditions. It was an appreciation that chemical substances are made up of particles—atoms and molecules—that helped scientists understand how this happens.**

### Factors affecting reactions
Chemical particles react when they collide with each other with enough energy, and any factor that increases the chance of this happening will increase the rate of the chemical reaction.

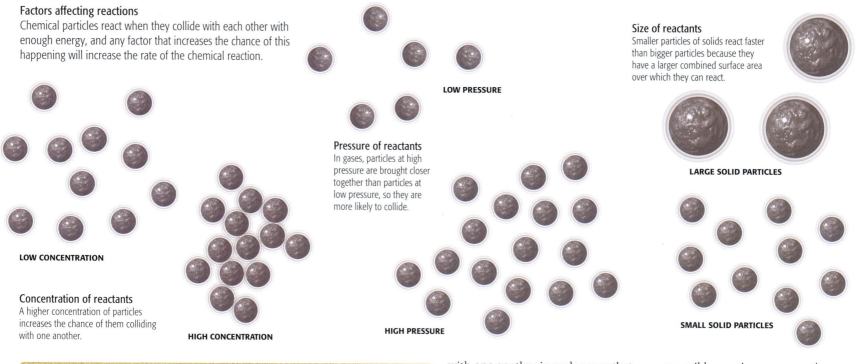

**LOW PRESSURE**

### Size of reactants
Smaller particles of solids react faster than bigger particles because they have a larger combined surface area over which they can react.

**LARGE SOLID PARTICLES**

### Pressure of reactants
In gases, particles at high pressure are brought closer together than particles at low pressure, so they are more likely to collide.

**LOW CONCENTRATION**

### Concentration of reactants
A higher concentration of particles increases the chance of them colliding with one another.

**HIGH CONCENTRATION**

**HIGH PRESSURE**

**SMALL SOLID PARTICLES**

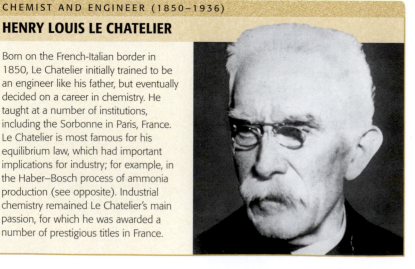

**CHEMIST AND ENGINEER (1850–1936)**

## HENRY LOUIS LE CHATELIER

Born on the French-Italian border in 1850, Le Chatelier initially trained to be an engineer like his father, but eventually decided on a career in chemistry. He taught at a number of institutions, including the Sorbonne in Paris, France. Le Chatelier is most famous for his equilibrium law, which had important implications for industry; for example, in the Haber–Bosch process of ammonia production (see opposite). Industrial chemistry remained Le Chatelier's main passion, for which he was awarded a number of prestigious titles in France.

with one another in such a way that they can engage together in bonding (see pp.288–89). Just how many of them succeed in doing this over time determines the speed of the reaction.

If there is a high concentration of particles, they are more likely to collide, so the reaction is faster than with particles at low concentration. Gases at high pressure will give rise to faster reactions than those at low pressure, and solids with larger surface areas will react more quickly than those with smaller surface areas.

Temperature also has an effect on reactions. In 1877 Austrian physicist Ludwig Boltzmann explained that as temperature rises, particles become more energetic and move faster: the

reversible reactions are sometimes controllable; if a reversible reaction results in a decrease in pressure, then increasing the pressure of the starting materials will push the reaction forwards. In the 20th century, the German chemist Max Trautz and then, independently, the British chemist William Lewis brought together ways in which particle behavior affects reactions in what became known as collision theory.

### Catalysts

By the time collision theory was accepted, Latvian chemist Wilhelm Ostwald was laying the foundation of another principle affecting reaction rates. He found that metal could encourage the production of nitric acid from ammonia. The metal was unaffected by, but increased the rate of, the reaction. Half a century earlier Swedish chemist Jöns Berzelius had coined a term for this: catalysis. Catalysts often speed up reactions by bringing particles closer together, effectively lowering the activation energy in doing so.

The action of catalysts can speed up reactions that would otherwise take millions of years. Biological catalysts are known as enzymes and these are responsible for most of the chemical processes such as digestion (pp.216–17) that take place in the body. Catalysts have also had an important role in facilitating processes for industry, such as the Haber–Bosch process (see below).

**Effect of acid rain on stone**
The face of this statue has been obliterated by acid rain because the detail of the features had a large surface area, encouraging a reaction between acid and limestone.

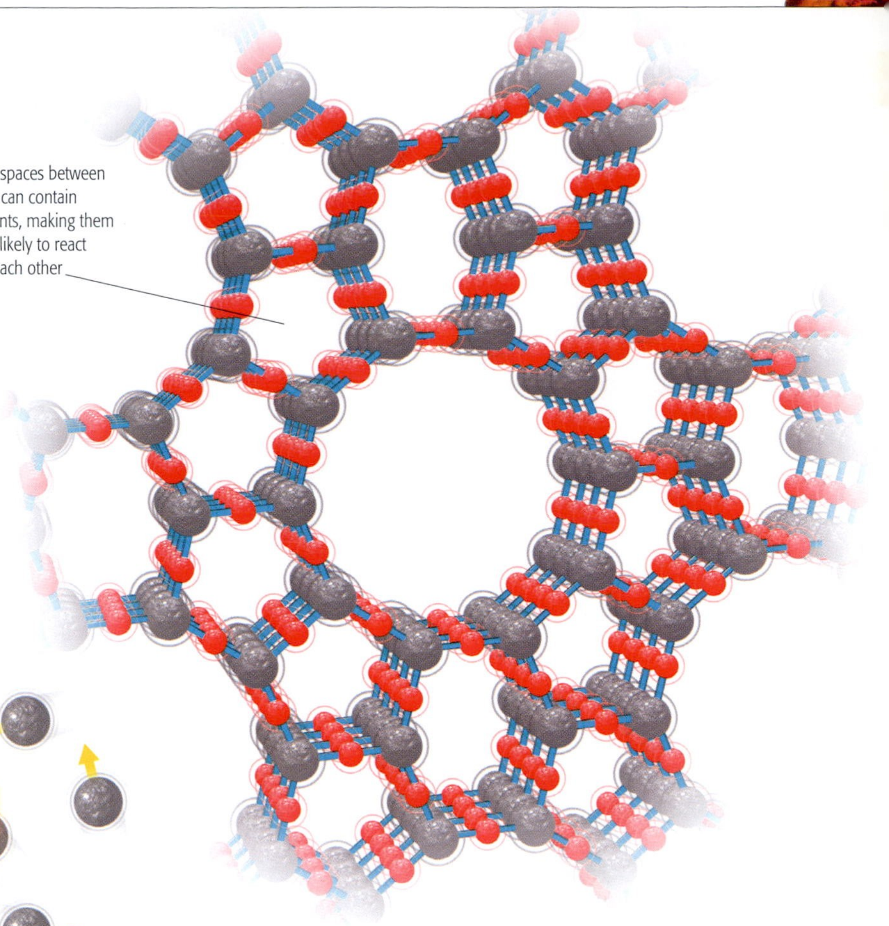

Small spaces between pores can contain reactants, making them more likely to react with each other

**Zeolite catalyst**
Zeolite is a mineral with a very porous structure, and is used widely as a catalyst in the petrochemical industry.

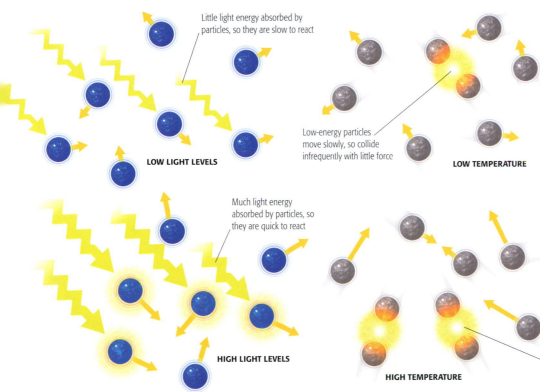

Little light energy absorbed by particles, so they are slow to react

**LOW LIGHT LEVELS**

Low-energy particles move slowly, so collide infrequently with little force

**LOW TEMPERATURE**

Much light energy absorbed by particles, so they are quick to react

**HIGH LIGHT LEVELS**

High-energy particles move quickly, so collide frequently with force

**HIGH TEMPERATURE**

**Light levels**
Some particles can absorb light energy, making them more reactive at high light levels than they are at low light levels.

**Temperature of reactants**
As temperature increases, particles move faster so collide more frequently than they do at low temperatures.

> "Wherever we look, the **work** of the **chemist** has **raised** the **level** of our **civilization**."
>
> JOHN CALVIN COOLIDGE, 30TH US PRESIDENT, 1924

**AFTER**

The beginnings of atomic theory developed in the 19th century were supported by new discoveries from physics in the 1900s.

**ATOMIC THEORY VINDICATED**
In 1905 German theoretical physicist **Albert Einstein** demonstrated how **Brownian motion**—the visible movement of particles in gas and liquid—provided a **theoretical proof of atomic theory**, which underpinned the behavior of chemical reactions.

**RISE OF CHEMICAL THERMODYNAMICS**
The tendency of chemicals to **react** depends on the amount of **energy** they contain, with some chemical particles being more "energetic" than others. In 1923 American chemists **Gilbert Lewis** and **Merle Randall** determined the **energy content** of many chemical substances, which enabled scientists to calculate the energy changes that take place during reactions.

**SEE ALSO »**
pp.286–87 STRUCTURE OF THE ATOM
pp.288–89 CHEMICAL BONDS

**IN PRACTICE**

## HABER–BOSCH PROCESS

The industrial production of ammonia was made possible by the process invented by German chemists Fritz Haber and Carl Bosch in 1909, using principles of equilibrium developed by French-Italian chemist Henry Le Chatelier (see opposite). Gaseous nitrogen and hydrogen are mixed together at a pressure of 200 atmospheres and at high temperatures—between 572° F (300° C) and 1,022° F (550° C). This encourages them to react with each other to make ammonia over an iron catalyst. Any nitrogen and hydrogen that does not react is recycled; this results in around 98 percent conversion. The Haber–Bosch process was an important breakthrough because it was a way of producing ammonia on an industrial scale for use in nitrogenous fertilizer for agriculture, synthetic fibres, and commercial explosives. It was also a way of harnessing elemental nitrogen, which is otherwise very unreactive (see pp.278–79).

## BEFORE

By the end of the Middle Ages scholars had succeeded in isolating some of the most important acids used today.

**AQUA REGIA AND THE PHILOSOPHER'S STONE**

The study of acids began when the 8th-century Persian alchemist **Geber** isolated **hydrochloric acid** by mixing common salt (sodium chloride) with **sulphuric acid**. Geber also found that **gold**, which is not affected by any other acids, would **dissolve on contact** with a mixture of hydrochloric acid and nitric acid. This mixture, which became known as **aqua regia** ("royal water"), was associated with the **philosopher's stone**, a mythical tool that alchemists believed could **turn base metals into gold**.

**PURE SULPHURIC ACID**

The 9th-century Persian physician **Rhazes** produced **pure sulphuric acid** by heating a combination of metal sulphates. The **vapors of water** and **sulphur trioxide** reacted to form the acid in much the same way acid rain forms.

**« SEE ALSO**
*pp.234–35* CHEMICAL REACTIONS
*pp.236–37* SPEEDING UP REACTIONS

# Acids and Bases

**That some acids and certain bases are dangerously corrosive was known—and even exploited—by the scholars of antiquity, but it would take centuries of painstaking scientific scrutiny before the exact chemical natures of acids and bases were finally revealed by a Swedish chemist.**

The sharp, stinging properties of lemon juice typify a class of natural substances that are familiar in everyday life. Organic, plant-derived acids are prominent in vinegar and sour fruit, while inorganic ones—some of which are extremely corrosive—are mainly of mineral origins.

Acids can be tempered by a class of substances called bases. However, bases can also be corrosive in high concentrations.

### Properties of acids and bases

Chemically, acids and bases are characterized by what happens when they are mixed, a feature recognized by French chemist Guillaume-François Rouelle in 1754. When mixed, acids and bases neutralize each other,

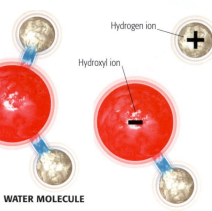

Hydrogen ion

Hydroxyl ion

**WATER MOLECULE**

forming a different type of substance called a salt. Salts come in millions of different forms, depending on the acids and bases mixed. Common table salt (sodium chloride), for example, is formed by mixing hydrochloric acid and the base sodium hydroxide. No

**Water splitting**

Pure water is neutral because it can split (dissociate) into equal numbers of positively charged hydrogen ions (H+) and negatively charged hydroxyl ions (OH-). Acids have more hydrogen ions, and bases more hydroxyl ions.

matter which acids or bases are mixed, the chemical reaction also forms water, and it is this that gives the biggest clue as to the true chemical identities of acids and bases.

### Chemical identities

In the 19th century various chemists including the German chemist Justus von Liebig suggested correctly that hydrogen was the key element in all acids. Then in 1884 the Swedish chemist Svante Arrhenius (see p.234) provided the first precise definitions of acids and bases.

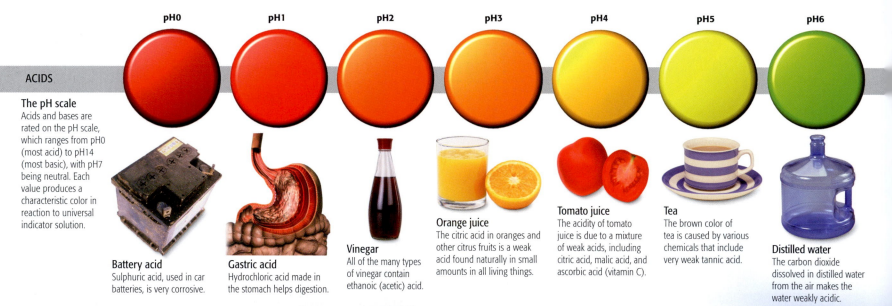

| pH0 | pH1 | pH2 | pH3 | pH4 | pH5 | pH6 |

ACIDS

**The pH scale**
Acids and bases are rated on the pH scale, which ranges from pH0 (most acid) to pH14 (most basic), with pH7 being neutral. Each value produces a characteristic color in reaction to universal indicator solution.

**Battery acid**
Sulphuric acid, used in car batteries, is very corrosive.

**Gastric acid**
Hydrochloric acid made in the stomach helps digestion.

**Vinegar**
All of the many types of vinegar contain ethanoic (acetic) acid.

**Orange juice**
The citric acid in oranges and other citrus fruits is a weak acid found naturally in small amounts in all living things.

**Tomato juice**
The acidity of tomato juice is due to a mixture of weak acids, including citric acid, malic acid, and ascorbic acid (vitamin C).

**Tea**
The brown color of tea is caused by various chemicals that include very weak tannic acid.

**Distilled water**
The carbon dioxide dissolved in distilled water from the air makes the water weakly acidic.

**"...acids and bases are substances that are capable of splitting off or taking up hydrogen ions, respectively."**

JOHANNES BRØNSTED, 1923

◁ **Acid rain**
Polluting sulphur dioxide gas from industry reacts with oxygen and water droplets in the air to form sulphuric acid. The acid acidifies the rain, damaging trees and limestone buildings.

**Reducing soil acidity**
Many crops grow best in soil with a pH of about 6.5. The addition of limestone to neutralize overly acidic soils can improve plant growth and encourage nutrient-releasing bacteria to thrive.

## Testing for acids and bases

Litmus paper is used to determine whether a liquid is either an acid or a base. The paper turns red (as shown to the right) if the liquid is acidic; the paper turns blue if the liquid is a base.

The significance of an acid's hydrogen is apparent when it forms water at neutralization. A molecule of water contains two hydrogen atoms and one oxygen atom. When an acid reacts with a base, the acid provides a positively charged hydrogen atom—a hydrogen ion (H⁺), or proton—and the base provides the remaining part of the water molecule in the form of a negatively charged hydroxyl ion: OH⁻.

According to Arrhenius, then, it was the provision of hydrogen and hydroxyl ions that defined acids and bases respectively. Stronger acids and bases release higher concentrations of their respective hydrogen and hydroxyl ions; it is the reaction of these with other materials that explains the corrosive actions of these chemicals. However, Arrhenius's definitions are applicable only for acids and bases that dissolve in water, which not all do. Therefore in 1923 the Danish chemist Johannes Brønsted and the English chemist Martin Lowry each independently suggested that the definitions of acids and bases be changed. They each defined an acid as a chemical that donates hydrogen ions (protons), and a base as one that accepts them. These remain the accepted definitions.

## Measuring acidity

Following Arrhenius, scientists realized that acidity was so important in the natural world that it should be quantified. German chemist Hans Friedenthal suggested a system classifying the concentration of

**ACID** A chemical substance that releases hydrogen ions (protons). The stronger the acid, the higher the concentration of hydrogen ions released.

**BASE** A chemical substance that can accept hydrogen ions released by acids, often reacting with them to form water.

BIOCHEMIST (1868–1939)

### SØREN SØRENSEN

The son of a Danish farmer, Sørensen entered the University of Copenhagen to study medicine, switching later to chemistry. He invented the pH scale while director of the Chemistry Department at the Carlsberg Laboratory in Copenhagen (a facility supported by the brewing company). He also developed an electrode to measure pH, as well as indicator solutions; both methods are used today. Sørensen won many awards and honors for his contribution to science, which also included medical research into diabetes and epilepsy.

hydrogen ions, but there was an enormous variation over the range—the most basic solutions had a hydrogen ion concentration that was around a hundred trillion times lower than the most acidic ones.

Then in 1909 a Danish chemist, Søren Sørensen, came up with a much easier, logarithmic scale that ranged from 0 (most acid) through 7 (neutral) to 14 (most basic). Each step up this simple scale represents a hydrogen ion concentration 10 times lower than the previous one. Sørensen's scale became known as the pH scale, which stands for *pondus hydrogenii*: "potential of hydrogen." At neutral pH7 the concentration of hydrogen ions is equal to the concentration of hydroxyl ions.

Universal indicator solution is now widely used to test pH, passing from red when most acid through orange, yellow, green (neutral), and blue to purple when most basic (below).

pH7 pH8 pH9 pH10 pH11 pH12 pH13 pH14 — BASES

**Cow's milk** While cow's milk is neutral when it is fresh, it becomes weakly acidic when bacteria in it produce lactic acid.

**Seawater** While mainly a solution of sodium chloride, seawater contains small quantities of other weakly basic chemicals.

**Toothpaste** Some of the chemicals in toothpaste are weak bases, such as sodium bicarbonate.

**Antacid tablets** When dissolved in water, antacid tablets produce a basic solution that raises the pH of the digestive juices in the stomach.

**Ammonia** Dissolved in water, ammonia forms a basic solution called ammonium hydroxide.

**Household bleach** The strong base sodium hypochlorite is the agent in household bleach.

**Cement** Made by heating limestone with clay, cement is a mix of highly basic chemicals.

**Drain cleaner** Most drain cleaners contain caustic soda—corrosive, concentrated sodium hydroxide.

**AFTER**

**Highly corrosive acids have a variety of industrial uses from removing corrosion from iron and steel to etching glass.**

**POLISHING SILICON CHIPS**
Nitric acid, sulphuric acid, hydrochloric acid, and hydrofluoric acid are all used in their purest forms to remove silicon dioxide from the surfaces of the wafers of silicon crystal that are key components of **computer microchips**.

**PICKLING AND ETCHING**
Sulphuric acid, hydrofluoric acid, and hydrochloric acid are all also commonly used to **remove rust** and **iron oxide** from sheets of iron and steel—a process called pickling. **Hydrofluoric acid**, which **dissolves glass**, is also commonly used to etch or frost the surface of glass.

SEE ALSO ≫
*pp.240–41* MASS PRODUCTION OF CHEMICALS

<< **BEFORE**

The practical applications of chemistry began with medieval Islamic scholars, who transformed alchemy into science.

### ORIGINS IN ALCHEMY
In the 9th century the Islamic scholar **Geber** managed to **synthesize numerous chemical substances**, including mineral (inorganic) acids and some elements. He was able to use many of these techniques to make perfumes and drugs. A century later the alchemist **Rhazes** << 52–53 developed **methods of distillation** and also devised a way of **producing sulphuric acid**.

### ALUM AND INDUSTRIAL CHEMISTRY
Alum generally refers to a variety of aluminium-based substances known since antiquity. It has been used in **numerous industrial processes**, including the production of cosmetics and as a **dye-fixer**. Alum can be extracted from clay or shale, and in Yorkshire, England, alum manufacture was **Britain's first chemical industry**.

<< SEE ALSO
*pp.52–53* ALCHEMY
*pp.56–57* GUNPOWDER AND FIRE WEAPONS
*pp.150–51* ORGANIC CHEMISTRY

# Mass Production of Chemicals

**The products of chemistry have wide-ranging applications in manufacturing and design, agriculture, medicine, engineering, and warfare. Sometimes useful products can be extracted from plants or other natural sources, but the principles of chemistry have also paved the way for synthesis in the laboratory and the factory.**

Early chemists developed ways of purifying key ingredients from natural mixtures, but if the extraction and purification was complicated, they sometimes tried to synthesize them. This could simply be a matter of discovering the right conditions for a reaction to happen—such as the conversion of nitrogen and hydrogen into ammonia —and once that was done synthesis could often be achieved in bulk. But in other cases, such as with the antibiotic penicillin (see pp.310–11), the chemical target was more complex and involved many stages: fermentation of raw ingredients followed by synthesis of the final product. Even so, a chemist who could devise a practical way of making a much-demanded product more cheaply than extracting it from natural sources could reap great financial rewards.

## The color purple
Occasionally synthesis happened by accident. In 1856 William Perkin was an 18-year-old student at London's Royal College of Chemistry. Inspired by his mentor, August van Hofmann, he tried to produce quinine in his homemade laboratory. This was an expensive natural drug, much in demand for treating malaria, and a substance called aniline was a likely precursor. But instead of making quinine, he ended up with a purple dye. Perkin named it mauveine, and it was the first organic dye to be made artificially.

At the time dyes were costly to extract from natural products, and Perkin quickly recognized the value of his discovery. Against the encouraging background of the Industrial Revolution, he patented his dye and set up factories for its commercial production. Perkin grew rich on his accidental discovery.

## Farming revolution
Perkin sold his factories at the start of the 20th century, just as the chemical dye industry was starting to expand in Germany. Here, too, a different breakthrough was taking place. Fritz Haber had succeeded in producing ammonia at relatively low temperatures from hydrogen and nitrogen using an iron catalyst. There was great demand for ammonia to make artificial fertilizers: until then, much of the nitrogen used to promote crop growth was supplied in the form of nitrates mined in places such as South America.

As populations boomed during the Industrial Revolution, demand for food spiraled. With the onset of World War I the shortage of nitrogen became a serious crisis in Germany, because only the Allies had access to South American nitrogen. Carl Bosch, an

### The structure of mauveine
The molecular structure of mauveine was not known with certainty until 1994. Mauveine dye is actually a mixture of four related aromatic compounds, which differ only in the number and placement of methyl groups (groupings of three hydrogen atoms and one carbon atom).

CHEMIST (1838–1907)

## WILLIAM PERKIN

Perkin was a British chemist with an entrepreneurial spirit. The family business he built on the discovery of the dye mauveine established his reputation and brought him great wealth. Perkin was one of the first chemists to make his scientific discoveries commercially viable. But he continued with his research too, discovering new dyes and successfully synthesizing perfumes. In 1906—the year of his knighthood—Perkin visited America, where he was additionally honored with the first Perkin Medal, which subsequently became the country's highest award for industrial chemistry.

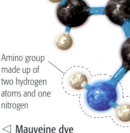

Nitrogen atom

Hydrogen atom

Rings made up of six carbon atoms

Carbon atom

Amino group made up of two hydrogen atoms and one nitrogen

Methyl group made up of three hydrogen atoms and one carbon atom

◁ **Mauveine dye**
One of the earliest samples of the first synthetic dye, mauveine, paved the way for a new dye industry. Until then, the natural source of purple dye was the mucus of a marine snail from the Mediterranean region, but its extraction made it expensive.

The successful synthesis of chemical products was **advanced** by a greater understanding of their structures and chemical reactions.

### SYNTHETIC DRUGS

Willow-tree-based **herbal remedies** containing salicylic acid compounds had been used for centuries but it was chemists at the Bayer drug and dye company that developed the **active ingredient** into a medicine called **aspirin 312–13 »** in 1899. Aspirin proved a successful painkiller—the first of many drugs based on natural products to be synthesized by chemists.

### CATALYZING CHEMICAL INDUSTRY

Cracking is a process that **splits hydrocarbons** from distilled crude oil into lighter products, such as **petroleum**. In 1937 the French engineer **Eugene Houdry** devised a better way of cracking petroleum that laid the foundations for modern oil refineries.

INVENTION

### DYNAMITE

Nobel's dynamite (patented 1867) made nitroglycerine safer to handle. The liquid explosive is mixed with an absorbent material, such as kieselguhr (a porous type of clay), and the mixture formed into a stick, which is then wrapped in paper. It is detonated via an electrical cable. Dynamite is still used for mining and construction, but the military use more stable explosives.

> "I intend to leave after my death a large **fund** for the **promotion** of the **peace idea**..."
> ALFRED NOBEL

### Chemical industry during World War I

A female employee working at the Brunner Mond and Company chemical works in Northwich, Manchester, England, feeds an ash furnace in September 1918.

industrial chemist, developed Haber's method into a process that could deliver ammonia in megaton amounts. Today around a third of humanity is sustained by crops grown using fertilizer made using ammonia obtained by the Haber-Bosch process.

### Explosive chemistry

In World War I much of the ammonia from the Haber-Bosch process was used for making explosives. An explosive is a substance that reacts to produce a huge volume of hot gas. The reaction spreads rapidly through the material. In gunpowder this reaction is started by a flame, but in other explosives it is initiated by a shock wave. One such explosive, the highly unstable liquid nitroglycerine, was discovered in 1847 by an Italian chemist, Ascanio Sobrero. A student of his, Alfred Nobel, made it safer to handle after a nitroglycerine explosion killed his younger brother Emil in 1864. He patented his new product as Dynamite, and made a fortune. Nobel left much of his wealth to fund the prizes that carry his name.

### Explosives factory

Because of the instability of the materials used in the production of nitroglycerine, this processing plant near Cengio, Italy, used earth embankments to separate the buildings and to surround the whole complex.

# The Spread of Disease

**Disease can devastate whole populations. Since antiquity humans have sought to understand how disease spreads and why it affects some people but not others. Scientists call it epidemiology, and in the last 150 years its findings have had far-reaching implications for public health.**

## PHYSICIAN (1818–1865)

### IGNAZ SEMMELWEIS

The Hungarian physician Ignaz Semmelweis successfully used one of the first hygiene regimens to reduce the spread of disease. In 1847 he was working at a maternity clinic in Vienna, where he instructed staff to wash their hands in disinfectant between performing autopsies and attending to patients. As a result patient deaths due to puerperal fever, caused by infections contracted during childbirth, dropped dramatically. Despite this, Semmelweis's theory was ridiculed and he was dismissed. He never recovered and died in a mental asylum.

As far back in history as ancient Roman times the spread of disease was reduced with the installation of the first advanced sewerage system, but scholars of the time did not understand why. This knowledge did not come until the 19th century and the advent of a new type of scientist—the epidemiologist.

It was during the 19th century that the vague notion, passed on from the ancient Greeks, that contagious diseases were caused by "noxious vapors" was finally replaced. This advance came in part through the work of scientists such as Louis Pasteur (see pp.244–45), and also the work of early epidemiologists such as William Budd, an English doctor. Budd championed the idea that disease was spread from person to person, rather than being caused by "bad air." His idea was based on observations of an outbreak of typhoid in his village in

### Cholera bacterium
This false-color image shows the *Vibrio* bacterium, which causes cholera. If ingested by humans, the bacterium causes diarrhea and rapid dehydration. Left untreated, cholera can kill in less than 24 hours.

Devon, in 1839, in which he noticed the disease spread quickly from sufferers to their caregivers. Today we know that contagious diseases are caused by microscopic organisms called pathogens that pass from person to person, carried in water, through the air, or transmitted during sexual contact.

### Waterborne disease
William Budd continued to study the spread of typhoid and, in 1847, he made the suggestion that the disease was transmitted through infected water. He went on to make a similar case for cholera the following year. Cholera is an example of an epidemic disease: one that sweeps rapidly from place to place. In the 19th century, this intestinal disease was spreading through Europe, and the epidemic that reached London in 1854 claimed over 10,000 lives. It took a physician named John Snow to help stop it. Snow demonstrated a connection between incidence of the disease and sewage-contaminated drinking water when he traced an outbreak to a public water pump in Broad Street, London. He saw to it that the pump was put out of action and local cases plummeted. Snow's influence went on to improve public health in London and beyond, prompting many to consider him the "Father of Epidemiology."

### Airborne disease
Another way in which disease spread was brought to the public's attention by the work of the German physician Robert Koch. On March 24, 1882 Koch announced that he had found the microbe responsible for tuberculosis, which had ravaging effects on the body and was responsible for nearly one-quarter of all deaths in early 19th-century England. Koch's discoveries revealed that this disease was transmitted by water droplets in the air, and could spread quickly in overcrowded slums through coughing, sneezing, and spitting. Personal hygiene became an important public concern and sufferers were quarantined.

**« BEFORE**

Before the understanding in the 19th century that contagious diseases consisted of live particles, popular belief held that disease was caused by bad gases.

### ORIGINS OF THE MIASMA THEORY
The Miasma theory of **bad gases spreading disease** probably originated in Ancient Greece, where **Hippocrates « 24–25** had already distinguished between diseases that spread freely and those that are restricted to a particular region.

### DISEASE SPREAD BY PARTICLES
In Persia, the physician **Avicenna** advocated **quarantine** in his medical encyclopedia, *The Canon of Medicine*, to reduce the spread of disease—suspecting **contamination by "bodily secretions."** However, **Ibn Khatima**, an Andalusian physician attributed the spread of **bubonic plague** in the 14th century to particles that could be carried on clothes.

### PERSONAL HYGIENE
In 1546 **Girolamo Fracastoro**, an Italian physician, implied that disease-causing particles were living. On this basis he advocated a **regimen of personal hygiene**.

**« SEE ALSO**
*pp.72–73* THE HUMAN BODY REVEALED

**CANON OF MEDICINE**

### The Broad Street Epidemic
This caricature shows people around a water pump contaminated with cholera. John Snow traced an outbreak to one such pump in Broad Street by plotting the occurrence of the disease in buildings in the neighborhood.

**Protecting troops from influenza in 1918**
In an attempt to prevent the spread of influenza among the armed forces during World War I, troops were issued with gauze masks.

Another airborne disease, influenza, is caused by a microbe that is constantly mutating; this results in periodic epidemics, as people's natural defenses are caught unprepared. A particularly devastating strain of the virus caused a pandemic (an epidemic of continental or global scale) in 1918–19, aided by the movements of World War I troops, and by sailors traveling on international trade routes.

## Sexually transmitted disease

While some diseases can be contained by public hygiene measures or effective quarantine, for others spread is more a matter of personal behavior. One such disease is syphilis, which can be unwittingly passed on during sexual contact before its ravaging symptoms develop many years later. The first recorded outbreak of the disease occurred during the Italian War of 1494. French soldiers invading Naples became infected and the disease then spread to France on their return home.

By the 19th century the focus of epidemologists was on finding a cure for syphilis and many other infections.

∇ **Spread of influenza in the US**
During the global influenza pandemic of 1918, the disease was initially confined to coastal areas in the US, but quickly spread inland. The disease infected much of the country and killed over 650,000 people in less than four weeks (shown by increasing prevalence of purple areas on the maps below).

**WEEK ONE**

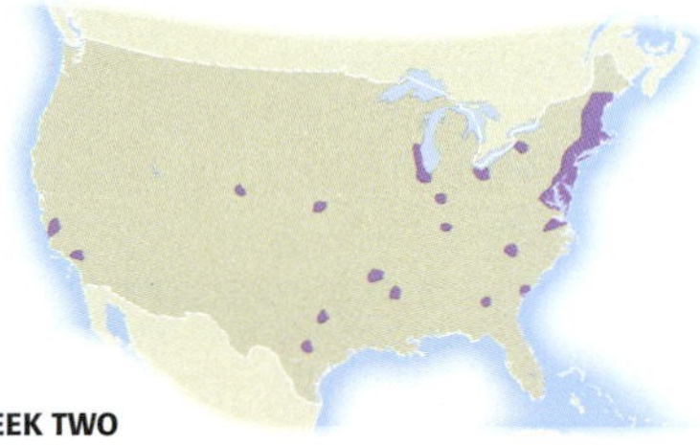

**WEEK TWO**

**WEEK THREE**

**WEEK FOUR**

These cures would come with the development of both antibiotics and sulfonamide drugs (see pp.312–13) in the 20th century. These medicines killed many bacteria outright and so diminished the spread of disease.

### AFTER ≫

Better understanding of the spread of disease in the 20th century led to more effective control measures.

#### MATHEMATICS AND SPREAD OF DISEASE
British physician **Ronald Ross** (1857–1932) applied mathematics to epidemiology. He developed **mathematical models** in order to study the **spread of malaria** in India and provided a method for calculating the rate of the spread of disease and controlling the economics of the disease (the material and work force required to track the epidemic). This approach enabled the global elimination of smallpox in 1977 **258–59 ≫**.

#### EPIDEMIOLOGY OF CANCER
British epidemiologist **Richard Doll** pioneered methods that link disease to **environmental factors**. His research team was the first to confirm the link between smoking and lung cancer.

SEE ALSO ≫
pp.248–49 IMMUNIZATION AND VACCINATION
pp.408–09 DISEASE CHALLENGES

**« BEFORE**

The earliest references to bacteria and viruses were about their effects in causing disease or spoiling food.

**HIDDEN LIFE**
Early humans were aware of the effects of bacteria and viruses long before they could see them—they **spoiled foods** but also allowed production of **vinegar** and **cheese**. Early scholars (Arab Ibn Khatima in the 14th century and Fracastoro in the 16th century) proposed that epidemics of diseases were caused by **invisible "seedlike" particles**, but the truth of these claims was not realized until the 19th century.

**« SEE ALSO**
*pp.242–43* SPREAD OF DISEASE

**IN PRACTICE**
## GRAM STAINING

Developed by Danish biologist Hans Gram in 1884, Gram staining identifies different types of bacteria: Gram-positive bacteria with thick walls stain purple and Gram-negative ones with thinner walls and an outer oily layer stain pink. Many Gram-negative bacteria are pathogenic (disease-causing) due to the presence of an outer oily layer.

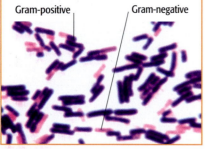

Gram-positive          Gram-negative

**Typical bacterium shapes**

The shape of a bacterium depends on the cell-wall protein filaments inside the cell and is determined by its DNA. It affects mobility and the bacterium's ability to infect other organisms.

**Rods**
One of the most abundant and most studied forms of bacteria, *Escherichia coli* is a bacillus.

**Spheres**
This form of bacterium occurs singly, in pairs (*diplococci*), chains (*streptococci*), or in groups (*staphylococci*).

**Corkscrews**
A form of bacteria that has a flagellum, which helps them move in a twisting fashion.

**Commas**
This form of bacteria, including the cholera bacterium, is comma shaped due to the flagellum.

---

Despite their microscopic size, bacteria have a profound influence on the rest of life on Earth. They recycle resources and affect the atmosphere, but also cause diseases. Viruses cause disease too, by attacking living cells and sabotaging the machinery inside.

### The smallest living things
A bacterium is a small cell with no nucleus—a prokaryotic organism. Bacteria come in a range of shapes, but their basic structure is quite uniform. They reproduce by binary fission—a bacterium splits into two cells, with each receiving a copy of the parent cells' genetic material (see pp.256–57). Bacteria can also join and exchange small rings of DNA called plasmids by way of conjugation—this explains the rapid spread of antibiotic resistance (see pp.408–409 and opposite). Dutch microbiologist Antonie van Leeuwenhoek (see pp.94–95) first observed bacteria in 1674 and in 1828 German biologist Christian Ehrenberg named them from the Greek for "little stick." Later that century, another German biologist Ferdinand Cohn observed a greater diversity of shapes, and classified them into rods, spheres, commas, and spirals—a system still used today (see below).

### Good bacteria, bad bacteria
Bacteria are an essential part of the cycle of life because they feed on dead matter, decomposing it and recycling many of the chemical constituents. Bacteria themselves have astonishing chemical diversity and can perform many processes that are impossible in plants and animals. In the 1880s Russian microbiologist Sergei Winogradsky discovered that some bacteria get energy from minerals, making food in a process that mirrors photosynthesis (see pp.256–57).

**CHEMIST (1822–95)**
## LOUIS PASTEUR

Pasteur is one of the founding fathers of microbiology: his work on microorganisms led to pasteurization of milk and the first vaccines. Despite a modest background in working-class France, Pasteur became one of the country's greatest scientists. After his death, his remains were interred beneath the Institut Pasteur, named in his honor.

# Bacteria and Viruses

**A bacterium is the smallest living cell. As many as 1,000 bacteria could comfortably line up across just one period on this page and still have room to spare. Smaller still are viruses, infectious particles that can multiply only inside host cells of living bacteria, animals, or plants.**

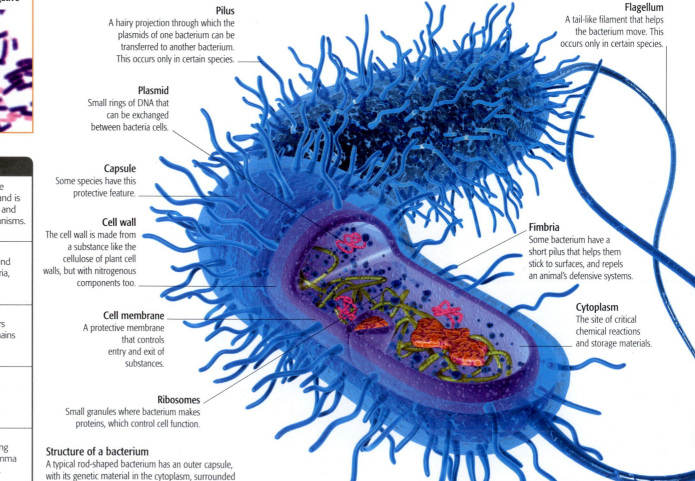

**Pilus**
A hairy projection through which the plasmids of one bacterium can be transferred to another bacterium. This occurs only in certain species.

**Plasmid**
Small rings of DNA that can be exchanged between bacteria cells.

**Capsule**
Some species have this protective feature.

**Cell wall**
The cell wall is made from a substance like the cellulose of plant cell walls, but with nitrogenous components too.

**Cell membrane**
A protective membrane that controls entry and exit of substances.

**Ribosomes**
Small granules where bacterium makes proteins, which control cell function.

**Flagellum**
A tail-like filament that helps the bacterium move. This occurs only in certain species.

**Fimbria**
Some bacterium have a short pilus that helps them stick to surfaces, and repels an animal's defensive systems.

**Cytoplasm**
The site of critical chemical reactions and storage materials.

**Structure of a bacterium**
A typical rod-shaped bacterium has an outer capsule, with its genetic material in the cytoplasm, surrounded by a cell membrane, and overlaid with a cell wall.

## Spreading antibiotic resistance

Recent years have seen antibiotic resistance rise. It can spread by exchange of DNA between conjugating bacteria.

**1 Duplicating plasmid**
Plasmids (rings of DNA) help bacteria make enzymes that enable the bacteria to resist drugs called antibiotics. Bacteria replicate these plasmids.

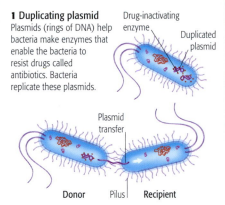

Drug-inactivating enzyme
Duplicated plasmid

Plasmid transfer

Donor    Pilus    Recipient

**2 Plasmid transfer** The plasmid copy is passed from the donor, through a pilus, to the recipient bacterium. This process is known as conjugation.

Drug-inactivating enzymes

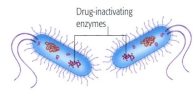

**3 Drug-resistant strains** Recipient bacteria inherit a gene that makes new enzymes. This process produces populations of antibiotic-resistant bacteria.

## 7 BILLION is the number of bacteria that can live in a human mouth—equivalent to more than the number of people living on the whole planet.

Instead of using light as an energy source, they use ammonia released from decomposers, changing it into nitrates, so making energy for themselves and nitrogen available to plants (see pp.278–79). However, some bacteria can be harmful not only to plants but also to humans, causing diseases such as tuberculosis, pneumonia, and tetanus.

## Discovery of viruses

Viruses differ from bacteria in that they are not cells. In fact, a virus is nothing more than a particle of genetic material surrounded by a protein casing (see right). In 1898 Dutch microbiologist Martinus Beijerinck first became aware of viruses while studying mosaic disease on tobacco plants. These tiny particles were finally seen by German physician Helmut Ruska and his team, in 1939, using the electron microscope that had been invented just eight years earlier.

## Cycle of a virus

A virus is not made of cells, so has no cellular "machinery" to reproduce or perform metabolic processes on its own. It can only complete its cycle by invading and taking over a living host cell.

**Communities of bacteria**
The chemical versatility of bacteria is demonstrated by stromatolites—films of photosynthetic bacteria that bind rock particles into mounds. These examples are found in coastal waters off Australia.

**Genetic material**
Viruses contain either DNA or RNA.

**Capsid**
Protein shell built from highly ordered protein subunits called capsomeres.

**Structure of a virus**
Much smaller than a bacterium, a virus is not a cell—it is a capsid (shell) of protein, which envelopes genetic material.

**Receptor molecules**
These help the virus bind to the host cell.

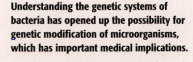

Understanding the genetic systems of bacteria has opened up the possibility for genetic modification of microorganisms, which has important medical implications.

### VACCINES FROM BACTERIA TOXINS
The harmful effects of disease-causing bacteria are caused by **poisonous substances called toxins**. Many toxins are neutralized by the body's immune system and have been used to develop **harmless forms** called **toxoids**, which are the **basis of vaccines** that protect against diseases such as tetanus and diphtheria. Tetanus and diphtheria vaccines were routinely combined with pertussis (whooping cough) vaccination from the 1940s.

### SEE ALSO »
*pp.246–47* NATURAL DEFENSES
*pp.248–49* IMMUNIZATION AND VACCINATION
*pp.310–11* THE DISCOVERY OF PENICILLIN

### Typical virus shapes
There are many types of virus. The varying shape or make-up of viruses may assist in the replication cycle by helping to penetrate a host cell or resist attack by the host defense system.

**Icosahedral**
Most animal viruses are roughly spherical, with at least 12 units making up the protein capsid.

**Spiral-helical**
Some viruses have capsid units arranged in a helix around the length of genetic material.

**Complex**
These viruses do not come in just one shape, but have variations to the capsid, such as a "tail."

**Enveloped**
These viruses coat themselves in host-cell membrane as a disguise from the host immune system.

**Virus**

**1 Attachment**
Virus capsid (shell) binds to cell membrane of specific type of cell.

Host-cell cytoplasm

Host-cell nucleus

**New virus**

**6 Release**
New virus is released from dying cell.

**2 Penetration**
Virus enters the cell, where its capsid breaks down to release the genetic material inside.

**3 Replication**
Genetic material of virus enters host nucleus.

**5 Assembly**
New capsids assemble around viral genetic material to make a new virus.

**4 Synthesis**
Virus instructs production of new viral DNA.

The interior of the human body encourages the growth of many different types of microbes. Once inside, certain kinds of bacteria and viruses breed (see pp.244–45) and can cause harm. Such disease-causing organisms are known as pathogens. Their ubiquity means that invasion is inevitable, but the body puts up an impressive defense.

### Barriers and fighters

Human skin forms a natural barrier to attackers: its outer layers are made of hardened, dead cells, which are lubricated by impervious oil. If this barrier becomes torn, a self-healing process automatically starts. A substance dissolved in the exposed blood, called fibrinogen—with the help of another blood component called platelets—rapidly solidifies into fibers, which are known as fibrin. These trap the precious blood cells within the body, slowing the bleeding and eventually stopping it entirely. Over the coming days, as the skin is repaired beneath the wound, the drying scab is pushed to the surface until it eventually falls away from the repaired skin.

Inside the body, a defending army of cells is mobilized to deal with invasions at the source of infection: phagocytes (pathogen-eating cells) attack microbes. Russian microbiologist Ilya Mechnikov discovered them in 1882 when he observed mobile cells in the larvae of starfishes. To put this idea to the test, he inserted small thorns into the larvae, and, the next day, he observed that the thorns were surrounded by mobile cells. However, a focused attack by phagocytes, no matter how effective, gives no long-term protection against a disease. Long-term immunity requires a much more sophisticated type of cell.

### Understanding immunity

The body can distinguish its own cells from alien ones, such as bacteria, by recognizing chemical markers, known as antigens, on their surfaces. Any cell with an unfamiliar, "non-self," marker is targeted for attack by phagocytes. In

> **PHAGOCYTES** White blood cells that protect the body by swallowing and digesting bacteria or harmful particles.
>
> **LYMPHOCYTES** White blood cells that can release antibodies to provide long-term immunity.

substance dissolved in the blood, something that is undoubtedly secreted by blood cells.

### Targeting specific invaders

In 1891 microbiologist Paul Ehrlich identified the protective substances in blood and called them antibodies. They are produced by white blood cells called lymphocytes, which are coated in antibody molecules with shapes that fit particular antigens. The result is that any invading microbe is likely to find itself stuck via their antigens to these lymphocytes. When this happens the lymphocytes are triggered to divide and release their surface antibodies, which circulate through the bloodstream, binding and disabling any microbes with the right-shaped antigens. This accounts for the observations of Shibasaburo and Behring that serum can enhance immunity. The system is efficient because a specific microbe will only

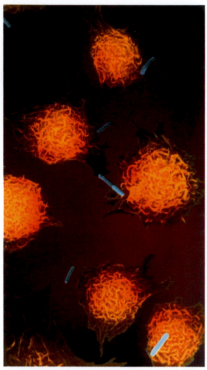

**RED AND WHITE BLOOD CELLS**

<< SEE ALSO
pp.242–43 THE SPREAD OF DISEASE
pp.244–45 BACTERIA AND VIRUSES

<< **BEFORE**

Immunity is gained naturally (by contracting a disease) or artificially (by vaccination) << 248–49. Immunity was appreciated from antiquity, but not fully understood.

**IMMUNITY RECOGNIZED**
The concept of immunity is an old one. In 430 BCE Greek historian Thucydides recorded that in Athens **people surviving the plague** were not attacked a **second time**. Al-Razi, a medieval Islamic physician, observed a similar thing with measles and smallpox.

**IDENTIFYING BLOOD CELLS**
**White blood cells**—distinguished from more abundant red blood cells by their prominent nuclei—were first seen by **Renaissance microscopists**, but their roles in defense would not be appreciated for over 200 years.

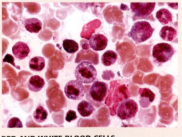
Phagocytes: blood cells that swallow bacteria
Around 70 percent of white blood cells are phagocytes. They contain little sacs of digestive fluid, which they use to destroy bacteria that they swallow.

# Natural Defenses

**The body is under constant attack from microorganisms, which can cause damage and disease. However, a natural, effective system of defenses barricades their entry, and swiftly disables any that get through. This system is so sophisticated that it remembers attackers, and makes the body immune from future invasion.**

Blood clotting is fast, but, while the wound is open, bacteria can infect the exposed area. Elsewhere in the body, where the skin barrier is necessarily compromised—such as the eyes and mouth—this invasion is a constant threat, even though tears and other fluids contain natural antiseptics, which stave off some attackers.

1890 Japanese bacteriologist Kitasato Shibasaburo and German physiologist Emil von Behring injected the antigens of tetanus bacteria into animals, making them immune to the disease. They discovered that serum (a fluid component of blood) extracted from these animals could be injected into others to make these immune too. Shibasaburo and Behring had discovered humoral immunity: immunity that depends on a

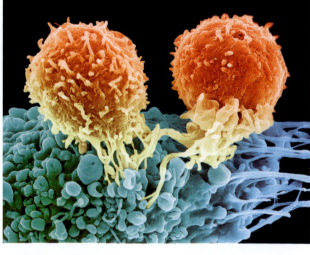
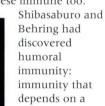

**Killer cells**
These specialized lymphocytes (shown here in orange) were discovered in the 1970s. They bond to diseased cells, such as those of cancerous tumors or infected with viruses, and kill them by making them autodestruct.

stimulate the lymphocytes that are able to release the exact type of antibody to target that particular microbe, an idea first explained in the 1950s by an Australian scientist called Frank Burnet in his Clonal Selection Theory.

### Remembering the invader

Burnet also explained long-term immunity. He discovered that some of the lymphocytes stimulated in the first invasion do not release their antibodies as part of the attack, but are held back in reserve. They become memory cells, and can remain in circulation in the body for months or even years. It is the presence of memory cells for a particular pathogen that makes the body immune to the disease caused by it. A second invasion "awakens" the memory cells: they proliferate more quickly and release more antibodies. The result is that this second response staves off the damaging effects of the disease so the symptoms may not develop.

## Immune response

The body's ability to kill harmful invading microbes involves a chain of defensive white blood cells, triggered by microbe-swallowing phagocytes, and culminating in an army of different lymphocytes.

## THE SECONDARY IMMUNE RESPONSE

Initial exposure to a pathogen (foreign particle) stimulates a primary immune response—the slow production of a moderate amount of antibodies. But a second exposure causes a bigger, faster secondary immune response—production of enough antibodies to destroy the invader without causing symptoms.

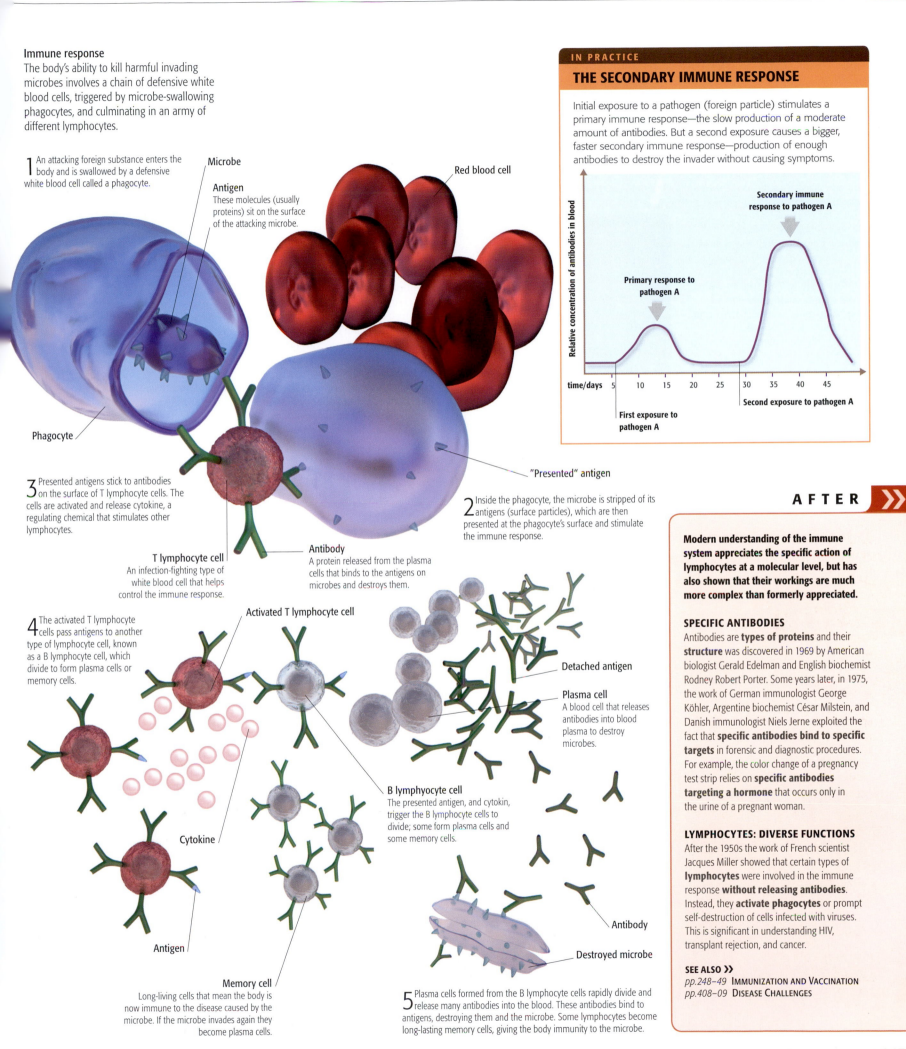

1 An attacking foreign substance enters the body and is swallowed by a defensive white blood cell called a phagocyte.

**Microbe**

**Antigen**
These molecules (usually proteins) sit on the surface of the attacking microbe.

**Red blood cell**

**Secondary immune response to pathogen A**

**Primary response to pathogen A**

Relative concentration of antibodies in blood

time/days   5   10   15   20   25   30   35   40   45

**First exposure to pathogen A**

**Second exposure to pathogen A**

**Phagocyte**

3 Presented antigens stick to antibodies on the surface of T lymphocyte cells. The cells are activated and release cytokine, a regulating chemical that stimulates other lymphocytes.

"Presented" antigen

2 Inside the phagocyte, the microbe is stripped of its antigens (surface particles), which are then presented at the phagocyte's surface and stimulate the immune response.

**T lymphocyte cell**
An infection-fighting type of white blood cell that helps control the immune response.

**Antibody**
A protein released from the plasma cells that binds to the antigens on microbes and destroys them.

4 The activated T lymphocyte cells pass antigens to another type of lymphocyte cell, known as a B lymphocyte cell, which divide to form plasma cells or memory cells.

**Activated T lymphocyte cell**

**Detached antigen**

**Plasma cell**
A blood cell that releases antibodies into blood plasma to destroy microbes.

**B lymphyocyte cell**
The presented antigen, and cytokin, trigger the B lymphocyte cells to divide; some form plasma cells and some memory cells.

**Cytokine**

**Antigen**

**Antibody**

**Destroyed microbe**

**Memory cell**
Long-living cells that mean the body is now immune to the disease caused by the microbe. If the microbe invades again they become plasma cells.

5 Plasma cells formed from the B lymphocyte cells rapidly divide and release many antibodies into the blood. These antibodies bind to antigens, destroying them and the microbe. Some lymphocytes become long-lasting memory cells, giving the body immunity to the microbe.

Modern understanding of the immune system appreciates the specific action of lymphocytes at a molecular level, but has also shown that their workings are much more complex than formerly appreciated.

### SPECIFIC ANTIBODIES

Antibodies are **types of proteins** and their **structure** was discovered in 1969 by American biologist Gerald Edelman and English biochemist Rodney Robert Porter. Some years later, in 1975, the work of German immunologist George Köhler, Argentine biochemist César Milstein, and Danish immunologist Niels Jerne exploited the fact that **specific antibodies bind to specific targets** in forensic and diagnostic procedures. For example, the color change of a pregnancy test strip relies on **specific antibodies targeting a hormone** that occurs only in the urine of a pregnant woman.

### LYMPHOCYTES: DIVERSE FUNCTIONS

After the 1950s the work of French scientist Jacques Miller showed that certain types of **lymphocytes** were involved in the immune response **without releasing antibodies**. Instead, they **activate phagocytes** or prompt self-destruction of cells infected with viruses. This is significant in understanding HIV, transplant rejection, and cancer.

SEE ALSO »
pp.248-49 IMMUNIZATION AND VACCINATION
pp.408-09 DISEASE CHALLENGES

A body is immune to a disease when its defense system has been primed to fight it. It can do this because white blood cells called lymphocytes release substances called antibodies, which disable harmful agents (see pp.246–47). This process happens naturally when a body contracts a disease. Medicine can help in two ways. First, passive immunization involves injecting the antibody directly and is done when immediate treatment is needed, against for example fast-acting food poisoning, such as botulism, or to treat the bites of a venomous animal. Second, active immunization involves injecting a "trigger" to "fool" the immune system. This trigger, called an antigen, is sometimes just an innocuous molecule from the outer coat of the microbe (bacteria or virus). If this molecule is isolated and

**Early vaccination tools**
Edward Jenner was the first doctor to vaccinate his patients against smallpox. His tools, right, consisted of small sharp blades, which were used to pierce the patient's skin so the vaccine could be applied.

injected, the army of lymphocytes is still activated even though the disease itself does not break out. In fact, sometimes the trigger doesn't even have to come from the dangerous microbe itself.

**Discovery of vaccinations**
English doctor Edward Jenner's experiments famously induced immunity to smallpox, not with a smallpox trigger—but with cowpox (see pp.156–57). He called the process "vaccination" from the Latin *vacca* for cow, a term that has stuck ever since.

At the time disease was thought to be caused by "bad gases" or similar vague notions. It wasn't until the late 19th century when the work of French chemist Louis Pasteur and German physician Robert Koch showed that disease is caused by living particles. Pasteur began experiments to develop immunity triggers—which he called vaccines—for diseases, and the direction of his research came about by accident. In his laboratory, a neglected culture of cholera bacteria became weakened because the culture medium they were growing in was "rejected" so that they no longer produced the disease, but they still triggered immunity. It spurred a flurry of work. Pasteur successfully produced a vaccine for cholera in 1879 and for anthrax in 1881. Perhaps his most celebrated success came with his work on rabies. Pasteur and French physician Emile Roux produced a vaccine, but it had been tested only

on animals. Pasteur risked prosecution by trying it on a boy who had been attacked by a rabid dog. The boy never developed rabies and Pasteur was fêted as a hero.

**Safer vaccines**
The vaccines of Jenner and Pasteur were "live" vaccines: they contained living microbes, but far milder than the ones

> **IMMUNIZATION** Any process by which the body is made immune to a particular disease.
>
> **VACCINATION** The administration of a vaccine (containing an antigen) in order to trigger immunity to a particular disease.
>
> **INOCULATION** The administration of a mild pathogen, in order to trigger immunity or for another purpose, such as producing a disease for study.

# Immunization and Vaccination

**There are few other areas of medical science that have had as great an impact on the health of humankind as that of vaccination: medicine creates immunity to disease where previously there was none. The first vaccination protected people against smallpox, and since then vaccines have been developed to immunize people against many more diseases, saving millions of lives.**

## BEFORE

**The principle of vaccination can be traced back to medieval times, but the procedure was not without risk.**

**EARLY PRACTICES OF VARIOLATION**
**Variolation** was **deliberate inoculation** with a mild form of the smallpox virus in order to **induce smallpox immunity << 156–57**. This was practiced in early China and India and was attempted in Europe by the Greek physician Giacomo Pylarini in 1701. Lady Montagu, wife of the British ambassador in Constantinople in 1716 was so impressed by its effects that she had her own children variolated, and later encouraged others to do the same.

**SMALLPOX SUFFERER**

**<< SEE ALSO**
*pp.242–43* SPREAD OF DISEASE
*pp.244–45* BACTERIA AND VIRUSES
*pp.246–47* NATURAL DEFENSES

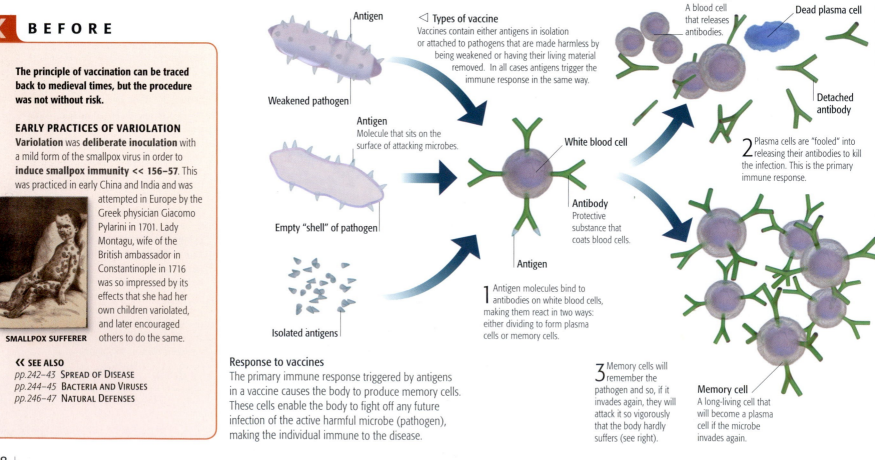

Antigen

◁ **Types of vaccine**
Vaccines contain either antigens in isolation or attached to pathogens that are made harmless by being weakened or having their living material removed. In all cases antigens trigger the immune response in the same way.

Weakened pathogen

**Antigen**
Molecule that sits on the surface of attacking microbes.

Empty "shell" of pathogen

Isolated antigens

White blood cell

**Antibody**
Protective substance that coats blood cells.

Antigen

**1** Antigen molecules bind to antibodies on white blood cells, making them react in two ways: either dividing to form plasma cells or memory cells.

**Plasma cell**
A blood cell that releases antibodies.

**Dead plasma cell**

**Detached antibody**

**2** Plasma cells are "fooled" into releasing their antibodies to kill the infection. This is the primary immune response.

**3** Memory cells will remember the pathogen and so, if it invades again, they will attack it so vigorously that the body hardly suffers (see right).

**Memory cell**
A long-living cell that will become a plasma cell if the microbe invades again.

**Response to vaccines**
The primary immune response triggered by antigens in a vaccine causes the body to produce memory cells. These cells enable the body to fight off any future infection of the active harmful microbe (pathogen), making the individual immune to the disease.

**Childhood vaccination**
Programs for vaccinating children have been in place since the mid 20th century and have played a key role in controlling infectious diseases. Here, Native American Shosone children line up for vaccines in Wyoming in 1955.

IN PRACTICE

## SMALLPOX ERADICATION PROGRAM

In 1959 the World Health Organization (WHO) of the United Nations launched an intensive program to eradicate smallpox. Immunizing everyone would have been impracticable, so instead the strategy was one of ring immunization: cases were isolated by vaccination of neighbors, so containing the disease, and preventing it from spreading. Smallpox was declared eradicated on December 9, 1979 and is the only disease that has been completely wiped out by humanity.

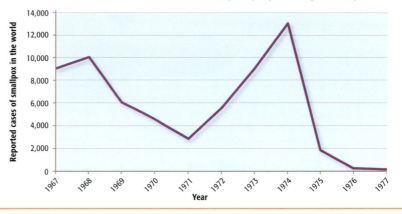

that inflicted terrible disease. Pasteur's work led to more vaccines being made in the 20th century, and the realization that live microbes were not always needed: that the triggers for immunity were specific molecules—the antigens—which could be isolated from harmful microbes to make vaccines even safer.

### Vaccination programs
If enough people in a population are immune to a disease, it is difficult for the disease to spread. This idea, called "herd immunity", is behind government initiatives that have made vaccination compulsory—such as the 1853 English law for smallpox vaccination. This has drastically reduced the incidence of infectious disease in many parts of the world. However, it was not until the middle of the 20th century that the World Health Organization was able to oversee a vaccination program (see above) that ultimately led to the complete eradication of smallpox in 1979.

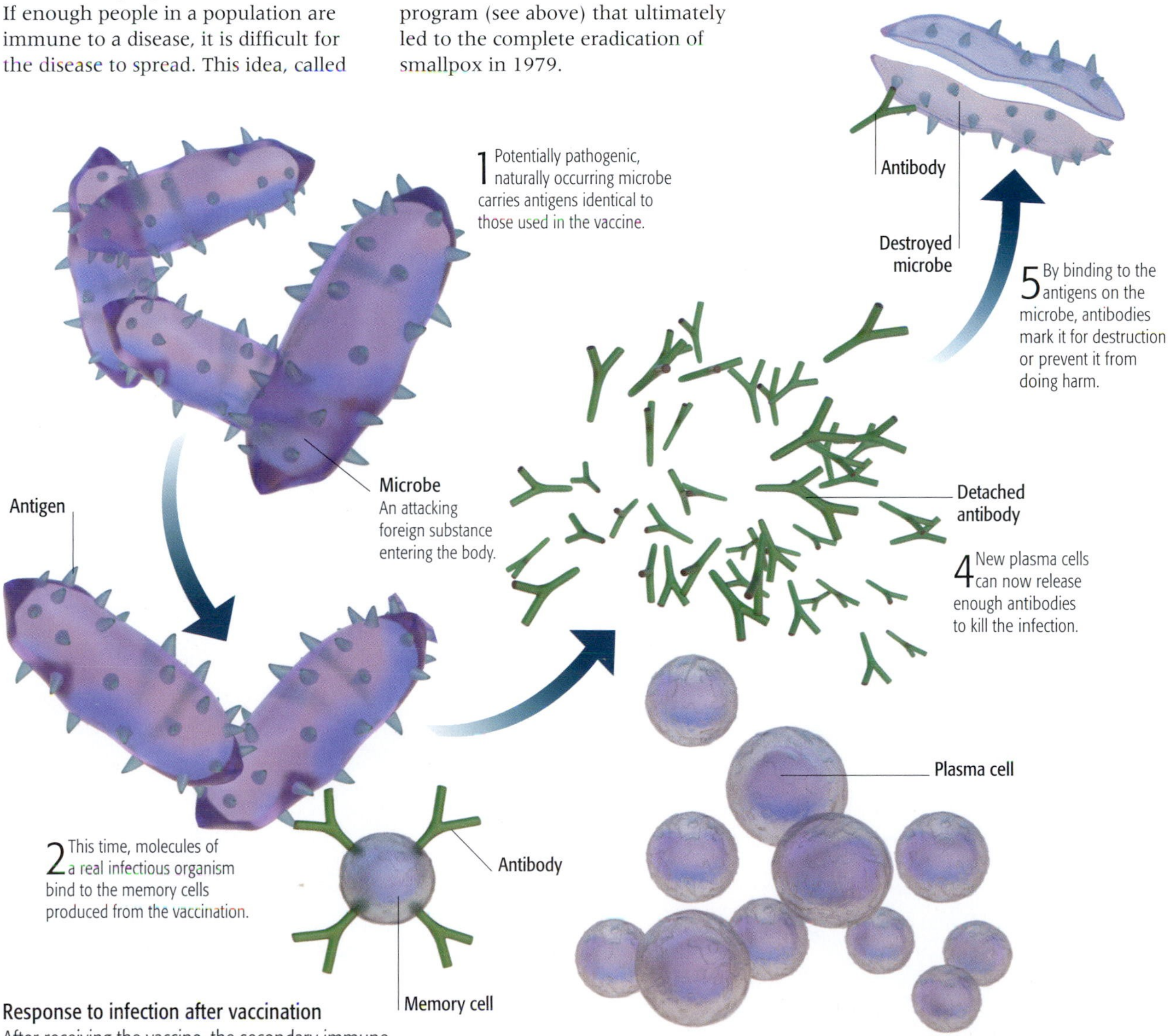

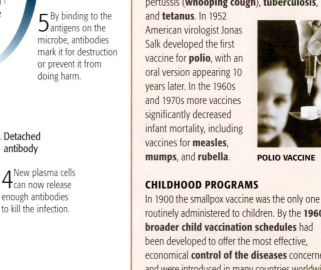

1 Potentially pathogenic, naturally occurring microbe carries antigens identical to those used in the vaccine.

**Antibody**

**Destroyed microbe**

5 By binding to the antigens on the microbe, antibodies mark it for destruction or prevent it from doing harm.

**Antigen**

**Microbe**
An attacking foreign substance entering the body.

**Detached antibody**

4 New plasma cells can now release enough antibodies to kill the infection.

2 This time, molecules of a real infectious organism bind to the memory cells produced from the vaccination.

**Antibody**

**Plasma cell**

**Memory cell**

### Response to infection after vaccination
After receiving the vaccine, the secondary immune response will be more effective when the body is infected naturally by the pathogen. This means that the pathogen will be destroyed before it can cause disease.

3 New plasma cells are formed from the memory cells being triggered by antigens.

### AFTER

Organizations around the world are fighting to combat the spread of disease with the development of new vaccination programs.

#### NEW VACCINES
The start of the 20th century saw the development of vaccines for **diphtheria**, pertussis (**whooping cough**), tuberculosis, and **tetanus**. In 1952 American virologist Jonas Salk developed the first vaccine for **polio**, with an oral version appearing 10 years later. In the 1960s and 1970s more vaccines significantly decreased infant mortality, including vaccines for **measles**, **mumps**, and **rubella**.

**POLIO VACCINE**

#### CHILDHOOD PROGRAMS
In 1900 the smallpox vaccine was the only one routinely administered to children. By the **1960s broader child vaccination schedules** had been developed to offer the most effective, economical **control of the diseases** concerned and were introduced in many countries worldwide.

#### NEW CONTROVERSIES
**Objections** to vaccination have been based on **perceived health concerns**. For example, in the UK in 1998 a published report linked the measles-mumps-rubella combined vaccine with **autism**. Subsequently the claim was discredited by a number of authorities, but it caused many parents to **opt out of this vaccine** for their children.

SEE ALSO »
pp.408–09 DISEASE CHALLENGES

« BEFORE

Early humans had no artificial light until they learned to make fire. From then on they built fires at cave mouths to provide light (and warmth) for the early part of the night.

**NATURAL LIGHT**

The first people did their work **during the hours of daylight**, so they had far more usable hours in summer than in winter. Hunting during the winter months would have been difficult and dangerous; many animals are most easily hunted in the evening, but pursuing them would have entailed a dark walk home, in danger from predators with **better night vision**.

**OIL LAMPS**

Once humans began cooking, they had access to animal fats. They soon began to make **simple tallow lamps**, which emitted a weak, smoky, yellow flame but provided light throughout the night.

**EARLY EGYPTIAN OIL LAMP**

**LIGHT FROM CANDLES**

Animal fat was also used to make **candles**. The ancient Egyptians are thought to be the first to use them; however, it was the Romans who introduced **the wick** that runs down the center. It was not until the Middle Ages that **beeswax** started to be used in the production of candles instead of animal fat.

« SEE ALSO
pp.114–15 THE NATURE OF LIGHT
pp.116–17 SPLITTING AND BENDING LIGHT
pp.164–65 ELECTRIC CURRENT

# Artificial Light

Until 200 years ago the best artificial light was candlelight, and even today many millions of people still rely upon simple oil lamps. The widespread use of indoor and outdoor lighting across the developed world became possible as scientists harnessed the powers of gas and then electricity, and invented new forms for their use—such as the lightbulb.

Living without any artificial light can be restricting, so the development of oil lamps and tallow candles made life much easier for early humans. The invention of gas lighting at the end of the 18th century marked a significant breakthrough in lighting technology, as did incandescent electric lamps at the end of the 19th century.

## Oil and gaslight

Oil will not burn easily at low temperatures, and it can be lit with a match only if it has a wick: a piece of absorbent material—such as string or tape—that soaks up the oil. When a match is put to the wick, the heat from the flame turns a little oil into vapor, which catches fire.

Coal gas was thought to have no real use until William Murdoch used it for lighting his house in Cornwall around 1799. Two years later he lit all of the Soho Foundry in

### Adjustable lamps

Still the only source of light for many people, oil lamps of some form have been used since 70,000 BCE. This 19th-century lamp has a fuel reservoir in the base, a knob for adjusting the wick, and a glass chimney to protect the flame.

PHYSICIST AND CHEMIST (1828–1914)

## JOSEPH SWAN

Born in Sunderland, northern England, in 1828, Joseph Swan earned his living by running a pharmacy in nearby Newcastle, and did his own scientific research in his spare time. He invented the dry photographic plate and several other photographic processes. Swan also produced the world's first artificial filament. He did this by dissolving blotting paper in zinc chloride solution, and then using a syringe to squirt this solution into alcohol. By carefully carbonizing this filament in the absence of air, he managed to produce the world's first successful lightbulb.

### Edison's bulb

Although Edison was not the first to create a lightbulb, his model was more commercially viable because it was long-lasting and formed part of a whole lighting system.

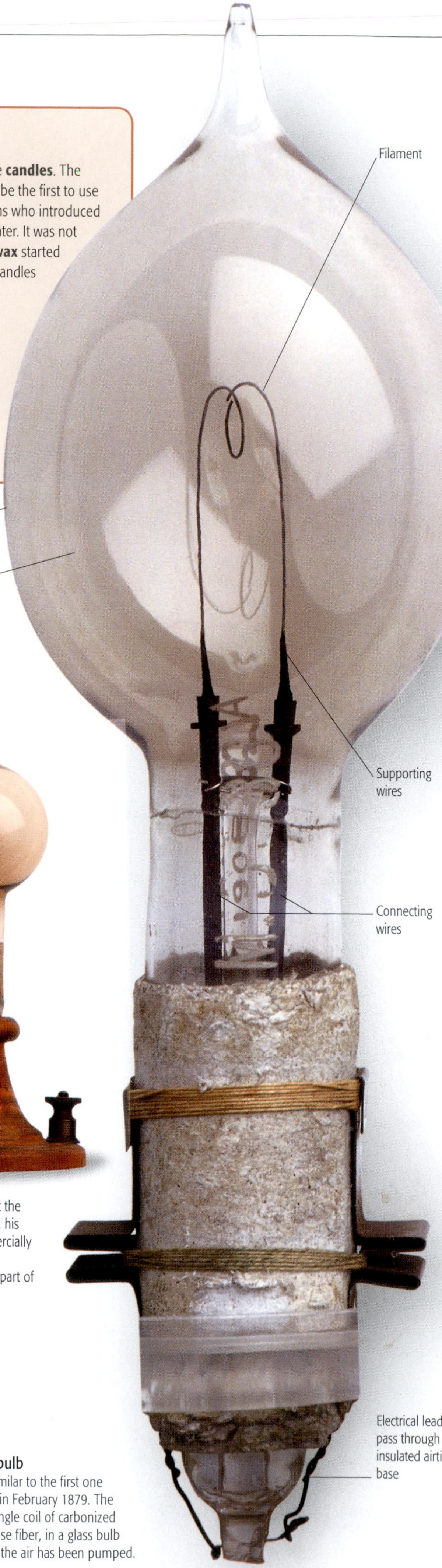

### Swan's lightbulb

This lamp is similar to the first one demonstrated in February 1879. The filament is a single coil of carbonized artificial cellulose fiber, in a glass bulb from which all the air has been pumped.

Filament

Thin glass bulb

Mixture of inert gases at low pressure

Supporting wires

Connecting wires

Electrical leads pass through insulated airtight base

# "I have not **failed**, I have found **10,000 ways** that do not work."

THOMAS ALVA EDISON, PHYSICIST, C.1910

Birmingham, and gradually gas works and gaslight spread across Britain, both in the streets and in homes. The bare gas flame was a dim flickering yellow until 1890, when it was transformed into bright white by the gas mantle—a fishnet of metal oxide placed over the gas. But by then electricity had started to become a viable alternative.

## The electrical arc lamp

The electric battery was invented by Volta in 1799 (see pp.162–63) and people all over Europe were soon experimenting with this new kind of electricity. Humphry Davy, working in London in around 1810, invented the arc lamp using two carbon rods, each connected to one terminal of a large battery. Davy touched the electrodes together, shorting the circuit, and then separated them by a millimeter or so. The electric current jumped the gap between the two electrodes, creating a plasma that glowed brightly.

Unfortunately this type of carbon arc is difficult to manage, since the electrodes continually burn away, and its extreme brightness can cause eye damage. Arc lamps were used, however, for some street lighting and lighthouses.

## Inventing the lightbulb

Many scientists noticed that passing an electric current through a wire made it heat up and glow, and that this could form the basis of an "incandescent"

lamp. The problem was that when the wire reached a dull red heat it would either melt or burn out. Two scientists competed to create the first successful lightbulb: the American Thomas Alva Edison (see pp. 264–65) and the British chemist Joseph Wilson Swan.

Swan won the race in February 1879, with a lamp that had a filament of pure carbon, contained in a vacuum within a glass bulb.

## Fluorescence

Fluorescent tubes first appeared at the end of the 19th century. They work by discharging electricity through a gas at low pressure; this produces ultraviolet light, which makes the phosphorescent coating of the tube glow with a bright white light.

Fluorescent lamps are about six times more efficient than incandescent lamps. Compact fluorescent lamps (CFLs) have a shorter tube, folded up to make a neat package.

As pressure grows to reduce energy consumption, there will be further evolution of sources of artificial light.

### LIMELIGHT

In the 1820s British scientist Goldsworthy Gurney invented a blowpipe that burned hydrogen gas in oxygen, creating an extremely hot flame—it had a temperature of around 3,500° F (2,000° C). He then discovered that by heating a piece of chalk or limestone within the flame he could make an intense white light. This "limelight" was used for stage lighting in theaters—hence the expression "in the limelight"; theater spotlights are called "limes" to this day. The high temperature excites the electrons within the calcium atoms, making them emit an intense white light.

**Neon street signs in Hong Kong**
Fluorescent tubes containing neon gas are frequently used for street signs, because they can be made in any shape and color, are weather-resistant and long lasting. Future possibilities include energy-efficient fluorescents, electrodeless lamps, and light-emitting diodes (LEDs).

## AFTER

Scientists and engineers are continually looking for new and more efficient ways of turning electricity into light, and at the same time to improve its color.

### LIGHT-EMITTING DIODES (LEDS)

LEDs are energy-efficient devices that produce light through semiconductor diodes. They first appeared in the 1960s, and initially emitted only a red, dim light. They are now very bright, work more efficiently than fluorescent lamps, and do not get hot.

### SAVING ENERGY

Incandescent lamps—traditional lightbulbs—waste some 90 percent of their energy as heat, so there is increasing pressure to save energy by using CFLs or LEDs instead.

**LED**

SEE ALSO >>
pp.252–53 GENERATING ELECTRICITY
pp.364–65 LASERS AND HOLOGRAMS

## The fluorescent lamp

The glass tube is filled with an inert gas—such as argon—at low pressure. Current flowing through the gas vaporizes the mercury. Mercury atoms that collide with electrons become excited and then emit photons of ultraviolet light as they fall back to their original state (see p.287). The phosphor coating on the inside of the tube converts this ultraviolet light to white visible light.

Glass tube

Electron

Inert gas, such as argon, neon, krypton, or xenon, at very low pressure (less than one percent of atmospheric pressure)

Phosphor coating on inside of glass tube

Contact pins to plug into socket and carry current from lighting circuit

"Ballast" or "choke" to regulate the current; without it the fluorescent tube would carry more and more current and burn out

Tungsten electrode, coated with a mixture of barium, strontium, and calcium oxides

Mercury

Mercury vapor exists as single atoms

# Generating Electricity

**Current electricity was invented in 1799, but it remained a laboratory curiosity for much of the 19th century. The invention of practical electric lighting, however, followed by other electrical machines, created a rapidly growing demand for electricity. Most of the world's supply is now generated in large power stations.**

## BEFORE

**Transmission of electricity on an industrial scale required metals such as aluminium and copper: excellent and inexpensive conductors.**

### INSULATORS

All electricity would have been lost from cables underground or under the sea without **gutta percha**—a natural latex material related to rubber—whose insulating properties were discovered in 1842.

**« SEE ALSO**
*pp.164–65* ELECTRIC CURRENT
*pp.166–67* ELECTROMAGNETISM
*pp.170–71* MICHAEL FARADAY

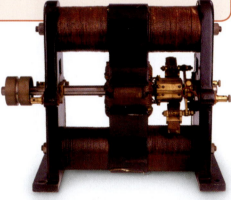

**The Gramme dynamo**
The first machine to generate power commercially was built in 1871 by Belgian engineer Zénobe Gramme. It used a series of armature coils.

The first electric currents came from batteries (see pp.162–63), but scientists realized a continuous process was needed for protracted use. The first device to provide this was Michael Faraday's dynamo, invented in 1831 at London's Royal Institution. This contains a coil of wire that rotates within a magnetic field, which makes an electric current flow in the wire.

### Direct and alternating current

The dynamo produces direct current (DC), but when the principles of alternating current (AC) were established in the late 1880s (see pp.164–65), dynamos were replaced by alternators. These use a rotating magnet inside a coil of wire or a coil of wire in a magnetic field (below) to produce AC. Motor vehicles use alternators to convert some of the energy produced by the internal combustion engine (see pp.254–55) into electrical energy.

The challenge for electricity producers is to find effective ways of generating the rotary motion that can run alternators and produce electricity. Hydroelectric plants and windmills do it directly, with spinning blades driven by water or wind. In the UK Parsons (see right) developed steam turbines in the 1880s that allowed production of cheap and plentiful electriciy. This is the most common form of generation today.

### Universal generation

In the early days there was much debate about whether it would be better to use DC or AC. The great inventor Thomas Edison championed DC, while the Serbian-born American Nikola Tesla favored the AC system.

In 1886 the first AC general-purpose generators were built in Massachusetts, and at the Rome-Cerchi Italian power station. Two years later Tesla introduced the idea of greatly stepping up the voltage for transmission (in order to reduce power loss), and then stepping it down again for use at lower voltages.

### Power stations

About 80 percent of the world's energy is produced in large power stations. Heat is generated by burning fossil fuels or by allowing controlled fission of radioactive elements; the heat is used to boil water and make super-heated steam. At high pressure this drives steam turbines, which use dozens of blades to extract kinetic (motion) energy from the steam (see right).

### Transmission of electricity

Electricity can be transmitted across great distances through power lines, generally suspended from pylons. There is some energy loss on the way, since resistance in the wires causes energy to be converted to heat. This is minimized by transmitting at high voltage.

**ENGINEER (1854–1931)**

### SIR CHARLES PARSONS

Born in London, Parsons was the third son of the third Earl of Rosse, an astronomer. He worked for several engineering firms in the north of England developing the turbine engine and electricity generator, before setting up his own company in 1889. Within 10 years he had produced a megawatt turbine and a turbine-powered boat *Turbinia*, which was faster than any ship in the Royal Navy. Steam turbines now produce most of the world's electricity.

## AFTER

**The demand for electricity is rising; many countries are seeking clean, new, renewable ways of generating electricity.**

### FUEL CELLS

Fuel cells are **clean sources of energy** that generate electricity locally in small amounts.

### EXPANDING USE OF ELECTRICITY

Threats of global warming **414–15 »** have intensified efforts to tap renewable energy sources, such as wind and wave power.

**SEE ALSO »**
*pp.324–25* FISSION AND FUSION
*pp.416–17* RENEWABLE ENERGY

## How an alternator works

Alternators produce current either by spinning a coil of wire through a fixed magnetic field or by spinning a magnet through a fixed coil. Current is generated as the wire intersects the magnetic field.

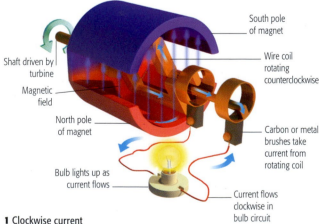

South pole of magnet
Shaft driven by turbine
Wire coil rotating counterclockwise
Magnetic field
North pole of magnet
Carbon or metal brushes take current from rotating coil
Bulb lights up as current flows
Current flows clockwise in bulb circuit

**1 Clockwise current**
As one side of the coil sweeps past the north pole of the magnet a current is induced through the coil, and in this case clockwise through the bulb circuit.

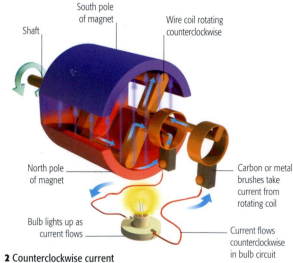

South pole of magnet
Shaft
Wire coil rotating counterclockwise
North pole of magnet
Carbon or metal brushes take current from rotating coil
Bulb lights up as current flows
Current flows counterclockwise in bulb circuit

**2 Counterclockwise current**
Half a turn later, the other side of the coil sweeps past the north pole, and the current is again induced into the coil, but in the opposite direction, so the current runs counterclockwise in the bulb circuit.

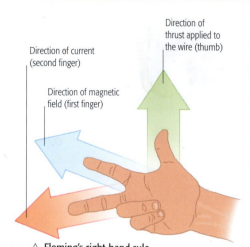

Direction of thrust applied to the wire (thumb)
Direction of current (second finger)
Direction of magnetic field (first finger)

**△ Fleming's right-hand rule**
Hold out your right hand with thumb up, first finger at 90° to it, and second finger pointing back to you at 90°: they show directions.

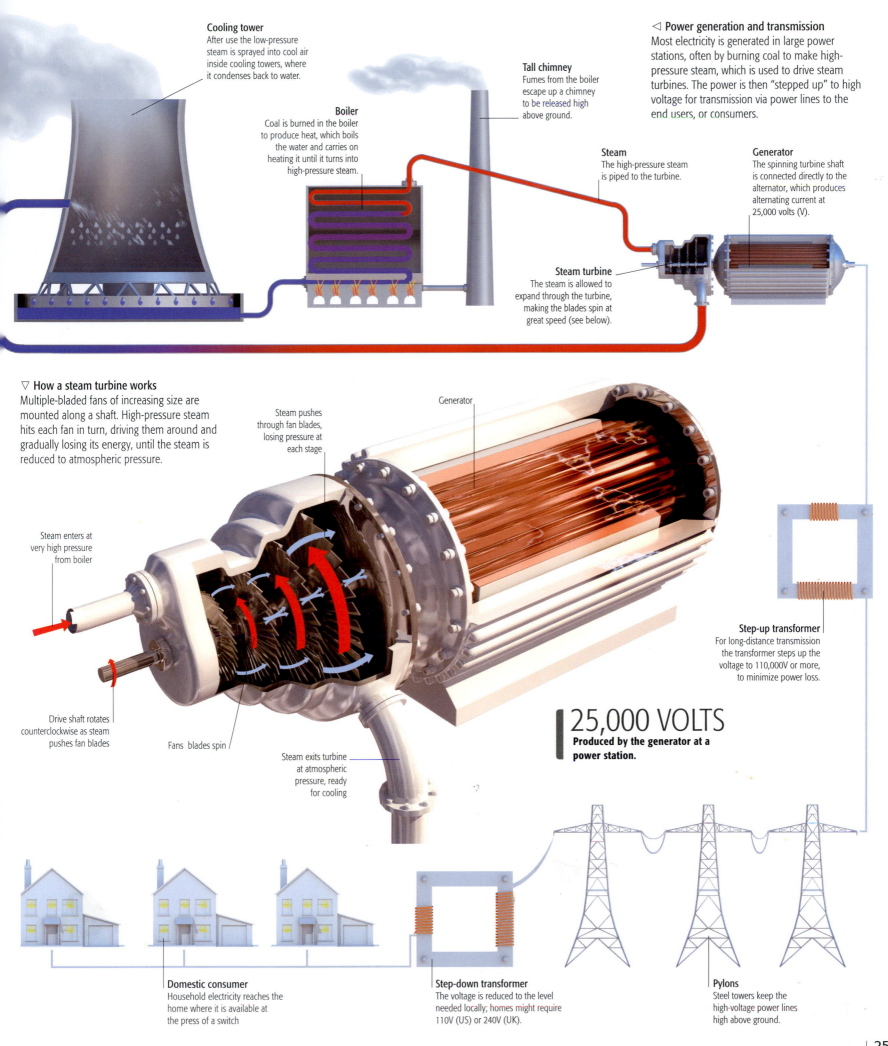

**Cooling tower**
After use the low-pressure steam is sprayed into cool air inside cooling towers, where it condenses back to water.

**Tall chimney**
Fumes from the boiler escape up a chimney to be released high above ground.

**Boiler**
Coal is burned in the boiler to produce heat, which boils the water and carries on heating it until it turns into high-pressure steam.

**Steam**
The high-pressure steam is piped to the turbine.

**Generator**
The spinning turbine shaft is connected directly to the alternator, which produces alternating current at 25,000 volts (V).

◁ **Power generation and transmission**
Most electricity is generated in large power stations, often by burning coal to make high-pressure steam, which is used to drive steam turbines. The power is then "stepped up" to high voltage for transmission via power lines to the end users, or consumers.

**Steam turbine**
The steam is allowed to expand through the turbine, making the blades spin at great speed (see below).

▽ **How a steam turbine works**
Multiple-bladed fans of increasing size are mounted along a shaft. High-pressure steam hits each fan in turn, driving them around and gradually losing its energy, until the steam is reduced to atmospheric pressure.

Steam pushes through fan blades, losing pressure at each stage

Generator

Steam enters at very high pressure from boiler

Drive shaft rotates counterclockwise as steam pushes fan blades

Fans blades spin

Steam exits turbine at atmospheric pressure, ready for cooling

**Step-up transformer**
For long-distance transmission the transformer steps up the voltage to 110,000V or more, to minimize power loss.

# 25,000 VOLTS
**Produced by the generator at a power station.**

**Domestic consumer**
Household electricity reaches the home where it is available at the press of a switch

**Step-down transformer**
The voltage is reduced to the level needed locally; homes might require 110V (US) or 240V (UK).

**Pylons**
Steel towers keep the high-voltage power lines high above ground.

# The **Internal Combustion Engine**

**The internal combustion engine replaced steam as the driving force behind industrial expansion, and put the world on wheels. These engines are at the heart of the automobile, as well as providing power for machines from lawn mowers to aircraft. The first engines appeared in the 19th century, but the technology actually had its roots in a much earlier age.**

There are an estimated 700 million cars on today's roads. Almost all, as well as most trucks, trains, motorcycles, aircraft, boats, and generators, are powered by an invention with its origins in the 13th century—the internal combustion engine.

The principle behind the internal combustion engine is straightforward enough: take a small amount of a high-energy liquid fuel, ignite it, and use the energy released to move a piston a little bit. Do this hundreds of times a minute, and you move a piston a lot. All you have to do now is connect your piston to some wheels, and you are on the move.

### Early combustion engines

The roots of combustion engines can be traced back, via the gunpowder-fueled designs of Leonardo da Vinci, Dutch physicist Christiaan Huygens, and British polymath Sir Samuel Morland, to the genius of 13th-century Arab inventor Aljazari. His 1206 *Book of Knowledge of Ingenious Mechanical Devices* included the first known description of the crankshaft, a device that is central to the combustion engine. It would be another 600 years, however, before an actual combustion engine was created.

In 1806 Swiss inventor François Isaac de Rivaz built the first working model, powered by a mixture of hydrogen and oxygen, and he knew what he wanted to use it for; just a year later Rivaz had built a wooden automobile around his revolutionary engine. The machine, however, was not commercially successful. Mass production of internal combustion engines would not start in earnest until the mid-19th century.

### Taking to the road

The turning point came with the invention of the four-stroke, petrol-driven engine by German inventor Nikolaus Otto. This was the first engine able to burn fuel efficiently in the

△ **An early production line**
Production line workers assemble motor cars at the Ford Motor Company factory in Dearborn, Michigan. Henry Ford paid his workers enough for them to afford to buy the cars they were making—a ready-made market.

## BEFORE

Some of the components and concepts associated with internal combustion engines have developed over several centuries.

**ENGINES IN THEORY**
Centuries before the first internal combustion engines, **Leonardo da Vinci** imagined a compressionless engine that used **gunpowder as a fuel**. He sketched his ideas in 1509 but, as far as we know, his machine was never built.

**HUYGENS'S GUNPOWDER ENGINE**
Having never seen Leonardo's drawings, in 1666 Dutch genius **Christiaan Huygens** tried to build an engine that **pushed a piston upward** when **gunpowder was ignited under it**, but he had no materials that could take the strain.

**STEAM FOLLIES**
The **very first automobiles** were steam-driven military tractors invented by French engineer **Nicolas Cugnot** in 1769. They had three wheels and reached speeds of **2½ mph (4 kph)**. Cugnot's machines met a premature end when one of his patrons died and the other was exiled.

**≪ SEE ALSO**
*pp.168–69* THE ELECTRIC MOTOR
*pp.170–71* MICHAEL FARADAY

## The Otto four-stroke cycle
Powering nearly all modern cars and trucks, this process, named after its 1876 inventor Nikolaus Otto, takes place at the heart of the internal combustion engine.

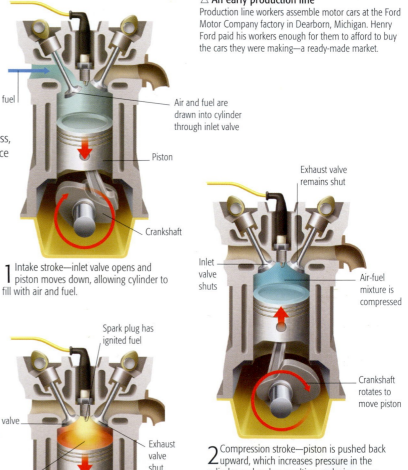

Air and fuel

Air and fuel are drawn into cylinder through inlet valve

Piston

Crankshaft

**1** Intake stroke—inlet valve opens and piston moves down, allowing cylinder to fill with air and fuel.

Exhaust valve remains shut

Inlet valve shuts

Air-fuel mixture is compressed

Crankshaft rotates to move piston

**2** Compression stroke—piston is pushed back upward, which increases pressure in the cylinder and makes resulting explosion more powerful.

Exhaust valve opens as piston rises

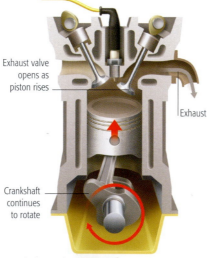

Exhaust

Crankshaft continues to rotate

**4** Exhaust stroke—when piston reaches the bottom, exhaust valve opens. As piston rises again, it pushes waste gases out into exhaust pipe.

Spark plug has ignited fuel

Inlet valve shut

Exhaust valve shut

Piston at top of its stroke

Crankshaft continues to rotate

**3** Combustion stroke—at top of piston's stroke, a spark plug fires, igniting fuel and causing an explosion. This shoots piston downward again.

piston chamber, making it practically and commercially viable. Designed as a stationary engine, in which the action of the pistons is maintained as back-and-forth motion, it was soon adapted into an automobile engine, complete with crankshaft to produce circular motion. Finally, here was an engine that could power an automobile.

The man who achieved this feat was Karl Benz, who later founded Mercedes Benz; for this reason he is generally regarded as the inventor of the automobile. The three-wheeled Benz Patent Motorwagen, built in 1885 and first sold in 1888, was the first automobile designed to generate its own power, as opposed to being a motorized stage coach or horse carriage. The other major pioneer at this time was Gottlieb Daimler, who patented the prototype of the modern gas engine and made the world's first four-wheeled cars.

### Diesel's influence

There is one other man who made a major contribution—Rudolf Diesel. Fascinated by the developments in

### ENGINEER (1858–1913)
## RUDOLF DIESEL

In 1889, French-German inventor Rudolf Diesel was a frustrated, if well-respected, refrigerator engineer. By the end of the 1890s he had created the Diesel engine and was a millionaire. He had a keen interest in social theories as well as engineering, and conceived of his engine as an energy-efficient machine, using local fuels, for craftsmen and artisans who could not afford the steam engines used by large companies.

engine technology, Diesel set about creating his own version, which did not require a spark to ignite the fuel. His endeavors almost came to an end when, in 1894, his new "diesel engine" exploded, almost killing him. To Diesel, however, this proved that his idea worked. His first successful engine fired into life in 1897, and all diesel engines in use today are refined versions of his original concept.

Diesel engines owe their success mainly to their efficiency. Unlike gas engines, a diesel engine compresses air

and then injects fuel into the chamber. This means that, while gas engines compress at a ratio of about 10:1, diesel engines have a compression ratio as high as 25:1. The higher compression ratio leads to better fuel efficiency.

Diesel engines are now used to power nearly all heavy machinery, including generators, factory machines, trucks, buses, other large vehicles, and ships. In addition, diesel is becoming ever more popular as a fuel for cars—more than 40 percent of the new cars bought today have a diesel engine.

The internal combustion engine radically changed not only the way we travel but our entire economic system.

**MASS PRODUCTION**
With internal combustion came the car, and with that **Henry Ford** and revolutionary ideas about factory production. The **mass-production line**, put to famous effect with the Ford Model T, **increased outputs** and **reduced prices**.

**THE IMPORTANCE OF OIL**
Rudolf Diesel's first engines ran on peanut oil, but it was **fossil fuels** that quickly took precedence. In 1900 the world produced about **100 million barrels of oil**. In 2000 it produced **25 billion barrels**.

**TAKING TO THE SKY**
Daimler's engine was the first with a **power-to-weight ratio** high enough to enable **powered flight**, which was achieved just 18 years later.

SEE ALSO »
pp.290–91  TAKING FLIGHT
pp.338–39  ROCKET
          PROPULSION

OIL PUMPJACKS, CALIFORNIA

**Model T Ford**
Created in 1908, the Ford Model T, or Tin Lizzie, was the first affordable car the world had ever seen. Ford designed his "universal car" so that it would be easy for anyone to run and maintain.

# The Nature of Sound

**Breakthroughs in the understanding of sound have often owed as much to the work of musicians as to the experiments of scientists, but once Isaac Newton had identified sound as a longitudinal wave in 1687, scientists of the 18th and 19th centuries soon discovered how to measure its lengths and frequencies.**

The principle that sound is a wave transmitted through the air from a vibrating object to the human ear was universally accepted from the 17th century onward. However, the nature of sound waves remained obscure. Most early scientists assumed that sound waves, like the vibrating strings that were often used to produce them, were transverse waves—waves whose vibrations oscillate at right angles to the direction of their travel.

## Longitudinal waves

This view was abandoned when Newton described the motion of sound in *Philosophiae Naturalis Principia Mathematica* (1687) as a moving region of high and low pressure "nodes"—a longitudinal compression wave. He calculated a theoretical value of 979 ft (298 m) a second for the speed of sound. The English scientist William Derham found this to be only slightly inaccurate when he tested it a few years later.

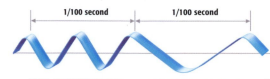

**HIGH–FREQUENCY WAVE**  **LOW–FREQUENCY WAVE**

1/100 second   1/100 second

**Frequency and pitch**
Pitch is dependent simply on the frequency of the sound, and therefore on its wavelength, because higher pitches equate to higher frequencies, which equate to shorter wavelengths.

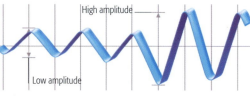

High amplitude

Low amplitude

**Loudness**
The amplitude of the sound wave is equivalent to its intensity, or the pressure it exerts. But the perceived "loudness" of a sound heard by the human ear depends not only on the wave's intensity, but also on its frequency and pitch.

> The loudest sound ever recorded was the explosive eruption of the Krakatoa volcano off Indonesia in 1883. Estimated at 180 decibels (see AFTER), it was heard worldwide.

## Measuring sound

Derham calculated the speed of sound through air to be 1,116 ft (348 m) a second. Today the accepted average speed of sound through air at sea level is 1,115 ft (340 m) a second, though it varies with air humidity.

Derham's method, which has often been repeated, was to send a colleague to various distant locations with orders to fire a gun at a certain time. To this end, they synchronized their watches. Watching through a telescope from a church tower, Derham was able to time the delay between the flash of the gun and the arrival of the sound waves it triggered over the various distances.

**Range of hearing in different animals**
The frequency, or pitch, of sound is measured in hertz (Hz). While we can typically detect sound waves with frequencies between 20Hz and 20,000Hz, bats and dolphins can hear much higher frequencies. Elephants and some other animals can hear and communicate through low-frequency "infrasound."

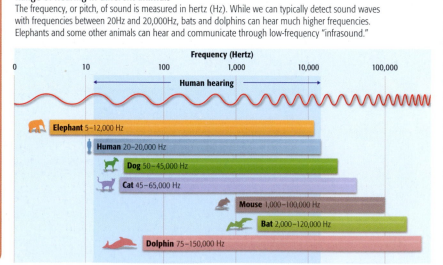

**Frequency (Hertz)**

0   10   100   1,000   10,000   100,000

Human hearing

Elephant 5–12,000 Hz
Human 20–20,000 Hz
Dog 50–45,000 Hz
Cat 45–65,000 Hz
Mouse 1,000–100,000 Hz
Bat 2,000–120,000 Hz
Dolphin 75–150,000 Hz

The basic physics of sound were now understood, but there remained the problem of relating the perceived qualities of sound to the properties of the waves themselves.

In the 1630s the French philosopher priest Marin Mersenne had shown that the frequency of the sound made by a vibrating string was twice that of the sound made by a string vibrating an octave lower. This proved that the pitch of a sound was predominantly due to the frequency of its vibrations in air.

Although there was at first no way of measuring the exact frequencies of sound waves, it was possible to figure out the wavelength of the vibrations within various instruments. Organ pipes, for instance, contained either a half-wavelength or quarter-wavelength of sound depending on their design. Since the frequency of a wave is its speed divided by its wavelength, it was then relatively simple to find the frequency of the waves being produced.

## Sound patterns

In the 1780s German musician and physicist Ernst Chladni developed an ingenious way of measuring sound waves directly. He used sound to vibrate a thin metal plate with a fine powder on its surface, and found that the powder collected into various elegant patterns as the vibrations interfered with one another. He then developed formulae for figuring out the properties of the sound from the patterns that formed. By observing the way that closed organ tubes filled with various gases produced notes of different pitch, Chladni was also able to calculate the speed of sound in these gases.

## Wavelengths and frequencies

The physics of sound advanced rapidly from the early 19th century onward. That the Doppler effect—the change in pitch experienced by someone as a noisy object approaches and then passes them (see opposite)—was due to a

## BEFORE

**The exact nature of sound has fascinated people since the earliest times.**

### EARLY WAVE THEORY
The ancient Greek philosopher **Aristotle** developed a theory of sound as a wave in the 4th century BCE.

### THE MATHEMATICS OF MUSIC
**Galileo Galilei** and **Marin Mersenne** rediscovered the **laws of vibrating strings** in the 17th century; Mersenne developed mathematical **equations** for calculating the frequency of waves produced by different types of vibration.

### IN A VACUUM
Using an air pump made for him by his assistant **Robert Hooke**, 17th-century scientist **Robert Boyle** demonstrated that a bell is inaudible in a vacuum. This proved that **sound needs a medium** such as air to pass through.

**BOYLE'S BELL IN A VACUUM**

### SEE ALSO
*pp.36–37* ARISTOTLE
*pp.96–97* DISCOVERY OF THE VACUUM
*pp.98–99* ROBERT BOYLE

**Faster than the speed of sound**
Ernst Mach first described how a projectile breaking the sound barrier builds up a shockwave in front of it, creating a "sonic boom." At this point water vapor condenses in the low-pressure region behind the shock, shown here around the plane's wing.

## AFTER »

The 20th century saw major advances in measuring and recording sound.

### SOUND UNITS
The "**bel**" and "**decibel**"—units for measuring **sound intensity**—were devised by telephone engineers in the 1920s as a convenient means of comparing sound levels.

### RECORDING SOUND
The ability to **record** and **reproduce sound** has also led to a deeper understanding of its other physical properties and their relation to human perception.

### SEE ALSO »
pp.260–61 TELEGRAPH TO TELEPHONE
pp.266–67 CAPTURING SOUND
pp.268–69 RADIO AND RADIO
WAVES

▷ **Sound harmonics**
Harmonics are the key to music. Related waves with frequencies that are whole-number multiples of a single fundamental frequency, they can be overlaid on each other for a harmonious effect.

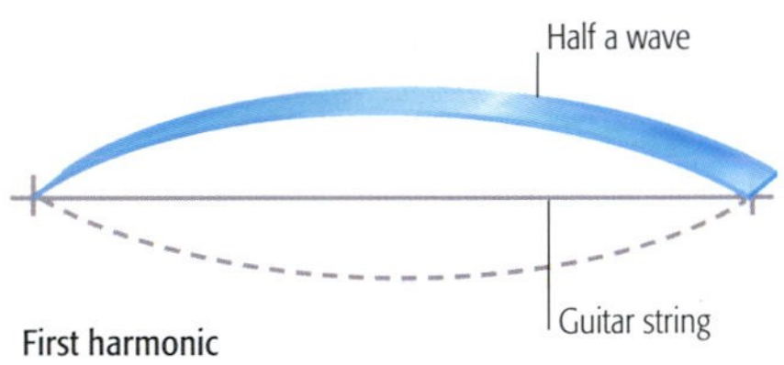

Half a wave

Guitar string

**First harmonic**
An instrument's fundamental, lowest frequency (longest wavelength) is its first harmonic. For the first harmonic of a guitar string, the wave is twice as long as the string.

Whole wave

**Second harmonic**
The second harmonic is twice the frequency and half the wavelength of the first harmonic. For the second harmonic of a guitar string, the wave is the same length as the string.

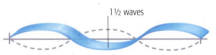

1½ waves

**Third harmonic**
The third harmonic has three times the frequency and one-third the wavelength of the first harmonic. For the third harmonic of a guitar string, the string is equal to 1½ waves.

change in the frequency of the sound wave was predicted by Austrian physicist Christian Doppler in 1842. Dutch scientist Buys Ballot proved it three years later with musicians playing a calibrated note on a passing train.

In 1866 the German scientist August Kundt developed an ingenious means of precisely studying sound waves: he used a sealed transparent tube in which fine powder such as talc collected at the nodes of a sound wave vibrating within it. This allowed the wavelength to be measured directly, and led to the most precise calculations of the speed of sound ever made.

## Supersonic predictions
One of the most intriguing discoveries about sound came from the Austrian Ernst Mach, who correctly described the shock wave generated when objects move at supersonic (faster than sound) speeds. In 1877 Mach predicted many of the effects aircraft would encounter when they broke the so-called sound barrier in the middle of the next century.

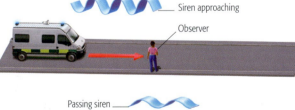

**1 Vehicle approaches**
Frequency of sound waves passing observer is increased; siren sounds higher to observer.

Siren approaching

Observer

**2 Vehicle passing**
Sound waves have "normal" frequency and are regular in pitch.

Passing siren

Observer

**3 Vehicle retreats**
Frequency of sound waves reaching observer is reduced; pitch sounds lower to the observer.

Siren moving away

Observer

### The Doppler effect
The most familiar version of the Doppler effect is the distortion of sound waves experienced when a speeding emergency vehicle approaches, passes, and retreats from a stationary observer.

IN PRACTICE
## ACOUSTICS

For centuries architects knew that a building's internal shape affects the way that sound reverberates—its acoustics—without understanding how. Then in 1895 a talented young Harvard physicist, Wallace Clement Sabine, developed a formula for accurately calculating the sound-absorbing properties of a building, such as London's Royal Albert Hall (shown here), in terms of its volume, shape, and materials.

## Across the spectrum

The electromagnetic spectrum forms a "continuum" of radiation waves with a potentially infinite range of wavelengths. All these forms of radiation have the same speed (the speed of light), and their frequency increases as their wavelength decreases. High-frequency waves carry more energy than lower-frequency ones, and are normally generated by hotter or otherwise more violent processes. Earth's atmosphere blocks many forms of radiation from space, only allowing through radio waves, visible light, the near-infrared, and some ultraviolet.

## Radio waves

The longest radio waves are absorbed in Earth's upper atmosphere by the electrically charged ionosphere. Those with shorter wavelengths are absorbed by atmospheric carbon dioxide and water vapor.

## Radio window

Intermediate radio waves with wavelengths between a few centimeters and about 10 meters, and with frequencies extending from six to 30,000 megahertz (MHz), pass through Earth's atmosphere and reach the surface.

## Microwaves

The shortest form of radio waves, with wavelengths between 1 cm and 100 µm, are known as microwaves. They are widely used in communications and microwave ovens.

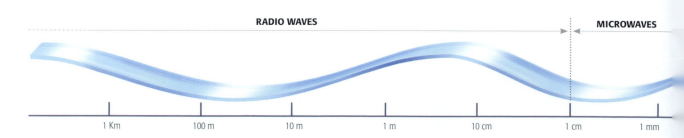

RADIO WAVES | MICROWAVES

1 Km    100 m    10 m    1 m    10 cm    1 cm    1 mm

« **BEFORE**

Since the Middle Ages, people had noticed that lenses produced images with colored fringes. Newton was the first person to see this as potentially significant.

### THE COLOR OF LIGHT

There was much debate about the **nature of light** in the late 1600s Europe. **Huygens** put forward a **wave theory**, which was countered by **Newton's particle theory**. Newton had also shown that "pure" sunlight is in fact a mix of different colors « 114–15.

### AN ELECTROMAGNETIC WAVE

**Thomas Young's** "double slit" experiment « 114–15 reopened the debate around 1800, and pushed the consensus back toward a **wave interpretation**, setting the stage for **Maxwell's** wave-based theory of light. Scientists continued to speculate on human vision and the perception of color « 166–67.

« **SEE ALSO**

pp.114–15 THE NATURE OF LIGHT
pp.116–17 SPLITTING AND BENDING LIGHT
pp.166–67 ELECTROMAGNETISM

## Herschel's experiment

In order to study the temperature of various colours, William Herschel split apart a band of light. When he moved a thermometer into the invisible region beyond the red, he found that the temperature shot up sharply.

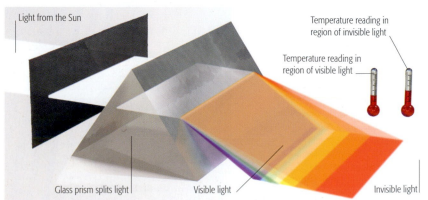

Light from the Sun

Temperature reading in region of invisible light

Temperature reading in region of visible light

Glass prism splits light    Visible light    Invisible light

# Electromagnetic Spectrum

**Throughout the 19th century scientists discovered previously unknown types of radiation within the electromagnetic spectrum, beyond visible light. As they uncovered radio waves, infrared, ultraviolet, X-rays, and gamma rays, they reached a new understanding of the nature of radiation itself.**

Before 1800 nobody even suspected the existence of invisible forms of "light" beyond the extremes of the visible spectrum. The discovery was an unexpected result of an experiment carried out in 1800 by William Herschel, the German-born British astronomer already famous for his discovery of the planet Uranus.

## Color and heat

During his observations, Herschel noticed that differently colored filters seemed to transmit different amounts of heat, and he began to suspect that the colors might have different temperatures associated with them.

As he carried out his measurements, he found that the temperatures increased noticeably at the red end of the spectrum compared with the blue, and so he decided to measure the apparently empty region where the spectrum's red light faded into invisibility. To his surprise, this turned out to be the hottest region of all.

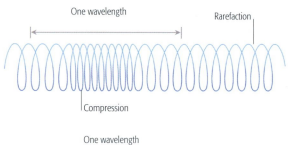

One wavelength    Rarefaction

Compression

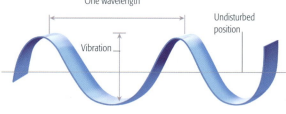

One wavelength

Vibration    Undisturbed position

He concluded that the Sun emitted invisible, very hot rays, and named this new type of radiation "calorific rays."

## Chemical rays

Herschel's discovery inspired German chemist Johann Wilhelm Ritter, who found another type of invisible radiation beyond the violet end of the spectrum. Ritter is known today for his invention of modern electroplating techniques, but it was his investigations into the way that silver salts darken when exposed to light—crucial to the eventual development of photography (see p.262)—that proved key to the new discovery.

In 1801 Ritter performed a similar experiment to Herschel's, to see how the different colors of light affected silver salts. He found that, in this case,

they darkened more rapidly in blue light than in red. Following the same line of reasoning as Herschel, Ritter tested the salts' reaction when placed beyond the violet end of the spectrum, and found that this was stronger still. Ritter's new radiation became known as "chemical rays," and their ability to stimulate chemical reactions was soon being widely used.

## Energy travels in waves

Intriguing though they were, the new rays did not immediately offer further insights into the nature of light—the physics of the time really had nothing to say on the possibility of radiations invisible to the human eye. The context for these discoveries would have to wait until later in the 19th century, and Scottish physicist James

## Longitudinal wave

A longitudinal wave is much like the waves that travel along a spring—it forms regions of compression and rarefaction (decrease in density) parallel to the direction in which it travels. Sound waves are longitudinal, and some early physicists thought that light was too.

## Transverse wave

In a transverse wave, the direction of the disturbances caused by the wave is at right angles to its direction of motion—as in water waves and light. The shape of the wave determines properties such as wavelength and amplitude.

**Infrared**
Emitted by warm objects, infrared radiation is mostly absorbed in Earth's atmosphere by gases such as carbon dioxide and water vapor.

**Optical window**
Wavelengths measured in billionths of a meter pass straight through Earth's atmosphere. Animal eyesight has evolved to be sensitive to radiations in this "window".

**Utraviolet**
The wavelengths of ultraviolet rays are shorter than visible light. Most ultraviolet rays from the Sun are blocked by the atmospheric ozone layer.

**X-rays**
With wavelengths from 10 nanometers (nm) down to 0.01 nm, X-rays carry large amounts of harmful energy. Fortunately Earth's atmosphere acts as a good barrier to them.

**Gamma rays**
Even shorter than X-rays, gamma rays have an extreme potential for penetrating materials and damaging tissues. As with X-rays, our atmosphere prevents them reaching Earth's surface.

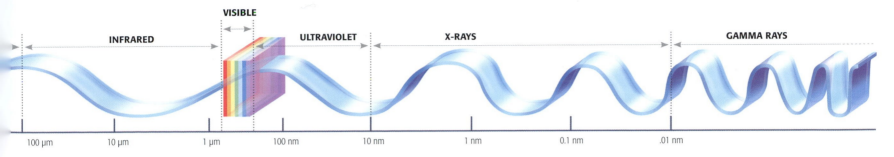

VISIBLE

INFRARED | ULTRAVIOLET | X-RAYS | GAMMA RAYS

100 μm | 10 μm | 1 μm | 100 nm | 10 nm | 1 nm | 0.1 nm | .01 nm

**The ultraviolet world**
Many flowers fluoresce and are far more strongly marked when viewed in ultraviolet light. This evolutionary adaptation aids pollination by attracting insects that can see in the ultraviolet part of the spectrum.

Clerk Maxwell's discovery of the equations linking electric and magnetic effects into a single phenomenon of electromagnetism (see pp.166–67).

Maxwell's work described light as a "transverse" wave, where particles oscillate up and down about their individual "home" positions (see opposite). He also suggested that the wave had interlinked electric and magnetic elements. However, he erroneously assumed that such a wave would require a medium in which to propagate (as sound and water waves do), and the search for this "luminiferous (light-carrying) ether" became a major concern for late 19th-century physics.

### A continuous spectrum

Maxwell's equations demonstrated how the different properties of light arose from the changing wavelengths of the electromagnetic waves. They explained why shorter-wavelength blue light was more susceptible to diffraction than the longer-wavelength red light. They also showed that the spectrum is continuous, and established that higher-energy objects and events emit waves of higher frequencies and shorter, bluer wavelengths.

Maxwell's equations made it clear that there was no physical reason why electromagnetic waves could not be many times longer or shorter than the known spectrum. In fact, the wavelengths of visible light, infrared, and ultraviolet all proved to be in the order of nanometers (billionths of a meter), but the first immediate proof of Maxwell's theories came nine years after his death, with Heinrich Hertz's demonstration of microwaves. These were radio waves with longer wavelengths than infrared (see p.268).

### X-rays and gamma rays

Despite these realizations, Wilhelm Röntgen's 1895 discovery of X-rays (see p.294) still came unexpectedly, and the rays were not automatically identified as a new form of electromagnetic radiation. Nor were gamma rays (see p.298), which were discovered five years later. Within a hundred years, the rainbow spectrum of light had been opened out to reveal a huge range of new radiations, each with its own strange properties. The quest to put these newly discovered waves into use would be a driving force throughout the following century.

**BREAKTHROUGH**

## SPECTROSCOPY

As scientists improved their ability to split light into a spectrum in the 19th century—in a technique known as spectroscopy—it became possible to identify finer details. The spectrum of sunlight was found to be crossed with many dark "absorption lines" at specific colors, while light from heated laboratory samples was found to be a series of bright "emission lines" of certain colors. This led to studying the spectral "signature" of chemical elements in gaseous form. Each element has is its own unique signature.

**Emission lines of helium**
The spectrum comprises red, yellow, cyan (two lines), and blue (three lines).

**The Swift satellite**
This multi-wavelength observatory is designed to study the puzzling "gamma ray bursts" occasionally seen across vast expanses of space. Observatories like this take advantage of being above the Earth's atmosphere to detect radiations that do not penetrate it.

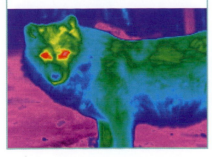
**AFTER**

Maxwell's theories allowed scientists to develop their understanding of light, and led to new theories and discoveries.

**LIGHT AS A WAVE AND A PARTICLE**
Ultraviolet fails to follow the pattern of **Maxwell's equations**. This ultimately led to the development of **quantum physics 314–17 》**.

**LASER TECHNOLOGY**
Scientists discovered a means of producing **"coherent light,"** with the peaks and troughs of all its waves "in step" in the 1950s. This is the key to **laser technology 364–65 》**.

**IMPROVED IMAGING**
Earth is constantly bombarded with invisible radiations. **Radio telescopes** and **satellite observatories** now use these radiations to help us learn more about the secrets of the Universe.

**SEE ALSO 》**
*pp.314–17* QUANTUM REVOLUTION
*pp.364–65* LASERS AND HOLOGRAMS

**LASER SPEED GUN**

## « BEFORE

Long-distance messaging has existed since ancient times, in the form of smoke signals, burning torches, waving flags, and mirrored light reflections.

### SIGNAL FIRES
One of the earliest means of communicating over long distances was to use fire. American Indians were able to send complex signals by controlling the **smoke** from a fire, while in other countries **chains of fires** within sight of one another were used to pass messages.

### NAPOLEONIC SEMAPHORE
In the 1790s Frenchman **Claude Chappe** devised a semaphore system that used **telegraph towers** with signaling arms mounted on the roof. This device spread widely across Europe.

A CLAUDE CHAPPE SIGNAL TOWER

**« SEE ALSO**
*pp.252–53* GENERATING ELECTRICITY
*pp.256–57* NATURE OF SOUND
*pp.258–59* ELECTROMAGNETIC SPECTRUM

Early methods of sending signals instantaneously over long distances were known as telegraphy, from the Greek for "distant writing." There were several problems with the various optical telegraph systems developed by the early 1800s: they were slow and limited in the detail they could convey, they required a direct line of sight between the stations, and they were heavily reliant on the weather. Semaphore stations could only operate in daylight, and even signal fires and heliographs (messages sent by mirrors) were useless in fog or low cloud.

## Signaling with electricity
Electricity offered an obvious means of communication, since it propagated over long distances almost instantaneously, but the problems of producing a working system based on electrostatic discharges initially proved

laboratory curiosity for another two decades or more. Three key discoveries were needed to make the telegraph and all later wired telecommunication systems possible. The first was the

**Women working at a telephone exchange**
In 1882 London had around 35 operators; by 1910 New York employed 6,000. The operators plugged cords into a panel: one for the incoming call, and a second for the receiving party's local circuit. Long-distance calls were connected to a trunk circuit between exchanges.

# Telegraph to Telephone

**Leading on from the line-of-sight telegraphs of the 1800s, the Victorian age saw a rapid development in methods of long-distance communication, making full use of the growing knowledge and use of electricity to develop more efficient telegraph and telephone devices.**

Movable metal armature
Spring holding up armature so that movable and fixed contacts are separated
Electromagnet (wire wrapped around an iron core)

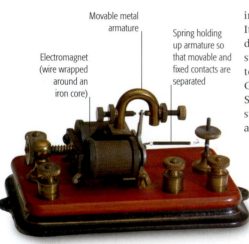

**Armature telegraph relay**
Joseph Henry's relay machine was used to telegraph messages across long distances. When the coil is energized, it attracts the armature, pulls it down, closes the contacts, and completes the circuit.

## SCIENTIST AND INVENTOR (1847–1922)
### ALEXANDER GRAHAM BELL

Born in Edinburgh, Scotland, Bell moved to Canada with his family at the age of 23, and later settled in Boston, Massachusetts. His family had a long tradition of teaching linguistics and elocution, and Bell too became a distinguished teacher in this field. However, he combined it with a passion for experimentation, and persuaded his father-in-law, a wealthy Boston lawyer, to back his plans for an improved telegraph. In 1874, he employed an electrical engineer, Thomas Watson, who did much to make the telephone a reality.

insurmountable. It was only after the Italian scientist Alessandro Volta discovered a means of generating steady current in 1799 that an electric telegraph became a real possibility. German experimentalist Samuel von Sömmerring demonstrated a working system just a few years later in 1809, although the telegraph would remain a

# 130,000 POLES
**held up the 14,000 miles (22,500 km) of copper wire used to make connections from New York to San Francisco.**

galvanometer, invented by the Dane Hans Christian Ørsted in 1820. This made the strength of an electric current visible through the deflection of an electric needle. The second was English scientist William Sturgeon's electromagnet (invented in 1825, and considerably improved in the US by Joseph Henry in 1828), which provided a means of multiplying the magnetic force generated by a comparatively feeble current. Using these two inventions alone, the great German scientists Karl Friedrich Gauss and Wilhelm Weber were able to build the first operational telegraph system in 1833. The final piece of the jigsaw, invented by Henry in 1835, was the relay: a device capable of boosting a weak signal. so that it could travel down a much greater length of wire.

## Telegraphy in action
In 1837 US inventors Samuel Morse and Alfred Vail devised the system that would eventually become the standard: the Morse Code. This entailed sending pulses of current of different lengths

("dots and dashes") down a single wire. The code was then easily transcribed into letters. A series of high-profile demonstrations secured US government backing for the system, and it was rapidly adopted, spreading widely across the continental United States. In 1866, Europe and America were linked via the first transatlantic cables.

Further innovations were rapid: the addition of a telegraph "sounder" that

**An early telegraph machine, 1837**
Samuel Morse's original telegraph machines were electromagnetic. They involved molded type with built-in dots and dashes, which was able to make or break the electrical circuit to send coded messages.

generated sound from the current pulses made the Morse system even easier to use, since it freed the operator's eyes for transcribing the message. Teletype machines soon automated the transcription process, and by 1843 Alexander Bain had even invented a primitive version of the modern fax machine for transmitting pictures through telegraph signals.

## Birth of the telephone

The story of the telephone's early development is a complex one, confused by rival claims, but the patent for the first telephone system was awarded to Scottish-born US inventor

Alexander Graham Bell in 1876. Bell was being funded to work on a system for sending multiple telegraph messages simultaneously using different signal frequencies (using a technique known as multiplexing).

In order to convert sound into an electrical signal, Bell's telephone used a thin membrane that vibrated when sound was directed toward it, moving a rod of magnetized iron in front of an electromagnet. At the other end of the wire, the signal passed into a receiver electromagnet, causing a thin metallic disk to vibrate and produce sound waves. Within a short time news of the invention had spread around the world.

## AFTER

The arrival of radio waves and wireless transmission caused another telecommunication revolution in the 20th century.

### WIRELESS COMMUNICATION

Wireless broadcasting of a single signal to a large number of receivers caught on rapidly, as did **wireless telegraphy** for long-range communications such as those between ships.

**EARLY CELL PHONE**

### CELL PHONES

Personal wireless telephony has become a widespread reality more than a century after Bell's invention, either by means of **cell phones** that rely on land-based radio masts, or **satellite phones** that operate via orbiting satellites.

**SEE ALSO >>**
pp.268–69 RADIO AND RADIO WAVES
pp.368–69 ARTIFICIAL SATELLITES
pp.378–79 THE INTERNET

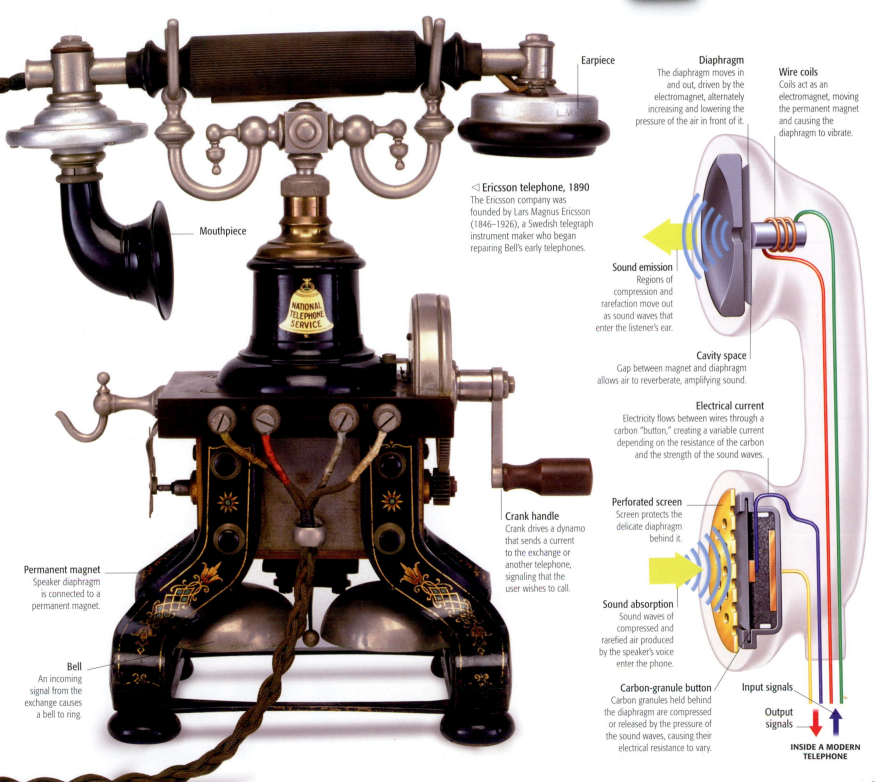

Earpiece

**Mouthpiece**

◁ **Ericsson telephone, 1890**
The Ericsson company was founded by Lars Magnus Ericsson (1846–1926), a Swedish telegraph instrument maker who began repairing Bell's early telephones.

**Diaphragm**
The diaphragm moves in and out, driven by the electromagnet, alternately increasing and lowering the pressure of the air in front of it.

**Wire coils**
Coils act as an electromagnet, moving the permanent magnet and causing the diaphragm to vibrate.

**Sound emission**
Regions of compression and rarefaction move out as sound waves that enter the listener's ear.

**Cavity space**
Gap between magnet and diaphragm allows air to reverberate, amplifying sound.

**Electrical current**
Electricity flows between wires through a carbon "button," creating a variable current depending on the resistance of the carbon and the strength of the sound waves.

**Perforated screen**
Screen protects the delicate diaphragm behind it.

**Crank handle**
Crank drives a dynamo that sends a current to the exchange or another telephone, signaling that the user wishes to call.

**Sound absorption**
Sound waves of compressed and rarefied air produced by the speaker's voice enter the phone.

**Permanent magnet**
Speaker diaphragm is connected to a permanent magnet.

**Bell**
An incoming signal from the exchange causes a bell to ring.

**Carbon-granule button**
Carbon granules held behind the diaphragm are compressed or released by the pressure of the sound waves, causing their electrical resistance to vary.

**Input signals**

**Output signals**

**INSIDE A MODERN TELEPHONE**

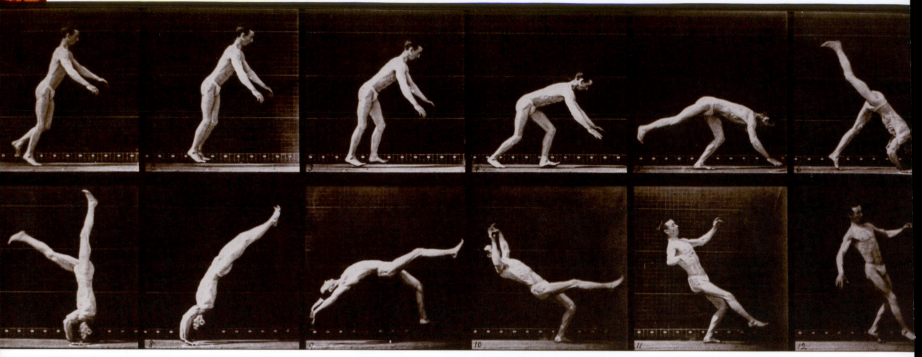

**Motion study**
This series of motion study photographs was taken by the British photographer Eadweard Muybridge in 1887. Muybridge made his name when he photographed horses in motion. He proved that all the animal's hooves leave the ground and are tucked under the body when galloping.

# Photography

**From its early beginnings as the preserve of dedicated enthusiasts, photography was to become crucial in transforming our view of the world. Geographic expeditions, wars, everyday human society, and flora and fauna could all be recorded, leading to a much greater understanding of our shared environment.**

## « BEFORE

**While the way in which light rays react when passing through a small hole was understood more than 2,000 years ago, it was not until the 19th century that the phenomenon was translated into a way of making a permanent image.**

### FIRST PRINCIPLES
In the 5th century BCE the Chinese philosopher **Mozi** discovered that **light rays passing through a pinhole** into a darkened room would create an **inverted image** on the surface opposite the hole. In the 4th century BCE **Aristotle** was also aware of the principle. **Alhazen** wrote about the pinhole camera in 1021 CE in the *Book of Optics*.

### LIGHT INTO ART
In the 16th century **Giovanni Battista della Porta** (1538–1615) described the use of the **camera obscura** ("darkened room") as a drawing aid. Artists including **Canaletto** and **Vermeer** are thought to have used the camera obscura to help them draw more accurately.

### PRINTING THE IMAGE
In Britain in 1802, **Thomas Wedgwood** and **Sir Humphry Davy** presented a paper on the results of their work in the field of photography. They had experimented with **creating images on light-sensitive materials**, but they were unable to make the images permanent.

**« SEE ALSO**
*pp.60–61* **ALHAZEN**

The 19th-century pioneers of photography often worked in isolation, experimenting with different ways of making light-sensitive surfaces. In Brazil, Antoine Florence covered glass plates with gum arabic (a natural adhesive taken from the acacia tree) and soot, while in England Thomas Wedgwood and Humphry Davy immersed paper and leather in silver nitrate. In France, Joseph Nicéphore Niepce tried similar methods; they all came up against the same problem: the light-sensitive surface continued to darken and the image was lost.

## Fixing the image
In about 1826 Niepce exposed a pewter plate coated with bitumen of Judea, a light-sensitive form of natural tar, to daylight for a few hours. After rinsing the plate with a mixture of lavender oil and petroleum oil,

**An early daguerreotype**
This photograph of Le Sueur's statue of Charles I on Whitehall, London, was taken in 1839. The image has been reversed to show the scene as it existed, since daguerreotypes only produce reversed views.

he found he had captured a negative image which, when exposed to iodine, was partially reversed to a positive —a process he called heliography.
In 1833 Louis-Jacques-Mandé Daguerre discovered that an iodine-coated silver plate would develop a

latent image when exposed to light. The image could be made visible by treating it with mercury fumes and then, crucially, fixed with a solution of table salt dissolved in hot water. The British chemist John Herschel also managed to fix a photographic image.

### DESIGNER (1789–1851)
## LOUIS DAGUERRE

Born in Val-d'Oise, France, Daguerre was a noted designer of theater sets and dioramas, for which he used a camera obscura as a drawing aid. His experiments in fixing a photographic image came from the desire to speed up his work rather than the interest in the science of photography that drove other pioneers in the field. Daguerre took little interest in further explorations in photography and retired the year after his process was publicized.

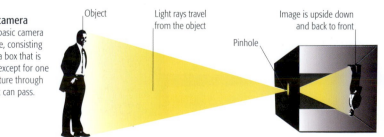

**Pinhole camera**
The most basic camera is a pinhole, consisting merely of a box that is light-tight except for one small aperture through which light can pass.

Object — Light rays travel from the object — Image is upside down and back to front — Pinhole

## How black-and-white film works

The film consists of a celluloid base on to which the other layers are coated. The image is formed in the emulsion layer, which contains the light-sensitive silver halides, a chemical mixture that usually includes silver bromide, silver iodide, and silver chloride. Until the film is developed, the image is latent.

SCRATCH-RESISTANT COATING
EMULSION
CELLULOID
ANTI-CURL LAYER
ANTI-HALATION LAYER

Anti-halation layer blocks out light

## How color negative film works

Silver halides in the emulsion layers are mixed with spectral sensitizers to make them respond to different colors of the spectrum. The silver halides react to light in the same way as in black-and-white film, but the image is converted to color by dye couplers in each layer.

SCRATCH-RESISTANT COATING
BLUE-SENSITIVE EMULSION LAYER
YELLOW FILTER
GREEN-SENSITIVE EMULSION LAYER
MAGENTA INTERLAYER
RED-SENSITIVE EMULSION LAYER
CELLULOID
ANTI-CURL AND -HALATION LAYER

**Digital cameras continue to develop, offering higher-quality images with each new model. Cameras enable us to find out more about the Universe.**

### PHOTOGRAPHING THE UNIVERSE
Cameras sent into space on probes and satellites send back **images of our galaxy and others**, extending our knowledge about stars **328–29 ▷▷** and possible **life on other planets**.

### PRESERVING THE RAIN FORESTS
In 2012 a **64-megapixel camera** costing £1m is to be launched on the **satellite Amazonia-1 368–69 ▷▷** to record illegal logging in the Amazon basin and the Congo.

**SEE ALSO ▷▷**
pp.378–79 THE INTERNET
pp.396–97 SPACE PROBES AND TELESCOPES

## Digital SLR (single-lens reflex)
The light comes into the lens of the camera and is reflected off the mirror and upward. It hits the top and side of the prism before passing through the viewfinder. The lower mirror flies upward on exposure to allow light to hit the sensor.

Built-in pop-up flash unit

Light is reflected around the pentaprism to the viewfinder

Built-in focusing motor

Zoom lens groups

Path of light through the camera

Circuit block for connectors

Sub-mirror for auto-focus detection

Mirror reflects light up toward the prism

still current today. Although it could not capture as much detail as the daguerrotype, the calotype process enabled the multiple reproduction of the image—the photograph as it became universally recognized.

Eleven years later Frederick Scott Archer's wet collodion process, using glass coated with silver salts, largely took over from the calotype and daguerrotype processes, since its greater sensitivity to light much reduced exposure times. In 1871, English physician Richard Maddox introduced gelatin-coated dry plates that gave faster exposure times still.

### The arrival of film
In the 1880s flexible nitrate-based plastic roll film was developed by George Eastman, along with the first mass-market cameras loaded with enough film to take 100 photographs. Once the customer had exposed the film the camera was returned to Kodak, where the film was developed and prints made. These were then sent back to the customer along with the camera, loaded

with fresh film. The first color transparency film was produced by Kodak in 1935. This was followed, in 1941, by color negative film that could be purchased from most pharmacies. Photography was no longer exclusive to professionals.

### The digital era
Digital imaging technology was first used by television in the 1950s, and was later used for mapping the Moon. The first digital camera for the professional market was the Nikon F3 in 1991, but it was not until the 21st century that they gained dominance over film in this area. For consumers, the digital camera was immediately successful, allowing instant gratification and the ability to email digitized images across the Internet.

**1 MILLION Number of 35 mm cameras produced by Leica up to 1961. Over three decades the Leica became a favorite of many important photographers, including Henri Cartier-Bresson.**

His cyanotype process, using iron salts to produce a blue image, was inexpensive and simple, making it commercial and popular.

In 1840 William Fox-Talbot developed the calotype, the first method to use the negative–positive process that is

**c.1840 Fox Talbot's camera** from which he produced the first photographs using the negative–positive process.

**c.1885** The London Stereoscopic Company led the market in both **stereoscopic cameras and stereo cards**, which became a craze both sides of the Atlantic.

**1925–36** The **Leica A** series introduced small, portable 35 mm film cameras offering high-quality images for professional photographers.

**1969** Thirteen modified **Hasselblad 500E** cameras were taken on the Apollo Moon landings.

**1980 Nikon** introduced the F3. In 1991, with a Kodak digital back, it became the first digital camera for professional use.

**1984** The **Nikonos V** was the latest in the range of the legendary underwater cameras from Nikon. It remained in production until 2001.

**2008 Cell phones** had become so popular that 9 billion were in use worldwide, causing some manufacturers of digital cameras to cease production.

| 1840 | 1860 | 1880 | 1900 | 1920 | 1940 | 1960 | 1980 | 2000 |
|---|---|---|---|---|---|---|---|---|

### Development of cameras
Originally bulky pieces of equipment with a basic mechanism, cameras have become highly sophisticated and so small they can be incorporated in a mobile phone.

**c.1840** The world's first widely available **daguerreotype camera** (below).

**1871** The **gelatin dry plate** was introduced by Richard Maddox.

**1915** The **Minnigraph**, made in Germany, was the first European camera to use 35 mm cine film, producing half-frame 18 x 24 mm negatives.

**1957–75** The **Hasselblad** 500C (left) was the first of the company's V series, which became the industry standard for professionals.

**1963** Kodak released the **Instamatic**, an inexpensive snapshot camera. By 1970, sales topped 50 million.

**1990** The **Polaroid Spectra Pro** offered instant results along with multiple exposures and self-timed sequential pictures.

**2008 Canon's PowerShot SX200** IS joined the travel-zoom camera market, aimed at travelers wanting a lightweight digital model.

INVENTOR AND ENTREPRENEUR **Born 1847** Died 1931

# Thomas Edison

## "**Genius** is one percent **inspiration** and 99 percent **perspiration**."

THOMAS EDISON

Thomas Alva Edison is considered to be the greatest inventor of his time. Among his world record 1,093 patents were key inventions that ushered in the modern world. He was one of the first men to realize the true potential of electricity and harnessed it to bring electric lights, sounds, and even moving pictures into our homes.

Edison was born in Milan, a port on the Huron River in Ohio. By the age of seven, his insatiable curiosity had worn his teacher's patience so far that his mother decided that home schooling was the only option. Convinced that her son's idiosyncrasies resulted from a remarkable brain, she gave "little Al" the self-belief to change the world.

### A young entrepreneur

Edison's voracious inquisitiveness was soon joined by entrepreneurship. After selling newspapers, snacks, and candy on the local railways, at the age of 12 he started his own business selling fruit and vegetables. The newspaper he started at 14, the *Weekly Herald*, soon had 300 commuting subscribers.

One day in 1862 Edison saved the three-year old son of J.U. MacKenzie, a Michigan station master, from an oncoming boxcar. In return MacKenzie agreed to teach him Morse code and telegraphy. By the age of 16 Edison had produced his first invention, an "automatic repeater" to transmit signals between unmanned stations.

Having moved to Boston to work as a telegrapher for Western Union, Edison gave up full-time employment in 1869

to become an inventor. By June he had produced an electric vote recorder. However, his "beautifully constructed" machine met a distinctly lukewarm reception—it was simply too quick for the politicians' liking. From then on, Edison would waste no more time on inventions that no one wanted.

### A succession of patents

After moving to New York, Edison patented several more improvements in telegraphy, culminating in the development of a quadruplex telegraph that could send two messages in each direction simultaneously.

By 1876 Edison was able to open a laboratory in New Jersey. Here, he and his staff would develop several different inventions at any one time. One example was a telephone transmitter that improved the clarity and volume of the voices transmitted over Alexander Graham Bell's nascent system of communications.

It was Edison's work on telephones that led to his first major invention: the phonograph. Realizing that sounds could be recorded as indentations on a rapidly moving piece of paper, he soon produced a machine that used a needle to score grooves on a tinfoil-coated cylinder. He was justifiably delighted when it replayed his own voice saying "Mary had a little lamb." Despite initial enthusiasm, however, the phonograph was destined to remain a virtual

novelty for another 10 years, by which time Edison's attention had shifted to other even more ambitious projects.

### Lighting the world

On November 15, 1878 the inventor formed the Edison Electric Light Co., with the intention of producing, not just an incandescent bulb, but a lighting system that could be extended throughout an entire city. The success of Edison's system was due not only to the development of British inventor Joseph Swan's long-lasting lightbulb (see p.250), thanks to the use of a carbonized filament, but

**Edison's 1877 phonograph**
Sounds entering the central mouthpiece caused a diaphragm to vibrate, making indentations with a stylus on tinfoil wrapped around the rotating drum. A second stylus and diaphragm could play back the recording.

**Thomas Edison**
Photographed c.1900, Edison examines motion picture film threaded through one of his film projectors. His first projector, known as a projecting kinetoscope, was patented in 1887. The Edison Company also produced film footage for home entertainment.

also to the invention of a host of other crucial components: the parallel circuit; an improved dynamo; an underground conductor network; various devices to maintain constant voltage; safety fuses; and light sockets with on–off switches. By 1882 the world's first commercial

**Menlo Park**
Edison stands in his chemical research laboratory at Menlo Park, New Jersey. The electric light bulb and many other patented inventions were developed in electrical laboratories on the same site.

lighting system—400 lamps in total—had been installed on Pearl Street in lower Manhattan. A year later 513 customers were using 10,300 lamps. The electricity that powered them was provided by arguably Edison's most important invention—the first commercially viable power station. Electricity would eventually bring light to millions of homes worldwide.

## Man of extraordinary vision

Edison was not the first to invent a lightbulb, or the power station, but he was the man who understood how to put all the components together and demonstrate to the world the possibilities afforded by electricity.

By the end of his life Edison held 1,093 patents for his inventions and innovations. He returned to the phonograph, developing the most advanced sound systems of his day, invented the alkali storage battery, and also helped to kick-start the age of motion pictures with the kinetoscope, one of the world's first projectors of moving images. "I never quit until I get what I'm after," he once said.

TIMELINE

**February 11, 1847** Thomas Alva Edison is born in Milan, Ohio.

**1859** Starts work on the Grand Trunk Railroad, selling newspapers and confectionery.

**1863–67** Works as a telegraph operator in various cities of the Midwest.

**1868** Moves to Boston to work for Western Union and invents an automatic vote recorder.

**January 1869** Becomes a full-time inventor.

**December 25, 1871** Marries one of his employees, Mary Stillwell.

**1874** Invents the quadruplex telegraph for Western Union.

**1876** Moves to Menlo Park, New Jersey, and constructs his first full-scale research laboratories, designed to perform electrical, chemical, and mechanical experiments.

**December 7, 1877** Demonstrates the phonograph at the offices of Scientific American.

**November 15, 1878** Forms Edison Electric Light Co.

**December 31, 1879** Demonstrates the first fully-functioning incandescent lighting system at the Menlo Park laboratories.

**EDISON'S INCANDESCENT LIGHT BULB**

**September 1882** Opens the first permanent central power station and public lighting system, on Pearl Street. He joined forces with Bergmann & Company to market electrical equipment; the company eventually becomes General Electric.

**1884** Mary (Stillwell) Edison dies.

**1886** Marries Mina Miller.

**1887** Moves into a new laboratory in West Orange, New Jersey.

**1888** Becomes embroiled in the "war of the currents" with George Westinghouse and Nikola Tesla. His promotion of direct current (DC) systems is ultimately unsuccessful.

**1893** Demonstrates the kinetograph and kinetoscope.

**1896** Introduces the home phonograph.

**1899** Unsuccessfully promotes concrete for cheap housing.

**1900** The first modern research laboratory opens at

**WAX MUSIC CYCLINDER MADE BY EDISON**

General Electric. Begins working on a battery for use in electric cars.

**1915** Employed to lead naval investigations into new military technologies.

**1920s** With deteriorating health, he continues his experiments at home. Spends much effort trying to find an alternative to rubber.

**October 18, 1931** Dies in Llewellyn Park, New Jersey. Lights are dimmed across the US.

**East River Power Plant**
The world's first power station at Pearl Street, New York City, burned down in 1890. It was succeeded in 1901 by the Waterside plant on Fourteenth Street and the East River; its capacity of 120,000 watts was 10 times that of Pearl Street.

# Capturing Sound

**During the second half of the 19th century inventors competed to build the first commercially viable machine that was capable of recording and reproducing sound. The recording industry that emerged was transformed during the 20th century by the invention of magnetic tape and digitally encoded sound.**

In 1857 Frenchman Edouard Leon-Scott patented a device that was capable of recording in visible form the frequency and shape of sound waves. The "phonautograph" used a hornlike receiver to focus waves on to a membrane, vibrating a fine bristle that scratched patterns on a glass plate covered in carbon. It was an ingenious and useful invention, but it would be another 20 years before someone successfully recorded and reproduced sound.

## First sound recording

The first person to record sound was US inventor Thomas Alva Edison (see pp.264–65), who demonstrated his "phonograph" device late in 1877. Edison's system used the pressure of sound waves to create up-and-down motions in a stylus, which in turn created indentations in a tinfoil sheet wrapped around a rotating cylinder and guided by spiral grooves. After a recording had been made, the stylus was allowed to ride along the grooves again, and the resulting vibrations recreated the original sound.

Edison's phonograph was an immediate sensation, but had many limitations. Perhaps the most obvious problem was that each recording was singular, while each time the cylinder

Tin cylinder with sound recorded as indentations

Handle to rotate cylinder

Combined stylus and loudspeaker apparatus

**An Edison phonograph**
The first cylinder phonographs were comparatively compact, with small speaker horns and a clockwork mechanism or even a hand crank to rotate the cylinder at a constant speed.

was played back, the wave pattern was worn down a little bit and the sound quality deteriorated.

## Early improvements

Wax cylinders, introduced in the late 1880s, were more durable than Edison's tinfoil originals, and could more easily be produced in greater numbers. Mechanical devices were used to copy a groove pattern onto several cylinders at the same time.

However, in 1889 Emile Berliner invented the "gramophone," which used flat disks instead of cylinders; these were easier to store, were more durable, and allowed for easier sound reproduction. The disks had a single spiral groove from the center to the edge, and each disk could simply be "pressed" on a metallic master to reproduce the grooves. Sound was recorded as a side-to-side movement rather than an up-and-down one. By the 1920s gramophone disks had almost completely replaced cylinder recordings.

## Electronic recording

Sound recording was a mechanical process until the 1920s, when advances in valve technology led to electrical recording, in which changing current was used to drive an electromagnetic recording head. Now microphones, which transformed the oscillation of sound waves into an electrical signal, could be used in recording. Amplifier circuits removed limits on the volume of playback, allowing louder sound.

Electronic technology also opened the way for a new recording medium: magnetic tape. The principle of capturing a varying wave pattern by lining up the iron oxide particles on a wire or tape had been established in the 1890s, but the microphone made it possible to record sound waves in this format. The first commercial systems appeared in the US during the 1940s. Magnetic tape and electronic

**The gramophone**
Early gramophones relied on a large speaker horn to amplify the sound from a vibrating diaphragm naturally. In order to achieve sufficient volume, the stylus was held down onto the recorded track with considerable force, causing rapid deterioration of the stylus and disk.

« BEFORE

**Before sound could be recorded, it was first necessary to understand its nature.**

### SOUND WAVES
Work done on the properties of sound from the 1600s onward had established that sound was a **wave of compression** carried through the air, which **stimulated the eardrum** by causing it to vibrate.

### EARLY STORED MUSIC
Automated musical instruments that could **play back musical notes** stored on a rotating cylinder or paper punchcard first appeared in Baghdad, Iraq, as early as the 9th century.

« SEE ALSO
pp.256–57 THE NATURE OF SOUND
pp.260–61 TELEGRAPH TO TELEPHONE
pp.264–65 THOMAS EDISON

**The telegraphone**
The earliest magnetic recording system, pioneered by Danish engineer Valdemar Poulsen in 1899, used wire rather than a magnetic tape to record sound signals.

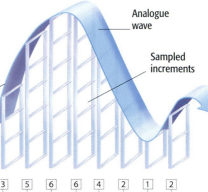

## The digital age

Modern recording techniques involve a complex sequence of procedures to digitize sound waves and reduce the information to binary values, which are then stored on a compact disc or other medium..

**Analogue wave**

**Sampled increments**

### 1 Analogue wave is sampled
An electronic circuit measures the intensity of the sound waves many thousands of times per second, with a frequency of at least 44.1 kHz.

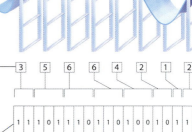

### 2 Sampled number values
The sampling measurements are converted to number values; there may be 65,000 to 17 million different levels, depending on the number of digital "bits" in the final recording.

3 5 6 6 4 2 1 2

1 1 1 0 1 1 1 0 1 1 0 1 0 0 1 0 1 1 0

### 3 Conversion to binary
The stream of numbers is converted to binary code; 16 digital "1s" or "0s" can represent 65,000 different sound levels, while 24 can represent as many as 17 million.

### 4 Ups and downs
The binary stream of data is now converted into a continuous spiral of "pits" and "flats" (raised areas) on the surface of a metal master disc.

### 5 Surface of a CD
The master recording is used to manufacture CDs, in which the flats become pits and vice versa. The entire disc is covered with a thin layer of reflective aluminium, and another of transparent protective plastic.

### 6 Sound from lasers
A CD player projects a focused laser beam onto the disc's spiral track. The light reflects back to a sensor only when it strikes a flat, recreating a changing binary signal that is used to reconstruct and ultimately reproduce the music.

recording together allowed the introduction of many new techniques. These included editing and mixing multiple tracks, addition of electronic effects, and reduction of background noise. Stereophonic sound—which reproduced the effect of different elements of a recording coming from different directions—soon became a recording-industry standard.

## The digital revolution

In the early 1980s the advent of digital technology made it possible to "sample" the strength of a signal thousands of times a second and convert this into a value that could be digitally encoded, usually in binary code. Digital recording allows many identical copies to be made, but some believe the sound quality is not quite as good as an analogue recording.

## MOVIE SOUND

In 1927 *The Jazz Singer* was the first ever "talkie," gramophone disks playing in synch with the film projection. But by the 1930s movies were using "optical recording," which involved printing a "soundtrack" of varying density in a strip along one side of the film. Light shining through the soundtrack hit a photoelectric sensor on the other side, which generated an electrical signal that reproduced the original sound waves.

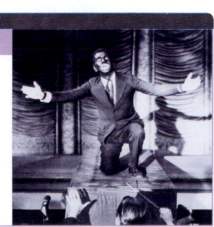

**Al Jolson in "The Jazz Singer"**

## AFTER ⟫

**Sound recording spearheaded the way for the digitization of nearly every form of data; an advance made possible only by vast increases in computer processing power.**

### DIGITAL VERSATILE DISCS
The DVD was invented in the 1990s as a high-capacity version of the CD. These discs proved especially useful for storing movies in **high-quality digital format**. Some movies were made in digital format from the 1980s.

### PIRACY
The music industry failed to anticipate the boom in home computing, which allowed users to **copy CDs and DVDs directly**, reproduce them, and even transfer them to compact files for **illegal sharing** over the Internet.

**SEE ALSO ⟫**
*pp.268–69* RADIO AND RADIO WAVES
*pp.378–79* THE INTERNET

### BEFORE

The radio revolution of the late 1800s was the culmination of a century of scientific and technological developments.

**ELECTROMAGNETIC WAVES**
The electromagnetic-induction effects discovered in the early 19th century were rapidly put to use, but the electrical "**induction field**" around a moving current was forgotten. No one realized at the time that another part of the same electric field interacted with the magnetic field to produce **electromagnetic waves**.

**TELEGRAPH MACHINE**

**TELEGRAPH AND TELEPHONE**
The invention of telegraphy in the 19th century was limited by the need for **cable connections** between transmitter and receiver. But it still found a wide variety of applications, culminating in **Alexander Graham Bell's** development of the **telephone** in the 1870s.

**« SEE ALSO**
*pp.256–57* THE NATURE OF SOUND
*pp.260–61* TELEGRAPH TO TELEPHONE

# Radio and Radio Waves

**The discovery of radio waves in the late 19th century was the start of a revolution in communications technology. Radio waves provided a means for the wireless transmission of information, allowing almost instantaneous contact over long distances without telephone wires or telegraph cables.**

In 1864 Scottish physicist James Clerk Maxwell predicted the existence of electromagnetic waves beyond the visible spectrum (see pp.258–59).

Fifteen years later, in 1879, London-born American David Edward Hughes reported that sparks from an isolated "transmitter" circuit appeared to be affecting an unconnected telephone system several hundred meters away. However, even when he demonstrated the effect to the British Royal Society and to the Post Office, experts dismissed the effect as a "standard" case of electromagnetic induction.

Then in 1883 Irish physicist George Francis Fitzgerald described how a current oscillating back and forth through a conductor could theoretically be used to generate low-frequency, long-wavelength electromagnetic waves. Five years later, in 1888, German scientist Heinrich Hertz became the first person to generate such waves—radio waves.

### Producing radio waves
Hertz developed the "dipole antenna," which used the motions of electrons oscillating through a long wire, triggered by a sudden discharge from an

electrical capacitor, to generate radio waves. The waves triggered a current in a second, isolated antenna, which produced a spark across a small gap. Hertz's experiments proved Maxwell and Fitzgerald right, and the basic unit of frequency is named in his honor.

### The race for radio
The first person to develop a practical application for radio waves was Nikola Tesla, who built antennae capable of transmitting relatively high-frequency radio waves over quite long distances. Throughout the 1890s Tesla conducted a series of pioneering demonstrations of "wireless telegraphy." His system was capable of sending electrical pulses similar to those used in cable telegraphy.

**Crystal set**
This popular early radio used a long antenna to receive signals, a wire coil to tune the frequency, and a silicon crystal to extract the sound waves. With no amplification, users had to listen through an earpiece.

**Radio transmission and reception**
For radio transmission, a low-frequency sound signal is combined with a high-frequency carrier signal in one of two ways: Amplitude Modulation (AM) or Frequency Modulation (FM).

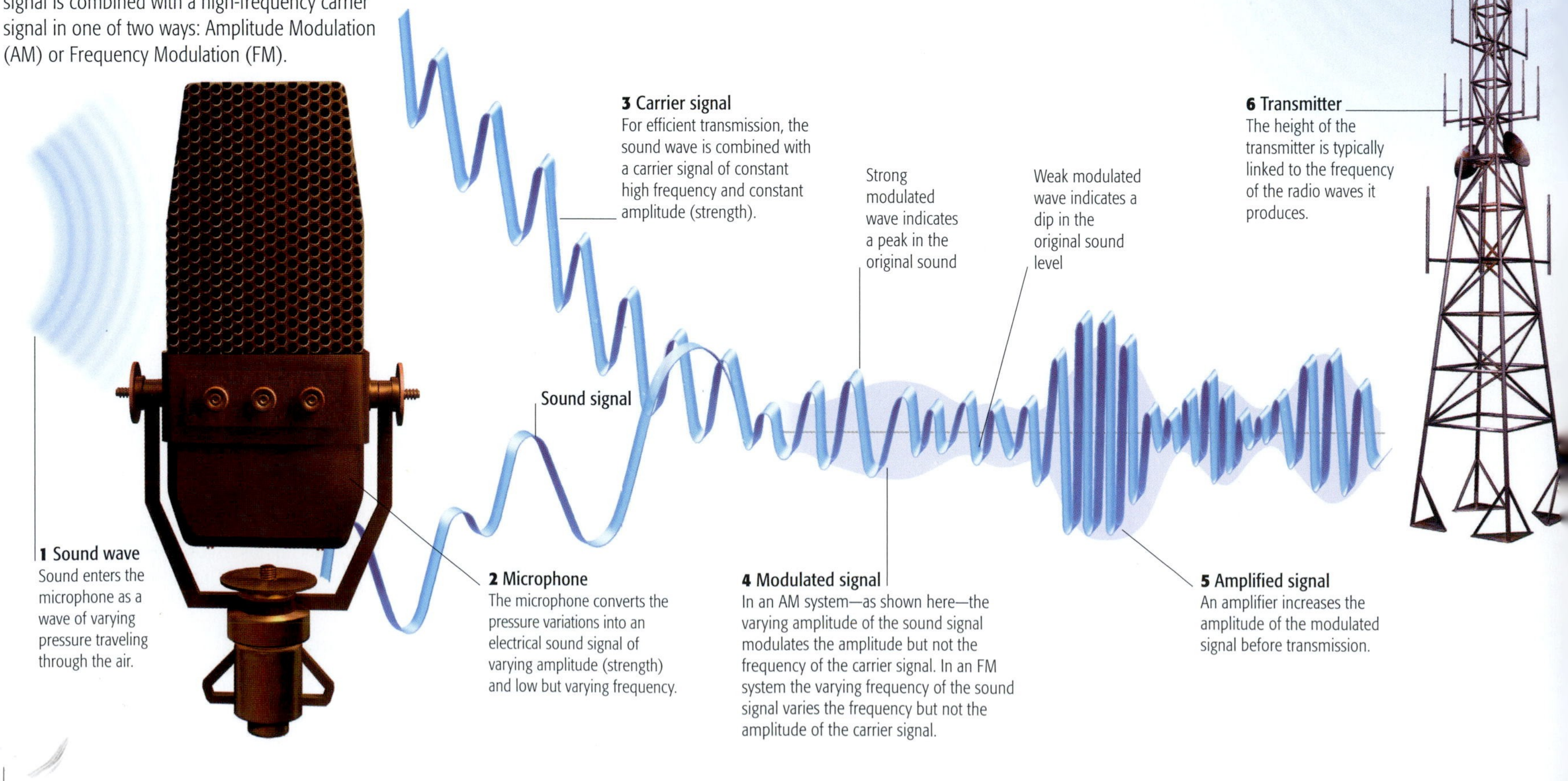

**3 Carrier signal**
For efficient transmission, the sound wave is combined with a carrier signal of constant high frequency and constant amplitude (strength).

Strong modulated wave indicates a peak in the original sound

Weak modulated wave indicates a dip in the original sound level

**6 Transmitter**
The height of the transmitter is typically linked to the frequency of the radio waves it produces.

Sound signal

**1 Sound wave**
Sound enters the microphone as a wave of varying pressure traveling through the air.

**2 Microphone**
The microphone converts the pressure variations into an electrical sound signal of varying amplitude (strength) and low but varying frequency.

**4 Modulated signal**
In an AM system—as shown here—the varying amplitude of the sound signal modulates the amplitude but not the frequency of the carrier signal. In an FM system the varying frequency of the sound signal varies the frequency but not the amplitude of the carrier signal.

**5 Amplified signal**
An amplifier increases the amplitude of the modulated signal before transmission.

**INVENTOR (1856–1943)**

## NIKOLA TESLA

Serbian inventor and engineer Nikola Tesla had a huge influence on the development of early electricity generation and radio engineering. Spending much of his life in the United States, he made breakthroughs in radio technology as well as developing and promoting the alternating current (AC) power system. His ingenuity was not matched by business acumen, however, and he failed to benefit from many of his inventions. Through the early 20th century he advocated apparently unlikely projects such as wireless energy transmission and projected energy weapons, and ended his life poor and ostracized.

for easier detection. Such valves remained a key component of radio equipment until transistors replaced them in the 1950s.

### The birth of broadcast

Canadian inventor Reginald Fessenden solved many of the remaining problems, including those of sound transmission. It was difficult to transmit the relatively low-frequency waves of audible sound on relatively high-frequency radio waves. To overcome this, Fessenden invented the "heterodyne system," which mixed the signal wave current generated from a microphone with a high-frequency carrier wave, so that the signal modulated the strength of the carrier. His first attempts at sound transmission were hampered by the inability to generate a stable, high-frequency carrier wave, but by 1906 the introduction of valves and improvements to electrical generators allowed the first such transmissions.

The "amplitude modulation" (AM) system was improved on in the 1930s by American inventor Edwin H. Armstrong's frequency modulation (FM) system, which is less prone to interference, but both systems are still in use.

The 20th century saw new applications for radio technology both inside and outside the field of wireless communication.

### CAVITY MAGNETRON

**Radar** became a practical reality in the 1930s after the invention of the cavity magnetron—a source of **high-powered** short-wavelength radio waves (microwaves).

### MICROWAVES

In the 1950s **microwaves** became the chief means of **point-to-point** radio communication, and they were key to the development of communication satellites.

### CONFIRMING THE BIG BANG

In 1964 Americans Arno Penzias and Robert Wilson, using a specialized **radio telescope**, detected the "**cosmic microwave background radiation**", supporting the Big Bang theory of the origin of the universe. Conventional radio telescopes have been used by astronomers since the 1930s.

**RADIO TELESCOPE**

SEE ALSO »
pp.320–21 THE BIG BANG
pp.368–69 ARTIFICIAL SATELLITES

Italian-Irish inventor Guglielmo Marconi developed Tesla's system by introducing new components, including the "coherer"—essentially an evacuated tube containing iron filings. This device could generate varying electrical resistance depending on the strength of the radio signals received, and the alternating current (AC) could be detected by a galvanometer.

The need for a more sensitive means of detecting signals inspired British physicist John Ambrose Fleming to develop the first "valve diode" in 1904. This turned the oscillating signals generated in an antenna back into direct current (DC)

Spreading radio waves

Motion of electrons

**7 Transmission**
Within the antenna, the amplified signal drives the movement of electrons, creating varying electromagnetic waves that carry the original information within them.

**Valve and transistor**
Valves and transistors both offer means of rectifying, extracting, and amplifying a sound signal. Although now miniaturized on to integrated circuits, transistors still lie at the heart of modern electronics.

**12 Speaker**
The sound signal drives the oscillations of a speaker, which reproduce the original sound.

**8 Aerial**
As waves from the transmitter encounter a receiving aerial, they produce a slight induced electric current.

**9 Selected frequency**
In order to select a frequency range from the many influencing the antenna, a radio has a tuning circuit. This selects only frequencies that match the "natural frequency" (time to charge and discharge) of an adjustable capacitor.

**11 Amplification**
The weak sound signal passes through an amplifier to boost its strength.

Sound signal

**10 Extracted signal**
An electronic circuit "extracts" the sound signal from the carrier signal in one of several ways, typically using a valve or transistor.

Redundant carrier signal

Sound

Early ideas about respiration and metabolism were hampered by a lack of chemical understanding, but some scholars appreciated the roles of food and oxygen.

## BURNING FOOD

In medieval Islam the physician **Ibn al-Nafis** gave an early description of **metabolism** as a **chemical process inside the body**, but muddled it with references to spontaneous generation. Later metabolism was compared with "burning" food: in the 1500s Italian polymath **Leonardo da Vinci** compared **animal nutrition** with the **burning of a candle**.

## LIFE-GIVING AIR

In 1605 the Polish alchemist **Michał Sędziwój** proposed the existence of a **"food of life"** in **air**—this was more than a century before the discovery of **oxygen**. Shortly afterward in the mid-1600s English chemist **John Mayow** suggested that **the body extracted a component of air**—which he called "spiritus nitro-aereus"—that was necessary for life.

**« SEE ALSO**
pp.90–91  CIRCULATION OF BLOOD
pp.214–15  ANIMAL AND PLANT CELLS
pp.216–17  DIGESTION

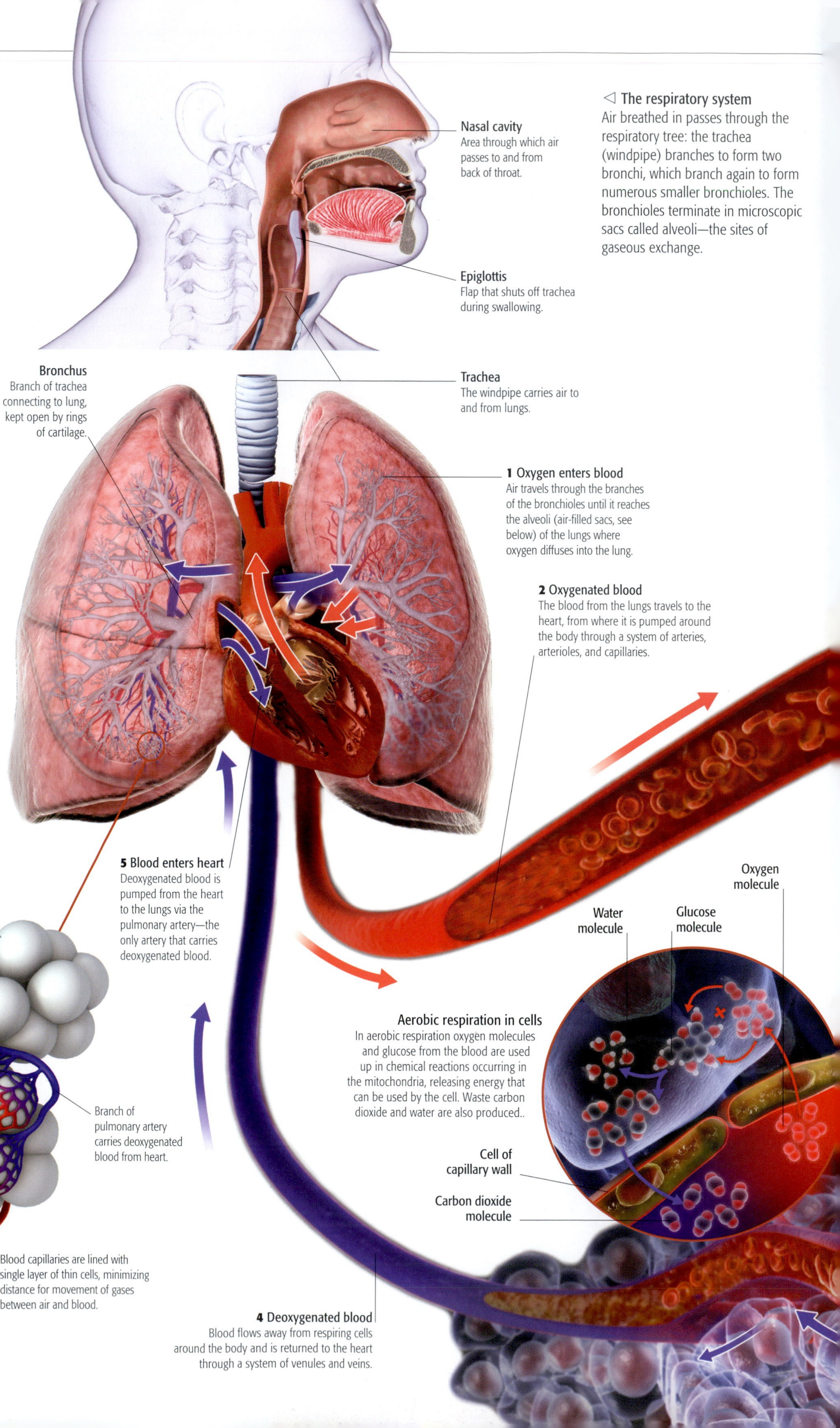

**Nasal cavity**
Area through which air passes to and from back of throat.

**Epiglottis**
Flap that shuts off trachea during swallowing.

**◁ The respiratory system**
Air breathed in passes through the respiratory tree: the trachea (windpipe) branches to form two bronchi, which branch again to form numerous smaller bronchioles. The bronchioles terminate in microscopic sacs called alveoli—the sites of gaseous exchange.

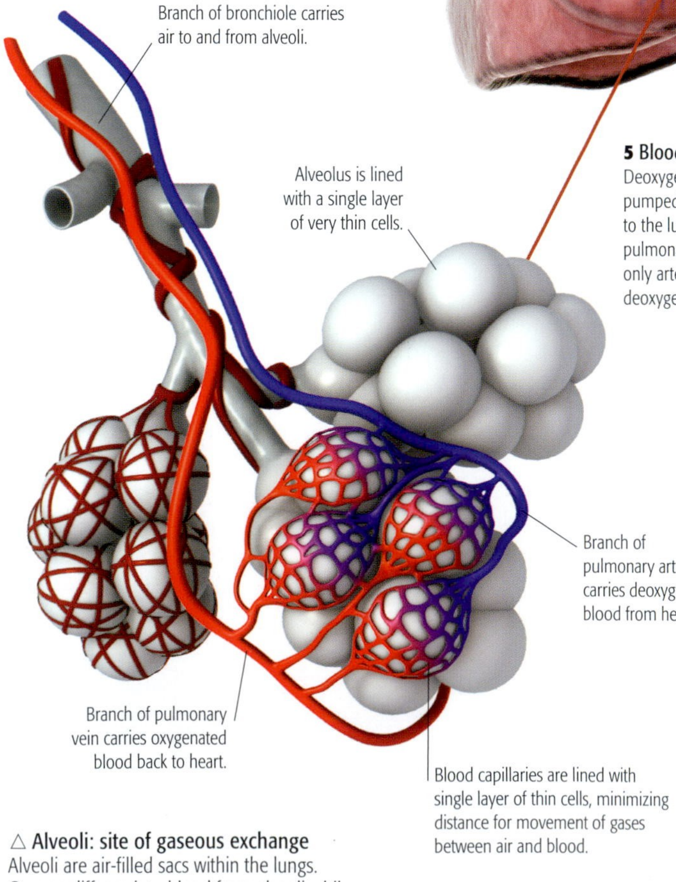

**Bronchus**
Branch of trachea connecting to lung, kept open by rings of cartilage.

**Trachea**
The windpipe carries air to and from lungs.

**1 Oxygen enters blood**
Air travels through the branches of the bronchioles until it reaches the alveoli (air-filled sacs, see below) of the lungs where oxygen diffuses into the lung.

**2 Oxygenated blood**
The blood from the lungs travels to the heart, from where it is pumped around the body through a system of arteries, arterioles, and capillaries.

**5 Blood enters heart**
Deoxygenated blood is pumped from the heart to the lungs via the pulmonary artery—the only artery that carries deoxygenated blood.

Branch of bronchiole carries air to and from alveoli.

Alveolus is lined with a single layer of very thin cells.

Branch of pulmonary artery carries deoxygenated blood from heart.

Branch of pulmonary vein carries oxygenated blood back to heart.

Blood capillaries are lined with single layer of thin cells, minimizing distance for movement of gases between air and blood.

**△ Alveoli: site of gaseous exchange**
Alveoli are air-filled sacs within the lungs. Oxygen diffuses into blood from alveoli while carbon dioxide passes into alveoli from blood —both through thin capillary and alveolar walls.

**4 Deoxygenated blood**
Blood flows away from respiring cells around the body and is returned to the heart through a system of venules and veins.

**Oxygen molecule**

**Water molecule**

**Glucose molecule**

**Aerobic respiration in cells**
In aerobic respiration oxygen molecules and glucose from the blood are used up in chemical reactions occurring in the mitochondria, releasing energy that can be used by the cell. Waste carbon dioxide and water are also produced..

**Cell of capillary wall**

**Carbon dioxide molecule**

# Breathing and Respiration

**All living organisms burn fuel like engines, although the fuel they burn is sugar and fat. The energy released from this process is used for movement, growth, and a host of other functions that sustain life. Breathing and respiration are the fundamental characteristics of life that enable this energy release to take place.**

Breathing is the process that brings fresh air into the lungs and allows oxygen in air to pass into the bloodstream for transport to body cells, where it is used in respiration to release energy from food.

## Breathing
When the body inhales, the intercostal muscles between the ribs and the diaphragm both contract, moving the ribs up and out, and flattening the diaphragm so that air flows into the lungs. When the body exhales, these muscles relax, moving the ribs down and in, and making the diaphragm dome-shaped so that air is pushed out of the lungs.

The breathing process enables an exchange of gases within the lungs. Air enters minute air-filled sacs called alveoli, then oxygen diffuses into the blood via millions of microscopic capillaries, which permeate the lungs. Waste carbon dioxide diffuses in the opposite direction to be expelled in air from the body on exhalation.

The blood then carries oxygen to the rest of the cells in the body. In the mid-1800s German physiologist Felix Hoppe-Seyler discovered how this happens—oxygen is picked up by the blood flowing through lung capillaries by binding to hemoglobin (a red iron-containing pigment) found inside red blood cells. When blood reaches other cells in the body where oxygen levels are much lower, hemoglobin gives up the oxygen to the cells.

## Aerobic and anaerobic respiration
Respiration releases heat and keeps mammals and birds warm-blooded. Respiratory energy is also used for cell activities: building protein and DNA,

**Red blood cells**
Disk-shaped cells have oxygen attached to their hemoglobin.

**3 Respiring cells**
As blood cells reach capillaries surrounded by respiring cells, oxygen bound to their hemoglobin is freed for use in aerobic respiration.

Matrix contains enzymes to initiate energy release

Crista is site where most of the energy is released

Inner membrane folded into cristae

Outer membrane

△ **Mitochondrion: site of aerobic respiration**
Mitochondria are found in the cytoplasm of cells. Each capsule is bounded by a double membrane: the inner is folded into cristae to maximize the surface area for enzymes, which catalyze (speed up) the reactions of aerobic respiration; the outer is a smooth boundary.

dividing cells, or pumping substances in and out of cells (see pp.214–15). After food is digested in the gut (see pp.216–17), glucose is carried in the blood to cells. Each glucose molecule is divided into two, and transferred into bean-shaped capsules called mitochondria (see above), where they react with oxygen in aerobic respiration, releasing energy in the form of adenosine triphosphate (ATP).

Anaerobic respiration (also known as fermentation) was first recognized in microorganisms; in 1838 French physicist Cagniard de la Tour discovered that a yeast culture produces alcohol (a biproduct of anaerobic respiration) when supplied with glucose. Anaerobic respiration also occurs in humans: when the demands of vigorous activity reach the limits of cardiovascular

performance, extra energy can be extracted without oxygen through breaking down molecules including sugar. Lactic acid, rather than alcohol is the byproduct of this process in humans.

In 1897 German chemist Eduard Büchner found that whole cells were not needed for fermentation—extracts of yeast could ferment glucose too. He had found enzymes: the chemical catalysts vital to respiration.

AFTER »

**Advances in the 20th century revealed more about the chemical reactions of respiration, and improved understanding of potentially fatal lung diseases.**

### MAKING RESPIRATION MORE EFFICIENT
In 1903 Danish physician **Christian Bohr** showed that the unloading of **oxygen from hemoglobin** is encouraged by **high levels of carbon dioxide**, meaning that as more of this waste is produced in **exercise**, more oxygen becomes available for respiration in a self-regulating process.

### KEY CYCLE OF ENERGY RELEASE
In 1937 the German-British physician **Hans Krebs** discovered the Krebs cycle, or the citric acid cycle, which demonstrated the key sequence of **metabolic chemical reactions** that are central to processes of energy release within the **mitochondria**. This discovery won him a Nobel Prize in 1953.

### SMOKING AND LUNG CANCER
In 1996 scientists demonstrated experimentally that a **chemical in tobacco smoke** called **benzopyrene** disrupted a part of DNA that normally stops lung cells from dividing too quickly; this disruption can result in a **cancerous tumor**. It was the first conclusive proof of a **link between smoking and lung cancer**.

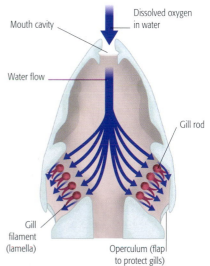

Dissolved oxygen in water

Mouth cavity

Water flow

Gill rod

Gill filament (lamella)

Operculum (flap to protect gills)

**Breathing in fish**
Aquatic organisms have gills instead of lungs. Water enters the mouth of a fish and flows over gill filaments, which are packed with blood capillaries to exchange oxygen and carbon dioxide between water and blood.

# The Five Senses

**Sensation is a fundamental characteristic of life: animals use their senses to take in information about the environment around them. Sensory cells called receptors are specialized to detect factors such as light, sound, pressure, taste, and smell.**

Sense organs contain receptor cells that can detect changes in our environment and so gather information about the world around us. Different kinds of stimuli, such as light, sound, and pressure are detected by specific receptors. The strength of a stimulus determines the frequency of electrical impulses that travel along sensory nerves to the brain. Detection of stimuli is usually beneficial: the pleasurable sensation of taste is the promise of nourishment. Even pain (or fear of it) can help us avoid potentially damaging stimuli, such as burning-hot objects.

### Detecting light

The eye focuses light onto the retina with its cornea and lens (see right), producing an inverted image that is later corrected by the brain. In 1856 German physician Hermann von Helmholtz proved how the eye focuses light: ciliary muscles encircling the lens control its shape, contracting to bend light from near objects or relaxing to adjust it from objects farther away. In the same decade Heinrich Müller showed that light is detected by photoreceptor cells (rods and cones) in the retina. These cells contain rhodopsin (a light-absorbing pigment), which converts light into electrical signals that are relayed to the brain, for processing into visual images.

### Detecting chemicals

The mechanisms of our senses of taste and smell are the same: substances reach the sensory parts of the mouth and nose (whether as odor or food) to stimulate receptor cells, and the brain interprets this as flavor (see below).

**BEFORE**

The ancient Greek idea that sensation was a separate process from thought promoted studies on how the eyes and ears could pick up signals of light and sound.

**DETECTING THE SIGNAL**
During the Renaissance many still agreed with the Roman physician **Galen** that the **eye's lens** was the **site** of **visual sensation**. In 1583 Swiss physician **Felix Platter** proved that vision was possible even after disconnecting the lens and correctly surmised that the **retina** was the part of the eye **stimulated by light.**

**ANATOMY IN MINIATURE**
Revealing the **structure of the ear** was a challenge to early anatomists. Andreas Vesalius **«« 72–73** advised that the inner ear should be removed during dissection to allow close scrutiny; Italians **Eustachi** and **Fallopio** dissected its parts—**identifying** the **eardrum, tiny bones**, and its intricate **labyrinth of tubes** for the first time.

**«« SEE ALSO**
*pp.220–21* THE NERVOUS SYSTEM
*pp.222–23* THE BRAIN

Choroid
Blood-rich layer that stops internal reflection of light in the eyeball.

Superior rectus
Small muscle that is used for moving the eyeball.

Iris
Ring of muscles that relax or contract to let more or less light through the central pupil.

Sclera
Outer opaque (white) wall of the eyeball.

Pupil
An opening in the iris, through which light passes to the lens.

Cornea
Transparent layer to let light into the eye. It is coated with the thinner protective conjunctival membrane.

Lens
Soft transparent disc that focuses light onto the retina.

Ciliary muscle
Ring of muscle that changes the shape of the lens to focus light.

Sensory cell of nerve going to brain

Olfactory bulb

Bone

Mucus-secreting gland

Chemoreceptor

Surface of olfactory epithelium where chemicals of odors bind

Airflow

Odor molecule

**NERVE PATHWAYS FOR TASTE AND SMELL**

◁ **How we detect smell**
When odorous chemicals are breathed in, they dissolve in the mucus lining of the nasal cavities. Here, chemoreceptor sensory cells have surface proteins, which bind to specific types of chemicals. Once stimulated, the cells transmit nerve impulses along nerves to the brain.

Larger papillae contain taste buds and openings of salivary ducts

Smaller taste buds are concentrated at the edges and tip

Sensory nerves from taste and touch receptors carry nerve signals to the brain

△ **How we detect taste**
The tongue has a surface coating of taste buds, each of which contains around 100 chemoreceptor cells. When these are stimulated, they fire electrical impulses along sensory nerve fibers to the brain.

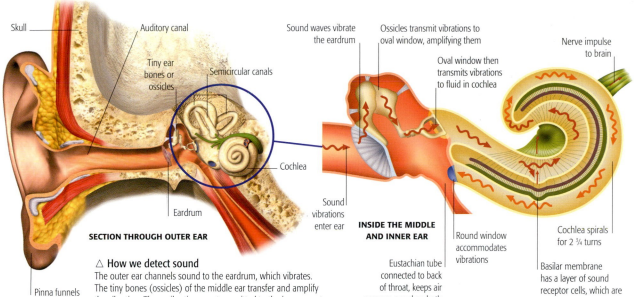

◁ **How we detect light**
The eye focuses light onto a spot of the retina called the fovea. Here, cone cell receptors are stimulated to send nerve impulses along the optic nerve to the brain, where a sharp, color image results.

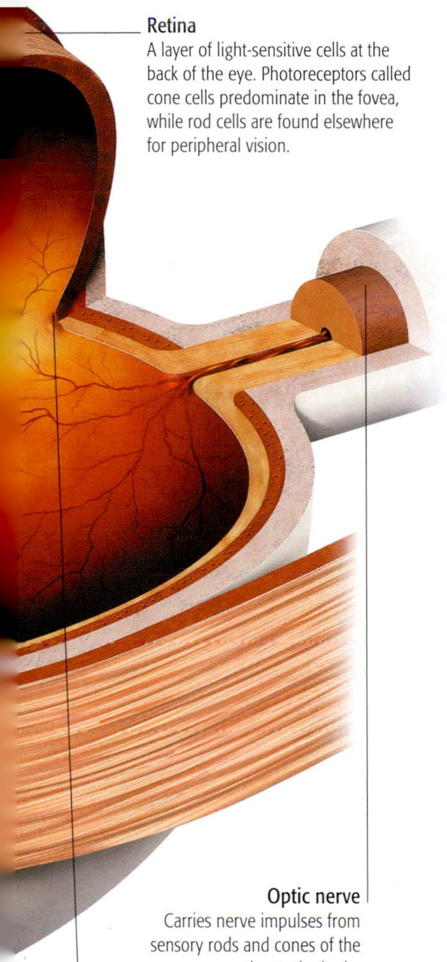

Skull
Auditory canal
Tiny ear bones or ossicles
Semicircular canals
Sound waves vibrate the eardrum
Ossicles transmit vibrations to oval window, amplifying them
Oval window then transmits vibrations to fluid in cochlea
Nerve impulse to brain
Cochlea
Eardrum
Sound vibrations enter ear
Round window accommodates vibrations
Cochlea spirals for 2 ¾ turns

**SECTION THROUGH OUTER EAR**

**INSIDE THE MIDDLE AND INNER EAR**

Eustachian tube connected to back of throat, keeps air pressure equal on both sides of eardrum

Basilar membrane has a layer of sound receptor cells, which are stimulated to send nerve impulses to the brain

Pinna funnels sound waves

△ **How we detect sound**
The outer ear channels sound to the eardrum, which vibrates. The tiny bones (ossicles) of the middle ear transfer and amplify the vibration. These vibrations are transmitted to the inner ear at the oval window. Disruption of fluid in the cochlea then stimulates receptor cells to send signals to the brain.

**Retina**
A layer of light-sensitive cells at the back of the eye. Photoreceptors called cone cells predominate in the fovea, while rod cells are found elsewhere for peripheral vision.

## Detecting sound
Receptors in the ear are stimulated by vibration of sound waves. Such vibrations are transmitted and amplified through a system of first air-filled and then fluid-filled cavities. In the cochlea, layers of receptor cells with hairlike projections bend in response to the vibration, firing off signals to the brain (see above). Semicircular canals in the inner ear are responsible for balance: as the head tilts and moves, the flow of liquid stimulates hair cells to send impulses to the brain.

### INVENTION
## HEARING AID

Devices based on the technology developed in the 1920s and 1930s by American physicist Harvey Fletcher allow people with hearing loss to experience sounds when wearing such a hearing aid. Those worn behind the ear transmit amplified sound into the ear canal. Some devices produce the amplified sound within the canal itself, thereby minimizing distortion. Others are surgically implanted to stimulate the inner ear via the skull, bypassing the ear canal entirely.

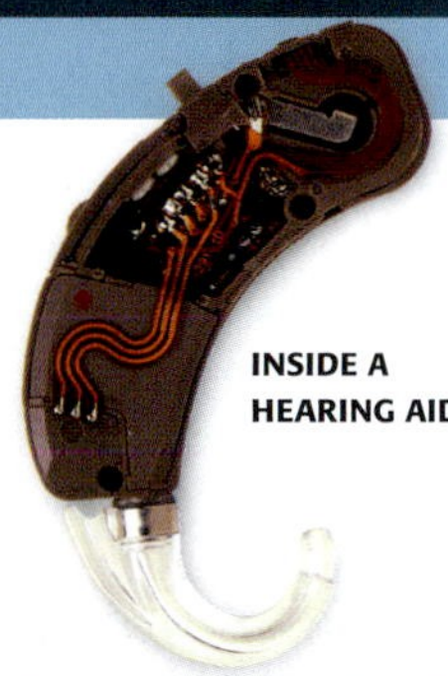

**INSIDE A HEARING AID**

**Optic nerve**
Carries nerve impulses from sensory rods and cones of the retina to the brain.

**Fovea**
Spot of the retina immediately behind the lens packed with cone cells for sharp, color vision.

> " Our **sight** is the **most perfect** and **delightful** of our **senses**. "

JOSEPH ADDISON, ENGLISH ESSAYIST, 1712

**AFTER**

Disorders of sense organs and the brain can interfere with the vision or hearing of sufferers. Advances in understanding the organs and how such disorders arise are important in opening up new treatments.

### EYE SURGERY
Modern **ophthalmic surgery** now allows for the successful **treatment** of many **disorders** that hinder vision. The most common procedure is the **surgery on cataracts**: cloudy lenses can be replaced with plastic implants. **Glaucoma**, where raised pressure affects the optic nerve, is treated by removal of excess fluid from the eye.

### TRANSLATING THE SIGNALS
In the 1950s Japanese scientist **Ichiji Tasaki** and his colleagues started to study the way **sound detection** within the inner ear was **translated** into **electrical signals**, which are then interpreted by the brain. Such work laid the **foundation of audiology** for the diagnosis and treatment of **hearing disorders**.

**SEE ALSO »**
pp.406–07 MODERN SURGICAL PROCEDURES

Sensory nerve fiber endings respond to pain, temperature and pressure

Meissner's corpuscle responds to light pressure

Sensory nerve fibers transmit impulses through the nervous system

Pacinian corpuscle responds to firmer, prolonged pressure

▷ **How we detect pressure**
The skin has many functions, including sensory reception. Seen here in cross section, it has different receptor cells at varying depths, which detect touch, pressure, pain, and changes in temperature.

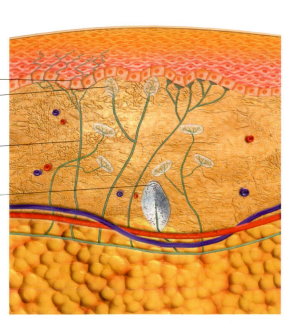

### UNIQUE ANIMAL SENSES

Many animals are able to sense stimuli that humans cannot, sometimes involving unique types of receptor cells.

**Heightened sensitivity to chemicals**
Many animals can detect exceptionally low concentrations of chemicals. For example, moths can detect minute levels of a sex pheromone.

**Detection of ultraviolet**
Many insects and birds have eyes specialized for vision in ultraviolet light.

**Polarized light**
Bees detect polarized light for orientation.

**Pressure detection**
Fish have a lateral line of hair cells on their body, that helps detect water pressure and navigate while swimming.

**Echolocation**
Bats and dolphins detect reflected sounds to find objects and to navigate.

**Electroreception**
Certain species of fish have the ability to detect electric fields, which helps to find food, perceive predators, and interact with others.

## BEFORE

The first physiologists recognized that the body was regulated and that imbalance was associated with disease, but many continued to rely on supernatural explanations.

### ORIGINS OF THE STUDY OF HORMONES
In 200 BCE Chinese physicians used **urine extracts** for medicinal use, and their therapeutic effects may have been caused by the presence of hormones. Later, medieval **Islamic physicians** recognized the symptoms of certain hormone imbalances: **Avicenna** noted the sweet-tasting urine of **diabetics** and **al-Jurjani** described the **goiter** of thyroid disorder. But hormones occur in such tiny amounts that techniques for detecting them were only found in the 19th century.

**GOITER SUFFERER**

### VITALISM
Until the mid 1800s it was thought that the body was governed by a **"vitalist force"** and its functions **could not be explained by science**. But chemist **Friedrich Wöhler** synthesized the human waste product **urea** in 1828 **‹‹ 22–23,** showing that an organic body product could be made artificially, and starting a shift away from vitalism.

**‹‹ SEE ALSO**
pp.90–91 CIRCULATION OF THE BLOOD
pp.216–17 DIGESTION

## PHYSIOLOGIST (1813–78)
### CLAUDE BERNARD

A French physiologist, Bernard is often considered the "father of physiology" (the study of processes within living organisms). He used rigorous experimental work to make discoveries about the liver, pancreas, and nervous system, and established the principle of biological self-regulation. He insisted that physiology obeyed normal scientific laws as opposed to vitalism. His experiments on living animals were controversial, but resulted in breakthroughs in medicine. After his death he was the first scientist to be honored with a public funeral in France.

The internal environment of the human body needs to be regulated in various ways, including the control of body temperature, sugar levels, and water levels. The control system for many factors is centered on a part of the brain called the hypothalamus, which is linked to the pituitary gland—the master gland for hormone secretion. Hormones are molecules secreted into the blood from endocrine glands, to bring about a specific response.

### Controlling body temperature
In the 18th century Scottish surgeon John Hunter discovered that so-called "warm-blooded" mammals and birds were homeotherms—they maintain a constant internal body temperature whatever the external environment. The hypothalamus monitors blood

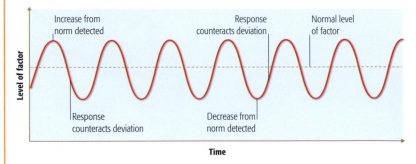

### IN PRACTICE
### HOMEOSTASIS

Homeostasis refers to the processes that regulate the internal environment of an organism. When a factor deviates from the norm, the deviation is detected by a part of the body called a receptor, which then communicates with an effector to counteract the change. For example, a slight increase in temperature causes receptors to stimulate the sweat glands—the effectors—to release sweat for cooling (by evaporation).

*Increase from norm detected*
*Response counteracts deviation*
*Normal level of factor*
*Level of factor*
*Response counteracts deviation*
*Decrease from norm detected*
*Time*

# Regulating the Body

**The vital functions of the human body operate only under a specific set of conditions, and so it is critical that these are regulated and controlled correctly, from the tiniest cells up to the largest organ systems. This regulation is precise, even though the body's external surroundings may change from moment to moment.**

temperature, and sends nerve impulses to various parts of the body to correct any changes. Some muscles in the skin can alter the diameter of blood vessels to help the body lose more or less heat; other muscles cause the body to shiver, or make hairs stand on end to trap warm air by the skin. These processes are part of homeostasis (see above).

### The study of hormones
In the following century, in 1849, German physiologist Arnold Berthold found he could reduce the aggressive behavior of cockerels by castrating them. His deduction that chemicals from the organs stimulated this change in behavior signified the birth of endocrinology—the study of hormones.

### Controlling sugar and water levels
Toward the end of the 19th century, English physiologist Edward Sharpey-Schafer resolved how the body controls

its blood sugar levels: these are regulated by a balance of the hormones insulin and glucagon (both released by the pancreas), which target cells of the liver to lower or raise glucose levels of the blood as required (see below).

At the start of the next century, in 1917, English physician Arthur Cushny found that urine was formed in the kidney by filtering waste products and

reabsorbing useful ones (see right). The hypothalamus detects falls in the blood's water content, and the correcting anti-diuretic hormone is released from the pituitary gland. This hormone targets microscopic tubules, called nephrons, in the kidneys, causing them to reabsorb more water when the body is dehydrated, so less water is lost in urine.

**Regulating body temperature**
Mammals maintain a constant body temperature. The yellow/orange colors show the hottest areas—the temperature of the mouse is higher than its surroundings.

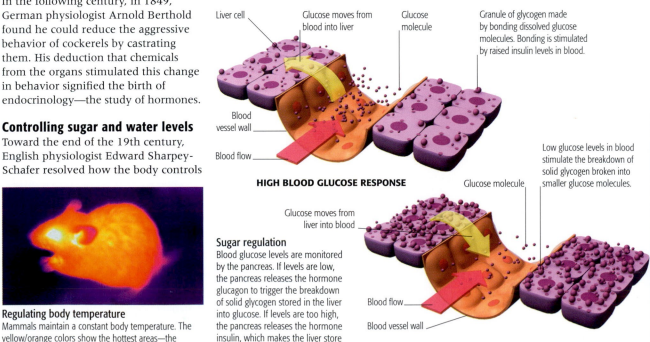

Liver cell
Glucose moves from blood into liver
Glucose molecule
Granule of glycogen made by bonding dissolved glucose molecules. Bonding is stimulated by raised insulin levels in blood.

Blood vessel wall
Blood flow

**HIGH BLOOD GLUCOSE RESPONSE**

Low glucose levels in blood stimulate the breakdown of solid glycogen broken into smaller glucose molecules.

Glucose molecule

Glucose moves from liver into blood

### Sugar regulation
Blood glucose levels are monitored by the pancreas. If levels are low, the pancreas releases the hormone glucagon to trigger the breakdown of solid glycogen stored in the liver into glucose. If levels are too high, the pancreas releases the hormone insulin, which makes the liver store more glycogen.

Glucose molecule
Blood flow
Blood vessel wall

**LOW BLOOD GLUCOSE RESPONSE**

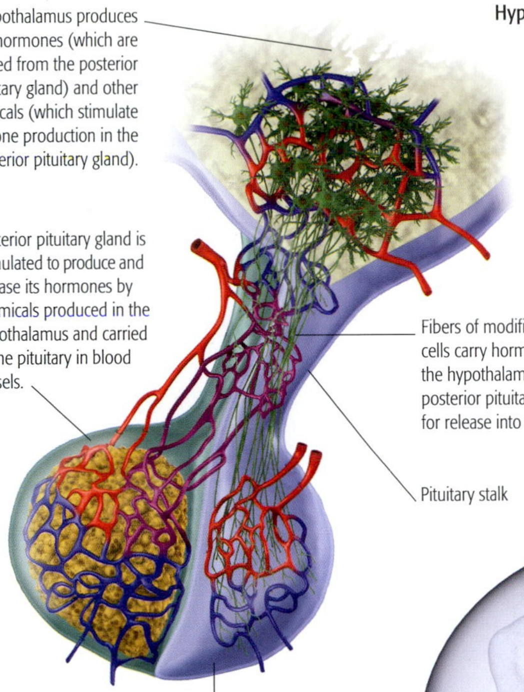

Hypothalamus produces hormones (which are released from the posterior pituitary gland) and other chemicals (which stimulate hormone production in the anterior pituitary gland).

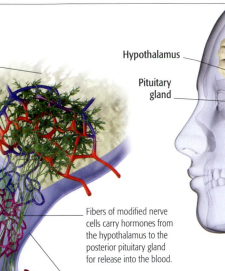

Anterior pituitary gland is stimulated to produce and release its hormones by chemicals produced in the hypothalamus and carried to the pituitary in blood vessels.

Fibers of modified nerve cells carry hormones from the hypothalamus to the posterior pituitary gland for release into the blood.

Pituitary stalk

Posterior pituitary gland releases hormones, which are produced in the hypothalamus and reach the pituitary along fibers of modified nerve cells.

### The pituitary gland
This master gland produces a range of different hormones, including anti-diuretic hormone (ADH) from the posterior lobe to stimulate reabsorption of water in the kidneys and, from the anterior lobe, the sex hormones FSH and LH, which regulate the function of the ovaries in women or testes in men.

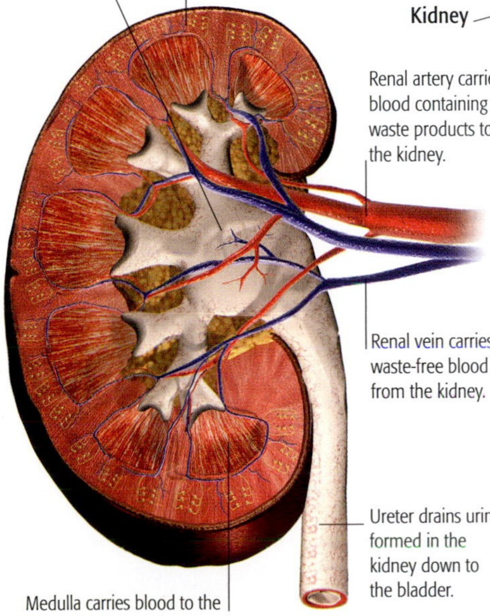

Renal pelvis receives urine from millions of collecting ducts and empties it into the ureter.

The cortex contains billions of microscopic tubules called nephrons, which reabsorb useful substances from the filtrate derived from blood.

**Kidney**

Renal artery carries blood containing waste products to the kidney.

Renal vein carries waste-free blood away from the kidney.

Ureter drains urine formed in the kidney down to the bladder.

Medulla carries blood to the cortex and drains urine from the cortex through collecting ducts.

### Kidney function
The kidney removes waste products from the blood and regulates its salt levels by controlling the volume of water lost in urine—this is determined by levels of ADH produced by the pituitary (see above).

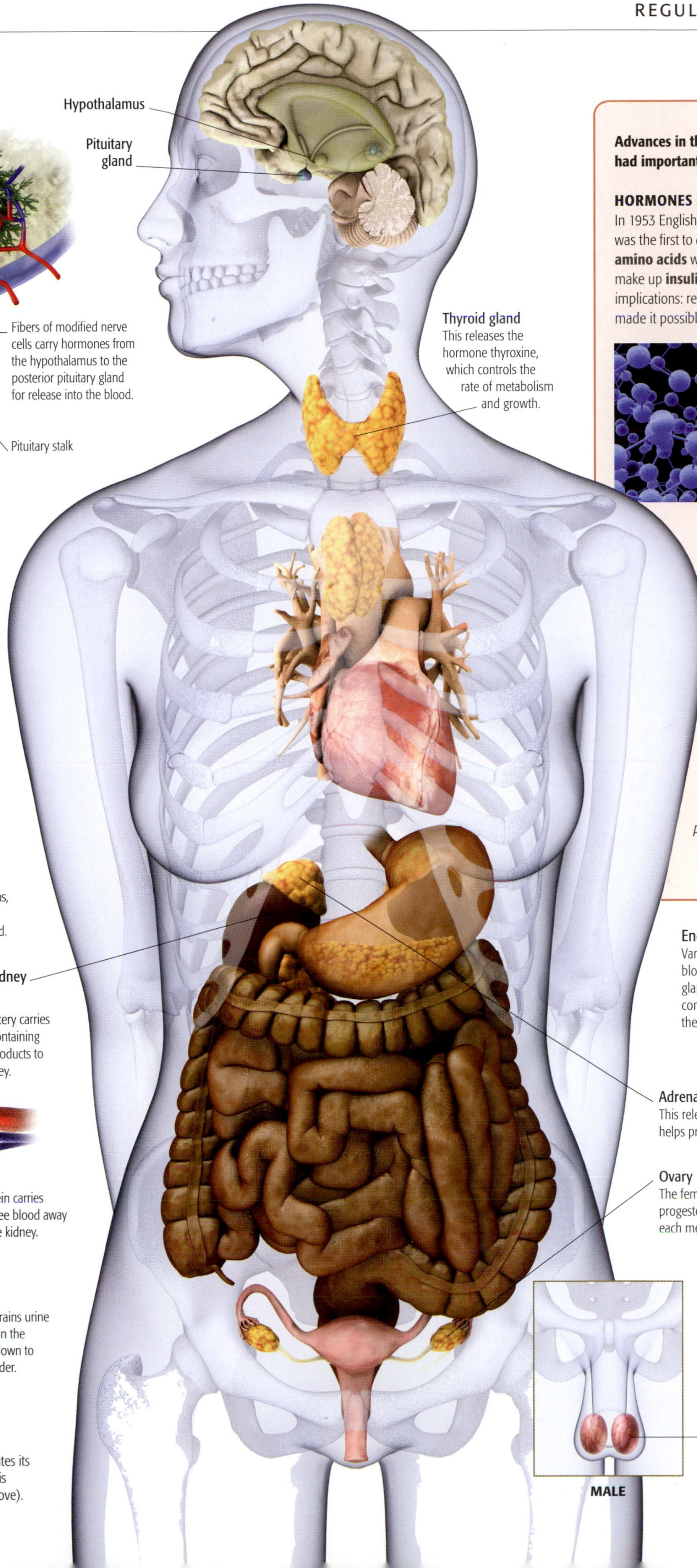

Hypothalamus

Pituitary gland

**Thyroid gland**
This releases the hormone thyroxine, which controls the rate of metabolism and growth.

**Advances in the study of hormones have had important medical implications.**

#### HORMONES AS MOLECULES
In 1953 English biochemist **Frederick Sanger** was the first to determine the exact sequence of **amino acids** within the chains of protein that make up **insulin**. Sanger's work had significant implications: revealing the structure of insulin made it possible for chemists to synthesize it.

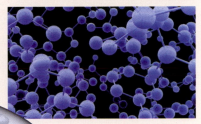

**INSULIN MOLECULE**

#### KIDNEY DIALYSIS
The kidney dialysis machine, invented by Dutch physician **Willem Kolff** in the 1940s, treats kidney failure by **removing impurities from the blood**, but it cannot replicate the kidney's hormone production. Surgical transplant is an alternative solution.

SEE ALSO »
pp.406–07 MODERN SURGICAL PROCEDURES
pp.446–50 REFERENCE SECTION

**Endocrine (hormone) glands**
Various hormones are released into the bloodstream from specific endocrine glands around the body. Many are controlled by the pituitary gland below the brain.

**Adrenal gland**
This releases the hormone adrenaline, which helps prepare the body for physical exertion.

**Ovary**
The female sex organ releases estrogen and progesterone and produces a ripe egg with each menstrual cycle.

**Testes**
The male sex organ releases testosterone for regulation of male characteristics, including sperm production.

**MALE**

# Animal Behavior

**It is an animal's brain that makes complex behavior possible. The bigger the brain, the more sophisticated the behavior becomes, resulting in a repertoire of activities such as finding food, raising families, and even—in humans—attaining a degree of creativity that forms the basis of art and science itself.**

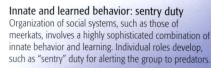
**Innate and learned behavior: sentry duty**
Organization of social systems, such as those of meerkats, involves a highly sophisticated combination of innate behavior and learning. Individual roles develop, such as "sentry" duty for alerting the group to predators.

**Innate behavior: courting rituals**
The Siamese fighting fish above has an elaborate innate courtship ritual, which culminates in the male fish wrapping his body around the female.

## « BEFORE

**Before the innate–learned idea emerged in the mid-1800s, understanding of animal behavior was limited.**

**ANIMAL AS MACHINE**
French philosopher Descartes, in the 17th century, took a **mechanical view** of animal activity by proposing **behavior** was **driven** by a **stimulus–response system**, but without considering that it could be modified by learning.

**ACQUIRED BEHAVIOR**
In the early 1800s French naturalist **Jean-Baptiste Lamarck** devised an **evolutionary theory « 200–201**, which suggested that animals in different environments would **adopt different habits**, causing them to acquire new **inherited characteristics** during their life.

**« SEE ALSO**
*pp.200–01* HOW EVOLUTION WORKS
*pp.220–21* THE NERVOUS SYSTEM
*pp.222–23* THE BRAIN

Ethology—the science of animal behavior—was pioneered in the 19th century by English biologist Douglas Spalding. He recognized that behavior has two components: an innate one that is inherited and a learned one honed by experience. At least some of all animal behavior is innate, determined by genes that dictate the hardwiring of the nervous system. Mammals and birds are capable of sophisticated learning because they have enlarged cerebrums (part of the brain that retains information in memory —see pp.222–23). This means that their behavior can be modified by learning.

## Innate behavior

This type of behavior tends to be stereotyped, meaning that all individuals (often of a certain sex or age) of a species perform it in much the same way. Innate behavior is often a straightforward response to a stimulus, such as when earthworms move into shade to avoid the drying effects of the sun.

However, in the early 1900s more complex instincts were uncovered. Austrian ethologist Konrad Lorenz studied instinctive behavior in birds and reinforced the idea of "imprinting"—a type of learning that occurs at a particular age. He found that chicks can become imprinted on a surrogate mother (even a human) during a critical period after hatching. Just as light is a stimulus to an earthworm, so a surrogate mother is to a chick.

Dutch ethologist Nicolaas Tinbergen studied the complex courtship ritual of the stickleback fish and found that the red throat of the male acts as a stimulus to initiate courtship. In these three cases, the behavior is beneficial to the animal.

## Learned behavior

This type of behavior also varies in levels of complexity. The simplest type of learning is habituation, whereby an animal learns to ignore an insignificant repetitive stimulus. For example, birds learn to disregard car sounds in cities—they do not waste time and energy in responding to them—unlike birds in the wild that may desert their breeding territories in response to any unfamiliar noise.

Conditioning is another form of simple learning, where one stimulus

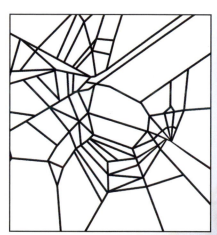
**Changing innate behavior**
The innate web-building behavior of spiders is fixed by the nervous system, but can be experimentally altered by treating the spider with drugs, as here with caffeine, which interfere with the behavior of nerve cells.

**Learned behavior: feeding methods**
Japanese macaques have learned that they can clean food in water. Potato washing was introduced by a female macaque in 1953, her family imitated her behaviour.

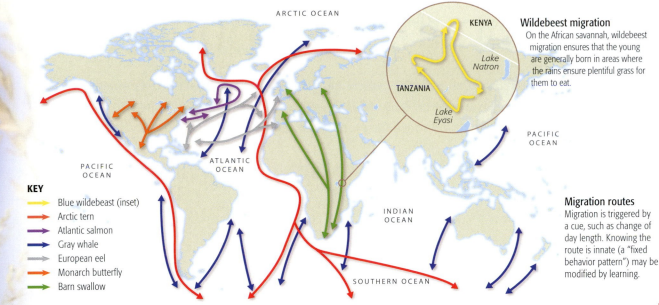

ARCTIC OCEAN

KENYA

**Wildebeest migration**
On the African savannah, wildebeest migration ensures that the young are generally born in areas where the rains ensure plentiful grass for them to eat.

*Lake Natron*

TANZANIA

*Lake Eyasi*

PACIFIC OCEAN

PACIFIC OCEAN

ATLANTIC OCEAN

INDIAN OCEAN

**KEY**
- → Blue wildebeast (inset)
- → Arctic tern
- → Atlantic salmon
- → Gray whale
- → European eel
- → Monarch butterfly
- → Barn swallow

SOUTHERN OCEAN

**Migration routes**
Migration is triggered by a cue, such as change of day length. Knowing the route is innate (a "fixed behavior pattern") may be modified by learning.

**3,500 MILES** (5,600 kilometers) of a connected hive of billions of Argentine ants, stretching from Portugal to Italy, the largest supercolony ever recorded.

is associated with another. A Russian physiologist, Ivan Pavlov, discovered simple conditioning with his experiments on dogs in the 1890s (see below). Another form of conditioning was discovered 30 years later by American pyschologist B.F. Skinner. In a laboratory, he found that an animal who accidentally triggered a switch to deliver food would learn from this experience, so the switch-flicking

> "We **animals** are the **most complicated things** in the known **universe**."
>
> RICHARD DAWKINS, BIOLOGIST, 1986

would become reinforced because of promise of food. In the wild, a bird that stumbles across food by flicking over leaves will learn to associate leaf-flicking with the promise of food, so this behavior is reinforced.

For some time, and largely due to the influence of American psychologist Edward Thorndike, most apparently intelligent animal behavior was thought to be explained by a process of trial and error. But then a German psychologist proposed something different. In his studies of captive chimpanzees, Wolfgang Köhler found that solving problems (such as stacking crates in order to reach bananas) required more insight. This kind of

behavior is the most sophisticated of all—it is behavior with purpose and guided by inspiration. Problem solving explains the superior skills involved in the most basic of biological functions, including acquiring food, avoiding predators, and raising offspring. It also encourages the use of tools and tool modification—something that reaches its pinnacle of achievement in the human species.

**Conditioning in birds: tool use**
The woodpecker finch, from the Galapagos Islands, is one of the few bird species that can control a tool—it uses a thin spine to probe crevices for food. Learning to master this tool is a sophisticated form of conditioning, whereby the promise of food reinforces behavior.

**AFTER** ≫

**In the 1970s the science of ethology (animal behavior) became increasingly influenced by genetics and ecology.**

**EVOLUTIONARY GAMES**
In 1976 English evolutionary biologist Richard Dawkins published *The Selfish Gene*, which stressed that the **evolution** of **organisms** is driven by **genes**: an individual body is a "community" of working genes and the behavior of the body maximizes the reproduction of these genes.

**SOCIOBIOLOGY**
American biologist Edward O. Wilson was instrumental in urging that **animal behavior could evolve by natural selection**, and that this could even lead to the **evolution of complex societies**. The nonbreeding drones in a beehive, for example, can pass on their genes indirectly by helping the queen, with whom they share genetic material.

**SEE ALSO** ≫
*pp.332–33* ECOLOGY AND ECOSYSTEMS
*pp.334–35* CONSERVATION BIOLOGY

**INNATE BEHAVIOR** Pre-set by genes, this is often the predominant type of behavior seen in animals with simple nervous systems. Because it is inherited, it can evolve by the process of natural selection.

**LEARNED BEHAVIOR** This behavior is modified by experience so is not stereotyped. It is most sophisticated in animals with well-developed nervous systems. Learned behavior is not inherited, but the potential to learn is.

**BREAKTHROUGH**

**PAVLOV'S DOG**

Ivan Pavlov's experiments on dogs in the 1890s—initially aimed at investigating digestion—proved to be a landmark in the understanding of animal behavior. He was the first scientist to demonstrate classical conditioning—where an animal learns to associate one stimulus (such as the sound of a bell) with another (food). Pavlov measured the degree of salivation by dogs in response to these stimuli. He found that if food was presented in association with a sound, then eventually the dogs would learn to associate the two stimuli, and salivate in response to sound alone.

# Cycles in the Biosphere

**Compared with Earth's total mass of nonliving matter, the biosphere is a comparatively thin veneer on its surface, comprising the living world on the land and in the oceans. In the biosphere life interacts with the atmosphere above and the soil below through the processes of respiration, photosynthesis, death, and decay.**

Organisms do not live in isolation from the environment around them. Organisms affect the air when they extract oxygen and excrete carbon dioxide during respiration (see pp.270–71). And every living thing is reduced to its separate chemical components during the inexorable advance of decay after its death. Organisms are made up of the same sorts of atoms that exist in the nonliving world and these atoms shuffle from life to nonlife as chemical processes recycle them through the biosphere. Such ideas were first proposed by a Russian geochemist called Vladimir Vernadsky (see above), who popularized the idea of the biosphere in his book of the same name, first published in 1926. One element is abundant in all living bodies: carbon. Carbon atoms can bond into a network with other atoms, forming complex organic molecules, which are the basis of life itself. Animals eat food, and in doing so convert one complex molecule into another. But plants make food from simple inorganic materials: carbon and oxygen from carbon dioxide in the air, hydrogen from water, and nitrogen from nitrates in the soil.

**Effect of combustion on carbon cycle**
A forest fire, such as at Puebla, Mexico (see below), causes carbon to be released into the air. But the ash returns minerals to soil, stimulating regrowth in plants, which absorb carbon dioxide as they photosynthesize.

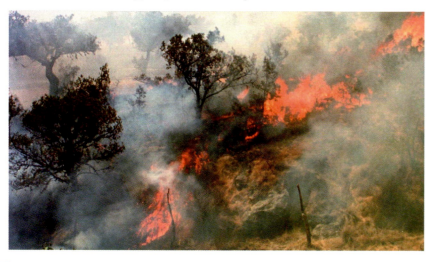

### The carbon cycle

The carbon atoms that make up biological molecules are returned to the atmosphere as carbon dioxide whenever an organism respires. Decay involves the same process: decomposers obtain food and energy from dead material. Plants recapture this carbon when they make food—using the energy of the sun in the process of photosynthesis (see pp.154–55) to convert carbon dioxide and water into sugars, fats, and protein. In doing so, they effectively make the carbon cycle complete.

There is a natural balance between the carbon locked in living bodies and the carbon of the atmosphere and other nonliving reservoirs, such as chalk. But if combustion speeds the rate of carbon emission, and this is not matched by a corresponding increase in photosynthesis, then carbon dioxide levels rise. A great deal of carbon has become locked into organic matter—the so-called carbon sinks—over extraordinary periods of time: hundreds of years for long-lived trees, and millions for coal and oil deposits. But combustion can happen in an instant, causing rapid rises in atmospheric carbon dioxide.

### The nitrogen cycle

A significant amount of organic material contains nitrogen, mostly in the form of protein and DNA. However, even though about 70 percent of the air is pure nitrogen, it is inaccessible to most living organisms in this form. Plants rely on the action of various types of bacteria and decomposers to convert nitrogen in dead matter first into ammonia, then into nitrites, and finally into nitrates. Animals obtain their nitrogen in the food they eat (see opposite). Nitrogen in the atmosphere can only be "fixed" into more usable forms naturally by nitrogen-fixing bacteria or the action of lightning.

> **"Living matter is the most powerful geological force."**
> VERNADSKY, THE CHEMICAL STRUCTURE OF THE BIOSPHERE AND ITS SURROUNDINGS, 1965

**Decomposition**
The breakdown of dead organic matter in compost happens because bacteria and fungi feed on this material and digest it. In doing so, they release chemicals, such as nitrates, which plants absorb for growth.

GEOCHEMIST (1863–1945)

## VLADIMIR VERNADSKY

Born of Russian–Ukrainian parents, Vernadsky effectively founded a new discipline of earth science that combined geology, chemistry, and biology. After graduating from St. Petersburg University, Russia, he worked as a mineralogist before writing *The Biosphere*, in which he identified the influence that life, including humans, has on cycles of matter. Late in his career he advocated use of atomic energy and also founded the Ukrainian Academy of Science.

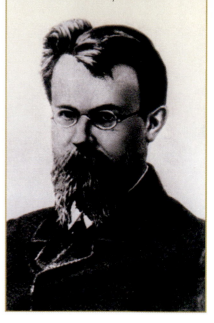

## « BEFORE

**The elements of carbon and nitrogen were characterized by chemists in the 1700s.**

**IDENTIFYING THE CYCLES**
The **carbon cycle was identified** in the **18th century**, culiminating in 1796, when Swiss pastor **Jean Senebier** showed that inorganic carbon dioxide was taken up by plants in photosynthesis. Meanwhile, French chemist **Antoine Lavoisier** discovered that **much of the gas in air was inert—nitrogen**. By 1785 French chemist **Claude Berthollet** found that **nitrogen was a component of ammonia**, but processes of the cycle were not understood for another 100 years.

**123 BILLION TONS of carbon used in photosynthesis and absorbed by the oceans per year.**

**6 BILLION TONS of atmospheric carbon—the net increase per year due to natural and human activities.**

**« SEE ALSO**
*pp.146–47* Discovery of Gases
*pp.154–55* How Plants Work
*pp.270–71* Breathing and Respiration

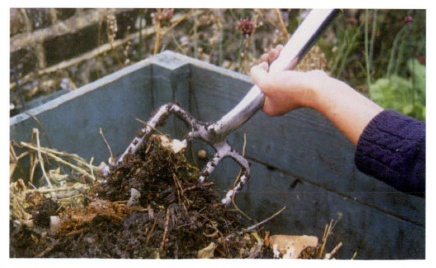

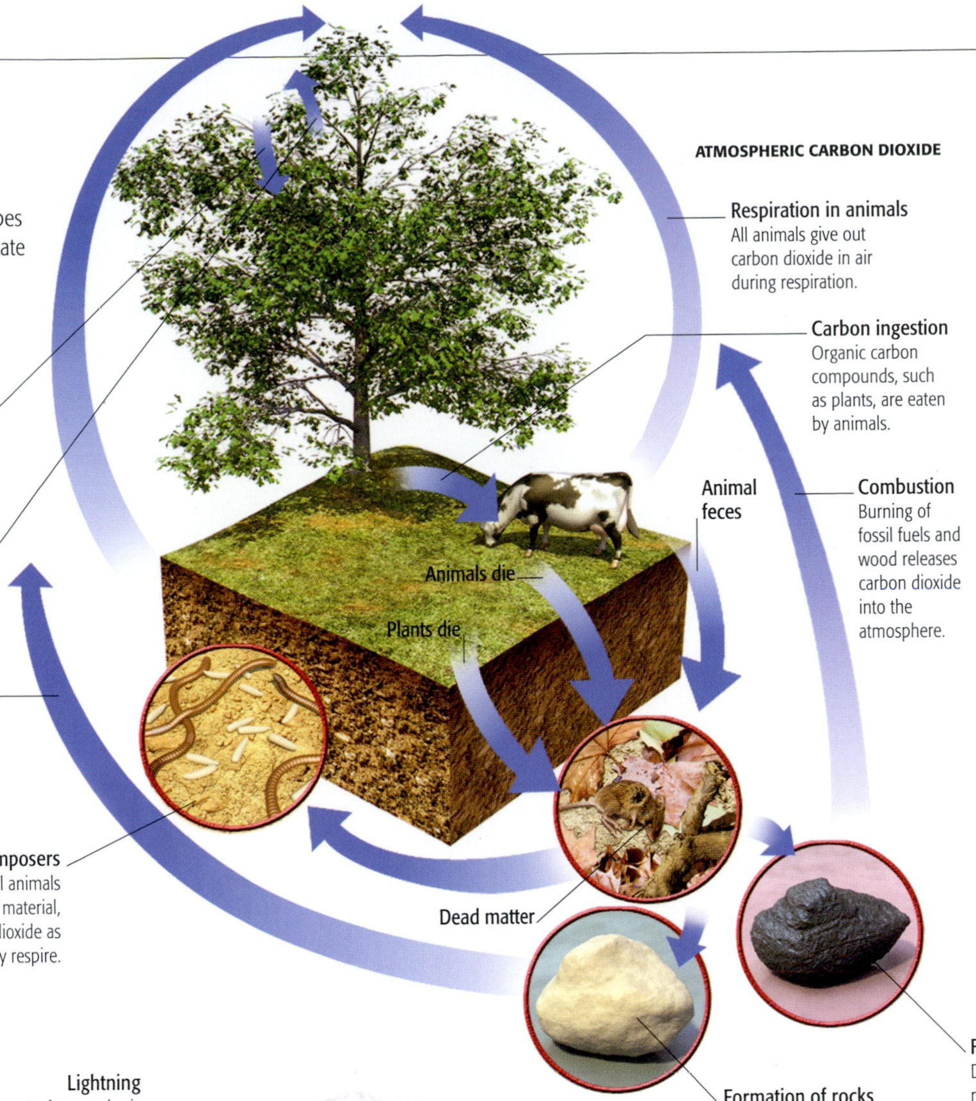

▷ **The carbon cycle**
Consisting of a series of complex processes, the carbon cycle describes the way in which carbon atoms rotate around plants, animals (including humans), rocks, and the atmosphere.

**ATMOSPHERIC CARBON DIOXIDE**

**Respiration in animals**
All animals give out carbon dioxide in air during respiration.

**Carbon ingestion**
Organic carbon compounds, such as plants, are eaten by animals.

**Photosynthesis**
Carbon dioxide used by plants as a raw material for photosynthesis.

**Respiration in plants**
Carbon dioxide is given out by plants as a waste product of respiration.

**Weathering and volcanic activity**
Processes that release carbon dioxide from carbonates in rocks.

**Respiration in decomposers**
Bacteria, fungi, and soil animals feed on dead material, releasing carbon dioxide as they respire.

Animals die

Plants die

Animal feces

**Combustion**
Burning of fossil fuels and wood releases carbon dioxide into the atmosphere.

Dead matter

**Formation of rocks**
Dead matter forms sediments, which over millions of years are compacted to make rocks, such as carbonate (chalk).

**Formation of fossil fuels**
Dead matter compacted in certain rock types altered over millions of years to form fossil fuels.

**AFTER** »

20th-century science has revealed the importance of biosphere cycles and how they can be affected by human activity.

**THE GREENHOUSE EFFECT**
In 1824 French physicist **Joseph Fourier** found that **certain atmospheric gases can absorb infrared radiation** and warm the surface of the planet. This natural phenomenon became known as **the greenhouse effect 416** » and is important in sustaining life on the planet. A number of gases are involved: **water vapor** is the most significant, but **carbon dioxide**, **methane**, and **ozone** also contribute.

**GLOBAL WARMING**
Since the middle of the 20th century, the **average temperature of Earth's surface has been rising** due to an increase in concentration of greenhouse gases in the air—most notably **carbon dioxide** from combustion of fossil fuels and **methane** from livestock.

**SEE ALSO** »
*pp.414–15* GLOBAL WARMING
*pp.418–19* TACKLING CLIMATE CHANGE

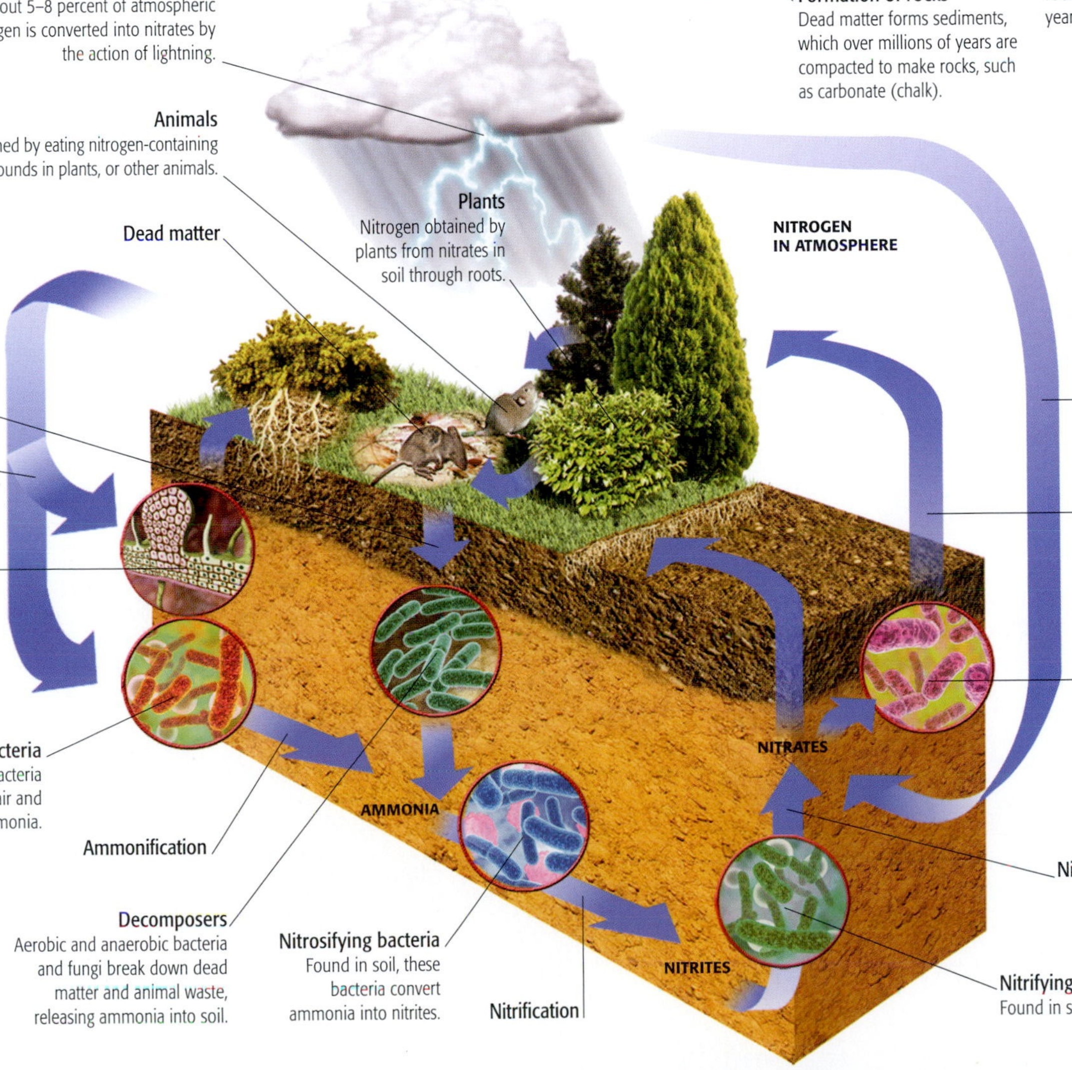

**Lightning**
About 5–8 percent of atmospheric nitrogen is converted into nitrates by the action of lightning.

**Animals**
Nitrogen obtained by eating nitrogen-containing compounds in plants, or other animals.

**Dead matter**

**Plants**
Nitrogen obtained by plants from nitrates in soil through roots.

**NITROGEN IN ATMOSPHERE**

◁ **The nitrogen cycle**
This cycle describes the processes whereby atmospheric nitrogen enters the soil and becomes part of living organisms before being returned to the atmosphere. Bacteria are key in converting nitrogen into a usable form.

**Decomposition**

**Biological fixation**

**Symbiotic nitrogen-fixing bacteria**
Found in root nodules of plants, such as legumes. Bacteria extract nitrogen from air and convert it into ammonia.

**Nitrogen-fixing bacteria**
Found in soil, these bacteria extract nitrogen from air and convert it into ammonia.

**Ammonification**

**Decomposers**
Aerobic and anaerobic bacteria and fungi break down dead matter and animal waste, releasing ammonia into soil.

**Nitrosifying bacteria**
Found in soil, these bacteria convert ammonia into nitrites.

**Nitrification**

**AMMONIA**

**NITRITES**

**NITRATES**

**Atmospheric fixation**

**Denitrification**

**Denitrifying bacteria**
Found in soil, these bacteria convert nitrates into atmospheric nitrogen and release into air.

**Nitrification**

**Nitrifying bacteria**
Found in soil, bacteria convert nitrites into nitrates.

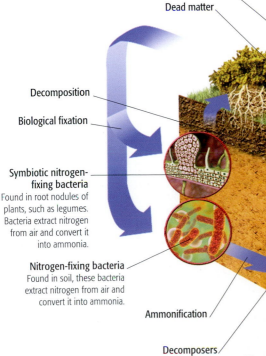

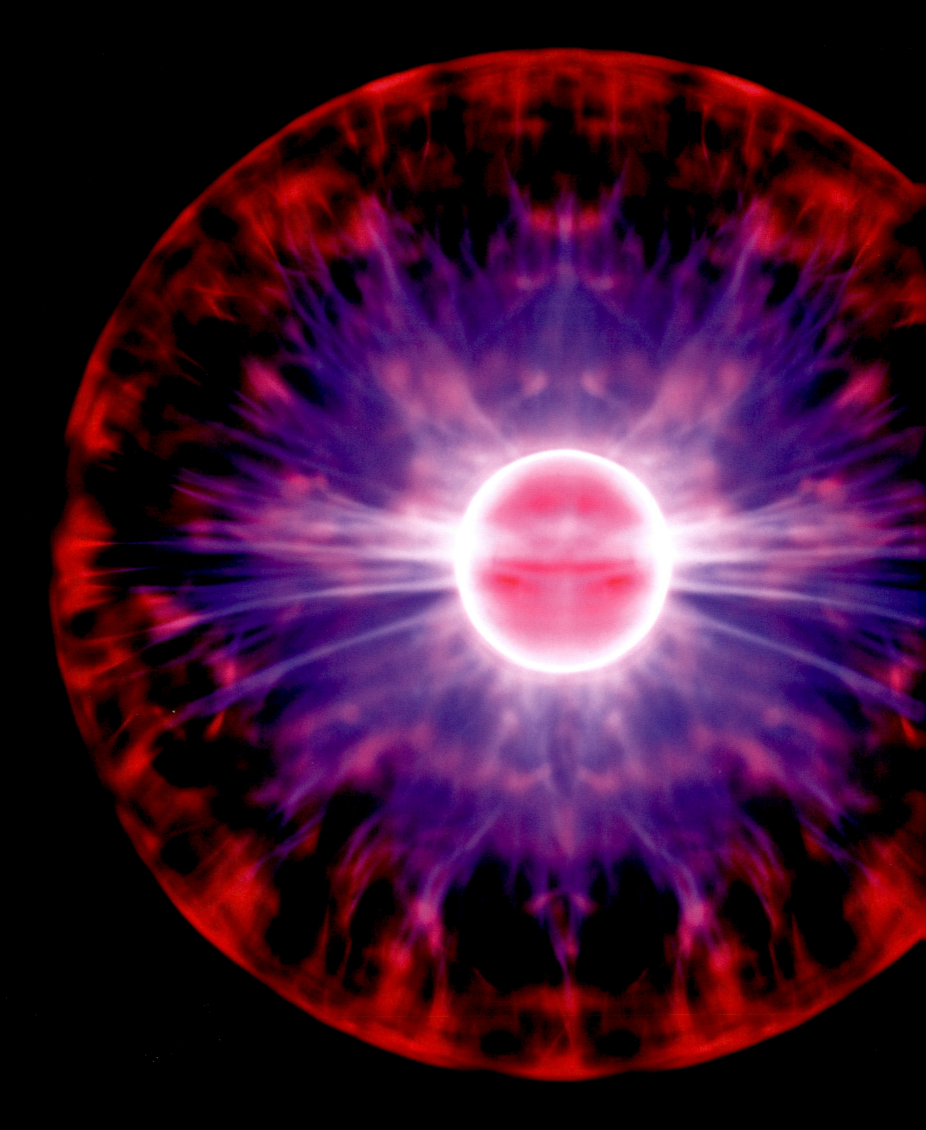

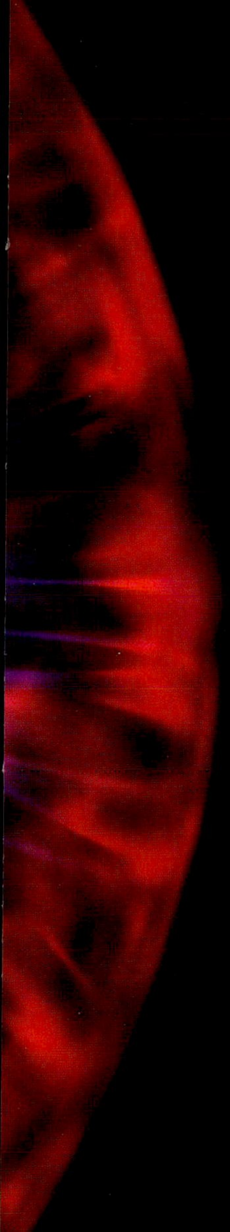

# THE ATOMIC AGE

## 1890–1970

Just when it seemed in 1890 that science had discovered all there was to know, the atom was broken open to reveal a dizzying array of smaller particles along with unimaginable amounts of energy. Two wholly new kinds of science—quantum theory and Einstein's relativity—were called into being to try to understand it all.

## 1890

**1891**
Otto Lilienthal makes the first of 2,000 hang-glider flights.

◀ Becquerel plate

**1892**
Hans Driesch finds what we now call stem cells. Vacuum flask invented by James Dewar.

▲ Otto Lilienthal in flight

**1893**
Alfred Werner discovers the octahedral structure of cobalt complexes. Ernst Mach publishes *The Science of Mechanics*.

Vacuum flask ▶

**1896**
Antoine Henri Becquerel finds radiation from uranium salts. Louis Benoist discovers the ionizing effects of X-rays.

**1897**
J.J. Thomson discovers the electron, using vacuum tubes. Paul Ehrlich identifies antibodies.

**1898**
Marie and Pierre Curie discover that an ore of uranium can turn air into a conductor of electricity, and coin the term "radioactive."

▼ The first X-ray

**1895**
Wilhelm Röntgen discovers X-rays. Eugen Warming finds that unrelated plants evolve similar strategies to cope with extremes of climate.

**1899**
Ernest Rutherford and Frederick Soddy discover alpha and beta particles. Bayer manufactures aspirin.

Marie Curie ▶

## 1900

**1900**
Max Planck proposes a new model for the energy released by a black body. Karl Landsteiner discovers four blood groups.

**1901**
Ernest Rutherford discovers that disintegrating atoms release radioactivity. Alois Alzheimer first describes the form of dementia that bears his name.

Brain of patient with ▶
Alzheimer's disease

▼ J.J. Thomson's vacuum tube

▼ The Stanley Steamer

**1902**
Walter Sutton and Theodor Boveri show that genes occur on chromosomes. Léon de Bort identifies the troposphere and the stratosphere. The Stanley Steamer, a steam-powered car, goes into production.

**1904**
J.J. Thomson proposes his "plum pudding" model of the atom. The first transatlantic radio message transmitted. Ambrose Fleming makes first electronic valve. The first geothermal generator opens in Italy.

**1903**
The first controlled, powered flight to carry people made by Orville and Wilbur Wright.

Plum pudding model ▶
of the atom

**1905**
William Bateson coins the term "genetics." Albert Einstein produces four scientific papers on quantum theory, Brownian motion, relativity, and the photoelectric effect.

▲ One of Einstein's books

**1906**
First voice transmissions carried out, using the AM system. Richard Oldham discovers "P" waves, which travel through Earth.

**1907**
Leo Baekeland produces Bakelite, the first commercial plastic. Guglielmo Marconi sets up the first commercial transatlantic radio communications. Bertram Boltwood develops a technique for dating rocks.

**1908**
Hans Geiger and Ernest Marsden perform the gold foil test, proving that atoms have a nucleus. Cellophane first produced.

▲ The gold foil test

**1909**
Korbinian Brodmann defines 52 areas of the cerebral cortex in the brain. Soren Sorensen develops the pH scale. The Burgess Shale fossil site found in Canada. Louis Blériot flies across the English Channel.

The startling discovery that the atom is not the smallest particle spurred scientists to explore the extraordinary "quantum" rules that apply at the smallest scale. In one direction, research led in 1945 to the unleashing in the atom bomb of the extraordinary energy bound within the atomic nucleus. In another, it led in the 1950s to the revelation of the structure of DNA, the amazing molecule inside every living cell that carries all life's instructions as a chemical code. At the other end of the scale, Einstein's theories overturned the accepted view of space and time—just as Hubble revealed that our galaxy is but one of many galaxies in a gigantic, ever-expanding universe, flung out from its origin long ago in the "Big Bang."

# 1910

# 1920

## 1910
Paul Ehrlich discovers that salvarsan kills syphilis bacteria.

## 1911
Ernest Rutherford proposes that electrons orbit the nucleus of an atom. Thomas Hunt Morgan discovers genes are on chromosomes.

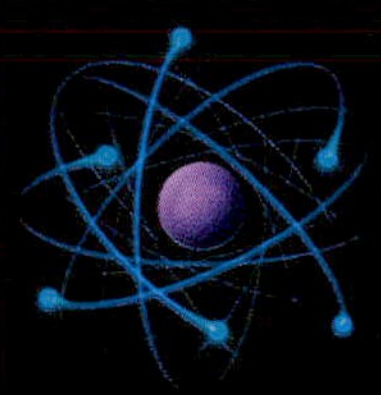

## 1912
Alfred Wegener proposes the theory of continental drift. Max von Laue proves X-rays are short-wave electromagnetic radiation. Vesto Slipher measures redshifts of galaxies.

▲ Ernest Rutherford's atomic model

▼ Niels Bohr's model of the atom

## 1913
Arthur Homes publishes *The Age of the Earth*, in which he estimates that Earth is around 1.6 million years old. Niels Bohr proposes the Bohr model of the atom.

▼ Albert Einstein

## 1915
Albert Einstein's *General Theory of Relativity* published.

## 1916
Harlow Shapley measures the Milky Way. Lennart von Post studies pollen grains to record the distribution of plant species.

▼ Cross section of kidney

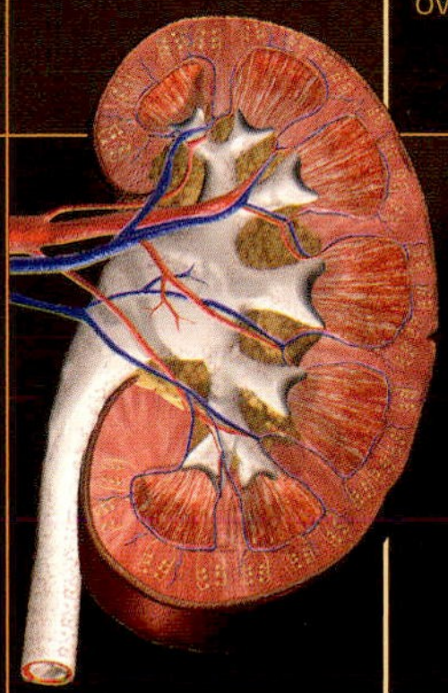

## 1917
Arthur Cushny finds how urine is formed in the kidneys. The Hooker telescope is completed and is the world's largest telescope until 1948.

## 1918
Sonar systems are used by the US and British navies. Ernest Rutherford discovers protons and neutrons. The Enigma cipher machine built.

◄ Enigma cipher machine

## 1919
Arthur Eddington finds supporting evidence for Einstein's general theory of relativity, by measuring stars during a solar eclipse.

▼ Solar eclipse

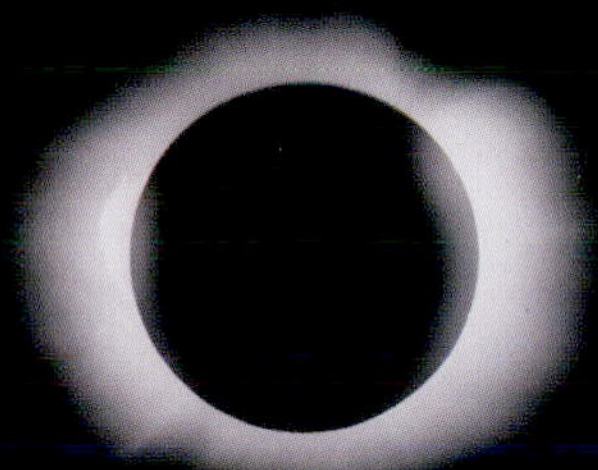

## 1920
Hans Winkler coins the term "genome." Frederick Hisaw, Harry Fevold, and Samuel Leonard find that hormones from the pituitary gland cause changes in the ovaries.

## 1921
Edward Mellanby discovers Vitamin D and shows that its absence causes rickets.

## 1923
Edwin Hubble uses the Cepheid variable stars in the Andromeda galaxy to prove that it is outside the Milky Way.

Andromeda galaxy ►

## 1924
Edwin Hubble discovers there are galaxies in every direction. Arthur Eddington discovers the relationship between mass and luminosity in stars.

## 1925
The Mid-Atlantic ridge is discovered. Wolfgang Pauli proposes his exclusion principle.

Mid-ocean ridge ►

## 1926
First successful rocket launch by Robert Goddard. Vladimir Vernadsky publishes *The Biosphere*, stating that all matter rotates in natural cycles. John Logie Baird gives the first demonstration of a working television system. Vinyl made.

## 1927
Werner Heisenberg publishes the uncertainty principle of quantum physics. The quartz clock invented. Farnsworth demonstrates a practical TV-picture scanning system.

Werner Heisenberg ►

▼ *Penicillium* mold

## 1928
Lidwell and Edgar Booth develop an artificial pacemaker for the heart. Baird Television broadcasts the first transatlantic television signal.

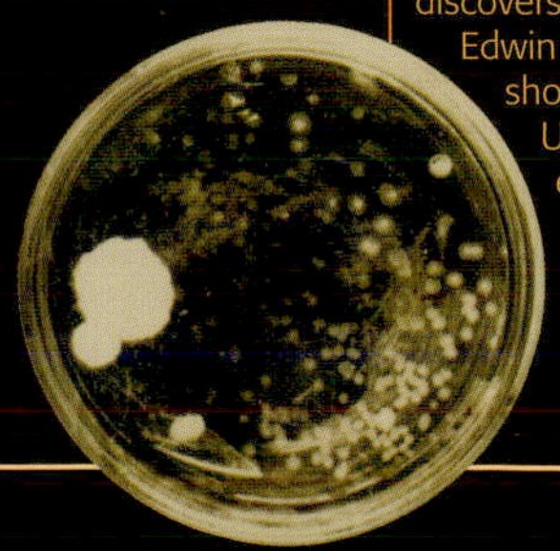

## 1929
Alexander Fleming discovers penicillin. Edwin Hubble shows that the Universe is expanding.

**1930**
Paul Dirac publishes *The Principles of Quantum Mechanics*. Neoprene and polyester first made.

**1931**
The electron microscope invented.

▼ Splitting uranium-235

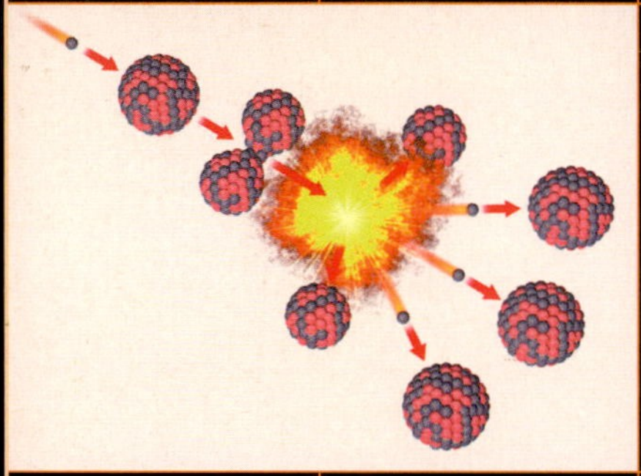

**1935**
Arthur Tansley develops the concept of an ecosystem. Charles Richter devises the Richter magnitude scale to measure earthquakes. Nylon first made.

▲ Charles Richter

**c.1940**
New layers in the atmosphere are discovered—the mesosphere and the ionosphere.

The atmosphere ▶

▲ Germany's V-2 rocket

**1937**
Grote Reber builds the first radio telescope. The first shipborne radar systems available.

**1938**
Otto Hahn discovers an isotope of barium produced by the breakdown of uranium.

**1941**
Herman Kalckar discovers that adenosine triphosphate (ATP) is the main energy-transferring molecule inside cells. Styrofoam (expanded polystyrene) first produced.

**1944**
Germany's V-2 rockets, the first ballistic missiles, fall on London.

**1932**
Cockcroft and Walton perform the first artificial nuclear fission. James Chadwick proves the existence of neutrons in an atom.

**1934**
Kodak produces the first color transparency film. Irène Joliot-Curie and Frédéric Joliot create the first artificial radioactivity.

▼ Linus Pauling

**1939**
Helmut Ruska first sees a virus using an electron microscope. The first jet aircraft has its maiden flight. Linus Pauling publishes *The Nature of the Chemical Bond*. The Cygnus A galaxy is discovered by its radio output.

**1945**
Dorothy Hodgkin discovers the structure of penicillin. The Trinity Test is the first artificial nuclear explosion.

Dorothy Hodgkin ▶

**1933**
Thomas Morgan Hunt wins the Nobel Prize for discovering the role of chromosomes in heredity. Fritz Zwicky infers that dark matter must exist in the Universe.

▼ Human chromosomes

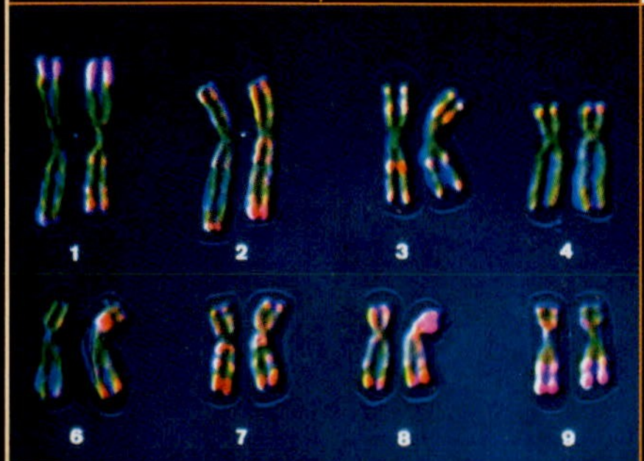

▼ Structure of a virus

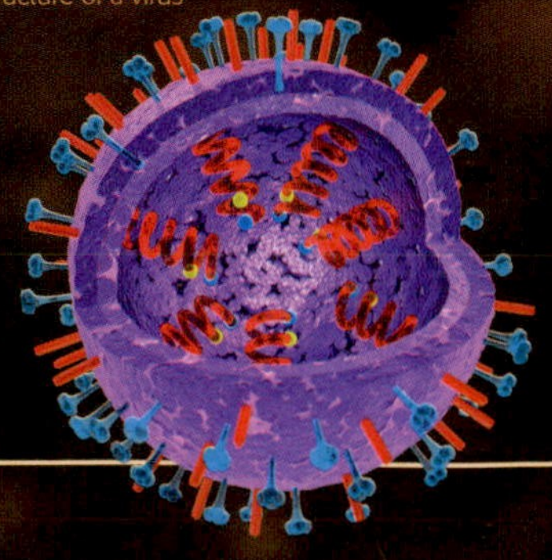

**1942**
The first fission reactor, Chicago Pile-1, goes into operation as part of the Manhattan Project.

Manhattan Project ▶

**1946**
Genetic recombination demonstrated.

**1948**
George Gamow and Ralph Alpher predict the proportions of light elements produced by the birth of the Universe.

**1943**
Oswald Avery proves that DNA is the carrier of genes in cells. The Colossus code-breaking machine and first electronic computer is built.

**1947**
The transistor is invented. Dennis Gabor develops theory of the hologram.

Security hologram ▶

> **"The stellar system is a swarm of stars isolated in space. It drifts through the Universe as a swarm of bees drifts through the summer air."**
>
> EDWIN HUBBLE, *The Realm of the Nebulae*, 1936

## 1950

## 1960

**1950**
Ocean floor is now mapped. 24-hour weather prediction launched. The first internal transplant operation—of a kidney. Introduction of the Turing test for machine intelligence. Fred Hoyle coins the term "Big Bang."

▲ Map of the ocean floor

**1955**
Radiometric dating confirms Earth is 4.55 billion years old.

**1960**
SI (*Système International*) Units system introduced. Harry Hess develops the theory of seafloor spreading. First laser built. TIROS-1, the first weather satellite, launched.

▼ Ocean floor spreading

**1967**
Jocelyn Bell and Antony Hewish discover the first pulsar, a type of neutron star. Kornberg, Goulian, and Sinsheimer synthesize DNA molecule. Christian Bernard performs the first heart transplant.

**1952**
Jonas Salk develops the first vaccine for polio. First artificial heart valve implanted. The first hydrogen bomb is tested over the Pacific Ocean.

Children being ▶ vaccinated against polio

**1957**
*Sputnik 1*, the first satellite, launched by the Soviet Union. James Lovelock invents the first version of the ECD, which measures traces of chemicals in the atmosphere.

▼ *Sputnik 1*

▲ Pulsar

**1968**
First amniocentesis. Richard Feynman develops his theory of partons.

▼ Richard Feynman

**1953**
James Watson and Francis Crick explain the double-helix structure of DNA. Frederick Sanger determines the sequence of amino acids within the chains of protein that make up insulin.

◀ DNA double helix

**1961**
Soviet cosmonaut Yuri Gagarin becomes the first person to orbit Earth in a spacecraft. World Wildlife Fund (WWF) established.

**1962**
Telstar 1, the first active communications satellite, launched.

▼ The butterfly effect

**1958**
The transistor is first demonstrated by Jack Kilby. F.C. Stewart grows carrots from carrot cells, the first case of artificial cloning.

Early transistor ▶

**1963**
Edward Lorenz coins the term "butterfly effect" to describe chaotic mathematical systems.

**1964**
The first black hole—Cygnus X-1—is discovered.

**1969**
Neil Armstrong and Buzz Aldrin become the first astronauts on the Moon. A laser measures the distance between Earth and the Moon. Dorothy Hodgkin reveals the structure of insulin.

Apollo 11 astronaut ▶

▼ Insulin molecule

**1954**
Townes and Schawlow invent the maser. Absolute zero is defined as −273.15° Celsius.

**1959**
Edelman and Porter discover the structure of antibodies as types of proteins. The far side of the Moon photographed by the Soviet Luna 3 probe.

**1965**
Leonard Hayflick discovers that abnormal cells can keep dividing, causing cancerous tumors. Alexei Leonov is the first man to walk in space. Quasars are discovered.

◀ Cancerous cells

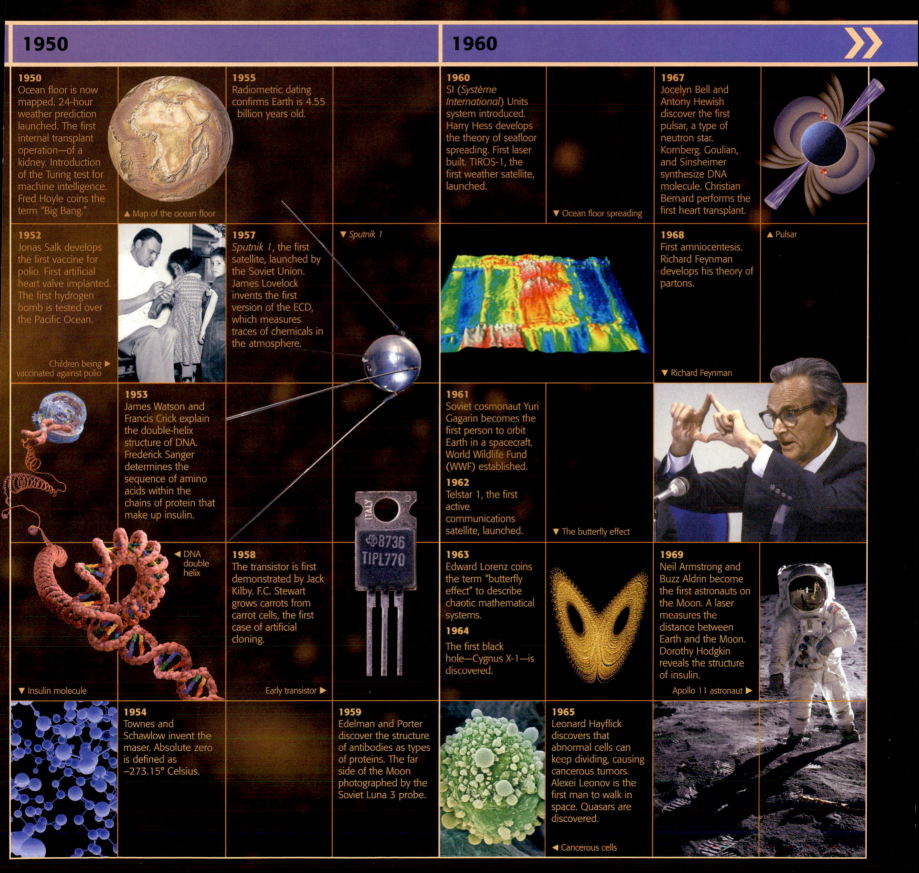

<< BEFORE

Throughout the 19th century scientists worked to piece together the mysteries of the elements and their chemical reactions.

**ATOMIC THEORY**
**John Dalton's atomic theory** of matter **<< 138–39**, published in the early 1800s, proved very successful in explaining both physical phenomena and chemical processes, and led to a **revolution in chemistry** in the 19th century.

**THE PERIODIC TABLE**
**Mendeleev's** periodic table **<<230–33** revealed that **elements fell neatly into groups** with similar chemical reactions and apparent patterns to their masses. **There were gaps** in the table as many elements had not yet been discovered.

<< SEE ALSO
pp.138–39 THE NATURE OF MATTER
pp.232–33 THE PERIODIC TABLE
pp.234-35 CHEMICAL REACTIONS

**Static electrons**
Thomson recognized the existence of negatively charged electrons within the atom, but thought that they were static.

**Nucleus**

**Cloudlike body**
The body of the atom was thought to be cloudlike.

**Orbiting electrons**

**Thomson's "Plum Pudding" model (1904)**
In J.J. Thomson's model of the atom, negatively charged electrons were dotted randomly, like plums in a pudding, through a positively charged sphere that contained most of the atom's mass. There is no nucleus.

**Rutherford's model (1911)**
Ernest Rutherford proposed that an atom consisted of a small, dense core of positively charged particles (the nucleus), around which negatively charged electrons orbited, like moons around a planet.

# Structure of the Atom

At the beginning of the 20th century J.J. Thomson's discovery of the electron revolutionized our view that the atom was the basic unit of matter and gave rise to the first model of the atom. Over the next 30 years further insights into atomic structure led to the development of other more complex models.

PHYSICIST (1871–1937)

## ERNEST RUTHERFORD

New Zealand-born chemist Lord Ernest Rutherford is often known as the "father" of nuclear physics, both for his own pivotal discoveries and for his influence over later generations. The son of an immigrant farmer from Britain, he became Professor of Physics at the University of Manchester in 1907, and won the 1908 Nobel Prize for chemistry for his groundbreaking proof that radioactivity was caused by the disintegration of atoms. He won many awards, gaining a life peerage in 1931.

In 1897 British physicist Joseph John Thomson began an exhaustive series of experiments into cathode rays that would lead to a revolution in our understanding of the atom. Using a type of cathode ray tube called a Crookes Tube (see pp.292–93), he was able to test how the rays behaved in electric and magnetic fields.

As the rays passed through these fields, they bent toward a positively charged electric plate, so Thomson realized they must be negatively charged particles generated by the cathode itself (rather than electromagnetic radiation as others had proposed). He called these particles "corpuscles", but they later became known as electrons.

The particles had several other characteristics: they always appeared to have the same charge, and they had a mass less than one thousandth of a hydrogen atom. It seemed clear that the particles were coming from inside the atoms of the cathode—proving that the atoms themselves were made up of smaller parts hidden within them.

## From plum pudding to nucleus
Thomson's discoveries won him the Nobel Prize for physics in 1906, but this was just the beginning of the

investigation. Since it seemed impossible to extract positive charge from an atom, and yet atoms as a whole were known to be electrically neutral, Thomson developed his own theory of the atom to explain the anomaly. He suggested that the negative electrons were moving around inside a uniform sphere of positive

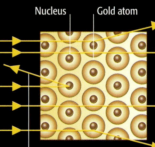

**Nucleus** **Gold atom**

**Most alpha particles** pass straight through the atom or are slightly deflected by the nucleus.

**DETAIL OF GOLD FOIL**

**Scintillation screen**
A screen coated in zinc sulphide emits light when struck by alpha particles, revealing their path through the gold foil.

A few particles hit the nucleus head-on and rebound.

**Slit in lead sheet**
This slit ensures the alpha particles reach the foil as a narrow beam, and filters out particles from other directions.

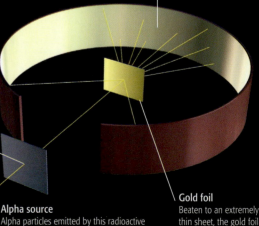

**Alpha beam**
Alpha particles are beamed on to the gold foil.

**Gold foil**
Beaten to an extremely thin sheet, the gold foil is just a few atoms thick.

**Alpha source**
Alpha particles emitted by this radioactive source spread out in all directions.

**Lead block**
This encases and shields the radium, which produces the rays.

**The gold leaf experiment**
The ingenious experiment developed by Rutherford and his students, Geiger and Marsden, used alpha particles to probe the properties of individual atoms, revealing that they are mostly made up of empty space.

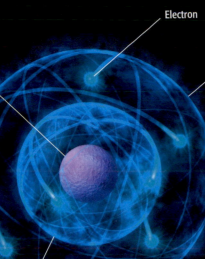

Nucleus

Electron

Second orbital level or "shell"

**First orbital level or "shell"**
The "shell" with the lowest energy level contains a maximum of two electrons .

**Bohr's model (1913)**
To explain how atoms emitted and absorbed light, Niels Bohr developed a model in which the orbits of electrons were fixed by atomic "shells" that each contained a fixed number of atoms.

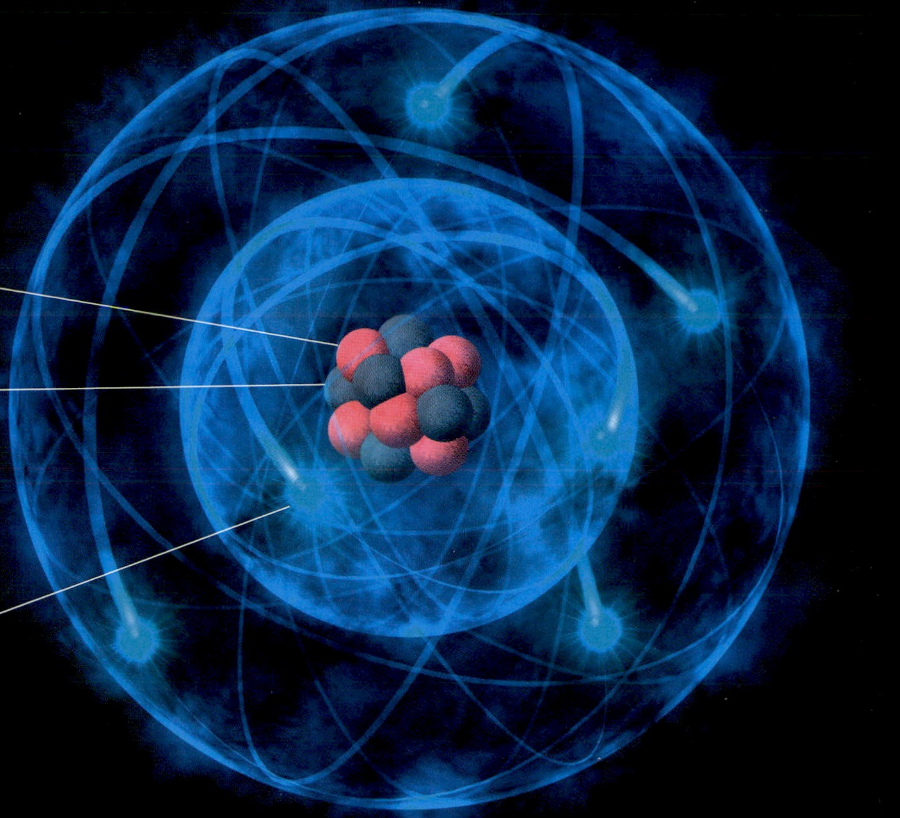

**Proton**
The number of positively charged protons determines an element's atomic number.

**Neutron**
Uncharged neutrons with the same mass as a proton help to make up the atom's overall mass.

**Electron**
The number of negatively charged electrons balances the number of protons in a neutral atom.

**Nucleus of protons and neutrons**
The discovery of protons and neutrons within the nucleus further refined Bohr's model. The number of protons determines an element's chemical properties.

charge that accounted for most of the atom's mass. The particles were like plums in a cake or pudding, so his model became known as the "plum pudding" model.

In 1909, while working under Ernest Rutherford, Hans Geiger and Ernest Marsden tested this idea. They fired alpha particles from a radium source at a thin sheet of gold leaf and measured the paths taken by the particles. The plum pudding model suggested the rays would only be slightly deflected, but in fact some were deflected by more than 90° and others rebounded from the foil. Rather than being like plum puddings, atoms actually concentrated their mass and positive charge in a small area at the center—the nucleus. In 1911 Rutherford proposed that electrons orbited the nucleus of an atom like moons around a planet.

### Electrons in fixed orbits

Another of Rutherford's students, Danish physicist Niels Bohr, used the new ideas emerging from quantum mechanics to explain what kept the electrons in orbit. The Bohr model suggested that the electrons were confined in fixed orbits by atomic "shells," and that the chemical properties of an element were governed by the number of electrons in the outer shell. He also said that the amount of energy an electron had was closely linked to how far it was from the nucleus of the atom. Electrons, he suggested, could absorb or emit radiation of specific wavelengths and colors in order to move between inner and outer shells, or orbitals.

### Nuclear structure

There were still some problems however. The patterns in the reactions of elements (signified by their atomic numbers) did not always match the patterns in their masses. In 1913 Henry Moseley showed that atomic number was a real property that indicated the positive electric charge in the nucleus. But if this was not always proportional to its mass, how were the two properties decided? Rutherford solved part of the problem in 1918 by identifying particles with a positive electric charge that was equal and opposite to the electron's negative charge. These particles were soon named protons, but they did not account for an atom's entire mass—the remainder seemed to lack any electric charge at all. Rutherford named these particles neutrons. It was another 14 years before the neutron's existence was finally proven, by James Chadwick, who won the Nobel Prize for his work in 1935.

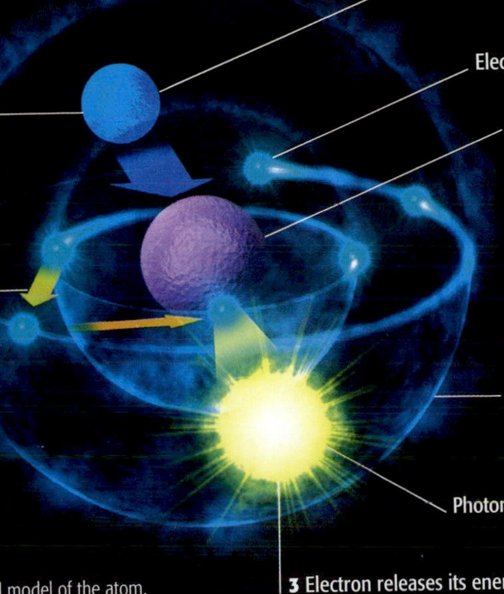

**1 Collision**
A particle carrying energy—either another atom, an electron, or even a photon of high-energy radiation—collides with the atom.

**2 Electron jumps**
An orbiting electron is given a boost of energy and jumps into an empty space within the outer higher-energy shell.

Particle

Electron

Nucleus

Electron shell

Photon of light

**How atoms emit light**
According to Bohr's electron-shell model of the atom, each electron has an energy determined both by the shell within which it orbits and the electron's distance from the nucleus. Various events can make electrons jump between shells, making atoms emit and absorb light of very specific wavelengths.

**3 Electron releases its energy**
If there is an empty space in the lower-energy shell below it, the electron drops back into it, losing the energy it had gained and releasing a photon, a particle of light.

Understanding the basic units of atomic structure led to further advances that would revolutionize the 20th century.

**NUCLEAR PHYSICS**
**Nuclear fission 324–25 »** was discovered by **Hahn, Meitner,** and **Strassmann** in 1938, leading ultimately to nuclear power stations and the fission (atomic) bomb.

**PARTICLES TO WAVES**
By 1926 electrons were no longer thought to act as particles but as waves. This led to the development of a more abstract **mathematical model of the atom**. It was devised by the quantum physicist **Erwin Schrödinger.** Instead of saying where an electron actually was, he used **"electron clouds"** (seen here in yellow and blue) to predict where the electrons were most likely to appear.

**ELECTRON CLOUDS OF CARBON**

**‹‹ BEFORE**

Once the idea of matter being made of small particles became accepted, there had to be some mechanism to hold them together.

**ROMAN IDEAS**
Around 60 BCE the Roman poet **Titus Lucretius Carus**, who is more commonly known simply as Lucretius, wrote an **epic philosophical poem** called *De rerum natura* ("On the nature of the Universe"). In this poem, which ranged over fundamental topics such as the **mind, spirit, atoms, and the Universe**, Lucretius suggested that the **world**, including people and their minds, was **made entirely of atoms**. These **tiny spheres** were **identical in substance**, but varied in shape and size. He believed that these atoms were in **perpetual motion** at great speeds, with random "swerves" to provide humans with free will, and that they had little **"hooks" to hold them together**—"tightly packed and closely joined."

**LUCRETIUS**

**‹‹ SEE ALSO**
pp.22–23 ELEMENTS OF LIFE
pp.138–39 THE NATURE OF MATTER
pp.286–87 STRUCTURE OF THE ATOM

# Chemical Bonds

Atoms, the smallest particles of elements, are held together by five major types of bond: covalent, ionic, metallic, and hydrogen bonds, and van der Waals forces. All of these bonds are formed by the movements of electrons, but the electrons are shared or transferred in various ways, according to the nature of the atoms.

Covalent bonds often form between light atoms; molecules of hydrogen, oxygen, chlorine, and water are all held together by covalent bonds. A hydrogen atom has just one electron whizzing around the nucleus, in a region called an orbital. When two hydrogen atoms combine to form a molecule—$H_2$—they share the electrons, so that each hydrogen nucleus has a pair of electrons in an orbital.

The electrons orbiting the nucleus of an atom are arranged in a number of "shells" (see below). Compounds are likely to be stable if the shells of their atoms are full, or complete. When atoms form covalent bonds they share electrons to achieve this stability. An atom's innermost shell is full when occupied by two electrons, so hydrogen, with a single electron, forms one bond to become stable, as in $H_2$. Helium, with two electrons, forms no bonds at all,

since its shell is already full. The second shell needs eight electrons to be complete; so carbon, with four electrons in its second shell, forms four covalent bonds, as in $CH_4$ (methane); nitrogen, with five, forms three bonds (see below); oxygen, with six, forms two, as in $H_2O$; and chlorine, with seven electrons, forms one bond, as in HCl (hydrogen chloride).

## Ionic bonds

Mineral salts such as sodium chloride, NaCl (common salt), do not have shared electrons. Instead the atoms lose or gain electrons to complete their shells. Sodium has one electron in its outer shell. It can achieve a filled shell by losing this electron to become a positively charged sodium ion, $Na^+$. Meanwhile the chlorine atom can fill its outer shell by gaining one electron and becoming a negatively charged chloride ion, $Cl^-$. The sodium atom

gives an electron to the atom of chlorine, so that both atoms have complete shells, and the substance becomes $Na^+Cl^-$. Its crystals do not contain molecules, but millions of alternating ions arranged in a regular, 3-D pattern and held together by the attraction between the positive and negative charges.

Hydrogen atom in bond has slight positive charge

Electrostatic attraction forms a hydrogen bond

The oxygen atom in the bond takes more than an equal share of the pair of electrons, and so is slightly negative

### Electron shells
Central nucleus

One complete inner shell and four extra electrons in outer shell

Two complete inner shells and one extra electron in outer shell

**CARBON ATOM**

**SODIUM ATOM**

Atoms are stabilized when the nucleus is surrounded by a filled "shell" of electrons—two in the first shell, eight in the second, and eight in the third.

Each of these two oxygen atoms now has a full outer shell—four of its own electrons and four shared

### Double bonds
Double (covalent) bonds are formed between two atoms when both have two spare electrons ready to pair. This is the structure of an oxygen molecule, $O_2$ or O=O.

### Hydrogen bonds
Hydrogen bonds are weak electrical attractions between hydrogen atoms in an OH (oxygen–hydrogen) or NH (nitrogen–hydrogen) group in one molecule and an oxygen or nitrogen atom in another. Here they are holding together water molecules ($H_2O$).

**CHEMIST (1829–96)**

## AUGUST FRIEDRICH KEKULÉ

Born in Darmstadt, Germany, August Kekulé was educated in Germany, Paris, Switzerland, and London before becoming professor at the University of Bonn. He studied the valency of elements (the number of bonds their atoms can form), and in 1857 he announced that carbon has a valency of 4. Kekulé later became the primary originator of the idea of chemical structure. His most dramatic discovery was that the structure of benzene, $C_6H_6$, is a symmetrical ring of carbon atoms. Kekulé allegedly "saw" this arrangement after having a dream of a snake biting its own tail—a common symbol in many Eastern cultures.

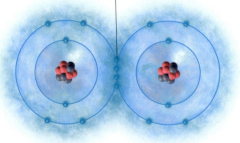

When bonded with three hydrogen atoms, the nitrogen atom has eight electrons in its outer shell—two of its own and six shared with hydrogen atoms

Hydrogen nucleus

Central nucleus of atom

Hydrogen atom shares two electrons with nitrogen atom

### Covalent bonds
Covalent bonds are formed mainly between light elements sharing electrons. In this representation of the ammonia molecule ($NH_3$) there are covalent bonds between the larger nitrogen atom and each of the smaller hydrogen atoms.

Hydrogen atom

## Metallic bonds and hydrogen bonds

Atoms in solid metals sit in a regular, 3-D pattern, and share the loose outer electrons from all the atoms. It is like a grid of positive metal ions bathed in a sea of loose electrons that can move freely through the grid (see below). Metals are good conductors, since these free electrons can travel easily through the metal carrying electrical current or heat.

Covalent compounds with OH and NH groups often have hydrogen bonds between their molecules. Oxygen is more "electronegative" than hydrogen; it takes more than its fair share of the electrons that form a bond between them, giving the oxygen atom a slight negative charge and the hydrogen a slight positive charge. The opposing charges result in attractions between neighboring molecules, especially in water, where a hydrogen atom on one molecule forms a hydrogen bond with the oxygen atom on another molecule.

## Van der Waals forces

Substances made of lone atoms, such as neon and argon, or small molecules, such as nitrogen and oxygen, tend to be gases with low boiling points. The only forces between the lone atoms or the separate molecules are van der Waals forces. These forces are named after the Dutch scientist Johannes Diderik van der Waals, who, in 1873, was one of the first to postulate their existence. Van der Waals forces are much weaker than any other type of bond, and are rather like a slight stickiness between the particles.

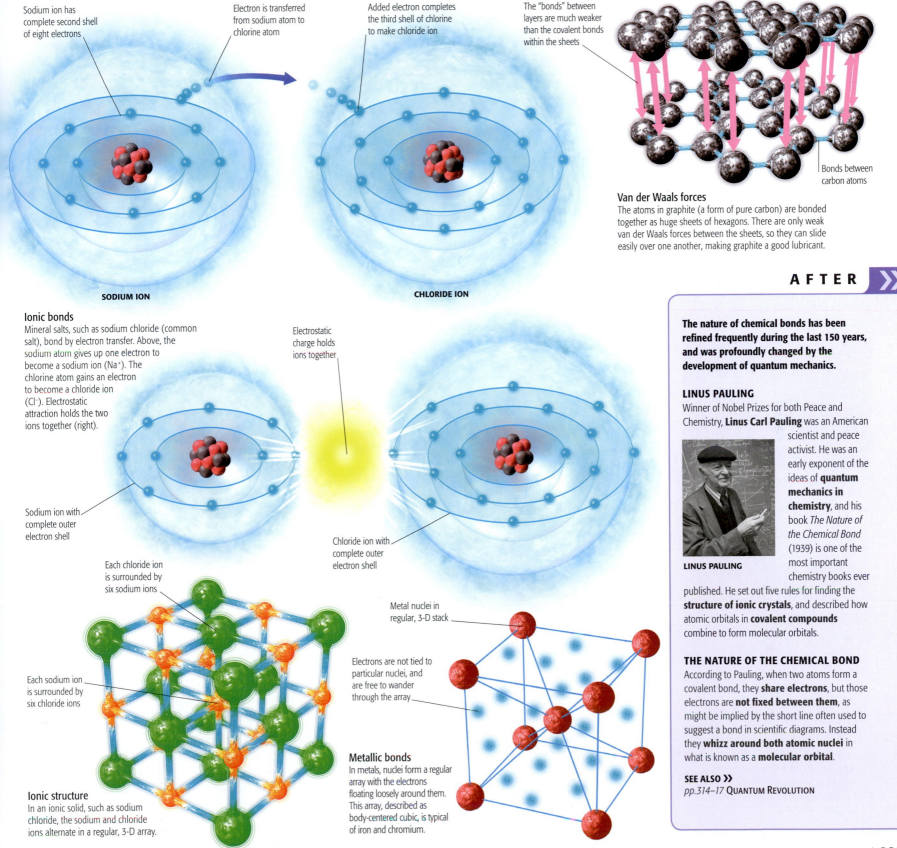

Sodium ion has complete second shell of eight electrons

Electron is transferred from sodium atom to chlorine atom

Added electron completes the third shell of chlorine to make chloride ion

The "bonds" between layers are much weaker than the covalent bonds within the sheets

Bonds between carbon atoms

**SODIUM ION**

**CHLORIDE ION**

### Van der Waals forces
The atoms in graphite (a form of pure carbon) are bonded together as huge sheets of hexagons. There are only weak van der Waals forces between the sheets, so they can slide easily over one another, making graphite a good lubricant.

### Ionic bonds
Mineral salts, such as sodium chloride (common salt), bond by electron transfer. Above, the sodium atom gives up one electron to become a sodium ion ($Na^+$). The chlorine atom gains an electron to become a chloride ion ($Cl^-$). Electrostatic attraction holds the two ions together (right).

Electrostatic charge holds ions together

Sodium ion with complete outer electron shell

Chloride ion with complete outer electron shell

Each chloride ion is surrounded by six sodium ions

Each sodium ion is surrounded by six chloride ions

Metal nuclei in regular, 3-D stack

Electrons are not tied to particular nuclei, and are free to wander through the array

### Ionic structure
In an ionic solid, such as sodium chloride, the sodium and chloride ions alternate in a regular, 3-D array.

### Metallic bonds
In metals, nuclei form a regular array with the electrons floating loosely around them. This array, described as body-centered cubic, is typical of iron and chromium.

≪ BEFORE

Early attempts at flight were in hot-air balloons and gliders. Often optimistic, they were in some cases involuntary.

**EARLY GLIDER**
When he became a Buddhist, the brutal 6th-century CE Chinese emperor **Kao Yang** "released" all prisoners by making them jump from a high tower with makeshift bamboo wings. One of them, Yuan Huang-Thou, managed to **glide around 1½ miles (2.4 km)**, and land. The emperor later had him starved to death.

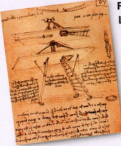

**FLYING MACHINE**
**Leonardo da Vinci** was an artist with a technological mind and a vivid imagination. Among his notebooks were several superb sketches of **flying machines**.

ONE OF LEONARDO'S FLYING MACHINES

**HOT-AIR BALLOONS**
The first hot-air balloon was built by French paper manufacturers and brothers **Jacques-Etienne** and **Joseph-Michel Montgolfier**, and demonstrated in June 1783. The second flight, in September 1783, carried a sheep, a duck, and a chicken. The third, in November, carried a doctor and an army officer 6 miles (9 km) over Paris.

MONTGOLFIER BALLOON TAKE OFF

≪ SEE ALSO
pp.254-55 THE INTERNAL COMBUSTION ENGINE

# Taking Flight

**People have always longed to fly like birds. The first heavier-than-air machines limped off the ground during the 19th century, but by the end of the 20th century military aircraft were essential to every major power, and cheap flights carried millions of people to vacations destinations all over the world.**

In order to get off the ground, a flying machine must have more lift than weight. It can achieve this either by being less dense than the surrounding air, or by having the power to push suitably shaped wings through the air at high speed. Balloons and airships are lighter than air; helicopters pull themselves up with huge rotors; other aircraft rely on the lift from their wings.

## Pioneers of aviation
Inspired by the Montgolfier Brothers (see BEFORE), in the 1790s George Cayley investigated aerodynamics, watching crows, and using a whirling-arm machine to study lift and drag at his home in Yorkshire, England. In 1853 he achieved a manned flight with a glider, volunteering his coachman to be the world's first test pilot. He said that he would be able to power his aircraft "if he could get a hundred horsepower into a pint pot." This was achieved with the internal combustion engine (see pp.254–55), invented some 30 years after his death in 1857.

## Powered flight
The first powered flight was achieved by John Stringfellow, who flew a model aircraft with a 10 ft (3 m) wingspan inside a lace mill in Chard, Somerset, in 1848. He used a miniature steam engine of his own design; this was probably the only successful steam-powered aircraft.

The first controlled, powered, man-carrying flight was achieved on

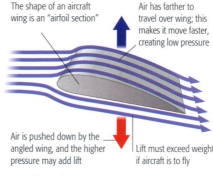

The shape of an aircraft wing is an "airfoil section"

Air has farther to travel over wing; this makes it move faster, creating low pressure

Air is pushed down by the angled wing, and the higher pressure may add lift

Lift must exceed weight if aircraft is to fly

**How lift works**
The wing is tilted upward at the "angle of attack." Moving through the air, it pushes air downward, and by Newton's third law an equal force pushes the wing up. The streamlined airfoil shape keeps the air flow "laminar," which reduces drag, and may contribute lift.

December 17, 1903 by Orville and Wilbur Wright, owners of a bicycle shop in Dayton, Ohio. They chose to take their "Wright Flyer" or "Flyer 1" to Kitty Hawk, North Carolina, because they were informed by the US Weather Bureau that Kitty Hawk had some of the steadiest winds in the country, and they wanted a reliable headwind to assist takeoff.

They had experimented extensively with gliders and box kites, finding out how to achieve both lift and control. Their successful aircraft was essentially a box kite, fitted with a gas engine.

Once the Wright Brothers had shown the way, others followed. In 1909 Louis Blériot flew across the English

**Wright Brothers' test flight**
One of a series of test flights by Orville Wright at the US Army base Fort Myer, Virginia, in 1908. On September 17, a propeller broke, causing a serious crash in which Orville was injured and his passenger killed.

**AVIATION PIONEER (1848–96)**

## OTTO LILIENTHAL

A dedicated champion of flight, Otto Lilienthal, from Anklam in Pomerania (now in Germany), made a series of hang gliders, and starting in 1891 used them for more than 2,000 flights, taking off from the top of an artificial hill he built near Berlin.

Given a reasonable headwind, he could hold the machine stationary, and shout instructions to a photographer on the ground below.

Unfortunately he made one flight too many; on August 9, 1896 he crashed and broke his back. The next day, before he died, he said "Small sacrifices must be made."

## Aircraft development
Getting off the ground was difficult in the 19th century, but once the internal combustion engine arrived in the 1880s, technological advances came fast.

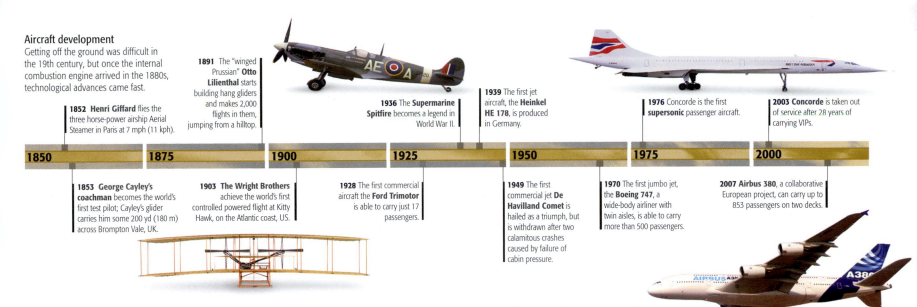

**1852** Henri Giffard flies the three horse-power airship Aerial Steamer in Paris at 7 mph (11 kph).

**1891** The "winged Prussian" **Otto Lilienthal** starts building hang gliders and makes 2,000 flights in them, jumping from a hilltop.

**1936** The **Supermarine Spitfire** becomes a legend in World War II.

**1939** The first jet aircraft, the **Heinkel HE 178**, is produced in Germany.

**1976** Concorde is the first **supersonic** passenger aircraft.

**2003 Concorde** is taken out of service after 28 years of carrying VIPs.

| 1850 | 1875 | 1900 | 1925 | 1950 | 1975 | 2000 |
|---|---|---|---|---|---|---|

**1853** George Cayley's **coachman** becomes the world's first test pilot; Cayley's glider carries him some 200 yd (180 m) across Brompton Vale, UK.

**1903** The Wright Brothers achieve the world's first controlled powered flight at Kitty Hawk, on the Atlantic coast, US.

**1928** The first commercial aircraft the **Ford Trimotor** is able to carry just 17 passengers.

**1949** The first commercial jet **De Havilland Comet** is hailed as a triumph, but is withdrawn after two calamitous crashes caused by failure of cabin pressure.

**1970** The first jumbo jet, the **Boeing 747**, a wide-body airliner with twin aisles, is able to carry more than 500 passengers.

**2007** Airbus 380, a collaborative European project, can carry up to 853 passengers on two decks.

> "To invent an **airplane** is nothing. To **build one** is something. But **to fly** is **everything**."
>
> OTTO LILIENTHAL (1848–96)

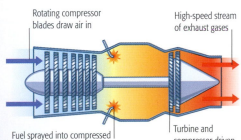

Channel, a distance of around 25 miles (40 km), and in 1927 Charles Lindberg flew the 3,000 miles (5,000 km) across the Atlantic Ocean.

Such record-breaking flights continue to this day; in 1986 Dick Rutan and Jeana Yeager flew around the world in a single flight, a record-breaking 25,000 miles (40,000 km) without refueling. Meanwhile, human-powered flight was launched by Gossamer Condor, which won the Kremer Prize by flying a 1 mile (1.6 km) figure-of-eight course on August 23, 1977. Two years later the human-powered Gossamer Albatross crossed the English Channel.

### The jet engine
In 1928 air cadet Frank Whittle wrote an essay, "Future developments in aircraft design," predicting that to fly faster, aircraft would have to fly much higher where the air is thin and drag is reduced. To do that, he said, would require a different sort of engine—a jet engine. No one paid much attention, and when he went ahead and built a prototype the UK government refused to back it; so the technology passed to the US, and German engineers also ran with the idea. The first jet aircraft, the Heinkel He 178, flew in 1939, but was

**6 MILLION** The total number of parts in a Boeing 747–400 airliner.

**565 MPH (910 KPH)** The usual cruising speed of a 747–400.

### AFTER »

Aircraft design continues to evolve and has led to changes in the way we live.

**NASA'S HELIOS**
In 1999 NASA launched an unmanned, **solar-powered aircraft**, Helios, with a wingspan of 247 ft (75.3 m). In 2001 Helios achieved a **world altitude record** for propeller-driven aircraft of 96,863 ft (29,524 m). Designed as a vehicle for atmospheric research and as a communications platform, she was brought down by turbulence over the Pacific Ocean in June 2003.

**GROWTH IN TRAVEL AND TOURISM**
Since the **Wright Brothers** made their first unsteady hops in 1903 the travel industry has taken to the air with enthusiasm. In 2006 scheduled airlines offered more than 3 billion seats on 28 million flights and on average carried more than **6 million passengers every day**.

**AFFORDABLE FLIGHTS LEAD TO CROWDED AIRPORTS**

SEE ALSO »
pp.338–39 ROCKET PROPULSION

Rotating compressor blades draw air in

High-speed stream of exhaust gases

Fuel sprayed into compressed air burns continuously

Turbine and compressor driven around by hot gases

### How a jet engine works
Air is pulled in at the front by a fan, mixed with fuel and ignited, and blasts out behind through a nozzle. The force pushing the exhaust gases out of the back is matched by an equal force pushing the aircraft forward.

## **‹‹ BEFORE**

Scientists had been working with vacuums for nearly 200 years before the invention of the vacuum tube.

### FIRST VACUUMS
The work of **Evangelista Torricelli** in 1643 allowed people to realize that vacuums **‹‹ 96–97** were possible and might even exist. The space above the **column of mercury** in Torricelli's barometer seemed completely empty (in fact, it is filled with very low-pressure mercury vapor).

### VACUUM PUMPS
In 1650 Otto von Guericke produced the first **vacuum pump**, which he used in a dramatic demonstration in his home town of Magdeburg **‹‹ 96–97**. He fit two hemispheres together and pumped out most of the air from inside. The hemispheres held together so strongly that **they could not be separated**, even by two teams of horses.

**MAGDEBURG SPHERE**

### ‹‹ SEE ALSO
*pp.96–97* DISCOVERY OF THE VACUUM
*pp.98–99* ROBERT BOYLE
*pp.286–87* STRUCTURE OF THE ATOM

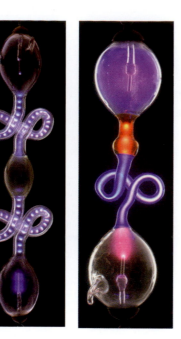

In 1838 English scientist Michael Faraday put metal electrodes inside either end of a sealed glass tube from which he had removed much of the air. The air pressure inside the tube was around a hundredth that of the air outside. He connected the electrodes to a high-voltage electricity supply and noticed a strange glow in the tube: electricity was flowing between the electrodes and "exciting" the low-pressure air to produce light.

Nearly 20 years later, in 1855, German scientific-instrument maker Heinrich Geissler made a vacuum pump that could achieve much lower pressures, so the tubes were

### Early vacuum tubes
Geissler made his first evacuated tubes in 1855, and noticed their strange glow in 1857. The colored light is caused by cathode rays hitting air molecules inside the tube.

# Vacuum Tubes

**During the 19th century experiments with electricity in glass vacuum tubes led to the discovery of cathode rays, electrons, and X-rays. The tubes themselves enabled the development of sound amplification and many broadcasting systems, such as radio and television. They were even essential in building the early computers.**

Maltese cross mask placed in the path of the rays

Cathode

Negative terminal

Positive terminal (anode)

Shadow cast by the cross on the back of the tube

### Crookes Tube
In William Crookes' Maltese cross experiment of 1879, the cross in the path of the cathode rays cast a crisp shadow on the phosphor screen. The experiment proved that cathode rays travel in straight lines.

"evacuated" (nearly completely emptied). He was also a talented glassblower, and managed to find a way to seal platinum wires inside these intricately designed Geissler Tubes.

### Cathode rays
One of Geissler's colleagues, German physicist Julius Plücker, experimented with the tubes. He found that the glow inside a tube vanished at one ten-thousandth atmospheric pressure because there was virtually no air to excite. But electricity was still flowing across the vacuum between the two electrodes. Plücker also noticed that the glass at one end was glowing. It was always the end farthest from the cathode (the electrode connected to the negative terminal) that glowed. Something was emanating from the

cathode and traveling the entire length of the tube to hit the glass opposite. In 1876 German physicist Eugen Goldstein named this "cathode rays."

Objects placed between the two electrodes cast shadows, suggesting that cathode rays travel in straight lines. The fact that the rays came from the cathode and traveled toward the positive electrode (the anode) suggested that they were carrying negative electric charge.

### Discovery of the electron
In 1875 English scientist William Crookes managed to make tubes with even better vacuums, and his tube—

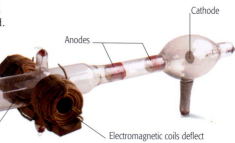

Cathode

Anodes

Electromagnetic coils deflect moving electrons magnetically

Calibrations to show deflection

Vacuum tube

Metal plates deflect electrons electrostatically

### Discovery of the electron
J.J. Thomson used this vacuum tube to investigate the nature of cathode rays. Some people suspected that they were streams of particles, but Thomson figured out the electric charge and mass of the particles, by measuring to what extent magnetic and electric fields deflected them.

INVENTION

## CATHODE RAY TUBE (CRT) TELEVISION TUBES

Television sets in the 20th century used the cathode ray tube (CRT), invented in 1897 by German scientist Karl Ferdinand Braun. At the back of the tube a "gun" fires high-speed electrons. The front of the tube forms a screen, the inside of which is coated with phosphors—metal compounds that glow when the electron beam hits. The beam varies in strength, and is scanned from side to side by the electromagnets around the tube. This builds up a picture on the screen, line by line. A color CRT has three beams and three different-colored phosphors.

Electron beams fired from three guns

Vacuum allows electrons to move freely

Inside of screen coated with phosphors

the Crookes Tube—became the standard apparatus for investigating cathode rays. Crookes found that the rays could make a paddle wheel turn when it was placed between the electrodes. This suggested that the rays were streams of particles.

In 1897 English physicist Joseph John (J.J.) Thomson used a vacuum tube to calculate the mass of these particles and the amount of electric charge they carried. He found that they were tiny—less than a thousandth the mass of even the lightest atoms. Thomson had discovered the electron.

Another key discovery that was made using vacuum tubes was the existence of X-rays (see pp.294–95). Cathode rays sometimes cause the emission of these penetrating, invisible rays.

### Rapidly developing technology

It was found that the cathode rays could be deflected by magnetic fields. This, combined with the fact they could make a glow on the end of the tube,

led to the development of the cathode ray tube (see above), and ultimately both television and radar.

In 1904 the world's first electronic component—Ambrose Fleming's "diode valve"—used the fact that electrons travel in only one direction in vacuum tubes. In 1907 American engineer Lee de Forest added a metal grid between the two electrodes. By supplying a small electrical signal to the grid he could control the large electric current passing through the valve. His device, the Audion, was the world's first amplifier. It was used first in radio, and later in television and the film industry. Subsequent configurations of valves eventually led to the consumer electronics industry.

Valves can also be used as electronic switches: when there is no charge on the grid, current flows across the valve (the device is "on"). By supplying a negative charge, the electron flow stops (turning it"off").

**Vacuum tube valves in computing**
Early computers such as the ENIAC (Electronic Numerical Integrator and Computer), 1946, used thousands of vacuum tube valves to perform operations and calculations. ENIAC had 17,468 valves.

AFTER »

Vacuum tubes are still used to produce X-rays. But their role in electronic circuits and televisions has been superseded.

### THE TRANSISTOR
Transistors **366–67 »** were invented in 1947. They do the same job as valves—amplifying and switching—but they are smaller, cheaper, more reliable, and use much less power.

### FLAT-SCREEN TELEVISION
Televisions with cathode ray tubes are now being replaced by plasma display panels (PDPs). In a PDP, cells between two glass panels contain an inert gas mixture. This is electrically turned into a plasma that excites phosphors to emit light.

SEE ALSO »
*pp.294–95* The Discovery of X-rays
*pp.366–67* Microchip Technology

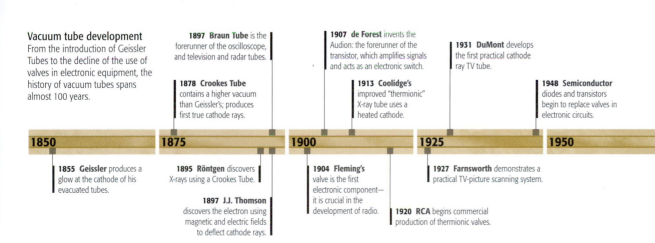

**Vacuum tube development**
From the introduction of Geissler Tubes to the decline of the use of valves in electronic equipment, the history of vacuum tubes spans almost 100 years.

**1897 Braun Tube** is the forerunner of the oscilloscope, and television and radar tubes.

**1878 Crookes Tube** contains a higher vacuum than Geissler's; produces first true cathode rays.

**1907 de Forest** invents the Audion: the forerunner of the transistor, which amplifies signals and acts as an electronic switch.

**1913 Coolidge's** improved "thermionic" X-ray tube uses a heated cathode.

**1931 DuMont** develops the first practical cathode ray TV tube.

**1948 Semiconductor** diodes and transistors begin to replace valves in electronic circuits.

| 1850 | 1875 | 1900 | 1925 | 1950 |
|---|---|---|---|---|

**1855 Geissler** produces a glow at the cathode of his evacuated tubes.

**1895 Röntgen** discovers X-rays using a Crookes Tube.

**1897 J.J. Thomson** discovers the electron using magnetic and electric fields to deflect cathode rays.

**1904 Fleming's** valve is the first electronic component—it is crucial in the development of radio.

**1920 RCA** begins commercial production of thermionic valves.

**1927 Farnsworth** demonstrates a practical TV-picture scanning system.

# The **Discovery** of X-rays

**The discovery of X-rays in 1895 came as a complete surprise, encouraging a fresh wave of investigations into the nature of electromagnetic radiation, instantly paving the way for a variety of new life-saving technologies, and leading to novel ways of exploring the structure of materials.**

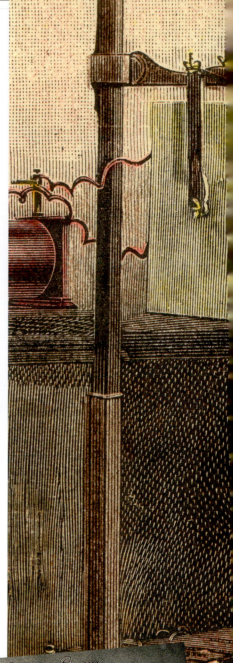

No one suspected X-rays existed before German physicist Wilhelm Röntgen stumbled across them in his laboratory, but their discovery was to revolutionize imaging techniques across the sciences.

### Unblockable light

In 1895 Röntgen was investigating the properties of various types of vacuum tube, and the mysterious "cathode rays" that they produced. He erected a cardboard screen to block out the effects of visible light from the tube, so as to isolate the effects of the cathode rays. But as he began experimenting, he noticed that a nearby fluorescent screen had begun to glow, despite the heavy cardboard. It could not be due to light, nor to the cathode rays (since they emerged from the tube in just one direction); it had to be something new.

### Putting X-rays to work

Röntgen threw himself into the study of his new rays, naming them "X" rays after the mathematical term for something unknown. Two months later he published the first account of them, sparking a sensation.

One of Röntgen's first steps had been to test the power of the rays to penetrate different objects. He found that they were blocked by metal and other dense materials, but not by paper

#### Early Crookes tubes

In the 1850s British scientist Sir William Crookes experimented with passing electrical discharges through gases at very low pressures in glass tubes, which came to be called Crookes tubes. These Crookes tubes show discharges through air at different pressures.

---

**PHYSICIST (1845–1923)**

### WILHELM CONRAD RÖNTGEN

Wilhelm Röntgen was born in Lennep, Germany, in 1845. Unexceptional at school, he had a natural aptitude for making mechanical devices, but after being unfairly expelled from technical school he gained a place to study physics at the University of Utrecht. Here he studied under the theoretical physicists Rudolf Clausius and August Kundt, who were huge influences on his work. Röntgen initially focused on the conductivity of crystals, but he was best known for the discovery of X-rays, winning the Nobel Prize in 1901.

---

or human skin. He realized that the rays affected photographic plates, devised a simple X-ray camera, and used it to make an image of the bones in his wife's hand. This new technique for looking at the inside of the human body had obvious medical applications, and X-rays were soon being recommended for the diagnosis of a huge range of problems. In 1896 French physicist Louis Benoist discovered the rays' ionizing effect, finding that they left air electrically charged as they passed and, as early as 1900, this ability to irradiate matter was adapted to treat skin cancers.

### Recognizing the dangers

At around the same time physicians and experimenters began to realize that the overenthusiastic use of X-rays could cause problems, damaging skin cells, leaving radiation burns, and even triggering cancers. From the 1900s onward X-rays began to be used with more caution and in smaller doses when creating diagnostic images of the body. However, it was not until the second half of the 20th century that the full risks of X-rays were

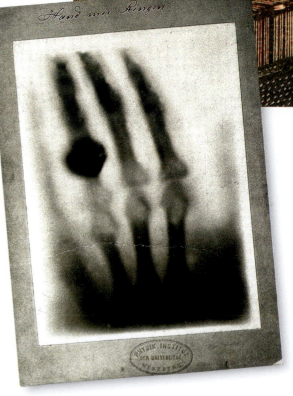

#### The first medical X-ray

One of the most famous images in photographic history, this X-ray of the hand of Röntgen's wife, Anna, clearly shows her bones and wedding ring. The image was taken in December 1895, and immediately revealed the medical implications: fractured bones were being X-rayed by January 1896.

---

<< **BEFORE**

**The 17th and 18th centuries saw breakthroughs in the understanding of light and radiation.**

#### TYPES OF LIGHT

**Isaac Newton** proved in 1666 that sunlight was made up of different colors. In 1800 **William Herschel** discovered infrared radiation, and in 1801 **Johann Ritter** discovered ultraviolet radiation.

#### MAXWELL'S PREDICTION

**James Clerk Maxwell** described light as an **electromagnetic wave** in the 1850s, and it became clear that the wavelengths of infrared radiation were longer than visible light, and those of ultraviolet radiation were shorter.

#### EARLY VACUUM TUBES

**Sir William Crookes** proved cathode rays existed by making early versions of the **cathode ray tube**. The discovery of X-rays would not have been possible without them.

<< SEE ALSO
*pp.114–15* THE NATURE OF LIGHT

X-rays allowed scientists to delve deeper into structures ranging from the human body to black holes in space.

### THE STRUCTURE OF DNA

X-ray crystallography offered clues to the double-helix **structure of DNA**, which in turn gave rise to the modern revolution in **genetics** and **biotechnology**.

### MEDICAL DIAGNOSES

X-rays are still used today in **medical imaging** techniques, such as computerized tomography (CT) scans.

### X-RAY ASTRONOMY

Orbiting **telescopes** are used to study the X-ray radiation that is produced by some of the most **energetic events** and objects in the Universe, including **black holes**.

### AIRPORT SECURITY

X-ray machines are used to **scan baggage** at airports, revealing the contents on a monitored computer screen.

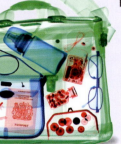

**SEE ALSO ⟩⟩**
pp.296–97 MARIE CURIE
pp.396–97 SPACEPROBES
AND TELESCOPES

LUGGAGE SCANNED AT AIRPORT

▷ **Medical X-ray machine**
The penetrating power of X-rays makes them well suited for studying the human body. Body tissues absorb X-rays to varying degrees, depending on their density, and so appear as different shades—from black through to white—on radiographic film.

**2 Free electrons**
The electrons are emitted from the heated filament and accelerated into a metal target—the anode —producing high-energy photons.

**1 Electron production**
A heated filament in the cathode ray tube generates high-energy electrons.

**3 Stream of photons**
The photons pass out of the cathode ray tube, channeled through a lead cylinder.

**4 Low-energy filters**
The photons pass through filters that block the low-energy rays, so that only the high-energy X-rays go through toward the patient. They are focused onto one particular part of the patient's body.

**5 Body imaging**
Dense tissues, such as bone, absorb X-rays almost completely and appear white on radiographic film; soft tissues absorb X-rays less well and appear gray.

**Perforated filter**
The filter screens out scattered X-rays that would blur the image.

**X-ray plate**
The plate is coated with silver halide emulsion, which blackens when exposed to X-rays.

**Screen glow**
A fluorescent screen emits light when struck by X-rays.

**6 Unabsorbed X-rays**
The areas where X-rays pass through the body will show as black on the X-ray.

**Image of bone**

△ **Early chest X-ray**
Wilhelm Röntgen prepares a patient for an X-ray in 1896. X-rays were quickly adopted by doctors as a diagnostic tool, and the first radiological journal— *Archives of Clinical Skiagraphy*—was documenting progress in print by April 1896.

recognized and rigorous guidelines were introduced to minimize exposure.

There were still questions about the true nature of X-rays. It was not until 1912, when Max von Laue of Munich University successfully proved that they were very short-wavelength electromagnetic radiation, that they began to be understood. Von Laue showed that the rays were subject to diffraction, by passing them between the narrow planes of aligned atoms in a crystal. His experiments also showed that the structure of the crystal itself affected the pattern of the diffracted X-rays, thereby offering a potential means of probing the internal arrangement of materials on the atomic scale.

### Continuing developments

In the 1940s the X-rays' extremely short wavelengths were used in microscopes to produce images of detail far smaller than that detectable with visible light. The rays' ability to penetrate bodies has continued to be developed in medicine through new forms of computer tomography, and in engineering to look deep into materials for indications of strain or degradation before there are any outward signs.

PHYSICIST AND CHEMIST  **Born 1867**  **Died 1934**

# Marie Curie

## "**Nothing** in life is to be **feared**, it is **only** to be **understood**."

MARIE CURIE

A world-famous scientist and remarkable role model for women, Marie Curie was the first woman to win a Nobel prize and the first person to be awarded two Nobel prizes in different sciences. She was a pioneering scientist who introduced the world to the marvels of radioactivity and worked selflessly to promote the use of radium to alleviate human suffering. By a sad irony it was her groundbreaking work, and the exposure to dangerous radiation that went with it, that ultimately caused her death.

Marie Curie was born Maria Sklodowska in 1867 in Warsaw, at that time part of the Russian Empire. Her father Wladyslaw was a teacher of physics and mathematics, whose pro-Polish

sentiments lost him his well-paid job as a school principal; her mother ran a girls' boarding school, but resigned from her position after Maria's birth. Life was hard for the Sklodowska family, and the eldest daughter, Zosia, died of typhus when Maria was eight. Two years later her mother, who had been suffering from tuberculosis, also died.

Maria and her siblings inherited a strong passion for education from their parents. Maria's brother, Josef, went to medical school in Warsaw, but women were not admitted to the Russian-run university. Maria and her elder sister, Bronya, agreed that Maria would work as a governess to help pay for Bronya's medical studies in Paris. In return, once Bronya was married, she invited Maria to Paris so that she could continue her studies, and in 1891 Maria enrolled at the College de Sorbonne. It was on arrival in France that Maria became a naturalized French citizen and registered her name as Marie.

### Radioactive notebooks
Marie and Pierre Curie meticulously recorded their observations in notebooks. Marie Curie's notebooks, now held in the National Library, Paris, remain radioactive to this day and carry a health warning.

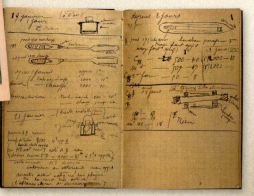

### Nobel prize awards
This picture of Pierre and Marie Curie working together in their laboratory appeared on the cover of Le Petit Parisien in 1901. Two years later they won the Nobel prize for physics.

## Marie Curie at work

Taken in 1925, this picture shows Marie Curie working in her laboratory at the College de Sorbonne, Paris, where she became the first woman professor.

## A scientific partnership

Marie was awarded a degree in physics in 1893 and stayed at the College de Sorbonne for another year to obtain a second degree in mathematical sciences. It was during this year that she met Pierre Curie, a young physics teacher at the Ecole Supérieure. Like Marie, Pierre was also an idealist, wishing to devote his life to scientific research. Their shared passion for physics brought them closer together, and in 1895 they married.

Marie carried out her postgraduate research in Paris and became fascinated by the radiation from uranium that Henri Becquerel had discovered, she went on to coin the term "radioactivity" for the phenomenon. Marie showed that the element thorium was radioactive, so uranium was not unique. Marie also discovered that two different uranium ores were more radioactive than uranium itself, and deduced that the ores must contain some other more active material. Pierre was so intrigued that he gave up his own research and joined her, and together they discovered two new radioactive elements: polonium and radium. Marie chose the names: polonium after her native Poland, and radium because of its intense radioactivity.

## Triumphs and tragedies

Anxious to share her discoveries, Marie published her findings before presenting her doctoral thesis, but even so was beaten in the race by English chemist Gerhard Schmidt. However,

in 1903 Marie and Pierre were jointly awarded the Nobel prize for physics with Becquerel for their work in the field of radioactivity.

As a result, Pierre was made professor at the College de Sorbonne, and established a research laboratory with Marie as its director. With the birth of their daughters Irène and Eve in 1897

> "Marie Curie is, of all **celebrated** beings, the only one whom **fame has not corrupted**."
>
> ALBERT EINSTEIN, 1934

and 1904, life was going well for the Curies, but in 1906 disaster struck: Pierre was knocked down and killed by a horse-drawn carriage. Marie was left to raise two children and to continue

## Converted radiological unit

Marie Curie drives a car that had been converted into a radiological unit during World War I. These vehicles were called "Little Curies." She and her daughter, Irène, also trained the nurses that operated the radiographic equipment.

the partnership's work alone. She was offered Pierre's professorship, and became the first woman professor at the College de Sorbonne. Marie finally isolated pure radium metal in 1910. For her exceptional work on the extraction and the properties of this element she was awarded the 1911 Nobel prize in chemistry.

Unaware of the dangers of handling radioactive substances, carrying samples and storing them in her desk, Marie at this time began to show signs of radiation sickness. It was also this year that she was accused in the press of breaking up the marriage of another physicist with an affair. This played on the sexism of the period, which was reinforced when she was refused membership of the Académie des Sciences despite her unprecedented two Nobel prizes.

## The spread of knowledge

During World War I Marie put her knowledge to practical use by training nurses in radiography and running a fleet of mobile radiology units, nicknamed Little Curies; she funded these with the money from her Nobel prizes. Marie later made the first experiments in the use of radioactivity in the treatment of cancer.

After the war, she toured the US twice to raise funds for her research. On her return to Paris, she continued working at the University and was made head of the Pasteur Institute.

Marie Curie's exposure to radiation finally took its toll and she died of leukemia on July 4, 1934. Marie and Pierre's hard work was continued by their daughter Irène and her husband Frédéric Joliot, and their legacy lives on in today's nuclear medicine, physics, and chemistry.

### FRENCH CHEMIST (1897–1956)
## IRENE JOLIOT-CURIE

The older daughter of Marie and Pierre Curie, Irène helped her mother run mobile hospitals equipped with primitive radiographic equipment, during World War I. After the war she completed her doctorate on polonium and met and married Frédéric Joliot, with whom she worked on atomic nuclei. In 1934 they succeeded in creating radioactive isotopes from other elements, an achievement that won them the Nobel prize for chemistry in 1935. Her work led to her contracting leukemia and she died at 58 on March 17, 1956.

### TIMELINE

- **Nov 7, 1867** Maria Sklodowska is born in Warsaw, Poland, then part of the Russian Empire.
- **June 12, 1883** Graduates from high school.
- **1885** Becomes a governess to help family finances and pay for sister Bronya's studies.
- **1891** Goes to Paris to study at the Sorbonne at Bronya's insistence.
- **July 1893** Graduates top of her year from the Sorbonne. A scholarship from Poland allows her to continue studying.
- **1894** Meets Pierre Curie in the spring; in July she is awarded her degree in mathematical sciences at the Sorbonne.
- **July 26, 1895** Marries Pierre.
- **1896** Henri Becquerel discovers radiation from uranium salts, for which Marie Curie later coins the term radioactivity.
- **Sept 12, 1897** Pierre and Marie's first daughter, Irène, is born.
- **April 1898** Professor Lippmann presents Marie's paper to the Académie des Sciences showing thorium emits radiation in the same way as uranium; in July she and Pierre announce the discovery of polonium, and on 26 December they announce the existence of radium.
- **1903** Marie receives her Doctorate in Science from the University of Paris. On December 10, Marie and Pierre are jointly awarded the Nobel prize for physics with Becquerel.

**NOBEL PRIZE, 1903**

- **December 6, 1904** The Curies' second daughter, Eve, is born. Around this time Marie is showing first signs of radiation sickness.
- **April 19, 1906** Pierre is run over by a horse-drawn carriage and killed. Marie is offered his professorship at the Sorbonne, but only becomes a full professor in 1908.
- **1911** Marie is awarded a second Nobel prize, for chemistry, for her discovery of radium and polonium.
- **1911** Marie is the victim of a hostile press campaign after revelations of an alleged affair with physicist Paul Langevin.
- **1914** The Institut de Radium (now the Institut Curie) is completed, and Marie is appointed director of the physics and chemistry laboratories.
- **1914** At the outbreak of World War I, Marie donates her and Pierre's Nobel medals for the war effort. She trains radiology nurses and runs mobile radiography units.
- **1921** Tours the US with her two daughters to raise funds for research.
- **1929** Makes a second tour of the US, raising funds for the Radium Institute in Warsaw (now the Marie Skłodowska-Curie Institute of Oncology), inaugurated in 1932 with her sister Bronya as director.
- **July 4, 1934** Marie Curie dies of leukemia in the Sancellemoz Sanatorium in Passy, France, and is buried next to Pierre in Sceaux. Their ashes are moved to the Paris Panthéon in 1995.

# Radiation and Radioactivity

**The discovery that some elements were radioactive—emitting energy and particles as their nuclei changed from one form to another—triggered a revolution in late 19th-century physics. Soon it would offer a means to probe the structure of atoms and, eventually, a way of tapping the enormous energies of the nucleus.**

**Uranium-238**
With a half-life of 4.47 billion years, uranium-238 decays through release of an alpha particle, losing two protons and two neutrons to produce thorium-234.

**Thorium-234**
As a neutron in the nucleus transforms into a proton, thorium-234 undergoes beta decay. It has a half-life of 24 days. The atom emits an electron and a neutrino as it is transmuted into protactinium-234.

**Protactinium-234**
With a half-life of 6.7 hours, protactinium-234 decays again through beta emission. Because two neutrons have now been converted to protons, the product of this decay is uranium-234.

**Uranium-234**
More stable than its two predecessors in the decay chain, uranium-234 has a half-life of 245,500 years. It undergoes alpha decay to produce thorium-230.

**Thorium-230**
Also relatively stable, thorium-230 decays through alpha emission to create radium-226, with a half-life of 75,380 years.

**Radium-226**
With a half-life of 1,602 years, radium-226 changes into radon-222 through alpha decay.

**Radon-222**
Less stable than its predecessor, radon-22 has a half-life of 3.8 days, and decays through another alpha emission into polonium-218.

**Alpha particle**

**Polonium-218**
With a half-life of just 3.1 minutes, this isotope decays rapidly. Almost all polonium-218 atoms undergo an alpha emission to produce lead-214, but a small fraction (0.02 percent) undergoes beta decay and follows a slightly different decay chain at this point.

**△ Radioactive decay chain: uranium decay**
All radioactive materials typically decay through a series of alpha and beta emissions, often producing a succession of other radioactive isotopes with different properties, before eventually reaching a stable form. One of the best-known chains is the decay of uranium-238.

Radioactivity is one of the great chance discoveries in science. In 1896 French scientist Henri Becquerel (see opposite) was investigating whether the recently discovered X-rays might be linked to the way in which some phosphorescent materials glowed after exposure to bright light. In preparation for an experiment, he wrapped a sample of phosphorescent uranium salts in black cloth and placed

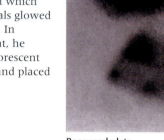

**Becquerel plate**
This photographic plate shows one of Becquerel's early experiments. The bottom section shows where he placed a metal cross between the uranium salts and the plate, successfully blocking some of the radiation.

**‹‹ BEFORE**

Throughout the 18th and 19th centuries evidence for the atomic nature of matter started to accumulate, but the secrets of atomic structure and the distinctions between elements remained a mystery.

**ATOMIC STRUCTURE**
Dmitri Mendeleev's periodic table ‹‹ 230–31, developed in the 1860s, revealed patterns that seemed to relate the atomic mass of elements to their properties. However, many elements seemed to have masses that were multiples of hydrogen, while other atomic masses had awkward fractions (ultimately explained by the existence of chemically identical isotopes present in various proportions).

**RÖNTGEN'S X-RAYS**
In late 1895 Wilhelm Röntgen discovered the new form of radiation known as X-rays ‹‹ 294–95, triggering a frenzy of speculation into how they might be used.

**‹‹ SEE ALSO**
*pp.190–91* DATING THE EARTH
*pp.286–87* STRUCTURE OF THE ATOM
*pp.296–97* MARIE CURIE

**△ Alpha decay**
Release of an alpha particle removes two protons and two neutrons from the atomic nucleus, so its mass drops by four "atomic mass units", while its "atomic number" drops by two.

**Alpha particle**

**◁ Alpha and beta decay**
The two types of radioactivity that produce particles also alter the structure of the radioactive atoms themselves, transmuting one element into another.

**△ Beta decay**
Most beta decay involves a neutron in the atomic nucleus spontaneously changing into a proton. The atom's mass is unaltered but its atomic number increases by one, and an electron is emitted.

**Beta particle**

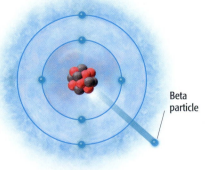

**▽ Penetration of radiation**
The ability of different materials to block radiation gives a clue to the nature of radioactive particles or rays—large alpha particles are more easily blocked by materials than smaller beta particles, while high-energy gamma rays have the most penetrating power.

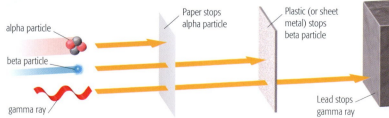

alpha particle

beta particle

gamma ray

Paper stops alpha particle

Plastic (or sheet metal) stops beta particle

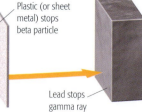

Lead stops gamma ray

PHYSICIST (1852–1908)

## ANTOINE HENRI BECQUEREL

Born into an illustrious French family of scientists, Henri Becquerel studied at the École Polytechnique in Paris, where he specialized in investigating the properties of light, such as polarization and phosphorescence. In 1892 he became the third member of his family to occupy the Chair in Physics at the famous Paris Museum of Natural History, and it was here that he made his landmark discovery. The SI unit of radioactivity, equivalent to one nuclear decay per second, is named the becquerel in his honor.

it between photographic plates. However, the plates became "fogged" even before the uranium was exposed to the light. He later discovered that even non-phosphorescent compounds of uranium caused the plates to fog.

At first, Becquerel suspected that his "rays" were similar to X-rays, but when he found that they could be deflected by a magnetic field he realized that they were, in fact, charged particles. What was more, the passage of the rays through air could turn air molecules into electrically charged ions. For this reason, the new phenomenon was termed "ionizing radiation."

Becquerel's discovery earned him the 1903 Nobel Prize for Physics. He shared it with his former student Marie Curie

(see pp.296–97) and her husband Pierre, who coined the term "radioactivity."

**HALF-LIFE** This is the period in which half of a given sample of a radioactive isotope will have decayed into another form. After one half-life, half remains undecayed; after two, one quarter; and so on. Half-lives range from fractions of a second to billions of years.

### New elements
The Curies found that uranium ore was often more radioactive than the pure metal. This led them to think that there were other radioactive materials mixed with it, and they went on to extract the new elements polonium and radium. The Curies used ionization of air to detect radioactive substances; the more electrically charged the air became, the more radioactive the substance.

### Alpha, beta, and gamma
In 1899 New Zealand-born physicist Ernest Rutherford and English-born Frederick Soddy identified two distinct types of radioactive particle, which Rutherford named "alpha" and "beta."

Alpha particles were relatively heavy and produced a strong ionizing effect, but traveled only a short distance in air and could be stopped by a sheet of paper. A few years later, they were identified as positively charged ions of the element helium. Beta particles were much lighter and less ionizing, but could travel through several millimeters of metal sheeting. They were eventually identified as lone negatively charged electrons, identical to those found in the structure of all atoms.

In 1900 French chemist Paul Villard discovered a third type of ionizing radiation—gamma rays—that could penetrate up to 8 in (20 cm) of iron. These proved to be high-energy radiation similar to X-rays, but with shorter wavelengths.

In 1901 Rutherford and Soddy found the atoms in a sample of radioactive thorium transforming into radium. They identified a pattern in the rate of a material's radioactive decay, which Rutherford termed "half-life." Soddy established that certain radioactive atoms had nonradioactive equivalents that were chemically identical but differed in mass. He called such atoms "isotopes." By 1905 Rutherford had figured out that radioactive atoms followed a "decay chain," transforming through different isotopes until they reached a stable, nonradioactive form.

Understanding radioactivity paved the way for harnessing the forces within the atom.

### SUBATOMIC STRUCTURE
In 1909 **Ernest Rutherford** and his students used alpha particles to probe the structure of the atom **286–87 »**. It took the arrival of **quantum physics**, however, to explain patterns of radioactive decay between elements.

### NUCLEAR FISSION
In 1919 Rutherford transformed **nitrogen into oxygen** by bombarding it with alpha particles. In 1934 **Irene Joliot-Curie** (daughter of Marie and Pierre) and her husband **Frédéric** successfully produced the first **artificial radioactive isotopes**.

### MODERN APPLICATIONS
Modern-day uses of radioactivity range from weapons to power stations and **medical diagnostics**.

SEE ALSO »
pp.324–25 FISSION AND FUSION
pp.406–07 MODERN SURGICAL PROCEDURES

**IN PRACTICE**

### PET SCANNING

Positron emission tomography (PET) is a means of imaging body tissues using radioactivity. The subject ingests material dosed with a short-lived radioactive isotope that is absorbed into specific tissues. Positrons (positively charged "antimatter" electrons) emitted by this material are "annihilated" by contact with electrons in the body, releasing gamma rays that are measured by detectors.

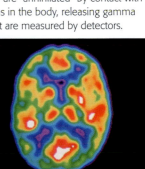

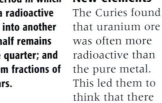

Beta particle

**Detecting radiation**
The Geiger counter is a radiation detector that contains an inert gas and two electrodes with a high voltage difference between them. Current flows only when a radioactive particle enters the detector through the window at one end, ionizing the gas within and briefly allowing it to conduct electricity.

**Lead-214**
This isotope undergoes beta decay, with a half-life of 26.8 minutes, producing bismuth-214.

**Bismuth-214**
An isotope with a half-life of 19.9 minutes, nearly all bismuth-214's atoms undergo beta decay to produce polonium-214, but 0.02 percent take a different route through alpha decay to become thallium-210.

**Polonium-214**
With an extremely short half-life of 0.16 milliseconds, polonium-214 decays by alpha emission to produce lead-210.

**Lead-210**
Beta emission is how lead-210 decays to produce bismuth-210. Its half-life is 22.3 years.

**Bismuth-210**
All but a tiny fraction of this decays by beta emission. It has a half-life of just a little over 5 days.

**Polonium-210**
This has a half-life of 138 days.

**Lead-206**
This stable isotope marks the end of the decay chain. All the various "side branches" of the decay chain also eventually reach lead-206.

# Einstein's Equation

In 1905 Albert Einstein published a scientific paper with the unpromising title "Does the Inertia of a Body Depend upon its Energy Content?" In it, he brought the tools of his special theory of relativity to bear on the question of objects traveling close to the speed of light. In the process he produced the most famous equation of all time: $E = mc^2$.

The three components of Einstein's equation are $E$, the energy content of a body, $m$, its mass, and $c$, the speed of light in a vacuum. The early 19th century had already seen a major change in the understanding of energy, with the idea that it was always "conserved" and could be transferred through electromagnetic waves. Similarly, advances in chemistry a few decades earlier had led to the idea that matter and mass could not be created or destroyed, and neither could energy.

The significance of $c$ had first been identified by James Clerk Maxwell in the 1860s. His famous equations had shown that an electromagnetic wave would travel through vacuum at a fixed speed—today measured as 299,792,458 meters per second. The fact that his predicted speed closely matched early measurements of the speed of light spurred the realization that light itself was such a wave.

Later physicists had difficulty in accepting this fact. By the late 19th century the motion of light, and objects traveling at speeds close to that of light, were subject to much research and debate, leading to discoveries including "time dilation,"

an apparent slowing down of time for objects measured at speeds close to that of light, and the "Lorentz-Fitzgerald contraction"—an apparent reduction in length of such objects.

However, it took the genius of Albert Einstein to show that the behavior of light and the distortion of objects at so-called relativistic speeds could be explained by a theory based on two simple assumptions—his special theory of relativity.

When Einstein applied his theory to an object approaching the speed of light, he discovered another twist. While there was no theoretical limit on the amount of energy that could be supplied to an object to drive it forward, it would eventually reach a point where it could not accelerate any further. Instead, it seemed, the energy would be poured into the object's mass, causing this to increase in line with his famous equation. Einstein soon showed that the same result applied in other circumstances, and that all mass and energy, even in objects at rest, were theoretically interchangeable. This breakthrough led to greater understanding of the forces in atoms, the power sources of stars, and the origins of the Universe.

**Relativistic kinetic energy**
This early derivation of Einstein's equation shows the kinetic energy of a moving body—the additional terms on the bottom line of the equation are the "Lorentz-Fitzgerald contraction"—a formula that explains many of the distortions that arise at speeds close to that of light.

> "Who could imagine that this **simple law** has plunged the conscientiously thoughtful physicist into the **greatest intellectual difficulties?**"
>
> EINSTEIN, "RELATIVITY," 1916

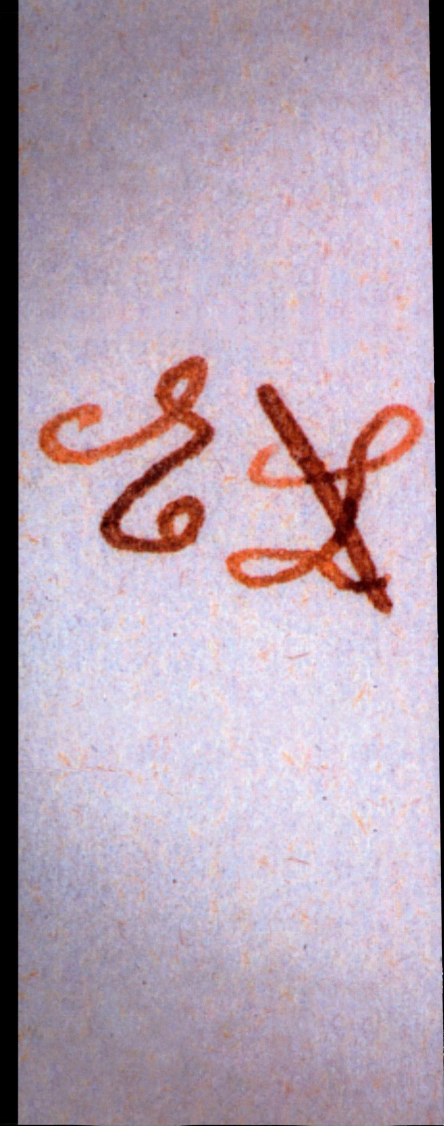

$$= \frac{mc^2}{\sqrt{1 - \frac{q^2}{c^2}}}$$

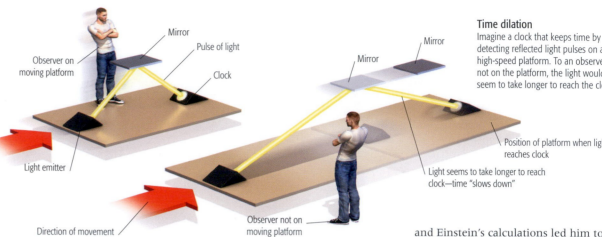

Einstein's first theory of relativity, which he proposed in 1905, is known as the special theory. It is built on two assumptions. The first, set out by Galileo as early as the 17th century, is the principle of relativity—that physical laws should be the same for all observers. The second, predicted by Maxwell's equations (see p.259), was that the speed of light is a constant, regardless of the motions of the light source and the observer.

Einstein began by considering how these rules would affect two observers in different "inertial reference frames" (environments free of acceleration or

**Time dilation**
Imagine a clock that keeps time by detecting reflected light pulses on a high-speed platform. To an observer not on the platform, the light would seem to take longer to reach the clock.

*Observer on moving platform*
*Mirror*
*Pulse of light*
*Clock*
*Light emitter*
*Direction of movement of platform*
*Mirror*
*Mirror*
*Position of platform when light reaches clock*
*Light seems to take longer to reach clock—time "slows down"*
*Observer not on moving platform*

# Theories of Relativity

**Einstein's special and general theories of relativity resolved many unexplained features of the Universe, and predicted others. Based on simple assumptions, they showed that space and time are intimately linked dimensions that can be stretched, warped, and even traded off against one another.**

deceleration). He realized that if these observers were to experience gravity, mechanics, and the speed of light in the same way, then many phenomena would, of necessity, seem to behave differently when one observer looked at the other reference frame. Events that seemed to occur simultaneously for one observer, for example, might look to the other as if they were separated by a substantial period of time. This effect, called the "relativity of simultaneity," shows that something very strange is happening—when two

observers are in motion relative to each other, each will say time is running differently for the other (see above).

In everyday experience the effect is negligible, but the greater the relative speed of objects, the more obvious it is. If an observer on Earth could view a rocket moving at a substantial fraction of the speed of light relative to him (a relativistic speed), then time on board the rocket would seem to have slowed down, or dilated. The rocket would also seem to have contracted in a way that matches the Lorentz-Fitzgerald contraction (see p.300). But those on board the rocket would say it was Earth that was slowed and distorted.

There are several ways to interpret these effects, but all of them rely on an important realization—that space and

time are intrinsically linked in a "continuum," and our perception of objects in this continuum depends on relative motion. Einstein's former tutor, Hermann Minkowski, did much to explain the properties of this four-dimensional "space-time."

## Mass-energy equivalence
One result of Einstein's theories was a rewriting of the laws of momentum and energy. While it is theoretically possible for a rocket to continue accelerating forever, it will never actually be able to move faster than the speed of light. In normal movement, an object's kinetic energy (energy of motion) is defined as $\frac{1}{2}mv^2$, where $m$ is mass and $v$ is velocity—but if the rocket's velocity has an unbreakable limit in $c$, the speed of light, then where does the additional energy from acceleration "go" at relativistic speeds?

Einstein realized that if energy was to be conserved, it must in fact be used to increase the object's mass, $m$. Energy and mass, then, are interchangeable,

and Einstein's calculations led him to the simple, celebrated equation that revealed for the first time the true nature of their relationship: $E = mc^2$.

## Generalizing relativity
In the decade following publication of the special theory, Einstein continued to work on the consequences of the theory, extending the principles he had already established to look at "non-inertial" situations—those involving acceleration and deceleration.

To do this, Einstein focused on the nature of gravity. He showed that a gravitational field is physically equivalent to a frame of reference that is under constant acceleration. Since beams of light will seem to bend in rapid acceleration, he predicted that they would also bend in strong gravitational fields. He found that strong gravity would cause distortions of time, mass, and appearance—large objects warp space-time around them.

Einstein used the theory to solve a long-running astronomical mystery. The elliptical orbit of Mercury, the closest planet to the Sun, "precesses" or rotates around the Sun at a faster rate than it should according to classical Newtonian physics. Einstein's theory explained it perfectly as a result of Mercury's passage through the distorted space-time around the Sun. Despite this success, however, it was not until 1919 that Arthur Eddington's solar eclipse expedition to Principe, off West Africa, demonstrated the accuracy of the theory: he showed that starlight really did bend in the Sun's gravity.

## BEFORE

**While the physics of the early 1900s had a great deal of predictive power, it left some fundamental problems to be solved.**

### THE NATURE OF GRAVITY
**Newtonian mechanics** was good for describing everyday interactions between objects, while his **theory of gravity** seemed to give accurate predictions for the behavior of larger bodies such as planets and stars.

### THE SPEED OF LIGHT
The discovery that the **speed of light is constant** and independent of relative motion posed awkward questions about how **objects might behave** in extreme circumstances and at speeds **approaching the speed of light**.

**The constancy of the speed of light**
Light from a moving source travels at the same speed as light from a stationary source—so the speed of light from a car's headlights does not increase when the vehicle accelerates, or decrease when it slows down.

**"Put your hand on a hot stove for a minute, and it seems like an hour. Sit with a pretty girl for an hour, and it seems like a minute. That's relativity."**

ALBERT EINSTEIN (1879–1955)

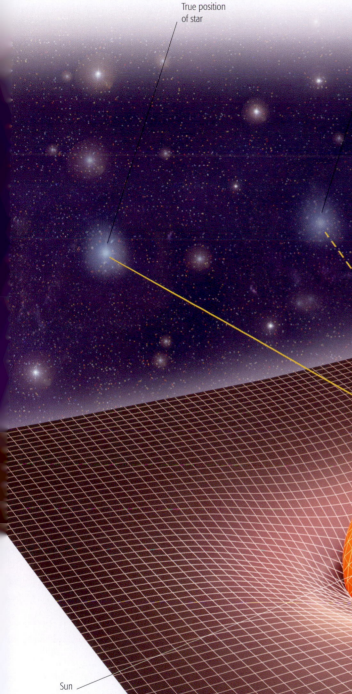

True position of star

Apparent position of star to observers on Earth, who assume light has traveled in a straight line

Sun

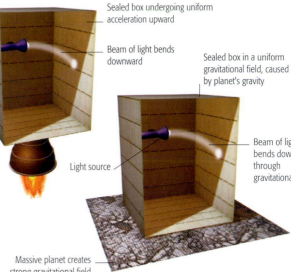

Sealed box undergoing uniform acceleration upward

Beam of light bends downward

Light source

Massive planet creates strong gravitational field

Sealed box in a uniform gravitational field, caused by planet's gravity

Beam of light bends downward through gravitational field

### Thought experiment with light
The equivalence of acceleration and gravity is the cornerstone of general relativity. This experiment shows that, just as extreme acceleration should deflect the path of light beams, so should extreme gravitational fields.

## AFTER »

Since their publication, Einstein's theories have played a vital part in areas such as nuclear fission and fusion, and the evidence supporting them continues to mount.

### TIME DILATION
This effect has been proved by **flying atomic clocks** at high speed around the world, and comparing the time that passes for them with that registered by similar devices on the ground.

### GRAVITATIONAL LENS
The **Hubble Space Telescope** has used **"gravitational lensing,"** caused as light bends around galaxies, to detect **dark matter**.

### SEE ALSO »
pp.304–05 ALBERT EINSTEIN
pp.398–99 DARK UNIVERSE
pp.402–03 STRING THEORY

Distortion of space-time caused by the Sun's mass deflects light from distant star

Telescope on Earth

Space-time around the Sun is warped by the Sun's mass, creating a so-called gravitational well

Two-dimensional rubber sheet represents three-dimensional space—dents in sheet represent distortions of space-time

### The proof of Einstein's theory
Arthur Eddington's measurements of stars near the Sun during the solar eclipse of 1919 showed that their positions were distorted in accordance with Einstein's theory—a result of the Sun's mass deflecting their light.

### Distorted space-time
Demonstrations of relativity often reduce the three dimensions of space to a two-dimensional "sheet" to visualize how large masses distort space-time. In this case, a star such as the Sun creates a "dent" in space. The path of light from distant stars and galaxies is also deflected, causing them to change their apparent positions.

PHYSICIST **Born 1879** Died 1955

# Albert Einstein

> "The **whole of science** is nothing more than a **refinement of everyday thinking**."

ALBERT EINSTEIN, "PHYSICS AND REALITY," 1936

German-born physicist Albert Einstein was at the heart of two great revolutions in 20th-century physics. Among his important contributions were his special and general theories of relativity and the discovery of the quantum nature of light. In later years his reputation and personality transcended scientific limits, transforming him into a political figure and popular icon.

### Early schooling

Born and raised in Germany, Einstein showed remarkable perception from an early age. He read widely and had mastered Euclidean geometry by the age of 12. However, he rebelled against the unimaginative educational system of his time. Fate intervened when his father's business failed in 1894 and the family moved to Italy in search of work. Einstein eventually completed his schooling in Switzerland and graduated from Zürich Polytechnic in 1900. He joined the Swiss patent office in Bern in 1902. Although Einstein had few personal contacts in the scientific community, his job at the patent office immersed him in the research being done on electromagnetism and other phenomena.

### Scientific papers published

In his spare time Einstein was working on the series of scientific papers that would make his reputation. His working methods, often based on intuitive "thought experiments," led to a series of remarkable breakthroughs. Four scientific papers, published in the leading journal *Annalen der Physik* during 1905, brought Einstein to the world's notice in a single year. The first paper—on the photoelectric effect—was to become a keystone of quantum theory. The second investigated the issue of Brownian motion.

◁ **Theories of relativity**
Einstein's theories were first published in detailed papers in German-language scientific journals, but he later wrote several book-length explanations for a more general readership.

▷ **Bern apartment**
While he worked at the Bern patent office, Einstein lived with his wife Mileva and their first son in the second-floor apartment of Kramgasse 49. Today the apartment is maintained by the Albert Einstein Society, with further exhibits on the third floor.

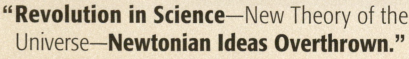

> "**Revolution in Science**—New Theory of the Universe—**Newtonian Ideas Overthrown**."

HEADLINE FROM "THE TIMES," NOVEMBER 7, 1919

The third was perhaps the boldest of all, a special theory of relativity (see pp302–303), which investigated the relative motion of objects moving at constant velocity. Einstein's fourth landmark paper went on to show how relativity implied that mass and energy were both aspects of the same property, linked by the iconic equation $E = mc^2$.

## General theory of relativity

The significance of Einstein's special theory of relativity was not immediately recognized; in fact, his explanations of the photoelectric effect (see p.314) and Brownian motion caused more of a stir. Undeterred, Einstein pressed on to develop a general theory of relativity, showing that strong gravitational fields could distort space and time. He published his completed version of the general theory of relativity in 1915. Although World War I was in progress, the theory reached the scientific community via his publisher in neutral Holland.

## Later achievements

After the publication of the general theory of relativity, Einstein wrote several more papers that proved

### Einstein the violinist
At his mother's insistence, Einstein learned the violin from the age of six. Although he disliked the lessons, he would later play duets with physicist Max Planck in Berlin.

### Albert Einstein
Although the iconic images of Einstein are from his later years at Princeton, he was only 26 years of age when he rose to scientific prominence through the revolutionary work of his *annus mirabilis*.

influential in the developing field of quantum physics (although he later clashed with Niels Bohr over the implications of the Copenhagen interpretation (see p.317). In 1917 he formulated a "cosmological constant" to explain why the Universe did not collapse under its own gravity. This was a precursor to the modern discovery of what is now termed "dark energy" (see p.398). Most of his later work, however, went into a doomed attempt at developing a "unified field theory"—a theory that would explain both electromagnetism and gravitation.

Einstein's fame meant that he had less time to devote to his work,

### Solvay Conference
In 1911 Belgian industrialist Ernest Solvay invited eminent physicists to a conference discussing "radiation and the quanta." Chaired by Hendrik Lorentz, the attendees included Marie Curie, Max Planck, Henri Poincaré—and Einstein, the youngest to be invited.

especially after he moved to Princeton with his second wife in 1933. His socialist views and Jewish heritage led him to speak out against the rise of the Nazis in Germany during the 1930s, and in 1939 he co-wrote a letter to US President Franklin D. Roosevelt warning about the potential threat from a German-built nuclear weapon.

After the World War II he was involved in the formation of the state of Israel, supported the campaign for black civil rights in the US, and became an energetic advocate of nuclear disarmament.

On 17 April 1955 Einstein was rushed to hospital with an aortic aneurysm. He died the next day, and doctors later (and without permission) removed his brain in the hope that one day they might learn what made him so intelligent. Five decades on, the source of Einstein's genius remains a mystery.

---

### ASTROPHYSICIST (1882–1944)

## ARTHUR STANLEY EDDINGTON

Arthur Eddington is best known today for his confirmation of Einstein's general theory of relativity. In 1919 he took measurements on the Atlantic island of Principe during a total solar eclipse, showing that massive objects such as the Sun distort space around them. In his own right, he developed the first detailed models of the internal structure of stars, recognizing that, through most of their lifetimes, stars obey a simple relationship linking their mass with their luminosity. He also supported models of an expanding Universe, but virulently attacked suggestions of "extreme" stars such as white dwarves.

---

### TIMELINE

**March 14, 1879** Albert Einstein is born in Ulm, southern Germany. His parents move to Munich, where his father Hermann establishes an electrical engineering business.

**1892** Begins studying at the Luitpold Gymnasium (high school) in Munich.

**1894** Hermann Einstein's business fails, and he takes the family to Italy in search of work. Einstein at first remains in Munich to continue his studies, but joins his family in 1895.

**1896** With his father's consent, he renounces German citizenship to avoid military service. He begins to study mathematics at the Zürich Polytechnic.

**1901** Graduates and takes Swiss citizenship. He seeks work as a teacher, without success.

**1903** After finding work at the patent office in Bern, Einstein marries Mileva Maric.

**1905** During his *annus mirabilis*, Einstein writes four landmark scientific papers and is awarded his doctorate from the University of Zürich.

**1915** Announces the completed version of his general theory of relativity.

**EINSTEIN'S FIRST WIFE AND CHILDREN**

**1917** Publishes papers that propose the existence of a "cosmological constant" (an error, as he later admits) and raise the possibility of stimulated emission—the principle behind the laser.

**1919** Results from Arthur Eddington's expedition to measure positions of stars during a total solar eclipse provide support for the general theory of relativity. Einstein gains international fame. In the same year he completes his divorce from his first wife and marries his cousin, Elsa Löwenthal.

**1922** Receives the 1921 Nobel Prize in Physics, awarded in recognition of his explanation for the photoelectric effect.

**1933** Takes up the offer of a position at the Institute for Advanced Study, and moves with Elsa to Princeton, New Jersey. He is also involved in the formation of the International Rescue Committee, founded to assist victims of Nazi persecution in Germany.

**1939** A letter from Einstein to US President Franklin D. Roosevelt warns of the prospect that Nazi Germany may be developing a nuclear weapon.

**1940** Einstein takes US citizenship.

**1947** Writes an article proposing that the US should hand over all its nuclear weapons to the United Nations as a deterrent against nuclear proliferation.

**EINSTEIN MEMORIAL, ULM, GERMANY**

**April 18, 1955** Dies in a hospital in Princeton, as a result of heart failure caused by an aortic aneurysm.

## Making new cells

The body grows as its cells go through the cell cycle—growth and DNA replication during interphase, four phases of mitosis, and a phase called cytokinesis, when the cell divides to form two daughter cells. Cells then respond to chemical cues to form tissues and organs.

Centrosomes with centriole pairs

Cell nucleus

**1 Interphase**
The cell grows and replicates its DNA, which is contained in the nucleus as long, loose threads of chromatin. This is the substance that forms chromosomes – the part of the cell that carries genes.

Spindle develops from centrosome microtubules

Chromosome

**2 Early prophase**
Chromatin condenses to form chromosomes. Each chromosome is made up of two identical joined parts, called chromatids.

Spindle

**3 Late prophase**
The nucleus of the cell fragments. The microtubules from the spindle attach to the chromatids.

Chromatids

**7 Cytokinesis**
Cell divides completely to form two daughter cells, which enter new interphase.

**6 Telophase**
Chromosomes regroup at either end of the cell forming two nuclei.

**5 Anaphase**
Chromatids separate entirely to form new chromosomes.

**4 Metaphase**
Chromatids align along the center of the cell.

Chromatids

Nucleus

Chromosomes become less condensed

Spindle microtubules shorten

Daughter chromosomes

**BEFORE**

In the 1800s new cellular theories emerged about how life was formed.

**CELLS FROM CELLS**
The idea that organisms are made up of units called cells began with the **Cell Theory** of biologist **Theodor Schwann** and botanist **Matthias Schleiden** in 1838 **‹‹ 214–15**. German doctor **Rudolf Virchow** asserted that **cells could only form from existing cells** and, together with the experiments by chemist **Louis Pasteur**, this was important in rejecting the idea that **life can form spontaneously ‹‹ 226–27**.

**NUCLEI FROM NUCLEI**
In 1876 Polish-German botanist **Eduard Strasburger** suggested that nuclei only arose from **division of existing nuclei,** based on his studies of cells in a pollinated flower. This helped focus attention on the role of the **cell nucleus** in **growth**, **development**, and **reproduction**.

**‹‹ SEE ALSO**
*pp.152–53* PLANT LIFE CYCLES
*pp.204–05* LAWS OF INHERITANCE
*pp.214–15* ANIMAL AND PLANT CELLS
*pp.226–27* HUMAN REPRODUCTION

# How Cells Divide

**Cells are the living units of an organism, and their division is the fundamental process that causes growth and the reproduction of life itself. Each cell contains a package of genetic material, and precise cell division ensures that daughter cells inherit all the necessary information to control their own functions.**

As a plant or animal grows, its cells undergo a cycle of enlargement whereby they replicate their DNA in preparation for cell division, when copies of genetic information are distributed into two daughter cells.

## Two types of division

There is a surprisingly large quantity of DNA packed inside the nucleus of a cell and, after its replication, it is intricately manipulated at cell division. First, the molecules of DNA must coil, and coil again, to condense them into manageable structures, which are known as chromosomes. An array of fibers guides them so that two separate bundles end up at opposite poles of the cell. Then the cell splits between these

bundles to make two daughter cells. During body growth, these cells are genetically identical: this type of cell division is called mitosis (see above).

But division to produce sex cells (sperm and eggs) is more complex. It halves the number of chromosomes and produces genetically different cells.

> " The **nucleus** has to take care of the **inheritance** of the heritable characters. "
>
> ERNST HEINRICH HAECKEL, FROM "GENERELLE MORPHOLOGIE," 1866

This type of cell division is meiosis (see opposite), which occurs only in sexual reproduction. It shuffles genes and is partly responsible for variety in offspring.

## Understanding mitosis

Around the 1850s a new type of stain called aniline was being used to reveal the structure of cells. Its alkaline property made it stick to the acidic material in a cell nucleus. In a paper of 1878, German biologist Walther Flemming called this material

**Mitosis in plant cells**
In plants, mitosis occurs in areas called meristems, behind the root tip. In this micrograph of a root tip the actively dividing meristem cells are visible in darker green.

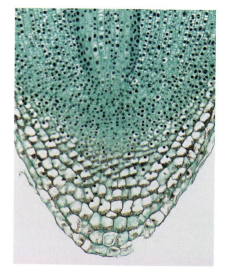

"chromatin" (substance containing DNA) and used his skills with the microscope to study cell division in the gills of a salamander. He discovered the thickening of the chromatin at the start of mitosis and traced the movement of its more solid threads through the rest of the process. Another German scientist, Heinrich Waldeyer-Hartz, studied Flemming's mitotic threads, and in 1888 he named them "chromosomes." Flemming himself had identified the principal stages of mitosis, which are recognized today (see opposite). He saw that, once formed, chromosomes are aligned along the middle of the cell, before two opposite groups move to separate poles. Although Flemming found that the

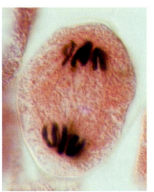

### Moving chromosomes
During anaphase of cell division an arrangement of fibers called a spindle rapidly pulls sets of chromosomes to opposing ends of the cell prior to division.

chromatin material doubled before division, he did not realize that each aligned chromosome contained two replicas of genetic material (DNA), known as chromatids, and that each of these splits apart before movement to the poles.

### Understanding meiosis
German professor Oscar Hertwig was one of the earliest embryologists. He studied reproduction in the sea urchin —a marine animal whose sex cells and fertilization can be easily observed on a microscope slide. In 1876 he saw that fertilization involved the sperm

penetrating the egg, and deduced that the cells' nuclear material had to halve in order to make sperm and eggs. German biologist August Weismann (see right) went further by correctly suggesting that chromosome number halved in the formation of germ cells (sperm and eggs) from body cells and that new combinations form at fertilization. He also deduced that two successive divisions were needed in meiosis: the first to separate maternal and paternal sets of chromosomes, and the second to split their replicated chromatids (see below). His conclusions had great influence on genetic ideas in the 1900s.

**MITOSIS** Conserves chromosome number and results in genetically identical cells.

**MEIOSIS** Halves chromosome number and results in genetically different cells.

## Making sex cells
During meiosis, two successive divisions produce sex cells (sperm or eggs). Each division follows the same phases as mitosis (see opposite), but in meiosis, four genetically different daughter cells are produced from the parent cell as opposed to two genetically identical ones.

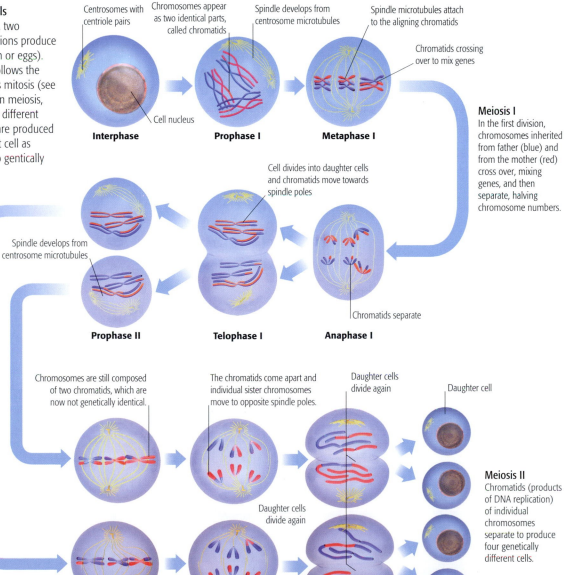

Centrosomes with centriole pairs

Chromosomes appear as two identical parts, called chromatids

Spindle develops from centrosome microtubules

Spindle microtubules attach to the aligning chromatids

Chromatids crossing over to mix genes

Cell nucleus

**Interphase**  **Prophase I**  **Metaphase I**

**Meiosis I**
In the first division, chromosomes inherited from father (blue) and from the mother (red) cross over, mixing genes, and then separate, halving chromosome numbers.

Cell divides into daughter cells and chromatids move towards spindle poles

Spindle develops from centrosome microtubules

**Prophase II**  **Telophase I**  **Anaphase I**

Chromatids separate

Chromosomes are still composed of two chromatids, which are now not genetically identical.

The chromatids come apart and individual sister chromosomes move to opposite spindle poles.

Daughter cells divide again

Daughter cell

Daughter cells divide again

**Meiosis II**
Chromatids (products of DNA replication) of individual chromosomes separate to produce four genetically different cells.

**Metaphase II**  **Anaphase II**  **Telophase II**  **Cytokinesis**

### AFTER »

The discovery of systems that control the rate of cell division and restrict the ways in which they become specialized had enormous medical implications.

### TREATMENT OF CANCEROUS CELLS
In 1965 American professor **Leonard Hayflick** discovered that normal cells can only **divide a certain number of times**. However, genetic disorders that interfere with this constraint can make cells **effectively immortal** and **cause cancerous tumors**. Today research into cancer treatment focuses on ways of targeting such cells with medications.

**CANCEROUS CELLS**

### CLONING AND STEM CELLS
In the 1890s scientists developed the idea that most embryo cells lose the ability to become specialized into different tissues as they grow older. **Early embryo cells** (known as **stem cells**) can be **cloned** and each has the potential to form organs or even whole individuals—this is being exploited to provide **body parts for transplantation 390-91 »**.

## ‹‹ BEFORE

The foundations of genetics were laid in the 19th century with the discovery of chromosomes and the principles of heredity.

### DISCOVERY OF CHROMOSOMES
Swiss botanist **Karl von Nägeli** may have been the first to see **chromosomes** in 1842, but the first unequivocal illustration of chromosomes during **mitosis** (cell division) ‹‹ **306** was made in 1849 by German botanist **Wilhelm Hofmeister**.

### PRINCIPLES OF HEREDITY
In 1866 Austrian monk and scientist **Gregor Mendel** ‹‹ **204–205** established the **principles of heredity** in pea plants by showing that characteristics were determined by **discrete particles** (later to be called genes), which were **inherited in definite proportions**.

### ‹‹ SEE ALSO
pp.204–05 Laws of Inheritance
pp.214–15 Animal and Plant Cells
pp.306–07 How Cells Divide

Each cell of an organism contains a fixed number of DNA threads, which carry genetic information. At cell division (see pp.306–307), replicated threads shorten and thicken to form visible chromosomes, and it was their movements that gave scientists the first clues to their importance.

## Chromosomes as genetic carriers
In 1900 Dutch botanist Hugo de Vries studied the evening primrose plant, and concluded that inheritance was due to particles, which he called "pangenes" (later changed to "gene"). Meanwhile, German

**Human X and Y chromosomes**
All cells in females have a pair of X chromosomes, while cells in males have one Y (left) and one X (right) chromosome. The larger X chromosome carries many more genes than the Y.

### BIOLOGIST (1866–1945)
### THOMAS HUNT MORGAN

Born in Kentucky, Thomas Hunt Morgan worked as a professor at Bryn Mawr College and Columbia University where he started breeding experiments using *Drosophila* fruit flies. In 1933 he became the first scientist to be awarded the Nobel Prize for genetics—he discovered the roles played by chromosomes in heredity (see below). He later established a division of research at the California Institute of Technology, which attracted the talents of some of the best geneticists of the day, including seven future Nobel prize winners.

> ## "Genes are the atoms of heredity."
> SEYMOUR BENZER, AMERICAN GENETICIST, SEPTEMBER 15, 1960

# Chromosomes and Inheritance

Chromosomes were first seen inside cells in the mid 1800s, but their roles as packages of inherited information that determine the characteristics of a living body were not appreciated until half a century later. Two botanists uncovered the work of a little-known Austrian monk, Gregor Mendel, and in so doing they ushered in a new age of genetics; one that, with the help of technology, is now developing at a furious pace.

### IN PRACTICE
### PRENATAL TESTS

An embryo or fetus can be tested for abnormalities by examining cell samples from the amniotic fluid (amniocentesis) or the placenta (chorionic villus sampling). Chromosomal abnormalities such as Down syndrome can be diagnosed in this way, and tests may be offered when there is a family history of genetic disorders or chromosomal anomalies. This micrograph shows the chromosomes of a female with Down syndrome; the 21st set has three instead of the normal two chromosomes.

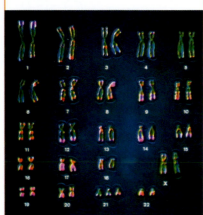

botanist Carl Correns discovered the principles of heredity in hawkweed. But both acknowledged a debt to an Austrian monk and scientist Gregor Mendel (see p.204), who had the same results with pea plants 30 years earlier.

Two scientists developed de Vries' and Correns' discoveries, and established that it is chromosomes that carry the genetic particles. German biologist Theodor Boveri studied fertilization in worms and sea urchins and showed that each chromosome regulated

embryo formation in a different way. A young student called Walter Sutton was doing the same thing with grasshoppers in the US, finding that chromosome behavior at reproduction mirrored the behavior of Mendel's "particles."

## Genes and sex chromosomes
Another American geneticist, Thomas Hunt Morgan (see above), was cautious of accepting Mendel's theory too soon. He began to do his own experiments in

the 1920s, but combined them with chromosome studies. To do this, he needed a variable organism with a small number of chromosomes to simplify analysis—he found it in the common *Drosophila* fruit fly. Morgan crossbred flies, just as Mendel had done

**1,442** The highest known number of chromosomes in an organism (*Ophioglossum*, adder's-tongue fern). Such a number is the result of chromosome multiplication over time.

**Male or female?**
The sex of a baby is determined by the father. Half his sperm carry X chromosomes, and if one of these fertilizes the egg, the result is a girl. The other half carry Y chromosomes, which produce boys. There is an equal chance of a girl or a boy at each fertilization.

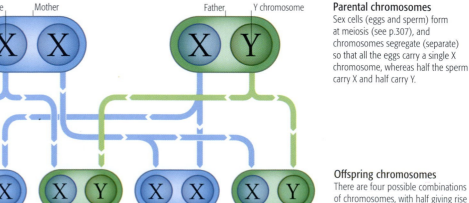

**Parental chromosomes**
Sex cells (eggs and sperm) form at meiosis (see p.307), and chromosomes segregate (separate) so that all the eggs carry a single X chromosome, whereas half the sperm carry X and half carry Y.

**Offspring chromosomes**
There are four possible combinations of chromosomes, with half giving rise to girls and the other half boys.

**Fruit-fly breeding experiments**
Breeding experiments with fruit flies suggested that genes that determine characteristics (such as red eye color, above) are carried on the sex chromosomes.

with pea plants, and from his two experiments (see below), he concluded that the eye-color gene resided on one of the chromosomes that are involved in sex determination.

## Human chromosomes

Humans have more chromosomes than fruit flies, and counting them proved difficult. Eventually in 1955 Chinese scientist Joe Hin Tjio established that human cells have 46 chromosomes. Today these can be stained so that their individualities can be observed.

In 1959 French geneticist Jérôme Lejeune found an abnormality in the chromosomes of a condition that had been identified nearly 100 years earlier by an English physician John Down. Sufferers of what is now known as Down syndrome have an additional copy of chromosome 21 in their cells.

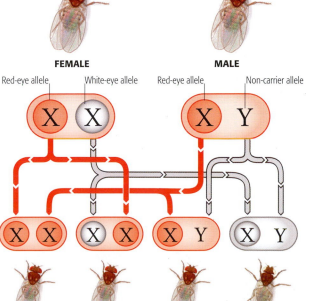

**Fruit-fly chromosomes**
This is a microscopic image of stained fruit-fly chromosomes. The banding corresponds to high and moderate densities of DNA. Characteristics that are often inherited together are determined by genes that sit close together on a chromosome.

**AFTER** »

New techniques in cell biology meant that the genetic material of different organisms could be clearly defined. Later biochemical analysis revealed the sequence of genes on chromosomes—genomes.

**THE DOUBLE HELIX STRUCTURE OF DNA**
In 1953 American geneticist **James Watson** and English geneticist **Francis Crick** established that **deoxyribonucleic acid (DNA)** had a **double helical** molecular structure **346–47** ». The ideas, later supported by experiment, **explained how chromosomes** (and therefore genetic material) **could replicate** during reproduction of life **348–49** ».

**GENOMES: SEQUENCES OF GENES**
In 1988, under the leadership of James Watson, scientists began to **catalogue the locations of genes on chromosomes** for different species.

The first bacterial **genome** (gene sequence) was published in 1995, and the first for an animal (a worm called *Caenorhabditis*) in 1998. In 2006 the gene sequence of the last **human chromosome** was completed.

**CAENORHABDITIS WORM**

**SEE ALSO** »
pp.346–47 THE STRUCTURE OF DNA
pp.348–49 THE GENETIC CODE
pp.386–87 GENE TECHNOLOGY

## Thomas Hunt Morgan's first fruit-fly experiment

When Morgan crossed a white-eyed male fruit fly with a red-eyed female fruit fly, all the subsequent offspring had red eyes, which suggested that the red-eye allele (variety of gene) is dominant and the white-eye allele is recessive.

**1** In the parental generation red-eyed females have two identical alleles on their X chromosomes. White-eyed males have the white-eye allele on their X chromosome, but their Y chromosome has no eye-color gene.

FEMALE WITH RED EYES

Red-eye allele     Eye color

MALE WITH WHITE EYES

White-eye allele     Non-carrier allele

**2** Sex cells form at meiosis (cell division, see p.307), and chromosomes segregate: the eggs all have a single red-eye X allele, while half the sperm have a single white-eye X and half have Y.

X X     X Y

**3** Eggs with red-eye alleles on their X chromosomes are fertilized with either sperm carrying the white-eye allele on the X chromosome or sperm with Y. All the resulting flies have red eyes because the red-eye allele is dominant.

X X     X Y     X X     X Y

FEMALE     MALE     FEMALE     MALE

## Thomas Hunt Morgan's second fruit-fly experiment

When Morgan crossed flies from the first generation, one-quarter of the offspring had white eyes, as predicted by Mendel's Laws (see p.205). The fact that all white-eyed flies were male suggested that the gene for eye color is on the sex (X) chromosomes.

**1** Flies from the first generation (see bottom left) were crossed. Females carry one red- and one white-eye allele on their X chromosomes in every cell. The males have a red-eye allele on only their single X chromosome.

FEMALE     MALE

Red-eye allele     White-eye allele     Red-eye allele     Non-carrier allele

**2** Segregation of the chromosomes at meiosis (cell division) means half the eggs have red-eye allele and half have white-eye allele. Half the sperm have red-eye X allele and half have Y.

X X     X Y

**3** Sex cells fuse at fertilization to produce four possible combinations of chromosomes and alleles in offspring. All females will have red eyes, but half the males will have red eyes and half white.

X X     X X     X Y     X Y

FEMALE     FEMALE     MALE     MALE

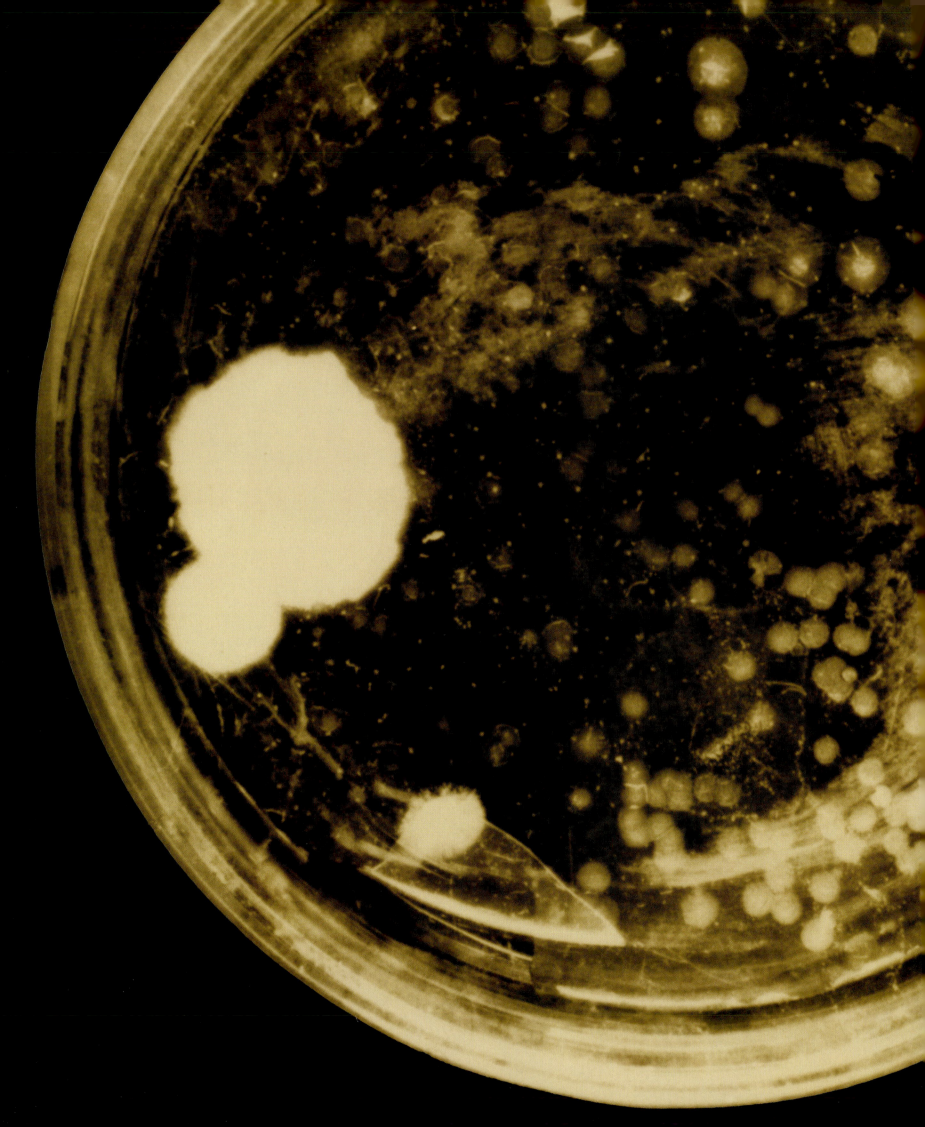

# The Discovery of Penicillin

**The discovery of the antibiotic penicillin remains one of the greatest breakthroughs ever in the fight against disease. It happened by chance in 1928, when Scottish bacteriologist Alexander Fleming noticed something unusual in a laboratory dish. But it was 15 years before the wonder drug could be produced on a large scale. Fleming and two other scientists shared the 1945 Nobel prize for medicine for the discovery.**

Fleming was a noted researcher. In 1923 he identified the enzyme lysozyme—an antibacterial found in tears, mucus, and saliva, which attacks some of the substances in bacterial cell walls. Five years later, working in St. Mary's Hospital, London, he was studying staphylococci—the bacteria responsible for many infectious diseases. He had been away for a few days, and was clearing some used laboratory dishes that contained bacterial colonies. In one dish, he noticed that the gel used for growing bacteria was infected by mold spores, and the area around it was free of bacteria. This is an example of antibiosis—a form of chemical warfare in which one microorganism produces substances (antibiotics) that destroy another. Fleming kept the dish. He identified the mold as *Penicillium* and confirmed that it killed the bacteria by secreting a previously unknown substance, which he called penicillin.

Fleming published his findings in 1929. Initial prospects for exploiting them seemed mixed. The mold was difficult to cultivate, and produced minute amounts of penicillin, which was also difficult to store. Its effects seemed promising, though: in 1930 it was used to cure eye infections. At this stage the drug was only applied externally; Fleming doubted that it would survive internally.

Nearly a decade later at the University of Oxford, Australian pharmacologist Howard Florey and his colleague, German-born chemist Ernst Chain, were researching natural antibacterials and read Fleming's report. Florey and Chain had the chemical expertise to work on the new antibiotic. By 1940 they had enough penicillin to test on mice. The first human to be treated was Albert Alexander. He was suffering from a life-threatening infection and was given a penicillin injection in February 1941. He improved rapidly, but stocks ran out and he died.

With Britain at war, it was difficult to produce the new drug. American drug companies became involved and, while in 1942 there was enough to treat 10 patients, by 1945 output rose to over 600 billion doses a year. Since then many other antibiotics have been developed.

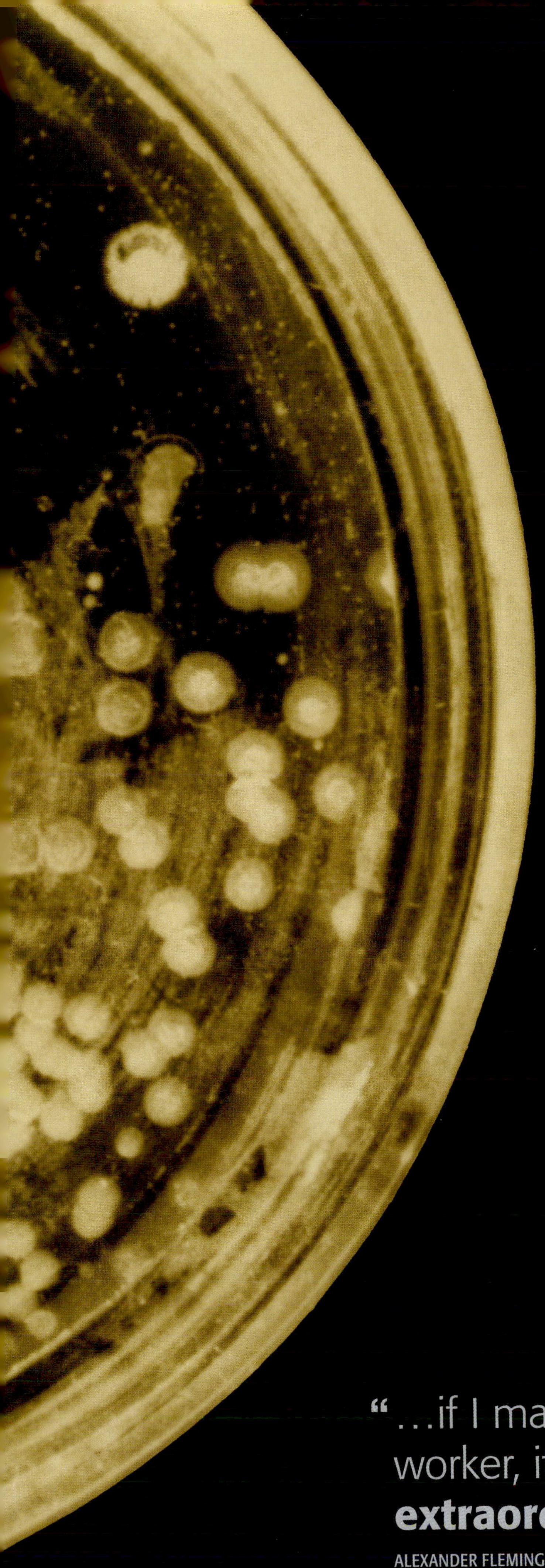

**The past preserved**
A contact print of Fleming's original laboratory dish shows a scattering of bacterial colonies, together with a large growth of *Penicillium* mold, far left, surrounded by an area clear of bacteria. Fleming commented that it was "rather dried up," 25 years after the dish famously caught his eye.

" …if I may offer **advice** to the young laboratory worker, it would be this—**never neglect** an **extraordinary appearance** or happening."

ALEXANDER FLEMING, LECTURE DELIVERED AT HARVARD UNIVERSITY, 1945

**BEFORE**

Until the 19th century, remedies were based almost entirely on natural substances—a tradition that still holds true in herbal medicine across the world.

**EARLY CURES AND REMEDIES**

The Swiss-born alchemist and botanist **Paracelsus ‹‹ 70–71** investigated the uses of chemicals as medicines. He recognized that **poisons** could be beneficial in small doses. He also advocated **learning by experiment**—a heresy in 16th-century medicine, which had held true since classical times.

PARACELSUS

**DIET AND HEALTH**

In 1747 the British naval physician **James Lind ‹‹ 218–19** identified a way of preventing **scurvy**, a disease now known to be caused by a deficiency of vitamin C; he gave sufferers citrus juice. **Vitamin C** was finally isolated in 1932.

**IMMUNIZATION**

After successful trials performed by **Edward Jenner ‹‹ 156–57**, immunization proved to be a powerful weapon in the fight against **smallpox ‹‹ 242–43**. However, it failed to stop killer diseases of the 19th century and had no effect on opportunistic infections such as **pneumonia**.

**‹‹ SEE ALSO**
pp.70–71    Renaissance Medicine and Surgery
pp.228–29   Safer Surgery
pp.244–45   Bacteria and Viruses
pp.310–11   The Discovery of Penicillin

Until the 1800s pure medicines were almost unknown. Instead, most medicines—or remedies—came from a wide range of different ingredients. For example, laudanum, or tincture of opium, was used to reduce pain, and willow bark for headaches and fever. Extracts of foxglove leaves helped to relieve dropsy or edema (fluid retention), one of the symptoms of heart failure.

Chemists began to examine these plant-based medicines, isolating and purifying the substances that made them work. The first of these drugs, morphine—a powerful painkiller—was isolated

**Willow bark**
Acetylsalicylic acid, a substance derived from willow bark, was isolated in 1853. About forty years later, Germany's Bayer laboratories patented the compound, and marketed it as one of the world's most successful drugs—aspirin.

from the opium poppy in 1805 by the German pharmacist Friedrich Sertürner. Codeine came from the same plant in 1832. Foxglove extracts turned out to have active ingredients that improved heart function (cardiac glycosides): digoxin was isolated in 1875. These drugs could be administered in standardized doses and were reliable.

**Antibiotics and synthetics**

With the discovery of penicillin in 1928 (see pp.310–11), antibiotics joined the list of drugs derived from natural sources. Unlike plant-based remedies, they are produced by molds or microorganisms. Penicillin was followed by others such as streptomycin, isolated in 1943, and then tetracyclines in the late 1940s.

# The Development of Medicines

**The last 200 years have seen an explosive growth in medicines, as plant-based remedies have given way to specific biochemicals with precise effects. Some of the greatest breakthroughs have come from chance observations, which have then been put to medical use.**

**Antibiotics in action on a bacterium**
This group of drugs work by blocking vital chemical processes in bacteria, without harming any human cells. They are categorized into five classes, or groups, which reflects the way they act on the cell. Some antibiotics stop bacteria from copying their DNA, or inhibit protein synthesis, which prevents bacteria from multiplying. Other antibiotics act on the cell walls.

**1 Rifamycin**
The antibiotic blocks production of messenger RNA from DNA, killing bacteria by preventing essential protein production.

**Messenger RNA**

**Bacterium DNA**

**2 Quinolones**
This group of antibiotics inhibit DNA replication and transcription, which prevents bacteria from multiplying.

**Ribosome**

**3 Chloramphenicol, erythromycin, tetracyclines, streptomycin**
These drugs bind permanently to bacterial ribosomes, which normally build proteins from the instructions held in messenger RNA. Without proteins, the bacterial cell rapidly dies.

**Flagellum**
Tail-like filament that helps bacterium move.

**Bacterium cell wall**

**Pilus**
Hairy projections through which healthy bacterium exchange genetic material.

**Bacterial protein**
Protein synthesis is essential for bacterium to survive.

**Outer capsule**

**Plasma membrane**

**Plasmid ring**

**4 Polymixin**
This drug binds to the plasma membrane around bacterial cells, causing it to become permeable. Osmotic pressure drives water into the bacterium, eventually killing it.

**Peptoglycan sheet**
A meshlike layer that forms part of the bacterial cell walls.

**5 Penicillins, cephalosporins, vancomycin**
The most widely used antibiotics, these prevent bacteria from forming the peptoglycan sheet in the cell walls. Without complete cell walls, bacteria cannot grow or multiply.

## PAUL EHRLICH

Paul Ehrlich spent most of his career in Berlin, where he studied blood cells with the help of recently discovered synthetic stains. Using these dyes, he discovered the blood-brain barrier—a sheet of cells that selectively shields the brain and spinal cord. Ehrlich realized that stains might lead the way to antibacterial drugs, and began the search for what he called "magic bullets." In 1909 his Japanese assistant, Sahachiro Hata, tested their compound number 606 (Salvarsan), and found that it was deadly to syphilis bacteria. Besieged by demands for the drug, Ehrlich was forced to make it available commercially before any further tests.

### Agonist drugs in action
Most cell activity is controlled by natural chemical messengers that fit into receptors on a cell's surface. If activity is inhibited, an agonist drug can mimic one of these messengers, triggering the receptors, which in turn steps up a process inside the cell.

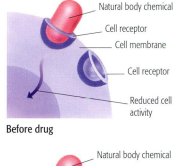

Natural body chemical
Cell receptor
Cell membrane
Cell receptor
Reduced cell activity

Before drug

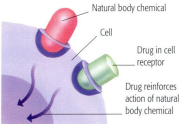

Natural body chemical
Cell
Drug in cell receptor
Drug reinforces action of natural body chemical

After agonist drug

### Antagonist drugs in action
Unlike agonists, these drugs compete with chemical messengers for receptor sites, without triggering the receptors. Once the antagonist drug is in position, the receptors remain blocked. As a result, a process in the cell slows down or stops.

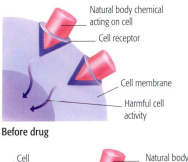

Natural body chemical acting on cell
Cell receptor
Cell membrane
Harmful cell activity

Before drug

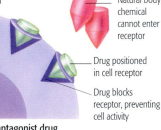

Cell
Natural body chemical cannot enter receptor
Drug positioned in cell receptor
Drug blocks receptor, preventing cell activity

After antagonist drug

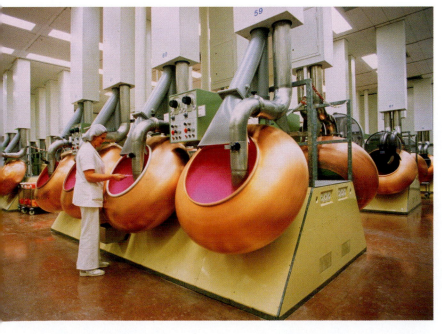

Many of the antibiotics used today are semi-synthetic, meaning that they have been treated to alter their chemical composition to aid absorption or reduce the risk of resistance. Fully synthetic drugs represent a further step along this path, by being entirely man-made.

Synthetic drug production began in the 19th century, using the sticky black sludge known as coal tar. Coal tar contains thousands of carbon-based chemicals, and had already yielded synthetic dyes (see pp.240–41). During the late 1800s chemists discovered its medicinal potential. Among the earliest drugs synthesized from this source were phenacetin, introduced in 1887, and acetaminophen in 1893.

### Chemical drug therapy
During the late 1800s synthetic dyes also found uses in microscopy, as coloring agents or stains. Stains are absorbed in varying amounts by different tissues, making structures easier to see under a microscope: some are strongly absorbed by bacteria, but only weakly by human cells. German immunologist Paul Ehrlich realized that this might lead to a way of targeting drugs. In 1910 he officially announced Salvarsan—a substance that killed syphilis bacteria—the first chemical therapy for a major infectious disease.

Nearly two decades later the German chemist Gerhard Domagk found that a dye called prontosil, which the body converts into a powerful compound called sulphanilamide, was highly effective against streptococci bacteria. In 1940 researchers discovered why. Sulphanilamide resembles a bacterial nutrient and takes its place in bacterial cells. Deprived of a key nutrient, the cells die. This "key-and-lock" concept has been broadened to explain how most drugs work. Agonist drugs

### Sugaring the pill
A technician prepares a pigmented coating during the last step in drug manufacture. Coatings identify drugs and preserve them. Once the drug is swallowed, the coating liberates the drug in a controlled way.

enhance the rate of particular processes in cells, while others (known as antagonists) depress them (see left).

### Boom times
Since World War II, pharmaceuticals have become a multi-billion dollar industry. Thousands of new drugs have been synthesized ranging from pain relievers (analgesics) to highly complex molecules such as insulin analogues—hormone replacements used in the treatment of diabetes.

Many challenges remain. One of the greatest is cancer, which exists in many forms, and complicates the task of drug research. General chemotherapy targets all rapidly dividing cells. Other drugs work more specifically: for example, tamoxifen blocks specific receptor sites on breast cancer cells.

**AFTER** ≫

Drug development requires a detailed knowledge of molecular structure. Recent advances in computer-modeling techniques have helped with this process.

#### ANTIREJECTION DRUGS
Produced by a soil-dwelling fungus tested in 1969, **cyclosporin** was found to **suppress the immune system**. In 1978 it was used in transplant surgery to prevent organ rejection.

**CYCLOSPORIN**

#### ANTIVIRAL DRUGS
The development of antiviral drugs **lagged behind antibiotics**, because viruses replicate inside living body cells. One of the earliest was **acyclovir**, approved for use in the US in 1982, which is highly effective against **herpes viruses**. In the 1980s research centered on HIV. The **first licensed HIV drug**, AZT, entered use in 1985.

SEE ALSO ≫
pp.406–07 MODERN SURGICAL PROCEDURES
pp.408–09 DISEASE CHALLENGES

IN PRACTICE

### DRUG ADMINISTRATION

Oral drug administration to a patient is straightforward, but some drugs are weakened by stomach acids. From the early 1900s the problem was addressed by giving pills resistant coatings, which release the drug after its transit through the stomach. Other methods of administration include suppositories, which are inserted into the rectum or vagina where they dissolve to release the drug, and giving drugs by injection. Disposable syringes first entered use in the 1950s.

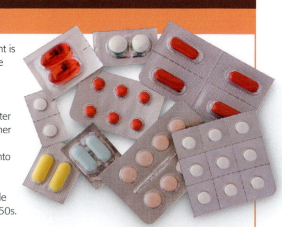

By the late 19th century an increasing amount was known about the structure of the world on the atomic level, but this was revealing flaws in the traditional understanding of many physical problems.

**THE MYSTERY OF LIGHT**
Following the debate between **Newton** ⟪ **110–11** and **Huygens** in the 17th century, and **Thomas Young's** experiments in the early 19th ⟪ **114–15**, light was widely accepted as a form of **wave**, but the medium that carried it

remained elusive. It had been thought that this medium was a substance called "**æther,**" but the Michelson-Morley experiment of 1887 showed **no evidence for æther's existence**.

**ATOMIC STRUCTURE**
Great leaps had been made in explaining the **internal structure of the atom**, from **Thomson's** discovery of electrons ⟪ **286** to **Rutherford's** identification of the atomic nucleus. However, there were major questions about **how electrons behaved** inside the atom.

**SPECTRAL QUESTIONS**
Studies of the **emission of light** from atoms and molecules revealed emission lines with very specific **wavelengths** and **energies**, unique to each atom. Their origins clearly lay **within the atom**, but no one was completely sure how the precise energies were defined.

⟪ **SEE ALSO**
*pp.258–59* ELECTROMAGNETIC SPECTRUM
*pp.286–87* STRUCTURE OF THE ATOM

PHYSICIST (1858–1947)

**MAX PLANCK**

Born into an intellectual family in Kiel, Germany, Max Planck's work in physics led to him being hailed as the "father" of quantum physics. After studying in Munich and Berlin, he became a professor at Kiel in 1885, and then at Berlin University. It was here that he carried out his work on the black body problem, and later welcomed Einstein as a friend and collaborator. After World War I he rejected the Copenhagen interpretation (see p.317) of quantum physics.

# Quantum Revolution

The development of quantum theory in the early 20th century resolved several major puzzles in the physics of the time, and also revealed the behavior of subatomic particles to be totally unlike anything seen in the everyday "macroscopic" world. The implications of the quantum revolution are still being felt today.

Quantum theory began as little more than a mathematical trick. By the beginning of the 20th century there were two rival theories to explain the phenomenon known as the "black body radiation curve." Many people are aware that when light hits objects, darker objects become warmer than

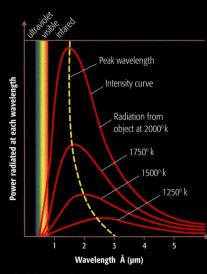

**Black body radiation curve**
The curve for the way in which incandescent objects emit radiation at different wavelengths reveals a peak that shifts to shorter wavelengths at higher temperatures, but it cannot be explained using classical physics.

pale-colored objects; this is because dark objects absorb more of the light's electromagnetic radiation. A black body is an idealized object that absorbs every bit of electromagnetic radiation that hits it. This also makes it a pure emitter of radiation. At any given temperature, the body emits a range of electromagnetic vibrations at various

wavelengths; this range can be plotted as a curve. The equation for one of these early theories described it accurately at longer wavelengths and lower energies, while the other described it accurately at shorter wavelengths and higher energies. The problem was that neither theory could match the overall distribution of radiation measured from such a body in the laboratory.

**Planck's idea**
Then, in 1900, German physicist Max Planck came up with an ingenious new model. He suggested that, because the energy released by a black body must ultimately come from vibrations within the body itself, it would be released only at well-defined and separated, or "discrete," wavelengths. His reasoning was that the energy came from oscillations of individual atoms, and the oscillations, rather like those of a violin string, would only have certain harmonic "modes." Radiation was released when an atom changed its mode of oscillation, and

**Photoelectric effect**
Photons of some forms of electromagnetic radiation provide sufficient power to a metal plate to release electrons from its atoms.

so tended to have distinct wavelengths. For any given temperature, very small or large changes would be less common than those in the middle of the range, so the distribution of wavelengths would form the shape of the black body radiation curve.

Planck's idea set the stage for the revolution to come, and its success in solving the black body problem was widely welcomed, but at the time no one believed it was saying anything very new about the nature of light or matter. It was only thanks to a young research scientist called Albert Einstein (see pp.304–05), who was attracted to the implications of Planck's theory, that the next important step was taken.

**Einstein's quanta of light**
One of Einstein's four great scientific papers of 1905 set out to resolve another contemporary problem—the photoelectric effect. While black body radiation involved the emission of light when a black body was heated (frequently by electricity), the photoelectric effect involved the emission of electricity when light illuminated a charged surface.

The problem was that the electric current produced was not dependent on the intensity of light as much as on its wavelength—for example, a brilliant source of red light might illuminate the surface without any effect, while a

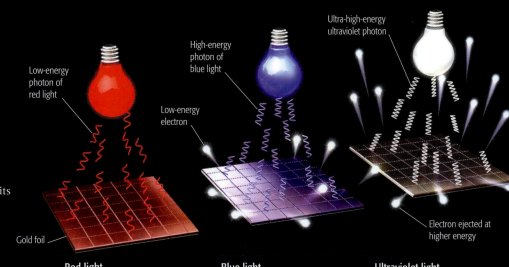

**Red light**
If the photons striking the surface do not have enough energy, the metal does not produce any electrons, however intense the light.

**Blue light**
Blue light is more energetic than red, so its photons have sufficient energy to allow electrons to escape from the surface.

**Ultraviolet light**
The number of electrons released is proportional to the intensity of the light and the number of photons striking the metal.

**1 Individual electrons** striking the screen fall roughly where one would expect from classical physics.

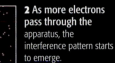

**2 As more electrons pass through the** apparatus, the interference pattern starts to emerge.

**3 Over time a complex pattern** of light and dark bands, formed by constructive and destructive interference of the waves, builds up on the screen.

## Quantum dual-slit experiment

When electrons or other particles are passed through a dual-slit apparatus similar to that of Thomas Young, classical physics would lead us to expect two narrow bands of "illumination." But because these particles have wavelike properties, they will actually interfere and produce a banded effect.

Particle

Particle source

Pair of narrow parallel slits

Fringe pattern on detector screen

wavelengths emitted when certain substances were heated in a laboratory. By modeling the structure of the atom, with electrons orbiting in discrete "shells" whose distance from the nucleus determined their energy, Bohr was able to explain the "emission spectra" of atoms in terms of photons given off as electrons jumped between different orbits (see p.286–87).

By the late 1910s most physicists were starting to accept the quantum model of light. One other advantage was the fact that it did away with the need for a

> "Anyone who is **not shocked** by quantum theory has **not understood it**."
>
> NEILS BOHR, 1958

supposed light-transmitting "ether" in the vacuum of space, therefore meshing neatly with that other great theory of the 20th century— relativity (see pp.300–01). However, Bohr's quantum model of the atom lacked a theoretical foundation, and so far could only accurately predict the emissions from hydrogen, the simplest atom. »

---

feeble source of ultraviolet light would cause a current to flow. If light was a continuous wave, then the effect of either source should be the same.

Einstein's solution was to take Planck's theory one step further, and suggest that light itself was "quantized"—broken down into discrete packets, or "quanta," each with its own wavelike properties of frequency, wavelength, and energy—what we now call photons. The release of an electron from the charged surface, Einstein said, required the absorption of a photon with the right wavelength

and energy, so it didn't matter how many infrared photons bombarded the surface—since the photons did not have a short enough wavelength and high enough energy, no charge would be emitted.

### From Bohr to de Broglie

It took some time for the full implications of this explanation to sink in. One of the first people to grasp that it was saying something fundamental about both the nature of light and the nature of atoms was Danish physicist Niels Bohr, who by 1913 had used it to crack another longstanding problem of light and atoms. This was the precise and unique nature of the light

## SCANNING TUNNELING MICROSCOPE

The scanning tunneling microscope is an ingenious means of studying the surface of materials at an atomic level. It relies on the principle of quantum tunneling (see left), in which the wavefunction of electrons allows them to overcome an apparently insuperable barrier. In this image, a scanner with an extremely sharp probe is brought close to iron atoms (in blue on the image below) on

a copper surface. A voltage applied between the probe and the material causes electrons to flow through the gap. Measurements of the current flow as the scanner is drawn across the surface allow the surface to be mapped out to a precision of less than 0.1 nanometers. The probe can even be used to manipulate atoms on the material.

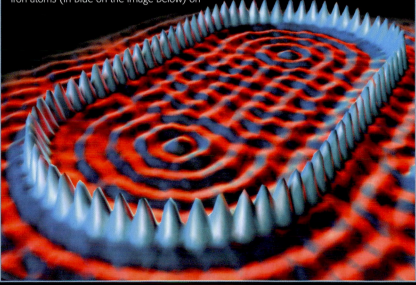

### The classical world

In classical physics, a particle cannot escape from within a "well" without gaining enough energy to get over the walls and out of the other side. This should make phenomena such as radioactive alpha decay impossible.

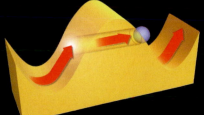

### Quantum tunneling

But the uncertainty principle (see over) makes it possible for the particle to "borrow" energy from its surroundings for a brief instant, allowing it to "tunnel" through the barrier and out of the other side.

THEORETICAL PHYSICIST (1901–76)

## WERNER HEISENBERG

Physicist Werner Heisenberg was born in Würzburg, Germany, and studied at the universities of Munich and Göttingen. Aware of Heisenberg's fascination with atomic physics, his tutor arranged for him to meet Niels Bohr in 1922, marking the beginning of a fruitful relationship. He played a crucial role in developing "matrix mechanics," needed to calculate electron energy levels, as well as the Copenhagen interpretation (see opposite). In World War II he worked on the German nuclear energy project, and afterward became the first director of the Max Planck Institute for Physics.

**Superposition**
The cat exists in a "superposition" of two different states, both alive and dead. As long as the box is not opened, there is a 50 percent chance the cat is still alive.

» But this was only the beginning of the story. The next chapter opened in 1924, when a young French physicist called Louis Victor de Broglie submitted a PhD thesis in which he made a highly daring suggestion: if light could have the properties of both a wave and a particle, then might matter have the properties of both a particle and a wave?

De Broglie backed up his proposal that "wave-particle duality" might extend beyond light to matter itself with a theoretical means of calculating the "wavelength" of a particle. Such a wavelength turned out to be inversely proportional to the mass of the particle in question, and becomes vanishingly small for any object above the atomic scale. But on the subatomic level the wavelength was measurable, and by 1927 scientists at two separate laboratories had shown that electrons diffracted and interfered with one another in the same way as photons of light.

### Exclusion and entanglement
Meanwhile, two young Austrian physicists added their ideas. The first was Wolfgang Pauli, who put Bohr's model of the hydrogen atom orbitals on a firmer footing with his "exclusion principle" of 1925. Reasoning that the overall "quantum state" of a particle was described by certain properties, each with a fixed number of possible discrete values, Pauli declared that it was impossible for two particles in the same system to occupy the same quantum state at the same time.

In order to explain the pattern of electron shells that was apparent from the periodic table (see pp.230–31), Pauli realized that the electrons must be defined by four quantum numbers,

**1 Particles are independent**
In this example, a laser beam is fired at two unentangled particles.

**2 Particles become entangled**
Action of the laser beam forces the particles into an entangled state.

**3 Particles operate as one**
The particles can be separated by any distance, but remain entangled. Manipulating one particle affects the other instantaneously.

**4 The bond is broken**
However, a manipulation that reveals the state of the particles causes entanglement to break down.

**5 All is revealed**
Once the link between the particles is broken, the state of each of them becomes evident once again.

Laser beam

**Quantum entanglement**
Quantum entanglement involves separating a pair of particles that are known to be in different quantum states because of the Pauli exclusion principle. This makes it possible to manipulate one particle instantaneously by altering the other one, and to extract information about both particles at the same time.

**Schrödinger's cat**
Schrödinger used this "thought experiment" to point out a problem in the Copenhagen interpretation: while the box remained unopened, the cat would hover in an indeterminate state between life and death.

which added a further property to the three already recognized. This property came to be known as "spin." It became clear that spin values separated two distinct families of fundamental particle: "fermions," with spins that have half-integer values, and "bosons," with spins that are whole-number values or zero. All matter particles are fermions, and obey the exclusion principle, while bosons, which do not obey this principle, are "force-carrying"

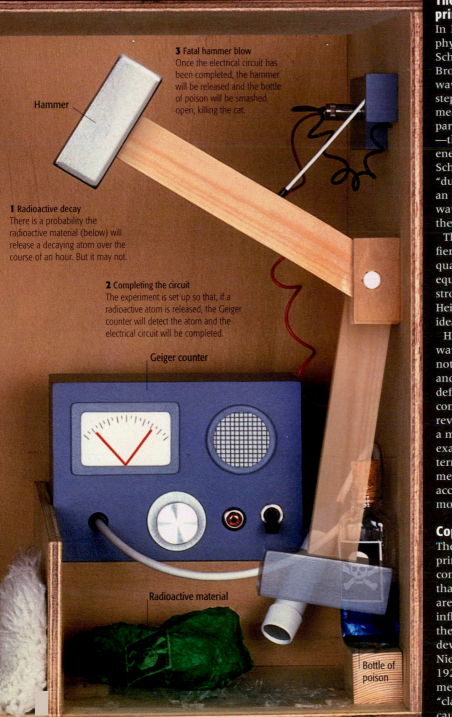

**3 Fatal hammer blow**
Once the electrical circuit has been completed, the hammer will be released and the bottle of poison will be smashed open, killing the cat.

Hammer

**1 Radioactive decay**
There is a probability the radioactive material (below) will release a decaying atom over the course of an hour. But it may not.

**2 Completing the circuit**
The experiment is set up so that, if a radioactive atom is released, the Geiger counter will detect the atom and the electrical circuit will be completed.

Geiger counter

Radioactive material

Bottle of poison

## The uncertainty principle

In 1926 Austrian physicist Erwin Schrödinger took de Broglie's theories of wave-particle duality a step further by devising a means of calculating a particle's "wavefunction" —the distribution of its energy in space. Schrödinger argued that "duality" was not really an issue, and that the wave alone represented the reality.

This approach sparked fierce debates about the meaning of quantum theory, but Schrödinger's equation also allowed one of his strongest opponents, Werner Heisenberg, to develop another important idea —the "uncertainty principle."

Heisenberg realized that the wavefunction meant a particle could not be "localized" to a point in space and at the same time have a well-defined wavelength; a wave that is concentrated at a specific point cannot reveal its wavelength, and a wave with a measurable wavelength cannot be exactly located in space. In particle terms, the more accurately one measures a particle's position, the less accurately one can know what its momentum is.

## Copenhagen and beyond

The interpretation of the uncertainty principle and the wavefunction remain controversial, with deep implications that are as much philosophical as they are scientific. Perhaps the most influential approach to the problem is the "Cophenhagen interpretation," developed in Denmark by Heisenberg, Niels Bohr, and others in the late 1920s. They believed that the act of measurement and interaction with a "classical" (i.e. large-scale) system caused the wavefunction of a particle to collapse, and that the very nature of the measurement influenced the way in which the wavefunction collapsed.

This explains, for example, why it is possible to devise an experiment that measures either the wavelike or the particle-like properties of an electron or photon, but not both at the same time. Taken to extremes, it implies that no subatomic event is decided on until it is observed—an idea that led Schrödinger to invent the "Schrödinger's cat" thought experiment (see left).

However, the Copenhagen interpretation is far from being the only way to resolve the uncertainty principle. Other ideas include the "ensemble interpretation," which states that quantum physics only really

**Electron cloud**
Thanks to quantum physics, modern chemists know that the atomic structure of elements such as helium is in fact composed of clouds of probability rather than particles in distinct orbits.

applies to large samples of particles; the "objective collapse" theory, in which the wavefunction collapses regardless of whether it is measured or not; and the "Bohn interpretation," in which particles always have discrete positions, but these are guided by the wavefunction.

However, perhaps the best-known alternative is the "many worlds" interpretation, in which the different possible results of the collapse all come true in mutually unobservable "histories"— effectively an infinite number of universes representing every possible quantum outcome.

## AFTER »

Although many of the findings of quantum theory seem obscure, they have actually had some important implications for the modern world.

### ATOMIC ENERGY
Understanding of the **atom's deep structure** and the quantum principles behind **radioactive decay** has led to the development of **nuclear fission 322–23 »** for both bombs and power stations. In the future, this same knowledge may open the way for clean and unlimited energy from **nuclear fusion**.

**NUCLEAR POWER STATION**

### UNDERSTANDING LASERS
Quantum physics also underlies **laser technology 364–65 »**, in which cascades of intense, synchronized, or "coherent" **photons** are generated in **tightly focused beams**. Lasers are used for a variety of medical, scientific, and manufacturing applications.

### QUANTUM COMPUTING
In the future, it is hoped that quantum concepts such as **entanglement** of quantum bits (**qubits**) and **superposition** of multiple states may be used to solve **complex problems** rapidly, beyond the ability of traditional computers, heralding a new generation of **superfast computers**. So far quantum computing **experiments** have been limited to manipulating small amounts of data, but the potential is **vast**.

particles, such as photons. Since Pauli's four quantum numbers defined the orbital shell and energy level in which an electron moved, they provided a neat theoretical explanation of how electrons distribute themselves around the nucleus of an atom, the transitions they make between energy levels, and therefore their emission spectra and even chemical reactions. The exclusion principle therefore underpins all of modern chemistry, and has other significant consequences—most notably the phenomenon of "entanglement," by which information in one place can effectively be transferred to another instantaneously. However, while the exclusion

> "The **more success** the quantum theory has, the **sillier it looks**."
>
> ALBERT EINSTEIN TO HEINRICH ZANGER, 1912

principle was a huge advance in the understanding of subatomic particles, other events would soon alter the way in which these particles were themselves considered.

**A universe of galaxies**
An estimated 10,000 galaxies have been caught on camera in this Hubble Space Telescope image. The patch of sky viewed is just one-tenth the diameter of the full Moon and appears largely empty in ground-based images.

# The **Expanding** Universe

**At the start of the 20th century, astronomers believed the Universe was a finite size and that it contained only one galaxy, the Milky Way. By 1930 the American astronomer Edwin Hubble had shown that the Universe contains many galaxies all moving away from us.**

For decades astronomers had observed a number of fuzzy-looking objects they termed "nebulae," which were believed to be part of the Milky Way. Hubble observed these in the 1920s using the 100-inch Hooker telescope, on Mount Wilson, California, then the largest telescope in the world.

Hubble established distances to the nebulae by observing Cepheid stars (usually giant yellow stars) within them. It had been shown that the regular variation in brightness of these stars could be used to determine distance. His calculations revealed that most nebulae were outside the Milky Way, which led him to suggest these were independent galaxies and that the Universe was millions of times bigger than previously thought.

Hubble and other astronomers, notably American Vesto Slipher, had also been looking at dark lines in the light spectra of these galaxies. Slipher had found the lines were displaced toward the red end of the spectrum, a displacement known as "red shift." This implied that galaxies were speeding away from us and that the universe was expanding.

In 1929 Hubble noted a relationship between red shift and distance; the more distant the galaxy, the faster it is receding. This is now known as "Hubble's Law." Hubble's initial value for the expansion rate, now known as the Hubble Constant, was around 300 miles (500 km)/sec per 3.26 million light years. This was soon proved wrong since it indicated the Universe was younger than Earth. The figure is now believed to be around 45 miles (72 km)/sec per 3.26 million light years.

Hubble's findings proved what earlier astronomers had theorized about. He opened the way for astronomers to speculate on how expansion began and, in time, to develop the Big Bang theory of the origin of the Universe.

### Red shift

Dark lines within an object's light spectrum shift depending on whether the object is moving away from or toward us. When a galaxy is moving away the wavelengths are stretched out (red shift). If it is moving closer the wavelengths are squashed up (blue shift).

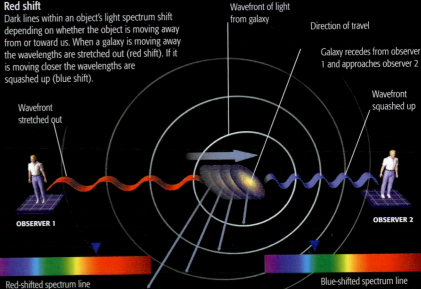

Wavefront of light from galaxy

Direction of travel

Galaxy recedes from observer 1 and approaches observer 2

Wavefront stretched out

Wavefront squashed up

OBSERVER 1

OBSERVER 2

Red-shifted spectrum line

Blue-shifted spectrum line

## "The **history** of **astronomy** is a history of receding **horizons**."

EDWIN HUBBLE, AMERICAN ASTRONOMER, 1889–1953

« **BEFORE**

Cosmology—the study of the origin and structure of the Universe—did not exist as a scientific subject before the 20th century.

### THEORIES OF CREATION

Traditionally, **different cultures** and religions had their own views of the creation of the Universe. Many believe that it was created by a god.

### BELIEF IN ONE GALAXY

Before 1920 people thought the Universe consisted of one huge galaxy and that the solar system lay near its center. Astronomers used **photography and improved telescopes**

to prove both these facts wrong during the early decades of the 20th century.

« SEE ALSO
*pp.74–75* THE SUN-CENTERED UNIVERSE
*pp.318–19* THE EXPANDING UNIVERSE

**"GOD OVERLOOKING HIS EARTH-CENTERED UNIVERSE," FROM THE LUTHER BIBLE, 1534**

# The Big Bang

**Scientists questioned the origin of the Universe for the first time in the 20th century. Could it really have existed in one form, and forever—or was there a starting point, and a process of continual change? The evidence pointed more and more toward the Big Bang, and a Universe only 13.7 billion years old.**

Until around 100 years ago the Universe was thought not to change over time. But during the 20th century scientists collected evidence proving that the Universe is not static, but is in a constant state of expansion and change. Its expansive state is linked to an explosive event known as the Big Bang.

### Origin of the Big Bang theory

The seeds for the new view of the Universe were planted in 1916 when Albert Einstein published his theory of general relativity, from which the Dutch astronomer Willem de Sitter developed an imaginary expanding Universe. The idea became a reality in 1929, when Edwin Hubble (see p.341) showed the Universe is expanding, which means that it was once smaller and denser. It was, however, Belgian Georges

Lemâitre's suggestion, in 1931, that the Universe's origin was a primeval "cosmic egg" that exploded, creating an expanding universe, which served as the first Big Bang model.

### The evidence mounts

In the 1940s Fred Hoyle, together with Americans Hermann Bondi and Thomas Gold, put forward an alternative view of the Universe. The "Steady State" theory proposed that the Universe appears the same in every location and at all times. It has no beginning or end, and matter is continuously created.

In 1948 George Gamow outlined how the relative proportions of the hydrogen and helium in today's Universe could be produced in a Big Bang Universe. The case for the Big Bang strengthened after 1955, when Martin Ryle showed that the distant, older radio galaxies were

more numerous and densely packed than those nearby. This disproved a vital characteristic of the hypothetical Steady State Universe, in which density would always stay the same.

Direct evidence for the Big Bang came a decade later. George Gamow had predicted that a remnant of the Universe's initial radiation would remain: the microwave background that would have permeated all space, before starting to cool. This was detected in 1964 by Arno Penzias and Robert Woodrow Wilson, confirming that the Universe was intensely hot in the past.

### The start of the Universe

We now think that the Big Bang occurred 13.7 billion years ago, starting our universe. It produced all energy, matter, and space, and marked the recognized start of time. At the outset

**Origin of the Universe**
The Big Bang produced all the energy, matter, and space in the Universe.

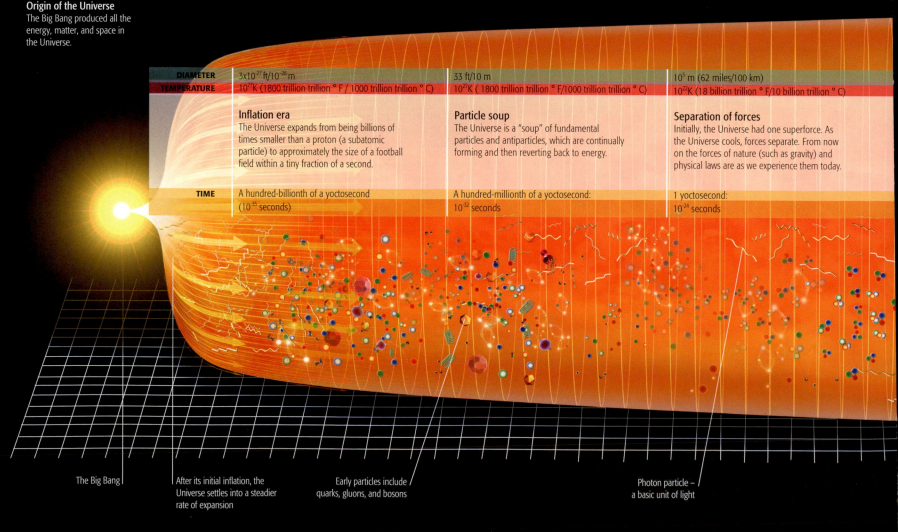

| DIAMETER | $3\times10^{27}$ ft/$10^{-26}$ m | 33 ft/10 m | $10^5$ m (62 miles/100 km) |
|---|---|---|---|
| TEMPERATURE | $10^{27}$K (1800 trillion trillion ° F / 1000 trillion trillion ° C) | $10^{27}$K ( 1800 trillion trillion ° F/1000 trillion trillion ° C) | $10^{22}$K (18 billion trillion ° F/10 billion trillion ° C) |

**Inflation era**
The Universe expands from being billions of times smaller than a proton (a subatomic particle) to approximately the size of a football field within a tiny fraction of a second.

**Particle soup**
The Universe is a "soup" of fundamental particles and antiparticles, which are continually forming and then reverting back to energy.

**Separation of forces**
Initially, the Universe had one superforce. As the Universe cools, forces separate. From now on the forces of nature (such as gravity) and physical laws are as we experience them today.

| TIME | A hundred-billionth of a yoctosecond ($10^{-35}$ seconds) | A hundred-millionth of a yoctosecond: $10^{-32}$ seconds | 1 yoctosecond: $10^{-24}$ seconds |
|---|---|---|---|

The Big Bang

After its initial inflation, the Universe settles into a steadier rate of expansion

Early particles include quarks, gluons, and bosons

Photon particle – a basic unit of light

AFTER

## FRED HOYLE

Sir Fred Hoyle was one of the greatest scientific thinkers of the 20th century. As the main exponent of the Steady State theory, this Englishman single handedly stimulated the worldwide development of cosmological research in the 1950s and 1960s. His demonstration of the ways in which heavier elements are made in stars was equally significant. He coined the phrase "Big Bang" while making a radio broadcast in early 1950.

**300,000 YEARS AFTER THE BIG BANG (BEFORE STARS)**

Hydrogen 76% — Helium 24% — Trace of lithium

**TODAY, AFTER MANY CYCLES OF STAR BIRTH AND DEATH**

Hydrogen 74% — Helium 23%

Oxygen 1%
Carbon 0.5%
Neon 0.5%
Iron 0.1%
Nitrogen 0.1%
Traces of other elements

the Universe was dense, hot, and contained pure energy. It rapidly ballooned in size, before settling to a steadier rate of expansion. It consisted of tiny particles of energy, which became tiny particles of matter within seconds. Within minutes the Universe had turned almost entirely into the nuclei of hydrogen and helium atoms.

### The first stars

Over the next 300,000 years or so the Universe expanded and cooled. When the temperature had dropped enough, the first atoms formed and the Universe became a transparent mix of hydrogen and helium, in a 3:1 ratio. This matter was unevenly distributed and, over millions of years, density irregularities

### Composition of the Universe
The early Universe consisted of hydrogen and helium, with a trace of lithium. Today it still consists mainly of hydrogen and helium, along with other chemical elements that were created in the stars.

formed into clumps, giving birth to the first stars and galaxies, possibly as early as 200 million years after the Big Bang.

The first stars were vast, and made almost entirely of hydrogen and helium. Nuclear reactions deep within their cores produced new chemical elements such as carbon, oxygen, and silicon which were dispersed into space after the stars' death. The Universe's chemical mix slowly changes as successive stars produce elements in this way, or as they die explosively.

The Big Bang theory continues to be refined. Yet, the biggest question of all—"what came before?"—remains completely unanswered.

### EXPLORING THE COSMIC BACKGROUND
In 1992 the Cosmic Background Explorer satellite (COBE) recorded tiny variations in the intensity of the **cosmic microwave background**, from which present-day galaxies and clusters grew.

### HADRON COLLIDER
Physicists are using the world's **largest scientific instrument**, the Large Hadron Collider **400–01 »**, to study **particle collisions** that would have occurred immediately after the Big Bang.

> **COSMIC MICROWAVE BACKGROUND RADIATION**
> The dying heat of the Big Bang. It has a temperature of –454° F (–270° C). It cools as the Universe expands.

SEE ALSO »
pp.324–25 FISSION AND FUSION
pp.398–99 DARK UNIVERSE

| 60 billion miles/100 billion km | 1,000 light years | 100 million light years |
| 10¹³K (18 trillion ° F/10 trillion ° C) | 10⁶K (180 million ° F/100 million ° C) | 3000K (4900° F/2700° C) |

**First protons and neutrons**
The Universe has cooled enough for particles to combine, producing the first protons and neutrons. During the next three minutes these combine to form atomic nuclei of hydrogen and helium.

**Opaque era**
For the next 300,000 years or so, matter particles are in a continual state of interaction with photon particles. The movement of the photons is restricted and the Universe appears "foggy."

**Matter era**
Protons and atomic nuclei begin to capture electrons and form the first atoms. The photons now travel freely through the Universe as radiation, making the Universe transparent.

1 microsecond:
10⁻⁶ seconds (1 millionth of a second)

200 seconds

300,000 years

Quarks combine with gluons to make protons and neutrons

Helium-3 nuclei and other atomic nuclei are formed by colliding protons and neutrons

Electron, a subatomic particle with negative charge

Helium-4 nucleus

Hydrogen atom (one proton and one electron): nine times more numerous than any other atom

Helium atom (two protons, two neutrons, two electrons)

# The First Atom Bomb

**On July 16, 1945, a huge fireball rose into the skies above the Alamogordo Test Range in the New Mexico desert. The "Trinity Test" was the culmination of decades of research on atomic structure and the nature of radioactivity. With energy equivalent to 16,000 tons of TNT, this was the first artificial nuclear explosion—herald of a new technology that would transform the world, for good or ill.**

As early as 1919 physicist Ernest Rutherford had shown that it was possible to trigger "transmutations" from one element to another artificially, by bombarding a material with alpha particles. It was clear that "splitting the atom" (see pp.324–25) could release enormous energy, but scientists had no idea how this might be done in practice on a large scale.

Then in 1932 came confirmation of the existence of the neutron, followed by the realization that bombardment with high-energy neutrons was a powerful way to trigger radioactive decay and release further neutrons in the process. In 1933 Hungarian-born physicist Leo Szilard suggested that, in the right conditions, a densely packed "critical mass" of radioactive material could undergo a "chain reaction," with neutrons from one decay event triggering others and potentially releasing a huge amount of energy.

Certain heavy elements were known as potential sources of radioactive emissions, and in the mid 1930s experiments by the Joliot-Curies and Enrico Fermi successfully created artificial radioactivity by bombarding stable heavy elements with alpha particles. In 1938 German physicist Otto Hahn discovered an isotope of barium produced by the breakdown of uranium. His former student Lise Meitner and her nephew Otto Frisch showed how this could be the result of "nuclear fission"—the breaking apart of heavy isotopes to form much lighter ones. This fission of heavy elements clearly offered the potential for Szilard's chain reaction.

On the brink of World War II eminent physicists urged US President Franklin D. Roosevelt to obtain this technology before Germany did. The resulting Manhattan Project brought together many of the finest minds in 20th-century physics—under the scientific direction of American physicist J. Robert Oppenheimer—to figure out the practicalities of nuclear reactors and weaponry, and the means of refining and concentrating "fissile" material. The test detonation was watched in secrecy by fewer than 300 people, but people all around the world were to see images of the "mushroom cloud" over Hiroshima, Japan, just three weeks later.

> **"**I am become **death**, the **destroyer of worlds.**"
>
> J. ROBERT OPPENHEIMER

**The Gadget**
The first offspring of the Manhattan Project was known as "the Gadget," shown here with US physicist Norris Bradbury. This makeshift-looking device was covered in conventional explosives that, when detonated, compressed the core of plutonium to reach a critical density.

## BEFORE

The exploitation of nuclear energy was the culmination of several breakthroughs in the understanding of atoms.

### ATOMIC STRUCTURE
In the early 1900s **Ernest Rutherford** and his team began to explore the **internal structure of atoms**. Frederick Soddy identified the existence of **isotopes ≪ 299** and noted that heavier isotopes tended to be radioactive.

### NEUTRON BREAKTHROUGH
**James Chadwick's** 1932 discovery of **neutrons ≪ 287** provided the means to **manufacture radioactive elements**.

**≪ SEE ALSO**
*pp.286–87* STRUCTURE OF THE ATOM
*pp.298–99* RADIATION AND RADIOACTIVITY
*pp.322–23* THE FIRST ATOM BOMB

# Fission and Fusion

There are two possible ways to extract energy from atomic nuclei—either by fission (splitting them apart) or by fusion (joining them together). Fusion holds the enticing promise of unlimited clean energy, if it can be harnessed in a controlled way, but so far only nuclear fission has been used in producing energy.

The principles of nuclear fission were figured out through the 1930s and put to the test during World War II. The bombs that were detonated over Hiroshima and Nagasaki relied on an uncontrolled fission "chain reaction"—a phenomenon first identified by Hungarian-born scientist Leo Szilard as early as 1933. The principle was simple. Within a dense enough sample of material (a "critical mass"), high-energy neutrons escaping from one nuclear decay would trigger another, releasing further high-energy neutrons, and so on, causing all the material to decay instantaneously with an enormous burst of energy.

## Fission versus fusion
Fission releases energy because the mass of the fragmented products from each decay is slightly less than the mass of the original atom—the lost mass is converted directly into energy in accordance with Einstein's equation $E = mc^2$ (see p.300). As physicists learned more about the structure of particular atomic nuclei, it became clear that this release of energy occurred only in the fission of elements heavier than iron. Lighter nuclei, in contrast, absorb energy when they are split, but the process of joining them together—nuclear fusion— has the potential to release larger amounts of energy than fission. Fusion is far harder to achieve than fission, since nuclei can only be forced together at tremendous

temperatures and pressures. In nature this happens in the cores of stars. However, after World War II scientists striving to develop more powerful weapons for the encroaching Cold War realized they had already built a means of producing such conditions—the fission bomb. The hydrogen bomb, first tested over the Pacific Ocean in the 1950s, relies in essence on

channeling the explosive force, radiation, and neutrons from a small fission "trigger" into a secondary charge of fusion fuel in order to create a reaction.

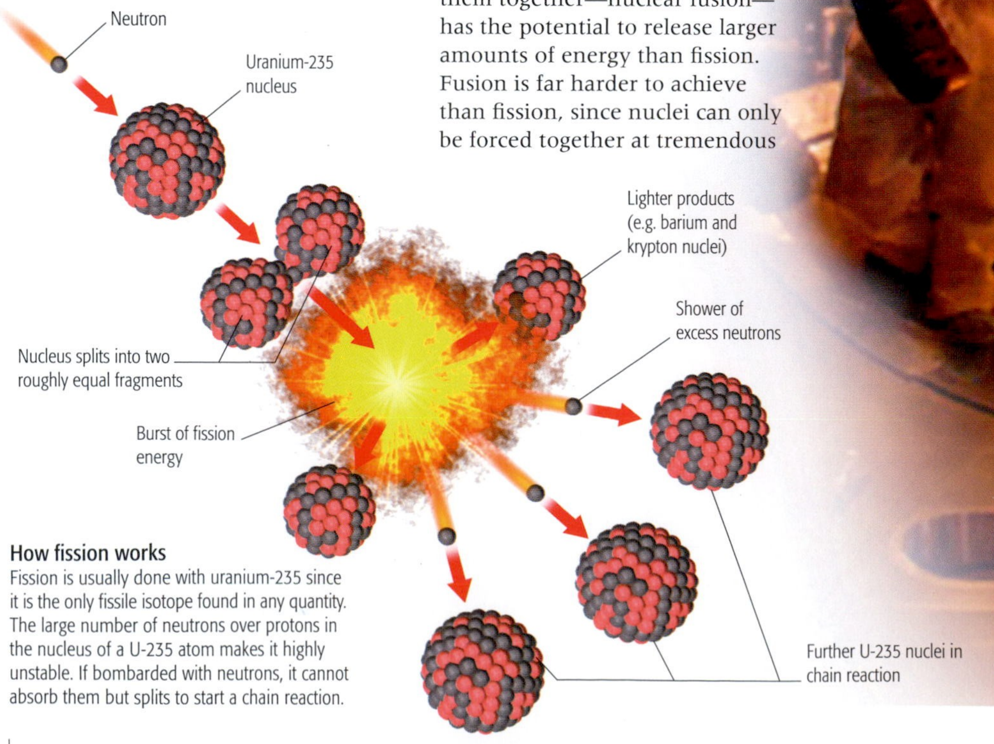

**Cockroft–Walton generator**
In 1932 British physicists John Cockroft and Ernest Walton used this early form of particle accelerator to bombard lithium with high-energy protons, splitting it to produce helium—the first artificial nuclear fission.

Neutron

Uranium-235 nucleus

Nucleus splits into two roughly equal fragments

Burst of fission energy

Lighter products (e.g. barium and krypton nuclei)

Shower of excess neutrons

Further U-235 nuclei in chain reaction

**How fission works**
Fission is usually done with uranium-235 since it is the only fissile isotope found in any quantity. The large number of neutrons over protons in the nucleus of a U-235 atom makes it highly unstable. If bombarded with neutrons, it cannot absorb them but splits to start a chain reaction.

## Putting fission to work

The first fission reactor, "Chicago Pile-1," began operation as part of the Manhattan Project in December 1942. By the mid-1950s nuclear power plants, able to extract energy from controlled reactions and convert it into electricity, were running in both the US and the Soviet Union.

### Nuclear fusion test facility

Attempts at fusion usually involve a doughnut-shaped "torus," such as this one at the KSTAR facility in South Korea, which compresses and heats fusion fuel until conditions mimic those in the core of the Sun.

Regulated nuclear fission involves a core containing pellets of enriched radioactive fuel, usually moderated by an intervening material (such as water or graphite) that slows the movement of the "fast neutrons" that are released, making them more likely to interact with other atoms and sustain the chain reaction. Control rods made from neutron-absorbing materials are inserted or removed to manage the flow of neutrons and the rate of the chain reaction. Power plants are typically designed so that the reaction will be automatically halted in the event of a catastrophic failure. They usually extract energy by allowing the

reaction to heat water in the reactor pile, and transferring that heat to spin the blades of steam turbines.

### Fusion power

While fission has proved effective in generating energy, attempts to harness fusion have been less successful. The usual method is to heat and compress the fusion fuel using electrical and magnetic fields, but producing the right conditions for controlled fusion tests modern science and engineering to the limit. So far none of the prototype reactors has been able to produce more energy than it consumes. Translating the process to an industrial scale presents further challenges. However, the potential for fusion power, which can generate energy from common elements with none of fission's harmful waste products, is ample for the foreseeable future.

## IN PRACTICE

### USS *ENTERPRISE*

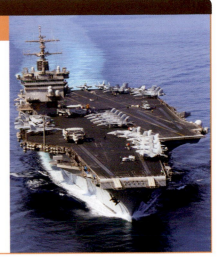

Almost as soon as they became feasible, nuclear reactors were adopted by both the US and Soviet navies. Building compact reactors was a major challenge, but their efficiency and low fuel use made them ideal power sources both above the waves and below during the tense years of the Cold War. They radically reduced the need for refueling and allowed ships and nuclear-armed submarines to stay at sea for months at a time. Perhaps the most impressive of all nuclear surface ships is the USS *Enterprise*: commissioned in 1961 and still in service today, she carries eight nuclear reactors.

Deuterium nucleus:
1 proton, 1 neutron

Fusion releases energy

Excess neutron

Excess energy

Tritium nucleus:
1 proton,
2 neutrons

Helium nucleus:
2 protons,
2 neutrons

### How fusion works

Fusion is easier to achieve with heavier isotopes of hydrogen such as deuterium and tritium. When forced together, these merge to form helium nuclei, releasing a burst of energy in the process.

**The Sun**
The core of the Sun can reach temperatures of 15 million °C and densities of 165 tons per cubic meter—enough to trigger nuclear fusion.

## AFTER

Although controversial, nuclear power is likely to be with us for years to come.

### CAUSES FOR CONCERN

Nuclear power stations produce **radioactive waste** from fission reactors, that remains radioactive for centuries, and is a hazard to transport and store. Another problem is that **accidents can be deadly**. The 1957 fire in the British reactor at Windscale, the 1979 escape of radioactive gas from the Three Mile Island plant in Pennsylvania, and the 1986 explosion at Chernobyl in Ukraine, all increased **public fear** about the safety of nuclear power.

### CLIMATE CHANGE

Nuclear fission is a viable **alternative to fossil fuels** as an energy source, so could help **cut global warming**. Some nuclear reactors can be used to "breed" radioactive nuclei, potentially making the process **self-sustaining** and reducing nuclear waste.

**TRANSPORTATION OF NUCLEAR WASTE**

SEE ALSO >>
pp.414–15 GLOBAL WARMING
pp.416–17 RENEWABLE ENERGY

PHYSICIST **Born 1918** Died 1988

# Richard Feynman

## "I think I can safely say that **nobody** understands **quantum mechanics**."

RICHARD FEYNMAN, FROM "IN CHARACTER OF PHYSICAL LAW," 1964

The American physicist Richard Feynman combined an astonishing aptitude for solving impenetrable problems in physics with being a brilliant teacher and communicator. He is credited with the development of quantum electrodynamics—the theory of how electrically charged particles (such as electrons) interact with photons (minuscule packages of electromagnetic radiation). He also made major contributions in other areas of physics, including superfluidity and the development of the quark model of particle physics.

### Early life and career
Feynman was born into a Jewish family in New York City in 1918. From an early age he showed a strong interest in scientific matters. In his school years he enjoyed tinkering with radios, clocks, and any other gadgets he could lay his hands on. He obtained a degree in physics from

MIT (the Massachusetts Institute of Technology). While completing a PhD in the same subject at Princeton, Feynman was approached in 1941 to work on the Manhattan Project—the secret development of a weapon based on nuclear fission (an "atomic bomb") at Los Alamos, New Mexico. He initially refused, but then agreed to participate, worried by thoughts of what might happen should Nazi Germany first develop such a weapon. Much of his time on the project was involved in supervising computational work, although he also developed the formula for calculating the yield of a fission bomb.

Throughout his time at Los Alamos, Feynman gained respect from many of the top minds in physics because of his openness to new ideas and willingness to speak his mind, argue with distinguished scientists such as Niels Bohr, and challenge ingrained ideas.

### The Manhattan Project
Assigned to the top-secret Manhattan Project, Feynman conducted bomb research at the remote National Laboratory site at Los Alamos, New Mexico. He and the many other personnel involved were accommodated in the site's sprawling trailer park.

Richard P. Feynman

### The postwar years
After World War II, in the fall of 1945, Feynman was appointed professor of theoretical physics at Cornell University. Eventually he was to

**Richard Feynman**
This 1959 photograph of Feynman, in characteristically animated mood, was taken during one of his lectures at the California Institute of Technology (Caltech), Pasadena.

## FEYNMAN DIAGRAMS

The Feynman diagram is a pictorial method that he invented to describe the possible interactions between subatomic particles (see pp.382–83). Feynman diagrams are widely used today by particle physicists.

The diagrams represent fermions (matter particles and their antiparticles, such as electrons and positrons), as straight lines. Bosons ("messenger" particles), such as photons, are depicted by means of wavy lines. In each diagram, time is represented along one axis and space along another axis. Each point of intersection indicates an interaction between two particles, but is also a type of shorthand notation for a series of (often highly complicated) equations. The diagrams represent particle interactions in a similar way that curves on ordinary graphs can represent mathematical equations.

One application of the diagrams has been to calculate the probabilities of various particle interactions happening in particle accelerators such as the Large Hadron Collider, built near Geneva, Switzerland.

**Examples of Feynman diagrams**
The top diagram represents electromagnetic repulsion: two electrons (e-) approach, exchange a photon, and then move apart. In the lower diagram, an electron (e-) and positron (e+) annihilate to produce a photon, which turns into a muon (μ-) and antimuon (μ+).

return to developing ideas in quantum electrodynamics, or QED. This research continued after he became professor of theoretical physics at the California Institute of Technology (Caltech) in 1950.

### Mid-life researches

QED is the most precise and successful theory in physics and is important to the deep understanding of phenomena involving electromagnetic radiation and electrons. This includes all of chemistry but also such questions as why the sky is blue, why diamonds are hard, and how the retina in the eye works. While at Caltech, Feynman also worked to explain superfluidity—a phenomenon in which liquid helium, when cooled close to absolute zero, is able to flow without viscosity. In 1965 Feynman was awarded (jointly with

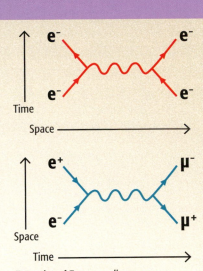

**Shuttle Commission public hearing**
Feynman uses his hands to illustrate a point about the functioning of O-ring seals during the public hearing into the causes of the Space Shuttle *Challenger* disaster, in which the seven crew members died.

Julian Schwinger and Sin-Itiro Tomonaga) a Nobel Prize for his work on quantum electrodynamics.

In the late 1960s Feynman developed a model of the subatomic world called the theory of "partons," which was a precursor of the modern theory of quarks. Meanwhile, he gained fame for brilliant and entertaining lectures.

### Later years

In the 1970s and early 1980s Feynman helped introduce nanotechnology (creation of devices at the molecular scale), and pioneered the development of quantum computing, in which the switches, or bits, in a computer exist in more than one state simultaneously.

He died on February 15, 1988 in Los Angeles as a result of abdominal cancer. His last words are said to have been either "I'd hate to die twice. It's so boring" or "This dying is boring."

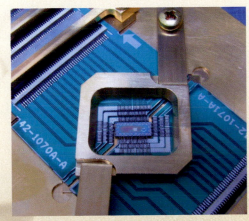

**Microchip from a quantum computer**
Digital computers work by manipulating bits that are limited to one of two states: 0 and 1, or off and on. In quantum computers off and on occur simultaneously, making their processing speed millions of times faster.

- **May 11, 1918** Richard Feynman is born in New York City. In childhood he is fascinated with science and puzzle solving.

- **1939** Graduates in physics from MIT (Massachusetts Institute of Technology).

- **1942** Completes PhD at Princeton University in New Jersey. Marries Arline Greenbaum, even though she has been diagnosed as having tuberculosis and his family oppose the marriage.

- **1942–45** Joins the Manhattan Project (to develop the first nuclear bomb). As a junior physicist, Feynman is not central to the work.

- **June 16, 1945** Suffers a personal tragedy as his wife Arline dies as a result of her illness.

- **1945–50** Appointed professor of theoretical physics at Cornell University (New York State), where he returns to developing the theory of quantum electrodynamics (QED).

- **1950** Becomes professor of theoretical physics at the California Institute of Technology (Caltech), in Pasadena, where in addition to continuing his work on QED, and inventing Feynman diagrams, he helps further develop the theory of superfluidity.

**FEYNMAN LECTURES**

- **1952** Is briefly married to Mary Louise Bell.

- **1960** Marries Gweneth Howarth. (They have a son in 1962 and adopt a daughter in 1968.)

- **1961–63** Delivers a series of lectures that become *The Feynman Lectures on Physics*, a textbook first published in 1964.

- **1965** Awarded (jointly with Julian Schwinger and Sin-Itiro Tomonaga) the Nobel Prize in Physics for his theory of quantum electrodynamics.

- **1968** Develops the theory of partons—hypothetical hard particles inside the protons and neutrons in the center of atoms. His work in this area contributes significantly to the modern understanding of quarks.

- **1970s** Gives many public lectures, such as an address on the topic of Cargo Cult Science.

- **1979** Has successful surgery for stomach cancer.

- **1982** Conceives quantum computing, in which the effects of quantum mechanics are harnessed in the working of computers.

- **1985** Publishes *QED: The Strange Theory of Light and Matter*.

- **1986** Serves on a panel that investigates the causes of the explosion of the Space Shuttle *Challenger*. His written account of the experience reveals a lack of harmony between NASA's engineers and executives.

**CHALLENGER DISASTER**

- **February 15, 1988** Dies in Los Angeles as a result of cancer.

**« BEFORE**

In premodern times astronomers viewed the stars simply as a backdrop to the motions of the Sun, Moon, and planets.

### CREATION THEORY

For thousands of years people had generally accepted that the Universe was **created by God**. In the 17th century Irish Archbishop James Ussher claimed God made the Universe in six days and even gave an exact date for the creation. According to Ussher, the Universe began on Sunday evening, **October 23, 4004 BCE**, making the Sun less than 6,000 years old.

### SUN'S ENERGY

The question of how the Sun generated its heat and stayed **luminous** for so long was a physical and intellectual problem. In 1854 the German physician and physicist **Hermann von Helmholtz** calculated that the Sun generated heat by shrinking 246 ft (75 m) per year.

### STARLIGHT

In 1666 **Isaac Newton split light** into a spectrum of colors **« 110–11**. In 1863 **William Huggins** studied the spectra of stars and identified their **chemical elements**, proving that stellar atmospheres consist of the same elements found on Earth.

#### Life cycle of a solar-mass star
All stars form from vast cold and dense "interstellar" clouds (mainly molecular hydrogen). The series of images below follows the formation and life cycle of a Sun-like star. The mass of the star will determine the length of its life cycle, with smaller stars lasting longer.

# The Life Cycle of Stars

**From the late 18th century astronomers began to question the nature of the Sun and the stars. By the early 1900s they had established that there is a range of star types, including red giants and white dwarfs. But these types are actually stages in a star's life cycle—each star moves from one type to another as it ages.**

The groundwork for our present understanding of the nature and evolution of stars started over a century ago, when astronomers in North America and Europe independently classified stars according to their spectra. They split the light of an individual star into its spectrum (its rainbow band of colors) and analyzed this to establish a range of stellar properties such as temperature and luminosity.

### Giants and dwarfs
The US Harvard *Annals* of 1897 included Antonia Maury's classification of stars, and this caught the attention of Danish astronomer Ejnar Hertzsprung. From 1905 he used the data to establish a relationship between stellar properties. His work was independently corroborated by the American astronomer Henry Norris Russell in 1910.

Both astronomers showed that when the surface temperature of stars is plotted against their luminosity on a graph, well-defined groupings appear. They found that stars divide into two main groups: dwarfs and giants.

In fact, about 90 percent of stars are dwarfs—they make a band that sweeps diagonally across the graph, known as "the main sequence." The remainder form other groups: red giants, supergiants, and white dwarfs. The

#### Star-forming nebula
New stars are forming within this small portion of the Eagle Nebula, which is illuminated by nearby, hot young stars. The Eagle Nebula is an immense starbirth cloud in one of the spiral arms of the Milky Way galaxy.

graph these astronomers used—now generally referred to as the H–R (Hertzsprung-Russell) diagram (see p.431)—is the world's most important astronomical diagram.

### Energy and luminosity
The H–R diagram revolutionized the study of stellar evolution, and it is still a fundamental tool in our understanding of the stars, their lives,

ages, and composition. It allowed astronomers to theorize on stellar evolution, but they still lacked an understanding of the stars' internal structure and mechanism. They knew a star's energy was produced according to Einstein's equation ($E = mc^2$), but they did not know how.

Two English astronomers provided new information in the 1920s. Cecilia Payne (later Payne-Gaposchkin) showed that stars are made primarily of hydrogen and helium, although her work was not seen as important at the time. But Arthur Eddington's suggestion—that a star's energy could come from nuclear fusion reactions

1 A cloud of hydrogen begins to collapse owing to a trigger, such as a collision with another cloud. A pressure wave pushes the cloud in, so it becomes unstable and fragments.

2 A fragment of cloud collapses as its gravity pulls material into the core, where the density, pressure, and temperature build up. It is now a protostar.

3 Infalling gas picks up speed, creating a rotating disk of stellar material from which two jets of gas are expelled. The protostar's center becomes dense and hot enough for fusion.

4 The protostar's outward pressure and its inpulling gravity balance each other out. The protostar has stabilized and is now a main sequence star.

5 Gas and dust unused in the star-making process surround the star. They cool and condense to form tiny pieces that join together to form planets.

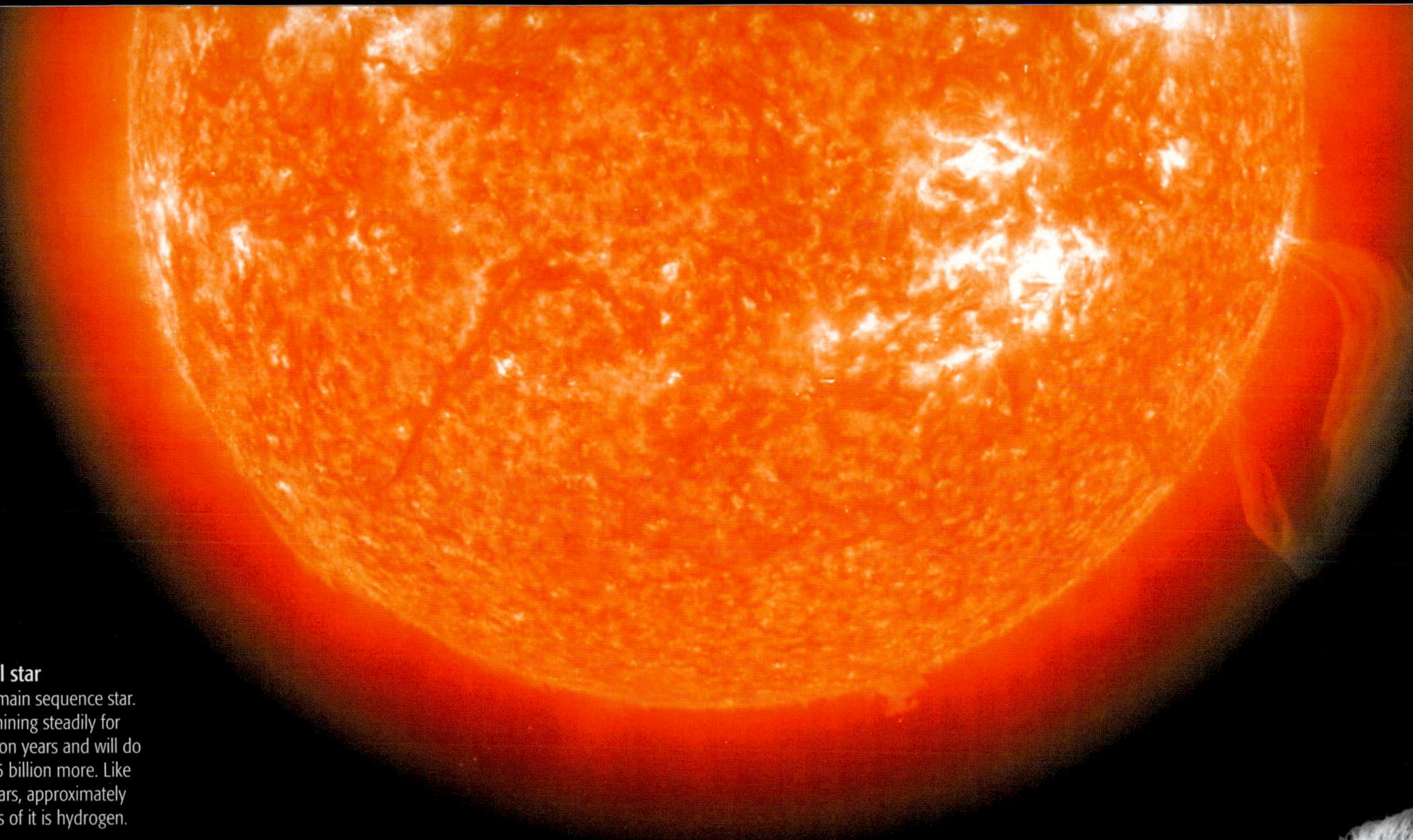

## Earth's local star

The Sun is a main sequence star. It has been shining steadily for around 5 billion years and will do so for about 5 billion more. Like most other stars, approximately three-quarters of it is hydrogen.

deep within the star itself, where hydrogen converts to helium—had an immediate impact.

More important still was Eddington's formalization of the relationship between the mass (amount of material) of a main sequence star and its luminosity (energy output). By knowing a star's luminosity it is possible to calculate its mass—it is this property, more than any other, that dictates the pattern and length of a star's life. The more massive the star, the shorter its life. Eddington's discovery was the key to a true understanding of the life cycle of stars.

»

"The **museum** of the sky contains a...**range of exhibits**, and we ...wonder what is the **origin**...?"

JAMES HOPWOOD JEANS, ASTRONOMER, 1931

ASTRONOMER (1873–1967)

### EJNAR HERTZSPRUNG

Born in Denmark, Hertzsprung trained as a chemical engineer but turned his interest in astronomy into a full-time career at the age of 29. He spent 10 years at Potsdam Observatory in Germany, followed by 25 at Leiden Observatory in the Netherlands— becoming its director—before retiring to Denmark, where he worked into his nineties. His name lives on in the Hertzsprung-Russell diagram (see p.429), but he also greatly contributed to the study of binary stars, Cepheid variable stars, and open star clusters.

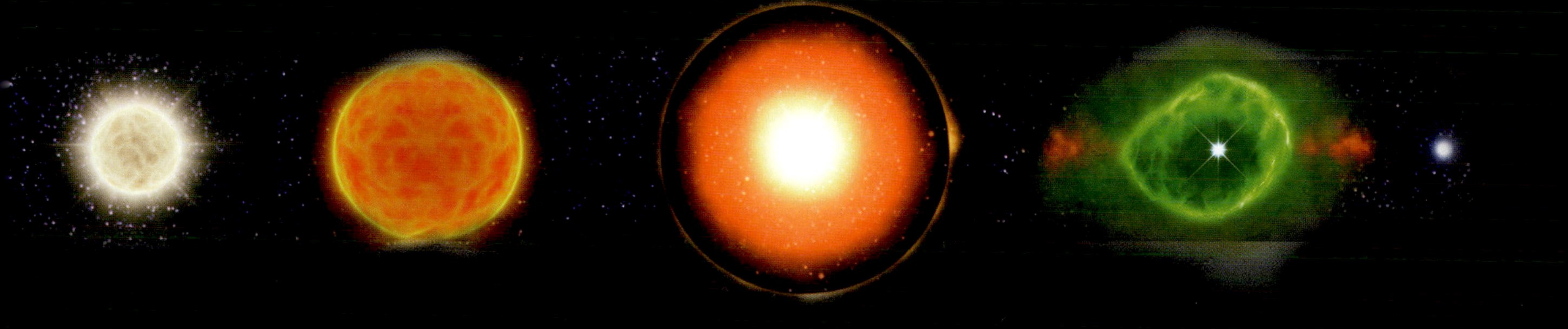

6 The main sequence star is stable and shines steadily. As it continues to convert hydrogen into helium in this longest stage, it gradually becomes larger and more luminous.

7 When the core has been converted to helium, it collapses: the core gets hotter and the outer layers are pushed out. The star becomes a red giant, up to 100 times larger than before.

8 The core is so hot that nuclear reactions convert helium to heavier elements. As the core becomes hotter still, the star's surface is blown off and starts to form a planetary nebula.

9 The dying star is surrounded by a planetary nebula—a glowing shell of gas and dust that slowly disperses. The remains shrink to form a white dwarf.

10 All that is left is a white dwarf. The nuclear fuel has been spent and this Earth-sized star fades until eventually it will be a cold, dead black dwarf star.

During the 1950s a host of scientists, including Fred Hoyle, worked on the nuclear reactions inside stars. They established how energy is produced; explained how increasingly heavy elements are formed in the reactions; and outlined star formation and the stellar life cycle.

## The stellar life cycle

Stars are vast balls of spinning, hot, luminous, mainly hydrogen gas, which start life in clouds of gas and dust. Their light and heat is produced by nuclear reactions, and mass lost in the process is converted into energy. Stars like the Sun, and those up to about eight times the Sun's mass, eventually become red giants; then they evolve into planetary nebulae, and finally Earth-sized white dwarfs. White dwarfs are extremely common. Over 97 percent of all stars in the Milky Way will eventually become white dwarfs.

## Higher-mass stars

Stars made of more than about eight times the Sun's mass die in a more spectacular fashion. The star suddenly explodes and temporarily becomes a more brilliant star: a supernova. The gas and dust blasted away forms into a supernova remnant. Supernovae are rare; each galaxy has one supernova every 200 years or so. The original star's core remains after the explosion and its fate depends on its mass.

## 100 BILLION

**100** BILLION ° C is the temperature reached by a dying star just before it explodes and becomes a supernova. Only some stars do this so supernovae are rare.

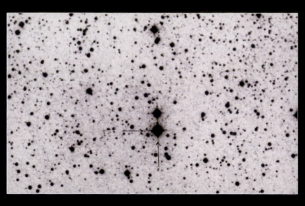

In 1932 Sir James Chadwick discovered the neutron (the subatomic particle). One year later, astronomers Wilhelm Baade and Fritz Zwicky suggested that the core remains of a supernova would be an extremely small, dense star made completely of neutrons: a neutron star. These city-sized stars are the smallest,

### The black hole Cygnus X-1

The largest black dot on this image is a blue supergiant star. It has a less massive but invisible companion: a black hole that is slowly pulling material from the supergiant.

densest stars we can detect. In 1967 Jocelyn Bell and Antony Hewish discovered the first pulsar, which turned out to be a rotating neutron star.

In 1939 Robert Oppenheimer and George Volkoff showed there is an upper limit for a stable neutron star of about three times the Sun's mass—beyond this the star continues to collapse and forms a black hole. These

A nebula of gas and dust surrounds the central pulsar

A wind of particles from the pulsar ploughs into the surrounding nebula, creates a shock wave, and forms the inner ring

The rapidly rotating pulsar

One of the two beams of radiation that flow away from the star

### Supernova core

The Crab Nebula pulsar, discovered in 1968, is the core of a supernova. It rotates around 30 times per second, surrounded by stellar material thrown off in the explosion. This image is the central detail of the Crab Nebula (see right).

### Supernova remnant

The Crab Nebula is the remains of a supernova that was seen from Earth in 1054. Strands of glowing gas were pushed off by the supernova explosion. The remains of the original star (pulsar) is seen as a white dot in the center.

## SUBRAHMANYAN CHANDRASEKHAR

Born in Lahore (then in India), the third of ten children, Chandrasekhar spent most of his working life at the University of Chicago. In 1983 he won the Nobel Prize for Physics for his work on stellar evolution and structure. The "Chandrasekhar Limit" states that the upper limit of a white dwarf is 1.4 times the mass of the Sun. This is important when looking at stars that initially have a mass of more than about eight times that of the Sun. If the mass of the star's core at the end of its life exceeds this limit, the core will produce a neutron star or a black hole.

are so densely compacted that even light cannot escape them. Their existence was first suggested by John Mitchell in 1783, but largely forgotten. In 1939 Oppenheimer proposed that a black hole could be produced in a supernova explosion, but it remained a theoretical concept until Cygnus X-1 was identified as one in the early 1970s. Even after that, it was several more years before most astronomers accepted Cygnus X-1 as a black hole.

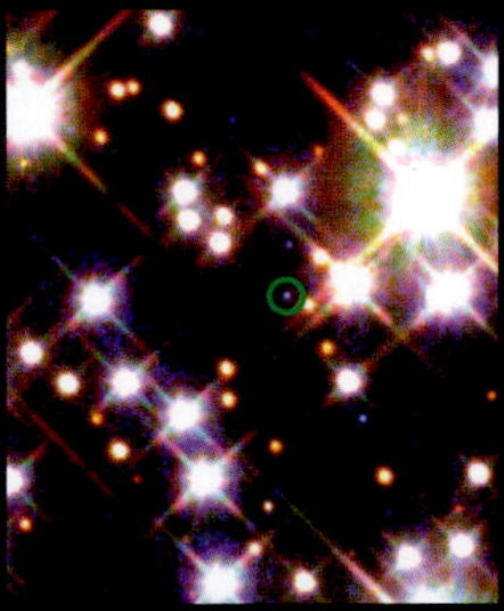

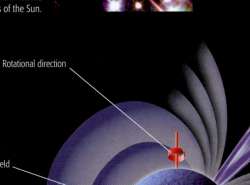

**White dwarf**
A dim white dwarf (circled) is outshone by the other stars in the cluster NGC 6397. It is the core remains of a star that was originally up to eight times the mass of the Sun.

### Stellar recycling
Higher-mass stars produce elements such as carbon, oxygen, silicon, and progressively heavier elements up to iron, either through nuclear reactions or during supernovae explosions. Discarded material moves off into space, eventually joining with other interstellar material to form the clouds that give birth to new stars. Supernovae are vitally important in seeding the Universe with elements heavier than iron—elements that are essential for terrestrial planet production, and the generation and evolution of life forms.

### Stellar cycle
Material discarded by previous generations of stars becomes part of a cloud that in turn gives birth to a new generation of stars.

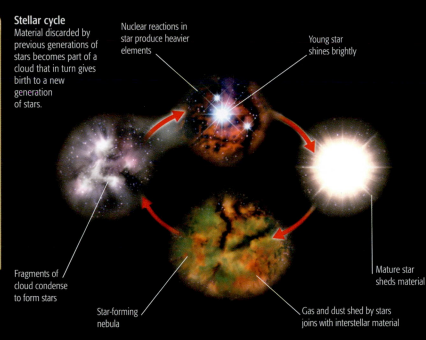

Nuclear reactions in star produce heavier elements

Young star shines brightly

Mature star sheds material

Gas and dust shed by stars joins with interstellar material

Star-forming nebula

Fragments of cloud condense to form stars

"Stars are born, **live**—often for billions of years—and **die**…sometimes in a…**spectacular manner**."

CARL SAGAN, ASTRONOMER, 1978

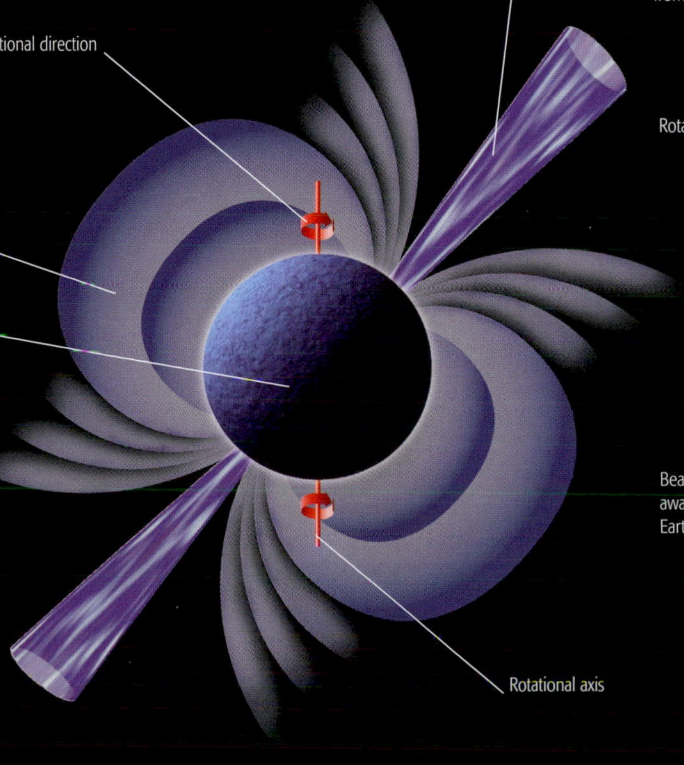

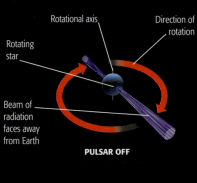

Rotational axis
Direction of rotation
Rotating star
Beam of radiation faces away from Earth
**PULSAR OFF**

Rotational axis
Direction of rotation
Beam aligned with Earth
**PULSAR ON**

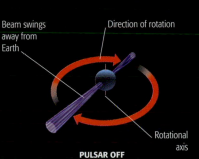

Beam swings away from Earth
Direction of rotation
Rotational axis
**PULSAR OFF**

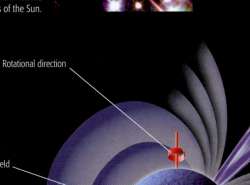

Beam of radiation
Rotational direction
Magnetic field
Neutron star
Rotational axis

**Pulsar**
Pulsars are rapidly rotating neutron stars which have very strong magnetic fields. Beams of radiation are emitted from their magnetic-polar regions, and we can detect these if Earth is within the path of a beam.

AFTER »

Astronomers continue to learn about the stellar world as space-age advances transform the studies of stellar objects.

**BROWN DWARFS**
The existence of **brown dwarf stars** (and their name) was suggested by **Jill Tarter** in 1975; the first was found in 1988. They have a small mass and a temperature too low for fusion reactions.

**SUPERNOVA EXPLOSION**
In 1987 the supernova SN1987A exploded in the nearby **Large Magellanic Cloud galaxy**. Astronomers were able to test their predictions of the theory of supernova explosions.

**SPACE TELESCOPES**
The **X-ray telescope** Chandra (launched 1999), and the **infrared telescope** Spitzer (launched 2003), collect radiation from stars.

**FAR FUTURE**
In the far future stars will have no usable **hydrogen and helium** left for nuclear reactions, so the stellar universe will be very much fainter than it is now.

SEE ALSO »
pp.340–41 GALAXIES, CLUSTERS, AND SUPERCLUSTERS
pp.394–95 INSIDE THE SOLAR SYSTEM

## BEFORE

Much ecological thinking has its origins in studies of natural history, which flourished particularly during the age of exploration.

### EARLIEST ECOLOGICAL IDEAS

The medieval Arabic scholar **al-Jahiz** wrote the *Book of Animals*—an encyclopedia of **zoology** that included numerous ideas that presaged later biological theories, including **natural selection** and the concept of the **food chain**.

### EXPLORERS AND NATURALISTS

The first naturalists to make **accurate observations** of the living world laid the foundations of **field ecology**: Englishman **Joseph Banks** famously **recorded many new species** on explorer Captain James Cook's first voyage across the Pacific (1768–71). In England, **Gilbert White** is widely considered to be one of the first ecologists because of his **observations of the animals and plants** at his home in Selborne.

### « SEE ALSO

*pp.200–01* How Evolution Works
*pp.202–03* Charles Darwin
*pp.278–79* Cycles in the Biosphere

# Ecology and Ecosystems

**Millions of different species of organisms live on Earth. Each individual species is the product of a long evolutionary process, which has been profoundly influenced both by the conditions and by the resources of each and every species' surroundings.**

In the conclusion to his famous book *On the Origin of Species* in 1859, English naturalist Charles Darwin (see pp. 202–03) contemplated: "an entangled bank, clothed with many plants… birds… insects… and worms… these elaborately constructed forms, so different from each other, and dependent on each other in so complex a manner, have all been produced by laws acting around us."

Darwin's description foreshadowed the rise of ecology: a new type of science that would seek to explain the scene that he saw on the "entangled bank." Ecology is the study of the distribution and abundance of organisms, and of their interactions both with their environment and with each other. It examines each individual organism in all levels of biological hierarchy: population, community, ecosystem, and biosphere.

The term ecosystem encompasses not only living organisms but also nonliving physical features of an environment, all of which function together as an independent unit.

### Effect of environment on organisms

A living organism affects, and is affected by, the world around it. Environmental conditions, such as temperature, determine whether an organism can live and breed. In 1895 Danish botanist Eugen Warming proved this by showing that unrelated plants coexisting in the same regions had evolved strategies to cope with temperature extremes.

Struggle for survival, as outlined by Charles Darwin, also plays a key role in shaping natural groups of organisms. Living things use up resources from the environment (food, oxygen, water, and salts). If the number of individuals increases, competition for resources becomes greater, which means that each individual gets a smaller share and the weaker produce fewer offspring.

### Interactions between organisms

A population is a group of individuals of the same kind of organism, united by an underlying genetic bond: all belong to the same species. A habitat ordinarily harbors numerous coexisting species. The genetic difference between species is reflected by their different capabilities and the roles that they play. In order for species to coexist, each must occupy a subtly different role—its niche—or competition for resources would make

### Rain forest pyramid of biomass

Chemical energy flows through a food web, from photosynthetic plants (producers) to animals (consumers). Each organism in this web gives off heat energy—and loses chemical waste—meaning that the amount of energy available to successively higher levels of the food chain diminishes. Consequently, there is always a lower combined weight of top-level predators than organisms at lower levels.

Jaguar — **TERTIARY CONSUMERS: TOP PREDATORS**

Motmot bird   Tarantula   Coati — **SECONDARY CONSUMERS: CARNIVORES AND OMNIVORES**

Tapir   Butterfly   Mouse   Agouti — **PRIMARY CONSUMERS: HERBIVORES**

Elephant ear   Aroid vine   Acacia tree   Coccoloba tree   Passion flower vine   Strangler fig — **PRODUCERS: PHOTOSYNTHETIC PLANTS**

## BOTANIST (1871–1955)

### ARTHUR TANSLEY

Tansley pioneered the study of plant ecology in England, and was responsible for developing the concept of the ecosystem. Tansley became the first president of the British Ecological Society, which he helped to establish, an organization that gained an international reputation for ecological science. He also helped set up and lead the Council for the Promotion of Field Studies as well as the Nature Conservancy Council in Britain. Following World War I Tansley further developed his interest in psychology, too, spending some time being mentored by Sigmund Freud.

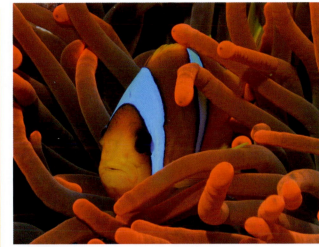

**Mutualism**
Clownfish live among the tentacles of sea anemones and are resistant to their sting. Both species benefit from such a relationship: the clownfish is protected from predators and the anemone eats leftover food brought in by the clownfish.

**Parasitism**
*Rafflesia* is the world's largest single flower, capable of reaching a diameter of 3 ft 3 in (1 m). It is a parasite that grows on tropical vines. The relationship is beneficial only to the parasite, which has no leaves and so cannot make its own food via photosynthesis.

one drive the other to extinction. Russian scientist Georgii Gause proved this competitive exclusion theory in 1932 when, in experimental cultures of microorganisms, one species consistently drove another to extinction.

Interactions between species were defined in the mid 1900s by two brothers, American ecologists Eugene and Howard Odum, who classified species based on whether inter-relationships were beneficial or harmful to individuals. Thus, some interactions benefit one organism and harm another, such as predation (the predator hunts and feeds on the prey) or parasitism (the parasite takes from the host). Competition is harmful to both; mutualism is beneficial to both (see above).

### Ecosystems: cycles and energy flow
In 1935 English botanist Arthur Tansley (see left) developed the concept of an ecosystem—how organisms of a community interact with the nonliving world within a unified system. In 1926 Russian geochemist Vladimir Vernadsky articulated the idea that all matter is rotated as part of natural cycles (see pp.278–79), meaning that all organisms add waste to their surroundings and most take oxygen from it. At the same time the ecosystem is fuelled by energy from the Sun. Developing this idea, Tansley, along with American botanist Frederic Clements, proposed that there was a continual flow of energy through a food chain, from the power of the Sun to the release of heat by every living thing. Their theory was later elaborated into the idea of the food web (see opposite), where energy does not follow a simple linear chain, but passes through a complex network of interactions.

If an ecosystem is destroyed, for instance, by fire (see below), volcanic eruption, or landslide, there follows a process of succession. Early stages of this process usually give rise to fast-growing, well-dispersed species, but as succession continues, more competitive species may replace them.

> ## "Nature undisturbed seems like a symphony whose harmonies arise from variation and change."
> DANIEL B. BOTKIN, AMERICAN ECOLOGIST AND WRITER, DISCORDANT HARMONIES, 1990

## AFTER

As ecology became a quantitative discipline, scientists developed mathematical methods to help explain this complex subject.

### CONTROLLING POPULATIONS
In the 1920s two mathematicians, American **Alfred Lotka** and Italian **Vito Volterra**, devised an equation to describe the **relationship between the population sizes of predators and prey**. This equation helped scientists to predict how **populations change over time** and how different species can interact. Scientists use such equations to develop **strategies to control** crop pests or parasites that cause disease.

### CONCEPTS OF PRODUCTIVITY
The productivity of a plant or animal is a measure of how much **living material is built up during growth**, in relation to the amount of **material the organism takes from the environment**. Productivity assessments are important in helping agriculturalists balance the cost of maintaining crops or livestock, and the potential returns in terms of food production.

**SEE ALSO »**
*pp.334–35* CONSERVATION BIOLOGY

### RELATIONSHIP BETWEEN PREDATOR AND PREY

Interactions between predators and prey result in oscillations in population sizes of both species. As more prey are born, there is more food for predators, so more predator offspring survive; the increase in numbers of predators then reduces the number of prey. The result is that the pattern of predator numbers is slightly out of phase with the pattern for prey.

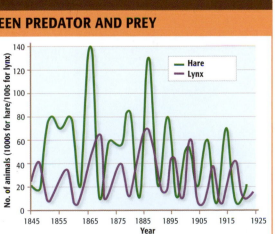

Legend:
— Hare
— Lynx

(y-axis: No. of animals (1000s for hare/100s for lynx); x-axis: Year, 1845–1925)

**Succession after destruction**
Habitat destroyed in the Shoshone Fire of 1988 in the Yellowstone National Park, Wyoming, showed recovery about 10 years later. Combustion enriches the soil and encourages growth of plants, such as this fireweed (right).

**BEFORE**

Concern for degradation of the environment has ancient origins, but concerted efforts to recognize and address many issues only began in the 20th century.

**ORIGINS OF ENVIRONMENTALISM**
Many **Islamic scholars** during the **Middle Ages** voiced their **concern regarding environmental issues** related to pollution and habitat destruction. Prominent among these was the philosopher **Al-Kindi**, who wrote, during the 9th century CE, about the **contaminating effects** of pollutants on **water supplies**, **air**, and **soil**.

**« SEE ALSO**
*pp.332–33* ECOLOGY AND ECOSYSTEMS

**CATEGORIES OF THREAT**

Scientists at WWF (World Wildlife Fund) and IUCN (International Union for the Conservation of Nature) have amassed sufficient information on the population ecology of many species to allow them to be ranked in terms of conservation priority, from "least concern" to "critically endangered." Below are listed a selection of species in each of the recognized categories of threat. The rankings have been important in guiding the priorities of these—and other—organizations.

**Least concern**
Kirk's Dik-dik (*Madoqua kirkii*)
House Sparrow (*Passer domesticus*)

**Near threatened**
European Otter (*Luta lutra*)
Manta Ray (*Manta birostris*)

**Vulnerable**
Hawaiian Goose (*Branta sandvicensis*)
Polar Bear (*Ursus maritimus*)

**Endangered**
Giant Panda (*Ailuropoda melanoleuca*)
Pink Pigeon (*Nesoenas mayeri*)

**Critically endangered**
Javan Rhinoceros (*Rhinoceros sondaicus*)
Spix's Macaw (*Cyanopsitta spixii*)

**Extinct in the wild**
Père David's Deer (*Elaphurus davidianus*)
Scimitar-horned Oryx (*Oryx dammah*)

**Extinct**
Golden Toad (*Incilius* or *Bufo periglenes*)
Thylacine (*Thylacinus cynocephalus*)

# Conservation Biology

**As humans explored and exploited, Earth was found not to be limitless, and living things were exhaustible: species formed by millions of years of evolution could be obliterated forever. Over the last century, however, there has been a growing awareness of humanity's practical and moral duty to conserve nature.**

In North America, in the 19th century, huge migratory flocks of the Passenger Pigeon could take days to pass by overhead. The species was a victim of hunting on a massive scale because it damaged crops, and its meat, being cheap and plentiful, was eaten by slaves. By the time scientists realized that such enormous flocks were vital for successful breeding, it was already too late; the last Passenger Pigeon died in captivity in 1914. Such an example showed that even the most common species could be driven to extinction.

The need for conservation biology was stimulated by events of the 20th century. The World Wildlife Fund (WWF) was established in 1961. Since then its focus has been to conserve those of Earth's habitats and species that are most in danger.

## Earth's wild places
In the 19th century German naturalist Alexander von Humboldt saw that climate determined the distribution of vegetation types such as rain forest, desert, and grassland, which are known as biomes. Although these biomes were structurally similar in different parts of the world, their species were very different. English ornithologist Philip Sclater recognized six regions according to bird distribution. British naturalist Alfred Russel Wallace (see pp.200–201) refined the system to take account of animals other than birds. In 1878, he wrote about deforestation and foresaw environmental issues to come.

Earth's wild places provide food, medicines, and the raw materials of technology, but ecology was teaching that if these resources were used

**Back from the brink**
Some species, such as Przewalski's Horse, have been exterminated in the wild, but intensive management and breeding of captive stocks have saved them from total extinction.

faster than they grew back, then they would run out entirely. In 1948, an organization, later known as the International Union for the Conservation of Nature (IUCN), was set up both to develop ways of using resources sustainably and to examine how protected areas could be representative of global living diversity. IUCN and WWF scientists have since identified over 600 ecoregions around the world with distinct ecological identities (see right).

## Conserving habitat
In 1914 the creation of a dam in Central America flooded an area of the Panama Canal Zone. Forest survived only on the hilltops, becoming islands in (what was then) the largest human-made lake in existence. One such island—Barro Colorado—became a nature reserve; its research station documented changes to the rain forest following its isolation. Although much of the food web remained intact, top predators became extinct, since they needed a big area to sustain a viable population.

> "We **shan't save all** we should like to, **but we shall save** a great deal **more than** if we had **never tried**."
> SIR PETER SCOTT, WWF FOUNDER, 1961

In the 1960s American ecologists Robert MacArthur and Edward O. Wilson looked more closely at island ecology and found that smaller, more isolated islands generally supported fewer species. The implications were profound: any patch of habitat surrounded by hostile landscape was effectively an island. It meant that one large nature reserve could preserve more biodiversity than many smaller ones of the same total area.

## Conserving species
Biologists have uncovered some places in the world with such an astonishing variety of life that they called them biodiversity "hot spots." Against a background of habitat destruction, hot spots such as the forests of Colombia or Madagascar serve to guide priorities in conservation. Media technology—combined with the efforts of WWF and bodies like it—has raised public and government awareness of these wild places. It means that there is now a greater global incentive to conserve Earth's natural resources than ever before—and the scientific know-how to keep a great deal of biodiversity intact.

**Seeding the land**
Dropping seed bombs (clods of soil containing live plant material) is an ingenious method for introducing vegetation to cleared land, and may offer a way of helping restore degraded habitats.

### Tropical forest
High temperature and humidity of the tropics favor diversification—half of the world's animals and plants are found here.

### Temperate forest
Extending from the tropics to the poles, broad-leaved deciduous forests eventually give way to needle-leaved coniferous forests.

### Coniferous forest
In cold regions close to the poles conifers predominate. Their needlelike leaves ensure survival in areas where water is scarce.

### Polar and tundra
With very cold winter extremes and little rainfall, the tundra has low-growing vegetation, such as mosses, grasses, and lichens.

### Biomes around the world
Scientists have many hundreds of biome descriptions for the myriad habitats of the world. Here is a simple breakdown of biogeographical regions, each of which nurtures unique wildlife as well as unique conservation challenges.

### Desert
Found mainly in Africa, Australia, and Asia, deserts have extremely low rainfall and endure huge extremes of temperature.

### Savannah
Grassland with small or sparse trees, found in South America, Africa, and Asia, is often associated with highly seasonal rainfall.

### Temperate grassland
Mostly found in North America and Eurasia, grasses thrive on cropping and so sustain a variety of herbivorous animals.

### Ocean
Covering 72 percent of Earth's surface, this diverse environment ranges from species-rich coastal reefs to vast unexplored depths.

## TARGETING SPECIES AND HABITATS

Popular species, such as tigers, can be "ambassadors" for drawing attention to the plight of certain habitats, such as forests in Asia. But some habitats are important because they contain biologically unique forms. Even though they may be less familiar, these distinctive forms of animals or plants are important pieces of biodiversity. Below is a selection of such species.

| Habitat | Distinctive species | Comments |
|---|---|---|
| Central Pacific forests of western North America | Tailed Frog (*Ascaphus*)—two species | Distinctive from all other frogs: has internal fertilization and cannot vocalize. Related to frogs of New Zealand |
| Northland temperate forests of New Zealand | Tuatara (*Sphenodon*)—two species | Sole surviving form of a reptile group from 200 million years ago: almost a "living fossil" |
| Madagascar dry and lowland forests and spiny thicket | Mesites (*Mesitornis*)—two species (*Monias*)—one species | Pheasantlike birds, dubiously related to doves, all endangered and numbers declining |
| Amazon basin forests | Hoatzin (*Opisthocomus*)—one species | Large, bizarre tree-dwelling birds; the chicks have claws on their wings |
| Eastern Australian rain forests | Platypus (*Ornithorhynchus*)—one species | Sole member of one of the two families of egg-laying mammals |
| Chilean temperate forests | Monito del Monte (*Dromiciops*)—one species | Small marsupial—actually more closely related to those in Australia than those of the rest of South America |

## AFTER

Modern conservation biology is a science that depends as much on genetics and reproductive function as it does ecology.

### BIOTECHNOLOGY AND CONSERVATION
Advances in reproductive biology are important in assisting conservation efforts. **Cloning** of **endangered species of plants** can be important in **increasing stocks** for distribution among botanic gardens, minimizing the risk of stock destruction by disease.

**SEE ALSO »**
*pp.390–91* HOW CLONING WORKS
*pp.414–15* GLOBAL WARMING

**« BEFORE**

Natural rubber has been used since ancient times, but it was not until the 19th century that it became a truly useful material.

## VULCANIZATION OF RUBBER

The **rubber tree** is native to South and Central America, where the ancient Mesoamericans used it for various purposes, including making balls for games, as far back as 1600 BCE. Ordinary natural rubber is **prone to snapping**, **stretches too much**, and is **sticky**. In 1839 American inventor **Charles Goodyear vulcanized rubber** with heat and sulphur, making it **tougher**, **harder wearing**, **less elastic**, and **less sticky**. Huge rubber plantations were created in Africa and Asia to meet demand.

CHARLES
GOODYEAR

**« SEE ALSO**
pp.150–51 ORGANIC CHEMISTRY
pp.240–41 MASS PRODUCTION
OF CHEMICALS

# The Age of Plastics

Until the early 20th century natural materials were used in construction, engineering, and the home. But as more and more ways of making inexpensive synthetic polymers emerged the situation changed radically, so that by the end of the century we were living in what has been called the Age of Plastics.

People have used natural polymers for millennia—a polymer is a chemical compound made up of long-chain molecules (see below). Natural polymers occur in plants and wood. As chemists began to understand natural polymers, the idea of trying to mimic them emerged. This led to the production of many useful synthetic materials, including plastics, which have had great impact in many areas.

## The first synthetic polymer

In the 1830s it was found that treating starch or wood fibers with nitric acid could nitrate the cellulose, and in 1856 English inventor Alexander Parkes treated these nitrocellulose materials with solvent. Thus he invented the first synthetic polymer, named Parkesine, which was an early form of celluloid: a class of materials that resemble animal horn, which are waterproof, but also elastic. Parkes intended the material to be used as waterproofing for fabrics, but commercial success eluded him.

## Practical plastics

In the 1860s American inventor John Wesley Hyatt was looking for a replacement for ivory in billiard balls, and experimented with nitrocellulose and camphor to make celluloid in 1872. While highly flammable and easy to decompose, this material was also easy to mold and shape, making it the first thermoplastic—polymer that is liquid when heated, and solid when cooled (see below left). Celluloid was the material of choice for film in the early years of photography (see pp.262–63) and cinema, in spite of its notorious flammability.

By 1909 Belgian chemist Leo Baekeland had accidentally discovered that mixing phenol and formaldehyde with powdered wood produced a resinlike material he called Bakelite, which was the first thermosetting plastic (polymer that is soft until set solid by heat, see below left). Bakelite became the design material of the Art Deco

Chemical bond

▷ **Structure of starch, a natural polymer**
Starch, a natural polymer, is made up of lots of glucose molecules strung together in chains. Scientists made different long-chain molecules when manufacturing synthetic polymers.

Carbon atom

Oxygen atom

Hydrogen atom

Long chains of polymers of high molecular weight

Cross-links between long chains of polymers form when heated

**THERMOPLASTIC POLYMER CHAINS**

**THERMOSETTING PLASTIC POLYMER CHAINS**

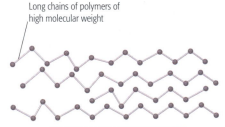

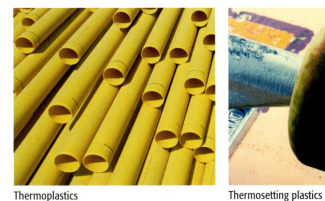

## Thermoplastics
Polymers that are liquid when heated and solid when cooled, thermoplastics are often used for pipes since they can be blow-molded to make hollow tubes with no joints, and can withstand high gas and liquid pressures.

## Thermosetting plastics
Thermosetting polymers can be molded and then irreversibly set by heating into solid objects, such as this skateboard wheel, which is hard-wearing yet has some of the useful properties of rubber.

## Plastics: an industrial revolution
From the mid-19th century chemists experimented with natural substances to produce semi-synthetic plastics. Wholly synthetic materials soon followed, culminating in rapid progress in polymer production in the 20th century.

**1887** American clergyman Hannibal Goodwin unsuccessfully files a patent claim for **celluloid film**. Two years later Eastman Kodak succeeds with its own claim.

**1909** Belgian chemist Leo Baekeland unveils **Bakelite** at the February meeting of the New York branch of the American Chemical Society.

**1926** Researcher Waldo Semon invents **vinyl**—flexible, plasticized PVC, the first commercially useful form of PVC—while working for the B.F. Goodrich Company in Ohio.

**1935** DuPont patents **nylon**. Three years later a nylon toothbrush is launched. Stockings follow in 1940.

| 1860 | 1870 | 1880 | 1890 | 1900 | 1910 | 1920 | 1930 | 1940 |
|---|---|---|---|---|---|---|---|---|

**1862** Alexander Parkes unveils **Parkesine**, the first synthetic plastic—made by treating cellulose with nitric acid—at the London International Exhibition.

**1872** German chemist Eugen Baumann makes the first **polyvinyl chloride (PVC)**, but not until the 1920s is the material made flexible enough to be commercially useful.

**1894** English chemist Charles Cross and his colleagues Edward Bevan and Clayton Beadle patent **viscose rayon**, a semi-synthetic silklike fiber made from cellulose.

**1912** Swiss chemist Jacques Brandenberger patents **cellophane**, a semi-synthetic polymer still used in packaging.

**1933** Eric Fawcett and Reginald Gibson of ICI in Northwich, England find a practical way to synthesize **polythene**.

## "Well, it was kind of an **accident**, because **plastic** is not what I meant to **invent**..."

LEO BAEKELAND, BELGIAN CHEMIST, ON INVENTING BAKELITE

▷ **Plastics in the home**
Plastics transformed modern living, as can be seen in this 1955 advertisement for Formica, a heat-resistant polymer developed for worktops.

movement owing to its durability and ability to be shaped. With the addition of colorful dyes, it lent itself to many practical uses, from light switches to lamps, and telephone handsets to pipe stems.

### Synthetic clothing fibers
The Age of Plastics truly began in the DuPont chemical company's laboratory, run by American chemist Wallace Carothers (see below). In 1930 his

**Bulletproof Kevlar**
Kevlar is a synthetic fiber that was developed by Polish-American chemist Stephanie Kwolek during the 1960s. Its incredible strength means that it is used to make bulletproof and stabproof vests.

team made an artificial rubber, dubbed neoprene, and also discovered the starting point for the synthesis of an artificial clothing fiber. In 1934 Carothers' team succeeded at this when they developed nylon, which was the perfect substitute for expensive, fragile silk. Nylon became symbolic of wartime strife in the US with the rush to acquire it in World War II, when new supplies of silk could not be imported from Asia.

### Modern plastics
Once chemists began making polymers, more and more followed. In 1938 American chemist Roy Plunkett accidentally discovered the "nonstick" material polytetrafluoroethylene (PTFE), or Teflon, widely used to coat metal cooking pans, and also used to make highly airtight and inert seals for uranium enrichment equipment during the building of the atomic bomb in World War II. In 1965 Polish-American chemist Stephanie Kwolek from

DuPont invented a lightweight but bulletproof polymer, Kevlar, a modified form of nylon also used in stabproof vests. A porous form of PTFE emerged

> **4 TRILLION** The number of disposable plastic bags made globally each year. This presents a huge disposal problem, which only reusable, recyclable bags may solve.

later as the water-repellent, but breathable, clothing material Gore-Tex, bringing the history of polymers full circle to Parkes's dream of plastic waterproofing.

### The price of polymers
Our ever-growing worldwide demand for polymers has come at a price. They are made from raw materials that come from oil—a finite resource—and, not being biodegradable, they pose a huge waste problem. Ironically however,

---

**IN PRACTICE**

### RECYCLING PLASTICS

Domestic plastic waste once went to landfill sites like the one here. Now it is increasingly being recycled. But first it must be sorted. This is mainly done by hand, but automatic sorting using X-ray fluorescence, electrostatics, flotation, and infrared and near-infrared spectroscopy is increasingly being used.

---

**AFTER** »

As oil supplies worldwide dwindle, the long-term future of plastics may lie in mining landfill sites or in bioplastics.

**MINING PLASTICS**
Experiments are underway in the US to **mine landfill sites** and **recycle used plastics** as "raw materials" for new plastics. A major problem, though, is the buildup of methane gas from decomposing organic waste buried with everything else at old landfill sites.

**BIOPLASTICS**
**Biodegradable and compostable bioplastics** are derived from renewable natural biomass sources such as corn starch and vegetable oil. They can be used for nondisposable as well as disposable products.

---

**CHEMIST (1896–1937)**

### WALLACE CAROTHERS

First born of four children to a teacher father, American Wallace Carothers worked as an instructor at Harvard before joining the DuPont chemical company's fledgling research laboratory, where he led a team in the production of many important materials, including synthetic rubber (neoprene) and nylon. His life was blighted by depression and alcoholism—he reputedly carried cyanide tablets with him. He married Helen Sweetman in 1936, but took his own life after the death of a favorite sister in 1937, so never saw his daughter nor the plastics revolution his pioneering research spawned.

---

**1937** The Dow Chemical Company introduces **polystyrene** —developed in Germany by BASF—to the US.

**1954** Italian chemist Giulio Natta makes **polypropylene**, a tough plastic widely used in many products.

**1950**

**1960**

**1941** Kinetic Chemicals Inc, a joint venture between DuPont and General Motors, patents **Teflon**—PTFE.

**1966** Stephanie Kwolek of DuPont patents **Kevlar**, first used in the 1970s in tires.

## « BEFORE

The rocket was used as a weapon for a long time before it its potential as a means of propulsion became clear.

### ROCKET FIREWORKS
The first rockets were **Chinese fireworks** « 56–57, which were used for celebrations and in warfare, to set fires and spread panic, from around the 13th century. They often veered off target and lacked force.

### NEWTON'S LAWS
In the late 1600s, Newton's laws of motion clarified the principle of **action and reaction** « 104–05 that allow a rocket to operate. Explosive gases

**EARLY CHINESE ROCKETS**

igniting at the base of a rocket are forced out in one direction, and this action creates a reaction—**an equal and opposite force**—that pushes the body of the rocket in the other direction. The rocket does not need to push against any surrounding medium, so it can **produce thrust even in a vacuum.**

### CONGREVE'S ROCKETS
British soldiers encountered **metal-cased rockets** when fighting in India during the late 18th century. William Congreve modified the design, and his rockets were widely used through the early 1800s. Improvements to artillery saw rockets fall out of use as weapons, but they continued to be used in rescues at sea and for scientific research.

**« SEE ALSO**
pp.290–91 TAKING FLIGHT

# Rocket Propulsion

**The development of rockets has been linked both to warfare and space exploration, as nations vied for military advantage. The key lay in unlocking the secrets of rocket propulsion, which required an inspired application of Newtonian physics, tiered stages, and unusual fuel combustion.**

Although people had fantasized for centuries about journeying into space, the origins of modern rocket propulsion can be traced to Konstantin Tsiolkovskii (1857–1935), a Russian schoolteacher. He realized that Newton's principle of action and reaction (see pp.104–05) would enable a rocket to generate thrust even beyond the atmosphere, provided it carried with it both fuel and an "oxidizer" (a chemical with which the fuel could react).

Tsiolkovskii built many models and made a series of breakthroughs, including the idea of using a series of separate "stages" for a rocket. Each stage

**Konstantin Tsiolkovskii**
The Russian pioneer of rocket propulsion built many models to demonstrate his various designs, but he never took the next step toward building a working rocket engine.

could have its own fuel tanks and engines, in order to minimize the amount of useless weight carried all the way to space. However, it was left to others to take the next step.

### Rocket development
The two pioneers of practical rocketry were the American Robert Goddard and Germany's Hermann Oberth. Goddard built the first flying liquid-fueled rocket, powered by the

combustion of gasoline and liquid oxygen, and he conducted a series of increasingly ambitious flights from 1926 onward. Oberth produced a number of widely read books and gathered a group of

**Space Shuttle launch**
The Space Shuttle system uses five rocket engines for launch. Three on the back of the orbiter spacecraft burn liquid fuel from the external fuel tank (they are burning too fiercely to be visible here), while two boosters, one on either side of the external tank, burn solid fuel.

**Germany's V-2 rocket**
The "Vengeance weapon," developed by Wernher von Braun and his German rocket team, was designed for rapid launch from a portable platform.

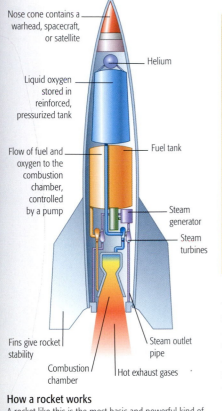

Nose cone contains a warhead, spacecraft, or satellite

Helium

Liquid oxygen stored in reinforced, pressurized tank

Fuel tank

Flow of fuel and oxygen to the combustion chamber, controlled by a pump

Steam generator

Steam turbines

Fins give rocket stability

Steam outlet pipe

Combustion chamber

Hot exhaust gases

### How a rocket works
A rocket like this is the most basic and powerful kind of engine. It is powered by burning either liquid or gas fuel which chemically reacts and forms hot gas. The gas is ejected at high pressure from the chamber, providing the thrust that propels the rocket.

ambitious young engineers and enthusiasts around him, who soon began making their own launches. Liquid-fueled rockets had significant benefits in terms of their thrust: it was possible to "throttle" the engine and even turn it off and on.

### Political and military use
By around 1930 rocket societies were flourishing in the US, Germany, and the Soviet Union, but the Nazi party in Germany directed German resources toward producing missiles rather than spacecraft. In 1944 the first "ballistic missiles," Germany's V-2 "vengeance weapons," fell on London.

The V-2 revealed that German rocketry was far ahead of its rivals, and the US and Soviet Union raced to develop their own ballistic missiles—as potential delivery systems for nuclear weapons. Two spaceflight enthusiasts led these projects: Wernher von Braun, the German architect of the V-2, who moved to the US at the end of the war, and Sergei Korolev in the Soviet Union. In the early 1950s they both argued that a presence in space would be an advantage to their political masters, and

### ROBERT GODDARD

Massachusetts-born physicist Robert Goddard became fascinated with space travel after reading H.G. Wells' science-fiction novel *The War of the Worlds* at the age of 16. He did much of his work on the development of liquid-fueled rockets while teaching at Clark University, and journalists delighted in mocking what they saw as his outlandish ideas. This attitude changed after Goddard's first successful rocket launch in 1926, and increased publicity and finance from farsighted backers—such as aviation pioneer Charles Lindbergh—allowed him to conduct increasingly ambitious experiments up to his death in 1945.

work began on projects to launch the first satellites. Korolev's team took an early lead in the "space race" thanks to their more powerful rockets (initially developed to carry their more primitive nuclear weapons, which were heavier than their US equivalents). In the 1960s the US fought back in the race to reach the Moon, and von Braun masterminded the construction of the largest rocket ever built, the Saturn V.

### Fuel developments
Fuel sources for rocket propulsion have evolved for a variety of reasons. Many liquid rocket stages use a mix of liquid oxygen and kerosene, but the V-2 and

many later ballistic missiles use nitrogen tetroxide and hydrazine—an oxidizer and propellant combination that can be stored for long periods at normal temperatures. High-powered engines such as those on the Space Shuttle use liquid oxygen with pure liquid hydrogen, which produces enormous thrust per pound of fuel, but is very difficult to handle and store, requiring extremely low temperatures.

Although liquid rockets are successful, solid rockets still have their place; the Space Shuttle's booster rockets use a rubberlike chemical compound that incorporates both fuel and oxidizer in a single material, lit by an ignition spark.

> " One of the surprising things was… the **lack** of very **loud roar**."
>
> ROBERT GODDARD, PHYSICIST, MARCH 16, 1926

A number of alternative versions to the traditional multistage rocket have been proposed, each with different advantages.

### ALTERNATE PROPULSION SYSTEMS
Nuclear-powered rockets propel themselves forward through small **nuclear explosions**. While powerful and fuel-efficient, the dangers associated with this technology led to it being largely abandoned in the 1960s.

### PRIVATE VENTURE SPACE ROCKETS
Since the 1990s there have been many attempts to develop a "**Single-Stage to Orbit**" (SSTO), a reuseable vehicle that would fly to orbit and then return to Earth intact (without having jettisoned any "stage"). These vehicles, or "**spaceplanes**," use conventional or modified aircraft technology to reach the edge of space, and then rockets for the last boost into orbit.

SPACESHIPONE "SPACEPLANE"

### SPACE EXPLORATION
Overcoming Earth's gravitational pull requires the powerful thrust of a rocket engine, but once in orbit other technologies become available. **Ion engines** use **solar power** to **split atoms** of fuel into charged ions that are ejected from the spacecraft at high speed. Solar sails use the pressure of light from the Sun itself.

SEE ALSO »
pp.370–71 MOON LANDING
pp.372–73 MANNED SPACE TRAVEL

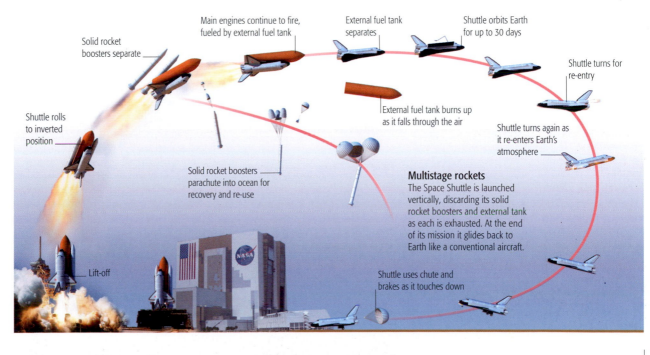

Solid rocket boosters separate

Main engines continue to fire, fueled by external fuel tank

External fuel tank separates

Shuttle orbits Earth for up to 30 days

Shuttle turns for re-entry

Shuttle rolls to inverted position

External fuel tank burns up as it falls through the air

Shuttle turns again as it re-enters Earth's atmosphere

Solid rocket boosters parachute into ocean for recovery and re-use

Lift-off

### Multistage rockets
The Space Shuttle is launched vertically, discarding its solid rocket boosters and external tank as each is exhausted. At the end of its mission it glides back to Earth like a conventional aircraft.

Shuttle uses chute and brakes as it touches down

# Galaxies, Clusters, and Superclusters

**In the 20th century, astronomers made unprecedented progress in understanding the nature and structure of the Universe. They discovered that it consists of billions of galaxies, which exist in clusters, and that these clusters also form groups, known as superclusters. The superclusters make thread- and sheet-like forms, which contain tens of thousands of galaxies and are the Universe's largest structures.**

At the beginning of the 20th century, Earth and the rest of the solar system were thought to lie at the center of a vast stellar system, called the Milky Way. Astronomers were divided about the nature and size of this galaxy, and the extent of the Universe itself.

### Galaxies beyond our own

In 1916 the American astronomer Harlow Shapley measured the size of the Milky Way and found it to be much larger than previously thought. This led him to believe in the one-galaxy universe. Others believed that the Universe consists of many galaxies, each existing like an "island" in space.

Observations made by Edwin Hubble on the huge Hooker telescope on Mount Wilson, California, were to prove one of these theories correct. The Hooker was the world's largest telescope from its first use in 1917 until

1948; its 100 in (254 cm) mirror marked a move away from the lens-based refractors of the late 19th century to mirror-based reflectors.

Hubble first detected the Cepheid stars (see p.319) in the Andromeda Galaxy in 1923, and used these to determine Andromeda's distance. By establishing the Cepheids's luminosity and apparent magnitude (brightness from Earth), he was able to calculate the stars' and the galaxy's distance from Earth, and so he proved that Andromeda Galaxy lies outside the Milky Way. By 1924 Hubble had found that there are galaxies in every

**Interacting galaxies**
Two galaxies, known together as Arp 87, interact as they swing past each other. Galaxies form when groups of stars join together; they grow and change through interacting with other galaxies.

direction. Today it is believed by scientists that the Universe consists of at least 125 billion galaxies.

### Spiral galaxies

A galaxy is a vast system of stars with massive amounts of interstellar gas and dust, all held together by gravity. Each contains hundreds of billions of stars. Galaxies

## BEFORE

The earliest astronomers observed that we are surrounded by stars. Countless more stars were seen for the first time in 1610, when Galileo turned the newly invented telescope on to the milky band of light that straddles Earth's sky.

### EARLY OBSERVATIONS

The first recorded observation of galaxies beyond our own was made by Persian astronomer **Al Sufi** in the late 10th century. He saw Andromeda and the **Large Magellanic Cloud**, but did not understand their true nature.

### MILKY WAY MODEL

After a telescopic study of the stars in the 1780s, William Herschel drew up the **first model of the system of stars** around us—unbeknown to him, this was the galaxy of the Milky Way. He discovered that we are immersed in a roughly disk-shaped stellar system.

### ASTROPHOTOGRAPHY

Immediately after its invention, photography was used to **record astronomical objects**. The earliest images included the Moon (1840); the stars (1850); and **the Milky Way** (1889).

**« SEE ALSO**
*pp.76–79* PLANETARY MOTION
*pp.318–19* THE EXPANDING UNIVERSE
*pp.328–31* THE LIFE CYCLE OF STARS

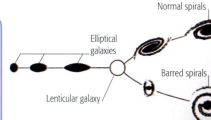

Normal spirals

Elliptical galaxies

Barred spirals

Lenticular galaxy

**The Hubble sequence**
This diagram shows Hubble's classification of galaxy types. Spirals are disk-shaped galaxies with arms of stars. Lenticular galaxies are disk-shaped, while elliptical galaxies range from spherical to squashed-egg shape.

**IMAGINARY GALAXY**

**SPIRAL REALITY**

**Spiral arms**
In an imaginary galaxy (top) stars follow aligned orbits. Stars closest to the center travel fastest; the more distant, the slower they travel. In a real spiral the orbits are not aligned, causing traffic jams of stars (bottom). Stars move in and out of them as they orbit.

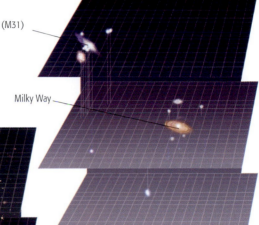

◁ **Andromeda Galaxy**
This is the closest large galaxy to Earth. It is a spiral and a member of the Local Group. By showing that it lies outside the Milky Way, Hubble vastly increased the size of the known universe.

Andromeda Galaxy (M31)

Milky Way

**Local Group**
Our Milky Way Galaxy is a part of the Local Group of galaxies. This also contains the Andromeda and Triangulum spiral galaxies. Many of the cluster's less massive and smaller galaxies orbit around them.

Dense core of cluster containing many large galaxies

**Rich cluster of galaxies**
Clusters containing many massive elliptical galaxies as well as hundreds of dwarf ellipticals occupy the same space as poor clusters such as the Local Group (right).

**13 BILLION** The number of years ago that the first galaxies formed.

**220 MILLION** The number of years the Sun takes to make one orbit of the Milky Way.

range in width from a few thousand to more than one million light years (a light year [ly] is the distance that light travels in one year: 5,878 billion miles or 9,460 billion km). Hubble studied the structure of the galaxies from 1925–53, and he devised a classification system based on shape and composition that is still used today, with some modifications. He recognized shapes such as spirals, barred spirals, ellipticals, and lenticulars, but there are also irregulars which have little or no structure.

The Milky Way was classified as a spiral galaxy throughout the 20th century, but is now widely believed to be a barred spiral. It measures 100,000 ly across and less than 4,000 ly deep, and it contains between 200 and 500 billion stars.

### New types of galaxy
In the middle of the 20th century, astronomers started to use telescopes that collected wavelengths other than visible light, which led to the discovery of more galaxies. The first discovered by its radio output (and so called a "radio galaxy") was Cygnus A, in 1939. Soon other types of galaxies

were discovered: Seyferts (1943), quasars (1963), and blazars (1978). These, together with radio galaxies, are known as active galaxies, because they emit more energy than comes from their stars. The energy source is material falling into a supermassive black hole at each galaxy's center. The four types of galaxy may actually be one type of object seen from different angles.

### Clusters and superclusters
When Shapley was studying wide-angle, deep-sky photographs in the 1930s he noticed that certain areas of

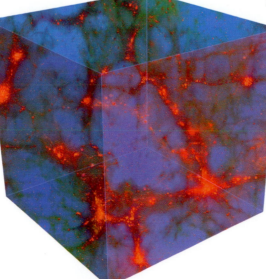

the sky contain more galaxies than others. This is because galaxies aren't randomly scattered in space; they exist in clusters.

The Milky Way is part of a poor cluster; these clusters have few members and are sometimes referred to as groups. Ours is the Local Group; it consists of more than 40 galaxies in a space stretching over 10 million ly. Rich clusters contain hundreds or thousands packed into a similar volume.

### The Great Wall
An astronomer in the US, George Abell, catalogued about 4,000 clusters. He showed that they exist together in superclusters in the form of flat sheets and filaments separated by huge voids. The first sheet, the Great Wall, was found in 1989. The largest structure known—The Sloan Great Wall—was revealed when over 200,000 galaxies were plotted on a 3-D map in 2003.

**Superclusters**
A computer-generated model reveals galactic superclusters (red) separated by huge voids (blue). The superclusters each contain thousands of galaxies, while the voids appear to hold nothing at all.

## AFTER »

The huge number of galaxies and stars discovered in the course of the 20th century diminished the importance of the Milky Way and underlined the possibility of life existing elsewhere in the Universe.

### CHARGED COUPLED DEVICE
From 1979 an electronic chip known as the **CCD** (charged coupled device) began to be used for **recording astronomical images**. This device, which is nearly 100 percent efficient when converting light into electronic signals, is now routinely used in everyday digital cameras.

### HUBBLE SPACE TELESCOPE
The Hubble Space Telescope was launched into orbit around Earth in 1990. It peers into space without the distortion of the Earth's atmosphere, **recording tens of thousands of new galaxies**.

### SEE ALSO »
pp.368–69 ARTIFICIAL SATELLITES
pp.394–95 INSIDE THE SOLAR SYSTEM
pp.398–99 DARK UNIVERSE

## BEFORE

The use of codes and ciphers dates back thousands of years, to the ancient Greeks and Romans and possibly before.

**EARLY CODES AND CIPHERS**
Ancient **Babylonian tablets** have **recipes encrypted** on them, and the ancient **Greeks and Romans** made much use of simple ciphers, often **for military purposes**. In ancient India, the *Kama Sutra* suggested ciphers as a way for **lovers to communicate secretly**.

**MORE COMPLEX CIPHERS**
In the 9th century the aristocratic Arab polymath **Al-Kindi** invented the first **polyalphabetic ciphers**, which were much more secure. In the 15th century, European scholars **Leone Battista Alberti** and **Johannes Trithemius** devised even more complex polyalphabetic ciphers.

**AMERICAN CIVIL WAR**
In 1862 **the Confederate Signal Service Bureau** adopted a letter-substitution cipher disk. In the American Civil War of 1861–65 **both sides devised ciphers** for secrecy.

**« SEE ALSO**
*pp.174–75* CALCULATING AND COMPUTING
*pp.260–61* TELEGRAPH TO TELEPHONE

# Codes and Ciphers

**Keeping information secret has been important to governments, diplomats, spies, and traders for thousands of years. Secret communication is vital in military campaigns, for example, and codes were developed so that messages could not be understood if they were intercepted by enemies.**

Specific phrases such as "attack at dawn" can be transmitted simply with a prearranged code such as "fly with the lark," but more complex communications require a system that encodes the basic units of the message in a particular way using a cipher. Encryption, the process of translating a message using a cipher, involves substituting letters or words of the original text (the plaintext) with another letter or symbol to produce the encrypted text (the ciphertext). The message can then be decrypted if the recipient has a key to the cipher.

### Codes and code breaking

Early ciphers, which date back at least as far as the ancient Babylonians, used simple substitution of letters in the plaintext to produce the ciphertext.

But messages coded in this way were vulnerable to cryptanalysis, or code breaking. The breakthrough in cryptanalysis came in the 9th century with frequency analysis—an invention often credited to the Arab polymath Al-Kindi. In any language, certain

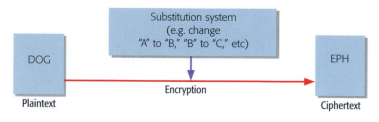

Substitution system
(e.g. change
"A" to "B," "B" to "C," etc)

DOG
Plaintext

Encryption

EPH
Ciphertext

**How a substitution code works**
One of the simplest types of encryption, a substitution code works by substituting letters in the plaintext in a systematic way (such as by shifting them by one letter, as above) to produce the ciphertext.

letters and letter combinations occur more frequently than others; for example, in English the letters E, T, A, and O are the most common. By analyzing the frequency of letters in the ciphertext, the corresponding letters in the plaintext can usually be deduced.

To try to stay ahead of cryptanalysis, cryptographers from Al-Kindi onward worked on cipher systems that used several alphabets—polyalphabetic systems. But as both encryption and decryption became more complex, code breaking depended increasingly

on knowing the key. This is where mathematicians and technicians stepped in, to develop complex mathematics and machines for cryptography and cryptanalysis.

### Mechanical and electronic cipher machines

Simple mechanical cipher devices such as cipher disks and wheels, which consist of rings inscribed with letters and numbers that can be aligned according to the cipher, had existed since the ancient Romans and continued to be used into the 20th century.

Much more complex, mathematically based ciphers demanded more complicated machines, although the principle was still the same: the system should be secure even when everything is known about it except the key. During World War I cryptographers increased the security of cipher systems with the one-time pad—a unique key used only once and then discarded—rather than a single re-use key for all messages. Machines were invented with a keyboard operating an intricate set of rotors, and these were further developed with the

### Bolton cipher wheel
Devised in the 19th century and based on the cipher disk of Leone Battista Alberti, the Bolton cipher wheel comprised concentric disks with two sets of letters and numbers that could be rotated. The movable aperture was used to encrypt and decrypt individual letters and numbers: the letter "I" becomes "A" in the example above.

Concentric ring of numbers

Movable aperture for encrypting and decrypting

Concentric ring of letters

Turning knob

Reading aperture

### Kryha cipher machine
Used in World War II by German diplomats, the Kryha cipher machine had two rings, each with an alphabet. Encryption involved finding the plaintext letter on the outer ring and using the adjacent letter on the inner ring for the ciphertext letter.

Inner cover (in open position)

Concentric rings, each with an alphabet

### Enigma cipher machine
First built in 1918 and used by German forces during World War II, the Enigma was an electromechanical cipher machine that used rotors and plug connections to encrypt messages. The rotor settings and plug connections could be altered to produce almost unbreakable polyalphabetic codes.

Metal cover fits over rotor cylinders

Viewing window shows code letters

Coding rotor

Lightboard

Keys for typing in messages

Plugboard setting altered regularly to change cipher

Klappe schließen

**Colossus code-breaking machine**
One of the earliest electronic digital computers, the Colossus played a key role in deciphering coded German messages during World War II. The first one was built in 1943 and filled an entire room.

## AFTER

Computer-based systems have entirely replaced mechanical devices for encryption, but even these are not totally secure and so new methods have been developed.

### QUANTUM CRYPTOGRAPHY

First proposed in the 1970s and developed in the 1980s, quantum cryptography relies on **quantum mechanical effects** for secure communication. Essentially it uses **quantum states** (usually of photons—quanta of light) to **encode the key**. The key is then sent to the intended recipients. Because the key is encoded in quantum states, any **interception can be detected** and a new key sent. When a key arrives safely, **messages are transmitted** using **normal encryption** methods.

**SEE ALSO >>**
pp.344–45 ALAN TURING
pp.378–79 THE INTERNET

advent of electronic technology into devices, such as the extraordinarily complex German Enigma machine, that were capable of producing almost unbreakable ciphers.

As cryptography made increasing use of new technology, so too did cryptanalysis. Breaking the German Enigma and other ciphers was vital to the allied victory in World War II, and a huge investment in terms of time, money, and people was made to achieve this difficult end. The work also stimulated the development of electronic code-breaking machines such as the Bombe and Colossus, the forerunners of modern computers.

### Modern encryption

With the rise of the computer not only did cryptography become increasingly mathematical, but more nonmilitary

uses were found for it. While still used for transmitting military and government secrets, encryption is now also common for protecting data in business and personal life. Public key encryption (see below) lets us access coded information, and protects it from misuse in daily transactions, such as using ATMs and Internet shopping.

**MP3 encoding**
Music downloads are encoded in a digital audio format (MP3) that converts the music into binary form and also compresses the music data to make the MP3 files small enough for easy transmission and storage on an MP3 player.

### IN PRACTICE

## PUBLIC KEY ENCRYPTION

The basis of most Internet communications, public key encryption uses two huge prime numbers and their product. A public key, consisting of one of the prime numbers and their product is freely available, but the private key – the other prime number – remains the secret of its owner. Because of the huge size of the numbers, it is not feasible to calculate the private key, even knowing the public key. Therefore anybody can encrypt a message using the public key but only the owner of the private key can decrypt it.

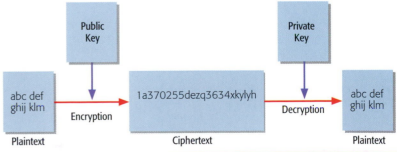

MATHEMATICIAN AND CRYPTANALYST **Born 1912** Died 1954

# Alan Turing

## "**We can only see** a short distance ahead, but we can **see plenty** there that **needs to be done**."

ALAN TURING, "COMPUTING MACHINERY AND INTELLIGENCE", 1950

A pioneer in the fields of computing and artificial intelligence, Alan Turing was one of the founders of modern computer science. His work on algorithms led to the concept of an abstract "machine" that was fundamental to the development of computers, as well as a test that considered in a practical way how far a machine can be said to be "intelligent" (see pp.380–81). Later, his interests diversified into chemistry and developmental biology.

Turing was born into a middle-class family in London, England, but he was brought up by friends when his parents returned to India, where his father worked as a civil servant. He attended Sherborne School, a public school in Dorset, where he voraciously pursued his interests in mathematics and science through a guided reading programme, although he neglected his classics. He was accepted on a degree course at King's College, Cambridge, to study mathematics.

**Turing the athlete**
Turing, who ran as a relief from stress, reached world-class Marathon standards. In a 1948 cross-country race he finished ahead of Tom Richards, who won the silver medal in the 1948 Olympics.

### The Turing machine
At Cambridge, Turing studied for the mathematical tripos. After graduating he followed Max Newman's course on the foundations of mathematics, producing a dissertation entitled *On the Gaussian Error Function*. The dissertation earned Turing a fellowship at King's College, where he worked on probability theory; he later studied under Alonzo Church at Princeton University. In a paper published in 1936 he proposed an abstract device that mathematically modelled the operation of a computing machine, reading instructions from a tape. Later known as a "Turing machine", this described the working of a computer before the technology existed to build such a machine.

### Cracking codes
Gaining his PhD at Princeton in 1938, Turing then returned to Britain, where he was approached by the Government Code and Cypher School to work on deciphering codes of the German Enigma machine (see pp.342–43). At first he worked for them only part time, while starting the construction of a mechanical computing device back at Cambridge, but after the outbreak of World War II he moved to Bletchley Park, headquarters of the Code

**Alan Turing**
Photographed here in 1951, Turing is widely recognized as the visionary of the general-purpose computer. Not only was he responsible for the creation of the first practical prototype, he also foresaw the limitations that computers would theoretically encounter.

## ELECTRONIC STORED-PROGRAM COMPUTER

A stored-program computer stores not only data but also its programmed instructions within its writable random access memory. Between 1946 and 1947 Turing formulated designs for a large computer of this type, the ACE, but its development was postponed. The Pilot ACE, a smaller model based on Turing's designs, was first demonstrated in 1950. It superseded earlier computers that were either not fully electronic or used only read-only memory.

**Bletchley Park**
The urgent wartime task at Bletchley Park was to break the German Enigma cypher, used for military and intelligence communications, and convey decoded information to the Allied military. About 10,000 people were involved.

**The Turing–Welchman Bombe**
To decode Enigma transmissions, Turing calculated that a machine capable of replicating 60 German Enigma machines would be required. Conceived with the help of Gordon Welchman and built by Harold "Doc" Keen, the machine, named the Bombe, was installed at Bletchley Park in August 1940. In all, about 200 bombes were built.

## TIMELINE

- **23 June 1912** Alan Turing is born in Maida Vale, London. His father, who works for the Indian Civil Service, wants Alan to be brought up in England. When he is about one year old his parents return to India, leaving him with friends of the family in Hastings.

- **1926** Takes the Common Entrance Examination for entry to public school, and is accepted at Sherborne School in Dorset.

- **February 1930** His best school-friend, Christopher Morcom, dies of bovine tuberculosis. Turing, greatly affected by his death, questions his own religious faith and becomes an atheist.

- **1931** Fails to win a scholarship to Trinity College, Cambridge, but is accepted by his second choice, King's College, to study mathematics.

- **1935** Elected Fellow of King's College for his dissertation in probability theory, in which he proves the central limit theorem.

- **1936** Publishes *On Computable Numbers, with an Application to the Entscheidungsproblem*. In this paper he introduces the abstract machine, later called a "Turing machine".

- **1936** Travels to the US and studies at Princeton University under Alonzo Church, later gaining a PhD in mathematical logic.

- **1939** After war begins, Turing works at the Government Code and Cypher School at Bletchley Park. He helps develop the "Bombe" to decode messages from the Enigma machine.

- **1942–43** Goes to the US to liaise with American experts on cryptanalysis.

- **1945** Is awarded the OBE for his war services.

- **1945** Designs a computer for the National Physical Laboratory in London. Because of its size and complexity, work on its construction is delayed until after Turing has left the laboratory.

- **1947** Returns to Cambridge for a sabbatical year. He explores interests such as neurology and physiology, and becomes interested in athletics.

- **1948** Is offered a readership at the Mathematics Department of the University of Manchester, where he works in the computing laboratory.

- **1950** Publishes *Computing Machinery and Intelligence* in the journal *Mind*, studying the fundamental questions of computing and artificial intelligence and introducing the "Turing test" to determine whether a machine can be called "intelligent".

- **1951** Is elected a Fellow of London's Royal Society.

- **1952** Opts for a course of hormone treatment following a trial for homosexuality.

- **7 June 1954** Dies at his home in Wilmslow, Cheshire, of cyanide poisoning. The ruling that he took his own life is generally accepted.

and Cypher School, to devote himself full time to code-breaking. With his colleague, Gordon Welchman, he developed the "Bombe", a computing machine based on work done by Polish cryptographers, and successfully broke the Enigma code. Subsequent, more complex German codes were also eventually deciphered, and he was awarded the OBE in 1945.

### Stored-program computer

Turing was invited to work at the National Physical Laboratory in London, where he used his experience of Colossus, the code-breaking computer (see p.343), to design an Automatic Computing Engine (ACE). This would have been the first-ever stored-program computer, but its construction was delayed and it did not run its first program until 1950.

Disillusioned by the delay, he accepted a readership at Manchester University in 1948. Here he studied the question of how far a machine can be considered capable of "thinking". He devised an experiment to help determine whether a computer could convince an interrogator in conversation that it was in fact human. The "Turing test", as it became known, is still used today to ascertain a machine's ability to show human-like intelligence.

### Studies in morphogenesis

During the early 1950s he returned to subjects he had pursued at Cambridge a few years before. He was fascinated by neurology and physiology, and he sought to apply his mathematical theories to biology, which culminated in a paper on morphogenesis (the development of form and pattern in living organisms) in 1952.

### Trial and suicide

However, in that same year, during the investigation of a break-in at his home, he naively admitted to a homosexual affair with the intruder. He was tried and convicted for homosexuality, which was a serious offence at that time, and offered the choice of either a prison sentence or hormone treatment. Rather than face jail, he underwent a series of oestrogen injections, which he took with characteristic good humour. The government no longer gave him security clearance for decoding work, although this freed him to concentrate on his Manchester projects.

In June 1954 he was found dead at his home, apparently poisoned by cyanide on a half-eaten apple. His death was ruled suicide, but his mother maintained that he had accidentally poisoned himself with chemicals he was using to silver-plate cutlery.

> ## "Machines take me by surprise with **great frequency**."
> ALAN TURING, "COMPUTING MACHINERY AND INTELLIGENCE", 1950

**TURING MEMORIAL STATUE AT BLETCHLEY**

# The **Structure** of **DNA**

**It was the ultimate biological question: how does life spring from life? By the early 1950s scientists had unraveled many of the body's inner workings, including the discovery that DNA carried genetic information, but the structure of this molecule eluded them. But on a spring day in England, in a year that a young woman was crowned Queen Elizabeth II and Edmund Hillary conquered Everest, the mystery was solved.**

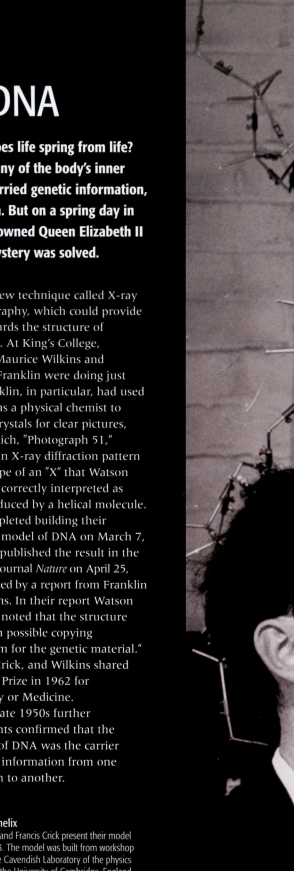

New kinds of biologists, their roots in physics, emerged in the 1950s from a world rocked by war. At the University of Cambridge, England, Francis Crick was one such scientist. He was joined by a precocious, young American called James Watson, who arrived at the Cavendish Laboratory in Cambridge for postdoctoral research on viruses. Neither Watson nor Crick was officially sanctioned to work on deoxyribonucleic acid (DNA), but both had been inspired by tantalizing reports questioning its structure.

Proof that DNA was the molecule responsible for carrying genetic information had already come in 1943 with an experiment by the American medical researcher Oswald Avery. However, most scientists suspected that his conclusions were flawed; they thought that protein, rather than DNA, had to be responsible since it had the complexity necessary. By the early 1950s opinion was still divided among scientists, but Watson and Crick were convinced that DNA was the true genetic material. The prospect of unlocking the secrets of DNA had been opened up by an exciting new technique called X-ray crystallography, which could provide clues towards the structure of molecules. At King's College, London, Maurice Wilkins and Rosalind Franklin were doing just that. Franklin, in particular, had used her skills as a physical chemist to produce crystals for clear pictures, one of which, "Photograph 51," revealed an X-ray diffraction pattern in the shape of an "X" that Watson and Crick correctly interpreted as being produced by a helical molecule. They completed building their structural model of DNA on March 7, 1953 and published the result in the scientific journal *Nature* on April 25, accompanied by a report from Franklin and Wilkins. In their report Watson and Crick noted that the structure "suggests a possible copying mechanism for the genetic material." Watson, Crick, and Wilkins shared the Nobel Prize in 1962 for Physiology or Medicine.

By the late 1950s further experiments confirmed that the structure of DNA was the carrier of genetic information from one generation to another.

**The double helix**
James Watson and Francis Crick present their model of DNA in 1953. The model was built from workshop materials in the Cavendish Laboratory of the physics department of the University of Cambridge, England.

**Rosalind Franklin**
Franklin tragically died from ovarian cancer in 1958. The Nobel Prize rules prevent posthumous nominations, so she was never fully rewarded for her crucial X-ray diffraction image of DNA.

Every living organism is made up of a complex substance called protein. It is central to the structural fabric of cells and, in the form of enzymes, it controls the chemical reactions of life.

By the late 19th century many scientists already thought that the cell nucleus held the secret of inheritance. A phosphoric-acid-rich substance had been identified in cell nuclei in 1869. In 1937 the British physicist William Astbury produced X-ray diffraction patterns that showed this substance to be deoxyribonucleic acid (DNA), which had a regular structure, quite unlike the tangled form of proteins.

By 1953 James Watson and Francis Crick had confirmed that this structure was a double helix, which enabled the fundamental process known as replication to occur (see pp.346–47). Scientists knew that protein

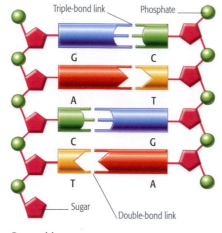

**Base pairing**
The bases of nucleotides of DNA can pair up only in two configurations because their sizes and shapes fit. Each pair has one large base and one small, adhered by either three bonds—guanine (G) with cytosine (C)—or two bonds—A (adenine) with T (thymine).

# The Genetic Code

**The characteristics of organisms are established by differences in their chemical makeup. One of these chemicals, DNA, carries the instructions for building another important substance—protein. It is the remarkable variety of proteins that accounts for the great diversity of life processes.**

**BEFORE**

The idea of genes as particles of inheritance originated with Gregor Mendel in 1866, but that the genes were composed of DNA would not be widely appreciated until the 1900s.

**CONFLICTING IDEAS ABOUT HEREDITY**
In the early 1800s French zoologist **Jean-Baptiste Lamarck** proposed a theory that characteristics acquired during the life of an organism could be passed on. In 1859 **Darwin's** evolutionary theory **≪ 198–99** corrected the principle of this view. But it was not until the 20th century that **Gregor Mendel's Laws of Inheritance ≪ 306–307** were combined with studies of **DNA** and **protein** production to prove that acquired characteristics could not be inherited.

**DARWIN'S FINCHES**

**≪ SEE ALSO**
*pp.308–309* CHROMOSOMES AND INHERITANCE

production was influenced by DNA, but they did not know how and so began an exciting new line of enquiry.

## Determining characteristics
DNA provides the instructions for the formation of protein (see opposite). As an organism grows, its cells divide in such a way that every cell has a copy of the DNA instructions for making protein; although some cells use these instructions selectively. In 1902 English physician Archibald Garrod anticipated a link between genetic material and protein. He recognized that some inherited diseases upset the balance of chemical reactions: some proteins were not made properly because the instructions were faulty. He called them "inborn errors of metabolism."

In 1941 American geneticists George Beadle and Edward Tatum found they could induce similar errors in bread mold. Their work confirmed that proteins—specifically enzymes (protein that speed reactions)—were affected. Beadle and Tatum identified that sections of DNA—the gene—provided the codes to make enzymes and was later modified to include all proteins.

Every organism produces thousands of proteins, each encoded by thousands of different genes packaged in every cell of the body. DNA's ability to copy itself and replicate means that the genes in all cells of the same body are identical, but vary from one body to another. Likewise the protein encoded differs, so their characteristics vary too.

## Cracking the code
DNA and proteins are long-chain molecules (polymers) made up of repeated building blocks. By 1929 Phoebus Levene, a Russian–American biochemist, discovered that DNA was made up of units called nucleotides, which consist of one sugar, one phosphate, and one base. The base can be one of four types: adenine, cytosine, thymine, or guanine. Proteins, on the other hand, have 20 different types of building blocks called amino acids.

In the aftermath of the double-helix breakthrough biologists began to look at how the nucleotide sequence in DNA determined the amino-acid chain of a protein. In 1961 Francis Crick and South African biologist Sydney Brenner confirmed earlier research and demonstrated that a set of three

**Cell**
All plant and animal cells have a nucleus.

**Nucleus**
This is the cell control center and contains the DNA in the form of chromosomes.

**Cytoplasm**
The part of a cell in which protein synthesis takes place.

**Chromosome**
Tight package of DNA that develops at cell division, formed to make DNA copies easier to manipulate.

**Chromatin thread**
Consists of DNA coiled around proteins called histones.

**Packaging of DNA**
Lengths of DNA double helix are organized into threads that solidify to form chromosomes. These unravel between cell divisions at certain points, enabling the DNA molecules to expose their information for making protein and to allow replication.

**Histone protein**
This acts as a scaffold for the huge quantity of DNA to coil around.

**Sugar-phosphate "backbones"**
A double helix of DNA consists of two chains of nucleotides. The "backbones" are made up of units of sugar and phosphates either side of a core of paired bases.

nucleotide bases (or triplet) encodes one amino acid. Crick and Brenner found that a virus stopped making protein if they inserted (or deleted) one or two nucleotides from its DNA; but if three nucleotides were inserted, the virus could still make protein.

In the same year, American geneticist Marshall Nirenberg and German biochemist J. Heinrich Matthaei cracked the first part of the code. They made a chain of a single nucleotide, which produced a protein composed of identical amino acids. Over the next 10 years all 20 amino acids were decoded.

## Making proteins
Protein is made on granules called ribosomes in a cell's cytoplasm. So somehow, instructions from the genes must "move out" of the nucleus. Spanish biochemist Severo Ochoa focused on a substance related to DNA called ribonucleic acid (RNA). First identified in 1909, RNA is found in both the nucleus and the cytoplasm of cells, and transpired to be the "messenger" molecule (mRNA). The final link came in 1964 when American biochemist Robert Holley cracked the structure of smaller transfer RNA (tRNA) molecules

that brought the amino-acid building blocks to the ribosome for assembly into proteins. By understanding protein synthesis, molecular biologists had finally unveiled the blueprint of life itself.

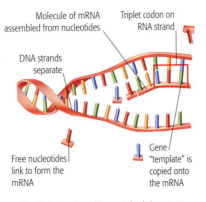
Molecule of mRNA assembled from nucleotides
Triplet codon on RNA strand
DNA strands separate
Free nucleotides link to form the mRNA
Gene "template" is copied onto the mRNA

**1** The first stage in making proteins is known as "transcription". Part of the DNA double helix unwinds to expose a length of one strand called a gene. A molecule of mRNA is assembled along the other strand, building a complementary "copy" of the gene sequence or gene code.

### Building proteins
Protein synthesis occurs within all cells and has two distinct stages. Transcription takes place inside the nucleus and translation outside it.

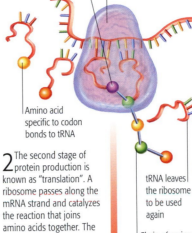

Amino acid brought to the ribosome by transfer RNA (tRNA) molecule
Ribosome straddles two base triplets to accommodate two amino acids
Amino acid specific to codon bonds to tRNA
tRNA leaves the ribosome to be used again
Chain of amino acids develops

**2** The second stage of protein production is known as "translation". A ribosome passes along the mRNA strand and catalyzes the reaction that joins amino acids together. The mRNA provides the information and the tRNA brings amino acids to build up the chain. Once complete, the chain folds into a protein.

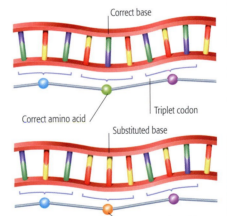
More amino acids added till code fully translated
Protein folds up into a shape determined by its sequence of amino acids

With the understanding of the chemical basis of inheritance came the controversial possibility of manipulating genetic material.

### MAPPING GENES
A genome is a complete **catalogue of the genes** of a cell or organism. The first draft of the **human genome 410–11 »**, published in 2001, provides a greater understanding of **genetic diseases**.

### DNA-CUTTING ENZYMES
Scientists developed tools to isolate genes for further study. Bacterial DNA exists free in the cytoplasm. Bacteria contain chemicals called restriction enzymes, which the cells use to **"cut"** DNA and **exchange genetic material**; a process that can be replicated in the laboratory.

**SEE ALSO »**
*pp.386–87* GENE TECHNOLOGY
*pp.410–11* THE HUMAN GENOME

Adenine–thymine link
Guanine–cytosine link

Correct base
Correct amino acid
Triplet codon
Substituted base
Incorrect amino acid

△ **Gene mutation**
DNA replicates when cells divide, and normally makes an exact copy of itself. Errors in the replication process can result in alterations in the base-pair sequence of a gene, so that a different amino acid is substituted, producing a different variety of protein.

**Complementary base pairs**
Pairs of nucleotides are arranged in a specific order that is replicated before cell division. The sequences of three bases (the gene code or triplet codon) determine the manufacture of specific proteins.

**Lack of pigment**
A pigment called melanin normally occurs in skin, hair, and the iris of the eye. Lack of pigment, known as albinism, is caused by a faulty variant of a gene that carries the pigment-making code for melanin.

MICROBIOLOGIST (1927–)
### SYDNEY BRENNER
Brenner grew up in South Africa, but was educated at Oxford University. In 1953 he saw Watson and Crick's model of the DNA double helix, which inspired his work on the genetic code. He made groundbreaking discoveries in animal development using a roundworm as his model species. He shared the Nobel Prize for Medicine in 2002 and also founded the Molecular Sciences Institute in California.

# Chaos Theory

**Until recently the apparently random, unpredictable behavior of systems such as weather was hard to reconcile with the traditional idea of a world governed by universal physical laws. Chaos theory can now explain these phenomena, with surprising and often beautiful results.**

For centuries scientists explained seemingly chaotic phenomena, such as weather and turbulence in fluids, as complex but ultimately predictable according to the laws that Newton and his contemporaries had proposed. This situation changed in the latter half of the 20th century, when a theory of "deterministic chaos," now better known simply as chaos theory, emerged to explain the underlying patterns of behavior.

## Unpredictability

The beginnings of chaos theory lay in the discoveries of Jacques Hadamard and Henri Poincaré, who were both working in the field of differential equations and dynamical systems (systems of variables that interact and change with time). Poincaré noticed that even tiny changes in the initial conditions of a system often resulted in large—and unpredictable—changes in outcome; in other words, his results exhibited chaotic behavior. Although he published his findings as early as 1908, they were largely ignored by the scientific community, who were more concerned with the hot topics of the day: relativity (see pp.302–03) and quantum theory (see pp.314–17).

It was not until the 1960s that chaotic behavior entered mainstream science, largely due to the development of electronic computers. The discovery that sparked this interest was made by Edward Lorenz, who worked at the Massachusetts Institute of Technology in the notoriously unpredictable field of meteorology. Using a series of nonlinear equations, he ran models of atmospheric conditions through his computer. Following one such run, he set up a second run using his figures for the initial conditions, but this time rounded down from six decimal places

### MATHEMATICIAN (1854–1912)

### HENRI POINCARÉ

Mathematical physicist Henri Poincaré was born in Nancy, France. In 1879 he earned his doctorate from the University of Paris, for his thesis on the properties of functions defined by differential equations. In applying these ideas to celestial mechanics, he identified some of the concepts behind chaotic systems. Poincaré made important contributions to various mathematical fields, notably algebraic topology. He also wrote extensively on science for the public.

## BEFORE

**Before chaos theory, scientists thought every aspect of the Universe operated according to a "deterministic" pattern—if you knew the initial conditions, you could predict how the system would behave.**

(often wrongly attributed to Newton himself) that the **Universe** operates in an **entirely predictable** way, like clockwork.

### CLOCKWORK UNIVERSE

In 1686 **Isaac Newton** published his groundbreaking *Philosophiae Naturalis Principia Mathematica* ("Mathematical Principles of Natural Philosophy"), in which he proposed that **gravity was a universal force**, acting on every kind of matter from apples to planets. Newton's discoveries gave rise to the idea

**NEWTON'S MODEL OF PLANETARY ORBITS**

The smoke suddenly breaks into turbulent, disordered motion

The smoke initially rises in a predictable and regular flow

## Unpredictable behavior

One example of a chaotic system is smoke rising from a point (such as an extinguished candle). The smoke leaves the source with only minute differences in speed, direction, and other conditions. Beyond a certain point, the increase in these differences suddenly causes the flow to break down into turbulent eddies.

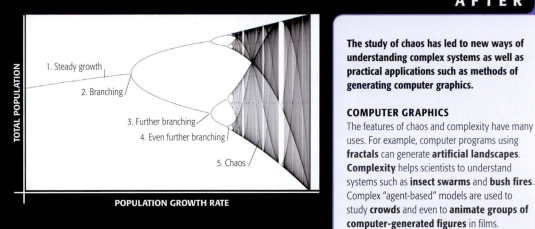

**Chaos over time**
Biologist Robert May showed how changes in just one parameter—the population growth rate—can lead to chaos. At first the population increases steadily. However, at a certain growth rate there is a sudden branching, with the population alternating between two levels in successive years; this leads to further splits, until chaos results.

TOTAL POPULATION

1. Steady growth
2. Branching
3. Further branching
4. Even further branching
5. Chaos

**POPULATION GROWTH RATE**

The study of chaos has led to new ways of understanding complex systems as well as practical applications such as methods of generating computer graphics.

## COMPUTER GRAPHICS

The features of chaos and complexity have many uses. For example, computer programs using **fractals** can generate **artificial landscapes**. **Complexity** helps scientists to understand systems such as **insect swarms** and **bush fires**. Complex "agent-based" models are used to study **crowds** and even to **animate groups of computer-generated figures** in films.

**COMPUTER-GENERATED FRACTAL LANDSCAPE**

to three; the results were wildly different. After repeating this experiment with a simplified model, Lorenz was convinced that even minor changes in initial conditions led to unpredictably different outcomes. But he also discovered that there was some kind of pattern to this unpredictability: plotting the results in a three-dimensional graph revealed a complex double spiral, later dubbed the "Lorenz attractor" (see below, right). Lorenz's discoveries were published in the *Journal of the Atmospheric Sciences* in 1963 under the title "Deterministic Nonperiodic Flow," but, like Poincaré's, they went unnoticed for some years.

### Attractors and fractals
At much the same time as Lorenz was finding a structure in unpredictable systems, mathematicians were exploring ways to model dynamical systems. Classical mechanics describes the behavior of such a system geometrically as an "attractor": a set of states into which a dynamical system evolves in time, with points that get close enough to the attractor, staying close even when they are slightly disturbed. In a steady state an attractor would be represented as a single point, in periodic cycles as a closed loop, and

### The Mandelbrot set
The Mandelbrot set is a group of complex numbers that produce striking images when represented graphically. The boundaries of the shapes show a self-similarity at all scales, as you zoom in, which is characteristic of fractality.

in several cycles as a torus (the shape of a ring doughnut). At the University of California at Berkeley, though, Stephen Smale discovered a new class of attractors—"strange attractors," which showed chaotic dynamics in the same way as Lorenz's double spiral.

Another mathematician, James Yorke, read Lorenz's paper in the 1970s and saw its mathematical importance. It tied in with his study of Robert May's work on population biology, in which nonlinear equations representing cyclical changes in population began to double before suddenly leading to unpredictable fluctuations—but within the "random" period, doubling would re-emerge. It was Yorke who first coined the term "chaos" in describing the study of this subject.

Similar discoveries were being made in other disciplines. Scientists including Mitchell Feigenbaum, investigating the phase transition from steady state to turbulence in fluid dynamics, found period-doubling patterns just like Yorke and May's. Mathematical economist Benoît Mandelbrot found the same in fluctuations of cotton prices.

What was most startling about the patterns of strange attractors and periodic order within disorder was that they had the same detailed structure at every level. Mandelbrot's rendition of the geometry of this phenomenon, aided by computer graphics, revealed recurring patterns at all scales of magnification in a class of complex geometric shapes he called fractals.

By the late 1970s chaos had become a recognized field of study, and centers for chaos and nonlinear studies were set up in the United States and the Soviet Union. It was considered a key part of the field of mathematics known as complexity, which includes study of both "organized" and "disorganized" systems and the interactions of their individual components.

Because of the almost simultaneous discovery of chaos theory in so many unrelated disciplines, it soon became clear that applications could be found in a variety of different fields where apparent disorder manifested itself. The mathematics of chaos theory could be used, for example, in astronomy, to investigate the structure of galaxies.

**BREAKTHROUGH**

## THE BUTTERFLY EFFECT

Edward Lorenz coined the phrase "the butterfly effect" to describe the main feature of chaotic systems, "sensitive dependence upon initial conditions"—in other words, even one tiny alteration to initial conditions, when repeated again and again, would quickly lead to large deviations. As Lorenz put it, "Does the flap of a butterfly's wing in Brazil set off a tornado in Texas?" The (coincidentally butterfly-shaped) model that he produced from his data, called a "Lorenz attractor," shows the chaotic process in motion—with sudden reversals, and never repeating the same path twice.

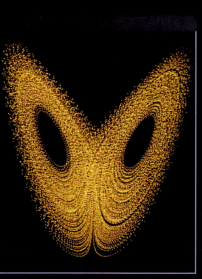

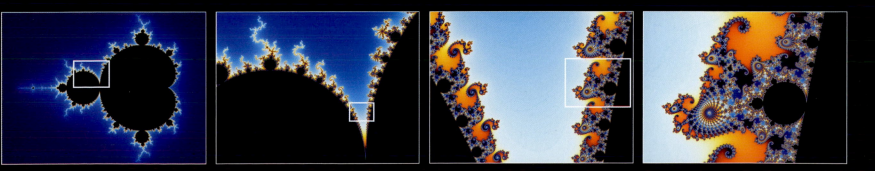

# The Structure of the Earth

**Until recently, knowledge about Earth's interior has been sparse, though many people have made some fanciful guesses. Since Earth's core is so inaccessible, scientists have had to use indirect means, notably the study of earthquake waves, to investigate its structure and understand its composition.**

## « BEFORE

Some past ideas about Earth's structure seem bizarre given current knowledge.

### WATER-FILLED EARTH
17th-century German scholar Athanasius Kircher produced drawings suggesting that Earth's interior contained **huge water-filled caverns**, rivers, and interlinked fiery chambers. In *Sacred Theory of the Earth*, 1681, English theologian Thomas Burnet also suggested that the planet contained large subterranean voids full of water.

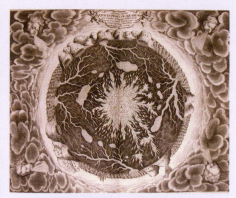

KIRCHER DRAWING OF EARTH'S INTERIOR

### HYPOTHESIS AND FACT
English scientist **Edmond Halley** proposed, in 1682, that Earth's interior consisted of a series of thin, **spherical shells**, with gas in the spaces between. In 1798 English chemist Henry Cavendish was the first to calculate Earth's density accurately to be 5.5 times that of water.

**« SEE ALSO**
*pp.38–39* Ancient Ideas of the World

▽ **Lava flow**
Lava is melted rock from the mantle that has reached and flowed on to Earth's surface via a volcanic vent. Here, thick lava can be seen flowing down the flank of Kilauea, Hawaii.

From the 19th century scientists realized that the best way to find out more about Earth's interior was to examine how earthquake waves pass through it. When an earthquake occurs, it produces shock waves, some of which travel through Earth's interior—these are known as body waves and can be detected when they arrive back at the surface. In the late 19th century Irish geologist Richard Oldham discovered two types of body wave—primary (P) waves, which can pass through solids and liquids, and secondary (S) waves, which can travel only through solids. The velocity of these waves varies with the density of the material they are passing through,

and they refract (change direction) when they encounter any sharp changes in density. In 1906 Oldham found that P-waves take longer than expected to travel straight through the earth and inferred that the planet must contain a dense core, which slows the waves as they pass (see below left). He also noted a seismic shadow zone on Earth's surface, where no P-waves were recorded, and concluded this must be due to waves being refracted by the core.

Later, in 1909, Croatian geophysicist Andrija Mohorovičić found that at various locations distant from the source identical sets of P- and S- waves arrived twice and deduced that one set

has traveled near the surface and another has traveled via a deeper, denser layer—the mantle.

## Probing the core
Over the next 30 years more information was gleaned about Earth's core from studying earthquake waves. In the 1920s British scientist Harold Jeffreys discovered that for every earthquake, there is a large shadow zone on the opposite side of Earth for S-waves and this implied that at least part of the core was liquid (see below center). A decade later Danish seismologist Inge Lehmann found that the core has inner and outer parts, by discovering a weak P-wave in the

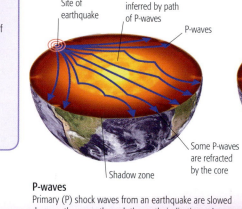

**P-waves**
Primary (P) shock waves from an earthquake are slowed down as they pass through the earth, indicating a dense core. The core also produces a surface shadow zone, where no P-waves are recorded.

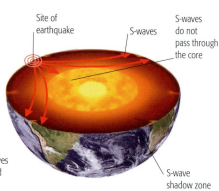

**S-waves**
No secondary (S) shock waves can be detected on the opposite side of the globe from an earthquake. This provided early evidence that at least the outer region of Earth's core is liquid.

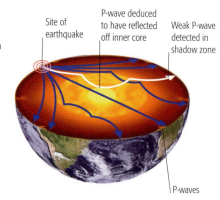

**Evidence of inner core**
Inge Lehmann detected a weak P-wave in the P-wave shadow zone following an earthquake. She called this P′ (P prime) and deduced that it must have been caused by a P- wave reflecting off the surface of an inner part of the core.

SEISMOLOGIST (1888–1993)

## INGE LEHMANN

Born in Copenhagen, Inge Lehmann studied mathematics at college, then became a seismologist (expert in earthquakes)—for 25 years, she was the only Danish seismologist. In 1936 she set out her argument that Earth's core has inner and outer parts in a paper called *P′*—reputedly the shortest title in the history of science. An enthusiastic traveler and skier, she spent several years in the US after her retirement in 1953, and collaborated with other scientists in studying Earth's upper mantle.

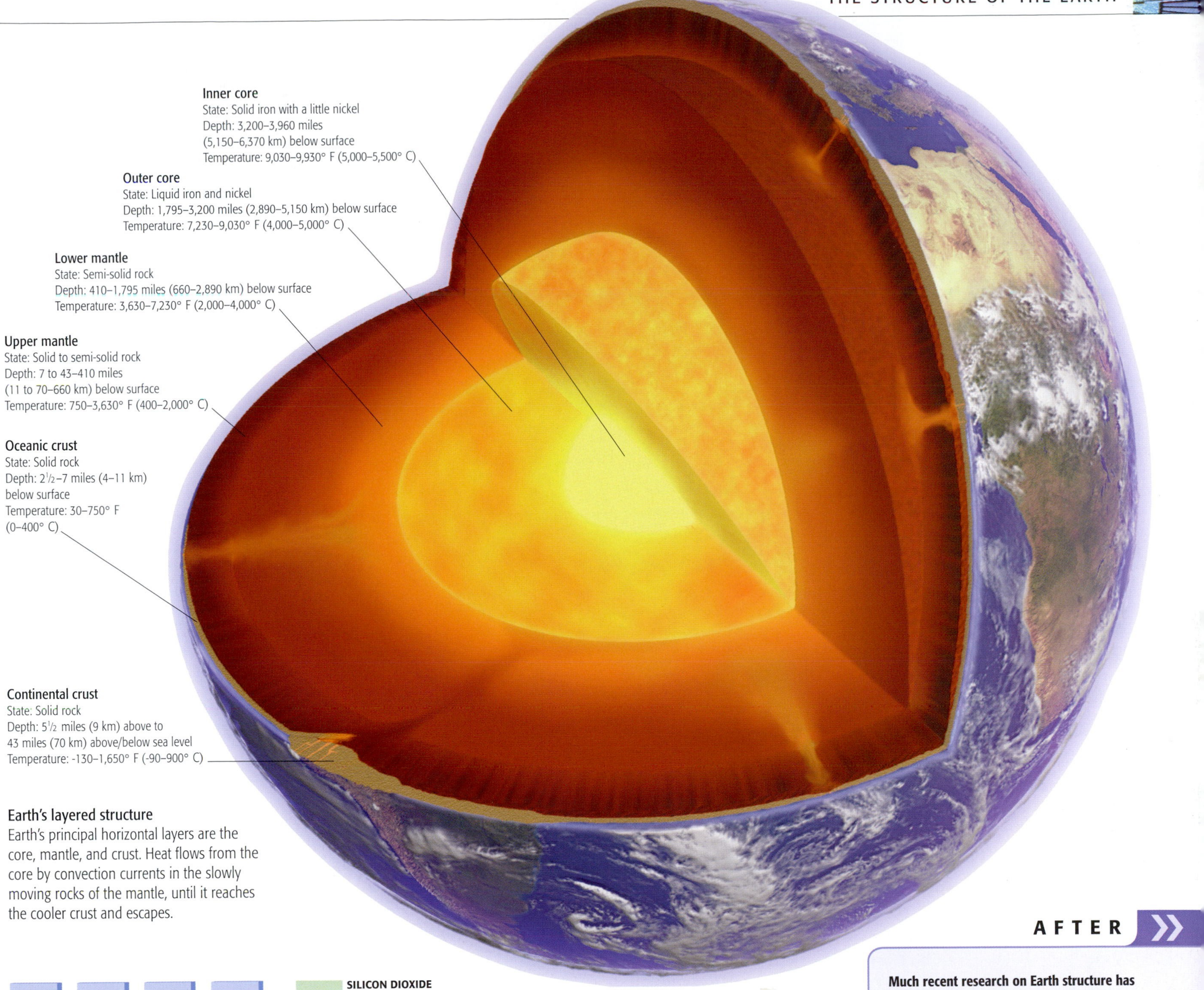

**Inner core**
State: Solid iron with a little nickel
Depth: 3,200–3,960 miles
(5,150–6,370 km) below surface
Temperature: 9,030–9,930° F (5,000–5,500° C)

**Outer core**
State: Liquid iron and nickel
Depth: 1,795–3,200 miles (2,890–5,150 km) below surface
Temperature: 7,230–9,030° F (4,000–5,000° C)

**Lower mantle**
State: Semi-solid rock
Depth: 410–1,795 miles (660–2,890 km) below surface
Temperature: 3,630–7,230° F (2,000–4,000° C)

**Upper mantle**
State: Solid to semi-solid rock
Depth: 7 to 43–410 miles
(11 to 70–660 km) below surface
Temperature: 750–3,630° F (400–2,000° C)

**Oceanic crust**
State: Solid rock
Depth: 2½–7 miles (4–11 km)
below surface
Temperature: 30–750° F
(0–400° C)

**Continental crust**
State: Solid rock
Depth: 5½ miles (9 km) above to
43 miles (70 km) above/below sea level
Temperature: -130–1,650° F (-90–900° C)

## Earth's layered structure
Earth's principal horizontal layers are the core, mantle, and crust. Heat flows from the core by convection currents in the slowly moving rocks of the mantle, until it reaches the cooler crust and escapes.

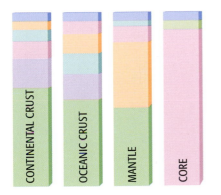

SILICON DIOXIDE
IRON AND IRON OXIDES
ALUMINIUM OXIDE
CALCIUM OXIDE
MAGNESIUM OXIDE
NICKEL OXIDE
OTHERS

**Chemical composition of Earth's layers**
The proportions of different chemical elements and compounds varies between Earth's layers. The core at the center is made almost entirely of metallic iron and nickel.

shadow zone, which she realized must have reflected off the inner part of the core (see opposite right).

## Current views of Earth's structure
Earth's interior is now regarded as being layered in two ways: chemically and mechanically. In chemical composition, it is differentiated into core, mantle, and two types of crust —continental and oceanic (see above). From a mechanical viewpoint, Earth's outer shell consists not just of the crust, but also the topmost layer of the upper mantle. Together these make a rigid and brittle layer called the lithosphere, which is now known to be broken up into plates (see pp.356–57). Beneath is a hot mantle layer, the asthenosphere, and then comes the rest of the mantle, and the outer and inner core.

## Theory of isostasy
Isostasy is an important idea that describes the condition of equilibrium between sections of the lithosphere and the underlying asthenosphere. It proposes that the less dense crust "floats" on the denser mantle like an iceberg floats in water. So, wherever continental crust has become thickened to produce belts of high mountains, the crustal rocks also extend a considerable distance downward—thus the Himalayas, with a maximum elevation of nearly 5½ miles (9 km), are underlain by a thick layer of continental crust 37-miles (60-km) thick displacing the underlying mantle.

**AFTER** »

Much recent research on Earth structure has focused on the properties of the core.

### INNER CORE ROTATION
In 1996 researchers at the Lamont-Doherty Earth Observatory in New York found that Earth's **solid inner core rotates freely within the fluid outer core**. Computer models suggest this is due to flows of material within the liquid outer core. These carry a magnetic field that "tugs" on the inner core, giving its rotation an extra boost.

### TWO-PART INNER CORE
In 2008 geologists at the University of Illinois announced that study of seismic waves passing through Earth's **solid inner core** suggests that it has **two parts**, inner and outer, with different texturing of iron crystals.

**SEE ALSO** »
*pp.356–57* PLATE TECTONICS
*pp.358–61* ACTIVE EARTH

# Moving Continents

**The development of plate tectonics—a synthesis of the earlier "continental drift" hypothesis with new discoveries, such as seafloor spreading—was the most important geological breakthrough of the 20th century.**

At the beginning of the 20th century, several topics in Earth science lacked a satisfactory explanation. One of these was how mountains form. Existing hypotheses, such as the idea that Earth contracts as it cools, leaving "wrinkles" on its surface, were unconvincing. Another anomaly to explain was how rocks on the tops of mountains could once have existed on the ocean floor. Nor could anyone come up with a good reason for why there are so many similarities between parts of continents that are separated by wide oceans—in their fossils, sequence of rock strata, and so on—or why deposits of coal, formed from tropical plants, exist in the frozen wastes of Antarctica. Finally, it was not clear why there were such heavy concentrations of earthquakes and volcanoes in some parts of the world.

In 1912 a novel idea, called continental drift, which proposed that continents had once been joined but had since split apart, seemed to provide a solution to some of the questions, but was rejected for lack of an explanatory mechanism. However, in subsequent decades, a series of discoveries in various areas of geology led to a new theory, called plate tectonics. This proposed that Earth's rigid outer shell is composed of several pieces, called plates, that are slowly shifted around as a result of large-scale processes occurring deep within the planet. This new theory explained not only how continents can separate or join together; it also accounted for inconsistencies in the fossil record and, through consideration of what happens at the plate boundaries, many other phenomena—such as how mountains form, and why earthquakes occur where they do. With its ability to explain so much, plate tectonics has revolutionized the way we look at Earth and the processes that occur at its surface.

### Plate motion
Plate tectonics proposes that Earth's surface is composed of separate plates, which are moved around by convection currents in Earth's interior. Interactions between the plates at their boundaries account for phenomena, such as the distribution of volcanoes.

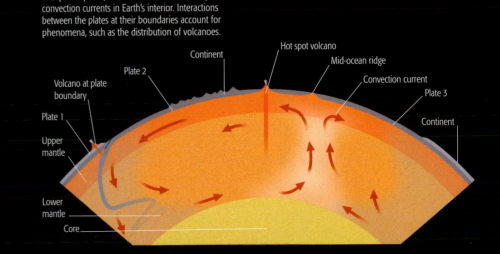

Volcano at plate boundary · Plate 2 · Continent · Hot spot volcano · Mid-ocean ridge · Convection current · Plate 3 · Plate 1 · Continent · Upper mantle · Lower mantle · Core

> **"The crust of the earth must be a shell floating on a fluid interior."**
>
> BENJAMIN FRANKLIN , 1782

**BEFORE** ≪

The idea that the continents were arranged differently in the past goes back centuries.

**COMPLEMENTARY COASTLINES**
Flemish cartographer **Abraham Ortelius**, in 1596, and English philosopher **Sir Francis Bacon** ≪ **69**, in 1620, noticed that the **shapes of continents** on either side of the Atlantic match, **as if they had once been joined.** Ortelius thought the Americas had been "torn away from Europe and Africa... by earthquakes and floods."

**THEORY OF LAND BRIDGES**
In 1885 Austrian geologist **Edward Suess** noted **similarities between plant fossils** in India, Africa, and South America, and suggested that all three may once have been **linked by wide land bridges,** forming a supercontinent, which he called Gondwanaland. He thought that the oceans had flooded these land bridges, thereby separating the continents.

≪ **SEE ALSO**
pp.184–85 HOW ROCKS FORM
pp.186–87 THE FOSSIL RECORD
pp.354–55 MOVING CONTINENTS

# Plate Tectonics

**Continental drift was a revolutionary new idea proposed in 1912. Originally it was dismissed, but by the 1960s geologists had combined many new discoveries, mainly about the ocean floor, into the groundbreaking theory of plate tectonics, which explained how continents really could move relative to each other.**

In 1912 German scientist Alfred Wegener proposed the new concept of continental drift, a forerunner of plate-tectonic theory later that century.

## Wegener's theory

Wegener suggested that hundreds of millions of years ago today's continents were joined in a single large continent, which had since broken up, with the land masses "drifting" apart. To back up this theory, he pointed to the coastlines of the Americas, which correspond in shape to those of Africa and Europe. Similar fossils were also found in these continents, as well as similar sequences of rock layers. The Appalachians of eastern North America, for example, matched the Scottish Highlands,

magnetic poles were in the past. By the 1950s studies such as these were bringing up anomalies that could only be explained if continents had moved—otherwise the magnetic poles would have had to have been in two places at once. In this way, interest in the idea of continental drift was rekindled.

## Theory of seafloor spreading

Building on an earlier idea by English geologist Arthur Holmes—that convection currents in Earth's mantle might carry large slabs of lithosphere (crust and top layer of mantle) along at the surface—in 1960 American geologist Harry Hess developed the theory of seafloor spreading. He

Earth's crust. These subduction regions are found at the edges of some oceans, where deep trenches form in the ocean

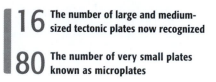

**16** The number of large and medium-sized tectonic plates now recognized

**80** The number of very small plates known as microplates

floor and frequent earthquakes occur. In the 1960s symmetrical patterns of magnetic stripes in rocks were found either side of mid-ocean ridges, confirming Hess's theory (see below).

## Putting the jigsaw together

The seafloor spreading theory quickly led, in the 1960s, to the idea that

**Continent separation**
The continental-drift hypothesis, now known to be correct, proposed that the continents were joined in the past in a supercontinent called Pangaea, and have since moved apart.

Eurasia beginning to split from North America

Supercontinent of Pangaea

**200 MILLION YEARS AGO**

India is in the middle of Indian Ocean

Africa has separated from the other southern continents

**75 MILLION YEARS AGO**

India has joined Eurasia

Atlantic has widened

Antarctica is isolated

**PRESENT DAY**

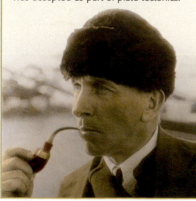

METEOROLOGIST (1880–1930)

**ALFRED WEGENER**

Born in Berlin, Germany, Alfred Wegener trained as an astronomer before becoming a meteorologist. He began collecting evidence for his continental drift theory in 1911, after reading about similarities between fossils on either side of the Atlantic. Sadly, he died on an expedition, decades before his theory was accepted as part of plate tectonics.

suggesting they had once been part of one mountain range. Wegener could not, however, explain how such large masses could move apart, so his idea was not developed.

## Mid-ocean ridges and magnetism

Over the next 40 years, there were two key breakthroughs. The first involved mapping the ocean floors. In 1925 a German expedition discovered the mid-Atlantic ridge, a range of mountains running down the middle of the Atlantic seafloor. Later, in the 1950s, American oceanographers Maurice Ewing and Bruce Heezen discovered similar ridges extending into the other oceans. It was soon realized that these ridges contain narrow rift valleys, where new seafloor is created. The second discovery concerned traces of Earth's magnetic field that are locked into rocks when they form. Sequences of rocks sometimes record changes in this magnetism, and by measuring rocks in different continents, it is possible to track where Earth's

proposed that new oceanic lithosphere is continually created at mid-ocean ridges, then moves away from them. Where continents are attached to this lithosphere, they move too. The creation of new oceanic lithosphere is balanced by activity at "subduction zones," where it is destroyed by descending toward the mantle under

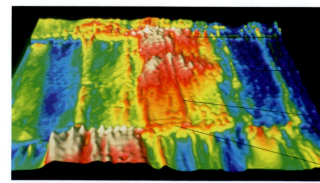

**Spreading of ocean floor**
Color-coded variations in Earth's magnetic field over millions of years are seen here in the rocks either side of a mid-ocean ridge (an underwater mountain range). The magnetic patterns are symmetrical on either side of the ridge—an observation that helped confirm the theory of seafloor spreading in the 1960s.

Red and yellow indicate magnetization of one polarity

Blue and green indicate magnetization of opposite polarity

Center of mid-ocean ridge (runs vertically)

Transform fault that has offset sections of ridge

Earth's surface consists of a jigsaw of "plates," which shift in relation to each other, driven by new plate creation at mid-ocean ridges. A new analysis of the degree of geometric fit between the coastlines of Africa and South America found that they matched even more closely than had been previously thought. In the late 1960s researchers

## Plates and boundaries

This map shows the main plates that form a jigsaw over Earth's surface, and their boundaries. The positioning of many of these boundaries was established in the late 1960s, by mapping where earthquakes took place around the world.

**MAP KEY**
—— Transform boundary
—— Divergent boundary
—— Convergent boundary
—— Deep-sea trench

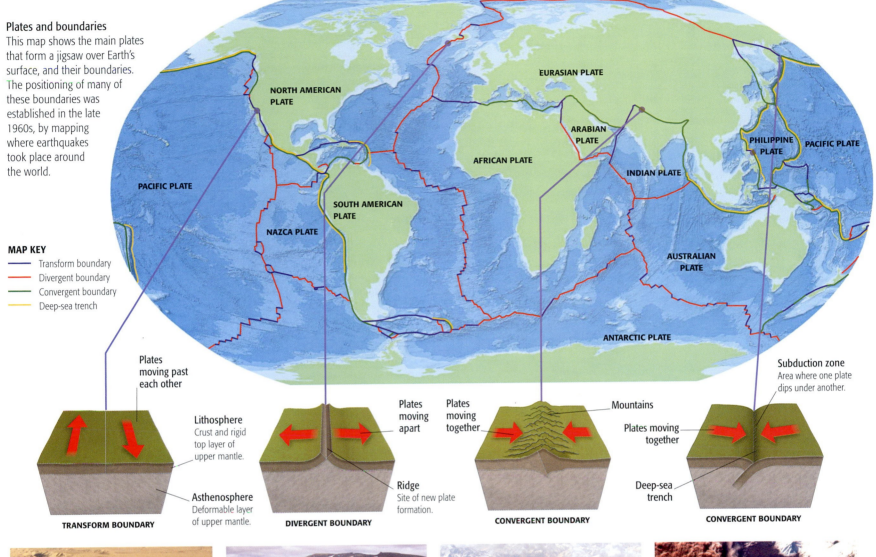

NORTH AMERICAN PLATE

EURASIAN PLATE

ARABIAN PLATE

AFRICAN PLATE

PHILIPPINE PLATE

PACIFIC PLATE

INDIAN PLATE

PACIFIC PLATE

SOUTH AMERICAN PLATE

NAZCA PLATE

AUSTRALIAN PLATE

ANTARCTIC PLATE

Plates moving past each other

**Lithosphere**
Crust and rigid top layer of upper mantle.

**Asthenosphere**
Deformable layer of upper mantle.

**TRANSFORM BOUNDARY**

Plates moving apart

**Ridge**
Site of new plate formation.

**DIVERGENT BOUNDARY**

Plates moving together

Mountains

**CONVERGENT BOUNDARY**

**Subduction zone**
Area where one plate dips under another.

Plates moving together

**Deep-sea trench**

**CONVERGENT BOUNDARY**

**San Andreas fault, US**
Slicing through California, this fault lies along a transform boundary where the Pacific and North American plates grind past each other, at a rate of approximately 2 in (5 cm) per year.

**Thingvellir, Iceland**
This crack in Iceland is part of a divergent boundary where the North American and Eurasian plates are moving apart. It marks a short section of the mid-Atlantic ridge, where it breaks the ocean surface.

**Himalayas, Asia**
The Himalayas were pushed up when two slabs of continental lithosphere began ploughing into each other. They started forming 70 million years ago when the Indian Plate collided with the Eurasian Plate.

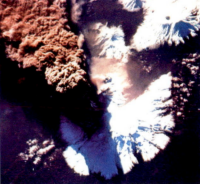

**Ring of Fire, northwest Pacific**
Volcanoes forming part of the Ring of Fire around the Pacific sit near convergent boundaries, where plates of oceanic lithosphere subduct (dip under) other plates. The area is affected by both volcanic activity and earthquakes.

**AFTER »**

constructed and tested models based on the idea that Earth's outer shell consisted of around a dozen moving plates. They traced the history of plate motions to indicate that a single large continent (Pangaea, see opposite) had once existed, as originally proposed by Wegener.

## Plate boundaries

The theory of plate tectonics was born, and within a few years it had gained wide acceptance. Subsequent study figured out the details of the theory. Three basic types of boundary were recognized: divergent (plates move apart), convergent (plates move toward each other), and transform (plates slide past each other). Based on both observation and theory, each type was associated with high levels of seismicity (earthquake occurrence). Global mapping of earthquake locations —which by the late 1960s was quite straightforward (see p.361)—played an important part in establishing where the plate boundaries are.

**Over the past decades, much work has gone into measuring exact plate movements and predicting what their future effects will be.**

**MONITORING PLATE MOVEMENT**
In 1987 NASA began to measure the **rates of plate motion** using radiotelescopes, which confirmed that plates typically **move a few inches (centimeters) a year** in relation to each other. The effects have now been projected into the future. Some 80 million years from now, Africa will collide with Italy as it moves northward, and later Australia will collide with Japan. In about 250 million years, all continents will plough into each other, **creating a new Pangaea** (supercontinent).

**SEE ALSO »**
pp.358–61 ACTIVE EARTH

# Active Earth

**Earth is a dynamic planet, which contains colossal amounts of pent-up energy. This energy is released at the surface—sometimes in slow, long-term processes, such as mountain building, but also in dramatic events, such as volcanic eruptions and earthquakes.**

## « BEFORE

**Throughout history, people have tried to explain how natural features form.**

### VOLCANIC ORIGINS
Aristotle (384–322 BCE) believed volcanoes to be caused by **underground winds** breaking Earth's surface, shattering the rocks so that they caught fire. In the 17th century German scholar **Athanasius Kircher** stated that they arose from subterranean **chambers of fiery matter**.

### EXPLAINING MOUNTAINS
The prolific Persian scholar **Avicenna** (981–1037 CE) developed a theory about how some **mountains** might have formed as a result of **violent earthquakes** that took **place in the past**.

**« SEE ALSO**
*pp.352–53* THE STRUCTURE OF THE EARTH
*pp.354–55* MOVING CONTINENTS
*pp.356–57* PLATE TECTONICS

For centuries scholars tried to understand a wide range of geological phenomena, from how mountains and rift valleys form to the causes of earthquakes and volcanoes. However, it was not until the development of plate tectonics (see pp.356–57) in the 20th century that a coherent explanation for all these phenomena, and other puzzling features of Earth's surface, was found.

## Mountain building
Before plate tectonics, many of the theories on how mountains formed were complex and somewhat unconvincing. Plate tectonics, on the other hand, provided a straightforward and convincing set of explanations. The theory showed how the uplift of areas of Earth's crust required

**Cooled lava**
A volcanic bomb, such as the example above, is a lump of molten lava that erupted from a volcano and then cooled and solidifed as it flew through the air.

for mountain formation results from stresses caused by plate movement. These stresses are greatest near plate boundaries and are most extreme where plate motion makes continents collide head-on (see bottom left). The stresses have two main effects on rocks in the crust: folding, in which rock strata (layers) buckle into wavelike shapes; and faulting, which occurs when blocks of crust fracture along fault planes, areas of stress (see opposite). Subsequent erosion of the uplifted crust produces mountains and valleys.

A separate process that can lead to mountain and valley formation is rifting (see opposite). If an area of continental crust thins out due to heat flowing up from Earth's interior, the crust and the underlying upper mantle on either side of the thinned

**Mountain formation**
Continents colliding at plate boundaries push gigantic quantities of folded and deformed crust upward, forming mountains, and downward, forming the mountains' roots.

**Continent–continent convergence**
When continents collide, the upper layers of mantle on both sides are forced down, one below the other.

**Altered crust**
When huge slabs of crust are forced together, faulting, folding, and deformation of rocks occur.

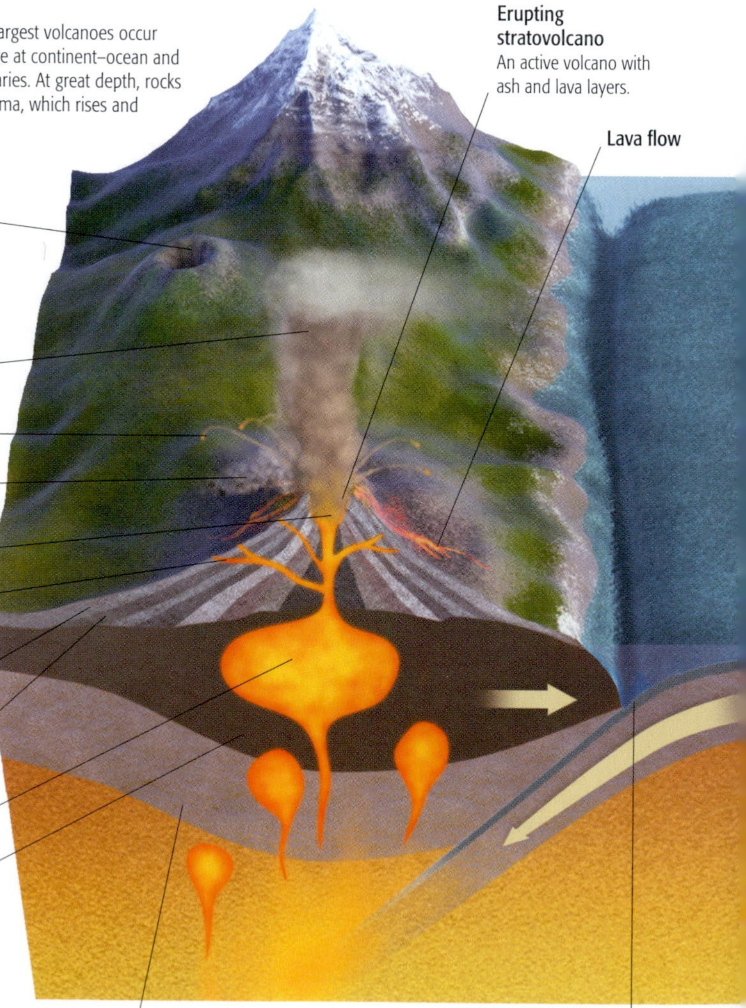

**Volcanic activity**
Many of the world's largest volcanoes occur where plates converge at continent–ocean and ocean–ocean boundaries. At great depth, rocks melt to produce magma, which rises and erupts at the surface.

**Dormant volcano**
Although presently inactive, a dormant volcano may still erupt in the future.

**Eruptive cloud of ash**

**Volcanic bombs**

**Pyroclastic (gas and rock) flow**

**Vent**

**Magma-filled fissure**

**Ash layer**

**Lava layer**

**Magma chamber**

**Continental crust**
From 12–25 miles (20–70 km) thick, continental crust is made of lighter rocks than oceanic crust.

**Erupting stratovolcano**
An active volcano with ash and lava layers.

**Lava flow**

**Top layer of upper mantle**
Along with the oceanic or continental crust, this rigid rock layer forms the solid lithosphere.

**Continent–ocean convergence**
Where continent collides with ocean, denser oceanic lithosphere moves beneath lighter continental lithosphere, forming a deep ocean trench.

zone begin to move apart, and a new plate boundary forms between them. Blocks of crust collapse along the new boundary, creating a rift valley with mountainous escarpments on either side, as in parts of Africa's Great Rift Valley. As rifting continues, the ocean floods in along the new boundary, along which magma (melted rock from Earth's mantle) continuously erupts. Over millions of years the eruptions produce a chain of underwater volcanoes and mountains known as a mid-ocean ridge.

## Volcano formation

When converging plates cause collisions involving oceanic lithosphere, volcanoes may be formed. Where dense oceanic lithosphere is subducted (dips down) beneath lighter continental lithosphere or another younger, less dense slab of oceanic lithosphere, a deep trench is created in the seafloor at the point of descent. As the oceanic lithosphere is driven down, water escapes from its top layer of oceanic crust and lowers the melting point of the mantle rocks that now surround it. The resulting magma rises up and erupts to form volcanoes.

At ocean–continent convergences, such as along the west coast of South America, plate movement results in the continent's edge becoming thickened, buckled, and forced up to form mountain ranges, such as the Andes, which have many active volcanoes. At ocean–ocean convergences, the rising magma creates a gentle arc of underwater volcanoes, which may rise out of the ocean as islands, such as the Indonesian volcanic arc.

## Volcanic hot spots

Volcanoes also occur away from plate boundaries at sites called hot spots. These result from plumes of hot mantle carrying heat energy up from Earth's interior (exactly why they occur in particular fixed spots is not known). Where a plume reaches the surface as magma it erupts and forms a volcano.

When hot spots are located beneath oceans, the magma initially erupts onto the seafloor, forming an underwater volcano. As more magma is added, the volcano grows and becomes an island. Over time the movement of the plate beneath the volcano gradually carries the volcano away from the hot spot, cutting it off from its magma source and making it extinct. However, as one volcano becomes extinct, another starts to grow over the hot spot. Whole chains of volcanic islands, such as the Hawaiian islands, can form in this way.

Hot spots also exist under continental areas, such as Yellowstone Park. There have been no eruptions here for thousands of years, yet a huge magma chamber sits beneath the park, powering its hot springs and geysers.

## Understanding earthquakes

At the start of the 20th century, little was known about earthquakes, although they were believed »

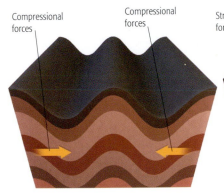

**Compressional forces**    **Compressional forces**

### Folding
As plates collide, prolonged compression, combined with heating, of a region of Earth's crust can lead to folding—buckling and deformation of its rock layers.

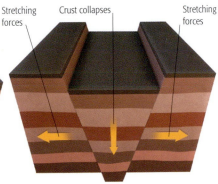

**Stretching forces**    **Crust collapses**    **Stretching forces**

### Rifting
Heat flowing from Earth's interior can cause regions of continental crust to thin and move apart. The intervening crust collapses, forming a rift valley.

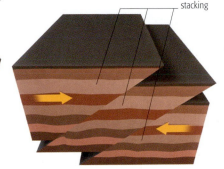

**Movement causes stacking**

### Faulting
Thrust faulting is a common type of faulting, and causes blocks of crust to stack up above one another and can cause great thickening of the crust locally.

**Mid-ocean ridge**
An underwater mountain range that has a rift (valley) along its spine. Rising magma forms new oceanic lithosphere, causing ocean floor to spread.

**Hot spot volcano**
Where a mantle hot spot occurs under an ocean, a volcanic island may arise—magma from Earth's interior erupts on the seafloor and eventually rises up above sea level.

**Volcanic island arc**
Rising magma where two areas of oceanic lithosphere meet may produce an arc of volcanic islands.

**Ocean-floor spreading**
Magma, originating from volcanic activity at the mid-ocean ridge, forms new lithosphere, which moves away from ridge, a few inches/centimeters a year.

**Extinct volcano**
Plate movement has carried this former volcano away from its magma source, making it extinct. It may have sunk as it cooled.

**Ocean–ocean convergence**
Where ocean collides with ocean, the denser section of oceanic lithosphere moves beneath the other and a deep ocean trench develops.

**Oceanic crust**
Less than 6 miles (10 km) thick, oceanic crust is mainly composed of the rocks basalt and gabbro.

**Kilauea volcano, Hawaii**
A spitting stream of fiery lava pours down the flank of Kilauea, one of five volcanoes that form the main island of Hawaii and the most active volcano on Earth.

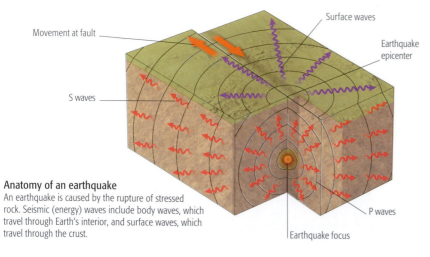

**Anatomy of an earthquake**
An earthquake is caused by the rupture of stressed rock. Seismic (energy) waves include body waves, which travel through Earth's interior, and surface waves, which travel through the crust.

IN PRACTICE

## LOCATING AN EARTHQUAKE FOCUS

The fact that P (primary) waves and S (secondary) waves travel at different speeds through the earth provides the basis for locating the focus of an earthquake. The difference in arrival times between the two types of wave is measured at a minimum of three stations. By calculating the distance from each station to the earthquake focus, its location can be found.

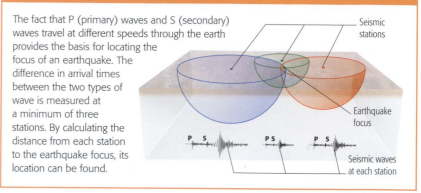

to emanate from specific underground spots. Improved seismometers (devices for detecting earthquakes) enabled scientists to study shock waves more closely. Study of these shock waves was also put to use in analyzing Earth's internal structure (see pp.352–53). In 1906 American scientist Harry Reid developed the "elastic rebound" theory to explain earthquake triggers. Pressure produces a buildup of energy in a region of crust. Eventually there is a massive release of energy as the rocks fracture, causing an earthquake, which propagates some distance through the stressed block of crust.

In the 1920s it was discovered that the centers of energy release of some quakes are concentrated near ocean trenches and that the depth of these centers increases with distance from the trench. This finding assisted the development of the plate tectonic theory (pp.356–57).

### Earthquake magnitude scales
In the 1930s Charles Richter devised the Richter magnitude scale to measure earthquakes. On this scale each step represents a 30-fold increase in the amount of energy released. Major earthquakes measure 7 or 8 on the Richter scale. However, seismologists now more frequently use the moment

**Earthquake damage, Kobe, Japan**
Officials inspect damage to the Hanshin expressway, an elevated freeway in Kobe, Japan, which collapsed during an earthquake in January 1995. The quake was caused by a rock fracture 16 miles (25 km) beneath the sea off Kobe and was responsible for more than 6,400 deaths.

**Place** Alaska, US
**Date** March 27, 1964
**Moment magnitude** 9.2
**Number of fatalities** 131

**Place** Messina, Italy
**Date** December 28, 1908
**Moment magnitude** 7.1 (estimated)
**Number of fatalities** 60,000–200,000

**Place** Izmit, Turkey
**Date** August 17, 1999
**Moment magnitude** 7.4
**Number of fatalities** 17,000–40,000

**Place** Kashmir, Pakistan
**Date** October 8, 2005
**Moment magnitude** 7.6
**Number of fatalities** 80,000 +

**Place** San Francisco, US
**Date** April 18, 1906
**Moment magnitude** 7.8 (estimated)
**Number of fatalities** 3,000+

**Place** Tangshan, China
**Date** July 28, 1976
**Moment magnitude** 8.2
**Number of fatalities** 240,000–250,000

**Place** Kanto, Japan
**Date** September 1, 1923
**Moment magnitude** 7.9
**Number of fatalities** 100,000–142,000

**Place** Mexico City, Mexico
**Date** September 19, 1985
**Moment magnitude** 8.0
**Number of fatalities** 5,000–45,000

**Place** Kobe, Japan
**Date** January 17, 1995
**Moment magnitude** 6.8
**Number of fatalities** 6,400+

### Earthquake zones
The distribution of earthquakes throughout the world closely follows plate boundaries. Highlighted here, in addition to the earthquake-prone regions, are the locations of 12 major earthquakes that have occurred since 1900, significant either because of the number of human fatalities, the economic damage caused, or their sheer size.

**Place** Valdivia, Chile
**Date** May 22, 1960
**Moment magnitude** 9.5
**Number of fatalities** 2,000–6,000
Most powerful earthquake recorded

**Place** Off Sumatra, Indonesia
**Date** December 26, 2004
**Moment magnitude** 9.1
**Number of fatalities** 225,000+
Deadliest tsunami ever recorded

**Place** Sichuan, China
**Date** May 12, 2008
**Moment magnitude** 7.9
**Number of fatalities** 69,000+

**AFTER**

magnitude scale. As a rule, earthquakes with magnitudes less than 5 cause little damage; those above 6 are destructive; and those above 8 are cataclysmic.

### Tectonics: the causes of earthquakes
In the 1960s and 1970s the plate tectonics theory helped to clarify the underlying causes of earthquakes considerably. The largest earthquakes

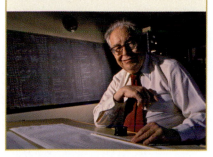

**SEISMOLOGIST (1900–85)**
### CHARLES RICHTER
Born in Ohio, Charles Richter spent most of his life in southern California, a region that suffers frequent and occasionally devastating earthquakes. In the 1930s, with German-born seismologist Beno Gutenberg, Richter developed the earthquake rating scale that bears his name. It is based on the strength of the shock waves recorded by seismometers, taking into account the distance of the instruments from the epicenter of an earthquake.

## 32,000 TIMES
more energy is released by an earthquake of magnitude 8 than by one of magnitude 5, or by one of magnitude 7 compared with an earthquake of magnitude 4.

tend to occur at convergent or transform plate boundaries, in areas where the plates cannot move past each other smoothly. Some, however, occur along faults that are located a little distance from, and often parallel to, plate boundaries, rather than being in the same place as them. These faults have usually developed over time, due to stresses in the crust caused by plate movements. Many shallow earthquakes (with a focus less than 230 feet/70 meters below Earth's surface) also occur along mid-ocean ridges, which are linked to rifting and the ocean floor spreading from the ridges (see pp.358–59). Mapping earthquake locations has helped geologists work out where the plate boundaries are.

### Effects of an earthquake
The effects of an earthquake derive from the violent shaking caused by the shock waves. Buildings not designed to withstand shaking are likely to collapse. Some types of soil can liquefy during a severe earthquake and lose all strength, so that buildings previously supported by them sink into the ground. Landslides are a common secondary effect when buildings collapse, as are fires, and

electrical and gas supplies are damaged. If the focus of a quake is beneath the ocean floor, it can cause a tsunami (see below), as chunks of oceanic crust are suddenly pushed upward.

### Earthquake prediction
Seismologists can predict where, but very rarely when, quakes are likely to occur. Indicators include ground uplift within the vicinity of known faults, which can indicate strain building up in the rocks. Another technique has been to look for "seismic gaps"—earthquake zones that have had little seismic activity for years, and where strain energy may be building up.

### The birth of a tsunami
Earthquakes below the seafloor are the most common cause of tsunamis. The tsunami waves increase in size as they encounter shallow water near the shore. They can grow from 15 ft (5 m) in height when traveling across the open ocean to up to 100 ft (30 m) or more in height when they reach the coast.

Much research over the last 50 years has gone into earthquake prediction and control.

#### CONTROL MEASURES
Since the 1960s scientists have known that **injecting high-pressure fluids** deep into the ground can lubricate faults and so trigger several small earthquakes before enough strain energy has built up to cause a large one. **Benefits versus risks** are still, however, being evaluated.

#### BUILDING PROTECTION
Knowledge of the effects of earthquakes on buildings, and the use of tremor-resistant design features, has meant that **skyscrapers expected to resist earthquakes** as large as moment magnitude 8.5 have been built in vulnerable cities, such as Los Angeles and Mexico City.

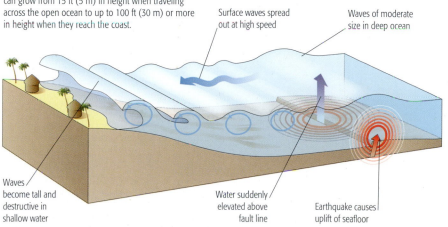

Surface waves spread out at high speed

Waves of moderate size in deep ocean

Waves become tall and destructive in shallow water

Water suddenly elevated above fault line

Earthquake causes uplift of seafloor

Agriculture is the practice of growing plants and crops and rearing animals as livestock to satisfy human requirements. Humans derive most of their needs from a limited number of plant and animal species—descendants of wild ancestors that have been domesticated since antiquity. Closely managed, the yield of crops and livestock can be maximized by providing them with good growing conditions while at the same time controlling pests and diseases. An understanding of ecological principles has found ways of protecting the land, without undermining the ability of agriculture to provide for populations.

## Domestication and breeding

Plant and animal species became domesticated when humans began raising them for their own use. The earliest selective breeding of useful varieties was purposeful, but occurred

# Agriculture

**Humans rely upon plants and animals as sources of food as well as for materials for clothing, construction, and even medicines. Science would show how crops and livestock could be managed to sustain the demand of growing populations and—ultimately—how to minimize damaging influences on the environment.**

« **BEFORE**

**Agriculture, much like medicine, was driven by the needs of civilization. Population growth over time has led to the the need to increase food production.**

### ORIGINS OF CROPS

**Domestication** of crops can be traced back to the birth of civilization in Africa and the Middle East. **Wheat** crops originated in the "Fertile Crescent" of the rivers Euphrates and Tigris, while **rice** and **corn** were staples in the Far East and the Americas. Intensive cultivation probably began in Mesopotamia, and Egypt around 5000 BCE.

### BEGINNINGS OF AGRICULTURAL SCIENCE

**Mechanized farming** methods and **selective breeding** increased agricultural productivity. In 18th-century England, **Jethro Tull** pioneered many farming techniques, including the design for the **seed drill** that mechanized seed sowing.

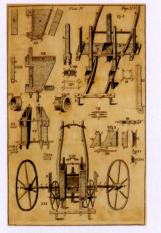

**JETHRO TULL'S SEED DRILL**

long before knowledge of genetics gave it a scientific basis. Scientists became aware that selective breeding was possible because of genetic variation, but the subsequent inbreeding reduced variation. Although domesticated crops can produce superior foods, they are inherently vulnerable as a result. In the 1840s a disease called potato blight devastated potato production in Ireland, causing widespread famine in a population dependent upon the crop. In 1908 American botanist George Shull showed that plant hybrids (combinations of different varieties) were more vigorous, and he used them to improve the quality of corn. By the 1920s geneticists had found a way of measuring heritability—the degree to which variation was due to genes. Selective plant and animal breeding works only when traits are genetically determined, so the understanding of heritability enables such traits to be identified with greater confidence.

## Improving crop yield

The Industrial Revolution of the 19th century was made possible at least partly by improved agricultural practices. This in turn brought mechanization to agriculture with the invention of machines like the tractor.

Meanwhile, a better understanding of soil nutrients and pest and disease control helped increase yields still further. Different crop species use nutrients in different ways. A farmer can ensure that the nutrient content of soil is managed effectively by rotating his crops (see right). This also prevents build-up of pests or diseases specific to any one type of plant.

**Good agricultural practice**
This aerial view of fields in Burgundy, France, in early summer shows a regimen that allows cultivation of a range of crops, including potatoes (foreground), winter wheat, winter barley, oilseed rape (from the cabbage family), and corn, as well as grass for livestock.

Adding fertilizer to the soil enhances plant growth. Traditional fertilizer, such as animal manure, releases nutrients in the soil slowly as it decomposes. It also improves water retention and opens the texture of heavy clay soils, encouraging root growth and nutrient-releasing bacteria. However, inorganic (artificial) fertilizer—in the form of pure nitrate or phosphate—works faster. In the 1800s phosphate-producing industries became possible with advances in chemistry. At the same time guano (bird dung)—collected from the South Pacific—proved to be a powerful fertilizer, and Chile supplied much of the world's agricultural nitrate. In 1908 Fritz Haber first filed his patent on the synthesis of ammonia from its elements. Haber and Carl Bosch went on to develop an economic way to make nitrogenous fertilizer. The invention—known as the Haber-Bosch process (see pp.240–41)—produced ammonia for conversion to

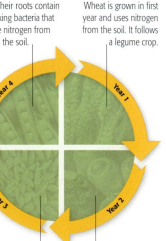

**Legumes**
Peas and beans replenish nitrogen. Their roots contain nitrogen-fixing bacteria that incorporate nitrogen from the air into the soil.

**First cereal crop**
Wheat is grown in first year and uses nitrogen from the soil. It follows a legume crop.

**Second cereal crop**
Barley is often grown in third year since it uses less nitrogen than wheat.

**Root vegetables**
Potatoes or turnips are grown in second year, and use potassium.

**Crop rotation**
This is an ancient regimen popularized in England in the 18th century by Charles Townshend. Crops are "cycled" around four fields over four years for effective management of soil nutrients. Nitrogen-"hungry" cereals are alternated with less demanding crops or those that replenish nutrients. Variation also prevents build-up of pests of diseases that affect plant families. Livestock farmers add an extra three years grazing into the cycle.

**Intensive olive farming**
Spain now produces about 60 percent of the world's olive oil. Intensive close planting of olive trees in the south—one of the hottest, driest parts of Europe—is possible only through an ambitious government-funded irrigation system.

nitrate and revolutionized productivity. Other areas of plant science were also finding new ways of improving yields. Cultivation in greenhouses enabled year-round supply of some crops. In 1938 American agriculturalists Dennis Hoagland and Daniel Arnon

> **ORGANIC FARMING** A type of agriculture that excludes artificial fertilizers and pesticides, as well as growth enhancers and genetically modified organisms.

developed an idea from William Gericke that allowed plants to be grown without soil. The technique —called hydroponics—would go on to make crop production possible for farmers with poor-quality land.
In 1939 chemistry offered agriculture something other than fertilizer. German chemist Paul Müller found that a chemical called DDT was a very effective insecticide. Until then pest control had used natural plant-based products. DDT was man-made—and proved much more effective. It was later found to be harmful and its use has since been banned.

### Sustainable development
The age of the pesticide was not universally welcomed. British botanist Albert Howard (1873–1947) rejected polluting pesticides and synthetic fertilizers after seeing traditional agricultural practices in India in the 1920s. He went on to pioneer chemical-free agricultural methods that became known as organic farming. A landmark book of 1962, *Silent Spring*, by American biologist Rachel Carson, warned of the polluting effects of pesticides and did much to change the public view. Today environmental attitudes have shifted as a result of Howard's and Carson's efforts. Pesticides are used, but they target specific pests and are degradable. In addition, there is a move to combine their use with biological pest control.

Developments in agriculture lie in making use of new capabilities of crop plants— either those developed by genetic modification or through new discoveries.

### NEW CROPS
The **winged bean** is a plant native to New Guinea, which exemplifies the potential of **wild species** to produce **new crops**. Flour can be made from its seeds, the pods are rich in vitamins, and its roots are high in protein. Other species include the salt-tolerant Salicornia and Tamarind Tree.

PSOPHOCARPUS TETRAGONOLOBUS, OR WINGED BEAN

### GENETIC MODIFICATION
**Selective breeding** and more advanced techniques can **alter plants**. It is possible to manipulate the genetic material in tissues to give plants characteristics such as **pest resistance**.

SEE ALSO »
pp.386–87 GENE TECHNOLOGY
pp.390–91 HOW CLONING WORKS

**Hydroponic lettuces**
Modern hydroponics allows crops such as lettuces to be cultivated on a very large scale in any climate. Plants are grown in solutions of nutrient that permeate a soilless inert medium, such as rock wool, perlite, or gravel.

BREAKTHROUGH
### BIOLOGICAL PEST CONTROL

Pest control by natural enemies of pests—their predators or parasites— prevents chemical pollution and a buildup of pesticide resistance. However, extensive research is needed to ensure that the introduced enemy does not become a pest. Biological control, for example, using predatory mites to kill spider mites (below) requires that some pest damage is tolerated. A compromise is often reached whereby biological pest control is used in combination with traditional techniques.

**Plant breeding**
Rice is the second most abundant cereal (grass) crop—after corn. As with many crops, varieties and hybrids have been bred from their wild ancestor (far left) to create plants with a bigger seed grain, or heavier crop (near left).

AGRICULTURALIST (1914–)
### NORMAN BORLAUG

American agronomist Borlaug is widely considered to be the founding father of the Green Revolution, which has seen the development of disease-resistant varieties of wheat. Following a period of research in plant pathology, he combined the results of selective breeding with improved agricultural techniques to increase the food productivity of Mexico, India, and Pakistan, averting famine in these countries. In 1970 he received the Nobel Peace Prize for his humanitarian efforts, and in 1986 he established the World Food Prize for individuals working to the same ends.

**Light show**
Coordinated displays of different colored laser beams creating spectacular patterns in the air are often used to entertain audiences at rock concerts and other events.

# Lasers and Holograms

**The word "laser" stands for Light Amplification by the Stimulated Emission of Radiation. It refers to the production of a narrow beam of light from a substance whose atoms have been "excited" artificially. Lasers have many uses, including delicate surgery, accurate measuring, and the creation of holograms.**

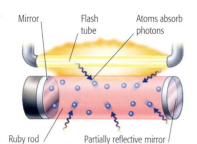

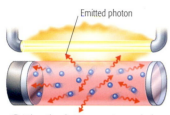

Mirror | Flash tube | Atoms absorb photons

Ruby rod | Partially reflective mirror

**1** Photons from a flash tube are absorbed by atoms in the ruby rod, raising the atoms' electrons to a higher energy level.

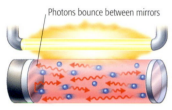

Emitted photon

**2** When the electrons spontaneously drop back to their normal energy level, the atoms emit photons in random directions.

Photons bounce between mirrors

**3** If an emitted photon hits another atom, the second atom emits a photon of the same energy and direction of travel.

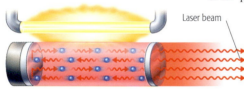

Laser beam

**4** The number of photons rapidly rises, until a stream of photons emerges from the partially reflective mirror as a laser beam.

### How a ruby laser works
When energy is fed into a ruby rod, the ruby atoms start to release photons of a specific wavelength. When these photons strike neighboring atoms, even more photons are emitted, causing a chain reaction. The photons bounce between two mirrors until the light is so intense that it emerges as a laser beam.

## BEFORE

Laser technology stemmed from research during the early years of the 20th century into how light and matter interact.

### EINSTEIN'S CONTRIBUTION
In 1917 **Albert Einstein ≪ 304–305** realized that a photon hitting an already excited atom might cause the emission of another photon going in the same direction as the first.

### THE MASER
In 1954 Charles Townes and Arthur Schawlow invented the **maser (Microwave Amplification by Stimulated Emission of Radiation)**, a predecessor of the laser. Masers are used today in devices such as atomic clocks.

### ≪ SEE ALSO
*pp.286–87* STRUCTURE OF THE ATOM
*pp.314–17* QUANTUM REVOLUTION

Theodore Maiman, an American physicist, built the first laser in 1960. When researching what to use as a lasing medium—the substance that is artificially excited in order to produce a laser beam—Maiman chose a ruby crystal, so his device was called a ruby laser. Since then, many different materials have been employed as lasing media. The first laser to use a gas (a helium–neon mixture) was also devised in 1960, and the carbon dioxide laser came soon after. Diode lasers, developed in the 1970s, were the first to produce a continuous beam rather than pulses of light. These lasers use semiconductor material as the medium, and they make up the majority of lasers in use today.

Laser beams vary in intensity, but all share certain properties. All the photons (packets of radiation) in a laser beam have the same wavelength, so laser light is monochromatic—it has just one color. The photons are also coherent, or in "phase," which means that all their wave peaks and troughs line up. Laser beams are usually highly directional and do not diffuse (spread out), so they can be focused on specific targets and projected over great distances.

### Lasers in everyday life
Industry employs lasers to cut through steel and other materials. Surgeons use lasers of much lower power in eye surgeries and cosmetic procedures. A laser beam cuts tissues more precisely than a scalpel, and the heat from the beam cauterizes severed blood vessels instantly, preventing bleeding. Lasers are also used to clear out deposits clogging arteries.

Lasers are ideal for making precise measurements, in areas ranging from surveying to astronomy. In 1969 a laser measured the distance to the Moon to within centimeters (see p.173). Low-cost diode lasers are used for retrieving data from compact discs (CDs), and also in bar-code readers, photocopiers, and printers. Other applications of lasers include scientific research, fiber-optic communication, guiding missiles, and holography.

### Holography
Pioneered by Hungarian engineer Dennis Gabor, a hologram is an image formed by the interference patterns created by the interaction of two laser light beams—one scattered from the subject of the hologram (usually a 3-D object), the other a "reference" beam. These patterns are recorded in a

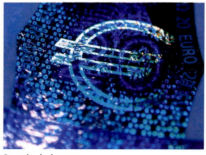

### Security hologram
A banknote hologram consists of a surface pattern on a plastic film. Light reflecting off an aluminium backing layer, and passing through the film, creates an image when viewed from the correct angle.

transparent film or plate. To view a hologram, the plate or film has to be illuminated either by a laser or by white light. The interference patterns diffract (scatter) the light waves in such a way that, when they reach a viewer's eyes, the subject seems to appear. Laser-viewed holograms give a monochrome image, but white light holograms can produce images in all colors. Holograms are often used as a security feature on banknotes, credit cards, and product packaging. They are also used for optical data storage, retrieval, and processing, and in instrument displays for aircraft pilots.

## AFTER

Research scientists are continuously finding practical and theoretical new applications for both lasers and holograms.

### GAMMA-RAY LASERS
In 2007 scientists fleetingly created a **new form of matter called positronium**. A combination of electrons and positrons, it is a **potential energy source for ultra-powerful gamma-ray lasers**, which might one day be used to **kick-start nuclear fusion reactors 324–25 ≫**.

### HOLOGRAPHIC DATA STORAGE
Holographic disks could become the **standard storage media of the future**. Disks capable of holding up to 4 terabytes (4,000 gigabytes) of data could be available within a few years.

### CURVED LASER BEAMS
Researchers in Arizona produced **curved** pulses of laser light in 2009. The beams leave an **arc of plasma** in the air. By firing lasers into thunderclouds, plasma arcs **could be used to guide lightning safely to the ground**. This would be useful to airplanes, which are particularly vulnerable to lightning strikes.

### SEE ALSO ≫
*pp.406–07* MODERN SURGICAL PROCEDURES

### ELECTRICAL ENGINEER (1900–79)
## DENNIS GABOR

In 1933 the Hungarian engineer Dennis Gabor fled from Nazi Germany, where he was working in electronics. After settling in Britain he joined an engineering company. In 1947 realizing that information about the 3-D shape of an object could be stored by comparing light from the object with a "reference" beam, he developed the idea of holography. Gabor had to wait until 1962, and the invention of lasers, to see his idea become reality. In 1971 he received the Nobel Prize for Physics for his invention of holography. In later life, Gabor wrote a book on social theory and the future of humanity.

« **BEFORE**

Before microchips, electronic devices were bulky and had to be assembled piece by piece.

**CIRCUIT BOARDS**
On pre-microchip circuit boards, all the **components** had to be **individually wired** and soldered together. Modern circuit boards are usually **printed** and components just plugged in.

**EARLY ELECTRONIC COMPUTERS**
Completed in 1946 the **first general purpose electronic computer** was ENIAC (Electronic Numerical Integrator And Computer), which was **made for the US army** primarily to carry out **ballistics calculations**. It weighed about 30 tons (27 tonnes) and filled an entire room.

**ENIAC COMPUTER**

**Microchips on circuit board**
The circuit boards of modern computers often contain several microchips, each with its own data processing function. A multicore microchip contains two or more microprocessors on the same microchip to increase processing power.

Try to imagine a world without microchips. The most obvious difference is that there would be no cell phones or personal computers. Without microchips, computers would be just too large and expensive. It is not just personal computers, however, that would be impracticable; an increasing number of everyday devices, from cars to toasters, use built-in computers and microchip technology. But what are microchips, and how has society come to rely on them?

**Circuits in miniature**
A microchip is a tiny electronic circuit in which all the components are housed on a single chip. Today most microchips, or integrated circuits, are made from silicon, but the first working chip, demonstrated by engineer Jack Kilby at Texas Instruments in 1958, was made from germanium (a chemical element with many similar properties to silicon).

Kilby's idea was to make all components from the same material so that they did not have to be wired together, thereby dramatically reducing the time and cost involved in manufacturing electronic devices. Kilby's chip allowed the electronics industry to realize the true potential of transistors—the semiconducting devices used to switch or amplify electrical signals in a circuit. Previously, engineers had painstakingly soldered hundreds of components to build complex circuits. But manufacturing them all from a single semiconducting material meant it was possible to cram thousands, and eventually millions, of transistors onto an area the size of a

**Early transistor**
Transistors are the basic building blocks of most electronic devices, acting as switches or amplifiers. Early transistors, such as the one shown at right, were quite large, but modern ones are so small that more than two billion of them can fit on a single microchip.

# Microchip Technology

**Often said to be the most important invention of the 20th century, the microchip has helped put computer technology into the palms of our hands. But keeping up with the rapid pace of progress means squeezing billions of components on to every chip.**

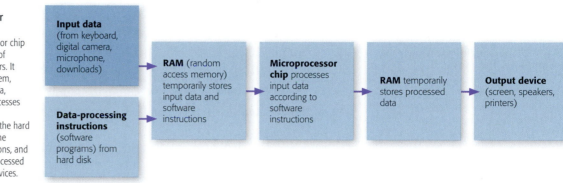

grain of rice. This paved the way for more compact and reliable electronics, including powerful computers, and in 2008 the electronics company IBM was able to announce that it had succeeded in packing more than two billion transistors onto a single chip.

## Uses of microchips

Integrated circuits have become ubiquitous in modern technology, finding their way into everything from cell phones and wristwatches to kitchen appliances such as dishwashers and refrigerators. Chips are also used in credit cards to store encrypted information, helping to protect users from fraud.

The proliferation of the microchip over the last half-century is due to the development of photolithography, a mass-production process that uses light to make chips from silicon wafers. Production processes have now advanced to a stage where it is possible to make as many as 100 microchips from a single 8 in (20 cm) square silicon wafer.

## Microchips and computers

In the 1980s the economies of "micro" production dramatically reduced the price of silicon chips, and this helped to usher in the era of personal computing. At the heart of every modern PC is a microprocessor chip, which is responsible for most of the machine's core information-processing functions (see right).

The breakneck speed at which microchip technology has advanced over the years has driven the construction of ever more powerful computers. This, in turn, has driven

### How a microprocessor chip works

The microprocessor chip is the workhorse of modern computers. It boots up the system, receives input data, retrieves and processes information and instructions from the hard disk, carries out the software instructions, and passes on the processed data to output devices.

**Input data** (from keyboard, digital camera, microphone, downloads)

**Data-processing instructions** (software programs) from hard disk

**RAM** (random access memory) temporarily stores input data and software instructions

**Microprocessor chip** processes input data according to software instructions

**RAM** temporarily stores processed data

**Output device** (screen, speakers, printers)

### Chip under the skin

Microchips can be used to keep track of pets. Implanted under the skin (in the neck of a cat in the X-ray above), a chip contains details of the pet owner's name and address, which can be read by a scanner if a wandering pet is found.

▷ **Microchip miniaturization**
The small size of microchips, and other components, makes it possible to pack immense computing power into relatively compact personal computers, such as the modern laptop shown in this X-ray.

the demand for ever smaller and more advanced microchips.

In 1965, Intel co-founder Gordon Moore described this trend in terms of the number of transistors on a microchip. Moore's Law, as it is now called, states that this number is likely to double every two years. But pushing up the number of transistors is becoming increasingly difficult. Having too many transistors on a single chip generates a lot of heat, which can prevent the tiny circuits from working properly. To overcome this problem, microchip developers are focusing their efforts on creating new types of chips—quantum chips, which rely on the physical state of individual subatomic particles to store and process data. Some nanotechnologists are also attempting to create wafers from potentially high-performance electronic materials such as graphene sheets—one-atom-thick sheets of carbon.

Battery

Port for external device (such as mouse)

Hard disk

Motherboard (main circuit board)

Central processing unit (CPU, or microprocessor)

Cooling fan

AFTER »

## BREAKTHROUGH

### INTEL'S FIRST MICROPROCESSOR CHIP

The microprocessor, or central processing unit (CPU), does most of the computation in a modern computer. The first ever commercially produced microprocessor chip, Intel's 4004, was introduced in 1971. By today's standards the 4004 was relatively simple, with just 2,300 transistors (compared with more than two billion on modern chips). Despite the ongoing trend, for miniaturization, the 4004 chip remains one of the smallest microprocessor designs that ever reached commercial production, measuring just over $1/8$ in (3 mm) by $1/6$ in (4 mm).

**New materials and new types of microchips could replace conventional silicon microchips in the future.**

#### PRINTABLE ELECTRONICS

**New semiconducting materials**, such as carbon nanotubes, could make it possible to **print electronics** in the same way that documents are printed using inkjet printers. Going one step further than microchips, these **electronic circuits** would be bendy, stretchy, and paper thin, allowing them to be **incorporated into clothing**, food packaging, and ultra-thin display screens.

#### QUANTUM CHIPS

**Quantum** computer chips hold the potential to be **much more powerful** than ordinary microchips. In 2006 scientists at the University of Michigan created the world's first quantum chip, in which the spin of an electron determined the value of a quantum bit, or qubit (the quantum equivalent of a bit—the basic unit of data storage in conventional computing). This could lead to even faster computers in the future.

SEE ALSO »
pp.378–79 THE INTERNET
pp.380–81 ARTIFICIAL INTELLIGENCE AND ROBOTICS

**‹‹ BEFORE**

For artificial satellites to become a reality, a number of scientific and technological boundaries had to be crossed.

**UNDERSTANDING ORBITS**
Planetary orbits were understood by German astronomer **Johannes Kepler** in 1608 **‹‹ 76–79**, but it took **Isaac Newton's theory of gravitation ‹‹ 108–09** to explain the forces behind them. Any orbit is simply an **elliptical path** on which, at every point, an object's tendency to continue in a straight line is precisely balanced by the inward force of **gravity**.

**DEVELOPING ROCKETS**
Launching any object into orbit requires a means of propulsion that will work **beyond Earth's atmosphere**. This technology only became a reality with the development of **liquid-fueled rockets** from the 1920s onward, and especially during World War II.

**‹‹ SEE ALSO**
*pp.38–39* ANCIENT IDEAS OF THE WORLD
*pp.136–37* NAVIGATING THE OCEANS
*pp.338–39* ROCKET PROPULSION

# Artificial Satellites

**Like the moon, artificial satellites orbit the Earth. Since the launch of Sputnik 1 in 1957, they have found a huge range of applications and have revolutionized telecommunications, astronomy, and our understanding of Earth.**

Scientists and space enthusiasts had been promoting the potential of artificial satellites for almost a century before they became a reality, in the late 1950s. However, it was political necessity, rather than pure science, that gave rise to the first satellite launches. In the lead-in to the International Geophysical Year of 1957–58, the United States announced its intention to launch a satellite to learn more about the near-Earth environment. The Soviet Union made a similar promise, but thanks to secrecy about their rocket program, this statement was widely dismissed. Then, on October 4, 1957, the Soviets shocked the world by announcing that they had launched their satellite. The 185 lb (84 kg) steel ball, called *Sputnik 1*, carried little more than a radio transmitter that sent out a steady beeping signal, which could be picked up by shortwave radio wherever it passed overhead; it had little scientific use, but was a publicity coup that demonstrated Soviet space superiority.

## Rivalry in space
The US struck back within months, launching *Explorer-1*—a scientific satellite carrying radiation detectors, which discovered the Van Allen radiation belts that encircle the earth. The Soviets, meanwhile, had already launched *Sputnik 2*, carrying a dog called Laika who briefly became the first living creature in orbit. The space race that would see astronauts walk on the Moon

Radio antenna 8–10 ft (2.4–2.9 m) long

△ **The first satellite**
*Sputnik 1* completed 1,440 96-minute orbits of Earth, traveling at 18,000 mph (29,000 kph) on an elliptical path that ranged in altitude between 133 and 583 miles (215 and 939 km). The capsule deployed its long radio antennae only after separating from the upper stage of its R-7 rocket.

▽ **Global village**
In conjunction with ground-based cell phone networks, communications satellites now allow almost instant communication between any two points on Earth. Direct satellite phones can even cut out the need for the ground-based network.

**3 Cosmat**
The comsat transmits the signal around the Earth to another satellite, which is visible from its location in geostationary orbit but out of sight from the original ground station.

Power supply

Solar panel

**2 Ground station**
The ground station beams the signal to a communications satellite ("comsat") in geostationary orbit above the equator, at a fixed location in the sky as seen from the station.

A network of interlocking hexagonal cells keeps phone in constant touch with antenna.

Receiver

**1 Cell Phone**
The signal is transmitted by a cell phone to a local antenna, which then relays it to a satellite ground station.

**4 Solar-powered comsat**
Another satellite transmits the signal to a solar-powered comsat on the network, and beams it down to a ground station in another place.

**5 Cell phone network**
The call is routed to the receiver's local antenna cell by the cell phone network.

Cell network

**6 Receiver's cell phone**
The signals are detected by the receiver's cell phone and it picks up the call.

Polar orbit used for Earth-observing satellites—rotating around Earth brings most areas into view

Highly elliptical orbit used for high-latitude communications satellites

Geostationary, or Clarke, orbit circles Earth once a day and remains above the same point on equator

Sun-synchronous orbit keeps the same angle relative to the Sun, so the ground below is always lit in the same way

Low Earth Orbit is used by manned spacecraft and space stations

◁ **Satellite orbits**
Depending on the function satellites have to fulfill, they can be launched into a wide variety of different orbits, ranging from a circular Low Earth Orbit to high, elliptical paths for interplanetary transfer (travel to other planets).

less than 12 years later had truly begun. As rocket technology became more reliable, scientists and engineers began to investigate uses for orbiting satellites. One of the most obvious was to observe Earth's atmosphere, and the first successful weather satellite, TIROS-1, was launched in 1960.

A major challenge was sending images back to Earth. Modern electronic methods make this relatively simple, but early satellites used traditional photographic film that was developed on board and then scanned and transmitted using a technique like that of a fax machine. Spy satellites, by contrast, carried telescopic cameras and film canisters that were later parachuted back through the atmosphere, allowing the retrieval of images.

### Remote sensing
Despite these advances, the idea of gathering information about Earth's surface from space came surprisingly late—a result of reports from the first astronauts that were then tested on the early Apollo missions. Multispectral cameras revealed detail about conditions on Earth. The first true "remote sensing" satellite, ERTS-1, was not launched until 1972. Today these orbiters use a variety of techniques to monitor every aspect of Earth and have led to a revolution in our understanding of the environment. In contrast, astronomical satellites take advantage of their situation beyond Earth's atmosphere to look out and reveal more about the Universe (see pp.412–13).

### The first communications satellites
As early as 1946, Arthur C. Clarke (see left) had pointed out the potential uses of a communications satellite ("comsat") in "geostationary" orbit— 22,236 miles (35,786 km) over the equator—where it would circle Earth in

precisely one day and so remain above exactly the same point on the surface. From here, the comsat would be a fixed point in the sky, allowing signals to be bounced between any two points on Earth where the satellite was within view. However, the rockets of the early Space Age did not have the power to put large objects into orbit so high up, so instead early comsats such as *Telstar 1* of 1962 operated in lower orbits, where they remained above the horizon for brief periods of time. They were equipped with transponders that allowed them to receive a signal as they passed over a ground station, and send it back to Earth over another area.

**At work in orbit**
From the 1980s the space shuttle has allowed the retrieval of faulty satellites for repairs in orbit or back on Earth. Here astronauts maneuver the *Intelsat VI* satellite into the cargo bay of the space shuttle *Endeavour*.

**WRITER (1917–2008)**

## ARTHUR C. CLARKE

British science and science-fiction author Arthur C. Clarke was hailed as the "prophet of the Space Age" for his knack of predicting technological advances. Many of his predictions concerned how humankind can survive in space. The idea of using geostationary satellites for telecommunications was published in an essay in 1945. It became a reality within 20 years and there are now more than 300 satellites in Clarke orbits.

> **AFTER**

Satellites continue to have a huge effect on the world around us, and may have even more applications in the future.

**GPS**
One of the most recent revolutions has been the spread of **Global Positioning System (GPS)** receivers. This system is based on contacting three or four satellites in low orbits, which send out **precise time signals**; the GPS receiver then works out the user's position by measuring the time delay between satellites.

**FUTURE PROJECTS**
Direct **satellite telephony systems** such as Iridium now allow anyone with a handset to make a call directly via satellite, cutting out reliance on a ground-based phone network and providing **truly global coverage**.

**SEE ALSO** »
pp.372–73 MANNED SPACE TRAVEL
pp.378–79 THE INTERNET
pp.396–97 SPACEPROBES AND TELESCOPES
pp.398–99 DARK UNIVERSE

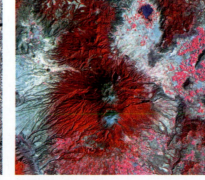

**Satellite image of Manhattan and Brooklyn**
Satellite photos of Earth from space, such as this image of New York, have revolutionized the way we see our planet, with applications that range from weather forecasting and mapmaking to surveillance.

**Multispectral imaging**
Photographs of an area in a variety of wavelengths can reveal information about mineral composition and land use. This image of the Colima Volcano in Mexico shows the extent of ash deposits and solidified lava flows.

400,000 people working on the Apollo program, the Moon landing was a result of the US government's decision to put a man on the Moon ahead of the Soviet Union.

The Apollo missions were born in the aftermath of Yuri Gagarin's historic first orbit of Earth in April 1961. President John F. Kennedy, concerned about being repeatedly embarrassed by the Soviets in the "Space Race," demanded a realistic space goal that might allow the US to overtake their Soviet rivals. At the time, the Soviets' technological lead was so great that the prospect of a lunar landing was the only target that NASA (National Aeronautics and Space Administration) officials believed they could achieve.

Enormous resources were thrown at a project that stamped its mark on the country throughout the 1960s. While astronauts practiced space maneuvers such as rendezvous docking, and spacewalking, a series of robot probes traveled to the Moon —crash landing, conducting orbital surveys, and ultimately making "soft" landings in order to learn as much about Earth's satellite as possible.

At the same time engineers were finalizing a flight plan called "lunar orbit rendezvous." This would involve a number of docking and undocking procedures in the Moon's orbit, which would reduce the weight of the spacecraft. But the project still required the most powerful rocket ever built, and Wernher von Braun, the man behind the German World War II V-2 rocket, would make the three-stage Saturn V his masterpiece.

After years of planning, Apollo 11 launched from Florida on July 16, 1969. From orbit around Earth, a brief burn of the upper-stage rocket engine put it on course for the Moon. The streamlined Command and Service Module (CSM) then broke free of the upper stage, turned around, and extracted the ungainly Lunar Module (LM) from its resting place within the rocket shroud. As the linked modules swung behind the Moon, a brief burst of the CSM's own engines put them into the Moon's orbit. Neil Armstrong and Buzz Aldrin boarded the LM for the journey to the surface, while Michael Collins stayed on the orbiting CSM.

As the LM drifted toward its landing site, the retrorockets slowed its descent and it dropped on to the lunar soil, thus realizing perhaps the greatest scientific and engineering achievement of modern times.

**One giant leap for mankind**
Apollo 11 astronaut Buzz Aldrin walks on the Moon. Aldrin and mission commander Neil Armstrong spent about two and a quarter hours on the Moon's surface collecting samples of lunar materials, taking photographs of their environs, and carrying out scientific experiments.

"Houston, **Tranquility Base** here. **The *Eagle*** has landed."

NEIL ARMSTRONG, THE LUNAR MODULE TOUCHES DOWN ON THE MOON, 1969

## « BEFORE

Some of the most important developments associated with spaceflight evolved from advances and experimental work in military aviation, as well as in rocketry.

### PRESSURE SUITS
The development of **jet engines** during World War II meant aircraft could reach **higher altitudes** and greater speeds. Pressure suits capable of protecting the human body from **low air pressures** became a necessity.

### ROCKET PROPULSION
Test aircraft in the 1950s ascended to the **very edge of space** using rocket propulsion before gliding back to Earth.

### SATELLITES AND THE SPACE RACE
In October 1957 the launch of the *Sputnik 1* satellite marked the **beginning of the space age**. Putting human beings into space was the next logical step.

**«SEE ALSO**
*pp.290–91* TAKING FLIGHT
*pp.338–39* ROCKET PROPULSION
*pp.368–69* ARTIFICIAL SATELLITES

The era of manned spaceflight began on April 12, 1961, with the Soviet Union's announcement that cosmonaut Yuri Gagarin was in orbit above Earth.

Putting a person into space is a bigger challenge than orbiting a satellite, largely because of the need to keep that person alive and bring them safely back to Earth. Early Soviet spacecraft used a pressurized steel ball to hold in an atmosphere against the vacuum of space, while NASA's vehicles relied on stronger but lighter materials. The need to carry a pressurized capsule, with oxygen, water, and other supplies, increased the weight of the spacecraft. The more powerful rockets that the Soviets used gave them an early advantage in counteracting this problem. By contrast, NASA's early launch vehicles were incapable of reaching orbit—the first Mercury flight in May 1961 was a "suborbital" hop, and it was only in 1962 that John Glenn became the first US astronaut to circle the earth.

**136** The number of missions made to date by the space shuttle.

**437** DAYS The longest time spent in space on a single flight by Valeri Polyakov in 1994.

### "**Exploration** is not a choice, really; it's an **imperative**."
MICHAEL COLLINS, ASTRONAUT ON GEMINI AND APOLLO MISSIONS (1930–)

meanwhile, aimed to develop a large, reusable space transport—the space shuttle (see p.339). This system consisted of an airplanelike "orbiter" installed with rocket engines for a vertical ascent, a large external fuel tank, and a pair of external solid rocket boosters. Once the orbiter reached space, its forward cabin could house a crew of up to seven in relative comfort for two weeks or more, while the cargo bay that made up most of its length could carry satellites and space probes into orbit, retrieve them for repairs in space or back on Earth, or house Spacelab—a European-built pressurized laboratory module that could be adapted for scientific experiments. When the mission

was over, a brief engine burn brought the orbiter spiraling back to Earth, making a gliderlike descent to land like a conventional aircraft.

### Cooperation in space
The shuttle was heralded as a new age of cheap access to space, but continued problems (including the

# Manned Space Travel

**Keeping human beings alive in space, and bringing them safely back to Earth, were much greater challenges than sending up satellites. From the first cramped capsules, space vehicles—and even spacesuits—have evolved into safe, comfortable places for sustained work and exploration.**

COSMONAUT (1932–68)

### YURI GAGARIN

Russia's first cosmonaut was born to peasant parents on a collective farm near Smolensk—a humble background that may have influenced his selection as the pilot of *Vostok 1*. He studied engineering at college, then trained as a fighter pilot and was selected for cosmonaut training in 1960. He became a national hero following his 108-minute, single-orbit flight. However, politics meant he was now far too precious to risk on another mission. He died in a jet crash during training for a planned return to space.

Returning to Earth safely was, and remains, another challenge. Various solutions have been found to counteract the heat generated during high-speed atmospheric re-entry (see opposite). Earth's atmosphere slows the spacecraft down and parachutes ensure a gentle landing. NASA opted for the safety of a splashdown at sea, while Soviet spacecraft came down over land, with a final blast from braking "retrorockets" just before reaching the ground.

### The Space Race and beyond
The technology that took astronauts to the Moon in the late 1960s was in place, but the spacecraft were small, cramped, and expensive because they could not be reused. After NASA's victory in the Space Race with the Apollo Moon landings (see pp.370–71), the two powers took different approaches to manned flight. The Soviet Union developed large space stations that could remain in orbit for several years. Crews traveled on Soyuz spacecraft used purely for transfer, and automated ferries brought in new supplies. NASA,

**Zero gravity experiments**
The "Spacelab" extended the craft's functions as a laboratory. Here the crew of the *Endeavour* shuttle work on materials and life sciences experiments during a 1992 mission.

**April 12, 1961** The **Vostok Descent Module** (left) is used by Yuri Gagarin to become the first person to orbit Earth and descend.

**June 16, 1963** Valentina Tereshkova becomes the first female cosmonaut.

**September 29, 1977** Launch of **Salyut 6**, the first in a new wave of "modular" Soviet space stations.

**December 24, 1968** The crew of **Apollo 8** become the first people to orbit the Moon.

| 1955 | 1960 | 1965 | 1970 | 1975 |

**Timeline of man in space**
Many of the greatest advances in manned spaceflight occurred within a single decade—the 1960s.

**February 20, 1962** **John Glenn** becomes the first American astronaut to orbit Earth.

**July 20, 1969** **Neil Armstrong** and **Buzz Aldrin** become the first astronauts to walk on the Moon during the *Apollo 11* mission.

### Physical damage

Long-term spaceflight has adverse effects on the body. For example, red blood cells lessen in number and grow distorted, as shown in these images, while bones lose minerals and become weaker.

**NORMAL RED BLOOD CELLS**  **DISTORTED RED BLOOD CELLS**

tragic loss of two spacecraft and their crews) meant that it never achieved its potential. Since the 1990s political changes have brought about a new spirit of cooperation between Russia and the US, including the construction of the International Space Station. Meanwhile, NASA has plans for a return to the Moon and, eventually, a manned flight to Mars.

IN PRACTICE

## RE-ENTERING THE ATMOSPHERE

Re-entering the atmosphere at high speed creates tremendous friction that can heat a spacecraft exterior to around 2,900° F (1,600° C)—hot enough to melt steel. The traditional solution is to fit the underside of a spacecraft with an "ablative" heat shield that breaks away in small fragments during re-entry, carrying the excess heat away with it and safely protecting the hull and interior. The reusable space shuttle (right), in contrast, uses advanced ceramic tiles that remain in place but will not transfer heat from the outside to the inside. Damage and loss of the delicate tiles has been a problem.

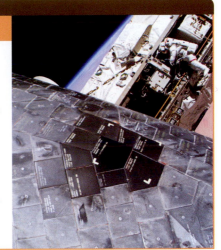

Helmet contains communications and ventilation systems

Gold-plated visor protects astronaut from the Sun's glare

Suit includes an airtight layer, insulation, and tough outer layers to provide protection

Insulated gloves with rubber fingertips to assist grip

Ports connect to air, water, and power supplies

Specially designed joints to allow easy movement

Overboots provide traction and protection

Oxygen purge system (for emergencies)

Communications equipment

Primary oxygen subsystem

Liquid transport loop (to cool astronaut's body)

Remote control unit

**The Apollo spacesuit and life-support system**
Spacesuits like this one worn during the Apollo Moon landings provide protection and life-support systems for astronauts away from the spacecraft. The Portable Life Support System (PLSS) is worn as a back pack.

**AFTER**

Recent years have seen more nations and companies engaging in manned spaceflight.

**CHINA IN SPACE**
In 2002 China became **the third nation to launch its own astronauts** into space, with the flight of *Shenzhou 5*. Although the Chinese space program is shrouded in secrecy, it is thought to involve ambitious plans for **missions to the Moon, Mars**, and even beyond.

**FUTURE MANNED SPACEFLIGHT**
**Commercial spaceflight** is also on the cusp of becoming a reality. Following the successful flights of the reusable SpaceShipOne in 2004, Virgin Galactic aims to offer this technology for suborbital trips. Other companies plan to enter orbit and even set up commercial space stations.

SEE ALSO >>
*pp.394–95* INSIDE THE SOLAR SYSTEM
*pp.396–97* SPACEPROBES AND TELESCOPES

**April 12, 1981** The **space shuttle** *Columbia* launches a new era of reusable space vehicles.

**November 20, 1998** Construction work begins on the **International Space Station**.

**October 15, 2003** **Yang Liwei** is the first Chinese astronaut to go into space.

| 980 | 1985 | 1990 | 1995 | 2000 | 2005 | 2010 |

**January 28, 1986** The Space Shuttle *Challenger* **explodes on launch** and is destroyed.

**1992** Russia and America announce their intention to work together on joint **shuttle–Mir missions**.

**1998** Final Mir space station mission.

**January 2004** US **President George Bush** announces plans for a return to the Moon by 2019.

**September 25, 2008** China counts down to its third space mission **Shenzhou 7**.

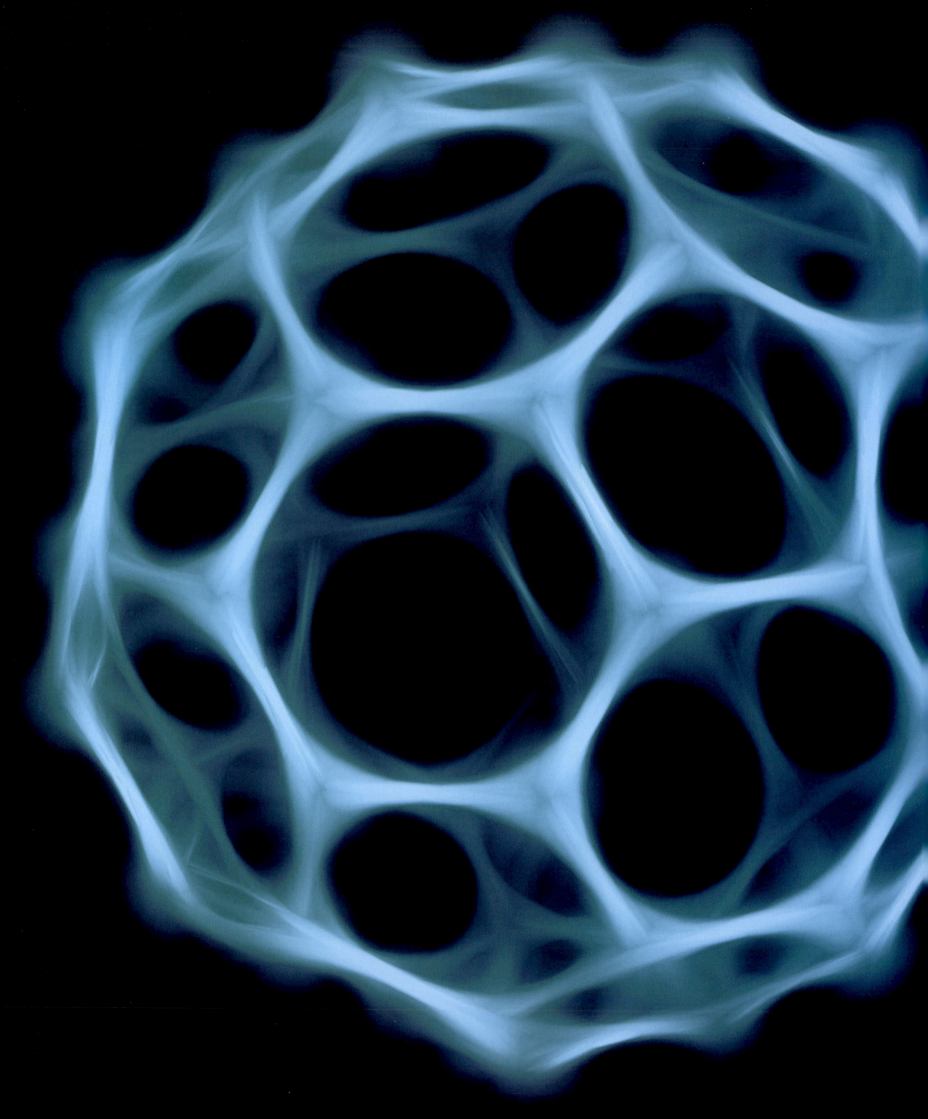

# 5

# THE INFORMATION AGE

## 1970 ONWARD

By the 1970s science and technology were in the ascendancy as astronauts landed on the Moon, and fertility experts produced the first test-tube baby. Meanwhile advances in electronics and computing enabled almost instant global communication—and people started to talk about the Information Age.

## 1970

## 1980

**1970**
First vaccine for rubella. String theory is born. Restriction enzymes, products of bacterial cells, discovered. The Soviet Venera probe lands on Venus.

▼ Intel 4004, Intel's first microprocessor chip

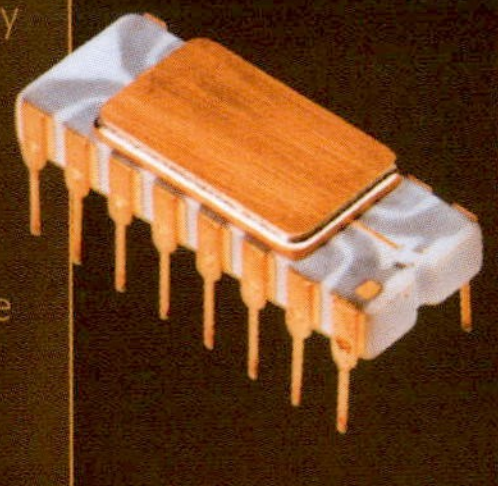

**1975**
Voyager 1 launched. The term "global warming" coined. Vera Rubin proposes that each galaxy has a halo of invisible dark matter.

Vera Rubin ▶

**1985**
ATZ, the first licensed HIV drug, launched. Buckminsterfullerine, a form of carbon, discovered.

Buckminsterfullerine ▶

**1971**
Intel introduces its first microprocessor chip (Intel 4004). Mariner 9 sends back pictures of the surface of Mars.

Surface of Mars ▶

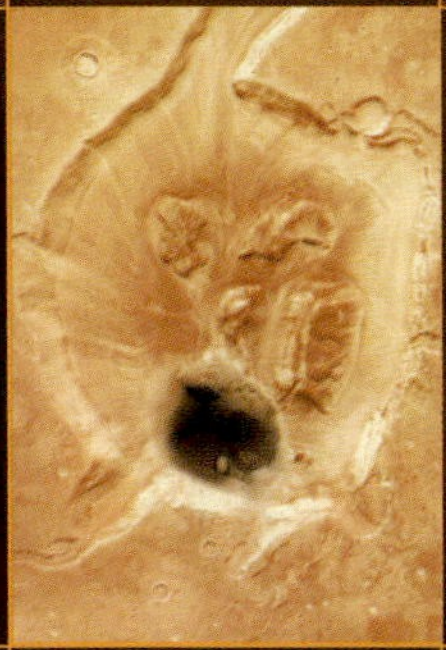

**1976**
Richard Dawkins publishes *The Selfish Gene*, stressing that the evolution of organisms is driven by genes. Concorde is the first supersonic passenger aircraft. The down quark discovered.

**1977**
Human growth hormone produced. Rings discovered around Uranus. Carl Woese proposes that there are three domains of life: two bacteria and one eukaryote. First MRI scan of a human.

▲ Wind farm

**1980**
The first wind farm built in the US.

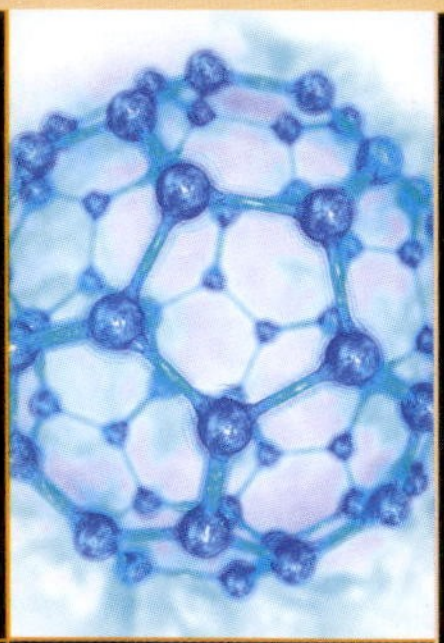

**1986**
Giotto photographs Halley's comet.

**1987**
NASA begins to measure tectonic plate movement. A supernova explodes in the galaxy the Large Magellanic Cloud.

Exploding supernova ▶

**1972**
The base pairs of a viral gene identified. The first remote sensing satellite (Landsat) launched. Moon rock is dated as 4.5 billion years old. Jacob Bekenstein suggests black holes have an entropy.

▼ Base pairs of a gene

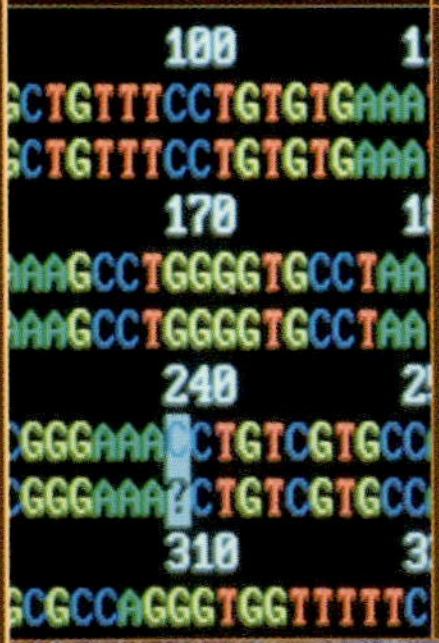

**1973**
Z and W bosons (subatomic particles) produced.

▼ Concorde

**1981**
The Space Shuttle first launched. First vaccine for Hepatitis B. The Scanning Tunneling Microscope (STM) developed.

Space Shuttle launch ▶

**1974**
First close-up of the surface of Mercury. Sheldon Glashow and Howard Georgi propose unifying the strong and electroweak interactions into a Grand Unified Theory.

▼ Close-up of Mercury

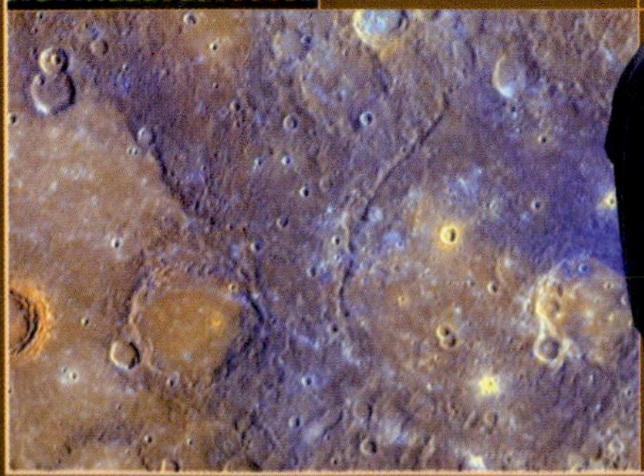

▼ James Lovelock

**1978**
Louise Brown, the first baby conceived by IVF, born. Venus examined by spacecraft radar. Human insulin produced.

**1979**
James Lovelock publishes *Gaia: A New Look at Life on Earth*. The Moment Magnitude scale introduced for measuring earthquakes.

**1982**
Richard Feynman conceives quantum computing.

**1983**
HIV identified as the cause of AIDS. The weak gauge boson discovered. The Infrared Astronomical Observatory launched.

**1984**
String theory developed further. Alec Jeffreys devises a genetic fingerprinting method. Kary Mullis invents the polymerase chain reaction.

▼ Genetic fingerprinting

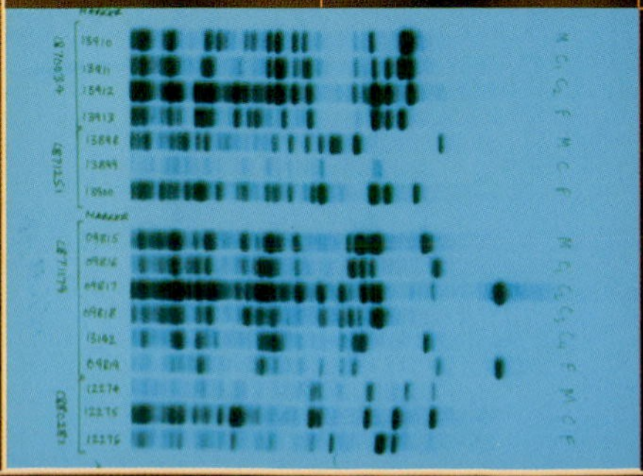

**1988**
Under James Watson, scientists begin to catalogue the locations of genes on chromosomes for different species.

**1989**
The first sheet of superclusters of galaxies, the Great Wall, is found.

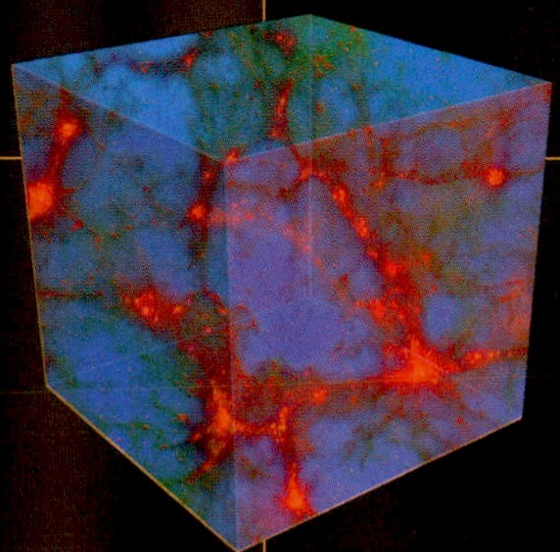

▲ Superclusters

The last 40 years have seen a strangely ambivalent view of science and technology emerging. On the one hand, science has continued to deliver astonishing advances in our understanding of the Universe. Space probes have landed on Mars and voyaged to the farthest reaches of the solar system. Microbiologists have mapped the genomes (the complete set of genes) of everything from nematode worms to human beings. And physicists believe they are on the verge of discovering the ultimate theory that will explain how every particle and force in the Universe interacts. On the other hand, science has been at the forefront of dire warnings of the consequences of technology and the damage it can do the environment.

## 1990

## 2000

**1990**
The Human Genome Project launched. The Hubble Space Telescope sent into orbit around Earth. Tim Berners-Lee invents the World Wide Web.

**1996**
Dolly, a sheep, is born: the first cloned animal.

Dolly, a cloned sheep ▶

**2000**
Genome of the fruit fly is sequenced. The Millennium Bridge, UK, is affected by resonance and is closed.

▲ Bike with frame made from carbon nanotubes

**1998**
The first genome for an animal (a worm) published. Google launched.

◀ Genome

**2001**
A zircon crystal is radiometrically dated as 4.4 billion years old, the oldest known object on Earth.

**2005**
A vaccine for the human papilloma virus developed. Bikes with frames made of carbon nanotubes are used in the Tour de France. Pluto is reclassified as a dwarf planet. The Huygens probe lands on Titan.

▲ Zircon crystal

**1991**
The Compton Gamma-Ray Observatory launched. Work on stringy black holes leads to a new understanding of how different versions of string theory are related.

**1992**
The first child conceived by intracytoplasmic sperm injection is born. The first planet orbiting another star discovered.

**1993**
The first smartphone, designed by IBM, goes on sale.

▲ Chandra space telescope

**2002**
The first artificial virus produced "from scratch."

**2006**
The gene sequence of the last human chromosome completed by the Human Genome Project.

▲ Mobile internet access

**1999**
The Chandra X-ray space telescope launched. Brian Greene publishes *The Elegant Universe*, explaining the string theory of particle physics.

▼ Brian Greene

**2003**
The Spitzer infrared space telescope launched.

Spitzer telescope ▶

**2008**
The number of mobile web users exceeds the number of PC web users. The Large Hadron Collider, the world's largest particle accelerator, is set up in Switzerland.

◀ Large Hadron Collider

**1995**
Top quarks (the most massive known elementary particles) are discovered. The first bacterial genome published. The first extrasolar planet around a sunlike star observed.

◀ Evidence of top quarks

**2004**
The Cassini space probe sent into orbit around Saturn. Gravity probe B experiment by NASA to test Einstein's general theory of relativity.

Gravity probe B ▶

**2009**
Researchers at the University of Arizona produce curved pulses of laser light.

# The Internet

## BEFORE

When telephone lines and developments in computer technology were united, the Internet age was born.

### TELECOMMUNICATIONS NETWORKS

A **communications revolution** occurred during the late 19th and early 20th centuries when telegraph and **telephone networks** connected all continents except Antarctica. When computers arrived in the 1960s it made sense to send electronic computer information via established telecommunications networks. In 1965 scientists **connected two computers** between California and Massachusetts over dial-up telephone lines. But the telephone line's circuit switching couldn't cope, leading to the invention of packet switching.

### PERSONAL COMPUTERS

The **Internet** existed in the 1970s but remained a tool for universities and governments until the rise of **personal home computers** in the 1980s and 1990s. At the same time **user-friendly software** was developed, which led to the **World Wide Web** and the explosion of Internet growth.

### SEE ALSO

### INVENTOR (1955–)
## TIM BERNERS-LEE

Tim Berners-Lee studied physics at Oxford University and by 1980 was writing software at CERN, the European Laboratory for Particle Physics in Geneva. Dreaming of improved communications between scientists, by 1990 he invented the software for building web pages and linking them to others via the Internet. In the CERN cafeteria he came up with a name for his system: World Wide Web.

**From blogging to googling, downloading to spamming, the Internet has spawned strange new words and a cyberworld that has altered how we communicate, work, and play. With this technological triumph geographical barriers have been transcended and anywhere in the world is just a click away.**

During the 1960s US-government computer scientists wondered how to keep US communications running following a potential nuclear attack. Working for the Advanced Research Project Agency (ARPA) they invented packet switching to break up messages and route them via several network pathways to ensure the message arrived. The first network, ARPA-net, was running by 1969, marking the dawn of the Internet—the global entirety of internetworked computers.

A group of computers that share information is a computer network. Like road networks, the world has lots of computer networks interconnected to each other over multiple routes. Computers communicate with packets of digital information sent via wires or radio waves to a router or modem, channeling the signals down a telephone line or television cable to a small local network called an ISP (Internet Service Provider). ISPs connect to bigger networks via a backbone of fiber-optic cables spanning land and sea, allowing the digital packets potentially to reach any device online. Like a postal address, all Internet devices have a unique IP (Internet Protocol) address so that the packets are traceable and know where to go.

## The World Wide Web

The World Wide Web uses the Internet to link billions of documents together. Like a vast digital library, web content is broken down into sites ("books") and pages. Every page is written using HTML (Hypertext Markup Language), a special computer language allowing pages (and websites) to be designed for particular functions, from selling groceries to emailing. HTML is difficult to read, so programs called web browsers were developed to display web pages. Any web page can be linked to any other page on the web using hyperlinks— often highlighted in blue—comprising HTTP (Hypertext Transfer Protocol) information. HTTP information gives each page a unique web address called a URL (Uniform Resource Locator) so it can be located. After clicking on a link or typing in a URL, the personal computer sends a digital request to a server—a large storage computer that hosts websites—via the Internet to view the HTML file that contains the relevant web page. The server then sends the file back to the personal computer, which displays the page.

## 1.6

**BILLION:** The estimated number of people across the world who surf the Internet every month. That's about one in four of the world's total estimated 6.7 billion inhabitants.

**Mobile internet access**
The vision of a new, mobile entertainment and information age has been realized with smartphone devices like this iPhone, which offers Internet access and multimedia entertainment on the move. The total number of mobile web users exceeded PC web users for the first time in 2008.

**Internet traffic**
These colorful interconnected threads show the Internet routing paths from an ISP network in Switzerland. When networks contact each other, many routes may be taken. If one fails, the information can take another path to reach its intended target. The blue lines show the busiest paths, pink and red less so, with yellow representing the least traffic.

## SEARCH ENGINES

Search engines use software robots, called spiders or crawlers, which follow every hyperlink and compile a huge list of words found on individual web pages. An index, like a book index, is then created using a ranking system. Different search engines use different ranking systems; some are based on how many times a particular word appears, while others assign a weight to words. Google™ ranks pages by the number of times other sites link to a particular page.

## Internet services
Hundreds of billions of emails are estimated to pass through cyberspace daily, while instant messaging and Internet phones like Skype™ combine with webcams to provide immediate interactions between people anywhere. Social networking sites, chat rooms, and online dating services offer new ways of establishing relationships, and

million of gamers explore virtual worlds. Bloggers can publish personal views online, and traditional media deliver immediate global news via their websites. Thanks to auction sites, buying and selling almost anything is possible, while digital music and video and fast broadband connections make songs and films easily obtained through downloading and streaming sites.

## Internet security
The Internet makes hacking possible, so secure connections are vital. Firewalls monitor traffic that enters and leaves a user's computer, blocking unauthorized access. Junk emails (spam) may contain viruses that infect computers and often delete files. ISPs scan emails for viruses, which can also be blocked using anti-virus software. Sending data via WiFi potentially allows unauthorized access, but signals can be encrypted using a system called WEP (Wired Equivalent Privacy). Sites requiring security of information like bank details and passwords also use encryption, signified by "https" in the URL. Filters built into the web browsers can restrict access to sites that may contain malicious software or undesirable content.

**Techno teaching tool**
With a seemingly infinite mine of information, the Internet has entered the classroom. Undoubtedly a useful educational resource, learning to use the Internet is also an important skill for children to develop in this modern information age.

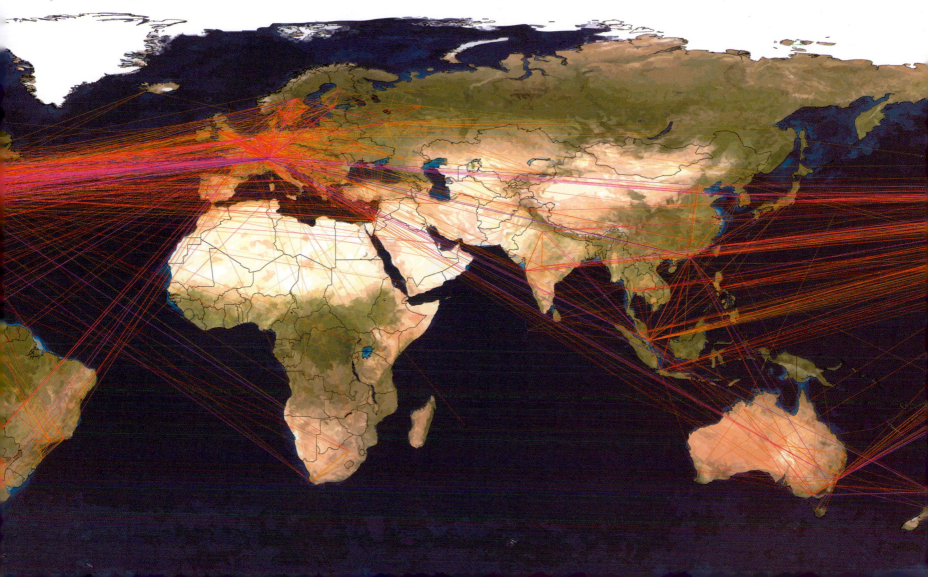
**Internet anywhere**
Even in the world's most remote places, including the polar regions, the Internet can be accessed without phone lines or cables by using special equipment that enables Internet devices to transmit and receive data via orbiting satellites. These systems are, however, expensive.

From fiber-optic networks to a new Internet system, the need for speed is driving the Internet into the future.

### FIBER-OPTIC NETWORKS
Broadband speeds may put dial-up connections to shame, but soon, broadband is still comparatively slow. Most **broadband** is **ADSL** (Asymmetric Digital Subscription Line), which still uses copper phone lines, so transfer speeds are limited. The newer technology of ADSL2+ can achieve download speeds up to 24 mbps (megabits per second) but also uses copper phone lines. In contrast, a network using **fiber-optic cable** is significantly faster, with potential speeds of 100 mbps.

### THE GRID
Since the Internet evolved from a jumble of cables and routers originally designed for telephone calls, an entirely new Internet, called **the Grid,** has been built from **dedicated fiber-optic cables** by the people who invented the World Wide Web at CERN. Currently used only for scientific purposes, it could one day be in all our homes, promising an **entertainment revolution** with speeds 10,000 times faster than current connections, and enabling us to talk to friends and relatives via holographic images.

# Artificial Intelligence and Robotics

**Today robots build walls, perform heart surgery, and even make scientific discoveries. In the future the field of artificial intelligence promises works of creative genius, safer cars, and more faithful companions. However, replicating human intelligence and complex human emotions and interactions in machines turns out to be more difficult than science-fiction films often suggest.**

## « BEFORE

The word "robot" was not in use until the 1900s, but people have been trying to make humanlike mechanical figures for centuries.

### EARLY MECHANICAL AUTOMATA

In the 13th century the Islamic engineer **Aljazari** wrote *The Book of Knowledge of Ingenious Mechanical Devices* detailing some of the earliest examples of automata—**humanoid machines**. Scientists recently reconstructed some of these, including **musical figures** designed to entertain party guests by banging drums and cymbals. In the 18th and 19th centuries **mechanical dolls and animals** amused European audiences, and the Japanese made automata that served tea.

**MAGICIAN AUTOMATON**

### EARLY ELECTRONIC ROBOTS

In World War II the British scientist **William Grey Walter** created the first **electronic autonomous robots**. Named "Elmer" and "Elsie," they were **tortoises with perspex shells** and moved around in response to light. Walter designed their **control systems** based on a greatly simplified version of the **human brain**.

### « SEE ALSO

pp.174–75 CALCULATING AND COMPUTING
pp.366–67 MICROCHIP TECHNOLOGY

Artificial intelligence (AI) is a branch of computer science that deals with machine intelligence. When we think of robots we often think of machines that resemble people, but these make up only a tiny proportion of what scientists would call robots. According to most definitions, a robot can be any machine or device that performs tasks usually performed by humans.

## From science fiction to science fact

Science-fiction writers used the idea of humanoid robots as early as the 1920s, before the age of electronic computers. In 1950, before AI even existed, the British mathematician Alan Turing (see pp.344–45) laid the groundwork for the field when he described a test for machine intelligence—the so-called Turing test (see opposite).

The 1968 film *2001: A Space Odyssey* showed a computer called HAL conversing intelligently with humans. In the real world no machine then could match HAL's intellect. Two years earlier, however, the German-born computer scientist Joseph Weizenbaum created a program called ELIZA that convinced many people they were talking to a real person via a computer rather than a computer itself. What ELIZA was actually doing was generating responses based on key words inputted by the participants.

## Machine intelligence

Language is one aspect of human culture that marks out our intelligence, and computers have not yet been able to match our skill with words.

**Industrial robot**
There are more than a million robots working in industry worldwide. Many are used in the car and food industries and are programmed to perform repetitive tasks such as welding (above), assembling, or packaging.

Something computers are better at, however, is data crunching—they are experts at memorizing huge quantities of data and performing calculations that would take us years. This makes them excellent problem solvers—and formidable opponents when it comes to games such as chess.

Emotional and social intelligence cannot be as easily programmed and these more complex attributes are yet to be developed in machines. In a first step in 2000, Danish scientists built Feelix, a robot that could convey six different emotions—anger, happiness, sadness, fear, surprise, and a "neutral" emotion—through its facial expressions in response to physical contact.

Creativity is another trait considered to be uniquely human. However, there are computer programs that can emulate creativity by, for example, composing music in the style of a famous composer such as Vivaldi.

Machines can also learn. Rather than programming in all the necessary information, machines can be given basic programs that allow them to learn by trial and error. They can even

**Humanoid robot**
ASIMO is one of the world's most famous humanlike robots. Created by car manufacturer Honda, the latest model is about about 5 ft (1.6 m) tall, knows when to recharge its own batteries, and is equipped with cameras for eyes to stop it from colliding with people and objects.

run experiments and make discoveries—and in 2009 a robot called Adam, built by UK researchers, became the first machine to do so. It ran a series of experiments that confirmed the function of a specific yeast gene. When the robot's work was checked by scientists, the experimental results were verified.

## Humans and machines

For many of us, trust in machine capabilities stops at human health and safety. Yet computer programs are regularly used to pilot planes to safety, and competitors in the DARPA Grand

### BREAKTHROUGH

## COMPUTER BEATS HUMAN

In 1996 Russian world chess champion Gary Kasparov played six rounds against Deep Blue, a chess-playing computer built by IBM, and won 4-2—human triumphed over machine. However, in a rematch a year later Kasparov lost to an updated version of Deep Blue by 3½-2½, the first time a computer had beaten a world chess champion in match play. Deep Blue won because it was able to calculate a vast number of moves at each turn and select the best. Unlike Kasparov, Deep Blue was also completely unaffected by the emotional pressure of the match.

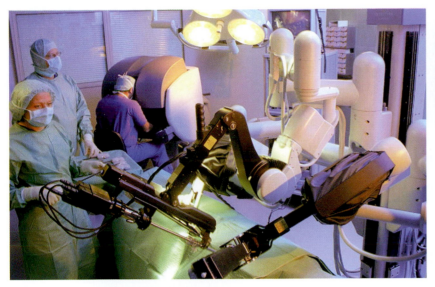

**Robotic vacuum cleaners**
Robotics is being applied to domestic appliances, such as vacuum cleaners, which use infrared sensors to guide themselves around furniture and operate entirely without human control.

**Robotic surgery**
Using remotely controlled robot arms to perform certain delicate surgical procedures reduces the risk of infection and the chance of mistakes caused by unsteady hands. The close-up cameras also give a much clearer, more detailed view of the area being operated on.

The next steps in AI will be some of the most difficult, as scientists try to develop machines that can respond to human emotions and, incredibly, evolve.

**ROBOT PARTNERS**
Finding the perfect partner can be problematic, but it is possible that in the future we will have our partners **custom made**. However, **robot partners** would need to be able to detect and respond to complex emotional cues and it may be many years before we learn how to program **emotional intelligence** into machines.

**ROBOT EVOLUTION**
An **artificial species** that could evolve would need artificial DNA. In 2005 South Korean scientists described the basis for **artificial chromosomes**: sections of computerized code that they programmed into **virtual robots**. The robots existed only inside a computer but because of differences in their code each responded to its virtual world as a distinct individual.

**SEE ALSO »**
*pp.386–87* GENE TECHNOLOGY
*pp.392–93* NANOTECHNOLOGIES

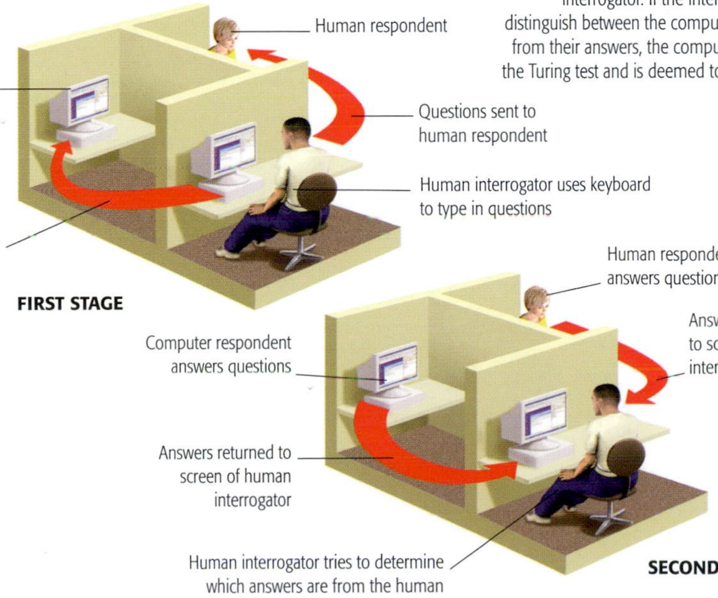

Challenge (a race for self-driving vehicles) have proved that intelligent cars can successfully negotiate traffic jams. Robots are also beginning to be used for surgical operations, even for difficult procedures such as heart surgery. These surgical robots are not automata, but act as extensions of the surgeon's hands. They have cameras to relay views of the surgery site to the human surgeon, who controls the cameras and instruments held by the robot's arms from a remote console.

Many robots are designed to carry out tasks that are too dangerous, difficult, or unpleasant for humans to do. For example, scientists are developing robots that can lift and manipulate heavy objects for use in the construction industry. Robots fitted with cameras and manipulating devices are being used by the emergency services and military for tasks such as dangerous search-and-rescue operations and defusing bombs.

ASIMO

HONDA

**Turing test**
In this test of artificial intelligence, a human interrogator first puts questions to a computer and another person, both of which are isolated from each other and from the interrogator. The computer and human respondent return their answers to the interrogator. If the interrogator cannot distinguish between the computer and person from their answers, the computer has passed the Turing test and is deemed to be intelligent.

Computer respondent

Human respondent

Questions sent to human respondent

Human interrogator uses keyboard to type in questions

Questions sent to computer respondent

**FIRST STAGE**

Computer respondent answers questions

Human respondent answers questions

Answers returned to screen of human interrogator

Answers returned to screen of human interrogator

Human interrogator tries to determine which answers are from the human and which from the computer

**SECOND STAGE**

**‹‹ BEFORE**

The concept of subatomic particles only came into existence in 1897 with the discovery of the first particle of this type—the electron. Further particles were discovered during the 20th century.

## LIGHT AS PARTICLES

Though light has wavelike properties, the idea that it consists of particles was first considered by **Isaac Newton ‹‹ 114**, put forward by **Einstein** in 1905, and widely accepted by the 1920s. Small packets, or **quanta**, of light radiation came to be known as **photons**.

## A SUBATOMIC ZOO

Following the discovery of the **proton** in 1910 and the **neutron** in 1932, scientists identified a whole **"zoo" of new particles**, many of which are extremely short-lived. The first few were found by studying what happens when **cosmic rays**—highly energetic particles from space—collide with matter on Earth. These included the **positron**, which is a positively charged version of the electron, and the **muon**, which is like an electron but with more mass. Closely following were the discoveries of **mesons** and **neutrinos**.

## PARTICLE TRACKS

In the early 1950s the American physicist **Donald Glaser** invented the **bubble chamber**. This is a tank of superheated liquid through which **charged particles** travel, making the liquid boil in trails behind them. The bubble chamber made visible the **tracks of scores of mysterious particles** sprayed out when known particles were smashed together in machines called particle accelerators.

**‹‹ SEE ALSO**

*pp.286–87* Structure of the Atom
*pp.314–17* Quantum Revolution
*pp.320–21* The Big Bang

### Evidence for the top quark
Top quarks—the most massive known elementary particles—were discovered at the Fermi National Accelerator Laboratory (Fermilab), US, in 1995. They are created in proton–antiproton collisions and decay almost instantly into other particles. The colored lines in this image show the tracks of top quarks in the moments before they decay.

### Helium nucleus
A helium nucleus contains two protons and normally (as here) two neutrons. They are held together by what is known as the "nuclear force." Physicists now regard this as a type of "leakage" of the strong nuclear force that binds quarks together inside protons and neutrons.

# Subatomic Particles

**Subatomic particles include the basic building blocks of matter and various force-carrier particles. Studying them has helped unravel the nature of matter and energy and, thus, the origins of the universe.**

In 1932 the only known subatomic particles were protons, neutrons, electrons, and photons, but over the next 30 years many new ones were found. They are studied alongside the four basic forces of nature: gravity, the weak nuclear interaction, the strong nuclear force, and electromagnetism.

## Quarks and the standard model
By the mid-1960s, physicists realized that a new model of the subatomic world was needed to explain the new particles, their interactions, and their relationships with the fundamental forces. This led to a new theory in which the particles were organized into a mathematically based classification plan. Developing into what became known as the standard model, this theory also helped to explain how all the forces except gravity operated.

Central to the theory was an idea that some particles, including protons and neutrons, consisted of subunits, termed quarks. Originally, quarks were just a

### Subatomic particles in a helium atom
When the protons and neutrons inside atoms, such as this helium atom, were first discovered, it was not realized that they had an internal structure. Now it is known that they contain smaller particles called quarks. However, the electrons in atoms are truly elementary—they possess no substructure.

## CLASSIFICATION OF SUBATOMIC PARTICLES

Physicists distinguish between elementary particles, which have no substructure, and composite particles, which are composed of smaller structures. They also divide particles into fermions, which are building blocks of matter, and bosons, which are primarily force-carrier particles.

Antiparticles normally have the opposite electric charge to their corresponding particles, although with antineutrons it is their quarks that differ. It is possible that antineutrinos are identical to neutrinos.

| | ELEMENTARY PARTICLES | | COMPOSITE PARTICLES (also called hadrons) | ELEMENTARY ANTIPARTICLES (examples) | | COMPOSITE ANTIPARTICLES (examples) |
|---|---|---|---|---|---|---|
| **FERMIONS** | **QUARKS** up $+\frac{2}{3}$ down $-\frac{1}{3}$ — charm $+\frac{2}{3}$ strange $-\frac{1}{3}$ — top $+\frac{2}{3}$ bottom $-\frac{1}{3}$ | **LEPTONS** electron $-1$ electron neutrino — muon $-1$ muon neutrino — tau particle $-1$ tau neutrino | **BARYONS** made of 3 quarks, bound by gluons — proton $+1$ 1 down and 2 up quarks — neutron 2 down and 1 up quark — lambda particle 1 down, 1 up, and 1 strange quark — plus numerous others | **ANTIQUARKS** up antiquark $-\frac{2}{3}$ — down antiquark $+\frac{1}{3}$ | **ANTILEPTONS** positron (anti-electron) $+1$ — electron antineutrino | **ANTIBARYONS** antiproton $-1$ 1 down and 2 up antiquarks — antineutron 2 down and 1 up antiquark |
| **BOSONS** | **GAUGE BOSONS** force-carrier particles photon — $+1$ $-1$ W+, W–, and Z bosons gluon | | **MESONS** quark–antiquark pairs positive pion $+1$ 1 up quark and 1 down antiquark — negative kaon $-1$ 1 strange quark and 1 up antiquark — plus numerous others | **ANTIGAUGE BOSONS** $+1$ $-1$ W+ and W– bosons are antiparticles of each other | | **ANTIMESONS** negative pion $-1$ 1 down quark and 1 up antiquark |

**Inside a neutron**
A neutron consists of one "up" quark (with a charge of $+\frac{2}{3}$) and two "down" quarks (each with a charge of $-\frac{1}{3}$), held together by the strong nuclear force. A neutron has a total charge of $+\frac{2}{3} - \frac{1}{3} - \frac{1}{3} = 0$.

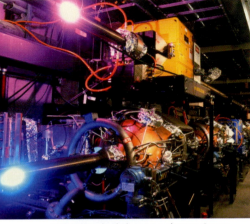

Up quark

Gluon

Down quark

**Electron**
Electrons are themselves elementary particles, each with a charge of −1. In a helium atom there are two electrons, each moving around within a spherical region, or "shell," that surrounds the nucleus.

**Inside a proton**
A proton consists of two "up" quarks (each with a charge of $+\frac{2}{3}$) and one "down" quark ($-\frac{1}{3}$), held together by the strong nuclear force. A proton has a total charge of $+\frac{2}{3} + \frac{2}{3} - \frac{1}{3} = +1$.

mathematical abstraction, but scientists came to recognize that they were real—elementary particles that bind to form composite particles. In the 1970s it became known that the force holding quarks together was the strong nuclear force and that this force was mediated by other massless particles called gluons. At first quarks were thought to come in three types or "flavors"— "up," "down," and "strange"—but three more "flavors" are now known: "charm," "top," and "bottom."

## Leptons and gauge bosons
As well as quarks, two more types of elementary particle were recognized: leptons and gauge bosons.

Leptons are particles of low mass that are unaffected by the strong nuclear force but respond to the weak nuclear interaction. There are six types—the electron, muon, and tau particle, and their associated neutrinos. The latter are uncharged particles, with a minuscule mass, that often travel at the speed of light and are created by radioactive decay processes, nuclear reactions, and cosmic-ray collisions.

Bosons differ from all of the other subatomic particles – collectively known as fermions – in a property called "spin". This can be thought of as a particle's internal angular momentum. Bosons have spin values that are either equal

## "Who ordered that?"

ISIDORE RABI, 1936, ON LEARNING OF THE DISCOVERY OF THE MUON

to 0 or to an integer (1, 2, and so on). Fermions, which include all the building blocks of matter, such as quarks, electrons, and protons, have spins with half-integer values (such as ½ or 1½).

Gauge bosons are "carrier" particles for the four forces. They include photons, which are carriers for the electromagnetic force; gluons, which

are carriers for the strong nuclear force; and W and Z bosons, which mediate the weak nuclear interaction. Gravitons are thought to mediate gravity.

## Further bosons
Other types of boson have been found or proposed. One is the pion, which mediates the attraction between protons and neutrons in atomic nuclei. One that has been proposed is the

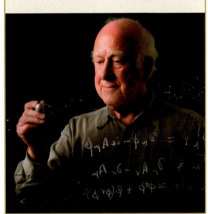

**PEP-II, Stanford Linear Accelerator Center**
Light beams show the paths of electrons (blue) and positrons (pink). These particles are sent in opposite directions and forced to collide, forming large quantities of B mesons and anti-B mesons.

Higgs boson, which is thought to give other particles mass. The existence of this boson, it is said, will be the ultimate test of the Standard Model. So far, it has not been detected, but this may be because it has a mass way beyond the capacity of current particle accelerators to generate.

## Antiparticles
For most subatomic particles, there are corresponding antimatter particles, or antiparticles. These may be produced in radioactive decay and in cosmic-ray collisions in Earth's atmosphere, and artificially in particle accelerators.

Most particles and their antiparticles are identical except for opposite electrical charges. Certain neutrally charged particles are also known to have antiparticles. In some (such as neutrons and antineutrons), the quarks making up the particle and its antiparticle have opposite charges. In others (such as photons and Z bosons), a particle is its own antiparticle. When a particle and its antiparticle meet, they may destroy each other, producing energy in the form of radiation (photons). A major question in current Big Bang research is why there are more particles than antiparticles in the Universe.

PHYSICIST (1929–)

## PETER HIGGS

Born in Newcastle, England, Peter Higgs studied physics at King's College London. In 1960 he began work at the University of Edinburgh, and from 1980 to 1996 he was professor of theoretical physics there. Higgs is best known for his involvement in developing a theory that explains the origin of the mass of elementary particles, particularly the W and Z gauge bosons. According to this theory, particles obtain their masses by interacting with an energetic field that permeates space, called the Higgs field. He also predicted that this field should produce its own type of particle, which has become known as the Higgs boson.

## AFTER

Particle physicists predict the existence of hitherto unobserved particles, and they will almost certainly continue to find them.

**LARGE HADRON COLLIDER (LHC)**
The LHC in Switzerland is the biggest and highest-energy **particle accelerator** built to date. Physicists anticipate that it has the best chance yet of **confirming the existence of the Higgs boson** and other predicted new particles.

**MAGNET BEING FITTED IN THE LHC**

**DARK MATTER**
A major goal for physicists and cosmologists is to discover what particles, if any, constitute **"dark" matter**, which forms a **high proportion** of the **mass of the Universe**. It might possibly consist of neutrinos, or perhaps some hitherto unknown types of particles, such as neutralinos.

CHEMIST AND CRYSTALLOGRAPHER **Born 1910** Died 1994

# Dorothy Hodgkin

## "I was captured for life by chemistry and by crystals."

DOROTHY HODGKIN, ON FIRST WATCHING CRYSTALS GROW IN THE LABORATORY

The chemist Dorothy Hodgkin was one of the few scientists of her generation who, in addition to being incredibly smart, was also very much loved—some of her colleagues found her almost saintly. She was a pioneer in the technique of X-ray crystallography, and is best known today for her research in proteins. She was the first British woman to receive the Nobel Prize, and her work has been fundamental to our understanding of how molecules function in living systems. She was also a remarkably able teacher.

### Early training
The eldest of four sisters, Hodgkin was born in Cairo, Egypt, but was sent to school in England where she spent much of her childhood away from her parents. Having developed a passionate interest in crystals at the age of 10, she fought to be allowed to study science with the boys at school, and this she duly achieved. In 1928 she was accepted at Somerville College at Oxford University to study

chemistry; the college would prove to be the mainstay of her academic career. In 1932 Hodgkin moved to Cambridge University to work with John Desmond Bernal, extending her Oxford work on sterols and helping him make the first X-ray diffraction studies of pepsin, a crystalline protein. However, the opportunity of a temporary fellowship back at her old college was too good to turn down, and she would stay at Somerville until her retirement in 1977.

After she was given some insulin crystals by her friend Robert Robinson, she began to research the structure of insulin, creating an X-ray laboratory in a corner of Oxford's Natural History Museum, a place usually better known for its dinosaur and mineral collections.

In 1937 she married historian Thomas Hodgkin and between 1938 and 1946 the couple had three children. Sadly, Dorothy developed rheumatoid arthritis at the age of just 28 due to an infection following the birth of her first child, Luke. Her hands became swollen and painful, but she never let the condition prevent her from working.

### Researches into penicillin
Hodgkin's work on insulin was forced to take a back seat when Australian pathologist Howard Florey and his colleagues successfully isolated penicillin. This group of antibiotics had long been a subject of controversy, and Florey and his team invited Hodgkin to complete the picture by revealing its structure. She duly obliged in 1945, describing the atoms in three dimensions and in the process settling many long-standing arguments between organic

**A woman of achievement**
This plaque acknowledges Hodgkin's attendance at the Sir John Leman Grammar School in Suffolk, England, where she, along with one other girl, Norah Pusey, was allowed to join the boys for chemistry classes.

**Dorothy Hodgkin** △
At a young age Hodgkin was given a copy of *Concerning the Nature of Things* by Sir William Henry Bragg, a founder of X-ray crystallography. The book sparked her interest in gaining data by analyzing the intensity of X-rays diffracted from planes in crystals.

**Molecular structure of vitamin B$_{12}$** ▷
Using X-ray crystallography, Hodgkin was able to determine the structure of vitamin B$_{12}$ in 1957. This ball-and-spoke model was made in 1958 for a research group at Cambridge University led by Sir Alexander Todd, who had been trying to figure out the vitamin's structure by chemical means.

## BREAKTHROUGH

### NOBEL PRIZE IN CHEMISTRY

In 1964 Dorothy Hodgkin was awarded the Nobel Prize in chemistry "for her determinations by X-ray techniques of the structures of important biochemical substances." She was only the third woman chemist to be so honored, after Marie Curie and Irène Joliot-Curie.

At Oxford University Hodgkin had pioneered the technique of using X-ray crystallography to determine the three-dimensional structure of biomolecules. A form of the same technique was employed by Rosalind Franklin at King's College, London, to produce the X-ray diffraction pictures of DNA that Watson and Crick at Cambridge University used to reveal the structure of DNA (see p.346).

Hodgkin took part in meetings in 1946 that led to the foundation of the International Union of Crystallography. X-ray crystallography is still widely used by scientists requiring knowledge of a molecule's structure in order to gain a clear understanding of its function.

chemists. At the time penicillin was the largest molecule to be described using X-ray techniques. In recognition of her work, in 1947 Hodgkin was elected to the Royal Society in London.

### Work on vitamin $B_{12}$

In the academic world, paleontologists tend to be shown fossils all the time, while ornithologists are often brought preserved birds for investigation. In Hodgkin's case, scientific friends would regularly bring crystals that might interest her. The chemist E. Lester Smith was one such friend who gave Hodgkin some cherry-red crystals of vitamin $B_{12}$. Characteristically, Hodgkin became determined to fathom out their structure and, using an array of complex computer models, it was not long before she had her answer.

In time more awards followed, most notably the Nobel Prize in chemistry in 1964—a remarkable achievement for a woman in the field of science. A year later she was made a member of the

Order of Merit. This institution of only 24 members had been founded by Edward VII in 1902 and is considered Britain's highest honor in science, the arts, and public life. Before Hodgkin, the only woman to have achieved this accolade was the pioneering nurse and writer Florence Nightingale.

first given a small sample of crystalline insulin by Robert Robinson. Analyzing this complex protein had been extremely difficult, requiring advanced X-ray diffraction techniques and high-speed computing. Hodgkin was careful to give due credit to the international team of young

> "She will be remembered as a **great chemist,** a saintly, gentle and tolerant lover of people, and as a **devoted protagonist** of world peace."
>
> M. F. PERUTZ, REMEMBERING HODGKIN IN "THE INDEPENDENT" NEWSPAPER, 1994

However, by Hodgkin's own admission, her greatest achievement did not come until 1969, when long years of patient work were finally rewarded by her discovery of the structure of insulin; this was some 35 years after she was

researchers who had helped her achieve her goal. When she retired from Somerfield College in 1977, however, Hodgkin decided to turn her insight to a completely different field.

### Advocate for peace

Enjoying the respect of the scientific community, Hodgkin wanted to exert her influence to help secure peace in international relations. She had long been a member of Science for Peace, a group dedicated to ending war, and in 1953 had been denied a visa to the US because the group had communist members. In 1976, though, these restrictions had been lifted and she was able to travel the world freely as a passionate advocate for peace.

Crystals never ceased to fascinate Hodgkin, however, and she continued her tours of American universities, discussing proteins, insulin, and the history of crystallography—right up until her death in England in 1994.

### Distinguished company

Photographed in 1985, Hodgkin stands next to Stephen Hawking, Lucasian Professor of Mathematics. Hodgkin, like Hawking today, was both a Nobel Laureate and a member of the prestigious Royal Society.

**WALLPAPER BASED ON STRUCTURE OF INSULIN**

## BEFORE

Gene technology became a practical possibility once the chemical basis of genetics was understood.

**PROTEIN SYNTHESIS**
As the result of numerous collaborative experiments in the 1950s, biologists demonstrated that the **inherited material was DNA**, and that it exerted its influence on the characteristics of organisms by instructing them to make **specific combinations of proteins**. The potential use of **genetic engineering** lay in the promise of transferring genes between organisms to **change these proteins**.

**GENE EXCHANGE IN BACTERIA**
**Joshua Lederberg**'s discovery that bacteria could **exchange genes** explained observations that bacteria changed their characteristics when mixed with others. They can transfer genes by passing small rings of DNA called **plasmids** between them, known as **conjugation**.

BACTERIA UNDERGOING CONJUGATON

« SEE ALSO
pp.244–45 BACTERIA AND VIRUSES
pp.348–49 THE GENETIC CODE

BIOLOGIST (1925–2008)

## JOSHUA LEDERBERG

Joshua Lederberg was an American biologist who was distinguished by his contributions in many areas of science. In 1958 he won the Nobel Prize for Medicine after discovering that bacteria exchanged genes in plasmids. He led the Genetics Department at Stanford University, California, and was a key scientific advisor to the US government, as well as being involved in NASA's space exploration—researching the potential extraterrestrial spread of microbes.

# Gene Technology

**The realization that the characteristics of living organisms were determined by sections of molecules called genes opened up the possibility of altering them in a way that went far beyond the breeding programs of plants and animals. Any organism could be endowed with features that would make it more useful.**

In the 1950s scientists established that inherited characteristics are determined by sections of DNA in cells called genes (pp.308–309). Genetic engineering involves artificially transferring genes from one organism to another—usually a completely different species—so that characteristics become altered in the resulting genetically modified organism or GMO. Theory became a reality in 1970 with the discovery of restriction enzymes. Bacteria produce these enzymes to cut up the DNA of invading viruses, but unlike conventional digestive enzymes they cut the DNA at precise positions. This ability was to prove critical in their use in genetic engineering.

**70 PERCENT** the proportion of therapeutic insulin currently used worldwide that has been produced by genetic engineering.

### The benefits of genetic engineering
From the 1970s scientists became increasingly aware of the potential benefits of genetic engineering. With the knowledge that genes determined proteins came the realization that an organism's protein production could be altered by transplanting genes.

Back in 1955 English biochemist Frederick Sanger had been the first to decipher the complete amino acid sequence of a protein (that of insulin). Animals produce insulin to control blood-sugar levels; if insulin cannot be made, a disease called diabetes results. Diabetes can be controlled with insulin treatment. At that time, the only source of insulin was from the pancreases of cattle or pigs and, while purification methods were improving, there was a risk that animal version could be contaminated with viruses. Genetic engineering offered the possibility of producing insulin from bacteria, which reduced this risk; furthermore, such engineering paved the way for modifying "higher" organisms, such as plants, to improve food production.

### The first GM bacteria
In 1973 American biochemists Herbert Boyer and Stanley Cohen started to use restriction enzymes to cut threads of DNA. Because these enzymes cut bacterial DNA at specific points, they could be used to cut out those precise fragments of DNA that contained genes—a process called gene splicing. Boyer and Cohen also knew that

bacteria routinely exchanged plasmids, so they used restriction enzymes to cut the rings open and another enzyme—ligase—to bond the gene fragments into the loops. In this way plasmids could be used as carriers, or vectors, for the inserted genes. The "new" genes would be transferred from one bacterium to another and replicated whenever the modified bacteria reproduced.

Using this technique they produced the human hormone somatostatin in 1977 and then followed that with human insulin in 1978. Boyer established a company for the commercial production of these and other proteins; in 1982 "man-made" insulin was the first genetically engineered product approved by the US Food and Drug Administration.

### Further developments
The sequencing work of Sanger and others paved the way for scientists to assemble genes from their component building blocks, removing the complex task of isolating the proteins they code

**Luminescent fish**
The first genetically modified pet to be made available to the public, this tiny fish—known as a Glofish—was produced by inserting genes for bioluminescence from jellyfish into zebra fish.

for first. Most therapeutic insulin and the blood-clotting factor used to treat the disorder hemophilia is made this way.

Restriction enzymes proved useful in other areas. In 1984 English geneticist Alec Jeffreys realized that the number and size of the DNA fragments they produce could characterize a person's genetic identity. "Genetic fingerprinting" is now used in forensic science and to establish family relationships.

**USES FOR GENE TECHNOLOGY**

| SUBJECT | POSSIBLE USES |
|---|---|
| **Plants** | Improving nutritional content, for example, golden rice contains a gene that produces a precursor of vitamin A. |
| | Lengthening shelf life by changing gene that ripens fruit, preventing overripening and bruising, for example, long-life FlavrSavr tomatoes. |
| | Adding pesticide and/or herbicide resistance to crops, such as soy, so that farmers can use fewer chemicals without reducing the yield. |
| | Ensuring good harvest by adding genes for pest and/or disease resistance, for example, virus-resistant papaya and insecticide-producing corn. |
| | Improving hardiness, such as adding ice resistance to crops normally grown in warmer climates to enable cultivation in colder areas. |
| **Animals** | Improving nutritional content of food, for example, growth-enhanced tilapia fish or transgenic chickens engineered for their high-protein eggs. |
| | Developing experimental models, such as mice engineered to develop cancer or mice with missing genes, to demonstrate gene function. |
| | Altering aesthetics, for example, zebra fish modified to create bioluminescent Glofish (see above). |
| **Chemical products created by genetic engineering** | Proteins for treating conditions such as diabetes (insulin), hemophilia (blood-clotting factor), and hepatitis B (interferon). |
| | Products that arise from protein activity, such as aspartame used as low-calorie sweetener in food. |
| **Humans** | Gene therapy using somatic GM is in its infancy but in theory it can replace a faulty gene with a functional one in the affected part of the body; possible disease candidates are cystic fibrosis and immunodeficiency. |
| | Germ-line GM could modify sex cells. It is widely considered unethical and almost universally illegal. |

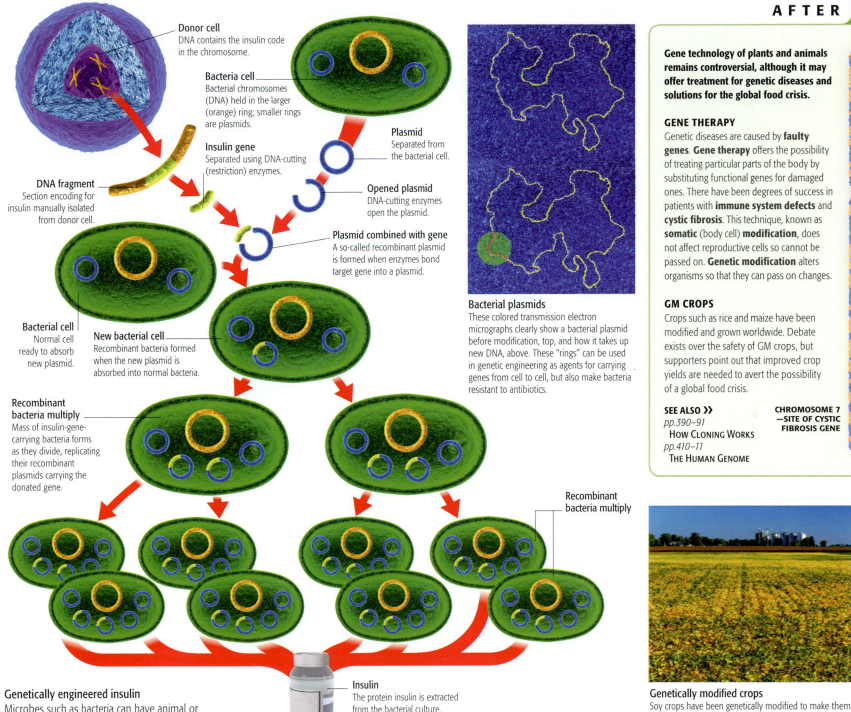

**Donor cell**
DNA contains the insulin code in the chromosome.

**Bacteria cell**
Bacterial chromosomes (DNA) held in the larger (orange) ring; smaller rings are plasmids.

**Insulin gene**
Separated using DNA-cutting (restriction) enzymes.

**Plasmid**
Separated from the bacterial cell.

**DNA fragment**
Section encoding for insulin manually isolated from donor cell.

**Opened plasmid**
DNA-cutting enzymes open the plasmid.

**Plasmid combined with gene**
A so-called recombinant plasmid is formed when enzymes bond target gene into a plasmid.

**Bacterial cell**
Normal cell ready to absorb new plasmid.

**New bacterial cell**
Recombinant bacteria formed when the new plasmid is absorbed into normal bacteria.

**Recombinant bacteria multiply**
Mass of insulin-gene-carrying bacteria forms as they divide, replicating their recombinant plasmids carrying the donated gene.

**Recombinant bacteria multiply**

**Insulin**
The protein insulin is extracted from the bacterial culture.

**Bacterial plasmids**
These colored transmission electron micrographs clearly show a bacterial plasmid before modification, top, and how it takes up new DNA, above. These "rings" can be used in genetic engineering as agents for carrying genes from cell to cell, but also make bacteria resistant to antibiotics.

**AFTER** »

Gene technology of plants and animals remains controversial, although it may offer treatment for genetic diseases and solutions for the global food crisis.

### GENE THERAPY

Genetic diseases are caused by **faulty genes**. **Gene therapy** offers the possibility of treating particular parts of the body by substituting functional genes for damaged ones. There have been degrees of success in patients with **immune system defects** and **cystic fibrosis**. This technique, known as **somatic** (body cell) **modification**, does not affect reproductive cells so cannot be passed on. **Genetic modification** alters organisms so that they can pass on changes.

### GM CROPS

Crops such as rice and maize have been modified and grown worldwide. Debate exists over the safety of GM crops, but supporters point out that improved crop yields are needed to avert the possibility of a global food crisis.

SEE ALSO »
pp.390–91
HOW CLONING WORKS
pp.410–11
THE HUMAN GENOME

**CHROMOSOME 7 —SITE OF CYSTIC FIBROSIS GENE**

**Genetically modified crops**
Soy crops have been genetically modified to make them resistant to herbicides, allowing farmers to exercise better control over weeds that grow alongside the soy.

## Genetically engineered insulin
Microbes such as bacteria can have animal or plant genes inserted into them so that they can be modified to produce a useful protein. The donated gene that instructs the manufacture of the protein is first isolated and then bonded into a plasmid ring—a molecule that is naturally absorbed by another bacteria.

## The future of GM
Gene technology has evolved to make more ambitious modifications possible. For example, naturally occurring plasmids in some bacteria can be modified to suppress certain genes and to carry other useful genes into plant cells instead. Modification of animals necessitates manipulation of embryos. GM animals, such as insects and mice, have mainly been used to further understanding of gene function, for example, genes that cause cancer.

A key area of gene technology concerns the production of crops that are genetically modified to resist pests or herbicides or to change qualities. Such GM products first became commercially available in the 1990s, the first being the FlavrSavr tomato. Most GM crops are grown in North America but they are increasingly sown in developing countries, such as Brazil and India.

Not everyone is convinced about the safety of such GM crops and research is ongoing to assess the implications for human health. Organizations such as Greenpeace campaign for an embargo on such foods until their safety has been undeniably proven.

**BREAKTHROUGH**

## GENETIC FINGERPRINTING

This technique, also known as DNA profiling or DNA testing, compares samples of DNA from different people. Restriction enzymes "digest" each DNA sample, generating a pattern of bands that looks like a bar code; the bands correspond to clusters of DNA fragments. Although 99.9 percent of DNA sequences are the same in every individual, each person does have a unique pattern. Similarities usually show family relationships.

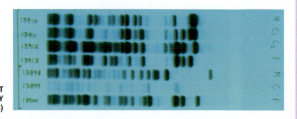

**GENETIC FINGERPRINT OF A FAMILY (TOP FOUR STRIPS)**

# In Vitro Fertilization (IVF)

**On July 25, 1978, at Oldham General Hospital, near Manchester in England, Lesley Brown gave birth to a healthy baby girl. This was the result of a groundbreaking technique that had been developed to help women with blocked or missing fallopian tubes conceive. The birth of Louise Brown was a particular cause for celebration: she was the world's first baby who had been successfully conceived by IVF.**

Lesley Brown had blocked fallopian tubes, which meant that sperm could not reach her eggs for conception—but in the late 1970s scientists developed a procedure that offered her hope. In vitro (in glass) fertilization involved a woman receiving fertility medicines in order to stimulate ovulation, retrieving eggs from the woman's ovaries, and introducing sperm and egg together in a laboratory Petri dish. Scientists provided the right conditions to encourage fertilization and monitored the first few days of embryological development. The tiny embryo—no more than a ball of cells—is then implanted back into the woman's uterus for a natural pregnancy.

For nearly 20 years before Louise Brown's birth, the physiologist Robert Edwards had been perfecting the medium needed for fertilization at the University of Cambridge, England. Edwards had managed to nurture human eggs to the critical stage where they were receptive, and in 1969 he witnessed the first human in vitro fertilization: in his petri dish he saw clear evidence that sperm had successfully penetrated an egg.

Edwards developed the clinical side of his work in collaboration with a surgeon, Patrick Steptoe, who used a probe to retrieve eggs from the ovaries of volunteering patients. Together they solved other technical problems: they used drugs to control the menstrual cycle of patients so that they knew when eggs were ready for extraction, and they perfected a method of returning embryos to the uterus. During much of their work Edwards and Steptoe endured condemnation from critics who called them "dehumanizing." However, the critics were silenced when Louise Brown was born and mother and baby's condition was described as excellent.

Since the birth of Louise Brown the techniques pioneered by Edwards and Steptoe have been adopted and refined by hospitals around the world. Drugs based on natural hormones have been developed to stimulate egg production and fertilization itself can be assisted using ICSI (intracytoplasmic sperm injection), a method where a single sperm is injected directly into the egg.

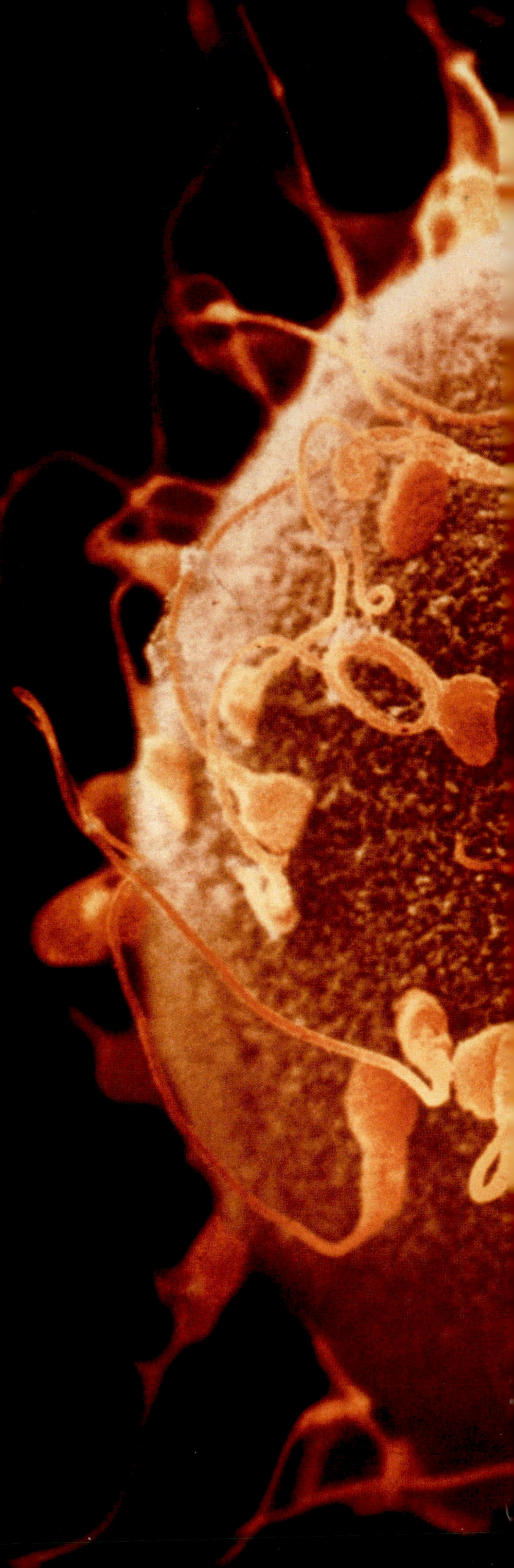

**The moment of fertilization**
This false-color image was produced by a scanning electron microscope and shows numerous sperm cells clustered around an egg. As soon as one penetrates the coat around the egg will harden, barring entry to the others.

## "I don't feel any **more special than anyone else.**"

LOUISE BROWN, FIRST "TEST TUBE" CONCEPTION

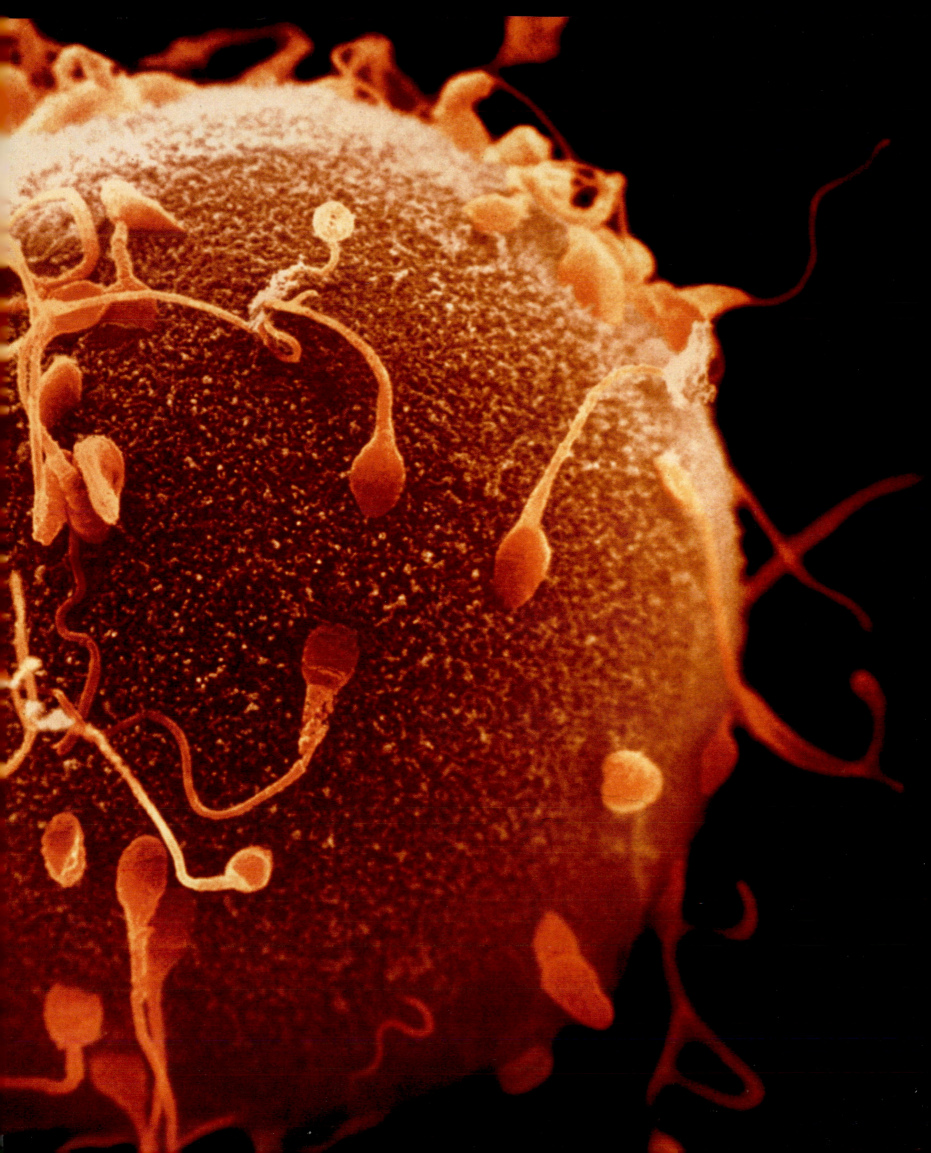

> **Cloning by embryo fragmentation**
This technology mimics the natural production of identical multiple births. All embryo cells are genetically identical. If the embryo is fragmented at the stem-cell stage, each fragment will develop into an identical embryo.

< **Cell-tissue cloning**
Genetically modified plants, such as the petroleum plants shown here, are tissue cloned. Balls of tissue—called calluses—are grown in a sterile nutrient medium.

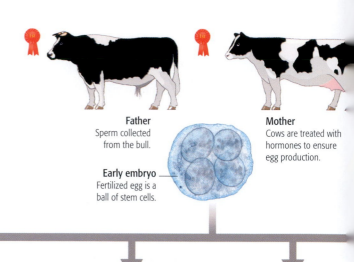

**Father**
Sperm collected from the bull.

**Mother**
Cows are treated with hormones to ensure egg production.

**Early embryo**
Fertilized egg is a ball of stem cells.

# How Cloning Works

**The production of genetically identical individuals, known as cloning, is an entirely natural action in some organisms. But the process—as controversial as it is misunderstood—can be artificial too, and in medicine and other applications has the potential to impact us all.**

**Stem cells**
Individual cells from the divided embryo.

Clones are organisms or cells that are genetically identical. This means that the sequence of bases in their DNA is identical. The term was coined by the British biologist J.B.S. Haldane in 1963. Whole body clones are produced naturally when an embryo splits to produce identical (monozygotic) twins

or when plants branch to produce rooted offspring. The latter is known as asexual reproduction.

As a body grows, the packages of DNA in every cell are replicated and, since copying mistakes are very rare, almost all the cells of a human being are genetic clones. When a body matures, cells become specialized in their functions as they switch off certain genes and form muscle, bone, or other tissue. Only cells in the early embryo—called stem cells—have the ability to form any body part. In 1952 American biologists Robert Briggs and Thomas King found that they could successfully produce artificial clones of

**Implantation**
Each embryo is implanted into a separate foster mother.

**New embryos**
Genetically identical embryos developed from the stem cells.

**Identical calves**
The calves are genetically identical, and, like normal offspring, each has half the genes from the mother and half from the father.

<< **BEFORE**

**Cloning technology arose from breakthroughs in understanding the development and genetics of embryos.**

**MOLECULE OF DNA REPLICATING ITSELF**

### DISCOVERY OF STEM CELLS
In 1891 the German biologist **Hans Driesch** discovered that cells of early sea urchin embryos could **generate any other cell** in the body.

### REPLICATING DNA
In 1953 **James Watson** and **Francis Crick** proposed DNA's double-helix structure. In 1958 its **self-replicating** mechanism was confirmed by **Matthew Meselson** and **Franklin Stahl**.

<< SEE ALSO
pp.348–49 THE GENETIC CODE

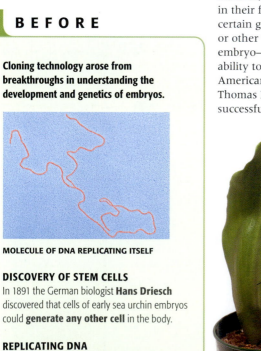

**Plant cuttings**
This method relies on a plant's natural ability to reproduce asexually. Cuttings from the parent plant are inserted into soil and kept humid.

Parent plant

Leaf sections rooting in soil

tadpoles by transplanting embryo nuclei into eggs. But when they tried it using the nuclei of older body cells they failed. It led them to conclude (wrongly) that the genetic potential of cells diminished as a body grew older.

### Cloning plants artificially
A plant's tendency to fragment means that cloning can simply be a matter of taking cuttings; especially using the growing tips containing meristems, where cell division is concentrated. These can be placed in soil and will naturally form clone plants. Scientists can induce cloning from non-meristematic parts by manipulating the growth-enhancing chemicals and nutrients (see above left). Such cloning was first achieved in 1958 when biologist

F.C. Steward grew carrot plants from carrot cells—the first case of artificial cloning from fully differentiated cells. This experiment was also the first indication that Briggs and King had been wrong about the genetic potential of older body cells.

### Cloning animals
Over the next 20 years scientists reported cloning of animals by manipulating cells, but many claims were flawed. A breakthrough came in 1984 when the Danish biologist Steen Willadsen fused cells from an early lamb embryo with unfertilized eggs that had had their nuclei removed. He nurtured the embryos in the laboratory, then transplanted them into the uteruses of surrogate mothers. Two lambs died at birth, but one survived to become the first mammal to be produced this way. Willadsen joined an American bioengineering company working on a project

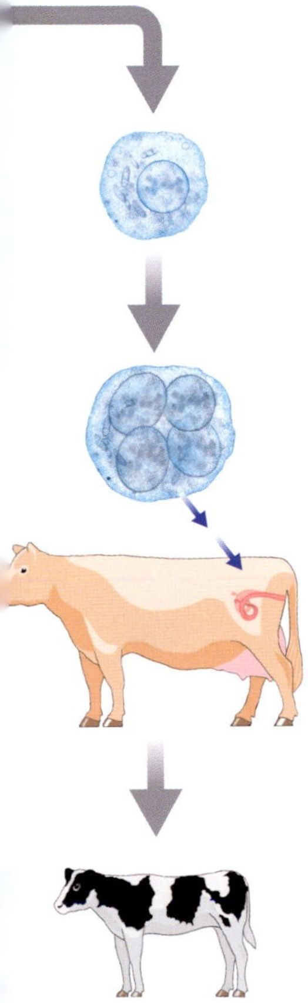

attempting to clone prize cattle. At the same time, a team consisting of Neal Frist, Randal Prather, and Willard Eyestone refined a similar technique that had become known as embryo fragmentation (see opposite). This technique, later patented by the funding cattle company, offered the potential to produce more animals with desirable characteristics. Despite early promise, few live births resulted and the industry collapsed. Only in recent years has it been revived—but with a new focus on cloning cattle to make human proteins for medical use.

Ian Wilmut of the Roslin Institute in Scotland wanted to use older, mature body cells (called somatic cells) as the source of genetic material, because they were more easily manipulated. In 1996 Wilmut and Keith Campbell (an expert in the cycle of cell growth and division) found that the nuclei of skin cells—they used udder cells—could be transferred into eggs provided they were synchronized into the same state as the egg by starvation. The fusion of the egg and nucleus was completed by electric shock. This technique is known as somatic nuclear transfer (see below). Of 277 udder cells used, 29 grew into embryos, but only one survived.

## Cloning tissue and molecules

Whole-body cloning has never been the sole aim of this kind of technology. At least potentially, any living material containing self-replicating genetic material can be cloned—even raw DNA. In 1984 American scientist Kary Mullis developed a technique of molecular "cloning," or copying, DNA, which became known as polymerase chain reaction (PCR). A DNA sample, DNA building blocks (nucleotides), and an enzyme called polymerase are mixed together and a cycle of

temperature changes allows successive stages of the replication process. As a result large amounts of DNA can be produced, or copied, from a tiny sample, which has proved invaluable in biological and forensic analysis.

Tissue cloning also offers the prospect of producing body parts for use in transplantation (see pp.406–407). Organ transplants are complicated by the possibility of rejection: genetically different organs are attacked by the recipient's immune system. Tissue cloning offers the possibility of using genetically identical organs, eliminating the risk of rejection entirely.

## STEM CELLS

These are cells that can form body tissues. The cells in a fertilized egg can form any body cell, see below. When the embryo becomes a hollow ball, the cells become pluripotent: they can form all structures except the placenta. Another stem cell has been identified in bone marrow and the umbilical cord, which can make any blood cell.

### Early embryo cells

Cells in a fertilized egg or early embryo that is no more than a ball of cells are totipotent, meaning they can form into any of a huge variety of cells in the body. These are also called precursor cells.

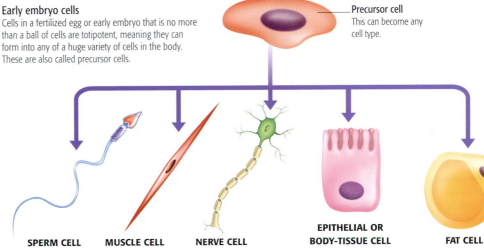

**Precursor cell**
This can become any cell type.

SPERM CELL    MUSCLE CELL    NERVE CELL    EPITHELIAL OR BODY-TISSUE CELL    FAT CELL

## AFTER »

Cloning technology has the potential to have groundbreaking applications in medicine and conservation.

### TREATMENT FOR PARKINSON'S DISEASE
In 2008 stem cell scientists in New York reported how **therapeutic cloning** had been used to cure Parkinson's disease in mice.

### SAVING ENDANGERED SPECIES
Noah, a baby gaur—a rare Indian Ox—was one of the **first endangered species** born by reproductive cloning in 2001. He later died but the technology is developing. Cloning of rare plants is also being undertaken by botanical gardens using **tissue culture** techniques.

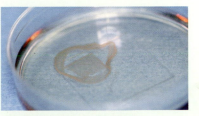

**SKIN SHEET CULTURE BEING GROWN IN A LABORATORY DISH**

SEE ALSO »
pp.410–11 THE HUMAN GENOME

### Dolly the sheep
The first mammal clone to be produced by somatic nuclear transfer—Dolly—was born on July 5, 1996. Of 277 udder cells used, 29 grew into embryos, but only one embryo developed into a fully formed lamb.

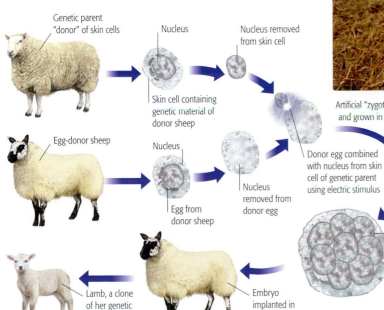

Genetic parent "donor" of skin cells

Nucleus

Nucleus removed from skin cell

Skin cell containing genetic material of donor sheep

Egg-donor sheep

Nucleus

Nucleus removed from donor egg

Egg from donor sheep

Artificial "zygote" created and grown in laboratory

Donor egg combined with nucleus from skin cell of genetic parent using electric stimulus

Cell division of zygote creates embryo, genetically identical to skin-cell nucleus

Lamb, a clone of her genetic parent

Embryo implanted in a foster sheep

### Somatic nuclear transfer
Dolly the sheep was genetically identical to the donor sheep (top left in the diagram). The nucleus of the donor's skin cell was inserted into an egg taken from a different ewe (center) that had had its nucleus removed. Whereas an egg has only one set of chromosomes, the skin-cell nucleus has a double set of chromosomes—just like a fertilized egg. Replication of the skin cell's DNA produced an embryo that is genetically identical.

# Nanotechnologies

**Conceived in the 1950s, nanotechnology is engineering on a scale invisible to the naked eye. By manipulating individual molecules and atoms, nanoengineers are able to make materials with unique properties that promise enormous developments across every industry.**

In December 1959, at the annual meeting of the American Physical Society, physicist Richard Feynman (see pp.326–27) laid down two challenges: the first was to take the words from the page of a book and write them on an area 25,000 times smaller, and the second was to produce a rotating electric motor smaller than a cube with sides of 0.4 mm. He offered a reward of $1,000 for each challenge. British engineer William McLellan claimed the prize for the second challenge the following year, but the first was not met for more than 25 years. Tom Newman of Stanford University in the US obliged in 1986 by reducing the first page of *A Tale of Two Cities* to nanoproportions using an electron beam machine.

## Small but powerful

In nature almost everything happens at the nanoscale—that is 100 nanometers and below, one nanometer (nm) being one millionth of a millimeter. But until the Scanning Tunneling Microscope (STM) was invented in the 1980s this miniature world was invisible to our eyes. Now we have microscopes capable of tracking the intimate details of our cells' inner workings. As a result, scientists in every field, from medicine to materials science, have been inspired to start working on much smaller scales.

However, particles begin to act very strangely once taken down to the nanoscale. Otherwise innocuous materials, such as metal powders and even flour, start exploding. Strange subatomic effects come into play when particles are this small, changing the electrical behavior and reactivity of the materials. It is this sort of unpredictable behavior that has led some to question the safety of using nanoparticles. Nevertheless, there are many problems for which nanotechnology might be the only solution—for instance, in keeping up with the current trend for ever-more powerful computers or delivering drugs to diseased cells in the human body.

## Using nanoparticles

The power of nanotechnology has long been within our grasp: we just have not realized it. As early as the 1920s car tires were treated with a material called carbon black to cut down on wear and tear; no one had realized that the same carbon particles being used were nano sized. For centuries, gold nanoparticles have featured in production of red and blue stained glass—another odd property of nanoparticles is that they do not necessarily appear the same color as their larger counterparts.

Since Newman met Feynman's challenge, nanotechnology has grown into a science in its own right. In 1985 buckminsterfullerine, a form

« BEFORE

**Nanotechnology could not exist until scientists could see individual atoms and molecules and had advanced knowledge of how matter behaves at the nanoscale.**

**SCANNING TUNNELING MICROSCOPE**
Not until the development of the **Scanning Tunneling Microsope (STM)** by Swiss researchers **Gerd Binnig** and **Heinrich Rohrer** in 1981 were scientists able to see things at a **single atom level**.

**CLUSTER PHYSICS**
The investigation of properties acquired by materials as they increase in size from just a few atoms is known as **cluster physics**. The science arose—in the early 1980s—because it was realized that the **behavior of materials alters with mass** and can be **unpredictable** at very small size, when **subatomic quantum effects** become more prevalent.

« SEE ALSO
pp.326–27 RICHARD FEYNMAN

**How small is small?**
Small objects can be seen with an ordinary microscope, but the nanoworld is so small it must be magnified many tens of thousands of times to be visible.

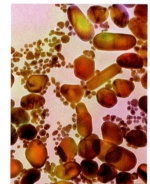

**Colloidal gold**
Gold nanoparticles, also known as colloidal gold, have many uses, ranging from staining glass to treating arthritis. Scientists also think they could be useful in detecting cancer cells.

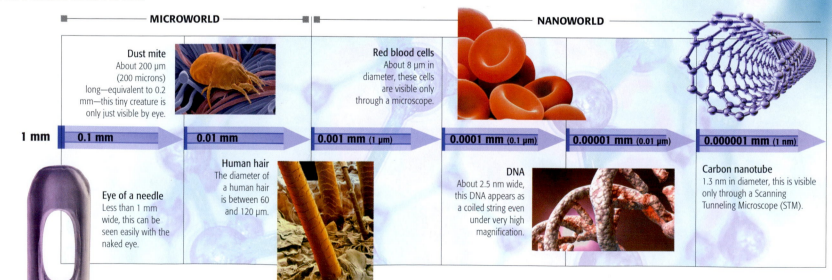

| MICROWORLD | | | NANOWORLD | | |
|---|---|---|---|---|---|

**Dust mite**
About 200 μm (200 microns) long—equivalent to 0.2 mm—this tiny creature is only just visible by eye.

**Red blood cells**
About 8 μm in diameter, these cells are visible only through a microscope.

| 1 mm | 0.1 mm | 0.01 mm | 0.001 mm (1 μm) | 0.0001 mm (0.1 μm) | 0.00001 mm (0.01 μm) | 0.000001 mm (1 nm) |
|---|---|---|---|---|---|---|

**Eye of a needle**
Less than 1 mm wide, this can be seen easily with the naked eye.

**Human hair**
The diameter of a human hair is between 60 and 120 μm.

**DNA**
About 2.5 nm wide, this DNA appears as a coiled string even under very high magnification.

**Carbon nanotube**
1.3 nm in diameter, this is visible only through a Scanning Tunneling Microscope (STM).

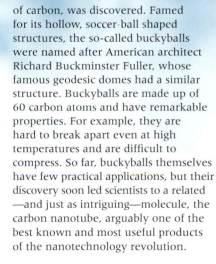

**Carbon buckyball**
Now recognized as a form of the element carbon, buckminsterfullerine has 60 atoms arranged in a regular pattern of pentagons (five sides) and hexagons (six sides). It occurs naturally in candle soot, but is now produced commercially by other means.

of carbon, was discovered. Famed for its hollow, soccer-ball shaped structures, the so-called buckyballs were named after American architect Richard Buckminster Fuller, whose famous geodesic domes had a similar structure. Buckyballs are made up of 60 carbon atoms and have remarkable properties. For example, they are hard to break apart even at high temperatures and are difficult to compress. So far, buckyballs themselves have few practical applications, but their discovery soon led scientists to a related —and just as intriguing—molecule, the carbon nanotube, arguably one of the best known and most useful products of the nanotechnology revolution.

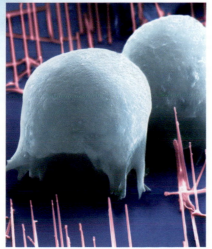

**Stem cells on nanowire**
This Scanning Electron Micrograph (SEM) shows mouse embryonic stem cells growing on vertical silicon nanowires. These nanowires relay electrical signals to and between the stem cells growing on them.

## IN PRACTICE
### CARBON NANOTUBE

In 2005 bikes with frames made of carbon nanotubes were used in the Tour de France. Apart from making the bikes very strong and rigid, the use of nanotubes made them extremely light. Each bike frame weighed less than 2.2 lb (1 kg). This technology is now being used to produce other sports equipment, such as tennis rackets and golf clubs.

Molecular models reveal a structure similar to a roll of chicken wire. It is extraordinary then to find that these tiny tubes are about 100 times stronger than steel. They are also extremely light and can be conducting or semiconducting —attractive properties for electronic engineers who want to create cheap solar panels and touch screens.

In biology, tiny fluorescent crystals called quantum dots are already used to track the movements of cells and molecules in minute detail. In 2005 scientists showed some crystals could latch on to viruses that cause respiratory infections in children, raising hopes of developing new diagnostic tests. The crystals were able to recognize and "stick" to structures in the virus's outer coating. Quantum dots could also revolutionize the computer industry, being used to increase data storage and speed up data retrieval.

### Technology for everything
Despite facilities springing up at most universities, the potential of nanotechnology remains largely untapped. There are also concerns about its safety. Nanoparticles are already used in dirt-repellent paints, sunscreens, and sports equipment, but there are difficulties in producing nanoparticles consistently on a commercial scale.

Medical advances have given rise to a number of novel medical devices, including nanomagnets. These have been successfully inserted into mouse blood cells to provide detailed images of blood vessels using Magnetic Resonance Imaging (MRI).

Nanomachines, including motors made from DNA (genetic material) and nanoscale computers that can perform simple calculations, could form the basis for more sophisticated nanoscale devices. Engineers are now working on self-propelled nanorobots that may one day be able to perform operations without the need for major surgery.

### NANOTECHNOLOGY APPLICATIONS

Few people realize that nanotechnology is already being used in almost every industry. From energy to electronics, manufacturing to medicine, nanotechnology is proving its worth and providing solutions to problems once thought impossible to solve.

**Consumer goods**
Clothing and other textiles; food; home appliances; optics; sportswear and equipment; cosmetics

**Construction**
Paint; surface treatments and protection

**Manufacturing**
Aerospace; vehicle manufacturing; print and packaging; refineries

**Medicine and health**
Drug delivery; diagnostics; tissue engineering; skin care

**Chemistry and the environment**
Filtration; catalysis (speeding up of chemical reactions); chemical sensors; water quality

**Electronics**
Memory storage; computing

**Energy**
Fuels; solar cells; batteries

> "It is a **staggeringly small world** that is below."
>
> RICHARD P. FEYNMAN, AMERICAN PHYSICIST, 1959

### AFTER

Nanotechnology is still very much work in progress, and some of the most promising nanoparticles have yet to find their niche.

#### UNLIMITED POTENTIAL
In the future, nanotechnology could yield super-strong **bone implants**, ultra-high speed **computer chips**, and artificial, touch-sensitive "skins" for **humanoid robots**. Nanodevices circulating in the bloodstream could be used to monitor human health from within the body. Superstrong but light materials could **revolutionize the construction industry** while others may provide an economic and efficient means of harnessing energy.

« **BEFORE**

Early astronomers studied the positions and movements of the planets and stars. Later scientists explored their characteristics.

**EARLY TELESCOPIC OBSERVATIONS**
From 1609 onward, **astronomers** began to study the **planets** with **telescopes** « 84–85. In 1659 **Christiaan Huygens** recorded a dark marking on Mars and two of Jupiter's cloud belts.

**ASTRONOMICAL PHOTOGRAPHY**
From the mid-19th century **photography** was used to **record astronomical data**; it captures more detail than the human eye.

**SPACE MISSIONS**
**Mariner 2** became the first spacecraft to visit a planet when it flew by **Venus** in December 1962.

« **SEE ALSO**
*pp.74–75* THE SUN-CENTERED UNIVERSE

# Inside the Solar System

At the beginning of the space age in the 1960s, astronomers had identified all of the major and many of the minor solar system bodies, and had established their basic properties. Since then unexpected discoveries have been made using Earth- and space-based telescopes, and spacecraft sent to explore them.

The scale of the solar system was estimated accurately for the first time more than 200 years ago. Since then astronomers have been refining measurements, such as the size of the solar system, and the size, mass, and spin of the planets. They have also established other characteristics of the planets, such as their surface temperatures, chemical compositions, and internal structures.

## A closer view
Orbiting satellites allow us to monitor the entire earth, identifying large-scale trends in its oceans, land, and atmosphere. Spacecraft have visited other planets – such as Venus and Mars—since the 1960s, but they were not able to provide detailed images until the missions of the 1970s. These images revealed what the planets look like up close and enabled scientists to piece together the geological histories of the rocky inner planets (Mercury, Venus, and Mars). They also helped monitor the atmospheres of the giant outer planets (Jupiter, Saturn, Uranus, and Neptune).

## The rocky planets
The first close-up of any planetary surface was of Mercury in 1974. Its cratered world is reminiscent of Earth's moon, which had been explored by manned and unmanned spacecraft in the 1960s.

Venus was examined by spacecraft radar in 1978, when its permanent dense cloud was penetrated to reveal a world of low-lying volcanic plains with hundreds of seemingly dormant volcanoes. Its carbon-dioxide-rich atmosphere traps heat in the way that

### Planetary properties
The composition of the surface and internal structure of the bodies in our solar system. These have been studied through space probes and telescopes, and, for the Moon, manned landings.

Mercury · Venus · Earth · Moon · Mars · Jupiter · Saturn · Uranus · Neptune

**The Sun**
Sun · Photosphere · Convective zone · Radiative zone · Core
The Sun generates energy in its highly dense core, which is then released through its surface.

**Mercury**
Silicate rock crust · Rocky, silicate mantle · Iron core
Mercury is rich in metals compared to the other rocky planets; its high density indicates a huge iron core.

**Venus**
Silicate crust · Rocky mantle · Molten outer core · Solid iron-nickel inner core
Venus's similar size and density to Earth suggest it has a comparable internal structure.

**Earth**
Silicate-rock mantle · Rocky crust · Molten outer core · Solid iron-nickel inner core
Earth's surface crust is broken into plates that move against and away from each other.

**The Moon**
Granitelike rocky crust · Rocky mantle · Possible metallic core
Moon's density is similar to that of the Earth's mantle—from which it was originally formed.

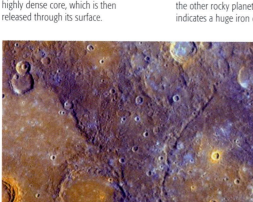

**Cratered Mercury**
On October 8, 2008 the Messenger spacecraft photographed the planet's cratered terrain. Most of the craters were formed when the planet was bombarded by asteroids more than 3.5 billion years ago.

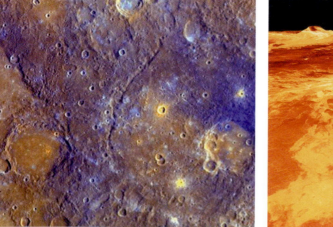

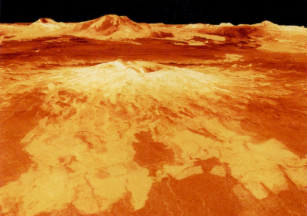

**Volcanic Venus**
The bright volcano in the foreground is Sapas Mons, one of Venus's largest shield volcanoes. On the horizon is Maat Mons, Venus's tallest volcano. The pair grew as successive lava flows accumulated.

**The far side of the Moon**
Spacecraft have provided images of the far side of the Moon—the side never seen from Earth. This view of the Daedalus Crater was taken during the Apollo 11 manned mission in July 1969.

## BREAKTHROUGH

## DWARF PLANETS

Improved technology has allowed scientists to detect ever-smaller objects in the solar system such as Pluto's three moons (Pluto at right). In 1992 they started finding icy rock objects beyond Neptune in a region called the Kuiper Belt. It was here, in 2005, that they discovered Eris, an object larger than Pluto. This led to Pluto's reclassification from a planet to a new category of object: the dwarf planet. These are roughly spherical and orbit the Sun in a region populated by other objects, as in the Kuiper Belt.

glass traps heat in a greenhouse, making Venus the hottest of all eight planets in the solar system.

Mars, which has deep canyons and the highest volcanoes in the solar system, is the only planet whose surface spacecraft have driven over. The first successful Mars rover was NASA's Sojourner, which landed on Mars on July 4, 1997. The images sent back to Earth by this and other Mars exploration missions have revealed Mars's rock-strewn plains, dry river beds, and ancient floodplains. These features are evidence that water once flowed across Mars's surface, while today it exists only as ice or vapor.

## Giant planets

Scientists have found out more about the four giant planets from six spacecraft missions since 1973. It is now known, for example, that Jupiter has a turbulent atmosphere and huge thunderstorms. More about the complex nature of Saturn's rings has also been revealed.

In 1977 astronomers unexpectedly discovered rings around Uranus. Spacecraft later found rings around Jupiter and Neptune too, and tens of relatively small moons orbiting the four giant worlds. They also showed that the largest moons, such as Jupiter's Io with its active volcanoes, are worlds in their own right.

Our knowledge of the solar system continues to grow through analysis of collected data. We are still learning more about individual objects, and of the history and future of the system as a whole.

### ONGOING EXPLORATION

**New Horizons**, launched in 2006, will be the first craft to visit **Pluto** when it arrives in 2015. The spacecraft **Rosetta**, launched in 2004, will be the first to travel alongside a **comet** as it orbits the Sun, starting in 2014.

### OTHER PLANETARY SYSTEMS

The solar system is not unique. The first **planet orbiting another star** was detected in 1992, and in 1995 astronomers discovered the first planet around a sunlike star. More than 340 are now known.

### SEE ALSO ❯❯

pp.396–97 SPACEPROBES AND TELESCOPES
pp.398–99 DARK UNIVERSE
pp.424–25 REFERENCE: ASTRONOMY

" This is **the time** when humans have begun to sail the **sea of space**."

CARL SAGAN, ASTRONOMER, 1980

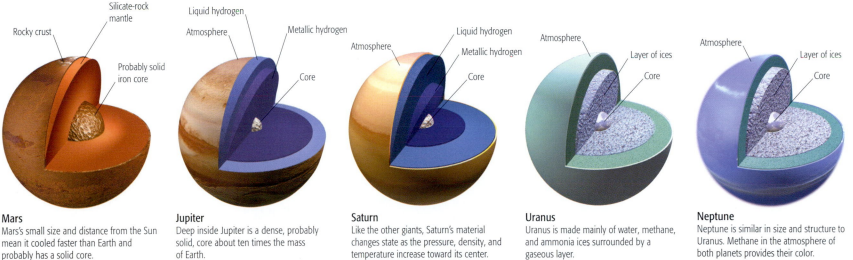

**Mars**
Rocky crust — Silicate-rock mantle — Probably solid iron core
Mars's small size and distance from the Sun mean it cooled faster than Earth and probably has a solid core.

**Jupiter**
Atmosphere — Liquid hydrogen — Metallic hydrogen — Core
Deep inside Jupiter is a dense, probably solid, core about ten times the mass of Earth.

**Saturn**
Atmosphere — Liquid hydrogen — Metallic hydrogen — Core
Like the other giants, Saturn's material changes state as the pressure, density, and temperature increase toward its center.

**Uranus**
Atmosphere — Layer of ices — Core
Uranus is made mainly of water, methane, and ammonia ices surrounded by a gaseous layer.

**Neptune**
Atmosphere — Layer of ices — Core
Neptune is similar in size and structure to Uranus. Methane in the atmosphere of both planets provides their color.

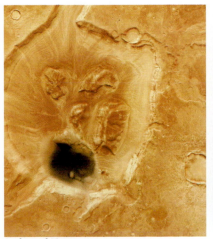

**Surface of Mars**
Spacecraft images supply evidence of Mars's watery past. The floor of this depression in Mamers Valles was shaped by flowing water, which also carried the darker rock into the depression.

**Galileo spacecraft near Jupiter**
The Galileo craft made an in-depth study of Jupiter and its system in the 1990s. A wok-shaped probe released into Jupiter's atmosphere returned data on its temperature, pressure, and composition.

**The ringed world of Saturn**
All four giants have rings, but Saturn's are the only ones obvious from Earth. Its seven rings consist of ringlets, which are made of individual pieces and lumps of dirty water-ice that orbit the planet.

## « BEFORE

As the limitations of earthbound astronomy became more apparent, pioneering experiments paved the way for studying the solar system from above the earth.

### DEPTH OF VIEW

**Earth-based telescopes**, such as Leviathan in Ireland, lacked the resolving power to show the planets' surfaces in detail, or to solve mysteries such as the origin of the Moon's craters.

**LEVIATHAN TELESCOPE, 1845**

### ATMOSPHERIC ANALYSIS

In the 1930–40s US and Soviet scientists launched **short-range rockets** that carried radiation detectors above the atmosphere, revealing that Earth was bombarded with radiation from space.

« SEE ALSO
*pp.368–69* ARTIFICIAL SATELLITES

## IN PRACTICE
### GRAVITATIONAL SLINGSHOTS

Exploring the solar system in a reasonable timeframe presents some difficult challenges, but since the 1970s NASA engineers have used an ingenious technique to shorten journey times without the need for additional fuel. A "gravitational slingshot" involves directing the spacecraft to head straight for a planet, swinging around it, and leaving in the same direction as the planet is traveling, picking up a significant speed boost in the process. The technique was first used to adjust the orbit of the Mariner 10 probe, which flew past Venus on its way to Mercury in 1973, but its most famous application was for the Voyager probes, which took advantage of a rare planetary alignment to tour all the gas giants in the 1970s and 1980s (see right).

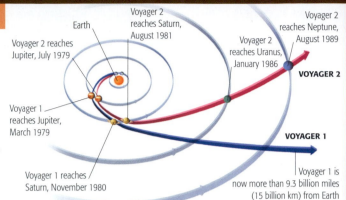

Earth

Voyager 2 reaches Jupiter, July 1979

Voyager 2 reaches Saturn, August 1981

Voyager 2 reaches Uranus, January 1986

Voyager 2 reaches Neptune, August 1989

**VOYAGER 2**

Voyager 1 reaches Jupiter, March 1979

Voyager 1 reaches Saturn, November 1980

**VOYAGER 1**

Voyager 1 is now more than 9.3 billion miles (15 billion km) from Earth

### Cassini spaceprobe
The Cassini Orbiter went into orbit around Saturn in 2004. It carried instruments to study the planet, its environment and satellites, and also deployed Huygens, a small lander that sent back images from Saturn's mysterious moon Titan.

Low-gain antenna (1 of 2)

Radar bay

Fields and particles pallet

Huygens Titan probe

Radioisotope thermoelectric generator (1 of 3)

Remote sensing pallet

Engine (1of 2)

36 ft (11 m) magnetometer boom

# Space Probes and Telescopes

**The space age has shown us astounding phenomena, both within our solar system and beyond. Space probes have journeyed to the Moon, all the major planets, and several smaller worlds, while orbiting telescopes have brought stunning clarity to our deep views of the Universe.**

It is little wonder that the Moon was the first object to be targeted by what we would now call a space probe. Lying on average just under 250,000 miles (400,000 km) away, it is on our cosmic doorstep and modifications to some of the earliest satellite launchers brought it within reach. Although early US and Soviet missions ran far from smoothly, a landmark came on October 7, 1959, when Luna 3 passed behind the Moon and sent back pictures of its unseen far side. Later moon probes surveyed its entire surface—literally testing the ground in preparation for the manned Apollo landings (see pp.370–71)—but

it is only since the 1990s that probes have returned to the Moon to study its geology in detail.

A series of Soviet failures in the 1960s allowed NASA to make the first flybys of both Venus and Mars. It was the Soviet Venera 7 probe that finally made it to the surface of Venus in 1970 and sent back data that revealed surface temperatures of up to 860° F (460° C) and atmospheric pressures 100 times those on Earth. In these conditions landers on Venus do not last long, and our most detailed look at the planet so far has

come from radar-mapping satellites in orbit (see p.369).

The first flybys to Mars revealed the heavily cratered southern region, disappointing scientists who had hoped to find a more hospitable world. It was only in 1971 that Mariner 9 revealed the towering volcanoes and riverlike channels of the northern hemisphere. Attempts to find life on Mars with the Viking missions

### Early probe
At 20 in (51 cm) tall, NASA's tiny, conical Pioneer 4 probe, launched in 1959, had just enough room to carry a camera and a radiation detector.

of 1976 failed. Further missions, including orbiting surveyors and robot rovers, have enabled scientists to study the surface in detail.

Beyond Mars lie the asteroid belt and the giant planets. Each of these worlds has been visited by at least one Voyager mission (see above). In addition, Jupiter and Saturn, with their moons, have been studied by the Galileo and Cassini orbiters.

Smaller bodies in the solar system are just as interesting. Halley's Comet, for example, was subjected to an armada of visiting spaceprobes during its 1986

A number of ambitious new probes and observatories are being planned for the next decade, and some are already en route to their destinations.

**PLANET DETECTOR**
In 2009 NASA launched **Kepler**, a telescope equipped with extremely sensitive detectors to **find possible planets** crossing the face of distant stars.

**BEYOND HUBBLE**
Development of the **Hubble Space Telescope's successor**, the 6.5-meter James Webb Space Telescope, is under way, although the telescope is unlikely to launch before about 2013.

**MISSIONS TO MARS**
Exploration of Mars is continuing—NASA has **several missions currently active**, and there are plans to launch new rovers, orbiters, and eventually a mission to bring rocks from the Red Planet back to Earth.

**THE EDGES OF THE SOLAR SYSTEM**
The **New Horizons mission** is on a high-speed route to Pluto and the mysterious worlds of the Kuiper Belt **beyond Neptune**, while the **DAWN** mission is on its way to explore the largest bodies in the asteroid belt.

SEE ALSO »
*pp.398–99* DARK UNIVERSE

◁ **Chandra Space Telescope**
The Chandra X-ray telescope is one of a series of NASA satellites designed to observe the sky at various wavelengths. Because X-rays will pass straight through conventional mirrors, it uses a series of curving metal surfaces to bring the rays to a focus on its detectors.

△ **Combined image**
This composite image of X-rays from Chandra and optical light data from Hubble Space Telescopes shows the largest region of star formation in a nearby galaxy, M33.

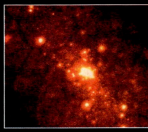
X-RAY IMAGE

VISIBLE LIGHT

MID-INFRARED (IRAS)

MID-INFRARED (SPITZER)

△ **Same galaxy, different wavelengths**
Different types of telescope show different features, as in these images of galaxy M31. They range from those using X-rays, which detect very hot objects, to infrared telescopes, which pick up faint radiation from objects too cool to shine in visible light.

visit to the inner solar system, while more probes have targeted asteroids and other comets.

**Space telescopes**
Telescopes in orbit around Earth have revealed details of some of the most distant objects in the universe. Space-based astronomy allows observation at any time of day or night, and it is not affected by weather or atmospheric turbulence. Detailed images using X-rays, infrared, and gamma rays are only possible above Earth's atmosphere. Such images reveal objects undetectable with ground-based telescopes. The first such telescopes included the Orbiting

Astronomical Observatory (OAO-2), an ultraviolet instrument launched in 1968, and IRAS, the InfraRed Astronomical Observatory, launched in 1983. The Hubble Space Telescope was launched in 1990 and has sent back countless images (see pp.318–19), leading to important observations about the Universe. Since the 1990s NASA has launched much larger and more ambitious "Great Observatories," including the Compton Gamma-Ray Observatory (1991), the Chandra X-ray Telescope (1999), and the Spitzer Infrared Space Telescope (2003).

"**Space** isn't remote at all. It's only an **hour's drive** away if your car could go **straight upwards**."
SIR FRED HOYLE, BRITISH ASTRONOMER AND COSMOLOGIST (1979)

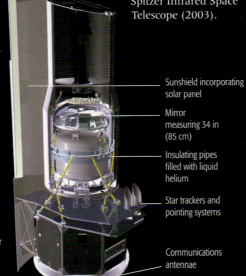

**Spitzer telescope**
This infrared telescope, launched in 2003, detects infrared radiation. Its telescope mirror focuses light on three infrared detectors and shows faint stars and planets forming.

Sunshield incorporating solar panel

Mirror measuring 34 in (85 cm)

Insulating pipes filled with liquid helium

Star trackers and pointing systems

Communications antennae

ASTROPHYSICIST (1914–97)
**LYMAN SPITZER**

One of the first scientists to recognize the potential of space-based astronomy was US astrophysicist Lyman Spitzer. Born in Ohio, Spitzer was head of his department at Princeton University by the age of 33. He made a name for himself through pioneering studies of the interstellar medium—the gas and dust that lies between the stars—and later of plasma physics and nuclear fusion. He first described the benefits of a space-based optical telescope in 1946, and in 1965 NASA asked him to help plan what became the Hubble Space Telescope. With his deft lobbying of politicians and scientists, he ensured that the project became a reality.

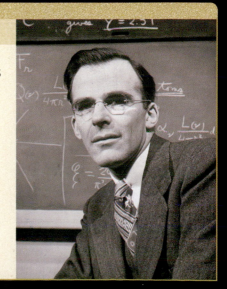

## « BEFORE

The Universe was previously thought to contain no more than we can see, or at least detect directly in other ways—such as by collecting X-rays or infrared radiation.

### UNIVERSE MADE BY GOD

One fundamental belief for many past astronomers was that the Universe was made by God. This being true, they also supposed that God would not make unnecessary things. Therefore the idea that God made most of the Universe out of materials we cannot "see" was regarded as highly unlikely.

### EINSTEIN'S THEORIES

Albert Einstein introduced his **general theory of relativity « 300–01** in 1916, when it was believed the Universe was static. He realized that **a completely static Universe would collapse** under the influence of gravity and become a single huge central mass, so he introduced a constant to his equation: a repulsion that essentially kept the galaxies apart. Dark energy is a similar repulsion.

**« SEE ALSO**
*pp.318–19* THE EXPANDING UNIVERSE
*pp.396–97* SPACEPROBES AND TELESCOPES

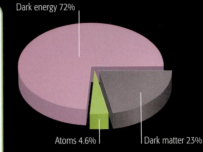

Dark energy 72%

Atoms 4.6%    Dark matter 23%

**Composition of the Universe**
Data (accurate to two digits) collected by the Wilkinson Microwave Anisotropy Probe (WMAP) spacecraft suggest that atoms—the building blocks of stars and planets—make up a tiny proportion of the Universe.

# Dark Universe

In recent decades astronomers have known that the Universe consists of much more than the known stars and galaxies. Recent findings suggest that in fact 95 percent of the ever-expanding Universe is made up of two completely unknown elements: dark matter and dark energy.

Ordinary matter consists of atoms, which form the planets, stars, and galaxies. It is known as "baryonic matter" by astronomers, and it makes up less than five percent of the Universe. The next largest constituent of the Universe is also a form of matter, known as "dark matter," because it emits no light and we don't know what

it consists of. The remaining 72 percent of the Universe is not matter, but "dark energy," whose nature is unknown.

The idea that the Universe contains unaccountable material first arose in 1933. Fritz Zwicky, a Swiss astronomer who had been studying the Coma Cluster of galaxies, concluded that each galaxy was many times more massive than inferred by their stellar content.

### Evidence for dark matter

During the 1970s Vera Rubin discovered that stars orbiting in the outer part of a spiral galaxy travel as fast as those close to the center. This was contrary to

Newton's laws of gravity, and implied that each galaxy had a halo of invisible matter whose gravitational force affected the outer stars. In 1970 Rubin showed this to be true of the Andromeda Galaxy; by 1985 she had examined 60 spiral galaxies and realized it was a general phenomenon.

Physicists think that dark matter might be a form of elementary particle. The suggested name is a WIMP, a Weakly Interacting Massive Particle. WIMPs can pass through normal matter and the hope is that every now and then a WIMP will collide with the nucleus of an atom and physicists will be able to detect the recoil. Normal cosmic rays produce similar effects, so experiments in search of WIMPs are performed in mines at least one kilometer ($^2/_3$ mile) below Earth's surface. No WIMPS have yet been found.

By the 1990s astronomers accepted that the Universe was expanding at an ever-slower rate, due to the influence

### Searching for dark matter

The world's most powerful telescopes have all at some time been part of the search for dark matter. Surveys by the Sloan Foundation Telescope, New Mexico, have also helped to prove the existence of dark energy.

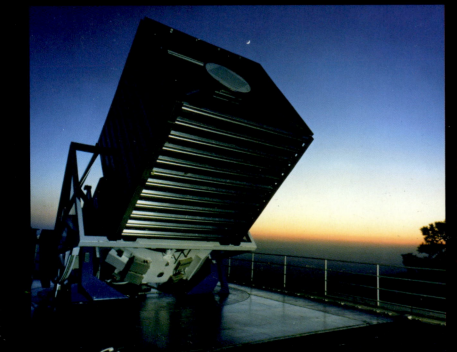

### Dark matter distribution

A computer simulation created by an international group of astronomers in 2005 shows the large-scale distribution of dark matter within the Universe. The pattern echoes that produced by the large-scale distribution of the light from galaxies. Yellow denotes the densest matter, then red, blue, and black.

ASTRONOMER (1928–)

## VERA RUBIN

Born in Philadelphia, PA, Vera Rubin was fascinated by the stars as a young girl. She became a research astronomer, and her discoveries have altered our understanding of the Universe. Rubin's work on the orbital motion of stars around the center of galaxies convinced astronomers that dark matter exists. She has received many prestigious awards, including the National Medal of Science—America's highest scientific award—in 1993.

## GRAVITATIONAL LENSING

The presence of matter in space can cause light rays to be deflected. This effect is known as gravitational lensing. The light from galaxies lying directly beyond a galaxy cluster (pp.340–41), for instance, is "bent" by the cluster. The cluster works like a giant lens, deflecting the light. From our position on Earth, the light paths from a more distant galaxy appear to arrive from different directions, because we are receiving multiple distorted images of the distant galaxy. Astronomers can use the effect of gravitational lensing to determine the mass of a galaxy cluster, figure out where the mass is distributed within the cluster, and assess the amount of dark matter it contains. Gravitational lensing has also been used to confirm the theory that the expansion of the Universe is accelerating (pp.318–19).

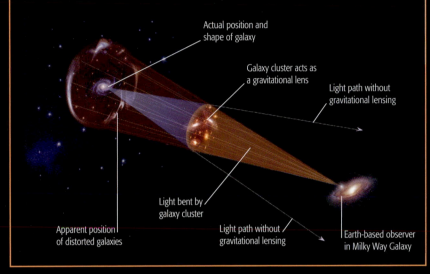

Actual position and shape of galaxy

Galaxy cluster acts as a gravitational lens

Light path without gravitational lensing

Light bent by galaxy cluster

Apparent position of distorted galaxies

Light path without gravitational lensing

Earth-based observer in Milky Way Galaxy

> "The **Universe** is made mostly of **dark matter** and **dark energy**, and we don't know what either of them is."

SAUL PERLMUTTER, ASTROPHYSICIST, 1999

of gravity. But observations of supernovae in distant galaxies contradicted this: they were found to be fainter, and hence farther away, than would be the case if the expansion of the Universe were slowing down. This indicates that the expansion of the Universe is actually speeding up.

### Dark energy's influence

More recent observations have corroborated this theory. It seems the gravitational attraction which would slow the expansion rate is being overcome by a repulsion; a sort of "anti-gravity." Detailed observations of the supernovae indicate that the Universe was slowing down until about 5–6 billion years ago, but then the deceleration somehow switched to acceleration. It is thought that the density of matter in the young Universe may initially have been high enough for gravity to dominate, but as matter thinned out, the repulsive influence took over. This influence is attributed to an unknown substance that is distributed relatively uniformly in space, now known as "dark energy."

**Gravitational lensing in action**
Thin arclike features are visible in this colored image of galaxy cluster Abell 2218. They are the result of gravitational lensing and are distorted images of galaxies that are some 5–10 times farther away than the cluster.

**Galaxies in the Coma Cluster**
The Coma Cluster consists of more than 3,000 galaxies. While studying the movements of these galaxies in 1933, Fritz Zwicky found that the cluster contains much more mass than its visible galaxies suggest. The unexplained mass was later shown to be dark matter.

Dark matter must have influenced the formation of the large-scale structure of the Universe, with its superclusters and voids.

### FUTURE OF THE UNIVERSE

The presence of dark energy **will affect the future of our universe**. If its influence continues in the same way, the acceleration of the **expansion of the Universe will continue**. Should its influence increase, the Universe could disintegrate in a "big rip." If dark energy's repulsion turns to attraction, it will contract in a "big crunch."

### ONGOING ENQUIRY

Although **no dark matter particles have been detected yet**, scientists are searching for them using increasingly sensitive underground detectors. They are also looking into space for evidence of the by-products of collisions between dark matter particles, and are attempting to **create dark matter particles** in particle accelerators such as the Large Hadron Collider.

### RADICAL ALTERNATIVE

One suggestion to explain the **accelerating expansion of the Universe** is that time itself is slowing down, and our time dimension slowly turning into a new space dimension. This would explain why the far-distant, ancient stars appear as though they are accelerating.

## BEFORE

The idea of forces of nature, or fundamental forces, goes back at least to the ancient Greeks. For most of history, people have thought there were two or three such forces.

### THREE FORCES OF NATURE
In the 16th century the recognized fundamental forces were **magnetism**, **electricity**, and **gravity**. The English scientist **William Gilbert** ≪ **80–81** studied magnetism and electricity, believing they were separate phenomena.

### ELECTROMAGNETISM
In 1819 the Danish physicist **Hans Christian Ørsted** ≪ **166** discovered that **electric currents caused magnetic fields**. He noticed that a compass needle deflected from magnetic north when the electric current from a battery he was using was switched on and off. In the 1820s the French physicist **André-Marie Ampère** showed that wires carrying electric currents could attract or repel each other.

### FARADAY'S WORK
English scientist **Michael Faraday** ≪ **170–71** explored how electricity and magnetism were connected. He developed applications of the link between the two, such as **electric motors**.

≪ **SEE ALSO**
*pp.314–17* QUANTUM REVOLUTION
*pp.320–21* THE BIG BANG
*pp.382–83* SUBATOMIC PARTICLES

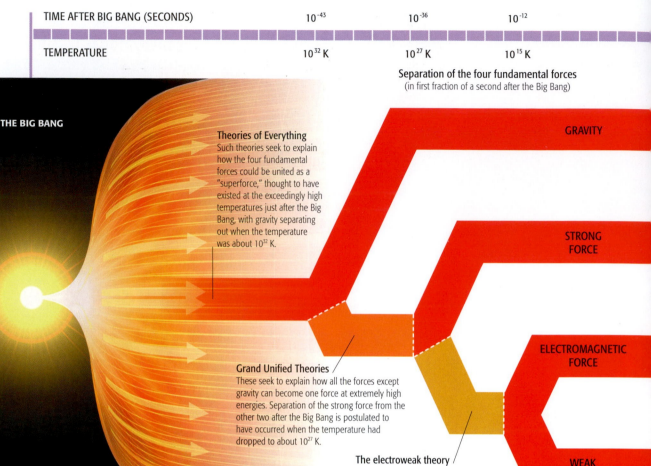

TIME AFTER BIG BANG (SECONDS)   $10^{-43}$   $10^{-36}$   $10^{-12}$

TEMPERATURE   $10^{32}$ K   $10^{27}$ K   $10^{15}$ K

**Separation of the four fundamental forces**
(in first fraction of a second after the Big Bang)

THE BIG BANG

GRAVITY

STRONG FORCE

ELECTROMAGNETIC FORCE

WEAK INTERACTION

**Theories of Everything**
Such theories seek to explain how the four fundamental forces could be united as a "superforce," thought to have existed at the exceedingly high temperatures just after the Big Bang, with gravity separating out when the temperature was about $10^{32}$ K.

**Grand Unified Theories**
These seek to explain how all the forces except gravity can become one force at extremely high energies. Separation of the strong force from the other two after the Big Bang is postulated to have occurred when the temperature had dropped to about $10^{27}$ K.

**The electroweak theory**
This implies that the weak interaction and the electromagnetic force are unified at high energies in the electroweak force. These two forces are thought to have separated when the temperature had dropped to about $10^{15}$ K.

# Grand Unified Theory

**Grand Unified Theory (GUT) seeks to unite three of the four fundamental forces of nature: electromagnetism, the strong nuclear force, and the weak nuclear interaction. It is not to be confused with the ongoing search for a "Theory of Everything" (see pp.402–403), which seeks to include the fourth fundamental force, gravity.**

Around the turn of the 20th century it seemed to scientists that the electromagnetic (EM) force, which causes unlike charges to attract and like charges to repel, together with gravity, could be used to explain everything in physics.

The existence of atomic nuclei was discovered in 1908; but, by the late 1920s, some puzzling facts had emerged about these nuclei. For one thing, they contained only positively charged particles (protons) and uncharged ones (neutrons), and it was hard to explain why the protons did not fly apart, repulsed by the EM force. Another question that required explanation was why neutrons sometimes suddenly turn into protons with the emission of electrons, producing the phenomenon known as beta radioactive decay.

Consequently, in the 1930s, two new forces were proposed: the strong force, which holds protons and neutrons together in the nucleus, and the weak interaction, which (among other effects) causes beta radioactive decay. Hence, four fundamental forces, or interactions, were now recognized.

### Role of messenger particles
At this time, little was known about how the four forces actually work. But over subsequent decades, a model was developed in which the forces are transmitted by different "messenger" particles called gauge bosons. By the middle of the 20th century quantum mechanics had established that particles can appear ephemerally in the subatomic world, in accordance with Heisenberg's Uncertainty Principle (see pp.316–17), and this is what happens with the messenger bosons. They are "virtual" particles that appear, take part in an interaction, and then disappear. The distance over which they can act depends on how much energy has to be "borrowed" from the vacuum of the Universe to create them. The more energy used, the shorter a particle's life

**THE FOUR FUNDAMENTAL FORCES OF NATURE**

The four fundamental forces differ in their messenger particles, the elementary particles that they affect, their relative strengths (calculated here at a range of $10^{-18}$m), and the range over which they operate.

| FORCE OR INTERACTION | | ELEMENTARY PARTICLES AFFECTED | MESSENGER PARTICLES | RELATIVE STRENGTH | RANGE IN METERS |
|---|---|---|---|---|---|
| GRAVITY | | All particles with mass | Gravitons (hypothetical) | $10^{-41}$ | Infinite |
| STRONG | | Quarks, gluons | Gluóns | 25 | $10^{-15}$ |
| ELECTRO-MAGNETIC | | All electrically charged particles | Photons | 1 | Infinite |
| WEAK | | Quarks, leptons | W and Z bosons | 0.8 | $10^{-18}$ |

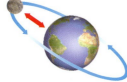

**ATTRACTION OF CELESTIAL OBJECTS**

**DOWNHILL MOVEMENT**

**BINDING OF QUARKS IN PROTONS AND NEUTRONS**

**BINDING OF ATOMIC NUCLEI**

**NUCLEAR ENERGY**

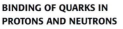

**ATTRACTION OF UNLIKE CHARGES**

**ATOMIC AND MOLECULAR STRUCTURES**

**CHEMICAL REACTIONS**

**MAGNETISM AND ELECTRICITY**

**LIGHT AND HEAT RADIATION**

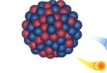

**BETA RADIOACTIVE DECAY**

**The four fundamental forces of nature**
Physicists theorize that the four forces were once unified in a single force that split in the first fraction of a second after the Big Bang. After splitting, the four forces exerted different effects, producing separate phenomena such as the examples shown above.

### THEORETICAL PHYSICIST (1926–96)

## ABDUS SALAM

Born in Pakistan, Abdus Salam studied mathematics and physics at the University of Cambridge, and became a physics professor at Imperial College, London. There he worked on establishing the electroweak theory, for which he was awarded a share in the 1979 Nobel Prize for physics. Abdus Salam was the first Muslim Nobel Laureate in science, and to date is the only Pakistani Nobel Laureate. His other achievements included setting up a center in Trieste, Italy, to promote science in the developing world.

span and distance range. In the case of the EM force, the virtual particles are photons (tiny packets of electromagnetic radiation). The way photons act in transmitting the EM force became the basis of the successful theory of Quantum Electrodynamics (QED), developed by Richard Feynman (see pp.326–27) and others.

## The electroweak theory

Physicists suspected that the weak interaction might also be mediated by messenger particles, although it was established that these would have to possess relatively high mass (which would also account for the short range of the force). During the 1960s it was realized that at very high energies these messenger particles could appear as real, long-lived particles that would act in the same way as photons. Under these conditions, the EM force and the weak interaction would become a single unified force. This was the basis on which Abdus Salam (see above) and the US physicists Sheldon Glashow and Steven Weinberg independently developed the electroweak theory, which explained not only how the weak interaction works, but also how

**Hadron Collider**
One use of the Large Hadron Collider (LHC), at CERN in Switzerland, is to test Grand Unification Theories. The collider has magnets to keep particles on the correct paths; here, a magnet is being moved into place.

it could be unified with the EM force. The electroweak theory postulated the existence of messenger particles (called W+, W-, and Z bosons) for the weak interaction, with different messengers being exchanged in different instances of the interaction. It also predicted the Higgs boson—a particle that gives other particles mass. Later, in 1983, the W and Z bosons were detected in particle accelerators, confirming the validity of the theory, although the Higgs boson so far remains undetected.

## Strong force and grand unification

Originally the strong force was thought to exist between nucleons (protons and neutrons), but following the discovery that these particles consist of sub-units called quarks (see pp.382–83), the strong force came to be seen as one that primarily binds quarks together within nucleons, while a "leakage" of the force binds together the nucleons.

But could the strong force be united with the electroweak force? Because of the success of the electroweak theory and QED, physicists looked to develop a similar theory that could explain how the strong force operates. This led in the 1970s to the theory of quantum chromodynamics (QCD)—the proposal that quarks, too, interact by means of messenger bosons called gluons. QCD is based on the idea that every quark can exist in any of three so-called colors, with the quarks inside nucleons constantly changing "color" as they interact with the gluons to produce the strong force.

Regarding unification of the strong force (as defined by QCD) with the electroweak force, various GUTs have been proposed, but none is entirely satisfactory. Any such theory would need to postulate the existence of very high-mass bosons (dubbed X-bosons) that could interconvert quarks and leptons (see pp.382–83). A problem in testing any GUT is that the creation of these bosons would require far higher energies than can at present be created on Earth. But attempts are being made with a new particle accelerator—the Large Hadron Collider—to recreate the conditions that existed in the first fraction of a second after the Big Bang, in the hope of proving the existence of the Higgs boson and other particles.

## AFTER »

**Beyond a Grand Unified Theory is the goal of unifying all four fundamental forces in an ultimate "Theory of Everything."**

### THINKING IN STRINGS

At present, **gravity** is most exactly described by Einstein's theory of general relativity, while the explanation of the other three forces is based on the theory of **quantum mechanics**, and these two theories appear in some respects to contradict each other. One of the best known attempts at unifying all four forces is **string theory 402–403 »**. In string theory, **elementary particles** are treated as

**infinitesimally thin, stringlike objects** rather than dimensionless points. But string theory is very much in its infancy.

### A FIFTH FORCE?

**Dark energy** is a hypothetical form of energy that permeates all of space and increases the rate at which the Universe is expanding. As such, it is like a "fifth force" that **opposes the action of gravity**. So far, not much at all is known about dark energy.

**SEE ALSO »**
*pp.402–03* STRING THEORY

## « BEFORE

The ideas that led to the emergence of string theory started to appear in the 1960s and 1970s, as scientists looked for a way to incorporate gravity with electromagnetism and the strong and weak nuclear forces.

### WHEN FOUR BECAME THREE
The first step toward **unifying** the four forces of nature (gravity, electromagnetism, and strong and weak nuclear forces) came in the 1960s. · Physicists found a mathematical package, now known as the **electroweak theory,** describes the **electromagnetic force** and the **weak nuclear force** in one set of equations.

### POINTING THE WAY
In 1979 **Abdus Salam**, **Sheldon Glashow**, and **Steven Weinberg** were awarded the **Nobel Prize in Physics** for their discovery of the electroweak theory. This theory was important not only in its own right, but because it pointed the way toward **further unification**, with the goal of finding **one set of equations** to describe all four forces.

« SEE ALSO
*pp.302–03* Theories of Relativity
*pp.314–17* Quantum Revolution
*pp.400–01* Grand Unified Theory

The search for a unified theory of physics is the quest to unite the two great theories of 20th-century physics: the general theory of relativity (see p.302) and quantum mechanics (see p.314). The general theory of relativity relates gravity to the structure of space and time, treating them as a single four-dimensional entity, space-time. Quantum mechanics describes the other three forces of nature: electromagnetism and the strong and weak nuclear forces (see pp.401–02).

A unified theory must describe gravity as well as the other three forces in quantum terms. This idea implies that space-time itself must be quantized (packaged) into discrete lumps. From

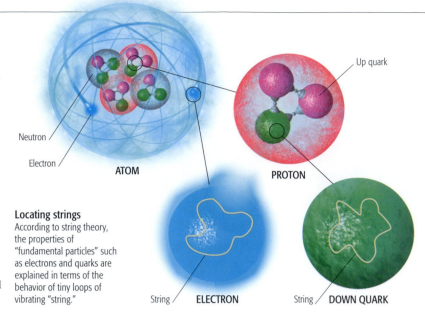

**Locating strings**
According to string theory, the properties of "fundamental particles" such as electrons and quarks are explained in terms of the behavior of tiny loops of vibrating "string."

# String Theory

**In particle physics fundamental entities such as electrons are defined as mathematical points. In string theory this idea is replaced by the concept of one-dimensional entities ("strings") or multi-dimensional entities ("membranes"), whose vibrations produce the properties formerly associated with "particles."**

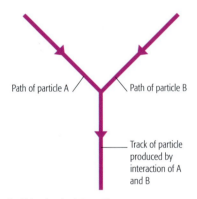

**Particle physics interaction**
If two particles with zero size meet and interact at a point, the forces operating between them become infinite, because these forces are divided by zero. This is a sign that there is something wrong with this description of particle interactions.

the weakness of gravity compared with the other three forces, physicists know that the scale on which this happens is incredibly tiny—only around $10^{-35}$m. The motivation for string theory came from attempts to mathematically model the behavior of space-time on this miniscule scale.

### Loops and threads
Although physicists had previously considered the possibility of treating fundamental entities as strings, string theory only took off in the mid 1980s, when John Schwarz of the US and Michael Green of the UK set out to include gravity in the package. They developed the idea that fundamental entities such as electrons and quarks are outward manifestations of "strings" vibrating on the quantum gravity scale. Properties such as mass are explained as different vibrations of the same kind of string. An electron is a piece of string vibrating in a certain way, while a quark is an identical piece of string vibrating in a different way.

At first, the physicists thought of the strings as open-ended, like a wiggly line; but they soon developed an improved version of the theory, in which the strings are tiny loops, like stretched elastic bands. It would take about $10^{20}$ such strings, side by side, to span the diameter of a proton.

### Good vibrations
One of the motivations for studying string theory is that it is free from infinities, which plague theories of

**String interaction**
Unlike two particles with zero size, two loops of vibrating string can meet and merge with one another to make a loop vibrating in a different way (a different particle), without any problems arising with infinities.

fundamental particles. The problem is essentially that forces such as the electromagnetic force are inversely proportional to distance (or the square of distance); this means that the "self interaction" of a particle with zero size would be 1 divided by zero, which is infinity. Since strings do not exist at a point, they don't have this problem.

String theory caused great excitement in the mid 1980s because it turned out that the mathematical description of

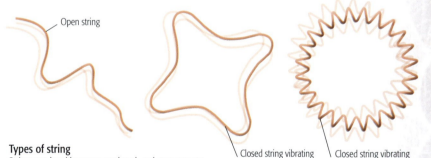

**Types of string**
Strings can be either open or closed, and can resonate at different frequencies. The frequencies of their vibrations give particles their different properties.

### THEORETICAL PHYSICIST (1963–)
## BRIAN GREENE

American physicist Brian Greene is one of the leading string theorists today. Greene was born in New York City in 1963, and has worked at Harvard, Oxford, Cornell, and Columbia universities. His special interest is applying ideas from string theory to cosmological problems, including the origin of the Universe and the nature of black holes. Greene is the most eloquent spokesman for the idea of string theory.

His book *The Elegant Universe* was nominated for the Pulitzer Prize in non-fiction in 2000, and was later made into a television special.

»

## BREAKTHROUGH

## COMPACTIFICATION

"Extra" dimensions are essential in defining the properties of strings, but they are thought of as hidden by "compactification." The usual example of compactification is to think of a hose pipe. Viewed from a distance, it looks like a one-dimensional line; but look closer and you see that it is made of a two-dimensional sheet (which might ripple in interesting ways) wrapped around the third dimension. The same trick can be used mathematically to wrap up any number of dimensions, shrinking them and leaving only the familiar four dimensions visible.

**Hidden dimensions**
A flat sheet is rolled, made into a ring and shrunk to the size of a full stop, concealing its original form.

these tiny loops of vibrating string automatically includes a description of gravity. The equations were set up to describe the other forces and particles of nature, and physicists were surprised to find that a description of another kind of particle, technically known as a spin-2 boson, fell out of the equations. At first they were baffled; then they realized that this was the particle associated with the gravitational field, in the same way that the photon is the particle associated with the electromagnetic field (see pp.382–83).

### Beyond the fourth dimension

The snag about string theory is that it requires many more dimensions than the familiar four of space-time. In order

**The latest development of string theory, known as M-theory, unites different versions of string theory in one package.**

#### WHEN SIX BECOME ONE

In the mid 1990s string theory seemed to be in trouble. There were **five different versions** of string theory, plus a **sixth theory called supergravity**, and they seemed unrelated. But in 1995 **Edward Witten** of the Institute for Advanced Study in New Jersey, caused a stir when he proposed that all six of these theories could be described in **one mathematical package**, which he called **M-theory**. (The "M" has variously been said to stand for Matrix, Mystery, Master, Mother, Magic, or even Murky.) It is as if the six string theories were the six sides of a die, which look different but are actually connected. Which one you see depends on how you roll the die. Could this be the long-sought **theory of everything**?

to produce all the variety of forces and particles from the vibrations of a single kind of string or membrane, the vibrations have to take place in many different dimensions. In one version of string theory, a closed loop of string has 10-dimensional waves—which describe fermions (see pp.382–83)—running around it one way, and 26-dimensional waves—which describe bosons—running around it the other way. Similar processes are needed with membranes.

So why is it that we only perceive three dimensions of space and one of time? The answer may be that the extra dimensions are rolled up very small, or "compactified" (see above). Strings may be thought of as loops embedded in higher-dimensional membranes, with harmonics on the vibrating string translating into properties such as mass and charge.

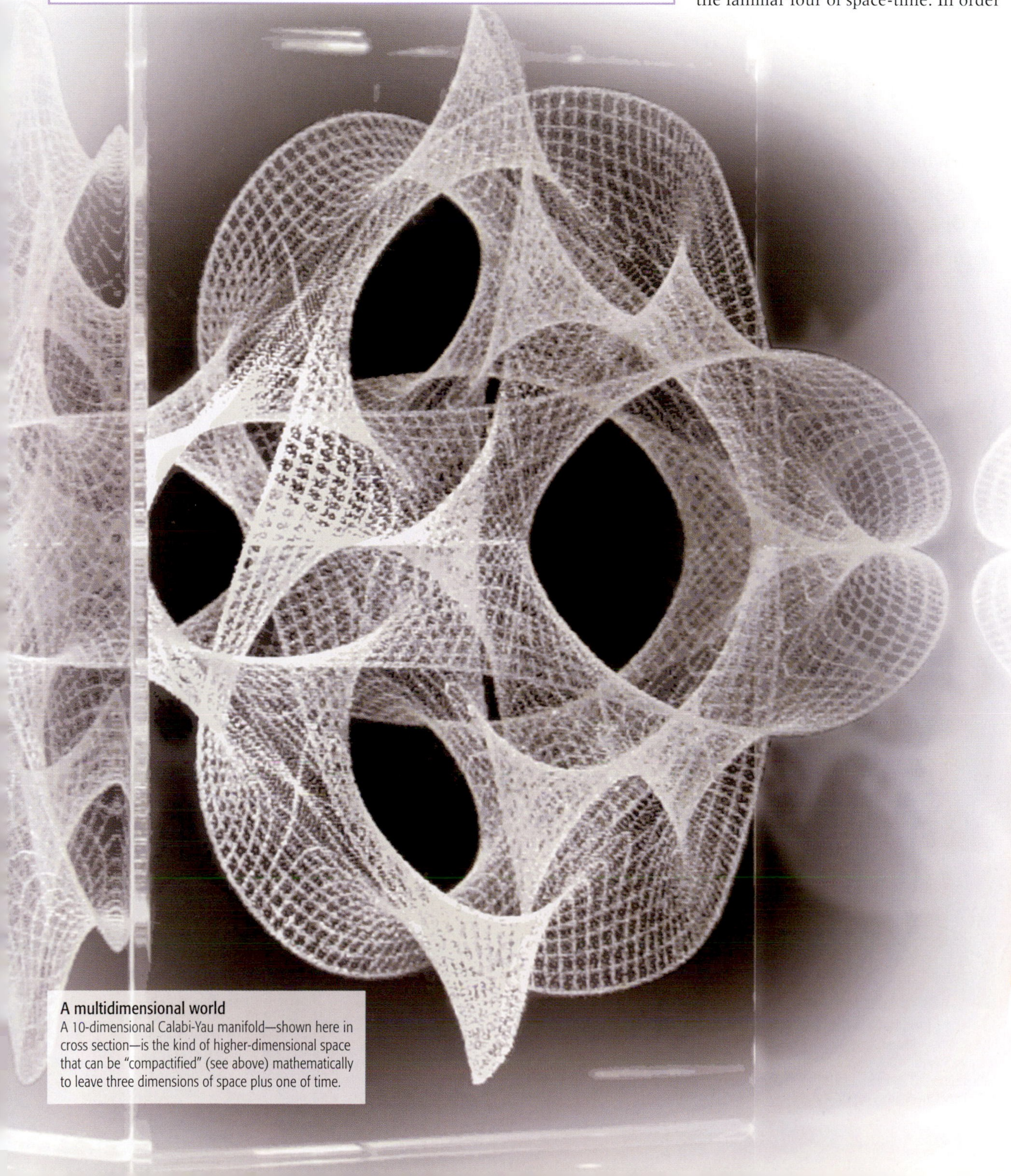

**A multidimensional world**
A 10-dimensional Calabi-Yau manifold—shown here in cross section—is the kind of higher-dimensional space that can be "compactified" (see above) mathematically to leave three dimensions of space plus one of time.

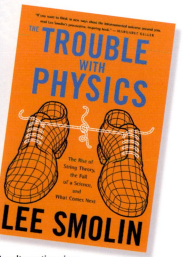

**An alternative view**
Some physicists worry that too much effort goes into string theory, compared with other areas of research. American physicist Lee Smolin speaks forcefully for this minority in *The Trouble with Physics*, published in 2006.

The first internal image of the body dates from 1895, when Wilhelm Röntgen made an X-ray photo of his wife's hand (see p.294). Within weeks, X-ray images made their debut in medical diagnostics, a role that they have maintained ever since.

X-ray technology is constantly improving and it is ideally suited for imaging solid structures within the body, such as teeth and bones. They can also be used to identify small regions of unusually dense tissue, which may be a sign of cancer. However, standard X-rays have one major limitation: because they reduce three-dimensional objects to a flat image, dense parts of the body, such as the skull and rib cage

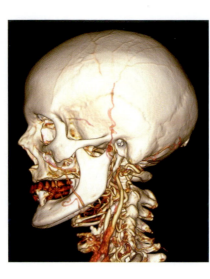

◁ CT skull
The astounding clarity of CT scans is demonstrated by this 3-D image of a man's skull. In this mode, the image has been processed by edge-detection software, giving the bones a lifelike appearance.

# Body Imaging

**By gathering data in different ways, modern imaging techniques have revolutionized our ability to look inside the human body. In addition to revealing internal organs, they can often show body processes at work—an important aid in diagnosing disorders and disease.**

## ◀◀ BEFORE

**Before medical imaging became available, doctors relied mainly on observation of symptoms and signs to make diagnoses.**

### SYMPTOMS AND SIGNS

A **symptom** is a **subjective** condition noticed by the patient. A **sign** is something **noticed by a doctor**, possibly visual such as change in skin color, but not by the patient.

**DOCTORS RELIED ON OBSERVATIONS**

### ENDOSCOPY

**Light-conducting tubes**, or **endoscopes**, were the earliest means of looking inside the body. The first recorded use was in 1822, by **William Beaumont**, an American army surgeon. In the early 20th century **laparoscopes** with their own light source were used in **abdominal surgery**.

### X-RAYS

The earliest form of **noninvasive body** imaging, **X-rays** were discovered by accident in 1895 by **Wilhelm Röntgen**. His breakthrough led to a new medical discipline, **radiology**.

### ◀◀ SEE ALSO
*pp.294–95* THE DISCOVERY OF X-RAYS

mask less dense tissue inside them. The solution to this problem was first sketched out in the 1920s, although it did not become a practical possibility until the late 1960s, with the beginning of the electronic age. Known as computerized tomography, or CT scanning, it uses a rotating X-ray source that circles the body, passing X-rays through it to a detector on the opposite side. Instead of producing direct X-ray images, the scanner builds up a bank of data that can be processed to show specific regions in 3-D form. Using digital techniques, superficial structures can be subtracted, revealing the tissues underneath.

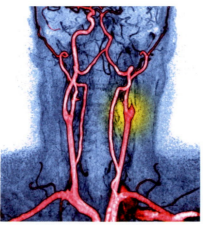

**Imaging blood vessels (angiography)**
This colored MRI angiogram shows a blocked artery in the neck of a stroke patient. Blood vessels can also be viewed using CT scanning, if they are first injected with a contrasting medium to highlight them on the CT scan.

## Magnetic imagery

CT scanning is one of several 3-D imaging techniques that are now widely used in medicine. The most widespread alternative, magnetic resonance imaging, or MRI, also co-evolved with the computer. Unlike CT scanning, MRI exposes the body to an intense magnetic field, created by a giant superconducting magnetic coil. Smaller electromagnets apply a much weaker variable field, and the scanner also emits pulses of radio waves. These radio pulses act on the body's hydrogen atoms in water, making them precess, or resonate. When each radio pulse ends, these atoms give off signals that the scanner detects. The pattern of these signals varies from one tissue type to another, because each contains different amounts of water.

Like CT scans, MRI scans provide a snapshot of the body at a single moment in time. A more recent development, called functional MRI, records real-time changes

## MRI cross-section
MRI scanning produces a high degree of contrast in soft tissues, which makes it ideal for examining the nervous system. This artificially colored scan shows the brain and spinal cord, along with the vertebral column, of a healthy adult.

### PET scan activity
By measuring energy use, PET scans can show the level of activity in the brain. This pair of artifically-colored scans shows normal brain activity (left) and reduced brain activity (right). Reduction of brain activity is often an indication of the onset of Alzheimer's, a degenerative disease of the brain.

in specific parts of the body, such as blood flow in the brain. Functional MRI is widely used in brain mapping, as it does not involve prolonged exposure to potentially harmful X-rays.

### Tracking tracers
With CT and MRI, a patient's symptoms are often used to decide which part of the body to scan. Positron emission tomography, or PET, is different: it can identify parts of the body that need further investigation. A radioactive substance, or tracer, is injected into the bloodstream. art of the scanner detects gamma rays, which are emitted when the tracer undergoes radioactive decay releasing positrons that immediately collide with neighboring electrons.

The chemical tracers used in PET are usually bioactive—in other words, they are taken up by body cells. One of the most important and commonly used tracers, fluorodeoxyglucose (FDG), has a structure similar to glucose, and, as such, is strongly absorbed by highly active cells, which use glucose in large amounts. These cells include neurons in the brain, and also rapidly dividing cells in cancerous tumors. FDG was first used on human volunteers in 1976. Since then, scanning with FDG has become an important tool for tracking down cancer, both in its initial stages and when it has seeded secondary tumors elsewhere in the body. PET is also used in the early diagnosis of Alzheimer's disease, because affected neurons take up less FDG than their healthy counterparts.

### Scanning with sound
Ultrasound scanning is one of the safest forms of imaging, with a proven track record dating back over 50 years. Unlike other forms of scanning, it can be done at the bedside, using a compact imaging unit and a handheld transducer or probe. It works by using high-pitched sound waves that are pulsed into the body. Wherever the sound meets a boundary between two different tissues, part of the sound energy is reflected and is detected by the probe to form the image on screen. Ultrasound has many uses, but its most common use is in assessing fetal development during pregnancy, when X-ray technology cannot be used.

The early ultrasound scanners created static images, but today's machines can show movement in real time by taking a series of scans in quick succession. A form called Doppler ultrasound can be used to monitor blood flow. Relative to the probe, the movement of blood cells causes a change in the pitch of the reflected sound. This is due to the Doppler effect—the same physical principle that changes the pitch of a siren when a fast-moving vehicle passes by (see p.257). Doppler ultrasound is rapid and noninvasive, making it very useful for identifying narrowed blood vessels or blood clots beneath the surface of the skin.

Imaging technology is set to continue its rapid growth and development, as data-processing speeds climb.

### SEARCH AND DESTROY
PET scanning locates cancers using **radioactive tracers**. Research is under way to combine imaging with **tumor destruction**, using tracers that work as cancer fighting drugs. **Search-and-destroy** tracers will seek out cancer cells, without harming healthy tissue.

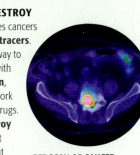

**PET SCAN OF CANCER IN LUMBAR VERTEBRA**

### ULTRA-FAST SCANS
Scanners are becoming faster and more precise. The newest CT scanners can complete a **chest scan in just half a second**, and MRI scanners operate hundreds of times faster than they did 20 years ago. Rapid scanning **reduces exposure to X-rays ❮❮ 294–95**. It also improves image quality, since the patient is less likely to move during the scan.

### COMPARE AND CONTRAST
In the future, imaging systems will be able to **compare scans** taken weeks or months apart **automatically**. Such systems will identify disorders that are still in their early stages.

**SEE ALSO** ❯❯
pp.406–07 MODERN SURGICAL PROCEDURES

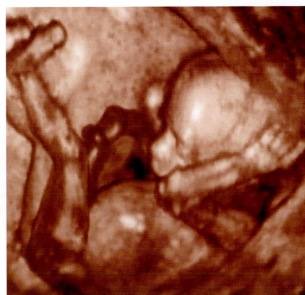

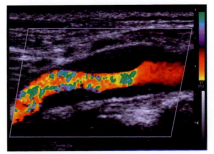

### Seeing moving blood
This Doppler ultrasound scan shows blood through a stricture, or narrowing, in an artery. The scanner detects changes in pitch of reflected sound waves, caused by differences in blood-flow rate. The flow is fastest when shown as red and slowest when green.

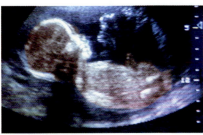

△ Fetal ultrasound
This colored ultrasound scan shows a six-month-old fetus. During pregnancy, ultrasound is used to check for abnormalities as well as the position of the fetus and placenta. This gives advance warning of complications, such as breech birth when a baby is born buttocks first.

◁ 3-D ultrasound
Seen in three dimensions, this scan shows a fetus at about the same age as the one above, but provides more external detail so physical disorders can be identified before birth. The fetus is lying from right to left across the field of view and the umbilical cord is visible.

**Keyhole surgery**
Surgery can have a traumatic effect on the body, but the smaller the incision, the quicker the patient recovers. In keyhole surgery, a fine probe called a laparoscope is guided through a tiny incision, or "keyhole." It carries surgical instruments and an optical system, which relays images to a computer screen.

The number of casualties during World War II triggered rapid improvement in medical procedures, especially surgery, which was helped by antibiotics and reliable blood transfusions. By the early 1950s significant progress was made in orthopedic surgery, which deals with bones and joints, and deep abdominal and thoracic surgery became routine.

### Organ transplants
The heart dominates the history of transplant surgery, but the kidney was the first internal organ to be successfully transplanted from one living person to another. This groundbreaking surgery was performed in 1954 by American surgeon

hundred heart transplants were performed. However, failure rates remained high until the introduction of the immunosuppressant drug cyclosporin in the 1980s. Tissue rejection could now be controlled; some patients have had "new" hearts for over 25 years.

### New parts for old
The idea of replacing body parts far predates transplant surgery. Until the 20th century, most of the parts were external appliances, such as wooden limbs and dentures made of bone or porcelain. Hip replacements were attempted in the 19th century, but the first versions to meet with any real success were created in the 1940s. In

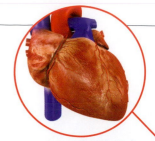

**Heart transplant**
Donor hearts have to be carefully matched to their recipients, to ensure that they are of appropriate size and tissue type. In most transplants, the donor heart is treated to stop it from beating, and then restarted once in place.

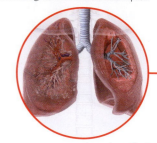

**Lung transplant**
Depending on the condition of the recipient, lungs may be transplanted partially, singly, or as a pair. More rarely, the heart and lungs may be replaced together in a combined transplant.

# Modern Surgical Procedures

**In the last 50 years average life expectancy in many developed countries has increased by up to a third. Some of this increase is due to new drugs and to improvements in public health, but in older people particularly, much is due to new medical procedures that repair or replace failing body parts.**

**Liver and pancreas transplants**
The liver is unusual in that it regenerates. Normally, it is transplanted intact, but a piece of healthy liver may be used, which regrows once it is in place. Pancreas transplants are often inserted alongside the existing organ.

## BEFORE

Throughout most of medical history, surgery was a treatment of last resort. Mortality rates were high until the introduction of anesthetics and antiseptics.

**REPLACEMENT BODY PARTS**
The use of **wooden limbs** dates back to the ancient Egyptians. Metal parts came later: the Danish astronomer **Tycho Brahe ≪ 77** had a **metal nose**, having lost his own in a duel.

**ANESTHETICS AND ANTISEPTICS**
In the mid 1800s the introduction of **anesthetics** gave surgeons longer to operate. The use of **antiseptics**, dating from 1860s, saw sharp improvements in survival rates.

**ANESTHESIA IN THE 19TH CENTURY**

**BLOOD TRANSFUSIONS**
In the opening years of the 20th century the **discovery of blood types** allowed safe transfusions after injuries and during surgery.

**≪ SEE ALSO**
*pp.228–29* SAFER SURGERY

Dr. Joseph Murray in Boston. He and his surgical team transplanted a kidney from Ronald Herrick to save the life of his dying twin brother, Richard. About 18,000 kidney transplants are performed each year in the US.

The kidney's blood supply is straightforward, making it a relatively simple organ to transplant. The liver's is more complex, which explains why the first successful transplant did not take place until 1967. But it was soon overtaken by the ultimate in organ transplants: that of a live human heart. First performed in South Africa on the December 3, 1967 by Dr. Christiaan Barnard, it was a technical success, but the recipient, Louis Washkansky died 18 days later after tissue rejection set in. In the following year, over a

recent years, new plastics and alloys have helped reduce friction and increase resistance to wear—a combination that is also valuable in replacement of knee joints.

Other implantable devices include those that improve hearing and vision, as well as those that play a life-saving role in improving heart function. The first artificial heart valve was implanted in 1952 (into a beating heart), triggering a wave of innovation. Implanted cardiac pacemakers, first used in 1958, monitor the heart's performance, intervening if they detect any unusual rhythms. Backup pumps, known as ventricular assist devices (see opposite), can take over the heart's role—a significant step toward the implantable artificial heart, which is already in development.

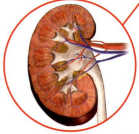

**Kidney transplant**
The body can survive with a single kidney, which allows transplants to be sourced from live donors. The recipient's kidneys are usually left in place, and the donor kidney inserted nearby, with a diverted blood supply.

### CHRISTIAAN BARNARD

Catapulted to stardom at the age of 45, the South African surgeon Christiaan Barnard won the race to transplant the human heart. His margin was narrow—American surgeons performed a heart transplant four weeks later—but in terms of publicity, it was a case of winner takes all. In December 1967 "The Man With Golden Hands" became the most famous surgeon in the world. Barnard attracted his share of critics, but his surgical talents were beyond doubt. He maintained an active interest in surgery at Cape Town's Groote Schuur Hospital, and was an outspoken critic of his country's apartheid regime.

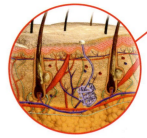

**Skin transplant**
Skin autografts—grafting a patient's own skin —are one of the oldest forms of transplant. It is widely used in plastic surgery, which reshapes living tissue after injury. More recently whole face transplants have become possible.

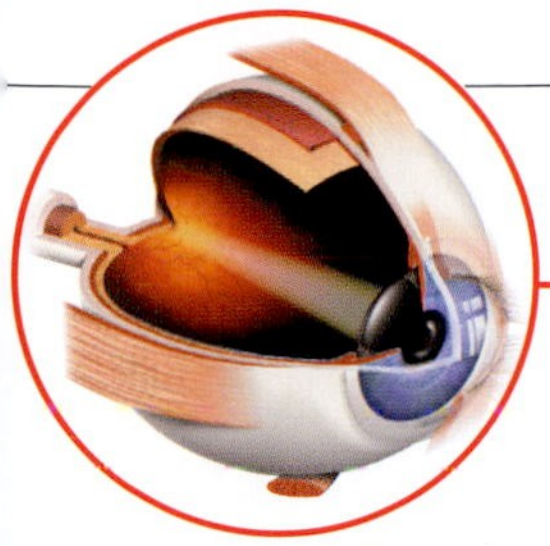

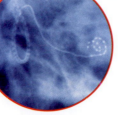

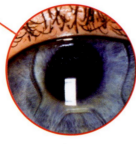

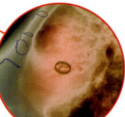

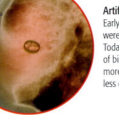

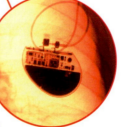

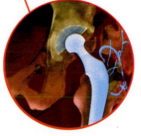

### Corneal transplant
First performed in the early 1900s this procedure replaces the thin film of tissue in front of the iris and pupil. It can restore vision if the original cornea is diseased or scarred.

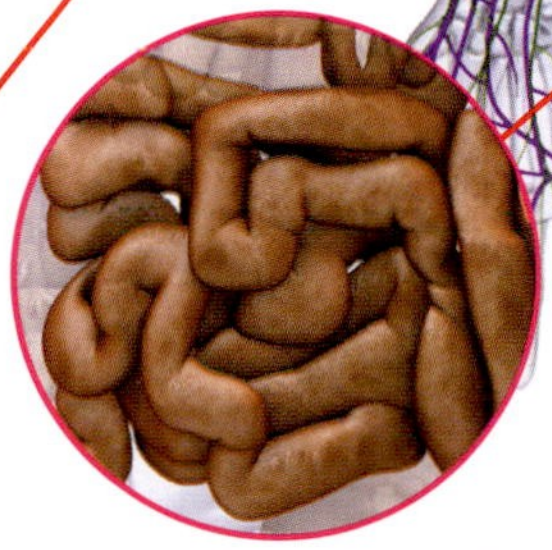

### Small intestine transplant
This rare form of surgery is used to treat cases of severe intestinal failure. A length of intestine is transplanted, and then connected to the blood supply so that it can absorb nutrients from food.

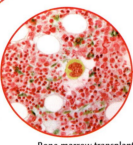

### Bone marrow transplant
Blood stem cells from bone marrow can be replaced in the treatment of blood diseases and cancer. The grafted stem cells may be from a donor, or they may be harvested from the patient before cancer-fighting therapy.

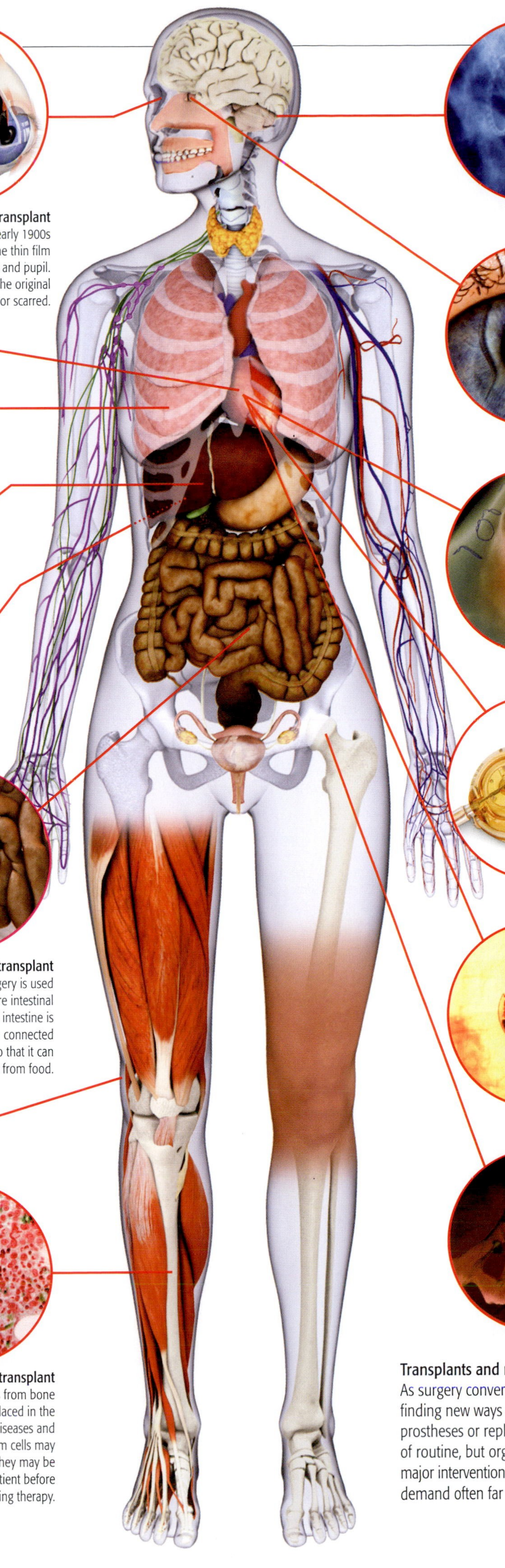

### Cochlear implant
This electronic device bypasses the outer ear, using an external microphone to stimulate auditory nerves in the cochlea of the inner ear.

### Artificial lens
A simple and effective treatment for eye disease such as cataracts, artificial lenses have been in use since the 1970s. New multifocal versions allow close or distant objects to be brought into focus.

### Artificial heart valve
Early replacement valves were entirely mechanical. Today, many include material of biological origin, which is more durable, and causes less damage to blood cells.

### Ventricular assist device (VAD)
This temporary backup pump helps one or both ventricles to expel blood from the heart. Such devices are normally used for patients awaiting a transplant.

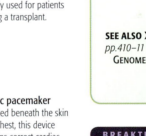

### Cardiac pacemaker
Implanted beneath the skin of the chest, this device maintains correct cardiac rhythms by using electrodes that are placed in one or more of the heart's chambers.

### Joint replacement
As the population ages, arthritic joints are routinely replaced with artificial implants. Hip replacements are the most common form of this replacement surgery, but others such as knee and elbow joints are possible.

### Transplants and replacements
As surgery converges with technology, medicine is constantly finding new ways to support ailing or aging bodies. Many prostheses or replacements, shown above, have become matters of routine, but organ transplants, shown on the left, are usually major interventions, carrying a degree of risk. With many transplants, demand often far outstrips the supply of donor organs.

## AFTER

**The divide between biology and technology will narrow even further as medicine restores and repairs defective body parts.**

### STEM-CELL RESEARCH
**Stem cells** have the potential to **differentiate** into almost all the tissues found in the adult body. In future they may **enable diseased organs** to be **restored** to a functioning state.

### TOTAL ARTIFICIAL HEART
Ventricular assist devices (VADs) currently provide a **temporary lifeline** for patients with severe heart failure. An **artificial heart** is in development as a permanent solution, which is expected to be ready by 2013.

### MYOELECTRIC LIMBS
Instead of being inert, **prosthetic limbs** will be powered and under direct nervous control. Prototype **"myoelectric" hands** that indicate how tightly an object is grasped already exist.

**MYOELECTRIC HAND**

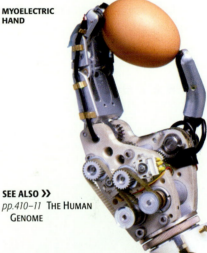

SEE ALSO »
pp.410–11 THE HUMAN GENOME

**BREAKTHROUGH**

## ADVANCED PACEMAKER

Normally the heartbeat is controlled by the heart's built-in pacemaker, which is influenced by nerves and hormones. But if a heart beats erratically or too slowly, then doctors can install a pacemaker to control the rhythm. When the device detects an anomaly in heart rate, it fires an electrical pulse that triggers the heart to correct its rhythm. Modern pacemakers can be programmed to deliver a shock if the heart goes into rapid, chaotic contraction (fibrillation), which can kill within minutes if not controlled.

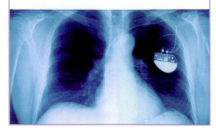

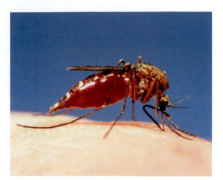

**Animal vectors**
Many infectious diseases, such as malaria, are spread indirectly via animal carriers, or "vectors". Malaria, carried by mosquitoes, is caused by a single-celled parasite that lives in blood. Although it can be treated and prevented, it still kills up to three million people every year.

▷ **Targeting HIV**
HIV multiplies by infecting specific cells in the immune system—most commonly the T4 lymphocytes. Each step in the virus's replication process involves chemical interactions that antiretroviral drugs can block or disrupt. By using drugs in varying combinations (see 1, 2, 3, and 4), patients can survive for many years without developing AIDS.

**Free HIV virus**

**Docking protein**
Antigen that helps lock virus on to the host cell CD4 receptors.

**2 Stopping reverse transcription**
The virus's genetic material (RNA) enters the cell. The enzyme reverse transcriptase frees it, and then reads it "backward," turning it into DNA. Transcriptase-inhibitor drugs bind to the enzyme or to the DNA strand to stop this process.

**T4 lymphocyte**

**CD4 receptors**

**Capsid**

**Viral RNA**

**Glycoprotein complexes**

**1 Inhibiting attachment and fusion**
The virus normally binds to a receptor (called a CD4 receptor) on the cell's surface, before fusing with its plasma membrane. Entry and fusion inhibitors stop this happening, so the virus cannot get access to the cell.

# Disease Challenges

**A burgeoning human population, together with rapid travel, have created ideal conditions for the spread of both new and existing forms of disease. Some have been contained successfully, but others threaten to become global epidemics or pandemics.**

**Viral DNA enters nucleus**

**Viral DNA**
This can combine with host's DNA.

**Transcription**
The instructions in the viral DNA are used to make viral proteins and new viral RNA within the cell nucleus. The viral DNA is copied and passed on whenever the infected cell divides.

## BEFORE

**Historically epidemics have often been linked with exploration and migration, and also with changes in population and land use.**

**EUROPEANS ARRIVING IN THE NEW WORLD**

### COLUMBIAN DISEASE EXCHANGE
Following **Columbus's** discovery of the New World in 1492, 90 percent of the population of American Indians died from diseases such as **measles** and **smallpox introduced by the Europeans**. Syphilis is the only major disease that traveled the other way.

### FIGHTING DISEASES WITH DRUGS
In the 1930s and 1940s the launch of **sulfa drugs** and **antibiotics** widened the variety of bacterial diseases that could be treated by drugs.

The World Health Organization (WHO) monitors dozens of diseases that cause major loss of life, or threaten to unless they are kept under control. Some of these, such as mosquito-borne malaria and tuberculosis (TB) are age-old enemies of humanity. Others are more recent, and are the result of viruses crossing the species barrier from animals to humans.

One disease, AIDS, or Acquired Immunodeficiency Syndrome, is caused by the Human Immunodeficiency Virus (HIV). It is thought to have evolved from Simian Immunodeficiency Virus, or SIV, which has existed in non-human primates in Africa for millions of years. Early in the 20th century SIV seems to have jumped from an infected chimpanzee to a human host and mutated into HIV—a form that can be transmitted from human to human. For decades HIV spread unnoticed, until a cluster of AIDS cases occurred in the US in the early 1980s. It has since spread around the world, and claimed over two million lives. HIV was finally unmasked as the causative agent of AIDS by the French virologist Luc Montagnier in 1983.

### Emerging epidemics
Once a disease outbreak occurs, many factors can affect its course. In some cases, the disease-causing organism is so deadly that it effectively contains itself by killing its host victims before it can spread too widely. This is typical of the Zaïre form of the Ebola hemorrhagic (bleeding) virus, which is

**4 Prevention of protein assembly**
Outside the nucleus viral proteins and RNA self-assemble, forming a capsid. This step requires enzymes called proteases, which cut up the proteins into smaller sub-units. Drugs called protease inhibitors interfere with this step.

**Budding**
With its proteins and RNA complete, the immature viral particle moves toward the exterior of the cell. It pushes through it, wrapping itself in a layer of cell membrane. The virus now enters the bloodstream.

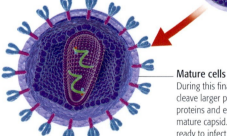

**Mature cells**
During this final stage, viral enzymes cleave larger proteins into smaller proteins and enzymes, creating a mature capsid. The virus is now ready to infect new host cells.

one of the most lethal disease agents known. First identified in the 1970s Ebola virus is restricted almost entirely to the African tropics, where outbreaks occur every few years.

The situation is different with respiratory diseases, such as flu and severe acute respiratory syndrome (SARS). These diseases spread via tiny droplets of mucus in the air, and initial symptoms are not debilitating enough to confine people to their homes. As a result, they can spread with extraordinary speed. During the 2009 pandemic of swine flu, the disease originated in Mexico. Air passengers then carried it around the world in a matter of days.

**3 Preventing integration**
If the viral DNA is created, the enzyme—integrase—inserts it into the host cell's DNA in the nucleus. Integrase inhibitors bind to the enzyme deactivating it.

Nuclear membrane

Host's DNA

**Cycle continues**
New HIV viruses attack and destroy other body cells.

Avian flu virus transmitted to pigs

Avian flu virus transmitted between wild birds

**Natural cycle**

Avian virus

New virus transmitted back to pigs

**New virus strain**

New virus transmitted to humans

**Pandemic cycle**

Interchange between birds and humans

New strain transmitted to other humans

**Jumping the species barrier**
Flu viruses infect a variety of birds and mammals. Pigs can harbor viruses adapted to birds and to humans, allowing new strains to emerge. If a new strain can be transmitted directly from person to person, an epidemic, or even pandemic, can occur.

Unlike flu viruses, HIV may exist in the body for several years without producing any visible symptoms. Moreover, during this latent period, it replicates at a phenomenal rate—in a 24-hour period, infected cells may release a total of 10 billion viruses into the blood. It can be passed on only through contact with body fluids.

### Fighting viral disease
Despite this dark backdrop, some remarkable successes have been achieved using vaccines (see pp.248–49). The smallpox virus, for example, was eradicated in 1979 (see pp.242–43), while the number of poliomyelitis (polio) cases has fallen by 99 percent in the last 20 years. Yellow fever is preventable, and a malaria vaccine may follow within the next 10 years. With flu, the outlook is more mixed. Vaccination programs based on known variants help control winter epidemics, but they cannot be stockpiled in advance against new ones. Antiviral drugs can inhibit development of the virus, although they do not destroy it.

**1 MILLION** The number of times faster HIV can evolve than the immune system cells it infects.

HIV, a retrovirus, remains an adversary with no vaccine in sight. Moreover, it targets cells (in particular T4 cells) in the immune system, the very weapon that the body uses to defend itself against attack. However, in the late 1980s, antiretroviral drugs became available that prevent HIV replicating in living cells. Today these

drugs are typically taken in combinations of three or four different types. This drug regimen targets different parts of the replication cycle (see opposite), minimizing resistance to any one particular drug—one of the greatest challenges in fighting disease.

### Drug resistance
Microorganisms reproduce rapidly, constantly evolving new chemical characteristics. If a single virus or bacterium develops drug resistance (see p.244), the drug will kill only the less-resistant forms.
Resistance can occur with almost any organism that causes infection or disease. For example, the bacterium *Staphylococcus aureus* is often found on the skin and in the lining of the nose, where it does little harm. However, the methicillin-resistant form, known as MRSA, is resistant to a wide range of antibiotics, making it a dangerous pathogen in surgical incisions.
Tuberculosis is even more problematic.

**Persistent shadows**
This chest X-ray shows the characteristic shadows caused by tuberculosis bacteria in the lungs. TB infections can be started by a single inhaled bacterium. Left untreated, they can gradually spread to other parts of the body.

In future, faster responses and antiviral drugs will play an important part in limiting the spread of infectious diseases.

PROTECTION AGAINST THE 2003 SARS OUTBREAK

#### OUTBREAK ALERTS
New disease outbreaks are monitored by the **World Health Organization** (WHO), through its **Epidemic and Pandemic Alert and Response** (EPR) system as well as by organizations such as Google.org. Part of this search is carried out automatically, by special computer programs that continuously search **news wires and websites**.

#### ANTIVIRAL DRUGS
**Neuraminidase inhibitors** prevent flu viruses **emerging from infected cells**. Future research into these and similar drugs may **combat viral infections** as effectively as antibiotics can now combat bacteria.

SEE ALSO »
pp.248–49 IMMUNIZATION AND VACCINATION

For the last 50 years treatment has involved combinations of up to four antibiotics, but new forms that are resistant to antibiotics continue to emerge. Extensively drug-resistant tuberculosis (XDR-TB) shrugs off almost all antibiotics, raising fears of epidemics. It has been identified in nearly 50 countries—an alarming statistic since tuberculosis spreads through the air.

EPIDEMIOLOGIST (1944–)

### LARRY BRILLIANT
Known first and foremost as an epidemiologist, American physician Larry Brilliant has been involved in many different campaigns to promote public health, particularly in the developing world. In the late 1970s, he was co-founder of a charitable foundation that combats eye disease and he took part in the World Health Organization's drive to eradicate smallpox. He helped to pioneer the use of the Internet as a social tool, creating one of the first online communities in 1985. From 2006 to 2008, he directed Google.org, a charitable foundation whose aims include predicting and preventing emerging diseases.

# The Human Genome

**The genome is the complete set of genetic (inherited) information stored in each of an organism's cells—the DNA. While DNA carries the codes that make the body work, it also carries the genetic variations that trigger disorders and disease.**

Virtually every cell in the human body contains up to 8 ft (2.5 m) of DNA, packed into nuclei as small as 5 micrometers across. Chromosomes package this DNA in a highly organized way. For most of a cell's life, its genetic material is diffuse and cannot be seen, but just before cell division (see pp.306–307) chromosomes duplicate and are easier to observe. In the 1950s the structure of DNA was discovered (see pp.346–47). In the 1960s the genetic code that it carries was identified (see pp.348–49), marking a huge change in genetics. Scientists could look beyond the structure of chromosomes to the molecular instructions they contain.

## BEFORE

Understanding that DNA held the key to inheritance and cracking the gene code within it paved the way for the genome.

### CHROMOSOMES AND GENE MAPS
Swiss botanist **Karl Wilhelm von Nägeli** first saw chromosomes in plants in the 1840s. By the early 1900s they were recognized as carrying genes. In 1911 American geneticist **Thomas Hunt Morgan** found that groups of genes in fruit flies can be passed on together. The first gene map was published by the American **Alfred Henry Sturtevant** in 1913.

### FROM GENES TO GENOMES
In 1972 Belgian scientist **Walter Fiers** identified the **sequence of base pairs** that code for a single viral gene. In 1976 the **first genome,** of the MS2 bacteriophage virus, was sequenced by Fiers.

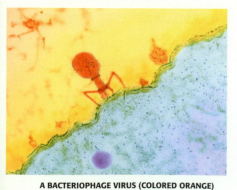

**A BACTERIOPHAGE VIRUS (COLORED ORANGE)**

« SEE ALSO
pp.348–49 THE GENETIC CODE
pp.386–87 GENE TECHNOLOGY

In 1972 an important milestone was reached when the sequence of base pairs that code for a single viral gene was identified (see below). The first DNA-based genome, that of the phi-X174 bacteriophage virus, was sequenced in 1977. Researchers had breached the divide separating simple viral genomes from more complex DNA-based genomes of fully living things.

### Reading the human genome
A bacterial genome was sequenced in 1995, followed by the first animal—the nematode worm *Caenorhabditis elegans*—in 1998 (see opposite). However, even before these were complete, the Human Genome Project was underway.

The human genome is not the largest in the living world, but its size meant that no single team of scientists could sequence it. Separate teams around the world worked on sections about 150,000 base pairs long. The sections were cloned and "read" using a technique called shotgun sequencing. This involves breaking the strands into random fragments, typically under 1,000 base pairs long. The fragments are "read," and the results analyzed to identify overlapping ends. The overlaps are then used to reassemble the fragments in the correct order, producing a sequence for the entire section. Then, the sections are assembled using a chromosome map.

The Human Genome Project was launched in 1990, and the first chromosome was sequenced in 1999. The project was completed in 2003—exactly 50 years after the discovery of the structure of DNA.

### Using the genome
Knowledge of the human genome throws light on how the body works, and on how humans have evolved. It also opens up new forms of medical treatment, based on genetic variations. The project has shown that about 90 percent of this variation consists of single-nucleotide polymorphisms or SNPs. Many SNPs have no apparent effect, but others affect susceptibility to disorders and diseases, as well as responses to medicinal drugs.

### The genome revealed
Each one of the body's cells contains the entire human genome. The genome includes all the information needed to form an individual body, as well as the information needed to control the sum of its chemical processes, or metabolism.

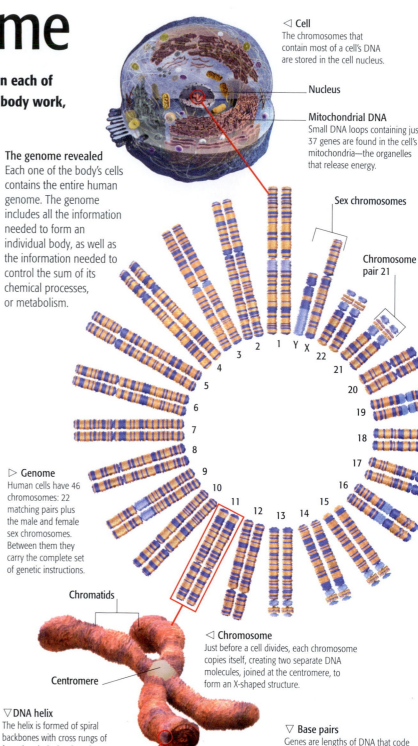

◁ **Cell**
The chromosomes that contain most of a cell's DNA are stored in the cell nucleus.

**Nucleus**

**Mitochondrial DNA**
Small DNA loops containing just 37 genes are found in the cell's mitochondria—the organelles that release energy.

**Sex chromosomes**

**Chromosome pair 21**

▷ **Genome**
Human cells have 46 chromosomes: 22 matching pairs plus the male and female sex chromosomes. Between them they carry the complete set of genetic instructions.

**Chromatids**

**Centromere**

◁ **Chromosome**
Just before a cell divides, each chromosome copies itself, creating two separate DNA molecules, joined at the centromere, to form an X-shaped structure.

▽ **DNA helix**
The helix is formed of spiral backbones with cross rungs of four chemicals that form base pairs—guanine only pairs with cytosine, and adenine pairs with thymine.

**Guanine– cytosine link**

**Adenine–thymine link**

**Chemical base G**

▽ **Base pairs**
Genes are lengths of DNA that code for specific proteins. Triplets of bases, called codons, specify the amino acids used to build each protein. Humans have about three billion base pairs.

**Base pair T - A**

▷ **Genetic sequence**
The order of the base pairs in the DNA helix provides the gene code.

**Chromosomal variation**
A geneticist studies a human karyotype. Humans have a set number of chromosomes. Unlike variations in genes, large-scale variations in chromosomes are rare.

People absorb and eliminate drugs at different rates: some undergo powerful immune reactions, while others do not respond. Currently, these responses are discovered by experience; in the future, SNP profiles may guide drug choice.

Decoding the entire human genome is slow and expensive, but in the last 15 years, the cost of reading specific parts of the genome has plunged. Companies offer genetic tests based on readings of over half a million SNPs. These tests can have far-reaching implications—particularly if they reveal untreatable conditions that have not yet expressed themselves.

## Close connections

The human genome is one of many that have been decoded. Others include species that are easy to breed in the laboratory and have short life cycles. Among them is the fruit fly *Drosophila melanogaster*, which was sequenced in 2000. This insect was used in early research (see pp.308–309), before the chemical nature of genes was understood. Other sequenced genomes include fungi, plants, protozoa, and our closest ancestor, the chimpanzee.

The chimpanzee's genome was released as a first draft in 2005. It contains 24 pairs of chromosomes—in humans, two ancestral pairs fused, reducing our count to 23. Humans differ from chimps in that there are fewer chromosomal rearrangements, and over 30 million pinpoint differences in our DNA. Some of the greatest divergences occur in the genes involved in speech and hearing. Across the genome as a whole, it is believed these rearrangements accumulate at a set rate. Geneticists can work backward to the point where the lines leading to chimps and humans split. Estimates put it between five and seven million years ago.

### CHROMOSOMES KNOWN TO CARRY DISEASES

| CHROMOSOME | NUMBER OF GENES (APPROXIMATELY) | EXAMPLES OF ASSOCIATED DISORDERS |
|---|---|---|
| 1 | 3,000 | Breast cancer, color vision deficiency, glaucoma |
| 2 | 1,400 | Congenital hypothyroidism, pulmonary hypertension |
| 3 | 1,300 | Cataracts, diabetes, night blindness, ovarian cancer |
| 4 | 1,300 | Hemophilia, Huntingdon's disease |
| 5 | 900 | Parkinson's disease, spinal muscular dystrophy |
| 6 | 1,200 | Ankylosing spondylitis, celiac disease |
| 7 | 1,150 | Cystic fibrosis, hemochromatosis |
| 8 | 800 | Cleft lip and palate, congenital hypothyroidism |
| 9 | 900 | Friedreich's ataxia, galactosemia |
| 10 | 800 | Breast cancer, Crohn's disease, heart defects, leukemia |
| 11 | 1,500 | Beta thalassemia, bladder and breast cancer, sickle cell |
| 12 | 1,100 | Parkinson's disease, phenylketonuria |
| 13 | 600 | Breast cancer, bladder cancer, retinoblastoma |
| 14 | 800 | Alzheimer's disease, multiple myeloma, deafness |
| 15 | 750 | Breast cancer, Marfan syndrome, Tay-Sachs disease |
| 16 | 850 | Crohn's disease, inflammatory bowel disease, leukemia |
| 17 | 1,200 | Bladder and breast cancer, deafness, leukemia |
| 18 | 300 | Porphyria, trisomy 18, skeletal abnormalities |
| 19 | 1,300 | Alzheimer's disease, breast cancer |
| 20 | 600 | Celiac disease, inherited immunodeficiency disease |
| 21 | 300 | Deafness, Down syndrome, leukemia |
| 22 | 500 | Methemoglobinemia, neurofibromatosis |
| X | 1,100 | Klinefelter's, triple X, and Turner's syndromes |
| Y | 100 | Y chromosome infertility |

The reading of the human genome provides a greater understanding of disease.

### DELIVERING GENE THERAPY

In future gene therapy could use **viruses** to **modify** the **genomes** of specific cells. Viruses can be used as carriers to deliver genes either that act on their own or that modify the effects of existing ones. This form of treatment has the potential to treat s**pecific genetic disorders**.

### INTERNATIONAL HAPMAP PROJECT

Launched in 2002 this international research project is a **global database screening** for **inherited diseases**. The map will record widespread **human haplotypes**—groups of single-nucleotide polymorphisms typically inherited as single units. Shared haplotypes can be used as markers, to pinpoint **genes that trigger** the particular disorders.

### EXPRESSED GENES

Cells vary in the genes that they put into action, or express. In future, identification of proteins coded by **expressed genes** will make it easier to identify specific kinds of cancerous cells.

### GENETIC MAPPING

By about 2020 genetic mapping will provide an early warning of inherited disorders in newborn babies, improving the scope for treatment.

**MAPPING OF NEWBORNS MAY BECOME ROUTINE**

**BIOLOGIST (1942–)**

## JOHN SULSTON

British biologist John Sulston became involved in gene sequencing in the 1980s—well before the introduction of automated techniques. His most important work involved *Caenorhabditis elegans*, a tiny transparent nematode worm just 0.03 in (1 mm) long. Unlike most animals, *C. elegans* has a precise number of cells—adult males have 1,031. Sulston mapped the origin of every cell from the egg to the adult stage. This "fate map" showed that over 10 cells are programmed to die during development, a process known as apoptosis. In 2002 John Sulston, Sydney Brenner and Robert Horvitz (see p.349) shared the Nobel Prize for Medicine for their work.

**BREAKTHROUGH**

## GENOME SIZE

Until the late 1980s geneticists believed that the human genome contained approximately 100,000 genes. The Human Genome Project showed that the true total is 20,000–25,000, far fewer than some superficially simpler forms of life. This is not restricted to humans. The largest known genome for an animal belongs to the marbled lungfish, with a genome size of 130 billion base pairs.

| ORGANISM | GENOME SIZE (BASE PAIRS) | NUMBER OF GENES | NUMBER OF CHROMOSOMES |
|---|---|---|---|
| Human *Homo sapiens* | 3.2 billion | 20,000–25,000 | 46 |
| Chimpanzee *Pan troglodytes* | 2.8 billion | 20,000 | 48 |
| Cow *Bos taurus* | 3 billion | 22,000 | 40 |
| House mouse *Mus musculus* | 2.6 billion | 25,000 | 40 |
| Fruit fly *Drosophila melanogaster* | 137 million | 13,600 | 8 |
| Roundworm *Caenorhabditis elegans* | 97 million | 19,000 | 12 |
| Bacterium *Escherichia coli* | 4.6 million | 32,00 | 1 |

INVENTOR AND ENVIRONMENTALIST **Born 1919**

# James Lovelock

> "**Life** does more than adapt to the Earth. It **changes** the **Earth** to its own purposes."
>
> JAMES LOVELOCK, 2000

James Lovelock, who trained in chemistry and medicine, is best known as the founder of Gaia theory, which describes the entire earth, made of both living and nonliving parts, as a single system. This gives the theory its alternative name, Earth System Science. His far-seeing and accurate predictions have transformed our understanding of the global environment.

## Early education

Born in 1919 in Letchworth, England, Lovelock moved to London six years later, attending the local public school in Brixton. He got a job in photographic chemistry and went to evening classes at the University of London. When World War II began Lovelock obtained grants to enable him to study chemistry at the University of Manchester. He graduated in 1941, and subsequently got a job with the National Institute for Medical Research (NIMR) in London.

Although he had been a conscientious objector, Lovelock gave up this status when he realized the appalling loss of life among merchant sailors bringing food to Britain. But his work at NIMR was deemed too important for him to be spared for active service, and he stayed there, working on studies of the way diseases spread and the effect of burns on flesh. Instead of using laboratory animals in the burns experiments, Lovelock used himself, burning his own arm repeatedly. In 1942 he married Helen Hyslop and they started a family.

After the war, Lovelock moved to the London School of Hygiene and Tropical Medicine, where he wrote up his work on air hygiene and was awarded a PhD in medicine from the University of London.

### James Lovelock
A career as chemist, physician, space scientist, and inventor led Lovelock to the Gaia theory, which suggests that everything on Earth is part of a single, interacting, living system.

---

**INVENTION**

## ELECTRON CAPTURE DETECTOR

Lovelock's career in invention began when he was a boy, when he produced a crystal set radio. His most famous invention, the electron capture detector (ECD), was designed to measure traces of chemicals in the atmosphere in minute quantities. The ECD has never been superseded, and the best modern versions can detect one part of a pollutant in one hundred thousand billion parts of air. Conventional chemical techniques can concentrate the pollutant a hundredfold, which enables the ECD to detect a trace of gas as small as one part in ten million billion. Although Lovelock's design is more than 50 years old, his ECD is still the most sensitive instrument of its kind.

> "People sometimes have the attitude that 'Gaia will look after us.' But that's totally wrong. If the concept means anything at all, Gaia will look after herself. And the best way for her to do that might well be to get rid of us."
>
> JAMES LOVELOCK, 1987

### The living planet
An alien visitor could immediately identify Earth as a living planet. Dead planets have inert atmospheres, in chemical equilibrium. In contrast, Earth has a chemically active atmosphere rich in oxygen.

But he was not an ordinary medical man. An inveterate tinkerer, he invented detectors to trace the movement of air around buildings.

### Journey into space
In 1961 Lovelock joined a NASA team designing instruments, first for lunar probes and then for the search for life on Mars. His family relocated to the US, but in 1963 returned to England, where he established himself as an independent scientific consultant. One of his consultancies was with the Jet Propulsion Laboratory in California; on a visit he had a sudden insight that led to the concept of Gaia. He realized that there was no need to go to Mars to look for signs of life. Life on Earth is responsible for the chemically reactive atmosphere, but a dead planet would have an inert atmosphere, so an infrared telescope would reveal whether the atmosphere of Mars is made of inert carbon dioxide. Astronomers in France soon found this to be the case.

This insight led Lovelock to the idea of Earth as a "living planet" in the sense of a superorganism, like a single living cell. But before developing Gaia theory, Lovelock was involved in another environmental saga.

### The hole in the sky
One of Lovelock's inventions is called the electron capture detector, or ECD (see opposite). In the early 1970s measurements using ECDs showed that man-made compounds known as CFCs, then widely used in refrigeration and spray cans, had spread throughout the atmosphere. Sunlight was breaking CFCs apart, releasing chlorine that was destroying the ozone layer over Antarctica, making what became known as "the hole in the sky." After CFCs were banned, the ozone layer, which shields us from ultraviolet radiation, has begun to recover. Meanwhile, Lovelock was developing his Gaia idea, which he published in a book in 1979. Responding to some criticism, he came up with Daisyworld, his greatest invention.

### Daisyworld
Lovelock explained how organisms acting in their own "selfish" interest can maintain an environment that benefits all life on a planet. In its simplest form, there are just two kinds of daisies on a planet. Black daisies flourish when it is cold, and white daisies flourish when it is hot. But black daisies absorb heat, and white daisies reflect heat away. So there is always a tendency for the temperature to be pushed toward a stable in-between value that suits both. Partly thanks to Daisyworld, Gaia theory is now established, although some prefer the term Earth System Science.

### Gaia in peril
Human activities such as destruction of the tropical rain forests are upsetting the natural mechanisms by which Gaian feedbacks maintain stable environmental conditions on Earth. Global warming is one result.

### TIMELINE

- **July 26, 1919** James Ephraim Lovelock is born in Letchworth, England. Spends his first six years with his grandparents there.
- **1928–38** Attends Brixton Grammar School. He hates formal education and learns about science from reading books.
- **1938–39** Works for a photographic chemist and takes at evening classes.
- **1941** Completes degree in chemistry at Manchester University.
- **1948** Awarded PhD in medicine.
- **1946-1951** Works at the Common Cold Research Unit at Harvard Hospital in Salisbury, Wiltshire.
- **1949** Travels on board the aircraft carrier HMS *Vengeance* in Arctic waters to perform air hygiene tests.
- **1955** Receives CIBA Foundation Award for research into aging.
- **1957** Invents first version of the electron capture detector (ECD).

**PROTOTYPE ECD**

- **1959** Awarded DSc (Doctor of Science) in biophysics by the University of London.
- **1961** Quits medical research, initially to work for NASA and then to become a freelance scientist.
- **1965** Has the flash of insight that living things are regulating the composition of the atmosphere.
- **Late 1960s** Detects the problem of the spread of CFCs and other man-made substances by measurements taken at the family's vacation home on the Beara Peninsula in the west of Ireland, far from any sources of pollution.
- **1971–72** Sails on the research ship *Shackleton* to measure CFC concentrations in the air of the southern hemisphere.
- **1973** Sails on the *Meteor* to measure air pollution in the North Atlantic and Caribbean.
- **1974** Elected a Fellow of the Royal Society.
- **1975** Publishes an article, "The Quest for Gaia," in *New Scientist*.
- **1979** Publishes first book, *Gaia: A New Look at Life on Earth*.
- **1981** Invents Daisyworld. Publishes *Gaia: The Practical Science of Planetary Medicine*.
- **1988** Publishes *Ages of Gaia*.
- **1989** His first wife, Helen, dies.
- **1990** Made a CBE. Awarded the Amsterdam Prize for the Environment.
- **1991** Marries Sandy Orchard.
- **1997** Awarded Japan's Blue Planet prize.
- **2000** Publishes *Homage to Gaia: The Life of an Independent Scientist*.
- **2004** Begins writing his most polemical book, *The Revenge of Gaia*, published in 2006.
- **2009** Publishes *The Vanishing Face of Gaia*.

**‹‹ BEFORE**

Suspicions of a link between changes in atmospheric chemistry and temperature have existed for over 150 years.

**GREENHOUSE GAS INVESTIGATION**
In 1859 Irish physicist **John Tyndall** set out to find which gases play a part in **trapping heat** in the atmosphere. He found the most important were water vapor and carbon dioxide ($CO_2$).

**EARLY WARMING WARNING**
At the turn of the 20th century the Swedish physicist **Svante Arrhenius** warned that **human activities** were raising $CO_2$ levels in the air at a rate that **might cause future global warming**.

**‹‹ SEE ALSO**
pp.206–07 ATMOSPHERIC MOVEMENT
pp.208–09 PREDICTING THE WEATHER
pp.210–11 STRUCTURE OF THE ATMOSPHERE

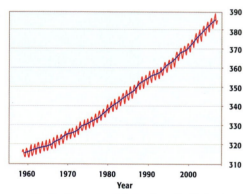

**The Keeling curve: rising carbon dioxide levels**
Keeling's graph shows that the level of carbon dioxide in the atmosphere has risen steadily since at least 1958. A small annual fluctuation (shown in red) occurs due to seasonal changes in carbon dioxide uptake by plants.

**GEOCHEMIST (1928–2005)**

## CHARLES DAVID KEELING

While working at the California Institute of Technology in the 1950s, American Charles Keeling developed a method for measuring levels of carbon dioxide ($CO_2$) in the air, which alerted the world to the relentless rise in atmospheric $CO_2$. He continued to measure atmospheric $CO_2$ for over 40 years. In 2002 he received the National Medal of Science, the highest award in the US for lifetime achievement in scientific research.

# Global Warming

**Rising carbon dioxide levels, caused partly by human emission of greenhouse gases, are causing an increase in the temperature of both the atmosphere and the oceans—a "global" warming. This is a massive environmental problem, which could have catastrophic consequences for the planet.**

The realization that carbon dioxide ($CO_2$) levels in the atmosphere are not only rising but might also cause disastrous warming first came to widespread scientific attention in the 1950s. By the late 1970s most climate scientists had come to accept the idea of the "greenhouse" effect (see below). The challenge that lay ahead was to ascertain the rate of current and potential future warming, and the consequences that this increase in temperature might have for the planet.

### Awareness of the greenhouse effect

In the 1950s studies of ocean chemistry made it clear that Earth's oceans would not be able to absorb as much atmospheric $CO_2$ as had once been thought. In 1958 American scientist Charles Keeling started to measure the levels of the gas in the atmosphere year by year. He found that levels were rising steadily, at the rate of a few percent per decade. By the mid-1970s, the graph had come to be known as Keeling's curve.

The so-called greenhouse effect had first been recognized in the mid 1800s. This phenomenon explains how gases such as $CO_2$ keep the atmosphere warm, but it soon became associated with a risk of overheating. Computer modeling in the 1970s allowed experts to predict what effect the rising greenhouse gas concentrations would have on Earth's temperature. The results showed a steady warming. Studies of temperature records going back for decades seemed to indicate

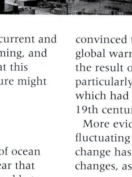

**Industrial emissions**
An acceleration in a long-term rising trend in global temperature coincided with increased industrial emissions of greenhouse gases in the 19th century, as seen here in Pittsburgh, PA, in the 1890s.

that this warming had already started, and that a distinct acceleration was occurring within a much longer-lived, more gradual warming trend. By the late 1970s most scientists became convinced that rising $CO_2$ levels and global warming were, at least partly, the result of human activities, particularly the burning of fossil fuels, which had increased rapidly in the 19th century.

More evidence for a link between fluctuating $CO_2$ levels and climate change has come from the study of past changes, as revealed, for example, by the analysis of samples of ice formed hundreds of thousands of years ago in the Antarctic. Measurements from air trapped in the ice can reveal what atmospheric $CO_2$ levels were, and other clues give an indication of air temperature around the time the ice originally formed. This information has shown that significant alterations in global temperatures have occurred in the past without human intervention, over surprisingly short time periods.

### Current warming

The latest estimates indicate that average global surface temperature will rise between 2° F (1.1° C) and 11.5° F (6.4° C) above current levels by the end of the 21st century. Uncertainty arises from different assumptions about the future emission of greenhouse gases. It has also been predicted that warming will not be spread evenly over the globe, but will be more pronounced in polar regions.

One of the main predicted effects of this warming is the melting of glaciers and ice sheets. Since the 1980s satellite

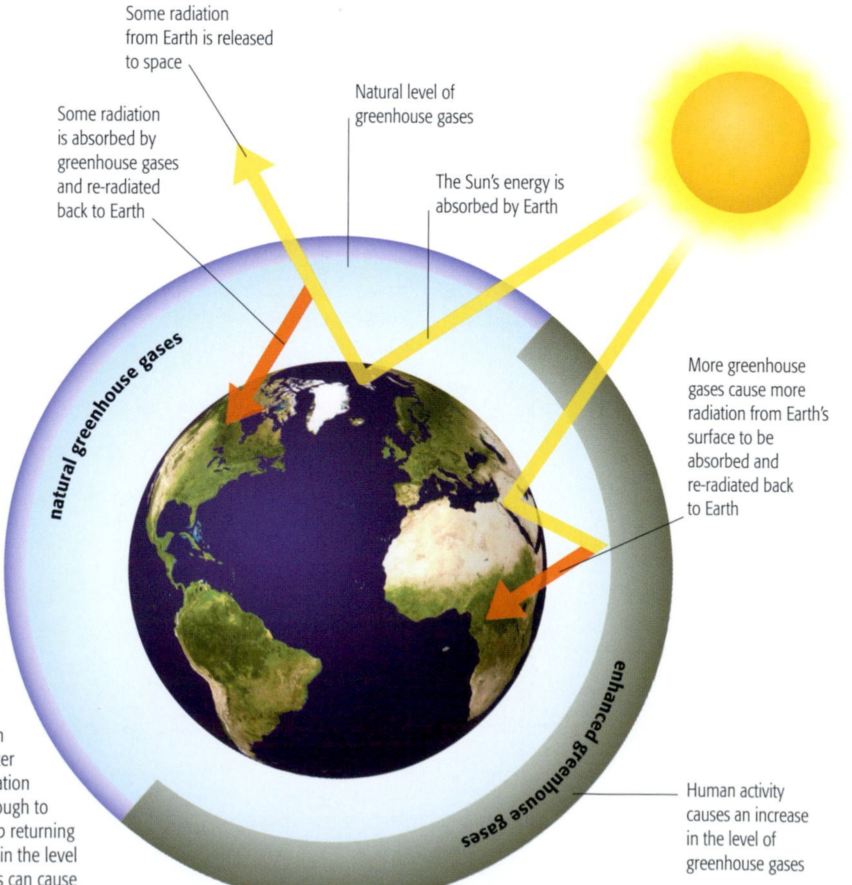

Some radiation from Earth is released to space

Natural level of greenhouse gases

Some radiation is absorbed by greenhouse gases and re-radiated back to Earth

The Sun's energy is absorbed by Earth

natural greenhouse gases

More greenhouse gases cause more radiation from Earth's surface to be absorbed and re-radiated back to Earth

enhanced greenhouse gases

Human activity causes an increase in the level of greenhouse gases

**Greenhouse effect**
Certain gases in the atmosphere, such as carbon dioxide, methane, and water vapor, allow short-wave radiation from the Sun to pass through to Earth's surface, but trap returning long-wave radiation. Any rise in the level of these greenhouse gases can cause overheating of the planet.

### Melting of glaciers

One of the effects of global warming is the melting of glaciers as seen here in this glacier in the Pyrenees. The ice was 33–49 ft (10–15 m) thick in 1933 (left) but had almost completely disappeared by 2004 (right).

**GLACIER DU LAC TOURRAT, FRANCE, 1933**

**GLACIER DU LAC TOURRAT, FRANCE, 2004**

observations have shown that this is already happening at a high rate—with a marked shrinking of glaciers in most parts of the world, including the Alps and Pyrenees in Europe (see above), the Himalayas in Asia (with increased risk of floods from glacier outbursts), and at the top of high mountains in the tropics, such as Kilimanjaro. In addition, an increased flow of glaciers and iceberg calving (the formation of an iceberg from a chunk of ice broken away from an ice shelf) have been observed in Greenland, several large ice shelves have disintegrated around the coast of Antarctica, and there has been a marked reduction in the extent of summer sea ice in the Arctic.

### Consequences of global warming

Global warming is predicted to raise sea levels worldwide (as past warmer periods in Earth's history have done), for two main reasons. First, ice melting from the Antarctic and Greenland ice sheets will raise the total amount of water held within the ocean basins. Second, as seawater warms, it expands slightly. Satellite measurements of Earth's surface indicate that sea level is currently rising at a rate of about $1/10$ in (3 mm) a year, but this is now expected to accelerate upward toward $1/3$ in (1 cm) a year. In 2009 the predicted rise in global sea level for all of the 21st century has been revised from around $1^{1}/_3$ ft (40 cm) upward to closer to 3 ft (1 m). This scale of rise is likely to displace tens of millions of people from low-lying coastal areas and islands, as well as cause more frequent and extensive floods.

Small changes in temperature can have a dramatic effect on climate. Future global warming is anticipated to increase the number and severity of extreme weather events such as storms and hurricanes, but in some areas may cause more frequent droughts with consequential effects on agricultural yields and an increased risk of famine. Some of the increased atmospheric $CO_2$ will be absorbed by the oceans. However, as $CO_2$ dissolves in water it slightly acidifies it, which may have consequences for many marine organisms—some may have problems making their calcium carbonate shells in a more acidic ocean. Other anticipated consequences of global warming include increased extinctions, and shifts in biomes (major ecological communities; see p.335), with marked differences between organisms that quickly adapt to change, and those that cannot respond quickly except in a negative way by dying off.

---

**The evidence that global warming is happening and, arguably, accelerating has been getting stronger for many years.**

#### DROUGHT AND BUSH FIRES

A government-commissioned report in Australia in 2008 warned that **climate change** could make **drought,** with accompanying **bush fires,** a near-permanent feature of the Australian environment.

**BUSH FIRES, QUEENSLAND, AUSTRALIA**

#### METHANE BURSTS

In December 2008, it was announced that large releases or "bursts" of **methane,** a greenhouse gas, have been occurring in the Arctic Tundra. These releases, of millions of tons of methane, have raised **fears of runaway climate change**.

**SEE ALSO »**
pp.416–17 RENEWABLE ENERGY
pp.418–19 TACKLING CLIMATE CHANGE

### Drought-affected land

During a severe drought that affected China in 2007, a farmer takes water from a dried-up pond to water his field near Yingtan, Jiangxi Province. Climate experts predict that global warming will increase the frequency of such droughts in the future

## « BEFORE

For centuries, renewable energy, such as water and wind, was the main or only form of energy that was available.

**EARLY WIND AND WATER POWER**
Early **windmills**, which had blades rotating around a vertical axis, were similar in structure to some in use today. They were first developed in **Persia and China** between the 6th and 10th centuries BCE. The wind power was used for grinding grain and pumping water **« 50–51**.

**DUTCH WINDMILLS**

First used by ancient Greeks, **water wheels** were **widely used in Europe** by the 16th century for grinding grain, sawing wood, raising water, and operating textile mills. Early versions of today's tidal power plants, called **tide mills**, also operated on coasts in medieval Europe.

**FIRST RENEWABLE ENERGY**
Some of the earliest large-scale **renewable energy** power plants started up around the **turn of the 20th century**. The world's first

**hydroelectric plant** began operating in Appleton, Wisconsin, in 1882: a water wheel on the Fox river generated hydroelectric power for two paper mills and a house. The **first geothermal generator** opened at the Larderello dry steam field in Italy in 1904 and in 1911, the first **geothermal power plant** was built there.

**« SEE ALSO**
*pp.50–51* Water and Wind Power
*pp.414–15* Global Warming

# Renewable Energy

**Renewable energy is generated from sources that are naturally replenished, such as wind, sunlight, rain, and geothermal heat. It also includes energy derived from recently dead plant material. With concerns about the rate of energy demand and global warming, the role of renewables is becoming increasingly significant.**

Renewables generate less carbon dioxide ($CO_2$) than fossil fuels. Using them is potentially a means of combating global warming. However, renewable energy facilities can be expensive to establish, and a reduction in $CO_2$ emissions may take years. Some forms provide only intermittent power (wind power, for example). The challenge now is to increase the amount of renewable energy produced and to get more out of that energy.

## Hydroelectric and wind power
Hydroelectric power is one of the most important forms of renewable energy in terms of its global contribution. It

harnesses the energy of flowing water using turbines, and is easily regulated. However, plant sites can be disruptive to aquatic ecosystems, upstream and

downstream. Also, potential sites for dams are already fully exploited in some areas. Although wind power currently provides less than 1 percent of global energy needs, it could provide much more, and it is one of the fastest-growing forms of renewable energy. To harness wind power fully would require

**Harnessing wind power**
With wind power production growing at a rate of over 20 percent a year, arrays of wind turbines such as these at Yucca Valley, California, are likely to become increasingly familiar.

**CONTRIBUTIONS TO WORLD ENERGY NEEDS**
- Coal 25.5%
- Oil 32%
- Gas 21%
- Renewables and nuclear energy 21.5% (see below)

**CONTRIBUTIONS FROM RENEWABLES AND NUCLEAR ENERGY**
- Hydro 26.5% (5.7% of whole)
- Biomass (wood and waste crops) 41% (8.8% of whole)
- Nuclear 23% (4.9% of whole)
- Other (renewable) 9.5% (2% of whole)

**CONTRIBUTIONS FROM OTHER RENEWABLES**
- Solar PV (photovoltaics) 3.2% (0.06% of whole)
- Other solar 35.4% (0.7% of whole)
- Geothermal 19.2% (0.4% of whole)
- Ocean power 0.2% (0.0004% of whole)
- Biofuels 13% (0.3% of whole)
- Wind 29% (0.6% of whole)

**Worldwide energy usage**
Renewables and nuclear energy supplied about 21.5 percent of global needs (top) in 2007. Of that, hydroelectric, nuclear, and biomass contributed the most (middle), with others making smaller contributions (bottom).

**Svatsengi geothermal power plant**
In Iceland, geothermal plants such as Svatsengi provide heating and hot water for 87 percent of homes and 26 percent of the nation's electricity. Most of the rest comes from hydroelectricity.

## BREAKTHROUGH

### PHOTOVOLTAIC CELLS

A photovoltaic, or solar, cell converts light from the Sun into electricity, based on the behavior of certain semiconductor substances, such as silicon and cadmium telluride, when exposed to sunlight. The main breakthrough occurred in 1954, when scientists at Bell Laboratories in the US accidentally discovered that silicon doped with certain impurities was sensitive to light. Within a year they had produced the first devices that could transform sunlight directly into power.

### Wave power

The world's first commercial wave farm is the Aguçadoura Wave Park off the coast of Portugal. This uses three Pelamis machines to convert the up-and-down movement of waves into electricity. With expansion to 28 units, the farm is expected to generate up to 21 megawatts, enough to power more than 10,000 homes.

huge areas of land and shallow sea areas to be covered in wind turbines. Currently, only the sites with reliable strong winds are considered in order to maintain a constant supply of energy. The expense of installing wind turbines is also currently high.

### Solar and geothermal power

The amount of solar energy reaching Earth's surface is vast, so this resource has perhaps the greatest potential. Solar energy collectors include photovoltaics (solar panels, see above), devices that concentrate solar radiation using mirrors or lenses, solar water heaters, and solar ovens. Once installed, they are cheap to run, and are quiet and unobtrusive. Disadvantages include a high initial set-up cost and an intermittent power supply, depending on the strength of the sunlight available.

Geothermal power plants exploit temperature differences between the Earth's surface and "hot spots" a few miles (kilometers) beneath. But, these exist only in a few places such as Iceland, parts of Japan, New Zealand, and California. Water is pumped at high pressure down a borehole into the hot zone, and then forced out of a second borehole. The hot water is converted into electricity by a steam turbine, or used directly to heat homes.

### Ocean power

Another form of renewable energy includes the generation of electricity from tidal currents, wave motion, or salinity gradient power, which exploits difference in salt levels between seas and rivers. At present, ocean power makes only a tiny contribution to total energy supply, but has great potential. Tidal power generation is predictable by nature, but is expensive to set up.

### Biomass and biofuels

Fuels such as wood and waste crops, which are burned to provide heat energy are known as biomass. These can be used to produce steam for making electricity, or to provide heat for homes. Burning wood is a major source of energy in the developing world. Biofuels—bioethanol, biodiesel, and other combustible substances derived from plant materials—can also be burned to produce heat energy or used directly as fuel for cars.

The combustion of biomass and biofuels releases $CO_2$ into the atmosphere, but, since the plants from which they are derived have absorbed

**1** The number of hours of sunlight needed, if harnessed effectively, to fulfil the energy demands of the world for a year.

Politicians and scientists agree on the need to step up renewable energy development for use in the future.

### OBAMA PLEDGE

In 2008 US President-elect Barack Obama pledged to spend **$150 billion over the next ten years** to support renewable energy. But the International Energy Agency projects that $26 trillion will be necessary to meet needs.

### OCEAN THERMAL ENERGY PLANT

There are plans to build a 10 megawatt **Ocean Thermal Energy Conversion** (OTEC) plant off Hawaii in 2013. This technology exploits the temperature difference between surface and deep ocean water to generate electricity.

### SOLAR UPDRAFT TOWERS

The **solar updraft tower** is a large structure like a greenhouse, built around the base of a tall chimney. Air, heated in the greenhouse by the Sun, rises up the chimney, **driving turbines for electricity**. One tower has been approved in Namibia and another is planned in Spain.

SEE ALSO ≫
pp.418–19 TACKLING CLIMATE CHANGE

$CO_2$ from the atmosphere when growing, overall their use is less polluting than the use of fossil fuels. One disadvantage of these fuels is the fact that land used for growing crops reduces the amount of land available for food crops. An additional drawback is that in some countries farmers are destroying rain forests to plant biofuel crops, which threatens wildlife and releases greenhouse gases by removing the trees that lock away $CO_2$.

# Tackling Climate Change

**To combat global warming and climate change, the rate at which carbon dioxide levels are increasing in the atmosphere needs to be reduced to zero, then will need to be reversed. People with ideas on how to achieve this difficult task fall into two main camps—there are those who favor novel "geo-engineering" solutions, and those who prefer more conventional approaches.**

To solve the problem of global warming, some believe that we need to start removing carbon dioxide ($CO_2$) from the atmosphere and locking it away (sequestration), while at the same time developing means of blocking sunlight from reaching Earth's surface—geo-engineering solutions. Others feel it is more important to concentrate on reducing the world's overall energy consumption and on switching from burning fossil fuels to renewable energy (see pp.416–17).

## Encouraging natural sequestration

Natural carbon "sinks" (places where carbon can naturally be stored long term) include forests, wood used as structural material, soil, and the oceans. So growing forests rather than destroying them, and using wood to build permanent structures rather than burning it, fall into this category. Public

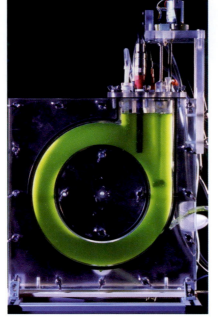

**Production of algae**
Culturing algae (in a scaled-up version of this device, left) could be used to capture carbon dioxide from the atmosphere and to produce biofuels (or, under certain conditions, hydrogen gas) for use as nonpolluting fuel.

awareness of the significance of $CO_2$ sinks has grown since the Kyoto Protocol in 1992, which promotes their use as a form of carbon offset.

One of the more radical suggestions for increasing natural carbon sequestration is to "fertilize" the oceans with iron (see right). Advocates say

> **CARBON SEQUESTRATION** The removal of carbon dioxide from the atmosphere followed by its long-term storage. Sequestration can either be natural, in "sinks" such as oceans or forests, or can involve artificial methods.

that it could solve half the climate change problem, and also might revive major fisheries and whale populations. But there is no certainty that it would work, and there are fears that it could harm marine life. Another idea, being investigated in Japan, is to farm huge amounts of a type of marine algae (see above), which makes small plates of calcium carbonate (limestone), absorbing $CO_2$ in the process. The algae could be harvested, compacted, and used as a building material.

## Artificial carbon sequestration

There are also methods of artificially capturing $CO_2$ (known as "carbon scrubbing") and storing it. $CO_2$ can be captured either from power stations or from anywhere in the atmosphere. This is expensive, but scientists in the US

are looking at how it might be achieved economically and, on a large scale, using "artificial trees." These would draw in air and pass it over a chemical that reacts with $CO_2$ to form a solution of sodium carbonate. This can later be converted back to $CO_2$ gas or into a solid salt, such as calcium carbonate, for storage. But many are unconvinced of the viability of this approach and feel it would be better just to plant more real trees. If the technology is adopted, the $CO_2$ or carbonate salt still has to be locked away. Possible methods include storing $CO_2$ as a gas in geological formations or as solid carbonate at the bottom of the oceans. It can also be converted naturally into minerals by peridotite rock. In theory, the peridotite rock in Oman alone could be used to absorb several billion tons of $CO_2$ every year. But, with all these methods, prime storage sites are likely to be used up quickly. Captured

> # "For each of our **actions** there are only **consequences**."
> JAMES LOVELOCK, ENVIRONMENTALIST (1919–)

**GEOCHEMIST (1931–)**

## WALLACE BROECKER

Professor at Columbia University, New York City, American Wallace Broecker is best known for his discovery of the role played by the oceans in triggering the sudden climate change in the past. He developed chemical tracer techniques for studying the behavior of carbon dioxide ($CO_2$) in the oceans and its interactions with the atmosphere. In 1975 he coined the term "global warming." A pioneer of geo-engineering, he believes in capturing and storing $CO_2$ as a possible solution.

**BREAKTHROUGH**

## OCEAN FERTILIZATION

The concept of ocean fertilization as a way of naturally sequestering (capturing and storing) carbon dioxide is based on the idea that there are areas of ocean where no phytoplankton currently grows because of a deficiency of iron in the seawater. If powdered iron were to be scattered over these areas, it could trigger massive plankton "blooms" (the light green areas in the ocean, right), which would absorb carbon dioxide from the atmosphere. Some of the plankton would clump together, forming heavy masses that would sink to the bottom of the ocean. Reaching there, some of it would become buried in sediments and ultimately be turned into rock.

**BEFORE**

The idea that humanity needs to stabilize the effect its activities are having on global climate goes back some 40 years.

**FEARS ABOUT FOSSIL FUELS**
Worries about use of fossil fuels started back in the 1970s, although initially these fears were centered on the **soaring price of crude oil** and **depletion of oil reserves**, rather than on the increase in carbon emissions and the effect that this might have on the climate.

**TANK FARM AT OIL REFINERY**

**EMISSIONS TRADING**
Administrative approaches such as **placing caps on greenhouse gas emissions** and international emissions trading were first introduced during the 1990s.

**SEE ALSO**
pp.414–15 GLOBAL WARMING
pp.416–17 RENEWABLE ENERGY

CO₂ might also be turned into fuel, using photosynthesizing devices known as "artificial leaves." Researchers in Italy have developed a solar-powered device that can turn CO₂ and water into hydrocarbon fuel. Producing and then burning this fuel is "carbon neutral"—it neither adds nor removes CO₂ from the atmosphere.

## Reducing energy consumption
Whatever solutions might be devised for mitigating carbon emissions, reducing world energy consumption still has to be one of the main parts of any anti-global warming strategy. Achieving this depends to a large extent on individuals. People can make their homes more energy efficient (see below). But there are other aspects to energy-efficient living. Whether switching to telecommuting rather than traveling to work, from using cars to bicycles or public transportation, switching from eating meat to vegetables, and using home appliances in a more disciplined way, everyone can make a significant contribution.

> **1 MILLION** fewer tons of carbon dioxide would be emitted if just 10 percent of the US's and Europe's workforce telecommuted once a week

### Eco-friendly house
Many improvements can be made to homes to reduce energy usage. These may be financially more expensive initially but over the long term would pay off and help the environment.

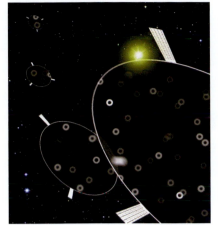

**Space sunshade**
A suggested method of partially blocking sunlight as a means of tackling global warming is depicted above. It involves stationing a "cloud" of tiny sunlight-diffusing disks in a stable orbit between Earth and the Sun.

Apart from solutions based on carbon sequestration (capturing and storage) and space sunshades, various other geo-engineering options have been suggested and may be employed in future.

### AEROSOLS
Tiny sulphate particles could be injected into the stratosphere at relatively low cost to **reflect the Sun's rays**. However, there would be a risk of ozone depletion and unknown consequences for the weather.

### REFLECTIVE CROPS
Some scientists believe that planting arable land with crop varieties whose **leaves reflect more light back** into the atmosphere could significantly cut regional summer temperatures.

### CLOUD SYNTHESIS
There are suggestions that **seawater could be atomized to create extra clouds**, which would reflect the Sun's rays. Again, there would be a risk of unknown weather effects.

**Wind power**
Wind turbines can produce electricity to supplement the supply from the local grid.

**Loft insulation**
A highly effective way of saving energy, loft insulation is particularly eco-friendly if it uses natural materials.

**Active solar power**
Solar energy collectors can be used to generate hot water and electricity.

**Rafter insulation**
In a lived-in roof space, ensuring that the ceiling is sufficiently insulated can be a key energy-saving measure.

**Green living roof**
This type of structure provides good insulation and a habitat for wildlife.

**Ground heating systems**
Various systems offer an alternative for space and water heating.

**Biomass boilers**
A carbon-neutral option that offers a viable alternative to conventional boilers.

**Rainwater storage**
Rainwater can be stored in tanks and used for purposes such as garden maintenance.

**Cavity wall insulation**
Insulating these walls can help reduce energy requirements further.

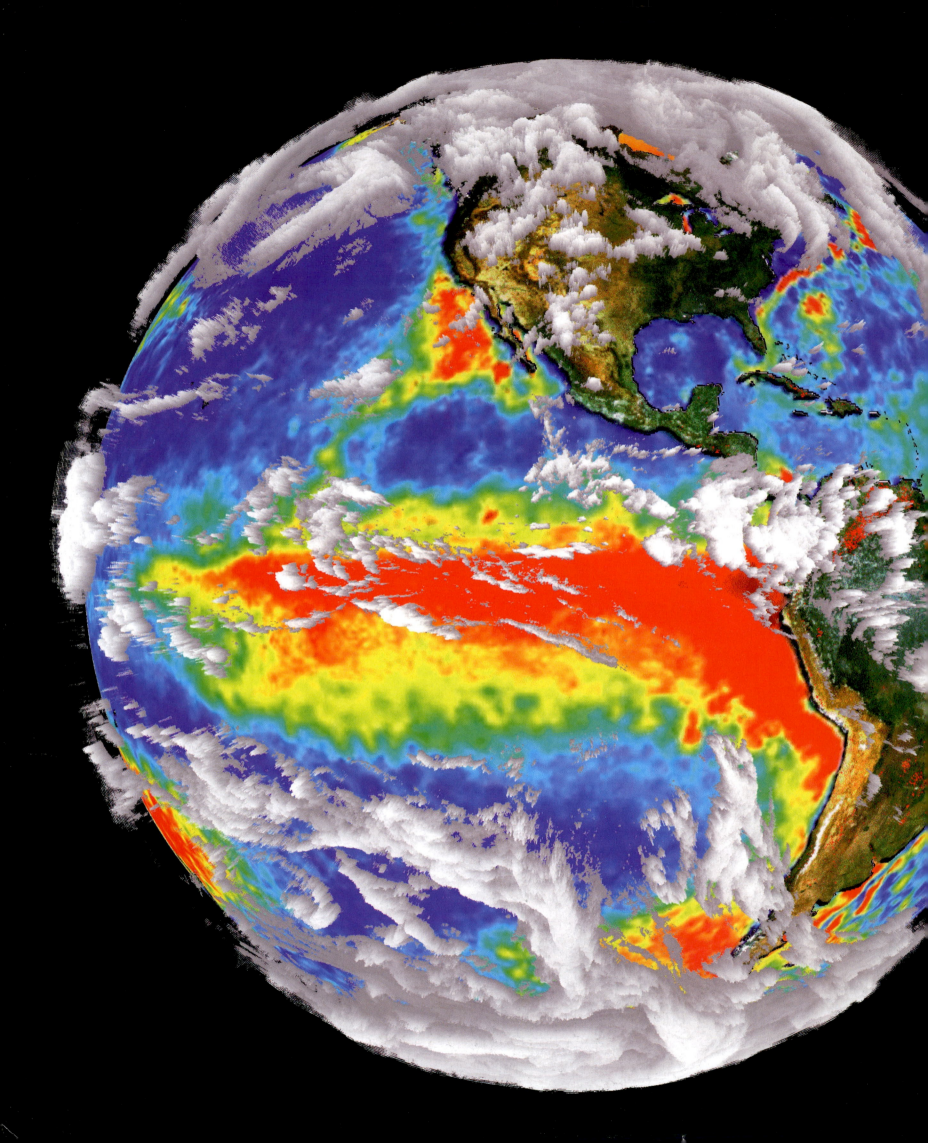

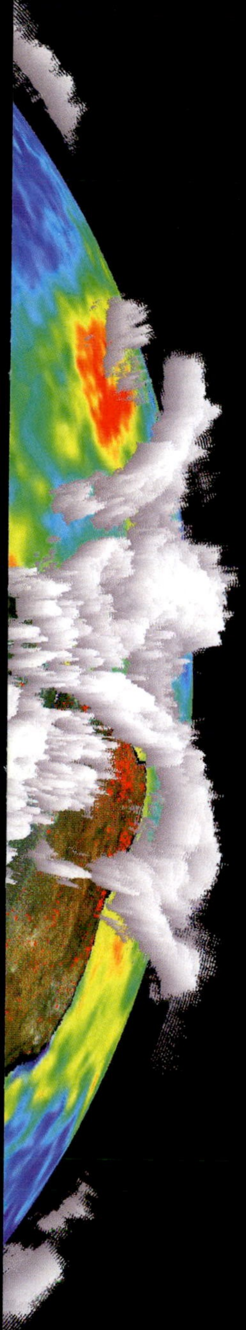

# REFERENCE

The following section provides quick access to facts and figures on the major fields of science covered in this book. Tables, laws, formulae, illustrations, and maps are included.

# Measurement

Historically, different people used independently formulated, and often imprecise, units of measurement—such as a grain of wheat. Standard measuring systems developed when commerce spread beyond local boundaries. International trade led to the introduction of units defining basic quantities, such as weight and length, that meant the same everywhere. Worldwide, metric measurement is now the most commonly used system.

## Base SI units

The SI (Système International d'Unités) metric system of measurement is founded on seven base units of independent physical quantities. All other physical units are obtained from these units.

| Unit | Symbol | Definition |
|---|---|---|
| meter | m | The meter is the length of a path traveled by light in a vacuum during a time interval of $1/299\,729\,458$ of a second. |
| kilogram | kg | The kilogram is the unit of mass equal to the mass of the international prototype kilogram. |
| second | s | The second is the duration of 9,192,631,770 periods of the radiation corresponding to the transition between the two hyperfine levels of the ground state of the cesium-133 atom. |

| Unit | Symbol | Definition |
|---|---|---|
| ampere | A | The ampere is that constant electric current which, if maintained in two straight parallel conductors of infinite length, of negligible cross section, and placed 1 meter apart in vacuum, would produce between these conductors a force equal to $2 \times 10^{-7}$ newton per meter. |
| kelvin | K | The kelvin, unit of thermodynamic temperature, is the fraction $1/273.16$ of the thermodynamic temperature of the triple point of water. |
| candela | cd | The candela is the luminous intensity, in a given direction, of a source that emits monochromatic radiation of frequency $540 \times 10^{-12}$ hertz and that has a radiant intensity in that direction of $1/683$ watt per steradian. |
| mole | mol | The mole is the amount of substance that contains as many elementary units as there are carbon atoms in 0.012 kilogram of carbon-12. |

## Supplementary and derived SI units

Two supplementary units are used in addition to the seven SI base units. Many other units, some examples of which are listed here, are derived from the SI base units using a system of quantity equations.

| Supplementary unit | Symbol | Definition |
|---|---|---|
| radian | rad | The unit of measurement of angle; it is the angle subtended at the center of a circle by an arc equal in length to the circle radius. |
| steradian | sr | The unit of measurement of solid angle; it is the solid angle subtended at the center of a circle by a spherical cap equal in area to the square of the circle radius. |

| Derived unit | Symbol | Definition |
|---|---|---|
| coulomb | C | A unit of electrical charge equal to the amount of charge transferred by a current of 1 ampere in 1 second. |
| farad | F | The capacitance of a capacitor that has an equal and opposite charge of 1 coulomb on each plate and a voltage difference of 1 volt between the plates. |
| henry | H | A unit of inductance in which an induced electromotive force of 1 volt is produced when the current is varied at the rate of 1 ampere per second. |
| hertz | Hz | The unit of frequency; 1 hertz has a periodic interval of 1 second. |
| joule | J | The energy exerted by a force of 1 newton acting to move an object through a distance of 1 meter. |
| lumen | lm | A unit of luminous flux equal to the amount of light given out through a solid angle of 1 steradian by a point source of 1 candela intensity radiating uniformly in all directions. |
| newton | N | A unit of force equal to the force that imparts an acceleration of 1 m/sec/sec to a mass of 1 kilogram. |
| ohm | Ω | A unit of electrical resistance equal to the resistance between two points on a conductor when a potential difference of 1 volt between them produces a current of 1 ampere. |
| pascal | Pa | A unit of pressure equal to 1 newton per square meter. |
| volt | V | A unit of potential equal to the potential difference between two points on a conductor carrying a current of 1 ampere when the power dissipated between the two points is 1 watt; equivalent to the potential difference across a resistance of 1 ohm when 1 ampere of current flows through it. |
| watt | W | A unit of power equal to 1 joule per second; the power dissipated by a current of 1 ampere flowing across a resistance of 1 ohm. |

## SI conversion factors

The table below gives the SI equivalents and conversion factors for units of measurement from other systems that are still in use. "SI equivalent" refers to one unit of the type named in the first column; the column headed "Reciprocal" gives the reverse conversion factors. For example: 1 yd = 0.914 m; 1 m = 1.093 yd.

| Unit | Symbol | Quantity | SI equivalent | SI unit | Reciprocal |
|---|---|---|---|---|---|
| acre | | area | 0.405 | hm² | 2.471 |
| ångström | Å | length | 0.1 | nm | 10 |
| astronomical unit | AU | length | 0.150 | Tm | 6.684 |
| atomic mass unit | amu | mass | $1.661 \times 10^{-27}$ | kg | $6.022 \times 10^{26}$ |
| bar | bar | pressure | 0.1 | MPa | 10 |
| barrel (US) = 42 US gal | bbl | volume | 0.159 | m³ | 6.290 |
| British thermal unit | btu | energy | 1.055 | kJ | 0.948 |
| calorie | cal | energy | 4.187 | J | 0.239 |
| cubic foot | cu ft | volume | 0.028 | m³ | 35.315 |
| cubic inch | cu in | volume | 16.387 | cm³ | 0.061 |
| cubic yard | cu yd | volume | 0.765 | m³ | 1.308 |
| curie | Ci | activity of radionuclide | 37 | GBq | 0.027 |
| degree = $1/90$ right angle | ° | plane angle | 0.0175 | rad | 57.296 |
| degree Celsius | °C | temperature | 1 | K | 1 |
| degree Fahrenheit | °F | temperature | 0.556 | K | 1.8 |
| degree Rankine | °R | temperature | 0.556 | K | 1.8 |
| dyne | dyn | force | 10 | μN | 0.1 |
| electronvolt | eV | energy | 0.160 | aJ | 6.241 |
| erg | erg | energy | 0.1 | μJ | 10 |
| fathom (6 ft) | | length | 1.829 | m | 0.547 |
| fermi | fm | length | 1 | fm | 1 |
| foot | ft | length | 30.48 | cm | 0.033 |
| foot per second | ft s⁻¹ | velocity | 0.305 | m s⁻¹ | 3.281 |
| | | | 1.097 | km h⁻¹ | 0.911 |
| gallon (UK) | gal | volume | 4.546 | dm³ | 0.220 |
| gallon (US) = 231cu in | gal | volume | 3.785 | dm³ | 0.264 |

**Continued »**

>> Continued

| Unit | Symbol | Quantity | SI equivalent | SI unit | Reciprocal |
|---|---|---|---|---|---|
| gauss | Gs, G | magnetic flux density | 100 | μT | 0.01 |
| grade = 0.01 rt angle | rt angle | plane angle | 0.0157 | rad | 63.662 |
| grain | gr | mass | 0.065 | g | 15.432 |
| hectare | ha | area | 1 | hm² | 1 |
| horsepower | hp | power | 0.746 | kW | 1.341 |
| inch | in | length | 2.54 | cm | 0.394 |
| kilogram-force | kgf | force | 9.807 | N | 0.102 |
| knot | | velocity | 1.852 | km h$^{-1}$ | 0.540 |
| light year | ly | length | $9.461 \times 10^{15}$ | m | $1.057 \times 10^{-16}$ |
| liter | l | volume | 1 | dm³ | 1 |
| Mach number | Ma | velocity | 1193.3 | km h$^{-1}$ | $8.380 \times 10^{-4}$ |
| maxwell | Mx | magnetic flux | 10 | nWb | 0.1 |
| micron | μ | length | 1 | μm | 1 |
| mile (nautical) | | length | 1.852 | km | 0.540 |
| mile (statute) | | length | 1.609 | km | 0.621 |
| miles per hour (mph) | mile h$^{-1}$ | velocity | 1.609 | km h$^{-1}$ | 0.621 |
| minute = $(1/60)°$ | ' | plane angle | $2.91 \times 10^{-4}$ | rad | 3437.75 |
| oersted | Oe | magnetic field strength | 0.079 | kA m$^{-1}$ | 12.6 |
| ounce (avoirdupois) | oz | mass | 28.349 | g | 0.035 |
| ounce (troy) = 480 gr | | mass | 31.103 | g | 0.032 |
| parsec | pc | length | 30857 | Tm | 0.0000324 |
| phot | ph | illuminance | 10 | klx | 0.1 |
| pint (UK) | pt | volume | 0.568 | dm³ | 1.760 |
| poise | P | viscosity | 0.1 | Pa s | 10 |
| pound | lb | mass | 0.454 | kg | 2.205 |
| pound force | lbf | force | 4.448 | N | 0.225 |
| pound force/in | | pressure | 6.895 | kPa | 0.145 |
| poundal | pdl | force | 0.138 | N | 7.233 |
| pounds per square inch | psi | pressure | $6.895 \times 10^3$ | kPa | 0.145 |
| rad | rad | absorbed dose | 0.01 | Gy | 100 |
| rem | rem | dose equivalent | 0.01 | Sv | 100 |
| right angle = $\pi/2$ rad | | plane angle | 1.571 | rad | 0.637 |
| röntgen | R | exposure | 0.258 | mC kg$^{-1}$ | 3.876 |
| second = $(1/60)'$ | " | plane angle | $4.85 \times 10^{-6}$ | mrad | $2.063 \times 10^{-5}$ |
| slug | | mass | 14.594 | kg | 0.068 |
| solar mass | M | mass | $1.989 \times 10^{30}$ | kg | $5.028 \times 10^{-31}$ |
| square foot | sq ft | area | 9.290 | dm² | 0.108 |
| square inch | sq in | area | 6.452 | cm² | 0.155 |
| square mile (statute) | sq mi | area | 2.590 | km² | 0.386 |
| square yard | sq yd | area | 0.836 | m² | 1.196 |
| standard atmosphere | atm | pressure | 0.101 | MPa | 9.869 |
| stere | st | volume | 1 | m³ | 1 |
| stilb | sb | luminance | 10 | kcd m$^{-2}$ | 0.1 |
| stokes | St | viscosity | 1 | cm² s$^{-1}$ | 1 |
| therm = $10^5$ btu | | energy | 0.105 | GJ | 9.478 |
| ton = 2240 lb | | mass | 1.016 | Mg | 0.984 |
| ton-force | tonf | force | 9.964 | kN | 0.100 |
| ton-force/sq in | | pressure | 15.444 | MPa | 0.065 |
| tonne | t | mass | 1 | Mg | 1 |
| torr, or mmHg | torr | pressure | 0.133 | kPa | 7.501 |
| X unit | | length | 0.100 | pm | 10 |
| yard | yd | length | 0.914 | m | 1.093 |

## SI prefixes

| Factor | Prefix | Symbol | Factor | Prefix | Symbol | Factor | Prefix | Symbol |
|---|---|---|---|---|---|---|---|---|
| $10^{18}$ | exa- | E | $10^2$ | hecto- | h | $10^{-9}$ | nano- | n |
| $10^{15}$ | peta- | P | $10^1$ | deca- | da | $10^{-12}$ | pico- | p |
| $10^{12}$ | tera- | T | $10^{-1}$ | deci- | d | $10^{-15}$ | femto- | f |
| $10^9$ | giga- | G | $10^{-2}$ | centi- | c | $10^{-18}$ | atto- | a |
| $10^6$ | mega- | M | $10^{-3}$ | milli- | m | | | |
| $10^3$ | kilo- | k | $10^{-6}$ | micro- | μ | | | |

## Converting common measures

### IMPERIAL TO METRIC

| Length | Multiply by |
|---|---|
| inches→millimeters | 25.4 |
| inches→centimeters | 2.54 |
| feet→meters | 0.3048 |
| yards→meters | 0.9144 |
| statute miles→kilometers | 1.6093 |
| nautical miles→kilometers | 1.852 |

| Area | Multiply by |
|---|---|
| square inches→square centimeters | 6.4516 |
| square feet→square meters | 0.0929 |
| square yards→square meters | 0.8361 |
| acres→hectares | 0.4047 |
| square miles→square kilometers | 2.5899 |

| Volume | Multiply by |
|---|---|
| cubic inches→cubic centimeters | 16.3871 |
| cubic feet→cubic meters | 0.0283 |
| cubic yards→cubic meters | 0.7646 |

| Capacity | Multiply by |
|---|---|
| UK fluid ounces→liters | 0.0284 |
| US fluid ounces→liters | 0.0296 |
| UK pints→liters | 0.5682 |
| US pints→liters | 0.4732 |
| UK gallons→liters | 4.546 |
| US gallons→liters | 3.7854 |

| Weight | Multiply by |
|---|---|
| ounces (avoirdupois)→grams | 28.3495 |
| ounces (troy)→grams | 31.1035 |
| pounds→ kilograms | 0.4536 |
| tons (long)→tonnes | 1.016 |

### METRIC TO IMPERIAL

| Length | Multiply by |
|---|---|
| millimeters→inches | 0.0394 |
| centimeters→inches | 0.3937 |
| meters→feet | 3.2806 |
| meters→yards | 1.0936 |
| kilometers→statute miles | 0.6214 |
| kilometers→nautical miles | 0.54 |

| Area | Multiply by |
|---|---|
| square centimeters→square inches | 0.155 |
| square meters→square feet | 10.764 |
| square meters→square yards | 1.196 |
| hectares→acres | 2.471 |
| square kilometers→square miles | 0.386 |

| Volume | Multiply by |
|---|---|
| cubic centimeters→cubic inches | 0.061 |
| cubic meters→cubic feet | 35.315 |
| cubic meters→cubic yards | 1.308 |

| Capacity | Multiply by |
|---|---|
| liters→UK fluid ounces | 35.1961 |
| liters→US fluid ounces | 33.8150 |
| liters→UK pints | 1.7598 |
| liters→US pints | 2.1134 |
| liters→UK gallons | 0.2199 |
| liters→US gallons | 0.2642 |

| Weight | Multiply by |
|---|---|
| grams→ounces (avoirdupois) | 0.0353 |
| grams→ounces (troy) | 0.0322 |
| kilograms→ pounds | 2.2046 |
| tonnes→tons (long) | 0.9842 |

### TEMPERATURE

| To convert | Equation |
|---|---|
| °Fahrenheit→°Celsius | −32, × 5, ÷ 9 |
| °Celsius→°Fahrenheit | × 9, ÷ 5, + 32 |
| Kelvin→°Celsius | − 273.16 |
| °Celsius→Kelvin | + 273.16 |

# Astronomy and Space
## The Solar System

The expanse of space known as the solar system contains the Sun together with the planets and their attendant moons, and countless rocky objects—such as asteroids and comets—that fall within its gravitational influence (see pp.182–83). This immense area measures 9,000 billion miles (15,000 billion km) in diameter and is divided roughly into three regions: the inner and outer planet groups and, in the farthest reaches, a vast comet cloud.

**SUN**

## The planets

Eight planets orbit the Sun. In the inner region of the solar system there are four rocky planets—Mercury, Venus, Earth, and Mars—each comprising a metallic core and a rocky mantle and crust. The four planets of the outer region are the gas giants Jupiter, Saturn, Uranus, and Neptune, which have rocky cores surrounded by liquid gases. A ninth planet, Pluto, on the outer reaches of the solar system, was reclassified as a dwarf planet in 2006.

| Planet | MERCURY | VENUS | EARTH | MARS | JUPITER | SATURN | URANUS | NEPTUNE |
|---|---|---|---|---|---|---|---|---|
| **Distance from Sun** millions of miles (km) | 36.0 (57.9) | 67.2 (108.2) | 93 (149.6) | 141.5 (227.9) | 483.3 (778.3) | 886 (1,427) | 1,782 (2,870) | 2,774 (4,497) |
| **Diameter at equator** miles (km) | 3,033 (4,879) | 7,523 (12,104) | 7,928 (12,756) | 4,222 (6,786) | 88,784 (142,984) | 74,914 (120,536) | 31,770 (51,118) | 30,757 (49,528) |
| **Mass** (Earth = 1) | 0.06 | 0.82 | 1 | 0.107 | 318 | 95 | 14.5 | 17 |
| **Volume** (Earth = 1) | 0.056 | 0.86 | 1 | 0.15 | 1,319 | 744 | 67 | 57 |
| **Surface temperature** °F (°C) | −356 to +800 (−180 to +430) | +896 (+480) | −158 to +133 (−70 to +55) | −248 to +77 (−120 to +25) | −238 (−150) | −292 (−180) | −353 (−214) | −364 (−220) |
| **Surface gravity** (Earth = 1) | 0.38 | 0.9 | 1 | 0.38 | 2.64 | 0.925 | 0.79 | 1.12 |
| **Time to orbit Sun** ("year") | 87.97 days | 224.7 days | 365.26 days | 686.98 days | 11.86 years | 29.46 yrs | 84.01 yrs | 164.8 yrs |
| **Time to turn 360°** ("day") | 58.65 days | 243.01 days | 23 h 56 m 4 s | 24 h 37 m 23 s | 9 h 55 m 30 s | 10 h 39 m | 17 h 14 m | 16 h 7 m |
| **Orbital speed** miles/s (km/s) | 29.7 (47.9) | 21.8 (35) | 18.5 (29.8) | 15 (24.1) | 8.1 (13.1) | 6 (9.6) | 4.2 (6.8) | 3.4 (5.4) |
| **Number of observed moons** | 0 | 0 | 1 | 2 | 63 | 60 | 27 | 13 |

## Planetary formation

The planets began to form about 4.6 billion years ago, each developing from a ring of dust and gases moving around the Sun (see pp.182–83). Rock and metallic particles clumped together to form the inner planets. In the colder outer areas of the spinning solar material, ice combined with solid debris to form the cores of the gas giants, which had enough gravity to attract large amounts of gas. The following data are based on one particular calculation of planetary formation.

| Planet | Made from | Mass of ring (Earth = 1) | Planet's present mass (Earth = 1) | Time to form in years |
|---|---|---|---|---|
| MERCURY | Rock, metal | 30 | 0.06 | 80,000 |
| VENUS | Rock, metal | 160 | 0.82 | 40,000 |
| EARTH | Rock, metal | 200 | 1.00 | 110,000 |
| MARS | Rock, metal | 200 | 0.11 | 200,000 |
| JUPITER | Rock, metal, snow, gas | 4,000 | 318.00 | 1 million |
| SATURN | Rock, metal, snow, gas | 400 | 95.16 | 9 million |
| URANUS | Rock, metal, snow, gas | 80 | 14.54 | 300 million |
| NEPTUNE | Rock, metal, snow, gas | 100 | 17.15 | 1 billion |

## Main Belt asteroids

Asteroids are small, mostly irregularly shaped bodies that orbit the Sun. The majority, around a billion, congregate in the Main Belt, an area between Mars and Jupiter. Most are only a few kilometers across. Ceres, the first asteroid to be discovered, in 1801, is now classified as a dwarf planet. Listed below are some asteroids that have been imaged in detail.

| Name | Size miles (km) | Orbital period (years) |
|---|---|---|
| Vesta | Diameter: 348 (560) | 3.63 |
| Mathilde | Length: 41 (66) | 4.31 |
| Ida | Length: 37 (60) | 4.84 |
| Eros | Length: 19.25 (31) | 1.76 |
| Gaspra | Length: 11.2 (18) | 3.29 |
| Annefrank | Length: 3.7 (6) | 3.29 |

# Kepler's laws of planetary motion

Formulated by German astronomer Johannes Kepler, these laws show that the orbits of the planets are elliptical, not circular, and that orbital speed is slower the further away a planet is from the Sun (see pp.76–78).

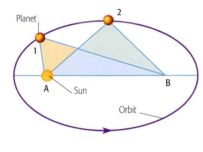

### First law
This states that a planet orbits the Sun in a path known as an ellipse, with the Sun at one focus. There are two focuses within an ellipse (**A** and **B**). On any particular ellipse, the total distance from one focus (**A**) to any point on the ellipse (**1** or **2**) and back to the other focus (**B**) is always the same.

### Second law
This describes how the speed of a planet changes as it orbits the Sun. The law states that a line from the Sun to a planet sweeps out equal areas in equal periods of time. This means that the planet moves faster when closer to the Sun and slower when farther away.

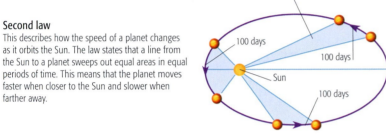

### Third law
Kepler's third law describes the mathematical relationship between the distances of planets from the Sun and their orbital periods. It states that the square of each planet's orbital period is directly proportional to the cube of its average distance from the Sun. This allows orbital period and distance to be calculated accurately for all planets.

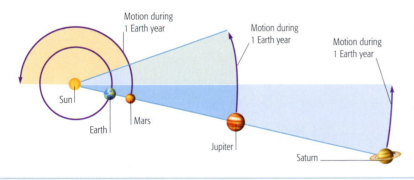

# Meteor showers

Meteors, the dust and debris from comets or asteroids, constantly enter Earth's upper atmosphere (see pp.118–19). As Earth orbits, it passes through these meteor streams or showers, which may be visible as trails of light. Some showers occur at fixed times of year and have been named after the constellations from which they appear to radiate.

| Name | Date | Constellation |
|---|---|---|
| Quadrantids | Jan 1–6 | Boötes |
| April Lyrids | April 19–24 | Lyra |
| Eta Aquarids | May 1–8 | Aquarius |
| Delta Aquarids | July 15–Aug 15 | Aquarius |
| Perseids | July 25–Aug 18 | Perseus |
| Orionids | Oct 16–27 | Orion |
| Taurids | Oct 20–Nov 30 | Taurus |
| Leonids | Nov 15–20 | Leo |
| Geminids | Dec 7–15 | Gemini |

# Main planetary moons

Besides Earth, most of the other solar system planets have their own orbiting moons. The exceptions are Mercury and Venus. Listed here are the major planetary moons. There are many others that have not been named.

| PLANET/ moons | Orbital period (days) | Distance from planet in miles (km) | | Diameter in miles (km) | |
|---|---|---|---|---|---|
| **EARTH** | | | | | |
| Moon | 27.32 | 238,000 | (384,000 ) | 2,155 | (3,476) |
| **MARS** | | | | | |
| Phobos | 0.31 | 582,700 | (937,800) | 17 | (27) |
| Deimos | 1.26 | 1,458,000 | (2,34,6000) | 9 | (15) |
| **JUPITER** | | | | | |
| Metis | 0.29 | 79,000 | (128,000) | 25 | (40) |
| Adrastea | 0.29 | 80,000 | (129,000) | 15 | (24) |
| Amalthea | 0.49 | 112,000 | (181,000) | 168 | (270) |
| Thebe | 0.67 | 138,000 | (222,000) | 60 | (100) |
| Io | 1.76 | 262,000 | (422,000) | 2,260 | (3,650) |
| Europa | 3.55 | 417,000 | (671,000) | 1,950 | (3,140) |
| Ganymede | 7.15 | 665,000 | (1,070,000) | 3,270 | (5,260) |
| Callisto | 16.68 | 1,170,000 | (1,883,000) | 3,000 | (4,800) |
| Leda | 238.72 | 6,900,000 | (11,100,000) | 12 | (20) |
| Himalia | 250.57 | 7,134,000 | (11,480,000) | 116 | (186) |
| Lysithea | 259.22 | 7,283,000 | (11,720,000) | 22 | (36) |
| Elara | 259.65 | 7,295,000 | (11,740,000) | 50 | (80) |
| Ananke | 631 | 13,174,000 | (21,200,000) | 19 | (30) |
| Carme | 692 | 14,044,000 | (22,600,000) | 25 | (40) |
| Pasiphae | 735 | 14,603,000 | (23,500,000) | 30 | (50) |
| Sinope | 758 | 14,727,000 | (23,700,000) | 22 | (36) |
| **SATURN** | | | | | |
| Pan | 0.57 | 83,000 | (134,000) | 6 | (10) |
| Atlas | 0.6 | 86,000 | (138,000) | 10 | (25) |
| Prometheus | 0.61 | 86,000 | (139,000) | 60 | (100) |
| Pandora | 0.62 | 88,000 | (142,000) | 60 | (100) |
| Epimetheus | 0.69 | 94,000 | (151,000) | 90 140 | (140) |
| Janus | 0.69 | 94,000 | (151,000) | 120 | (200) |
| Mimas | 0.94 | 116,000 | (186,000) | 240 | (390) |
| Enceladus | 1.37 | 148,000 | (238,000) | 310 | (500) |
| Calypso | 1.88 | 183,000 | (295,000) | 19 | (30) |
| Telesto | 1.88 | 183,000 | (295,000) | 19 | (30) |
| Tethys | 1.88 | 183,000 | (295,000) | 660 | (1,060) |
| Dione | 2.73 | 234,000 | (377,000) | 700 | (1,120) |
| Helene | 2.73 | 234,000 | (377,000) | 9 | (15) |
| Rhea | 4.51 | 327,000 | (527,000) | 950 | (1,530) |
| Titan | 15.94 | 759,000 | (1,222,000) | 3,200 | (5,150) |
| Hyperion | 21.27 | 920,000 | (1,481,000) | 300 | (480) |
| Iapetus | 14.72 | 2,212,000 | (3,560,000) | 910 | (1,460) |
| Phoebe | 550.48 | 8,047,000 | (12,950,000) | 137 | (220) |
| **URANUS** | | | | | |
| Miranda | 1.41 | 81,000 | (130,000) | 300 | (480) |
| Ariel | 2.52 | 119,000 | (191,000) | 720 | (1,160) |
| Umbriel | 4.14 | 165,000 | (266,000) | 730 | (1,170) |
| Titania | 8.7 | 271,000 | (436,000) | 980 | (1,580) |
| Oberon | 13.46 | 362,000 | (583,000) | 947 | (1,524) |
| **NEPTUNE** | | | | | |
| Triton | 5.87 | 221,000 | (355,000) | 1,681 | (2,705) |
| Nereid | 360.16 | 5,510,000 | (3,424,000) | 210 | (340) |

# Earth, Moon, and Sun

Earth is the third closest planet to the Sun after Mercury and Venus, and is the only planet known to support living organisms. It is orbited by one natural satellite, the Moon, which is one of the largest satellites in the solar system in comparison to its parent planet. Interactions between the Earth, Moon, and Sun are responsible for variations in Earth's atmosphere, night and day cycles, climate, seasons, and tidal patterns.

## Earth at a glance

Largest of the inner planets, Earth is unique in the solar system for having an oxygen-rich atmosphere, liquid as well as frozen surface water, and a crust of moving plates. Earth orbits the Sun at an average speed of 18.5 miles/s (29.8 km/s)—about 67,000 mph (108,000 kph).

**STRUCTURE**

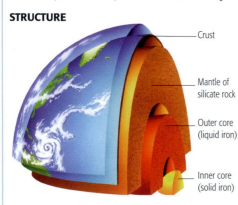

- Crust
- Mantle of silicate rock
- Outer core (liquid iron)
- Inner core (solid iron)

**SCALE**

Earth is about four times the size of its moon

**TILT, SPIN, AND ORBIT**

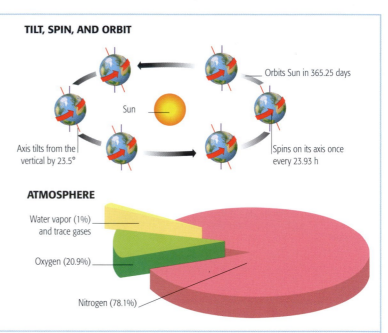

- Orbits Sun in 365.25 days
- Sun
- Axis tilts from the vertical by 23.5°
- Spins on its axis once every 23.93 h

**ATMOSPHERE**

- Water vapor (1%) and trace gases
- Oxygen (20.9%)
- Nitrogen (78.1%)

## Earth: vital statistics

| Diameter: miles (km) | 7,928 (12,756) |
|---|---|
| Average distance from Sun: miles (km) | 93 million (149.6 million) |
| Orbital speed around Sun: miles/s (km/s) | 18.5 (29.8) |
| Sunrise to sunrise | 23 hr 59m 4s |
| Mass | $5.98 \times 10^{24}$ kg |
| Volume | $1.08321 \times 10^{12}$ km³ |
| Average density (water = 1) | 5.52 |
| Surface gravity | 9.8m/s² |
| Average surface temperature | 59° F (15° C) |
| Number of moons | 1 |

## Earth's chemical composition

The chemical compounds that make up Earth occur in different proportions in each of its layers (see pp.352–53). Gravitational separation of material means that the deeper the layer is beneath the surface, the richer the concentrations of denser materials such as iron.

- Silicon dioxide
- Aluminium oxide
- Iron and iron oxides
- Calcium oxide
- Magnesium oxide
- Nickel oxide
- Others

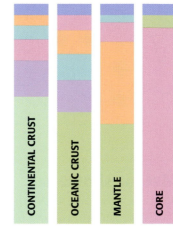

CONTINENTAL CRUST | OCEANIC CRUST | MANTLE | CORE

## Moon at a glance

The Moon is a barren, dusty sphere with a similar structure to that of the rocky planets. It has no atmosphere or liquid water. In the 27.32 days the Moon takes to orbit Earth, it makes a complete rotation on its axis.

**STRUCTURE**

- Crust of granitelike rock
- Rocky mantle
- Partly molten outer core
- Solid inner core

**TILT, SPIN, AND ORBIT**

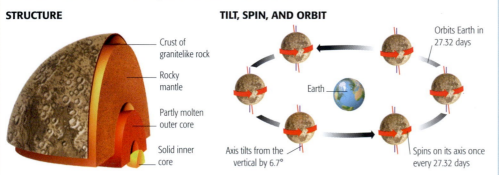

- Orbits Earth in 27.32 days
- Earth
- Axis tilts from the vertical by 6.7°
- Spins on its axis once every 27.32 days

## Moon: vital statistics

| Diameter: miles (km) | 2,160 (3,476) |
|---|---|
| Average distance from Earth: miles (km) | 238,900 (384,400) |
| Orbital speed around Earth: miles/s (km/s) | 0.63 (1.02) |
| New Moon to new Moon | 29.53 days |
| Mass (Earth = 1) | 0.01 |
| Volume (Earth = 1) | 0.02 |
| Average density (water = 1) | 3.34 |
| Surface gravity (Earth = 1) | 0.17 |
| Average surface temperature | -4° F (-20° C ) |

# Phases of the Moon

As the Moon orbits, the proportion of its Earth-facing hemisphere that is lit by the Sun changes in a cycle of phases. During the 29.5 days of the lunar month, the Moon's appearance ranges from a thin crescent to quarter, gibbous (more than half illuminated), and full face. When the Moon passes between the Sun and Earth, sunlight falls on its far side; the side facing Earth is in darkness and cannot be seen at all.

**CYCLE OF THE MOON'S PHASES**

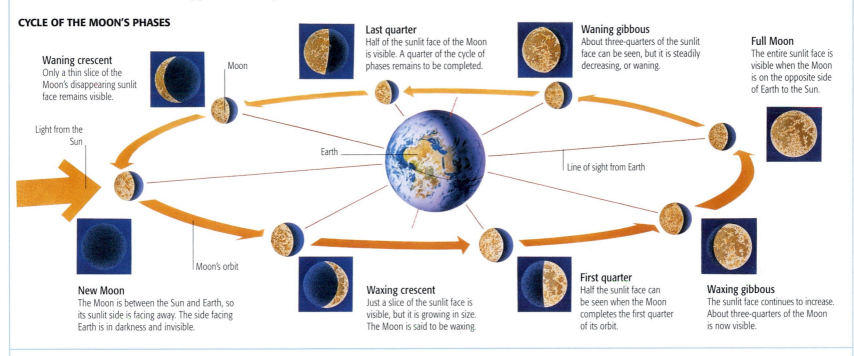

**Waning crescent**
Only a thin slice of the Moon's disappearing sunlit face remains visible.

**Moon**

**Last quarter**
Half of the sunlit face of the Moon is visible. A quarter of the cycle of phases remains to be completed.

**Waning gibbous**
About three-quarters of the sunlit face can be seen, but it is steadily decreasing, or waning.

**Full Moon**
The entire sunlit face is visible when the Moon is on the opposite side of Earth to the Sun.

**Light from the Sun**

**Earth**

**Line of sight from Earth**

**Moon's orbit**

**New Moon**
The Moon is between the Sun and Earth, so its sunlit side is facing away. The side facing Earth is in darkness and invisible.

**Waxing crescent**
Just a slice of the sunlit face is visible, but it is growing in size. The Moon is said to be waxing.

**First quarter**
Half the sunlit face can be seen when the Moon completes the first quarter of its orbit.

**Waxing gibbous**
The sunlit face continues to increase. About three-quarters of the Moon is now visible.

# Lunar eclipses 2010–15

A lunar eclipse occurs when the orbiting Moon passes through Earth's shadow and its brightness is dimmed to a varying degree (see p.79). In a penumbral eclipse, the Moon passes through the lighter, outer shadow; when part of the Moon passes through full shadow, that is a partial eclipse; in a total eclipse, the entire Moon passes through full shadow.

| Date | Type | Region of visibility | Date | Type | Region of visibility |
|---|---|---|---|---|---|
| June 26, 2010 | Partial | E. Asia, Australia, Pacific, Americas | May 25, 2013 | Penumbral | Americas, Africa |
| December 21, 2010 | Total | E. Asia, Australia, Pacific, Americas, Europe | October 18, 2013 | Penumbral | Americas, Europe, Africa, Asia |
| June 15, 2011 | Total | S. America, Europe, Africa, Asia, Australia | April 15, 2014 | Total | Australia, Pacific, Americas |
| December 10, 2011 | Total | Europe, E. Africa, Asia, Australia, Pacific, N. America | October 8, 2014 | Total | Asia, Australia, Pacific, Americas |
| June 4, 2012 | Partial | Asia, Australia, Pacific, Americas | April 4, 2015 | Total | Asia, Australia, Pacific, Americas |
| November 28, 2012 | Penumbral | Europe, E. Africa, Asia, Australia, Pacific, N. America | September 28, 2015 | Total | E. Pacific, Americas, Europe, Africa, W. Asia |
| April 25, 2013 | Partial | Europe, Africa, Asia, Australia | | | |

*Eclipse predictions by Fred Espenak, NASA/GSFC*

# The Sun: vital statistics

| | |
|---|---|
| **Mass (Earth = 1)** | 333,000 |
| **Surface temperature** | 9,900° F (5,500° C) |
| **Core temperature** | 27 million° F (15 million° C) |
| **Diameter: miles (km)** | 864,900 (1,392,000) |
| **Distance from Earth: miles (km)** | 93 million (149.6 million) |
| **Luminosity** | $3.9 \times 10^{26}$ megawatts |
| **Rotation period** | 25.4 days (at equator) |
| **Age** | 4.6 billion years |

# Solar composition

The Sun's outer layers are 73 percent hydrogen, 25 percent helium, and two percent other elements. In the superheated core, where more than 600 million tonnes of hydrogen are converted into helium every second, the amount of hydrogen is only about 34 percent, while the amount of helium is about 64 percent.

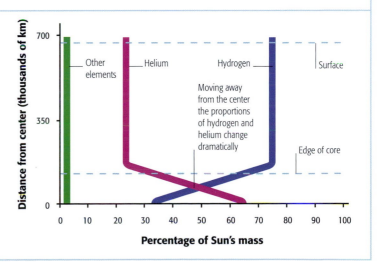

## Total solar eclipses 2010–20

In a total eclipse, the Moon completely obscures the Sun from viewers within a limited region (see pp.78–79). This region, known as the path of totality, can be predicted precisely.

| Date | Duration | Region of totality |
|---|---|---|
| July 11, 2010 | 5 minutes 20 seconds | S. Pacific, Easter Island, Chile, Argentina |
| November 13, 2012 | 4 minutes 2 seconds | N. Australia, S. Pacific |
| March 20, 2015 | 2 minutes 47 seconds | N. Atlantic, Faroe Islands, Svalbard |
| March 9, 2016 | 4 minutes 9 seconds | Sumatra, Borneo, Sulawesi, Pacific |
| August 21, 2017 | 2 minutes 40 seconds | N. Pacific, United States, S. Atlantic |
| July 2, 2019 | 4 minutes 33 seconds | S. Pacific, Chile, Argentina |
| December 14, 2020 | 2 minutes 10 seconds | S. Pacific, Chile, Argentina, S. Atlantic |

*Eclipse predictions by Fred Espenak, NASA/GSFC*

## Solar heating

Heat from the Sun in the form of infrared radiation warms Earth's atmosphere, oceans, and land. Because of Earth's curvature, solar heating is more intense in the tropical regions than at the poles.

# The stars

A star is a body formed when a cloud of dust and gas shrinks and collapses under its own gravity (see pp.330–31). The collapsed cloud heats up, setting off nuclear reactions that create energy and light. Stars are held together by gravity in groups called galaxies (see pp.340–41). Between galaxies lie the immensities of empty space. Our home galaxy, the Milky Way, is made up of 200–500 billion stars of varying types and brightness, of which the Sun is one.

## Magnitude of stars

Astronomers measure the luminosity, or brightness, of stars in magnitudes. The smaller the magnitude number, the brighter the star. The very brightest stars have negative magnitudes. Each step on the scale shown below represents an increase or decrease in brightness of 2.5 times. For comparison, the scale includes the planet Venus, which sometimes appears in the sky as a far brighter object than any star.

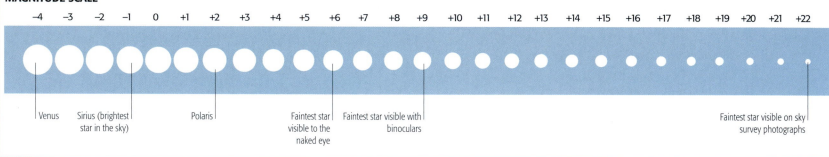

## Spectral types

When a star's light is split into different wavelengths, it produces a spectrum containing dark lines (absorption lines), the pattern of which reveals the star's composition, temperature, and color. Astronomers classify stars into seven spectral types: O, B, A, F, G, K, and M. O, blue-white, stars are the hottest; M, orange-red, the coolest. Each type has 10 subdivisions, numbered 0 to 9 (hotter to cooler). The Sun is type G2.

| Type O (40,000–29,000° C) | Type B (28,000–9,700° C) | Type A (9,600–7,200° C) | Type F (7,100–5,800° C) | Type G (5,700–4,700° C) | Type K (4,600–3,300° C) | Type M (3,200–2,100° C) |
|---|---|---|---|---|---|---|

# Average sizes of stars

There are enormous differences in the sizes of stars, which range from supergiants 300 times larger than the Sun to giant and dwarf stars. A star may end its life as a very small, dense neutron star, or as a black hole very much smaller than Earth.

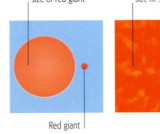

Supergiant: 10 times size of red giant

Red giant

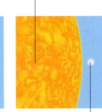

Red giant: 30 times size of Sun

Sun

Hydrogen-burning type B star: 7 times size of Sun

Sun

Hydrogen-burning type M star: 1/10 size of Sun

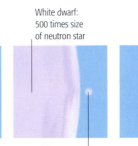

Sun: 100 times size of white dwarf

White dwarf

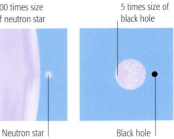

White dwarf: 500 times size of neutron star

Neutron star

Neutron star: 5 times size of black hole

Black hole

# Hertzsprung–Russell (H–R) diagram

Named after the astronomers who devised it, the H–R diagram is a graph that plots the temperature, luminosity, and size of stars (see pp.328–29). Most stars lie in a band called the main sequence, and are at roughly midlife. Older, cooler stars—red giants and supergiants—occupy the upper right side of the graph. Dying stars shrink to become tiny, hot white dwarfs and move to the bottom left of the graph.

## Main sequence stars

The main sequence runs diagonally from the brightest stars at top left to the faintest at bottom right. Main sequence stars burn hydrogen in nuclear reaction and convert it to helium. Stars spend about 90 percent of their lives on the main sequence, changing very little in luminosity or temperature while they are there. The Sun is a typical main sequence star.

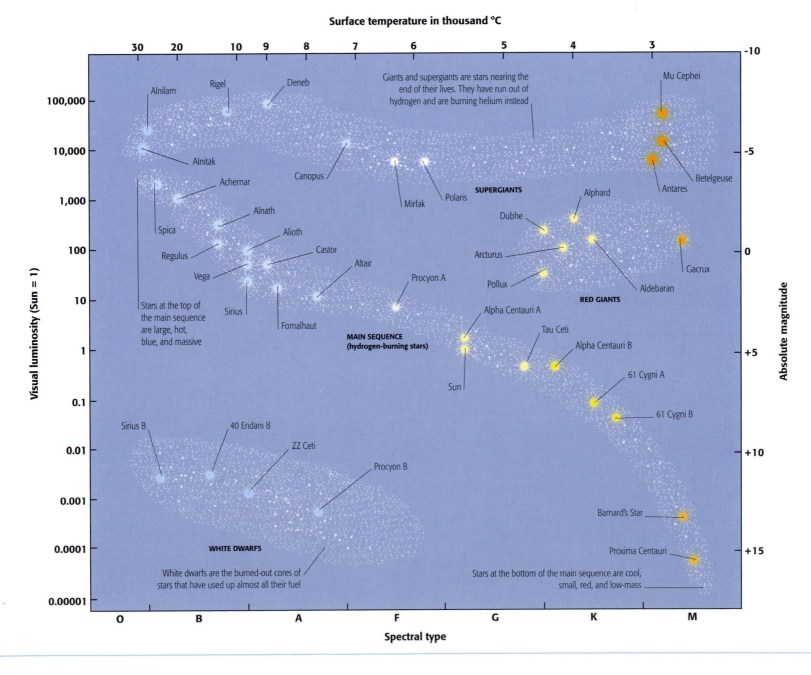

# Brightest stars

A star's brightness, or magnitude, is described in two ways. The apparent magnitude of a star is its brightness as seen from Earth. A star's absolute magnitude refers to the amount of light it actually emits. The stars listed below are shown in descending order of apparent magnitude.

| Name | Magnitude | | Distance from |
| | Apparent | Absolute | Sun (light years) |
| --- | --- | --- | --- |
| Sirius | -1.46 | +1.4 | 8.65 |
| Canopus | -0.73 | -4.6 | 1,200 |
| Alpha Centauri | -0.1 | +4.1 | 4.38 |
| Arcturus | -0.06 | -0.3 | 36 |
| Vega | +0.04 | +0.5 | 26 |
| Capella | +0.08 | -0.5 | 42 |
| Rigel | +0.10 | -7.0 | 900 |
| Procyon | +0.35 | +2.6 | 11.4 |
| Betelgeuse | +0.49 | -5.7 | 310 |
| Achernar | +0.51 | -2.5 | 117 |
| Hadar | +0.63 | -4.6 | 490 |
| Altair | +0.77 | +2.3 | 16 |
| Aldebaran | +0.85 | -0.7 | 69 |
| Acrux | +0.90 | -3.7 | 370 |
| Antares | +0.92 | -4.5 | 430 |
| Spica | +0.96 | -3.6 | 260 |
| Pollux | +1.15 | +1.0 | 35 |
| Fomalhaut | +1.16 | +1.9 | 23 |
| Deneb | +1.25 | -7.1 | 1,800 |
| Beta Crucis | +1.25 | -5.1 | 489 |
| Regulus | +1.35 | -0.7 | 85 |
| Adhara | +1.50 | -4.4 | 681 |

# Nearest stars and star systems

Stellar distances are measured in light years (the distance light travels in a year: 9,461 billion km). The following stars and systems (groups) of stars are among the Sun's nearest neighbors.

| Name of star or star system | Apparent magnitude | Spectral type | Distance (light years) |
| --- | --- | --- | --- |
| Alpha Centauri A, B, C | 0.1, 1.4, 11.0 | G2, K1, M5 | 4.4 |
| Barnard's Star | 9.5 | M5 | 5.9 |
| Lalande 21185 | 7.5 | M2 | 8.3 |
| Sirius A, B | -1.4, 8.5 | A0, white dwarf | 8.6 |
| Ross 154 | 10.4 | M4 | 9.7 |
| Epsilon Eridani | 3.7 | K2 | 10.5 |
| HD 217987 | 7.4 | M2 | 10.7 |
| Ross 128 | 11.1 | M4 | 10.9 |
| 61 Cygni A, B | 5.2, 6.1 | K5, K7 | 11.4 |
| Procyon | 0.4, 10.7 | F5, white dwarf | 11.4 |

# The constellations

Earlier peoples saw distinctive star patterns as mythical figures, and named them accordingly. Usually, the stars in a constellation are chance groupings and are not related. Currently, 88 constellations are listed, now recognized as regions of sky rather than as celestial images.

| Latin name | Common name | Latin name | Common name |
| --- | --- | --- | --- |
| Andromeda | Andromeda | Lacerta | Lizard |
| Antlia | Air Pump | Leo | Lion |
| Apus | Bird of Paradise | Leo Minor | Little Lion |
| Aquarius | Water Carrier | Lepus | Hare |
| Aquila | Eagle | Libra | Scales |
| Ara | Altar | Lupus | Wolf |
| Aries | Ram | Lynx | Lynx |
| Auriga | Charioteer | Lyra | Lyre |
| Boötes | Herdsman | Mensa | Table Mountain |
| Caelum | Chisel | Microscopium | Microscope |
| Camelopardalis | Giraffe | Monoceros | Unicorn |
| Cancer | Crab | Musca | Fly |
| Canes Venatici | Hunting Dogs | Norma | Level |
| Canis Major | Great Dog | Octans | Octant |
| Canis Minor | Little Dog | Ophiuchus | Serpent Bearer |
| Capricornus | Sea Goat | Orion | Orion, Hunter |
| Carina | Keel | Pavo | Peacock |
| Cassiopeia | Cassiopeia | Pegasus | Winged Horse |
| Centaurus | Centaur | Perseus | Perseus |
| Cepheus | Cepheus | Phoenix | Phoenix |
| Cetus | Whale | Pictor | Painter's Easel |
| Chamaeleon | Chameleon | Pisces | Fishes |
| Circinus | Compasses | Piscis Austrinus | Southern Fish |
| Columba | Dove | Puppis | Poop Deck |
| Coma Berenices | Berenice's Hair | Pyxis | Mariner's Compass |
| Corona Australis | Southern Crown | Reticulum | Net |
| Corona Borealis | Northern Crown | Sagitta | Arrow |
| Corvus | Crow | Sagittarius | Archer |
| Crater | Cup | Scorpius | Scorpion |
| Crux | Southern Cross | Sculptor | Sculptor |
| Cygnus | Swan | Scutum | Shield |
| Delphinus | Dolphin | Serpens | Serpent |
| Dorado | Swordfish | Sextans | Sextant |
| Draco | Dragon | Taurus | Bull |
| Equuleus | Foal | Telescopium | Telescope |
| Eridanus | River | Triangulum | Triangle |
| Fornax | Furnace | Triangulum Australe | Southern Triangle |
| Gemini | Twins | Tucana | Toucan |
| Grus | Crane | Ursa Major | Great Bear |
| Hercules | Hercules | Ursa Minor | Little Bear |
| Horologium | Clock | Vela | Sails |
| Hydra | Water Snake | Virgo | Virgin |
| Hydrus | Little Water Snake | Volans | Flying Fish |
| Indus | Indian | Vulpecula | Fox |

# Star maps

Around 6,000 stars are visible to the naked eye. These star maps show the main constellations as viewed from the northern and southern hemispheres. As the earth rotates, some stars appear to rise and set in the sky. Depending on the viewer's latitude, certain constellations may be observable only at certain times of year. The stars around the poles are circumpolar—they never set and are visible year round.

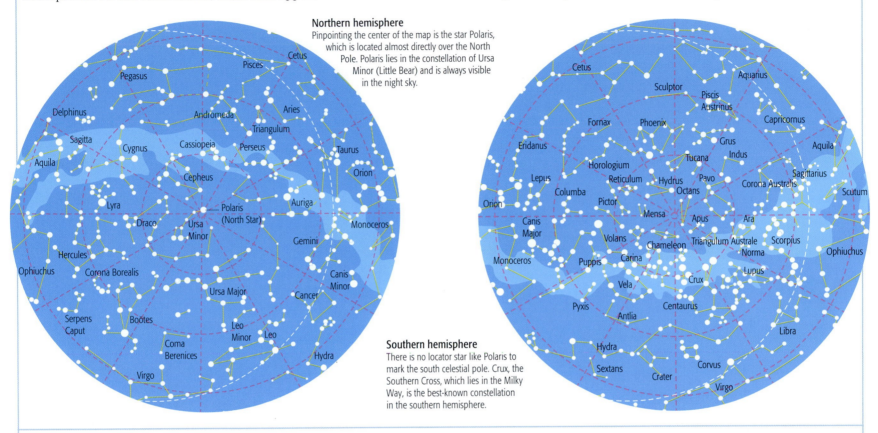

**Northern hemisphere**
Pinpointing the center of the map is the star Polaris, which is located almost directly over the North Pole. Polaris lies in the constellation of Ursa Minor (Little Bear) and is always visible in the night sky.

**Southern hemisphere**
There is no locator star like Polaris to mark the south celestial pole. Crux, the Southern Cross, which lies in the Milky Way, is the best-known constellation in the southern hemisphere.

# Local Group galaxies

The Local Group is a cluster of more than 30 galaxies that includes our home galaxy, the Milky Way (see p.341). Although the members of the group are scattered over a region of space measuring approximately 3 million light years, they are bound together by powerful gravitational forces. By far the largest galaxies in the Local Group are the Milky Way and the Andromeda Galaxy (which is just discernible to the naked eye).

| Name | Distance in light years | Diameter in light years | Luminosity in millions of Suns | Type | Name | Distance in light years | Diameter in light years | Luminosity in millions of Suns | Type |
|---|---|---|---|---|---|---|---|---|---|
| Milky Way | 0 | 100,000 | 14,000 | Spiral | NGC 185 | 2,000,000 | 6,000 | 110 | Elliptical |
| Sagittarius | 78,000 | 15,000 | 30 | Elliptical | Andromeda | 2,500,000 | 150,000 | 40,000 | Spiral |
| LMC | 160,000 | 30,000 | 2,000 | Irregular | M32 | 2,500,000 | 5,000 | 300 | Elliptical |
| SMC | 190,000 | 20,000 | 250 | Irregular | NGC 205 | 2,500,000 | 10,000 | 250 | Elliptical |
| Ursa Minor | 225,000 | 1,000 | 0.3 | Elliptical | Triangulum | 2,500,000 | 40,000 | 4,000 | Spiral |
| Draco | 248,000 | 500 | 0.3 | Elliptical | IC 1613 | 2,500,000 | 12,000 | 80 | Irregular |
| Sculptor | 250,000 | 1,000 | 1.5 | Elliptical | LGS 3 | 2,500,000 | 1,000 | 0.6 | Irregular |
| Carina | 280,000 | 500 | 0.4 | Elliptical | And I | 2,570,000 | 2,000 | 5 | Elliptical |
| Sextans | 290,000 | 1,000 | 0.8 | Elliptical | And III | 2,570,000 | 3,000 | 1 | Elliptical |
| Fornax | 430,000 | 3,000 | 20 | Elliptical | EGB0427+63 | 2,600,000 | 1,000 | 0.8 | Elliptical |
| Leo II | 750,000 | 500 | 1 | Elliptical | Tucana | 2,900,000 | 500 | 0.6 | Elliptical |
| Leo I | 880,000 | 1,000 | 10 | Elliptical | WLM | 3,000,000 | 7,000 | 30 | Irregular |
| Phoenix | 1,270,000 | 1,000 | 0.8 | Irregular | SagDIG | 3,700,000 | 5,000 | 2 | Irregular |
| NGC 6822 | 1,750,000 | 8,000 | 300 | Irregular | IC 10 | 4,000,000 | 6,000 | 1,000 | Irregular |
| And II | 1,910,000 | 2,000 | 5 | Elliptical | Pegasus | 5,800,000 | 7,000 | 50 | Irregular |
| NGC 147 | 1,920,000 | 10,000 | 80 | Elliptical | | | | | |

# Messier objects

The French astronomer and comet-hunter Charles Messier compiled a list of 110 nebulous objects to distinguish them from comets, with which they could be readily confused. Most Messier objects belong to one of three major categories: galaxies, star clusters, and nebulae. All of them are identified by the prefix M and a number; a few Messier objects have been given common names.

| Number/Name | Constellation | Type | Number/Name | Constellation | Type | Number/Name | Constellation | Type |
|---|---|---|---|---|---|---|---|---|
| M1 Crab Nebula | Taurus | Supernova remnant | M37 | Auriga | Star cluster | M74 | Pisces | Galaxy |
| M2 | Aquarius | Star cluster | M38 | Auriga | Star cluster | M75 | Sagittarius | Star cluster |
| M3 | Canes Venatici | Star cluster | M39 | Cygnus | Star cluster | M76 Little Dumbbell | Perseus | Nebula |
| M4 | Scorpius | Star cluster | M40 | Ursa Major | Double star | M77 | Cetus | Galaxy |
| M5 | Serpens | Star cluster | M41 | Canis Major | Star cluster | M78 | Orion | Nebula |
| M6 Butterfly Cluster | Scorpius | Star cluster | M42 Orion Nebula | Orion | Nebula | M79 | Lepus | Star cluster |
| M7 Ptolemy Cluster | Scorpius | Star cluster | M43 De Mairan's Nebula | Orion | Nebula | M80 | Scorpius | Star cluster |
| M8 Lagoon Nebula | Sagittarius | Nebula | M44 Beehive Cluster | Cancer | Star cluster | M81 Bode's Galaxy | Ursa Major | Galaxy |
| M9 | Ophiuchus | Star cluster | M45 Pleiades | Taurus | Star cluster | M82 Cigar Galaxy | Ursa Major | Galaxy |
| M10 | Ophiuchus | Star cluster | M46 | Puppis | Star cluster | M83 Southern Pinwheel Galaxy | Hydra | Galaxy |
| M11 Wild Duck Cluster | Scutum | Star cluster | M47 | Puppis | Star cluster | M84 | Virgo | Galaxy |
| M12 | Ophiuchus | Star cluster | M48 | Hydra | Star cluster | M85 | Coma Berenices | Galaxy |
| M13 Great Globular Cluster | Hercules | Star cluster | M49 | Virgo | Galaxy | M86 | Virgo | Galaxy |
| M14 | Ophiuchus | Star cluster | M50 | Monoceros | Star cluster | M87 Virgo A Galaxy | Virgo | Galaxy |
| M15 Pegasus Cluster | Pegasus | Star cluster | M51 Whirlpool Galaxy | Canes Venatici | Galaxy | M88 | Coma Berenices | Galaxy |
| M16 Eagle Nebula | Serpens | Nebula | M52 | Cassiopeia | Star cluster | M89 | Virgo | Galaxy |
| M17 Omega Nebula | Sagittarius | Nebula | M53 | Coma Berenices | Star cluster | M90 | Virgo | Galaxy |
| M18 | Sagittarius | Star cluster | M54 | Sagittarius | Star cluster | M91 | Coma Berenices | Galaxy |
| M19 | Ophiuchus | Star cluster | M55 | Sagittarius | Star cluster | M92 | Hercules | Star cluster |
| M20 Trifid Nebula | Sagittarius | Nebula | M56 | Lyra | Star cluster | M93 | Puppis | Star cluster |
| M21 | Sagittarius | Star cluster | M57 Ring Nebula | Lyra | Nebula | M94 | Canes Venatici | Galaxy |
| M22 | Sagittarius | Star cluster | M58 | Virgo | Galaxy | M95 | Leo | Galaxy |
| M23 | Sagittarius | Star cluster | M59 | Virgo | Galaxy | M96 | Leo | Galaxy |
| M24 | Sagittarius | Star cluster | M60 | Virgo | Galaxy | M97 Owl Nebula | Ursa Major | Nebula |
| M25 | Sagittarius | Star cluster | M61 | Virgo | Galaxy | M98 | Coma Berenices | Galaxy |
| M26 | Scutum | Star cluster | M62 | Ophiuchus | Star cluster | M99 | Coma Berenices | Galaxy |
| M27 Dumbbell Nebula | Vulpecula | Nebula | M63 Sunflower Galaxy | Canes Venatici | Galaxy | M100 | Coma Berenices | Galaxy |
| M28 | Sagittarius | Star cluster | M64 Blackeye Galaxy | Coma Berenices | Galaxy | M101 Pinwheel Galaxy | Ursa Major | Galaxy |
| M29 | Cygnus | Star cluster | M65 | Leo | Galaxy | M102 | not fully identified | – |
| M30 | Capricornus | Star cluster | M66 | Leo | Galaxy | M103 | Cassiopeia | Star cluster |
| M31 Andromeda Galaxy | Andromeda | Galaxy | M67 | Cancer | Star cluster | M104 Sombrero Galaxy | Virgo | Galaxy |
| M32 | Andromeda | Galaxy | M68 | Hydra | Star cluster | M105 | Leo | Galaxy |
| M33 Triangulum Galaxy | Triangulum | Galaxy | M69 | Sagittarius | Star cluster | M106 | Canes Venatici | Galaxy |
| M34 | Perseus | Star cluster | M70 | Sagittarius | Star cluster | M107 | Ophiuchus | Star cluster |
| M35 | Gemini | Star cluster | M71 | Sagitta | Star cluster | M108 | Ursa Major | Galaxy |
| M36 | Auriga | Star cluster | M72 | Aquarius | Star cluster | M109 | Ursa Major | Galaxy |
| | | | M73 | Aquarius | Star cluster | M110 | Andromeda | Galaxy |

## Galaxy statistics

Galaxies are classified into three main groups: elliptical (E), spiral (S), and barred spiral (SB). Variations are denoted by a, b, c, and so on. Some galaxies are irregular (Irr) in shape. Examples are listed here.

| Name | Constellation | Type | Distance in millions of light years |
|---|---|---|---|
| M105 | Leo | E0 | 38 |
| M59 | Virgo | E5 | 50 |
| Sombrero | Virgo | Sa | 50 |
| NGC 2841 | Ursa Major | Sb | 33 |
| Andromeda | Andromeda | Sb | 2.5 |
| Pinwheel | Ursa Major | Sc | 20 |
| Triangulum | Triangulum | Sc | 2.5 |
| Whirlpool | Canes Venatici | Sc | 20 |
| NGC 2859 | Leo Minor | SBa | 72 |
| NGC 5850 | Virgo | SBb | 100 |
| NGC 7479 | Pegasus | SBc | 110 |
| M82 | Ursa Major | Irr | 12 |
| Large Magellanic Cloud | Dorado | Irr | 0.16 |

## Important clusters of galaxies

Galaxies tend to form clusters, pulled together by huge gravitational forces (see p.341). Such clusters differ in density and pattern, some having a regular structure while others are more random groupings.

| Name | Distance in millions of light years | Size in millions of light years | Gas temperature (million °C) |
|---|---|---|---|
| Virgo | 50 | 11 | 30 |
| Fornax | 70 | 8 | – |
| Centaurus | 140 | 5 | 45 |
| Cancer | 210 | 11 | – |
| Perseus | 240 | 17 | 75 |
| Coma | 290 | 20 | 95 |
| Hercules | 490 | 15 | 45 |
| Abell 2256 | 760 | 10 | 85 |
| Corona Borealis | 940 | 8 | 100 |
| Gemini | 1,000 | 9 | – |

# Space exploration

Since 1957, man has explored the Universe from beyond Earth's atmosphere. Space probes have returned data from every planet in the solar system; people live on space stations for months at a time; and observatories both on Earth and in orbit provide information about near and deep space. Although space exploration is dominated by Russia and the United States, some 30 other countries now have their own space programs.

## Artificial satellites

Hundreds of artificial satellites orbit Earth (see pp.368–69) and most are launched into one of four main orbits. A low-Earth orbit is just above Earth's atmosphere. Satellites in polar orbit are often up to 500 miles (800 km) above Earth. In a highly elliptical orbit, a satellite is at much higher altitude at one end of its orbit than the other. In geostationary orbit, 23,500 miles (36,000 km) high, a satellite synchronizes with Earth's rotation and appears fixed in the sky.

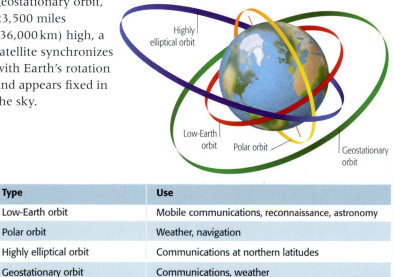

Highly elliptical orbit

Low-Earth orbit    Polar orbit    Geostationary orbit

| Type | Use |
|---|---|
| Low-Earth orbit | Mobile communications, reconnaissance, astronomy |
| Polar orbit | Weather, navigation |
| Highly elliptical orbit | Communications at northern latitudes |
| Geostationary orbit | Communications, weather |

## Major events in space exploration

| Mission | Date | Country | First |
|---|---|---|---|
| Sputnik 1 | October 1957 | USSR | Artificial satellite in space |
| Sputnik 2 | November 1957 | USSR | Animal in space (dog, Laika) |
| Vostok 1 | April 1961 | USSR | Human in space (Yuri Gagarin) |
| Mercury | February 1962 | USA | US manned space flight (John Glenn) |
| Voskhod 2 | March 1965 | USSR | Space walk (Alexsey Leonov) |
| Luna 9 | February 1966 | USSR | Soft landing on the Moon |
| Apollo 11 | July 1969 | USA | Human on the Moon (Neil Armstrong) |
| Venera 7 | December 1970 | USSR | Soft landing on Venus |
| Salyut 1 | April 1971 | USSR | Space station launched |
| Mars 3 | December 1971 | USSR | Soft landing on Mars |
| Pioneer 10 | December 1973 | USA | Flyby of Jupiter |
| Voyager 1 & 2 | August/September 1977 | USA | Probes launched to fly by Jupiter, Saturn, Uranus, Neptune |
| STS 31 | April 1990 | USA/ESA | Space telescope (Hubble) launch |
| Mercury Messenger | August 2004 | USA | Probe sent to Mercury |
| New Horizons | January 2006 | USA | Launch of probe to Pluto |

*(ESA = European Space Agency)*

# Earth Sciences
## Structure of Earth

Earth has taken about 4.6 billion years to develop. It is a remarkably varied world, especially on the thin outer crust that covers its rock mantle and metallic core (see pp.352–53). The processes that shaped Earth are ongoing. The influence of forces such as shifting tectonic plates is usually imperceptible, except over geological time, but occasionally tectonic movement causes sudden, violent events, including earthquakes.

## Geological timescale

Earth's timescale covers thousands of millions of years. Study of the rocks and fossils of the planet's crust has made it possible to divide this time into segments (see p.191). The divisions are constantly being refined.

| Era | Period | Epoch | Start (million years ago) |
|---|---|---|---|
| Cenozoic | Neogene | Holocene | 0.01 |
| | | Pleistocene | 1.8 |
| | | Pliocene | 5.3 |
| | | Miocene | 23 |
| | Paleogene | Oligocene | 34 |
| | | Eocene | 55 |
| | | Paleocene | 65 |
| Mesozoic | Cretaceous | | 145 |
| | Jurassic | | 199 |
| | Triassic | | 251 |
| Paleozoic | Permian | | 299 |
| | Carboniferous (divided into Lower Mississippian and Upper Pennsylvanian in United States) | | 359 |
| | Devonian | | 416 |
| | Silurian | | 433 |
| | Ordovician | | 488 |
| | Cambrian | | 542 |
| Precambrian | Formation of Earth | | 4,554 |

## Earth's minerals

Minerals are the basic materials of Earth's rocks. All of them have an internal crystal structure composed of atoms arranged in a regular, repeating, three-dimensional pattern. A free-growing crystal forms a geometric shape with flat planes, or faces, that are often arranged symmetrically. Patterns of crystal symmetry are classified into six main groups, or crystal systems, which are defined by the specific angles of their faces. These systems are: cubic; hexagonal/trigonal; monoclinic; orthorhombic; tetragonal; and triclinic.

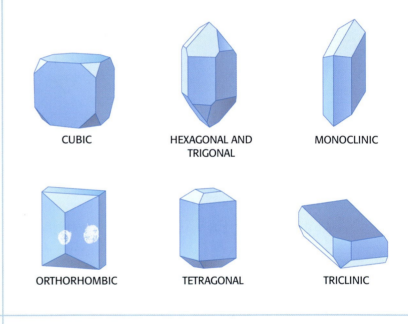

CUBIC · HEXAGONAL AND TRIGONAL · MONOCLINIC

ORTHORHOMBIC · TETRAGONAL · TRICLINIC

## Mineral classification

There are more than 4,000 known natural minerals, although only about 100 of these are abundant. Minerals are classified according to their chemical composition, and are commonly divided into the major groups listed below. Each grouping contains many minerals.

| Group | Approx. no. of minerals | Examples |
|---|---|---|
| Sulphides | 600 | Pyrite, galena |
| Silicates | 500 | Olivine, quartz, feldspar, garnet |
| Oxides and hydroxides | 400 | Chromite, hematite |
| Phosphates and vanadates | 400 | Apatite, carnotite |
| Sulphates | 300 | Anhydrite, barite, gypsum |
| Carbonates | 200 | Calcite, aragonite, dolomite |
| Halides | 140 | Fluorite, halite, sylvite |
| Borates and nitrates | 125 | Borax, colemanite, kernite, nitratine |
| Molybdates and tungstates | 42 | Wulfenite, wolframite |
| Native elements | 20 | Gold, platinum, copper, sulphur, carbon |

## Earth's rocks

Rocks are grouped according to the way they have formed. There are three main types—igneous, sedimentary, and metamorphic—common examples of which are listed here. Igneous rocks form from molten magma; sedimentary rocks accumulate from layers of sediment; metamorphism occurs when heat or pressure converts one rock type into another.

| Igneous | Sedimentary | Metamorphic |
|---|---|---|
| Syenite | Limestone | Slate |
| Trachyte | Dolomite | Phyllite |
| Granite | Evaporite | Schist |
| Rhyolite | Breccia | Gneiss |
| Diorite | Conglomerate | Hornfels |
| Andesite | Sandstone | Marble |
| Gabbro | Siltstone | Quartzite |
| Basalt | Mudstone | Migmatite |
| Peridotite | Shale | Amphibolite |
| Picrite | Coal | Mylonite |

# The "rock cycle"

All types of rock—igneous, metamorphic, and sedimentary—form continuously both within Earth and on its surface. Many of these rock materials are also "recycled" over geological time, with igneous and metamorphic rocks from Earth's interior being broken down by surface processes, deposited as sediments, and buried. The "rock cycle" (see below and pp.184–185) is a simplified illustration of some of these interlinked processes and transformations.

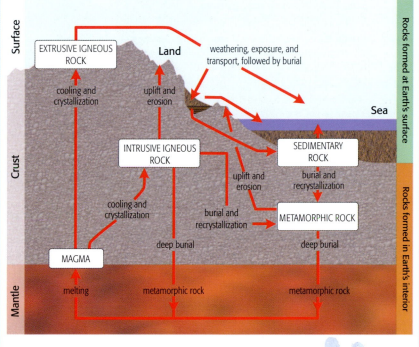

# Tectonic plates

Earth's outermost rocks are broken into seven major tectonic plates and about twelve smaller ones (see pp.356–57). Over time the plates have moved apart (diverged) and joined together (converged) to shift continents, open and close oceans, and form mountains. At transform boundaries, plates slide past one another along fault planes.

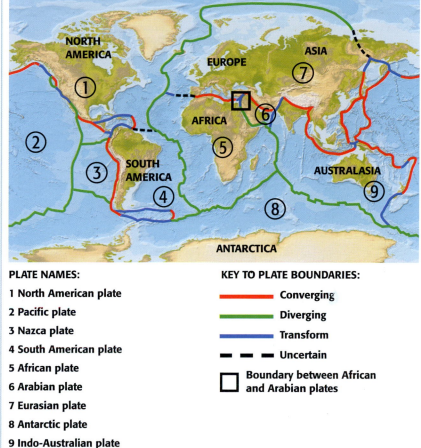

**PLATE NAMES:**

1 North American plate

2 Pacific plate

3 Nazca plate

4 South American plate

5 African plate

6 Arabian plate

7 Eurasian plate

8 Antarctic plate

9 Indo-Australian plate

**KEY TO PLATE BOUNDARIES:**

— Converging

— Diverging

— Transform

-- -- Uncertain

☐ Boundary between African and Arabian plates

# Earthquake zones

Movement of Earth's crust and tectonic plates produces stress and fracture of its rocks along fault planes. The buildup of stress in these brittle crustal rocks results in sudden failure and the release of massive amounts of energy as earthquakes (see pp.359–61). The shock waves are transmitted through the rocks, often over great distances; those of major quakes reach the other side of the world. Earth's major earthquake zones follow the boundaries where tectonic places move relative to one another.

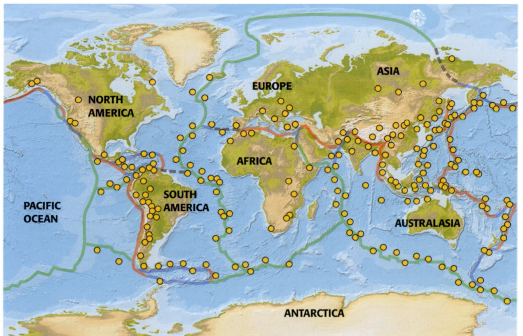

**KEY TO PLATE BOUNDARIES:**

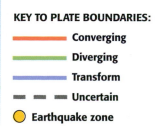

— Converging

— Diverging

— Transform

-- -- Uncertain

● Earthquake zone

## Earthquake severity scales

The size of an earthquake can be measured by magnitude and by intensity (see pp.360–61). Two traditionally used scales are the Mercalli Intensity Scale (which measures the effects of an earthquake on the surface, at 12 levels) and the Richter Scale (which measures the amount of energy released). However, in scientific use these have largely been superseded by the moment magnitude scale. Almost every day, there are hundreds of minor earthquakes around the world. Very large disturbances causing widespread damage are rare.

| RICHTER SCALE: TYPICAL RANGES | |
|---|---|
| <3.5 | Recordable, but not generally felt |
| 3.5–5.4 | Felt by many, but causes little or no damage |
| 5.5–6.0 | May be slight damage to well-constructed buildings; more severe damage to poorly constructed buildings |
| 6.1–6.9 | Can cause severe and widespread damage up to a range of 62.5 miles (100 km) |
| 7.0–7.9 | Major effects, causing severe damage over larger areas |
| >8.0 | Can cause devastation over several hundred miles (kilometers) |

| THE MERCALLI INTENSITY SCALE | | |
|---|---|---|
| I | Instrumental | No Earth movement felt; detectable only by seismograph |
| II | Feeble | Movement felt only by people at rest or on upper floors of buildings |
| III | Slight | Noticeable indoors, especially on upper floors; hanging objects swing |
| IV | Moderate | Felt indoors by most people and outdoors by a few; hanging objects swing; dishes, doors, and windows rattle; sensation of a heavy truck hitting the walls; parked cars rock |
| V | Rather strong | Felt by nearly all; sleeping people waken; doors swing; small objects move; dishes and glassware break; liquids may spill |
| VI | Strong | Felt by all; difficult to walk; furniture moves; objects fall from shelves and pictures from walls; plaster may crack; little structural damage |
| VII | Very strong | Difficult to stand; drivers feel cars shaking; furniture may break; bricks or chimney pots may fall; some structural damage depending on construction of building |
| VIII | Destructive | Car drivers cannot steer; considerable damage in poorly constructed buildings or tall structures such as factory chimneys; branches broken from trees; wet or sloping ground may crack; water levels in wells may be affected |
| IX | Ruinous | Considerable structural damage to all building types; houses may shift off their foundations; the ground cracks and reservoirs are damaged seriously |
| X | Disastrous | Most buildings destroyed; bridges and dams seriously damaged; large cracks in ground; major landslides, rail tracks bend |
| XI | Very disastrous | Few structures remain standing; pipelines and rail tracks destroyed |
| XII | Catastrophic | Total destruction; upheavals of ground and rock |

# The land and oceans

The topography of Earth's thin outer layer, or crust, is extremely varied, ranging from mountain peaks so high there is almost no oxygen, to plains, valleys, and deep-sea trenches. There are two distinct types of crust—continental and oceanic. The continental crust forms all the major landmasses and their shallow seas, but the greater part of Earth's surface is formed by the oceanic crust. Water is the predominant feature of the planet, with five major oceans and many smaller seas; and much of the land has been shaped by the powerful forces of rivers.

## Earth's surface

About 70 percent of Earth's total surface is covered in water. More than half (53.5 percent) of this water covered area lies at depths ranging between approximately 1¾–3¾ miles (3 and 6 km). About 20.9 percent of the land surface lies between sea level and 0.6 miles (1 km) in altitude. The highest point on Earth's surface is Mount Everest at 29,035 ft (8,850 m) above sea level; the lowest point is the bottom of the Mariana deep-sea trench, located in the floor of the Pacific Ocean, at a depth of 36,201 ft (11,034 m) below sea level.

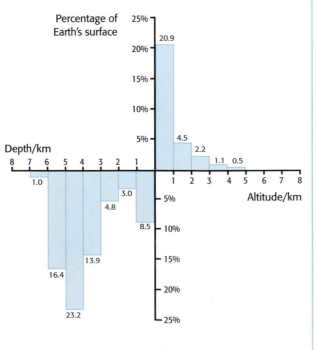

## Continents of the world

Most geographers recognize the seven continents named here, although some combine Europe and Asia as Eurasia, and regard North and South America as one American continent.

| Continent | Size sq miles (km²) | Estimated population |
|---|---|---|
| Asia | 17,179,000 (44,493,000) | 3,879,000,000 |
| Africa | 11,690,000 (30,970,000) | 877,500,000 |
| North America | 9,442,000 (24,454,000) | 501,500,000 |
| South America | 6,887,000 (17,838,000) | 379,500,000 |
| Antarctica | 6,000,000 (15,500,000) | None |
| Europe | 3,956,000 (10,245,000) | 727,000,000 |
| Australia/Oceania | 3,454,000 (8,945,000) | 32,000,000 |

# The world's longest rivers

The lengths given here are approximations, as exact measurement of a river is difficult. Multiple tributaries often make a river's source hard to pinpoint and the outflow may follow numerous channels.

| River | Length miles (km) | Location |
|---|---|---|
| Nile | 4,160 (6,690) | North and East Africa |
| Amazon | 4,080 (6,570) | South America |
| Mississippi–Missouri | 3,740 (6,020) | United States of America |
| Chang Jiang (Yangtze) | 3,720 (5,980) | China |
| Yenisey | 3,650 (5,870) | Russia |
| Amur | 3,590 (5,780) | Northeast Asia |
| Ob–Irtysh | 3,360 (5,410) | Russia |
| Parana | 3,030 (4,880) | South America |
| Huang He (Yellow) | 3,010 (4,840) | China |
| Congo | 2,880 (4,630) | Central Africa |
| Lena | 2,730 (4,400) | Russia |
| Mackenzie | 2,630 (4,240) | Northwest Canada |
| Mekong | 2,600 (4,180) | Southeast Asia |
| Niger | 2,550 (4,100) | West Africa |

# The world's highest mountain peaks

Fourteen peaks rise over 26,000 ft (8,000 m) above sea level, all of them located in Asia. The majority are in the Himalayas, the highest and one of the most recently formed mountain ranges on Earth.

| Mountain | Range | Location | Height feet (meters) |
|---|---|---|---|
| Everest | Himalayas | Nepal/Tibet | 29,035 (8,850) |
| K2 | Karakoram | Pakistan/China | 28,250 (8,611) |
| Kangchenjunga | Himalayas | Nepal/India | 28,169 (8,586) |
| Lhotse | Himalayas | Nepal/Tibet | 27,940 (8,516) |
| Makalu | Himalayas | Nepal/Tibet | 27,766 (8,463) |
| Cho Oyu | Himalayas | Nepal/Tibet | 26,906 (8,201) |
| Dhaulagiri | Himalayas | Nepal | 26,795 (8,167) |
| Manaslu | Himalayas | Nepal | 26,781 (8,163) |
| Nanga Parbat | Himalayas | Pakistan | 26,660 (8,125) |
| Annapurna | Himalayas | Nepal | 26,545 (8,091) |
| Gasherbrum I | Karakoram | Pakistan/China | 26,470 (8,068) |
| Broad Peak | Karakoram | Pakistan/China | 26,400 (8,047) |
| Gasherbrum II | Karakoram | Pakistan/China | 26,360 (8,035) |
| Shishna Pangma | Himalayas | Tibet | 26,289 (8,013) |

# Ocean currents

The water in the oceans is in constant motion, circulating in strong currents that flow both near the surface and, more slowly, at great depths (see pp.212–13). Surface currents are driven by winds but their patterns are modified by the effects of Earth's rotation on its axis, which causes circular water movements called gyres. These gyres rotate clockwise in the northern hemisphere and counterclockwise in the southern. Specific components of these gyres are called boundary currents. The boundary currents on the eastern side of oceans are predominantly cold and move toward the equator; those on the western sides tend to be warm and move away from the equator. The present global circulation pattern is also partly controlled by the configuration of the ocean basins and surrounding coastlines. In the past, the distribution of oceans and continents, and therefore current circulation, was very different.

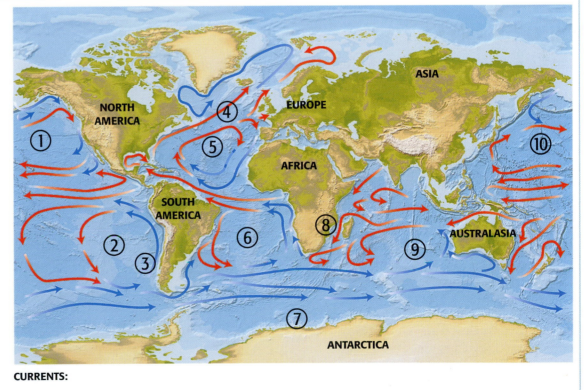

CURRENTS:

1 North Pacific gyre
2 South Pacific gyre
3 Humboldt current
4 Gulf Stream

5 North Atlantic gyre
6 South Atlantic gyre
7 Antarctic circumpolar current
8 Aghulas current

9 South Indian gyre
10 North Pacific gyre

→ Warm ocean currents
→ Cold ocean currents

## Major mountain ranges

The world's major mountain ranges, such as the Andes, Himalayas, Alps, and Rockies are comparatively young, having formed within the last few hundred million years. They are situated along the boundaries of tectonic plates that collided in the geological past (see pp.358–59). The ranges were formed as immense pressures caused by the convergence of the plates deformed and uplifted the rocks high above the surrounding land. This process of uplift is still continuing.

RANGES:

| | |
|---|---|
| 1 Alaska Range | 8 Drakensberg |
| 2 Rocky Mountains | 9 Ethiopian Highlands |
| 3 Appalachians | 10 Caucasus |
| 4 Andes | 11 Ural Mountains |
| 5 Pyrenees | 12 Tien Shan |
| 6 Atlas Mountains | 13 Himalayas |
| 7 Alps | 14 Great Dividing Range |

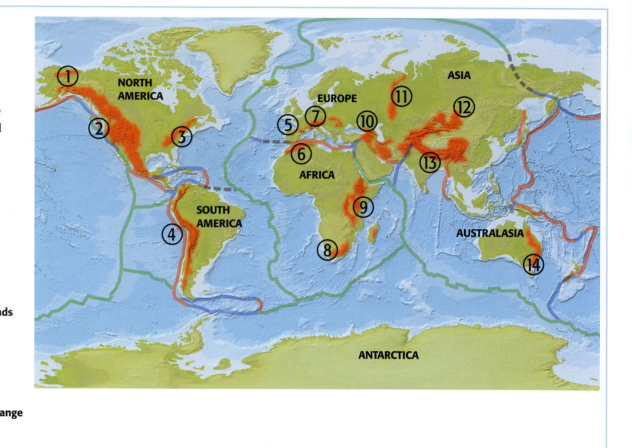

# Climate and weather

Earth's atmosphere, surface waters, and landmasses combine to produce the characteristic climates and weather patterns of regions around the world. In particular, oceans, which cover about 70 percent of Earth's surface, have a major role in the regulation of climates, transporting heat from the equatorial regions to higher, colder latitudes and releasing water vapor into the atmosphere.

## Climate regions

Earth's landmasses can be divided into climate regions, based on average temperatures and rainfall, and type of vegetation. Tropical areas are hot all year round and often very wet, while polar regions and the tops of high mountains are characterized by extreme cold. In the temperate regions between the poles and the tropics, climates are more moderate. Areas classified as desert usually have an annual rainfall of less than 10 in (25 cm). Climate is influenced not only by latitude but also by factors such as height above sea level and distance from the ocean.

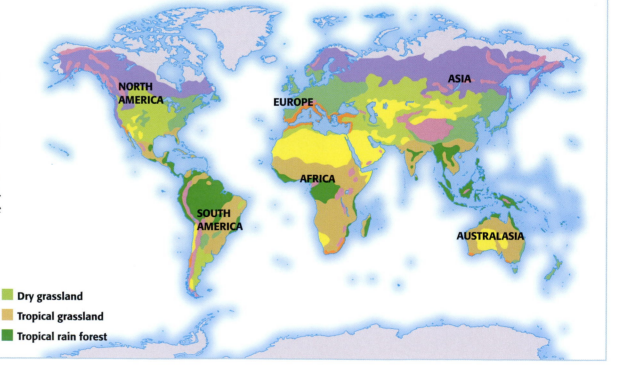

CLIMATE REGIONS:

| | | |
|---|---|---|
| Polar and tundra | Temperate forest | Dry grassland |
| Boreal forest | Mediterranean | Tropical grassland |
| Mountain | Desert | Tropical rain forest |

## Weather map symbols

The symbols shown below are an internationally recognized system used for classifying types of weather. They were devised by the World Meteorological Organization and represent varying intensities of precipitation and other conditions.

| Symbol | Description | Symbol | Description |
|---|---|---|---|
| ● | Intermittent light rain | ∼ | Freezing rain |
| ●● | Continuous light rain | ∼ | Freezing drizzle |
| ●●● | Intermittent moderate rain | R | Thunderstorm |
| ●●●● | Continuous moderate rain | ⊕ | Tornado |
| ●●●●● | Continuous heavy rain | ☰ | Fog |
| ▽ | Light rain shower | ∞ | Haze |
| ▽ | Moderate rain shower | Stationary front | Stationary front |
| ▽ | Heavy rain shower | Cold front | Cold front |
| ✳ | Snow | Occluded front | Occluded front |
| ❟ | Drizzle | Warm front | Warm front |

## The Beaufort wind force scale

Beaufort's scale was originally devised for sailors as a means of judging wind strength by observations of sea disturbance and the way sails, and therefore steerage, were affected. The scale was later modified for use on land, substituting features such as trees and cars.

| Beaufort no. | Wind speed mph (kph) | Description |
|---|---|---|
| 0 | 0–1 (0–2) | Calm—smoke rises vertically, air feels still |
| 1 | 1–3 (2–6) | Light air—smoke drifts |
| 2 | 4–7 (7–11) | Slight breeze—wind detectable on face, some leaf movement |
| 3 | 8–12 (12–19) | Gentle breeze—leaves and twigs move gently |
| 4 | 13–18 (20–29) | Moderate breeze—loose paper blows around |
| 5 | 19–24 (30–39) | Fresh breeze—small trees sway |
| 6 | 25–31 (40–50) | Strong breeze—difficult to use an umbrella |
| 7 | 32–38 (51–61) | High wind—whole trees bend |
| 8 | 39–46 (62–74) | Gale—twigs break off trees, walking into wind is difficult |
| 9 | 47–54 (75–87) | Severe gale—roof tiles blow away |
| 10 | 55–63 (88–101) | Whole gale—trees break and are uprooted |
| 11 | 64–74 (102–119) | Storm—damage is extensive, cars overturn |
| 12 | 75+ (120+) | Hurricane—widespread devastation |

## Where the wind blows

At Earth's poles, the air is at high pressure and low temperature; at the equator, air is at low pressure but higher temperature. This, together with the rotation of Earth on its axis, creates a predictable pattern of warm and cool winds around the globe (see pp.206–207). Continental landmasses and high mountains also influence wind patterns, such as the monsoon winds over southern Asia. On the equator there is an area known as the doldrums, once a trap for sailing ships, where there is very little wind. In certain latitudes, a combination of disturbances in wind speed and direction and high sea surface temperatures can create tropical cyclones, also known as hurricanes or typhoons. In a cyclone, the winds move in a spiral pattern and may reach speeds of more than 125 mph (200 kph).

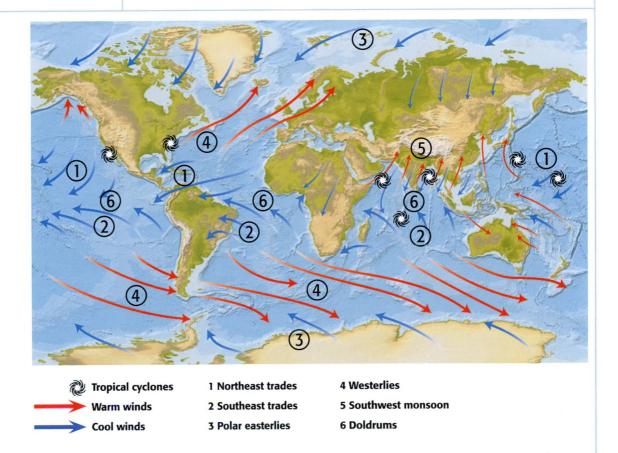

🌀 Tropical cyclones
→ Warm winds
→ Cool winds

1 Northeast trades
2 Southeast trades
3 Polar easterlies

4 Westerlies
5 Southwest monsoon
6 Doldrums

# Biology
## Defining life

Living things exist in many forms. Some are highly complex, multicellular organisms, readily identifiable as, for example, a tree, or a fish, or an elephant. Others are microscopic, consisting of no more than a single cell, and cannot be seen with the naked eye. Regardless of size, complexity, or internal chemistry, all things classified as living take in and use energy to grow and survive. Another feature that defines life is the ability to replicate.

## Scientific classification

Various systems are used to classify the world's living organisms (see pp.122–23). The most widely used system is that of the five kingdoms, the major divisions into which all lifeforms are placed according to their fundamental characteristics. Each kingdom contains subdivisions, or phyla, from which in turn further groupings are organized. The chart below shows the hierarchy from kingdom to species, the basic unit of scientific classification. Not all lifeforms are categorized easily and some are placed into intermediate groupings created to accommodate them.

### Classification levels

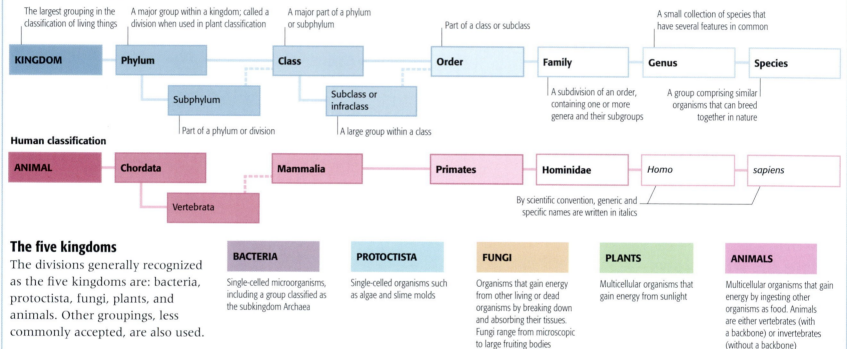

The largest grouping in the classification of living things — **KINGDOM**

A major group within a kingdom; called a division when used in plant classification — **Phylum**

**Subphylum** — Part of a phylum or division

A major part of a phylum or subphylum — **Class**

**Subclass or infraclass** — A large group within a class

Part of a class or subclass — **Order**

**Family** — A subdivision of an order, containing one or more genera and their subgroups

**Genus** — A group comprising similar organisms that can breed together in nature

A small collection of species that have several features in common — **Genus** — **Species**

### Human classification

**ANIMAL** — **Chordata** — **Vertebrata** — **Mammalia** — **Primates** — **Hominidae** — *Homo* — *sapiens*

By scientific convention, generic and specific names are written in italics

### The five kingdoms

The divisions generally recognized as the five kingdoms are: bacteria, protoctista, fungi, plants, and animals. Other groupings, less commonly accepted, are also used.

**BACTERIA**
Single-celled microorganisms, including a group classified as the subkingdom Archaea

**PROTOCTISTA**
Single-celled organisms such as algae and slime molds

**FUNGI**
Organisms that gain energy from other living or dead organisms by breaking down and absorbing their tissues. Fungi range from microscopic to large fruiting bodies

**PLANTS**
Multicellular organisms that gain energy from sunlight

**ANIMALS**
Multicellular organisms that gain energy by ingesting other organisms as food. Animals are either vertebrates (with a backbone) or invertebrates (without a backbone)

## Bacteria

These single-celled organisms (see pp.244–45) live in a wide range of habitats on land and in water, sometimes in extreme conditions. A few thousand species out of possibly millions of different bacteria have been named. Only a tiny fraction of them cause disease.

**BACTERIA**
Kingdom: Bacteria — **Species** *5,000 named*

**Subkingdom**
Archaea — **Species** *209*

## Fungi

This group includes single-celled organisms such as yeasts, as well as the familiar mushrooms and toadstools. Fungi absorb nutrients made by plants or animals. They reproduce by shedding spores.

**FUNGI**
Kingdom: Fungi — **7 Phyla** — **Species** *100,000*

## Protoctista

This kingdom contains simple organisms that usually consist of a single cell. There are at least 65,000 species of protoctista, most of which live in water. A selection of important phyla is shown here.

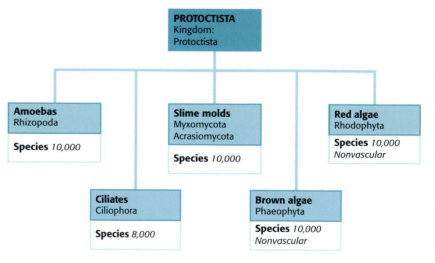

**PROTOCTISTA**
Kingdom: Protoctista

**Amoebas**
Rhizopoda
**Species** *10,000*

**Slime molds**
Myxomycota
Acrasiomycota
**Species** *10,000*

**Red algae**
Rhodophyta
**Species** *10,000*
*Nonvascular*

**Ciliates**
Ciliophora
**Species** *8,000*

**Brown algae**
Phaeophyta
**Species** *10,000*
*Nonvascular*

# The plant kingdom

Plants are organisms that produce their food by photosynthesis, using the green pigment chlorophyll in their leaves to absorb energy from sunlight (see pp.154–55). Chemical reactions combine energy with water and carbon dioxide to make glucose, which fuels growth. The plant kingdom contains over 400,000 species divided into two major groups: nonflowering plants that reproduce by shedding spores or naked seeds; and flowering plants that produce seeds in ovaries. Flowering plants are divided into two other main categories: monocotyledons and dicotyledons.

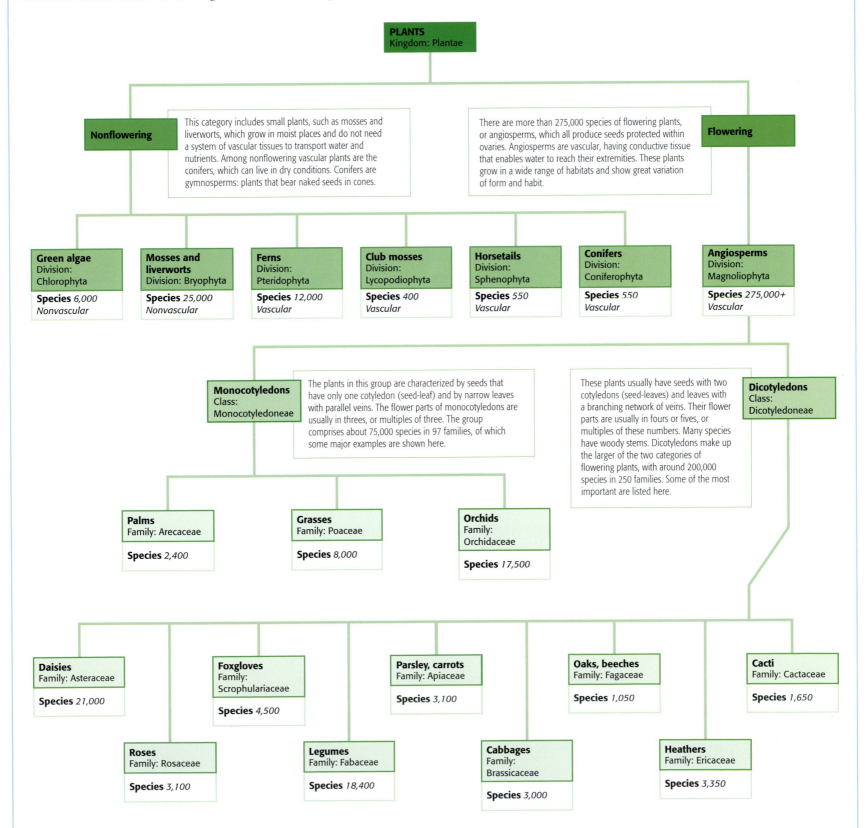

**PLANTS**
Kingdom: Plantae

**Nonflowering**
This category includes small plants, such as mosses and liverworts, which grow in moist places and do not need a system of vascular tissues to transport water and nutrients. Among nonflowering vascular plants are the conifers, which can live in dry conditions. Conifers are gymnosperms: plants that bear naked seeds in cones.

**Flowering**
There are more than 275,000 species of flowering plants, or angiosperms, which all produce seeds protected within ovaries. Angiosperms are vascular, having conductive tissue that enables water to reach their extremities. These plants grow in a wide range of habitats and show great variation of form and habit.

**Green algae**
Division: Chlorophyta
**Species** 6,000
*Nonvascular*

**Mosses and liverworts**
Division: Bryophyta
**Species** 25,000
*Nonvascular*

**Ferns**
Division: Pteridophyta
**Species** 12,000
*Vascular*

**Club mosses**
Division: Lycopodiophyta
**Species** 400
*Vascular*

**Horsetails**
Division: Sphenophyta
**Species** 550
*Vascular*

**Conifers**
Division: Coniferophyta
**Species** 550
*Vascular*

**Angiosperms**
Division: Magnoliophyta
**Species** 275,000+
*Vascular*

**Monocotyledons**
Class: Monocotyledoneae
The plants in this group are characterized by seeds that have only one cotyledon (seed-leaf) and by narrow leaves with parallel veins. The flower parts of monocotyledons are usually in threes, or multiples of three. The group comprises about 75,000 species in 97 families, of which some major examples are shown here.

**Dicotyledons**
Class: Dicotyledoneae
These plants usually have seeds with two cotyledons (seed-leaves) and leaves with a branching network of veins. Their flower parts are usually in fours or fives, or multiples of these numbers. Many species have woody stems. Dicotyledons make up the larger of the two categories of flowering plants, with around 200,000 species in 250 families. Some of the most important are listed here.

**Palms**
Family: Arecaceae
**Species** 2,400

**Grasses**
Family: Poaceae
**Species** 8,000

**Orchids**
Family: Orchidaceae
**Species** 17,500

**Daisies**
Family: Asteraceae
**Species** 21,000

**Foxgloves**
Family: Scrophulariaceae
**Species** 4,500

**Parsley, carrots**
Family: Apiaceae
**Species** 3,100

**Oaks, beeches**
Family: Fagaceae
**Species** 1,050

**Cacti**
Family: Cactaceae
**Species** 1,650

**Roses**
Family: Rosaceae
**Species** 3,100

**Legumes**
Family: Fabaceae
**Species** 18,400

**Cabbages**
Family: Brassicaceae
**Species** 3,000

**Heathers**
Family: Ericaceae
**Species** 3,350

# The animal kingdom

Animals are multicellular life-forms that obtain energy by eating food, including other living organisms or their remains. One of the defining features of an animal is mobility. Most are able to move from one place to another, or to move part of their body to feed or protect themselves. There are about two million known species of animals on Earth, but the true total may be as high as 10 million. Because of this enormous diversity, only a summary of the classification system for animals is given here, and the number of species in each phylum can be no more than an estimate. Any of the groupings is liable to change in response to shifts in current scientific opinion.

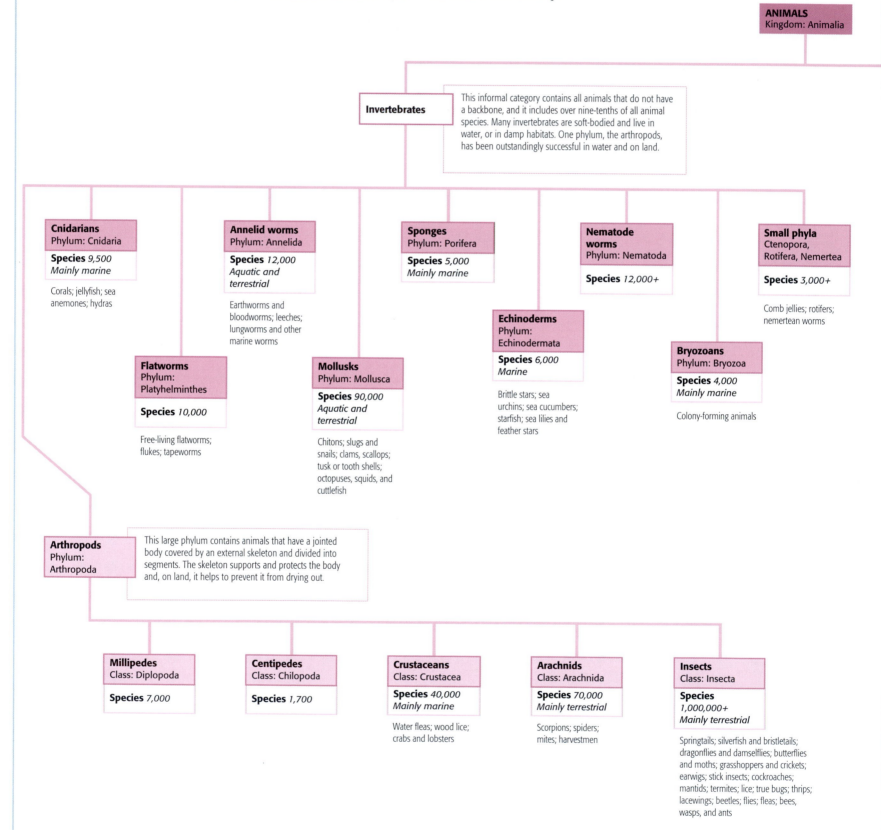

**ANIMALS**
Kingdom: Animalia

**Invertebrates**

This informal category contains all animals that do not have a backbone, and it includes over nine-tenths of all animal species. Many invertebrates are soft-bodied and live in water, or in damp habitats. One phylum, the arthropods, has been outstandingly successful in water and on land.

**Cnidarians**
Phylum: Cnidaria

**Species** 9,500
*Mainly marine*

Corals; jellyfish; sea anemones; hydras

**Annelid worms**
Phylum: Annelida

**Species** 12,000
*Aquatic and terrestrial*

Earthworms and bloodworms; leeches; lungworms and other marine worms

**Sponges**
Phylum: Porifera

**Species** 5,000
*Mainly marine*

**Nematode worms**
Phylum: Nematoda

**Species** 12,000+

**Small phyla**
Ctenopora, Rotifera, Nemertea

**Species** 3,000+

Comb jellies; rotifers; nemertean worms

**Echinoderms**
Phylum: Echinodermata

**Species** 6,000
*Marine*

Brittle stars; sea urchins; sea cucumbers; starfish; sea lilies and feather stars

**Flatworms**
Phylum: Platyhelminthes

**Species** 10,000

Free-living flatworms; flukes; tapeworms

**Mollusks**
Phylum: Mollusca

**Species** 90,000
*Aquatic and terrestrial*

Chitons; slugs and snails; clams, scallops; tusk or tooth shells; octopuses, squids, and cuttlefish

**Bryozoans**
Phylum: Bryozoa

**Species** 4,000
*Mainly marine*

Colony-forming animals

**Arthropods**
Phylum: Arthropoda

This large phylum contains animals that have a jointed body covered by an external skeleton and divided into segments. The skeleton supports and protects the body and, on land, it helps to prevent it from drying out.

**Millipedes**
Class: Diplopoda

**Species** 7,000

**Centipedes**
Class: Chilopoda

**Species** 1,700

**Crustaceans**
Class: Crustacea

**Species** 40,000
*Mainly marine*

Water fleas; wood lice; crabs and lobsters

**Arachnids**
Class: Arachnida

**Species** 70,000
*Mainly terrestrial*

Scorpions; spiders; mites; harvestmen

**Insects**
Class: Insecta

**Species** 1,000,000+
*Mainly terrestrial*

Springtails; silverfish and bristletails; dragonflies and damselflies; butterflies and moths; grasshoppers and crickets; earwigs; stick insects; cockroaches; mantids; termites; lice; true bugs; thrips; lacewings; beetles; flies; fleas; bees, wasps, and ants

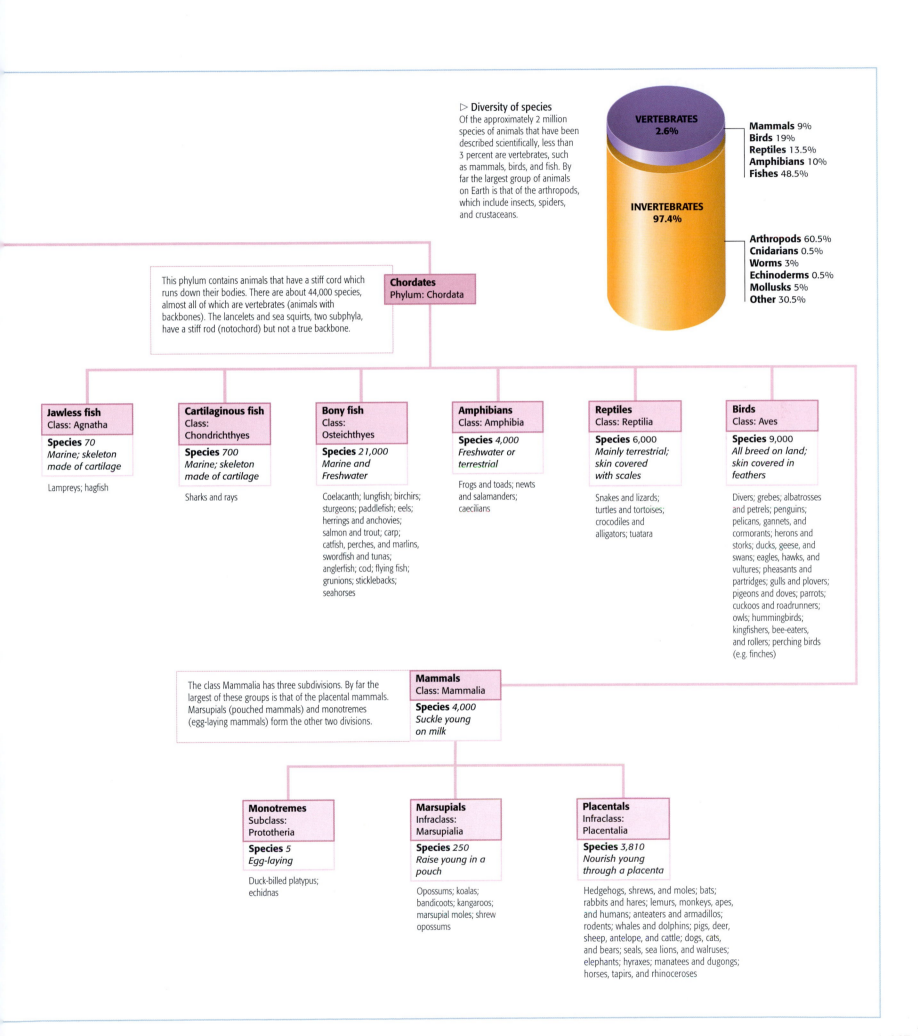

▷ Diversity of species
Of the approximately 2 million species of animals that have been described scientifically, less than 3 percent are vertebrates, such as mammals, birds, and fish. By far the largest group of animals on Earth is that of the arthropods, which include insects, spiders, and crustaceans.

**VERTEBRATES 2.6%**

**INVERTEBRATES 97.4%**

**Mammals** 9%
**Birds** 19%
**Reptiles** 13.5%
**Amphibians** 10%
**Fishes** 48.5%

**Arthropods** 60.5%
**Cnidarians** 0.5%
**Worms** 3%
**Echinoderms** 0.5%
**Mollusks** 5%
**Other** 30.5%

This phylum contains animals that have a stiff cord which runs down their bodies. There are about 44,000 species, almost all of which are vertebrates (animals with backbones). The lancelets and sea squirts, two subphyla, have a stiff rod (notochord) but not a true backbone.

**Chordates**
Phylum: Chordata

**Jawless fish**
Class: Agnatha

**Species** *70*
*Marine; skeleton made of cartilage*

Lampreys; hagfish

**Cartilaginous fish**
Class: Chondrichthyes

**Species** *700*
*Marine; skeleton made of cartilage*

Sharks and rays

**Bony fish**
Class: Osteichthyes

**Species** *21,000*
*Marine and Freshwater*

Coelacanth; lungfish; birchirs; sturgeons; paddlefish; eels; herrings and anchovies; salmon and trout; carp; catfish, perches, and marlins; swordfish and tunas; anglerfish; cod; flying fish; grunions; sticklebacks; seahorses

**Amphibians**
Class: Amphibia

**Species** *4,000*
*Freshwater or terrestrial*

Frogs and toads; newts and salamanders; caecilians

**Reptiles**
Class: Reptilia

**Species** *6,000*
*Mainly terrestrial; skin covered with scales*

Snakes and lizards; turtles and tortoises; crocodiles and alligators; tuatara

**Birds**
Class: Aves

**Species** *9,000*
*All breed on land; skin covered in feathers*

Divers; grebes; albatrosses and petrels; penguins; pelicans, gannets, and cormorants; herons and storks; ducks, geese, and swans; eagles, hawks, and vultures; pheasants and partridges; gulls and plovers; pigeons and doves; parrots; cuckoos and roadrunners; owls; hummingbirds; kingfishers, bee-eaters, and rollers; perching birds (e.g. finches)

The class Mammalia has three subdivisions. By far the largest of these groups is that of the placental mammals. Marsupials (pouched mammals) and monotremes (egg-laying mammals) form the other two divisions.

**Mammals**
Class: Mammalia

**Species** *4,000*
*Suckle young on milk*

**Monotremes**
Subclass: Prototheria

**Species** *5*
*Egg-laying*

Duck-billed platypus; echidnas

**Marsupials**
Infraclass: Marsupialia

**Species** *250*
*Raise young in a pouch*

Opossums; koalas; bandicoots; kangaroos; marsupial moles; shrew opossums

**Placentals**
Infraclass: Placentalia

**Species** *3,810*
*Nourish young through a placenta*

Hedgehogs, shrews, and moles; bats; rabbits and hares; lemurs, monkeys, apes, and humans; anteaters and armadillos; rodents; whales and dolphins; pigs, deer, sheep, antelope, and cattle; dogs, cats, and bears; seals, sea lions, and walruses; elephants; hyraxes; manatees and dugongs; horses, tapirs, and rhinoceroses

# The human species

The human body took a few million years to evolve once we branched away from our closest relatives, but essentially it works in the same way as that of any other mammal. The body's systems are made up of tissues and organs designed for specific functions. A system cannot work alone: each one depends on the others. For example, the skeletal system needs muscles for mobility, blood to provide oxygen and nutrients, and nerves for control.

## Human ancestors

The evolution of humans follows a complicated, many-branched family tree. We did not simply advance in an unbroken line, progressing from the first hominid to stand upright and walk on two legs to modern man.

At one time in the distant past, several closely related species of hominids co-existed. All of them except our direct ancestors died out, leaving *Homo sapiens* as the sole representative of the human species.

| Years ago | Key event | Evidence | Discovery site |
|---|---|---|---|
| 7 million | Existence of a possible hominid | The "Chad" skull (*Sahelanthropus*) | Chad, Central Africa |
| 6 million | Possible earliest date of upright, bipedal walking | Leg bones | Tugen Hills, Kenya |
| 5.8 million | Earliest evidence of first known hominid (*Ardipithecus kaddaba*) | Toe bone and other fragments | Middle Awash, Ethiopia |
| 4.4 million | Early hominid *Ardipithecus ramidus* believed to be in existence | Teeth and other fragments | Middle Awash, Ethiopia |
| 4.2 million | Remains of bipedal hominid (*Australopithecus anamensis*) found | Teeth and limb fragments | Lake Turkana, Kenya |
| 3.6 million | First clear evidence of bipedalism | Footprints of *Australopithecus afarensis* | Laetoli, northern Tanzania |
| 3.2 million | "Lucy" (*Australopithecus afarensis*) found | Partial skeleton of bipedal hominid | Hadar, Ethiopia |
| 3 million | Hominid (*Australopithecus africanus*) found Notable for powerful build of upper body | Skull and other fragments | 4 sites Southern Africa |
| 2.6 million | Hominid *Australopithecus aethiopicus* in existence | Skulls and other fragments | Ethiopia; northern Tanzania |
| 2.3 million | Possible direct ancestor of humans (*Homo habilis*) appears | Skull and other fragments | Olduvai Gorge, Tanzania |
| 2 million | *Australopithecus robustus* comes into existence | Jaw and skull fragments | Southern Africa |
| 1.8 million | *Homo erectus*, first hominid to leave Africa, emerges | Nearly complete skeleton | Kenya |
| 1.6 million | Earliest evidence of *Homo* genus ("Java Man") in Asia | Skull and limb bone | Java, Indonesia |
| 1.5 million | "Turkana Boy," most complete specimen of *Homo erectus*, found | Nearly complete skeleton | Southern Africa |
| 1.2 million | *Australopithecus robustus* known to be extinct in Africa | | |
| 800,000 | Oldest evidence of human life in Western Europe | Jawbone | near Burgos, northern Spain |
| 600,000 | Emergence of *Homo heidelbergensis* | Jawbone | near Heidelberg, Germany |
| 260,000 | Possible existence of *Homo sapiens* | Africa | |
| 250,000 | Emergence of *Homo neanderthalensis* | Skull, limb bones, and other fragments | Neander Valley, Germany |
| 160,000 | Existence of *Homo sapiens* confirmed | Skull fragments | Middle Awash, Ethiopia |
| 120,000 | *Homo sapiens* reaches southern Africa | Skeletons and stone tools | Kwazulu–Natal, South Africa |
| 100,000 | *Homo sapiens* reaches Near East | Skeletons at grave site | near Nazareth, Israel |
| 60,000 | *Homo sapiens* reaches China | Skull and teeth | Shaanxi and Guizhou Provinces, China |
| 55,000 | *Homo sapiens* reaches Australia | Skull | Lake Mungo, South Australia |
| 40,000 | Cro-Magnons appear in Europe | Skeletons | Dordogne region, southern France |
| 30,000 | *Homo neanderthalensis* disappears | | |
| 13,500 | *Homo sapiens* crosses into American continent | Skull | La Brea Tar Pits, Los Angeles, USA |
| 12,500 | *Homo sapiens* reaches South America | Artefacts | Monte Verde, Chile |

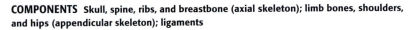

# Skeletal system

The skeleton is a strong but flexible framework that provides support for all other body parts, and protects delicate organs such as the brain and heart (see pp.224–25). Bone is made of living tissue; it is a reservoir of important minerals, especially calcium and phosphorus, and also makes new cells for the blood. The bones are arranged symmetrically on either side of the body into two main divisions: the axial and appendicular skeletons. The axial skeleton consists of 80 bones, including the skull,

spine, ribs, and breastbone. Attached to these bones is the appendicular skeleton, which has 126 bones and provides the body with mobility. Bones in this division, which include those of the arms and legs, have many joints. Also part of the appendicular skeleton are the shoulder blades and pelvis, the linking structures between the limbs and the core of the body.

**COMPONENTS** **Skull, spine, ribs, and breastbone (axial skeleton); limb bones, shoulders, and hips (appendicular skeleton); ligaments**

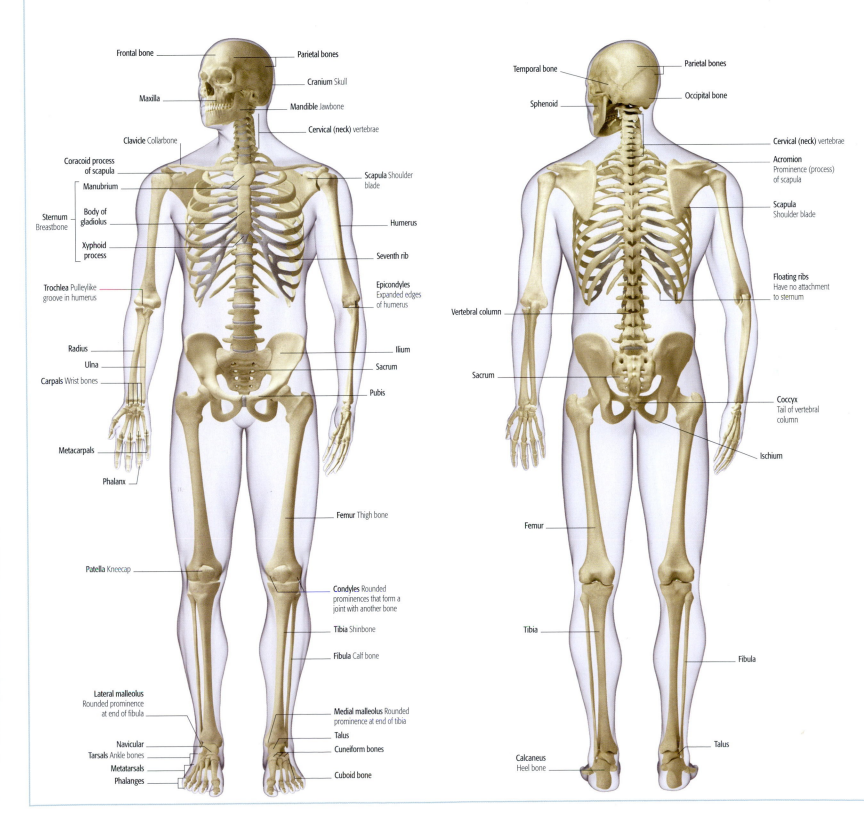

Frontal bone
Parietal bones
Cranium Skull
Maxilla
Mandible Jawbone
Cervical (neck) vertebrae
Clavicle Collarbone
Coracoid process of scapula
Scapula Shoulder blade
Manubrium
Body of gladiolus
Sternum Breastbone
Humerus
Xyphoid process
Seventh rib
Trochlea Pulleylike groove in humerus
Epicondyles Expanded edges of humerus
Radius
Ilium
Ulna
Sacrum
Carpals Wrist bones
Pubis
Metacarpals
Phalanx
Femur Thigh bone
Patella Kneecap
Condyles Rounded prominences that form a joint with another bone
Tibia Shinbone
Fibula Calf bone
Lateral malleolus Rounded prominence at end of fibula
Medial malleolus Rounded prominence at end of tibia
Talus
Navicular
Cuneiform bones
Tarsals Ankle bones
Metatarsals
Phalanges
Cuboid bone

Temporal bone
Parietal bones
Sphenoid
Occipital bone
Cervical (neck) vertebrae
Acromion Prominence (process) of scapula
Scapula Shoulder blade
Floating ribs Have no attachment to sternum
Vertebral column
Sacrum
Coccyx Tail of vertebral column
Ischium
Femur
Tibia
Fibula
Calcaneus Heel bone
Talus

# Muscular system

Muscles (see pp.224–25) account for about half the body's total weight. There are three types: skeletal, cardiac, and smooth muscle. Over 600 skeletal muscles make up the flesh of the body. They come in a variety of sizes—from large triangular slabs such as the deltoid in the shoulder, to thin slivers, such as the sartorius that runs from the hip to the inside knee. The muscles cover the skeleton in overlapping layers. Those just below the skin and its underlying fat are described as superficial and those beneath them are the deep muscles. They enable the body to move by contracting quickly and powerfully to pull on the bones, to which they are linked by cordlike tendons. In addition, they maintain a steady tension that sustains posture. The skeletal muscles work in groups, usually under voluntary control. Both the other types of muscles are involuntary and work automatically: cardiac muscle controls heart rhythm; and smooth muscle controls such actions as movement of food through the digestive tract.

**COMPONENTS  Skeletal muscles (attached to bones); smooth muscles within organs; tendons; cardiac muscle of heart**

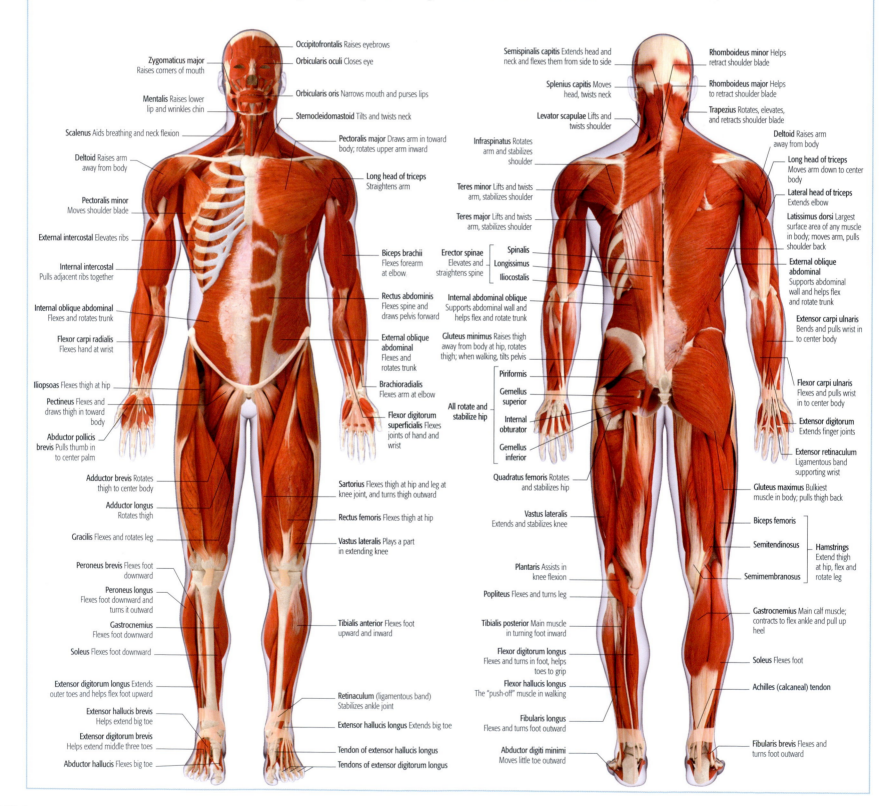

Occipitofrontalis Raises eyebrows

Zygomaticus major Raises corners of mouth

Orbicularis oculi Closes eye

Mentalis Raises lower lip and wrinkles chin

Orbicularis oris Narrows mouth and purses lips

Scalenus Aids breathing and neck flexion

Sternocleidomastoid Tilts and twists neck

Pectoralis major Draws arm in toward body; rotates upper arm inward

Deltoid Raises arm away from body

Long head of triceps Straightens arm

Pectoralis minor Moves shoulder blade

External intercostal Elevates ribs

Internal intercostal Pulls adjacent ribs together

Biceps brachii Flexes forearm at elbow

Rectus abdominis Flexes spine and draws pelvis forward

Internal oblique abdominal Flexes and rotates trunk

External oblique abdominal Flexes and rotates trunk

Flexor carpi radialis Flexes hand at wrist

Brachioradialis Flexes arm at elbow

Iliopsoas Flexes thigh at hip

Flexor digitorum superficialis Flexes joints of hand and wrist

Pectineus Flexes and draws thigh in toward body

Abductor pollicis brevis Pulls thumb in to center palm

Adductor brevis Rotates thigh to center body

Sartorius Flexes thigh at hip and leg at knee joint, and turns thigh outward

Adductor longus Rotates thigh

Rectus femoris Flexes thigh at hip

Gracilis Flexes and rotates leg

Vastus lateralis Plays a part in extending knee

Peroneus brevis Flexes foot downward

Peroneus longus Flexes foot downward and turns it outward

Gastrocnemius Flexes foot downward

Tibialis anterior Flexes foot upward and inward

Soleus Flexes foot downward

Extensor digitorum longus Extends outer toes and helps flex foot upward

Retinaculum (ligamentous band) Stabilizes ankle joint

Extensor hallucis brevis Helps extend big toe

Extensor hallucis longus Extends big toe

Extensor digitorum brevis Helps extend middle three toes

Tendon of extensor hallucis longus

Abductor hallucis Flexes big toe

Tendons of extensor digitorum longus

Semispinalis capitis Extends head and neck and flexes them from side to side

Rhomboideus minor Helps retract shoulder blade

Splenius capitis Moves head, twists neck

Rhomboideus major Helps to retract shoulder blade

Levator scapulae Lifts and twists shoulder

Trapezius Rotates, elevates, and retracts shoulder blade

Infraspinatus Rotates arm and stabilizes shoulder

Deltoid Raises arm away from body

Teres minor Lifts and twists arm, stabilizes shoulder

Long head of triceps Moves arm down to center body

Teres major Lifts and twists arm, stabilizes shoulder

Lateral head of triceps Extends elbow

Erector spinae Elevates and straightens spine — Spinalis — Longissimus — Iliocostalis

Latissimus dorsi Largest surface area of any muscle in body; moves arm, pulls shoulder back

External oblique abdominal Supports abdominal wall and helps flex and rotate trunk

Internal abdominal oblique Supports abdominal wall and helps flex and rotate trunk

Gluteus minimus Raises thigh away from body at hip, rotates thigh; when walking, tilts pelvis

Extensor carpi ulnaris Bends and pulls wrist in to center body

All rotate and stabilize hip — Piriformis — Gemellus superior — Internal obturator — Gemellus inferior

Flexor carpi ulnaris Flexes and pulls wrist in to center body

Extensor digitorum Extends finger joints

Extensor retinaculum Ligamentous band supporting wrist

Quadratus femoris Rotates and stabilizes hip

Gluteus maximus Bulkiest muscle in body; pulls thigh back

Vastus lateralis Extends and stabilizes knee

Biceps femoris

Semitendinosus

Hamstrings Extend thigh at hip, flex and rotate leg

Plantaris Assists in knee flexion

Semimembranosus

Popliteus Flexes and turns leg

Gastrocnemius Main calf muscle; contracts to flex ankle and pull up heel

Tibialis posterior Main muscle in turning foot inward

Flexor digitorum longus Flexes and turns in foot, helps toes to grip

Soleus Flexes foot

Flexor hallucis longus The "push-off" muscle in walking

Achilles (calcaneal) tendon

Fibularis longus Flexes and turns foot outward

Soleus Flexes foot

Abductor digiti minimi Moves little toe outward

Fibularis brevis Flexes and turns foot outward

# Nervous system

The body's prime communication and coordination network, the nervous system (see pp.220–21) has two main parts. The central nervous system (CNS) comprises the brain and the spinal cord. Nerves branch from the CNS to form the peripheral nervous system (PNS), which extends to every organ and tissue of the body. A third component, the autonomic nervous system (ANS), controls involuntary body functions such as heart rate.

**COMPONENTS** Brain; spinal cord; peripheral nerves; sense organs

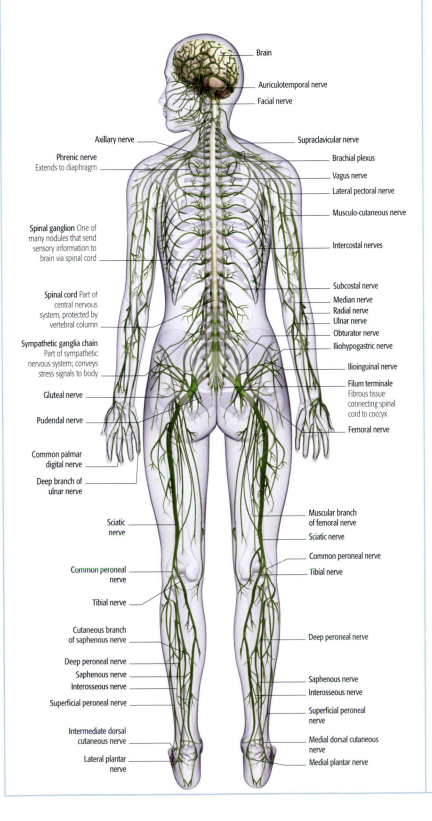

Brain

Auriculotemporal nerve

Facial nerve

Axillary nerve

Supraclavicular nerve

Phrenic nerve
Extends to diaphragm

Brachial plexus

Vagus nerve

Lateral pectoral nerve

Musculo-cutaneous nerve

Spinal ganglion One of many nodules that send sensory information to brain via spinal cord

Intercostal nerves

Spinal cord Part of central nervous system, protected by vertebral column

Subcostal nerve

Median nerve

Radial nerve

Ulnar nerve

Obturator nerve

Iliohypogastric nerve

Sympathetic ganglia chain Part of sympathetic nervous system; conveys stress signals to body

Ilioinguinal nerve

Filum terminale
Fibrous tissue connecting spinal cord to coccyx

Gluteal nerve

Pudendal nerve

Femoral nerve

Common palmar digital nerve

Deep branch of ulnar nerve

Sciatic nerve

Muscular branch of femoral nerve

Sciatic nerve

Common peroneal nerve

Common peroneal nerve

Tibial nerve

Tibial nerve

Cutaneous branch of saphenous nerve

Deep peroneal nerve

Deep peroneal nerve

Saphenous nerve

Saphenous nerve

Interosseous nerve

Interosseous nerve

Superficial peroneal nerve

Superficial peroneal nerve

Intermediate dorsal cutaneous nerve

Medial dorsal cutaneous nerve

Lateral plantar nerve

Medial plantar nerve

# Endocrine system

The body's endocrine system is composed of bodies of glandular tissue, such as the thyroid, and glands within certain organs, such as the testis, ovary, and heart. These glands secrete hormones—chemical messengers that regulate many body processes—including the breakdown of chemical substances in metabolism, fluid balance and urine production, growth and development, and sexual reproduction (see pp.274–75).

**COMPONENTS** Pituitary gland; pineal gland; hypothalamus; thyroid gland; thymus gland; heart; stomach; pancreas; intestines; adrenal glands; kidney; ovaries (female); testes (male)

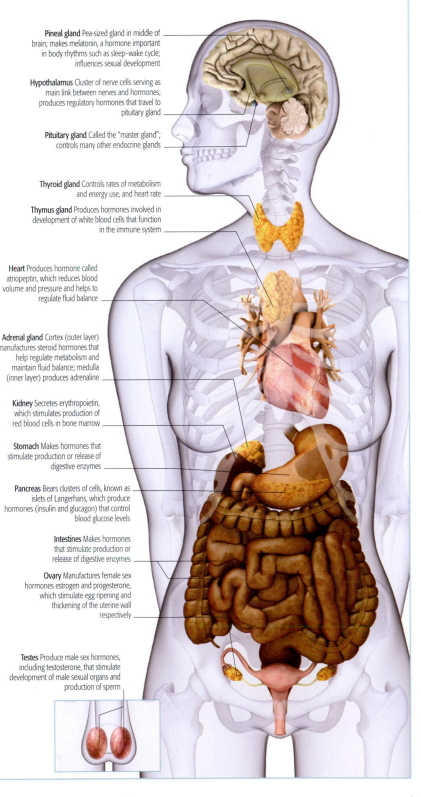

Pineal gland Pea-sized gland in middle of brain; makes melatonin, a hormone important in body rhythms such as sleep–wake cycle; influences sexual development

Hypothalamus Cluster of nerve cells serving as main link between nerves and hormones; produces regulatory hormones that travel to pituitary gland

Pituitary gland Called the "master gland"; controls many other endocrine glands

Thyroid gland Controls rates of metabolism and energy use, and heart rate

Thymus gland Produces hormones involved in development of white blood cells that function in the immune system

Heart Produces hormone called atriopeptin, which reduces blood volume and pressure and helps to regulate fluid balance

Adrenal gland Cortex (outer layer) manufactures steroid hormones that help regulate metabolism and maintain fluid balance; medulla (inner layer) produces adrenaline

Kidney Secretes erythropoietin, which stimulates production of red blood cells in bone marrow

Stomach Makes hormones that stimulate production or release of digestive enzymes

Pancreas Bears clusters of cells, known as islets of Langerhans, which produce hormones (insulin and glucagon) that control blood glucose levels

Intestines Makes hormones that stimulate production or release of digestive enzymes

Ovary Manufactures female sex hormones estrogen and progesterone, which stimulate egg ripening and thickening of the uterine wall respectively

Testes Produce male sex hormones, including testosterone, that stimulate development of male sexual organs and production of sperm

# Cardiovascular system

Also known as the circulatory system, the cardiovascular system comprises the heart, blood vessels—arteries, veins, capillaries, and many smaller vessels—and blood (see pp.90–91). The heart continuously pumps oxygen-rich blood round the body to all organs and tissues through a network of arteries. Deoxygenated blood returning to the heart through the veins carries away wastes produced by body cell processes.

**COMPONENTS   Heart; blood; major blood vessels (arteries and veins); minor blood vessels (arterioles, venules, and capillaries)**

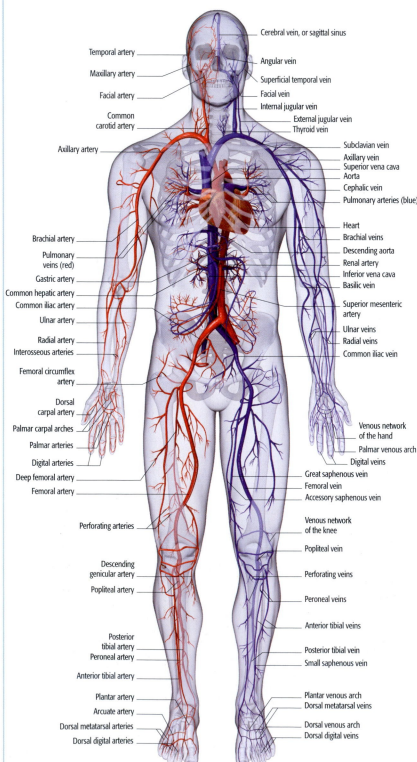

Cerebral vein, or sagittal sinus

Temporal artery

Maxillary artery

Facial artery

Common carotid artery

Axillary artery

Angular vein

Superficial temporal vein

Facial vein

Internal jugular vein

External jugular vein

Thyroid vein

Subclavian vein

Axillary vein

Superior vena cava

Aorta

Cephalic vein

Pulmonary arteries (blue)

Brachial artery

Pulmonary veins (red)

Gastric artery

Common hepatic artery

Common iliac artery

Ulnar artery

Radial artery

Interosseous arteries

Femoral circumflex artery

Heart

Brachial veins

Descending aorta

Renal artery

Inferior vena cava

Basilic vein

Superior mesenteric artery

Ulnar veins

Radial veins

Common iliac vein

Dorsal carpal artery

Palmar carpal arches

Palmar arteries

Digital arteries

Deep femoral artery

Femoral artery

Venous network of the hand

Palmar venous arch

Digital veins

Great saphenous vein

Femoral vein

Accessory saphenous vein

Perforating arteries

Venous network of the knee

Popliteal vein

Descending genicular artery

Popliteal artery

Perforating veins

Peroneal veins

Anterior tibial veins

Posterior tibial artery

Peroneal artery

Anterior tibial artery

Plantar artery

Arcuate artery

Dorsal metatarsal arteries

Dorsal digital arteries

Posterior tibial vein

Small saphenous vein

Plantar venous arch

Dorsal metatarsal veins

Dorsal venous arch

Dorsal digital veins

# Respiratory system

The respiratory tract carries oxygen into the body and removes carbon dioxide, the waste produced by cells (see pp.270–71). With each inhalation, air passes through passages of varying sizes until it reaches tiny air sacs deep inside the lungs. Here, vital oxygen diffuses across the thin walls of the sacs and into the bloodstream. In exchange, carbon dioxide passes into the lungs, from where it is exhaled. The respiratory system is powered by breathing muscles—the diaphragm and the intercostal muscles between the ribs. On inhalation, these contract to increase the space inside the chest cavity, which allows the lungs to expand and take in air. On exhalation, the breathing muscles return to their resting positions.

**COMPONENTS   Nasal passage; throat (pharynx); windpipe (trachea); lungs; major and minor lung airways (bronchi and bronchioles); diaphragm and other respiratory muscles**

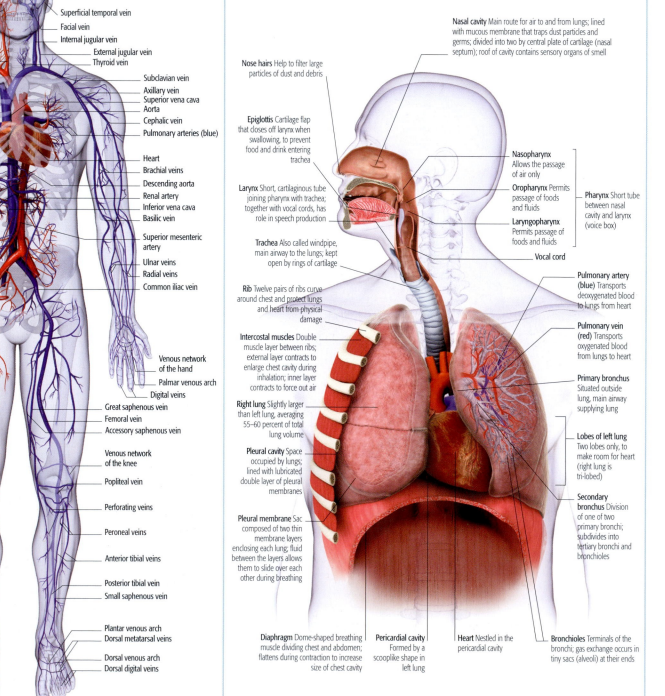

**Nasal cavity** Main route for air to and from lungs; lined with mucous membrane that traps dust particles and germs; divided into two by central plate of cartilage (nasal septum); roof of cavity contains sensory organs of smell

**Nose hairs** Help to filter large particles of dust and debris

**Epiglottis** Cartilage flap that closes off larynx when swallowing, to prevent food and drink entering trachea

**Larynx** Short, cartilaginous tube joining pharynx with trachea; together with vocal cords, has role in speech production

**Trachea** Also called windpipe, main airway to the lungs; kept open by rings of cartilage

**Nasopharynx** Allows the passage of air only

**Oropharynx** Permits passage of foods and fluids

**Laryngopharynx** Permits passage of foods and fluids

**Pharynx** Short tube between nasal cavity and larynx (voice box)

**Vocal cord**

**Rib** Twelve pairs of ribs curve around chest and protect lungs and heart from physical damage

**Intercostal muscles** Double muscle layer between ribs; external layer contracts to enlarge chest cavity during inhalation; inner layer contracts to force out air

**Right lung** Slightly larger than left lung, averaging 55–60 percent of total lung volume

**Pleural cavity** Space occupied by lungs; lined with lubricated double layer of pleural membranes

**Pleural membrane** Sac composed of two thin membrane layers enclosing each lung; fluid between the layers allows them to slide over each other during breathing

**Pulmonary artery (blue)** Transports deoxygenated blood to lungs from heart

**Pulmonary vein (red)** Transports oxygenated blood from lungs to heart

**Primary bronchus** Situated outside lung, main airway supplying lung

**Lobes of left lung** Two lobes only, to make room for heart (right lung is tri-lobed)

**Secondary bronchus** Division of one of two primary bronchi; subdivides into tertiary bronchi and bronchioles

**Diaphragm** Dome-shaped breathing muscle dividing chest and abdomen; flattens during contraction to increase size of chest cavity

**Pericardial cavity** Formed by a scooplike shape in left lung

**Heart** Nestled in the pericardial cavity

**Bronchioles** Terminals of the bronchi; gas exchange occurs in tiny sacs (alveoli) at their ends

# Lymphatic and immune system

The body has an internal defense mechanism, the immune system, which is based on specialized white blood cells and helps to provide protection against disease. The lymphatic system is an integral part of the immune system. Its active element is lymph, a watery fluid that circulates the body and drains through lymph nodes that filter out harmful organisms.

**COMPONENTS** White blood cells; antibodies; spleen; tonsils and adenoids; thymus gland; lymph fluid; lymph vessels, nodes, and ducts

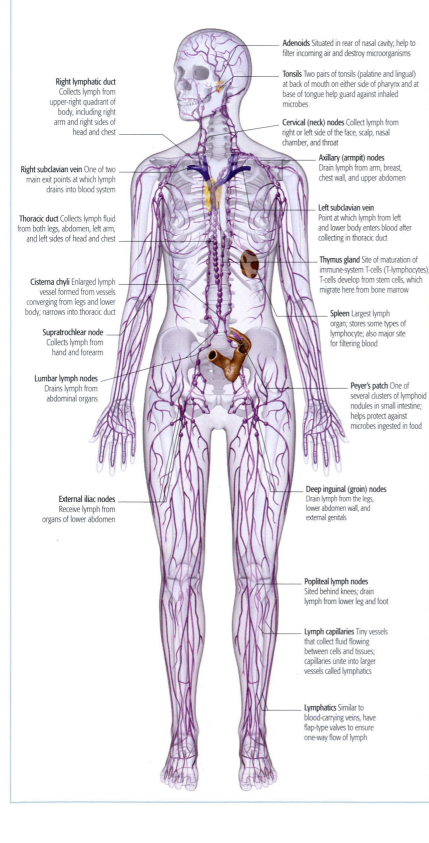

**Adenoids** Situated in rear of nasal cavity; help to filter incoming air and destroy microorganisms

**Tonsils** Two pairs of tonsils (palatine and lingual) at back of mouth on either side of pharynx and at base of tongue help guard against inhaled microbes

**Cervical (neck) nodes** Collect lymph from right or left side of the face, scalp, nasal chamber, and throat

**Axillary (armpit) nodes** Drain lymph from arm, breast, chest wall, and upper abdomen

**Left subclavian vein** Point at which lymph from left and lower body enters blood after collecting in thoracic duct

**Thymus gland** Site of maturation of immune-system T-cells (T-lymphocytes); T-cells develop from stem cells, which migrate here from bone marrow

**Spleen** Largest lymph organ; stores some types of lymphocyte; also major site for filtering blood

**Peyer's patch** One of several clusters of lymphoid nodules in small intestine; helps protect against microbes ingested in food

**Deep inguinal (groin) nodes** Drain lymph from the legs, lower abdomen wall, and external genitals

**Popliteal lymph nodes** Sited behind knees; drain lymph from lower leg and foot

**Lymph capillaries** Tiny vessels that collect fluid flowing between cells and tissues; capillaries unite into larger vessels called lymphatics

**Lymphatics** Similar to blood-carrying veins, have flap-type valves to ensure one-way flow of lymph

**Right lymphatic duct** Collects lymph from upper-right quadrant of body, including right arm and right sides of head and chest

**Right subclavian vein** One of two main exit points at which lymph drains into blood system

**Thoracic duct** Collects lymph fluid from both legs, abdomen, left arm, and left sides of head and chest

**Cisterna chyli** Enlarged lymph vessel formed from vessels converging from legs and lower body; narrows into thoracic duct

**Supratrochlear node** Collects lymph from hand and forearm

**Lumbar lymph nodes** Drains lymph from abdominal organs

**External iliac nodes** Receive lymph from organs of lower abdomen

# Digestive system

The long passageway known as the digestive tract starts at the mouth and continues through the esophagus, stomach, and intestines to the anus. Along its length, food is broken down and nutrients absorbed, while waste is formed into feces to be eliminated from the body. The liver, gallbladder, and pancreas are also associated with digestion (see pp.216–17).

**COMPONENTS** Mouth; throat (pharynx); esophagus; stomach; pancreas; liver; gallbladder; small intestine (duodenum, jejunum, and ileum); large intestine (colon and rectum); anus

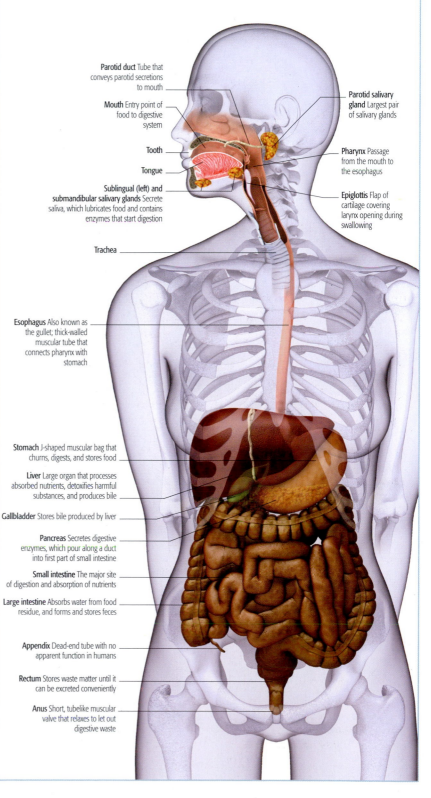

**Parotid duct** Tube that conveys parotid secretions to mouth

**Mouth** Entry point of food to digestive system

**Tooth**

**Tongue**

**Sublingual (left) and submandibular salivary glands** Secrete saliva, which lubricates food and contains enzymes that start digestion

**Trachea**

**Esophagus** Also known as the gullet; thick-walled muscular tube that connects pharynx with stomach

**Parotid salivary gland** Largest pair of salivary glands

**Pharynx** Passage from the mouth to the esophagus

**Epiglottis** Flap of cartilage covering larynx opening during swallowing

**Stomach** J-shaped muscular bag that churns, digests, and stores food

**Liver** Large organ that processes absorbed nutrients, detoxifies harmful substances, and produces bile

**Gallbladder** Stores bile produced by liver

**Pancreas** Secretes digestive enzymes, which pour along a duct into first part of small intestine

**Small intestine** The major site of digestion and absorption of nutrients

**Large intestine** Absorbs water from food residue, and forms and stores feces

**Appendix** Dead-end tube with no apparent function in humans

**Rectum** Stores waste matter until it can be excreted conveniently

**Anus** Short, tubelike muscular valve that relaxes to let out digestive waste

# Urinary system

The urinary tract is the body's filtering system. Circulating blood passes through the kidneys, which remove waste products and excess water and convert them into urine. Long tubes, called ureters, convey urine from the kidneys to the bladder for storage and excretion. As the urinary system filters the blood, it regulates the body's levels of fluids and chemicals.

**COMPONENTS** Kidneys; ureters; bladder; urethra

# Reproductive system

Of the body's major systems, the reproductive system (see pp.226–27) is the one that differs most between males and females. In both sexes the system starts to function at puberty, enabling the production of offspring by bringing together male and female sex cells. Hormones manufactured by the ovaries in women and the testes in men influence the development of sexual characteristics such as the shape of the body.

**COMPONENTS** Female: ovaries; fallopian tubes; uterus; vagina and external genitalia; breasts. Male: testes; spermatic ducts, seminal vesicles; urethra; penis; prostate and bulbourethral glands

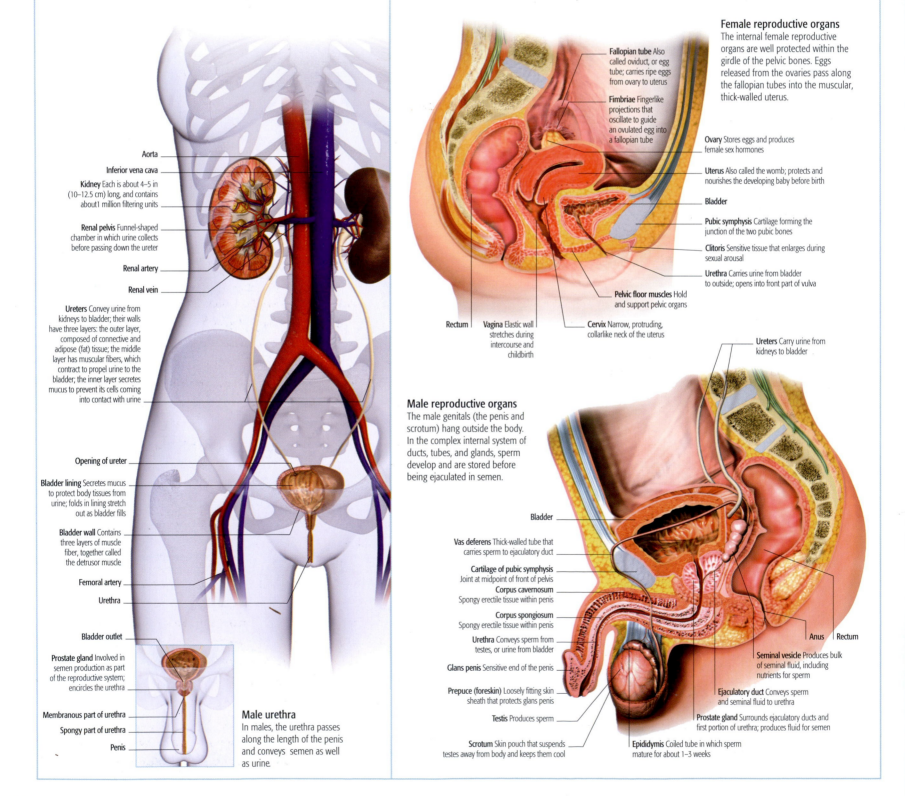

**Aorta**

**Inferior vena cava**

**Kidney** Each is about 4–5 in (10–12.5 cm) long, and contains about 1 million filtering units

**Renal pelvis** Funnel-shaped chamber in which urine collects before passing down the ureter

**Renal artery**

**Renal vein**

**Ureters** Convey urine from kidneys to bladder; their walls have three layers: the outer layer, composed of connective and adipose (fat) tissue; the middle layer has muscular fibers, which contract to propel urine to the bladder; the inner layer secretes mucus to prevent its cells coming into contact with urine

**Opening of ureter**

**Bladder lining** Secretes mucus to protect body tissues from urine; folds in lining stretch out as bladder fills

**Bladder wall** Contains three layers of muscle fiber, together called the detrusor muscle

**Femoral artery**

**Urethra**

**Bladder outlet**

**Prostate gland** Involved in semen production as part of the reproductive system; encircles the urethra

**Membranous part of urethra**

**Spongy part of urethra**

**Penis**

**Male urethra**
In males, the urethra passes along the length of the penis and conveys semen as well as urine.

### Female reproductive organs
The internal female reproductive organs are well protected within the girdle of the pelvic bones. Eggs released from the ovaries pass along the fallopian tubes into the muscular, thick-walled uterus.

**Fallopian tube** Also called oviduct, or egg tube; carries ripe eggs from ovary to uterus

**Fimbriae** Fingerlike projections that oscillate to guide an ovulated egg into a fallopian tube

**Ovary** Stores eggs and produces female sex hormones

**Uterus** Also called the womb; protects and nourishes the developing baby before birth

**Bladder**

**Pubic symphysis** Cartilage forming the junction of the two pubic bones

**Clitoris** Sensitive tissue that enlarges during sexual arousal

**Urethra** Carries urine from bladder to outside; opens into front part of vulva

**Pelvic floor muscles** Hold and support pelvic organs

**Rectum**

**Vagina** Elastic wall stretches during intercourse and childbirth

**Cervix** Narrow, protruding, collarlike neck of the uterus

**Ureters** Carry urine from kidneys to bladder

### Male reproductive organs
The male genitals (the penis and scrotum) hang outside the body. In the complex internal system of ducts, tubes, and glands, sperm develop and are stored before being ejaculated in semen.

**Bladder**

**Vas deferens** Thick-walled tube that carries sperm to ejaculatory duct

**Cartilage of pubic symphysis** Joint at midpoint of front of pelvis

**Corpus cavernosum** Spongy erectile tissue within penis

**Corpus spongiosum** Spongy erectile tissue within penis

**Urethra** Conveys sperm from testes, or urine from bladder

**Glans penis** Sensitive end of the penis

**Prepuce (foreskin)** Loosely fitting skin sheath that protects glans penis

**Testis** Produces sperm

**Scrotum** Skin pouch that suspends testes away from body and keeps them cool

**Anus** **Rectum**

**Seminal vesicle** Produces bulk of seminal fluid, including nutrients for sperm

**Ejaculatory duct** Conveys sperm and seminal fluid to urethra

**Prostate gland** Surrounds ejaculatory ducts and first portion of urethra; produces fluid for semen

**Epididymis** Coiled tube in which sperm mature for about 1–3 weeks

# Skin, hair, and nails

The body's outer coverings of skin, hair, and nails (together known as the integumentary system) provide first-line defense against physical injury and harmful microorganisms. Skin has two main layers. The top layer, or epidermis, is a tough waterproof tissue with a surface of dead cells that are constantly sloughed off. Beneath the epidermis is the dermis. This layer contains sweat glands, sebaceous glands that produce oily sebum to lubricate the skin, and sensory organs that keep the body informed about pressure, heat, and cold. The dermis also contains blood vessels that regulate body temperature. Below the epidermis and dermis is the subcutaneous fat layer, which acts as an insulator and shock absorber.

**COMPONENTS** **Skin; hair; nails; subcutaneous fat layer**

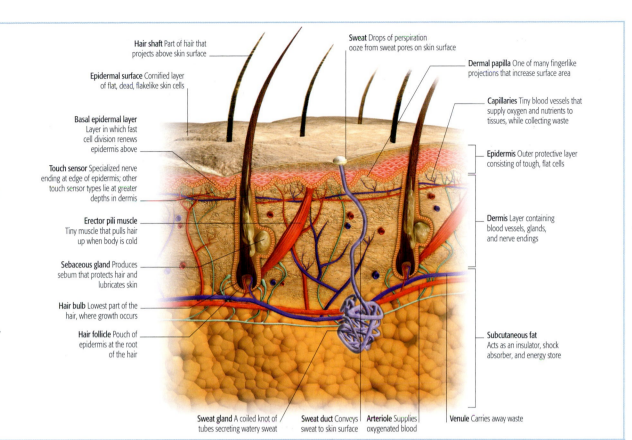

Hair shaft Part of hair that projects above skin surface

Sweat Drops of perspiration ooze from sweat pores on skin surface

Epidermal surface Cornified layer of flat, dead, flakelike skin cells

Dermal papilla One of many fingerlike projections that increase surface area

Basal epidermal layer Layer in which fast cell division renews epidermis above

Capillaries Tiny blood vessels that supply oxygen and nutrients to tissues, while collecting waste

Touch sensor Specialized nerve ending at edge of epidermis; other touch sensor types lie at greater depths in dermis

Epidermis Outer protective layer consisting of tough, flat cells

Erector pili muscle Tiny muscle that pulls hair up when body is cold

Dermis Layer containing blood vessels, glands, and nerve endings

Sebaceous gland Produces sebum that protects hair and lubricates skin

Hair bulb Lowest part of the hair, where growth occurs

Hair follicle Pouch of epidermis at the root of the hair

Subcutaneous fat Acts as an insulator, shock absorber, and energy store

Sweat gland A coiled knot of tubes secreting watery sweat

Sweat duct Conveys sweat to skin surface

Arteriole Supplies oxygenated blood

Venule Carries away waste

# Chromosome set

Each body cell contains genetic material (DNA) arranged on structures called chromosomes (see pp.308–309). There are 46 chromosomes in a full set. Of these, 22 are matching pairs, one chromosome of which is derived from the mother and one from the father. The 23rd pair are the sex chromosomes: XX signifying female and XY (shown here) male. When stained by chemicals, each chromosome shows a banding pattern that makes it possible for the locations of genes to be mapped (see pp.410–11).

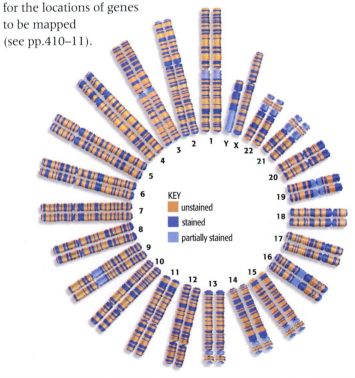

KEY
- unstained
- stained
- partially stained

| Chromosome No. | Genes | Examples of gene functions |
|---|---|---|
| 1 | 3,100 | Pancreatic secretion; tumor suppression; type of collagen; coagulation factor V (F5) |
| 2 | 1,900 | Red hair color; growth of bone and cartilage (BMPR2) |
| 3 | 2,000 | Rhodopsin retinal pigment (RHO); smell (olfactory receptors); DNA repair (MLH1) |
| 4 | 1,200 | Red hair color (HCL2); coagulation factor XI; dentin (DSPP) |
| 5 | 1,300 | Taste receptor; growth hormone receptor; embryonic development (NIPBL) |
| 6 | 1,500 | Control of light-sensing pigments (RDS); iron balance (HFE); immune system (HLA-B) |
| 7 | 1,500 | Type of collagen; color blindness (blue/yellow); growth-rate controlling factor |
| 8 | 1,000 | Fibroblast growth factor receptor 1; fat transportation and breakdown (LPL) |
| 9 | 1,100 | Blood group; removal of excess cholesterol (ABCA1) |
| 10 | 1,100 | Fibroblast growth factor receptor 2; tumor suppression (PTEN) |
| 11 | 1,800 | Albinism (OCA1); beta hemoglobin (HBB); parathyroid hormone; calcitonin |
| 12 | 1,400 | Interferon; type of collagen; regulate muscle proteins (PPP1R12A) |
| 13 | 550 | Cholesterol-lowering factor; coagulation factors VII and X |
| 14 | 1,300 | Thyroid-stimulating hormone factor; coagulation factor C |
| 15 | 950 | Brown and blue eye color (Bey 1/Bey 2); brown hair color (HC13); albinism (OCA2) |
| 16 | 1,100 | Alpha hemoglobin; skin color—melanin (MC1R); breakdown of fatty acids (MLYCD) |
| 17 | 1,500 | Growth hormone; tumor suppression (FLCN); urea production in liver (NAGS) |
| 18 | 400 | Controlling inflammation (MEFV); tumor suppression (SMAD4) |
| 19 | 1,700 | Green eye color (Gey 1); brown hair color (HCL1); protection of nerve cells (PRX) |
| 20 | 700 | Embryonic development (JAG1); nervous system function (PRNP) |
| 21 | 350 | Calcium channels in cardiac muscle and ears (KCNE1) |
| 22 | 700 | Tumor suppression (CHEK2); normal nerve function (NEFH) |
| X | 1,300 | Color blindness (red/green); maternal–fetal interface role; brain development (CDKL5) |
| Y | 300 | Sex determination; skeletal development (SHOX), also present on X chromosome |

# Sources and effects of hormones

The various glands of the hormonal system constitute a control and communications network that is complementary to the nervous system. However, instead of using nerve impulses, the glands secrete chemical messages (in the form of hormones) to affect other glands and tissues in various parts of the body. Hormones are carried in the bloodstream to their targets, where they exert their specific effects. The table below lists the hormones that are secreted by different parts of the body and describes their wide-ranging actions.

| Gland or hormone-secreting tissue | Hormone secreted | Effects |
| --- | --- | --- |
| Hypothalamus | Releasing or inhibiting hormones | Stimulate or suppress hormone secretion by pituitary gland |
| Pituitary gland | Growth hormone | Stimulates growth and metabolism |
| | Prolactin | Stimulates milk production after childbirth |
| | ACTH (adrenocorticotrophic hormone) | Stimulates hormone production by adrenal glands |
| | TSH (thyroid-stimulating hormone) | Stimulates hormone production by thyroid gland |
| | FSH (follicle-stimulating hormone); LH (luteinizing hormone) | Stimulate gonads (ovaries or testes) |
| | ADH (antidiuretic hormone) | Acts on kidneys to conserve water |
| | Oxytocin | Stimulates contractions of uterus during labor and milk letdown reflex in breast-feeding |
| | MSH (melanocyte-stimulating hormone) | Acts on skin to promote production of skin pigment (melanin) |
| Brain | Endorphins; enkephalins | Alleviate pain; boost mood |
| Thyroid gland | Thyroid hormones | Increase metabolic rate; affect growth |
| | Calcitonin | Lowers level of calcium in blood |
| Parathyroid glands | Parathyroid hormone | Increases level of calcium in blood |
| Thymus | Thymic hormone | Stimulates lymphocyte development |
| Heart | Atrial natriuretic factor | Lowers blood pressure |
| Adrenal glands | Adrenaline (epinephrine); noradrenaline (norepinephrine) | Prepare body for physical and mental stress |
| | Hydrocortisone | Affects metabolism |
| | Aldosterone | Regulates sodium and potassium excretion by kidneys |
| | Androgens | Affect growth and sex drive (in both males and females) |
| Kidneys | Renin | Regulates blood pressure |
| | Erythropoietin | Stimulates production of red blood cells |
| Pancreas | Insulin | Lowers blood sugar level |
| | Glucagon | Raises blood sugar level |
| Placenta | Chorionic gonadotrophin (HCG); estrogens; progesterone | Maintain pregnancy |
| Gastrointestinal tract | Gastrin; secretin; cholecystokinin | Regulate secretion of some digestive enzymes |
| Testes | Testosterone | Affects development of male secondary sexual characteristics and genital organs |
| Ovaries | Estrogens; progesterone | Affect development of female secondary sexual characteristics and genital organs; control menstrual cycle; maintain pregnancy |

# Functions of vitamins and minerals

Vitamins are organic substances that are mostly incorporated into coenzymes—molecules that assist and support enzymes in the control of metabolic processes. Regular vitamin intake is required because only a few vitamins can be manufactured in the body. Minerals are simple inorganic substances such as calcium, iron, chloride, and iodine. They are needed both for general metabolism and specialized uses, such as iron for hemoglobin in red blood cells.

| Function | Vitamins and minerals needed |
| --- | --- |
| Blood clotting | Vitamin K<br>Calcium<br>Iron |
| Blood cell formation and functioning | Vitamins $B_6$ and $B_{12}$<br>Vitamin E<br>Folic acid<br>Copper<br>Iron<br>Cobalt |
| Healthy skin and hair | Vitamin A<br>Vitamin $B_2$ (riboflavin)<br>Vitamin $B_3$ (niacin)<br>Vitamin $B_6$<br>Vitamin $B_{12}$<br>Biotin<br>Sulphur<br>Zinc |
| Heart functioning | Vitamin $B_1$ (thiamine)<br>Vitamin D<br>Inositol<br>Calcium<br>Potassium<br>Magnesium<br>Selenium<br>Sodium<br>Copper |
| Healthy teeth | Vitamins C and D<br>Calcium<br>Phosphorus<br>Fluorine<br>Magnesium<br>Boron |
| Healthy eyes | Vitamin A<br>Zinc |
| Bone formation | Vitamins A, C, and D<br>Fluorine<br>Calcium<br>Copper<br>Phosphorus<br>Magnesium<br>Boron |
| Muscle functioning | Vitamin $B_1$ (thiamine)<br>Vitamin $B_6$<br>Vitamin $B_{12}$<br>Vitamin E<br>Biotin<br>Calcium<br>Potassium<br>Sodium<br>Magnesium |

# Medical landmarks

The history of medicine is punctuated by breakthroughs that have radically changed our understanding of the human body and the processes of disease. In the 21st century the development of new treatments and technologies is accelerating rapidly, as scientists and the chemical industry make advances on an almost daily basis. The following are some of the medical events that have made a significant contribution to health.

| Date | Developments in diagnosis and prevention of disease |
|---|---|
| c.400 BCE | Disease concept introduced by the Greek physician Hippocrates. |
| c.1666 | Single-lens light microscope developed by the Dutch naturalist Antonie van Leeuwenhoek, who discovered microorganisms using it. |
| 1725 | Medical thermometer with a scale devised by Daniel Gabriel Fahrenheit. |
| 1796 | First smallpox vaccination given by the English physician Edward Jenner. The first true vaccine (comprising weakened microorganisms), against chicken cholera, was developed in 1880 by the French scientist Louis Pasteur. |
| 1816 | Stethoscope invented by the French physician René Laënnec. |
| 1850–1900 | The germ theory of disease proposed by the French scientist Louis Pasteur and developed by the German bacteriologist Robert Koch. |
| 1851 | Ophthalmoscope invented by the German scientist Herman von Helmholtz. |
| 1881 | The Czech–Austrian physician Samuel von Basch invented the sphygmomanometer to measure blood pressure. |
| 1887 | First electrocardiograph (ECG) described by the English physiologist Augustus Waller at St. Mary's Hospital, London. Dutch physiologist Willem Einthoven went on to develop the five phases of the ECG and won the Nobel Prize in 1924. |
| 1891 | Baby incubator introduced by the French doctor Alexandre Lion. |
| 1895 | X-rays discovered by the German physicist Wilhelm Röntgen. |
| 1898 | British doctor Ronald Ross proved that malaria is transmitted by mosquitoes. |
| 1901 | Hearing aid (electric) developed by the American inventor Miller Reese Hutchinson. |
| c.1932 | Transmission electron microscope (TEM) constructed by the German scientists Max Knoll and Ernst Ruska. |
| 1938 | First cardiac catheterization performed by George Peter Robb and Israel Steinberg in New York. |
| 1946 | Franz Greiter developed sunscreen after severe sunburn while climbing Piz Buin, a mountain on the Swiss–Austrian border. |
| 1950 | Amniocentesis to detect fetal disease developed by English doctor Douglas Bevis, but its significance went unrecognized for 10 years. |
| 1951 | Link discovered between smoking and lung cancer demonstrated by British scientists Richard Doll and Austin Bradford Hill. |
| 1957 | Fiber-optic endoscopy pioneered by the South African-born doctor Basil Hirschowitz at the University of Michigan. |
| 1958 | Ultrasound first used on pregnant women to examine the fetus by obstetrician and gynecologist Ian Donald, working in Scotland. |
| 1972 | CT scanner invented by the British engineer Godfrey Hounsfield and the South African-born physicist Alan Cormack of Tufts University, Massachusetts. |
| 1976 | Chorionic villus sampling developed by Chinese gynecologists to aid the early diagnosis of genetic disorders. |
| 1977 | First MRI scanner completed by Raymond Damadian and team in the US. |
| 1979 | The global eradication of smallpox was formally certified, following an intensive campaign launched by the World Health Organization in 1967. |
| 1985 | Rapid copying of DNA sequences using polymerase chain reaction developed by Kary Mullis of the Cetus Corporation, California. |
| 1995 | Chromosome sequence of *Haemophilus influenzae* identified. This was the first free-living organism to have its entire genome sequenced. |
| 2003 | Completion of the Human Genome Project by the International Human Genome International Consortium, following first draft produced in 2001. |
| 2006 | Cervical cancer vaccine developed by Scottish-born doctor Ian Frazer, and Chinese molecular virologist Jian Zhou, at the University of Queensland, Australia. |

| Date | Developments in surgery and drug treatment |
|---|---|
| 1545 | Basic surgical principles established by the French surgeon Ambroise Paré. |
| 1666 | Quinine (in form of Jesuits' bark) popularized by the British physician Thomas Sydenham for treating malaria. |
| 1785 | Use of digitalis to treat heart failure first described by the British physician William Withering. |
| 1805 | Morphine, extracted from opium, first used as pain relief by the German pharmacist Friedrich Sertürner. |
| 1842 | First surgery using general anesthesia performed by the American surgeon Crawford Long, who used ether. In 1845 the American dentist Horace Wells used nitrous oxide (laughing gas) as an anesthetic. In 1847 the British obstetrician James Simpson introduced chloroform anesthesia. |
| 1870 | Antiseptic surgery pioneered by the British surgeon Joseph Lister, who used a carbolic acid (phenol) spray to help prevent infection. |
| 1878 | Acetaminophen discovered by American chemistry professor Harmon Morse; its use was not realized until 70 years later. |
| 1899 | Aspirin first manufactured by German company Bayer. |
| 1901 | ABO blood groups discovered by the Austrian pathologist Karl Landsteiner, thereby establishing the basis for safe transfusions. |
| 1911 | Salvarsan introduced by the German bacteriologist Paul Ehrlich to treat syphilis. |
| 1920 | Adhesive bandages invented by Earle Dickson while working for Johnson & Johnson. |
| 1928 | British bacteriologist Alexander Fleming first recognized penicillin's antibacterial action. |
| 1935 | Sulphonamides' antibacterial action discovered by the German pharmacologist Gerhard Domagk. |
| 1943 | Kidney dialysis machine developed by the Dutch surgeon Willem Kolff. |
| 1951 | Oral contraceptive pill developed by the American doctors Gregory Pincus and John Rock, and the Austrian-born American chemist Carl Djerassi. |
| 1955 | First successful kidney transplant (between identical twins) performed by a team of American surgeons, led by Joseph Murray, of Harvard Medical School. |
| 1962 | First beta-blocker heart drug, nethalide (pronethalol), developed by scientists at Imperial Chemical Industries, England. |
| 1967 | First coronary artery bypass graft performed by Argentinian doctor René Géronimo Favaloro working in Cleveland, Ohio. |
| 1967 | First human heart transplant performed by the South African surgeon Christiaan Barnard at the Groote Schur Hospital, Cape Town. |
| 1969 | Scottish scientist David Jack discovered salbutamol to treat asthma. |
| 1976 | Coronary angioplasty introduced by Swiss surgeon Andreas Grüntzig at the University Hospital, Zurich. |
| 1978 | First "test-tube baby" Louise Brown, resulting from in vitro fertilization, born in the UK. British gynecologist Patrick Steptoe and the embryologist Robert Edwards developed the IVF techniques. |
| 1980 | German gynecologist Kurt Semm performed the first keyhole surgery. |
| 1982 | First genetically engineered medicine, human insulin, developed at Genentech, California. |
| 1986 | Zidovudine (originally called AZT) developed to treat HIV/AIDS by scientists at Burroughs Wellcome Research Laboratories, North Carolina. |
| 1998 | Viagra (sildenafil) introduced to treat erectile dysfunction after development by the Pfizer Corporation, US. |
| 2005 | First face transplant performed by French team led by Bernarad Devauchelle and Jean-Michel Dubernard on 38-year-old Isabelle Dinoire. |

# Chemistry

Chemistry is the study of the composition, structure, and properties of substances and the changes they undergo during chemical reactions. Organic chemistry studies the structures and reactions of carbon (organic) compounds; inorganic chemistry deals with the properties of all other elements; physical chemistry investigates the physical properties of chemical reactions; and biochemistry studies the compounds and reactions integral to life processes.

## Periodic table

The periodic table currently contains 117 elements, of which around 90 occur natually on Earth. The elements are structured in order of increasing atomic number, and are arranged in vertical "groups" and horizontal "periods" (see pp.232–33). Groupings of elements demonstrate close family resemblances, with reactive metallic elements on the left, through less reactive metals to reactive nonmetals, to barely reactive gases on the far right. Two rows of "inner transition elements" are positioned at the bottom of the table to avoid making the periods too long.

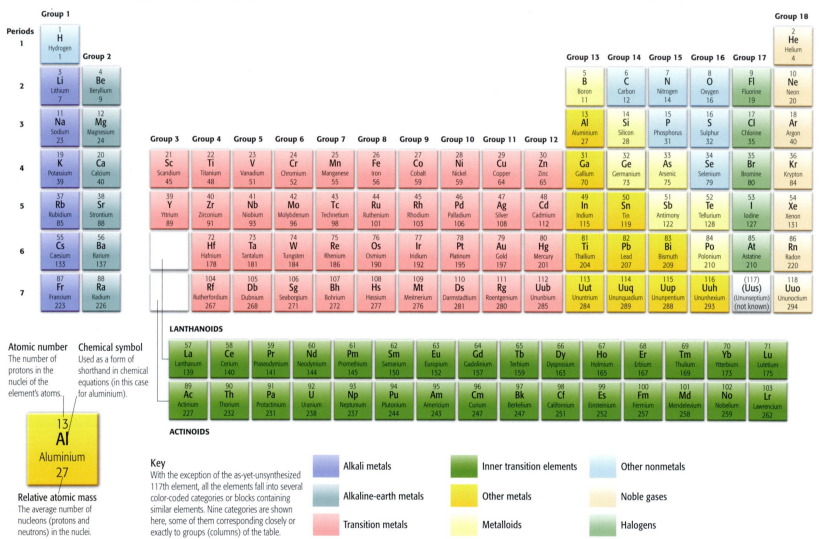

## Atomic mass

Atomic mass is the number of protons and neutrons in the nucleus of an atom. The greater the atomic mass of an atom, the smaller the atom is. Sulphur has an atomic weight of 32, and is 32 times heavier than hydrogen (atomic mass 1), 8 times heavier than helium (atomic mass 4), and twice as heavy as oxygen (atomic mass 16).

**1 Sulphur = 32 Hydrogen**
One sulphur atom has the same mass as 32 hydrogen atoms

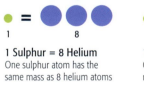

**1 Sulphur = 8 Helium**
One sulphur atom has the same mass as 8 helium atoms

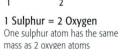

**1 Sulphur = 2 Oxygen**
One sulphur atom has the same mass as 2 oxygen atoms

# Table of Elements

This chart lays out the essential information for each of the known elements. It is sorted by atomic number—the number of protons in the nucleus of an element's atoms—and includes the commonly known symbol, atomic mass (number of protons and neutrons in the nucleus of an element's atoms) and valency (the number of chemical bonds formed by the atoms) for each given element.

| Atomic number | Element | Symbol | Atomic mass | Melting point °F | Melting point °C | Boiling point °F | Boiling point °C | Valency |
|---|---|---|---|---|---|---|---|---|
| 1 | Hydrogen | H | 1 | -434 | -259 | -253 | -423 | 1 |
| 2 | Helium | He | 4 | -458 | -272 | -269 | -452 | 0 |
| 3 | Lithium | Li | 7 | 354 | 179 | 2440 | 2990 | 1 |
| 4 | Beryllium | Be | 9 | 2341 | 1283 | 5400 | 3660 | 2 |
| 5 | Boron | B | 11 | 4170 | 2300 | 6620 | 4827 | 3 |
| 6 | Carbon | C | 12 | 6332 | 3500 | 8721 | 4827 | 2,4 |
| 7 | Nitrogen | N | 14 | -346 | -210 | -321 | -183 | 3,5 |
| 8 | Oxygen | O | 16 | -362 | -219 | -297 | -188 | 2 |
| 9 | Fluorine | F | 19 | -364 | -220 | -306 | -246 | 1 |
| 10 | Neon | Ne | 20 | -416 | -249 | -410 | 890 | 0 |
| 11 | Sodium | Na | 23 | 208 | 98 | 1634 | 1105 | 1 |
| 12 | Magnesium | Mg | 24 | 1202 | 660 | 2021 | 2467 | 2 |
| 13 | Aluminum | Al | 27 | 1220 | 1420 | 4473 | 2355 | 3 |
| 14 | Silicon | Si | 28 | 2588 | 44 | 4271 | 280 | 4 |
| 15 | Phosphorus | P | 31 | 111 | 113 | 536 | 445 | 3,5 |
| 16 | Sulphur | S | 32 | 235 | -101 | 832 | -34 | 2,4,6 |
| 17 | Chlorine | Cl | 35 | -150 | -189 | -29 | -186 | 1,3,5,7 |
| 18 | Argon | Ar | 40 | -308 | 64 | -303 | 754 | 0 |
| 19 | Potassium | K | 39 | 147 | 848 | 1389 | 1487 | 1 |
| 20 | Calcium | Ca | 40 | 1558 | 1541 | 2709 | 2831 | 2 |
| 21 | Scandium | Sc | 45 | 2806 | 1677 | 5128 | 3277 | 3 |
| 22 | Titanium | Ti | 48 | 3051 | 1917 | 5931 | 3377 | 3,4 |
| 23 | Vanadium | V | 51 | 3483 | 1903 | 6111 | 2642 | 2,3,4,5 |
| 24 | Chromium | Cr | 52 | 3457 | 1244 | 4788 | 2041 | 2,3,6 |
| 25 | Manganese | Mn | 55 | 2271 | 1539 | 3706 | 2750 | 2,3,4,6,7 |
| 26 | Iron | Fe | 56 | 2802 | 1495 | 4980 | 2877 | 2,3 |
| 27 | Cobalt | Co | 59 | 2723 | 1455 | 5211 | 2730 | 2,3 |
| 28 | Nickel | Ni | 59 | 2651 | 1083 | 4950 | 2582 | 2,3 |
| 29 | Copper | Cu | 64 | 1981 | 420 | 4680 | 907 | 1,2 |
| 30 | Zinc | Zn | 65 | 788 | 30 | 1665 | 2403 | 2 |
| 31 | Gallium | Ga | 70 | 86 | 937 | 4357 | 2355 | 2,3 |
| 32 | Germanium | Ge | 73 | 1719 | 817 | 4271 | 613 | 4 |
| 33 | Arsenic | As | 75 | 1503 | 217 | 1135 | 685 | 3,5 |
| 34 | Selenium | Se | 79 | 423 | -7 | 1265 | 59 | 2,4,6 |
| 35 | Bromine | Br | 80 | 19 | -157 | 138 | -152 | 1,3,5,7 |
| 36 | Krypton | Kr | 84 | -251 | 39 | -242 | 688 | 0 |
| 37 | Rubidium | Rb | 85 | 102 | 769 | 1270 | 1384 | 1 |
| 38 | Strontium | Sr | 88 | 1416 | 1522 | 2523 | 3338 | 2 |
| 39 | Yttrium | Y | 89 | 2772 | 1852 | 6040 | 4377 | 3 |
| 40 | Zirconium | Zr | 91 | 3366 | 2467 | 7911 | 4742 | 4 |
| 41 | Niobium | Nb | 93 | 4473 | 2610 | 8568 | 5560 | 3,5 |
| 42 | Molybdenum | Mo | 96 | 4730 | 2172 | 8811 | 4877 | 2,3,4,5,6 |
| 43 | Technetium | Tc | 98 | 3942 | 2310 | 8811 | 3900 | 2,3,4,6,7 |
| 44 | Ruthenium | Ru | 101 | 4190 | 1966 | 7052 | 3727 | 3,4,6,8 |
| 45 | Rhodium | Rh | 103 | 3571 | 1554 | 6741 | 2970 | 3,4 |
| 46 | Palladium | Pd | 106 | 2829 | 962 | 5378 | 2212 | 2,4 |
| 47 | Silver | Ag | 108 | 1764 | 321 | 4014 | 767 | 1 |
| 48 | Cadmium | Cd | 112 | 610 | 156 | 1413 | 2028 | 2 |
| 49 | Indium | In | 115 | 313 | 232 | 3680 | 2270 | 1,3 |
| 50 | Tin | Sn | 119 | 450 | 631 | 4118 | 1635 | 2,4 |
| 51 | Antimony | Sb | 122 | 1168 | 450 | 2975 | 990 | 3,5 |
| 52 | Tellurium | Te | 128 | 842 | 114 | 1814 | 184 | 2,4,6 |
| 53 | Iodine | I | 127 | 237 | -112 | 363 | 184 | 1,3,5,7 |
| 54 | Xenon | Xe | 131 | -170 | 29 | -161 | 671 | 0 |
| 55 | Cesium | Cs | 133 | 84 | 725 | 1240 | 1640 | 1 |
| 56 | Barium | Ba | 137 | | 1337 | | 2984 | 2 |

| Atomic number | Element | Symbol | Atomic mass | Melting point °F | Melting point °C | Boiling point °F | Boiling point °C | Valency |
|---|---|---|---|---|---|---|---|---|
| 57 | Lanthanum | La | 139 | 1690 | 921 | 6255 | 3457 | 3 |
| 58 | Cerium | Ce | 140 | 1470 | 799 | 6199 | 3426 | 3,4 |
| 59 | Praseodymium | Pr | 141 | 1708 | 931 | 6354 | 3512 | 3 |
| 60 | Neodymium | Nd | 144 | 1870 | 1021 | 5554 | 3068 | 3 |
| 61 | Promethium | Pm | 145 | 2134 | 1168 | 4892 | 2700 | 3 |
| 62 | Samarium | Sm | 150 | 1971 | 1077 | 4892 | 1791 | 2,3 |
| 63 | Europium | Eu | 152 | 1512 | 822 | 3256 | 1597 | 2,3 |
| 64 | Gadolinium | Gd | 157 | 2395 | 1313 | 2907 | 3266 | 3 |
| 65 | Terbium | Tb | 159 | 2473 | 1356 | 5911 | 3123 | 3 |
| 66 | Dysprosium | Dy | 163 | 2574 | 1412 | 5653 | 2562 | 3 |
| 67 | Holmium | Ho | 165 | 2685 | 1474 | 4644 | 2695 | 3 |
| 68 | Erbium | Er | 167 | 2784 | 1529 | 4883 | 2863 | 3 |
| 69 | Thulium | Tm | 169 | 2813 | 1545 | 5185 | 1947 | 2,3 |
| 70 | Ytterbium | Yb | 173 | 1506 | 819 | 3537 | 1194 | 2,3 |
| 71 | Lutetium | Lu | 175 | 3025 | 1663 | 2181 | 3395 | 3 |
| 72 | Hafnium | Hf | 178 | 4041 | 2227 | 6143 | 4602 | 4 |
| 73 | Tantalum | Ta | 181 | 5425 | 2996 | 8316 | 5427 | 3,5 |
| 74 | Tungsten | W | 184 | 6170 | 3410 | 9801 | 5660 | 2,4,5,6 |
| 75 | Rhenium | Re | 186 | 5756 | 3180 | 10220 | 5627 | 1,4,7 |
| 76 | Osmium | Os | 190 | 5510 | 3045 | 10161 | 5090 | 2,3,4,6,8 |
| 77 | Iridium | Ir | 192 | 4370 | 2410 | 9190 | 4130 | 3,4 |
| 78 | Platinum | Pt | 195 | 3222 | 1772 | 7466 | 3827 | 2,4 |
| 79 | Gold | Au | 197 | 1947 | 1064 | 6921 | 2807 | 1,3 |
| 80 | Mercury | Hg | 201 | -38 | -39 | 5080 | 357 | 1,2 |
| 81 | Thallium | Tl | 204 | 577 | 303 | 675 | 1457 | 1,3 |
| 82 | Lead | Pb | 207 | 622 | 328 | 2655 | 1744 | 2,4 |
| 83 | Bismuth | Bi | 209 | 520 | 271 | 3171 | 1560 | 3,5 |
| 84 | Polonium | Po | 210 | 489 | 254 | 2840 | 962 | 2,3,4 |
| 85 | Astatine | At | 210 | 572 | 300 | 1764 | 370 | 1,3,5,7 |
| 86 | Radon | Rn | 220 | -96 | -71 | 698 | -62 | 0 |
| 87 | Francium | Fr | 223 | 81 | 27 | -80 | 677 | 1 |
| 88 | Radium | Ra | 226 | 1292 | 700 | 1251 | 1200 | 2 |
| 89 | Actinium | Ac | 227 | 1922 | 1050 | 2190 | 3200 | 3 |
| 90 | Thorium | Th | 232 | 3182 | 1750 | 5792 | 4787 | 4 |
| 91 | Protactinium | Pa | 231 | 2907 | 1597 | 8649 | 4027 | 4,5 |
| 92 | Uranium | U | 238 | 2070 | 1132 | 7281 | 3818 | 3,4,5,6 |
| 93 | Neptunium | Np | 237 | 1179 | 637 | 6904 | 4090 | 2,3,4,5,6 |
| 94 | Plutonium | Pu | 244 | 1184 | 640 | 7394 | 3230 | 2,3,4,5,6 |
| 95 | Americium | Am | 243 | 1821 | 994 | 5850 | 2607 | 2,3,4,5,6 |
| 96 | Curium | Cm | 247 | 2444 | 1340 | 4724 | 3190 | 2,3,4 |
| 97 | Berkelium | Bk | 247 | 1922 | 1050 | 5774 | 710 | 2,3,4 |
| 98 | Californium | Cf | 251 | 1652 | 900 | 1310 | 1470 | 2,3,4 |
| 99 | Einsteinium | Es | 252 | 1580 | 860 | 2678 | 966 | 2,3 |
| 100 | Fermium | Fm | 257 | unknown | | unknown | | 2,3 |
| 101 | Mendelevium | Md | 258 | unknown | | unknown | | 2,3 |
| 102 | Nobelium | No | 259 | unknown | | unknown | | 2,3 |
| 103 | Lawrencium | Lr | 262 | unknown | | unknown | | 3 |
| 104 | Rutherfordium | Rf | 267 | unknown | | unknown | | unknown |
| 105 | Dubnium | Db | 268 | unknown | | unknown | | unknown |
| 106 | Seaborgium | Sg | 271 | unknown | | unknown | | unknown |
| 107 | Bohrium | Bh | 272 | unknown | | unknown | | unknown |
| 108 | Hassium | Hs | 277 | unknown | | unknown | | unknown |
| 109 | Meitnerium | Mt | 276 | unknown | | unknown | | unknown |
| 110 | Darmstadtium | Ds | 281 | unknown | | unknown | | unknown |
| 111 | Roentgenium | Rg | 280 | unknown | | unknown | | unknown |
| 112 | Ununbium | Uub | 285 | unknown | | unknown | | unknown |

# Atoms and molecules

An atom is the smallest part of an element that can exist as a stable entity. Atoms are the basic units of matter, and many bond together to form larger particles called molecules. Atoms within a molecule may be of the same chemical element, as in oxygen, or of different elements, as in water. Atoms are held together by strong chemical bonds (see pp.138–39) formed as electrons are shared or exchanged between the atoms.

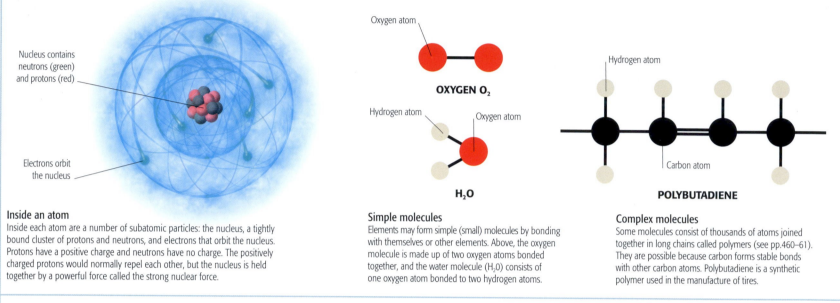

Nucleus contains
neutrons (green)
and protons (red)

Electrons orbit
the nucleus

Oxygen atom

**OXYGEN O$_2$**

Hydrogen atom

Oxygen atom

**H$_2$O**

Hydrogen atom

Carbon atom

**POLYBUTADIENE**

**Inside an atom**
Inside each atom are a number of subatomic particles: the nucleus, a tightly bound cluster of protons and neutrons, and electrons that orbit the nucleus. Protons have a positive charge and neutrons have no charge. The positively charged protons would normally repel each other, but the nucleus is held together by a powerful force called the strong nuclear force.

**Simple molecules**
Elements may form simple (small) molecules by bonding with themselves or other elements. Above, the oxygen molecule is made up of two oxygen atoms bonded together, and the water molecule (H$_2$0) consists of one oxygen atom bonded to two hydrogen atoms.

**Complex molecules**
Some molecules consist of thousands of atoms joined together in long chains called polymers (see pp.460–61). They are possible because carbon forms stable bonds with other carbon atoms. Polybutadiene is a synthetic polymer used in the manufacture of tires.

# Allotropes

The atoms of some elements can bond in a number of different ways to create forms called allotropes. For example, the element carbon has two common allotropes: diamond, where the carbon atoms are bonded together in a tetrahedral lattice arrangement; and graphite, where the carbon atoms are bonded together in hexagonal sheets. Allotropy refers only to different forms of the same element; allotropes do not exist in compounds.

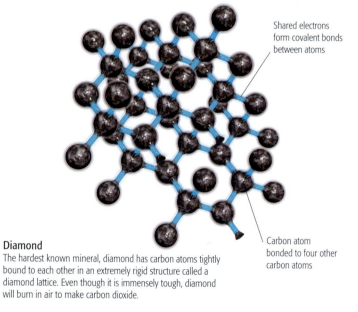

Shared electrons
form covalent bonds
between atoms

Carbon atom
bonded to four other
carbon atoms

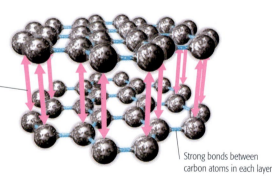

Bonds between layers
are weaker than those
within the layers

**Graphite**
This allotrope has strong bonds between the carbon atoms of each layer, but weak bonds between the different layers. The weak bonds allow the layers to move across one another, making graphite quite a soft material.

Strong bonds between
carbon atoms in each layer

**Diamond**
The hardest known mineral, diamond has carbon atoms tightly bound to each other in an extremely rigid structure called a diamond lattice. Even though it is immensely tough, diamond will burn in air to make carbon dioxide.

# The pH scale for acids and bases

The pH scale measures the strength of acids and bases (also known as alkalis). "pH" stands for potential of hydrogen, and it measures the concentration of hydrogen ions in a liquid. The scale ranges from pH0 (most acid) to pH14 (most basic), with pure water defining the neutral point in the scale at pH7. Each value produces a characterisic color in response to a universal indicator solution (see pp.238–39).

| pH | pH0 | pH1 | pH2 | pH3 | pH4 | pH5 | pH6 | pH7 | pH8 | pH9 | pH10 | pH11 | pH12 | pH13 | pH14 |
|---|---|---|---|---|---|---|---|---|---|---|---|---|---|---|---|
| Acid/alkali | stronger acid | | | weak acid | | | | Neutral | | weak alkali | | | | stronger alkali | |
| Substance | Battery acid = pH0 | | Orange juice = pH3 | | Tomato juice = pH4 | | | Water = pH7 | Seawater = pH8 | | toothpaste = pH9 | Ammonia = pH11 | | Drain cleaner = pH14 | |

# Chemical reactions

In any chemical reaction, the starting materials undergo bond breaking and/or bond making. The products of the reaction are different chemicals with different properties. There are several different types of chemical reaction (see pp.234–35).

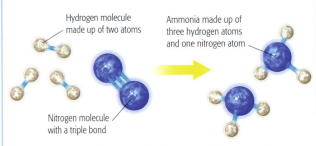

Hydrogen molecule made up of two atoms

Ammonia made up of three hydrogen atoms and one nitrogen atom

Nitrogen molecule with a triple bond

**HYDROGEN (H) + NITROGEN (N) = AMMONIA (NH$_3$)**

△ Synthesis
A synthesis reaction is when two or more simple compounds combine to form a more complicated one. In the above example, molecules of hydrogen and nitrogen bond together to form ammonia.

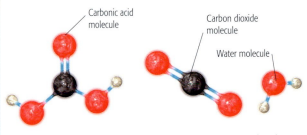

Carbonic acid molecule

Carbon dioxide molecule

Water molecule

**CARBONIC ACID (H$_2$CO$_3$) = CARBON DIOXIDE (CO$_2$) + WATER (H$_2$O)**

△ Decomposition
A decomposition reaction is the opposite of a synthesis reaction (see top), whereby a complex molecule breaks down into simpler ones. In the above reaction, carbonic acid (a weak acid) breaks down into carbon dioxide (CO$_2$) and water (H$_2$O) when warmed in water.

### Factors affecting reactions

| | |
|---|---|
| Concentration of reactants | A higher concentration of particles increases the chance of them colliding with one another, hence causing a reaction. |
| Pressure of reactants | In gases, particles at high pressure are brought closer together than particles at low pressure, so they are more likely to collide. |
| Temperature of reactants | As temperature increases, particles move faster and so collide more frequently than they do at low temperatures. |
| Light levels | Some particles can absorb light energy, making them more reactive at high light levels than they are at low light levels. |
| Size of reactants | Smaller particles of solids react faster than bigger particles because they have a larger combined surface area over which reactions can occur. |

# Fission and fusion

Fission occurs when the nuclei of radioactive atoms are split into fragments, emitting alpha (α) and beta (β) particles, gamma (γ) radiation, and masses of energy (see pp.298–99). Fusion occurs when, at very high temperatures (millions of degrees centigrade), some of the hydrogen atoms lose their electron shells and the unprotected nuclei collide with each other producing helium (see pp.324–25). When the collision occurs huge amounts of energy are released. The most striking natural example of nuclear fusion is the Sun.

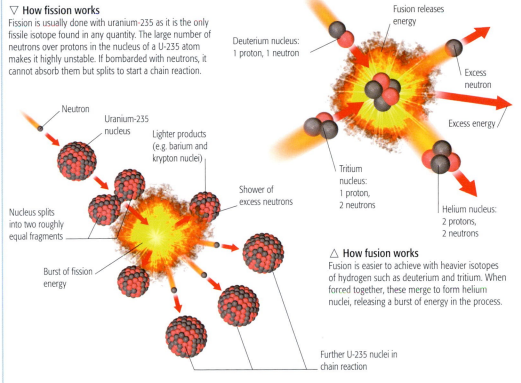

▽ How fission works
Fission is usually done with uranium-235 as it is the only fissile isotope found in any quantity. The large number of neutrons over protons in the nucleus of a U-235 atom makes it highly unstable. If bombarded with neutrons, it cannot absorb them but splits to start a chain reaction.

△ How fusion works
Fusion is easier to achieve with heavier isotopes of hydrogen such as deuterium and tritium. When forced together, these merge to form helium nuclei, releasing a burst of energy in the process.

# Radioactive half-life

Radioactive elements decay at different rates. A half-life is the time it takes for half of the original amount of a radioactive element to decay. Different elements emit different types of radiation when they decay—alpha (α) particles, beta (β) particles, and gamma (γ) rays. After two half-lives, a quarter the original sample will remain; after three half-lives one-eighth the original sample will remain, and so on. Carbon dating (see pp.190–91) uses carbon-14 to determine the age of organic material. Half lives of other materials are given in the table below.

| Element | Decay rate | Particle |
|---|---|---|
| Uranium-238 | 4,500 million years | α |
| Plutonium-239 | 24,400 years | α |
| Carbon-14 | 5,700 years | β |
| Radium-226 | 1,600 years | α |
| Strontium-90 | 28 years | β |
| Hydrogen-3 | 12.3 years | β |
| Cobalt-60 | 5.3 years | γ |
| Phosphorus-32 | 14.3 days | β |
| Iodine-131 | 8.1 days | β |
| Radon-222 | 3.8 days | α |
| Lead-214 | 26.8 minutes | β |
| Astatine-215 | 0.0001 seconds | α |

# Chemical bonds

The atoms within chemical compounds are held together by bonds, which form when atoms come into contact with each other and their electron shells interact (exchange electrons). The strongest bonds are ionic, covalent, or metallic (see pp.288–89), and they vary in strength according to the atoms involved; for example hydrogen forms strong bonds with many elements, but helium forms no bonds at all.

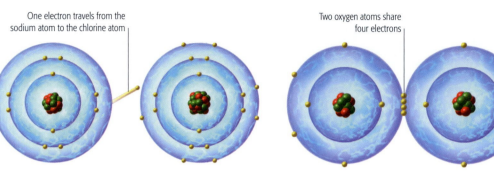

One electron travels from the sodium atom to the chlorine atom

Two oxygen atoms share four electrons

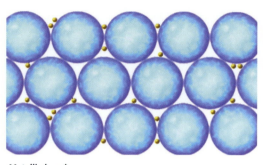

**Ionic bonds**
In ionic bonds, electrons are transferred from one atom to another. When sodium and chlorine combine to form sodium chloride (salt), sodium loses an electron and becomes positively charged; chlorine takes that electron and becomes negatively charged. Ionic bonds are difficult to break. Ionic compounds are usually solids with high melting points.

**Covalent bonds**
In a covalent bond, electrons are shared between two atoms. When two oxygen atoms bond together to form an oxygen molecule, they share four electrons—two from each oxygen atom. Other examples of covalent bonding are water ($H_2O$), and carbon dioxide ($CO_2$). Covalent compounds are usually liquids or gases with low melting points.

**Metallic bonds**
Metal atoms are bonded to each other through metallic bonding. In this type of bonding, all the atoms lose electrons, which float around in a common pool. The electrons in this pool can move around freely, which is why metals can transfer heat or electricity so well. If one part of the metal is heated, the electrons carry the heat quickly to other parts.

# Chemical nomenclature

The system of naming of chemical compounds is referred to as chemical nomenclature. Specific rules and conventions are used to avoid ambiguity between compounds, and to convey information about the structure of the compound. Prefixes are used to denote the number of atoms present, mono- indicating one atom, di- two atoms, and so on, while suffixes -ite and -ate indicate the presence of oxygen.

| Compound ending in | Description | Example |
| --- | --- | --- |
| –ide | Contains just the two elements in the name | Iron sulphide FeS |
| –ite | Contains oxygen as well as the other elements in the name | Iron sulphite $FeSO_3$ |
| –ate | Contains oxygen as well as the other elements in the name. Contains more oxygen than is found in –ites | Iron sulphate $FeSO_4$ |

| Prefix | Number of atoms in prefix | Example |
| --- | --- | --- |
| Mono- | 1 | Nitrogen monoxide NO |
| Di- | 2 | Nitrogen dioxide $NO_2$ |
| Tri- | 3 | Dinitrogen trioxide $N_2O_3$ |
| Tetra– | 4 | Dinitrogen tetroxide $N_2O_4$ |
| Penta- | 5 | Dinitrogen pentoxide $N_2O_5$ |

# Alkanes

Alkanes contain only hydrogen and carbon atoms, which are linked together by single bonds. Each carbon atom has 4 bonds and each hydrogen atom is linked to a carbon atom. Alkanes always have the suffix –ane. A selection of alkanes and their properties is listed below.

| Number of carbon atoms | Name | Physical state at 77° F (22° C) | Molecular formula |
| --- | --- | --- | --- |
| 1 | Methane | gas | $CH_4$ |
| 2 | Ethane | gas | $C_2H_6$ |
| 3 | Propane | gas | $C_3H_8$ |
| 4 | Butane | gas | $C_4H_{10}$ |
| 5 | Pentane | liquid | $C_5H_{12}$ |
| 6 | Hexane | liquid | $C_6H_{14}$ |
| 7 | Heptane | liquid | $C_7H_{16}$ |
| 8 | Octane | liquid | $C_8H_{18}$ |
| 9 | Nonane | liquid | $C_9H_{20}$ |
| 10 | Decane | liquid | $C_{10}H_{22}$ |

# Alkenes

Like alkanes, alkenes also contain only hydrogen and carbon atoms, but all contain at least one carbon–carbon double bond. This means that they are more reactive than alkanes. Alkenes always have the suffix –ene. A selection of alkenes and their properties is listed below.

| Carbon atoms in chain | Name | Physical state at 77° F (22° C) | Molecular formula |
| --- | --- | --- | --- |
| 2 | Ethene | gas | $C_2H_4$ |
| 3 | Propene | gas | $C_3H_6$ |
| 4 | Butene | gas | $C_4H_8$ |
| 5 | Pentene | liquid | $C_5H_{10}$ |
| 6 | Hexene | liquid | $C_6H_{12}$ |
| 7 | Heptene | liquid | $C_7H_{14}$ |
| 8 | Octene | liquid | $C_8H_{16}$ |
| 9 | Nonene | liquid | $C_9H_{18}$ |
| 10 | Decene | liquid | $C_{10}H_{20}$ |
| 11 | Undecene | liquid | $C_{11}H_{22}$ |

## Properties of common metals

About three-quarters of the known elements are metals. Most have high melting points and are good conductors of heat and electricity. Most are strong but malleable. Properties of common metals are listed below, with electronegativity referring to the tendency of the atoms in a stable molecule to attract electrons within covalent bonds.

| Metal | Electronegativity (0.7 to 4.0) | Atomic number | Melting point (°F) | Melting point (°C) | Boiling point (°F) | Boiling point (°C) |
|---|---|---|---|---|---|---|
| Aluminium (Al) | 1.6 | 13 | 1220 | 660 | 4473 | 2467 |
| Barium (Ba) | 0.9 | 56 | 1337 | 725 | 2984 | 1640 |
| Beryllium (Be) | 1.6 | 4 | 2341 | 1283 | 5400 | 2990 |
| Cesium (Cs) | 0.8 | 55 | 84 | 29 | 1240 | 671 |
| Calcium (Ca) | 1.0 | 20 | 1558 | 848 | 2709 | 1487 |
| Chromium (Cr) | 1.7 | 24 | 3457 | 1903 | 4788 | 2642 |
| Cobalt (Co) | 1.9 | 27 | 2723 | 1495 | 5211 | 2877 |
| Copper (Cu) | 1.9 | 29 | 1981 | 1083 | 4680 | 2582 |
| Iron (Fe) | 1.8 | 26 | 2802 | 1539 | 4980 | 2750 |
| Lead (Pb) | 1.8 | 82 | 622 | 328 | 3171 | 1744 |
| Lithium (Li) | 1.0 | 3 | 354 | 179 | 2440 | 1340 |
| Magnesium (Mg) | 1.3 | 12 | 1202 | 650 | 2021 | 1105 |
| Manganese (Mn) | 1.6 | 25 | 2271 | 1244 | 3706 | 2401 |
| Nickel (Ni) | 1.9 | 28 | 2651 | 1455 | 4950 | 2730 |
| Tin (Sn) | 2.0 | 50 | 232 | 119 | 4118 | 2270 |
| Zinc (Zn) | 1.7 | 30 | 788 | 420 | 1665 | 907 |

## Salt formation

When metals come into contact with acids, salts are formed along with hydrogen. This table describes what happens in a number of metal-acid reactions.

| Metal | Acid | Salt |
|---|---|---|
| Sodium | Hydrochloric acid | Sodium chloride |
| Potassium | Sulphuric acid | Potassium sulphate |
| Copper | Hydrochloric acid | Copper chloride |
| Zinc | Sulphuric acid | Zinc sulphate |
| Calcium | Hydrochloric acid | Calcium chloride |
| Sodium | Sulphuric acid | Sodium sulphate |

## Mohs scale of hardness

The mohs scale is a measure of the hardness (resistance to scratching) of minerals. It lists 10 minerals in ascending order: the higher the number, the harder the mineral. Each mineral in the scale scratches all those above it. All other materials can be compared to these minerals.

| Mineral | Composition | Simple Hardness Test | Hardness |
|---|---|---|---|
| Talc | $Mg_3Si_4O_{10}(OH)_2$ | Crushed by finger nail | 1 |
| Gypsum | $CaSO_4.2H_2O$ | Scratched by finger nail | 2 |
| Calcite | $CaCO_3$ | Scratched by copper coin | 3 |
| Fluorite | $CaF_2$ | Scratched by glass | 4 |
| Apatite | $Ca_5(PO_4)_3F$ | Scratched by penknife | 5 |
| Feldspar | $KAlSi_3O_8$ | Scratched by quartz | 6 |
| Quartz | $SiO_2$ | Scratched by steel file | 7 |
| Topaz | $Al_2SiO_4F_2$ | Scratched by corundum | 8 |
| Corundum | $Al_2O_3$ | Scratched by diamond | 9 |
| Diamond | C | | 10 |

## Organic compounds

Organic chemistry is concerned with compounds whose molecules contain carbon, the foundation of all living things. Carbon can combine with itself and other elements to form a huge variety of molecules (see pp.150–51).

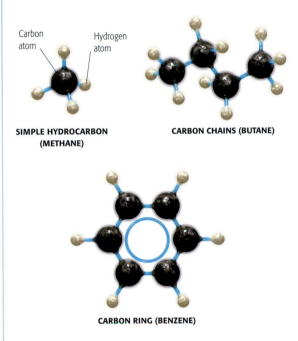

SIMPLE HYDROCARBON (METHANE)

CARBON CHAINS (BUTANE)

CARBON RING (BENZENE)

### Carbon compounds

Carbon atoms usually have a tetrahedral (pyramidal) arrangement of bonds, which means they can form complex branching chains and rings. This structural complexity allows carbon to form the structural and functional building blocks of all living cells. The typical shapes of a simple hydrocarbon (methane), a carbon chain (butane), and a carbon ring (benzene) are shown above.

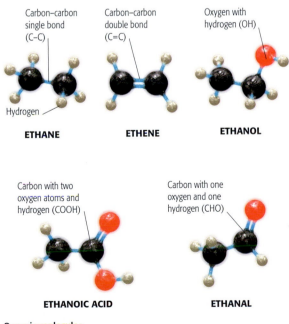

ETHANE

ETHENE

ETHANOL

ETHANOIC ACID

ETHANAL

### Organic molecules

The most common elements to bond with carbon are hydrogen, oxygen, and nitrogen. These elements form arrangements of atomic clusters called functional groups, which are attached to the carbon atoms and determine the chemical properties of each molecule. A selection of two-carbon molecules is shown above.

# Synthetic polymers

Polymers are large molecules made from thousands of identical smaller molecules, called "repeating units" or "monomers." They can be synthesized, in a process called polymerization. Synthetic polymers can be designed to have different features, such as flexibility, hardness, or stickiness, and used in substances such as plastics (see pp.336–37). The chart below shows the monomers, properties, and applications of some common synthetic polymers, although it does not faithfully represent their structure—usually long, twisting three-dimensional chains.

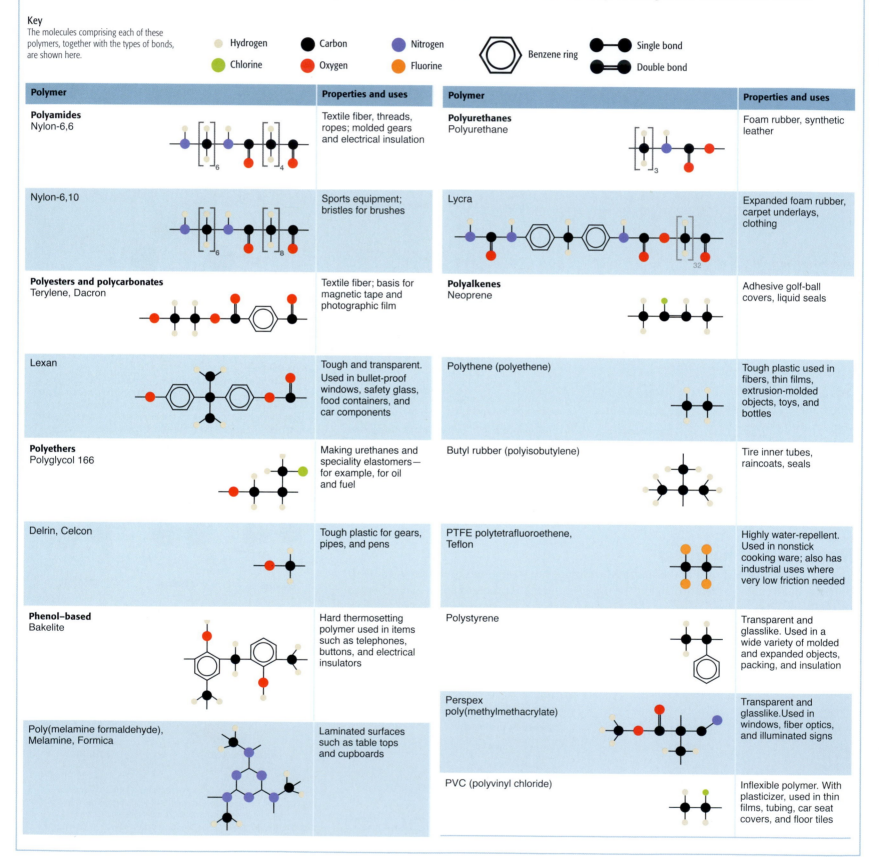

**Key**
The molecules comprising each of these polymers, together with the types of bonds, are shown here.

- Hydrogen
- Carbon
- Nitrogen
- Chlorine
- Oxygen
- Fluorine
- Benzene ring
- Single bond
- Double bond

| Polymer | Properties and uses | Polymer | Properties and uses |
|---|---|---|---|
| **Polyamides** Nylon-6,6 | Textile fiber, threads, ropes; molded gears and electrical insulation | **Polyurethanes** Polyurethane | Foam rubber, synthetic leather |
| Nylon-6,10 | Sports equipment; bristles for brushes | Lycra | Expanded foam rubber, carpet underlays, clothing |
| **Polyesters and polycarbonates** Terylene, Dacron | Textile fiber; basis for magnetic tape and photographic film | **Polyalkenes** Neoprene | Adhesive golf-ball covers, liquid seals |
| Lexan | Tough and transparent. Used in bullet-proof windows, safety glass, food containers, and car components | Polythene (polyethene) | Tough plastic used in fibers, thin films, extrusion-molded objects, toys, and bottles |
| **Polyethers** Polyglycol 166 | Making urethanes and speciality elastomers—for example, for oil and fuel | Butyl rubber (polyisobutylene) | Tire inner tubes, raincoats, seals |
| Delrin, Celcon | Tough plastic for gears, pipes, and pens | PTFE polytetrafluoroethene, Teflon | Highly water-repellent. Used in nonstick cooking ware; also has industrial uses where very low friction needed |
| **Phenol–based** Bakelite | Hard thermosetting polymer used in items such as telephones, buttons, and electrical insulators | Polystyrene | Transparent and glasslike. Used in a wide variety of molded and expanded objects, packing, and insulation |
| Poly(melamine formaldehyde), Melamine, Formica | Laminated surfaces such as table tops and cupboards | Perspex poly(methylmethacrylate) | Transparent and glasslike. Used in windows, fiber optics, and illuminated signs |
| | | PVC (polyvinyl chloride) | Inflexible polymer. With plasticizer, used in thin films, tubing, car seat covers, and floor tiles |

# Natural polymers

Polymers are abundant in the natural world. The most common is cellulose, which is formed from long chains of glucose repeating units. Other natural polymers, or biopolymers, include amber and starch. Polymers are even found in the human body, in the form of proteins, made from amino acids (see p.348), and DNA, made from nucleic acids. Some natural polymers can be adapted for different uses—for example, rubber can be modified, by a process called vulcanization, to make it tougher and more elastic.

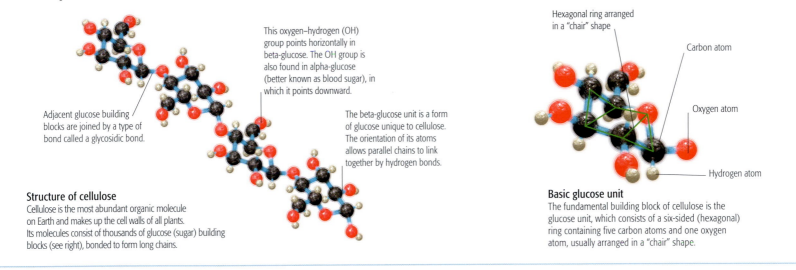

This oxygen–hydrogen (OH) group points horizontally in beta-glucose. The OH group is also found in alpha-glucose (better known as blood sugar), in which it points downward.

Adjacent glucose building blocks are joined by a type of bond called a glycosidic bond.

The beta-glucose unit is a form of glucose unique to cellulose. The orientation of its atoms allows parallel chains to link together by hydrogen bonds.

Hexagonal ring arranged in a "chair" shape

Carbon atom

Oxygen atom

Hydrogen atom

**Structure of cellulose**
Cellulose is the most abundant organic molecule on Earth and makes up the cell walls of all plants. Its molecules consist of thousands of glucose (sugar) building blocks (see right), bonded to form long chains.

**Basic glucose unit**
The fundamental building block of cellulose is the glucose unit, which consists of a six-sided (hexagonal) ring containing five carbon atoms and one oxygen atom, usually arranged in a "chair" shape.

# States of matter

Matter normally exists as a solid, liquid, or gas. Each state has particular features (see pp.140–43). Solids are fixed in shape and volume, and can be broken apart only by physical means; liquids and gases are fluids, which can flow freely and adapt to the shape of any space. Matter can be changed from one of these states to another, usually by heating ot cooling. There is also a fourth state, plasma, which comprises ionized gas particles that form filaments or beams of light when exposed to electric charge.

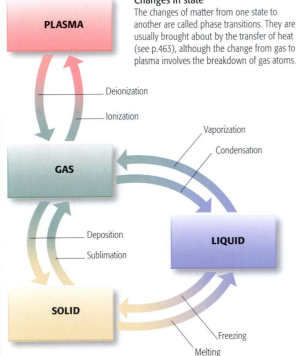

**Changes in state**
The changes of matter from one state to another are called phase transitions. They are usually brought about by the transfer of heat (see p.463), although the change from gas to plasma involves the breakdown of gas atoms.

PLASMA
Deionization
Ionization
GAS
Vaporization
Condensation
Deposition
Sublimation
LIQUID
SOLID
Freezing
Melting

**THE FOUR STATES**

| State | Features |
|---|---|
| Solid | Molecules are fixed in position by chemical bonds. Energy from heat makes them vibrate, but they stay in place. Their shape can be changed only by physical force. |
| Liquid | Molecules are more widely spaced and can move around (flow), but they are still held together by inter-particle attraction; as a result, liquids have a fixed volume. |
| Gas | Molecules move independently of one another; they can flow and scatter. Volume is not fixed. |
| Plasma | Particles comprise flowing negative electrons and positive ions. Plasma has no fixed shape or volume. It can conduct electricity, and responds strongly to magnetic fields. |

**FREEZING AND BOILING POINTS**

| Substance | Freezing point | | Boiling point | |
|---|---|---|---|---|
| | (°F) | (°C) | (°F) | (°C) |
| Water | 32 | 0 | 212 | 100 |
| Saltwater (brine | -5.8 | -21 | 226.4 | 108 |
| Alcohol (ethanol) | -173.2 | -114 | 172.9 | 78.3 |
| Gasoline | 140 | <-60 | 100–400 | 40–200 |
| Methane | 296.5 | -182.5 | 258.9 | -161.6 |
| Carbon dioxide | -68.1 | -55.6 | 109.3 | -78.5 |
| Acetic acid | 62.6 | 17 | 224.6 | 118.1 |
| Benzene | 41.9 | 5.5 | 176.4 | 80.2 |
| Acetone | -138.1 | -94.5 | 133.7 | 56.5 |
| Camphor | 355.6 | 179.8 | 399.2 | 204 |
| Chloroform | -82.3 | -63.5 | 143.6 | 62 |
| Nitrogen dioxide | 11.84 | -11.2 | 69.9 | 21.1 |
| Ammonia | -107.9 | -77.7 | -28.3 | -33.5 |
| Magnesium oxide | 5,165.6 | 2,852 | 6,512 | 3,600 |
| Sodium chloride | 1,473.8 | 801 | 2,669 | 1,465 |
| Calcium chloride | 1,421.6 | 772 | 3,515 | 1,935 |

# Physics

Physics is the study of the physical forces and qualities that govern the Universe, from the subatomic to the interstellar level. It provides definitions of matter, energy, force, and motion, and includes laws and principles that demonstrate the ways in which these phenomena relate to each other. There are various branches, including classical mechanics, electromagnetism, optics, and more recent disciplines such as quantum mechanics.

## Energy, work, and power

According to the First law of Thermodynamics (see opposite), energy cannot be created or destroyed. It can, however, be converted into other forms—for example, to do work. There are two main types of energy: potential energy, which is stored in an object, and kinetic energy, which an object possesses due to its motion.

### TYPES OF ENERGY

| Potential energy | Motion energy |
|---|---|
| Magnetic | Kinetic |
| Gravitational | Heat |
| Chemical | Light |
| Electrical | Sound |
| Elastic | |

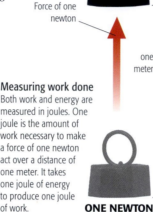

Force of one newton

one meter

**Measuring work done**
Both work and energy are measured in joules. One joule is the amount of work necessary to make a force of one newton act over a distance of one meter. It takes one joule of energy to produce one joule of work.

**ONE NEWTON**

### ENERGY FORMULAE

| Quantity | Description | Formula |
|---|---|---|
| Change in gravitational potential energy | mass × gravitational field strength × height difference | $\Delta E_p = mg\Delta h$ |
| Kinetic energy | $1/2$ mass × square of velocity | $E_k = 1/2\,mv^2$ |
| Weight | mass × gravitational field strength | $W = mg$ |
| Work done | force × distance moved in direction of force | $W = Fs$ |
| Power | $\dfrac{\text{work done (or energy transferred)}}{\text{time taken}}$ | $P = \dfrac{W}{t}$ |
| Efficiency | $\dfrac{\text{work output}}{\text{work input}} \times 100\%$ | $\dfrac{W_o}{W_i} \times 100\%$ |

## Mechanical forces

A force is a push or pull that makes an object move in a straight line or turn (see pp.112–13). Forces have a magnitude (size) and a direction, and are called vector quantities (unlike speed, which has only magnitude and is a scalar). They may act alone or together. We can harness forces by using machines to make work easier.

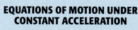

### MOTION FORMULAE

| Quantity | Description | Formula |
|---|---|---|
| Speed | $\dfrac{\text{distance}}{\text{time}}$ | $s = \dfrac{d}{t}$ |
| Time | $\dfrac{\text{distance}}{\text{speed}}$ | $t = \dfrac{d}{s}$ |
| Distance | speed × time | $d = st$ |
| Velocity | $\dfrac{\text{displacement (distance in a given direction)}}{\text{time}}$ | $v = \dfrac{s}{t}$ |
| Acceleration | $\dfrac{\text{change in velocity}}{\text{time taken for change}}$ | $a = \dfrac{(v-u)}{t}$ |
| Resultant force | mass × acceleration | $F = ma$ |
| Momentum | mass × velocity | $p = mv$ |

### TURNING FORCES

| Force | Description | Formula | Key |
|---|---|---|---|
| Moment of inertia | The equivalent of mass for an object rotating around an axis | $I = mr^2$ | $I$ = moment of inertia<br>$m$ = mass<br>$r^2$ = square of distance from axis |
| Angular velocity | The velocity of an object rotating around an axis | $\omega = \dfrac{\Delta\theta}{\Delta t}$ | $\omega$ = angular velocity<br>$\Delta\theta$ = angular displacement<br>$\Delta t$ = change in time |
| Angular momentum | The momentum of an object rotating around an axis | $L = I\omega$ | $L$ = angular momentum<br>$I$ = moment of inertia<br>$\omega$ = angular velocity |

### EQUATIONS OF MOTION UNDER CONSTANT ACCELERATION

These four equations are used to express constant acceleration in different ways:

$$v = u + at$$

$$s = \frac{(u+v)t}{2}$$

$$v^2 = u^2 + 2as$$

$$s = ut + 1/2\,at^2$$

$s$ = displacement
$u$ = original velocity
$v$ = final velocity
$a$ = acceleration
$t$ = time taken

### HOOKE'S LAW

$$F_s = -kx$$

$F_s$ = force of spring
$k$ = spring constant (indication of spring's stiffness)
$x$ = extension

**Mnemonic triangle**
The relationship between distance, speed, and time can be shown in various ways, as seen in the equations on the left. This triangle can help you to remember the relationships. The top element, $d$, can be divided by $s$ or $t$, and $s$ and $t$ can be multiplied together.

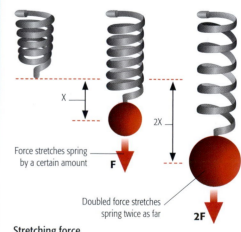

Force stretches spring by a certain amount  **F**

X

2X

Doubled force stretches spring twice as far  **2F**

**Stretching force**
Hooke's law states that the extension of a spring (or other stretchy object) is directly proportional to the force acting on it. This law is only true if the elastic limit of the object has not been reached. If the elastic limit has been reached, the object will not return to its original shape and may eventually break.

# Newton's laws

Newton's laws of motion demonstrate how forces make objects move, remain at rest, or interact with other objects and forces (see pp.104–05). Newton also formulated the law of universal gravitation to define the force of gravity, which holds together everything in the Universe.

| FIRST LAW | Unless disturbed by a force, an object will either stay still or travel at a constant speed in a straight line. |
|---|---|
| SECOND LAW | The force acting on a body is equal to its rate of change of momentum, which is the product of its mass and its acceleration. |
| THIRD LAW | For every action there is an equal and opposite reaction. |

**LAW OF UNIVERSAL GRAVITATION**

$$F = \frac{Gm_1 m_2}{r^2}$$

$F$ = force
$G$ = universal gravitational constant
$m_1$, $m_2$ = masses
$r^2$ = square of distances between masses

# Laws of thermodynamics

These laws are fundamental physical statements that describe what happens when a thermodynamic system goes through an energy change (see pp.180–81).

| ZEROTH LAW | If two bodies (distinct systems) are each in thermal equilibrium with a third body, these two bodies will also be in thermal equilibrium with each other. |
|---|---|
| FIRST LAW | Energy can be neither created nor destroyed. |
| SECOND LAW | The total entropy of an isolated system increases over time. |
| THIRD LAW | There is a theoretical minimum temperature at which the motion of the particles of matter would cease. |

# Heat transfer

Heat is a form of kinetic energy (see opposite) that causes atoms in a substance to vibrate. The more heat energy a substance contains, the more vigorously the atoms will vibrate. The vibrating atoms collide with each other, transferring heat from one to the next. If two objects of different temperatures come into contact, heat will be transferred from the warmer to the cooler one until the two hold the same amount of heat. There are three main forms of heat transfer—conduction, convection, and radiation (see p.179).

**HEAT FORMULAE**

| Quantity | Description | Formula | Key |
|---|---|---|---|
| Specific heat capacity | The amount of energy (measured in joules or calories) needed per unit mass (gram or kilogram) to raise the temperature of a given mass of a material by 1° Celsius. | $Q = cm\Delta T$ | $Q$ = energy<br>$c$ = specific heat capacity<br>$\Delta T$ = temperature change |
| Specific latent heat | The amount of energy absorbed or lost as a material changes from one state to another (e.g. from liquid to solid). | $E = mL$ | $E$ = energy<br>$m$ = mass<br>$L$ = specific latent heat |

**TYPES OF SPECIFIC LATENT HEAT**

| Type | Description |
|---|---|
| Specific latent heat of melting | The energy (in joules) needed to convert 1 kg of a material from solid to liquid without changing its temperature. |
| Specific latent heat of fusion | The energy needed to convert 1 kg of a material from liquid to solid without changing its temperature. |
| Specific latent heat of boiling (or vaporization) | The energy needed to convert 1 kg of a material from liquid to gas without changing its temperature. |

Hottest atoms vibrate most vigorously

Heat energy is transferred to adjacent atoms, making them vibrate more actively

## Conduction
Conduction occurs mainly in solid materials. Metals, such as iron bars, are very good conductors of heat energy, unlike nonmetals and gases. Atoms vibrate more when heated; if one end of an iron bar is heated, the atoms pass this heat energy on toward the cold end of the bar.

## Convection
Convection is the transfer of heat through fluids (moving liquids or gases). A sea breeze is a convection current. Air heated by land becomes less dense and rises; to replace it, cooler air is drawn in from over the sea. At night, the direction of the current is reversed.

## Radiation
Radiation is the transfer of heat through electromagnetic waves, such as from the Sun, or from the heating element in a toaster. Radiation is the only form of heat transfer that allows heat to pass through empty space.

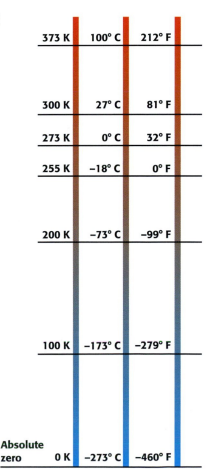

| Kelvin | Celsius | Fahrenheit |
|---|---|---|
| 373 K | 100° C | 212° F |
| 300 K | 27° C | 81° F |
| 273 K | 0° C | 32° F |
| 255 K | −18° C | 0° F |
| 200 K | −73° C | −99° F |
| 100 K | −173° C | −279° F |
| Absolute zero 0 K | −273° C | −460° F |

## Temperature
Temperature is a measure of how hot or cold an object or substance is. It actually shows how much kinetic (heat) energy something contains. Different scales are used to measure temperature, including the kelvin scale, on which 0 K is absolute zero—the point at which atoms in a substance have no heat energy, and do not vibrate at all.

# Pressure

Pressure is defined as the force applied to an object divided by the area over which the force is applied (see pp.100–01). It is a scalar quantity (one that has only magnitude or size), usually expressed in pascals (Pa), newtons (N) per m² (see p.469) or atmospheres (1 atmosphere being equivalent to 101.33kPa). Pressure varies according to the density of the body exerting the force. For example, atmospheric pressure on a climber decreases as he or she nears the summit of a mountain because there is less atmosphere the higher from sea level the climber ascends. In contrast, pressure increases with increasing depth in the ocean. For every 33 ft (10 m) of depth, the pressure increases by one atmosphere. Pressure in fluids also acts equally in all directions and can be transmitted to act in another place (see p.180). If a fluid is moving (for example, water flowing in a river), pressure is reduced if the rate of flow increases. This principle, named after Bernoulli, applies to all fluids.

Relative density is the ratio of the density of a substance to the density of a reference substance. For liquids or solids it is the ratio of the density to the density of fresh water at 4° C, which is given as 1g per cm³; substances less dense than this will float in water.

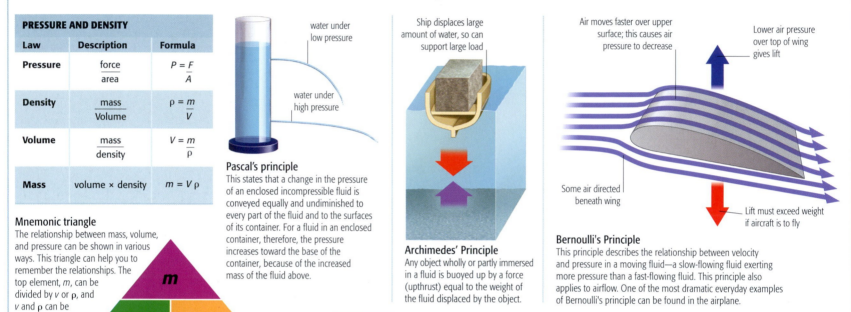

## PRESSURE AND DENSITY

| Law | Description | Formula |
|---|---|---|
| Pressure | force / area | $P = \dfrac{F}{A}$ |
| Density | mass / Volume | $\rho = \dfrac{m}{V}$ |
| Volume | mass / density | $V = \dfrac{m}{\rho}$ |
| Mass | volume × density | $m = V\rho$ |

**Mnemonic triangle**
The relationship between mass, volume, and pressure can be shown in various ways. This triangle can help you to remember the relationships. The top element, $m$, can be divided by $v$ or $\rho$, and $v$ and $\rho$ can be multiplied together.

water under low pressure

water under high pressure

**Pascal's principle**
This states that a change in the pressure of an enclosed incompressible fluid is conveyed equally and undiminished to every part of the fluid and to the surfaces of its container. For a fluid in an enclosed container, therefore, the pressure increases toward the base of the container, because of the increased mass of the fluid above.

Ship displaces large amount of water, so can support large load

**Archimedes' Principle**
Any object wholly or partly immersed in a fluid is buoyed up by a force (upthrust) equal to the weight of the fluid displaced by the object.

Air moves faster over upper surface; this causes air pressure to decrease

Lower air pressure over top of wing gives lift

Some air directed beneath wing

Lift must exceed weight if aircraft is to fly

**Bernoulli's Principle**
This principle describes the relationship between velocity and pressure in a moving fluid—a slow-flowing fluid exerting more pressure than a fast-flowing fluid. This principle also applies to airflow. One of the most dramatic everyday examples of Bernoulli's principle can be found in the airplane.

## ATMOSPHERIC AND WATER PRESSURE

The pressure of the air in our atmosphere, and of the water in lakes and oceans, varies with height and depth.

| Altitude/depth | Pressure (kPa)* |
|---|---|
| 65,000 ft (20,000 m) | 5.46 |
| 36,000 ft (11,000 m) | 22.69 |
| 24,600 ft (7,500 m) | 38.30 |
| Sea level: 0 ft (0 m) | 101.29 |
| -400 ft (-120 m) | 1,114.21 |
| -21,300 ft (-6,500 m) | 65,839.80 |
| -35,800 ft (-10,912 m) | 107,000.00 |

*kPa = kilopascals

## GAS LAWS

| Law | Description | Formula | Key |
|---|---|---|---|
| Avogadro's Law | At a constant temperature and pressure, volume is proportional to number of molecules. Equal volumes of different gases, in identical conditions of temperature and pressure, will contain equal numbers of molecules. | $V \propto n$ | $V$ = volume <br> $n$ = number of molecules <br> $\propto$ = proportional to |
| Boyle's Law | For a given mass of gas at a constant temperature, volume is inversely proportional to pressure. So, for example, if volume doubles, pressure halves. | $PV = $ constant | $P$ = pressure <br> $V$ = volume |
| Charles's Law | For a given mass of gas at a constant pressure, volume is directly proportional to absolute temperature (measured in Kelvin). | $V/T = $ constant | $V$ = volume <br> $T$ = (absolute) temperature |
| Gay-Lussac's Law | For a given mass of gas at a constant volume, pressure is directly proportional to absolute temperature (measured in Kelvin). | $P/T = $ constant | $P$ = pressure <br> $T$ = (absolute) temperature |
| Ideal gas law | An "ideal" gas is a hypothetical gas in which particles may collide but do not have attractive forces between them. The ideal gas law is a good approximation of the behavior of many gases in various conditions. | $pV = nRT$ | $n$ = number of moles <br> $R$ = universal gas constant <br> $T$ = absolute temperature |
| Dalton's Law of Partial Pressures | Applies to mixtures of two or more gases. States that total pressure of gaseous mixture equals sum of partial pressures of individual component gases. This law is used to figure out gas mixtures breathed by scuba divers. | $P = \Sigma p$ or $P_{total} = p1 + p2 + p3 \ldots$ etc | $P$ = total pressure <br> $p$ = individual partial pressures |

# Sound

Mechanical vibrations cause sound waves (see pp.256–57). These can travel through solids, liquids or gases, but cannot move through a vacuum. Sound has three elements: pitch, which increases with the frequency of the sound waves; loudness, which increases with the amplitude; and quality, which depends on the wave form. The speed of the waves depends on the medium through which they travel. The speed of sound varies with air density and temperature; in dry air at sea level, its speed is 1,125 ft (343 m) per second.

**SOUND-INTENSITY LEVEL**

| Source | Sound-intensity level (dB)* |
|---|---|
| Jet aircraft | 120 |
| Heavy machinery | 90 |
| Busy street | 70 |
| Conversation | 50 |
| Whisper | 20 |

*dB = decibels

**PROPERTIES OF SOUND WAVES**

Longitudinal

Cannot travel through empty space (vacuum)

Can travel through solids, liquids, and gases

Comprise back-and-forth oscillations of particles in a medium, e.g. air

Travel at about 1,125 ft/s (343 m/s) through air at sea level

### Frequency and amplitude

Wavelength is related to frequency: the lower the wave's frequency, the longer the wavelength. The energy of a sound wave is indicated by its amplitude: the higher the intensity, the greater the amplitude.

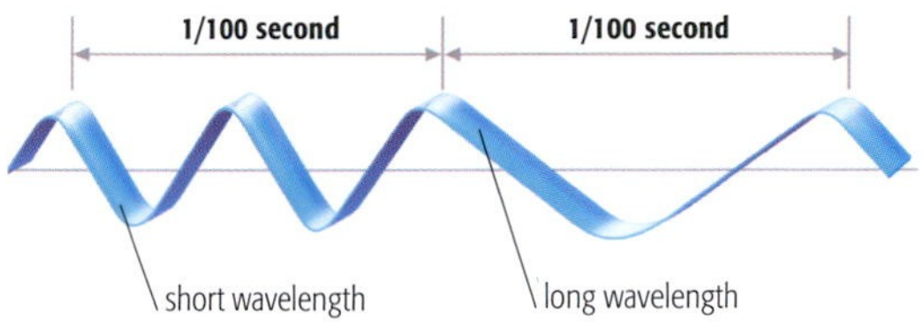

1/100 second    1/100 second

short wavelength    long wavelength

**HIGH–FREQUENCY WAVE    LOW–FREQUENCY WAVE**

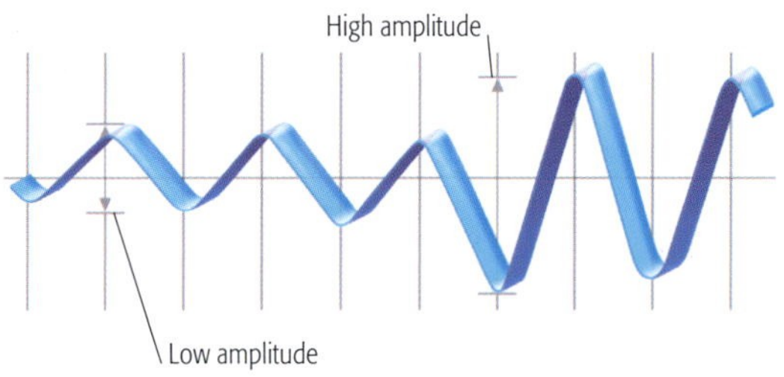

High amplitude

Low amplitude

**WAVE SPEED**

$$v = f\lambda$$

$v$ = speed    $f$ = frequency
$\lambda$ = wavelength

# Light

Light is a form of electromagnetic radiation that is visible to the human eye (see pp.114–17). It can only pass through certain solids and liquids, such as glass and water; however, unlike sound, it can travel through a vacuum. Light can be reflected from polished or shiny surfaces, such as water or a mirror, or it can be refracted through a medium such as glass —a quality used in the correction of near- and far-sightedness. White light can also be split to reveal its constituent parts—red, orange, yellow, green, blue, indigo, and violet light—by passing it through a prism.

**REFRACTION OF LIGHT**

| Feature | Description | Formula | Key |
|---|---|---|---|
| Refractive index | The ability of a material to refract (bend) light. | $n = \dfrac{c}{v}$ | $c$ = speed of light in vacuum $v$ = speed of light in medium |
| Snell's Law | Expresses relationship between angles of incidence and refraction in different media; shows what happens when light passes from one medium to another. | $n = \dfrac{\sin i}{\sin r}$ | $n$ = refractive index $i$ = angle of incidence $r$ = angle of refraction |

### Reflecting light

There are two laws that govern reflection. The first is that the incident ray (striking the boundary), the reflected ray (returning from the boundary), and the "normal" (the line drawn at right angles to the boundary) are in the same plane. The second is that the angle of incidence equals the angle of reflection. Images seen by reflection appear to be the same distance behind the reflecting surface as the object is in front of it.

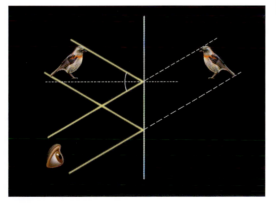

**PROPERTIES OF ELECTROMAGNETIC WAVES**

Transverse

Can travel through empty space (vacuum)

Can travel through some solids, liquids, and gases

Comprise side-to-side oscillations of electric and magnetic fields

Travel at about 186,000 miles/s (300,000 km/s) in a vacuum, slightly slower in air

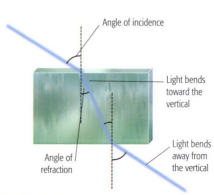

Angle of incidence

Light bends toward the vertical

Light bends away from the vertical

Angle of refraction

### Refracting light

Light bends toward the vertical when slowed on entering a denser medium, and bends away from it on exiting. The amount of bending is described quantitatively by Snell's Law.

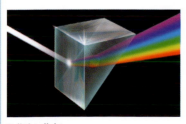

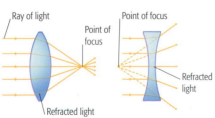

Ray of light

Point of focus

Point of focus

Point of focus

Refracted light

Refracted light

Refracted light

### Splitting light

White light passing through a prism is refracted to form a spectrum. This is because each of the colors is refracted at a slightly different angle. The blue light is "bent" most and the red light least. If the spectrum is passed through a second prism, white light can be restored (see p.116–17).

### Convex and concave lenses

Parallel light rays converge on passing through a convex lens and diverge on passing through a concave lens. The distance between the point of focus and the lens is the focal length. A highly curved lens has a smaller focal length and is more powerful than a less curved lens.

**EXAMPLES OF REFRACTIVE INDICES**

| Substance | Refractive index | Speed of light (m/s) |
|---|---|---|
| Air | 1.0 | 300,000,000 |
| Water | 1.33 | 225,000,000 |
| Perspex | 1.5 | 200,000,000 |
| Glass | 1.5 | 200,000,000 |
| Diamond | 2.4 | 120,000,000 |

# Magnetism and electricity

A magnet is any object or device that produces a magnetic field (see pp.80–81)—an area comprising lines of force that emanate from one end of a magnet, called the "north-seeking" (N) pole, and re-enter the magnet at the other end, the "south-seeking" (S) pole. This field can attract other objects, such as iron filings, with the potential to be magnetized. Certain substances, such as magnetite (rock containing iron oxide), are strongly magnetic, and Earth itself has a magnetic field. Magnetism has a close interrelationship with electricity—magnetic fields can be used to generate electric currents, and vice versa (see pp.166–67).

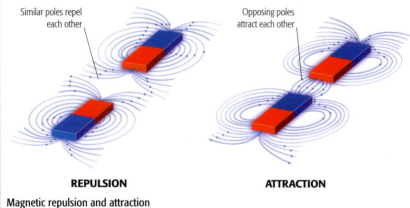

Similar poles repel each other

Opposing poles attract each other

**REPULSION**

**ATTRACTION**

**Magnetic repulsion and attraction**
If two magnets are placed with their N poles close together, the fields will repel each other. If the magnets are placed with opposing poles close together, the field from the N pole of one will be attracted to the S pole of the other.

| ELECTRICITY FORMULAE | | |
| --- | --- | --- |
| Quantity | Discription | Formula |
| Electric charge | current × time | $q = I\,t$ |
| Electrical energy | potential difference (or voltage) × current × time | $E = V\,I\,t$ |
| Electrical power | potential difference (or voltage) × current | $P = V\,I$ |

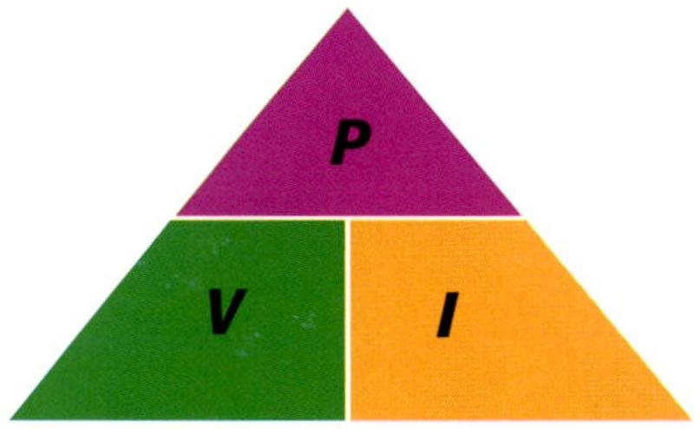

**Mnemonic triangle**
The relationship between electrical power, current, and potential difference (voltage) can be shown in various ways, as seen here. The top element, $P$, can be divided by $V$ or $I$, and $V$ and $I$ can be multiplied together.

# Maxwell's equations

These four equations or laws, formulated by James Clerk Maxwell, comprise a full description of classical electromagnetism, showing how electric and magnetic fields are produced, and how the rates of change of these fields are related to their sources (see p.167). For this reason, they are some of the most important equations used in science. In the process of figuring out these laws, Maxwell derived the electromagnetic wave equation, which demonstrated that light was a form of wave with interlinked electric and magnetic elements. This finding led to the discovery of radio waves and the understanding of other parts of the electromagnetic spectrum (see below).

| THE FOUR EQUATIONS | | |
| --- | --- | --- |
| Law | Statement | Application |
| Gauss's law for electricity | The electric flux through any closed surface is proportional to the total charge contained within that surface. | Used to calculate electric fields around charged objects. |
| Gauss's law for magnetism | For a magnetic dipole with any closed surface, the magnetic flux drawn inward toward the south pole will equal the flux directed outward from the north pole; the net flux will always be zero. | Describes sources of magnetic fields and shows that they will always be closed loops. |
| Faraday's law of induction | The induced electromotive force (emf) around any closed loop equals the negative of the rate of change of the magnetic flux through the area enclosed by the loop. | Describes how a changing magnetic field can generate an electric field; this is the operating principle for electric generators, inductors, and transformers. |
| Ampère's law with Maxwell's correction | In static electric field, the line integral of the magnetic field around any closed loop is proportional to the electric current that is flowing through the loop. | Used to calculate magnetic fields. Shows that magnetic current can be generated in two ways: by electric current and by changing electric fields. |

# Electromagnetic spectrum

The electromagnetic spectrum is a continuum comprising various forms of radiation wave. The waves travel at the same speed (the speed of light), but have different wavelengths. Their frequency, and the energy that they carry, increases as the wavelength decreases. Visible light is the part of the spectrum that we can perceive naturally; other forms can be detected by instruments. Some forms are blocked by Earth's atmosphere.

**Radio waves**
The longest radio waves are absorbed in Earth's upper atmosphere by the electrically charged ionosphere. Intermediate waves, with wavelengths between a few centimeters and about 10 meters, pass through the atmosphere and reach Earth's surface, while those with shorter wavelengths are absorbed by atmospheric carbon dioxide and water vapor.

**Microwaves**
The shortest form of radio waves, with wavelengths between 1 cm and 100 µm, are known as microwaves. These waves are widely used in communications as well as in microwave ovens.

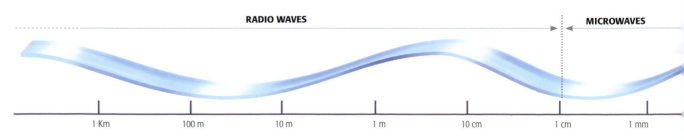

RADIO WAVES

MICROWAVES

1 Km   100 m   10 m   1 m   10 cm   1 cm   1 mm

# Electric currents and circuits

Electric charge can be made to travel in a flow, or current (see pp.164–65). This current can be generated by an electromotive force (emf) such as a battery cell, and directed around wire loops, or circuits, to provide power for electrical devices. There are two main ways to drive the flow around a circuit: by creating an "alternating" current (AC), which flows to and fro; or by producing a "direct" current (DC), which flows in one direction only (see p.252). The way current flows around a circuit is governed by particular laws (see below).

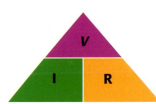

**Mnemonic triangle**
Ohm's Law can be illustrated by a triangle such as this one, which shows the relationship between electrical current, resistance, and potential difference (voltage). The top element, *V*, can be divided by *I* or *R*, and *I* and *R* can be multiplied together.

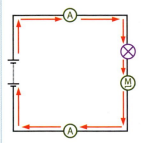

**Series circuit**
If a circuit has components connected in a single loop, the components are said to be "in series." In a series circuit, the same current flows through all of the components.

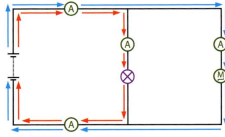

**Parallel circuit**
If a circuit includes two or more branches, the components are said to be "in parallel." In a parallel circuit, the components are connected so that the same voltage is supplied to each component.

**KEY**
→ Path of current around single circuit
→ Path of current around parallel branch

## KEY TO CIRCUIT SYMBOLS

| Symbol | |
|---|---|
| Switch (open) | |
| Switch (closed) | |
| Cell | |
| Battery | |
| Ammeter | |
| Voltmeter | |
| Junction of conductors | |
| Motor | |
| Power supply | |
| Lamp | |
| Fuse | |
| Fixed resistor | |
| Variable resistor | |
| Light dependent resistor | |
| Thermistor | |
| Diode | |
| Generator | |
| AC power supply | |

## ELECTRICITY AND CIRCUIT LAWS

| Law | Description | Formula | Key |
|---|---|---|---|
| Coulomb's Law | The force of attraction or repulsion between two charged particles is directly proportional to the product of the charges and inversely proportional to the distance between them. | $F = \dfrac{kq_1 q_2}{r^2}$ | $k$ = Coulomb's constant<br>$q_1$ and $q_2$ = electric (point) charges<br>$r^2$ = square of distance |
| Ohm's Law | This law expresses the relationships between voltage, resistance and current, which can be expressed in several ways. | $I = \dfrac{V}{R}$  $R = \dfrac{V}{I}$<br>$V = IR$ | $R$ = resistance<br>$I$ = current<br>$V$ = potential difference (voltage) |
| Kirchhoff's Current Law | The sum of the electric currents entering any junction in a circuit is equal to the sum of those leaving the junction. | $\Sigma I = 0$ | $\Sigma$ = summation symbol<br>$I$ = current |
| Kirchhoff's Voltage Law | The sum of the voltage changes around any closed loop path is zero. | $\Sigma V = 0$ | $\Sigma$ = summation symbol<br>$V$ = potential difference (voltage) |

**Infrared**
Emitted by warm objects, infrared radiation is mostly absorbed in Earth's atmosphere by gases such as carbon dioxide and water vapor.

**Optical window**
Wavelengths measured in billionths of a meter pass straight through Earth's atmosphere. Animal eyesight has evolved to be sensitive to radiations in this "window."

**Utraviolet**
The wavelengths of ultraviolet rays are shorter than visible light. Most ultraviolet rays from the Sun are blocked by the atmospheric ozone layer.

**X-rays**
With wavelengths from 10 nanometers (nm) down to 0.01 nm, X-rays carry large amounts of harmful energy. Fortunately Earth's atmosphere acts as a good barrier to them.

**Gamma rays**
Even shorter than X-rays, gamma rays have an extreme potential for penetrating materials and damaging tissues. As with X-rays, our atmosphere prevents them reaching Earth's surface.

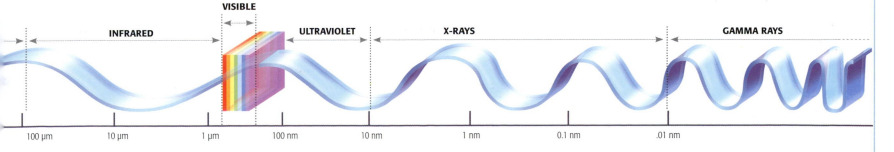

VISIBLE
INFRARED · ULTRAVIOLET · X-RAYS · GAMMA RAYS

100 µm · 10 µm · 1 µm · 100 nm · 10 nm · 1 nm · 0.1 nm · .01 nm

# Quantum mechanics

The branch of science that deals with minute, indivisible units of energy is known as quantum mechanics. Largely theoretical, it provided a means to answer questions that were beyond the realm of classical physics and has contributed to the development of many new ideas, including string theory and grand unified theory. Quantum mechanics has also been used to help explain fundamental forces and the behavior of electrons.

## The four fundamental forces

All matter in the Universe is subject to four basic forces—gravity, electromagnetism, and strong and weak nuclear forces (see pp.400–01). Each is associated with a particular subatomic particle.

### FORCE-CARRIER PARTICLES

Matter particles affected by a particular force produce and absorb specific force carriers. For example, strong forces are linked with gluons.

| Particle | Force | Relative strength | Range in (m) |
|---|---|---|---|
| Graviton | *Gravity | $10^{-41}$ | Infinite |
| Photon | Electromagnetic | 1 | Infinite force |
| Gluon | Strong force | 25 | $10^{-15}$ |
| W, Z bosons | Weak interaction | 0.8 | $10^{-18}$ |

\* hypothetical force particle

## Subatomic particles

Elementary particles are the fundamental constituents of everything in the Universe (see pp.382–83)—quarks and leptons are the basic components of matter, while the gauge bosons are the particles that transmit the fundamental forces. These particles may combine to form larger particles. Scientists think that every particle is twinned with an opposing "antiparticle." If paired, they can annihilate each other, producing particles of light (photons).

### SUBATOMIC PARTICLES: TYPES

| | Elementary particles | Composite particles (hadrons) |
|---|---|---|
| Fermions (matter particles) | **Quarks** Particles that make protons and neutrons. There are six "flavors": Up, Down, Charm, Strange, Top, Bottom.<br><br>**Leptons** A group of six particles, comprising electron, muon, and tau particles and their associated neutrinos. | **Baryons** Particles made up of three quarks. The best known are the proton (two Up and one Down particle) and the neutron (one Up and two Down particles) |
| Bosons (force-carrier particles) | **Gauge bosons** Particles associated with the four fundamental forces (see left). No gauge boson has yet been found for gravity, although the existence of a "graviton" has been hypothesized. | **Mesons** Particles made from a quark with an antiquark. There are many types, including the positive pion (Up quark with Down antiquark) and the negative kaon (Strange quark with Up antiquark). |

## Photoelectric effect

Einstein described how light can act like a particle or as a continuous wave. The photoelectric effect demonstrates the particle-like properties of light and involves the emission of electrons when light is shone on to a target (see pp.314–15). Changing the intensity, voltage, and/or wavelength of the light will affect the current and energy of electrons, as does changing the material of the target. For example, the shorter the wavelength of light, the more energy hits the target, and more electrons are ejected.

**Red, blue, and ultraviolet light**
When red light (left) is shone on to a metal surface, no electrons are ejected. Shorter-wavelength blue light (center) is more energetic than red, so does cause electrons to be ejected. Ultraviolet light (right) causes many electrons to be ejected at very high energy.

## Planck's constant

In 1900 Max Planck explained the peaks in "black body" curves (see p.314) by relating the temperature of a black body to the intensity of energy radiated by it at a given wavelength. He proposed that the peaks were areas where there were lots of atoms producing particles of light that, when added together, were greater in number than at either end. He divided the light into "packets," or quanta, and assigned each an energy ($E$) related to its frequency ($f$). This worked provided that $E = hf$, $h$ being a new constant in nature now called Planck's constant.

**Black body curves**
Plotted on a graph, black body curves at different temperatures have peaks in the same part of the spectrum with little radiation to either side. This phenomenon could not be explained by classical physics and so led to the quantum revolution.

*(graph axes: Power radiated at each wavelength vs. Wavelength Å (µm), values 1 2 3 4 5)*

## Einstein's theories of relativity

Albert Einstein stunned the scientific world by challenging Newton's accepted theory of gravitation with the publication of his Special Theory of Relativity in 1905, and his General Theory of Relativity in 1915 (see pp.302–03). The Special Theory showed that Newton's three Laws of Motion were only approximately correct, breaking down near the speed of light. The General Theory proved that Newton's Law of Gravitation didn't hold in the presence of very strong gravitational fields.

| THEORY | PROPOSITION |
|---|---|
| Special Theory of Relativity | 1) All physical laws are the same in all frames of reference in uniform motion with respect to one another.<br>2) The speed of light is a constant, regardless of the motions of the light source and the observer. |
| General Theory of Relativity | Space-time is curved: strong gravity causes distortions of time and mass, and large objects (such as stars) warp space-time around them. |

# Common equations

The following equations are commonly used in physics. Some of the units used to calculate these equations can be found in the metric and imperial measurement table on p.423.

| EQUATIONS | | |
|---|---|---|
| **Quantity** | **Statement** | **Formula** |
| **Gravitational force** | universal gravitational constant × mass$_1$ × mass$_2$ / distance$^2$ | $\dfrac{Gm_1 m_2}{r^2}$ |
| **Change in gravitational potential energy** | mass × gravitational field strength × height difference | $\Delta E_p = mg\Delta h$ |
| **Kinetic energy** | ½mass × velocity$^2$ | $E_k = \frac{1}{2}mv^2$ |
| **weight** | mass × gravitational field strength | $W = mg$ |
| **Work done** | force × distance moved in the direction of the force | $W = Fs$ |
| **Power** | $\dfrac{\text{work done}}{\text{time taken}}$ or $\dfrac{\text{energy transferred}}{\text{time taken}}$ | $P = \dfrac{W}{t}$ |
| **Efficiency** | $\dfrac{\text{work output}}{\text{work input}}$ × 100% | $\dfrac{W_o}{W_i}$ × 100% |
| **Speed** | $\dfrac{\text{distance moved}}{\text{time taken}}$ | $s = \dfrac{d}{t}$ |
| **Velocity** | $\dfrac{\text{displacement}}{\text{time}}$ | $v = \dfrac{s}{t}$ |
| **Acceleration** | $\dfrac{\text{change in velocity}}{\text{time taken for this change}}$ | $a = \dfrac{(v-u)}{t}$ |
| **Resultant force** | mass × acceleration | $F = ma$ |
| **Momentum** | mass × velocity | $mv$ |
| **Moment of inertia** | mass × radius from the axis$^2$ | $I = mr^2$ |
| **Density** | $\dfrac{\text{mass}}{\text{volume}}$ | $\rho = \dfrac{m}{V}$ |
| **Volume** | $\dfrac{\text{mass}}{\text{density}}$ | $V = \dfrac{m}{\rho}$ |
| **Mass** | volume × density | $m = V\rho$ |
| **Pressure** | $\dfrac{\text{force}}{\text{area}}$ | $P = \dfrac{F}{A}$ |
| **Refractive index** | $\dfrac{\text{speed of light in a vacuum}}{\text{speed of light in a medium}}$ | $n = \dfrac{c}{v}$ |
| **Wave speed** | frequency × wavelength | $v = f\lambda$ |
| **Electric charge** | current × time | $q = It$ |
| **Potential difference (voltage)** | current × resistance or $\dfrac{\text{energy transferred}}{\text{charge}}$ | $V = IR$   $V = \dfrac{W}{q}$ |
| **Resistance** | $\dfrac{\text{voltage}}{\text{current}}$ | $R = \dfrac{V}{I}$ |
| **Electrical energy** | potential difference (or voltage) × current × time | $E = VIt$ |
| **Electrical power (DC)** | potential difference (or voltage) × current | $P = VI$   $V = \dfrac{P}{I}$   $I = \dfrac{P}{V}$ |

# Common physical qualities

The following table shows some common physical properties and their symbols. The units by which they are measured are indicated by the SI units and their symbols.

| Physical quality | Symbol | SI unit | SI unit symbol |
|---|---|---|---|
| **Acceleration, deceleration** | $a$ | meter/second$^2$<br>kilometer/hour/second | m s$^{-2}$<br>km h$^{-1}$ s$^{-1}$ |
| **Angular velocity** | $\omega$ | radian/second | rad s$^{-1}$; s$^{-1}$ |
| **Density** | $\rho$ | kilogram/meter$^3$<br>kilogram/milliliter | kg m$^{-3}$<br>kg ml$^{-1}$ |
| **Electric charge** | $Q, q$ | coulomb | C |
| **Electric current** | $I, i$ | ampere (coulomb/second) | A (Cs$^{-1}$) |
| **Electrical energy** | – | megajoule<br>kilowatt-hour | MJ<br>kWh |
| **Electrical power** | $P$ | watt (joule/second) | W (J s$^{-1}$) |
| **Electromotive force (emf)** | $E$ | volt (watt/ampere) | V (W A$^{-1}$) |
| **Electrical conductance** | $S$ | siemen (ohm$^{-1}$) | A V$^{-1}$ |
| **Electrical resistance** | $R$ | ohm (volt/ampere) | $\Omega$ (V A$^{-1}$) |
| **Frequency** | $f$ | hertz (cycles/second) | Hz (s$^{-1}$) |
| **Force** | $F$ | newton (kilogram meter/second$^2$) | N (kg m s$^{-2}$) |
| **Gravitational intensity, field strength** | – | newton/kilogram | N kg$^{-1}$ |
| **Magnetic field strength** | $H$ | ampere/meter | A m$^{-1}$ |
| **Magnetic flux** | $\Phi$ | weber | Wb |
| **Magnetic flux density** | $B$ | tesla (weber/meter$^2$) | T (Wb m$^{-2}$) |
| **Mass** | $m$ | kilogram | kg |
| **Mechanical power** | $P$ | watt (joule/second) | W (J s$^{-1}$) |
| **Moment of inertia** | $I$ | kilogram meter$^2$ | Kg m$^2$ |
| **Momentum** | $p$ | kilogram meter/second | Kg m s$^{-1}$ |
| **Pressure** | $P$ | pascal (newton/meter$^2$) | Pa (N m$^{-2}$) |
| **Quantity of substance** | $n$ | mole | mol |
| **Specific heat capacity** | $C$ or $c$ | joule/kilogram/kelvin | J kg$^{-1}$ K$^{-1}$ |
| **Specific latent heats of fusion, vaporization** | $L$ | joule/kilogram | J kg$^{-1}$ |
| **Torque** | $\tau$ | newton meter | N m |
| **Velocity, speed** | $u, v$ | meter/second<br>kilometer/hour | m s$^{-1}$<br>km h$^{-1}$ |
| **Volume** | $V$ | meter$^3$<br>milliliter | m$^3$<br>ml |
| **Wavelength** | $\lambda$ | meter | m |
| **Weight** | $W$ | newton | N |
| **Work, energy** | $W$ | joule (newton meter) | J (N m) |

# Mathematics

Mathematics is a huge subject, encompassing the study of numbers, quantities, shapes, space, and patterns, as well as relationships between entities and operations that can be performed on them. Mathematics plays a fundamental role in all the sciences, and over the millennia it has developed its own set of symbols, theorems, and other key concepts, a selection of which are described in the following pages.

## Counting systems

Tallies have been used since prehistoric times to represent amounts and numbers, and from these, various counting systems have been developed. The Roman numeral system is based on seven letters of the alphabet whose combined values give a total. Our current system, based on Hindu–Arabic numerals, uses ten digits that can be arranged to represent larger numbers, while the binary system, which is used mainly in computing, uses just two digits: 1 and 0. Numerical equivalents for the different counting systems are shown here.

| Hindu–Arabic | Roman | Binary | Hindu–Arabic | Roman | Binary | Hindu–Arabic | Roman | Binary |
|---|---|---|---|---|---|---|---|---|
| 1 | I | 1 | 13 | XIII | 1101 | 70 | LXX | 1000110 |
| 2 | II | 10 | 14 | XIV | 1110 | 80 | LXXX | 1010000 |
| 3 | III | 11 | 15 | XV | 1111 | 90 | XC | 1011010 |
| 4 | IV | 100 | 16 | XVI | 10000 | 100 | C | 1100100 |
| 5 | V | 101 | 17 | XVII | 10001 | 200 | CC | 11001000 |
| 6 | VI | 110 | 18 | XVIII | 10010 | 300 | CCC | 100101100 |
| 7 | VII | 111 | 19 | XIX | 10011 | 400 | CD | 110010000 |
| 8 | VIII | 1000 | 20 | XX | 10100 | 500 | D | 111110100 |
| 9 | IX | 1001 | 30 | XXX | 11110 | 1,000 | M | 1111101000 |
| 10 | X | 1010 | 40 | XL | 101000 | 5,000 | $\overline{\text{V}}$ | 1001110001000 |
| 11 | XI | 1011 | 50 | L | 110010 | 10,000 | $\overline{\text{X}}$ | 10011100010000 |
| 12 | XII | 1100 | 60 | LX | 111100 | 100,000 | $\overline{\text{C}}$ | 11000011010100000 |

## Numerical equivalents

Percentages, decimals, and fractions are different ways of presenting a numerical value as a proportion of a given amount. For example, 10 percent (10%) has the equivalent value of the decimal 0.1 and the fraction 1/10. However, some values that are expressed precisely by fractions have unresolved decimal equivalents that contain recurring decimal numbers. For example, the fraction ⅓ has the decimal equivalent of 0.333 recurring, where the "3" repeats infinitely. In these instances decimal values are given to the first three decimal places below.

| % | Decimal | Fraction | % | Decimal | Fraction | % | Decimal | Fraction | % | Decimal | Fraction | % | Decimal | Fraction |
|---|---|---|---|---|---|---|---|---|---|---|---|---|---|---|
| 1 | 0.01 | 1/100 | 12½ | 0.125 | 1/8 | 24 | 0.24 | 6/25 | 36 | 0.36 | 9/25 | 49 | 0.49 | 49/100 |
| 2 | 0.02 | 1/50 | 13 | 0.13 | 13/100 | 25 | 0.25 | 1/4 | 37 | 0.37 | 37/100 | 50 | 0.50 | 1/2 |
| 3 | 0.03 | 3/100 | 14 | 0.14 | 7/50 | 26 | 0.26 | 13/50 | 38 | 0.38 | 19/50 | 55 | 0.55 | 11/20 |
| 4 | 0.04 | 1/25 | 15 | 0.15 | 3/20 | 27 | 0.27 | 27/100 | 39 | 0.39 | 39/100 | 60 | 0.60 | 3/5 |
| 5 | 0.05 | 1/20 | 16 | 0.16 | 4/25 | 28 | 0.28 | 7/25 | 40 | 0.40 | 2/5 | 65 | 0.65 | 13/20 |
| 6 | 0.06 | 3/50 | 16⅔ | 0.166 | 1/6 | 29 | 0.29 | 29/100 | 41 | 0.41 | 41/100 | 66⅔ | 0.666 | 2/3 |
| 7 | 0.07 | 7/100 | 17 | 0.17 | 17/100 | 30 | 0.30 | 3/10 | 42 | 0.42 | 21/50 | 70 | 0.70 | 7/10 |
| 8 | 0.08 | 2/25 | 18 | 0.18 | 9/50 | 31 | 0.31 | 31/100 | 43 | 0.43 | 43/100 | 75 | 0.75 | 3/4 |
| 8⅓ | 0.083 | 1/12 | 19 | 0.19 | 19/100 | 32 | 0.32 | 8/25 | 44 | 0.44 | 11/25 | 80 | 0.80 | 4/5 |
| 9 | 0.09 | 9/100 | 20 | 0.20 | 1/5 | 33 | 0.33 | 33/100 | 45 | 0.45 | 9/20 | 85 | 0.85 | 17/20 |
| 10 | 0.10 | 1/10 | 21 | 0.21 | 21/100 | 33⅓ | 0.333 | 1/3 | 46 | 0.46 | 23/50 | 90 | 0.90 | 9/10 |
| 11 | 0.11 | 11/100 | 22 | 0.22 | 11/50 | 34 | 0.34 | 17/50 | 47 | 0.47 | 47/100 | 95 | 0.95 | 19/20 |
| 12 | 0.12 | 3/25 | 23 | 0.23 | 23/100 | 35 | 0.35 | 7/20 | 48 | 0.48 | 12/25 | 100 | 1.00 | 1 |

# Mathematical signs and symbols

This table presents a selection of signs and symbols commonly used in mathematics and other branches of science. These generally represent abbreviations for entities, relationships between entities and/or values, and operations. Using signs and symbols, mathematicians and scientists can express complex equations and formulae in a standardized way that is universally understood.

| Symbol | Definition |
|---|---|
| $+$ | plus; positive |
| $-$ | minus; negative |
| $\pm$ | plus or minus; positive or negative; degree of accuracy |
| $\mp$ | minus or plus; negative or positive |
| $\times$ | multiplied by (6 x 4) |
| $\cdot$ | multiplied by (6·4); scalar product of two vectors (A·B) |
| $\div$ | divided by (6 ÷ 4) |
| $/$ | divided by; ratio of (6/4) |
| $—$ | divided by; ratio of (6/4) |
| $!$ | factorial (4! = 4 x 3 x 2 x 1) |
| $=$ | equals |
| $\neq$ | not equal to |
| $\equiv$ | identical with; congruent to |
| $\not\equiv$ | not identical with |
| $\triangleq$ | corresponds to |
| $:$ | ratio of (6:4) |
| $::$ | proportionately equals (1:2 :: 2:4) |
| $\approx$ | approximately equal to; equivalent to; similar to |
| $>$ | greater than |
| $\gg$ | much greater than |
| $\ngtr$ | not greater than |
| $<$ | less than |

| Symbol | Definition |
|---|---|
| $\ll$ | much less than |
| $\nless$ | not less than |
| $\geqslant, \geqq, \gtreqless$ | equal to or greater than |
| $\leqslant, \leqq, \lesseqgtr$ | equal to or less than |
| $\propto$ | directly proportional to |
| $(\ )$ | parentheses |
| $[\ ]$ | brackets |
| $\{\ \}$ | braces |
| $—$ | vinculum: division (a-b); chord of circle or length of line (AB); arithmetic mean (X) |
| $\exists$ | there exists |
| $\forall$ | for all |
| $\infty$ | infinity |
| $\rightarrow$ | approaches the limit |
| $\sqrt{\ }$ | square root |
| $\sqrt[3]{\ }\sqrt[4]{\ }$ | cube root, fourth root, etc. |
| $\%$ | per cent |
| $'$ | prime; minute(s) of arc; foot/feet |
| $''$ | double prime; second(s) of arc; inch/inches |
| $\cap$ | arc of circle |
| $°$ | degree of arc |
| $\angle, \angle^s$ | angle(s) |
| $\veebar$ | equiangular |
| $\perp$ | perpendicular |

| Symbol | Definition |
|---|---|
| $\parallel$ | parallel |
| $\bigcirc$ | circle |
| $\triangle$ | triangle |
| $\square$ | square |
| $\square$ | rectangle |
| $\square$ | parallelogram |
| $\cong$ | congruent to; identical with |
| $\therefore$ | therefore |
| $\because$ | because |
| $\underline{m}$ | measured by |
| $\Delta$ | increment |
| $\sum$ | summation |
| $\prod$ | product |
| $\int$ | integral |
| $\nabla$ | del: differential operator |
| $\cup$ | union |
| $\cap$ | intersection |
| $\in$ | is an element of |
| $\subset$ | strict inclusion |
| $\supset$ | contains |
| $\Rightarrow$ | implies |
| $\Leftarrow$ | implied by |
| $\Leftrightarrow$ | implies and is implied by |

# Platonic solids

Also known as convex regular polyhedrons, the five Platonic solids were known by the ancient Greeks and are named after the Greek philosopher Plato (see p.34). All of their faces are identical regular polygons. The same number of faces meet at every vertex (point where edges meet), and all angles at each vertex are equal. Their names come from their respective number of faces: 4, 6, 8, 12, 20.

**Regular tetrahedron**
Composed of equilateral triangles, the regular tetrahedron has 4 faces, 6 edges, and 4 vertices.

**Regular hexahedron (cube)**
The most easily recognizable of the Platonic solids, the cube consists of 6 square faces, 12 edges, and 8 vertices.

**Regular octahedron**
Composed of equilateral triangles, the regular octahedron has 8 faces, 12 edges, and 6 vertices.

**Regular dodecahedron**
Composed of regular pentagons, the regular dodecahedron has 12 faces, 30 edges, and 20 vertices.

**Regular icosahedron**
Composed of equilateral triangles, the regular icosahedron has 20 faces, 30 edges, and 12 vertices.

# Areas and volumes

The illustrations below show various two- and three-dimensional shapes, and the formulae for calculating their areas and volumes. In the formulae, two letters together means that they are multiplied together (for example, "ah" means "a" multiplied by "h"); the superscript $^2$ means squared (for example, "$a^2$" is "a" multiplied by "a"); the superscript $^3$ means cubed (for example, "$a^3$" is "a" multiplied by "a" multiplied by "a"); the symbol $\sqrt{}$ means square root; and the symbol $\pi$ represents pi, which is 3.142 to three decimal places.

## AREA OF TWO-DIMENSIONAL SHAPES

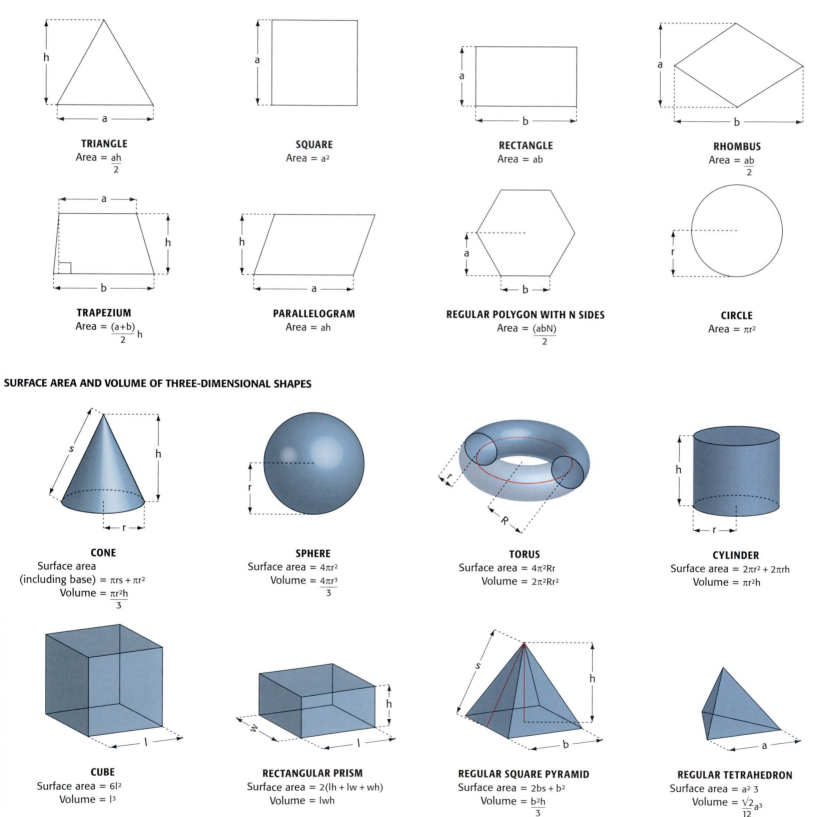

**TRIANGLE**
Area = $\dfrac{ah}{2}$

**SQUARE**
Area = $a^2$

**RECTANGLE**
Area = $ab$

**RHOMBUS**
Area = $\dfrac{ab}{2}$

**TRAPEZIUM**
Area = $\dfrac{(a+b)}{2}h$

**PARALLELOGRAM**
Area = $ah$

**REGULAR POLYGON WITH N SIDES**
Area = $\dfrac{(abN)}{2}$

**CIRCLE**
Area = $\pi r^2$

## SURFACE AREA AND VOLUME OF THREE-DIMENSIONAL SHAPES

**CONE**
Surface area
(including base) = $\pi rs + \pi r^2$
Volume = $\dfrac{\pi r^2 h}{3}$

**SPHERE**
Surface area = $4\pi r^2$
Volume = $\dfrac{4\pi r^3}{3}$

**TORUS**
Surface area = $4\pi^2 Rr$
Volume = $2\pi^2 Rr^2$

**CYLINDER**
Surface area = $2\pi r^2 + 2\pi rh$
Volume = $\pi r^2 h$

**CUBE**
Surface area = $6l^2$
Volume = $l^3$

**RECTANGULAR PRISM**
Surface area = $2(lh + lw + wh)$
Volume = $lwh$

**REGULAR SQUARE PYRAMID**
Surface area = $2bs + b^2$
Volume = $\dfrac{b^2 h}{3}$

**REGULAR TETRAHEDRON**
Surface area = $a^2\sqrt{3}$
Volume = $\dfrac{\sqrt{2}}{12}a^3$

# Factorials

The factorial of a number (denoted by the symbol !) is the product of all positive integers up to and including that number. For example, 4 factorial (written as 4!) = 4 x 3 x 2 x 1 = 24.

| N | N! | N | N! |
|---|---|---|---|
| 0 | 1 | 11 | 39916800 |
| 1 | 1 | 12 | 479001600 |
| 2 | 2 | 13 | 6227020800 |
| 3 | 6 | 14 | 87178291200 |
| 4 | 24 | 15 | 1307674368000 |
| 5 | 120 | 16 | 20922789888000 |
| 6 | 720 | 17 | 355687428096000 |
| 7 | 5040 | 18 | 6402373705728000 |
| 8 | 40320 | 19 | 121645100408832000 |
| 9 | 362880 | 20 | 2432902008176640000 |
| 10 | 3628800 | 21 | 51090942200000000000 |

# Compound interest

Compound interest is interest that is paid on both the principal amount of money invested and on any accumulated interest. The formula for calculating the total value of an invested sum after a period of compound interest if the interest is paid once a year is:

$$A = P(1+i)^n$$

$A$ = final amount of money accumulated, including interest
$P$ = principal amount of money invested
$i$ = rate of interest (expressed as a decimal) per period (per year for interest paid annually)
$n$ = number of periods for which the money is invested

For example, if $1000 was invested for three years at an interest rate of 5% per year:

P = $1000
i = 5% per year;  expressed as a decimal, 5% = 5/100 = 0.05
n = 3

Therefore the amount after three years, A = 1000 x $(1 + 0.05)^3$
$= 1000 \times 1.05^3$
$= 1000 \times 1.157625$
= $1157.62 (rounded down)

If the same amount ($1000) was invested for 5 years at 5% per year, the amount after that period would be:
1000 x $(1 + 0.05)^5$ = $1276.28

# Coefficients

In algebra a coefficient is a number that is the multiple of a variable in an equation. For example, in the equation $3x^2 + 7y - 4 = 0$, the coefficient of the variable $x^2$ is 3, and the coefficient of the variable y is 7; in the equation $3(x + y) = 0$, the coefficient of the variable $(x + y)$ is 3. If there is no number in front of a variable, the coefficient is 1; for example, in the equation x + 2 = 5, the coefficent of x is 1.

# Pascal's triangle

A triangular arrangement of numbers, Pascal's triangle is named after the 17th-century French mathematician Blaise Pascal (see p.196). In Pascal's triangle, each row has 1 as the first and last number. The other numbers are created by adding the two numbers that are to the left and right in the row above. Pascal's triangle is used mainly in algebra and probability.

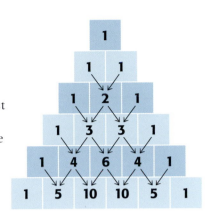

# Types of algebraic equations

Equations can be classified according to the highest power to which one or more of their variables is raised (see p.49). For example, in a quadratic equation at least one variable is squared (raised to the power of two).

| Equation | Common mathematical form | Description/uses |
|---|---|---|
| Linear | ax + by + c = 0 | No variable is raised to a power greater than one. Used in simple problems of addition, multiplication, and division. |
| Quadratic | $ax^2 + bx + c = 0$ | At least one of the variables is squared. Used in calculations involving area. |
| Cubic | $ax^3 + bx^2 + cx + d = 0$ | At least one of the variables is cubed. Used in calculations involving volume. |
| Quartic | $ax^4 + bx^3 + cx^2 + dx + e = 0$ | At least one of the variables is to the power of four. Used in computer graphics. |
| Quintic | $ax^5 + bx^4 + cx^3 + dx^2 + ex + f = 0$ | At least one of the variables is to the power of five. Solution of quintic equations was important to the development of abstract algebra. |

# Averages

An average is a single number that represents a group of numbers. The term "average" is usually used to refer to the arithmetic mean (often called simply the mean) but it may also refer to the median or mode.

| Measure | Definition | Example |
|---|---|---|
| Arithmetic mean (mean) | The sum of all the values in a group divided by the number of values. | To find the mean of 2, 4, 5, 5, 6, add together all of the values: 2 + 4 + 5 +5 + 6 =22. There are 5 values so the mean is 22 ÷ 5 = 4.4. |
| Median | The middle value of a group when the values are put in order. If there are two values in the middle, the median is the mean of those two values. | To find the median of 6, 5, 2, 4, 5, put the values in order: 2, 4, 5, 5, 6. The middle value—the median—is the third value along, which is 5. |
| Mode | The value that occurs most often in a group. It is possible to have more than one mode if more than one value appears "the most." | In the group of values 2, 4, 5, 5, 6, the value that appears most—the mode—is 5. In the group 2, 4, 4, 5, 5, 6, the values 4 and 5 both appear "the most" and so both are modes. |

# Basic trigonometry

The key formulae used in trigonometry are the sine (sin), cosine (cos), and tangent (tan) ratios. They can be used to find the angles in a right-angled triangle when at least two sides are known, or conversely, if an angle and the length of a side are known, these can be used to figure out lengths of the other sides. A useful way of recalling the formulae is by remembering the word "SOH-CAH-TOA."

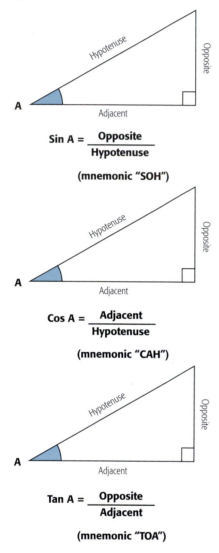

$$\text{Sin A} = \frac{\text{Opposite}}{\text{Hypotenuse}}$$

(mnemonic "SOH")

$$\text{Cos A} = \frac{\text{Adjacent}}{\text{Hypotenuse}}$$

(mnemonic "CAH")

$$\text{Tan A} = \frac{\text{Opposite}}{\text{Adjacent}}$$

(mnemonic "TOA")

# Conic sections

Also known as conics, conic sections are curves or shapes produced when a cone or double cone is intersected by a plane. The diagrams below show the shapes produced (in red) from various intersections.

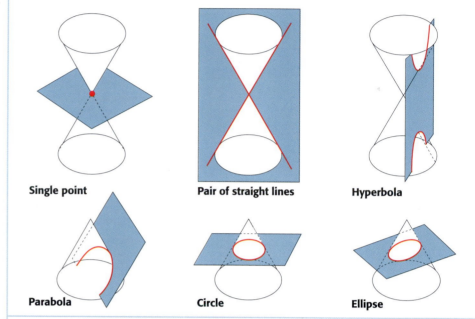

**Single point**  **Pair of straight lines**  **Hyperbola**

**Parabola**  **Circle**  **Ellipse**

# Squares, cubes, and roots

| Number | Square | Cube | Square root | Cube root | Number | Square | Cube | Square root | Cube root |
|---|---|---|---|---|---|---|---|---|---|
| 1 | 1 | 1 | 1.000 | 1.000 | 13 | 169 | 2,197 | 3.606 | 2.351 |
| 2 | 4 | 8 | 1.414 | 1.260 | 14 | 196 | 2,744 | 3.742 | 2.410 |
| 3 | 9 | 27 | 1.732 | 1.442 | 15 | 225 | 3,375 | 3.873 | 2.466 |
| 4 | 16 | 64 | 2.000 | 1.587 | 16 | 256 | 4,096 | 4.000 | 2.520 |
| 5 | 25 | 125 | 2.236 | 1.710 | 17 | 289 | 4,913 | 4.123 | 2.571 |
| 6 | 36 | 216 | 2.449 | 1.817 | 18 | 324 | 5,832 | 4.243 | 2.621 |
| 7 | 49 | 343 | 2.646 | 1.913 | 19 | 361 | 6,859 | 4.359 | 2.668 |
| 8 | 64 | 512 | 2.828 | 2.000 | 20 | 400 | 8,000 | 4.472 | 2.714 |
| 9 | 81 | 729 | 3.000 | 2.080 | 25 | 625 | 15,625 | 5.000 | 2.924 |
| 10 | 100 | 1,000 | 3.162 | 2.154 | 30 | 900 | 27,000 | 5.477 | 3.107 |
| 11 | 121 | 1,331 | 3.317 | 2.224 | 40 | 1,600 | 64,000 | 6.325 | 3.420 |
| 12 | 144 | 1,728 | 3.464 | 2.289 | 50 | 2,500 | 125,000 | 7.071 | 3.684 |

The square roots and cube roots are rounded to three decimal places.

# Pythagoras's theorem

Pythagoras's theorem states that in any right-angled triangle, the square of the length of the hypotenuse is equal to the sum of the squares of the other two sides (see p.35). It can be used to calculate the length of any side of such a triangle if the length of the other two sides are known. The formula is written $c^2 = a^2 + b^2$, where c is the length of the hypotenuse, and a and b are the lengths of the other two sides.

Although known to the Babylonians, tradition ascribes to Pythagoras himself the first proof. There are hundreds of proofs of Pythagoras's theorem. The one shown here uses algebra.

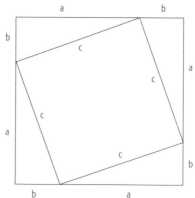

### Algebraic proof

The formula for Pythagoras's theorem, $c^2 = a^2 + b^2$, can be proved using this diagram, which features four right-angled triangles created when a square is positioned at an angle within a larger square. The following algebraic equations provide the proof:

The area of each of the the four triangles = ½ab

Therefore, the total area of the four triangles = 4(½ab) = 2ab

The area of the inner square = $c^2$

Therefore the total area of the large outer square = 2ab + $c^2$

But the total area of the outer square also = $(a + b)^2 = a^2 + 2ab + b^2$

Therefore 2ab + $c^2$ = $a^2$ + 2ab + $b^2$

Subtracting 2ab from both sides gives $c^2 = a^2 + b^2$

# Prime numbers

A prime number is any number that can only be exactly divided by 1 and itself without leaving a remainder. By definition, 1 is not a prime. The Greek mathematician Euclid proved that there is an infinite number of primes. There is no one formula for yielding every prime, and the largest prime numbers discovered by supercomputers run to many million of digits. Shown here are the first 918 prime numbers.

| 2 | 3 | 5 | 7 | 11 | 13 | 17 | 19 | 23 | 29 | 31 | 37 | 41 | 43 | 47 | 53 | 59 | 61 | 67 | 71 | 73 | 79 | 83 | 89 | 97 | 101 | 103 |
|---|---|---|---|----|----|----|----|----|----|----|----|----|----|----|----|----|----|----|----|----|----|----|----|----|-----|-----|
| 107 | 109 | 113 | 127 | 131 | 137 | 139 | 149 | 151 | 157 | 163 | 167 | 173 | 179 | 181 | 191 | 193 | 197 | 199 | 211 | 223 | 227 | 229 | 233 | 239 | 241 | 251 |
| 257 | 263 | 269 | 271 | 277 | 281 | 283 | 293 | 307 | 311 | 313 | 317 | 331 | 337 | 347 | 349 | 353 | 359 | 367 | 373 | 379 | 383 | 389 | 397 | 401 | 409 | 419 |
| 421 | 431 | 433 | 439 | 443 | 449 | 457 | 461 | 463 | 467 | 479 | 487 | 491 | 499 | 503 | 509 | 521 | 523 | 541 | 547 | 557 | 563 | 569 | 571 | 577 | 587 | 593 |
| 599 | 601 | 607 | 613 | 617 | 619 | 631 | 641 | 643 | 647 | 653 | 659 | 661 | 673 | 677 | 683 | 691 | 701 | 709 | 719 | 727 | 733 | 739 | 743 | 751 | 757 | 761 |
| 769 | 773 | 787 | 797 | 809 | 811 | 821 | 823 | 827 | 829 | 839 | 853 | 857 | 859 | 863 | 877 | 881 | 883 | 887 | 907 | 911 | 919 | 929 | 937 | 941 | 947 | 953 |
| 967 | 971 | 977 | 983 | 991 | 997 | 1009 | 1013 | 1019 | 1021 | 1031 | 1033 | 1039 | 1049 | 1051 | 1061 | 1063 | 1069 | 1087 | 1091 | 1093 | 1097 | 1103 | 1109 | 1117 | 1123 | 1129 |
| 1151 | 1153 | 1163 | 1171 | 1181 | 1187 | 1193 | 1201 | 1213 | 1217 | 1223 | 1229 | 1231 | 1237 | 1249 | 1259 | 1277 | 1279 | 1283 | 1289 | 1291 | 1297 | 1301 | 1303 | 1307 | 1319 | 1321 |
| 1327 | 1361 | 1367 | 1373 | 1381 | 1399 | 1409 | 1423 | 1427 | 1429 | 1433 | 1439 | 1447 | 1451 | 1453 | 1459 | 1471 | 1481 | 1483 | 1487 | 1489 | 1493 | 1499 | 1511 | 1523 | 1531 | 1543 |
| 1549 | 1553 | 1559 | 1567 | 1571 | 1579 | 1583 | 1597 | 1601 | 1607 | 1609 | 1613 | 1619 | 1621 | 1627 | 1637 | 1657 | 1663 | 1667 | 1669 | 1693 | 1697 | 1699 | 1709 | 1721 | 1723 | 1733 |
| 1741 | 1747 | 1753 | 1759 | 1777 | 1783 | 1787 | 1789 | 1801 | 1811 | 1823 | 1831 | 1847 | 1861 | 1867 | 1871 | 1873 | 1877 | 1879 | 1889 | 1901 | 1907 | 1913 | 1931 | 1933 | 1949 | 1951 |
| 1973 | 1979 | 1987 | 1993 | 1997 | 1999 | 2003 | 2011 | 2017 | 2027 | 2029 | 2039 | 2053 | 2063 | 2069 | 2081 | 2083 | 2087 | 2089 | 2099 | 2111 | 2113 | 2129 | 2131 | 2137 | 2141 | 2143 |
| 2153 | 2161 | 2179 | 2203 | 2207 | 2213 | 2221 | 2237 | 2239 | 2243 | 2251 | 2267 | 2269 | 2273 | 2281 | 2287 | 2293 | 2297 | 2309 | 2311 | 2333 | 2339 | 2341 | 2347 | 2351 | 2357 | 2371 |
| 2377 | 2381 | 2383 | 2389 | 2393 | 2399 | 2411 | 2417 | 2423 | 2437 | 2441 | 2447 | 2459 | 2467 | 2473 | 2477 | 2503 | 2521 | 2531 | 2539 | 2543 | 2549 | 2551 | 2557 | 2579 | 2591 | 2593 |
| 2609 | 2617 | 2621 | 2633 | 2647 | 2657 | 2659 | 2663 | 2671 | 2677 | 2683 | 2687 | 2689 | 2693 | 2699 | 2707 | 2711 | 2713 | 2719 | 2729 | 2731 | 2741 | 2749 | 2753 | 2767 | 2777 | 2789 |
| 2791 | 2797 | 2801 | 2803 | 2819 | 2833 | 2837 | 2843 | 2851 | 2857 | 2861 | 2879 | 2887 | 2897 | 2903 | 2909 | 2917 | 2927 | 2939 | 2953 | 2957 | 2963 | 2969 | 2971 | 2999 | 3001 | 3011 |
| 3019 | 3023 | 3037 | 3041 | 3049 | 3061 | 3067 | 3079 | 3083 | 3089 | 3109 | 3119 | 3121 | 3137 | 3163 | 3167 | 3169 | 3181 | 3187 | 3191 | 3203 | 3209 | 3217 | 3221 | 3229 | 3251 | 3253 |
| 3257 | 3259 | 3271 | 3299 | 3301 | 3307 | 3313 | 3319 | 3323 | 3329 | 3331 | 3343 | 3347 | 3359 | 3361 | 3371 | 3373 | 3389 | 3391 | 3407 | 3413 | 3433 | 3449 | 3457 | 3461 | 3463 | 3467 |
| 3469 | 3491 | 3499 | 3511 | 3517 | 3527 | 3529 | 3533 | 3539 | 3541 | 3547 | 3557 | 3559 | 3571 | 3581 | 3583 | 3593 | 3607 | 3613 | 3617 | 3623 | 3631 | 3637 | 3643 | 3659 | 3671 | 3673 |
| 3677 | 3691 | 3697 | 3701 | 3709 | 3719 | 3727 | 3733 | 3739 | 3761 | 3767 | 3769 | 3779 | 3793 | 3797 | 3803 | 3821 | 3823 | 3833 | 3847 | 3851 | 3853 | 3863 | 3877 | 3881 | 3889 | 3907 |
| 3911 | 3917 | 3919 | 3923 | 3929 | 3931 | 3943 | 3947 | 3967 | 3989 | 4001 | 4003 | 4007 | 4013 | 4019 | 4021 | 4027 | 4049 | 4051 | 4057 | 4073 | 4079 | 4091 | 4093 | 4099 | 4111 | 4127 |
| 4129 | 4133 | 4139 | 4153 | 4157 | 4159 | 4177 | 4201 | 4211 | 4217 | 4219 | 4229 | 4231 | 4241 | 4243 | 4253 | 4259 | 4261 | 4271 | 4273 | 4283 | 4289 | 4297 | 4327 | 4337 | 4339 | 4349 |
| 4357 | 4363 | 4373 | 4391 | 4397 | 4409 | 4421 | 4423 | 4441 | 4447 | 4451 | 4457 | 4463 | 4481 | 4483 | 4493 | 4507 | 4513 | 4517 | 4519 | 4523 | 4547 | 4549 | 4561 | 4567 | 4583 | 4591 |
| 4597 | 4603 | 4621 | 4637 | 4639 | 4643 | 4649 | 4651 | 4657 | 4663 | 4673 | 4679 | 4691 | 4703 | 4721 | 4723 | 4729 | 4733 | 4751 | 4759 | 4783 | 4787 | 4789 | 4793 | 4799 | 4801 | 4813 |
| 4817 | 4831 | 4861 | 4871 | 4877 | 4889 | 4903 | 4909 | 4919 | 4931 | 4933 | 4937 | 4943 | 4951 | 4957 | 4967 | 4969 | 4973 | 4987 | 4993 | 4999 | 5003 | 5009 | 5011 | 5021 | 5023 | 5039 |
| 5051 | 5059 | 5077 | 5081 | 5087 | 5099 | 5101 | 5107 | 5113 | 5119 | 5147 | 5153 | 5167 | 5171 | 5179 | 5189 | 5197 | 5209 | 5227 | 5231 | 5233 | 5237 | 5261 | 5273 | 5279 | 5281 | 5297 |
| 5303 | 5309 | 5323 | 5333 | 5347 | 5351 | 5381 | 5387 | 5393 | 5399 | 5407 | 5413 | 5417 | 5419 | 5431 | 5437 | 5441 | 5443 | 5449 | 5471 | 5477 | 5479 | 5483 | 5501 | 5503 | 5507 | 5519 |
| 5521 | 5527 | 5531 | 5557 | 5563 | 5569 | 5573 | 5581 | 5591 | 5623 | 5639 | 5641 | 5647 | 5651 | 5653 | 5657 | 5659 | 5669 | 5683 | 5689 | 5693 | 5701 | 5711 | 5717 | 5737 | 5741 | 5743 |
| 5749 | 5779 | 5783 | 5791 | 5801 | 5807 | 5813 | 5821 | 5827 | 5839 | 5843 | 5849 | 5851 | 5857 | 5861 | 5867 | 5869 | 5879 | 5881 | 5897 | 5903 | 5923 | 5927 | 5939 | 5953 | 5981 | 5987 |
| 6007 | 6011 | 6029 | 6037 | 6043 | 6047 | 6053 | 6067 | 6073 | 6079 | 6089 | 6091 | 6101 | 6113 | 6121 | 6131 | 6133 | 6143 | 6151 | 6163 | 6173 | 6197 | 6199 | 6203 | 6211 | 6217 | 6221 |
| 6229 | 6247 | 6257 | 6263 | 6269 | 6271 | 6277 | 6287 | 6299 | 6301 | 6311 | 6317 | 6323 | 6329 | 6337 | 6343 | 6353 | 6359 | 6361 | 6367 | 6373 | 6379 | 6389 | 6397 | 6421 | 6427 | 6449 |
| 6451 | 6469 | 6473 | 6481 | 6491 | 6521 | 6529 | 6547 | 6551 | 6553 | 6563 | 6569 | 6571 | 6577 | 6581 | 6599 | 6607 | 6619 | 6637 | 6653 | 6659 | 6661 | 6673 | 6679 | 6689 | 6691 | 6701 |
| 6703 | 6709 | 6719 | 6733 | 6737 | 6761 | 6763 | 6779 | 6781 | 6791 | 6793 | 6803 | 6823 | 6827 | 6829 | 6833 | 6841 | 6857 | 6863 | 6869 | 6871 | 6883 | 6899 | 6907 | 6911 | 6917 | 6947 |
| 6949 | 6959 | 6961 | 6967 | 6971 | 6977 | 6983 | 6991 | 6997 | 7001 | 7013 | 7019 | 7027 | 7039 | 7043 | 7057 | 7069 | 7079 | 7103 | 7109 | 7121 | 7127 | 7129 | 7151 | 7159 | 7177 | 7187 |

# Multiplication table

A multiplication table is a quick-reference tool that can be used to multiply two numbers. A range of numbers is aligned down the left hand side of the chart, starting a series of rows; another along the top of the chart, heads a series of columns. At the intersections of the rows and columns lies the products of the numbers at the far left and top respectively.

|    | 2 | 3 | 4 | 5 | 6 | 7 | 8 | 9 | 10 | 11 | 12 | 13 | 14 | 15 | 16 | 17 | 18 | 19 | 20 | 21 | 22 | 23 | 24 | 25 |
|----|---|---|---|---|---|---|---|---|----|----|----|----|----|----|----|----|----|----|----|----|----|----|----|----|
| 2 | 4 | 6 | 8 | 10 | 12 | 14 | 16 | 18 | 20 | 22 | 24 | 26 | 28 | 30 | 32 | 34 | 36 | 38 | 40 | 42 | 44 | 46 | 48 | 50 |
| 3 | 6 | 9 | 12 | 15 | 18 | 21 | 24 | 27 | 30 | 33 | 36 | 39 | 42 | 45 | 48 | 51 | 54 | 57 | 60 | 63 | 66 | 69 | 72 | 75 |
| 4 | 8 | 12 | 16 | 20 | 24 | 28 | 32 | 36 | 40 | 44 | 48 | 52 | 56 | 60 | 64 | 68 | 72 | 76 | 80 | 84 | 88 | 92 | 96 | 100 |
| 5 | 10 | 15 | 20 | 25 | 30 | 35 | 40 | 45 | 50 | 55 | 60 | 65 | 70 | 75 | 80 | 85 | 90 | 95 | 100 | 105 | 110 | 115 | 120 | 125 |
| 6 | 12 | 18 | 24 | 30 | 36 | 42 | 48 | 54 | 60 | 66 | 72 | 78 | 84 | 90 | 96 | 102 | 108 | 114 | 120 | 126 | 132 | 138 | 144 | 150 |
| 7 | 14 | 21 | 28 | 35 | 42 | 49 | 56 | 63 | 70 | 77 | 84 | 91 | 98 | 105 | 112 | 119 | 126 | 133 | 140 | 147 | 154 | 161 | 168 | 175 |
| 8 | 16 | 24 | 32 | 40 | 48 | 56 | 64 | 72 | 80 | 88 | 96 | 104 | 112 | 120 | 128 | 136 | 144 | 152 | 160 | 168 | 176 | 184 | 192 | 200 |
| 9 | 18 | 27 | 36 | 45 | 54 | 63 | 72 | 81 | 90 | 99 | 108 | 117 | 126 | 135 | 144 | 153 | 162 | 171 | 180 | 189 | 198 | 207 | 216 | 225 |
| 10 | 20 | 30 | 40 | 50 | 60 | 70 | 80 | 90 | 100 | 110 | 120 | 130 | 140 | 150 | 160 | 170 | 180 | 190 | 200 | 210 | 220 | 230 | 240 | 250 |
| 11 | 22 | 33 | 44 | 55 | 66 | 77 | 88 | 99 | 110 | 121 | 132 | 143 | 154 | 165 | 176 | 187 | 198 | 209 | 220 | 231 | 242 | 253 | 264 | 275 |
| 12 | 24 | 36 | 48 | 60 | 72 | 84 | 96 | 108 | 120 | 132 | 144 | 156 | 168 | 180 | 192 | 204 | 216 | 228 | 240 | 252 | 264 | 276 | 288 | 300 |
| 13 | 26 | 39 | 52 | 65 | 78 | 91 | 104 | 117 | 130 | 143 | 156 | 169 | 182 | 195 | 208 | 221 | 234 | 247 | 260 | 273 | 286 | 299 | 312 | 325 |
| 14 | 28 | 42 | 56 | 70 | 84 | 98 | 112 | 126 | 140 | 154 | 168 | 182 | 196 | 210 | 224 | 238 | 252 | 266 | 280 | 294 | 308 | 322 | 336 | 350 |
| 15 | 30 | 45 | 60 | 75 | 90 | 105 | 120 | 135 | 150 | 165 | 180 | 195 | 210 | 225 | 240 | 255 | 270 | 285 | 300 | 315 | 330 | 345 | 360 | 375 |
| 16 | 32 | 48 | 64 | 80 | 96 | 112 | 128 | 144 | 160 | 176 | 192 | 208 | 224 | 240 | 256 | 272 | 288 | 304 | 320 | 336 | 352 | 368 | 384 | 400 |
| 17 | 34 | 51 | 68 | 85 | 102 | 119 | 136 | 153 | 170 | 187 | 204 | 221 | 238 | 255 | 272 | 289 | 306 | 323 | 340 | 357 | 374 | 391 | 408 | 425 |
| 18 | 36 | 54 | 72 | 90 | 108 | 126 | 144 | 162 | 180 | 198 | 216 | 234 | 252 | 270 | 288 | 306 | 324 | 342 | 360 | 378 | 396 | 414 | 432 | 450 |
| 19 | 38 | 57 | 76 | 95 | 114 | 133 | 152 | 171 | 190 | 209 | 228 | 247 | 266 | 285 | 304 | 323 | 342 | 361 | 380 | 399 | 418 | 437 | 456 | 475 |
| 20 | 40 | 60 | 80 | 100 | 120 | 140 | 160 | 180 | 200 | 220 | 240 | 260 | 280 | 300 | 320 | 340 | 360 | 380 | 400 | 420 | 440 | 460 | 480 | 500 |
| 21 | 42 | 63 | 84 | 105 | 126 | 147 | 168 | 189 | 210 | 231 | 252 | 273 | 294 | 315 | 336 | 357 | 378 | 399 | 420 | 441 | 462 | 483 | 504 | 525 |
| 22 | 44 | 66 | 88 | 110 | 132 | 154 | 176 | 198 | 220 | 242 | 264 | 286 | 308 | 330 | 352 | 374 | 396 | 418 | 440 | 462 | 484 | 506 | 528 | 550 |

# Who's Who

**Agassiz, Louis 1835–1910**
See p.195.

**Al-Biruni 973-1048 CE**
Persian scholar and scientist who excelled in astronomy, mathematics, physics, medicine, and history, writing about them prolifically. Al-Biruni traveled to India, making a comprehensive study of its culture in *A History of India*. In his studies on astronomy he supported the theory of Earth's rotation on its axis and made accurate calculations of latitude and longitude. Al-Biruni's work in physics included an explanation of how natural springs were governed by the laws of hydrostatics, while in geography he radically suggested that the Indus valley had once been a sea basin.

**Al-Hazen 965–1039**
See pp.60–61.

**Al-Idrisi c.1099–1165**
Arab geographer, cartographer, and traveler, Al-Idrisi is best known for his *Tabula Rogeriana*, a remarkably accurate map of the world as it was then known. He prepared the map and its commentary in 1154 for King Roger II of Sicily after collating information from merchants and explorers. The map, which has south at the top, extends across all of Europe and Asia, although only northern Africa is shown. Al-Idrisi also created a huge, two-sided planisphere: one side showed the positions of the constellations, while the other showed where the countries of the world were.

**Al-Kwharizmi c.820**
See p.48.

**Alvarez, Luis Walter 1911–1988**
US experimental physicist who won the 1968 Nobel Prize for Physics for developing the liquid-hydrogen bubble chamber, which he used to identify many resonances (very short-lived subatomic particles that occur only in high-energy nuclear collisions). During World War II Alvarez invented the radar guidance system for aircraft landings and designed the detonators for the spherical implosives used on the Trinity and Nagasaki atomic bombs. In 1947 he helped put together the first proton linear accelerator. In 1980, with his son, Walter, he proposed the asteroid-impact theory to explain the extinction of the dinosaurs.

**Amontons, Guillaume 1663–1705**
French inventor and physicist who invented and perfected instruments for measuring temperature and pressure. In 1687 he invented an improved hygrometer, a device used to measure humidity. He improved the barometer and thermometer (1695), particularly for use at sea. He was the first researcher to discuss the theory of an absolute zero of temperature and proposed theories on the laws of friction.

**Ampère, André-Marie 1775–1836**
French mathematician and physicist who developed the mathematical theory of electromagnetism. His theory served as the basis for other fundamental ideas of the 19th century regarding electricity and magnetism, and inspired discoveries by scientists such as Faraday, Weber, Thomson, and Maxwell. The ampere, the unit of electric current, was named after him.

**Anaximander 611–547 BCE**
Greek philosopher. He drew a map of the entire known world and is credited as being the first scientific geographer. He invented the gnomon, used to determine the solstices, which became the basis for the development of sundials, and suggested that the heavens are a perfect sphere, with Earth floating freely at its center. He was also the first to try to determine the magnitude of the celestial bodies and how they are positioned in space.

**Anderson, Carl David 1905–91**
American physicist who opened up the field of particle physics and the study of the atom. He discovered the positron and muon, for which he was jointly awarded the Nobel Prize in Physics in 1936.

**Angstrom, Anders Jonas 1814–74**
Swedish physicist and spectroscopist who deduced in 1855 that a hot gas emits light at the same wavelengths at which it absorbs light when cooler. In 1861 Angstrom began to study the solar spectrum, proving the existence of hydrogen in the Sun and recording his measurements in units, subsequently named in his honor. He was also the first to examine the spectra of the aurora borealis.

**Appleton, Edward Victor 1892–1965**
English physicist who proved the existence of the ionosphere. From 1924 to 1936 Appleton was professor of physics at King's College, London, where he worked on experiments using radio waves. He determined the height at which the waves were reflected back to Earth—the ionosphere, and discovered another reflecting layer at approximately 150 miles (250 km), now called the Appleton layer. This work led directly to the development of radar technologies.

**Archimedes c.287–212 BCE**
Greek philosopher and inventor, and probably the greatest mathematician of the ancient world. He lived on the island of Sicily, wrote influential works on arithmetic, geometry, and mechanics, and famously, while taking a bath, solved a tricky problem concerning the king's crown—realizing that it would displace its own volume of water. This led him to investigate hydrostatics and discover Archimedes' Principle, which states that any object immersed in water will experience an upthrust equal to the weight of water displaced. He gave mathematical descriptions of levers and pulleys, devised machines to repel invading Roman ships, and may have invented the pump known as the Archimedes screw.

**Aristarchus of Samos c.310–230 BCE**
Greek astronomer noted as being the first to suggest that the earth revolves around the Sun. His only surviving treatise, *On the Magnitudes and Distances of the Sun and Moon*, details his geometric argument, based on observation, whereby he calculated that the Sun was about 20 times as far from Earth as the Moon, and 20 times the Moon's size. Although these estimates proved inaccurate, the fault lay in the inadequacies of the instruments he was using rather than in his reasoning, and his pioneering methods provided a platform for future astronomers and mathematicians.

**Aristotle 384–322 BCE**
See pp.37–37

**Arkwright, Richard 1733–92**
English inventor and industrialist, who in 1769 patented an automatic cotton-spinning machine which allowed an unskilled teenager to spin 96 threads simultaneously. Powered by a water wheel, it came to be called the water frame. He installed many of these machines in a purpose-built factory in Derbyshire in 1771, thus introducing the factory system and mass production, and driving forward the industrial revolution. He had to borrow money to build that first factory, but when he died in 1792 he left several factories, a town house in London, a castle in Derbyshire, and half a million pounds.

**Arrhenius, Svante August 1859–1927**
See p.234.

**Aston, Francis William 1877–1945**
English physicist and chemist who won the 1922 Nobel Prize for chemistry for his work on isotopes. He developed the mass spectrograph, which separates atoms or molecular fragments of different mass and measures those masses. This led to his discovery of a large number of naturally-occurring isotopes, and the formulation of his Whole Number Rule, which became important in the development of nuclear energy.

**Avicenna c.979–1037**
See p.25.

**Avogadro, Amedeo 1776–1856**
Italian physicist and mathematician most noted for his work on molecular theory and Avogadro's law. In 1811 he published work which made the distinction between the molecule and the atom. He suggested that equal volumes of all gases at the same temperature and pressure contain the same number of molecules. In honor of Avogadro's contributions to molecular theory, the number of molecules in one mole in the molecular weight in grams was named Avogadro's number.

**Babbage, Charles 1791–1871**
See p.174.

**Bacon, Francis 1561–1626**
English statesman who became a leader in natural philosophy and is regarded as the father of modern scientific methodology. Rejecting ancient beliefs that scientific truth could only be reached by authoritative discussion, Bacon tried to put natural science on a sound empirical foundation in *Novum Organum Scientiarum* (1620) and *The Advancement of Learning* (1623). His classification of the sciences later inspired the 18th-century French encylopedists, while his empiricism influenced 19th-century British philosophers of science. He died following an experiment on how freezing affected meat preservation: having braved a blizzard to stuff a chicken with snow, Bacon developed pneumonia and died.

**Bacon, Roger 1214–1294**
See p.57.

**Baekeland, Leo Hendrik 1863–1944**
Belgian-American chemist and inventor of Bakelite, whose first invention was Velox, an improved photographic paper that freed photographers from having to use sunlight for developing images. In 1899 George Eastman bought full rights to Velox for $1 million, and with this windfall Baekeland was able to invest in research and set up a company in 1909 to produce Bakelite, the first completely synthetic plastic, which could flow and be poured into molds before hardening.

**Baeyer, Adolf von 1835–1917**
German organic chemist who synthesized the dye indigo and determined its structure in the 1880s. For this and his other contributions to the synthetic dyestuffs industry, he was awarded the 1905 Nobel Prize in Chemistry.

**Baird, John Logie 1888–1946**
Scottish television pioneer. Baird began his television experiments in 1922 and obtained the world's first moving television image in 1925. He demonstrated the world's first color transmission in 1928, and made the first television program for the BBC. His mechanical scanning system, however, was less versatile than Marconi-EMI's electronic system, which was adopted by the BBC in 1936.

**Banks, Joseph 1743–1820**
English explorer and botanist who introduced many new plants to the West including the eucalyptus, and helped to establish the Royal Botanic Gardens at Kew, London. In 1768 Banks joined a Royal Society expedition to the South Pacific, led by Captain James Cook aboard HMS *Endeavour*. He collected numerous specimens en route, later writing a scientific account of the voyage and its discoveries. Banks was interested in the practical and commercial uses of plants and in introducing them into other countries. Around 75 species are named after him, as is a group of Pacific Islands. There was even a suggestion to name Australia "Banksia." He was a long-time president (1778–1820) of the Royal Society.

**Bardeen, John 1908–1991**
US physicist known for his work on semiconductors; he co-invented the transistor (1947), paving the way for modern electronics, and later helped develop a theory of superconductivity (1957). The transistor enabled the development of thousands of different electronic devices, while his developments in superconductivity are used in medical advances such as MRI. Bardeen won the Nobel Prize for each of these achievements in 1956 and 1972 respectively. He worked at Bell Laboratories from 1945–51 and was a professor of electrical engineering and physics at the University of Illinois until 1975.

**Barnard, Christian Neething 1922–2001**
See p.406.

**Baudot, Jean Maurice Emile, 1845–1903**
French telegraph engineer and inventor of the Baudot code. The telegraph service trained him in the Morse telegraph and the Hughes printing telegraph method, which inspired his own system. Baudot used synchronous distributors running at a constant speed, allowing several messages to be sent simultaneously. In June 1874 he patented his first printing telegraph. His apparatus was shown at the Paris Exposition Universelle (1878) and won him the gold medal.

**Baumé, Antoine 1728–804**
French chemist best known as the inventor of the Baumé scale hydrometer. He became professor of chemistry at the École de Pharmacie in 1752. He retired in 1780, but was ruined by the French Revolution and forced to return to commercial life. He improved many scientific instruments, among them the hydrometer, used to measure the specific gravity of liquids. Baumé devised an improved scale for the hydrometer which now bears his name.

**Becquerel, Antoine Henri 1852–1908**
See p.299.

**Behring, Emil Adolph von 1854–1917**
German physician who developed serum therapy. He discovered that the body produces antitoxins and developed treatments for diseases such as diphtheria. Together with Japanese bacteriologist Shibasaburo Kitasato he applied the serum therapy technique to tetanus. Behring also introduced early vaccination techniques against diphtheria and tuberculosis. He won the first Nobel Prize in Physiology or Medicine in 1901.

**Bell, Alexander Graham 1847–1922**
See p.260.

**Benz, Karl 1844–1929**
German inventor who, with Gottlieb Daimler, made the first practical gas-engined motor vehicle. In June 1886 Benz road-tested a three-wheeled car powered by a four-stroke single-cylinder engine with electric ignition. Two years later he was the first to offer gas-driven cars to sell to the public, so founding the motor industry.

**Berg, Paul 1926–**
American biochemist and molecular biologist who shared the 1980 Nobel Prize for Chemistry for studies on recombinant DNA. Berg developed recombinant DNA techniques which enable DNA to be spliced from different organisms and recombined. An important leap forward for scientific research, this has valuable commercial applications in fields such as agriculture, medicine, and industry. Berg has long been a leading figure in genetic engineering; he also played a key role in addressing its potential risks and devising guidelines for self-regulation.

**Bergius, Friedrich 1884–1949**
German chemist who made lubricants and synthetic fuels, such as gasoline, from coal. In 1910 Bergius established a private laboratory where he researched the hydrogenating effect of hydrogen on coal and heavy oils under high pressure. After WWI he managed to develop the hydrogenation of coal on a large scale in what is know as the Bergius process. In 1931 he shared the Nobel Prize in Chemistry with Carl Bosch for their invention of chemical high-pressure methods.

**Bernard, Claude 1813–1878**
See p.274.

**Berners-Lee, Tim 1955–**
See p.378.

**Bernoulli, Daniel 1700–1782**
Swiss mathematician and physicist who discovered that the pressure in a fluid decreases as the velocity of fluid flow increases. This became known as Bernoulli's principle and is fundamental to many engineering applications. Bernoulli also formulated a law enabling the velocity of a liquid to be measured, and contributed to probability theory. His publication *Hydrodynamica* (1738), in which he developed the theory of watermills, windmills, water pumps, and water propellers, was a milestone in the kinetic theory of fluids and gases. Bernoulli was the first to differentiate between hydrostatic and hydrodynamic pressure.

**Berthelot, Marcellin Pierre Eugène 1827–1907**
French chemist who, working on the synthesis of acetylene and other hydrocarbons, demonstrated that living processes are not essential to the production of organic chemicals. His research and publications made a significant contribution to the development of chemistry in the late 19th century.

**Berzelius, Jöns Jakob 1779–1848**
Swedish chemist considered one of the founders of modern chemistry, Berzelius is best known for his system of electrolysis. He produced the first accurate list of atomic weights and developed modern chemical symbols. He discovered selenium, cerium, and thorium, isolated the elements of zirconium, silicon, and titanium, developed classical analytical techniques, and investigated isomerism and catalysis. From 1807 to 1832 Berzelius was a professor of medicine and pharmacy at the Karolinska Institute and he became a member of the Royal Swedish Academy of Sciences in 1808.

**Best, Charles Herbert 1899–1978**
Canadian medical scientist who helped discover insulin in 1921, after which it became a standard treatment for diabetes. Best worked closely with Frederick Banting, and missed out on winning the Nobel Prize in Physiology or Medicine in 1923 to Banting and physiology professor John Macleod. Best went on to develop heparin, recognizing that it could be an important anticoagulant drug for preventing blood clotting.

**Bethe, Hans Albrecht 1906–2005**
German-born American physicist who, in 1938, made the first detailed calculations concerning the nuclear reactions that power the stars. He theorized that hydrogen is turned into helium by a series of nuclear reactions known as the carbon cycle. For his work, Bethe won the 1967 Nobel Prize in Physics. In the 1950s he joined a project to develop the hydrogen bomb. He later campaigned, alongside Albert Einstein, against nuclear testing.

**Biot, Jean Baptiste 1774–1862**
French physicist, mathematician, and astronomer who confirmed the reality of meteorites. He also worked on the relationship between electrical currents and magnetism. He discovered that the intensity of the magnetic field set up by a current flowing through a wire varies inversely with the distance from the wire. Known as Biot-Savart's law, it became fundamental to modern electromagnetic theory. A crater on the moon is named after him.

**Bjerknes, Vilhelm 1862–1951**
See p.208.

**Black, Joseph 1728–1799**
See pp.148–49.

**Blackett, Patrick Maynard Stuart 1897–1974**
English physicist best known for his work on cloud chambers, rock magnetism, and cosmic rays. In 1932, working with Giuseppe Occhialini, he devised a system of Geiger counters to investigate cosmic rays. In 1937 he went to the University of Manchester where he created a major international research laboratory. He played an important role in World War II, advising on military strategy, and in 1948 was awarded the Nobel Prize in Physics.

**Boas, Franz 1858–1942**
German-born US anthropologist known for applying scientific methods to the study of human cultures and societies, Boas is credited with playing a pivotal role in developing modern cultural anthropology. His concept was that all human races are equally capable of developing cultural forms; where differences exist it is because of culture rather than race or genetics. At Columbia University (1899–1942), Boas developed one of the leading departments of anthropology in the US. He specialized in Native American culture and languages and founded the *International Journal of American Linguistics*.

**Bode, Johann Elert 1747–1826**
German astronomer best known for Bode's law, or the Titus-Bode law, a rough rule which predicts the spacing of the planets in the solar system. As director of the Berlin Observatory he discovered several new star clusters and a comet. In 1801 he published the *Uranographia*, a celestial atlas that showed the positions of the stars and gave the planet Uranus its name.

**Bohr, Niels 1885–1962**
Danish physicist who carried out research into the atom and was the first to apply quantum theory, as established by Max Planck. In 1913, on the basis of Rutherford's theories, he published his model of atomic structure that still serves as the basis of the physical and chemical properties of the elements. He was awarded the Nobel Prize for Chemistry in 1922. During World War II he joined the Manhattan Project but later concentrated on the peaceful application of atomic physics and the political conflicts caused by the development of atomic weapons.

**Boole, George 1815–1864**
English teacher, mathematician, and philosopher who invented symbolic knowledge and formulated the rules that govern it, known as Boolean algebra. His most important work, *The Laws of Thought*, was published in 1854. Boole's ideas became the basis for the development of much modern abstract mathematics and computer science.

**Bosch, Carl 1874–1940**
German industrial chemist who developed the Haber-Bosch process for high-pressure synthesis of ammonia. This process became the main industrial procedure for nitrogen fixation and in 1931 he was jointly awarded the Nobel Prize for Chemistry with Friedrich Bergius. Bosch then studied how to make ammonia available for use in agriculture and industry, developing a means for the production of nitrogen fertilizers. This had a global impact on agriculture.

**Bose, Satyendranath 1894–1974**
Indian physicist best known for his work in quantum mechanics. In 1924 Albert Einstein arranged publication of an article Bose sent him, and developed Bose's work in the form of two Bose-Einstein theories. Bosons, which form half the fundamental particles in the Universe, are named after these two scientists.

**Boyle, Robert 1627–1691**
See pp.98–99.

**Brahe, Tycho 1546–1601**
Danish nobleman and astronomer known for his work in developing astronomical instruments and for his accurate and comprehensive planetary and astronomical observations. These included a thorough study of the solar system, proof that the orbit of the 1577 comet lay beyond the Moon, and accurate positions of more than 777 fixed stars. Incredibly, all were achieved before the invention of the telescope, and his work provided the gateway to future discoveries. Brahe is also known for his contributions to medicine. He famously lost the tip of his nose during a duel and wore a metal prosthetic.

**Bramah, Joseph 1748–1814**
See p.145.

**Broca, Paul 1824–1880**
See p.223.

**Broecker, Wallace 1931–**
See p.418.

**Brunel, Isambard Kingdom 1806–1859**
English engineer whose first project was helping his father to dig the first tunnel under a major river—the Thames in London. He designed the Clifton Suspension Bridge across the Avon Gorge in Bristol, and constructed the Great Western Railway from London all the way to Cornwall in the southwest, including all its tunnels, bridges, and viaducts. He also built three great ships—the *Great Western*, the first steamship built to carry passengers to America, the *Great Britain*, made of iron and equipped with the newly invented screw propeller, and the *Great Eastern*, which in 1866 laid the first successful cable across the Atlantic.

**Budd, William 1811–1880**
English doctor and epidemiologist whose studies on infectious diseases, such as typhoid, advanced contagion theory. Budd also advocated the use of disinfectant and the prevention of sewage entering the water supplies. Such preventative methods enabled Budd to significantly reduce the spread of cholera in Bristol.

**Burbank, Luther 1849–1926**
American botanist, horticulturalist, and pioneer in agricultural science noted for his plant-breeding experiments. Burbank developed more than 800 strains and varieties of plants. He helped transform plant breeding into a modern science, aiding the later study of genetics. His objective was to improve plant quality, so increasing the world's food supply. His Burbank potato was introduced to Ireland to counteract the blight epidemic. Burbank settled in Santa Rosa, California, where he established a nursery garden, greenhouse, and experimental farms that became internationally renowned.

**Cabrera, Blas c.1942–**
Spanish physicist at Stanford University, best known for his experiment in search of magnetic monopoles. In 1982 his detector recorded an event which bore the perfect signature hypothesized for a magnetic monopole. No similar event has since been recorded. The Cabrera Group at Stanford is currently involved in several research projects including the Cryogenic Dark Matter Search (CDMS) experiment. This seeks to identify the dominant form of matter in and around our galaxy, under the hypothesis that it comprises weakly interfacing massive particles (WIMPS). If WIMPS and therefore supersymmetry were discovered, the most important problems in elementary particle physics and cosmology would be simultaneously solved.

**Calmette, Albert (Leon Charles) 1863–1933**
French bacteriologist and immunologist best known for his work in public health with Camille Guérin. They developed the BCG (*bacillus* Calmette-Guérin) vaccine against tuberculosis, first trialed in 1922. He also produced the first antivenin for snake venom, the Calmette's serum.

**Cajal, Santiago Ramón y 1852–1934**
Spanish histologist, physician, and pathologist, Cajal is noted for his work on the structure of the central nervous system. Using a histological staining technique based on silver chromate developed by Camillo Golgi, Cajal identified the structure of individual neurons, which led him to conclude that nervous tissue is made of independent cells rather than a complex connecting network. He was joint-winner of the 1905 Nobel Prize for Physiology or Medicine.

**Cannizarro, Stanislao 1826–1910**
An Italian chemist, Cannizarro is remembered for his discovery of cyanamide, for obtaining alcohols from aldehydes (a process known as Cannizarro's reaction), and for distinguishing between molecular and atomic weights. He supported Avogadro's law—the idea that equal volumes of gas at the same pressure and temperature held equal numbers of molecules or atoms, and that equal volumes of gas could be used to calculate atomic weights—which led to a new understanding of chemistry. He held posts at various Italian universities including Alessandria, Palermo, and Rome, and was also politically active.

**Cannon, Walter Bradford 1871–1945**
American neurologist and physiologist who studied the regulation of hunger and thirst in animals and developed the concept of homeostasis (the ability of an organism to maintain its internal environment), popularizing it in *The Wisdom of the Body* (1932). With Philip Bard, he developed the Cannon-Bard theory to explain why people feel emotions first and then act upon them. Cannon was also the first to use X-rays in physiological studies; the barium meal is a modern derivative of this research. He investigated hemorrhagic and traumatic shock, worked on methods of storing blood, and in 1931 discovered sympathin, an adrenaline-like substance. He became a leading authority in the emerging field of psychosomatic medicine. Cannon was professor and chairman of the Department of Physiology at Harvard Medical School.

**Carnot, Nicolas Leonard Sadi 1796–1832**
French physicist and military engineer who gave the first successful theoretical account of heat engines, now known as the Carnot Cycle. His work laid the foundation for thermodynamics. Carnot's most important work, *Réflexions sur la Puissance Motrice du Feu* (Reflections on the Driving Power of Fire) published in 1824, provided the theoretical basis for the development of the steam engine and introduced the idea of the second law of thermodynamics. The Carnot Cycle demonstrates the law of entropy: an engine cannot convert all of its heat energy into mechanical energy; some always stays unused.

**Carson, Rachel Louise 1907–1964**
American marine biologist. Carson joined the US Bureau of Fisheries as a writer. In 1936 the Bureau hired her as a junior biologist and later she became the chief editor of all publications for the US Fish and Wildlife Service. In the 1940s Carson began to write books on marine life. In response to her most famous work, *Silent Spring* (1962), President Kennedy called for tests to be carried out on the chemicals mentioned in the book.

**Carver, George Washington 1860–1943**
African-American botanist and inventor who won acclaim for his contributions to agricultural chemistry. He researched and promoted alternative crops to cotton, such as peanuts and sweet potatoes. He also experimented to find new plant uses, making more than 300 products from peanuts alone. Most inspirational of all was how a gentle man born into slavery gained an education against the odds, graduating with a BS in Agriculture in 1894. Carver was later invited to teach at Tuskegee Institute, Alabama, where he remained for the rest of his life.

**Cassini, Giovanni Domenico 1625–1712**
Italian-born engineer and astronomer. He discovered a dark gap, 1,696 miles (2,730 km) wide, in Saturn's rings, caused by the gravitational influence of Saturn's inner satellites. The gap was named Cassini's division. He also discovered four satellites of Saturn and the Great Red Spot on Jupiter. The NASA *Cassini* spacecraft, named after him, is currently in orbit around Saturn.

**Cauchy, Augustin Louis, Baron 1789–1857**
French mathematician who laid the groundwork for rigor in analysis and researched differential equations, probability, and mathematical physics. A prolific science writer and one of the greatest modern mathematicians, he was elected to the Academy of Sciences in Paris at age 27.

**Cavendish, Henry 1731–1810**
See p.210.

**Caventou, Joseph Bienaimé 1795–1877**
French pharmacist and one of the earliest investigators of alkaloids. Working alongside Pierre-Joseph Pelletier he isolated a number of active ingredients from plants, including chlorophyll, strychnine, caffeine, and quinine sulphate, which became an important treatment for malaria. The Moon crater Caventou is named after him.

**Chadwick, James 1891–1974**
English physicist who won the Nobel Prize for Physics in 1935 for discovering the neutron. Chadwick worked on radioactivity with Ernest Rutherford and in 1932 proved Rutherford's prediction that a particle without electric charge, which Chadwick named a neutron, existed in the nucleus of an atom; he also calculated its mass. Chadwick's discovery advanced scientific experiment internationally, and scientists began to bombard all types of materials with neutrons; one of these was uranium, which resulted in nuclear fission. During World War II Chadwick worked on the Manhattan Project to develop the atomic bomb. He was knighted in 1945.

**Chamberlain, Owen 1920-2006**
American physicist who, with Emilio Segrè, won the Nobel Prize in Physics in 1959 for their discovery of the antiproton, the antimatter equivalent and negatively-charged mirror image of the proton. This discovery opened up a new field of physics and expanded the understanding of particle physics. He worked on the Manhattan Project, helping to develop the first atomic bomb.

**Chandrasekhar, Dr. Subrahmanyan 1910–95**
See p.331.

**Chappe, Claude 1763–1805**
French inventor who, with his brother Ignace, developed a mechanical telegraph system. Their practical semaphore design was demonstrated in 1792 when a message was passed between Paris and Lille. It eventually spanned all of France and parts of Europe.

**Chaptal, Jean-Antoine 1756–1832**
French technical chemist, industrialist, and statesman who discovered that adding sugar to unfermented wine increased the final alcohol level; this became known as chaptalization. Some of his achievements included improvements in the manufacture of sulphuric acid, saltpeter for gunpowder, wine, and dyeing. Focusing on technical chemistry, Chaptal supplemented Lavoisier's theoretical work, and his writings pioneered the application of chemical principles to industry and agriculture.

**Charles, Jaques Alexandre Cesar 1746–1823**
French chemist and physicist who invented the hydrogen balloon. On August 27, 1783, his first balloon ascended to a height of nearly 3,000 feet (914 m). In 1783 Charles and Ainé Roberts covered over 27 miles (43 km) in La Charlière. He is also known for his formulation of Charles's law (1787), which states that, at constant pressure, the volume occupied by a fixed mass of gas is directly proportional to the absolute temperature.

**Cherenkov, Pavel Alekseyevich 1904–1990**
Soviet physicist who shared the 1959 Nobel Prize for Physics with Igor Tamm and Illya Frank for the discovery and theoretical interpretation of Cherenkov radiation. As a research student, Cherenkov had observed that electrons produce a blue glow when passing through a transparent liquid at high velocity. This Cherenkov radiation, correctly interpreted by Tamm and Frank in 1937, led to the development of the Cherenkov Detector that became so useful in experimental nuclear and particle physics.

**Clarke, Arthur 1917–2008**
See p.369.

**Claude, Albert 1899–1983**
Belgian-American cell biologist and co-winner of the 1974 Nobel Prize for Physiology or Medicine for discoveries concerning the structural and functional organization of the living cell. During the 1930s and 40s Claude helped pioneer the use of an electron microscope for cellular study and was involved in developing differential centrifugation to separate various cell components. He discovered that the cell interior comprised a highly organized membranous network (the endoplasmic reticulum), dispelling the belief that it was merely a chaotic mass. Claude joined the Rockefeller Institute in 1929, and remained there for the rest of his career.

**Cockcroft, John Douglas 1897–1967**
English physicist and co-winner with Ernest Walton of the 1951 Nobel Prize for Physics for pioneering the use of particle accelerators to study atomic nuclei. While a professor at Cambridge, he and Walton designed the Cockcroft-Walton generator and used it to split lithium atoms by bombarding them with protons. This was one of the earliest experiments to artificially change the atomic nucleus of one element to a different nucleus. They then researched the splitting of other atoms, establishing the importance of accelerators like their generator as a tool for nuclear research. In 1946 he set up the British Atomic Energy Research Establishment with the aim of developing Britain's atomic power program. He was master of Churchill College, Cambridge, UK, from 1959 to 1967.

**Cohen, Stanley Norman 1935–**
American microbiologist known for pioneering new methods of combining and transplanting genes, which led to the field of genetic engineering. At Stanford University, US, from 1968, Cohen began researching how the genes of plasmids could make bacteria resistant to antibiotics. In 1972 he collaborated with Herbert Boyer and Paul Berg to devise ways of combining and transplanting genes, received the first patent for gene splicing, and in 1988 was awarded the National Medal of Science. Cohen remains at Stanford as professor of genetics and medicine; his primary field of research is molecular genetics.

**Compton, Arthur Holly 1892–1962**
American physicist and joint winner with CTR Wilson of the 1927 Nobel Prize in Physics for his discovery of the Compton Effect in which wave-lengths of X-rays increase when the rays collide with electrons in metals. This discovery provided the first evidence that X-rays can act like particles, confirming the theory that electromagnetic radiation can act as both a wave and a particle and helping to legitimize quantum theory. Compton worked on the Manhattan Project.

**Coolidge, William David 1873–1975**
US engineer, physical chemist, and inventor famous for his work on incandescent electric lighting and X-ray development. Coolidge conducted experiments that led to the use of tungsten filaments in light bulbs and invented the X-ray vacuum tube or Coolidge tube (1916). A major breakthrough in the field of radiology, it became the prototype of the modern X-ray tube. Coolidge also helped to develop the first successful submarine-detection system. He was the director of the General Electric Research Laboratory and a vice-president of the corporation, joining as a researcher in 1905.

**Cope, Edward Drinker 1840–1897**
American paleontologist and evolutionary theoretician who helped define American paleontology. Cope's contribution to knowledge of vertebrates of the Tertiary Period was invaluable. He discovered some 1,000 species of extinct vertebrates in the US, including the dinosaurs Camarasaurus and Coelophysis. Much of Cope's time was spent on expeditions across the American West. A competition with fellow paleontologist, Othniel Marsh, to discover the most American fossil dinosaurs, famously degenerated in to an ongoing feud dubbed the Bone Wars. Cope was one of the founders of the Neo-Lamarckian school of evolutionary thought, believing in the inheritance of acquired characteristics, rather than in natural selection.

**Copernicus, Nicolaus 1473–1543**
Polish polymath of the Renaissance, Copernicus combined studies as a mathematician, astronomer, and physician with a career as a scholar, cleric, civic leader, magistrate, diplomat, military commander, translator, and economist. In 1543, the year he died, he published his masterwork, De Revolutionibus Orbium Coelestium (On the Revolutions of the Celestial Spheres), in which he set out a fully predictive mathematical model of a heliocentric solar system, with Earth and all the planets revolving around the Sun, and all the bodies in the Universe moving in circular motions. The book had a crucial influence on the European Scientific Revolution.

**Coriolis, Gaspard-Gustave de 1792–1843**
French mathematician, mechanical engineer, and scientist, Coriolis dedicated his career to the study of friction, hydraulics, machine performance, and ergonomics. Teaching at the École Polytechnique, Paris, in 1829 he published his first paper, Du Calcul de l'Effet des Machines (On the Calculation of Mechanical Action), in which he introduced the important terms "work" and "kinetic energy." He is now best known for the Coriolis force, an apparent deflection of moving objects when viewed from a rotating reference frame, such as the surface of Earth.

**Cousteau, Jacques-Yves 1910–1997**
French explorer, ecologist, filmmaker, scientist, photographer, author, and nature presenter, Cousteau became known to millions through the television series he made from 1966 to 1994, such as The Undersea World of Jacques Cousteau (1968–76). In the 1940s he improved the aqualung underwater breathing apparatus, enabling his prolonged explorations underwater. His beautifully photographed and soundly researched television documentaries and over 50 books laid an increasing emphasis on the need to conserve marine life and habitats, and the televised explorations aboard his famous boat Calypso were pioneering in raising mass awareness of ecological issues.

**Crick, Francis 1916–2004**
English molecular biologist, physicist, and neuroscientist who, along with colleague James Watson, worked out the structure of deoxyribonucleic acid, or DNA when they were both carrying out research at Cambridge University. They discovered its double helix form, one that could duplicate itself, confirming theories that it carried life's heredity information. He then went on to find the link between DNA and the genetic code. His research spawned an entire industry of biotechnology, and earned Crick, Watson, and physicist Maurice Wilkins the Nobel Prize for Physiology or Medicine in 1962. In 1977 he became professor at the Salk Institute in San Diego and branched into neuroscience.

**Crookes, William 1832–1919**
See p.292.

**Curie, Marie 1867–1934**
See pp.296–297.

**Cuvier, Georges 1769–1832**
See p.186.

**Da Vinci, Leonardo 1452–1519**
Italian artist, musician, botanist, mathematician, and architect, Leonardo was the archetypal Renaissance man and his brilliant inventiveness was expressed in hundreds of detailed technical drawings. Leonardo set out to write the first systematic explanations of how machines work, but many of his inventions progressed no further than plans on paper. His submarine, clockwork road vehicle, hydraulic pumps, armored tank, helicopter and flapping-wing aircraft were never made. However, he greatly advanced the ideas of civil engineering and hydrodynamics for future generations

**Daguerre, Louis 1787–1851**
See p. 262.

**Dalton, John 1766–1844**
See p.138.

**Daoyuan, Li 469 or 476 527 CE**
A geographer during the Northern Wei Dynasty, Daoyuan traversed China's mountains and streams analyzing the topography. He discovered that the landscape had changed over time and set out to record a detailed geographical log, which he called Shui Jing Zhu, Commentary on the Waterways Classic, describing hills, lakes, cities, as well as hydrology and local climate.

**Darwin, Charles 1809–1882**
See pp.202–203.

**Davy, Humphry 1778–1829**
See p.170.

**Dawkins, Richard 1941–**
English evolutionary biologist, Dawkins taught zoology at the Universities of Oxford and California. As a student he was mentored by Niko Tinbergen, famous for his pioneering research on animal behavior. He was also influenced by the work of Crick and Watson pn DNA and genetics, and pushed his theories of evolutionary biology, the "selfish gene" and the term "meme" into mainstream scientific culture. He is a Fellow of the Royal Society, and remains an influential and controversial figure.

**de Betancourt, Agustín 1758–1824**
Spanish engineer whose work ranged from steam engines to balloons, urban planning, and structural engineering. De Betancourt was the first director of the Madrid School of Civil Engineering. On a visit to England in 1788 he met James Watt and designed a steam-powered pump and mechanical loom. He built an optical telegraph connection between Madrid and Aranjuez and in Russia he constructed Saint Petersburg's first bridge across Malaya Nevka.

**de Fermat, Pierre 1601–65**
See p.196.

**de Gimbernat, Anthoni 1734–1816**
Spanish anatomist and surgeon. In 1765 he became professor of anatomy at the Royal School of Surgery in Barcelona. He was a pioneer in ophthalmology, urology, and vascular surgery, and developed techniques for mending diaphragmatic hernias. He entered the king's medical cabinet in 1798, serving as personal physician to King Carlos III. In 1801 he was appointed First Royal Surgeon and president of all surgical schools in Spain.

**de la Cierva, Juna 1895–1936**
Spanish civil engineer and pilot, de la Cierva made a major contribution toward the practical take off and landing of aircraft. His autogyro of the 1920s had a freely turning rotor, and could take off and land in a short distance. He influenced the design of helicopters and is credited with establishing an understanding of rotor control and stalling.

**de la Place, Pierre 1749–1827**
See p.183.

**Delbrück, Max 1906–81**
German-American biologist who pioneered the study of molecular biology. Originally trained in astrophysics Delbrück became interested in biology after meeting Wolfgang Pauli and Neils Bohr. In the US Delbrück teamed up with Salvador Luria and, with Alfred Day Hershey, they won the Nobel Prize for Physiology or Medicine in 1969 for their work on the genetic structure of viruses.

**Descartes, René 1596–1650**
See p.103.

**de Vries, Hugo 1848–1935**
Dutch botanist and geneticist who is best known for his studies on plant mutation. De Vries rediscovered George Mendel's laws of heredity, hypothesized on the existence of pangenes, and experimented with osmosis, coining the term "isotonic." He became professor at the University of Amsterdam in 1878 and was awarded the Darwin Medal in 1906.

**Diesel, Rudolph 1858–1913**
See p.255.

**Diophantus of Alexandria c.200–c.284 CE**
Greek mathematician best known for his Arithmetica. He studied at the University of Alexandria in Egypt and wrote a series of books which set out to explore the riddle of algebraic equations. He is known as the father of algebra, though some claim this title should be shared with the Persian mathematician Al-Khwarizmi.

**Dirac, Paul 1902–1984**
Influential English theoretical physicist who won the Nobel Prize for physics in 1933 for his work on quantum theory. He applied Einstein's relativity theory to quantum mechanics and deduced that electrons must be spinning around their own axes and behaving like small magnets. Dirac predicted the existence of positrons and formulated the Dirac equation, a relativistic quantum mechanical wave equation that is important in modern theoretical physics. The equation demands the existence of antiparticles and actually predated their experimental discovery. Dirac held the Lucasian Chair of Mathematics at the University of Cambridge and spent the last years of his life at the University of Florida.

**Dolland, John 1706–1761**
English optician, the son of a Huguenot silk weaver, who found how to construct achromatic lenses by a combination of crown and flint glasses. He also developed the heliometer, a telescope used to find the angular distance between two stars. Awarded the Copley Medal from the Royal Society in 1758, he was elected as a Fellow in 1761 and was appointed optician to King George III.

**Doppler, Christian Johann 1803–53**
Austrian mathematician, famous for conceiving the Doppler Effect, which he presented at the Royal Bohemian Society in Prague, 1842. He theorized on the apparent change in the frequency of a light or sound wave when the observer and the source are moving relative to each other. In 1850 he was appointed as the first director of the new Institute of Physics at the Imperial University in Vienna.

**Duchenne, Guillaume 1806–1875**
See p.225.

**Eastman, George 1854–1932**
American philanthropist, inventor, and industrialist. Founder of the Eastman Kodak, he is remembered as the pioneer of snapshot photography.  The most significant innovations under Eastman's direction were the Kodak snapshot box camera (1888), the celluloid-based roll film (1889), and 16 mm ciné film (1923).

**Eddington, Arthur 1882–1944**
See p.305.

**Edison, Thomas 1847–1931**
See p.265.

**Ehrlich, Paul 1854–1915**
See p.313.

**Einstein, Arthur 1879–1955**
See pp.304–305.

**Empedocles c.495–c.435 BCE**
See p.22.

**Entralgo, Pedro Laín 1908–2001**
Spanish medical researcher and humanist, who wrote on the anthropology of hope.  He obtained the chair of the history of medicine at the University of Madrid in 1942, founded the Instituto Arnau de Vilanova of the History of Medicine in 1943, and was elected a member of the Royal National Academy of Medicine in 1946.

**Erastothenes 276–194 BCE**
See p.38.

**Esaki, Leo 1925–**
Japanese physicist who invented the Esaki tunnel diode, the first quantum electron device. In 1973 he was awarded the Nobel Prize in Physics for his work on electron tunneling in solids. Esaki was elected a Fellow of the American Academy of Arts and Sciences in May 1974, a Foreign Associate of the National Academy of Engineering (USA) in 1977, and a foreign member of the American Philosophical Society in 1991.

**Euclid c.330–c.260 BCE**
See p.34.

**Euler, Leonhard 1703–1783**
Swiss mathematician and physicist who created new areas of mathematical study, such as topology. He was a leading pioneer in astronomy, mechanics, and optics. Although totally blind toward the end of his life, his vast output was undiminished and he carried on performing astronomical calculations mentally.

**Eustachio, Bartolomeo c.1500–1574**
Italian founder of the science of human anatomy. He was the first to describe the anatomy of the teeth, identified the adrenal glands, and extended the knowledge of the inner ear. The eustachian tube linking the ear to the throat, to balance pressure on the eardrum, is named after him.

**Fabricius, Hieronymus 1537–1619**
Italian surgeon and anatomist. He was educated at the University of Padua where he studied under and succeeded Gabriel Fallopius as professor of anatomy in 1565. Fabricius's pioneering work on veins led to later knowledge of blood circulation. He identified the larynx as a vocal organ and elevated embryology to an independent science.

**Fahrenheit, Gabriel Daniel 1686–1736**
German physicist, engineer and instrument maker who made the first reliable thermometers. Fahrenheit used mercury successfully and introduced cylindrical bulbs instead of spherical ones. When Europe switched to Celsius, the Fahrenheit temperature scale was still widely used in the rest of the world.

**Fallopius, Gabriello 1523–1562**
Italian anatomist considered to be one of the most important of the 16th century. He spent much of his time working on the anatomy of the head, but is best known for identifying the reproductive organs of both men and women. The fallopian tube, which connects an ovary to the uterus, is named after him. In 1551 he took the Chair of anatomy and surgery at the University of Padua and held a professorship of botany.

**Faraday, Michael 1791–1867**
See pp.170–71.

**Fermi, Enrico 1901–54**
Italian physicist most noted for his development of the first nuclear reactor, and for his work on quantum theory and statistical mechanics. Fermi was elected professor of theoretical physics at the University of Rome in 1927, became the leading expert on neutrons, and was awarded the Nobel Prize for physics in 1938. He became professor at the Institute for Nuclear Studies of the University of Chicago in 1946, a position he held until his death.

**Ferrel, William 1817–1891**
See p.207.

**Feynman, Richard 1918–1988**
See pp.326–327.

**Fibonacci, Leonardo c.1170–c.1250**
See p.63.

**Fitzroy, Robert 1805–65**
English captain of HMS Beagle and meteorologist. He was awarded a Gold Medal by the Royal Geographic Society in 1837, and his achievement in nautical surveying and scientific navigation resulted in his election to the Fellowship of the Royal Society in 1851. He was Governor of New Zealand from 1843 to 1845 and became head of the British Meteorological Department in 1854. Promoted to Vice Admiral in 1863, he suffered from depression and committed suicide, leaving substantial debts.

**Fleming, Alexander 1881–1955**
Scottish bacteriologist who, in 1928, discovered the antibiotic penicillin. Working at St. Mary's Hospital in London, he noticed that penicillin mold growing on cultures of bacteria prevented the bacteria from multiplying. Penicillin has since saved millions of lives. He was knighted in 1944 and in 1945 shared the Nobel Prize for Physiology or Medicine with Ernest Chain and Sir Howard Florey, who developed the commercial production of penicillin.

**Fleming, Walter 1843–1905**
See p.306.

**Florey, Howard Walter 1898–1968**
Australian biochemist who, along with Ernst Chain, took up Alexander Fleming's neglected work on penicillium. They successfully isolated and purified penicillin. With mass production of the drug achieved in 1943, it was used in treating the war injured, saving countless lives. He shared the Nobel Prize for Physiology or Medicine with Chain and Fleming in 1945, and raised to the British peerage in 1965.

**Ford, Henry 1863–1947**
American inventor and industrialist who introduced the moving assembly line into motor manufacture, enabling full-scale mass-production.  He founded the Ford Motor Company in 1903. His first vehicle was the Model T Ford, introduced in 1908. The Model T remained in production almost unaltered for 15 years, during which time 15 million were built.

**Foucault, Jean Bernard Léon 1819–1868**
French physicist who was the first to show how a pendulum can demonstrate Earth's rotation. He is credited with naming the gyroscope, and measured the speed of light, showing how it travels more slowly through water than air. In 1855 he received the Copley Medal of the Royal Society and was made a Fellow in 1864.

**Fowler, William 1911–1955**
American nuclear astrophysicist who formulated a theory of element generation. Fowler won the Henry Norris Russell Lectureship of the American Astronomical Society in 1963 and the Eddington Medal in 1978. He shared the Nobel Prize for Physics in 1983 for his theoretical and experimental studies of nuclear reactions in the formation of the chemical elements in the universe. He received the national Medal of Science in 1974 and was awarded the Legion of Honor in 1989.

**Franklin, Benjamin 1706–1790**
See pp.160–161.

**Franklin, Rosalind 1920–58**
English scientist, who, along with Maurice Wilkins, took X-ray diffraction photographs of DNA and found that the DNA molecule had a corkscrewlike shape known as a helix. These results were put together by Francis Crick and James Watson to become the double helix model. Though less well-known than her colleagues, Franklin's crucial role in the discovery of DNA structure is now widely acknowledged.

**Fresnel, Augustin-Jean 1788–1827**
French physicist who built on the work of Thomas Young to extend the wave theory of light to a large range of optical uses. He studied the behavior of light, and is best known for the Fresnel lens, which used compound lenses instead of mirrors in lighthouses. He received scant recognition for his work during his lifetime, but became a member of the Royal Society in 1825, and was awarded the Rumford Medal by the Society in 1827.

**Gabor, Dennis 1900–79**
See p.365.

**Gagarin, Yuri 1934–68**
See p.372.

**Galen, Claudius c.130 CE–c.200 CE**
Greek physician, born in Pergamum in Asia Minor, who settled in Rome in 164. He was physician to five Roman emperors, and wrote extensively on physiology and anatomy. He based his writing on his knowledge of early medicine and his own experiments. Of the 500 treatises he wrote, 180 survive. Galen's arguments dominated medicine throughout the Middle Ages.

**Galilei, Galileo 1564–1642**
See pp.82–83.

**Galvani, Luigi 1737–1798**
See p.221.

**Gamow, George 1904–68**
American physicist, cosmologist and champion of the Big Bang theory of the Universe. He also predicted the existence of a background warmth in the Universe, left over from the Big Bang's fireball, a prediction confirmed in 1965. Renowned as a popular science writer, he joined the University of Colorado, US in 1956 as professor, and worked there until his death.

**Gassendi, Pierre 1592–1655**
French priest who achieved distinction as a scientist, astronomer, mathematician, and philosopher. He tried to revive the philosophy of Epicurus and reconcile it with Christianity. He is best known for his opposition to the ideas of Descartes. The lunar crater Gassendi is named after him.

**Gauss, Carl Friedrich 1777–1855**
German mathematician whose major work concerned number theory, or higher arithmetic. One of history's most influential mathematicians, Gauss worked on complex numbers, non-Euclidean geometry, electromagnetism, and the theory of errors, which led to the concept of the normal distribution curve. He was awarded the Royal Society's Copley Medal in 1838, and has a moon crater, ship, asteroid, and extinct volcano named after him.

**Gay-Lussac, Joseph-Louis 1778–1850**
See p.101.

**Geber, Jabir c.721–815 CE**
Islamic alchemist, astronomer, physicist, and philosopher. Also known by the full name of Abu Musa Jabir ibn Hayyan, he is described as the father of Arab chemistry. He turned alchemy into a science, invented many pieces of laboratory equipment, and identified chemical substances such as hydrochloric acid and nitric acid.

**Geiger, Hans 1882–1945**
German physicist best-known for developing the Geiger counter, an instrument for measuring and detecting nuclear radiation. In 1909, while working under the direction of Ernest Rutherford, Geiger and Ernest Marsden set up their famous gold foil experiment, the Geiger-Marsden experiment and discovered that atoms have a nucleus. In 1928 Geiger and Walther Müller created an improved Geiger counter with increased speed and sensitivity.

**Gilbert, William 1544–1603**
See p.81.

**Goddard, Robert H 1882–1945**
See p.339.

**Goeppert-Meyer, Marie 1906–72**
German-born American theoretical physicist. Most of her university career took place in Göttingen, where she took a doctorate in theoretical physics. In Chicago she became a professor at the Institute for Nuclear Studies, where she developed a model for the nuclear shell structure. During a time of sexism against women in science, she did most of her work as a volunteer, and in 1963 became only the third woman to win a Nobel Prize for Physics.

**Golgi, Camillo 1843–1926**
Italian scientist, pathologist, and physician who spent much of his career studying the central nervous system. In 1873 he published *On the Structure of the Brain Gray Matter*, in which he described the discovery of the "black reaction." This revolutionary tissue staining is still in use today and is named after him (Golgi staining or Golgi impregnation). Golgi established the Institute of General Pathology in 1881 and was jointly awarded the Nobel Prize for Physiology or Medicine in 1906 with Santiago Ramón y Cajal.

**Gould, Stephen Jay 1941–2002**
American paleontologist and evolutionary biologist. He is best known for his work with Niles Eldridge and their theory of punctuated equilibrium. He spent most of his career at Harvard University, where he campaigned against creationism and argued that science and religion should be kept as two distinct fields. He was posthumously awarded the prestigious Linnean Society of London's Darwin-Wallace Medal in 2008.

**Graaf, Regnier de 1641–73**
Dutch physician and anatomist who discovered the development of the ovarian follicles, later called Graafian follicles. He also gave exact descriptions of the testicles, and isolated and collected secretions of the gall bladder and pancreatic juices.

**Greene, Brian 1963–**
See p.402.

**Grew, Nehemiah 1641–1712**
English botanist, vegetable anatomist, and physiologist. His most popular and influential work, *The Anatomy of Plants* (1682) shows the microscopic detail of plant tissue. His work was contemporaneous with that of Marcello Malpighi and the two shared much of their research. He was elected Fellow of the Royal Society in 1671, and in 1684 was the first to describe sweat pores on the hands and feet.

**Guangqi, Xu 1562–1633**
Chinese astronomer, mathematician, and agricultural scientist who experimented with western irrigation methods and new agricultural practices. Working with Italian Jesuit Matteo Ricci, he translated several important western texts, including the first parts of Euclid's *Elements*, and began to reform the Chinese calendar. His collaboration with western scientists was hailed as the start of Chinese enlightenment.

**Guericke, Otto von 1602–86**
German inventor and philosopher who was the mayor of Magdeburg, Germany. In 1650 Guericke invented an air pump to help him create a partial vacuum. In 1663 he demonstrated the power of a vacuum with his Magdeburg Hemispheres to Emperor Ferdinand III. During public demonstrations, teams of horses would famously attempt to pull the large hemispheres apart. In 1660 he developed a machine that produced static electricity; the first electric generator, which led to later experiments with electricity. The Otto von Guericke University of Magdeburg, Germany, is named after him.

**Gutenberg, Johannes 1400–1468**
See p.58.

**Haber, Fritz 1868–1934**
German chemist who in 1908–1909 devised a method of making synthetic ammonia, a raw material essential in making explosives and fertilizers. His discovery led to a great expansion of the chemical industry, and for this he won the Nobel Prize in Chemistry in 1918. A patriotic German, he was given the rank of captain by the Kaiser and developed the use of poisonous gases in trench warfare.

**Hadamard, Jacques 1865–1963**
French mathematician best known for proving the prime number theorem, and the Hadamard three-circle theorem. He published *Psychology of Invention in the Mathematical Field* in 1945, demonstrating the use of introspection to describe mathematical thought processes. Hadamard is also associated with the Dreyfus Affair, a French political scandal in which Alfred Dreyfus, a Jewish army captain, was accused of treason. Dreyfus was Hadamard's brother-in-law, and Hadamard worked tirelessly to clear his name.

**Hadley, George 1685–1768**
English physicist and meteorologist who advanced the first accurate theory describing the trade winds and explained why they blew from the northeast in the northern hemisphere and the southeast in the southern hemisphere. Although a partial explanation had previously been offered by Edmond Halley (1686), he was unable to explain the associated north-south circulation pattern. In 1735 Hadley suggested that heated air at the equator rises and flows toward the poles, where it cools, descends, and flows back toward the equator; this pattern became known as the Hadley cell. However, Hadley's theory remained unacknowledged until 1793, when it was rediscovered by English scientist John Dalton.

**Haeckel, Ernst 1834–1919**
German biologist and artist best known for his now-discounted "biogenetic law" or theory of recapitulation. He was a supporter of Charles Darwin's theories and produced studies of marine organisms. Haeckel made the basic division of the animal kingdom into single-celled and multicellular creatures. He described and named thousands of new species and mapped a genealogical tree relating all forms of life. He was awarded the Linnean Society of London's Darwin-Wallace Medal in 1908.

**Hahn, Otto 1879–1968**
German chemist credited as the first person to demonstrate the deliberate fission of uranium atoms in 1938. Once nuclear fission had been accepted, Hahn was able to demonstrate its use in warfare and energy production. He was awarded the Nobel Prize in Chemistry in 1945 but could not attend the ceremony as he was interned at Farm Hall near Cambridge, suspected of working on the German nuclear energy project. Hahn had never worked on the program, and after the Americans dropped atomic bombs on Hiroshima and Nagasaki, Hahn became an outspoken advocate against nuclear weapons. In 1961 Pope John XXIII awarded him the Gold medal of the Papal Academy.

**Hales, Stephen 1677–1761**
See p.155

**Halley, Edmond 1656–1742**
See p.119.

**Harrison, John 1693–1776**
English carpenter and clockmaker who invented the marine chronometer crucial in establishing longitude at sea. Built in response to a prize offered by the British Government in 1714, it solved one of the greatest seagoing problems. Harrison built a succession of three chronometers, but it was with his fourth timepiece (H4) that Harrison intended to claim the prize. Despite its accuracy the Government was unconvinced, and Harrison had to prove the design could be replicated. Although Harrison eventually won some of the prize money, he remained bitter at his treatment.

**Harvey, William 1578–1657**
See p.90.

**Hawking, Stephen 1942–**
English physicist and expert on black holes, developing a single theory on the beginning and end of the Universe. Hawking is regarded as one of the leading contributors to the science of cosmology and his work is widely recognized. In 1979 he took the post of Lucasian Professor of Mathematics at Cambridge. Author of the bestselling *A Brief History of Time*, he has suffered from ALS (amyotrophic lateral sclerosis) since he was young. Confined to a wheelchair, he uses an electronic voice synthesizer to communicate. His many accolades include 12 honorary degrees, Fellowship of the Royal Society, and Membership of the US National Academy of Sciences.

**Heisenberg, Werner 1901–1976**
See p.316.

**Henry, Joseph 1797–1878**
See p.168.

**Herschel, William 1738–1822**
German-born English astronomer. He discovered Uranus in 1781 and developed a theory of nebulae and the evolution of stars. He also showed that the solar system moved through space, discovered infrared radiation, and constructed over 400 telescopes. Herschel coined the term "asteroid," and controversially believed that every planet was inhabited. He received the Copley Medal in 1781, was made a Fellow of the Royal Society, and knighted in 1816. Herschel helped to found the Astronomical Society of London in 1820, which later became the Royal Astronomical Society.

**Hertz, Heinrich 1857–1894**
German physicist whose experiments with electromagnetic waves led to the development of the wireless telegraph and the radio. All frequencies are measured in Hertz, meaning cycles per second. He progressed the electromagnetic theory of light formulated by English physicist James Clerk Maxwell in 1884, and proved that electricity can be transmitted in electromagnetic waves that travel at the speed of light and possess various other properties of light. Hertz was the first person to broadcast and receive radio waves, and to show that light was a form of electromagnetic radiation (1885-89).

**Hertzsprung, Henry Russell 1873–1967**
See p.329.

**Hilbert, David 1862–1943**
German mathematician whose many outstanding contributions span number theory, mathematical logic, differential equations, and the three-body problem. He invented a space-filling curve known as the Hilbert Curve. In his *Foundations of Geometry* he stated the first rigorous set of geometrical axioms. In a world meeting in Paris in 1900 he posed a list of 23 problems, now called Hilbert problems; this was considered the most influential speech ever given to mathematicians.

**Hipparchus 190–120 BCE**
See p.28.

**Hippocrates c.460–377 BCE**
Greek physician and teacher regarded as the father of clinical medicine. Hippocrates is credited with being the first physician to reject the idea that supernatural or divine forces caused illness. He based his medical practice on observations, study of the human body, and documentation of symptoms and treatments. He was the first physician to accurately describe many diseases and conditions, categorize illnesses according to severity (acute, chronic), and to use terms such as "relapse" and "convalescence." He founded the Hippocratic school of medicine on Cos. With his followers he taught medicine and set out rigorous codes of behavior, ethics, and good practice through the Hippocratic oath and some 60 medical treatises (the Hippocratic corpus).

**Hodgkin, Dorothy 1910–1994**
See pp.384–385.

**Hooke, Robert 1635–1703**
See pp.92–93.

**Hooker, Joseph Dalton 1817–1911**
See p.153.

**Hopper, Grace 1906–22**
American computer scientist who devised some of the first computer programs. Notably, Hopper was one of the first programmers of the Harvard Mark I computer. Her work on compilers led ultimately to the development of the business language COBOL. She was a United States Naval Officer and rose to the rank of Rear Admiral; the US Battleship USS Hopper was named after her. In 1973 she became the first woman to be made a Distinguished Fellow of the British Computer Society.

**Hoyle, Fred 1915–2001**
See p.321.

**Hubble, Edwin 1889–1953**
See p.340.

**Hutton, James 1726–1797**
See p.184.

**Huygens, Christiaan 1629–95**
See p.114.

**Ingenhousz, Jan 1730–1799**
Dutch-born English physician and scientist who discovered photosynthesis, whereby sunlight enables green plants to absorb carbon dioxide and release oxygen. Ingenhousz also invented an improved device for generating large quantities of static electricity (1766), made the first quantitative measurements of heat conduction in metal rods (1789), and was a discoverer of Brownian motion, describing the erratic movement of coal dust on the surface of alcohol (1785). He was a physician to the Austrian Empress, Maria Theresa, and worked in various European countries ranging from Britain to Switzerland.

**Jenner, Edward 1749–1823**
English scientist and medical practitioner, Edward Jenner demonstrated how inoculation with cowpox virus could safely provide immunity from the deadly viral disease of smallpox. Before Jenner's time, Europe had become aware that in the East people protected themselves from smallpox by treating themselves with material from smallpox survivors, but Jenner's "vaccination" with cowpox virus produced dependable immunization that carried no risk of smallpox infection from the vaccine itself. Thanks in large part to universal adoption of cowpox inoculation, smallpox was declared an eradicated disease by the World Health Organization in 1980.

**Joliot-Curie, Irène 1897–1956**
See p.297.

**Joule, James Prescott 1818–1889**
See p.177.

**Jun, Heo 1546–1615**
Korean court physician during the reign of King Seonjo of the Joseon Dynasty, Heo Jun based his herbal medicine on Chinese principles but rewrote the prescriptions, simplifying them and replacing expensive Chinese ingredients with native Korean equivalents. The greatest of his texts is *Dongui bogam*, in which he made remedies understandable to commoners by using simple hangui characters rather than Chinese hanja characters. Heo Jun was a holistic practitioner who would promote training of the mind and body and mind in preference to other forms of treatment.

**Kamil, Abu c.850–c.930**
Egyptian mathematician Abu Kamil, also known as al-Hasib al-Misri, the "calculator from Egypt," specialized in algebra. He was familiar with the works of Al-Khwarizmi, whom he acknowledged as inventor of all the principles of algebra. In his own *Book on Algebra*, Abu Kamil combined Greek geometry with Al-Khwarizmi's algebra and Babylonian methodology to solve geometrical problems. In his later *Book of Rare Things in the Art of Calculation*, Abu Kamil addressed solutions to indeterminate equations. Abu Kamil's books gained importance in the 13th century as the basis of Fibonacci's introduction of algebra to Europe.

**Kant, Immanuel 1724–1804**
German philosopher Immanuel Kant had a profound effect not only on mathematical and scientific thinking but also on other areas of philosophy, such as ethics and aesthetics. By addressing the question, "What can we know?" he worked to reconcile empiricism (we know only what our senses reveal) and rationalism (we know only what our minds can construct). Concluding that objective experience is actively constructed by the functioning of the human mind, he placed the human mind at the center of the acquisition of knowledge, drawing a definite boundary between physics (what we can know) and metaphysics (about which we can only speculate).

**Kapany, Narinder Singh 1927–**
Indian physicist whose 1955 doctoral research opened the gateway to fiber optics, with its myriad uses from data transmission to imaging. In addition to fiber-optics communications, his research and inventions have involved biomedical instrumentation, pollution monitoring, lasers, and solar energy; he has over 120 patents. Long based in the US, his eclectic career has also encompassed business, publishing, lecturing, farming, and academia.

**Keeling, Charles 1928–2005**
See p.416.

**Kekulé, Friedrich August 1829–1896**
See p.288.

**Kelvin, William Thompson, Lord 1824-1907**
See p.181.

**Kepler, Johannes 1571–1630**
See p.77.

**Khayyám, Omar  1048–1123 BCE**
Persian scholar, poet, and mathematician. In mathematics, he developed a geometric method for solving cubic equations by intersecting a parabola with a circle. Khayyam also contributed to calendar reform, compiled astronomical tables, and constructed important critiques of Euclid's theories of proportion and parallels. He is also known for his poem, *The Rubáiyát of Omar Khayyám*.

**Khorana, Har Gobind  1922–**
Indian-born US biochemist who shared the 1968 Nobel Prize in Physiology or Medicine with Marshall Nirenberg and Robert Holley for work on the interpretation of the genetic code and its role in protein synthesis. He later developed the first artificial gene. Such genes are used in laboratories for sequencing, cloning, and other forms of genetic engineering. In 1970 he became a professor of chemistry and biology at the Massachusetts Institute of Technology, remaining there until retirement in 2007.

**Koch, Robert  1843–1910**
German physician, whose discoveries facilitated the development of the first chemicals designed to attack specific bacteria. Koch isolated the bacilli for anthrax (1877), tuberculosis (1882), and cholera (1883). He also developed Koch's postulates of germ theory that sets out the criteria to be met to establish whether specific bacteria cause specific diseases. Koch won the 1905 Nobel Prize in Physiology or Medicine for his tuberculosis discoveries. He was appointed professor of hygiene at the University of Berlin in 1885 and was director of the Institute for Infectious Diseases, Berlin from 1891 to 1904.

**Kuhne, Wilhelm  1857–1945**
See p.216.

**LaPlace, Pierre-Simon  1749–1827**
See p.183.

**Lamarck, Jean-Baptiste  1749–1823**
French biologist and an early evolutionist, who believed that acquired characteristics are inheritable. Lamarck recognized that the environment helps shape the species that inhabit it and noted that Earth's surface undergoes constant transformation, creating significant changes over time. His Lamarckian theory of evolution suggested that living things evolved continuously upward, from dead matter to simple and then increasingly complex organisms. Although generally rejected during his lifetime, his theory of inheritance remained popular into the 20th century. Lamarck also made his mark as a botanical and zoological systematist and a founder of invertebrate paleontology.

**Laue, Max Von  1879–1960**
German physicist who discovered that X-rays could be diffracted by crystals, which opened the door to X-ray crystallography and the study of the structure of crystals. This marked the advent of solid-state physics which played a vital role in the development of modern electronics. He also supported Albert Einstein's theory of relativity and conducted research on superconductivity, quantum theory, and optics. Von Laue became director of the University of Berlin's Institute for Theoretical Physics in 1919 and director of the Max Planck Institute, Berlin in 1951.

**Lavoisier, Antoine Laurent  1743–94**
French chemist whose synthesis of existing chemical knowledge, sound methodology, and a quantitative approach revolutionized 18th-century chemistry. Lavoisier proved that matter is neither created nor destroyed during chemical changes, and discovered that air comprises two gases, naming them oxygen and azote (nitrogen). In demonstrating the role oxygen plays in combustion, Lavoisier laid the foundations for modern chemistry. He co-produced a system of chemical nomenclature (1787), wrote the first list of elements, and helped establish the metric system. He was guillotined during the French Revolution.

**Leavitt, Henrietta Swan  1868–1921**
Celebrated for her contributions to astronomy. Working at the Harvard College Observatory, Leavitt conducted groundbreaking work on Cepheid variable stars and stellar magnitudes. In 1912 she established a direct link between time span and brightness in Cepheid variables, pulsating stars that vary regularly in luminosity in periods ranging from days to months. Her discovery led to an invaluable method of determining the distances of stars and galaxies. Leavitt was also involved in determining the photographic magnitude of stars, developing the Harvard Standard of photographic measurements (1913).

**Le Chatelier  1850–1926**
See p.236.

**Lee, Tsung-Dao  1926–**
Chinese-born US physicist best known for his work in elementary particles, statistical mechanics, and astrophysics. He co-won the 1957 Nobel Prize for Physics for demonstrating violations of the principle of parity conservation, leading to important refinements in particle physics theory. He also created the Lee model (a solvable model of quantum field theory). Lee joined the Institute for Advanced Study, Princeton (1951–53) and was later professor of theoretical physics there (1960-63).

**Leeuwenhoek, Anton Van  1632–1723**
See p.94.

**Lehmann, Inge  1888–1993**
See p.352.

**Leibniz, Gottfried von  1646–1716**
German philosopher and mathematician. Leibniz published his discovery of differential and integral calculus (1684) which he made independently of Isaac Newton. He also invented the binary system, the foundation on which digital technology is based, and a calculating machine (1671). Leibniz made important contributions to metaphysics, logic, and physics, and anticipated ideas that emerged ater in geology, medicine, and psychology. Among his major works is *New Essays Concerning Human Understanding* (1704).

**Lilienthal, Otto  1848–1896**
See p.290.

**Lind, James  1716–1794**
See p.218.

**Linnaeus, Carolus  1707–1778**
See p.123.

**Lippershey, Hans  c.1570–c.1619**
Dutch glasses-maker credited with inventing the first practical telescope (1608). The States General of the Netherlands refused to award Lippershey a patent, but paid him with the proviso that the telescope was modified into a binocular device. Others recognized the telescope's importance in astronomy, however, and a lunar crater and planet have since been named after him. Lippershey is also credited with inventing the compound microscope, independently of others during the same period.

**Lister, Joseph  1828–1912**
See p.229.

**Locke, John  1632–1704**
English philosopher whose work influenced the development of political philosophy and empiricism. His most famous works include his *Essay Concerning Human Understanding* (1690) and *Two Treatises on Government*, published in 1690 concerning natural law and natural rights. Locke's ideas about liberty and the social contract influenced the US Declaration of Independence (1776).

**Lorentz, Hendrik Antoon  1853–1928**
Dutch physicist known for his work on the relationship between electricity, magnetism, and light. Lorentz shared the 1902 Nobel Prize in Physics with Pieter Zeeman for his theory of electromagnetic radiation based on the Zeeman effect. In 1904 he introduced the concept of local time (different times in different locations), creating transformation equations that paved the way for Einstein's theory of relativity. He also discovered, almost simultaneously with L V Lorenz, the relation between the refraction of light and the density of the translucent body (the Lorentz-Lorenz formula). From 1919 he led a study of projected movements of sea water caused by the reclamation of the Zuyderzee; his calculations contributed to the science of hydraulics.

**Lorenz, Edward  1917–2008**
American mathematician and meteorologist who first recognized the practical effect of chaos theory, influencing a wide range of basic sciences. In the early 1960s Lorenz realized that small differences in a dynamic system such as the atmosphere could trigger huge, often unpredictable changes. This led him to formulate the "butterfly effect." In 1991 Lorenz won the Kyoto Prize for establishing the theoretical basis of weather and climate predictability, as well as the basis for computer-aided atmospheric physics and meteorology. He was Professor Emeritus at Massachusetts Institute of Technology from 1987 until his death.

**Losada, Manuel  1929–**
Spanish biochemist and molecular biologist known for his research on the photosynthetic assimilation of nitrogen. Concerned with the practical aspects of scientific research, Losada has focused on the development of biological and biochemical systems that can transform solar energy into chemical energy, and the use of microalgae to produce valuable compounds, such as protein-rich biomass. Recently Losada has been involved in plant developmental biology at the Institute of Plant Biochemistry and Photosynthesis at the University of Seville, where he is director.

**Lovelace (Countess of), Ada  1815-1852**
English mathematician and daughter of poet Lord Byron, she was described as the first computer programmer. An acolyte of inventor and mathematician Charles Babbage, Lovelace devised programs for his analytical engine (1843). She detailed how it could be used to calculate Bernoulli numbers, foreseeing the functions of a machine such as a general-purpose computer and anticipating future developments, such as computer-generated music. A software language, Ada, was named after her in 1979.

**Lovelock, James  1919–**
See pp.412–413.

**Lyell, Charles  1797–1875**
See p.190.

**Malpighi, Marcello  1628-94**
Italian physiologist and biologist who, through his microscope studies of plant and animal tissue, founded the sciences of microscopic anatomy and histology. In 1661 he observed the anatomy of the frog lung. He named capillaries, scrutinized the silkworm, and made a major contribution to embryology. He discovered papillae (taste buds) and was the first to attempt a thorough study of the anatomical details of the brain. Fascinated by botany, he also discovered stomata, the pores of leaves. He was professor of the University of Bologna (1661-91) and nominated personal physician to Pope Innocent XII (1691).

**Malthus, Thomas Robert  1766-1834**
English clergyman and philosopher best remembered for his theories on population growth. He argued that natural growth in population tends to outstrip food supplies. Unless population growth could be restricted, overpopulation would be checked only by war or famine. His theory, the Malthusian growth model, had a major impact on political, economic and social thought. While berated by some, others, such as Darwin, were hugely influenced by him.

**Marconi, Guglielmo  1874–1937**
Italian physicist and inventor who proved radio communication was possible with the development of a successful wireless telegraph (1896). Marconi then sent the first wireless signal across the English Channel (1899) followed by the first wireless signal across the Atlantic (1902). This made headlines worldwide, paving the way for important developments in radio communications, broadcasting, and navigation services. Marconi won the 1909 Nobel Prize for Physics with German physicist Ferdinand Braun. He patented several new inventions between 1902 and 1912, including his magnetic detector, which became the standard wireless receiver for many years. He later worked on the development of shortwave wireless communication.

**Margalef i López, Ramón  1919–2004**
A Spanish founder of the modern science of ecology who made a huge contribution to limnology (study of lakes) marine biology, oceanography, and theoretical ecology. He became Spain's first professor of ecology (1967), setting up the department of ecology at the University of Barcelona. His book *The Theory of Information in Ecology* (1957) attracted international acclaim, as did his *Perspectives in Ecological Theory* (1968). Among his most important work were the application of information theory to ecological studies and the creation of mathematical models for population studies.

**Marsh, Othniel Charles  1831–99**
American paleontologist. Marsh became professor of paleontology at Yale University in 1866 and director of the Peabody Museum of Natural History. He led his students on field trips to the American West, and discovered many new dinosaur fossils. In 1871 his party discovered the first pterodactyl found in the United States. By the mid-1870s Marsh had an exceptional collection of early mammals.

**Maudslay, Henry  1771–1831**
See p.172.

**Maurey, Matthew  1888–1873**
See p.213.

**Maxim, Hiram Stevens  1840–1916**
Anglo-American inventor who perfected the first fully automatic machine gun. The Maxim gun, in which the recoil of each shot loaded and fired the next round, was introduced in 1888. He also built a large steam-powered flying machine, which almost took off. During his career Maxim took out 250 patents for various designs.

**Maxwell, James Clerk  1831–1879**
See p.167.

**May, Robert McCredie  1936–**
Australian theoretical physicist, zoologist, and ecologist. He trained as a theoretical physicist/applied mathematician and studied the way populations are structured. He pioneered work on chaos theory and ecological systems and devised methods of estimating the number of species on Earth. He holds a Professorship jointly at Oxford University and at Imperial College, London, is a former President of the Royal Society (who awarded him their Copley medal in 2007), a member of the UK Government's Climate Change Committee, and was made a life peer in 2001.

**Mayer, Julius Robert von  1814–78**
German physician and physicist who discovered the concept of energy conservation but was unable to prove it. He was the first to determine the mechanical equivalent of heat, but it took scientists years to confirm his findings, notably James Joule in 1847, leading to the First law of thermodynamics. Mayer was awarded the Copley Medal in 1871 from the Royal Society and was granted the honor of adding "von" to his name.

**McClintock, Barbara  1902–92**
American cytogeneticist, who for her groundbreaking work on the genetics of corn, earned a place among the leaders in genetics. She was elected to the National Academy of Sciences in 1944. Despite this, she met with prejudice in her department at the University of Missouri and was forced to leave. She kept her next appointment at the Carnegie Institute at Cold Spring Harbor for the remainder of her life. McClintock was awarded the Nobel Prize in Physiology or Medicine in 1983.

**Mead, Margaret  1901–78**
American social anthropologist noted for her pioneering studies on childhood, adolescence, and sexual diversity in Pacific communities. Her most famous work, *Coming of Age in Samoa* (1929), suggested that "civilized" peoples could learn much from "primitive" peoples. Her studies were controversial at the time. She was a prolific author who pioneered academic involvement in television by fronting documentaries. Mead was the first elected president for the American Association for Advancement of Science (1974).

**Meitner, Lise 1878–1968**
Swedish physicist who helped discover nuclear fission. In 1945 the Nobel Prize in Chemistry was awarded to Otto Hahn for the discovery of nuclear fission, overlooking the contribution of Lise Meitner, who collaborated with him and gave the first theoretical explanation of the fission process. This was partly rectified in 1966, when Hahn, Meitner, and Fritz Strassmann were awarded the US Fermi Prize.

**Mendel, Gregor 1822–1884**
See p.204.

**Mendeleev, Dimitri 1834–1907**
See p.233.

**Mercator, Gerardus 1512–94**
Flemish geologist, cartographer, and instrument maker. He produced his first globe in 1541. In 1569 he produced the first map of the world for navigational use, with longitude and latitude, using a map projection. He was the first to make a map of the British Isles and Flanders. The lunar crater Mercator is named after him.

**Michell, John 1724–93**
English philosopher and amateur astronomer. In 1760 Michell constructed a theory of earthquakes as wave motions in the interior of the earth. As an amateur astronomer, he was a talented telescope builder. In 1761 he was elected Fellow of the Royal Society. From 1762 to 1764, Michell held the Woodwardian Chair of Geology at Cambridge, and in 1767 he was appointed rector of St. Michael's Church of Thornhill, near Leeds, UK. While a rector he continued to do scientific work. He suggested the idea of black holes in 1784 and devised apparatus for measuring the density of the earth.

**Michelson, Albert Abraham 1852–1931**
American physicist whose experiment with Edward Morley in 1887 helped determine the speed of light. Their failure to detect any influence of the earth's motion on the velocity of light was the starting point for Einstein's theory of relativity. Regarded as a key finding in scientific history, the discovery was made on an instrument Michelson invented and is used today to measure the wavelengths of spectra. He was the first American to win a Nobel Prize for Physics in 1907. A crater on the Moon is named after him.

**Mohs, Friedrich 1773–1839**
German mineralogist who devised a scale to measure and specify the hardness of a mineral. In 1801 Mohs was given a job identifying minerals in a collection that belonged to a wealthy banker. The scale he devised to catalogue these minerals became known as the Mohs Hardness Scale. His approach was criticized by the mineralogical establishment of the time, but it ultimately brought him fame. In 1810 Mohs became a professor of mineralogy in Gratz and in 1826 professor of mineralogy at the University of Vienna.

**Morgan, Thomas Hunt 1866–1945**
See p.306.

**Morley, Edward 1838–1923**
American chemist and physicist who collaborated with Albert Michelson in 1887 on an experiment to discover the speed of light, leading to Einstein's theory of relativity. Morley also worked on the oxygen composition of the atmosphere, and the velocity of light in a magnetic field. He was president of the American Association for the Advancement of Science (1895) and president of the American Chemical Society (1899). The lunar crater Morley was named after him.

**Morse, Samuel Finlay Breese 1791–1872**
American artist and electrical engineer, best known for inventing the Morse Code, first demonstrated in 1838. In this code, short and long signals, dots and dashes, represent letters and numbers. Morse also introduced the relay, an electric switch that strengthened the signal at intervals, making long distance communication possible.

**Moseley, Harry 1887–1915**
English physicist whose law formed the basis of modern atomic and nuclear physics. Christened Henry, Moseley studied physics at Oxford University before joining Ernest Rutherford at Manchester University. Moseley provided an experimental basis for equating nuclear charge with what he called atomic number, making it possible to predict missing elements in the periodic table from discontinuities in the spectral series. He moved back to Oxford to continue his work, but enlisted in the Royal Engineers at the outbreak of World War I. He was killed during the Battle of Gallipoli.

**Murchison, Roderick Impey 1792–1871**
Scottish geologist who played a major role in establishing parts of the geographical time scale, notably the Silurian, Permian, and Devonian Periods. The development of this time scale, in the absence of any absolute dating of rocks, was regarded as one of the greatest achievements of 19th-century geologists. He was knighted in 1846 and made a baron in 1866. He spent his later years devoted to the Royal Geographical Society, which he had helped found. The crater Murchison on the Moon is named after him.

**Muybridge, Eadweard 1830–1904**
English photographer, known for his use of multiple cameras to capture motion. He became a successful photographer in the US, taking memorable pictures of the western landscape. In 1878 he used 24 cameras to film a horse galloping, *The Horse in Motion*, and began the transformation from stills to motion picture. He later returned to England where he published two popular books.

**Napier, John 1550–1716**
See p.88.

**Newcomen, Thomas 1663–1729**
English engineer and builder of the first steam engine to be put to practical use. It was erected in 1712 to pump water out of a coal mine near Dudley Castle, Staffordshire. During the next three-quarters of a century hundreds of Newcomen engines were installed in Britain, greatly increasing coal production. They were eventually superseded by James Watt's low-pressure condensing engines.

**Newton, Sir Isaac 1642–1727**
See pp.110–111.

**Nightingale, Florence 1820–1910**
English nurse who transformed hospital nursing in Britain and is regarded as the founder of modern nursing. During the Crimean War, despite official opposition, she organized a nursing service for the soldiers who named her "the Lady with the Lamp." She founded a training school for nurses at St. Thomas' Hospital in London in 1861, influencing many parts of the world. In 1907 she was the first woman to be awarded the Order of Merit. After her death she was immortalized in many films, statues, plays, and banknotes.

**Nobel, Alfred 1833–96**
Swedish inventor of dynamite and gelignite, of smokeless gunpowder and the detonator, and the establisher of the Nobel prizes. As a young man Nobel worked in the research laboratory of his father's explosives factory near Stockholm. After his young brother died when liquid explosive nitroglycerine blew up, Nobel tried to make the substance less sensitive. He succeeded in 1866 and called it dynamite. He later produced gelignite, which was even safer. Nobel also worked on synthetic rubber, leather, and silk. The synthetic element nobelium was named after him. He willed most of his large fortune to establishing the Nobel Prizes, annual awards for the year's most outstanding work for physics, chemistry, medicine, literature, and peace. First presented in 1901, they are awarded annually on December 10, the anniversary of his death.

**Ochoa, Severo 1905–93**
American biochemist best known for being the first to synthesize ribonucleic acid (RNA) outside the cell. He also discovered several important metabolic processes. The formation of RNA from appropriate nucleotides was later used for the synthesis of artificial RNA. Ochoa was jointly awarded the 1959 Nobel Prize for Physiology or Medicine with Arthur Kornberg, who synthesized deoxyribonucleic acid (DNA). Ochoa received the US National Medal of Science in 1979, and the asteroid 117435 Severochoa is named in his honor.

**Ørsted, Hans Christian 1777–1851**
Danish physicist and chemist who, in 1920, showed that when a magnetic compass is placed near a wire and electricity is passed through it, the compass needle moves. His findings resulted in intensive research throughout the scientific community in electrodynamics. The CGS unit of magnetic induction (oersted) is named after him, as was the first Danish satellite, which was launched in 1999.

**Ohm, Georg Simon 1789–1854**
German experimental physicist who established Ohm's law, which states that the electric current through a conductor is proportional to the potential difference across it. The current is equal to the voltage divided by the resistance. Ohm used thermocouples for this work, as they provided a steadier voltage than the batteries available at the time. The SI unit of electrical resistance, the ohm, is named after him. He was awarded the Copley Medal from the Royal Society in 1841, and became a Fellow of the Society in 1842.

**Olbers, Heinrich Wilhelm 1758–1840**
German astronomer who discovered the asteroids Pallas and Vesta in 1802 and 1807, and found a method of calculating the orbits of comets. He posed the question of why the night sky is so dark if stars are distributed infinitely throughout space, which became known as Olbers' paradox. He assumed that dust clouds block light from the stars. Later, astronomers realised that the total amount of energy and matter is too small to light up the night sky. Olbers became famous for his discoveries and in 1811 was elected to assist at the baptism of Napoleon II of France.

**Oort, Jan Hendrik 1900–92**
Dutch astronomer recognized as one of the greatest of the 20th century. Oort demonstrated that the Milky Way rotates, discovered that each of the stars in our galaxy was traveling independently through space, and found mysterious "dark matter" that makes up about 25 percent of the Universe. He is best known for his suggestion that the solar system is surrounded by a vast cloud of perhaps 100 billion comets, known as the Oort cloud. Oort received some of the highest awards in astronomy, including the 1966 Vetliesen Prize from Columbia University, US.

**Oppenheimer, (Julius) Robert 1904–67**
American theoretical physicist who developed electron-positron theory, the Oppenheimer-Phillips process, quantum tunneling, black holes, and cosmic rays. Oppenheimer was appointed director of the Manhattan Project in 1941. His efforts earned him the Presidential Medal of Merit in 1946. He was chairman of the General Advisory Committee of the Atomic Energy Commission, or AEC, from 1947 to 1952, but opposed the development of even more powerful bombs. In 1953, at the height of US anticommunist feeling, Oppenheimer was accused of having communist sympathies, but a decade later he was awarded the Fermi Award as a gesture of rehabilitation.

**Ostwald, Wilhelm 1853–1932**
German chemist who is regarded as the founder of modern physical chemistry. He investigated catalysts, the speed of chemical reactions, reversible reactions, and the way electrolytes split up and conduct electricity. Ostwald also established that indicators were weak acids whose ionized form differs in color from their undissociated or non-ionized form. In 1887 he accepted an invitation as professor of physical chemistry at Leipzig University where he remained until his retirement in 1906. His work was recognized by the Nobel committee who awarded him the Prize in Chemistry in 1909.

**Otto, Nikolaus August 1832–91**
German technician who pioneered the first practical four-stroke internal combustion engine. He was inspired by reports of the engine of the French engineer Etienne Lenoir, in which coal gas igniting inside a cylinder drove a piston. In 1872 Otto joined German engineer Eugene Langen to produce a stationary gas engine for industry. It was a commercial success. He employed the same four-stroke cycle in a new gas engine in 1876, after which it was known as the Otto cycle. The same four-stroke principle was later used by Karl Benz and Gottlieb Daimler to power the first automobiles.

**Oughtred, William 1575–1660**
English mathematician and Anglican minister renowned for his invention of an early form of the slide rule. He invented many new mathematical symbols, such as x for multiplication. His work influenced the chemist Robert Boyle, the architect Christopher Wren, and the mathematician-physicist Isaac Newton.

**Owen, Richard 1804–92**
English biologist and comparative anatomist. Owen is best remembered for coining the word dinosauria, meaning "Terrible Reptiles," and for his opposition to Darwin's theory of evolution by natural selection. In 1856 Owen became the first Superintendent of the British Museum's natural history departments. He campaigned for a new museum dedicated to natural history. Land was bought in South Kensington and in 1881 Owen's vision became reality. The famous building designed by Alfred Waterhouse is a lasting monument to Owen.

**Palade, George Emil 1912–2008**
Romanian biologist who pioneered modern cell biology, earning him the Nobel Prize in Physiology or Medicine in 1974. Palade's discoveries were later used in understanding diseases and in the protein production that is the basis of the biotechnology industry. In 1973 he moved to Yale, where he became the chairman of the new department of cell biology. In 1990 he became the first dean for scientific affairs at the School of Medicine at the University of California, San Diego.

**Paracelsus 1493–11541**
See p.71.

**Parsons, Sir Charles 1854–1931**
See p.252.

**Pascal, Blaise 1623–1662**
French mathematician and physicist whose greatest contribution to mathematics was his work on the theory of probability, developed from a problem sent to him by a noted gambler. In connection with this work he discovered the pattern of numbers known as Pascal's triangle. Pascal's other work included major discoveries in geometry and the invention of the first digital calculating machine. He established the basic principles of hydraulic pressure, and the SI unit of pressure, the pascal, is named after him. Deeply religious, he dedicated his later years to spiritual learning and wrote several philosophical essays.

**Pasteur, Louis 1822–1895**
See p.244.

**Pastor, Julio Rey 1888–1962**
Spanish mathematician who pioneered a modern approach to the history and teaching of mathematics. While making important contributions to projective geometry, he promoted the science of mathematics throughout Argentina and Spain; his textbooks and essays were widely publicized. A crater on the Moon is named after him, as well as several academic institutions in Spain.

**Pauli, Wolfgang 1900–58**
American theoretical physicist who in 1925 established the Pauli exclusion principle, which states that no two electrons in an atom can exist in exactly the same state, with the same quantum numbers. Pauli, and independently Arnold Sommerfeld, devised an atomic model that explained the electrical and thermal properties of metals. Pauli was the first to recognize the existence of neutrino, and helped to complete Dirac's quantum electrodynamics. He was awarded the Lorentz Medal in 1930.

**Pauling, Linus Carl 1901–94**
American chemist whose best-known work was on the structure of protein molecules. He found that fibrous proteins are helical, sometimes with several helices twisted together into a "super-helix." He also did research on the chemical action of antibodies. In the late 1940s Pauling was outspoken on the dangers of nuclear weapons. He was ostracized for his uncompromising stand, but went on to win two Nobel Prizes: the 1954 award for Chemistry and the 1962 Nobel Peace Prize. He devoted himself to fighting disease and educating the public about health issues.

**Pavlov, Ivan Petrovich 1849–1936**
Russian physiologist whose experiments on dogs led to the discovery of the conditioned reflex. His work has had a widespread influence on psychology. He made his findings in 1904 while studying dogs. Realizing that they salivated in anticipation of food, not just at the sight of it, Pavlov experimented with other stimuli. He developed the process of learning called classical conditioning, and worked out many of the basic laws underlying learning. He was awarded the Nobel Prize in Physiology or Medicine in 1904.

**Peral, Issac 1851–95**
Spanish sailor and inventor associated with the Peral submarine built for the Spanish Navy. In 1882 he was appointed professor of mathematical physics at the Escuela de Ampliación de Estudios de la Armada where he began work on designing a submarine. *El Peral*, built in Cadiz, was completed in 1888 and was designed for military use. In 1889 she became the first submarine to successfully fire Whitehead torpedoes while submerged. Internal politics, however, kept the Spanish Navy from pursuing the project. Discouraged, Peral left the navy in 1890.

**Perkin, William 1838–1907**
See p.240.

**Petit, Alexis Therese 1791–1820**
French physicist who, along with Pierre Dulong, discovered the Dulong-Petit Law. which states that for a solid element, the product of relative atomic mass and specific heat capacity is approximately constant. Petit's life was cut short by tuberculosis.

**Planck, Max 1858–1947**
See p.314.

**Plimsoll, Samuel 1824–98**
English politician and social reformer, best remembered for the Plimsoll line, which indicated how low or high a ship is resting in the water, serving as a warning against overloading. He directed his efforts against what he called "coffin ships," unseaworthy vessels, and attempted to push through laws which would secure better standards of safety on ships in the future.

**Poincaré, Henri 1854–1912**
See p.350.

**Posadillo, Gregorio Marañón y 1887–1960**
Spanish physician, writer, and philosopher. He specialized in endocrinology and became professor at the Complutense University in Madrid in 1931. He studied the emotional effects of adrenaline and was a pioneer in the development of cognitive theories of emotion. A prolific writer, he founded the Institute of Medical Pathology and was the first director of the Center of Biological Research in Madrid.

**Priestley, Joseph 1733–1804**
See p.146.

**Proust, Joseph-Louis 1754–1826**
French chemist best known for Proust's law. In 1785 Proust held teaching posts in Spain, where he began studying different types of sugar and was the first to identify glucose. Proust published his law of constant composition, which later evolved into the law of definite proportions (Proust's law). Most chemists followed the work of Claude Berthollet, who did not believe that substances always combine in constant and definite proportions as Proust did. A friendly rivalry ensued, and Proust was proved right.

**Ptolemy (Claudius Ptolemaeus) 90–168 CE**
Greek astronomer, author of the encyclopedic treatise the *Almagest* (*c.*150 CE), containing a summary of Greek astronomical thought. It is through him that the work of the mathematician Hipparchus survives. Ptolemy brought together the work of earlier Greek astronomers to produce the Ptolemaic system of the Universe, in which the Sun and planets orbit a centrally-placed Earth. This theory held for nearly 1,400 years before it was abandoned in favor of the Sun-centered system of the Universe put forward by Copernicus in 1543.

**Purkinje, Jan 1787–1869**
Czech physiologist who pioneered studies in vision, the function of the brain and heart, embryology, cells and tissue. He is best known for his discovery, in 1837, of Purkinje cells, large nerve cells with branching extensions found in the cerebellum. In 1823 he recognized that fingerprints can be used as a means of identification and in 1839 discovered Purkinje fibers, the fibrous tissue that conducts the pacemaker stimulus along the inside walls of the ventricles to all parts of the heart. Purkinje introduced the term plasma, and protoplasm, which is used to describe young animal embryos.

**Pythagoras c.580–500 BCE**
See pp.32–33.

**Rainwater, (Leo) James 1917–86**
American physicist who shared the 1975 Nobel Prize in Physics for his work identifying the asymmetrical shapes of certain atomic nuclei. During World War II Rainwater was involved with the Manhattan Project to build the atomic bomb. After the war he developed his theory that not all atomic nuclei were spherical, later confirmed by Neils Bohr and Ben Mottleson. He received the Ernest Orlando Lawrence Award for Physics in 1963 and was made professor of physics in 1982 at Columbia University, US.

**Raman, Chandrasekhar Venkata 1888–1970**
Indian physicist. Raman pioneered work on optics and the scattering of light, later called the Raman effect, and won the Nobel Prize for Physics in 1930. In 1949 he established the Raman Research Institute in Bangalore, where he worked until his death. He was elected a Fellow of the Royal Society in 1924, knighted in 1929, and in 1957 was awarded the Lenin Peace Prize.

**Ramsay, William 1852–1916**
English chemist who discovered the inert gases, beginning with argon in 1894. He isolated helium in 1895, neon, krypton, and xenon from liquid air in 1898, and in 1910 discovered the rare gas radon. His work had a great influence on atomic theory, and in recognition of this Ramsay was knighted in 1902 and awarded the Nobel Prize for Chemistry in 1904.

**Rees, Martin 1942–**
English cosmologist and astrophysicist who has made important contributions to the origin of cosmic microwave background radiation, galaxy clustering and formation, and was one of the first to theorize that vast black holes power quasars. He is Professor of Cosmology and Astrophysics at the University of Cambridge, was knighted in 1992, holds the honorary title of Astronomer Royal (1995), was appointed to the House of Lords in 2005, and made a member of the Order of Merit in 2007. He is a foreign associate of the National Academy of Sciences and the American Academy of Arts and Sciences, and has received many national and international awards in recognition of his work.

**Richter, Charles 1900–1985**
See p.361.

**Römer, Ole Christensen 1644–1710**
Danish astronomer best known for discovering light's finite velocity: the speed of light. Römer invented various scientific instruments including a micrometer, planetaria, and an alcohol thermometer that influenced Daniel Fahrenheit's researches. In his position as Royal Mathematician, he introduced the first national system for weights and measures in Denmark in 1683.

**Röntgen, Wilhelm 1845–1923**
See p.294.

**Ross, Ronald 1857–1932**
Indian-born English physician who discovered the life cycle of malarial parasites (*Plasmodium*) in mosquitoes. Ross was subsequently awarded the Nobel Prize for Physiology or Medicine in 1902. He spent much of his career initiating measures for the prevention of malaria. He was knighted in 1911 and in 1926 became Director of the Ross Institute and Hospital for Tropical Diseases in London, which was founded in his honor.

**Rubin, Vera 1928–**
See p.398.

**Rumford, Benjamin Thompson, Count 1753–1814**
See p.179.

**Rutherford, Ernest 1871–1937**
See p.286.

**Russell, Henry Norris 1877–1957**
American astronomer who helped establish modern theoretical astrophysics. He analyzed the physical conditions and chemical compositions of stellar atmospheres; his assertion that there is an abundance of hydrogen is now accepted as one of the basic facts of cosmology. He is best associated with the Hertzsprung-Russell diagram (1910), a graph in which the color or temperature of stars is plotted in relation to their brightness. Russell became president of the American Astronomical Society, the American Association for the Advancement of Science, and the American Philosophical Society. In 1921 he was awarded the gold medal of the Royal Astronomical Society.

**Sabin, Florence 1871–1953**
American medical scientist who was an important figure in her field, and a pioneer for women. In her research work Sabin made significant contributions to knowledge of the histology of the brain and the development of the lymphatic systems, and to the understanding of the pathology and immunology of tuberculosis. She became active in public health matters and modernized Colorado's health system, known as the Sabin health laws. She was the first woman to become a full professor at John Hopkins Medical School. She was also the first female president of the American Association of Anatomists.

**Salam, Abdus 1926–1996**
See p.401.

**Salas, Margarita 1938-**
Spanish pioneer in molecular biology. After working under Severo Ochoa in America, she returned to Spain to initiate research into molecular biology, a new field at the time. Married to the scientist Eladio Viñuela, both were responsible for boosting Spanish research in molecular biology and biochemistry. She is a member of the Royal Academy of Exact, Physical and Natural Sciences of Spain and is president of the Severo Ochoa Foundation. In 2007 she was named a member of the National Academy of Sciences in the United States, becoming the first Spanish woman to join that institution.

**Salk, Jonas Edward 1914–95**
American medical researcher who found a vaccine for polio. Benefiting from the work of Harvard's John Enders, who had developed a way to grow polio in test tubes, Salk began a human trial of the vaccine in 1955. When it was made public, he became famous overnight and polio was virtually eradicated. In 1963 Salk founded the Salk Institute for Biological Studies and in 1977 was awarded the US Presidential Medal of Freedom.

**Sanger, Frederick 1918–**
English biochemist whose research on the structure of insulin earned him the Nobel Prize for Chemistry in 1958, and which was a key influence on the work of Francis Crick in researching the nature of DNA. Sanger himself turned to DNA and by 1975 had developed the "dideoxy" method for sequencing DNA molecules, also known as "the Sanger method." He used this technique to develop the first fully sequenced DNA-based genome. This was important for the Human Genome Project and earned him a second Nobel Prize in Chemistry in 1980, which he shared with Walter Gilbert and Paul Berg.

**Santacilia, Jorge Juan y 1713–73**
Spanish mariner, mathematician, and scientist. In 1734 he accepted a mission to measure the length of a degree of the longitudinal meridian at the equator in South America and to determine the roundness of the earth. He found that Earth is not exactly spherical, but slightly flattened at the poles. His *Examen Maritimo* (1771), in which he applied Newtonian mechanics to naval engineering, is regarded as one of the greatest naval treatises of the 18th century.

**Schiaparelli, Giovanni Virginio 1835–1910**
Italian astronomer who in 1877 reported "canals" on Mars. Other astronomers thought the canals were artificial, though Schiaparelli had not claimed this. An asteroid and craters on Mars and the Moon are named after him.

**Schickard, Wilhelm 1592–1635**
German astronomer who invented one of the first calculating machines. In 1631 he was appointed professor of astronomy at the University of Tübingen. There his research also included mathematics and surveying. He was a renowned wood and copper engraver and he made significant advances in mapmaking. In 1623, long before Blaise Pascal and Gottfried Leibniz, Schickard invented a calculating machine which was used by Johannes Kepler. Known as the Speeding clock, it was used for calculating astronomical tables. Schickard died of bubonic plague; the lunar crater Schickard was named in his honor.

**Schrödinger, Erwin 1887–1961**
Austrian theoretical physicist who made significant contributions to quantum mechanics. He worked on wave mechanics, explaining the movement of an electron in an atom as a wave. The equation became known as Schrödinger's wave equation. After teaching at Berlin University, he left with the rise of Nazism and went to Oxford in 1933, where he learned that he had won the Nobel Prize with Paul Dirac. He emigrated to Ireland in 1940 to join the newly established Institute for Advanced Studies in Dublin. In 1956 Schrödinger returned to Vienna, where he later died of tuberculosis.

**Schwann, Theodor 1810–1882**
See p.215.

**Sedgwick, Adam 1785–1873**
English geologist, professor of geology at Cambridge University, who played a major role in establishing parts of the Geographical Time Scale, most notably the Cambrian and Devonian Periods. Sedgwick introduced the term Paleozoic for the era stretching from the Cambrian to the Permian Periods. He was an outspoken opponent of Darwin's theory of evolution, although the two remained friends. Sedgwick was awarded the Copley Medal in 1863 and the geological museum at Cambridge was renamed the Sedgwick Museum of Geology in his honor.

**Semmelweis, Ignaz 1818–1865**
See p.242.

**Servet, Miguel 1511–53**
Spanish theologian, cartographer, physician, and humanist. In his role as physician, Servet came to differ from Galen in pulmonary circulation. Galen had proposed that aeration of the blood took place in the heart and assigned the lungs a minor function. Servet concluded that transformation of the blood occurred in the lungs. He was the first to publish a modern understanding of pulmonary respiration.

**Shen Kuo 1031–1095**
Chinese astronomer, mathematician, and scholar who discovered magnetic declination, transforming navigation. Shen mentions his findings in his famous work, *Mengxi bitan*. His planetary observation project, which involved measuring planets' positions thrice nightly for five years, was on a scale unequaled in Europe for 500 years.

**Shizhen, Li 1518–93**
Chinese physician, regarded as one of the greatest in Chinese history. His courtesy name was Dongbi. He hailed from the Hubei Province of China, and lived during the Ming Dynasty. In his work as a doctor he became frustrated in the errors and omissions he found in textbooks and decided to compile a new comprehensive book. For this he traveled all over China, and experimented with many herbs and minerals. His *Bencao Gangmu* (Great Compendium of Herbs) was translated into several languages and introduced into Japan, and later Europe.

**Shockley, William Bradford 1910–89**
American who fathered the transistor and brought silicon to Silicon Valley. His co-invention of a "semiconductor sandwich" in 1947, with John Bardeen and Walter Brattain, was quickly adapted and radios of all shapes and sizes flooded the market. In 1956 they shared a Nobel Prize in Physics. Shockley set out to capitalize on his invention by developing a silicon transistor. He played a key role in the industrial development of the region at the base of the San Francisco Peninsula, US. In his later life, Shockley was a Professor at Stanford and controversially became an outspoken advocate of eugenics.

**Shoujing, Guo 1231–1316**
Chinese astronomer, mathematician, and engineer of the Yuan Dynasty, best known for the Shoushi Calendar. This calendar had 365.2425 days in a year which, remarkably, was only 26 seconds out from the time it takes Earth to go around the Sun. Shoujing also invented and improved astronomical instruments, including a tower sundial, raising its height and dramatically increasing its accuracy. He is renowned for building the artificial Kunming Lake in Beijing as a reservoir. The Asteroid 2012 Guo Shou-Jing is named after him.

**Simiao, Sun 581–682 CE**
Chinese physician and alchemist, hugely influential in the history of Chinese medicine. He wrote two important Chinese compilations of prescriptons and placed great emphasis on gynecology, pediatrics, and medical ethics, stipulating that all patients should be treated as equal. This doctrine was spread throughout China and acted as an early version of the Hippocratic oath.

**Skinner, BF 1904–90**
American psychologist who discovered practical applications for behaviorism in the form of programmed learning and psychotherapy. Skinner experimented on rats and pigeons, inventing a small box with a lever inside, the Skinner box, to deliver stimuli. The animal learned to associate pressing the lever with receiving a food reward. He founded a school for experimental analysis of behavior and is regarded as one of the leading psychologists of the 20th century.

**Smith, Adam 1723–90**
Scottish economist whose ideas formed the basis of the economic theories that became paramount in the 19th century, and have influenced economic thought ever since. The culmination of his life's work was the publication in 1776 of his book *An Inquiry Into the Nature and Causes of the Wealth of Nations*. He believed that wealth derived principally from labor; the more labor that was required to produce a commodity, the higher its value. His book also formed the basis of the policies of laissez faire and free trade, which became dominant in the 19th century.

**Smith, William 1769–1839**
English geologist who founded the science of stratigraphy. A surveyor and canal engineer who studied geology during his travels around Britain, Smith showed that rock strata could be correlated from locality to locality by examining their fossil contents. Smith's *Geographical Map of England and Wales, with Part of Scotland* (1815) was the first regional geographical map.

**Snell, Willebrord 1580–1626**
See p.116.

**Snow, John 1813–58**
English obstetrician who worked to convince other doctors that cholera was caused by drinking contaminated water. He published an article in 1839 outlining his theory, but doctors remained unconvinced and were certain that cholera was spread by breathing "vapors" in the atmosphere. Snow believed that sewage was the culprit, and when a major epidemic broke out with a water pump at its source, he had the pump handle removed. The outbreak ceased, but it was years before Snow was taken seriously. Scientists now consider Snow to be the pioneer of public health research in the field of epidemiology.

**Sorenson, Soren 1868–1939**
See p.239.

**Spallanzani, Lazzaro 1729–1799**
See p.226.

**Spemann, Hans 1869–1941**
German embryologist. Spemann studied how embryological cells become specialized and differentiated in the process of forming a complete organism. He developed a method for operating on objects often less than two millimeters in diameter, founding the techniques of modern microsurgery. His greatest contribution, for which he won the 1935 Nobel Prize in Physiology or Medicine, was his discovery of the "organizer" effect. Spemann found that when an area containing an "organizer" is transplanted into an undifferentiated host embryo, the transplanted area can induce the host to develop into an entirely new embryo.

**Spinoza, Baruch 1632–77**
Dutch-Jewish philosopher of rationalism, who advocated the intellectual love of God, whom he saw as the perfect being. Spinoza saw everything in the Universe as a manifestation of this perfect being. Since his view could not be identified with the Jewish God, he was expelled from his own community and was equally unpopular with Christians. Spinoza is best known for his *Ethics*, a monumental work which puts forward his vision, and is regarded as one of the greatest rationalists of 17th century philosophy, having a great impact on writers and scientists in the 20th century.

**Spitzer, Lyman 1914–1997**
See p.397.

**Sprengel, Christian Konrad 1750–1816**
German botanist whose studies of sex in plants helped establish the study of pollination ecology. Sprengel's observations were driven by a belief that everything in nature had a purpose, including every part of a plant. He discovered the role played by insects and wind in the cross-pollination of plants, and also described dichogamy, in which stigma and antlers on the same flower ripen at different times to ensure cross-fertilization. He published his findings in *The Newly Revealed Mystery of Nature in the Structure and Fertilization of Flowers* (1793), but his work failed to gain credence until it was later discovered by Charles Darwin.

**Steptoe, Patrick Christopher 1913–88**
English obstetrician and gynecologist whose technique for uniting an egg and a sperm in a Petri dish and implanting them into the mother's uterus led to breakthroughs in fertility treatment. The first "test tube" baby born using in vitro fertilization was Louise Brown in 1978. For their efforts, Steptoe and colleague Robert Edwards were both named Commanders of the British Empire, and in 1987 Steptoe was made a Fellow of the Royal Society.

**Stevens, Nettie Marie 1861–1912**
American biologist and geneticist who discovered the chromosomal basis of sex. By showing that the chromosomes known as "X" and "Y" were responsible for determining the sex of individuals, it ended a longstanding scientific debate as to whether sex was determined by heredity or other factors.

**Stringfellow, John 1799–1883**
English pioneer aviator under whose supervision the world's first powered flight took place in the UK in 1848. His model flying machine was powered by two contra-rotating propellers driven by one of Stringfellow's powerful steam engines. The first attempt took place indoors, but a lack of proper balance resulted in failure. The second attempt was more successful; the flying machine left a guide wire and flew straight and true for about 30 ft (10 m).

**Sturgeon, William 1783–1850**
English electrical engineer who in 1825 invented the electromagnet, which was essential in developing the telegraph, telephone, and electric motor. He left the army to become a teacher, and specialized in lecturing on electro-magnetism. His findings were published in several scientific magazines. In 1840 he became the Superintendent of the Royal Victoria Gallery of Practical Science in Manchester. Starved of funds, the gallery later shut down.

**Swammerdam, Jan 1637–80**
Dutch microscopist who made major discoveries in anatomy. His greatest contribution to biology was the understanding of insects and their development. Using dissections, he showed that the various stages of the growth of an insect, from egg to adult, are different forms of the same creature. He was the first to describe red blood cells, discovered valves in the lymphatic vessels, known as Swammerdam valves, described the ovarian follicles of mammals at the same time as physician Reinier de Graaf (1672), and devised new techniques for preserving and examining specimens.

**Swan, Joseph 1828–1914**
See p.250.

**Talbot, (William) Henry Fox 1800–77**
English chemist and photographer who made groundbreaking improvements to developing, fixing, and printing. Fox Talbot discovered that an image could be produced after a very short exposure to light and chemically developed into a useful negative. With the negative image it was possible to repeat the process of printing. Fox Talbot patented his process in 1841, and the following year was rewarded with a medal from the Royal Society.

**Tansley, Arthur 1871–1955**
See p.333.

**Tesla, Nikola 1856–1943**
See p.269.

**Thales c.625–546 BCE**
Greek philosopher, considered the first of the natural philosophers. He formed part of what became later known as the Miletus School with Anaximander and Anaximenes. He was the first to consider the nature of the elements and concluded that water was the fundamental principal of the Universe. He accurately predicted an eclipse of the Sun in 585 BCE, and realized that the Moon shone only by reflecting the Sun's light. The idea of geometry as a systematic science involving rigorous proofs arose from him. Five theorems are attributed to Thales.

**Thomson, Joseph John 1856–1940**
See p.140.

**Townes, Charles Hard 1915–**
American physicist and educator best known for inventing the maser and laser. At Columbia University, in New York, he continued his research into microwave physics and detected the first complex molecules in interstellar space. He was elected to the National Academy of Sciences in 1956, and jointly awarded the Nobel Prize in Physics in 1964. In 1967 he was appointed professor at the University of California and was chairman of the advisory committee for the first human landing on the Moon.

**Trevithick, Richard 1771–1833**
English engineer and pioneer of the successful high-pressure steam engine. This was the first type of steam engine to combine high power with low bulk and weight, making it more versatile than the Watt type that preceded it. Trevithick was the first engineer to design a boiler strong enough to withstand the new engine's high steam pressures. He made Britain's first self-propelled road vehicle in 1801and the first railway locomotive two years later. But Trevithick preferred to experiment with new ideas rather than develop his inventions and, as a result, he died penniless.

**Turing, Alan 1912–1954**
See pp.344–45.

**van der Waals, Johannes Diderik 1837–1923**
Dutch physicist who first proposed an equation of state for real gases to explain why ordinary gases do not obey the ideal gas laws, such as Boyle's law, at high pressures. He also put forward the Law of corresponding states, which served as a guide during experiments which ultimately led to the liquefaction of hydrogen and helium. For his achievements he won the Nobel Prize in Physics in 1910.

**van Helmont, Jan Baptista 1580–1644**
Flemish physiologist, physician, and chemist who was the first to postulate on the existence of gases distinct from air. He coined the word "gas," possibly from the word "chaos." He believed that the prime elements of the Universe were air and water, rejecting ancient ideas on the four elements. Van Helmont is also famous for his experiment on a willow tree which he planted and grew for five years to determine where plants gain their mass.

**Vega, Jurij 1754–1802**
Slovenian mathematician and physicist, author of the first complete table of logarithms. His book *Treasury of all Logarithms* published in 1794, became an essential aid in computations and teaching. Vega also made a name for himself when he calculated the mathematical constant to an accuracy of 140 decimal places, and is well known for his work on ballistics and astronomy. Vega's death was shrouded in mystery. His body was found in the River Danube; many suspect that he was murdered. A crater on the Moon is named after him.

**Verenadsky, Vladimir 1863–1945**
See p.278.

**Vesalius, Andreas 1514–1564**
See p.73.

**Virchow, Rudolf Karl Ludwig 1821–1902**
German physician and pathologist who popularized the medical rule that every cell originates from another cell. In *Cellular Pathology* he explained his discovery that disease occurs at a cellular level. He conducted medical research into the cause of pulmonary thromboembolism and named leukemia. Also well known for his anthropological work, Virchow wrote a study in 1885 which challenged the belief common at the time of Aryan or Nordic racial superiority, showing that Europeans belonged to a mixture of races.

**Volta, Alessandro 1745–1827**
Italian physician, best known for his work in electricity. In 1775 Volta devised the electrophorus, a device that produced a static electric charge. In 1799 he developed the voltaic pile, a forerunner of the electric battery, which produced a steady stream of electricity. Napoleon made him a Count in 1810, and in honor of his work, the electrical unit known as the volt was named after him.

**von Braun, Wernher 1912–77**
German rocket scientist and champion of space exploration. Von Braun was the leader of the team that developed the V-2 ballistic missile for the Germans during World War II. In 1955 he became an American citizen and worked for NASA where he was director of the Marshall Space Flight Center and the chief architect of the Saturn V launch vehicle that propelled the Apollo spacecraft to the Moon. He received the National Medal of Science in 1975.

**von Fraunhofer, Joseph 1787–1826**
German physicist best known for discovering the dark lines found in the Sun's spectrum. In 1813 von Fraunhofer independently rediscovered William Hyde Wollaston's dark lines in the solar spectrum, which are now known as Fraunhofer lines. He described a great number of the 500 or so lines he could see using self-designed instruments. Fraunhofer lines were later used to reveal the chemical composition of the Sun's atmosphere. Von Fraunhofer was honored with membership to the Bavarian Academy of Sciences and is considered the founder of the German optical industry. He was knighted in 1824.

**von Helmholtz, Hermann 1821–94**
German physician and physicist who made significant contributions to the understanding of the nervous system, the functions of the ear and eye, and in physics the concept of the conservation of energy. In 1851 Helmholtz revolutionized ophthalmology with the invention of the ophthalmoscope. The following year he successfully measured the speed of the nerve impulse. He moved to Berlin in 1871 to work as professor of physics, and attracted many promising students, among them Heinrich Hertz, who developed the propagation of electromagnetic oscillations in space.

**von Humboldt, Alexander 1769–1859**
German naturalist and explorer credited with being a founder of modern geography and who helped popularize science through his book *Kosmos*. Humboldt's expeditions to South and Central America (1799-1804) and Russia (1829), his studies, and the data he gathered radically changed western science in the 19th century. His interest in Earth's geomagnetic fields triggered one of the first examples of international scientific cooperation, the setting up of permanent observatories around the world. Humboldt made an invaluable contribution to comparative climatology and conducted pioneering work on the relationship between local geography and its flora and fauna. He also discovered and studied the Peru Current (originally called the Humboldt Current) off the west coast of South America.

**von Liebig, Justus 1803–73**
German chemist known for his work in organic, pharmacological, and agricultural chemistry. He pioneered the use of nitrogen-based fertilizers in agriculture, helped found Germany's new dye industry, and built a laboratory for chemical research which was copied worldwide. In later years he found a way to process beef extract from carcasses, which was given the trade name OXO.

**von Siemens, Werner 1816–92**
German electrical engineer who played a significant role in the development of the telegraph industry. Von Siemens made the world's first pointer telegraph and electric dynamo. Together with Johann Georg Halske, he set up a company, Telegraphen-Bau-Anstalt von Siemens & Halske, in Berlin in 1847. Siemens AG is now one of the largest electro-technological firms in the world. The SI unit of electrical conductance, the siemens, is named after him.

**Waksman, Selman Abraham 1888–1973**
Ukrainian-born American biochemist who discovered several important antibiotics. In 1939 Waksman and his colleagues undertook to identify soil organisms, producing soluble substances that might be useful in the control of infectious diseases, now known as antibiotics. Within a decade, 10 antibiotics were isolated and characterized, including actinomycin, streptomycin, and neomycin. Streptomycin was the first effective treatment for tuberculosis. Eighteen antibiotics were discovered under Waksman's general supervision. He won widespread recognition for his work, receiving the Lasker Award in 1948, the Legion of Honor in 1950, and the Nobel Prize in Physiology or Medicine in 1952.

**Wallace, Alfred Russell 1823–1913**
See p.201.

**Warburg, Otto Heinrich 1883–1970**
German biochemist, physiologist and medical doctor noted for his differentiation of a cancer cell from a normal, healthy cell. This was such a major finding he was awarded the Nobel Prize for Physiology or Medicine in 1931. Warburg studied chemistry under Emil Fischer in Berlin, and medicine under Ludolf von Krehl, before serving in the Prussian Horse Guards during World War I. In 1931 he also discovered the nature and mode of action of the respiratory enzyme. This discovery opened up new ways in the fields of cellular metabolism and cellular respiration.

**Watson, James Dewey 1928-**
American geneticist, who, with Francis Crick, discovered the double helix structure of the nucleic acid DNA in 1953. He shared a Nobel Prize in 1962 with Crick and Maurice Wilkins. He wrote highly regarded textbooks and, like Crick, was widely honored. He received the Lasker Award, Copley Medal, the Presidential Medal of Freedom (1961) and was made an Honorary Knight Commander in the Order of the British Empire. From 1988 to 1992 Watson took a leading role in the Human Genome Project.

**Watson-Watt, Robert 1892–1973**
English inventor of radar. His first interest was meteorology where he used short-wave radio to detect the location of thunderstorms. By combining this direction finding technique with the ranging capabilities of ionosondes, he devised a system that was capable of detecting airplanes. He called it Radio Detection and Ranging (RADAR). By the end of 1938 radar systems were in place along the south coast of Britain and were used during the Battle of Britain in 1940 to detect enemy aircraft. Watson-Watt was elected a Fellow of the Royal Society in 1941 and knighted in 1942 for his pivotal work in radar and its role in World War II.

**Watt, James 1892–1973**
See p.149.

**Weber, Wilhelm Eduard 1804–91**
German physicist, who, together with mathematician Carl Friedrich Gauss, invented the first electromagnetic telegraph. Weber made sensitive magnetometers to measure magnetic fields, and instruments to measure direct and alternating currents. The SI unit of magnetic flux, the weber, is named after him. Weber's later years were spent at the University of Göttingen, Germany, where he worked in electrodynamics and the electrical structure of matter. He developed the electrodynamometer and defined the absolute unit of electrical resistance. He was elected a member of the Royal Swedish Academy of Sciences in 1855 and received the Copley Medal in 1859.

**Wegener, Alfred 1880–1930**
See p.356.

**Weinberg, Steven 1933–**
American physicist who is best known for his work in unified field theory. Four forces were believed to drive the laws of physics: gravity, electromagnetism, the strong force (which holds an atom's nucleus together), and the weak force (which breaks an atom apart). Weinberg theorized that the electromagnetic and the weak forces are the same at extremely high energy levels. This was a major step in finding a single equation to explain all the matter and forces in nature. Weinberg, along with Sheldon Glashow and Abdus Salam, was awarded the Nobel Prize in Physics, 1979. Weinberg currently holds the Josey Regental Chair in Science at the University of Texas, where he is a member of the physics and astronomy departments. He is considered by many to be the preeminent living theoretical physicist.

**Weismann, August 1834–1914**
See p.309.

**Whipple, Fred Lawrence 1906–2004**
American expert on the solar system who in 1949 proposed the "dirty snowball" theory of the origin of the comets. This visualizes comets as balls of frozen gas mixed with particles of dust. During World War II he invented a device for cutting tinfoil into chaff, to confuse enemy radar. In recognition of this he was awarded a Certificate of Merit in 1948. Whipple, who organized the tracking networks for the first US satellites, also discovered that sporadic meteors, like shower meteors, are debris from comets. In 1955 he became director of the Smithsonian Astrophysical Observatory (SAO), where he remained until 1973. He was honored with the Gold Medal from the Royal Astronomical Society in 1983

**Whitworth, Joseph 1803–87**
English engineer, the pioneer of commercial manufacture of precision machine tools. Before Whitworth, engineers made their own precision machine tools. By 1850 Whitworth's lathes, drilling and planing machinery were world renowned for their accuracy. Whitworth also developed a series of standard sizes for components and gauges to be used by engineers. His name is still associated with standard screw-thread sizes; British Standard Whitworth. He designed the Whitworth Rifle to replace the Enfield, and a method for making artillery from soft steel which was less likely to shatter under the high pressure of firing. He was a Fellow of the Royal Society and a generous benefactor to educational institutions.

**Wilson, C.T.R. 1869–1959**
Scottish physicist and meteorologist, Charles Thomson Rees Wilson is best known for his invention of the cloud chamber, an instrument used for decades in the study of subatomic particles. Wilson, who studied physics at Cambridge University, spent two weeks in 1894 measuring the weather on Ben Nevis, Scotland's highest mountain. As he said in his Nobel banquet speech, "Morning after morning I saw the sun rise above a sea of clouds and the shadow of the hill on the clouds below surrounded by gorgeous colored rings." This inspired him to attempt cloud formation in the laboratory. In 1911 he demonstrated that a charged particle passing through supersaturated air produced a visible trail that could be photographed and studied. In 1927 Wilson received the Nobel Prize for his work in particle physics.

**Woese, Carl 1928–**
American microbiologist and physicist whose identification of the archaea changed the way life is classified on Earth. Before Woese most of biology had dealt with the larger creatures, which were organized into the three kingdoms of plants, animals, and fungi. By the 1970s there were two additional microbial kingdoms: the prokaryotes (or bacteria) and the eukaryotes. Woese discovered a new kingdom found in the so-called 16s mitochondral RNA. He redrew the taxonomic tree, which has proved invaluable to conservationists and ecologists. In 2006 he was made a Foreign Member of the Royal Society, and is currently professor of microbiology at the University of Illinois

**Wöhler, Friedrich 1800–1882**
See p.150.

**Wright, Wilbur & Orville 1867–1912/1871–1948**
American aviation pioneers who, inspired by the work of the German glider pioneer Otto Lilienthal, became the first men to achieve sustained and controlled flight in powered craft that were heavier than air. The Wright brothers, formerly bicycle manufacturers, designed a gas engine weighing about 170 lb (77 kg), making the aircraft sufficiently light and powerful. On December 17 1903 they made four epoch-making flights in North Carolina, US. In 1908 they flew for an hour and carried a passenger for the first time. Their breakthrough was the invention of "three-axis control," which enabled the pilot to steer the aircraft effectively and to maintain its equilibrium.

**Yalow, Rosalyn Sussman 1921–**
American medical physicist who collaborated with Soloman Berson to develop radioimmunoassay, a method which uses radioisotopes to measure substances in the blood that are present in the tiniest of quantities, such as hormones, vitamins, and enzymes. The technique revolutionized the treatment of conditions such as diabetes and infertility. In 1977 Yalow became only the second woman ever to win the Nobel Prize in Medicine.

**Yingxing, Song 1587–1666**
Chinese scientist and encyclopedist who lived during the late Ming Dynasty. He produced the *Tian Gong Kai Wu* (The Exploitation of the Works of Nature), considered one of China's technological classics, in 1637. It covered a wide range of subjects, from smelting, hydraulic devices and irrigation, silk, salt, sugar, ceramics, papermaking and ink, and transportation, with exquisitely detailed illustrations that helped explain Chinese production processes.

**Yukawa, Hideki 1907–81**
Japanese physicist who won the Nobel Prize in Physics in 1949 for his theory of elementary particles. His theory of how the nuclear force holds the nucleus together predicted the existence of the meson, a subatomic particle, which later became an important part of nuclear and high-energy physics. In 1955 Yukawa joined ten other leading scientists in signing the Russell-Einstein Manifesto, calling for nuclear disarmament.

**Zeiss, Carl 1816–88**
German optician famous for the company he founded, which bears his name. Zeiss became a noted lens maker in the 1840s when his lenses were mostly used for microscopes. When cameras were invented, his company began manufacturing high quality lenses for cameras, which soon gained an international reputation.

**Zhang Heng 78–139 CE**
See pp.54–55.

**Zhenning, Yang 1922–**
Chinese-American physicist who worked on statistical mechanics and symmetry principles, for which he received the Nobel Prize in Physics in 1957. Zhenning was an assistant to physicist Enrico Fermi at the University of Chicago. In 1965 he moved to New York's Stony Brook University, where he continued to work until 1999. He won the Rumford Prize in 1980 and the Albert Einstein Medal in 1995.

**Zhongjing, Zhang 150–219 CE**
Chinese herbal physician best known for his monumental work, *Shanghan Zabing Lun* (Treatise on Febrile Diseases Caused by Cold and Miscellaneous Diseases), considered one of the most influential works in the history of Chinese medicine. It contained new recipes and formulas, many of which are still used in traditional Chinese medicine.

**Zhongzhi, Zu 429–500 CE**
Chinese mathematician and astronomer who lived during the Song Dynasty of the Southern Dynasties. He produced the Daming Calendar in 465, calculated one year as 365.24281481 days, and derived two approximations of pi. He also achieved fame for his derivation and proof of the formula for the volume of a sphere.

**Ziegler, Karl 1888–1966**
German chemist who shared the 1963 Nobel Prize in Chemistry with the Italian Giulio Natta for developing organo-aluminium catalysts for the low-temperature polymerization of olefins, particularly in the production of ethylene. Ziegler conducted research into organo-metallic compounds, developing lead tetraethyl, the anti-knock additive used in petrol.

**Zu Chongzhi 429–500 CE**
Chinese astronomer, mathematician, and engineer who created the Daming calendar and found several close approximations for pi. His calendar provided a more accurate number of lunations per year and, for the first time in China, took into account the precession of the equinoxes. It was eventually adopted in 510 CE. Zu's work with pi (he calculated its numerical value as between 3.1415926 and 3.1415927) remained unsurpassed until the 15th century. Zu Chongzhi also calculated the volume of a sphere.

# Glossary

## A

**aberration** A distortion of an image created by a lens or mirror, because the light rays do not meet at the focus.

**absolute magnitude** A measure of the actual luminosity of a star.

**absolute scale** A scale of temperature, also known as the Kelvin scale, that begins at absolute zero. Its unit of measurement is the kelvin.

**absolute zero** The lowest possible temperature: 0° K or -459.67° F (-273.15° C).

**absorption** 1. The taking up of one substance by another. 2. The letting in of radiation (the opposite of emission).

**abyssal plain** The essentially flat floor of the deep ocean, typically 3 miles (5 km) below the surface.

**acceleration** The rate of change of velocity.

**acid** A compound containing hydrogen that splits up in water to give hydrogen ions. Acidic solutions have a pH less than 7.

**acid rain** Rain that has become acidic due to water in the air reacting with certain chemicals (such as sulphur dioxide, nitrogen oxide, and carbon dioxide) emitted as a result of human activities, such as power production and vehicle emissions, or natural processes, such as volcanic emissions.

**acoustics** 1. The study of sound. 2. How sound travels around a room. For example, a concert hall needs to have good acoustics.

**activation energy** The amount of energy needed to start a chemical reaction.

**active transport** The use of energy to move molecules across a biological cell membrane against the concentration gradient, such as the chemical energy of ATP.

**adaptation** The way in which an organism gradually changes over many generations to become better suited to its particular environment.

**additive** Any substance added in small amounts, especially to food or drink, to improve it or inhibit its deterioration, e.g. to change color or taste or increase shelf life.

**adhesion** The force of attraction between the atoms or molecules of two different substances.

**adhesive** A sticky substance, such as paste or glue, used to join two surfaces together.

**ADP** Adenosine diphosphate. A compound that is formed when ATP releases energy.

**advection fog** A type of fog that forms where warm, moist air moves over a colder surface. It is often known as sea fog.

**aerobic respiration** A type of respiration that requires oxygen, unlike anaerobic respiration.

**airfoil** The special shape of an aircraft wing. It is more curved on the top than on the bottom, and produces lift as it moves through the air.

**air resistance** The force that resists the movement of an object through the air.

**albedo** How much an object, especially a planet or moon, reflects the light that hits it: its reflecting power.

**alchemy** A medieval science that tried, among other things, to find a way of changing metals, such as lead, into gold.

**alga** (plural: **algae**) An informal grouping of organisms, most of which photosynthesize, that do not have real roots or stems. Seaweeds are probably the best known examples.

**aliphatic compound** An organic compound made up of chains, rather than rings, of carbon atoms.

**alkali** A base that dissolves in water.

**alkaline** Describes a solution with a pH greater than 7.

**allotropes** Different forms of the same element. For example, diamond and graphite are allotropes of carbon.

**alloy** A mixture of two or more metals, or of a metal and a nonmetal.

**alpha decay** A type of radioactive decay in which atomic nuclei emit alpha particles. An alpha particle consists of two protons and two neutrons bound together, and is identical to a helium nucleus.

**alternating current (AC)** An electric current whose direction reverses at regular intervals.

**alternator** An electric generator that produces an alternating current.

**alveolus** (plural: **alveoli**) A tiny air sac in the lungs.

**AM** Amplitude modulation. The transmission of a signal by changing the amplitude of the carrier wave.

**amalgam** An alloy of mercury and another metal, such as tin.

**amino acid** Any of a group of organic chemicals in which the molecules each contain one or more amino groups $(NH_2)$ and one or more carboxyl groups (COOH). Amino acids are the chemicals that make up proteins.

**ammeter** An instrument that measures electric current.

**ammonification** The conversion of nitrogen in rotting organic matter in the soil to ammonia by decomposers, such as bacteria and fungi.

**amniocentesis** A test during pregnancy in which a small amount of amniotic fluid is removed from inside the amniotic sac around the baby and examined for genetic, chromosome, and developmental abnormalities.

**amp (ampère)** The SI unit of electric current.

**amplifier** A device that changes and usually increases the amplitude of a signal, especially an electrical signal for audio equipment.

**amplitude** The size of a vibration, or the height of a wave, such as a sound wave.

**anabolism** A series of chemical reactions in living things, which builds up large molecules from small ones.

**anaerobic respiration** A type of respiration that does not require oxygen. It produces less energy than aerobic respiration, which does require oxygen.

**analogue** In computing, representing data in a directly measurable way, such as by a varying voltage or the position of a needle on a meter.

**anaphase** A stage in the natural division of living cells by mitosis. The paired chromosomes separate and begin moving to opposite ends of the cell.

**anatomy** The science that studies the structure of living things. Human anatomy is the study of the human body.

**andesite** Fine-grained brown or grayish volcanic rock.

**anesthesia** The medical inducement of pain relief by complete loss of sensation or consciousness, generally using drugs.

**angiogram** An X-ray of the blood vessels, usually taken to study blood flow.

**angle of incidence** The angle a light ray makes with the perpendicular to the surface it hits.

**angle of reflection** The angle a reflected light ray makes with the perpendicular to the reflecting surface.

**anion** A negatively charged ion.

**anode** A positive electrode.

**anodizing** The process of coating a metal object with a thin, protective layer of oxide, by electrolysis.

**anthracite** Hard coal that burns with hardly any flame or smoke.

**antiretroviral drugs** Drugs used for the treatment of retrovirus infections, principally HIV.

**antibiotic** A drug that is used to kill or inhibit the growth of bacteria that cause infections.

**antibodies** Special proteins produced in the body by white blood cells, to label foreign particles or antigens and stimulate the immune response.

**anticyclone** An area of high pressure air that often produces settled weather.

**antigen** Any substance that stimulates the body to produce antibodies and an immune response.

**antimicrobial** Capable of either killing microbes that cause infection or stopping them from multiplying.

**antioxidant** A compound added to foods and plastics, to prevent them from oxidizing and becoming stale or breaking down.

**antiparticles** A particle that is the same as a normal particle, except that it has an opposite electrical charge. If a particle and its antiparticle collide, they annihilate each other.

**antiseptic** An antimicrobial substance applied to the skin, to reduce the risk of infection.

**apogee** The point in the orbit of a moon or satellite at which it is farthest from the object it is orbiting.

**apparent magnitude** The brightness of a star as seen from Earth.

**arteries** Blood vessels that carry blood from the heart to other parts of the body.

**arthropod** An invertebrate, such as insect, spider, or centipede, that has an exoskeleton (external skeleton), a segmented body, and jointed appendages or feet.

**artificial insemination** The artificial insertion of sperm into the female reproductive tract to induce fertilization.

**artificial selection** The process by which humans change the genetic makeup of a species.

**aseptic technique** The performance of a medical or laboratory procedure under completely sterile conditions.

**asexual reproduction** Reproduction that involves only one parent.

**asteroid** A rocky body, also known as a minor planet or planetoid, that circles the Sun. Most asteroids are in the asteroid belt between Mars and Jupiter.

**asthenosphere** A soft layer of Earth's mantle.

**astrolabe** A historical astronomical instrument used by astronomers and sailors for locating the position of the Sun, Moon, planets, and stars.

**astronaut** A person who has been trained to be a crew member in a spacecraft.

**astronomy** The study of the stars, planets, and other bodies in space.

**atmosphere** The layer of gases that surrounds a planet.

**atmospheric pressure** The force per unit area exerted by the weight of the air above that area.

**atoll** A ring-shaped group of coral islands surrounding a lagoon, usually on top of a seamount.

**atom** The smallest part of an element that has the chemical properties of that element. An atom consists of a nucleus of protons and neutrons surrounded by orbiting electrons.

**atomic nucleus** The very dense region in the center of an atom, consisting of protons and neutrons.

**atomic number** The number of protons in the nucleus of an atom. All atoms of the same element have the same atomic number.

**atomic weight** Also known as the relative atomic mass, the ratio of the average mass of atoms of an element to 1/12 of the mass of an atom of carbon-12.

**ATP** Adenosine triphosphate. A nucleotide molecule used by organisms for storing and transporting energy.

**attractor** A point, state, or set of states in a dynamic system toward which the system tends to move.

**aurora borealis** A display of lights in the night sky in Arctic regions, caused by the impact of electrically-charged particles from the Sun with Earth's atmosphere.

**autoclave** A strong, steam-heated container that is used for sterilization at high temperature and pressure.

**autotrophic** A plant that makes its own food by photosynthesis.

**average** See mean, median, and mode.

**axis** 1. An imaginary line around which an object rotates. 2. The line along which rock bends in a fold. 3. A reference line on a graph. 4. The second cervical vertebra of the spine.

## B

**background radiation** 1. Low-intensity radiation emitted by radioactive substances in and around Earth. 2. Radiation (in the form of microwaves) detected in space that may have come from the Big Bang.

**bacteriophage** A parasitic virus that infects bacteria; also called a phage.

**bacterium** (plural: **bacteria**) A single-celled microscopic organism that does not have a membrane-enclosed nucleus or other organelles.

**barchan** A sand dune with a crest.

**basalt** A dark volcanic rock.

**base** 1. A compound that reacts with an acid to give water and a salt. 2. The number of different single-digit symbols used in a particular number system; e.g. binary has two digits, 0 and 1.

**base pairs** Two complementary chemical bases linked on either side of the double spiral of a DNA molecule. The order of base pairs spells out the DNA code.

**base triplet** See codon.

**batholith** A dome of igneous rock that solidifies in a huge underground mass.

**battery** A combination of two or more electrical cells connected together that produce electricity.

**Beaufort scale** A scale used to measure wind speed, ranging from 0 to 12 (calm to hurricane force).

**bell curve** The bell-shaped curve on a graph that represents the normal distribution of a set of data. The highest, middle point of the curve corresponds to the mean (average) and the curve slopes symmetrically downward on either side of this value.

**beta decay** A type of radioactive decay in which atomic nuclei emit beta particles (electrons or positrons).

**Big Bang** The theory that the Universe began with a massive explosion of matter. It is thought that everything in the Universe is still moving apart because of the explosion.

**binary number system** A method of representing numbers, typically for computing, in which only the digits 0 and 1 are used. Successive units are powers of 2.

**binocular vision** The ability of some animals to see objects in three dimensions, so that they are able to judge distance.

**binomial system** A system of giving an organism two names. The first name signifies the genus and the second name signifies the species.

**biodegradable** A substance that can decompose and become harmless naturally.

**biogas** A gas made when plant or animal waste rots down without the presence of air.

**biogenic** Produced by organisms.

**biology** The study of living things.

**biomass** 1. The total number of living organisms found in a given area. 2. Plant material, such as wood or peat, that is used as a source of energy.

**biome** A large ecosystem, such as a rain forest or desert.

**biosphere** The region of Earth and its atmosphere in which living things are found.

**biotechnology** The technology of manipulating living things on a microscopic level, often by modifying their genes.

**bit** In computing, a fundamental unit of information having just two possible values, as either of the binary digits 0 or 1.

**bituminous** Containing bitumen, a tarlike substance produced from petroleum.

**black body** A theoretical object that absorbs all the electromagnetic radiation that falls on it, and so appears black. At a particular temperature, a black body also radiates the maximum amount of energy possible at that temperature.

**black dwarf** The faded remains of a dead star. (See also white dwarf.)

**black hole** An extremely dense object in space. Its gravity is so strong that it pulls in anything around it, even light, so that it looks black.

**black ice** Thin, hard, transparent ice, especially on the surface of a road.

**blastocyst** A hollow ball of cells that is an early stage in the development of an embryo.

**blood type** Also known as blood group, any of several types into which blood can be classified based on the antigens on the surface of red blood cells.

**boiling point** The temperature at which a liquid becomes a gas.

**bond** The attraction between atoms, ions, or groups of atoms that holds them together in a molecule or solid.

**Boolean algebra** A symbolic system of algebra devised by George Boole that represents relationships between entities such as sets or objects. It can be used to solve logical problems and is fundamental to the way in which digital computers process data.

**boson** A class of subatomic particles that act as messengers carrying forces between other particles.

**brine** A strong solution of salt in water.

**brown dwarf** An object smaller than a star, but bigger than a planet, which never has quite enough mass to start hydrogen fusion in its core, so cannot shine.

**Brownian motion** The random movement of tiny particles in a liquid or a gas, caused by molecules colliding with them.

**bubble chamber** A vessel filled with a superheated transparent liquid (usually liquid hydrogen), that was once used in physics to track the movement of electrically-charged particles.
**buffer** 1. A solution that is resistant to changes in pH.
2. An electric circuit used to join two other circuits.
**butterfly effect** A characterization of how, in chaos theory, small random incidents can have massive effects, such as a butterfly flapping its wings in South America leading to a storm in Europe.
**byte** A unit of information storage and transmission in computing. A byte consists of eight bits. A kilobyte is a thousand ($10^3$) bytes; a megabyte is a million ($10^6$) bytes; and a gigabyte is a billion ($10^9$) bytes.

# C

**CAD** Computer-aided design.
**calorie** A unit of energy. The calorie used in food science is in fact a kilocalorie—1,000 calories.
**calculus** 1. The branch of mathematics that deals with infinitesimal changes. It comprises differential calculus, which is concerned with rates of change, and integral calculus, which can be used to calculate quantities such as areas and volumes.
2. Also called a stone, a hard mass that forms in the body, such as in the kidney, gallbladder, or urinary tract.
**camouflage** The color, markings, or body shape that help to hide an animal or plant in its surroundings.
**capacitance** The ability to store electric charge.
**capacitor** A device that is used to store electric charge temporarily.
**capillaries** Tiny blood vessels that carry blood to and from cells.
**capillary action** The movement of a liquid up or down due to attraction between its molecules and the molecules of the surrounding solid material.
**carbohydrate** Any of a group of organic compounds consisting of carbon, hydrogen, and oxygen that are a major dietary source of energy for animals. Carbohydrate-rich foods include sugar, potatoes, bread, and pasta.
**carbon** The chemical element with the symbol C and atomic number 6, which forms more compounds than any other element, including the important chemicals of life.
**carbon cycle** The continuous circulation of carbon, from the atmosphere (in carbon dioxide), through plants (trapped in carbohydrates by photosynthesis), then animals (that eat the plants), and finally back into the atmosphere (through respiration and decomposition).
**carbon scrubbing** A process for absorbing carbon dioxide from industrial waste, and spaceship and submarine life-support systems before it escapes into the air.
**carburized** The addition of carbon to the surface of iron and steel by heat treatment.
**carnivore** A meat eater.
**cartilage** Gristly connective tissue that makes up the soft parts of the vertebrate skeleton and is present in some joints. The skeletons of some fish, such as sharks and rays, are made entirely of cartilage.
**cartography** The science of map making.
**cast** A hollow in a rock, formed around a decomposed animal or plant. The cast is a mold in which minerals collect and solidify to form a fossil.
**catabolism** A series of chemical reactions in living things that break down large molecules into small ones, releasing energy in the process.
**catalyst** A chemical that speeds up a chemical reaction without being changed itself at the end of the reaction.
**catalytic converter** A device in a vehicle that uses a catalyst to change toxic exhaust gases into less harmful gases.
**catenation** The ability of a chemical element to form long chain polymer molecules.
**cathode** A negative electrode.
**cathode ray tube** A vaccum tube with a fluorescent screen at one end and an electron gun at the other. The electron gun produces a stream of electrons (cathode rays), which, when they hit the screen, make it glow. Cathode ray tubes were used in old-fashioned televisions.
**cation** A positively charged ion.
**cavitation** 1. The formation of small holes (cavities) in something, such as the formation of bubbles in carbonated drinks and around moving marine propellers.
2. The natural enlarging of cracks in rocks by compressed air.
**cecum** A pouch in an animal's intestines which may house cellulose-digesting bacteria.
**celestial body** A natural object in space, such as a planet or star.
**celestial pole** One of two points in the celestial sphere about which stars appear to revolve when seen from Earth.
**celestial sphere** The imaginary sphere in which the stars seem to lie when seen from Earth.
**cell** The smallest unit of an organism that can exist on its own.
**cell division** The process by which one cell splits to produce two cells, called daughter cells.
**cellulose** The carbohydrate that forms the walls of plant cells.
**Celsius** The temperature scale named after Anders Celsius, in which water normally freezes at 0° and boils at 100°.
**central nervous system** The collection of nerves in the brain and spinal cord that acts to control animals' bodies.
**center of mass** The point at which the mass of an object or a group of objects can be assumed to be concentrated.
**centrifugal force** (or **centrifugal effect**) The force that appears to push outward on a body that is moving in a circle.
**centrifuge** A device used to separate substances of different densities, by spinning them at high speed.
**centripetal force** The force that pulls inward on a body moving in a circle.
**Cepheid star** A star that varies in brightness with great regularity, and acts as a distance marker in space.
**ceramics** Objects made of clay or porcelain fired in a kiln.
**cerebellum** A part of the brain at the back of the skull. Its primary role is to control movement and maintain balance.
**cerebrum** The largest part of the brain in humans, responsible for most conscious thought and activity. In humans, it is divided into two cerebral hemispheres and surrounds most of the rest of the brain.
**cermet** A material made from ceramic and metal. Cermets can withstand very high temperatures.
**CERN** *Conseil Européen pour Recherches Nucléaires*. The research center of the European Organization for Nuclear Research in Geneva.
**CFC** Chlorofluorocarbon. Any of a series of gases based on an alkane (such as methane or ethane) and containing chlorine and fluorine. Once widely used as refrigerants and propellants in aerosols, CFCs cause depletion of the ozone layer if allowed to escape into the atmosphere.
**chain reaction** A chemical or nuclear reaction in which the product of one step triggers the next step, which in turn triggers the next step, and so on. For example, in nuclear fission reactions, neutrons from the splitting of atomic nuclei go on to cause other nuclei to split.
**chaos** A state of disorder. A chaotic system appears to behave randomly and unpredictably, even though it follows deterministic laws. Chaos theory describes dynamic systems whose behavior is extremely sensitive to the initial conditions, such as the weather.
**charged particle** A subatomic particle or ion that has an electric charge.
**chemical** Any substance that can change when it is joined or mixed with another substance.
**chemical bond** *See* bond.
**chemistry** The science that deals with the structure, composition, and properties of substances and the changes they undergo as a result of reactions with other substances.
**chemotherapy** The treatment of cancer with drugs that kill cancer cells (cytotoxic drugs).
**chlorophyll** The green pigment found in many plants that absorbs light to provide the energy for photosynthesis.
**chloroplasts** Tiny bodies in some plant cells that contain chlorophyll.
**chromatid** One of the two identical strands of a chromosome after replication.
**chromatography** A method of separating a mixture by running it through a medium such as filter paper. Different parts of the mixture will move through the medium at different speeds and travel a greater distance.
**chromosome** A structure made of DNA and protein that is found in cells. A chromosome contains the genetic information (in the form of genes) of a cell.
**chromosphere** A layer of gases in the Sun's atmosphere that shines red.
**cilium** (plural: **cilia**) A tiny "hair" that projects from the surface of many small organisms. In humans, cilia line tracts inside the body such as the respiratory tract.
**cipher** A code in which letters in the original text are substituted with other letters, numbers, or symbols.
**circuit** A path around which an electric current can flow.
**circulation** The movement around something. Blood circulation is the continuous flow of blood around the body. Atmospheric circulation is the continuous movement of air in the atmosphere.
**classification** The process of arranging items in groups, or the system of grouping used for a particular collection of objects, such as living organisms or stars.
**climate** The prevailing weather conditions in an area over a long period of time.
**clone** Two or more identical organisms that share exactly the same genes.
**cluster** 1. A group of stars or galaxies.
2. A particle consisting of several atoms bound together.
3. A grouping of occurrences of a disease in a localized area.
4. A group of linked computers that work together as a single unit.
**code** A system in which some units, such as sequences of chemicals, letters, numbers, or symbols, are substitutes for, represent, or provide instructions for other things.
**codon** A sequence of three adjacent nucleotide bases that forms part of the genetic code. Most codons spell out the code for making a particular amino acid or for stopping or starting protein synthesis.
**cohesion** The force of attraction between two particles of the same substance.
**coil** An electric device made of a coil of electric wire, used for converting voltages, creating a magnetic field or inducing an electric current.
**coke** A fuel made by baking coal. It is mostly carbon and gives off far more heat than coal.
**cold front** The boundary between masses of cold and warm air, where the cold air is advancing and sharply undercutting the warm air.
**collision boundary** *See* convergent boundary.
**colloid** A mixture made up of tiny particles of one substance that are dispersed in another but do not dissolve.
**colonization** The occupation of an area by a population or species.
**colony** A large group of the same species of organism that live together.
**coma** 1. A cloud of gas and dust around the center of a comet.
2. A state of unconsciousness.
**combustion** A chemical reaction (burning) in which a substance combines with oxygen, producing heat energy.
**comet** A ball of frozen gas and dust that travels around the Sun. Some of the dust streams out from the comet to make a "tail."
**commensalism** Where two or more organisms live together and neither causes harm to the other.
**community** A group of people or animals that live in the same place.
**commutator** A rotating device that periodically reverses the direction of an electric current.
**complexity theory** The study of the behavior of extremely complicated dynamic systems, particularly how structure and order can arise in complex, apparently chaotic systems.
**compound** A substance containing atoms of two or more elements.
**compressed air** Air that is squeezed and kept under pressure greater than normal atmospheric pressure.
**compression** 1. Being pressed together or bunched up.
2. The increase in density of a fluid.
**concave lens** A lens that curves inward.
**concentration** The strength of a solution, or the amount of solute that is dissolved in a certain amount of solvent.
**conception** The coming together of a male sperm and a female ovum to create a zygote.
**condensation** The change of a gas or vapor into a liquid.
**condenser** 1. A cooling device for reducing a gas or vapor to a liquid.
2. Another name for a capacitor.
**conduction** The movement of heat or electricity through a substance.
**conductor** A substance through which heat or electric current flows easily.
**cones** Light-sensitive cells in the retina of the eye of humans and some other mammals that make it possible to see colors.
**conjugation** The transfer of genetic material between bacteria by direct cell-to-cell contact.
**conjunctiva** The mucus membrane that covers the front of the eye and lines the inside of the eyelids.
**conservation** 1. The preservation of any process or object.
2. The principle that a total quantity remains constant, whatever changes happen within a system.
**constellation** One of 88 regions of the night sky. Each constellation contains a group of stars joined by imaginary lines to represent an image.
**continental drift** The slow movement of the continents around the world over millions of years, carried on the tectonic plates that make up Earth's surface.
**convection** The transfer of heat through a fluid by currents within the fluid.
**convergent boundary** The line along which two tectonic plates that are moving toward each other meet. Also known as a collision boundary.
**converging lens** *See* convex lens.
**convergent evolution** The way in which different species evolve similar features because they are subjected to similar environmental conditions and selection pressures.
**convex lens** A lens that curves outward.
**coordinate** Each of a group of numbers used to specify the position of a point.
**copper oxide** A compound of copper and oxygen. There are two types: copper(I) oxide ($Cu_2O$, also called cuprous oxide) and copper(II) oxide (CuO, also called cupric oxide). Copper(I) oxide occurs naturally as the mineral cuprite and was the first known semiconductor.
**coprolite** Fossilized dung.
**Coriolis effect** The deflection of winds and ocean currents by the rotation of the earth.
**corona** The plasma atmosphere of the Sun or another star extending thousands of kilometers into space.
**corrasion** The wearing away of a surface by the action of rocks carried in ice or water.
**corrosion** Chemical attack of the surface of a metal.
**cosine (cos)** In a right-angled triangle, the cosine of an angle is the ratio of the side adjacent to the angle to the hypotenuse (the longest side).
**cosmic rays** High-energy, electrically-charged particles from space originating either in the Sun (lower energy), supernovae (medium energy), or beyond the galaxy (high energy).
**cosmology** The science of the origin and development of the Universe.
**cotyledon** A simple leaf that forms part of a developing plant. Also called a seed-leaf.
**coulomb** The SI unit of electrical charge: the quantity of electricity transported in 1 second by a current of 1 ampere.
**covalent bond** A chemical bond formed by atoms sharing one or more electrons.
**CPU** The central processing unit of a computer.
**cracking** The process of splitting larger molecules into smaller ones, often by heating with a catalyst.
**cross fertilization** Fertilization of a plant with gametes from another plant of the same species.
**crucible** A container in which metals or other substances may be melted or heated to very high temperatures.

**crust** The thin rigid outer surface of Earth, made of rock.
**cryptanalysis/cryptography** The art of studying and deciphering coded messages.
**crystal** A solid whose constituent atoms, ions, or molecules are arranged in a regularly repeating pattern.
**crystal lattice** The repeating pattern of atoms or ions that forms a crystal.
**crystallogram** A pattern formed on a photographic plate by passing a beam of X-rays through a crystal.
**cuneiform** The wedge-shaped symbols of the writing system used in ancient Mesopotamia, Persia, and Ugarit.
**curve** A line on a graph that shows the varying relationship between two quantities.
**cyclone** Another name for a hurricane, especially for one over the Indian Ocean.
**cylinder** 1. A tube shape.
2. The chamber in an engine in which the piston slides up and down.
**cystic fibrosis** Hereditary disorder of the exocrine glands that causes excess mucus production, leading to clogging of the airways.
**cytokinesis** The division of the cytoplasm of a cell into two daughter cells at the end of meiosis or mitosis.
**cytoplasm** The contents of a cell, apart from the nucleus.

# D

**dark energy** The little understood, repulsive force that accounts for about three quarters of the Universe's mass-energy and acts in the opposite direction to gravity, making the Universe expand.
**dark matter** Invisible matter detectable only by its gravitational effects on ordinary matter, believed to account for about a quarter of the Universe's mass.
**decant** To separate a mixture of a solid and a liquid by allowing the solid to settle and pouring off the liquid.
**decay chain** The series of transformations through different isotopes which a radioactive atom goes through as it decays.
**decibel** A unit used to measure the loudness of a sound.
**decimal** Relating to a number system based on the number ten, the power ten, or tenths.
**decomposer** A tiny organism, such as a bacterium, that breaks down dead matter.
**decomposition** 1. Organic decay.
2. Breaking larger molecules into smaller ones.
**decryption** The decoding of a coded message.
**deep-sea trench** A long depression in the ocean floor that forms where one tectonic plate is subducted beneath another.
**defibrillator** A device for restoring the rhythmic beating of the heart by the administration of a controlled electric shock.
**denominator** The number below the line of a fraction, the divisor.
**density** The mass per unit volume of a substance.
**deposition** The natural laying down of loose material, such as sediment, carried by moving water, wind, or ice.
**depression** An area of low air pressure that often brings bad weather.
**dermis** The layer of tissue in the skin below the outer epidermis.
**desalination** The removal of salt from seawater.
**desertification** The formation of a desert.
**desiccate** To dry out a substance by removing water from it.
**desiccator** A sealed container used to desiccate substances and keep them dry.
**detector** In electronics, the circuit in a radio receiver that separates out the sound signal from a radio wave.
**detergent** A substance which, when added to water, helps remove grease, oil, and dirt.
**diagnosis** The identification of an illness from its symptoms and signs.
**dicotyledon** A flowering plant that has two cotyledons (seed-leaves).
**differentiation** 1. In mathematics, the process of finding the derivative of a function, which is a measure of how a function changes.
2. In biology, the process by which cells or tissues change to become more specialized, as occurs during embryonic development, for example.
**diffraction** The bending of waves around obstacles or the spreading out of waves when they pass through a narrow aperture, such as when light waves pass through a diffraction grating.
**diffusion** The mixing of two or more different substances because of the random movement of the molecules.
**digestion** The breaking down of food into simpler molecules that can be utilized by the body.
**digital** Representing, storing, or manipulating sound, video, images, text, or data by discrete values rather than continuously varying ones. The most common way of doing this is by using binary encoding, in which a series of 1s and 0s is used to handle the data.
**digital sound** Sound recorded as a series of discrete pulses rather than as a continuous wave.
**diode** An electronic component that lets electricity flow in one direction only.
**diploid cell** A cell containing the full complement of chromosomes.
**direct current (DC)** An electric current that flows in one direction only.
**discharge** 1. The release or conversion of stored energy.
2. The removal of electrical charge from an object.
3. The emission of a fluid from a body.
**disinfection** Making something clean and free from infection, especially with the use of germ-killing chemicals.
**displacement reaction** A chemical reaction in which one kind of atom or ion in a molecule is replaced by another.
**dissection** The methodical cutting up of a body or plant to study its internal anatomy.
**distillation** The purification of a liquid by heating it until it vaporizes, then collecting the vapor and condensing it back into a liquid.
**divergent boundary** The line along which two tectonic plates are moving away from each other.
**DNA** (deoxyribose nucleic acid) The large, double-helix-shaped molecule that makes up the chromosomes that are found in almost all cells. It carries the genetic information in all known living organisms and in some viruses.
**doldrums** Area along the equator where the trade winds meet and form an area where there is very little wind.
**domain** In physics, a small, discrete region of magnetism within a magnet.
**double helix** A double spiral, particularly used in relation to the DNA molecule.
**drag** The force that slows down an object as it travels through a liquid or gas.
**drone** 1. A remote-controlled flightless aircraft.
2. A male bee that does no work but is able to fertilize the queen.
**drought** A long period with no rain.
**ductile** Ability of a substance to stretch out into a wire without losing its strength.
**dwarf planet** A small, planetlike object that is big enough to have become rounded by its own gravity, but is not big enough to have cleared surrounding space of objects.
**dye** A substance that colors a material.
**dynamo** A generator that produces direct current.

# E

**easterlies** Prevailing winds that blow from the east.
**echo** A sound that is heard again because it bounces back off a solid object.
**eclipse** The shadow caused by a body blocking the light from another. *See* lunar eclipse and solar eclipse.
**ecliptic** The apparent path in the celestial sphere, through which the Sun passes during the course of a year.
**ecology** The study of the relationships between organisms and their environment.
**ecosystem** A distinct area in the biosphere, such as a lake or forest, that contains living things.
**effort** A force applied to move a load.
**egg** The female reproductive cell, or ovum.
**elasticity** The ability of a material to stretch and then return to its original shape.
**electrode** A solid electric conductor that collects or transmits electric current. Terminals in electrical components such as transistors and batteries are electrodes, as are the small disks attached to the body to monitor electrical activity of the heart in electrocardiography (ECG) and of the brain in electroencephalography (EEG).
**electric charge** The property of subatomic particles that makes them interact electromagnetically, either repelling or attracting each other.
**electric circuit** A complete loop of conducting material that carries an electric current and connects electrical elements, such as switches, resistors, and light bulbs.
**electric current** The flow of electrons or ions.
**electrical resistance** The tendency to restrict the flow of electricity.
**electrolysis** Chemical change in an electrolyte caused by an electric current flowing through it.
**electrolyte** A substance that conducts electricity when molten or in solution.
**electromagnetic spectrum** The complete range of electromagnetic radiation: gamma rays, X-rays, ultraviolet radiation, visible light, infrared radiation, microwaves, and radio waves.
**electromagnetism** The physics of the electromagnetic field created by the interaction of electricity and magnetism.
**electromotive force (emf)** The potential difference of a battery, cell, or generator. It pushes an electric current around a circuit.
**electron** A particle with a negative electric charge. Electrons orbit the nucleus of the atom in a "cloud." An electric current consists of a flow of electrons.
**electron cloud** A term used to describe the indeterminate position of an electron around the nucleus of an atom.
**electron gun** A device that produces a stream of electrons, called cathode rays. For use in a television, for example.
**electron micrograph** A magnified image of an object made by an electron microscope.
**electron microscope** A microscope that uses a beam of electrons to produce a magnified image of an object.
**electron shell** One of the layers in which electrons are arranged around the nucleus of an atom.
**electrophoresis** The separation of charged particles in a mixture.
**electroplating** Coating a metal surface electrolytically by passing an electric current through a liquid in which the metal is immersed.
**electroscope** An instrument that detects electric charge.
**electrostatic field** The field of force surrounding an electrically charged object.
**electroweak theory** The theory explaining the combined interaction of electromagnetism and the weak nuclear force on leptons, a group of subatomic particles that includes electrons.
**element** A substance that cannot be broken down into simpler substances.
**ellipse** An oval shape like a flattened circle, made by slicing across a cone at an angle.
**embryo** The first stage of development of newly conceived offspring. In humans, it covers the first eight weeks of pregnancy.
**emission lines** Bright lines in the spectrum of light emitted by a body.
**emulsifier** A substance that is used to make immiscible liquids blend.
**emulsion** Tiny particles of one liquid that are dispersed in another liquid.
**encryption** The process of coding a message.
**endangered species** A species of animal or plant that is at risk of becoming extinct, either because it is few in number or because of major threats to its habitat.
**endocrinology** The branch of medicine concerned with hormones and the endocrine glands.
**endometrium** The mucus membrane lining the uterus.
**endoplasmic reticulum** A system of membranes in a cell, on which chemical reactions take place.
**endoscope** An instrument for directly viewing inside the body. Endoscopes may be flexible or rigid, and comprise a light source, and a series of lenses or a miniature camera. Surgical instruments may be passed inside endoscopes to perform surgeries.
**endoscopy** A medical technique for seeing inside the body by inserting an endoscope. Surgery may also be performed during endoscopy.
**endoskeleton** The internal skeleton of a vertebrate.
**endosperm** The tissue in a seed that stores food.
**endothermic reaction** A reaction that results in the absorption of heat.
**energy** The capacity to do work.
**entanglement** In quantum physics, the linking of two particles as one, so that when the particles move apart a change in one instantly causes a change in the other.
**environment** The surroundings of an animal or plant.
**enzyme** A catalyst in living things that increases the speed of reaction in natural chemical processes.
**epidermis** The outer layer of the skin.
**equation** 1. A symbolic statement that the values of two mathematical expressions are equal to each other.
2. A symbolic representation of the changes that occur in a chemical reaction.
**equator** The imaginary circle around the middle of the earth, midway between the north and south poles.
**equilibrium** A state of physical or chemical balance.
**equinox** One of the two dates each year (about March 20/21 and September 22/23) on which the Sun is exactly above the equator and night and day are of equal length.
**erosion** The wearing away of Earth's surface due to the effects of weather, water, or ice.
**erythrocytes** Red blood cells.
**escape velocity** The minimum speed that a space rocket must reach to escape a planet's gravity.
**estivation** The deep sleep or immobility that some animals go into when the weather is very hot and dry.
**ethology** The study of animal behavior.
**eukaryotic cell** A cell with a nucleus.
**Eurasia** The combined land mass of Europe and Asia.
**eutrophication** Where an excess of nutrients, such as from fertilizers, seeps into water, causing the overgrowth of aquatic plants. This creates a shortage of oxygen in the water, killing animal life.
**evaporation** The changing of a liquid into a vapor as molecules escape from its surface.
**evolution** The change in organisms over time.
**evolve** To undergo evolution.
**excretion** The elimination of waste by organisms.
**exosphere** The outermost part of Earth's atmosphere, about 560 miles (900 km) above Earth's surface.
**exoskeleton** The hard outer skeleton of many invertebrates, such as insects.
**exothermic** A chemical reaction in which heat is released.
**exponent** The power, or superscript number to the right of a number, that indicates the number of times the number is to be multiplied by itself, as in $5^3 = 5 \times 5 \times 5$.
**exponential growth** Growth that accelerates with increasing speed: the larger something becomes, the faster its rate of growth.
**expressed gene** A gene that shows its effect in a living organism.
**extinction** The death of all the members of a species.

# F

**factorial** The product of an integer and all the integers smaller than it; e.g. factorial 5 (5!) is $5 \times 4 \times 3 \times 2 \times 1 = 120$.
**Fahrenheit** Temperature scale named after Gabriel Fahrenheit, in which water at sea level normally freezes at 32° and boils at 212°.
**family** In biological classification, the taxonomic group into which an order is divided.
**fault** A break in Earth's crust.
**faulting** The sudden shifting of a block of rock due to the disturbance of Earth's crust.
**fermentation** The process by which yeast converts sugars into alcohol.
**fermion** One of the group of subatomic particles that are associated with matter, such as electrons, quarks, and protons, rather than those that carry force, such as bosons.
**Ferrel cell** The circulation of the atmosphere in mid-latitudes that brings westerly winds at surface level and returning easterly winds at high level.
**fertilization** The joining of male and female gametes.
**fetus** The unborn, developing offspring of a mammal. In humans, it covers pregnancy after the first eight weeks.
**filigree** Delicate ornamental tracery in gold, silver, or coppery wire.
**filter** A device that removes solid material from a liquid.
**firmware** The small, fixed programs that control electronic items such as remote controls and calculators, as well as computer keyboards and robots.
**fluid** A substance that can flow, such as a gas or liquid.
**fluorescence** Light that is given off by certain atoms when they are hit by ultraviolet radiation.
**FM** Frequency modulation. The transmission of a signal by changing the frequency of the carrier wave, such as a radio wave.

**fold** A bend in layers of rock.
**folding** The slow crumpling of layers of rock due to the lateral compression of Earth's crust.
**food chain** A series of organisms, each of which is consumed by the next.
**food web** The system of food chains in an ecosystem.
**force** Something that changes the movement or shape of an object.
**force field** The area in which a force can be detected.
**forensic** The application of scientific methods to the study of crime.
**formula** (plural: **formulae**) 1. A set of chemical symbols that represent the make up of a substance.
2. A set of mathematical symbols expressing a rule, principle, or method for determining an answer.
**fossil** Evidence of past life such as animal and plant remains, eggs, and footprints.
**fossil fuel** A fuel, such as coal or oil, that has been formed over millions of years from the remains of living things.
**fractal** A curve or geometrical pattern in which similar shapes recur at different scales.
**fraction** 1. A numerical quantity that is not a whole number. A fraction is expressed as one number divided by another, as in ½, where the upper number is called the numerator and the lower one is called the denominator.
2. In chemistry, one of the portions into which a mixture is separated by a process such as distillation.
**fracture zone** The region on either side of divergent tectonic plate boundaries, in which rock is broken on a large scale by the movement of the plates.
**Fraunhofer lines** Dark lines in the Sun's spectrum caused by elements in the Sun's gases absorbing certain wavelengths of light.
**freezing point** The temperature at which a substance turns from liquid to solid.
**frequency** The number of waves that pass a point every second.
**friction** A force that resists or stops the movement of objects in contact with each other. The friction between an object and a fluid such as air or water is known as drag.
**front** The first part of an advancing mass of cold or warm air.
**fundamental particle** A basic subatomic particle, such as a quark, lepton, or boson, that is not thought to be made of smaller particles.
**fungus** (plural: **fungi**) Any of a group of eukaryotic organisms (those whose cells have nuclei) that reproduce by means of spores. The group includes molds, yeasts, smuts, and mushrooms.
**fuse** A safety device used in electrical circuits. It is a thin wire which melts if too much current passes through it.

# G

**Gaia hypothesis** The concept that all the living things and physical components of Earth interact to form a complex self-regulating system, like a huge "organism."
**galaxy** A large group of stars, dust, and gas, all loosely held together by gravity. Our galaxy is called the Milky Way.
**galvanize** To coat iron with zinc to protect it from rust.
**gamete** A reproductive cell, such as a sperm or egg.
**gametophyte** The gamete (reproductive cell)-producing stage in the life cycle of plants such as ferns and mosses.
**gamma rays** A form of electromagnetic radiation with a very short wavelength.
**ganglion** (plural: **ganglia**) A group of nerve cells enclosed in a casing of connective tissue.
**gas** An airlike substance that expands to fill any space it enters.
**gauge boson** Any of the subatomic particles that carry the four fundamental forces of nature—electromagnetism, gravity, the weak nuclear force, and the strong nuclear force.
**Geiger counter** An instrument used to detect and measure radioactivity.
**gene** The basic unit of heredity in living things, typically a segment of DNA or RNA that provides the coded instructions for a particular protein.
**gene code** The sequence of nucleotide bases on DNA that codes for a particular gene.
**gene map** A plot of the sequence of genes along an entire strand of DNA.
**generation** A group of individuals born or living at the same time.
**generator** A device that converts mechanical energy into electrical energy.
**genetic drift** The process of change in the overall genetic composition of a population as a result of random events rather than by natural selection. Genetic drift causes a change in the relative prevalence of alleles (variations of a gene that controls the same trait) over time.
**genetic engineering** The process of artificially modifying the characteristics of an organism by manipulating its genetic material.
**genetic fingerprinting** The analysis of a DNA sample to identify who it belongs to.
**genome** The complete set of genes for an organism.
**genotype** The genetic makeup of an organism.
**geochemistry** The study of the chemistry and composition of the earth.
**geology** The science that deals with rocks, minerals, and the physical structure of the earth.
**geomorphology** The study of the physical features on Earth's surface.
**geothermal** Relating to the internal heat of Earth, or energy produced from it.
**geothermal energy** Energy harnessed from the heat in the interior of Earth.
**germ** 1. A microbe that causes disease.
2. The part of an organism from which a new organism can develop.
**germination** The early stages in the growth of a seed.
**gibbous** The phase of the Moon between a half- and fully- illuminated disk.
**glacier** A slowly moving river of ice formed by the accumulation of hard-packed snow over many thousands of years.
**gland** An organ or group of cells that produces substances used by the body, such as hormones.
**glaze** A glassy substance fused on to the surface of pottery by the heat to protect it.
**global warming** An increase in the average temperature of Earth's atmosphere. The recent increase in global temperatures is at least partly due to the greenhouse effect.
**glucagon** A hormone that stimulates the liver to turn stored glycogen into glucose when blood sugar levels are low; the opposite effect of insulin.
**glucose** A simple sugar that is the main carbohydrate source of energy in most living cells.
**gluons** Particles within protons and neutrons that hold quarks together.
**glycogen** A stored form of glucose in animal cells, made mainly in the liver and muscles.
**granulation** The appearance of dark spots or granules in the Sun's photosphere caused by convection currents.
**gravitational force** The force of gravitational attraction between matter.
**gravitational lensing** The "bending" of light from a distant bright object in space, such as a quasar, around a massive object en route, due to the massive object's gravity.
**gravitational slingshot** The extra acceleration a space probe acquires by swinging close to a planet and using the planet's gravity to alter its path and speed.
**graviton** A hypothetical particle responsible for gravity.
**gravity** The force of attraction between any two masses. It attracts all objects toward Earth, giving them weight.
**great ocean conveyor** The deep circulation of ocean waters around the world, driven by differences in density resulting from salt concentration and temperature differences. Also known as thermohaline circulation.
**greenhouse effect** The way in which certain gases in Earth's atmosphere, especially carbon dioxide, trap heat. The buildup of these gases leads to global warming.
**greenhouse gases** Gases that increase the atmosphere's capacity for preventing the escape of heat. They include carbon dioxide and methane.
**grike** An enlarged crack in limestone that is produced as the rock gradually dissolves in rainwater.
**groynes** Low walls or fences built along the seashore to prevent coastal erosion.
**guttation** The loss of water from the surface of a plant as liquid rather than vapor.
**guyot** An undersea mountain with a flat top.
**gyre** The complete circulation of ocean currents around an ocean basin.
**gyroscope** A fast-spinning wheel whose axis tends to stay pointing in the same direction once the wheel is spinning. A gyrocompass uses a gyroscope to point to true north. Gyrocompasses are more accurate than magnetic compasses and are widely used in navigation.

# H

**habitat** The natural range of an animal or plant.
**Hadley cell** The circulation pattern of the atmosphere in the tropics, which brings north and southeasterly trade winds toward the equator at surface level and returns westerly winds at high level.
**half-life** 1. The time taken for radioactivity in a sample to drop to half of its original value.
2. The time taken for the concentration of a drug or other substance in an organism to naturally reduce to half its original level.
**haploid cell** A sex cell with half the number of chromosomes. See also diploid.
**hard water** Water that contains calcium and magnesium salts.
**hardware** The mechanism, wiring, and other physical components of a computer.
**harmonics** Waves whose frequencies are whole-number multiples of the basic frequency.
**heart** The hollow muscular organ in humans and other vertebrates that pumps blood around the body. Some invertebrates, such as mollusks and annelid worms, have a simpler organ that performs the same function.
**hemisphere** One half of a sphere. Earth is divided into the northern and southern hemispheres by the equator.
**hemoglobin** An iron-containing protein that is the oxygen-carrying component of red blood cells of vertebrates. It is also found in the tissues of some invertebrates.
**herbivore** An animal that only eats plants.
**heredity** The passing on of characteristics through the generations.
**hertz (Hz)** The SI unit of frequency. One hertz is one cycle per second.
**hibernation** The deep sleep or period of inactivity that some animals go into during the winter.
**hierarchy** An arrangement in ranks, according to factors such as complexity or power.
**high frequency** Short wavelength, rapid vibration.
**HIV** Human immunodeficiency virus, a retrovirus that causes the disease AIDS.
**hologram** A three-dimensional image created by the interference between two parts of a divided laser beam.
**homeostasis** The processes by which an animal keeps its internal environment (for example, temperature, blood pressure, and water balance) stable.
**hominid** Any member of the primate family Hominoidea, including humans.
**hormones** Chemical "messengers" that move around the bloodstream and control the functions of the body.
**hot-spot volcano** A volcano that forms away from tectonic plate boundaries above a hot convection current in Earth's mantle known as a mantle plume.
**H-R diagram** (Hertzsprung–Russell diagram) A graph that plots stars according to their absolute magnitude (real brightness), color, and temperature. Most stars fall along a line called the main sequence.
**html** An acronym for HyperText Markup Language, the main computer language used on the Internet.
**http** An acronym for HyperText Transfer Protocol, the call and response system used to link websites into the Internet.
**Hubble's law** The law stating that the distance of a galaxy is proportional to the speed at which it is moving away from us: the more distant the galaxy, the faster it is receding. This shows that the Universe is expanding.
**Human Genome Project** The worldwide science project completed in 2003 to map the entire sequence of genes in human DNA.
**humidity** The proportion of water vapor in the atmosphere.
**humorism** The ancient theory of the workings of the body according to the four humors or temperaments: sanguine, choleric, melancholic, and phlegmatic.
**hurricane** A huge, circular tropical storm in which there are wind speeds of 75 mph (120 km/h) or more.
**hydraulic** Describes a machine that operates by transferring pressure through a fluid.
**hydraulic pressure** The pressure created by a fluid pushed through a pipe.
**hydrocarbon** A chemical compound made up of hydrogen and carbon only.
**hydroelectricity** The generation of electricity by harnessing the energy in flowing water.
**hydrogen** The lightest and most abundant chemical element in the Universe. It has an atomic weight of just over 1 and makes up about 75% of the total mass of elements in the Universe.
**hydrometer** An instrument used to measure the density of a liquid.
**hydroponics** The process of growing plants without any soil in sand or gravel supplied with nutrients and water only.
**hygiene** Attention to cleanliness for the good of health.
**hypha** (plural: **hyphae**) One of many tiny threads that form the main body of a fungus.
**hypotenuse** The longest side of a right-angled triangle; the side opposite the right angle.

# I

**igneous rock** Rock that is formed when molten magma cools and solidifies.
**imaging** A general term for any techniques that produces images. Modern imaging techniques often rely on computers to process raw imaging data from devices such as sensors and body scanners (for example, CT scanners).
**immiscible** Describes two liquids that do not blend together, such as water and oil.
**immune system** The body's natural defense mechanisms, which react to foreign material and organisms with effects such as inflammation and antibody production.
**immunization** Priming the body's immune system to fight against future infections by inoculation.
**in vitro fertilization (IVF)** An artificial method of conception in which egg cells are fertilized by sperm outside the uterus, in vitro (in glass).
**indicator** A substance that shows the pH of a solution by its color.
**induction** 1. The process by which an electrical conductor becomes electrically charged when near a charged object.
2. The process by which a magnetizable object becomes magnetized when in a magnetic field.
3. The process by which an electric current is produced in a circuit by varying the magnetic field linked with the circuit.
**industrial plant** The land, buildings, and machinery used to perform an industrial process.
**inertia** The tendency of an object to remain at rest or to keep moving in a straight line until a force acts on it.
**infection** An illness or medical problem caused by disease-causing organisms, such as bacteria and viruses.
**infinity** A quantity, distance, time, or number that is immeasurably large.
**infrared radiation (IR)** A type of electromagnetic radiation with a wavelength just longer than that of visible light but shorter than that of microwaves. It is commonly experienced as heat.
**inheritance** The range of natural characteristics and potential passed on by parents or ancestors.
**inhibitor** A substance that slows down a chemical reaction.
**inoculation** The deliberate introduction of disease-causing organisms into the body in a mild or harmless form to stimulate the production of antibodies that will provide future protection against the disease.
**inorganic chemistry** The branch of chemistry that deals with all chemicals except the large number of organic compounds (those that contain carbon–hydrogen bonds).
**input** Information fed into a computer.
**insecticide** Any chemical that kills insects.
**insulator** A material that reduces or stops the flow of heat, electricity, or sound.
**insulin** A hormone released by the pancreas that regulates blood sugar levels by controlling the concentration of glucose in the blood.
**integrated circuit** A tiny electric circuit made of components built into the surface of a silicon chip.
**interference** The disturbance of signals caused where two or more waves meet.
**interglacial** The period of warmer weather between two ice ages.

**internal reflection** The reflection of light when it passes from a dense to a less dense medium, such as from glass to air.
**Internet** The electronic information network linking computers around the world.
**Internet service provider (ISP)** A company that provides access to the Internet.
**interphase** The phase of the cell cycle during which a cell is not undergoing cell division. Cells spend most of their time in interphase.
**interstellar space** Space between the stars.
**intestine** The lower part of the digestive tract, from the stomach to the anus, in humans and other mammals.
**invertebrate** An animal with no backbone.
**inverter** A device used to convert direct current into alternating current.
**invisible light** Light made of waves too short or too long for the human eye to see, including infrared and ultraviolet.
**ion** An atom or group of atoms that has lost or gained one or more electrons to become electrically charged.
**ionic compound** A chemical compound, such as sodium chloride, in which ions are held together in a lattice by ionic bonds.
**ionic bond** A chemical bond formed when one or more electrons are passed from one atom to another, creating two ions of opposite charge that attract each other.
**ionosphere** The part of the atmosphere, 30–250 miles (50–400 km) above Earth's surface, that reflects radio waves.
**iris** The colored part of the eye that surrounds and controls the size of the pupil.
**irradiation** The use of radiation to preserve food.
**irrational number** Any number that cannot be expressed as a simple fraction. Written as a decimal, irrational numbers have infinitely many digits after the decimal point.
**isobar** A line on a weather map that connects points with the same atmospheric pressure.
**isomer** A chemical compound with the same formula as another compound, but a different physical structure.
**isoseismal** A line on a map that connects points at which earthquake shocks are of equal intensity.
**isotope** One of two or more atoms with the same atomic number but with different numbers of neutrons.

# J

**jet propulsion** The pushing forward of an object by a stream of fluid or air.
**jet stream** Strong winds that circle Earth about 6 miles (10 km) above the surface.
**joint** 1. A structural line in sedimentary rocks.
2. An area of the body where bones meet.
**joule** The SI unit of work or energy, equal to the work done by a force of one newton moving one meter.

# K

**karyotype** 1. The characterization of the chromosomes of a species or individual in terms of the number, size, and structure of the chromosomes.
2. A diagram or image of an individual's total complement of chromosomes arranged in their assigned numerical order.
**Kelvin scale** *See* absolute scale.
**keratin** The protein that makes up hair, horns, hoofs, nails, and feathers.
**keyhole surgery** Surgery performed through a very small incision, using special instruments and an endoscope.
**kidney** One of a pair of internal organs that filter waste products and excess water from the blood.
**kinetic energy** The energy an object has because of its movement.
**kingdom** The highest level in the classification of life forms.
**Kyoto protocol** An international agreement on climate change that sets industrialized countries binding targets for reducing greenhouse gas (GHG) emissions.

# L

**laccolith** A mass of igneous rock that pushes the rock above into a dome shape.
**lactose** One of the sugars in milk.
**Lamarckism** The theory that evolution depends on the inheritance of characteristics acquired during an organism's life, named after Jean Baptiste Lamarck.
**larva** (plural: **larvae**) The second stage in the life of an insect, between the egg and the adult, such as a caterpillar.
**laser** (Light Amplification by the Stimulated Emission of Radiation). A device that emits an intense beam of light.
**laser surgery** Surgery performed with a laser beam, for example, reshaping the cornea to improve eyesight.
**latent heat** The heat needed to change a solid to a liquid or a liquid to a gas without a change of temperature.
**latitude** A measure of distance from the equator (the poles are at 90° latitude and the equator is at 0°). Lines of latitude are imaginary lines drawn around Earth, parallel with the equator.
**LDR** (Light-Dependent Resistor). A resistor whose resistance increases when the amount of light that hits it increases.
**leaching** The extraction of a soluble material from a mixture by passing a solvent through the mixture.
**LED** (Light-Emitting Diode). A diode that emits light when a current flows through it.
**lens** 1. Disk of glass ground to a precise shape so that it refracts light rays to a focus, to provide a sharp image or magnification.
2. The part of the eye that forms an image on the retina.
**lepton** A family of fundamental particles that are affected by electromagnetism, gravity, and the weak nuclear force, but not by the strong nuclear force, as quarks are.
**leucocyte** A general term for any white blood cell.
**lever** A rigid bar that pivots to move a load at one end when force is applied to the other.
**lift** The upward force produced by an aircraft's wings that keeps it airborne.
**ligament** A short, elastic band of fibers that connects two bones or cartilages at a joint.
**light year** The distance traveled by light in a year. It is equal to 5.9 million million miles (9.5 million million km).
**lignin** A polymer in the walls of the cells of trees and shrubs. It makes the plant woody.
**lithosphere** The layer of Earth that includes the crust and the upper mantle.
**logarithm** The power to which a base, such as 10, must be raised to produce a given number.
**longitude** A measure of distance around Earth, measured in degrees east or west of the prime meridian at Greenwich, London. Lines of longitude are imaginary lines drawn on Earth's surface between the poles.
**longitudinal wave** A wave in which the particles of the medium vibrate in the same direction as that in which the wave is traveling.
**low frequency** A radio frequency band with long waves that vibrate relatively slowly.
**luminosity** The amount of light given out by an object, such as a star.
**lunar eclipse** When the Moon moves into Earth's shadow so that it cannot be seen from Earth.
**lymphocytes** Any of a group of white blood cells that play a crucial role in immunity, by producing antibodies, for example.
**lymphatic system** A network of tubes and small organs that drains a fluid called lymph from the body's tissues into the bloodstream.

# M

**magma** Liquid, molten rock in Earth's mantle and crust. It cools to form igneous rock.
**magnetic field** The area around a magnet in which its effects are felt.
**magnetic poles** 1. The two regions of a magnet where he magnetic effect is strongest.
2. The two variable points on Earth where Earth's magnetic field is strongest and toward which a compass needle points.
**magnetism** The invisible force of attraction or repulsion produced by a magnetic field.
**magnetosphere** The magnetic field around a star or planet.
**magnification** The degree to which a lens or other optical device makes something appear larger than it is.
**main sequence** A term for stars that fall within the main diagonal band on the Hertzsprung–Russell diagram. Such stars generate heat and light by converting hydrogen to helium by nuclear fusion in their cores. Most stars are of this type.
**malleable** A term for metal that can be hammered or pressed into shape.
**mammal** Warm-blooded vertebrate animal that gives birth to young that feed on their mother's milk.
**manned maneuvering unit (MMU)** A backpack used by astronauts to move around in space.
**mantissa** The part of a logarithm after the decimal point.
**mantle** The large part of Earth's interior between the core and the crust, made of warm, often semi-molten rock.
**mass** The amount of matter in an object.
**mass wasting** The downslope movement of rock and weathered material in processes such as landslides and rockfalls.
**matter** Anything that has mass and occupies space.
**mean** The middle or expected value of a set, typically found by adding all the values together then dividing by the number of values in the set.
**meander** A looplike bend in a river.
**mechanical advantage** A measure of the effectiveness of a machine: the ratio of the force produced by the machine (output) to the force applied to it (input).
**median** 1. Situated along the middle of the body, in the plane that divides the body into right and left halves.
2. In a set of values, the middle value when all the values are arranged in order of size.
**meiosis** The type of cell division that results in daughter cells that have half the number of chromosomes of the parent cell. Meiotic division produces egg and sperm cells.
**melanin** A brown pigment that is found in the skin, hair, and eyes.
**melting point** The temperature at which a solid becomes liquid.
**membrane** A thin skin.
**meniscus** The curved upper surface of a liquid in a thin tube.
**menstrual cycle** The recurring cycle of changes in women and other female primates that prepares the body for reproduction, including ovulation and menstruation.
**Mercalli scale** A scale that is used to measure an earthquake's intensity.
**Mesolithic** The middle part of the Stone Age, ending typically in the development of farming, between 11,000 and 7,000 years ago in Europe.
**mesopause** The part of the atmosphere about 50 miles (80 km) above Earth's surface. It is the upper limit of the mesosphere.
**mesosphere** 1. The layer of Earth's mantle underneath the asthenosphere.
2. The layer of the atmosphere above the stratopause.
**metabolism** The sum total of chemical processes within a living organism.
**metal** Any of several elements that are usually shiny solids and good conductors of electricity and heat.
**metallic bond** A bond formed between metal atoms. The metal's electrons flow freely around the atoms.
**metamorphic rock** Rock that has been changed by great heat and pressure underground.
**metamorphosis** A change of form, such as from a caterpillar to a butterfly.
**metaphase** The stage of mitosis and meiosis, following prophase and preceding anaphase, during which the chromosomes are aligned across the middle of a cell.
**meteor** A tiny piece of dust or rock from space that burns up as it enters Earth's atmosphere, producing a streak of light.
**meteoroid** A small body of rock in space that will become a meteor if it enters Earth's atmosphere, and a meteorite if it reaches the ground.
**meteorite** A piece of rock or metal from space that enters Earth's atmosphere and reaches the ground without burning up.
**meteorology** The study of the weather.
**microchip** A small wafer of semiconducting material with an integrated circuit.
**microclimate** The climate of a small area, such as a valley.
**micrograph** A photograph taken using a microscope.
**microorganism** A tiny organism which can be seen only with the aid of a microscope.
**microprocessor** An integrated circuit that performs all the central processing tasks of a computer.
**microscope** An instrument that produces magnified images of very small objects.
**microwave** A type of electromagnetic radiation. Microwaves are very short radio waves.
**mid-ocean ridge** A ridge down the middle of the ocean floor, created by volcanic material erupting from the gap between diverging oceanic plates.
**migration** 1. The movement of animals to find food, a place to breed, or a more suitable climate.
2. The movement of one or more atoms, or of a chemical bond, from one position to another in a molecule .
3. The movement of ions between electrodes in electrolysis.
4. The change from one hardware or software system to another.
**mimicry** 1. Where a species of plant or animal evolves to look like another.
2. The copying of one individual's behavior by another.
**mineral** A naturally occurring inorganic solid.
**mineralogy** The study of minerals.
**mirage** An optical illusion produced by light bending through layers of air with different densities.
**miscible** Describes two or more liquids that can be blended together.
**mitochondrion** (plural: **mitochondria**) An organelle that produces energy for a cell.
**mitosis** Cell division in which the nucleus divides to produce two cells, each with the same number of chromosomes as the parent cell.
**mixture** A substance that contains two or more elements or compounds that are not combined chemically.
**mode** 1. Patterns of vibrations.
2. The value that occurs most frequently in a set of values.
**modem** A device for converting computer output and input into a form suitable for transmitting through telephone lines or optical fibers.
**modulation** The transmission of a signal by changing the characteristics of a radio wave (called the carrier wave).
**mole** The amount of a substance that contains the same number of atoms or molecules as there are in 0.4 oz (12 g) of carbon-12.
**molecule** The smallest unit of an element or compound, made up of at least two atoms.
**momentum** A quantity equal to a object's mass multiplied by its velocity.
**monocotyledon** A flowering plant with a single cotyledon (seed-leaf). *See also* dicotyledon.
**monomer** A molecule that can be bonded to other identical molecules to form a polymer.
**monsoon** A strong wind that changes direction according to the season, bringing torrential rain from the sea to areas such as India and Bangladesh.
**moon** A small body that orbits a planet.
**moraine** Rocks and debris that have been deposited by ice.
**mordant dyes** Dyes that need another chemical to be added to fix them to a fabric.
**mouse** In computing, a handheld device that is used to control a cursor on a computer monitor.
**MRI** (Magnetic Resonance Imaging). A noninvasive form of diagnostic medical imaging, in which a patient is placed in a strong magnetic field and the absorption and transmission of radio signals from the body is analyzed.
**multidimensions** The theoretical idea that there are many more than the standard three dimensions in space.
**muscle** A band of strong fibers in the body that is able to contract in response to nerve signals and thereby provide movement.
**mutation** A random change in the chromosomes of a cell.
**mutualism** Any close relationship between two species in which both benefit.
**myelin** A fatty material found around nerve fibers.
**myofibril** Stretchy threads found in muscle cells.

# N

**nacreous** Relating to the mother-of-pearl made inside shellfish such as oysters.
**nanometer (nm)** One billionth of a meter ($10^{-9}$m).
**nanotechnology** Minute technology on the atomic or molecular scale, less than 100 nanometers.
**natural logarithm (e)** A logarithm in which the base is the irrational number e (about 2.71828).

**natural selection** The biological process by which inheritable characteristics that increase an individual's chances of survival and reproduction are passed on to the next generation.
**nebula** (plural: **nebulae**) A visible cloud of dust and gas in space.
**nectar** A sugary liquid found in the flowers of some plants, on which pollinators feed.
**nematocyst** A long, coiled thread that shoots out of a stinging cell, such as in a sea anemone.
**nephron** One of the million or so minute purification and filtration units in the kidney.
**nerve** A fiber that carries electrochemical "messages" (nerve impulses) from one part of the body to another.
**neurone** A nerve cell.
**neutralize** To make an acid or alkali into a neutral solution that is neither acidic nor alkaline.
**neutron** A particle in the nucleus of an atom that has no electrical charge.
**neutron star** A small, very dense star made mostly of neutrons, formed by the gravitational collapse of a giant star.
**newton (N)** An SI unit of force.
**niche** The position that a living thing occupies in an ecosystem.
**nitrification** The conversion of ammonia in the soil into nitrites or nitrates.
**nitrogen** A colorless, odorless, unreactive gas that makes up the majority of the atmosphere and plays an important part in the life of plants and the soil.
**noble gases** Gases such as helium, argon, and neon that have a complete complement of electrons in their outer shell and are very unreactive.
**noctilucent cloud** A high, luminous cloud of ice crystals seen at night, especially in summer at high latitudes.
**nocturnal** A term used to describe a living organism that is active at night and sleeps during the day.
**nomenclature** A body or system of names.
**nuclear energy** 1. Energy released by an atomic nucleus as a result of nuclear fission, nuclear fusion, or radioactive decay. 2. Electricity generated by a nuclear power station.
**nuclear fission** A nuclear reaction in which the nucleus of an atom splits into two smaller nuclei, releasing energy.
**nuclear fusion** A nuclear reaction in which the nuclei of light atoms, such as hydrogen, fuse to form a heavier nucleus, releasing energy.
**nuclear reaction** A change in the nucleus of an atom.
**nucleolus** A small, dense, round body inside the nucleus of a cell.
**nucleus** 1. The central part of an atom, made up of protons and neutrons.
2. A structure found in most plant and animal cells that contains the genetic material of the cell.
**numerator** The number above the line in a fraction, indicating the number that is divided.
**nutrients** Substances in food that are used by organisms for growth, maintenance, and reproduction.
**nutrition** The processes in which an organism takes in food and uses it for growth and maintenance.

# O

**obelisk** A tapering, usually ancient, stone pillar.
**observatory** A building from which astronomers study space.
**occlusion** Where a cold front catches up with a warm front.
**ohm** (Ω) The SI unit of electrical resistance.
**okta scale** A scale for measuring cloud cover. One okta equals one-eighth cloud cover.
**omnivore** An animal that eats both plants and animals.
**Oort cloud** An immense spherical cloud surrounding the solar system and extending approximately three light years (about 30 trillion kilometers) from the Sun.
**opaque** Does not let light through.
**optical fibers** Thin glass fibers along which light travels. They are used in communications.
**optics** The branch of science that studies vision and the behavior of light, including how light is affected by lenses, mirrors, and other optical devices.
**orbit** The path of one body, such as a planet or satellite, around another body, such as a star or planet.
**ore** A naturally occurring rock from which metals can be extracted.
**organ** A group of tissues, usually grouped together in a discrete structure, that has a special function, such as the brain or the heart.
**organelle** Specialized structure that forms part of a plant or animal cell.
**organic** 1. A compound containing carbon.
2. Food production without the use of chemical fertilizers.
**organic chemistry** The chemistry of carbon compounds, mostly related to living things.
**organism** A living thing consisting of one or more cells.
**oscillation** A regular movement back and forth.
**oscillator** An instrument that produces an alternating current of known frequency.
**oscilloscope** An instrument that shows electrical signals on a screen.
**osmosis** The movement of water through a semi-permeable membrane from a weak solution to a more concentrated one.
**ossify** Turn to bone.
**output** Information from a computer.
**ovulation** The release of an ovum from the ovary about midway through the menstual cycle.
**ovum** The egg cell.
**oxidation** When a substance gains oxygen or loses hydrogen, or an atom loses electrons in a chemical reaction.
**oxide** A compound formed between an element and oxygen.
**oxidizing agent** A substance that causes the oxidation of another substance.
**oxygen** A colorless, odorless, reactive gas that is the third most abundant element in the Universe, makes up 21% of the atmosphere, and is essential for most forms of life on Earth.
**ozone** An unstable, toxic form of oxygen with three atoms in its molecules. It is present in the upper atmosphere and absorbs ultraviolet radiation. At ground level it is a respiratory irritant.

# P

**P wave** Primary wave. A fast-moving earthquake wave that alternately stretches and squeezes rocks as it moves.
**pacemaker** An electronic device implanted in the chest that delivers short electrical impulses via electrodes to regulate the heartbeat.
**palynology** The study of both living and fossil pollen and spores.
**pancreas** A gland behind the stomach that secretes digestive enzymes and hormones that regulate glucose levels.
**pandemic cycle** The now discredited idea that worldwide epidemics of diseases occur in cycles.
**parabolic dish** A specially shaped dish that collects and concentrates waves, such as radio waves.
**parallax** The apparent movement of objects against each other, such as the movement of nearby trees against background hills, as the observer moves.
**parallel circuit** A circuit in which there are at least two independent paths in a circuit to get back to the source.
**parasite** An organism that lives on and feeds off another organism, called the host, often until it destroys the host.
**particle** A tiny speck of matter.
**particle accelerator** A giant machine in which subatomic particles are accelerated around a tunnel by electromagnets, and smashed together at very high speeds.
**particle physics** The branch of physics that deals with subatomic particles.
**parthenogenesis** Reproduction without mating.
**pasteurization** The heating of food to destroy disease-carrying bacteria.
**pathogen** A microbe (microorganism) that causes disease.
**payload** The equipment carried into space by a spacecraft, such as a satellite.
**penumbra** A partial shadow, especially round the shadow of the Moon or Earth in an eclipse.
**perigee** The point in the orbit of a moon or satellite at which it is nearest the object it is orbiting.
**periodic table** A table of all the elements arranged in order of their atomic numbers.
**pesticide** A substance used to destroy insects and other pests of crops and domestic animals.
**PET** Positron Emission Tomography. A form of medical scan that creates an image by detecting positrons sent out by radioactive substances in the blood.
**petrochemical** Any chemical made from petroleum or natural gas.
**petrology** A science that deals with the study of rocks.
**pH** A measure of acidity or alkalinity of a solution.
**phagocyte** A type of body cell that engulfs and absorbs bacteria and other small particles.
**pharmacology** The branch of medicine concerned with drugs and their uses.
**phases** 1. The various visual aspects of a moon or planet, depending on how it is illuminated.
2. The three states in which matter occurs: solid, liquid, and gas or vapor.
**phenotype** The observable characteristics of an organism.
**pheromones** Chemical substances released by animals to communicate with one another by smell.
**phloem** Tissue that carries food in a plant.
**photocell** An electronic device that generates electricity when light falls on it, such as in a solar-powered calculator.
**photochromicity** The ability of an object, such as a lens, to darken or change color when exposed to light, and to return to its original color when the light is removed.
**photoelectric effect** The emission of electrons from the surfaces of some substances when light hits them.
**photon** The particle that makes up light and other electromagnetic radiation.
**photosphere** The visible surface of the Sun, which gives out almost all of its light.
**photosynthesis** The process by which plants make food from water and carbon dioxide, using energy from the Sun.
**physics** The science of the properties and nature of matter and the interactions of energy and matter.
**physiology** The study of how organisms work.
**phytoplankton** Tiny organisms that are part of plankton.
**pi** ($\pi$) The ratio of the circumference of a circle to its diameter, approximately 22 divided by 7, or about 3.14159.
**piezoelectric effect** The production of electricity by applying stress to certain crystals, such as quartz.
**pigment** A substance that gives color to a material, but unlike a dye, does not dissolve in it.
**piston** A sliding disk or short cylinder that is pushed up and down inside a cylinder to provide the power in an engine.
**pitch** The property of a sound that makes it high or low.
**placebo** An inactive substance given to a patient to compare its effects with a real drug.
**plaintext** Simple, unformatted text in a form that can be read by many different computer systems.
**Planck constant** (h) The ratio of the energy in one photon of electromagnetic radiation to its frequency. It is a fundamental constant in quantum physics.
**planet** A large body that orbits a star.
**planetesimals** Any small body circling the Sun from which planets are thought to have formed.
**plankton** Unicellular and juvenile organisms that live near the surface of the seas and inland waters.
**plant** Any organism that contains chlorophyll.
**plaque** A deposit on teeth where bacteria thrive.
**plasma** 1. The liquid part of the blood.
2. A hot, electrically charged gas, in which the electrons are free from their atoms.
**plasma membrane** The skin around a cell which regulates what molecules enter and what leave.
**plasmid** A normally circular strand of DNA in bacteria or protozoa.
**platelet** An irregular-shaped disc in the blood that releases chemicals to coagulate the blood.
**plate tectonics** The study of the plates that make up Earth's crust.
**Pleiades** The name of an open cluster of stars in the constellation of Taurus, one of the closest to Earth. Also known as M45 or the Seven Sisters.
**polar cell** The circulation of the atmosphere in polar regions in which cold descending air blows out from the poles at surface level, then returns at high level.
**polar reversal** The reversal of the direction of Earth's magnetic field.
**polarized light** Light in which the vibrations occur in only one plane.
**pollination** The deposition of pollen on a flower so that it can be fertilized and set seed.
**pollution** Dirtying or poisoning of the air, land, or water by substances such as waste chemicals from factories.
**polygon** A plane geometric figure with at least three straight sides and angles.
**polyhedron** A solid geometric figure with at least six flat faces.
**polymer** A substance made from long, chain-shaped molecules consisting of many identical monomer units.
**population** The number in a particular region, typically of inhabitants.
**potential difference** The difference in energy between two places in an electric field or circuit.
**potential energy** 1. Energy stored for use at a later time.
2. The stored energy that a body has because of its position or state.
**positron** Counterpart of an electron, with a positive charge.
**power** The rate of change of energy.
**precipitate** Tiny particles of solid in a liquid, made by a chemical reaction.
**predator** An animal that preys on others.
**preservation** The process of keeping something in its original state, or free from harm, erosion, or decay.
**pressure** Continual physical force pushing against a surface, or the force per unit area.
**prey** An animal that is hunted or eaten by another animal.
**prime number** Any positive that can be divided by itself and one to produce a whole number as the result.
**prism** 1. A solid geometric form with sides that form a parallelogram.
2. A prism-shaped block of glass, especially one with triangular sides, that is used to split white into the colors of the spectrum.
**probability** The likelihood of an event happening, normally expressed as a number between 0 and 1.
**processor speed** The rate at which the processor in a computer deals with data.
**program** A series of coded instructions to operate a computer.
**prokaryotic cell** A cell with no nucleus.
**prominence** A mass of glowing gas reaching out from the surface of the Sun.
**prosthetic** An artificial body part.
**protein** A substance found in foods such as meat, fish, cheese, and beans, which the body needs for growth and repair.
**protein synthesis** The process in which cells make proteins under instruction from DNA and RNA.
**proton** A particle in the nucleus of an atom that has a positive electric charge.
**pulley** A grooved wheel around which a rope or chain runs, used to lift heavy weights.
**pulsar** A dense star that emits regular pulses of radiation, usually radio waves.

# Q

**qualitative analysis** Finding out what a substance is made of.
**quantitative analysis** Finding out how much of each ingredient is in a substance.
**quantum chromodynamics (QCD)** The quantum physics theory that deals with subatomic particles involved in the strong or "color" interaction in the nucleus.
**quantum computing** A computing system, as yet hypothetical, that makes use of quantum phenomena, such as quantum entanglement.
**quantum electrodynamics (QED)** The quantum physics theory that deals with the interactions between electrons, positrons, and photons.
**quantum physics** The branch of science that deals with subatomic particles and energy interactions in terms of minute discrete energy packets called quanta.
**quantum theory** The theory that light and other electromagnetic radiation is made up of a stream of photons, each carrying a certain amount of energy.
**quarantine** A time or place of isolation for people or animals that may carry an infectious disease.
**quark** One of a group of small particles that make up protons, neutrons, and other subatomic particles.
**quasar** The brilliant core of a young galaxy, probably a disk of hot gas around a massive black hole.

# R

**radar** Radio detection and ranging. A way of detecting objects by sending out radio waves and collecting the "echoes."

**radiation** 1. An electromagnetic wave.
2. A stream of particles from a source of radioactivity. *See also* electromagnetic spectrum.

**radio antenna** An aerial or dish that picks up radio signals.

**radioactive** Emitting ionizing particles and radiation as atomic nuclei break up naturally.

**radioactive dating** A method of estimating the age of an object by measuring how much the radioactive isotopes in it have decayed.

**radioactive decay** The process in which unstable atomic nuclei emit ionizing particles and radiation as they break up.

**radioactive tracers** Substances that contain a radioactive atom to allow easier detection and measurement.

**radioactivity** The disintegration of the nuclei in an atom, causing radiation to be given off.

**radiometric rock dating** The process of finding an absolute age for rocks by detecting the stage of radioactive decay of particular isotopes in them.

**radiosonde** A package of instruments, including a miniature radio transmitter, carried into Earth's upper atmosphere by a weather balloon to gather meteorological information.

**RAM** Random access memory. Computer memory chips where information can be stored and retrieved.

**rarefaction** 1. Areas along a longitudinal wave, such as a sound wave, where the pressure and density of the molecules is decreased.
2. The lowering of the density of a gas.

**ratio** The proportional relation between two numbers.

**reactants** The substances that take part in a chemical reaction.

**reaction** 1. A force that is the same in magnitude, but opposite in direction to another force. Every force has a reaction.
2. Any change that alters the chemical properties of a substance or that forms a new substance.

**reactivity** The ability of a substance to take part in a chemical reaction.

**real image** An image formed where light rays focus. It can be seen on a screen. *Compare to* virtual image

**recycling** Using waste material again, thus saving resources and energy.

**red giant** A star near the end of its life, which has swollen in size and cooled.

**red shift** The stretching out of light (moving it toward the red end of the spectrum) from a galaxy that is moving away from Earth.

**reducing agent** A substance that causes the reduction of another substance.

**reduction** When a substance gains hydrogen or loses oxygen, or an atom gains electrons in a chemical reaction.

**reflection** The bouncing back of light, heat, or sound from a surface.

**reflex** An automatic reaction to something.

**refraction** The bending of light rays as they enter a different medium at an angle, such as from air to water.

**refractive index** The ratio of the speed of light in one medium to the speed of light in a second medium.

**relative density** The weight of a volume of a material divided by the weight of an equal volume of water.

**relativity** The description of space and time, energy and matter according to the theories of Einstein, which depend on the constancy of the speed of light.

**repoussé** The ancient art of decorating metal by hammering on the back of the piece.

**reproduction** The process of creating offspring.

**resistance** A measure of how much an electrical component opposes the flow of an electric current.

**resistor** A component in an electric circuit that opposes the flow of electricity.

**resonance** When the vibrations of an object become large because it is being made to vibrate at its "natural" frequency.

**respiration** The process in which oxygen is taken in by living things and used to break down food. Carbon dioxide and energy are produced.

**resultant** The overall force that results from two or more forces acting on an object.

**retrovirus** An RNA virus, such as HIV, the virus that causes AIDS. It replicates by inserting a DNA copy of its genes into the host cell.

**reverberation** Where an echo reaches a listener before the original sound has finished. It makes a sound seem to last longer.

**rheostat** A resistor whose resistance can be changed.

**ria** A long, narrow sea inlet caused by the flooding of a river valley.

**ribosomes** Tiny spherical bodies in the cytoplasm of cells, where proteins are made.

**Richter scale** The logarithmic scale for registering the magnitude of earthquake vibrations named after Charles F Richter.

**rifting** The opening up of a crack and the formation of a valley as a result of the separation of tectonic plates.

**RNA** Ribonucleic acid, a molecule related to DNA but that forms simpler, single, long strands of nucleotide bases. It plays a part in mediating between DNA and the rest of the cell.

**robot** A machine able to carry out a complex series of instructions according to a program.

**ROM** Read only memory. Computer memory in which information is stored permanently, so that it can be retrieved but not altered.

**router** An electronic networking device for forwarding data between computers and phone lines, often wirelessly, using radio waves.

# S

**S wave** Secondary wave, an earthquake wave that travels only through solid rock as lateral or horizontal waves.

**salt** 1. A compound formed from the reaction of an acid with a base.
2. The common name for sodium chloride.

**sap** The liquid that flows through a plant, carrying food and water.

**saprophyte** An organism, such as a fungus or bacterium, that lives on dead or decaying matter.

**satellite** An object that orbits a planet. There are natural satellites, such as a moon, and artificial satellites, such as a craft used to reflect radio signals.

**satellite dish** A dish-shaped aerial that receives signals broadcast from satellites.

**saturation** The point at which no more solute can be dissolved in a solution.

**scalar quantity** A quantity that has only magnitude, such as mass and time.

**SDSS** Sloan Digital Sky Survey. A large scale astronomical project to map space, which covered more than a quarter of the sky and created three-dimensional maps containing more than 930,000 galaxies and over 120,000 quasars.

**seamount** An isolated undersea mountain on the deep ocean floor.

**search engine** A program for locating and retrieving data and documents from a computer network, especially the Internet.

**secretion** The release of specific substances from plant and animal cells.

**sedimentary rock** Rock formed when fragments of material settle on the floor of a sea or lake, for example. The layers are cemented together over time by a process called lithification.

**sedimentation** The geological process in which loose material is laid down by water, wind and moving ice.

**seismic wave** A wave that travels through the ground, such as from an earthquake or explosion.

**seismology** The study of earthquakes.

**seismometer** A device for recording and measuring earthquake waves.

**selective breeding** The process of choosing particular domestic animals for breeding to encourage the development of particular traits over the generations.

**semiconductor** A substance that has a resistance somewhere between that of a conductor and an insulator.

**semipermeable membrane** A membrane that lets small molecules through, but stops large molecules.

**sequestration** Short for carbon sequestration. Any process that uses the natural world to reduce the amount of carbon dioxide in the atmosphere, including fertilizing the ocean with iron to stimulate plankton growth.

**sessile** 1. Animals that cannot move around as adults, such as sea anemones.
2. Plants with no stalks, such as algae.

**sex cell** *See* gamete.

**sextant** A navigational instrument designed to measure the altitude of an object, such as the Sun above the horizon at noon.

**sexual reproduction** Reproduction that involves the combination of male and female gametes.

**SI (Système International) unit** A unit in the international system of measures based on the meter, kilogram, second, ampere, kelvin, candela, and mole.

**sial** The silica- and aluminium-rich upper layer of Earth's crust.

**sickle-cell anemia** A severe hereditary disease in which a mutated form of hemoglobin distorts red blood cells into a "sickle" shape.

**silica** A white or colorless compound of silicon that occurs naturally, such as quartz.

**sima** The silica- and magnesium-rich lower layer of Earth's crust.

**sine (sin)** The sine of an angle is the ratio of the side opposite to that angle in a right-angled triangle to the hypotenuse.

**sinter** 1. A hard siliceous or calcareous deposit around mineral springs.
2. A hard material created by the coagulation of coal powder when it is heated.

**skeleton** The frame of bone and cartilage in vertebrates that supports the body and protects its organs.

**slag** The waste matter separated from metals in smelting or from ores during processing.

**slide rule** A ruler with a central sliding bar designed to make quick calculations using logarithms.

**smelting** The process of extracting metal from ores, using heat to melt the pure metal out.

**SNP** Single nucleotide polymorphism. A small genetic change, or variation that can occur within a person's DNA sequence.

**soft water** Water that is free of dissolved calcium and magnesium salts.

**software** The programs used by a computer.

**solar constant** The amount of heat energy from the Sun received per unit area of Earth's surface.

**solar eclipse** An eclipse in which the Moon passes between Earth and the Sun so that the Sun, or part of it, cannot be seen from Earth.

**solar flare** A sudden burst of radiation from the Sun.

**solar panel** An object that collects energy from the Sun and uses it to heat water, for example, or to produce electricity.

**solar system** The Sun, the planets that orbit the Sun, their moons, and the other bodies in space whose movements are affected by the Sun's gravity.

**solder** An alloy (often made of tin and lead) used to join metal surfaces together.

**solenoid** A cylindrical core of wire that becomes a magnet when an electric current is passed through it.

**solstice** Each of the two times each year in midsummer and midwinter (around June 21 and December 21), when the Sun reaches its highest or lowest point in the sky at noon.

**solubility** The ability of a solute to be dissolved.

**solute** The substance that dissolves in a solvent to form a solution.

**solution** A liquid in which another substance is mixed evenly and invisibly.

**solvent** A substance that can dissolve other substances, or a liquid in which other substances are dissolved to form a solution.

**sonar** Sound navigation and ranging. A means of detecting objects and of navigating underwater by sending out sound waves.

**somatic nuclear transfer** A laboratory technique for creating a fertilized ovum using a "somatic" cell (ordinary body cell) to create a clone of the organism.

**sonic boom** A loud, explosive noise made by the shockwave from an object that is traveling faster than the speed of sound.

**space age** The era of space travel.

**space probe** An unmanned spacecraft sent from Earth to investigate the solar system.

**space station** A spacecraft, big enough for people to live and work on, which orbits Earth.

**space-time** The three dimensions of space combined with time in a single continuum.

**species** A group of organisms that look alike and can breed only with one another.

**spectroscope** An optical instrument that divides the light given off by an object into its spectrum.

**spectroscopy** The branch of science that is concerned with studying the radiation spectra emitted and absorbed by different substances.

**SPECT** Single photon emission computed tomography. A medical imaging technique that scans the patient with gamma rays to build up a complete 3D picture.

**spectrum** (plural: **spectra**) A particular distribution of wavelengths and frequencies, such as the electromagnetic spectrum.

**specular reflection** When light bounces off a surface at exactly the same angle as that at which it hits it.

**speed** The rate at which something is moving.

**sperm** One of the male reproductive cells released in semen during ejaculation, which must enter an ovum for fertilization to take place.

**spinal cord** A bundle of nerves running from the brain down through the spine.

**standard model** The principle theoretical framework for particle physics, combining the theories on how three of the four fundamental forces interact (electromagnetism and the strong and weak nuclear forces) with 12 basic particles (six quarks and six leptons—the electrons, muon and tauon, and three neutrinos).

**star** A celestial body that releases energy from the nuclear reactions in its core.

**starch** A polymer found in plants that is an important part of the human diet.

**static electricity** An electric charge held on an object, caused by the gain or loss of electrons.

**stem cell** Basic, undifferentiated cells that can develop into any other kind of cell.

**sterilization** The removal of bacteria from an object.

**stoma** (plural: **stomata**) A tiny opening in a plant's leaf or stem through which gases and water vapor pass.

**stratigraphy** The study of rock layers.

**stratopause** The boundary between the stratosphere and the mesosphere in Earth's atmosphere.

**stratosphere** The part of Earth's atmosphere between the troposphere and the mesosphere.

**strong force** The force that binds quarks, gluons, and other particles together in the nuclei of atoms to form protons, neutrons, and other particles.

**strong interaction** *See* strong force.

**subatomic particle** A particle smaller than an atom, such as a proton or a neutron.

**subduction boundary** A boundary between two tectonic plates where one plate is being subducted or pushed beneath another.

**sublimation** When a solid turns straight from a solid into a gas without becoming a liquid first.

**submersible** A craft that can travel underwater.

**substance** Any kind of matter.

**succession** The process of change from one ecosystem to another, such as from grassland to woodland.

**sugars** A group of soluble, sweet-tasting carbohydrates.

**sunspot** A cooler patch on the Sun's surface that appears darker than its surroundings.

**super neutrino beam** A very intense neutrino beam produced by a high power accelerator.

**superconductor** A substance that has no electrical resistance.

**supernova** (plural: **supernovae**) The explosion of a very large star at the end of its life.

**supersonic** Faster than the speed of sound.

**surface tension** An effect that makes a liquid seem as though it has an elastic "skin." It is caused by cohesion between the surface molecules.

**suspension** A mixture of tiny particles of solid matter or globules of liquid.

**switch** A device that turns an electric current on or off.

**synapse** A junction of two nerve cells, or between a nerve cell and a muscle fiber or a gland.

**synthesis** The combining of separate parts or different theories to make a whole.

**synthetic** An artificially created chemical.

**synthesizer** A musical instrument that creates sound electronically.

# T

**tangent** 1. A straight line that just touches the outside of a circle or curve at a single point.
2. The tangent of an angle (tan) is the ratio of the side opposite that angle in a right-angled triangle to the side adjacent to it.
**taxonomy** The classification of living things.
**tectonics** The theory that Earth's surface is broken into a number of mobile slabs of rock called plates.
**telemedicine** A developing field of medicine in which problems are dealt with, or medical information transferred, over large distances using telecommunications technology.
**telophase** The final stage of cell division, in which two daughter cells are formed.
**temperate** Describes a climate that has mild summers and cool winters.
**temperature** A measure of how hot or cold something is.
**terawatt (TW)** A unit of power that is equal to one million watts or one thousand gigawatts.
**terminal** A connecting point on an electrical component.
**theorem** A concept that has been proved to be true or can be proved to be true.
**thermal** A current of rising hot air in the atmosphere.
**thermistor** A resistor whose resistance changes with temperature.
**thermoplastic** Plastics that can be continually reshaped whenever they are sufficiently heated.
**thermoset** Plastics that can only be shaped once by heating.
**thermosphere** The part of Earth's atmosphere between the mesosphere and the exosphere.
**three dimensions (3D)** Length, breadth, and depth.
**timbre** The quality of a musical sound.
**tissue** A group of similar cells that carry out the same function, such as muscle tissue, which can contract.
**tissue typing** The identification of antigens in tissue to minimize the possibility of rejection due to antigenic differences between a donor and recipient in organ transplantation.
**titration** A method of finding the concentration of a solution.
**tomography** Any technique for producing cross-sectional image ("slices") of parts of the body; for example, CT (computerized tomography) scanning.
**torque** A force that causes rotation. Known as "moment" by physicists.
**totipotency** The capability of a cell to develop into any kind of cell.
**toxicology** The study of toxic substances.
**trace elements** 1. Any element that occurs in very small concentrations.
2. Substances such as minerals, that are needed in only minute amounts by living things.
**trade wind** A wind that blows from the southeast or northeast all the year round in equatorial regions.
**transcription** The copying of sequences of genes from DNA to RNA.
**transform boundary** A boundary between tectonic plates where the plates are sliding horizontally past each other.
**transformer** A device that increases or decreases voltage.
**transfusion** The transfer of blood from a donor to a recipient.
**transistor** A semiconductor device that acts as an electronic switch, amplifier, or rectifier.
**translocation** The movement a segment of a chromosome from one location to another, either on the same chromosome or to another chromosome.
**translucent** Allows some light through, but is not transparent.
**transmission** Movement through from one place to another.
**transmutation** 1. The evolutionary change of one species into another.
2. Conversion of one element into another via a nuclear reaction.
**transparent** Allowing nearly all light through; is "see-through."
**transpiration** The giving off of water vapor through the stomata of a plant's leaves or stem.
**transplant** The taking and implanting of tissue or organs from one part of the body to another, or from a donor to a recipient.

**transverse wave** A wave in which the particles of the medium vibrate at right angles to the direction in which the wave is traveling.
**trigonometry** The branch of mathematics concerned with triangles.
**trophic level** The level at which an animal is in a food chain.
**tropical** Describes a climate that is hot, with periods of heavy rainfall.
**tropopause** The boundary in Earth's atmosphere between the troposphere and the stratosphere.
**troposphere** The lowest layer of Earth's atmosphere, immediately above the ground.
**tsunami** Giant sea waves started by earthquakes or other major disturbances.
**tunneling** Passing through an apparent barrier, as in quantum tunneling, in which a particle appears to jump from one side of a barrier to another.
**turbine** A machine that is made to rotate in order to drive a generator.
**typhoon** The name that is given to a hurricane in the Pacific Ocean.

# U

**UHF** Ultra-high-frequency radio waves.
**ultrasound** Sound with a frequency above that which the human ear can detect.
**ultraviolet (UV)** A type of electromagnetic radiation with a wavelength shorter than visible light.
**umbra** The dark central part of a shadow, where no light at all falls.
**uncertainty principle** The idea, formulated by Werner Heisenberg, that it is impossible to measure both the position and momentum of quantum objects, because the observation of one changes the other.
**Universe** All of space and everything it contains.
**uplift** The upward movement of a geological block, typically due to earthquakes or tectonic activity.
**Upper Paleolithic** The last period of the Old Stone Age, approximately 40,000 to 10,000 years ago.
**upthrust** A force pushing upward, opposite to the pull of gravity, such as buoyancy in water.
**urine** A yellowish waste fluid stored in the bladder of animals and discharged through the urethra.

# V

**vaccination** *See* inoculation.
**vacuole** A small fluid-filled sac in the cytoplasm of a cell.
**vacuum** A space in which there is no matter.
**vacuum tube** A gas-tight glass tube from which all gas has been sucked out to create a vacuum.
**valency** The number of chemical bonds that an atom can make with another atom.
**valve** 1. A device, flap, or membrane for controlling the flow of gas or fluid through an opening.
2. One of the two sides of the shell of a bivalve mollusk.
3. A vacuum or gas-filled device for controlling the flow of an electric current.
**variolation** The production of smallpox pustules.
**vector** An organism that transmits a disease.
**vector quantity** A quantity that has both magnitude and direction, such as a force.
**veins** Vessels that carry blood from all parts of the body back to the heart.
**velocity** Speed in a particular direction.
**ventricle** One of the two large cavities of the heart.
**Vernier scale** A small movable, graduated scale added to a larger scale for increased accuracy in precision instruments such as barometers; named after Pierre Vernier.
**vertebrate** An animal with a spine.
**vessel** A duct carrying blood or another fluid through plants or animals.
**VHF** Very-high-frequency radio waves.
**vibration** A quick back-and-forth movement. For example, an earthquake makes Earth vibrate; sound makes the air vibrate.
**virtual image** An image formed where light rays appear to be focused, as in a reflection in a mirror.
**virus** A microscopic particle that invades cells to reproduce, and may cause disease.
**viscosity** A measure of how easily a substance flows.

**viscous** Thick, sticky, and slow-flowing.
**visible light** Light in the range of wavelengths that can be seen by the human eye.
**vitalism** The theory that life depends on some other force than purely chemical and physical.
**vitamin** An organic compound found in foods that is essential for good health.
**volt** The unit of potential difference.
**voltaic cell** A device that produces electricity by chemical change.
**voltmeter** or **voltammeter** A device used to measure potential difference.
**volume** 1. The amount of space something takes up.
2. The loudness of a sound.
**vulcanization** The hardening of rubber by heating it with sulphur.

# W

**waning** Becoming smaller, as in the illuminated portion of the Moon visible from Earth.
**warm front** The boundary between masses of warm and cold air in which the warm air is advancing and gradually riding up over the cold air.
**waterspout** A column of water sucked up by a tornado moving over the sea.
**watt (W)** A unit of power (1 watt = 1 joule per second).
**wave** One of a series of repeated, often regular-moving variations in intensity or concentration, such as the compression of air created by sound as it moves.
**wavelength** The distance between successive wave crests.
**waxing** Growing, as in the illuminated portion of the Moon visible from Earth.
**weak force** The force in atomic nuclei involved in beta decay and affecting one of the four fundamental forces. Also known as the weak nuclear force or weak interaction.
**weak interaction** *See* weak force.
**weathering** The slow natural breakdown of rocks exposed to the weather.
**weight** The force with which a mass is pulled toward the center of Earth.
**westerlies** Prevailing winds blowing from the west.
**whirlwind** A column of air spinning rapidly in a funnel shape over land or water.
**white dwarf** The small, dense remains of a dead star.
**WIMPs** Weak Interacting Massive Particles. Hypothetical large particles that interact only via gravity, and may be prime candidates for dark matter.
**WMAP** Wilkinson Microwave Anisotropy Probe. A satellite space probe for measuring differences in the temperature of the Big Bang's remnant radiant heat across the full sky.
**WMO** World Meteorological Organization.
**work** The energy transferred when a force moves an object or changes its shape.
**World Wide Web (www)** A vast network on the Internet for gathering and exchanging data and documents through hypertext links.

# X

**X-ray** A type of electromagnetic radiation with a wavelength shorter than ultraviolet radiation.
**x-ray opaque dyes** Dyes that block X-rays. They are used in medical imaging to show structures that do not show up clearly in conventional plain X-rays.
**xylem** The tissue that carries water through a plant.

# Z

**zeolite** A natural or synthetic compound with an open structure that can act as a catalyst or a filter for individual molecules.
**zeugen** A ridge of hard rock formed by erosion.
**Zhou** The name of the dynasty of Chinese emperors who ruled from 11th-3rd centuries BCE, during which time philosophies such as Confucianism and Taoism were created, and Chinese writing originated.
**zodiac** The originally twelve (now thirteen) constellations that lie along the ecliptic, through which the Sun, Moon, and planets move.

**zooplankton** The tiny, often microscopic, animals in the sea that form part of plankton.
**zygote** A cell resulting from the fusion of two gametes, such as sperm and ovum at conception.

# Index

# Acknowledgments

**Dorling Kindersley** would like to thank the following people for their help in the preparation of this book: Polly Boyd and Nicky Twyman for proof reading; Jane Parker for the index; Juliet Jopson for additional consultancy on biology; Chris Bryan for additional consultancy on physics. Additional editorial help from Simon Tuite, Victoria Wiggins, Megan Hill, Miezan van Zyl, Nathan Joyce, David Summers, Peter Frances, Elizabeth Munsey, Cressida Malins, Ruth O'Rourke, Conor Kilgallon, Marcus Hardy, and Kajal Mistry. Additional design assistance from Maxine Pedliham, Francis Wong, Riccie Janus, Steve Woosnam-Savage, Dean Morris, and Anna Hall. DK Cartography: Paul Eames. DK images: Romaine Werblow, Rose Horridge, and Emma Shepherd.

**PICTURE CREDITS**

The publisher would like to thank the following for their kind permission to reproduce their photographs:

(Key: a-above; b-below/bottom; c-center; f-far; l-left; r-right; t-top)

Corner Tabs

**16-63 Alamy Images:** capt.digby (Ch 1). **68-123 Corbis:** Lawrence Manning (Ch 2). **130-279 Getty Images:** The Image Bank/Malcolm Piers (Ch 3). **286-373 Corbis:** Science Faction/William James Warren (Ch 4). **378-419 Science Photo Library:** Tony Craddock (Ch 5). **422-483 Corbis:** Zefa/Josh Westrich (Reference). **484-512 Corbis:** Science Faction/David Scharf (Endmatter).

**1 DK Images:** Courtesy of The Science Museum, London/Dave King. **2-3 Corbis:** Bettmann. **4-5 Science Photo Library:** Alexandros Alexakis. **6 Alamy Images:** nagelestock.com (tr). **Corbis:** Bettmann (tl, bl). **Getty Images:** Comstock (cla). **7 Alamy Images:** North Wind Picture Archives (br). **DK Images:** Courtesy of the Natural History Museum, London/Colin Keates (tl). **Getty Images:** LOOK/Konrad Wothe (bl); Time Life Pictures/Mansell (tr). **National Maritime Museum, Greenwich, London:** (ca). **8 Corbis:** Bettmann (bl); Hulton-Deutsch Collection (tl); Science Faction/Norbert Wu (br). **Getty Images:** National Geographic/James P. Blair (tr). **Science Photo Library:** Laguna Design (ca). **9 Chandra X-Ray Observatory:** X-Ray: NASA/CXC/CfA/R.Tuellmann et al.; Optical: NASA/AURA/STScl (br). **Corbis:** EPA/Jose Jacome (tr); Reuters/Stringer (bl). **Science Photo Library:** CNRI (tl); Kevin Curtis (ca); NASA/R.B. Husar (cra). **10-11 Science Photo Library:** CERN. **12 Getty Images:** Comstock. **14 akg-images:** James Morris (Astronomical Ceiling). **Alamy Images:** INTERFOTO Pressebildagentur (Herodotus Map); nagelestock.com (Callanish). **The Bridgeman Art Library:** Ashmolean Museum, University of Oxford (Abacus). **Corbis:** The Art Archive (Pythagoras); Christie's Images (Zhou Bronze); Dean Conger (Ziggurat); Reuters/Sergio Moraes (Eclipse); Sakamoto Photo Research Laboratory (Jomon Bowl); Werner Forman (Standard of Ur). **DK Images:** Courtesy of the Science Museum, London/Dave King (Wheel). **Getty Images:** The Bridgeman Art Library (Faience). **Wellcome Library, London:** (Acupuncture). **14-15 Corbis:** Edifice/Philippa Lewis (Background). **15 Alamy Images:** North Wind Picture Archives (Ptolemy Map); The Print Collector (Anatomy of the Eye, Archimedes). **Corbis:** Stefano Bianchetti (Fibonacci). **Getty Images:** Altrendo Travel (Windmills); The Bridgeman Art Library/Universitatsbibliothek, Gottingen (Gutenberg Bible). **NASA:** Marshall Space Flight Centre (Chinese Rocket). **Photo Scala, Florence:** Ministero Beni e Att. Culturali (Farnese Atlas). **Science & Society Picture Library:** (Chariot). **TopFoto.co.uk:** The British Library / HIP (Star Chart). **16 Corbis:** Dean Conger (bc); Peter Johnson (tc). **Getty Images:** Jan Rendek (c). **16-17 Corbis:** Ashley Cooper (c). **17 Alamy Images:** INTERFOTO Pressebildagentur (crb). **Corbis:** Burstein Collection (bc); Sakamoto Photo Research Laboratory (tl). **Getty Images:** The Bridgeman Art Library (tr). **18 DK Images:** Courtesy of The American Museum of Natural History/Lynton Gardiner (tl). **18-19 Corbis:** Christie's Images (cb). **19 akg-images:** Iraq Museum, Baghdad (tr). **Alamy Images:** Robin Weaver (br). **DK Images:** The British Museum, London/Nick Nicholls (tl). **Getty Images:** Gallo Images/Graeme Williams (c). **20 DK Images:** Courtesy of the Science Museum, London/Dave King (tl, tc, tr). **21 Corbis:** Werner Forman (b). **FOTOE:** (c). **22 Alamy Images:** Mary Evans Picture Library (bc). **Ancient Art & Architecture Collection:** Ronald Sheridan (tr). **Getty Images:** Hulton Archive (cl). **23 Mary Evans Picture Library:** AISA Media (tr). **TopFoto.co.uk:** The Granger Collection (l). **www.heikenwaelder.at:** (crb). **24 DK Images:** CONACULTA-INAH-MEX. Authorized reproduction by the Instituto Nacional de Antropologia e Historia/Michel Zabe (cl); The Science Museum, London/Adrian Whicher (br). **Wellcome Library, London:** (tr, bc). **25 Corbis:** Bettmann (tl); Historical Picture Archive (crb). **DK Images:** The Science Museum, London/David Exton (bl). **Wellcome Library, London:** (cra). **26 akg-images:** James Morris (cl). **Alamy Images:** nagelestock.com (tr). **Corbis:** Staffan Widstrand (crb). **Science Photo Library:** John Sanford (b). **27 Science Photo Library:** Gordon Garradd. **28 Alamy Images:** North Wind Picture Archives (tc). **Photo Scala, Florence:** Ministero Beni e Att. Culturali (bl). **Science Photo Library:** Royal Astronomical Society (tr). **TopFoto.co.uk:** The British Library / HIP (crb). **29 Alamy Images:** Wolfgang Kaehler (br). **Corbis:** K.M. Westermann (tr). **DK Images:** National Maritime Museum, London (c). **30 The Art Archive:** (br). **Corbis:** The Art Archive (ca). **Muséum des Sciences Naturelles, Brussels:** (clb). **30-31 Alamy Images:** The Print Collector (c). **31 iStockphoto.com:** (cra). **32 Corbis:** The Art Archive (r). **Science Photo Library:** Sheila Terry (bl). **33 akg-images:** (cl). **Science Photo Library:** Royal Astronomical Society (bc). **34 Corbis:** Free Agents Limited (cla). **Museum of the History of Science, University of Oxford:** Keiko Ikeuchi (tc). **Image courtesy History of Science Collections, University of Oklahoma Libraries; copyright the Board of Regents of the University of Oklahoma:** (bl). **35 Corbis:** Bettmann (tr). *Euclud's Elements* by Adelard of Bath, c.1309-1316 (br). **36-37 Corbis:** The Gallery Collection (b/Background). **37 Alamy Images:** North Wind Picture Archives (tl). **38 Alamy Images:** The London Art Archive (ca, bl). **Corbis:** Reuters/Sergio Moraes (tr). **38-39 Alamy Images:** INTERFOTO Pressebildagentur (b). **39 Alamy Images:** North Wind Picture Archives (tl). **40 Corbis:** Derek Croucher (cl). **DK Images:** The British Museum, London/Peter Hayman (br); Courtesy of the Charlestown Shipwreck and Heritage Centre, Cornwall/Alex Wilson (cb/Scissors). **Getty Images:** Time Life Pictures/Mansell (bl). **41 Alamy Images:** The Print Collector (t). **42 Getty Images:** Johner Images (br). **Rex Features:** Dragon News (cla). **Science & Society Picture Library:** (bl). **43 Getty Images:** The Bridgeman Art Library/Private Collection (tl, cra). **44-45 Alamy Images:** The Print Collector. **46 Alamy Images:** Martin Dalton (br). **Corbis:** Roman Soumar (cla). **47 Corbis:** Edifice/Philippa Lewis (b). **48 DK Images:** The British Museum, London/Peter Hayman (bl). **49 Getty Images:** Riser/Jeffrey Coolidge (l). **Science Photo Library:** Professor Peter Goddard (br). **50 Corbis:** Reuters/Daniel Aguilar (tr). **Getty Images:** AFP/Sanogo Issouf (clb); Altrendo Travel (b). **51 Corbis:** Dean Conger (br); Eye Ubiquitous/James Davis (crb). **DK Images:** Courtesy of the National Maritime Museum, London/James Stevenson and Tina Chambers (tl); Courtesy of The Science Museum, London (ftl). **Science & Society Picture Library:** (tc). **52 Corbis:** Bettmann (cr); Gianni Dagli Orti (cla). **Science Photo Library:** Sheila Terry (b). **53 Alamy Images:** Interfoto (tr). **The Bridgeman Art Library:** Derby Museum and Art Gallery, UK (bc). **Corbis:** Bettmann (c). **54 Alamy Images:** Dennis Cox (bl). **Science Photo Library:** (r). **55 China Tourism Photo Library:** (br). **Corbis:** EPA/Thomas Rensinghoff (tl). **DK Images:** The Science Museum, London/John Lepine (tc, c). **56 Corbis:** Christel Gerstenberg (bl). **NASA:** Marshall Space Flight Centre (c). **57 akg-images:** Bibliothèque Nationale (b). **Corbis:** (tc). Press Association Images: NULL/AP (cra). **58 Corbis:** (br); Gianni Dagli Orti (clb). **DK Images:** © British Library Board. All Rights Reserved (Picture number 1022251.611) (cla); Courtesy of the London College of Printing (c). **59 Corbis:** Bettmann (l). **Getty Images:** The Bridgeman Art Library/Universitatsbibliothek, Gottingen (c). **60 Alamy Images:** The Print Collector (br). **Getty Images:** Photographer's Choice/Sylvain Grandadam (bl). **Image courtesy History of Science Collections, University of Oklahoma Libraries; copyright the Board of Regents of the University of Oklahoma:** (cl). **Science Photo Library:** Sheila Terry (tr). **61 Alamy Images:** The Print Collector (cl); Eric Robison (br). **Corbis:** Angelo Hornak (cr). **Getty Images:** Robert Harding World Imagery/Guy Edwardes (tl). **62 Alamy Images:** The London Art Archive (br). **The Art Archive:** Bodleian Library, Oxford (*Pococke 375 folio 3v-4r*) (bl). **62-63 Wellcome Library, London:** (t). **63 The Bridgeman Art Library:** Giraudon/Musée Conde, Chantilly (cr). **Corbis:** Stefano Bianchetti (tr). **Getty Images:** Time Life Pictures/Mansell (br). **66 Alamy Images:** INTERFOTO Pressebildagentur (Paracelsus); The London Art Archive (Canon of Medicine). **Corbis:** Stefano Bianchetti (Copernicus World). **DK Images:** Courtesy of The Science Museum, London/Dave King (Microscope). **Image courtesy History of Science Collections, University of Oklahoma Libraries; copyright the Board of Regents of the University of Oklahoma:** (Anatomical Model). **Science Photo Library:** Jean-Loup Charmet (Sextant). **66-67 Wellcome Library, London:** (Background). **67 The Bridgeman Art Library:** The Science Museum, London (Napier Bones). **Corbis:** Bettmann (Gilbert, Newton, Galileo Telescope). **DK Images:** Courtesy of The Science Museum, London/Dave King (Newton Telescope). **Library Of Congress, Washington, D.C.:** (Spring). **Image courtesy History of Science Collections, University of Oklahoma Libraries; copyright the Board of Regents of the University of Oklahoma:** (Air Pump, Micrographia). **Science & Society Picture Library:** (Galileo Moon). **Science Photo Library:** Sheila Terry (Torricelli). **68 The Art Archive:** Dagli Orti/Musée Ampre Poleymieux (bl). *Novum Organum* by Francis Bacon, 1650 (tc). **69 Alamy Images:** INTERFOTO (t). **Getty Images:** Stock Montage (b). **70 Science & Society Picture Library:** (cb). **Wellcome Library, London:** (t). **71 Alamy Images:** INTERFOTO Pressebildagentur (br). **Corbis:** Bettmann (cr, bl). **DK Images:** The Science Museum, London/David Exton (cl, tc/Right). **Wellcome Library, London:** (tc/Left). **72 Alamy Images:** Mary Evans Picture Library (bl). **The Bridgeman Art Library:** Archives Charmet/Bibliothèque des Arts Decoratifs, Paris (bc). **DK Images:** Courtesy of the Museum of Natural History of the University of Florence, Zoology section 'La Specola'/Liberto Perugi (r). **73 The Art Archive:** Marc Charmet/Musée des Beaux Arts, Orléans (br). **Corbis:** Lawrence Manning (cra). **Image courtesy History of Science Collections, University of Oklahoma Libraries; copyright the Board of Regents of the University of Oklahoma:** (tc). **74-75 Corbis:** Stefano Bianchetti. **76 Science Photo Library:** Jean-Loup Charmet (bl). **TopFoto.co.uk:** National Pictures (c). **76-77 Corbis:** Stapleton Collection/ Philip Spruyt (c). **77 Corbis:** Bettmann (tr). **TopFoto.co.uk:** The Granger Collection (br). **78 Corbis:** Hulton-Deutsch Collection (c). **Science Photo Library:** Eckhard Slawik (tc). **78-79 Corbis:** Roger Ressmeyer (c). **79 Corbis:** Roger Ressmeyer (cr). **80 DK Images:** Courtesy of The Science Museum, London/Clive Streeter (bl). **81 Alamy Images:** Phil Degginger (bl). **Corbis:** Bettmann (tr). **DK Images:** National Maritime Museum, London/James Stevenson (c). **82 Alamy Images:** The Print Collector (ca). **Getty Images:** The Bridgeman Art Library (r). **Adam Hart-Davis:** (bl). **83 Corbis:** Stefano Bianchetti (tl). **DK Images:** Courtesy of The Science Museum, London/Dave King (cr). **84 Science & Society Picture Library:** (cl). **85 Corbis:** Roger Ressmeyer (crb). **DK Images:** Courtesy of The Science Museum, London/Dave King (l). **Adam Hart-Davis:** (tr). **NASA:** ESA, J. Hester and A. Loll (Arizona State University, ca). **86 Photo Scala, Florence:** (b). **87 Corbis:** Schlegelmilch (bc). **Getty Images:** Topical Press Agency/Monty Fresco Jnr (tl). **88 The Bridgeman Art Library:** Ashmolean Museum, University of Oxford (cl). **DK Images:** Courtesy of The Science Museum, London/Dave King (bc). **Image courtesy History of Science Collections, University of Oklahoma Libraries; copyright the Board of Regents of the University of Oklahoma:** (bl). **89 The Bridgeman Art Library:** The Science Museum, London (bl). **Corbis:** Kevin Schafer (tr). **Getty Images:** The Bridgeman Art Library/State Central Artilllery Museum, St Petersburg (br). **90 The Bridgeman Art Library:** The Trustees of the Weston Park Foundation (bl). **Corbis:** Bettmann (cla, tc). **Science Photo Library:** (bc). **91 Science Photo Library:** Steve Gschmeissner (crb). **92 Corbis:** Historical Picture Archive (r). **Image courtesy History of Science Collections, University of Oklahoma Libraries; copyright the Board of Regents of the University of Oklahoma:** (clb, bc). **93 Alamy Images:** Eric Nathan (crb). **DK Images:** Courtesy of The Science Museum, London/Dave King (tc). **Getty Images:** The Bridgeman Art Library/National Portrait Gallery, London (cra). **Library Of Congress, Washington, D.C.:** (bc). **94 DK Images:** Courtesy of The Science Museum, London/Dave King (cra, bl). **Image courtesy History of Science Collections, University of Oklahoma Libraries; copyright the Board of Regents of the University of Oklahoma:** (br). **Wellcome Library, London:** (cla). **95 Getty Images:** National Geographic/Paul Zahl (clb). **Photolibrary:** Phototake Science/Dr Gary D. Gaugler (cb). **Science Photo Library:** Steve Gschmeissner (bc); National Cancer Institute (tr); Omikron (br). **96 Corbis:** Bettmann (bl). **Science Photo Library:** Sheila Terry (br). **96-97 Getty Images:** Hulton Archive (t). **97 DK Images:** Courtesy of The Science Museum, London/Dave King (br). **98 Getty Images:** Hulton Archive (bc); Stock Montage (r). **iStockphoto.com:** Robin Brisco (clb). **99 Alamy Images:** Mary Evans Picture Library (tl). **Image courtesy History of Science Collections, University of Oklahoma Libraries; copyright the Board of Regents of the University of Oklahoma:** (bc, br). **100-101 Corbis:** Gianni Dagli Orti (c). **101 Corbis:** Bettmann (bc). **Getty Images:** Time Life Pictures/Mansell (tr). **102 The Bridgeman Art Library:** Private Collection (bl). **103 Getty Images:** Stock Montage (cr). **104 Alamy Images:** The Print Collector (tl). **DK Images:** Courtesy of the National Maritime Museum, London/Tina Chambers (cl). **Getty Images:** Popperfoto (ca). **105 NASA:** JPL (cra). **107 NASA:** JPL. **108-109 NASA:** Hubble Heritage Team (STScl/AURA) , R.G. French (Wellesley College) , J. Cuzzi (NASA/Ames) , L. Dones (SwRI) , and J. Lissauer (NASA/Ames) (t). **109 DK Images:** NASA (cl). **NASA:** JPL (tc). **110 Alamy Images:** Lebrecht Music and Arts Photo Library (r); The Print Collector (bl). **111 Alamy Images:** Neil McAllister (br). **Corbis:** Bettmann (tl); Walter Geiersperger (bc). **DK Images:** Courtesy of The Science Museum, London/Dave King (cla). **Getty Images:** Hulton Archive (tc). **Image courtesy History of Science Collections, University of Oklahoma Libraries; copyright the Board of Regents of the University of Oklahoma:** (c). **112 Corbis:** Frank Lukasseck (b). **113 Getty Images:** AFP/DSK (br); Photographer's Choice/Joe McBride (bl). **114 Corbis:** amanaimages/Kawai Kazuo (cl). **Image courtesy History of Science Collections, University of Oklahoma Libraries; copyright the Board of Regents of the University of Oklahoma:** (cra). **115 Getty Images:** LOOK/Konrad Wothe. **116 DK Images:** The British Museum, London/Nick Nicholls (cla). **Science Photo Library:** (bl). **117 Louise Thomas:** (tr, cra). **118 Getty Images:** Derke/ O'Hara (cl). **119 Corbis:** Bettmann (cla); Charles O'Rear (bc); Roger Ressmeyer (br). **Science Photo Library:** Detlev Van Ravenswaay (cb); Walter Pacholka, Astropics (tr). **120 The Art Archive:** Dagi Orti/Pharaonic Village, Cairo (cl). **The Bridgeman Art Library:** The Science Museum, London (c). **DK Images:** Courtesy of the National Maritime Museum, London/James Stevenson (tl). **Photo Scala, Florence:** (bl). **121 Alamy Images:** studiomode (cr). **Corbis:** Loop Images/Nadia Isakova (bc). **Science Photo Library:** National Physical Laboratory/Andrew Brookes (tr); Sheila Terry (c). **122 Corbis:** Visuals Unlimited (Kingdom/Bacteria, Kingdom/Fungi). **Getty Images:** Stone/Guy Edwardes (Family/Dogs); Visuals Unlimited/Carolina Biological (Kingdom/Protoctists). **123 DK Images:** Graham High at Centaur Studios - Model-maker/Dave King (bl/Dinosaur); Courtesy of the Linnean Society of London (cra). **Science Photo Library:** BSIP, Ermakoff (crb). **Wellcome Library, London:** (tc). **124 National Maritime Museum, Greenwich, London. 126 Alamy Images:** Lebrecht Music and Arts Photo Library (Newton). **The Art Archive:** (Newcomen). **Corbis:** Bettmann (Lind); Hulton-Deutsch Collection (Solar Eclipse). **DK Images:** Courtesy of The Science Museum, London/Clive Streeter (Leyden Jar). **Getty Images:** Derke/ O'Hara (Comet). **National Maritime Museum, Greenwich, London:** (Chronometer). **Wellcome Library, London:** (Hales). **126-127 Alamy Images:** North Wind Picture Archives (Background). **127 Corbis:** Gianni Dagli Orti (Balloon). **DK Images:** National Maritime Museum, London (Sextant); Courtesy of The Science Museum, London/Dave King (Watt Engine). **Getty Images:** James Sharples (Priestley). **National**

**Maritime Museum, Greenwich, London:** (H4). **Science Photo Library:** John A. Ey III (Lightning). **Wellcome Library, London:** (Linnaea Borealis, Jenner). **128 Corbis:** Bettmann (Gay-Lussac). **DK Images:** Courtesy of The Science Museum, London (Difference Engine); Courtesy of The Science Museum, London/Clive Streeter (Faraday Motor); The Science Museum, London (Dalton Atoms). **Science & Society Picture Library:** (Telegraph Machine). **Science Photo Library:** (Dalton Symbols). **128-129 Getty Images:** The Bridgeman Art Library (Background). **129 Corbis:** Louie Psihoyos (Archaeopteryx). **DK Images:** Courtesy of The Science Museum, London/Clive Streeter (Induction Motor); Courtesy of The Science Museum, London/Dave King (Edison Photograph); The Science Museum, London (Periodic Table). **Photolibrary:** Phototake Science/Dr Gary D. Gaugler (Anthrax Bacteria). **130-131 The Art Archive. 132 Alamy Images:** North Wind Picture Archives (cra). **Corbis:** Bettmann (br). **DK Images:** Courtesy of The Science Museum, London/Clive Streeter (tl, clb). **Science & Society Picture Library:** (tr). **133 Corbis:** Bettmann (tc). **DK Images:** Courtesy of The Science Museum, London/Dave King (b). **Getty Images:** The Bridgeman Art Library (tl). **134 National Maritime Museum, Greenwich, London. 136 The Trustees of the British Museum:** (tl). **National Maritime Museum, Greenwich, London:** (ca, bl). **137 DK Images:** National Maritime Museum, London (cl). **National Maritime Museum, Greenwich, London:** (bc). **Science Photo Library:** David Parker (crb). **138 Corbis:** Bettmann (cra); Michael Nicholson (cla). **DK Images:** The Science Museum, London (ca). **Science Photo Library:** (br); Sheila Terry (bc). **139 Corbis:** Bettmann (crb). **Science Photo Library:** Charles D. Winters (tr). **140 The Bridgeman Art Library:** The Royal Institution, London, UK (bl). **Corbis:** Terry W. Eggers (bc). **140-141 Corbis:** TongRo/ Beateworks (c). **141 DK Images:** Rough Guides/Eddie Gerald (tc). **Science Photo Library:** Richard Folwell (cl); Pasieka (br); Charles D. Winters (bc). **142 Alamy Images:** Peter Jordan (tl). **Getty Images:** Greg Pease (cl). **Science Photo Library:** Alex Bartel (cb). **142-143 Alamy Images:** Martin Shields (bc). **143 Corbis:** Moodboard (cl). **Science Photo Library:** Paul Rapson (tr). **144 Alamy Images:** Gabe Palmer (br). **Corbis:** Bettmann (cla); Richard Cummins (bc). **144-145 Still Pictures:** UNEP/S. Compoint (tc). **145 Alamy Images:** G.P. Bowater (cr). **The Art Archive:** Eileen Tweedy (tr). **Corbis:** Gilles Sabrié (br). **146 Getty Images:** James Sharples (bc). **TopFoto.co.uk:** The Granger Collection (cl). **147 The Bridgeman Art Library:** Roger-Viollet, Paris (tc). **Photo Scala, Florence:** HIP (tr). **148 Edinburgh University Library:** (crb). **Science Photo Library:** Sheila Terry (l). **149 Corbis:** Bettmann (tc). **Edinburgh University Library:** (cr, br). **Mary Evans Picture Library:** (bl). **150 Alamy Images:** Trip (cla). **Corbis:** Michael Nicholson (bl). **151 Construction Photography:** Andrew Holt (bc). **Corbis:** AgStock Images (cl). **152 The Bridgeman Art Library:** Archives Charmet/Bibliothèque des Arts Decoratifs, Paris (bl). **153 Corbis:** Hulton-Deutsch Collection (cra). **Getty Images:** ScienceFoto/G. Wanner (br). **Science Photo Library:** Philippe Psaila (crb). **154 Wellcome Library, London:** (bl). **155 Wellcome Library, London:** (br). **156-157 Wellcome Library, London. 158 Corbis:** Bettmann (bl). **DK Images:** Courtesy of The Science Museum, London/Clive Streeter (cb). **159 Science Photo Library:** John A. Ey III. **160 The Bridgeman Art Library:** Collection of the New York Historical Society (bl). **Getty Images:** The Bridgeman Art Library/Pennsylvania Academy of the Fine Arts, Philadelphia (r). **161 Corbis:** Bettmann (tr, crb); Visions of America/Joseph Sohm (bl). **DK Images:** Courtesy of The Science Museum, London/Dave King (cla). **Science Photo Library:** Sheila Terry (ca). **162-163 Corbis:** Stefano Bianchetti. **163 DK Images:** Courtesy of The Science Museum, London/Clive Streeter (bc). **165 DK Images:** Courtesy of The Science Museum, London/Clive Streeter (tr). **iStockphoto.com:** MH (crb). **166 DK Images:** Courtesy of The Science Museum, London/Clive Streeter (cl). **167 Corbis:** Reuters/Mark Ralston (br); Baldwin H. Ward & Kathryn C. Ward (tr). **DK Images:** Courtesy of The Science Museum, London/Clive Streeter (tl). **Science Photo Library:** Alex Bartel (cr). **168 Corbis:** Bettmann (bl); Car Culture (br). **DK Images:** Courtesy of The Science Museum, London/Clive Streeter (ca). **169 DK Images:** Courtesy of The Science Museum, London/Clive Streeter (bl). **170 Corbis:** Hulton-Deutsch Collection (r). **Science Photo Library:** Maria Platt-Evans (bl). **171 The Bridgeman Art Library:** The Royal Institution, London (cl). **Corbis:** (tl). **DK Images:** Courtesy of The Science Museum, London/Clive Streeter (cra). **Getty Images:** The Bridgeman Art Library/The Royal Institution, London (bc). **Science Photo Library:** (crb). **172 DK Images:** The British Museum, London/Alan Hills and Barbara Winter (bl); Courtesy of The Science Museum, London/Clive Streeter (r). **Science & Society Picture Library:** (bc). **173 Corbis:** Roger Ressmeyer (cr). **DK Images:** Courtesy of the Fire Investigation Unit at Acton Police Station, London/Andy Crawford (tl). **NASA:** (tc); Katherine Stephenson, Standford University and Lockheed Martin Corporation (br). **174 Corbis:** Hulton-Deutsch Collection (bc). **DK Images:** Courtesy of the National Maritime Museum, London/Tina Chambers (clb); Courtesy of The Science Museum, London (r). **175 Science & Society Picture Library:** (cl). **Science Photo Library:** NASA (cra). **176 Corbis:** Zefa/Paul Freytag (b/Chemical). **Getty Images:** Glowimages (b/Electrical); Iconica/Nicholas Eveleigh (b/Elastic); Photographer's Choice/Joe McBride (b/Gravitational); Stone/Influx Productions (cl). **176-177 Photolibrary:** Imagestate/Jose Fuste Raga (c). **177 Corbis:** Gary Braasch (b/Light); Pablo Corral (b/Sound); Reuters/Shaun Best (b/Kinetic). **Getty Images:** Hulton Archive (tc); Photographer's Choice/Eric Schnakenberg (b/Heat). **SOHO/EIT (ESA & NASA):** (b/Nuclear). **178 DK Images:** Alamy/Brand X Pictures. **179 Corbis:** Burstein Collection (cra); Zefa/Theo Allofs (bc). **180 Corbis:** (ca). **Science Photo Library:** Sheila Terry (cb). **Still Pictures:** Henry Groskinsky (c). **181 Corbis:** Underwood & Underwood (t). **183 Corbis:** Bettmann (tc). **184 Corbis:** EPA/Jose Jacome (bc). **DK Images:** Courtesy of the Natural History Museum, London/Colin Keates (fbr); Courtesy of the Natural History Museum, London/Harry Taylor (bc). **Getty Images:** National Geographic/Ralph Lee Hopkins (c); National Geographic/Jim Richardson (br). *Geological Sections of Strata in Thuringia* by Johann Gottlob Lehmann, Germany, 1759 (cl). **185 Alamy Images:** Scenics & Science (crb). **Corbis:** Science Faction/G. Brad Lewis (bl); Michael St. Maur Sheil (tr). **Science Photo Library:** Gregory Dimijian (c). **186 DK Images:** Courtesy of the Natural History Museum, London/Colin Keates (b/Carboniferous). **Getty Images:** Hulton Archive (cr). **187 Corbis:** James L. Amos (br); Reuters/Mike Segar (cr). **DK Images:** Courtesy of the Natural History Museum, London/Colin Keates (b/Triassic); Courtesy of the Natural History Museum, London/Harry Taylor (b/Quarternary). **188-189 Corbis:** Louie Psihoyos. **190 Corbis:** Hulton-Deutsch Collection (tr). **Getty Images:** The Bridgeman Art Library (clb). **191 Corbis:** Image Source (b). **Getty Images:** Stanford University (tc). **192 Science Photo Library:** Pasieka (cra). **192-193 Corbis:** Keren Su (b). **193 Alamy Images:** geogphotos (tr). **Corbis:** Kazuyoshi Nomachi (tl); Zefa/Micha Pawlitzki (ftr). **Science Photo Library:** Duncan Shaw (tc). **194 Corbis:** Karl-Heinz Haenel (tl); Adam Jones (cla). **195 Corbis:** Zefa/Theo Allofs (ca). **Image courtesy History of Science Collections, University of Oklahoma Libraries; copyright the Board of Regents of the University of Oklahoma:** (tr). **Science Photo Library:** Wayne Lawler (tl). **196 Alamy Images:** K-Photos (br). **Corbis:** Bettmann (bl); Michael Freeman (cla). **197 Corbis:** Rudy Sulgan (br). **Getty Images:** General Photographic Agency (tl); NOAA (cra). **198-199 The Natural History Museum, London. 200 Alamy Images:** Arco Images GmbH (ca, br); David Hosking (bc); imagebroker (clb). **Iggino Van Bael:** (cl). **Getty Images:** The Image Bank/Pete Turner (cra). **201 Alamy Images:** imagebroker (clb). **Iggino Van Bael:** (cl). **Corbis:** Martin Harvey (bl). **Getty Images:** London Stereoscopic Company (cla). **Photolibrary:** Herbert Kehrer (bc/Above). **Science Photo Library:** Christian Darkin (tr). **Still Pictures:** VISUM/Wolfgang Steche (bc/Below). **202 Corbis:** The Gallery Collection (l). **DK Images:** Courtesy of the Natural History Museum, London/Dave King (cra). **203 akg-images:** (ca). **Corbis:** Star Ledger/William Perlman (bc). **DK Images:** Courtesy of the Natural History Museum, London/Down House/Dave King (tr); Courtesy of the Natural History Museum, London/Dave King (crb). **Mary Evans Picture Library:** (cl). **Science Photo Library:** (tc). **204 Getty Images:** GAP Photos/FhF Greenmedia (tl); Time Life Pictures/Mansell (br). **The Natural History Museum, London:** (tr). **206 Corbis:** Bettmann (bc). **Science Photo Library:** Royal Astronomical Society (bl). **207 Corbis:** (br). **NOAA:** Central Library (tc). **208 Corbis:** Alinari Archives (cla); Jim Reed Photography (cra). **Getty Images:** Photographer's Choice/Harald Sund (tr). **Science Photo Library:** (bl); Detlev van Ravenswaay (br). **209 Corbis:** Eye Ubiquitous/Bryan Pickering (cra); James Randklev (tr). **NASA:** Image courtesy the QuikSCAT team at NASA's Jet Propulsion Laboratory (clb). **NOAA:** National Weather Service (NWS) Collection (bc). **210 Corbis:** Stefano Bianchetti (cl). **NASA:** Earth Sciences and Image Analysis Laboratory, Johnson Space Center (t). **Image courtesy History of Science Collections, University of Oklahoma Libraries; copyright the Board of Regents of the University of Oklahoma:** (bc). **211 Corbis:** Niall Benvie (c); Daniel J. Cox (tc); Darrell Gulin (cb); Ultimate Chase/Mike Theiss (ca). **Getty Images:** The Image Bank/Pete Turner (br). **NASA:** Courtesy the TOMS Science Team & the Scientific Visualization Studio, GSFC (cr). **212 NOAA:** Steve Nicklas, NOS, NGS (tr). **213 Corbis:** Medford Historical Society Collection (tl); Yogi, Inc. (tr). **Science Photo Library:** Lamont-Doherty Earth Observatory/W. Haxby (c). **214 Wellcome Library, London:** (tr). **215 Getty Images:** Hulton Archive (tl). **Science Photo Library:** Don W. Fawcett (crb). **216 Wellcome Library, London:** (cla). **217 Alamy Images:** The Print Collector (c). **218 Corbis:** Bettmann (tc). **DK Images:** Courtesy of The Science Museum, London/Clive Streeter (bl). **Wellcome Library, London:** (br). **219 Alamy Images:** PHOTOTAKE Inc. (cra). **220 Science Photo Library:** Kent Wood (br). **221 Getty Images:** Time Life Pictures/Mansell (tr). **223 The Bridgeman Art Library:** Archives Charmet/Mairie de Sainte-Foy-La-Grande (br). **Science Photo Library:** Pasieka (bl). **225 Corbis:** Hulton-Deutsch Collection (cra); Visuals Unlimited (tl, cla). **DK Images:** Courtesy of the Natural History Museum, London/Colin Keates (bl). **226 Alamy Images:** Mary Evans Picture Library (cla). *Essai de diotropique* by Nicolaus Hartsoeker, 1694 (bl). **227 Corbis:** Mediscan (tc). **228 Corbis:** Visuals Unlimited (br). **DK Images:** Courtesy of The Science Museum, London/Dave King (bl). **228-229 Wellcome Library, London:** (tc). **229 Corbis:** Bernardo Bucci (br); Hulton-Deutsch Collection (tr). **230-231 DK Images:** The Science Museum, London. **232 Lebrecht Music and Arts:** Interfoto (tl). **233 Getty Images:** Time & Life Pictures (cr). **234 Corbis:** Hulton-Deutsch Collection (c); Michael Nicholson (cla). **Science Photo Library:** Martyn F. Chillmaid (cra). **235 Science Photo Library:** Charles D. Winters (l). **236 Getty Images:** Roger Viollet (bc). **237 Archiv der Max-Planck-Gesellschaft, Berlin-Dahlem:** (bc). **Science Photo Library:** Cordelia Molloy (tc). **238 Corbis:** Lawrence Manning (c/Distilled Water). **Getty Images:** David Woodfall (bl). **238-239 Science Photo Library:** Bill Barksdale (bc). **239 Science Photo Library:** (tr); Andrew Lambert Photography (tl); Garry Watson (c/Ammonia ). **240 Science & Society Picture Library:** Science Museum/ SSPL (bl). **241 Alamy Images:** The Print Collector (br). **Imperial War Museum:** G P Lewis (tl). **Science Photo Library:** Charles D. Winters (crb). **242 Alamy Images:** INTERFOTO Pressebildagentur (cl). **Science Photo Library:** Dr Kari Lounatmaa (ca). **Wellcome Library, London:** (bl). **243 The Art Archive:** Culver Pictures (tc). **Science Photo Library:** (l). **244 Corbis:** Visuals Unlimited (cl, bl/Commas, bl/Rods, bl/Spheres). **Getty Images:** Edward Gooch (cra); Visuals Unlimited/Dr Fred Hossler (bl/Cork-Screws). **245 Corbis:** Australian Picture Library/Chris Boydell (tc); Visuals Unlimited (cr/Enveloped, cr/Icosahedral, cr/Spiral-Helical). **Getty Images:** Stone/Hans Gelderblom (cr/Complex). **246 Science Photo Library:** Herve Conge, ISM (bl); Steve Gschmeissner (bc); Dr Gopal Murti (tl). **248 Alamy Images:** Mary Evans Picture Library (bl). **Science & Society Picture Library:** (ca). **249 Getty Images:** Three Lions/Evans (tl). **Wellcome Library, London:** (cr). **250 Corbis:** Hulton-Deutsch Collection (bc); Roger Wood (tc). **DK Images:** Courtesy of The Science Museum, London/Dave King (cr); Courtesy of The Science Museum, London/Clive Streeter (cl, r). **251 Corbis:** Walter Hodges (c); Francis G. Mayer (tr). **252 DK Images:** Courtesy of The Science Museum, London/Clive Streeter (cl). **Science Photo Library:** Library of Congress/George Grantham Bain Collection (cra). **254 Getty Images:** Hulton Collection (cra). **255 Corbis:** Bettmann (tc). **Getty Images:** Car Culture (b); Riser/Jurgen Vogt (cra). **256 Science Photo Library:** Science Source (bl). **257 Corbis:** Arcaid/Richard Bryant (br). **Courtesy of U.S. Navy:** Photographer's Mate 3rd Class Jonathan Chandler (t). **259 Corbis:** Niall Benvie (cla/Left). **Getty Images:** Christopher Furlong (br). **NASA:** E/PO, Sonoma State University, Aurore Simonnet (cb). **Photolibrary:** Oxford Scientific/Owen Newman (cla/Right). **Science Photo Library:** Dept. of Physics, Imperial College (cr); T-Service (bl). **260 Alamy Images:** John Krstenansky (clb). **Corbis:** Leonard de Selva (cla). **Getty Images:** Stock Montage (bc). **Science & Society Picture Library:** (br). **261 Science & Society Picture Library:** (b). **262 Corbis:** Hulton-Deutsch Collection (t, br). **Science & Society Picture Library:** (c). **263 Canon UK Ltd:** (cla). **DK Images:** Canon UK Ltd (b/Canon); Judith Miller/Collectors Cameras (b/Stereoscopic); Courtesy of The Science Museum, London/Dave King (b/Fox Talbot). **Eugene Ilchenko:** (b/Hasselblad). **iStockphoto.com:** Oleg Prikhodko (b/Mobile Phones). **Arne List:** (b/Nikon F3). **264 DK Images:** Courtesy of The Science Museum, London/Dave King (bl). **264-265 Getty Images:** Hulton Archive (b). **265 Corbis:** (cb); Bettmann (tc). **DK Images:** Courtesy of The Science Museum, London/Dave King (cr, br). **266 Science & Society Picture Library:** (ca, bc). **267 Corbis:** Bettmann (bc). **DK Images:** Dave King (l). **268 DK Images:** Judith Miller/Otford Antiques and Collectors Centre (ca); Courtesy of The Science Museum, London/Clive Streeter (cla, bl). **269 Corbis:** (tc). **DK Images:** Courtesy of The Science Museum, London/Dave King (c/Left). **Getty Images:** Photographer's Choice/Cosmo Condina (cr). **274 Getty Images:** Hulton Archive/Sean Sexton (cla); Roger Viollet/Boyer (bl). **Science Photo Library:** Edward Kinsman (bc). **275 Science Photo Library:** Pasieka (cra). **276 FLPA:** Sunset (tl). **naturepl.com:** Miles Barton (bl). **Science Photo Library:** NASA (tr). **276-277 Getty Images:** Robert Harding World Imagery/Steve & Ann Toon (c). **277 Alamy Images:** INTERFOTO Pressebildagentur (br). **FLPA:** MInden Pictures/Tui de Roy (c). **278 Alamy Images:** RIA Novosti (cr). **Corbis:** Reuters (cl). **280 Science Photo Library:** Kevin Curtis. **282 Corbis:** Bettmann (Stanley Steamer); Hulton-Deutsch Collection (Lilienthal). **DK Images:** Courtesy of The Science Museum, London/Clive Streeter (Vacuum Tube). **Getty Images:** AFP (Curie); Daniel Berehulak (Einstein Book). **Science Photo Library:** (Becquerel); Pasieka (Alzheimer Brain). **Wellcome Library, London:** (First X-Ray). **282-283 Science Photo Library:** A. Barrington Brown (Background). **283 Corbis:** Bettmann (Heisenberg). **DK Images:** Courtesy of the Imperial War Museum, London/Geoff Dann (Enigma Machine). **Getty Images:** Keystone (Einstein). **Science Photo Library:** Robert Gendler (Andromeda Galaxy); Royal Astronomical Society (Solar Eclipse); St Mary's Hospital Medical School (Penicillin). **284 Alamy Images:** artpartner-images.com (Hologram). **Corbis:** (Manhattan Project); Bettmann (Pauling); Charles O'Rear (Richter). **Getty Images:** Popperfoto/Paul Popper (V-2). **National Portrait Gallery, London:** Peter Lofts Photography (Hodgkin). **Science Photo Library:** CNRI (Chromosomes). **284-285 Science Photo Library:** Robert Gendler (Background). **285 Corbis:** (Apollo 11). **Getty Images:** Three Lions/Evans (Vaccination); Time & Life Pictures/Ralph Morse (Feynman). **Science Photo Library:** Peter Arnold Inc./Prof. E. Lorenz (Lorenz Attractor); Steve Gschmeissner (Cancer Cell); Dr Ken MacDonald (Ocean Floor); Pasieka (Insulin). **286 Getty Images:** Topical Press Agency (bl). **287 Science Photo Library:** David Parker (crb). **288 Science Photo Library:** (bc); Sheila Terry (cla). **289 Corbis:** Bettmann (crb). **290 Alamy Images:** Dennis Hallinan (cla). **The Art Archive:** Museo del Prado, Madrid/Laurie Platt Winfrey (clb). **Corbis:** Hulton-Deutsch Collection (bc). **290-291 Corbis:** Underwood & Underwood (bc); EPA/Udo Weitz (br). **291 Corbis:** EPA/Ali Haider (t/Airbus). **DK Images:** Courtesy of the Imperial War Museum, Duxford/Andy Crawford (t/Concorde); Courtesy of The Shuttleworth Collection, Bedfordshire/Martin Cameron (t/Wright Flyer). **292 Corbis:** Hulton-Deutsch Collection (cra). **DK Images:** Courtesy of The Science Museum, London/Clive Streeter (bl/Insert, bl, br). **Getty Images:** Hulton Archive (cla). **Science & Society Picture Library:** (tc/Left, tc/Right). **293 Corbis:** Bettmann (tl).

**294 Corbis:** Bettmann (ca). **Science Photo Library:** Jean-Loup Charmet (cb). **Wellcome Library, London:** (br). **295 akg-images:** (tl). **Getty Images:** Nick Veasey (tr). **296 Alamy Images:** Mary Evans Picture Library (bl). **Getty Images:** AFP (r). **Wellcome Library, London:** (bc). **297 Alamy Images:** The Print Collector (t). **The Bridgeman Art Library:** Bibliotheque des Arts Decoratifs, Paris, France/ Archives Charmet (bl). **Corbis:** Ted Spiegel (cr). **298 Science Photo Library:** (tr). **299 Corbis:** Roger Ressmeyer (crb). **DK Images:** Courtesy of The Science Museum, London/Clive Streeter (cb). **Library Of Congress, Washington, D.C.:** (tc). **300-301 akg-images. 302 Corbis:** Lester Lefkowitz (bc). **303 Science Photo Library:** Royal Astronomical Society (bl). **304 Getty Images:** AFP/Jean-Pierre Clatot (bc); Daniel Berehulak (clb); Topical Press Agency (r). **305 Corbis:** Dave Bartruff (br); Hulton-Deutsch Collection (t). **Getty Images:** Keystone (cl). **Science & Society Picture Library:** (bc). **Science Photo Library:** American Institute of Physics/Emilio Segre Visual Archives (cra). **306 Science Photo Library:** M.I. Walker (br). **307 Corbis:** Bettmann (cra); Clouds Hill Imaging Ltd. (tl). **Science Photo Library:** Steve Gschmeissner (crb). **308 Corbis:** Bettmann (tr). **Science Photo Library:** Biophoto Associate (ca); CNRI (bl). **309 Getty Images:** Visuals Unlimited/Dr Dennis Kunkel (cr). **Photolibrary:** Oxford Scientific (tl). **Still Pictures:** Ed Reschke (tc). **310-311 Science Photo Library:** St Mary's Hospital Medical School. **312 Getty Images:** Hulton Archive (tl). **313 Corbis:** Bettmann (tl). **Getty Images:** Stone/Michael Rosenfeld (bl). **iStockphoto.com:** Richard Cano (br). **Science Photo Library:** Pasieka (cr). **314 Corbis:** Bettmann (cra). **315 Image originally created by IBM Corporation:** (br). **Dr Tonomura:** (t). **316 Corbis:** Bettmann (tc). **317 Corbis:** Macduff Everton (cr). **Science Photo Library:** ArSciMed (tc). **318-319 NASA:** ESA, S. Beckwith (STScI) and the HUDF Team. **320 The Bridgeman Art Library:** Bible Society, London, UK (cla). **321 Science Photo Library:** A. Barrington Brown (cla). **322-323 Corbis. 324 Science Photo Library:** Fermilab (cl). **324-325 Corbis:** EPA (c). **325 Getty Images:** Carsten Koall (br). **SOHO/EIT (ESA & NASA):** (cr). **Courtesy of U.S. Navy:** Lt. Scott Miller (tr). **326 Corbis:** (clb, bl). **Getty Images:** Hulton Archive/Joe Munroe (r). **327 Getty Images:** Time & Life Pictures/Ralph Morse (c). **NASA:** Kennedy Space Center (br). **Science & Society Picture Library:** (cra). **© D-Wave Systems Inc. All rights reserved:** (bl). **328 STSCI/ AURA/NASA:** ESA (cra). **329 Science Photo Library:** Emilio Segre Visual Archives / American Institute Of Physics (cr). **SOHO/EIT (ESA & NASA):** (t). **330 NASA:** ESA/ CXC/ STScI/ J. Hester and A. Loll (Arizona State Univ.) , R. Gehrz (Univ. Minn.) (br); HST/ASU/J. Hester and A. Loll (Arizona State Univ.), R. Gehrz (Univ. Minn.) (bl). **Science Photo Library:** Royal Greenwich Observatory (tr). **331 Science Photo Library:** NASA / ESA / STSCI / H.Richer,UBC (cl); Physics Today Collection / American Institute Of Physics (c). **332 Alamy Images:** Bob Gibbons (Acacia Tree). **Corbis:** Michael & Patricia Fogden (Tapir); Darrell Gulin (Aroid Vine); Steve Kaufman (Motmot); Frank Lane Picture Agency/Fritz Polking (Strangler Fig); Robert Pickett (Tarantula). **DK Images:** Courtesy of the Exmoor Zoo, Devon/Gary Ombler (Agouti). **Getty Images:** Digital Vision/Tom Brakefield (Jaguar); The Image Bank/Grant Fair (Butterfly); National Geographic/ Nicole Duplaix (Coccoloba Tree); Visuals Unlimited/Steve Maslowski (Mouse). **333 Corbis:** Daniel J. Cox (br); Raymond Gehman (bc). **Getty Images:** The Image Bank/A. & J. Visage (tr); Photographer's Choice/Sami Sarkis (tc). **The Natural History Museum, London:** (cla). **334 Corbis:** Frans Lanting (cra). **Kathryn Miller:** (bc, bc/Inset). **335 Corbis:** Darrell Gulin (ftr); Mark A. Johnson (fcr); Wolfgang Kaehler (tr); Frans Lanting (ftl); Frank Lukasseck (fcl); David Samuel Robbins (cl); Michel Setboun (cr). **Getty Images:** Photographer's Choice/Darrell Gulin (tl). **336 Alamy Images:** Joe Tree (cb). **Getty Images:** Andy Sotiriou (clb). **Science Photo Library:** Humanities & Social Sciences Library / New York Public Library (cla). **337 The Advertising Archives:** (tr). **akg-images:** PictureContact (crb). **Science Photo Library:** Edward Kinsman (clb). **TopFoto.co.uk:** The Granger Collection (bc). **338 Corbis:** (r). **Getty Images:** Popperfoto/ Paul Popper (bl). **NASA:** (tc). **Science Photo Library:** Ria Novosti (cl). **339 Corbis:** Bettmann (tc). **Courtesy of Virgin Galactic:** (cra). **340 NASA:** ESA/ The Hubble Heritage Team (STScI/ AURA, ca). **Science Photo Library:** Hale Observatories (br); Royal Astronomical Society (cb). **340-341 Science Photo Library:** Robert Gendler (c). **341 Science Photo Library:** Mark Garlick (bc). **342 DK Images:** Courtesy of the H. Keith Melton Collection (bc); Courtesy of the Royal Signals Museum, Blandford Camp, Dorset/Steve Gorton (bl). **343 Courtesy of Apple. Apple and the Apple logo are trademarks of Apple Computer Inc., registered in the US and other countries:** (cr). **DK Images:** Courtesy of the Imperial War Museum, London/Geoff Dann (l). **Science & Society Picture Library:** (tc). **344 Getty Images:** Time & Life Pictures/Life Magazine (r). **Science & Society Picture Library:** (bc). **TopFoto.co.uk:** The Granger Collection, New York (cl). **345 Alamy Images:** Alan Gallery (br). **Photolibrary:** Francis Firth Collection (tc). **Rex Features:** Andy Lauwers (ca). **346 Photolibrary:** Imagestate/Jewish Chronical (bl). **346-347 Science Photo Library:** A. Barrington Brown. **348 Alamy Images:** Mary Evans Picture Library (bl). **349 Science Photo Library:** Nancy Hamilton (cr); James King-Holmes (br). **350 Corbis:** Hulton-Deutsch Collection (cl). **DK Images:** Courtesy of the National Maritime Museum, London/Tina Chambers (bc). **Photolibrary:** Phillip Hayson (r). **351 Wolfgang Beyer:** (b). **Science Photo Library:** Peter Arnold Inc./Prof. E. Lorenz (crb); Gregory Sams (cra). **352 Corbis:** Frans Lanting (b). **Image courtesy History of Science Collections, University of Oklahoma Libraries; copyright the Board of Regents of the University of Oklahoma:** (cla). **Science Photo Library:** (br). **354-355 Corbis:** Roger Ressmeyer. **356 akg-images:** (bl). **Science Photo Library:** Dr Ken MacDonald (br). **357 Corbis:** Atlantide Phototravel (cl); Lloyd Cluff (fcl); EPA/ Olivier Matthys (cr). **NASA:** JPL (fcr). **358 DK Images:** Courtesy of the Natural History Museum, London/Colin Keates (tr). **359 Getty Images:** Stone/G. Brad Lewis (br). **360 Corbis:** TWPhoto (b). **361 Corbis:** Charles O'Rear (bl). **362** *Horse-Hoeing Husbandry* by Jethro Tull, 1752 (bl). **362-363 Corbis:** Sylvain Saustier (tc). **363 Alamy Images:** Bon Appetit (tr); Nigel Cattlin (br). **Corbis:** Michael Busselle (tc); Sygma/Micheline Pelletier (bc). **Getty Images:** Visuals Unlimited/Inga Spence (clb). **Photolibrary:** Animals Animals/ Patti Murray (fclb). **Science Photo Library:** Hank Morgan (cr). **364 Alamy Images:** Wolfgang Kaehler. **365 Alamy Images:** artpartner-images.com (c). **Corbis:** Bettmann (br). **366 Corbis:** Jerry Cooke (cla). **DK Images:** Clive Streeter (tr). **Getty Images:** Taxi/Lester Lefkowitz (b). **367 © Intel Corporation:** (bc). **Science Photo Library:** Gustoimages (cr). **369 Getty Images:** DigitalGlobe (bl); Hulton Archive (fbl). **Landsat 7 satellite image courtesy of NASA Landsat Project Science Office and USGS National Center for Earth Resources Observation Science:** (bc). **NASA:** (tr). **370-371 Corbis. 372 Corbis:** Hulton-Deutsch Collection (br). **Getty Images:** Popperfoto (bl). **372-373 Corbis:** NASA/ Roger Ressmeyer (c). **373 Corbis:** CCTV/ epa (bc); NASA/ Handout/ CNP (bl); Reuters (fbl). **NASA:** (cr). **Science Photo Library:** Steve Gschmeissner (tc); NASA (tr); D. Phillips (tl). **374 Science Photo Library:** Laguna Design. **376 Alamy Images:** Tim Cuff (Lovelock). **Corbis:** (Shuttle Launch). **© Intel Corporation:** (Intel 4004). **DK Images:** Courtesy of the Imperial War Museum, Duxford/Andy Crawford (Concorde). **ESA:** DLR/ FU Berlin (G. Neukum) (Mars). **Getty Images:** hemis.fr/Fred Derwal (Wind Farm). **NASA:** ESA/ CXC/ STScI/ J. Hester and A. Loll (Arizona State Univ.) , R. Gehrz (Univ. Minn.) (Supernova); Johns Hopkins University Applied Physics Laboratory/Carnegie Institution of Washington (Mercury). **Science Photo Library:** Emilio Segre Visual Archives / American Institute Of Physics (Rubin); Mark Garlick (Superclusters); David Parker (Genetic Fingerprinting); Philippe Plailly (Base Pairs of Gene). **376-377 Corbis:** Bettmann (Background). **377 Courtesy of Apple. Apple and the Apple logo are trademarks of Apple Computer Inc., registered in the US and other countries:** (iPhone). **Corbis:** EPA/Martial Trezzini (Hadron Collider); Karen Kasmauski (Dolly). **Getty Images:** AFP (Nanotubes); Stanford University (Zircon); Amy Sussman (Greene). **NASA:** (Chandra); JPL-Caltech/ K. Gordon (University of Arizona) (Spitzer Telescope); Katherine Stephenson, Standford University and Lockheed Martin Corporation (Gravity Probe B). **Science Photo Library:** Fermilab (Top Quarks). **378 Courtesy of Apple. Apple and the Apple logo are trademarks of Apple Computer Inc., registered in the US and other countries:** (tr). **Getty Images:** Andreas Rentz (bl). **379 Philippe Bourcier - sysctl.org:** (b). **Corbis:** Ariel Skelley (tc). **© Google Inc. Used with permission:** (tl). **Still Pictures:** Biosphoto/Vernay Frédéric/Polar Lys (ca). **380 akg-images:** Sotheby's (cla). **Corbis:** Gideon Mendel (tr). **Getty Images:** AFP/Stan Honda (bl). **381 Corbis:** EPA/Alberto Estevez (bl). **Getty Images:** Douglas McFadd (tc). **Intuitive Surgical, Inc.:** (tl). **382 Science Photo Library:** Fermilab (tc). **383 © CERN Geneva:** (cra). **Getty Images:** AFP/CERN (crb). **Science Photo Library:** David Parker (ca). **384 Alamy Images:** Adrian Muttitt (cr). **National Portrait Gallery, London:** Peter Lofts Photography (cl). **Science & Society Picture Library:** Science Museum (bc). **385 Corbis:** Clements/ Hulton-Deutsch Collection (tc). **TopFoto.co.uk:** Topham / PA (bl). **V&A Images, Victoria and Albert Museum:** (crb). **386 Getty Images:** Visuals Unlimited/Dr Dennis Kunkel (cla). **www.glofish.com:** (tr). **Science Photo Library:** © Estate of Francis Bello (bl). **387 Alamy Images:** AGStockUSA, Inc. (crb). **Science Photo Library:** David Parker (br); J. C. Revy, ISM (tc/ Above, tc/Below). **388-389 Science Photo Library:** D. Phillips. **390 Corbis:** Jonathan Blair (tl). **Science Photo Library:** Dr Gopal Murti (clb). **391 Corbis:** Karen Kasmauski (cb). **Science Photo Library:** Andy Crump (br); Professor Miodrag Stojkovic (tr). **392 Corbis:** Visuals Unlimited (b/ Needle). **Science Photo Library:** Steve Gschmeissner (b/ Human Hair); Laguna Design (b/Nanotube); David Mccarthy (c); Medical Rf.com (b/DNA, b/Red Blood Cells); Andrew Syred (b/Dust Mite). **393 Getty Images:** AFP (tr). **Science Photo Library:** Peidong Yang, Lawrence Berkeley National Laboratory (bc). **394 Corbis:** (br). **NASA:** Johns Hopkins University Applied Physics Laboratory/Carnegie Institution of Washington (bl). **Science Photo Library:** JPL / NASA (bc). **395 Corbis:** Roger Ressmeyer (bc). **ESA:** DLR/ FU Berlin (G. Neukum, bl). NASA: ESA, H. Weaver (JHU/APL) , A. Stern (SwRI), and the HST Pluto Companion Search Team (tc); JPL/ Space Science Institute (br). **396 Corbis:** (b). **Science Photo Library:** (cla). **Science & Society Picture Library:** Science Museum (cr). **397 Getty Images:** Al Fenn/ Time Life Pictures (br). **NASA:** (tl); Bill Schoening, Vanessa Harvey/ REU program/ NOAO/ AURA/ NSF (cl); Infrared Processing and Analysis Center, Caltech/ JPL (c, cr); JPL-Caltech/ K. Gordon (University of Arizona) (bl); ROSAT/ MPE (fcl); X-ray: CXC/ CfA/R. Tuellmann et al.; Optical: AURA/STScI (tc). **398 Science Photo Library:** Emilio Segre Visual Archives / American Institute Of Physics (br); Fermilab (bl). **398-399 Science Photo Library:** Volker Springel / Max Planck Institute For Astrophysics (tc). **399 Science Photo Library:** Celestial Image Co. (bc); W. Couch & R. Ellis (c). **401 Corbis:** Bettmann (tr); EPA/Martial Trezzini (bc). **402 Getty Images:** Amy Sussman (br). **403 Bathsheba Sculpture LLC:** (bl). **Houghton Mifflin Company:** (br). **404 Corbis:** Lester Lefkowitz (r). **Getty Images:** Roger Viollet/Harlingue (clb). **Photolibrary:** Phototake Science/Scott Camazine (tc). **Science Photo Library:** Newcastle Hospitals NHS Trust/ Simon Fraser (bc). **405 Corbis:** Mediscan (tc). **Photolibrary:** Nordic Photos/Chad Ehlers (crb). **Science Photo Library:** GE Medical Systems (bc); ISM (tr); Zephyr (bl). **406 Getty Images:** Express/Stan Meagher (bc). **Rex Features:** Olivier Pirard (tl). **Science Photo Library:** Jean-Loup Charmet (clb); Steve Gschmeissner (br). **407 Alamy Images:** Medical-on-Line (ca/Artificial Lens). **Corbis:** Charles O'Rear (cb/ Pacemaker, br). **Photolibrary:** AGE fotostock/Frank Siteman (c). **Science Photo Library:** (ca/Heart Valve); CC, ISM (tc); James King-Holmes (cr); Dr P. Marazzi (cb/Joint Replacement). **408 Corbis:** Historical Picture Archive (cl). **Getty Images:** Taxi/Geoff Du Feu (tl). **409 Corbis:** Reuters/ Kin Cheung (tr). **Getty Images:** Neilson Barnard (bc). **Science Photo Library:** Gustoimages (bl). **410 Science Photo Library:** Biozentrum, University of Basel (bl); Philippe Plailly (br). **411 Corbis:** Louis Quail (cr); Sygma/Ferdaus Shamim (bc). **Science Photo Library:** ICRF/James King-Holmes (tl). **412 Alamy Images:** Tim Cuff (r). **Science Photo Library:** Anthony Howarth (bc). **413 Getty Images:** Reportage/Daniel Beltra (bc). **NASA:** Image created by Reto Stockli with the help of Alan Nelson, under the leadership of Fritz Hasler (tl). **Science Photo Library:** Anthony Howarth (cra). **414 Corbis:** Bettmann (ca); Jim Sugar (bl). **415 Corbis:** Barry Lewis (cra); Reuters/Stringer (b). **Still Pictures:** Biosphoto/Ittel Jean-Frédéric & Association Moraine (tl, tr). **416 Corbis:** Benjamin Rondel (tc). **Getty Images:** hemis.fr/ Fred Derwal (c). **416-417 Corbis:** Paul A. Souders (b). **417 © Pelamis Wave Power Ltd:** (tc). **Science Photo Library:** Rosenfeld Images Ltd (cla). **418 Corbis:** George Steinmetz (clb). **NASA:** Provided by the SeaWiFS project/Goddard Space Flight Center, and ORBIMAGE (br). **Press Association Images:** AP Photo/Gregorio Borgia (cr). **Science Photo Library:** Pascal Goetgheluck (ca). **419 Illustration courtesy of UA Steward Observatory:** (tl). **420 Science Photo Library:** NASA/R.B. Husar.

Endpapers: **Image courtesy History of Science Collections, University of Oklahoma Libraries; copyright the Board of Regents of the University of Oklahoma.**

All other images © Dorling Kindersley
For further information see: www.dkimages.com

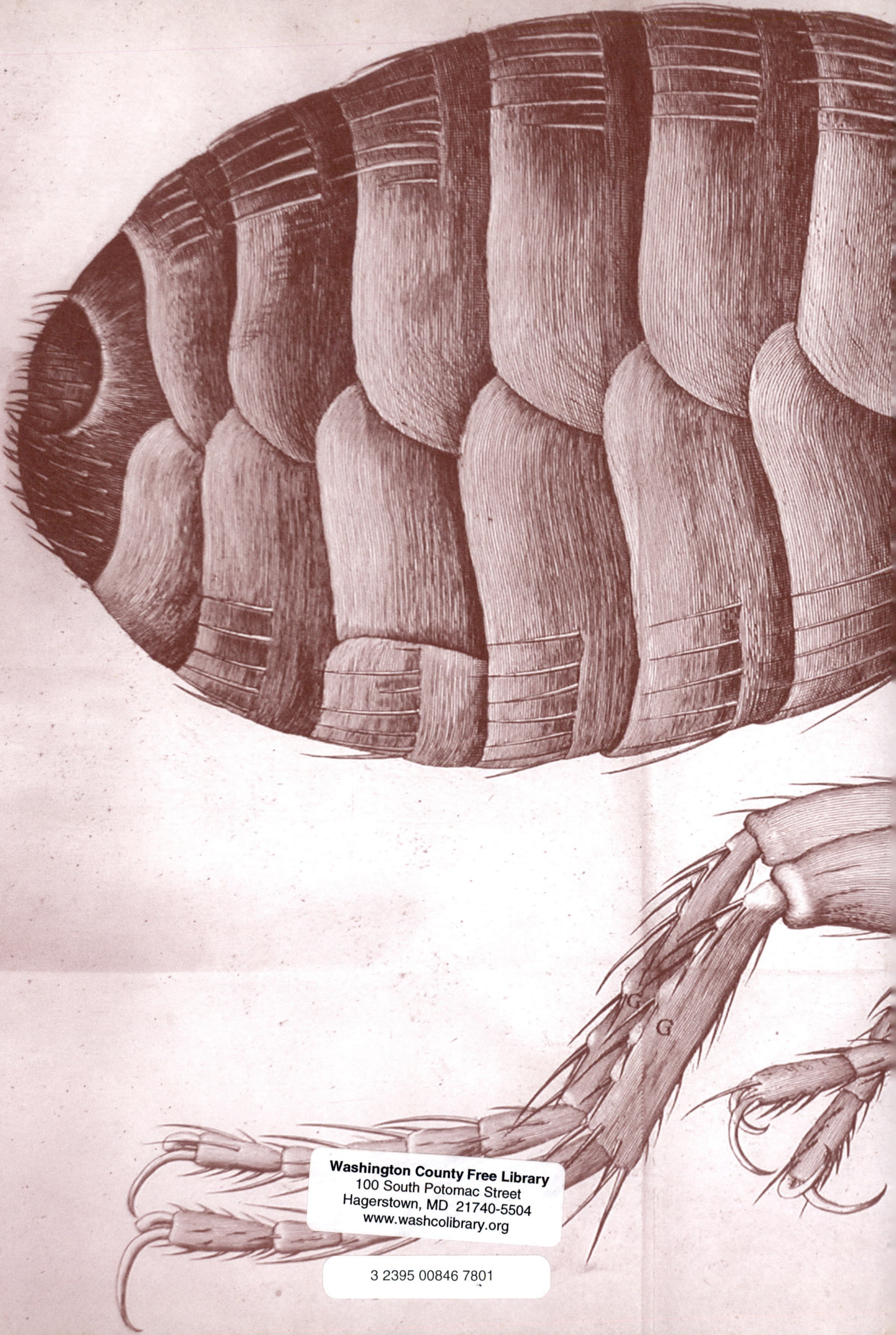
G
G